*Transport Processes
and Unit Operations*

CHRISTIE J. GEANKOPLIS

University of Minnesota

Transport Processes and Unit Operations

THIRD EDITION

PRENTICE HALL P T R
Englewood Cliffs, New Jersey 07632

Library of Congress Cataloging-in-Publication Data

```
Geankoplis, Christie J.
    Transport processes and unit operations / Christie J. Geankoplis.
  -- 3rd ed.
      p.   cm.
    Includes bibliographical references and index.
    ISBN 0-13-930439-8
    1. Transport theory.   I. Title.
  TP156.T7G4   1993
  660'.2842--dc20                                          92-43956
                                                               CIP
```

Editorial/production supervision: Mary P. Rottino
Cover design: Bruce Kenselaar
Manufacturing buyer: Mary E. McCartney
Acquisitions editor: Betty Sun

 © 1993, 1983, 1978 by Prentice Hall P T R
Prentice-Hall, Inc.
A Simon & Schuster Company
Englewood Cliffs, New Jersey 07632

The publisher offers discounts on this book when
ordered in bulk quantities. For more information,
contact:

Corporate Sales Department
PTR Prentice Hall
113 Sylvan Avenue
Englewood Cliffs, NJ 07632

Phone: 201-592-2863
Fax: 201-592-2249

Printed in the United States of America
20   19   18   17   16   15   14   13   12

ISBN 0-13-930439-8

Prentice-Hall International (UK) Limited, *London*
Prentice-Hall of Australia Pty. Limited, *Sydney*
Prentice-Hall Canada Inc., *Toronto*
Prentice-Hall Hispanoamericana, S.A., *Mexico*
Prentice-Hall of India Private Limited, *New Delhi*
Prentice-Hall of Japan, Inc., *Tokyo*
Simon & Schuster Asia Pte. Ltd., *Singapore*
Editora Prentice-Hall do Brasil, Ltda., *Rio de Janeiro*

Dedicated to the memory of my beloved mother, Helen,
for her love and encouragement

Contents

Preface   xi

PART 1
TRANSPORT PROCESSES: MOMENTUM, HEAT, AND MASS

Chapter 1   Introduction to Engineering Principles and Units   1

1.1	Classification of Unit Operations and Transport Processes	1
1.2	SI System of Basic Units Used in This Text and Other Systems	3
1.3	Methods of Expressing Temperatures and Compositions	5
1.4	Gas Laws and Vapor Pressure	7
1.5	Conservation of Mass and Material Balances	9
1.6	Energy and Heat Units	14
1.7	Conservation of Energy and Heat Balances	19
1.8	Graphical, Numerical, and Mathematical Methods	23

Chapter 2   Principles of Momentum Transfer and Overall Balances   31

2.1	Introduction	31
2.2	Fluid Statics	32
2.3	General Molecular Transport Equation for Momentum, Heat, and Mass Transfer	39
2.4	Viscosity of Fluids	43
2.5	Types of Fluid Flow and Reynolds Number	47
2.6	Overall Mass Balance and Continuity Equation	50
2.7	Overall Energy Balance	56
2.8	Overall Momentum Balance	69
2.9	Shell Momentum Balance and Velocity Profile in Laminar Flow	78
2.10	Design Equations for Laminar and Turbulent Flow in Pipes	83
2.11	Compressible Flow of Gases	101

Chapter 3   Principles of Momentum Transfer and Applications   **114**

3.1	Flow Past Immersed Objects and Packed and Fluidized Beds	114
3.2	Measurement of Flow of Fluids	127
3.3	Pumps and Gas-Moving Equipment	133
3.4	Agitation and Mixing of Fluids and Power Requirements	140
3.5	Non-Newtonian Fluids	153
3.6	Differential Equations of Continuity	164
3.7	Differential Equations of Momentum Transfer or Motion	170
3.8	Use of Differential Equations of Continuity and Motion	175
3.9	Other Methods for Solution of Differential Equations of Motion	184
3.10	Boundary-Layer Flow and Turbulence	190
3.11	Dimensional Analysis in Momentum Transfer	202

Chapter 4 **Principles of Steady-State Heat Transfer**   214

  4.1 Introduction and Mechanisms of Heat Transfer   214
  4.2 Conduction Heat Transfer   220
  4.3 Conduction Through Solids in Series   223
  4.4 Steady-State Conduction and Shape Factors   233
  4.5 Forced Convection Heat Transfer Inside Pipes   236
  4.6 Heat Transfer Outside Various Geometries in Forced Convection   247
  4.7 Natural Convection Heat Transfer   253
  4.8 Boiling and Condensation   259
  4.9 Heat Exchangers   267
 4.10 Introduction to Radiation Heat Transfer   276
 4.11 Advanced Radiation Heat-Transfer Principles   281
 4.12 Heat Transfer of Non-Newtonian Fluids   297
 4.13 Special Heat-Transfer Coefficients   300
 4.14 Dimensional Analysis in Heat Transfer   308
 4.15 Numerical Methods for Steady-State Conduction in Two   310
       Dimensions

Chapter 5 **Principles of Unsteady-State Heat Transfer**   330

  5.1 Derivation of Basic Equation   330
  5.2 Simplified Case for Systems with Negligible Internal Resistance   332
  5.3 Unsteady-State Heat Conduction in Various Geometries   334
  5.4 Numerical Finite-Difference Methods for Unsteady-State Conduction   350
  5.5 Chilling and Freezing of Food and Biological Materials   360
  5.6 Differential Equation of Energy Change   365
  5.7 Boundary-Layer Flow and Turbulence in Heat Transfer   370

Chapter 6 **Principles of Mass Transfer**   381

  6.1 Introduction to Mass Transfer and Diffusion   381
  6.2 Molecular Diffusion in Gases   385
  6.3 Molecular Diffusion in Liquids   397
  6.4 Molecular Diffusion in Biological Solutions and Gels   403
  6.5 Molecular Diffusion in Solids   408
  6.6 Numerical Methods for Steady-State Molecular Diffusion in
       Two Dimensions   413

Chapter 7 **Principles of Unsteady-State and Convective Mass Transfer**   426

  7.1 Unsteady-State Diffusion   426
  7.2 Convective Mass-Transfer Coefficients   432
  7.3 Mass-Transfer Coefficients for Various Geometries   437
  7.4 Mass Transfer to Suspensions of Small Particles   450
  7.5 Molecular Diffusion Plus Convection and Chemical Reaction   453
  7.6 Diffusion of Gases in Porous Solids and Capillaries   462
  7.7 Numerical Methods for Unsteady-State Molecular Diffusion   468
  7.8 Dimensional Analysis in Mass Transfer   474
  7.9 Boundary-Layer Flow and Turbulence in Mass Transfer   475

PART 2
UNIT OPERATIONS

Chapter 8 **Evaporation**   **489**

  8.1   Introduction   489
  8.2   Types of Evaporation Equipment and Operation Methods   491
  8.3   Overall Heat-Transfer Coefficients in Evaporators   495
  8.4   Calculation Methods for Single-Effect Evaporators   496
  8.5   Calculation Methods for Multiple-Effect Evaporators   502
  8.6   Condensers for Evaporators   511
  8.7   Evaporation of Biological Materials   513
  8.8   Evaporation Using Vapor Recompression   514

Chapter 9 **Drying of Process Materials**   520

  9.1   Introduction and Methods of Drying   520
  9.2   Equipment for Drying   521
  9.3   Vapor Pressure of Water and Humidity   525
  9.4   Equilibrium Moisture Content of Materials   533
  9.5   Rate of Drying Curves   536
  9.6   Calculation Methods for Constant-Rate Drying Period   540
  9.7   Calculation Methods for Falling-Rate Drying Period   545
  9.8   Combined Convection, Radiation, and Conduction
         Heat Transfer in Constant-Rate Period   548
  9.9   Drying in Falling-Rate Period by Diffusion and
         Capillary Flow   551
 9.10 Equations for Various Types of Dryers   556
 9.11 Freeze Drying of Biological Materials   566
 9.12 Unsteady-State Thermal Processing and Sterilization
         of Biological Materials   569

Chapter 10 **Stage and Continuous Gas–Liquid Separation Processes**   584

 10.1 Types of Separation Processes and Methods   584
 10.2 Equilibrium Relations Between Phases   586
 10.3 Single and Multiple Equilibrium Contact Stages   587
 10.4 Mass Transfer Between Phases   594
 10.5 Continuous Humidification Processes   602
 10.6 Absorption in Plate and Packed Towers   610
 10.7 Absorption of Concentrated Mixtures in Packed Towers   627
 10.8 Estimation of Mass Transfer Coefficients for Packed Towers   632

Chapter 11 **Vapor–Liquid Separation Processes**   640

 11.1 Vapor–Liquid Equilibrium Relations   640
 11.2 Single-Stage Equilibrium Contact for Vapor–Liquid System   642
 11.3 Simple Distillation Methods   644
 11.4 Distillation with Reflux and McCabe–Thiele Method   649
 11.5 Distillation and Absorption Tray Efficiencies   666
 11.6 Fractional Distillation Using Enthalpy–Concentration Method   669
 11.7 Distillation of Multicomponent Mixtures   679

Chapter 12 **Liquid–Liquid and Fluid–Solid Separation Processes** **697**

12.1 Introduction to Adsorption Processes 697
12.2 Batch Adsorption 700
12.3 Design of Fixed-Bed Adsorption Columns 701
12.4 Ion-Exchange Processes 708
12.5 Single-Stage Liquid–Liquid Extraction Processes 709
12.6 Equipment for Liquid–Liquid Extraction 715
12.7 Continuous Multistage Countercurrent Extraction 716
12.8 Introduction and Equipment for Liquid–Solid Leaching 723
12.9 Equilibrium Relations and Single-Stage Leaching 729
12.10 Countercurrent Multistage Leaching 733
12.11 Introduction and Equipment for Crystallization 737
12.12 Crystallization Theory 743

Chapter 13 **Membrane Separation Process** **754**

13.1 Introduction and Types of Membrane Separation Processes 754
13.2 Liquid Permeation Membrane Processes or Dialysis 755
13.3 Gas Permeation Membrane Processes 759
13.4 Complete-Mixing Model for Gas Separation by Membranes 764
13.5 Complete-Mixing Model for Multicomponent Mixtures 769
13.6 Cross-Flow Model for Gas Separation by Membranes 772
13.7 Countercurrent-Flow Model for Gas Separation by Membranes 778
13.8 Effects of Processing Variables on Gas Separation by Membranes 780
13.9 Reverse-Osmosis Membrane Processes 782
13.10 Applications, Equipment, and Models for Reverse Osmosis 788
13.11 Ultrafiltration Membrane Processes 791

Chapter 14 **Mechanical–Physical Separation Processes** **800**

14.1 Introduction and Classification of Mechanical–Physical Separation
Processes 800
14.2 Filtration in Solid–Liquid Separation 801
14.3 Settling and Sedimentation in Particle–Fluid Separation 815
14.4 Centrifugal Separation Processes 828
14.5 Mechanical Size Reduction 840

Appendix

Appendix A.1 Fundamental Constants and Conversion Factors 850
Appendix A.2 Physical Properties of Water 854
Appendix A.3 Physical Properties of Inorganic and Organic Compounds 864
Appendix A.4 Physical Properties of Foods and Biological Materials 889
Appendix A.5 Properties of Pipes, Tubes, and Screens 892

Notation **895**

Index **905**

Preface

In this third edition, the main objectives and the format of the first and second editions remain the same. The sections on momentum transfer have been greatly expanded, especially in the sections covering differential equations of momentum transfer. This now allows full coverage of the transport processes of momentum, heat, and mass transfer. Also, a section on adsorption and an expanded chapter on membrane processes have been added to the unit operations sections.

The field of chemical engineering involved with physical and physical–chemical changes of inorganic and organic materials, and to some extent biological materials, is overlapping more and more with the other process engineering fields of ceramic engineering, process metallurgy, agricultural food engineering, wastewater treatment (civil) engineering, and bioengineering. The principles of momentum, heat, and mass transport and the unit operations are used in these processing fields.

The principles of momentum transfer and heat transfer have been taught to all engineers. The study of mass transfer has been limited primarily to chemical engineers. However, engineers in other fields have become more interested in mass transfer in gases, liquids, and solids.

Since chemical and other engineering students must study so many topics today, a more unified introduction to the transport processes of momentum, heat, and mass transfer and to the applications of unit operations is provided. In this text the principles of the transport processes are covered first, and then the unit operations. To accomplish this, the text is divided into two main parts.

PART 1: Transport Processes: Momentum, Heat, and Mass

This part, dealing with fundamental principles, includes the following chapters: 1. Introduction to Engineering Principles and Units; 2. Principles of Momentum Transfer and Overall Balances; 3. Principles of Momentum Transfer and Applications; 4. Principles of Steady-State Heat Transfer; 5. Principles of Unsteady-State Heat Transfer; 6.

Principles of Mass Transfer; and 7. Principles of Unsteady-State and Convective Mass Transfer.

PART 2: *Unit Operations*

This part, on applications, covers the following unit operations: 8. Evaporation; 9. Drying of Process Materials; 10. Stage and Continuous Gas–Liquid Separation Processes (humidification, absorption); 11. Vapor–Liquid Separation Processes (distillation); 12. Liquid–Liquid and Fluid–Solid Separation Processes (adsorption, ion exchange, extraction, leaching, crystallization); 13. Membrane Separation Processes (dialysis, gas separation, reverse osmosis, ultrafiltration); 14. Mechanical–Physical Separation Processes (filtration, settling, centrifugal separation, mechanical size reduction).

In Chapter 1 elementary principles of mathematical and graphical methods, laws of chemistry and physics, material balances, and heat balances are reviewed. Many, especially chemical engineers, may be familiar with most of these principles and may omit all or parts of this chapter.

A few topics, involved primarily with the processing of biological materials, may be omitted at the discretion of the reader or instructor: Sections 5.5, 6.4, 8.7, 9.11, and 9.12. Over 230 example or sample problems and over 500 homework problems on all topics are included in the text. Some of the homework problems are concerned with biological systems, for those readers who are especially interested in that area.

This text may be used for a course of study using any of the following five suggested plans. In all plans, Chapter 1 may or may not be included.

1. *Study of transport processes of momentum, heat, and mass and unit operations.* In this plan, most of the complete text covering the principles of the transport processes in Part 1 and the unit operations in Part 2 are covered. This plan could be applicable primarily to chemical engineering and also to other process engineering fields in a one and one-half year course of study at the junior and/or senior level.

2. *Study of transport processes of momentum, heat, and mass and selected unit operations.* Only the elementary sections of Part 1 (the principles chapters—2, 3, 4, 5, 6, and 7) are covered, plus selected unit operations topics in Part 2 applicable to a particular field in a two-semester or three-quarter course. Those in wastewater treatment engineering, food process engineering, and process metallurgy could follow this plan.

3. *Study of transport processes of momentum, heat, and mass.* The purpose of this plan in a two-quarter or two-semester course is to obtain a basic understanding of the transport processes of momentum, heat, and mass transfer. This involves studying sections of the principles chapters—2, 3, 4, 5, 6, and 7 in Part 1—and omitting Part 2, the applied chapters on unit operations.

4. *Study of unit operations.* If the reader has had courses in the transport processes of momentum, heat, and mass, Chapters 2 through 7 can be omitted and only the unit operations chapters in Part 2 studied in a one-semester or two-quarter course. This plan could be used by chemical and certain other engineers.

5. *Study of mass transfer.* For those such as chemical or mechanical engineers who have had momentum and heat transfer, or those who desire only a background in mass transfer in a one-quarter or one-semester course, Chapters 6, 7, and 10 would be covered. Chapters 9, 11, 12, and 13 might be covered optionally, depending on the needs of the reader.

The SI (Système International d'Unités) system of units has been adopted by the scientific community. Because of this, the SI system of units has been adopted in this text for use in the equations, example problems, and homework problems. However, the most important equations derived in the text are also given in a dual set of units, SI and English, when different. Many example and homework problems are also given using English units.

Christie J. Geankoplis

PART 1

Transport Processes: Momentum, Heat, and Mass

CHAPTER 1

Introduction to Engineering Principles and Units

1.1 CLASSIFICATION OF UNIT OPERATIONS AND TRANSPORT PROCESSES

1.1A Introduction

In the chemical and other physical processing industries and the food and biological processing industries, many similarities exist in the manner in which the entering feed materials are modified or processed into final materials of chemical and biological products. We can take these seemingly different chemical, physical, or biological processes and break them down into a series of separate and distinct steps called *unit operations*. These unit operations are common to all types of diverse process industries.

For example, the unit operation *distillation* is used to purify or separate alcohol in the beverage industry and hydrocarbons in the petroleum industry. Drying of grain and other foods is similar to drying of lumber, filtered precipitates, and rayon yarn. The unit operation *absorption* occurs in absorption of oxygen from air in a fermentation process or in a sewage treatment plant and in absorption of hydrogen gas in a process for liquid hydrogenation of oil. Evaporation of salt solutions in the chemical industry is similar to evaporation of sugar solutions in the food industry. Settling and sedimentation of suspended solids in the sewage and the mining industries are similar. Flow of liquid hydrocarbons in the petroleum refinery and flow of milk in a dairy plant are carried out in a similar fashion.

The unit operations deal mainly with the transfer and change of energy and the transfer and change of materials primarily by physical means but also by physical–chemical means. The important unit operations, which can be combined in various sequences in a process and which are covered in Part 2, of this text, are described next.

1.1B Classification of Unit Operations

1. *Fluid flow.* This concerns the principles that determine the flow or transportation of any fluid from one point to another.
2. *Heat transfer.* This unit operation deals with the principles that govern accumulation and transfer of heat and energy from one place to another.

1

3. *Evaporation.* This is a special case of heat transfer, which deals with the evaporation of a volatile solvent such as water from a nonvolatile solute such as salt or any other material in solution.
4. *Drying.* In this operation volatile liquids, usually water, are removed from solid materials.
5. *Distillation.* This is an operation whereby components of a liquid mixture are separated by boiling because of their differences in vapor pressure.
6. *Absorption.* In this process a component is removed from a gas stream by treatment with a liquid.
7. *Membrane separation.* This process involves the separation of a solute from a fluid by diffusion of this solute from a liquid or gas through a semipermeable membrane barrier to another fluid.
8. *Liquid–liquid extraction.* In this case a solute in a liquid solution is removed by contacting with another liquid solvent which is relatively immiscible with the solution.
9. *Adsorption.* In this process a component of a gas or a liquid stream is removed and adsorbed by a solid adsorbent.
10. *Liquid–solid leaching.* This involves treating a finely divided solid with a liquid that dissolves out and removes a solute contained in the solid.
11. *Crystallization.* This concerns the removal of a solute such as a salt from a solution by precipitating the solute from the solution.
12. *Mechanical–physical separations.* These involve separation of solids, liquids, or gases by mechanical means, such as filtration, settling, and size reduction, which are often classified as separate unit operations.

Many of these unit operations have certain fundamental and basic principles or mechanisms in common. For example, the mechanism of diffusion or mass transfer occurs in drying, membrane separation, absorption, distillation, and crystallization. Heat transfer occurs in drying, distillation, evaporation, and so on. Hence, the following classification of a more fundamental nature is often made into transfer or transport processes.

1.1C Fundamental Transport Processes

1. *Momentum transfer.* This is concerned with the transfer of momentum which occurs in moving media, such as in the unit operations of fluid flow, sedimentation, and mixing.
2. *Heat transfer.* In this fundamental process, we are concerned with the transfer of heat from one place to another; it occurs in the unit operations heat transfer, drying, evaporation, distillation, and others.
3. *Mass transfer.* Here mass is being transferred from one phase to another distinct phase; the basic mechanism is the same whether the phases are gas, solid, or liquid. This includes distillation, absorption, liquid–liquid extraction, membrane separation, adsorption, and leaching.

1.1D Arrangement in Parts 1 and 2

This text is arranged in two parts:

Part 1: Transport Processes: Momentum, Heat, and Mass. These fundamental principles are covered extensively in Chapters 1 to 7 to provide the basis for study of unit operations.

Part 2: Unit Operations. The various unit operations and their applications to process areas are studied in Part 2 of this text.

There are a number of elementary engineering principles, mathematical techniques, and laws of physics and chemistry that are basic to a study of the principles of momentum, heat, and mass transfer and the unit operations. These are reviewed for the reader in this first chapter. Some readers, especially chemical engineers, agricultural engineers, civil engineers, and chemists, may be familiar with many of these principles and techniques and may wish to omit all or parts of this chapter.

Homework problems at the end of each chapter are arranged in different sections, each corresponding to the number of a given section in the chapter.

1.2 SI SYSTEM OF BASIC UNITS USED IN THIS TEXT AND OTHER SYSTEMS

There are three main systems of basic units employed at present in engineering and science. The first and most important of these is the *SI* (Système International d'Unités) *system*, which has as its three basic units the meter (m), the kilogram (kg), and the second (s). The others are the English foot (ft)–pound (lb)–second (s), or *fps, system* and the centimeter (cm)–gram (g)–second (s), or *cgs, system*.

At present the SI system has been adopted officially for use exclusively in engineering and science, but the older English and cgs systems will still be used for some time. Much of the physical and chemical data and empirical equations are given in these latter two systems. Hence, the engineer should not only be proficient in the SI system but must also be able to use the other two systems to a limited extent.

1.2A SI System of Units

The basic quantities used in the SI system are as follows: the unit of length is the meter (m); the unit of time is the second (s); the unit of mass is the kilogram (kg); the unit of temperature is the kelvin (K); and the unit of an element is the kilogram mole (kg mol). The other standard units are derived from these basic quantities.

The basic unit of force is the newton (N), defined as

$$1 \text{ newton (N)} = 1 \text{ kg} \cdot \text{m/s}^2$$

The basic unit of work, energy, or heat is the newton-meter, or joule (J).

$$1 \text{ joule (J)} = 1 \text{ newton} \cdot \text{m (N} \cdot \text{m)} = 1 \text{ kg} \cdot \text{m}^2/\text{s}^2$$

Power is measured in joules/s or watts (W).

$$1 \text{ joule/s (J/s)} = 1 \text{ watt (W)}$$

The unit of pressure is the newton/m² or pascal (Pa).

$$1 \text{ newton/m}^2 \text{ (N/m}^2) = 1 \text{ pascal (Pa)}$$

[Pressure in atmospheres (atm) is not a standard SI unit but is being used during the transition period.] The standard acceleration of gravity is defined as

$$1 \text{ } g = 9.80665 \text{ m/s}^2$$

A few of the standard prefixes for multiples of the basic units are as follows: giga (G) $= 10^9$, mega (M) $= 10^6$, kilo (k) $= 10^3$, centi (c) $= 10^{-2}$, milli (m) $= 10^{-3}$, micro (μ) $= 10^{-6}$, and nano (n) $= 10^{-9}$. The prefix c is not a preferred prefix.

Temperatures are defined in kelvin (K) as the preferred unit in the SI system. However, in practice, wide use is made of the degree Celsius (°C) scale, which is defined by

$$t°C = T(K) - 273.15$$

Note that $1°C = 1$ K and that in the case of temperature difference,

$$\Delta t°C = \Delta T \text{ K}$$

The standard preferred unit of time is the second (s), but time can be in nondecimal units of minutes (min), hours (h), or days (d).

1.2B   CGS System of Units

The cgs system is related to the SI system as follows:

$$1 \text{ g mass (g)} = 1 \times 10^{-3} \text{ kg mass (kg)}$$

$$1 \text{ cm} = 1 \times 10^{-2} \text{ m}$$

$$1 \text{ dyne (dyn)} = 1 \text{ g} \cdot \text{cm/s}^2 = 1 \times 10^{-5} \text{ newton (N)}$$

$$1 \text{ erg} = 1 \text{ dyn} \cdot \text{cm} = 1 \times 10^{-7} \text{ joule (J)}$$

The standard acceleration of gravity is

$$g = 980.665 \text{ cm/s}^2$$

1.2C   English fps System of Units

The English system is related to the SI system as follows:

$$1 \text{ lb mass (lb}_m) = 0.45359 \text{ kg}$$

$$1 \text{ ft} = 0.30480 \text{ m}$$

$$1 \text{ lb force (lb}_f) = 4.4482 \text{ newtons (N)}$$

$$1 \text{ ft} \cdot \text{lb}_f = 1.35582 \text{ newton} \cdot \text{m (N} \cdot \text{m)} = 1.35582 \text{ joules (J)}$$

$$1 \text{ psia} = 6.89476 \times 10^3 \text{ newton/m}^2 \text{ (N/m}^2)$$

$$1.8°F = 1 \text{ K} = 1°C \text{ (centigrade or Celsius)}$$

$$g = 32.174 \text{ ft/s}^2$$

The proportionality factor for Newton's law is

$$g_c = 32.174 \text{ ft} \cdot \text{lb}_m/\text{lb}_f \cdot \text{s}^2$$

The factor g_c in SI units and cgs units is 1.0 and is omitted.

In Appendix A.1, convenient conversion factors for all three systems are tabulated. Further discussions and use of these relationships are given in various sections of the text.

This text uses the SI system as the primary set of units in the equations, sample problems, and homework problems. However, the important equations derived in the text are given in a dual set of units, SI and English, when these equations differ. Some example problems and homework problems are also given using English units. In some cases, intermediate steps and/or answers in example problems are also stated in English units.

1.2D Dimensionally Homogeneous Equations and Consistent Units

A dimensionally homogeneous equation is one in which all the terms have the same units. These units can be the base units or derived ones (for example, kg/s² · m or Pa). Such an equation can be used with any system of units provided that the same base or derived units are used throughout the equation. No conversion factors are needed when consistent units are used.

The reader should be careful in using any equation and always check it for dimensional homogeneity. To do this, a system of units (SI, English, etc.) is first selected. Then units are substituted for each term in the equation and like units in each term canceled out.

1.3 METHODS OF EXPRESSING TEMPERATURES AND COMPOSITIONS

1.3A Temperature

There are two temperature scales in common use in the chemical and biological industries. These are degrees Fahrenheit (abbreviated °F) and Celsius (°C). It is often necessary to convert from one scale to the other. Both use the freezing point and boiling point of water at 1 atmosphere pressure as base points. Often temperatures are expressed as absolute degrees K (SI standard) or degrees Rankine (°R) instead of °C or °F. Table 1.3-1 shows the equivalences of the four temperature scales.

The difference between the boiling point of water and melting point of ice at 1 atm is 100°C or 180°F. Thus, a 1.8°F change is equal to a 1°C change. Usually, the value of -273.15°C is rounded to -273.2°C and -459.7°F to -460°F. The following equations can be used to convert from one scale to another.

$$°F = 32 + 1.8(°C) \qquad \text{(1.3-1)}$$

$$°C = \frac{1}{1.8}(°F - 32) \qquad \text{(1.3-2)}$$

$$°R = °F + 460 \qquad \text{(1.3-3)}$$

$$K = °C + 273.15 \qquad \text{(1.3-4)}$$

TABLE 1.3-1. *Temperature Scales and Equivalents*

	Centigrade	Fahrenheit	Kelvin	Rankine	Celsius
Boiling water	100°C	212°F	373.15 K	671.7°R	100°C
Melting ice	0°C	32°F	273.15 K	491.7°R	0°C
Absolute zero	-273.15°C	-459.7°F	0 K	0°R	-273.15°C

1.3B Mole Units, and Weight or Mass Units

There are many methods used to express compositions in gases, liquids, and solids. One of the most useful is molar units, since chemical reactions and gas laws are simpler to express in terms of molar units. A mole (mol) of a pure substance is defined as the amount of that substance whose mass is numerically equal to its molecular weight. Hence, 1 kg mol of methane CH_4 contains 16.04 kg. Also, 1.0 lb mol contains 16.04 lb_m.

The mole fraction of a particular substance is simply the moles of this substance divided by the total number of moles. In like manner, the weight or mass fraction is the mass of the substance divided by the total mass. These two compositions, which hold for gases, liquids, and solids, can be expressed as follows for component A in a mixture:

$$x_A(\text{mole fraction of } A) = \frac{\text{moles of } A}{\text{total moles}} \qquad (1.3\text{-}5)$$

$$w_A(\text{mass or wt fraction of } A) = \frac{\text{mass } A}{\text{total mass}} \qquad (1.3\text{-}6)$$

EXAMPLE 1.3-1.   Mole and Mass or Weight Fraction of a Solution
A container holds 50 g of water (B) and 50 g of NaOH (A). Calculate the weight fraction and mole fraction of NaOH. Also, calculate the lb_m of NaOH (A) and H_2O (B).

Solution: Taking as a basis for calculation $50 + 50$ or 100 g of solution, the following data are calculated:

Component	G	Wt Fraction	Mol Wt	G Moles	Mole Fraction
H_2O (B)	50.0	$\frac{50}{100} = 0.500$	18.02	$\frac{50.0}{18.02} = 2.78$	$\frac{2.78}{4.03} = 0.690$
NaOH (A)	50.0	$\frac{50}{100} = 0.500$	40.0	$\frac{50.0}{40.0} = 1.25$	$\frac{1.25}{4.03} = 0.310$
Total	100.0	1.00		4.03	1.000

Hence, $x_A = 0.310$ and $x_B = 0.690$ and $x_A + x_B = 0.310 + 0.690 = 1.00$. Also, $w_A + w_B = 0.500 + 0.500 = 1.00$. To calculate the lb_m of each component, Appendix A.1 gives the conversion factor of 453.6 g per 1 lb_m. Using this,

$$\text{lb mass of } A = \frac{50 \text{ g } A}{453.6 \text{ g } A/lb_m \ A} = 0.1102 \ lb_m \ A$$

Note that the g of A in the numerator cancels the g of A in the denominator, leaving lb_m of A in the numerator. The reader is cautioned to put all units down in an equation and cancel those appearing in the numerator and denominator. In a similar manner we obtain 0.1102 lb_m B (0.0500 kg B).

The analyses of solids and liquids are usually given as weight or mass fraction or weight percent, and gases as mole fraction or percent. Unless otherwise stated, analyses of solids and liquids will be assumed to be weight (mass) fraction or percent, and of gases to be mole fraction or percent.

1.3C Concentration Units for Liquids

In general, when one liquid is mixed with another miscible liquid, the volumes are not additive. Hence, compositions of liquids are usually not expressed as volume percent of a component but as weight or mole percent. Another convenient way to express concentrations of components in a solution is *molarity*, which is defined as g mol of a component per liter of solution. Other methods used are kg/m^3, g/liter, g/cm^3, lb mol/cu ft, lb_m/cu ft, and lb_m/gallon. All these concentrations depend on temperature, so the temperature must be specified.

The most common method of expressing total concentration per unit volume is density, kg/m^3, g/cm^3, or lb_m/ft^3. For example, the density of water at 277.2 K (4°C) is 1000 kg/m^3, or 62.43 lb_m/ft^3. Sometimes the density of a solution is expressed as *specific gravity*, which is defined as the density of the solution at its given temperature divided by the density of a reference substance at its temperature. If the reference substance is water at 277.2 K, the specific gravity and density of the substance are numerically equal.

1.4 GAS LAWS AND VAPOR PRESSURE

1.4A Pressure

There are numerous ways of expressing the pressure exerted by a fluid or system. An *absolute pressure* of 1.00 atm is equivalent to 760 mm Hg at 0°C, 29.921 in. Hg, 0.760 m Hg, 14.696 lb force per square inch (psia), or 33.90 ft of water at 4°C. *Gage pressure* is the pressure above the absolute pressure. Hence, a pressure of 21.5 lb per square inch gage (*psig*) is 21.5 + 14.7 (rounded off), or 36.2 psia. In SI units, 1 psia = 6.89476×10^3 pascal (Pa) = 6.89476×10^3 newtons/m^2. Also, 1 atm = 1.01325×10^5 Pa.

In some cases, particularly in evaporation, one may express the pressure as inches of mercury vacuum. This means the pressure as inches of mercury measured "below" the absolute barometric pressure. For example, a reading of 25.4 in. Hg vacuum is 29.92 − 25.4, or 4.52 in. Hg absolute pressure. Pressure conversion units are given in Appendix A.1.

1.4B Ideal Gas Law

An *ideal gas* is defined as one that obeys simple laws. Also, in an ideal gas the gas molecules are considered as rigid spheres which themselves occupy no volume and do not exert forces on one another. No real gases obey these laws exactly, but at ordinary temperatures and pressures of not more than several atmospheres, the ideal laws give answers within a few percent or less of the actual answers. Hence, these laws are sufficiently accurate for engineering calculations.

The *ideal gas law* of Boyle states that the volume of a gas is directly proportional to the absolute temperature and inversely proportional to the absolute pressure. This is expressed as

$$pV = nRT \tag{1.4-1}$$

where p is the absolute pressure in N/m^2, V the volume of the gas in m^3, n the kg mol of the gas, T the absolute temperature in K, and R the gas law constant of 8314.3 kg·m^2/kg mol·s^2·K. When the volume is in ft^3, n in lb moles, and T in °R, R has a value of 0.7302 ft^3·atm/lb mol·°R. For cgs units (see Appendix A.1), $V = cm^3$, $T = K$, $R = 82.057 \ cm^3$·atm/g mol·K, and n = g mol.

In order that amounts of various gases may be compared, *standard conditions of temperature and pressure* (abbreviated STP or SC) are arbitrarily defined as 101.325 kPa (1.0 atm) abs and 273.15 K (0°C). Under these conditions the volumes are as follows:

$$\text{volume of 1.0 kg mol (SC)} = 22.414 \text{ m}^3$$

$$\text{volume of 1.0 g mol (SC)} = 22.414 \text{ L (liter)}$$

$$= 22\,414 \text{ cm}^3$$

$$\text{volume of 1.0 lb mol (SC)} = 359.05 \text{ ft}^3$$

EXAMPLE 1.4-1. Gas-Law Constant

Calculate the value of the gas-law constant R when the pressure is in psia, moles in lb mol, volume in ft^3, and temperature in °R. Repeat for SI units.

Solution: At standard conditions, $p = 14.7$ psia, $V = 359 \text{ ft}^3$, and $T = 460 + 32 = 492°R$ (273.15 K). Substituting into Eq. (1.4-1) for $n = 1.0$ lb mol and solving for R,

$$R = \frac{pV}{nT} = \frac{(14.7 \text{ psia})(359 \text{ ft}^3)}{(1.0 \text{ lb mol})(492°R)} = 10.73 \frac{\text{ft}^3 \cdot \text{psia}}{\text{lb mol} \cdot °R}$$

$$R = \frac{pV}{nT} = \frac{(1.01325 \times 10^5 \text{ Pa})(22.414 \text{ m}^3)}{(1.0 \text{ kg mol})(273.15 \text{ K})} = 8314 \frac{\text{m}^3 \cdot \text{Pa}}{\text{kg mol} \cdot \text{K}}$$

A useful relation can be obtained from Eq. (1.4-1) for n moles of gas at conditions p_1, V_1, T_1, and also at conditions p_2, V_2, T_2. Substituting into Eq. (1.4-1),

$$p_1 V_1 = nRT_1$$

$$p_2 V_2 = nRT_2$$

Combining gives

$$\frac{p_1 V_1}{p_2 V_2} = \frac{T_1}{T_2} \tag{1.4-2}$$

1.4C Ideal Gas Mixtures

Dalton's law for mixtures of ideal gases states that the total pressure of a gas mixture is equal to the sum of the individual partial pressures:

$$P = p_A + p_B + p_C + \cdots \tag{1.4-3}$$

where P is total pressure and p_A, p_B, p_C, ... are the partial pressures of the components A, B, C, ... in the mixture.

Since the number of moles of a component is proportional to its partial pressure, the mole fraction of a component is

$$x_A = \frac{p_A}{P} = \frac{p_A}{p_A + p_B + p_C + \cdots} \tag{1.4-4}$$

The volume fraction is equal to the mole fraction. Gas mixtures are almost always represented in terms of mole fractions and not weight fractions. For engineering purposes, Dalton's law is sufficiently accurate to use for actual mixtures at total pressures of a few atmospheres or less.

EXAMPLE 1.4-2. Composition of a Gas Mixture

A gas mixture contains the following components and partial pressures: CO_2, 75 mm Hg; CO, 50 mm Hg; N_2, 595 mm Hg; O_2, 26 mm Hg. Calculate the total pressure and the composition in mole fraction.

Solution: Substituting into Eq. (1.4-3),

$$P = p_A + p_B + p_C + p_D = 75 + 50 + 595 + 26 = 746 \text{ mm Hg}$$

The mole fraction of CO_2 is obtained by using Eq. (1.4-4).

$$x_A(CO_2) = \frac{p_A}{P} = \frac{75}{746} = 0.101$$

In like manner, the mole fractions of CO, N_2, and O_2 are calculated as 0.067, 0.797, and 0.035, respectively.

1.4D Vapor Pressure and Boiling Point of Liquids

When a liquid is placed in a sealed container, molecules of liquid will evaporate into the space above the liquid and fill it completely. After a time, equilibrium is reached. This vapor will exert a pressure just like a gas and we call this pressure the *vapor pressure* of the liquid. The value of the vapor pressure is independent of the amount of liquid in the container as long as some is present.

If an inert gas such as air is also present in the vapor space, it will have very little effect on the vapor pressure. In general, the effect of total pressure on vapor pressure can be considered as negligible for pressures of a few atmospheres or less.

The vapor pressure of a liquid increases markedly with temperature. For example, from Appendix A.2 for water, the vapor pressure at 50°C is 12.333 kPa (92.51 mm Hg). At 100°C the vapor pressure has increased greatly to 101.325 kPa (760 mm Hg).

The *boiling point* of a liquid is defined as the temperature at which the vapor pressure of a liquid equals the total pressure. Hence, if the atmospheric total pressure is 760 mm Hg, water will boil at 100°C. On top of a high mountain, where the total pressure is considerably less, water will boil at temperatures below 100°C.

A plot of vapor pressure P_A of a liquid versus temperature does not yield a straight line but a curve. However, for moderate temperature ranges, a plot of $\log P_A$ versus $1/T$ is a reasonably straight line, as follows.

$$\log P_A = m\left(\frac{1}{T}\right) + b \tag{1.4-5}$$

where m is the slope, b is a constant for the liquid A, and T is the temperature in K.

1.5 CONSERVATION OF MASS AND MATERIAL BALANCES

1.5A Conservation of Mass

One of the basic laws of physical science is the *law of conservation of mass*. This law, stated simply, says that mass cannot be created or destroyed (excluding, of course, nuclear or atomic reactions). Hence, the total mass (or weight) of all materials entering any process must equal the total mass of all materials leaving plus the mass of any materials accumulating or left in the process.

$$\text{input} = \text{output} + \text{accumulation} \tag{1.5-1}$$

In the majority of cases there will be no accumulation of materials in a process, and then the input will simply equal the output. Stated in other words, "what goes in must come out." We call this type of process a *steady-state process.*

$$\text{input} = \text{output (steady state)} \qquad (1.5\text{-}2)$$

1.5B Simple Material Balances

In this section we do simple material (weight or mass) balances in various processes at steady state with no chemical reaction occurring. We can use units of kg, lb_m, lb mol, g, kg mol, etc., in our balances. The reader is cautioned to be consistent and not to mix several units in a balance. When chemical reactions occur in the balances (as discussed in Section 1.5D), one should use kg mol units, since chemical equations relate moles reacting. In Section 2.6, overall mass balances will be covered in more detail and in Section 3.6, differential mass balances.

To solve a material-balance problem it is advisable to proceed by a series of definite steps, as listed below.

1. *Sketch a simple diagram of the process.* This can be a simple box diagram showing each stream entering by an arrow pointing in and each stream leaving by an arrow pointing out. Include on each arrow the compositions, amounts, temperatures, and so on, of that stream. All pertinent data should be on this diagram.
2. *Write the chemical equations involved (if any).*
3. *Select a basis for calculation.* In most cases the problem is concerned with a specific amount of one of the streams in the process, which is selected as the basis.
4. *Make a material balance.* The arrows into the process will be input items and the arrows going out output items. The balance can be a total material balance in Eq. (1.5-2) or a balance on each component present (if no chemical reaction occurs).

Typical processes that do not undergo chemical reactions are drying, evaporation, dilution of solutions, distillation, extraction, and so on. These can be solved by setting up material balances containing unknowns and solving these equations for the unknowns.

> ### *EXAMPLE 1.5-1. Concentration of Orange Juice*
> In the concentration of orange juice a fresh extracted and strained juice containing 7.08 wt % solids is fed to a vacuum evaporator. In the evaporator, water is removed and the solids content increased to 58 wt % solids. For 1000 kg/h entering, calculate the amounts of the outlet streams of concentrated juice and water.
>
> *Solution:* Following the four steps outlined, we make a process flow diagram (step 1) in Fig. 1.5-1. Note that the letter *W* represents the unknown

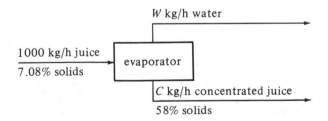

FIGURE 1.5-1. *Process flow diagram for Example 1.5-1.*

amount of water and C the amount of concentrated juice. No chemical reactions are given (step 2). Basis: 1000 kg/h entering juice (step 3).

To make the material balances (step 4), a total material balance will be made using Eq. (1.5-2).

$$1000 = W + C \qquad (1.5\text{-}3)$$

This gives one equation and two unknowns. Hence, a component balance on solids will be made.

$$1000\left(\frac{7.08}{100}\right) = W(0) + C\left(\frac{58}{100}\right) \qquad (1.5\text{-}4)$$

To solve these two equations, we solve Eq. (1.5-4) first for C since W drops out. We get $C = 122.1$ kg/h concentrated juice.

Substituting the value of C into Eq. (1.5-3),

$$1000 = W + 122.1$$

and we obtain $W = 877.9$ kg/h water.

As a check on our calculations, we can write a balance on the water component.

$$1000\left(\frac{100 - 7.08}{100}\right) = 877.9 + 122.1\left(\frac{100 - 58}{100}\right) \qquad (1.5\text{-}5)$$

Solving,

$$929.2 = 877.9 + 51.3 = 929.2$$

In Example 1.5-1 only one unit or separate process was involved. Often, a number of processes in series are involved. Then we have a choice of making a separate balance over each separate process and/or a balance around the complete overall process.

1.5C Material Balances and Recycle

Processes that have a recycle or feedback of part of the product into the entering feed are sometimes encountered. For example, in a sewage treatment plant, part of the activated sludge from a sedimentation tank is recycled back to the aeration tank where the liquid is treated. In some food-drying operations, the humidity of the entering air is controlled by recirculating part of the hot wet air that leaves the dryer. In chemical reactions, the material that did not react in the reactor can be separated from the final product and fed back to the reactor.

EXAMPLE 1.5-2. Crystallization of KNO_3 and Recycle

In a process producing KNO_3 salt, 1000 kg/h of a feed solution containing 20 wt % KNO_3 is fed to an evaporator, which evaporates some water at 422 K to produce a 50 wt % KNO_3 solution. This is then fed to a crystallizer at 311 K, where crystals containing 96 wt % KNO_3 are removed. The saturated solution containing 37.5 wt % KNO_3 is recycled to the evaporator. Calculate the amount of recycle stream R in kg/h and the product stream of crystals P in kg/h.

Solution: Figure 1.5-2 gives the process flow diagram. As a basis we shall use 1000 kg/h of fresh feed. No chemical reactions are occurring. We can

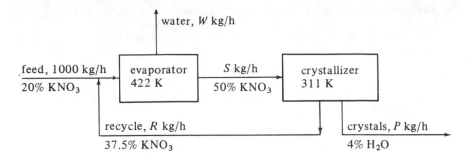

FIGURE 1.5-2. *Process flow diagram for Example 1.5-2.*

make an overall balance on the entire process for KNO_3 and solve for P directly.

$$1000(0.20) = W(0) + P(0.96) \qquad (1.5\text{-}6)$$

$$P = 208.3 \text{ kg crystals/h}$$

To calculate the recycle stream, we can make a balance around the evaporator or the crystallizer. Using a balance on the crystallizer since it now includes only two unknowns, S and R, we get for a total balance,

$$S = R + 208.3 \qquad (1.5\text{-}7)$$

For a KNO_3 balance on the crystallizer,

$$S(0.50) = R(0.375) + 208.3(0.96) \qquad (1.5\text{-}8)$$

Substituting S from Eq. (1.5-7) into Eq. (1.5-8) and solving, $R = 766.6$ kg recycle/h and $S = 974.9$ kg/h.

1.5D Material Balances and Chemical Reaction

In many cases the materials entering a process undergo chemical reactions in the process so that the materials leaving are different from those entering. In these cases it is usually convenient to make a molar and not a weight balance on an individual component such as kg mol H_2 or kg atom H, kg mol CO_3^- ion, kg mol $CaCO_3$, kg atom Na^+, kg mol N_2, and so on. For example, in the combustion of CH_4 with air, balances can be made on kg mol of H_2, C, O_2, or N_2.

EXAMPLE 1.5-3. Combustion of Fuel Gas
A fuel gas containing 3.1 mol % H_2, 27.2% CO, 5.6% CO_2, 0.5% O_2, and 63.6% N_2 is burned with 20% excess air (i.e., the air over and above that necessary for complete combustion to CO_2 and H_2O). The combustion of CO is only 98% complete. For 100 kg mol of fuel gas, calculate the moles of each component in the exit flue gas.

Solution: First, the process flow diagram is drawn (Fig. 1.5-3). On the

Chap. 1   Introduction to Engineering Principles and Units

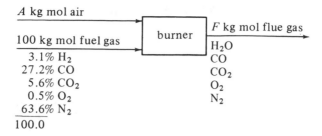

FIGURE 1.5-3.   *Process flow diagram for Example 1.5-3.*

diagram the components in the flue gas are shown. Let A be moles of air and F be moles of flue gas. Next the chemical reactions are given.

$$CO + \tfrac{1}{2}O_2 \rightarrow CO_2 \tag{1.5-9}$$

$$H_2 + \tfrac{1}{2}O_2 \rightarrow H_2O \tag{1.5-10}$$

An accounting of the total moles of O_2 in the fuel gas is as follows:

mol O_2 in fuel gas = $(\tfrac{1}{2})27.2(CO) + 5.6(CO_2) + 0.5(O_2) = 19.7$ mol O_2

For all the H_2 to be completely burned to H_2O, we need from Eq. (1.5-10) $\tfrac{1}{2}$ mol O_2 for 1 mol H_2 or $3.1(\tfrac{1}{2}) = 1.55$ total mol O_2. For completely burning the CO from Eq. (1.5-9), we need $27.2(\tfrac{1}{2}) = 13.6$ mol O_2. Hence, the amount of O_2 we must add is, theoretically, as follows:

mol O_2 theoretically needed = $1.55 + 13.6 - 0.5$ (in fuel gas)

$$= 14.65 \text{ mol } O_2$$

For a 20% excess, we add 1.2(14.65), or 17.58 mol O_2. Since air contains 79 mol % N_2, the amount of N_2 added is (79/21)(17.58), or 66.1 mol N_2.

To calculate the moles in the final flue gas, all the H_2 gives H_2O, or 3.1 mol H_2O. For CO, 2.0% does not react. Hence, 0.02(27.2), or 0.54, mol CO will be unburned.

A total carbon balance is as follows: inlet moles C = 27.2 + 5.6 = 32.8 mol C. In the outlet flue gas, 0.54 mol will be as CO and the remainder of 32.8 − 0.54, or 32.26, mol as CO_2.

For calculating the outlet mol O_2, we make an overall O_2 balance.

$$O_2 \text{ in} = 19.7 \text{ (in fuel gas)} + 17.58 \text{ (in air)} = 37.28 \text{ mol } O_2$$

O_2 out = $(3.1/2)$ (in H_2O) + $(0.54/2)$ (in CO) + 32.26 (in CO_2) + free O_2

Equating inlet O_2 to outlet, the free remaining $O_2 = 3.2$ mol O_2. For the N_2 balance, the outlet = 63.6 (in fuel gas) + 66.1 (in air), or 129.70 mol N_2. The outlet flue gas contains 3.10 mol H_2O, 0.54 mol CO, 32.26 mol CO_2, 3.20 mol O_2, and 129.7 mol N_2.

In chemical reactions with several reactants, the limiting reactant component is defined as that compound which is present in an amount less than the amount necessary for it to react stoichiometrically with the other reactants. Then the percent completion of a reaction is the amount of this limiting reactant actually converted, divided by the amount originally present, times 100.

1.6 ENERGY AND HEAT UNITS

1.6A Joule, Calorie, and Btu

In a manner similar to that used in making material balances on chemical and biological processes, we can also make energy balances on a process. Often a large portion of the energy entering or leaving a system is in the form of heat. Before such energy or heat balances are made, we must understand the various types of energy and heat units.

In the SI system energy is given in joules (J) or kilojoules (kJ). Energy is also expressed in btu (British thermal unit) or cal (calorie). The g calorie (abbreviated cal) is defined as the amount of heat needed to heat 1.0 g water 1.0°C (from 14.5°C to 15.5°C). Also, 1 kcal (kilocalorie) = 1000 cal. The btu is defined as the amount of heat needed to raise 1.0 lb water 1°F. Hence, from Appendix A.1,

$$1 \text{ btu} = 252.16 \text{ cal} = 1.05506 \text{ kJ} \tag{1.6-1}$$

1.6B Heat Capacity

The *heat capacity* of a substance is defined as the amount of heat necessary to increase the temperature by 1 degree. It can be expressed for 1 g, 1 lb, 1 g mol, 1 kg mol, or 1 lb mol of the substance. For example, a heat capacity is expressed in SI units as $J/kg \text{ mol} \cdot K$; in other units as $cal/g \cdot °C$, $cal/g \text{ mol} \cdot °C$, $kcal/kg \text{ mol} \cdot °C$, $btu/lb_m \cdot °F$, or $btu/lb \text{ mol} \cdot °F$.

It can be shown that the actual numerical value of a heat capacity is the same in mass units or in molar units. That is,

$$1.0 \text{ cal/g} \cdot °C = 1.0 \text{ btu/lb}_m \cdot °F \tag{1.6-2}$$

$$1.0 \text{ cal/g mol} \cdot °C = 1.0 \text{ btu/lb mol} \cdot °F \tag{1.6-3}$$

For example, to prove this, suppose that a substance has a heat capacity of 0.8 $btu/lb_m \cdot °F$. The conversion is made using 1.8°F for 1°C or 1 K, 252.16 cal for 1 btu, and 453.6 g for 1 lb_m, as follows:

$$\text{heat capacity} \left(\frac{\text{cal}}{g \cdot °C} \right) = \left(0.8 \frac{\text{btu}}{lb_m \cdot °F} \right) \left(252.16 \frac{\text{cal}}{\text{btu}} \right) \left(\frac{1}{453.6 \text{ g/lb}_m} \right) \left(1.8 \frac{°F}{°C} \right)$$

$$= 0.8 \frac{\text{cal}}{g \cdot °C}$$

The heat capacities of gases (sometimes called *specific heat*) at constant pressure c_p are functions of temperature and for engineering purposes can be assumed to be independent of pressure up to several atmospheres. In most process engineering calculations, one is usually interested in the amount of heat needed to heat a gas from one temperature t_1 to another at t_2. Since the c_p varies with temperature, an integration must be performed or a suitable mean c_{pm} used. These mean values for gases have been obtained for T_1 of 298 K or 25°C (77°F) and various T_2 values, and are tabulated in Table 1.6-1 at 101.325 kPa pressure or less as c_{pm} in $kJ/kg \text{ mol} \cdot K$ at various values of T_2 in K or °C.

EXAMPLE 1.6-1. Heating of N_2 Gas
The gas N_2 at 1 atm pressure absolute is being heated in a heat exchanger. Calculate the amount of heat needed in J to heat 3.0 g mol N_2 in the

TABLE 1.6-1. *Mean Molar Heat Capacities of Gases Between 298 and TK (25 and T°C) at 101.325 kPa or Less (SI Units: $c_p = kJ/kg\ mol \cdot K$)*

T(K)	T(°C)	H_2	N_2	CO	Air	O_2	H_2O	CO_2	CH_4	SO_2
298	25	28.86	29.14	29.16	29.19	29.38	33.59	37.20	35.8	39.9
373	100	28.99	29.19	29.24	29.29	29.66	33.85	38.73	37.6	41.2
473	200	29.13	29.29	29.38	29.40	30.07	34.24	40.62	40.3	42.9
573	300	29.18	29.46	29.60	29.61	30.53	34.39	42.32	43.1	44.5
673	400	29.23	29.68	29.88	29.94	31.01	35.21	43.80	45.9	45.8
773	500	29.29	29.97	30.19	30.25	31.46	35.75	45.12	48.8	47.0
873	600	29.35	30.27	30.52	30.56	31.89	36.33	46.28	51.4	47.9
973	700	29.44	30.56	30.84	30.87	32.26	36.91	47.32	54.0	48.8
1073	800	29.56	30.85	31.16	31.18	32.62	37.53	48.27	56.4	49.6
1173	900	29.63	31.16	31.49	31.48	32.97	38.14	49.15	58.8	50.3
1273	1000	29.84	31.43	31.77	31.79	33.25	38.71	49.91	61.0	50.9
1473	1200	30.18	31.97	32.30	32.32	33.78	39.88	51.29	64.9	51.9
1673	1400	30.51	32.40	32.73	32.76	34.19	40.90	52.34		

Mean Molar Heat Capacities of Gases Between 25 and T°C at 1 atm Pressure or Less (English Units: $c_p = btu/lb\ mol \cdot °F$)

T(°C)	H_2	N_2	CO	Air	O_2	NO	H_2O	CO_2	HCl	Cl_2	CH_4	SO_2	C_2H_4	SO_3	C_2H_6
25	6.894	6.961	6.965	6.972	7.017	7.134	8.024	8.884	6.96	8.12	8.55	9.54	10.45	12.11	12.63
100	6.924	6.972	6.983	6.996	7.083	7.144	8.084	9.251	6.97	8.24	8.98	9.85	11.35	12.84	13.76
200	6.957	6.996	7.017	7.021	7.181	7.224	8.177	9.701	6.98	8.37	9.62	10.25	12.53	13.74	15.27
300	6.970	7.036	7.070	7.073	7.293	7.252	8.215	10.108	7.00	8.48	10.29	10.62	13.65	14.54	16.72
400	6.982	7.089	7.136	7.152	7.406	7.301	8.409	10.462	7.02	8.55	10.97	10.94	14.67	15.22	18.11
500	6.995	7.159	7.210	7.225	7.515	7.389	8.539	10.776	7.06	8.61	11.65	11.22	15.60	15.82	19.39
600	7.011	7.229	7.289	7.299	7.616	7.470	8.678	11.053	7.10	8.66	12.27	11.45	16.45	16.33	20.58
700	7.032	7.298	7.365	7.374	7.706	7.549	8.816	11.303	7.15	8.70	12.90	11.66	17.22	16.77	21.68
800	7.060	7.369	7.443	7.447	7.792	7.630	8.963	11.53	7.21	8.73	13.48	11.84	17.95	17.17	22.72
900	7.076	7.443	7.521	7.520	7.874	7.708	9.109	11.74	7.27	8.77	14.04	12.01	18.63	17.52	23.69
1000	7.128	7.507	7.587	7.593	7.941	7.773	9.246	11.92	7.33	8.80	14.56	12.15	19.23	17.86	24.56
1100	7.169	7.574	7.653	7.660	8.009	7.839	9.389	12.10	7.39	8.82	15.04	12.28	19.81	18.17	25.40
1200	7.209	7.635	7.714	7.719	8.068	7.898	9.524	12.25	7.45	8.94	15.49	12.39	20.33	18.44	26.15
1300	7.252	7.692	7.772	7.778	8.123	7.952	9.66	12.39							
1400	7.288	7.738	7.818	7.824	8.166	7.994	9.77	12.50							
1500	7.326	7.786	7.866	7.873	8.203	8.039	9.89	12.69							
1600	7.386	7.844	7.922	7.929	8.269	8.092	9.95	12.75							
1700	7.421	7.879	7.958	7.965	8.305	8.124	10.13	12.70							
1800	7.467	7.924	8.001	8.010	8.349	8.164	10.24	12.94							
1900	7.505	7.957	8.033	8.043	8.383	8.192	10.34	13.01							
2000	7.548	7.994	8.069	8.081	8.423	8.225	10.43	13.10							
2100	7.588	8.028	8.101	8.115	8.460	8.255	10.52	13.17							
2200	7.624	8.054	8.127	8.144	8.491	8.277	10.61	13.24							

Source: O. A. Hougen, K. W. Watson, and R. A. Ragatz, *Chemical Process Principles*, Part I, 2nd ed. New York: John Wiley & Sons, Inc., 1954. With permission.

following temperature ranges:
   (a) 298–673 K (25–400°C)
   (b) 298–1123 K (25–850°C)
   (c) 673–1123 K (400–850°C)

Solution: For case (a), Table 1.6-1 gives c_{pm} values at 1 atm pressure or less and can be used up to several atm pressures. For N_2 at 673 K, $c_{pm} = 29.68$ kJ/kg mol · K or 29.68 J/g mol · K. This is the mean heat capacity for the range 298–673 K.

$$\text{heat required} = M \text{ g mol} \left(c_{pm} \frac{J}{\text{g mol} \cdot K} \right)(T_2 - T_1)K \qquad \textbf{(1.6-4)}$$

Substituting the known values,

$$\text{heat required} = (3.0)(29.68)(673 - 298) = 33\,390 \text{ J}$$

For case (b), the c_{pm} at 1123 K (obtained by linear interpolation between 1073 and 1173 K) is 31.00 J/g mol · K.

$$\text{heat required} = 3.0(31.00)(1123 - 298) = 76\,725 \text{ J}$$

For case (c), there is no mean heat capacity for the interval 673–1123 K. However, we can use the heat required to heat the gas from 298 to 673 K in case (a) and subtract it from case (b), which includes the heat to go from 298 to 673 K plus 673 to 1123 K.

$$\text{heat required (673–1123 K)} = \text{heat required (298–1123 K)}$$
$$- \text{ heat required (298–673)} \qquad \textbf{(1.6-5)}$$

Substituting the proper values into Eq. (1.6-5),

$$\text{heat required} = 76\,725 - 33\,390 = 43\,335 \text{ J}$$

On heating a gas mixture, the total heat required is determined by first calculating the heat required for each individual component and then adding the results to obtain the total.

The heat capacities of solids and liquids are also functions of temperature and independent of pressure. Data are given in Appendix A.2, Physical Properties of Water; A.3, Physical Properties of Inorganic and Organic Compounds; and A.4, Physical Properties of Foods and Biological Materials. More data are available in (P1).

EXAMPLE 1.6-2. Heating of Milk
Rich cows' milk (4536 kg/h) at 4.4°C is being heated in a heat exchanger to 54.4°C by hot water. How much heat is needed?

Solution: From Appendix A.4 the average heat capacity of rich cows' milk is 3.85 kJ/kg · K. Temperature rise, $\Delta T = (54.4 - 4.4)°C = 50 \text{ K}$.

$$\text{heat required} = (4536 \text{ kg/h})(3.85 \text{ kJ/kg} \cdot K)(1/3600 \text{ h/s})(50 \text{ K}) = 242.5 \text{ kW}$$

The enthalpy, H, of a substance in J/kg represents the sum of the internal energy plus the pressure–volume term. For no reaction and a constant-pressure process with a change in temperature, the heat change as computed from Eq. (1.6-4) is the difference in enthalpy, ΔH, of the substance relative to a given temperature or base point. In other units, $H = \text{btu/lb}_m$ or cal/g.

1.6C Latent Heat and Steam Tables

Whenever a substance undergoes a change of phase, relatively large amounts of heat changes are involved at a constant temperature. For example, ice at 0°C and 1 atm pressure can absorb 6013.4 kJ/kg mol. This enthalpy change is called the *latent heat of fusion*. Data for other compounds are available in various handbooks (P1, W1).

When a liquid phase vaporizes to a vapor phase under its vapor pressure at constant temperature, an amount of heat called the *latent heat of vaporization* must be added. Tabulations of latent heats of vaporization are given in various handbooks. For water at 25°C and a pressure of 23.75 mm Hg, the latent heat is 44 020 kJ/kg mol, and at 25°C and 760 mm Hg, 44 045 kJ/kg mol. Hence, the effect of pressure can be neglected in engineering calculations. However, there is a large effect of temperature on the latent heat of water. Also, the effect of pressure on the heat capacity of liquid water is small and can be neglected.

Since water is a very common chemical, the thermodynamic properties of it have been compiled in steam tables and are given in Appendix A.2 in SI and in English units.

EXAMPLE 1.6-3. Use of Steam Tables

Find the enthalpy change (i.e., how much heat must be added) for each of the following cases using SI and English units.

(a) Heating 1 kg (lb$_m$) water from 21.11°C (70°F) to 60°C (140°F) at 101.325 kPa (1 atm) pressure.

(b) Heating 1 kg (lb$_m$) water from 21.11°C (70°F) to 115.6°C (240°F) and vaporizing at 172.2 kPa (24.97 psia).

(c) Vaporizing 1 kg (lb$_m$) water at 115.6°C (240°F) and 172.2 kPa (24.97 psia).

Solution: For part (a), the effect of pressure on the enthalpy of liquid water is negligible. From Appendix A.2,

$$H \text{ at } 21.11°C = 88.60 \text{ kJ/kg} \quad \text{or} \quad \text{at } 70°F = 38.09 \text{ btu/lb}_m$$

$$H \text{ at } 60°C = 251.13 \text{ kJ/kg} \quad \text{or} \quad \text{at } 140°F = 107.96 \text{ btu/lb}_m$$

$$\text{change in } H = \Delta H = 251.13 - 88.60 = 162.53 \text{ kJ/kg}$$

$$= 107.96 - 38.09 = 69.87 \text{ btu/lb}_m$$

In part (b), the enthalpy at 115.6°C (240°F) and 172.2 kPa (24.97 psia) of the saturated vapor is 2699.9 kJ/kg or 1160.7 btu/lb$_m$.

$$\text{change in } H = \Delta H = 2699.9 - 88.60 = 2611.3 \text{ kJ/kg}$$

$$= 1160.7 - 38.09 = 1122.6 \text{ btu/lb}_m$$

The latent heat of water at 115.6°C (240°F) in part (c) is

$$2699.9 - 484.9 = 2215.0 \text{ kJ/kg}$$

$$1160.7 - 208.44 = 952.26 \text{ btu/lb}_m$$

1.6D Heat of Reaction

When chemical reactions occur, heat effects always accompany these reactions. This area where energy changes occur is often called *thermochemistry*. For example, when HCl is neutralized with NaOH, heat is given off and the reaction is exothermic. Heat is absorbed in an endothermic reaction. This heat of reaction is dependent on the chemical nature of each reacting material and product and on their physical states.

For purposes of organizing data we define a standard heat of reaction ΔH^0 as the change in enthalpy when 1 kg mol reacts under a pressure of 101.325 kPa at a temperature of 298 K (25°C). For example, for the reaction

$$H_2(g) + \tfrac{1}{2}O_2(g) \rightarrow H_2O(l) \qquad (1.6\text{-}6)$$

the ΔH^0 is -285.840×10^3 kJ/kg mol or -68.317 kcal/g mol. The reaction is exothermic and the value is negative since the reaction loses enthalpy. In this case, the H_2 gas reacts with the O_2 gas to give liquid water, all at 298 K (25°C).

Special names are given to ΔH^0 depending upon the type of reaction. When the product is formed from the elements, as in Eq. (1.6-6), we call the ΔH^0, *heat of formation* of the product water, ΔH_f^0. For the combustion of CH_4 to form CO_2 and H_2O, we call it *heat of combustion, ΔH_c^0*. Data are given in Appendix A.3 for various values of ΔH_c^0.

EXAMPLE 1.6-4. Combustion of Carbon
A total of 10.0 g mol of carbon graphite is burned in a calorimeter held at 298 K and 1 atm. The combustion is incomplete and 90% of the C goes to CO_2 and 10% to CO. What is the total enthalpy change in kJ and kcal?

Solution: From Appendix A.3 the ΔH_c^0 for carbon going to CO_2 is -393.513×10^3 kJ/kg mol or -94.0518 kcal/g mol, and for carbon going to CO is -110.523×10^3 kJ/kg mol or -26.4157 kcal/g mol. Since 9 mol CO_2 and 1 mol CO are formed,

$$\text{total } \Delta H = 9(-393.513) + 1(-110.523) = -3652 \text{ kJ}$$

$$= 9(-94.0518) + 1(-26.4157) = -872.9 \text{ kcal}$$

If a table of heats of formation, ΔH_f^0, of compounds is available, the standard heat of the reaction, ΔH^0, can be calculated by

$$\Delta H^\circ = \sum \Delta H_{f\,(\text{products})}^0 - \sum \Delta H_{f\,(\text{reactants})}^0 \qquad (1.6\text{-}7)$$

In Appencix A.3, a short table of some values of ΔH_f is given. Other data are also available (H1, P1, S1).

EXAMPLE 1.6-5. Reaction of Methane
For the following reaction of 1 kg mol of CH_4 at 101.32 kPa and 298 K,

$$CH_4(g) + H_2O(l) \rightarrow CO(g) + 3H_2(g)$$

calculate the standard heat of reaction ΔH^0 at 298 K in kJ.

Solution: From Appendix A.3, the following standard heats of formation are obtained at 298 K:

	ΔH_f^0 (kJ/kg mol)
$CH_4(g)$	-74.848×10^3
$H_2O(l)$	-285.840×10^3
$CO(g)$	-110.523×10^3
$H_2(g)$	0

Note that the ΔH_f^0 of all elements is, by definition, zero. Substituting into Eq. (1.6-7),

$$\Delta H^0 = [-110.523 \times 10^3 - 3(0)] - (-74.848 \times 10^3 - 285.840 \times 10^3)$$

$$= +250.165 \times 10^3 \text{ kJ/kg mol} \qquad (\text{endothermic})$$

1.7 CONSERVATION OF ENERGY AND HEAT BALANCES

1.7A Conservation of Energy

In making material balances we used the law of conservation of mass, which states that the mass entering is equal to the mass leaving plus the mass left in the process. In a similar manner, we can state the *law of conservation of energy*, which says that all energy entering a process is equal to that leaving plus that left in the process. In this section elementary heat balances will be made. More elaborate energy balances will be considered in Sections 2.7 and 5.6.

Energy can appear in many forms. Some of the common forms are enthalpy, electrical energy, chemical energy (in terms of ΔH reaction), kinetic energy, potential energy, work, and heat inflow.

In many cases in process engineering, which often takes place at constant pressure, electrical energy, kinetic energy, potential energy, and work either are not present or can be neglected. Then only the enthalpy of the materials (at constant pressure), the standard chemical reaction energy (ΔH^0) at 25°C, and the heat added or removed must be taken into account in the energy balance. This is generally called a *heat balance*.

1.7B Heat Balances

In making a heat balance at steady state we use methods similar to those used in making a material balance. The energy or heat coming into a process in the inlet materials plus any net energy added to the process is equal to the energy leaving in the materials. Expressed mathematically,

$$\sum H_R + (-\Delta H^0_{298}) + q = \sum H_p \qquad (1.7\text{-}1)$$

where $\sum H_R$ is the sum of enthalpies of all materials entering the reaction process relative to the reference state for the standard heat of reaction at 298 K and 101.32 kPa. If the inlet temperature is above 298 K, this sum will be positive. ΔH^0_{298} = standard heat of the reaction at 298 K and 101.32 kPa. The reaction contributes heat to the process, so the negative of ΔH^0_{298} is taken to be positive input heat for an exothermic reaction. q = net energy or heat added to the system. If heat leaves the system, this item will be negative. $\sum H_p$ = sum of enthalpies of all leaving materials referred to the standard reference state at 298 K (25°C).

Note that if the materials coming into a process are below 298 K, $\sum H_R$ will be negative. Care must be taken not to confuse the signs of the items in Eq. (1.7-1). If no chemical reaction occurs, then simple heating, cooling, or phase change is occurring. Use of Eq. (1.7-1) will be illustrated by several examples. For convenience it is common practice to call the terms on the left-hand side of Eq. (1.7-1) input items, and those on the right, output items.

> ### EXAMPLE 1.7-1. Heating of Fermentation Medium
> A liquid fermentation medium at 30°C is pumped at a rate of 2000 kg/h through a heater, where it is heated to 70°C under pressure. The waste heat water used to heat this medium enters at 95°C and leaves at 85°C. The average heat capacity of the fermentation medium is 4.06 kJ/kg · K, and that for water is 4.21 kJ/kg · K (Appendix A.2). The fermentation stream and the wastewater stream are separated by a metal surface through which heat is transferred and do not physically mix with each other. Make a complete heat balance on the system. Calculate the water flow and the amount of heat added to the fermentation medium assuming no heat losses. The process flow is given in Fig. 1.7-1.

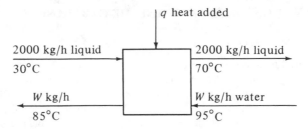

FIGURE 1.7-1. *Process flow diagram for Example 1.7-1.*

Solution: It is convenient to use the standard reference state of 298 K (25°C) as the datum to calculate the various enthalpies. From Eq. (1.7-1) the input items are as follows.

Input items. $\sum H_R$ of the enthalpies of the two streams relative to 298 K (25°C) (note that $\Delta t = 30 - 25°C = 5°C = 5$ K):

$$H(\text{liquid}) = (2000 \text{ kg/h})(4.06 \text{ kJ/kg} \cdot \text{K})(5 \text{ K})$$

$$= 4.060 \times 10^4 \text{ kJ/h}$$

$$H(\text{water}) = W(4.21)(95 - 25) = 2.947 \times 10^2 \ W \text{ kJ/h} \qquad (W = \text{kg/h})$$

$$(-\Delta H_{298}^0) = 0 \qquad \text{(since there is no chemical reaction)}$$

$$q = 0 \qquad \text{(there are no heat losses or additions)}$$

Output items. $\sum H_P$ of the two streams relative to 298 K (25°C):

$$H(\text{liquid}) = 2000(4.06)(70 - 25) = 3.65 \times 10^5 \text{ kJ/h}$$

$$H(\text{water}) = W(4.21)(85 - 25) = 2.526 \times 10^2 \ W \text{ kJ/h}$$

Equating input to output in Eq. (1.7-1) and solving for W,

$$4.060 \times 10^4 + 2.947 \times 10^2 \ W = 3.654 \times 10^5 + 2.526 \times 10^2 \ W$$

$$W = 7720 \text{ kg/h water flow}$$

The amount of heat added to the fermentation medium is simply the difference of the outlet and inlet liquid enthalpies.

$$H(\text{outlet liquid}) - H(\text{inlet liquid}) = 3.654 \times 10^5 - 4.060 \times 10^4$$

$$= 3.248 \times 10^5 \text{ kJ/h (90.25 kW)}$$

Note in this example that since the heat capacities were assumed constant, a simpler balance could have been written as follows:

$$\text{heat gained by liquid} = \text{heat lost by water}$$

$$2000(4.06)(70 - 30) = W(4.21)(95 - 85)$$

Then, solving, $W = 7720$ kg/h. This simple balance works well when c_p is constant. However, when the c_p varies with temperature and the material is a gas, c_{pm} values are only available between 298 K (25°C) and t K and the simple method cannot be used without obtaining new c_{pm} values over different temperature ranges.

EXAMPLE 1.7-2   *Heat and Material Balance in Combustion*

The waste gas from a process of 1000 g mol/h of CO at 473 K is burned at 1 atm pressure in a furnace using air at 373 K. The combustion is complete and 90% excess air is used. The flue gas leaves the furnace at 1273 K. Calculate the heat removed in the furnace.

Solution: First the process flow diagram is drawn in Fig. 1.7-2 and then a material balance is made.

$$CO(g) + \tfrac{1}{2}O_2(g) \rightarrow CO_2(g)$$

$$\Delta H^0_{298} = -282.989 \times 10^3 \text{ kJ/kg mol}$$

$$\text{(from Appendix A.3)}$$

$$\text{mol CO} = 1000 \text{ g mol/h} = \text{moles } CO_2$$

$$= 1.00 \text{ kg mol/h}$$

$$\text{mol } O_2 \text{ theoretically required} = \tfrac{1}{2}(1.00) = 0.500 \text{ kg mol/h}$$

$$\text{mol } O_2 \text{ actually added} = 0.500(1.9) = 0.950 \text{ kg mol/h}$$

$$\text{mol } N_2 \text{ added} = 0.950\frac{0.79}{0.21} = 3.570 \text{ kg mol/h}$$

$$\text{air added} = 0.950 + 3.570 = 4.520 \text{ kg mol/h} = A$$

$$O_2 \text{ in outlet flue gas} = \text{added} - \text{used}$$

$$= 0.950 - 0.500 = 0.450 \text{ kg mol/h}$$

$$CO_2 \text{ in outlet flue gas} = 1.00 \text{ kg mol/h}$$

$$N_2 \text{ in outlet flue gas} = 3.570 \text{ kg mol/h}$$

For the heat balance relative to the standard state at 298 K, we follow Eq. (1.7-1).

Input items

$$H \text{ (CO)} = 1.00(c_{pm})(473 - 298) = 1.00(29.38)(473 - 298) = 5142 \text{ kJ/h}$$

(The c_{pm} of CO of 29.38 kJ/kg mol·K between 298 and 473 K is obtained from Table 1.6-1.)

$$H \text{ (air)} = 4.520(c_{pm})(373 - 298) = 4.520(29.29)(373 - 298) = 9929 \text{ kJ/h}$$

$$q = \text{heat added, kJ/h}$$

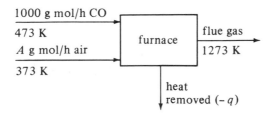

FIGURE 1.7-2.   *Process flow diagram for Example 1.7-2.*

(This will give a negative value here, indicating that heat was removed.)

$$-\Delta H^0_{298} = -(-282.989 \times 10^3 \text{ kJ/kg mol})(1.00 \text{ kg mol/h}) = 282\,990 \text{ kJ/h}$$

Output items

$$H(CO_2) = 1.00(c_{pm})(1273 - 298) = 1.00(49.91)(1273 - 298) = 48\,660 \text{ kJ/h}$$

$$H(O_2) = 0.450(c_{pm})(1273 - 298) = 0.450(33.25)(1273 - 298) = 14\,590 \text{ kJ/h}$$

$$H(N_2) = 3.570(c_{pm})(1273 - 298) = 3.570(31.43)(1273 - 298) = 109\,400 \text{ kJ/h}$$

Equating input to output and solving for q,

$$5142 + 9929 + q + 282\,990 = 48\,660 + 14\,590 + 109\,400$$
$$q = -125\,411 \text{ kJ/h}$$

Hence, heat is removed: $-34\,837$ W.

Often when chemical reactions occur in the process and the heat capacities vary with temperature, the solution in a heat balance can be trial and error if the final temperature is the unknown.

EXAMPLE 1.7-3. *Oxidation of Lactose*
In many biochemical processes, lactose is used as a nutrient, which is oxidized as follows:

$$C_{12}H_{22}O_{11}(s) + 12O_2(g) \rightarrow 12CO_2(g) + 11H_2O(l)$$

The heat of combustion ΔH^0_c in Appendix A.3 at 25°C is -5648.8×10^3 J/g mol. Calculate the heat of complete oxidation (combustion) at 37°C, which is the temperature of many biochemical reactions. The c_{pm} of solid lactose is 1.20 J/g·K, and the molecular weight is 342.3 g mass/g mol.

Solution: This can be treated as an ordinary heat-balance problem. First, the process flow diagram is drawn in Fig. 1.7-3. Next, the datum temperature of 25°C is selected and the input and output enthalpies calculated. The temperature difference $\Delta t = (37 - 25)$°C $= (37 - 25)$ K.

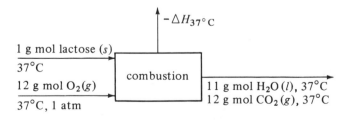

FIGURE 1.7-3. *Process flow diagram for Example 1.7-3.*

Input items

$$H(\text{lactose}) = (342.3 \text{ g})\left(c_{pm} \frac{J}{g \cdot K}\right)(37 - 25) \text{ K} = 342.3(1.20)(37 - 25)$$

$$= 4929 \text{ J}$$

$$H(O_2 \text{ gas}) = (12 \text{ g mol})\left(c_{pm} \frac{J}{g \text{ mol} \cdot K}\right)(37 - 25) \text{ K}$$

$$= 12(29.38)(37 - 25) = 4230 \text{ J}$$

(The c_{pm} of O_2 was obtained from Table 1.6-1.)

$$-\Delta H_{25}^0 = -(-5648.8 \times 10^3)$$

Output items

$$H(H_2O \text{ liquid}) = 11(18.02 \text{ g})\left(c_{pm} \frac{J}{g \cdot K}\right)(37 - 25) \text{ K}$$

$$= 11(18.02)(4.18)(37 - 25) = 9943 \text{ J}$$

(The c_{pm} of liquid water was obtained from Appendix A.2.)

$$H(CO_2 \text{ gas}) = (12 \text{ g mol})\left(c_{pm} \frac{J}{g \text{ mol} \cdot K}\right)(37 - 25) \text{ K}$$

$$= 12(37.45)(37 - 25) = 5393 \text{ J}$$

(The c_{pm} of CO_2 is obtained from Table 1.6-1.)

$\Delta H_{37°C}$:
Setting input = output and solving,

$$4929 + 4230 + 5648.8 \times 10^3 = 9943 + 5393 - \Delta H_{37°C}$$

$$\Delta H_{37°C} = -5642.6 \times 10^3 \text{ J/g mol} = \Delta H_{310 \text{ K}}$$

1.8 GRAPHICAL, NUMERICAL, AND MATHEMATICAL METHODS

1.8A Graphical Integration

Often the mathematical function $f(x)$ to be integrated is too complex and we are not able to integrate it analytically. Or in some cases the function is one that has been obtained from experimental data, and no mathematical equation is available to represent the data so that they can be integrated analytically. In these cases, we can use graphical integration.

Integration between the limits $x = a$ to $x = b$ can be represented graphically as shown in Fig. 1.8-1. Here a function $y = f(x)$ has been plotted versus x. The area under the curve $y = f(x)$ between the limits $x = a$ to $x = b$ is equal to the integral. This area is then equal to the sum of the areas of the rectangles as follows.

$$\int_{x = a}^{x = b} f(x) \, dx = A_1 + A_2 + A_3 + A_4 + A_5 \qquad (1.8\text{-}1)$$

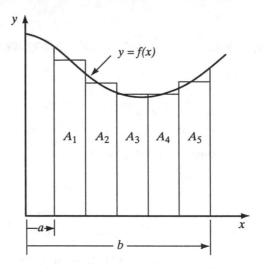

FIGURE 1.8-1. *Graphical integration of $\int_{x=a}^{x=b} f(x)\, dx$.*

1.8B Numerical Integration and Simpson's Rule

Often it is desired or necessary to perform a numerical integration by computing the value of a definite integral from a set of numerical values of the integrand $f(x)$. This, of course, can be done graphically, but if data are available in large quantities, numerical methods suitable for the digital computer are desired.

The integral to be evaluated is as follows:

$$\int_{x=a}^{x=b} f(x)\, dx \tag{1.8-2}$$

where the interval is $b - a$. The most generally used numerical method is the parabolic rule often called *Simpson's rule*. This method divides the total interval $b - a$ into an even number of subintervals m, where

$$m = \frac{b - a}{h} \tag{1.8-3}$$

The value of h, a constant, is the spacing in x used. Then, approximating $f(x)$ by a parabola on each subinterval, Simpson's rule is

$$\int_{x=a}^{x=b} f(x)\, dx = \frac{h}{3}\, [f_0 + 4(f_1 + f_3 + f_5 + \cdots + f_{m-1})$$

$$+ 2(f_2 + f_4 + f_6 + \cdots + f_{m-2}) + f_m] \tag{1.8-4}$$

where f_0 is the value of $f(x)$ at $x = a$, f_1 the value of $f(x)$ at $x = x_1, \ldots, f_m$ the value of $f(x)$ at $x = b$. The reader should note that m must be an even number and the increments evenly spaced. This method is well suited for digital computation.

PROBLEMS

1.2-1. *Temperature of a Chemical Process.* The temperature of a chemical reaction was found to be 353.2 K. What is the temperature in °F, °C, and °R?

Ans. 176°F, 80°C, 636°R

1.2-2. *Temperature for Smokehouse Processing of Meat.* In smokehouse processing of sausage meat, a final temperature of 155°F inside the sausage is often used. Calculate this temperature in °C, K, and °R.

1.3-1. *Molecular Weight of Air.* For purposes of most engineering calculations, air is assumed to be composed of 21 mol % oxygen and 79 mol % nitrogen. Calculate the average molecular weight.

Ans. 28.9 g mass/g mol, lb mass/lb mol, or kg mass/kg mol

1.3-2. *Oxidation of CO and Mole Units.* The gas CO is being oxidized by O_2 to form CO_2. How many kg of CO_2 will be formed from 56 kg of CO? Also, calculate the kg of O_2 theoretically needed for this reaction. (*Hint:* First write the balanced chemical equation to obtain the mol O_2 needed for 1.0 kg mol CO. Then calculate the kg mol of CO in 56 kg CO.)

Ans. 88.0 kg CO_2, 32.0 kg O_2

1.3-3. *Composition of a Gas Mixture.* A gaseous mixture contains 20 g of N_2, 83 g of O_2, and 45 g of CO_2. Calculate the composition in mole fraction and the average molecular weight of the mixture.

Ans. Average mol wt = 34.2 g mass/g mol, 34.2 kg mass/kg mol

1.3-4. *Composition of a Protein Solution.* A liquid solution contains 1.15 wt % of a protein, 0.27 wt % KCl, and the remainder water. The average molecular weight of the protein by gel permeation is 525 000 g mass/g mol. Calculate the mole fraction of each component in solution.

1.3-5. *Concentration of NaCl Solution.* An aqueous solution of NaCl has a concentration of 24.0 wt % NaCl with a density of 1.178 g/cm³ at 25°C. Calculate the following.
(a) Mole fraction of NaCl and water.
(b) Concentration of NaCl as g mol/liter, lb_m/ft^3, lb_m/gal, and kg/m³.

1.4-1. *Conversion of Pressure Measurements in Freeze Drying.* In the experimental measurement of freeze drying of beef, an absolute pressure of 2.4 mm Hg was held in the chamber. Convert this pressure to atm, in. of water at 4°C, μm of Hg, and Pa. (*Hint:* See Appendix A.1 for conversion factors.)

Ans. 3.16×10^{-3} atm, 1.285 in. H_2O, 2400 μm Hg, 320 Pa

1.4-2. *Compression and Cooling of Nitrogen Gas.* A volume of 65.0 ft³ of N_2 gas at 90°F and 29.0 psig is compressed to 75 psig and cooled to 65°F. Calculate the final volume in ft³ and the final density in lb_m/ft^3. [*Hint:* Be sure to convert all pressures to psia first and then to atm. Substitute original conditions into Eq. (1.4-1) to obtain n, lb mol.]

1.4-3. *Gas Composition and Volume.* A gas mixture of 0.13 g mol NH_3, 1.27 g mol N_2, and 0.025 g mol H_2O vapor is contained at a total pressure of 830 mm Hg and 323 K. Calculate the following.
(a) Mole fraction of each component.
(b) Partial pressure of each component in mm Hg.
(c) Total volume of mixture in m³ and ft³.

1.4-4. *Evaporation of a Heat-Sensitive Organic Liquid.* An organic liquid is being evaporated from a liquid solution containing a few percent nonvolatile dissolved solids. Since it is heat-sensitive and may discolor at high temperatures, it will be evaporated under vacuum. If the lowest absolute pressure that can be obtained in the apparatus is 12.0 mm Hg, what will be the temperature of evaporation in K? It will be assumed that the small amount of solids does not affect the vapor pressure, which is given as follows:

$$\log P_A = -2250\left(\frac{1}{T}\right) + 9.05$$

where P_A is in mm Hg and T in K.

Ans. $T = 282.3$ K or $9.1°C$

1.5-1. Evaporation of Cane Sugar Solutions. An evaporator is used to concentrate cane sugar solutions. A feed of 10 000 kg/d of a solution containing 38 wt % sugar is evaporated, producing a 74 wt % solution. Calculate the weight of solution produced and amount of water removed.

Ans. 5135 kg/d of 74 wt % solution, 4865 kg/d water

1.5-2. Processing of Fish Meal. Fish are processed into fish meal and used as a supplementary protein food. In the processing the oil is first extracted to produce wet fish cake containing 80 wt % water and 20 wt % bone-dry cake. This wet cake feed is dried in rotary drum dryers to give a "dry" fish cake product containing 40 wt % water. Finally, the product is finely ground and packed. Calculate the kg/h of wet cake feed needed to produce 1000 kg/h of "dry" fish cake product.

Ans. 3000 kg/h wet cake feed

1.5-3. Drying of Lumber. A batch of 100 kg of wet lumber containing 11 wt % moisture is dried to a water content of 6.38 kg water/1.0 kg bone-dry lumber. What is the weight of "dried" lumber and the amount of water removed?

1.5-4. Processing of Paper Pulp. A wet paper pulp contains 68 wt % water. After the pulp was dried, it was found that 55% of the original water in the wet pulp was removed. Calculate the composition of the "dried" pulp and its weight for a feed of 1000 kg/min of wet pulp.

1.5-5. Production of Jam from Crushed Fruit in Two Stages. In a process producing jam (C1), crushed fruit containing 14 wt% soluble solids is mixed in a mixer with sugar (1.22 kg sugar/1.00 kg crushed fruit) and pectin (0.0025 kg pectin/1.00 kg crushed fruit). The resultant mixture is then evaporated in a kettle to produce a jam containing 67 wt% soluble solids. For a feed of 1000 kg crushed fruit, calculate the kg mixture from the mixer, kg water evaporated, and kg jam produced.

Ans. 2222.5 kg mixture, 189 kg water, 2033.5 kg jam

1.5-6. Drying of Cassava (Tapioca) Root. Tapioca flour is used in many countries for bread and similar products. The flour is made by drying coarse granules of the cassava root containing 66 wt % moisture to 5% moisture and then grinding to produce a flour. How many kg of granules must be dried and how much water removed to produce 5000 kg/h of flour?

1.5-7. Processing of Soybeans in Three Stages. A feed of 10 000 kg of soybeans is processed in a sequence of three stages or steps (E1). The feed contains 35 wt % protein, 27.1 wt % carbohydrate, 9.4 wt % fiber and ash, 10.5 wt % moisture, and 18.0 wt % oil. In the first stage the beans are crushed and pressed to remove oil, giving an expressed oil stream and a stream of pressed beans containing 6% oil. Assume no loss of other constituents with the oil stream. In the second step the pressed beans are extracted with hexane to produce an extracted meal stream containing 0.5 wt % oil and a hexane-oil stream. Assume no hexane in the extracted meal. Finally, in the last step the extracted meal is dried to give a dried meal of 8 wt % moisture. Calculate:
(a) Kg of pressed beans from the first stage.
(b) Kg of extracted meal from stage 2.
(c) Kg of final dried meal and the wt % protein in the dried meal.

Ans. (a) 8723 kg, (b) 8241 kg, (c) 7816 kg, 44.8 wt % protein

1.5-8. Recycle in a Dryer. A solid material containing 15.0 wt % moisture is dried so that it contains 7.0 wt % water by blowing fresh warm air mixed with recycled air over the solid in the dryer. The inlet fresh air has a humidity of 0.01 kg water/kg dry air, the air from the drier that is recycled has a humidity of 0.1 kg water/kg dry air, and the mixed air to the dryer, 0.03 kg water/kg dry air. For a feed of 100 kg solid/h fed to the dryer, calculate the kg dry air/h in the fresh air, the kg dry air/h in the recycle air, and the kg/h of "dried" product.

> **Ans.** 95.6 kg/h dry air in fresh air, 27.3 kg/h dry air in recycle air, and 91.4 kg/h "dried" product

1.5-9. Crystallization and Recycle. It is desired to produce 1000 kg/h of $Na_3PO_4 \cdot 12H_2O$ crystals from a feed solution containing 5.6 wt % Na_3PO_4 and traces of impurity. The original solution is first evaporated in an evaporator to a 35 wt% Na_3PO_4 solution and then cooled to 293 K in a crystallizer, where the hydrated crystals and a mother liquor solution are removed. One out of every 10 kg of mother liquor is discarded to waste to get rid of the impurities, and the remaining mother liquor is recycled to the evaporator. The solubility of Na_3PO_4 at 293 K is 9.91 wt %. Calculate the kg/h of feed solution and kg/h of water evaporated.

> **Ans.** 7771 kg/h feed, 6739 kg/h water

1.5-10. Evaporation and Bypass in Orange Juice Concentration. In a process for concentrating 1000 kg of freshly extracted orange juice (C1) containing 12.5 wt % solids, the juice is strained, yielding 800 kg of strained juice and 200 kg of pulpy juice. The strained juice is concentrated in a vacuum evaporator to give an evaporated juice of 58% solids. The 200 kg of pulpy juice is bypassed around the evaporator and mixed with the evaporated juice in a mixer to improve the flavor. This final concentrated juice contains 42 wt % solids. Calculate the concentration of solids in the strained juice, the kg of final concentrated juice, and the concentration of solids in the pulpy juice bypassed. (*Hint:* First, make a total balance and then a solids balance on the overall process. Next, make a balance on the evaporator. Finally, make a balance on the mixer.)

> **Ans.** 34.2 wt % solids in pulpy juice

1.5-11. Manufacture of Acetylene. For the making of 6000 ft^3 of acetylene (CHCH) gas at 70°F and 750 mm Hg, solid calcium carbide (CaC_2) which contains 97 wt % CaC_2 and 3 wt % solid inerts is used along with water. The reaction is

$$CaC_2 + 2H_2O \rightarrow CHCH + Ca(OH)_2 \downarrow$$

The final lime slurry contains water, solid inerts, and $Ca(OH)_2$ lime. In this slurry the total wt % solids of inerts plus $Ca(OH)_2$ is 20%. How many lb of water must be added and how many lb of final lime slurry is produced? [*Hint:* Use a basis of 6000 ft^3 and convert to lb mol. This gives 15.30 lb mol C_2H_2, 15.30 lb mol $Ca(OH)_2$, and 15.30 lb mol CaC_2 added. Convert lb mol CaC_2 feed to lb and calculate lb inerts added. The total lb solids in the slurry is then the sum of the $Ca(OH)_2$ plus inerts. In calculating the water added, remember that some is consumed in the reaction.]

> **Ans.** 5200 lb water added (2359 kg), 5815 lb lime slurry (2638 kg)

1.5-12. Combustion of Solid Fuel. A fuel analyzes 74.0 wt % C and 12.0% ash (inert). Air is added to burn the fuel, producing a flue gas of 12.4% CO_2, 1.2% CO, 5.7% O_2, and 80.7% N_2. Calculate the kg of fuel used for 100 kg mol of outlet flue gas and the kg mol of air used. (*Hint:* First calculate the mol O_2 added in the air, using the fact that the N_2 in the flue gas equals the N_2 added in the air. Then make a carbon balance to obtain the total moles of C added.)

1.5-13. Burning of Coke. A furnace burns a coke containing 81.0 wt % C, 0.8% H, and the rest inert ash. The furnace uses 60% excess air (air over and above that needed to burn all C to CO_2 and H to H_2O). Calculate the moles of all components in the flue gas if only 95% of the carbon goes to CO_2 and 5% to CO.

1.5-14. Production of Formaldehyde. Formaldehyde (CH_2O) is made by the catalytic oxidation of pure methanol vapor and air in a reactor. The moles from this reactor are 63.1 N_2, 13.4 O_2, 5.9 H_2O, 4.1 CH_2O, 12.3 CH_3OH, and 1.2 $HCOOH$. The reaction is

$$CH_3OH + \tfrac{1}{2}O_2 \rightarrow CH_2O + H_2O$$

A side reaction occurring is

$$CH_2O + \tfrac{1}{2}O_2 \rightarrow HCOOH$$

Calculate the mol methanol feed, mol air feed, and percent conversion of methanol to formaldehyde.

> **Ans.** 17.6 mol CH_3OH, 79.8 mol air, 23.3% conversion

1.6-1. Heating of CO_2 Gas. A total of 250 g of CO_2 gas at 373 K is heated to 623 K at 101.32 kPa total pressure. Calculate the amount of heat needed in cal, btu, and kJ.

> **Ans.** 15 050 cal, 59.7 btu, 62.98 kJ

1.6-2. Heating a Gas Mixture. A mixture of 25 lb mol N_2 and 75 lb mol CH_4 is being heated from 400°F to 800°F at 1 atm pressure. Calculate the total amount of heat needed in btu.

1.6-3. Final Temperature in Heating Applesauce. A mixture of 454 kg of applesauce at 10°C is heated in a heat exchanger by adding 121 300 kJ. Calculate the outlet temperature of the applesauce. (*Hint:* In Appendix A.4 a heat capacity for applesauce is given at 32.8°C. Assume that this is constant and use this as the average c_{pm}.)

> **Ans.** 76.5°C

1.6-4. Use of Steam Tables. Using the steam tables, determine the enthalpy change for 1 lb water for each of the following cases.
  (a) Heating liquid water from 40°F to 240°F at 30 psia. (Note that the effect of total pressure on the enthalpy of liquid water can be neglected.)
  (b) Heating liquid water from 40°F to 240°F and vaporizing at 240°F and 24.97 psia.
  (c) Cooling and condensing a saturated vapor at 212°F and 1 atm abs to a liquid at 60°F.
  (d) Condensing a saturated vapor at 212°F and 1 atm abs.

> **Ans.** (a) 200.42 btu/lb_m, (b) 1152.7 btu/lb_m, (c) -1122.4 btu/lb_m,
> (d) -970.3 btu/lb_m, -2256.9 kJ/kg

1.6-5. Heating and Vaporization Using Steam Tables. A flow rate of 1000 kg/h of water at 21.1°C is heated to 110°C when the total pressure is 244.2 kPa in the first stage of a process. In the second stage at the same pressure the water is heated further, until it is all vaporized at its boiling point. Calculate the total enthalpy change in the first stage and in both stages.

1.6-6. Combustion of CH_4 and H_2. For 100 g mol of a gas mixture of 75 mol % CH_4 and 25% H_2, calculate the total heat of combustion of the mixture at 298 K and 101.32 kPa, assuming that combustion is complete.

1.6-7. Heat of Reaction from Heats of Formation. For the reaction

$$4NH_3(g) + 5O_2(g) \rightarrow 4NO(g) + 6H_2O(g)$$

calculate the heat of reaction, ΔH, at 298 K and 101.32 kPa for 4 g mol of NH_3 reacting.

> **Ans.** ΔH, heat of reaction $= -904.7$ kJ

1.7-1. Heat Balance and Cooling of Milk. In the processing of rich cows' milk, 4540 kg/h of milk is cooled from 60°C to 4.44°C by a refrigerant. Calculate the heat removed from the milk.

> **Ans.** Heat removed $= 269.6$ kW

1.7-2. Heating of Oil by Air. A flow of 2200 lb$_m$/h of hydrocarbon oil at 100°F enters a heat exchanger, where it is heated to 150°F by hot air. The hot air enters at 300°F and is to leave at 200°F. Calculate the total lb mol air/h needed. The mean heat capacity of the oil is 0.45 btu/lb$_m$·°F.

Ans.   70.1 lb mol air/h, 31.8 kg mol/h

1.7-3. Combustion of Methane in a Furnace. A gas stream of 10 000 kg mol/h of CH$_4$ at 101.32 kPa and 373 K is burned in a furnace using air at 313 K. The combustion is complete and 50% excess air is used. The flue gas leaves the furnace at 673 K. Calculate the heat removed in the furnace. (*Hint:* Use a datum of 298 K and liquid water at 298 K. The input items will be the following: the enthalpy of CH$_4$ at 373 K referred to 298 K; the enthalpy of the air at 313 K referred to 298 K; $-\Delta H_c^0$, the heat of combustion of CH$_4$ at 298 K which is referred to liquid water; and q, the heat added. The output items will include: the enthalpies of CO$_2$, O$_2$, N$_2$, and H$_2$O gases at 673 K referred to 298 K; and the latent heat of H$_2$O vapor at 298 K and 101.32 kPa from Appendix A.2. It is necessary to include this latent heat since the basis of the calculation and of the ΔH_c^0 is liquid water.)

1.7-4. Preheating Air by Steam for Use in a Dryer. An air stream at 32.2°C is to be used in a dryer and is first preheated in a steam heater, where it is heated to 65.5°C. The air flow is 1000 kg mol/h. The steam enters the heater saturated at 148.9°C, is condensed and cooled, and leaves as a liquid at 137.8°C. Calculate the amount of steam used in kg/h.

Ans.   450 kg steam/h

1.7-5. Cooling of Cans of Potato Soup After Thermal Processing. A total of 1500 cans of potato soup undergo thermal processing in a retort at 240°F. The cans are then cooled to 100°F in the retort before being removed from the retort by cooling water, which enters at 75°F and leaves at 85°F. Calculate the lb of cooling water needed. Each can of soup contains 1.0 lb of liquid soup and the empty metal can weighs 0.16 lb. The mean heat capacity of the soup is 0.94 btu/lb$_m$·°F and that of the metal can is 0.12 btu/lb$_m$·°F. A metal rack or basket which is used to hold the cans in the retort weighs 350 lb and has a heat capacity of 0.12 btu/lb$_m$·°F. Assume that the metal rack is cooled from 240°F to 85°F, the temperature of the outlet water. The amount of heat removed from the retort walls in cooling from 240 to 100°F is 10 000 btu. Radiation loss from the retort during cooling is estimated as 5000 btu.

Ans.   21 320 lb water, 9670 kg

1.8-1. Graphical Integration and Numerical Integration Using Simpson's Method. The following experimental data of $y = f(x)$ were obtained.

x	$f(x)$	x	$f(x)$
0	100	0.4	53
0.1	75	0.5	60
0.2	60.5	0.6	72.5
0.3	53.5		

It is desired to determine the integral

$$A = \int_{x=0}^{x=0.6} f(x)\,dx$$

(a) Do this by a graphical integration.
(b) Repeat using Simpson's numerical method.

Ans.   (a) $A = 38.55$; (b) $A = 38.45$

1.8-2. *Graphical and Numerical Integration to Obtain Wastewater Flow.* The rate of flow of wastewater in an open channel has been measured and the following data obtained:

Time (min)	Flow (m³/min)	Time (min)	Flow (m³/min)
0	655	70	800
10	705	80	725
20	780	90	670
30	830	100	640
40	870	110	620
50	890	120	610
60	870		

(a) Determine the total flow in m³ for the first 60 min and also the total for 120 min by graphical integration.

(b) Determine the flow for 120 min using Simpson's numerical method.

Ans. (a) 48 460 m³ for 60 min, 90 390 m³ for 120 m

REFERENCES

(C1) CHARM, S. E. *The Fundamentals of Food Engineering*, 2nd ed. Westport, Conn.: Avi Publishing Co., Inc., 1971.

(E1) EARLE, R. L. *Unit Operations in Food Processing.* Oxford: Pergamon Press, Inc., 1966.

(H1) HOUGEN, O. A., WATSON, K. M., and RAGATZ, R. A. *Chemical Process Principles*, Part I, 2nd ed. New York: John Wiley & Sons, Inc., 1954.

(O1) OKOS, M. R. M.S. thesis. Ohio State University, Columbus, Ohio, 1972.

(P1) PERRY, R. H., and GREEN, D. *Perry's Chemical Engineers' Handbook*, 6th ed. New York: McGraw-Hill Book Company, 1984.

(S1) SOBER, H. A. *Handbook of Biochemistry, Selected Data for Molecular Biology*, 2nd ed. Boca Raton, Fla.: Chemical Rubber Co., Inc., 1970.

(W1) WEAST, R. C., and SELBY, S. M. *Handbook of Chemistry and Physics*, 48th ed. Boca Raton, Fla.: Chemical Rubber Co., Inc., 1967–1968.

CHAPTER 2

Principles of
Momentum Transfer
and Overall Balances

2.1 INTRODUCTION

The flow and behavior of fluids is important in many of the unit operations in process engineering. A fluid may be defined as a substance that does not permanently resist distortion and, hence, will change its shape. In this text gases, liquids, and vapors are considered to have the characteristics of fluids and to obey many of the same laws.

In the process industries, many of the materials are in fluid form and must be stored, handled, pumped, and processed, so it is necessary that we become familiar with the principles that govern the flow of fluids and also with the equipment used. Typical fluids encountered include water, air, CO_2, oil, slurries, and thick syrups.

If a fluid is inappreciably affected by changes in pressure, it is said to be *incompressible*. Most liquids are incompressible. Gases are considered to be *compressible* fluids. However, if gases are subjected to small percentage changes in pressure and temperature, their density changes will be small and they can be considered to be incompressible.

Like all physical matter, a fluid is composed of an extremely large number of molecules per unit volume. A theory such as the kinetic theory of gases or statistical mechanics treats the motions of molecules in terms of statistical groups and not in terms of individual molecules. In engineering we are mainly concerned with the bulk or macroscopic behavior of a fluid rather than with the individual molecular or microscopic behavior.

In momentum transfer we treat the fluid as a continuous distribution of matter or as a "continuum". This treatment as a continuum is valid when the smallest volume of fluid contains a large enough number of molecules so that a statistical average is meaningful and the macroscopic properties of the fluid such as density, pressure, and so on, vary smoothly or continuously from point to point.

The study of *momentum transfer*, or *fluid mechanics* as it is often called, can be divided into two branches: *fluid statics*, or fluids at rest, and *fluid dynamics*, or fluids in motion. In Section 2.2 we treat fluid statics; in the remaining sections of Chapter 2 and in Chapter 3, fluid dynamics. Since in fluid dynamics momentum is being transferred, the term "momentum transfer" or "transport" is usually used. In Section 2.3 momentum transfer is related to heat and mass transfer.

2.2 FLUID STATICS

2.2A Force, Units, and Dimensions

In a static fluid an important property is the pressure in the fluid. Pressure is familiar as a surface force exerted by a fluid against the walls of its container. Also, pressure exists at any point in a volume of a fluid.

In order to understand *pressure*, which is defined as force exerted per unit area, we must first discuss a basic law of Newton's. This equation for calculation of the force exerted by a mass under the influence of gravity is

$$F = mg \qquad \text{(SI units)}$$

$$F = \frac{mg}{g_c} \qquad \text{(English units)} \tag{2.2-1}$$

where in SI units F is the force exerted in newtons $N(kg \cdot m/s^2)$, m the mass in kg, and g the standard acceleration of gravity, $9.80665 \, m/s^2$.

In English units, F is in lb_f, m in lb_m, g is $32.1740 \, ft/s^2$, and g_c (a gravitational conversion factor) is $32.174 \, lb_m \cdot ft/lb_f \cdot s^2$. The use of the conversion factor g_c means that g/g_c has a value of $1.0 \, lb_f/lb_m$ and that $1 \, lb_m$ conveniently gives a force equal to $1 \, lb_f$. Often when units of pressure are given, the word "force" is omitted, such as in $lb/in.^2$ (psi) instead of $lb_f/in.^2$. When the mass m is given in g mass, F is g force, $g = 980.665$ cm/s^2, and $g_c = 980.665$ g mass $\cdot$ cm/g force $\cdot s^2$. However, the units g force are seldom used.

Another system of units sometimes used in Eq. (2.2-1) is that where the g_c is omitted and the force ($F = mg$) is given as $lb_m \cdot ft/s^2$, which is called *poundals*. Then $1 \, lb_m$ acted on by gravity will give a force of 32.174 poundals $(lb_m \cdot ft/s^2)$. Or if 1 g mass is used, the force ($F = mg$) is expressed in terms of dynes $(g \cdot cm/s^2)$. This is the centimeter–gram–second (cgs) systems of units.

Conversion factors for different units of force and of force per unit area (pressure) are given in Appendix A.1. Note that always in the SI system, and usually in the cgs system, the term g_c is not used.

EXAMPLE 2.2-1. *Units and Dimensions of Force*

Calculate the force exerted by 3 lb mass in terms of the following.
   (a) Lb force (English units).
   (b) Dynes (cgs units).
   (c) Newtons (SI units).

Solution: For part (a), using Eq. (2.2-1),

$$F \text{ (force)} = m \frac{g}{g_c} = (3 \, lb_m)\left(32.174 \, \frac{ft}{s^2}\right)\left(\frac{1}{32.174 \, \dfrac{lb_m \cdot ft}{lb_f \cdot s^2}}\right) = 3 \, lb \text{ force } (lb_f)$$

For part (b),

$$F = mg = (3 \, lb_m)\left(453.59 \, \frac{g}{lb_m}\right)\left(980.665 \, \frac{cm}{s^2}\right)$$

$$= 1.332 \times 10^6 \, \frac{g \cdot cm}{s^2} = 1.332 \times 10^6 \, dyn$$

As an alternative method for part (b), from Appendix A.1,

$$1 \text{ dyn} = 2.2481 \times 10^{-6} \text{ lb}_f$$

$$F = (3 \text{ lb}_f)\left(\frac{1}{2.2481 \times 10^{-6} \text{lb}_f/\text{dyn}}\right) = 1.332 \times 10^6 \text{ dyn}$$

To calculate newtons in part (c),

$$F = mg = \left(3 \text{ lb}_m \times \frac{1 \text{ kg}}{2.2046 \text{ lb}_m}\right)\left(9.80665 \frac{\text{m}}{\text{s}^2}\right)$$

$$= 13.32 \frac{\text{kg} \cdot \text{m}}{\text{s}^2} = 13.32 \text{ N}$$

As an alternative method, using values from Appendix A.1,

$$1 \frac{\text{g} \cdot \text{cm}}{\text{s}^2}(\text{dyn}) = 10^{-5} \frac{\text{kg} \cdot \text{m}}{\text{s}^2}(\text{newton})$$

$$F = (1.332 \times 10^6 \text{ dyn})\left(10^{-5} \frac{\text{newton}}{\text{dyn}}\right) = 13.32 \text{ N}$$

2.2B Pressure in a Fluid

Since Eq. (2.2-1) gives the force exerted by a mass under the influence of gravity, the force exerted by a mass of fluid on a supporting area or force/unit area (pressure) also follows from this equation. In Fig. 2.2-1 a stationary column of fluid of height h_2 m and constant cross-sectional area A m^2, where $A = A_0 = A_1 = A_2$, is shown. The pressure above the fluid is P_0 N/m^2; that is, this could be the pressure of the atmosphere above the fluid. The fluid at any point, say h_1, must support all the fluid above it. It can be shown that the forces at any given point in a nonmoving or static fluid must be the same in all directions. Also, for a fluid at rest, the force/unit area or pressure is the same at all points with the same elevation. For example, at h_1 m from the top, the pressure is the same at all points shown on the cross-sectional area A_1.

The use of Eq. (2.2-1) will be shown in calculating the pressure at different vertical points in Fig. 2.2-1. The total mass of fluid for h_2 m height and density ρ kg/m^3 is

$$\text{total kg fluid} = (h_2 \text{ m})(A \text{ m}^2)\left(\rho \frac{\text{kg}}{\text{m}^3}\right) = h_2 A \rho \text{ kg} \qquad (2.2\text{-}2)$$

FIGURE 2.2-1.   *Pressure in a static fluid.*

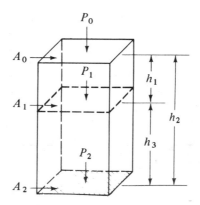

Substituting into Eq. (2.2-2), the total force F of the fluid on area A_1 due to the fluid only is

$$F = (h_2\, A\rho \text{ kg})(g \text{ m/s}^2) = h_2\, A\rho g\, \frac{\text{kg}\cdot\text{m}}{\text{s}^2}\ (\text{N}) \qquad (2.2\text{-}3)$$

The pressure P is defined as force/unit area:

$$P = \frac{F}{A} = (h_2\, A\rho g)\frac{1}{A} = h_2\,\rho g \text{ N/m}^2 \quad \text{or} \quad \text{Pa} \qquad (2.2\text{-}4)$$

This is the pressure on A_2 due to the mass of the fluid above it. However, to get the total pressure P_2 on A_2, the pressure P_0 on the top of the fluid must be added.

$$P_2 = h_2\,\rho g + P_0 \text{ N/m}^2 \quad \text{or} \quad \text{Pa} \qquad (2.2\text{-}5)$$

Equation (2.2-5) is the fundamental equation to calculate the pressure in a fluid at any depth. To calculate P_1,

$$P_1 = h_1\rho g + P_0 \qquad (2.2\text{-}6)$$

The pressure difference between points 2 and 1 is

$$P_2 - P_1 = (h_2\,\rho g + P_0) - (h_1\rho g + P_0) = (h_2 - h_1)\rho g \qquad \text{(SI units)}$$

$$P_2 - P_1 = (h_2 - h_1)\rho\,\frac{g}{g_c} \qquad\qquad\qquad\qquad\qquad \text{(English units)} \qquad (2.2\text{-}7)$$

Since it is the vertical height of a fluid that determines the pressure in a fluid, the shape of the vessel does not affect the pressure. For example, in Fig. 2.2-2, the pressure P_1 at the bottom of all three vessels is the same and equal to $h_1\rho g + P_0$.

EXAMPLE 2.2-2. *Pressure in Storage Tank*
A large storage tank contains oil having a density of 917 kg/m³ (0.917 g/cm³). The tank is 3.66 m (12.0 ft) tall and is vented (open) to the atmosphere of 1 atm abs at the top. The tank is filled with oil to a depth of 3.05 m (10 ft) and also contains 0.61 m (2.0 ft) of water in the bottom of the tank. Calculate the pressure in Pa and psia 3.05 m from the top of the tank and at the bottom. Also calculate the gage pressure at the tank bottom.

Solution: First a sketch is made of the tank, as shown in Fig. 2.2-3. The pressure $P_0 = 1$ atm abs $= 14.696$ psia (from Appendix A.1). Also,

$$P_0 = 1.01325 \times 10^5 \text{ Pa}$$

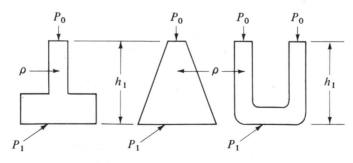

FIGURE 2.2-2. *Pressure in vessels of various shapes.*

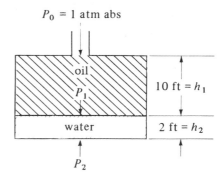

FIGURE 2.2-3. *Storage tank in Example 2.2-2.*

From Eq. (2.2-6) using English and then SI units,

$$P_1 = h_1 \rho_{\text{oil}} \frac{g}{g_c} + P_0 = (10 \text{ ft})\left(0.917 \times 62.43 \frac{\text{lb}_m}{\text{ft}^3}\right)\left(1.0 \frac{\text{lb}_f}{\text{lb}_m}\right)\left(\frac{1}{144 \text{ in.}^2/\text{ft}^2}\right)$$

$$+ 14.696 \text{ lb}_f/\text{in.}^2 = 18.68 \text{ psia}$$

$$P_1 = h_1 \rho_{\text{oil}} g + P_0 = (3.05 \text{ m})\left(917 \frac{\text{kg}}{\text{m}^3}\right)\left(9.8066 \frac{\text{m}}{\text{s}^2}\right) + 1.0132 \times 10^5$$

$$= 1.287 \times 10^5 \text{ Pa}$$

To calculate P_2 at the bottom of the tank, $\rho_{\text{water}} = 1.00 \text{ g/cm}^3$ and

$$P_2 = h_2 \rho_{\text{water}} \frac{g}{g_c} + P_1 = (2.0)(1.00 \times 62.43)(1.0)(\tfrac{1}{144}) + 18.68$$

$$= 19.55 \text{ psia}$$

$$= h_2 \rho_{\text{water}} g + P_1 = (0.61)(1000)(9.8066) + 1.287 \times 10^5$$

$$= 1.347 \times 10^5 \text{ Pa}$$

The gage pressure at the bottom is equal to the absolute pressure P_2 minus 1 atm pressure.

$$P_{\text{gage}} = 19.55 \text{ psia} - 14.696 \text{ psia} = 4.85 \text{ psig}$$

2.2C Head of a Fluid

Pressures are given in many different sets of units, such as psia, dyn/cm², and newtons/m², as given in Appendix A.1. However, a common method of expressing pressures is in terms of head in m or feet of a particular fluid. This height or head in m or feet of the given fluid will exert the same pressure as the pressures it represents. Using Eq. (2.2-4), which relates pressure P and height h of a fluid and solving for h, which is the head in m,

$$h(\text{head}) = \frac{P}{\rho g} \text{ m} \quad \text{(SI)}$$

$$\textbf{(2.2-8)}$$

$$h = \frac{P g_c}{\rho g} \text{ ft} \quad \text{(English)}$$

EXAMPLE 2.2-3. Conversion of Pressure to Head of a Fluid
Given the pressure of 1 standard atm as 101.325 kN/m² (Appendix A.1), do
as follows.

    (a) Convert this pressure to head in m water at 4°C.

    (b) Convert this pressure to head in m Hg at 0°C.

Solution: For part (a), the density of water at 4°C in Appendix A.2 is 1.000
g/cm³. From A.1, a density of 1.000 g/cm³ equals 1000 kg/m³. Substituting
these values into Eq. (2.2-8),

$$h(\text{head}) = \frac{P}{\rho g} = \frac{101.325 \times 10^3}{(1000)(9.80665)}$$

$$= 10.33 \text{ m of water at } 4°C$$

For part (b), the density of Hg in Appendix A.1 is 13.5955 g/cm³. For
equal pressures P from different fluids, Eq. (2.2-8) can be rewritten as

$$P = \rho_{\text{Hg}}\, h_{\text{Hg}}\, g = \rho_{\text{H}_2\text{O}}\, h_{\text{H}_2\text{O}}\, g \qquad\qquad \textbf{(2.2-9)}$$

Solving for h_{Hg} in Eq. (2.2-9) and substituting known values,

$$h_{\text{Hg}}(\text{head}) = \frac{\rho_{\text{H}_2\text{O}}}{\rho_{\text{Hg}}}\, h_{\text{H}_2\text{O}} = \left(\frac{1.000}{13.5955}\right)(10.33) = 0.760 \text{ m Hg}$$

2.2D   Devices to Measure Pressure and Pressure Differences

In chemical and other industrial processing plants it is often important to measure and
control the pressure in a vessel or process and/or the liquid level in a vessel. Also, since
many fluids are flowing in a pipe or conduit, it is necessary to measure the rate at which
the fluid is flowing. Many of these flow meters depend upon devices to measure a
pressure or pressure difference. Some common devices are considered in the following
paragraphs.

1. Simple U-tube manometer.   The U-tube manometer is shown in Fig. 2.2-4a. The

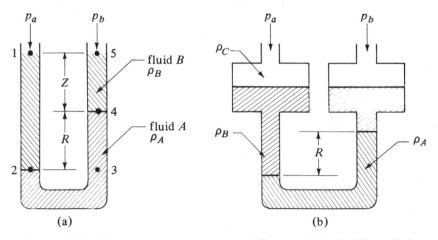

FIGURE 2.2-4.   *Manometers to measure pressure differences: (a) U tube; (b) two-fluid*
U tube.

pressure p_a N/m^2 is exerted on one arm of the U tube and p_b on the other arm. Both pressures p_a and p_b could be pressure taps from a fluid meter, or p_a could be a pressure tap and p_b the atmospheric pressure. The top of the manometer is filled with liquid B, having a density of ρ_B kg/m^3, and the bottom with a more dense fluid A, having a density of ρ_A kg/m^3. Liquid A is immiscible with B. To derive the relationship between p_a and p_b, p_a is the pressure at point 1 and p_b at point 5. The pressure at point 2 is

$$p_2 = p_a + (Z + R)\rho_B g \text{ N/m}^2 \qquad (2.2\text{-}10)$$

where R is the reading of the manometer in m. The pressure at point 3 must be equal to that at 2 by the principles of hydrostatics.

$$p_3 = p_2 \qquad (2.2\text{-}11)$$

The pressure at point 3 also equals the following:

$$p_3 = p_b + Z\rho_B g + R\rho_A g \qquad (2.2\text{-}12)$$

Equating Eq. (2.2-10) to (2.2-12) and solving,

$$p_a + (Z + R)\rho_B g = p_b + Z\rho_B g + R\rho_A g \qquad (2.2\text{-}13)$$

$$p_a - p_b = R(\rho_A - \rho_B)g \qquad \text{(SI)}$$

$$\qquad (2.2\text{-}14)$$

$$p_a - p_b = R(\rho_A - \rho_B)\frac{g}{g_c} \qquad \text{(English)}$$

The reader should note that the distance Z does not enter into the final result nor do the tube dimensions, provided that p_a and p_b are measured in the same horizontal plane.

EXAMPLE 2.2-4. Pressure Difference in a Manometer

A manometer, as shown in Fig. 2.2-4a, is being used to measure the head or pressure drop across a flow meter. The heavier fluid is mercury, with a density of 13 6 g/cm^3, and the top fluid is water, with a density of 1.00g/cm^3. The reading on the manometer is $R = 32.7$ cm. Calculate the pressure difference in N/m^2 using SI units.

Solution: Converting R to m,

$$R = \frac{32.7}{100} = 0.327 \text{ m}$$

Also converting ρ_A and ρ_B to kg/m^3 and substituting into Eq. (2.2-14),

$$p_a - p_b = R(\rho_A - \rho_B)g = (0.327 \text{ m})[(13.6 - 1.0)(1000 \text{ kg/m}^3)](9.8066 \text{ m/s}^2)$$

$$= 4.040 \times 10^4 \text{ N/m}^2 \text{ (5.85 psia)}$$

2. *Two-fluid U tube.* In Fig. 2.2-4b a two-fluid U tube is shown, which is a sensitive device to measure small heads or pressure differences. Let A m^2 be the cross-sectional area of each of the large reservoirs and a m^2 be the cross-sectional area of each of the tubes forming the U. Proceeding and making a pressure balance as for the U tube,

$$p_a - p_b = (R - R_0)\left(\rho_A - \rho_B + \frac{a}{A}\rho_B - \frac{a}{A}\rho_C\right)g \qquad (2.2\text{-}15)$$

where R_0 is the reading when $p_a = p_b$, R is the actual reading, ρ_A is the density of the heavier fluid, and ρ_B is the density of the lighter fluid. Usually, a/A is made sufficiently small to be negligible, and also R_0 is often adjusted to zero; then

$$p_a - p_b = R(\rho_A - \rho_B)g \qquad \text{(SI)}$$
$$p_a - p_b = R(\rho_A - \rho_B)\frac{g}{g_c} \qquad \text{(English)} \qquad \textbf{(2.2-16)}$$

If ρ_A and ρ_B are close to each other, the reading R is magnified.

EXAMPLE 2.2-5. *Pressure Measurement in a Vessel*
The U-tube manometer in Fig. 2.2-5a is used to measure the pressure p_A in a vessel containing a liquid with a density ρ_A. Derive the equation relating the pressure p_A and the reading on the manometer as shown.

Solution: At point 2 the pressure is

$$p_2 = p_{\text{atm}} + h_2 \rho_B g \ \text{N/m}^2 \qquad \textbf{(2.2-17)}$$

At point 1 the pressure is

$$p_1 = p_A + h_1 \rho_A g \qquad \textbf{(2.2-18)}$$

Equating $p_1 = p_2$ by the principles of hydrostatics and rearranging,

$$p_A = p_{\text{atm}} + h_2 \rho_B g - h_1 \rho_A g \qquad \textbf{(2.2-19)}$$

Another example of a U-tube manometer is shown in Fig. 2.2-5b. This device is used in this case to measure the pressure difference between two vessels.

3. Bourdon pressure gage. Although manometers are used to measure pressures, the most common pressure-measuring device is the mechanical Bourdon-tube pressure gage. A coiled hollow tube in the gage tends to straighten out when subjected to internal pressure, and the degree of straightening depends on the pressure difference between the inside and outside pressures. The tube is connected to a pointer on a calibrated dial.

4. Gravity separator for two immiscible liquids. In Fig. 2.2-6 a continuous gravity separator (decanter) is shown for the separation of two immiscible liquids A (heavy liquid) and B (light liquid). The feed mixture of the two liquids enters at one end of the separator vessel and the liquids flow slowly to the other end and separate into two distinct layers. Each liquid flows through a separate overflow line as shown. Assuming

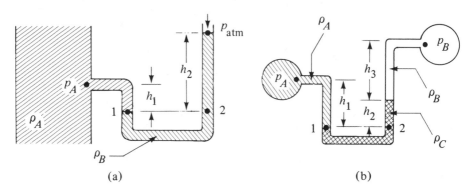

(a)   (b)

FIGURE 2.2-5. *Measurements of pressure in vessels: (a) measurement of pressure in a vessel, (b) measurement of differential pressure.*

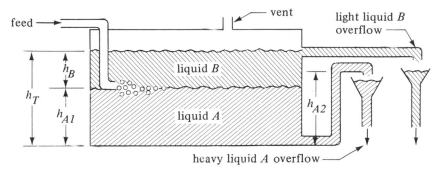

FIGURE 2.2-6. *Continuous atmospheric gravity separator for immiscible liquids.*

the frictional resistance to the flow of the liquids is essentially negligible, the principles of fluid statics can be used to analyze the performance.

In Fig. 2.2-6, the depth of the layer of heavy liquid A is h_{A1} m and that of B is h_B. The total depth $h_T = h_{A1} + h_B$ and is fixed by position of the overflow line for B. The heavy liquid A discharges through an overflow leg h_{A2} m above the vessel bottom. The vessel and the overflow lines are vented to the atmosphere. A hydrostatic balance gives

$$h_B \rho_B g + h_{A1} \rho_A g = h_{A2} \rho_A g \qquad (2.2\text{-}20)$$

Substituting $h_B = h_T - h_{A1}$ into Eq. (2.2-20) and solving for h_{A1},

$$h_{A1} = \frac{h_{A2} - h_T \rho_B/\rho_A}{1 - \rho_B/\rho_A} \qquad (2.2\text{-}21)$$

This shows that the position of the interface or height h_{A1} depends on the ratio of the densities of the two liquids and on the elevations h_{A2} and h_T of the two overflow lines. Usually, the height h_{A2} is movable so that the interface level can be adjusted.

2.3 GENERAL MOLECULAR TRANSPORT EQUATION FOR MOMENTUM, HEAT, AND MASS TRANSFER

2.3A General Molecular Transport Equation and General Property Balance

1. Introduction to transport processes. In molecular transport processes in general we are concerned with the transfer or movement of a given property or entity by molecular movement through a system or medium which can be a fluid (gas or liquid) or a solid. This property that is being transferred can be mass, thermal energy (heat), or momentum. Each molecule of a system has a given quantity of the property mass, thermal energy, or momentum associated with it. When a difference of concentration of the property exists for any of these properties from one region to an adjacent region, a net transport of this property occurs. In dilute fluids such as gases where the molecules are relatively far apart, the rate of transport of the property should be relatively fast since few molecules are present to block the transport or interact. In dense fluids such as liquids the molecules are close together and transport or diffusion procedes more slowly. The molecules in solids are even more close-packed than in liquids and molecular migration is even more restricted.

2. *General molecular transport equation.* All three of the molecular transport processes of momentum, heat or thermal energy, and mass are characterized in the elementary sense by the same general type of transport equation. First we start by noting the following:

$$\text{rate of a transfer process} = \frac{\text{driving force}}{\text{resistance}} \tag{2.3-1}$$

This states what is quite obvious—that we need a driving force to overcome a resistance in order to transport a property. This is similar to Ohm's law in electricity, where the rate of flow of electricity is proportional to the voltage drop (driving force) and inversely proportional to the resistance.

We can formalize Eq. (2.3-1) by writing an equation as follows for molecular transport or diffusion of a property:

$$\psi_z = -\delta \frac{d\Gamma}{dz} \tag{2.3-2}$$

where ψ_z is defined as the flux of the property as amount of property being transferred per unit time per unit cross-sectional area perpendicular to the z direction of flow in amount of property/s · m², δ is a proportionality constant called diffusivity in m²/s, Γ is concentration of the property in amount of property/m³, and z is the distance in the direction of flow in m.

If the process is at steady state, then the flux ψ_z is constant. Rearranging Eq. (2.3-2) and integrating,

$$\psi_z \int_{z_1}^{z_2} dz = -\delta \int_{\Gamma_1}^{\Gamma_2} d\Gamma \tag{2.3-3}$$

$$\psi_z = \frac{\delta(\Gamma_1 - \Gamma_2)}{z_2 - z_1} \tag{2.3-4}$$

A plot of the concentration Γ versus z is shown in Fig. 2.3-1a and is a straight line. Since the flux is in the direction 1 to 2 of decreasing concentration, the slope $d\Gamma/dz$ is negative and the negative sign in Eq. (2.3-2) gives a positive flux in the direction 1 to 2. In Section 2.3B the specialized equations for momentum, heat, and mass transfer will be shown to be the same as Eq. (2.3-4) for the general property transfer.

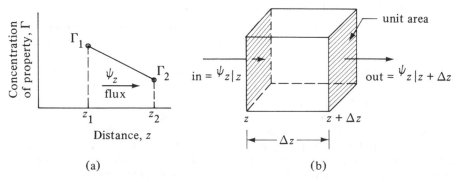

FIGURE 2.3-1. *Molecular transport of a property: (a) plot of concentration versus distance for steady state, (b) unsteady-state general property balance.*

EXAMPLE 2.3-1. Molecular Transport of a Property at Steady State
A property is being transported by diffusion through a fluid at steady state.
At a given point 1 the concentration is 1.37×10^{-2} amount of property/m^3
and 0.72×10^{-2} at point 2 at a distance $z_2 = 0.40$ m. The diffusivity
$\delta = 0.013$ m^2/s and the cross-sectional area is constant.
  (a) Calculate the flux.
  (b) Derive the equation for Γ as a function of distance.
  (c) Calculate Γ at the midpoint of the path.

Solution: For part (a) substituting into Eq. (2.3-4),

$$\psi_z = \frac{\delta(\Gamma_1 - \Gamma_2)}{z_2 - z_1} = \frac{(0.013)(1.37 \times 10^{-2} - 0.72 \times 10^{-2})}{0.40 - 0}$$

$$= 2.113 \times 10^{-4} \text{ amount of property/s} \cdot \text{m}^2$$

For part (b), integrating Eq. (2.3-2) between Γ_1 and Γ and z_1 and z and
rearranging,

$$\psi_z \int_{z_1}^{z} dz = -\delta \int_{\Gamma_1}^{\Gamma} d\Gamma \qquad (2.3\text{-}5)$$

$$\Gamma = \Gamma_1 + \frac{\psi_z}{\delta}(z_1 - z) \qquad (2.3\text{-}6)$$

For part (c), using the midpoint $z = 0.20$ m and substituting into Eq. (2.3-6),

$$\Gamma = 1.37 \times 10^{-2} + \frac{2.113 \times 10^{-4}}{0.013}(0 - 0.2)$$

$$= 1.045 \times 10^{-2} \text{ amount of property/m}^3$$

3. General property balance for unsteady state.   In calculating the rates of transport in a
system using the molecular transport equation (2.3-2), it is necessary to account for the
amount of this property being transported in the entire system. This is done by writing a
general property balance or conservation equation for the property (momentum, thermal
energy, or mass) at unsteady state. We start by writing an equation for the z direction
only, which accounts for all the property entering by molecular transport, leaving, being
generated, and accumulating in a system shown in Fig. 2.3-1b, which is an element of
volume $\Delta z(1)$ m^3 fixed in space.

$$\begin{pmatrix} \text{rate of} \\ \text{property in} \end{pmatrix} + \begin{pmatrix} \text{rate of generation} \\ \text{of property} \end{pmatrix}$$

$$= \begin{pmatrix} \text{rate of} \\ \text{property out} \end{pmatrix} + \begin{pmatrix} \text{rate of accum-} \\ \text{ulation of property} \end{pmatrix} \qquad (2.3\text{-}7)$$

The rate of input is $(\psi_{z|z}) \cdot 1$ amount of property/s and the rate of output is
$(\psi_{z|z+\Delta z}) \cdot 1$, where the cross-sectional area is 1.0 m^2. The rate of generation of the
property is $R(\Delta z \cdot 1)$, where R is rate of generation of property/s $\cdot$ m^3. The accumulation
term is

$$\text{rate of accumulation of property} = \frac{\partial \Gamma}{\partial t}(\Delta z \cdot 1) \qquad (2.3\text{-}8)$$

Substituting the various terms into Eq. (2.3-7),

$$(\psi_{z|z}) \cdot 1 + R(\Delta z \cdot 1) = (\psi_{z|z + \Delta z}) \cdot 1 + \frac{\partial \Gamma}{\partial t}(\Delta z \cdot 1) \qquad (2.3\text{-}9)$$

Dividing by Δz and letting Δz go to zero,

$$\frac{\partial \Gamma}{\partial t} + \frac{\partial \psi_z}{\partial z} = R \qquad (2.3\text{-}10)$$

Substituting Eq. (2.3-2) for ψ_z into (2.3-10) and assuming that δ is constant,

$$\frac{\partial \Gamma}{\partial t} - \delta \frac{\partial^2 \Gamma}{\partial z^2} = R \qquad (2.3\text{-}11)$$

For the case where no generation is present,

$$\frac{\partial \Gamma}{\partial t} = \delta \frac{\partial^2 \Gamma}{\partial z^2} \qquad (2.3\text{-}12)$$

This final equation relates the concentration of the property Γ to position z and time t.

Equations (2.3-11) and (2.3-12) are general equations for the conservation of momentum, thermal energy, or mass and will be used in many sections of this text. The equations consider here only molecular transport occurring and the equations do not consider other transport mechanisms such as convection, and so on, which will be considered when the specific conservation equations are derived in later sections of this text for momentum, energy, or mass.

2.3B Introduction to Molecular Transport

The kinetic theory of gases gives us a good physical interpretation of the motion of individual molecules in fluids. Because of their kinetic energy the molecules are in rapid random movement, often colliding with each other. Molecular transport or molecular diffusion of a property such as momentum, heat, or mass occurs in a fluid because of these random movements of individual molecules. Each individual molecule containing the property being transferred moves randomly in all directions and there are fluxes in all directions. Hence, if there is a concentration gradient of the property, there will be a net flux of the property from high to low concentration. This occurs because equal numbers of molecules diffuse in each direction between the high-concentration and low-concentration regions.

1. Momentum transport and Newton's law. When a fluid is flowing in the x direction parallel to a solid surface, a velocity gradient exists where the velocity v_x in the x direction decreases as we approach the surface in the z direction. The fluid has x-directed momentum and its concentration is $v_x \rho$ momentum/m^3, where the momentum has units of kg $\cdot$ m/s. Hence, the units of $v_x \rho$ are (kg $\cdot$ m/s)/m^3. By random diffusion of molecules there is an exchange of molecules in the z direction, an equal number moving in each direction ($+z$ and $-z$ directions) between the faster-moving layer of molecules and the slower adjacent layer. Hence, the x-directed momentum has been transferred in the z direction from the faster- to the slower-moving layer. The equation for this transport of momentum is similar to Eq. (2.3-2) and is Newton's law of viscosity written as follows for constant density ρ:

$$\tau_{zx} = -v \frac{d(v_x \rho)}{dz} \qquad (2.3\text{-}13)$$

where τ_{zx} is flux of x-directed momentum in the z direction, $(kg \cdot m/s)/s \cdot m^2$; v is μ/ρ, the momentum diffusivity in m^2/s; z is the direction of transport or diffusion in m; ρ is the density in kg/m^3; and μ is the viscosity in $kg/m \cdot s$.

2. Heat transport and Fourier's law. Fourier's law for molecular transport of heat or heat conduction in a fluid or solid can be written as follows for constant density ρ and heat capacity c_p.

$$\frac{q_z}{A} = -\alpha \frac{d(\rho c_p T)}{dz} \qquad (2.3\text{-}14)$$

where q_z/A is the heat flux in $J/s \cdot m^2$, α is the thermal diffusivity in m^2/s, and $\rho c_p T$ is the concentration of heat or thermal energy in J/m^3. When there is a temperature gradient in a fluid, equal numbers of molecules diffuse in each direction between the hot and the colder region. In this way energy is transferred in the z direction.

3. Mass transport and Fick's law. Fick's law for molecular transport of mass in a fluid or solid for constant total concentration in the fluid is

$$J^*_{Az} = -D_{AB} \frac{dc_A}{dz} \qquad (2.3\text{-}15)$$

where J^*_{Az} is the flux of A in kg mol $A/s \cdot m^2$, D_{AB} is the molecular diffusivity of the molecule A in B in m^2/s, and c_A is the concentration of A in kg mol A/m^3. In a manner similar to momentum and heat transport, when there is a concentration gradient in a fluid, equal numbers of molecules diffuse in each direction between the high- and the low-concentration region and a net flux of mass occurs.

Hence, Eqs. (2.3-13), (2.3-14), and (2.3-15) for momentum, heat, and mass transfer are all similar to each other and to the general molecular transport equation (2.3-2). All equations have a flux on the left-hand side of each equation, a diffusivity in m^2/s, and the derivative of the concentration with respect to distance. All three of the molecular transport equations are mathematically identical. Thus, we state we have an analogy or similarity among them. It should be emphasized, however, that even though there is a mathematical analogy, the actual physical mechanisms occurring can be totally different. For example, in mass transfer two components are often being transported by relative motion through one another. In heat transport in a solid, the molecules are relatively stationary and the transport is done mainly by the electrons. Transport of momentum can occur by several types of mechanisms. More detailed considerations of each of the transport processes of momentum, energy, and mass are presented in this and succeeding chapters.

2.4 VISCOSITY OF FLUIDS

2.4A Newton's Law and Viscosity

When a fluid is flowing through a closed channel such as a pipe or between two flat plates, either of two types of flow may occur, depending on the velocity of this fluid. At low velocities the fluid tends to flow without lateral mixing, and adjacent layers slide past one another like playing cards. There are no cross currents perpendicular to the direction of flow, nor eddies or swirls of fluid. This regime or type of flow is called *laminar flow*. At higher velocities eddies form, which leads to lateral mixing. This is called *turbulent flow*. The discussion in this section is limited to laminar flow.

A fluid can be distinguished from a solid in this discussion of viscosity by its behavior when subjected to a stress (force per unit area) or applied force. An elastic solid deforms by an amount proportional to the applied stress. However, a fluid when subjected to a similar applied stress will continue to deform, i.e., to flow at a velocity that increases with increasing stress. A fluid exhibits resistance to this stress. Viscosity is that property of a fluid which gives rise to forces that resist the relative movement of adjacent layers in the fluid. These *viscous forces* arise from forces existing between the molecules in the fluid and are of similar character as the *shear forces* in solids.

The ideas above can be clarified by a more quantitative discussion of viscosity. In Fig. 2.4-1 a fluid is contained between two infinite (very long and very wide) parallel plates. Suppose that the bottom plate is moving parallel to the top plate and at a constant velocity Δv_z m/s faster relative to the top plate because of a steady force F newtons being applied. This force is called the *viscous drag*, and it arises from the viscous forces in the fluid. The plates are Δy m apart. Each layer of liquid moves in the z direction. The layer immediately adjacent to the bottom plate is carried along at the velocity of this plate. The layer just above is at a slightly slower velocity, each layer moving at a slower velocity as we go up in the y direction. This velocity profile is linear, with y direction as shown in Fig. 2.4-1. An analogy to a fluid is a deck of playing cards, where, if the bottom card is moved, all the other cards above will slide to some extent.

It has been found experimentally for many fluids that the force F in newtons is directly proportional to the velocity Δv_z in m/s, to the area A in m^2 of the plate used, and inversely proportional to the distance Δy in m. Or, as given by Newton's law of viscosity when the flow is laminar,

$$\frac{F}{A} = -\mu \frac{\Delta v_z}{\Delta y} \tag{2.4-1}$$

where μ is a proportionality constant called the *viscosity* of the fluid, in Pa·s or kg/m·s. If we let Δy approach zero, then, using the definition of the derivative,

$$\tau_{yz} = -\mu \frac{dv_z}{dy} \quad \text{(SI units)} \tag{2.4-2}$$

where $\tau_{yz} = F/A$ and is the shear stress or force per unit area in newtons/m^2 (N/m^2). In the cgs system, F is in dynes, μ in g/cm·s, v_z in cm/s, and y in cm. We can also write

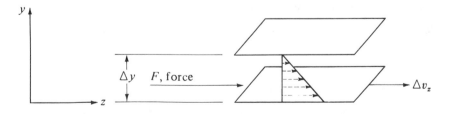

FIGURE 2.4-1. *Fluid shear between two parallel plates.*

Chap. 2   Principles of Momentum Transfer and Overall Balances

Eq. (2.2-2) as

$$\tau_{yz} g_c = -\mu \frac{dv_z}{dy} \qquad \text{(English units)} \qquad (2.4\text{-}3)$$

where τ_{yz} is in units of lb_f/ft^2.

The units of viscosity in the cgs system are $g/cm \cdot s$, called *poise* or centipoise (cp). In the SI system, viscosity is given in $Pa \cdot s \, (N \cdot s/m^2$ or $kg/m \cdot s)$.

$$1 \text{ cp} = 1 \times 10^{-3} \text{ kg/m} \cdot s = 1 \times 10^{-3} \text{ Pa} \cdot s = 1 \times 10^{-3} \text{ N} \cdot s/m^2 \text{ (SI)}$$

$$1 \text{ cp} = 0.01 \text{ poise} = 0.01 \text{ g/cm} \cdot s$$

$$1 \text{ cp} = 6.7197 \times 10^{-4} \text{ lb}_m/ft \cdot s$$

Other conversion factors for viscosity are given in Appendix A.1. Sometimes the viscosity is given as μ/ρ, kinematic viscosity, in m^2/s or cm^2/s, where ρ is the density of the fluid.

EXAMPLE 2.4-1. Calculation of Shear Stress in a Liquid
Referring to Fig. 2.4-1, the distance between plates is $\Delta y = 0.5$ cm, $\Delta v_z = 10$ cm/s, and the fluid is ethyl alcohol at 273 K having a viscosity of 1.77 cp (0.0177 g/cm · s).
    (a) Calculate the shear stress τ_{yz} and the velocity gradient or shear rate dv_z/dy using cgs units.
    (b) Repeat, using lb force, s, and ft units (English units).
    (c) Repeat, using SI units.

Solution: We can substitute directly into Eq. (2.4-1) or we can integrate Eq. (2.4-2). Using the latter method, rearranging Eq. (2.4-2), calling the bottom plate point 1, and integrating,

$$\tau_{yz} \int_{y_1=0}^{y_2=0.5} dy = -\mu \int_{v_1=10}^{v_2=0} dv_z \qquad (2.4\text{-}4)$$

$$\tau_{yz} = \mu \frac{v_1 - v_2}{y_2 - y_1} \qquad (2.4\text{-}5)$$

Substituting the known values,

$$\tau_{yz} = \mu \frac{v_1 - v_2}{y_2 - y_1} = \left(0.0177 \frac{g}{cm \cdot s}\right) \frac{(10 - 0) \text{ cm/s}}{(0.5 - 0) \text{ cm}}$$

$$= 0.354 \frac{g \cdot cm/s^2}{cm^2} = 0.354 \frac{dyn}{cm^2} \qquad (2.4\text{-}6)$$

To calculate the shear rate dv_z/dy, since the velocity change is linear with y,

$$\text{shear rate} = \frac{dv_z}{dy} = \frac{\Delta v_z}{\Delta y} = \frac{(10 - 0) \text{ cm/s}}{(0.5 - 0) \text{ cm}} = 20.0 \text{ s}^{-1} \qquad (2.4\text{-}7)$$

For part (b), using lb force units and the viscosity conversion factor from Appendix A.1,

$$\mu = 1.77 \text{ cp}(6.7197 \times 10^{-4} \text{ lb}_m/ft \cdot s)/cp$$

$$= 1.77(6.7197 \times 10^{-4}) \text{ lb}_m/ft \cdot s$$

Integrating Eq. (2.4-3),

$$\tau_{yz} = \frac{\mu \; \frac{lb_m/ft \cdot s}{g_c \; \frac{lb_m \cdot ft}{lb_f \cdot s^2}}} \frac{(v_1 - v_2)ft/s}{(y_2 - y_1) \; ft} \tag{2.4-8}$$

Substituting known values into Eq. (2.4-8) and converting Δv_z to ft/s and Δy to ft, $\tau_{yz} = 7.39 \times 10^{-4} \; lb_f/ft^2$. Also, $dv_z/dy = 20 \; s^{-1}$.

For part (c), $\Delta y = 0.5/100 = 0.005$ m, $\Delta v_z = 10/100 = 0.1$ m/s, and $\mu = 1.77 \times 10^{-3}$ kg/m·s $= 1.77 \times 10^{-3}$ Pa·s. Substituting into Eq. (2.4-5),

$$\tau_{yz} = (1.77 \times 10^{-3})(0.10)/0.005 = 0.0354 \; N/m^2$$

The shear rate will be the same at $20.0 \; s^{-1}$

2.4B Momentum Transfer in a Fluid

The shear stress τ_{yz} in Eqs. (2.4-1)-(2.4-3) can also be interpreted as a *flux of z-directed momentum in the y direction*, which is the rate of flow of momentum per unit area. The units of momentum are mass times velocity in kg·m/s. The shear stress can be written

$$\tau_{yz} = \frac{kg \cdot m/s}{m^2 \cdot s} = \frac{momentum}{m^2 \cdot s} \tag{2.4-9}$$

This gives an amount of momentum transferred per second per unit area.

This can be shown by considering the interaction between two adjacent layers of a fluid in Fig. 2.4-1 which have different velocities, and hence different momentum, in the z direction. The random motions of the molecules in the faster-moving layer send some of the molecules into the slower-moving layer, where they collide with the slower-moving molecules and tend to speed them up or increase their momentum in the z direction. Also, in the same fashion, molecules in the slower layer tend to retard those in the faster layer. This exchange of molecules between layers produces a transfer or flux of z-directed momentum from high-velocity to low-velocity layers. The negative sign in Eq. (2.4-2) indicates that momentum is transferred down the gradient from high- to low-velocity regions. This is similar to the transfer of heat from high- to low-temperature regions.

2.4C Viscosities of Newtonian Fluids

Fluids that follow Newton's law of viscosity, Eqs. (2.4-1)–(2.4-3), are called *Newtonian fluids*. For a Newtonian fluid, there is a linear relation between the shear stress τ_{yz} and the velocity gradient dv_z/dy (rate of shear). This means that the viscosity μ is a constant and independent of the rate of shear. For non-Newtonian fluids, the relation between τ_{yz} and dv_z/dy is not linear; i.e., the viscosity μ does not remain constant but is a function of shear rate. Certain liquids do not obey this simple Newton's law. These are primarily pastes, slurries, high polymers, and emulsions. The science of the flow and deformation of fluids is often called *rheology*. A discussion of non-Newtonian fluids will not be given here but will be included in Section 3.5.

The viscosity of gases, which are Newtonian fluids, increases with temperature and is approximately independent of pressure up to a pressure of about 1000 kPa. At higher pressures, the viscosity of gases increases with increase in pressure. For example, the viscosity of N_2 gas at 298 K approximately doubles in going from 100 kPa to about 5×10^4 kPa (R1). In liquids, the viscosity decreases with increasing temperature. Since liquids are essentially incompressible, the viscosity is not affected by pressure.

TABLE 2.4-1. *Viscosities of Some Gases and Liquids at 101.32 kPa Pressure*

Gases				Liquids			
Substance	Temp., K	Viscosity $(Pa \cdot s)10^3$ or $(kg/m \cdot s) \ 10^3$	Ref.	Substance	Temp., K	Viscosity $(Pa \cdot s)10^3$ or $(kg/m \cdot s) \ 10^3$	Ref.
Air	293	0.01813	N1	Water	293	1.0019	S1
CO_2	273	0.01370	R1		373	0.2821	S1
	373	0.01828	R1	Benzene	278	0.826	R1
CH_4	293	0.01089	R1				
				Glycerol	293	1069	L1
SO_2	373	0.01630	R1	Hg	293	1.55	R2
				Olive oil	303	84	E1

In Table 2.4-1 some experimental viscosity data are given for some typical pure fluids at 101.32 kPa. The viscosities for gases are the lowest and do not differ markedly from gas to gas, being about 5×10^{-6} to 3×10^{-5} Pa·s. The viscosities for liquids are much greater. The value for water at 293 K is about 1×10^{-3} and for glycerol 1.069 Pa·s. Hence, there is a great difference between viscosities of liquids. More complete tables of viscosities are given for water in Appendix A.2, for inorganic and organic liquids and gases in Appendix A.3, and for biological and food liquids in Appendix A.4. Extensive data are available in other references (P1, R1, W1, L1). Methods of estimating viscosities of gases and liquids when experimental data are not available are summarized elsewhere (R1). These estimation methods for gases at pressures below 100 kPa are reasonably accurate, with an error within about ±5%, but the methods for liquids are often quite inaccurate.

2.5 TYPES OF FLUID FLOW AND REYNOLDS NUMBER

2.5A Introduction and Types of Fluid Flow

The principles of the statics of fluids, treated in Section 2.2, are almost an exact science. On the other hand, the principles of the motions of fluids are quite complex. The basic relations describing the motions of a fluid are the equations for the overall balances of mass, energy, and momentum, which will be covered in the following sections.

These overall or macroscopic balances will be applied to a finite enclosure or control volume fixed in space. We use the term "overall" because we wish to describe these balances from outside the enclosure. The changes inside the enclosure are determined in terms of the properties of the streams entering and leaving and the exchanges of energy between the enclosure and its surroundings.

When making overall balances on mass, energy, and momentum we are not interested in the details of what occurs inside the enclosure. For example, in an overall balance average inlet and outlet velocities are considered. However, in a differential balance the velocity distribution inside an enclosure can be obtained with the use of Newton's law of viscosity.

In this section we first discuss the two types of fluid flow that can occur: laminar and turbulent flow. Also, the Reynolds number used to characterize the regimes of flow is considered. Then in Sections 2.6, 2.7, and 2.8 the overall mass balance, energy balance, and momentum balance are covered together with a number of applications. Finally, a discussion is given in Section 2.9 on the methods of making a shell balance on an element to obtain the velocity distribution in the element and pressure drop.

2.5B Laminar and Turbulent Flow

The type of flow occurring in a fluid in a channel is important in fluid dynamics problems. When fluids move through a closed channel of any cross section, either of two distinct types of flow can be observed according to the conditions present. These two types of flow can be commonly seen in a flowing open stream or river. When the velocity of flow is slow, the flow patterns are smooth. However, when the velocity is quite high, an unstable pattern is observed in which eddies or small packets of fluid particles are present moving in all directions and at all angles to the normal line of flow.

The first type of flow at low velocities where the layers of fluid seem to slide by one another without eddies or swirls being present is called *laminar flow* and Newton's law of viscosity holds, as discussed in Section 2.4A. The second type of flow at higher velocities where eddies are present giving the fluid a fluctuating nature is called *turbulent flow.*

The existence of laminar and turbulent flow is most easily visualized by the experiments of Reynolds. His experiments are shown in Fig. 2.5-1. Water was allowed to flow at steady state through a transparent pipe with the flow rate controlled by a valve at the end of the pipe. A fine steady stream of dye-colored water was introduced from a fine jet as shown and its flow pattern observed. At low rates of water flow, the dye pattern was regular and formed a single line or stream similar to a thread, as shown in Fig. 2.5-1a. There was no lateral mixing of the fluid, and it flowed in streamlines down the tube. By putting in additional jets at other points in the pipe cross section, it was shown that there was no mixing in any parts of the tube and the fluid flowed in straight parallel lines. This type of flow is called *laminar* or *viscous flow.*

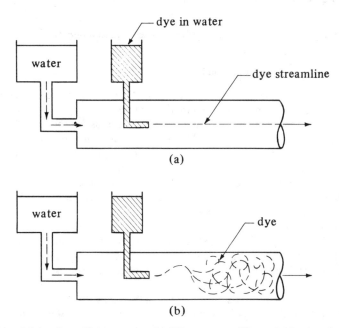

(a)

(b)

FIGURE 2.5-1. *Reynolds' experiment for different types of flow : (a) laminar flow. (b) turbulent flow.*

Chap. 2   Principles of Momentum Transfer and Overall Balances

As the velocity was increased, it was found that at a definite velocity the thread of dye became dispersed and the pattern was very erratic, as shown in Fig. 2.5-1b. This type of flow is known as turbulent flow. The velocity at which the flow changes is known as the *critical velocity*.

2.5C Reynolds Number

Studies have shown that the transition from laminar to turbulent flow in tubes is not only a function of velocity but also of density and viscosity of the fluid and the tube diameter. These variables are combined into the Reynolds number, which is dimensionless.

$$N_{Re} = \frac{Dv\rho}{\mu} \tag{2.5-1}$$

where N_{Re} is the Reynolds number, D the diameter in m, ρ the fluid density in kg/m^3, μ the fluid viscosity in Pa $\cdot$ s, and v the average velocity of the fluid in m/s (where average velocity is defined as the volumetric rate of flow divided by the cross-sectional area of the pipe). Units in the cgs system are D in cm, ρ in g/cm^3, μ in g/cm $\cdot$ s, and v in cm/s. In the English system D is in ft, ρ in lb$_m$/ft^3, μ in lb$_m$/ft $\cdot$ s, and v in ft/s.

The instability of the flow that leads to disturbed or turbulent flow is determined by the ratio of the kinetic or inertial forces to the viscous forces in the fluid stream. The inertial forces are proportional to ρv^2 and the viscous forces to $\mu v/D$, and the ratio $\rho v^2/(\mu v/D)$ is the Reynolds number $Dv\rho/\mu$. Further explanation and derivation of dimensionless numbers is given in Section 3.11.

For a straight circular pipe when the value of the Reynolds number is less than 2100, the flow is always laminar. When the value is over 4000, the flow will be turbulent, except in very special cases. In between, which is called the *transition region*, the flow can be viscous or turbulent, depending upon the apparatus details, which cannot be predicted.

EXAMPLE 2.5-1 Reynolds Number in a Pipe
Water at 303 K is flowing at the rate of 10 gal/min in a pipe having an inside diameter (ID) of 2.067 in. Calculate the Reynolds number using both English units and SI units.

Solution: From Appendix A.1, 7.481 gal = 1 ft^3. The flow rate is calculated as

$$\text{flow rate} = \left(10.0 \frac{\text{gal}}{\text{min}}\right)\left(\frac{1 \text{ ft}^3}{7.481 \text{ gal}}\right)\left(\frac{1 \text{ min}}{60 \text{ s}}\right) = 0.0223 \text{ ft}^3/\text{s}$$

$$\text{pipe diameter, } D = \frac{2.067}{12} = 0.172 \text{ ft}$$

$$\text{cross-sectional area of pipe} = \frac{\pi D^2}{4} = \frac{\pi(0.172)^2}{4} = 0.0233 \text{ ft}^2$$

$$\text{velocity in pipe, } v = \left(0.0223 \frac{\text{ft}^3}{\text{s}}\right)\left(\frac{1}{0.0233 \text{ ft}^2}\right) = 0.957 \text{ ft/s}$$

From Appendix A.2 for water at 303 K (30°C),

$$\text{density, } \rho = 0.996(62.43) \text{ lb}_m/\text{ft}^3$$

$$\text{viscosity, } \mu = (0.8007 \text{ cp})\left(6.7197 \times 10^{-4} \frac{\text{lb}_m/\text{ft} \cdot \text{s}}{\text{cp}}\right)$$

$$= 5.38 \times 10^{-4} \text{ lb}_m/\text{ft} \cdot \text{s}$$

Substituting into Eq. (2.5-1),

$$N_{Re} = \frac{Dv\rho}{\mu} = \frac{(0.172 \text{ ft})(0.957 \text{ ft/s})(0.996 \times 62.43 \text{ lb}_m/\text{ft}^3)}{5.38 \times 10^{-4} \text{ lb}_m/\text{ft} \cdot \text{s}}$$

$$= 1.905 \times 10^4$$

Hence, the flow is turbulent. Using SI units,

$$\rho = (0.996)(100 \text{ kg/m}^3) = 996 \text{ kg/m}^3$$

$$D = (2.067 \text{ in.})(1 \text{ ft/12 in.})(1 \text{ m/3.2808 ft}) = 0.0525 \text{ m}$$

$$v = \left(0.957 \frac{\text{ft}}{\text{s}}\right)(1 \text{ m/3.2808 ft}) = 0.2917 \text{ m/s}$$

$$\mu = (0.8007 \text{ cp})\left(1 \times 10^{-3} \frac{\text{kg/m} \cdot \text{s}}{\text{cp}}\right) = 8.007 \times 10^{-4} \frac{\text{kg}}{\text{m} \cdot \text{s}}$$

$$= 8.007 \times 10^{-4} \text{ Pa} \cdot \text{s}$$

$$N_{Re} = \frac{Dv\rho}{\mu} = \frac{(0.0525 \text{ m})(0.2917 \text{ m/s})(996 \text{ kg/m}^3)}{8.007 \times 10^{-4} \text{ kg/m} \cdot \text{s}} = 1.905 \times 10^4$$

2.6   OVERALL MASS BALANCE AND CONTINUITY EQUATION

2.6A   Introduction and Simple Mass Balances

In fluid dynamics fluids are in motion. Generally, they are moved from place to place by means of mechanical devices such as pumps or blowers, by gravity head, or by pressure, and flow through systems of piping and/or process equipment. The first step in the solution of flow problems is generally to apply the principles of the conservation of mass to the whole system or to any part of the system. First, we will consider an elementary balance on a simple geometry, and later we shall derive the general mass-balance equation.

Simple mass or material balances were introduced in Section 1.5, where

$$\text{input} = \text{output} + \text{accumulation} \qquad (1.5\text{-}1)$$

Since, in fluid flow, we are usually working with rates of flow and usually at steady state, the rate of accumulation is zero and we obtain

$$\text{rate of input} = \text{rate of output (steady state)} \qquad (2.6\text{-}1)$$

In Fig. 2.6-1 a simple flow system is shown where fluid enters section 1 with an

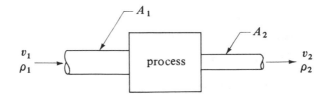

FIGURE 2.6-1.   *Mass balance on flow system.*

Chap. 2   Principles of Momentum Transfer and Overall Balances

average velocity v_1 m/s and density ρ_1 kg/m^3. The cross-sectional area is A_1 m^2. The fluid leaves section 2 with average velocity v_2. The mass balance, Eq. (2.6-1), becomes

$$m = \rho_1 A_1 v_1 = \rho_2 A_2 v_2 \qquad (2.6\text{-}2)$$

where $m =$ kg/s. Often, $v\rho$ is expressed as $G = v\rho$, where G is mass velocity or mass flux in kg/s·m^2. In English units, v is in ft/s, ρ in lb$_m$/ft^3, A in ft^2, m in lb$_m$/s, and G in lb$_m$/s·ft^2.

EXAMPLE 2.6-1.   Flow of Crude Oil and Mass Balance

A petroleum crude oil having a density of 892 kg/m^3 is flowing through the piping arrangement shown in Fig. 2.6-2 at a total rate of 1.388×10^{-3} m^3/s entering pipe 1.

The flow divides equally in each of pipes 3. The steel pipes are schedule 40 pipe (see Appendix A.5 for actual dimensions). Calculate the following using SI units.

   (a) The total mass flow rate m in pipe 1 and pipes 3.
   (b) The average velocity v in 1 and 3.
   (c) The mass velocity G in 1.

Solution: From Appendix A.5, the dimensions of the pipes are as follows: 2-in. pipe: D_1 (ID) $= 2.067$ in., cross-sectional area

$$A_1 = 0.02330 \text{ ft}^2 = 0.02330(0.0929) = 2.165 \times 10^{-3} \text{ m}^2$$

$1\frac{1}{2}$-in. pipe: D_3 (ID) $= 1.610$ in., cross-sectional area

$$A_3 = 0.01414 \text{ ft}^2 = 0.01414(0.0929) = 1.313 \times 10^{-3} \text{ m}^2$$

The total mass flow rate is the same through pipes 1 and 2 and is

$$m_1 = (1.388 \times 10^{-3} \text{ m}^3/\text{s})(892 \text{ kg/m}^3) = 1.238 \text{ kg/s}$$

Since the flow divides equally in each of pipes 3,

$$m_3 = \frac{m_1}{2} = \frac{1.238}{2} = 0.619 \text{ kg/s}$$

For part (b), using Eq. (2.6-2) and solving for v,

$$v_1 = \frac{m_1}{\rho_1 A_1} = \frac{1.238 \text{ kg/s}}{(892 \text{ kg/m}^3)(2.165 \times 10^{-3} \text{ m}^2)} = 0.641 \text{ m/s}$$

$$v_3 = \frac{m_3}{\rho_3 A_3} = \frac{0.619}{(892)(1.313 \times 10^{-3})} = 0.528 \text{ m/s}$$

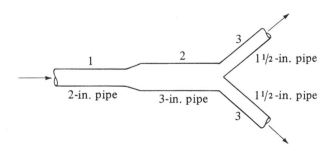

FIGURE 2.6-2.   *Piping arrangement for Example 2.6-1.*

For part (c),

$$G_1 = v_1 \rho_1 = \frac{m_1}{A_1} = \frac{1.238}{2.165 \times 10^{-3}} = 572 \ \frac{\text{kg}}{\text{s} \cdot \text{m}^2}$$

2.6B Control Volume for Balances

The laws for the conservation of mass, energy, and momentum are all stated in terms of a system, and these laws give the interaction of a system with its surroundings. A *system* is defined as a collection of fluid of fixed identity. However, in flow of fluids, individual particles are not easily identifiable. As a result, attention is focused on a given space through which the fluid flows rather than to a given mass of fluid. The method used, which is more convenient, is to select a control volume, which is a region fixed in space through which the fluid flows.

In Fig. 2.6-3 the case of a fluid flowing through a conduit is shown. The control surface shown as a dashed line is the surface surrounding the control volume. In most problems part of the control surface will coincide with some boundary, such as the wall of the conduit. The remaining part of the control surface is a hypothetical surface through which the fluid can flow, shown as point 1 and point 2 in Fig. 2.6-3. The control-volume representation is analogous to the open system of thermodynamics.

2.6C Overall Mass-Balance Equation

In deriving the general equation for the overall balance of the property mass, the law of conservation of mass may be stated as follows for a control volume where no mass is being generated.

$$\begin{pmatrix} \text{rate of mass output} \\ \text{from control volume} \end{pmatrix} - \begin{pmatrix} \text{rate of mass input} \\ \text{from control volume} \end{pmatrix}$$

$$+ \begin{pmatrix} \text{rate of mass accumulation} \\ \text{in control volume} \end{pmatrix} = 0 \ (\text{rate of mass generation})$$

$$(2.6\text{-}3)$$

We now consider the general control volume fixed in space and located in a fluid flow field, as shown in Fig. 2.6-4. For a small element of area dA m^2 on the control surface, the rate of mass efflux from this element $= (\rho v)(dA \cos \alpha)$, where $(dA \cos \alpha)$ is the area dA projected in a direction normal to the velocity vector $\mathbf{v}$, α is the angle between the velocity vector $\mathbf{v}$ and the outward-directed unit normal vector $\mathbf{n}$ to dA, and ρ is the

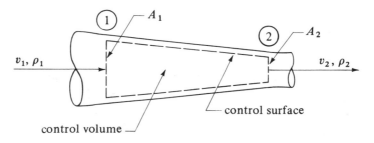

FIGURE 2.6-3.   *Control volume for flow through a conduit.*

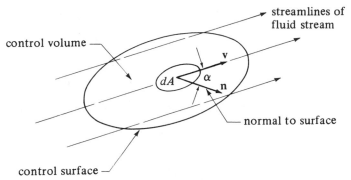

streamlines of
fluid stream

control volume

dA α $\mathbf{n}$

normal to surface

control surface

FIGURE 2.6-4.   *Flow through a differential area dA on a control surface.*

density in kg/m^3. The quantity ρv has units of $kg/s \cdot m^2$ and is called *a flux* or *mass velocity G*.

From vector algebra we recognize that $(\rho v)(dA \cos \alpha)$ is the scalar or dot product $\rho(\mathbf{v} \cdot \mathbf{n}) \, dA$. If we now integrate this quantity over the entire control surface A we have the net outflow of mass across the control surface, or the net mass efflux in kg/s from the entire control volume V:

$$\begin{pmatrix} \text{net mass efflux} \\ \text{from control volume} \end{pmatrix} = \iint_A v\rho \cos \alpha \, dA = \iint_A \rho(\mathbf{v} \cdot \mathbf{n}) \, dA \qquad (2.6\text{-}4)$$

We should note that if mass is entering the control volume, i.e., flowing inward across the control surface, the net efflux of mass in Eq. (2.6-4) is negative since $\alpha > 90°$ and $\cos \alpha$ is negative. Hence, there is a net influx of mass. If $\alpha < 90°$, there is a net efflux of mass.

The rate of accumulation of mass within the control volume V can be expressed as follows.

$$\begin{pmatrix} \text{rate of mass accumulation} \\ \text{in control volume} \end{pmatrix} = \frac{\partial}{\partial t} \iiint_V \rho \, dV = \frac{dM}{dt} \qquad (2.6\text{-}5)$$

where M is the mass of fluid in the volume in kg. Substituting Eqs. (2.6-4) and (2.6-5) into (2.6-3), we obtain the general form of the overall mass balance.

$$\iint_A \rho(\mathbf{v} \cdot \mathbf{n}) \, dA + \frac{\partial}{\partial t} \iiint_V \rho \, dV = 0 \qquad (2.6\text{-}6)$$

The use of Eq. (2.6-6) can be shown for a common situation for steady-state one-dimensional flow, where all the flow inward is normal to A_1 and outward normal to A_2, as shown in Fig. 2.6-3. When the velocity v_2 leaving (Fig. 2.6-3) is normal to A_2, the angle α_2 between the normal to the control surface and the direction of the velocity is 0° and $\cos \alpha_2 = 1.0$. Where v_1 is directed inward, $\alpha_1 > \pi/2$, and for the case in Fig. 2.6-3, α_1 is 180° ($\cos \alpha_1 = -1.0$). Since α_2 is 0° and α_1 is 180°, using Eq. (2.6-4),

$$\iint_A v\rho \cos \alpha \, dA = \iint_{A_2} v\rho \cos \alpha_2 \, dA + \iint_{A_1} v\rho \cos \alpha_1 \, dA$$

$$= v_2 \rho_2 A_2 - v_1 \rho_1 A_1 \qquad (2.6\text{-}7)$$

For steady state, $dM/dt = 0$ in Eq. (2.6-5), and Eq. (2.6-6) becomes

$$m = \rho_1 v_1 A_1 = \rho_2 v_2 A_2 \qquad \textbf{(2.6-2)}$$

which is Eq. (2.6-2), derived earlier.

In Fig. 2.6-3 and Eqs. (2.6-3)–(2.6-7) we were not concerned with the composition of any of the streams. These equations can easily be extended to represent an overall mass balance for component i in a multicomponent system. For the case shown in Fig. 2.6-3 we combine Eqs. (2.6-5), (2.6-6), and (2.6-7), add a generation term, and obtain

$$m_{i2} - m_{i1} + \frac{dM_i}{dt} = R_i \qquad \textbf{(2.6-8)}$$

where m_{i2} is the mass flow rate of component i leaving the control volume and R_i is the rate of generation of component i in the control volume in kg per unit time. (Diffusion fluxes are neglected here or are assumed negligible.) In some cases, of course, $R_i = 0$ for no generation. Often it is more convenient to use Eq. (2.6-8) written in molar units.

EXAMPLE 2.6-2. Overall Mass Balance in Stirred Tank

Initially, a tank contains 500 kg of salt solution containing 10% salt. At point (1) in the control volume in Fig. 2.6-5, a stream enters at a constant flow rate of 10 kg/h containing 20% salt. A stream leaves at point (2) at a constant rate of 5 kg/h. The tank is well stirred. Derive an equation relating the weight fraction w_A of the salt in the tank at any time t in hours.

Solution: First we make a total mass balance using Eq. (2.6-7) for the net total mass efflux from the control volume.

$$\iint_A v\rho \cos \alpha \, dA = m_2 - m_1 = 5 - 10 = -5 \text{ kg solution/h} \qquad \textbf{(2.6-9)}$$

From Eq. (2.6-5), where M is total kg of solution in control volume at time t,

$$\frac{\partial}{\partial t} \iiint_V \rho \, dV = \frac{dM}{dt} \qquad \textbf{(2.6-5)}$$

Substituting Eqs. (2.6-5) and (2.6-9) into (2.6-6), and then integrating,

$$-5 + \frac{dM}{dt} = 0$$

$$\int_{M=500}^{M} dM = 5 \int_{t=0}^{t} dt \qquad \textbf{(2.6-10)}$$

$$M = 5t + 500 \qquad \textbf{(2.6-11)}$$

Equation (2.6-11) relates the total mass M in the tank at any time t.

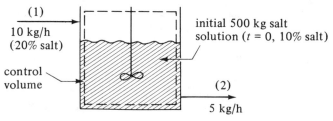

(1)

10 kg/h
(20% salt)

initial 500 kg salt
solution ($t = 0$, 10% salt)

control
volume

(2)

5 kg/h

FIGURE 2.6-5. *Control volume for flow in a stirred tank for Example 2.6-2.*

Next, making a component A salt balance, let w_A = weight fraction of salt in tank at time t and also the concentration in the stream m_2 leaving at time t. Again using Eq. (2.6-7) but for a salt balance,

$$\iint_A v\rho \cos \alpha \, dA = (5)w_A - 10(0.20) = 5w_A - 2 \text{ kg salt/h} \quad \textbf{(2.6-12)}$$

Using Eq. (2.6-5) for a salt balance,

$$\frac{\partial}{\partial t} \iiint_V \rho \, dV = \frac{d}{dt}(Mw_A) = \frac{M \, dw_A}{dt} + w_A \frac{dM}{dt} \text{ kg salt/h} \quad \textbf{(2.6-13)}$$

Substituting Eqs. (2.6-12) and (2.6-13) into (2.6-6),

$$5w_A - 2 + M\frac{dw_A}{dt} + w_A\frac{dM}{dt} = 0 \quad \textbf{(2.6-14)}$$

Substituting the value for M from Eq. (2.6-11) into (2.6-14), separating variables, integrating, and solving for w_A,

$$5w_A - 2 + (500 + 5t)\frac{dw_A}{dt} + w_A\frac{d(500 + 5t)}{dt} = 0$$

$$5w_A - 2 + (500 + 5t)\frac{dw_A}{dt} + 5w_A = 0$$

$$\int_{w_A = 0.10}^{w_A} \frac{dw_A}{2 - 10w_A} = \int_{t=0}^{t} \frac{dt}{500 + 5t}$$

$$-\frac{1}{10}\ln\left(\frac{2 - 10w_A}{1}\right) = \frac{1}{5}\ln\left(\frac{500 + 5t}{500}\right) \quad \textbf{(2.6-15)}$$

$$w_A = -0.1\left(\frac{100}{100 + t}\right)^2 + 0.20$$

$$\textbf{(2.6-16)}$$

Note that Eq. (2.6-8) for component i could have been used for the salt balance with $R_i = 0$ (no generation).

2.6D Average Velocity to Use in Overall Mass Balance

In solving the case in Eq. (2.6-7) we assumed a constant velocity v_1 at section 1 and constant v_2 at section 2. If the velocity is not constant but varies across the surface area, an average or bulk velocity is defined by

$$v_{av} = \frac{1}{A}\iint_A v \, dA \quad \textbf{(2.6-17)}$$

for a surface over which v is normal to A and the density ρ is assumed constant.

EXAMPLE 2.6-3. Variation of Velocity Across Control Surface and Average Velocity
For the case of imcompressible flow (ρ is constant) through a circular pipe of radius R, the velocity profile is parabolic for laminar flow as follows:

$$v = v_{max}\left[1 - \left(\frac{r}{R}\right)^2\right] \quad \textbf{(2.6-18)}$$

where v_{max} is the maximum velocity at the center where $r = 0$ and v is the velocity at a radial distance r from the center. Derive an expression for the average or bulk velocity v_{av} to use in the overall mass-balance equation.

Solution: The average velocity is represented by Eq. (2.6-17). In Cartesian coordinates dA is $dx\ dy$. However, using polar coordinates which are more appropriate for a pipe, $dA = r\ dr\ d\theta$, where θ is the angle in polar coordinates. Substituting Eq. (2.6-18), $dA = r\ dr\ d\theta$, and $A = \pi R^2$ into Eq. (2.6-17) and integrating,

$$v_{av} = \frac{1}{\pi R^2} \int_0^{2\pi} \int_0^R v_{max} \left[1 - \left(\frac{r}{R}\right)^2 \right] r\ dr\ d\theta$$

$$= \frac{v_{max}}{\pi R^4} \int_0^{2\pi} \int_0^R (R^2 - r^2) r\ dr\ d\theta$$

$$= \frac{v_{max}}{\pi R^4} (2\pi - 0) \left(\frac{R^4}{2} - \frac{R^4}{4}\right) \qquad \text{(2.6-19)}$$

$$v_{av} = \frac{v_{max}}{2} \qquad \text{(2.6-20)}$$

In this discussion overall or macroscopic mass balances were made because we wish to describe these balances from outside the enclosure. In this section on overall mass balances, some of the equations presented may have seemed quite obvious. However, the purpose was to develop the methods which should be helpful in the next sections. Overall balances will also be made on energy and momentum in the next sections. These overall balances do not tell us the details of what happens inside. However, in Section 2.9 a shell momentum balance will be made to obtain these details, which will give us the velocity distribution and pressure drop. To further study these details of the processes occurring inside the enclosure, differential balances rather than shell balances can be written and these are discussed in other later Sections 3.6 to 3.9 on differential equations of continuity and momentum transfer, Sections 5.6 and 5.7 on differential equations of energy change and boundary-layer flow, and Section 7.5B on differential equations of continuity for a binary mixture.

2.7 OVERALL ENERGY BALANCE

2.7A Introduction

The second property to be considered in the overall balances on a control volume is energy. We shall apply the principle of the conservation of energy to a control volume fixed in space in much the same manner as the principle of conservation of mass was used to obtain the overall mass balance. The energy-conservation equation will then be combined with the first law of thermodynamics to obtain the final overall energy-balance equation.

We can write the first law of thermodynamics as

$$\Delta E = Q - W \qquad \text{(2.7-1)}$$

where E is the total energy per unit mass of fluid, Q is the heat *absorbed* per unit mass of fluid, and W is the work of all kinds done per unit mass of fluid *upon* the surroundings.

In the calculations, each term in the equation must be expressed in the same units, such as J/kg (SI), btu/lb$_m$, or ft · lb$_f$/lb$_m$ (English).

Since mass carries with it associated energy because of its position, motion, or physical state, we will find that each of these types of energy will appear in the energy balance. In addition, we can also transport energy across the boundary of the system without transferring mass.

2.7B Derivation of Overall Energy-Balance Equation

The entity balance for a conserved quantity such as energy is similar to Eq. (2.6-3) and is as follows for a control volume.

rate of entity output − rate of entity input

$$+ \text{ rate of entity accumulation} = 0 \qquad \textbf{(2.7-2)}$$

The energy E present within a system can be classified in three ways.

1. *Potential energy zg* of a unit mass of fluid is the energy present because of the position of the mass in a gravitational field g, where z is the relative height in meters from a reference plane. The units for zg for the SI system are m · m/s^2. Multiplying and dividing by kg mass, the units can be expressed as (kg · m/s^2) · (m/kg), or J/kg. In English units the potential energy is zg/g_c in ft · lb$_f$/lb$_m$.
2. *Kinetic energy $v^2/2$* of a unit mass of fluid is the energy present because of translational or rotational motion of the mass, where v is the velocity in m/s relative to the boundary of the system at a given point. Again in the SI system the units of $v^2/2$ are J/kg. In the English system the kinetic energy is $v^2/2g_c$ in ft · lb$_f$/lb$_m$.
3. *Internal energy U* of a unit mass of a fluid is all of the other energy present, such as rotational and vibrational energy in chemical bonds. Again the units are in J/kg or ft · lb$_f$/lb$_m$.

The total energy of the fluid per unit mass is then

$$E = U + \frac{v^2}{2} + zg \qquad \text{(SI)}$$

$$E = U + \frac{v^2}{2g_c} + \frac{zg}{g_c} \qquad \text{(English)} \qquad \textbf{(2.7-3)}$$

The rate of accumulation of energy within the control volume V in Fig. 2.6-4 is

$$\begin{pmatrix} \text{rate of energy accumulation} \\ \text{in control volume} \end{pmatrix} = \frac{\partial}{\partial t} \iiint\limits_{V} \left(U + \frac{v^2}{2} + zg \right) \rho \; dV \qquad \textbf{(2.7-4)}$$

Next we consider the rate of energy input and output associated with mass in the control volume. The mass added or removed from the system carries internal, kinetic, and potential energy. In addition, energy is transferred when mass flows into and out of the control volume. Net work is done by the fluid as it flows into and out of the control volume. This pressure–volume work per unit mass fluid is pV. The contribution of shear work is usually neglected. The pV term and U term are combined using the definition of enthalpy, H.

$$H = U + pV \qquad \textbf{(2.7-5)}$$

Hence, the total energy carried with a unit mass is $(H + v^2/2 + zg)$.

For a small area dA on the control surface in Fig. 2.6-4, the rate of energy efflux is $(H + v^2/2 + zg)(\rho v)(dA \cos \alpha)$, where $(dA \cos \alpha)$ is the area dA projected in a direction normal to the velocity vector $\mathbf{v}$ and α is the angle between the velocity vector $\mathbf{v}$ and the outward-directed unit normal vector $\mathbf{n}$. We now integrate this quantity over the entire control surface to obtain

$$\binom{\text{net energy efflux}}{\text{from control volume}} = \iint\limits_{A} \left(H + \frac{v^2}{2} + zg \right)(\rho v) \cos \alpha \, dA \qquad \text{(2.7-6)}$$

Now we have accounted for all energy associated with mass in the system and moving across the boundary in the entity balance, Eq. (2.7-2). Next we take into account heat and work energy which transfers across the boundary and is not associated with mass. The term q is the heat transferred per unit time across the boundary to the fluid because of a temperature gradient. Heat absorbed by the system is positive by convention.

The work $\dot{W}$, which is energy per unit time, can be divided into $\dot{W}_S$, purely mechanical shaft work identified with a rotating shaft crossing the control surface, and the pressure–volume work, which has been included in the enthalpy term H in Eq. (2.7-6). By convention, work done by the fluid upon the surroundings, i.e., work out of the system, is positive.

To obtain the overall energy balance, we substitute Eqs. (2.7-4) and (2.7-6) into the entity balance Eq. (2.7-2) and equate the resulting equation to $q - \dot{W}_S$.

$$\iint\limits_{A} \left(H + \frac{v^2}{2} + zg \right)(\rho v) \cos \alpha \, dA + \frac{\partial}{\partial t} \iiint\limits_{V} \left(U + \frac{v^2}{2} + zg \right) \rho \, dV = q - \dot{W}_S \qquad \text{(2.7-7)}$$

2.7C   Overall Energy Balance for Steady-State Flow System

A common special case of the overall or macroscopic energy balance is that of a steady-state system with one-dimensional flow across the boundaries, a single inlet, a single outlet, and negligible variation of height z, density ρ, and enthalpy H across either inlet or outlet area. This is shown in Fig. 2.7-1. Setting the accumulation term in Eq. (2.7-7) equal to zero and integrating.

$$H_2 m_2 - H_1 m_1 + \frac{m_2 (v_2^3)_{av}}{2v_{2\,av}} - \frac{m_1 (v_1^3)_{av}}{2v_{1\,av}} + gm_2 z_2 - gm_1 z_1 = q - \dot{W}_S \qquad \text{(2.7-8)}$$

For steady state, $m_1 = \rho_1 v_{1\,av} A_1 = m_2 = m$. Dividing through by m so that the equation is on a unit mass basis,

$$H_2 - H_1 + \frac{1}{2} \left[\frac{(v_2^3)_{av}}{v_{2\,av}} - \frac{(v_1^3)_{av}}{v_{1\,av}} \right] + g(z_2 - z_1) = Q - W_S \qquad \text{(SI)} \qquad \text{(2.7-9)}$$

The term $(v^3)_{av}/(2v_{av})$ can be replaced by $v_{av}^2/2\alpha$, where α is the kinetic-energy velocity correction factor and is equal to $v_{av}^3/(v^3)_{av}$. The term α has been evaluated for various flows in pipes and is $\frac{1}{2}$ for laminar flow and close to 1.0 for turbulent flow. (See Section 2.7D.) Hence, Eq. (2.7-9) becomes

$$H_2 - H_1 + \frac{1}{2\alpha} (v_{2\,av}^2 - v_{1\,av}^2) + g(z_2 - z_1) = Q - W_S \qquad \text{(SI)}$$

$$\text{(2.7-10)}$$

$$H_2 - H_1 + \frac{1}{2\alpha g_c} (v_{2\,av}^2 - v_{1\,av}^2) + \frac{g}{g_c} (z_2 - z_1) = Q - W_S \qquad \text{(English)}$$

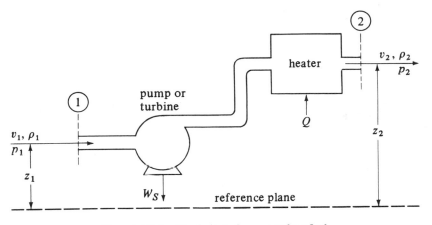

FIGURE 2.7-1. *Steady-state flow system for a fluid.*

Some useful conversion factors to be used are as follows from Appendix A.1:

$$1 \text{ btu} = 778.17 \text{ ft} \cdot \text{lb}_f = 1055.06 \text{ J} = 1.05506 \text{ kJ}$$

$$1 \text{ hp} = 550 \text{ ft} \cdot \text{lb}_f/\text{s} = 0.7457 \text{ kW}$$

$$1 \text{ ft} \cdot \text{lb}_f/\text{lb}_m = 2.9890 \text{ J/kg}$$

$$1 \text{ J} = 1 \text{ N} \cdot \text{m} = 1 \text{ kg} \cdot \text{m}^2/\text{s}^2$$

2.7D Kinetic-Energy Velocity Correction Factor α

1. Introduction.   In obtaining Eq. (2.7-8) it was necessary to integrate the kinetic-energy term,

$$\text{kinetic energy} = \iint\limits_{A} \left(\frac{v^2}{2}\right)(\rho v) \cos \alpha \, dA \qquad (2.7\text{-}11)$$

which appeared in Eq. (2.7-7). To do this we first take ρ as a constant and $\cos \alpha = 1.0$. Then multiplying the numerator and denominator by $v_{av} A$, where v_{av} is the bulk or average velocity and noting that $m = \rho v_{av} A$, Eq. (2.7-11) becomes

$$\frac{\rho}{2} \iint\limits_{A} (v^3) \, dA = \frac{\rho v_{av} A}{2 v_{av} A} \iint\limits_{A} (v^3) \, dA = \frac{m}{2 v_{av}} \frac{1}{A} \iint\limits_{A} (v^3) \, dA \qquad (2.7\text{-}12)$$

Dividing through by m so that Eq. (2.7-12) is on a unit mass basis,

$$\left(\frac{1}{2 v_{av}}\right) \frac{1}{A} \iint\limits_{A} (v^3) \, dA = \frac{(v^3)_{av}}{2 v_{av}} = \frac{v^2_{av}}{2 \alpha} \qquad (2.7\text{-}13)$$

where α is defined as

$$\alpha = \frac{v^3_{av}}{(v^3)_{av}} \qquad (2.7\text{-}14)$$

and $(v^3)_{av}$ is defined as follows:

$$(v^3)_{av} = \frac{1}{A} \iint\limits_A (v^3) \, dA \qquad \text{(2.7-15)}$$

The local velocity v varies across the cross-sectional area of a pipe. To evaluate $(v^3)_{av}$ and, hence, the value of α, we must have an equation relating v as a function of position in the cross-sectional area.

2. Laminar flow. In order to determine the value of α for laminar flow, we first combine Eqs. (2.6-18) and (2.6-20) for laminar flow to obtain v as a function of position r.

$$v = 2v_{av}\left[1 - \left(\frac{r}{R}\right)^2\right] \qquad \text{(2.7-16)}$$

Substituting Eq. (2.7-16) into (2.7-15) and noting that $A = \pi R^2$ and $dA = r \, dr \, d\theta$ (see Example 2.6-3), Eq. (2.7-15) becomes

$$(v^3)_{av} = \frac{1}{\pi R^2} \int_0^{2\pi} \int_0^R \left[2v_{av}\left(1 - \frac{r^2}{R^2}\right)\right]^3 r \, dr \, d\theta$$

$$= \frac{(2\pi)2^3 v_{av}^3}{\pi R^2} \int_0^R \frac{(R^2 - r^2)^3}{R^6} r \, dr = \frac{16v_{av}^3}{R^8}\int_0^R (R^2 - r^2)^3 r \, dr \qquad \text{(2.7-17)}$$

Integrating Eq. (2.7-17) and rearranging,

$$(v^3)_{av} = \frac{16v_{av}^3}{R^8} \int_0^R (R^6 - 3r^2 R^4 + 3r^4 R^2 - r^6) r \, dr$$

$$= \frac{16v_{av}^3}{R^8}\left(\frac{R^8}{2} - \frac{3}{4}R^8 + \frac{1}{2}R^8 - \frac{1}{8}R^8\right)$$

$$= 2v_{av}^3 \qquad \text{(2.7-18)}$$

Substituting Eq. (2.7-18) into (2.7-14),

$$\alpha = \frac{v_{av}^3}{(v^3)_{av}} = \frac{v_{av}^3}{2v_{av}^3} = 0.50 \qquad \text{(2.7-19)}$$

Hence, for laminar flow the value of α to use in the kinetic-energy term of Eq. (2.7-10) is 0.50.

3. Turbulent flow. For turbulent flow a relationship is needed between v and position. This can be approximated by the following expression:

$$v = v_{max}\left(\frac{R-r}{R}\right)^{1/7} \qquad \text{(2.7-20)}$$

where r is the radial distance from the center. This Eq. (2.7-20) is substituted into Eq. (2.7-15) and the resultant integrated to obtain the value of $(v^3)_{av}$. Next, Eq. (2.7-20) is substituted into Eq. (2.6-17) and this equation integrated to obtain v_{av} and $(v_{av})^3$. Combining the results for $(v^3)_{av}$ and $(v_{av})^3$ into Eq. (2.7-14), the value of α is 0.945. (See Problem 2.7-1 for solution.) The value of α for turbulent flow varies from about 0.90 to 0.99. In most cases (except for precise work) the value of α is taken to be 1.0.

2.7E Applications of Overall Energy-Balance Equation

The total energy balance, Eq. (2.7-10), in the form given is not often used when appreciable enthalpy changes occur or appreciable heat is added (or subtracted) since the kinetic- and potential-energy terms are usually small and can be neglected. As a result, when appreciable heat is added or subtracted or large enthalpy changes occur, the methods of doing heat balances described in Section 1.7 are generally used. Examples will be given to illustrate this and other cases.

EXAMPLE 2.7-1. Energy Balance on Steam Boiler
Water enters a boiler at 18.33°C and 137.9 kPa through a pipe at an average velocity of 1.52 m/s. Exit steam at a height of 15.2 m above the liquid inlet leaves at 137.9 kPa, 148.9°C, and 9.14 m/s in the outlet line. At steady state how much heat must be added per kg mass of steam? The flow in the two pipes is turbulent.

Solution: The process flow diagram is shown in Fig. 2.7-2. Rearranging Eq. (2.7-10) and setting $\alpha = 1$ for turbulent flow and $W_S = 0$ (no external work),

$$Q = (z_2 - z_1)g + \frac{v_2^2 - v_1^2}{2} + (H_2 - H_1) \qquad (2.7\text{--}21)$$

To solve for the kinetic-energy terms,

$$\frac{v_1^2}{2} = \frac{(1.52)^2}{2} = 1.115 \text{ J/kg}$$

$$\frac{v_2^2}{2} = \frac{(9.14)^2}{2} = 41.77 \text{ J/kg}$$

Taking the datum height z_1 at point 1, $z_2 = 15.2$ m. Then,

$$z_2 g = (15.2)(9.80665) = 149.1 \text{ J/kg}$$

From Appendix A.2, steam tables in SI units, H_1 at 18.33°C = 76.97 kJ/kg, H_2 of superheated steam at 148.9°C = 2771.4 kJ/kg, and

$$H_2 - H_1 = 2771.4 - 76.97 = 2694.4 \text{ kJ/kg} = 2.694 \times 10^6 \text{ J/kg}$$

Substituting these values into Eq. (2.7-21),

$$Q = (149.1 - 0) + (41.77 - 1.115) + 2.694 \times 10^6$$

$$Q = 189.75 + 2.694 \times 10^6 = 2.6942 \times 10^6 \text{ J/kg}$$

Hence, the kinetic-energy and potential-energy terms totaling 189.75 J/kg are negligible compared to the enthalpy change of 2.694×10^6 J/kg. This 189.75 J/kg would raise the temperature of liquid water about 0.0453°C, a negligible amount.

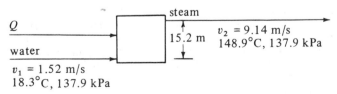

FIGURE 2.7-2. *Process flow diagram for Example 2.7-1.*

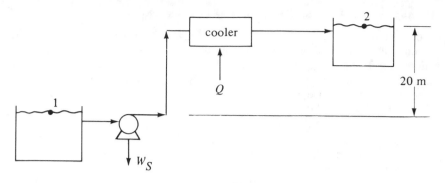

FIGURE 2.7-3. *Process flow diagram for energy balance for Example 2.7-2.*

EXAMPLE 2.7–2. Energy Balance on a Flow System with a Pump

Water at 85.0°C is being stored in a large, insulated tank at atmospheric pressure as shown in Fig. 2.7-3. It is being pumped at steady state from this tank at point 1 by a pump at the rate of 0.567 m³/min. The motor driving the pump supplies energy at the rate of 7.45 kW. The water passes through a heat exchanger, where it gives up 1408 kW of heat. The cooled water is then delivered to a second, large open tank at point 2, which is 20 m above the first tank. Calculate the final temperature of the water delivered to the second tank. Neglect any kinetic-energy changes since the initial and final velocities in the tanks are essentially zero.

Solution: From Appendix A.2, steam tables, H_1 (85°C) = 355.90 × 10³ J/kg, ρ_1 = 1/0.0010325 = 968.5 kg/m³. Then, for steady state,

$$m_1 = m_2 = (0.567)(968.5)(\tfrac{1}{60}) = 9.152 \text{ kg/s}$$

Also, $z_1 = 0$ and $z_2 = 20$ m. The work done by the fluid is W_S, but in this case work is done on the fluid and W_S is negative.

$$W_S = -(7.45 \times 10^3 \text{ J/s})(1/9.152 \text{ kg/s}) = -0.8140 \times 10^3 \text{ J/kg}$$

The heat added to the fluid is also negative since it gives up heat and is

$$Q = -(1408 \times 10^3 \text{ J/s})(1/9.152 \text{ kg/s}) = -153.8 \times 10^3 \text{ J/kg}$$

Setting $(v_1^2 - v_2^2)/2 = 0$ and substituting into Eq. (2.7-10),

$$H_2 - 355.90 \times 10^3 + 0 + 9.80665(20 - 0)$$
$$= (-153.8 \times 10^3) - (-0.814 \times 10^3)$$

Solving, $H_2 = 202.71 \times 10^3$ J/kg. From the steam tables this corresponds to $t_2 = 48.41$°C. Note that in this example, W_S and $g(z_2 - z_1)$ are very small compared to Q.

EXAMPLE 2.7-3. Energy Balance in Flow Calorimeter

A flow calorimeter is being used to measure the enthalpy of steam. The calorimeter, which is a horizontal insulated pipe, consists of an electric heater immersed in a fluid flowing at steady state. Liquid water at 0°C at a rate of 0.3964 kg/min enters the calorimeter at point 1. The liquid is vaporized completely by the heater, where 19.63 kW is added and steam leaves point 2 at 250°C and 150 kPa absolute. Calculate the exit enthalpy H_2 of the steam if the liquid enthalpy at 0°C is set arbitrarily as 0. The kinetic-energy changes are small and can be neglected. (It will be assumed that pressure has a negligible effect on the enthalpy of the liquid.)

Solution: For this case, $W_S = 0$ since there is no shaft work between points 1 and 2. Also, $(v_2^2/2\alpha - v_1^2/2\alpha) = 0$ and $g(z_2 - z_1) = 0$. For steady state, $m_1 = m_2 = 0.3964/60 = 6.607 \times 10^{-3}$ kg/s. Since heat is added to the system,

$$Q = +\frac{19.63 \text{ kJ/s}}{6.607 \times 10^{-3} \text{ kg/s}} = 2971 \text{ kJ/kg}$$

The value of $H_1 = 0$. Equation (2.7-10) becomes

$$H_2 - H_1 + 0 + 0 = Q - 0$$

The final equation for the calorimeter is

$$H_2 = Q + H_1 \tag{2.7-22}$$

Substituting $Q = 2971$ kJ/kg and $H_1 = 0$ into Eq. (2.7-22), $H_2 = 2971$ kJ/kg at 250°C and 150 kPa, which is close to the value from the steam table of 2972.7 kJ/kg.

2.7F Overall Mechanical-Energy Balance

A more useful type of energy balance for flowing fluids, especially liquids, is a modification of the total energy balance to deal with mechanical energy. Engineers are often concerned with this special type of energy, called *mechanical energy*, which includes the work term, kinetic energy, potential energy, and the flow work part of the enthalpy term. Mechanical energy is a form of energy that is either work or a form that can be directly converted into work. The other terms in the energy-balance equation (2.7-10), heat terms and internal energy, do not permit simple conversion into work because of the second law of thermodynamics and the efficiency of conversion, which depends on the temperatures. Mechanical-energy terms have no such limitation and can be converted almost completely into work. Energy converted to heat or internal energy is lost work or a loss in mechanical energy which is caused by frictional resistance to flow.

It is convenient to write an energy balance in terms of this loss, $\sum F$, which is the sum of all frictional losses per unit mass. For the case of steady-state flow, when a unit mass of fluid passes from inlet to outlet, the batch work done by the fluid, W', is expressed as

$$W' = \int_{V_1}^{V_2} p \, dV - \sum F \qquad (\sum F > 0) \tag{2.7-23}$$

This work W' differs from the W of Eq. (2.7-1), which also includes kinetic- and potential-energy effects. Writing the first law of thermodynamics for this case, where ΔE becomes ΔU,

$$\Delta U = Q - W' \tag{2.7-24}$$

The equation defining enthalpy, Eq. (2.7-5), can be written as

$$\Delta H = \Delta U + \Delta pV = \Delta U + \int_{V_1}^{V_2} p \, dV + \int_{p_1}^{p_2} V \, dp \tag{2.7-25}$$

Substituting Eq. (2.7-23) into (2.7-24) and then combining the resultant with Eq. (2.7-25), we obtain

$$\Delta H = Q + \sum F + \int_{p_1}^{p_2} V \, dp \qquad (2.7\text{-}26)$$

Finally, we substitute Eq. (2.7-26) into (2.7-10) and $1/\rho$ for V, to obtain the overall mechanical-energy-balance equation:

$$\frac{1}{2\alpha} [v_{2 \text{ av}}^2 - v_{1 \text{ av}}^2] + g(z_2 - z_1) + \int_{p_1}^{p_2} \frac{dp}{\rho} + \sum F + W_S = 0 \qquad (2.7\text{-}27)$$

For English units the kinetic- and potential-energy terms of Eq. (2.7-27) are divided by g_c.

The value of the integral in Eq. (2.7-27) depends on the equation of state of the fluid and the path of the process. If the fluid is an incompressible liquid, the integral becomes $(p_2 - p_1)/\rho$ and Eq. (2.7-27) becomes

$$\frac{1}{2\alpha} (v_{2 \text{ av}}^2 - v_{1 \text{ av}}^2) + g(z_2 - z_1) + \frac{p_2 - p_1}{\rho} + \sum F + W_S = 0 \qquad (2.7\text{-}28)$$

EXAMPLE 2.7-4. Mechanical-Energy Balance on Pumping System
Water with a density of 998 kg/m³ is flowing at a steady mass flow rate through a uniform-diameter pipe. The entrance pressure of the fluid is 68.9 kN/m² abs in the pipe, which connects to a pump which actually supplies 155.4 J/kg of fluid flowing in the pipe. The exit pipe from the pump is the same diameter as the inlet pipe. The exit section of the pipe is 3.05 m higher than the entrance, and the exit pressure is 137.8 kN/m² abs. The Reynolds number in the pipe is above 4000 in the system. Calculate the frictional loss $\sum F$ in the pipe system.

Solution: First a flow diagram is drawn of the system (Fig. 2.7-4), with 155.4 J/kg mechanical energy added to the fluid. Hence, $W_S = -155.4$, since the work done by the fluid is positive.

Setting the datum height $z_1 = 0$, $z_2 = 3.05$ m. Since the pipe is of

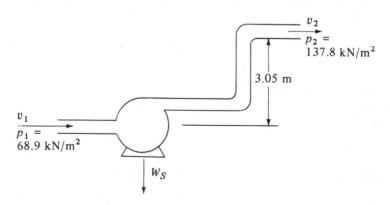

FIGURE 2.7-4. *Process flow diagram for Example 2.7-4.*

constant diameter, $v_1 = v_2$. Also, for turbulent flow $\alpha = 1.0$ and

$$\frac{1}{2(1)} (v_2^2 - v_1^2) = 0$$

$$z_2 g = (3.05 \text{ m})(9.806 \text{ m/s}^2) = 29.9 \text{ J/kg}$$

Since the liquid can be considered incompressible, Eq. (2.7-28) is used.

$$\frac{p_1}{\rho} = \frac{68.9 \times 1000}{998} = 69.0 \text{ J/kg}$$

$$\frac{p_2}{\rho} = \frac{137.8 \times 1000}{998} = 138.0 \text{ J/kg}$$

Using Eq. (2.7-28) and solving for $\sum F$, the frictional losses,

$$\sum F = -W_s + \frac{1}{2\alpha} (v_1^2 - v_2^2) + g(z_1 - z_2) + \frac{p_1 - p_2}{\rho} \qquad \textbf{(2.7-29)}$$

Substituting the known values, and solving for the frictional losses,

$$\sum F = -(-155.4) + 0 - 29.9 + 69.0 - 138.0$$

$$= 56.5 \text{ J/kg} \left(18.9 \frac{\text{ft} \cdot \text{lb}_f}{\text{lb}_m} \right)$$

EXAMPLE 2.7-5. Pump Horsepower in Flow System

A pump draws 69.1 gal/min of a liquid solution having a density of 114.8 lb_m/ft^3 from an open storage feed tank of large cross-sectional area through a 3.068-in.-ID suction line. The pump discharges its flow through a 2.067-in.-ID line to an open overhead tank. The end of the discharge line is 50 ft above the level of the liquid in the feed tank. The friction losses in the piping system are $\sum F = 10.0$ ft-lb force/lb mass. What pressure must the pump develop and what is the horsepower of the pump if its efficiency is 65% ($\eta = 0.65$)? The flow is turbulent.

Solution: First, a flow diagram of the system is drawn (Fig. 2.7-5). Equation (2.7-28) will be used. The term W_s in Eq. (2.7-28) becomes

$$W_S = -\eta W_p \qquad \textbf{(2.7-30)}$$

where $-W_S$ = mechanical energy actually delivered to the fluid by the

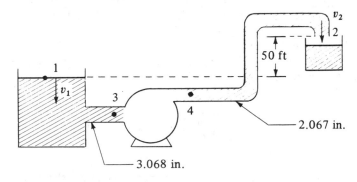

FIGURE 2.7-5.   *Process flow diagram for Example 2.7-5.*

pump or net mechanical work, η = fractional efficiency, and W_p is the energy or shaft work delivered to the pump.

From Appendix A.5, the cross-sectional area of the 3.068-in. pipe is 0.05134 ft^2 and of the 2.067-in. pipe, 0.0233 ft^2. The flow rate is

$$\text{flow rate} = \left(69.1 \frac{\text{gal}}{\text{min}}\right)\left(\frac{1 \text{ min}}{60 \text{ s}}\right)\left(\frac{1 \text{ ft}^3}{7.481 \text{ gal}}\right) = 0.1539 \text{ ft}^3/\text{s}$$

$$v_2 = \left(0.1539 \frac{\text{ft}^3}{\text{s}}\right)\left(\frac{1}{0.0233 \text{ ft}^2}\right) = 6.61 \text{ ft/s}$$

$v_1 = 0$, since the tank is very large. Then $v_1^2/2g_c = 0$. The pressure $p_1 = 1$ atm and $p_2 = 1$ atm. Also, $\alpha = 1.0$ since the flow is turbulent. Hence,

$$\frac{p_1}{\rho} - \frac{p_2}{\rho} = 0$$

$$\frac{v_2^2}{2g_c} = \frac{(6.61)^2}{2(32.174)} = 0.678 \frac{\text{ft} \cdot \text{lb}_f}{\text{lb}_m}$$

Using the datum of $z_1 = 0$, we have

$$z_2 \frac{g}{g_c} = (50.0)\frac{32.174}{32.174} = 50.0 \frac{\text{ft} \cdot \text{lb}_f}{\text{lb}_m}$$

Using Eq. (2.4-28), solving for W_S, and substituting the known values,

$$W_S = z_1 \frac{g}{g_c} - z_2 \frac{g}{g_c} + \frac{v_1^2}{2g_c} - \frac{v_2^2}{2g_c} + \frac{p_1 - p_2}{\rho} - \sum F$$

$$= 0 - 50.0 + 0 - 0.678 + 0 - 10 = -60.678 \frac{\text{ft} \cdot \text{lb}_f}{\text{lb}_m}$$

Using Eq. (2.7-30) and solving for W_p,

$$W_p = -\frac{W_S}{\eta} = \frac{60.678}{0.65} \frac{\text{ft} \cdot \text{lb}_f}{\text{lb}_m} = 93.3 \frac{\text{ft} \cdot \text{lb}_f}{\text{lb}_m}$$

$$\text{mass flow rate} = \left(0.1539 \frac{\text{ft}^3}{\text{s}}\right)\left(114.8 \frac{\text{lb}_m}{\text{ft}^3}\right) = 17.65 \frac{\text{lb}_m}{\text{s}}$$

$$\text{pump horsepower} = \left(17.65 \frac{\text{lb}_m}{\text{s}}\right)\left(93.3 \frac{\text{ft} \cdot \text{lb}_f}{\text{lb}_m}\right)\left(\frac{1 \text{ hp}}{550 \text{ ft} \cdot \text{lb}_f/\text{s}}\right)$$

$$= 3.00 \text{ hp}$$

To calculate the pressure the pump must develop, Eq. (2.7-28) must be written over the pump itself between points 3 and 4 as shown on the diagram.

$$v_3 = \left(0.1539 \frac{\text{ft}^3}{\text{s}}\right)\left(\frac{1}{0.05134 \text{ ft}^2}\right) = 3.00 \text{ ft/s}$$

$$v_4 = v_2 = 6.61 \text{ ft/s}$$

Since the difference in level between z_3 and z_4 of the pump itself is negligible, it will be neglected. Rewriting Eq. (2.7-28) between points 3 and 4 and substituting known values ($\sum F = 0$ since this is for the piping system),

$$\frac{p_4 - p_3}{\rho} = z_3 \frac{g}{g_c} - z_4 \frac{g}{g_c} + \frac{v_3^2}{2g_c} - \frac{v_4^2}{2g_c} - W_s - \sum F \qquad \text{(2.7-31)}$$

$$= 0 - 0 + \frac{(3.00)^2}{2(32.174)} - \frac{(6.61)^2}{2(32.174)} + 60.678 - 0$$

$$= 0 - 0 + 0.140 - 0.678 + 60.678 = 60.14 \frac{\text{ft} \cdot \text{lb}_f}{\text{lb}_m}$$

$$p_4 - p_3 = \left(60.14 \frac{\text{ft} \cdot \text{lb}_f}{\text{lb}_m} \right) \left(114.8 \frac{\text{lb}_m}{\text{ft}^3} \right) \left(\frac{1}{144 \text{ in.}^2/\text{ft}^2} \right)$$

$$= 48.0 \text{ lb force/in.}^2 \text{ (psia pressure developed by pump)} \quad \text{(331 kPa)}$$

2.7G Bernoulli Equation for Mechanical-Energy Balance

In the special case where no mechanical energy is added ($W_S = 0$) and for no friction ($\sum F = 0$), then Eq. (2.7-28) becomes the Bernoulli equation, Eq. (2.7-32), for turbulent flow, which is of sufficient importance to deserve further discussion.

$$z_1 g + \frac{v_1^2}{2} + \frac{p_1}{\rho} = z_2 g + \frac{v_2^2}{2} + \frac{p_2}{\rho} \qquad \text{(2.7-32)}$$

This equation covers many situations of practical importance and is often used in conjunction with the mass-balance equation (2.6-2) for steady state.

$$m = \rho_1 A_1 v_1 = \rho_2 A_2 v_2 \qquad \text{(2.6-2)}$$

Several examples of its use will be given.

EXAMPLE 2.7-6. Rate of Flow from Pressure Measurements

A liquid with a constant density ρ kg/m^3 is flowing at an unknown velocity v_1 m/s through a horizontal pipe of cross-sectional area A_1 m^2 at a pressure p_1 N/m^2, and then it passes to a section of the pipe in which the area is reduced gradually to A_2 m^2 and the pressure is p_2. Assuming no friction losses, calculate the velocity v_1 and v_2 if the pressure difference $(p_1 - p_2)$ is measured.

Solution: In Fig. 2.7-6, the flow diagram is shown with pressure taps to measure p_1 and p_2. From the mass-balance continuity equation (2.6-2), for constant ρ where $\rho_1 = \rho_2 = \rho$,

$$v_2 = \frac{v_1 A_1}{A_2} \qquad \text{(2.7-33)}$$

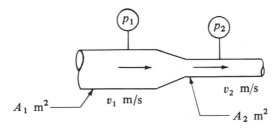

FIGURE 2.7-6. *Process flow diagram for Example 2.7-6.*

For the items in the Bernoulli equation (2.7-32), for a horizontal pipe,

$$z_1 = z_2 = 0$$

Then Eq. (2.7-32) becomes, after substituting Eq. (2.7-33) for v_2,

$$0 + \frac{v_1^2}{2} + \frac{p_1}{\rho} = 0 + \frac{v_1^2 A_1^2 / A_2^2}{2} + \frac{p_2}{\rho} \qquad \textbf{(2.7-34)}$$

Rearranging,

$$p_1 - p_2 = \frac{\rho v_1^2 [(A_1 / A_2)^2 - 1]}{2} \qquad \textbf{(2.7-35)}$$

$$v_1 = \sqrt{\frac{p_1 - p_2}{\rho} \frac{2}{[(A_1/A_2)^2 - 1]}} \qquad \text{(SI)}$$

$$\qquad \textbf{(2.7-36)}$$

$$v_1 = \sqrt{\frac{p_1 - p_2}{\rho} \frac{2g_c}{[(A_1/A_2)^2 - 1]}} \qquad \text{(English)}$$

Performing the same derivation but in terms of v_2,

$$v_2 = \sqrt{\frac{p_1 - p_2}{\rho} \frac{2}{1 - (A_2/A_1)^2}} \qquad \textbf{(2.7-37)}$$

EXAMPLE 2.7-7.   Rate of Flow from a Nozzle in a Tank

A nozzle of cross-sectional area A_2 is discharging to the atmosphere and is located in the side of a large tank, in which the open surface of the liquid in the tank is H m above the center line of the nozzle. Calculate the velocity v_2 in the nozzle and the volumetric rate of discharge if no friction losses are assumed.

Solution: The process flow is shown in Fig. 2.7-7, with point 1 taken in the liquid at the entrance to the nozzle and point 2 at the exit of the nozzle. Since A_1 is very large compared to A_2, $v_1 \cong 0$. The pressure p_1 is greater than 1 atm (101.3 kN/m²) by the head of fluid of H m. The pressure p_2, which is at the nozzle exit, is at 1 atm. Using point 2 as a datum, $z_2 = 0$ and $z_1 = 0$ m. Rearranging Eq. (2.7-32),

$$z_1 g + \frac{v_1^2}{2} + \frac{p_1 - p_2}{\rho} = z_2 g + \frac{v_2^2}{2} \qquad \textbf{(2.7-38)}$$

Substituting the known values,

$$0 + 0 + \frac{p_1 - p_2}{\rho} = 0 + \frac{v_2^2}{2} \qquad \textbf{(2.7-39)}$$

Solving for v_2,

$$v_2 = \sqrt{\frac{2(p_1 - p_2)}{\rho}} \qquad \text{m/s} \qquad \textbf{(2.7-40)}$$

FIGURE 2.7-7.   *Nozzle flow diagram for Example 2.7-7.*

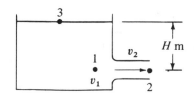

Since $p_1 - p_3 = H\rho g$ and $p_3 = p_2$ (both at 1 atm),

$$H = \frac{p_1 - p_2}{\rho g} \qquad \text{m} \tag{2.7-41}$$

where H is the head of liquid with density ρ. Then Eq. (2.4-40) becomes

$$v_2 = \sqrt{2gH} \tag{2.7-42}$$

The volumetric flow rate is

$$\text{flow rate} = v_2 A_2 \qquad \text{m}^3/\text{s} \tag{2.7-43}$$

To illustrate the fact that different points can be used in the balance, points 3 and 2 will be used. Writing Eq. (2.7-32),

$$z_2 g + \frac{v_2^2}{2} + \frac{p_2 - p_3}{\rho} = z_3 g + \frac{v_3^2}{2} \tag{2.7-44}$$

Since $p_2 = p_3 = 1$ atm, $v_3 = 0$, and $z_2 = 0$,

$$v_2 = \sqrt{2gz_3} = \sqrt{2gH} \tag{2.7-45}$$

2.8 OVERALL MOMENTUM BALANCE

2.8A Derivation of General Equation

A momentum balance can be written for the control volume shown in Fig. 2.6-3, which is somewhat similar to the overall mass-balance equation. Momentum, in contrast to mass and energy, is a vector quantity. The total linear momentum vector $\mathbf{P}$ of the total mass M of a moving fluid having a velocity of $\mathbf{v}$ is

$$\mathbf{P} = M\mathbf{v} \tag{2.8-1}$$

The term $M\mathbf{v}$ is the momentum of this moving mass M enclosed at a particular instant in the control volume shown in Fig. 2.6-4. The units of $M\mathbf{v}$ are kg·m/s in the SI system.

Starting with Newton's second law we will develop the integral momentum-balance equation for linear momentum. Angular momentum will not be considered here. Newton's law may be stated: The time rate of change of momentum of a system is equal to the summation of all forces acting on the system and takes place in the direction of the net force.

$$\sum \mathbf{F} = \frac{d\mathbf{P}}{dt} \tag{2.8-2}$$

where $\mathbf{F}$ is force. In the SI system $\mathbf{F}$ is in newtons (N) and $1\ \text{N} = 1\ \text{kg·m/s}^2$. Note that in the SI system g_c is not needed, but it is needed in the English system.

The equation for the conservation of momentum with respect to a control volume can be written as follows:

$$\begin{pmatrix} \text{sum of forces acting} \\ \text{on control volume} \end{pmatrix} = \begin{pmatrix} \text{rate of momentum} \\ \text{out of control volume} \end{pmatrix} - \begin{pmatrix} \text{rate of momentum} \\ \text{into control volume} \end{pmatrix}$$

$$+ \begin{pmatrix} \text{rate of accumulation of momentum} \\ \text{in control volume} \end{pmatrix} \tag{2.8-3}$$

This is in the same form as the general mass-balance equation (2.6-3), with the sum of the forces as the generation rate term. Hence, momentum is not conserved, since it is generated by external forces on the system. If external forces are absent, momentum is conserved.

Using the general control volume shown in Fig. 2.6-4, we shall evaluate the various terms in Eq. (2.8-3), using methods very similar to the development of the general mass balance. For a small element of area dA on the control surface, we write

$$\text{rate of momentum efflux} = \mathbf{v}(\rho v)(dA \cos \alpha) \tag{2.8-4}$$

Note that the rate of mass efflux is $(\rho v)(dA \cos \alpha)$. Also, note that $(dA \cos \alpha)$ is the area dA projected in a direction normal to the velocity vector $\mathbf{v}$ and α is the angle between the velocity vector $\mathbf{v}$ and the outward-directed-normal vector $\mathbf{n}$. From vector algebra the product in Eq. (2.8-4) becomes

$$\mathbf{v}(\rho v)(dA \cos \alpha) = \rho \mathbf{v}(\mathbf{v} \cdot \mathbf{n}) \, dA \tag{2.8-5}$$

Integrating over the entire control surface A,

$$\begin{pmatrix} \text{net momentum efflux} \\ \text{from control volume} \end{pmatrix} = \iint_A \mathbf{v}(\rho v) \cos \alpha \, dA = \iint_A \rho \mathbf{v}(\mathbf{v} \cdot \mathbf{n}) \, dA \tag{2.8-6}$$

The net efflux represents the first two terms on the right-hand side of Eq. (2.8-3).

Similarly to Eq. (2.6-5), the rate of accumulation of linear momentum within the control volume V is

$$\begin{pmatrix} \text{rate of accumulation of momentum} \\ \text{in control volume} \end{pmatrix} = \frac{\partial}{\partial t} \iiint_V \rho \mathbf{v} \, dV \tag{2.8-7}$$

Substituting Equations (2.8-2), (2.8-6), and (2.8-7) into (2.8-3), the overall linear momentum balance for a control volume becomes

$$\sum \mathbf{F} = \iint_A \rho \mathbf{v}(\mathbf{v} \cdot \mathbf{n}) \, dA + \frac{\partial}{\partial t} \iiint_V \rho \mathbf{v} \, dV \tag{2.8-8}$$

We should note that $\sum \mathbf{F}$ in general may have a component in any direction, and the $\mathbf{F}$ is the force the surroundings exert on the control-volume fluid. Since Eq. (2.8-8) is a vector equation, we may write the component scalar equations for the x, y, and z directions.

$$\sum F_x = \iint_A v_x \rho v \cos \alpha \, dA + \frac{\partial}{\partial t} \iiint_V \rho v_x \, dV \qquad \text{(SI)}$$

$$\tag{2.8-9}$$

$$\sum F_x = \iint_A v_x \frac{\rho}{g_c} v \cos \alpha \, dA + \frac{\partial}{\partial t} \iiint_V \frac{\rho}{g_c} v_x \, dV \qquad \text{(English)}$$

$$\sum F_y = \iint_A v_y \rho v \cos \alpha \, dA + \frac{\partial}{\partial t} \iiint_V \rho v_y \, dV \tag{2.8-10}$$

$$\sum F_z = \iint_A v_z \rho v \cos \alpha \, dA + \frac{\partial}{\partial t} \iiint_V \rho v_z \, dV \tag{2.8-11}$$

The force term $\sum F_x$ in Eq. (2.8-9) is composed of the sum of several forces. These are given as follows.

1. *Body force.* The body force F_{xg} is the x-directed force caused by gravity acting on the total mass M in the control volume. This force, F_{xg}, is Mg_x. It is zero if the x direction is horizontal.
2. *Pressure force.* The force F_{xp} is the x-directed force caused by the pressure forces acting on the surface of the fluid system. When the control surface cuts through the fluid, the pressure is taken to be directed inward and perpendicular to the surface. In some cases part of the control surface may be a solid, and this wall is included inside the control surface. Then there is a contribution to F_{xp} from the pressure on the outside of this wall, which is typically atmospheric pressure. If gage pressure is used, the integral of the constant external pressure over the entire outer surface can be automatically ignored.
3. *Friction force.* When the fluid is flowing, an x-directed shear or friction force F_{xs} is present, which is exerted on the fluid by a solid wall when the control surface cuts between the fluid and the solid wall. In some or many cases this frictional force may be negligible compared to the other forces and is neglected.
4. *Solid surface force.* In cases where the control surface cuts through a solid, there is present force R_x, which is the x component of the resultant of the forces acting on the control volume at these points. This occurs in typical cases when the control volume includes a section of pipe and the fluid it contains. This is the force exerted by the solid surface on the fluid.

The force terms of Eq. (2.8-9) can then be represented as

$$\sum F_x = F_{xg} + F_{xp} + F_{xs} + R_x \tag{2.8-12}$$

Similar equations can be written for the y and z directions. Then Eq. (2.8-9) becomes for the x direction,

$$\sum F_x = F_{xg} + F_{xp} + F_{xs} + R_x$$
$$= \iint_A v_x \, \rho v \cos \alpha \, dA + \frac{\partial}{\partial t} \iiint_V \rho v_x \, dV \tag{2.8-13}$$

2.8B Overall Momentum Balance in Flow System in One Direction

A quite common application of the overall momentum-balance equation is the case of a section of a conduit with its axis in the x direction. The fluid will be assumed to be flowing at steady state in the control volume shown in Fig. 2.6-3 and also shown in Fig. 2.8-1. Equation (2.8-13) for the x direction becomes as follows since $v = v_x$.

$$\sum F_x = F_{xg} + F_{xp} + F_{xs} + R_x = \iint_A v_x \, \rho v_x \cos \alpha \, dA \tag{2.8-14}$$

Integrating with $\cos \alpha = \pm 1.0$ and $\rho A = m/v_{\mathrm{av}}$,

$$F_{xg} + F_{xp} + F_{xs} + R_x = m \frac{(v_{x2}^2)_{\mathrm{av}}}{v_{x2 \, \mathrm{av}}} - m \frac{(v_{x1}^2)_{\mathrm{av}}}{v_{x1 \, \mathrm{av}}} \tag{2.8-15}$$

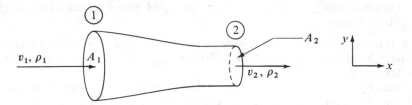

FIGURE 2.8-1.   *Flow through a horizontal nozzle in the x direction only.*

where if the velocity is not constant and varies across the surface area,

$$(v_x^2)_{av} = \frac{1}{A} \iint_A v_x^2 \, dA \qquad (2.8\text{-}16)$$

The ratio $(v_x^2)_{av}/v_{x\,av}$ is replaced by $v_{x\,av}/\beta$, where β, which is the momentum velocity correction factor, has a value of 0.95 to 0.99 for turbulent flow and $\frac{3}{4}$ for laminar flow. For most applications in turbulent flow, $(v_x)_{av}^2/v_{x\,av}$ is replaced by $v_{x\,av}$, the average bulk velocity. Note that the subscript x on v_x and F_x can be dropped since $v_x = v$ and $F_x = F$ for one directional flow.

The term F_{xp}, which is the force caused by the pressures acting on the surface of the control volume, is

$$F_{xp} = p_1 A_1 - p_2 A_2 \qquad (2.8\text{-}17)$$

The friction force will be neglected in Eq. (2.8-15), so $F_{xs} = 0$. The body force $F_{xg} = 0$ since gravity is acting only in the y direction. Substituting F_{xp} from Eq. (2.8-17) into (2.8-15), replacing $(v_x^2)_{av}/v_{x\,av}$ by v/β (where $v_{x\,av} = v$), setting $\beta = 1.0$, and solving for R_x in Eq. (2.8-15),

$$R_x = mv_2 - mv_1 + p_2 A_2 - p_1 A_1 \qquad (2.8\text{-}18)$$

where R_x is the force exerted by the solid on the fluid. The force of the fluid on the solid (reaction force) is the negative of this or $-R_x$.

EXAMPLE 2.8-1.   *Momentum Velocity Correction Factor β for Laminar Flow*
The momentum velocity correction factor β is defined as follows for flow in one direction where the subscript x is dropped.

$$\frac{(v^2)_{av}}{v_{av}} = \frac{v_{av}}{\beta} \qquad (2.8\text{-}19)$$

$$\beta = \frac{(v_{av})^2}{(v^2)_{av}} \qquad (2.8\text{-}20)$$

Determine β for laminar flow in a tube.

Solution: Using Eq. (2.8-16),

$$(v^2)_{av} = \frac{1}{A} \iint_A v^2 \, dA \qquad (2.8\text{-}21)$$

Chap. 2   Principles of Momentum Transfer and Overall Balances

Substituting Eq. (2.7-16) for laminar flow into Eq. (2.8-21) and noting that $A = \pi R^2$ and $dA = r\,dr\,d\theta$, we obtain (see Example 2.6-3)

$$(v^2)_{av} = \frac{1}{\pi R^2} \int_0^{2\pi} \int_0^R \left[2v_{av}\left(1 - \frac{r^2}{R^2}\right) \right]^2 r\,dr\,d\theta$$

$$= \frac{(2\pi)2^2 v_{av}^2}{\pi R^2} \int_0^R \frac{(R^2 - r^2)^2}{R^4}\, r\,dr \qquad \text{(2.8-22)}$$

Integrating Eq. (2.8-22) and rearranging,

$$(v^2)_{av} = \frac{8v_{av}^2}{R^6}\left(\frac{R^6}{2} - \frac{R^6}{2} + \frac{R^6}{6} \right) = \tfrac{4}{3}v_{av}^2 \qquad \text{(2.8-23)}$$

Substituting Eq. (2.8-23) into (2.8-20), $\beta = \tfrac{3}{4}$.

EXAMPLE 2.8-2. Momentum Balance for Horizontal Nozzle

Water is flowing at a rate of 0.03154 m³/s through a horizontal nozzle shown in Fig. 2.8-1 and discharges to the atmosphere at point 2. The nozzle is attached at the upstream end at point 1 and frictional forces are considered negligible. The upstream ID is 0.0635 m and the downstream 0.0286 m. Calculate the resultant force on the nozzle. The density of the water is 1000 kg/m³.

Solution: First, the mass flow and average or bulk velocities at points 1 and 2 are calculated. The area at point 1 is $A_1 = (\pi/4)(0.0635)^2 = 3.167 \times 10^{-3}$ m² and $A_2 = (\pi/4)(0.0286)^2 = 6.424 \times 10^{-4}$ m². Then,

$$m_1 = m_2 = m = (0.03154)(1000) = 31.54 \text{ kg/s}$$

The velocity at point 1 is $v_1 = 0.03154/(3.167 \times 10^{-3}) = 9.96$ m/s and $v_2 = 0.03154/(6.424 \times 10^{-4}) = 49.1$ m/s.

To evaluate the upstream pressure p_1 we use the mechanical-energy balance equation (2.7-28) assuming no frictional losses and turbulent flow. (This can be checked by calculating the Reynolds number.) This equation then becomes, for $\alpha = 1.0$,

$$\frac{v_1^2}{2} + \frac{p_1}{\rho} = \frac{v_2^2}{2} + \frac{p_2}{\rho} \qquad \text{(2.8-24)}$$

Setting $p_2 = 0$ gage pressure, $\rho = 1000$ kg/m³, $v_1 = 9.96$ m/s, $v_2 = 49.1$ m/s, and solving for p_1,

$$p_1 = \frac{(1000)(49.1^2 - 9.96^2)}{2} = 1.156 \times 10^6 \text{ N/m}^2 \qquad \text{(gage pressure)}$$

For the x direction, the momentum balance equation (2.8-18) is used. Substituting the known values and solving for R_x,

$$R_x = 31.54(49.10 - 9.96) + 0 - (1.156 \times 10^6)(3.167 \times 10^{-3})$$

$$= -2427 \text{ N}(-546 \text{ lb}_f)$$

Since the force is negative, it is acting in the negative x direction or to the left. This is the force of the nozzle on the fluid. The force of the fluid on the solid is $-R_x$ or $+2427$ N.

2.8C   Overall Momentum Balance in Two Directions

Another application of the overall momentum balance is shown in Fig. 2.8-2 for a flow system with fluid entering a conduit at point 1 inclined at an angle of α_1 relative to the

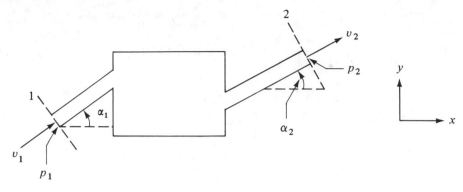

FIGURE 2.8-2. *Overall momentum balance for flow system with fluid entering at point 1 and leaving at 2.*

horizontal x direction and leaving a conduit at point 2 at an angle α_2. The fluid will be assumed to be flowing at steady state and the frictional force F_{xs} will be neglected. Then Eq. (2.8-13) for the x direction becomes as follows for no accumulation:

$$F_{xg} + F_{xp} + R_x = \iint\limits_A v_x \rho v \cos \alpha \, dA \tag{2.8-25}$$

Integrating the surface (area) integral,

$$F_{xg} + F_{xp} + R_x = m \frac{(v_2^2)_{av}}{v_{2\,av}} \cos \alpha_2 - m \frac{(v_1^2)_{av}}{v_{1\,av}} \cos \alpha_1 \tag{2.8-26}$$

The term $(v^2)_{av}/v_{av}$ can again be replaced by v_{av}/β with β being set at 1.0. From Fig. 2.8-2, the term F_{xp} is

$$F_{xp} = p_1 A_1 \cos \alpha_1 - p_2 A_2 \cos \alpha_2 \tag{2.8-27}$$

Then Eq. (2.8-26) becomes as follows after solving for R_x:

$$R_x = m v_2 \cos \alpha_2 - m v_1 \cos \alpha_1 + p_2 A_2 \cos \alpha_2 - p_1 A_1 \cos \alpha_1 \tag{2.8-28}$$

The term $F_{xg} = 0$ in this case.

For R_y the body force F_{yg} is in the negative y direction and $F_{yg} = -m_t g$, where m_t is the total mass fluid in the control volume. Replacing $\cos \alpha$ by $\sin \alpha$, the equation for the y direction becomes

$$R_y = m v_2 \sin \alpha_2 - m v_1 \sin \alpha_1 + p_2 A_2 \sin \alpha_2 - p_1 A_1 \sin \alpha_1 + m_t g \tag{2.8-29}$$

EXAMPLE 2.8-3. Momentum Balance in a Pipe Bend

Fluid is flowing at steady state through a reducing pipe bend, as shown in Fig. 2.8-3. Turbulent flow will be assumed with frictional forces negligible. The volumetric flow rate of the liquid and the pressure p_2 at point 2 are known as are the pipe diameters at both ends. Derive the equations to calculate the forces on the bend. Assume that the density ρ is constant.

Solution: The velocities v_1 and v_2 can be obtained from the volumetric flow rate and the areas. Also, $m = \rho_1 v_1 A_1 = \rho_2 v_2 A_2$. As in Example 2.8-2, the mechanical-energy balance equation (2.8-24) is used to obtain the upstream pressure, p_1. For the x direction Eq. (2.8-28) is used for the mo-

Chap. 2 *Principles of Momentum Transfer and Overall Balances*

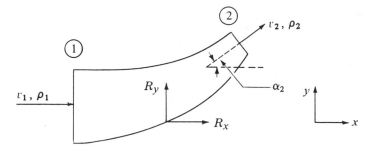

FIGURE 2.8-3. *Flow through a reducing bend in Example 2.8-3.*

mentum balance. Since $\alpha_1 = 0°$, $\cos \alpha_1 = 1.0$. Equation (2.8-28) becomes

$$R_x = mv_2 \cos \alpha_2 - mv_1 + p_2 A_2 \cos \alpha_2 - p_1 A_1 \qquad \text{(SI)}$$

$$R_x = \frac{m}{g_c} v_2 \cos \alpha_2 - \frac{m}{g_c} v_1 + p_2 A_2 \cos \alpha_2 - p_1 A_1 \qquad \text{(English)}$$

(2.8-30)

For the y direction the momentum balance Eq. (2.8-29) is used where $\sin \alpha_1 = 0$.

$$R_y = mv_2 \sin \alpha_2 + p_2 A_2 \sin \alpha_2 + m_t g \qquad \text{(SI)} \qquad \text{(2.8-31)}$$

where m_t is total mass fluid in the pipe bend. The pressures at points 1 and 2 are gage pressures since the atmospheric pressures acting on all surfaces cancel. The magnitude of the resultant force of the bend acting on the control volume fluid is

$$|\mathbf{R}| = \sqrt{R_x^2 + R_y^2} \qquad \text{(2.8-32)}$$

The angle this makes with the vertical is $\theta = \arctan (R_x/R_y)$. Often the gravity force F_{yg} is small compared to the other terms in Eq. (2.8-31) and is neglected.

EXAMPLE 2.8-4. *Friction Loss in a Sudden Enlargement*
A mechanical-energy loss occurs when a fluid flows from a small pipe to a large pipe through an abrupt expansion, as shown in Fig. 2.8-4. Use the momentum balance and mechanical-energy balance to obtain an expression for the loss for a liquid. (*Hint :* Assume that $p_0 = p_1$ and $v_0 = v_1$. Make a mechanical-energy balance between points 0 and 2 and a momentum balance between points 1 and 2. It will be assumed that p_1 and p_2 are uniform over the cross-sectional area.)

FIGURE 2.8-4. *Losses in expansion flow.*

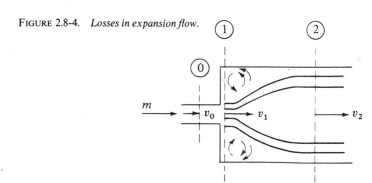

Solution: The control volume is selected so that it does not include the pipe wall, so R_x drops out. The boundaries selected are points 1 and 2. The flow through plane 1 occurs only through an area of A_0. The frictional drag force will be neglected and all the loss is assumed to be from eddies in this volume. Making a momentum balance between points 1 and 2 using Eq. (2.8-18) and noting that $p_0 = p_1$, $v_1 = v_0$, and $A_1 = A_2$,

$$p_1 A_2 - p_2 A_2 = mv_2 - mv_1 \tag{2.8-33}$$

The mass-flow rate is $m = v_0 \rho A_0$ and $v_2 = (A_0/A_2)v_0$. Substituting these terms into Eq. (2.8-33) and rearranging gives us

$$v_0^2 \frac{A_0}{A_2}\left(1 - \frac{A_0}{A_2}\right) = \frac{p_2 - p_1}{\rho} \tag{2.8-34}$$

Applying the mechanical-energy-balance equation (2.7-28) to points 1 and 2,

$$\frac{v_0^2 - v_2^2}{2} - \sum F = \frac{p_2 - p_1}{\rho} \tag{2.8-35}$$

Finally, combining Eqs. (2.8-34) and (2.8-35),

$$\sum F = \frac{v_0^2}{2}\left(1 - \frac{A_0}{A_2}\right)^2 \tag{2.8-36}$$

2.8D Overall Momentum Balance for Free Jet Striking a Fixed Vane

When a free jet impinges on a fixed vane as in Fig. 2.8-5 the overall momentum balance can be applied to determine the force on the smooth vane. Since there are no changes in elevation or pressure before and after impact, there is no loss in energy and application of the Bernoulli equation shows that the magnitude of the velocity is unchanged. Losses due to impact are neglected. The frictional resistance between the jet and the smooth vane is also neglected. The velocity is assumed to be uniform throughout the jet upstream and downstream. Since the jet is open to the atmosphere, the pressure is the same at all ends of the vane.

In making a momentum balance for the control volume shown for the curved vane in Fig. 2.8-5a, Eq. (2.8-28) is written as follows for steady state, where the pressure terms are zero, $v_1 = v_2$, $A_1 = A_2$, and $m = v_1 A_1 \rho_1 = v_2 A_2 \rho_2$:

$$R_x = mv_2 \cos \alpha_2 - mv_1 + 0 = mv_1(\cos \alpha_2 - 1) \tag{2.8-37}$$

Using Eq. (2.8-29) for the y direction and neglecting the body force,

$$R_y = mv_2 \sin \alpha_2 - 0 = mv_1 \sin \alpha_2 \tag{2.8-38}$$

Hence, R_x and R_y are the force components of the vane on the control volume fluid. The force components on the vane are $-R_x$ and $-R_y$.

EXAMPLE 2.8-5.   Force of Free Jet on a Curved, Fixed Vane
A jet of water having a velocity of 30.5 m/s and a diameter of 2.54×10^{-2} m is deflected by a smooth, curved vane as shown in Fig. 2.8-5a, where $\alpha_2 = 60°$. What is the force of the jet on the vane? Assume that $\rho = 1000$ kg/m^3.

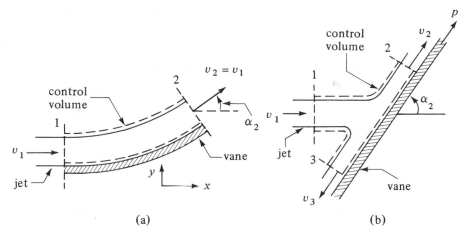

FIGURE 2.8-5. *Free jet impinging on a fixed vane: (a) smooth, curved vane, (b) smooth, flat vane.*

Solution: The cross-sectional area of the jet is $A_1 = \pi(2.54 \times 10^{-2})^2/4 = 5.067 \times 10^{-4}$ m². Then, $m = v_1 A_1 \rho_1 = 30.5 \times 5.067 \times 10^{-4} \times 1000 = 15.45$ kg/s. Substituting into Eqs. (2.8-37) and (2.8-38),

$$R_x = 15.45 \times 30.5 \,(\cos 60° - 1) = -235.6 \text{ N}(-52.97 \text{ lb}_f)$$

$$R_y = 15.45 \times 30.5 \sin 60° = 408.1 \text{ N}(91.74 \text{ lb}_f)$$

The force on the vane is $-R_x = +235.6$ N and $-R_y = -408.1$ N. The resultant force is calculated using Eq. (2.8-32).

In Fig. 2.8-5b a free jet at velocity v_1 strikes a smooth, inclined flat plate and the flow divides into two separate streams whose velocities are all equal ($v_1 = v_2 = v_3$) since there is no loss in energy. It is convenient to make a momentum balance in the p direction parallel to the plate. No force is exerted on the fluid by the flat plate in this direction; i.e., there is no tangential force. Then, the initial momentum component in the p direction must equal the final momentum component in this direction. This means $\sum F_p = 0$. Writing an equation similar to Eq. (2.8-26), where m_1 is kg/s entering at 1 and m_2 leaves at 2 and m_3 at 3,

$$\sum F_p = 0 = m_2 v_2 - m_1 v_1 \cos \alpha_2 - m_3 v_3$$

$$0 = m_2 v_1 - m_1 v_1 \cos \alpha_2 - m_3 v_1 \qquad (2.8\text{-}39)$$

By the continuity equation,

$$m_1 = m_2 + m_3 \qquad (2.8\text{-}40)$$

Combining and solving,

$$m_2 = \frac{m_1}{2}(1 + \cos \alpha_2), \qquad m_3 = \frac{m_1}{2}(1 - \cos \alpha_2) \qquad (2.8\text{-}41)$$

The resultant force exerted by the plate on the fluid must be normal to it. This means the resultant force is simply $m_1 v_1 \sin \alpha_2$. Alternatively, the resultant force on the fluid can be calculated by determining R_x and R_y from Eqs. (2.8-28) and 2.8-29) and then using Eq. (2.8-32). The force on the bend is the opposite of this.

2.9 SHELL MOMENTUM BALANCE AND VELOCITY PROFILE IN LAMINAR FLOW

2.9A Introduction

In Section 2.8 we analyzed momentum balances using an overall, macroscopic control volume. From this we obtained the total or overall changes in momentum crossing the control surface. This overall momentum balance did not tell us the details of what happens inside the control volume. In the present section we analyze a small control volume and then shrink this control volume to differential size. In doing this we make a shell momentum balance using the momentum-balance concepts of the preceding section, and then, using the equation for the definition of viscosity, we obtain an expression for the velocity profile inside the enclosure and the pressure drop. The equations are derived for flow systems of simple geometry in laminar flow at steady state.

In many engineering problems a knowledge of the complete velocity profile is not needed, but a knowledge of the maximum velocity, average velocity, or the shear stress on a surface is needed. In this section we show how to obtain these quantities from the velocity profiles.

2.9B Shell Momentum Balance Inside a Pipe

Engineers often deal with the flow of fluids inside a circular conduit or pipe. In Fig. 2.9-1 we have a horizontal section of pipe in which an incompressible Newtonian fluid is flowing in one-dimensional, steady-state, laminar flow. The flow is fully developed; i.e., it is not influenced by entrance effects and the velocity profile does not vary along the axis of flow in the x direction.

The cylindrical control volume is a shell with an inside radius r, thickness Δr, and length Δx. At steady state the conservation of momentum, Eq. (2.8-3), becomes as follows: sum of forces acting on control volume = rate of momentum out − rate of momentum into volume. The pressure forces become, from Eq. (2.8-17),

$$\text{pressure forces} = pA\,|_x - pA\,|_{x+\Delta x} = p(2\pi r\;\Delta r)\,|_x - p(2\pi r\;\Delta r)\,|_{x+\Delta x} \qquad \textbf{(2.9-1)}$$

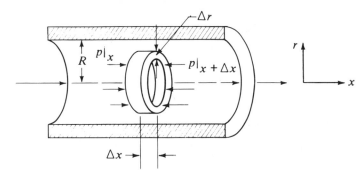

FIGURE 2.9-1. *Control volume for shell momentum balance on a fluid flowing in a circular tube.*

Chap. 2   Principles of Momentum Transfer and Overall Balances

The shear force or drag force acting on the cylindrical surface at the radius r is the shear stress τ_{rx} times the area $2\pi r \, \Delta x$. However, this can also be considered as the rate of momentum flow into the cylindrical surface of the shell as described by Eq. (2.4-9). Hence, the net rate of momentum efflux is the rate of momentum out − rate of momentum in and is

$$\text{net efflux} = (\tau_{rx} \, 2\pi r \, \Delta x)\big|_{r+\Delta r} - (\tau_{rx} \, 2\pi r \, \Delta x)\big|_r \tag{2.9-2}$$

The net convective momentum flux across the annular surface at x and $x + \Delta x$ is zero, since the flow is fully developed and the terms are independent of x. This is true since v_x at x is equal to v_x at $x + \Delta x$.

Equating Eq. (2.9-1) to (2.9-2) and rearranging,

$$\frac{(r\tau_{rx})\big|_{r+\Delta r} - (r\tau_{rx})\big|_r}{\Delta r} = \frac{r(p\big|_x - p\big|_{x+\Delta x})}{\Delta x} \tag{2.9-3}$$

In fully developed flow, the pressure gradient $(\Delta p/\Delta x)$ is constant and becomes $(\Delta p/L)$, where Δp is the pressure drop for a pipe of length L. Letting Δr approach zero, we obtain

$$\frac{d(r\tau_{rx})}{dr} = \left(\frac{\Delta p}{L}\right) r \tag{2.9-4}$$

Separating variables and integrating,

$$\tau_{rx} = \left(\frac{\Delta p}{L}\right)\frac{r}{2} + \frac{C_1}{r} \tag{2.9-5}$$

The constant of integration C_1 must be zero if the momentum flux is not infinite at $r = 0$. Hence,

$$\tau_{rx} = \left(\frac{\Delta p}{2L}\right) r = \frac{p_0 - p_L}{2L} r \tag{2.9-6}$$

This means that the momentum flux varies linearly with the radius, as shown in Fig. 2.9-2, and the maximum value occurs at $r = R$ at the wall.

Substituting Newton's law of viscosity,

$$\tau_{rx} = -\mu \frac{dv_x}{dr} \tag{2.9-7}$$

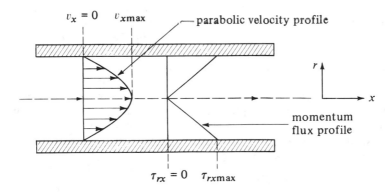

FIGURE 2.9-2. *Velocity and momentum flux profiles for laminar flow in a pipe.*

into Eq. (2.9-6), we obtain the following differential equation for the velocity:

$$\frac{dv_x}{dr} = -\frac{p_0 - p_L}{2\mu L} r \tag{2.9-8}$$

Integrating using the boundary condition that at the wall, $v_x = 0$ at $r = R$, we obtain the equation for the velocity distribution.

$$v_x = \frac{p_0 - p_L}{4\mu L} R^2 \left[1 - \left(\frac{r}{R}\right)^2 \right] \tag{2.9-9}$$

This result shows us that the velocity distribution is parabolic as shown in Fig. 2.9-2.

The average velocity $v_{x\,av}$ for a cross section is found by summing up all the velocities over the cross section and dividing by the cross-sectional area as in Eq. (2.6-17). Following the procedure given in Example 2.6-3, where $dA = r\,dr\,d\theta$ and $A = \pi R^2$,

$$v_{x\,av} = \frac{1}{A} \iint_A v_x\,dA = \frac{1}{\pi R^2} \int_0^{2\pi} \int_0^R v_x r\,dr\,d\theta = \frac{1}{\pi R^2} \int_0^R v_x\,2\pi r\,dr \tag{2.9-10}$$

Combining Eqs. (2.9-9) and (2.9-10) and integrating,

$$v_{x\,av} = \frac{(p_0 - p_L)R^2}{8\mu L} = \frac{(p_0 - p_L)D^2}{32\mu L} \tag{2.9-11}$$

where diameter $D = 2R$. Hence, Eq. (2.9-11), which is the *Hagen–Poiseuille equation*, relates the pressure drop and average velocity for laminar flow in a horizontal pipe.

The maximum velocity for a pipe is found from Eq. (2.9-9) and occurs at $r = 0$.

$$v_{x\,max} = \frac{p_0 - p_L}{4\mu L} R^2 \tag{2.9-12}$$

Combining Eqs. (2.9-11) and (2.9-12), we find that

$$v_{x\,av} = \frac{v_{x\,max}}{2} \tag{2.9-13}$$

2.9C Shell Momentum Balance for Falling Film

We now use an approach similar to that used for laminar flow inside a pipe for the case of flow of a fluid as a film in laminar flow down a vertical surface. Falling films have been used to study various phenomena in mass transfer, coatings on surfaces, and so on. The control volume for the falling film is shown in Fig. 2.9-3a, where the shell of fluid considered is Δx thick and has a length of L in the vertical z direction. This region is sufficiently far from the entrance and exit regions so that the flow is not affected by these regions. This means the velocity $v_z(x)$ does not depend on position z.

To start we set up a momentum balance in the z direction over a system Δx thick, bounded in the z direction by the planes $z = 0$ and $z = L$, and extending a distance W in the y direction. First, we consider the momentum flux due to molecular transport. The rate of momentum out-rate of momentum in is the momentum flux at point $x + \Delta x$ minus that at x times the area LW.

$$\text{net efflux} = LW(\tau_{xz})|_{x+\Delta x} - LW(\tau_{xz})|_x \tag{2.9-14}$$

The net convective momentum flux is the rate of momentum entering the area $\Delta x W$ at

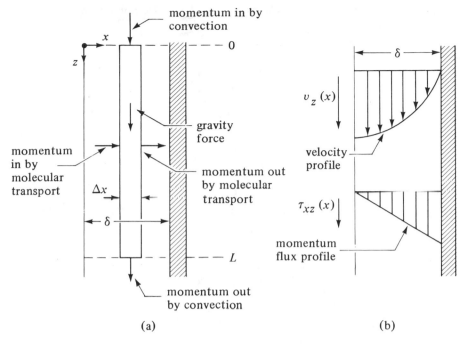

FIGURE 2.9-3. *Vertical laminar flow of a liquid film : (a) shell momentum balance for a control volume Δ x thick; (b) velocity and momentum flux profiles.*

$z = L$ minus that leaving at $z = 0$. This net efflux is equal to 0 since v_z at $z = 0$ is equal to v_z at $z = L$ for each value of x.

$$\text{net efflux} = \Delta x W v_z(\rho v_z)|_{z=L} - \Delta x W v_z(\rho v_z)|_{z=0} = 0 \qquad \textbf{(2.9-15)}$$

The gravity force acting on the fluid is

$$\text{gravity force} = \Delta x W L(\rho g) \qquad \textbf{(2.9-16)}$$

Then using Eq. (2.8-3) for the conservation of momentum at steady state,

$$\Delta x W L(\rho g) = L W(\tau_{xz})|_{x+\Delta x} - L W(\tau_{xz})|_{x} + 0 \qquad \textbf{(2.9-17)}$$

Rearranging Eq. (2.9-17) and letting $\Delta x \to 0$,

$$\frac{\tau_{xz}|_{x+\Delta x} - \tau_{xz}|_{x}}{\Delta x} = \rho g \qquad \textbf{(2.9-18)}$$

$$\frac{d}{dx} \tau_{xz} = \rho g \qquad \textbf{(2.9-19)}$$

Integrating using the boundary conditions at $x = 0$, $\tau_{xz} = 0$ at the free liquid surface and at $x = x$, $\tau_{xz} = \tau_{xz}$,

$$\tau_{xz} = \rho g x \qquad \textbf{(2.9-20)}$$

This means the momentum-flux profile is linear as shown in Fig. 2.9-3b and the maximum value is at the wall. For a Newtonian fluid using Newton's law of viscosity,

$$\tau_{xz} = -\mu \frac{dv_z}{dx} \tag{2.9-21}$$

Combining Eqs. (2.9-20) and (2.9-21) we obtain the following differential equation for the velocity:

$$\frac{dv_z}{dx} = -\left(\frac{\rho g}{\mu}\right)x \tag{2.9-22}$$

Separating variables and integrating gives

$$v_z = -\left(\frac{\rho g}{2\mu}\right)x^2 + C_1 \tag{2.9-23}$$

Using the boundary condition that $v_z = 0$ at $x = \delta$, $C_1 = (\rho g/2\mu)\delta^2$. Hence, the velocity distribution equation becomes

$$v_z = \frac{\rho g \delta^2}{2\mu}\left[1 - \left(\frac{x}{\delta}\right)^2\right] \tag{2.9-24}$$

This means the velocity profile is parabolic as shown in Fig. 2.9-3b. The maximum velocity occurs at $x = 0$ in Eq. (2.9-24) and is

$$v_{z\,max} = \frac{\rho g \delta^2}{2\mu} \tag{2.9-25}$$

The average velocity can be found by using Eq. (2.6-17).

$$v_{z\,av} = \frac{1}{A}\iint\limits_{A} v_z\, dA = \frac{1}{W\delta}\int_0^W \int_0^\delta v_z\, dx\, dy = \frac{W}{W\delta}\int_0^\delta v_z\, dx \tag{2.9-26}$$

Substituting Eq. (2.9-24) into (2.9-26) and integrating,

$$v_{z\,av} = \frac{\rho g \delta^2}{3\mu} \tag{2.9-27}$$

Combining Eqs. (2.9-25) and (2.9-27), we obtain $v_{z\,av} = (2/3)v_{z\,max}$. The volumetric flow rate q is obtained by multiplying the average velocity $v_{z\,av}$ times the cross-sectional area δW.

$$q = \frac{\rho g \delta^3 W}{3\mu} \text{ m}^3/\text{s} \tag{2.9-28}$$

Often in falling films, the mass rate of flow per unit width of wall Γ in kg/s·m is defined as $\Gamma = \rho \delta v_{z\,av}$ and a Reynolds number is defined as

$$N_{Re} = \frac{4\Gamma}{\mu} = \frac{4\rho \delta v_{z\,av}}{\mu} \tag{2.9-29}$$

Laminar flow occurs for $N_{Re} < 1200$. Laminar flow with rippling present occurs above a N_{Re} of 25.

EXAMPLE 2.9-1. Falling Film Velocity and Thickness
An oil is flowing down a vertical wall as a film 1.7 mm thick. The oil density is 820 kg/m^3 and the viscosity is 0.20 Pa·s. Calculate the mass flow rate per

unit width of wall, Γ, needed and the Reynolds number. Also calculate the average velocity.

Solution: The film thickness is $\delta = 0.0017$ m. Substituting Eq. (2.9-27) into the definition of Γ,

$$\Gamma = \rho \delta v_{z\,av} = \frac{(\rho\delta)\rho g\delta^2}{3\mu} = \frac{\rho^2\delta^3 g}{3\mu}$$

$$= \frac{(820)^2(1.7 \times 10^{-3})^3(9.806)}{3 \times 0.20} = 0.05399 \text{ kg/s} \cdot \text{m} \tag{2.9-30}$$

Using Eq. (2.9-29),

$$N_{Re} = \frac{4\Gamma}{\mu} = \frac{4(0.05399)}{0.20} = 1.080$$

Hence, the film is in laminar flow. Using Eq. (2.9-27),

$$v_{z\,av} = \frac{\rho g\delta^2}{3\mu} = \frac{820(9.806)(1.7 \times 10^{-3})^2}{3(0.20)} = 0.03873 \text{ m/s}$$

2.10   DESIGN EQUATIONS FOR LAMINAR AND TURBULENT FLOW IN PIPES

2.10A   Velocity Profiles in Pipes

One of the most important applications of fluid flow is flow inside circular conduits, pipes, and tubes. Appendix A.5 gives sizes of commercial standard steel pipe. Schedule 40 pipe in the different sizes is the standard usually used. Schedule 80 has a thicker wall and will withstand about twice the pressure of schedule 40 pipe. Both have the same outside diameter so that they will fit the same fittings. Pipes of other metals have the same outside diameters as steel pipe to permit interchanging parts of a piping system. Sizes of tubing are generally given by the outside diameter and wall thickness. Perry and Green (P1) give detailed tables of various types of tubing and pipes.

When fluid is flowing in a circular pipe and the velocities are measured at different distances from the pipe wall to the center of the pipe, it has been shown that in both laminar and turbulent flow, the fluid in the center of the pipe is moving faster than the fluid near the walls. These measurements are made at a reasonable distance from the entrance to the pipe. Figure 2.10-1 is a plot of the relative distance from the center of the pipe versus the fraction of maximum velocity v'/v_{max}, where v' is local velocity at the given position and v_{max} the maximum velocity at the center of the pipe. For viscous or laminar flow, the velocity profile is a true parabola, as derived in Eq. (2.9-9). The velocity at the wall is zero.

In many engineering applications the relation between the average velocity v_{av} in a pipe and the maximum velocity v_{max} is useful, since in some cases only the v_{max} at the center point of the tube is measured. Hence, from only one point measurement this relationship between v_{max} and v_{av} can be used to determine v_{av}. In Fig. 2.10-2 experimentally measured values of v_{av}/v_{max} are plotted as a function of the Reynolds numbers $Dv_{av}\rho/\mu$ and $Dv_{max}\rho/\mu$.

The average velocity over the whole cross section of the pipe is precisely 0.5 times the maximum velocity at the center as given by the shell momentum balance in Eq. (2.9-13) for laminar flow. On the other hand, for turbulent flow, the curve is somewhat flattened in the center (see Fig. 2.10-1) and the average velocity is about 0.8 times the

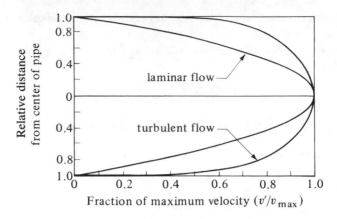

FIGURE 2.10-1. *Velocity distribution of a fluid across a pipe.*

maximum. This value of 0.8 varies slightly, depending upon the Reynolds number, as shown in the correlation in Fig. 2.10-2. (*Note :* See Problem 2.6-3, where a value of 0.817 is derived using the $\frac{1}{7}$-power law.)

2.10B Pressure Drop and Friction Loss in Laminar Flow

1. Pressure drop and loss due to friction.   When the fluid is in steady-state laminar flow in a pipe, then for a Newtonian fluid the shear stress is given by Eq. (2.4-2), which is rewritten for change in radius dr rather than distance dy, as follows.

$$\tau_{rz} = -\mu \frac{dv_z}{dr} \tag{2.10-1}$$

Using this relationship and making a shell momentum balance on the fluid over a

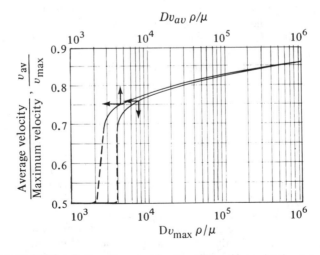

FIGURE 2.10-2.   *Ratio v_{av}/v_{max} as a function of Reynolds number for pipes.*

cylindrical shell, the Hagen–Poiseuille equation (2.9-11) for laminar flow of a liquid in circular tubes was obtained. A derivation is also given in Section 3.6 using a differential momentum balance. This can be written as

$$\Delta p_f = (p_1 - p_2)_f = \frac{32\mu v(L_2 - L_1)}{D^2} \tag{2.10-2}$$

where p_1 is upstream pressure at point 1, N/m^2; p_2, pressure at point 2; v is average velocity in tube, m/s; D is inside diameter, m; and $(L_2 - L_1)$ or ΔL is length of straight tube, m. For English units, the right-hand side of Eq. (2.10-2) is divided by g_c.

The quantity $(p_1 - p_2)_f$ or Δp_f is the pressure loss due to skin friction. Then, for constant ρ, the friction loss F_f is

$$F_f = \frac{(p_1 - p_2)_f}{\rho} = \frac{N \cdot m}{kg} \quad \text{or} \quad \frac{J}{kg} \quad \text{(SI)}$$

$$\tag{2.10-3}$$

$$F_f = \frac{ft \cdot lb_f}{lb_m} \quad \text{(English)}$$

This is the mechanical-energy loss due to skin friction for the pipe in $N \cdot m/kg$ of fluid and is part of the $\sum F$ term for frictional losses in the mechanical-energy-balance equation (2.7-28). This term $(p_1 - p_2)_f$ for skin friction loss is different from the $(p_1 - p_2)$ term, owing to velocity head or potential head changes in Eq. (2.7-28). That part of $\sum F$ which arises from friction within the channel itself by laminar or turbulent flow is discussed in Sections 2.10B and in 2.10C. The part of friction loss due to fittings (valves, elbows, etc.), bends, and the like, which sometimes constitute a large part of the friction, is discussed in Section 2.10F. Note that if Eq. (2.7-28) is applied to steady flow in a straight, horizontal tube, we obtain $(p_1 - p_2)/\rho = \sum F$.

One of the uses of Eq. (2.10-2) is in the experimental measurement of the viscosity of a fluid by measuring the pressure drop and volumetric flow rate through a tube of known length and diameter. Slight corrections for kinetic energy and entrance effects are usually necessary in practice. Also, Eq. (2.10-2) is often used in the metering of small liquid flows.

EXAMPLE 2.10-1. Metering of Small Liquid Flows

A small capillary with an inside diameter of 2.22×10^{-3} m and a length 0.317 m is being used to continuously measure the flow rate of a liquid having a density of 875 kg/m^3 and $\mu = 1.13 \times 10^{-3}$ Pa·s. The pressure-drop reading across the capillary during flow is 0.0655 m water (density 996 kg/m^3). What is the flow rate in m^3/s if end-effect corrections are neglected?

Solution: Assuming that the flow is laminar, Eq. (2.10-2) will be used. First, to convert the height h of 0.0655 m water to a pressure drop using Eq. (2.2-4),

$$\Delta p_f = h\rho g = (0.0655 \text{ m})\left(996 \frac{kg}{m^3}\right)\left(9.80665 \frac{m}{s^2}\right)$$

$$= 640 \text{ kg} \cdot m/s^2 \cdot m^2 = 640 \text{ N/m}^2$$

Substituting the following values into Eq. (2.10-2) of $\mu = 1.13 \times 10^{-3}$ Pa·s,

$L_2 - L_1 = 0.317$ m, $D = 2.22 \times 10^{-3}$ m, and $\Delta p_f = 640$ N/m², and solving for v,

$$\Delta p_f = \frac{32\mu v(L_2 - L_1)}{D^2} \tag{2.10-2}$$

$$640 = \frac{32(1.13 \times 10^{-3})(v)(0.317)}{(2.22 \times 10^{-3})^2}$$

$$v = 0.275 \text{ m/s}$$

The volumetric rate is then

$$\text{volumetric flow rate} = v\pi \frac{D^2}{4} = \frac{0.275(\pi)(2.22 \times 10^{-3})^2}{4}$$

$$= 1.066 \times 10^{-6} \text{ m}^3/\text{s}$$

Since it was assumed that laminar flow is occurring, the Reynolds number will be calculated to check this.

$$N_{Re} = \frac{Dv\rho}{\mu} = \frac{(2.22 \times 10^{-3})(0.275)(875)}{1.13 \times 10^{-3}} = 473$$

Hence, the flow is laminar as assumed.

2. Use of friction factor for friction loss in laminar flow. A common parameter used in laminar and especially in turbulent flow is the *Fanning friction factor*, f, which is defined as the drag force per wetted surface unit area (shear stress τ_s at the surface) divided by the product of density times velocity head or $\frac{1}{2}\rho v^2$. The force is Δp_f times the cross-sectional area πR^2 and the wetted surface area is $2\pi R\, \Delta L$. Hence, the relation between the pressure drop due to friction and f is as follows for laminar and turbulent flow.

$$f = \frac{\tau_s}{\rho v^2/2} = \frac{\Delta p_f \pi R^2}{2\pi R\, \Delta L} \bigg/ \frac{\rho v^2}{2} \tag{2.10-4}$$

Rearranging, this becomes

$$\Delta p_f = 4 f \rho \frac{\Delta L}{D} \frac{v^2}{2} \qquad \text{(SI)}$$
$$\tag{2.10-5}$$
$$\Delta p_f = 4 f \rho \frac{\Delta L}{D} \frac{v^2}{2g_c} \qquad \text{(English)}$$

$$F_f = \frac{\Delta p_f}{\rho} = 4f \frac{\Delta L}{D} \frac{v^2}{2} \qquad \text{(SI)}$$
$$\tag{2.10-6}$$
$$F_f = 4f \frac{\Delta L}{D} \frac{v^2}{2g_c} \qquad \text{(English)}$$

For laminar flow only, combining Eqs. (2.10-2) and (2.10-5),

$$f = \frac{16}{N_{Re}} = \frac{16}{Dv\rho/\mu} \tag{2.10-7}$$

Equations (2.10-2), (2.10-5), (2.10-6), and (2.10-7) for laminar flow hold up to a Reynolds number of 2100. Beyond that at an N_{Re} value above 2100, Eqs. (2.10-2) and (2.10-7) do not hold for turbulent flow. For turbulent flow Eqs. (2.10-5) and (2.10-6), however, are used extensively along with empirical methods of predicting the friction factor f, as discussed in the next section.

EXAMPLE 2.10-2.   Use of Friction Factor in Laminar Flow

Assume the same known conditions as in Example 2.10-1 except that the velocity of 0.275 m/s is known and the pressure drop Δp_f is to be predicted. Use the Fanning friction factor method.

Solution: The Reynolds number is, as before,

$$N_{Re} = \frac{Dv\rho}{\mu} = \frac{(2.22 \times 10^{-3} \text{ m})(0.275 \text{ m/s})(875 \text{ kg/m}^3)}{1.13 \times 10^{-3} \text{ kg/m} \cdot \text{s}} = 473$$

From Eq. (2.10-7) the friction factor f is

$$f = \frac{16}{N_{Re}} = \frac{16}{473} = 0.0338 \quad \text{(dimensionless)}$$

Using Eq. (2.10-5) with $\Delta L = 0.317$ m, $v = 0.275$ m/s, $D = 2.22 \times 10^{-3}$ m, $\rho = 875 \text{ kg/m}^3$,

$$\Delta p_f = 4 f \rho \frac{\Delta L}{D} \frac{v^2}{2} = \frac{4(0.0338)(875)(0.317)(0.275)^2}{(2.22 \times 10^{-3})(2)} = 640 \text{ N/m}^2$$

This, of course, checks the value in Example 2.10-1.

When the fluid is a gas and not a liquid, the Hagen–Poiseuille equation (2.10-2) can be written as follows for laminar flow:

$$m = \frac{\pi D^4 M(p_1^2 - p_2^2)}{128(2RT)\mu(L_2 - L_1)} \quad \text{(SI)}$$

$$m = \frac{\pi D^4 g_c M(p_1^2 - p_2^2)}{128(2RT)\mu(L_2 - L_1)} \quad \text{(English)} \tag{2.10-8}$$

where m = kg/s, M = molecular weight in kg/kg mol, T = absolute temperature in K, and $R = 8314.3$ N $\cdot$ m/kg mol $\cdot$ K. In English units, $R = 1545.3$ ft $\cdot$ lb$_f$/lb mol $\cdot$ °R.

2.10C   Pressure Drop and Friction Factor in Turbulent Flow

In turbulent flow, as in laminar flow, the friction factor also depends on the Reynolds number. However, it is not possible to predict theoretically the Fanning friction factor f for turbulent flow as it was done for laminar flow. The friction factor must be determined empirically (experimentally) and it not only depends upon the Reynolds number but also on surface roughness of the pipe. In laminar flow the roughness has essentially no effect.

Dimensional analysis also shows the dependence of the friction factor on these factors. In Sections 3.11 and 4.14 methods of obtaining the dimensionless numbers and their importance are discussed.

A large number of experimental data on friction factors of smooth pipe and pipes of varying degrees of equivalent roughness have been obtained and the data correlated. For design purposes to predict the friction factor f and, hence, the frictional pressure drop of round pipe, the friction factor chart in Fig. 2.10-3 can be used. It is a log–log plot of f

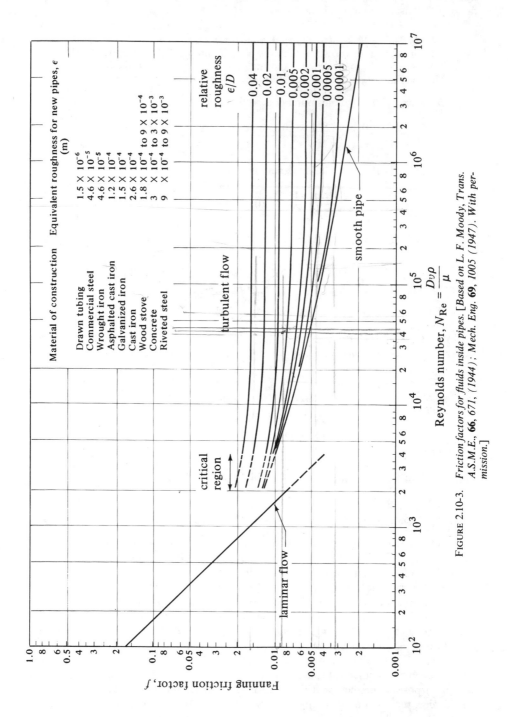

FIGURE 2.10-3. *Friction factors for fluids inside pipes.* [*Based on L. F. Moody, Trans. A.S.M.E.,* **66**, *671, (1944); Mech. Eng.* **69**, *1005 (1947). With permission.*]

Material of construction	Equivalent roughness for new pipes, ϵ (m)
Drawn tubing	1.5×10^{-6}
Commercial steel	4.6×10^{-5}
Wrought iron	4.6×10^{-5}
Asphalted cast iron	1.2×10^{-4}
Galvanized iron	1.5×10^{-4}
Cast iron	2.6×10^{-4}
Wood stove	1.8×10^{-4} to 9×10^{-4}
Concrete	3×10^{-4} to 3×10^{-3}
Riveted steel	9×10^{-4} to 9×10^{-3}

relative roughness ϵ/D

0.04
0.02
0.01
0.005
0.002
0.001
0.0005
0.0001

turbulent flow

smooth pipe

critical region

laminar flow

Reynolds number, $N_{\text{Re}} = \dfrac{Dv\rho}{\mu}$

Fanning friction factor, f

versus N_{Re}. This friction factor f is then used in Eqs. (2.10-5) and (2.10-6) to predict the friction loss Δp_f or F_f.

$$\Delta p_f = 4f\rho \frac{\Delta L}{D} \frac{v^2}{2} \quad \text{(SI)}$$

$$\Delta p_f = 4f\rho \frac{\Delta L}{D} \frac{v^2}{2g_c} \quad \text{(English)}$$
(2.10-5)

$$F_f = \frac{\Delta p_f}{\rho} = 4f \frac{\Delta L}{D} \frac{v^2}{2} \quad \text{(SI)}$$

$$F_f = 4f \frac{\Delta L}{D} \frac{v^2}{2g_c} \quad \text{(English)}$$
(2.10-6)

For the region with a Reynolds number below 2100, the line is the same as Eq. (2.10-7). For a Reynolds number above 4000 for turbulent flow, the lowest line in Fig. 2.10-3 represents the friction factor line for smooth pipes and tubes, such as glass tubes, and drawn copper and brass tubes. The other lines for higher friction factors represent lines for different relative roughness factors, ε/D, where D is the inside pipe diameter in m and ε is a roughness parameter, which represents the average height in m of roughness projections from the wall (M1). On Fig. 2.10-3 values of the equivalent roughness for new pipes are given (M1). The most common pipe, commercial steel, has a roughness of $\varepsilon = 4.6 \times 10^{-5}$ m $(1.5 \times 10^{-4}$ ft).

The reader should be cautioned on using friction factors f from other sources. The Fanning friction factor of f in Eq. (2.10-6) is the one used here. Others use a friction factor that may be 4 times larger.

EXAMPLE 2.10-3. Use of Friction Factor in Turbulent Flow
A liquid is flowing through a horizontal straight commercial steel pipe at 4.57 m/s. The pipe used is commercial steel, schedule 40, 2-in. nominal diameter. The viscosity of the liquid is 4.46 cp and the density 801 kg/m³. Calculate the mechanical-energy friction loss F_f in J/kg for a 36.6-m section of pipe.

Solution: The following data are given: From Appendix A.5, $D = 0.0525$ m, $v = 4.57$ m/s, $\rho = 801$ kg/m³, $\Delta L = 36.6$ m, and

$$\mu = (4.46 \text{ cp})(1 \times 10^{-3}) = 4.46 \times 10^{-3} \text{ kg/m} \cdot \text{s}$$

The Reynolds number is calculated as

$$N_{Re} = \frac{Dv\rho}{\mu} = \frac{0.0525(4.57)(801)}{4.46 \times 10^{-3}} = 4.310 \times 10^4$$

Hence, the flow is turbulent. For commercial steel pipe from the table in Fig. 2.10-3, the equivalent roughness is 4.6×10^{-5} m.

$$\frac{\varepsilon}{D} = \frac{4.6 \times 10^{-5} \text{ m}}{0.0525 \text{ m}} = 0.00088$$

For an N_{Re} of 4.310×10^4, the friction factor from Fig. 2.10-3 is $f = 0.0060$. Substituting into Eq. (2.10-6), the friction loss is

$$F_f = 4f \frac{\Delta L}{D} \frac{v^2}{2} = \frac{4(0.0060)(36.6)(4.57)^2}{(0.0525)(2)} = 174.8 \frac{\text{J}}{\text{kg}} \left(58.5 \frac{\text{ft} \cdot \text{lb}_f}{\text{lb}_m} \right)$$

In problems involving the friction loss F_f in pipes, F_f is usually the unknown with the diameter D, velocity v, and pipe length ΔL known. Then a direct solution is possible as in Example 2.10-3. However, in some cases, the friction loss F_f is already set by the available head of liquid. Then if the volumetric flow rate and pipe length are set, the unknown to be calculated is the diameter. This solution is by trial and error since the velocity v appears in both N_{Re} and f, which are unknown. In another case, with the F_f being again already set, the diameter and pipe length are specified. This is also by trial and error, to calculate the velocity. Example 2.10-4 indicates the method to be used to calculate the pipe diameter with F_f set. Others (M2) give a convenient chart to aid in these types of calculations.

EXAMPLE 2.10-4. Trial-and-Error Solution to Calculate Pipe Diameter

Water at 4.4°C is to flow through a horizontal commercial steel pipe having a length of 305 m at the rate of 150 gal/min. A head of water of 6.1 m is available to overcome the friction loss F_f. Calculate the pipe diameter.

Solution: From Appendix A.2 the density $\rho = 1000$ kg/m^3 and the viscosity μ is

$$\mu = (1.55 \text{ cp})(1 \times 10^{-3}) = 1.55 \times 10^{-3} \text{ kg/m} \cdot \text{s}$$

friction loss $F_f = (6.1 \text{ m}) g = (6.1)(9.80665) = 59.82$ J/kg

$$\text{flow rate} = \left(150 \, \frac{\text{gal}}{\text{min}} \right) \left(\frac{1 \text{ ft}^3}{7.481 \text{ gal}} \right) \left(\frac{1 \text{ min}}{60 \text{ s}} \right) (0.028317 \text{ m}^3/\text{ft}^3)$$

$$= 9.46 \times 10^{-3} \text{ m}^3/\text{s}$$

$$\text{area of pipe} = \frac{\pi D^2}{4} \text{ m}^2 \quad (D \text{ is unknown})$$

$$\text{velocity } v = (9.46 \times 10^{-3} \text{ m}^3/\text{s}) \left(\frac{1}{\pi D^2/4 \text{ m}^2} \right) = \frac{0.01204}{D^2} \text{ m/s}$$

The solution is by trial and error since v appears in N_{Re} and f. Assume that $D = 0.089$ m for first trial.

$$N_{Re} = \frac{Dv\rho}{\mu} = (0.089) \frac{0.01204(1000)}{(0.089)^2 (1.55 \times 10^{-3})} = 8.730 \times 10^4$$

For commercial steel pipe and using Fig. 2.10-3, $\varepsilon = 4.6 \times 10^{-5}$ m. Then,

$$\frac{\varepsilon}{D} = \frac{4.6 \times 10^{-5} \text{ m}}{0.089 \text{ m}} = 0.00052$$

From Fig. 2.10-3 for $N_{Re} = 8.730 \times 10^4$ and $\varepsilon/D = 0.00052$, $f = 0.0051$. Substituting into Eq. (2.10-6),

$$F_f = 59.82 = 4f \frac{\Delta L}{D} \frac{v^2}{2} = \frac{4(0.0051)(305)}{D(2)} \frac{(0.01204)^2}{D^4}$$

Solving for D, $D = 0.0945$ m. This does not check the assumed value of 0.089 m.

For the second trial D will be assumed as 0.0945 m.

$$N_{Re} = (0.0945) \frac{0.01204}{(0.0945)^2} \frac{1000}{1.55 \times 10^{-3}} = 8.220 \times 10^4$$

$$\frac{\varepsilon}{D} = \frac{4.6 \times 10^{-5}}{0.0945} = 0.00049$$

From Fig. 2.10-3, $f = 0.0052$. It can be seen that f does not change much with N_{Re} in the turbulent region.

$$F_f = 59.82 = \frac{4(0.0052)(305)}{D(2)} \frac{(0.01204)^2}{D^4}$$

Solving, $D = 0.0954$ m or 3.75 in. Hence, the solution checks the assumed value of D closely.

2.10D Pressure Drop and Friction Factor in Flow of Gases

The equations and methods discussed in this section for turbulent flow in pipes hold for incompressible liquids. They also hold for a gas if the density (or the pressure) changes less than 10%. Then an average density, ρ_{av} in kg/m^3, should be used and the errors involved will be less than the uncertainty limits in the friction factor f. For gases, Eq. (2.10-5) can be rewritten as follows for laminar and turbulent flow:

$$(p_1 - p_2)_f = \frac{4f \, \Delta L G^2}{D2\rho_{av}} \tag{2.10-9}$$

where ρ_{av} is the density at $p_{av} = (p_1 + p_2)/2$. Also, the N_{Re} used is DG/μ, where G is kg/m$^2 \cdot$ s and is a constant independent of the density and velocity variation for the gas. Equation (2.10-5) can also be written for gases as

$$p_1^2 - p_2^2 = \frac{4f \, \Delta L G^2 RT}{DM} \qquad \text{(SI)}$$

$$\tag{2.10-10}$$

$$p_1^2 - p_2^2 = \frac{4f \, \Delta L G^2 RT}{g_c DM} \qquad \text{(English)}$$

where R is 8314.3 J/kg mol $\cdot$ K or 1545.3 ft $\cdot$ lb$_f$/lb mol $\cdot$ °R and M is molecular weight.

The derivation of Eqs. (2.10-9) and (2.10-10) applies only to cases with gases where the relative pressure change is small enough so that large changes in velocity do not occur. If the exit velocity becomes large, the kinetic-energy term, which has been omitted, becomes important. For pressure changes above about 10%, compressible flow is occurring and the reader should refer to Section 2.11. In adiabatic flow in a uniform pipe, the velocity in the pipe cannot exceed the velocity of sound.

EXAMPLE 2.10-5. Flow of Gas in Line and Pressure Drop

Nitrogen gas at 25°C is flowing in a smooth tube having an inside diameter of 0.010 m at the rate of 9.0 kg/s $\cdot$ m^2. The tube is 200 m long and the flow can be assumed to be isothermal. The pressure at the entrance to the tube is 2.0265×10^5 Pa. Calculate the outlet pressure.

Solution: The viscosity of the gas from Appendix A.3 is $\mu = 1.77 \times 10^{-5}$ Pa $\cdot$ s at $T = 298.15$ K. Inlet gas pressure $p_1 = 2.0265 \times 10^5$ Pa, $G = 9.0$ kg/s $\cdot$ m^2, $D = 0.010$ m, $M = 28.02$ kg/kg mol, $\Delta L = 200$ m, and $R = 8314.3$ J/kg mol $\cdot$ K. Assuming that Eq. (2.10-10) holds for this case and that the pressure drop is less than 10%, the Reynolds number is

$$N_{Re} = \frac{DG}{\mu} = \frac{0.010(9.0)}{1.77 \times 10^{-5}} = 5085$$

Hence, the flow is turbulent. Using Fig. 2.10-3, $f = 0.0090$ for a smooth tube.

Substituting into Eq. (2.10-10),

$$p_1^2 - p_2^2 = \frac{4f\,\Delta LG^2RT}{DM}$$

$$(2.0265 \times 10^5)^2 - p_2^2 = \frac{4(0.0090)(200)(9.0)^2(8314.3)(298.15)}{0.010(28.02)}$$

$$4.1067 \times 10^{10} - p_2^2 = 0.5160 \times 10^{10}$$

Solving, $p_2 = 1.895 \times 10^5$ Pa. Hence, Eq. (2.10-10) can be used since the pressure drop is less than 10%.

2.10E Effect of Heat Transfer on Friction Factor

The friction factor f given in Fig. 2.10-3 is given for isothermal flow, i.e., no heat transfer. When a fluid is being heated or cooled, the temperature gradient will cause a change in physical properties of the fluid, especially the viscosity. For engineering practice the following method of Sieder and Tate (P1, S3) can be used to predict the friction factor for nonisothermal flow for liquids and gases.

1. Calculate the mean bulk temperature t_a as the average of the inlet and outlet bulk fluid temperatures.
2. Calculate the N_{Re} using the viscosity μ_a at t_a and use Fig. 2.10-3 to obtain f.
3. Using the tube wall temperature t_w, determine μ_w at t_w.
4. Calculate ψ for the case occurring below.

$$\psi = \left(\frac{\mu_a}{\mu_w}\right)^{0.17} \qquad \text{(heating) } N_{Re} > 2100 \qquad\qquad (2.10\text{-}11)$$

$$\psi = \left(\frac{\mu_a}{\mu_w}\right)^{0.11} \qquad \text{(cooling) } N_{Re} > 2100 \qquad\qquad (2.10\text{-}12)$$

$$\psi = \left(\frac{\mu_a}{\mu_w}\right)^{0.38} \qquad \text{(heating) } N_{Re} < 2100 \qquad\qquad (2.10\text{-}13)$$

$$\psi = \left(\frac{\mu_a}{\mu_w}\right)^{0.23} \qquad \text{(cooling) } N_{Re} < 2100 \qquad\qquad (2.10\text{-}14)$$

5. The final friction factor is obtained by dividing the f from step 2 by the ψ from step 4.

Hence, when the liquid is being heated, ψ is greater than 1.0 and the final f decreases. The reverse occurs on cooling the liquid.

2.10F Friction Losses in Expansion, Contraction, and Pipe Fittings

Skin friction losses in flow through straight pipe are calculated by using the Fanning friction factor. However, if the velocity of the fluid is changed in direction or magnitude, additional friction losses occur. This results from additional turbulence which develops because of vortices and other factors. Methods to estimate these losses are discussed below.

1. Sudden enlargement losses. If the cross section of a pipe enlarges very gradually, very little or no extra losses are incurred. If the change is sudden, it results in additional losses due to eddies formed by the jet expanding in the enlarged section. This friction loss can be calculated by the following for turbulent flow in both sections. This Eq. (2.8-36) was derived in Example 2.8-4.

$$h_{ex} = \frac{(v_1 - v_2)^2}{2\alpha} = \left(1 - \frac{A_1}{A_2}\right)^2 \frac{v_1^2}{2\alpha} = K_{ex} \frac{v_1^2}{2\alpha} \frac{J}{kg} \qquad (2.10\text{-}15)$$

where h_{ex} is the friction loss in J/kg, K_{ex} is the expansion-loss coefficient and $= (1 - A_1/A_2)^2$, v_1 is the upstream velocity in the smaller area in m/s, v_2 is the downstream velocity, and $\alpha = 1.0$. If the flow is laminar in both sections, the factor α in the equation becomes $\frac{1}{2}$. For English units the right-hand side of Eq. (2.10-15) is divided by g_c. Also, $h = ft \cdot lb_f/lb_m$.

2. Sudden contraction losses.   When the cross section of the pipe is suddenly reduced, the stream cannot follow around the sharp corner, and additional frictional losses due to eddies occur. For turbulent flow, this is given by

$$h_c = 0.55 \left(1 - \frac{A_2}{A_1}\right) \frac{v_2^2}{2\alpha} = K_c \frac{v_2^2}{2\alpha} \frac{J}{kg} \qquad (2.10\text{-}16)$$

where h_c is the friction loss, $\alpha = 1.0$ for turbulent flow, v_2 is the average velocity in the smaller or downstream section, and K_c is the contraction-loss coefficient (P1) and approximately equals $0.55(1 - A_2/A_1)$. For laminar flow, the same equation can be used with $\alpha = \frac{1}{2}$(S2). For English units the right side is divided by g_c.

TABLE 2.10-1.   *Friction Loss for Turbulent Flow Through Valves and Fittings*

Type of Fitting or Valve	Frictional Loss, Number of Velocity Heads, K_f	Frictional Loss, Equivalent Length of Straight Pipe in Pipe Diameters, L_e/D
Elbow, 45°	0.35	17
Elbow, 90°	0.75	35
Tee	1	50
Return bend	1.5	75
Coupling	0.04	2
Union	0.04	2
Gate valve		
Wide open	0.17	9
Half open	4.5	225
Globe valve		
Wide open	6.0	300
Half open	9.5	475
Angle valve, wide open	2.0	100
Check valve		
Ball	70.0	3500
Swing	2.0	100
Water meter, disk	7.0	350

Source : R. H. Perry and C. H. Chilton, *Chemical Engineers' Handbook,* 5th ed. New York: McGraw-Hill Book Company, 1973. With permission.

3. *Losses in fittings and valves.* Pipe fittings and valves also disturb the normal flow lines in a pipe and cause additional friction losses. In a short pipe with many fittings, the friction loss from these fittings could be greater than in the straight pipe. The friction loss for fittings and valves is given by the following equation:

$$h_f = K_f \frac{v_1^2}{2} \tag{2.10-17}$$

where K_f is the loss factor for the fitting or valve and v_1 is the average velocity in the pipe leading to the fitting. Experimental values for K_f are given in Table 2.10-1 for turbulent flow (P1) and in Table 2.10-2 for laminar flow.

As an alternative method, some texts and references (B1) give data for losses in fittings as an equivalent pipe length in pipe diameters. These data, also given in Table 2.10-1, are presented as L_e/D, where L_e is the equivalent length of straight pipe in m having the same frictional loss as the fitting, and D is the inside pipe diameter in m. The K values in Eqs. (2.10-15) and (2.10-16) can be converted to L_e/D values by multiplying the K by 50 (P1). The L_e values for the fittings are simply added to the length of the straight pipe to get the total length of equivalent straight pipe to use in Eq. (2.10-6).

4. *Frictional losses in mechanical-energy-balance equation.* The frictional losses from the friction in the straight pipe (Fanning friction), enlargement losses, contraction losses, and losses in fittings and valves are all incorporated in the $\sum F$ term of Eq. (2.7-28) for the mechanical-energy balance, so that

$$\sum F = 4f \frac{\Delta L}{D} \frac{v^2}{2} + K_{ex} \frac{v_1^2}{2} + K_c \frac{v_2^2}{2} + K_f \frac{v_1^2}{2} \tag{2.10-18}$$

If all the velocities, v, v_1, and v_2, are the same, then factoring, Eq. (2.10-18) becomes, for this special case,

$$\sum F = \left(4f \frac{\Delta L}{D} + K_{ex} + K_c + K_f \right) \frac{v^2}{2} \tag{2.10-19}$$

The use of the mechanical-energy-balance equation (2.7-28) along with Eq. (2.10-18) will be shown in the following examples.

EXAMPLE 2.10-6. *Friction Losses and Mechanical-Energy Balance*
An elevated storage tank contains water at 82.2°C as shown in Fig. 2.10-4. It is desired to have a discharge rate at point 2 of 0.223 ft³/s. What must be the height H in ft of the surface of the water in the tank relative to the

TABLE 2.10-2. *Friction Loss for Laminar Flow Through Valves and Fittings (K1)*

Type of Fitting or Valve	Frictional Loss, Number of Velocity Heads, K_f Reynolds Number					
	50	*100*	*200*	*400*	*1000*	*Turbulent*
Elbow, 90°	17	7	2.5	1.2	0.85	0.75
Tee	9	4.8	3.0	2.0	1.4	1.0
Globe valve	28	22	17	14	10	6.0
Check valve, swing	55	17	9	5.8	3.2	2.0

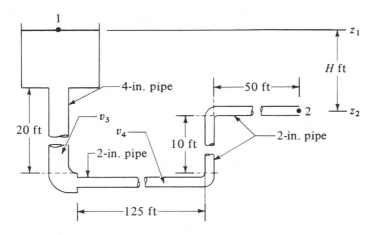

FIGURE 2.10-4. *Process flow diagram for Example 2.10-6.*

discharge point? The pipe used is commercial steel pipe, schedule 40, and the lengths of the straight portions of pipe are shown.

Solution: The mechanical-energy-balance equation (2.7-28) is written between points 1 and 2.

$$z_1 \frac{g}{g_c} + \frac{v_1^2}{2\alpha g_c} + \left(\frac{p_1}{\rho_1} - \frac{p_2}{\rho_2}\right) - W_S = z_2 \frac{g}{g_c} + \frac{v_2^2}{2\alpha g_c} + \sum F \quad \text{(2.10-20)}$$

From Appendix A.2, for water, $\rho = 0.970(62.43) = 60.52$ lb$_m$/ft^3 and $\mu = 0.347$ cp $= 0.347(6.7197 \times 10^{-4}) = 2.33 \times 10^{-4}$ lb$_m$/ft·s. The diameters of the pipes are

For 4-in. pipe: $D_3 = \dfrac{4.026}{12} = 0.3353$ ft; $A_3 = 0.0884$ ft^2

For 2-in. pipe: $D_4 = \dfrac{2.067}{12} = 0.1722$ ft; $A_4 = 0.02330$ ft^2

The velocities in the 4-in. and 2-in. pipe are

$$v_3 = \frac{0.223 \text{ ft}^3/\text{s}}{0.0884 \text{ ft}^2} = 2.523 \text{ ft/s} \quad \text{(4-in. pipe)}$$

$$v_4 = \frac{0.223}{0.02330} = 9.57 \text{ ft/s} \quad \text{(2-in. pipe)}$$

The $\sum F$ term for frictional losses in the system includes the following: (1) contraction loss at tank exit, (2) friction in the 4-in. straight pipe, (3) friction in 4-in. elbow, (4) contraction loss from 4-in. to 2-in. pipe, (5) friction in the 2-in. straight pipe, and (6) friction in the two 2-in. elbows. Calculations for the six items are as follows.

1. Contraction loss at tank exit. From Eq. (2.10-16) for contraction from

A_1 to A_3 cross-sectional area since A_1 of the tank is very large compared to A_3,

$$K_c = 0.55\left(1 - \frac{A_3}{A_1}\right) = 0.55(1 - 0) = 0.55$$

$$h_c = K_c \frac{v_3^2}{2g_c} = 0.55 \frac{(2.523)^2}{2(32.174)} = 0.054 \text{ ft} \cdot \text{lb}_f/\text{lb}_m$$

2. *Friction in the 4-in. pipe.* The Reynolds number is

$$N_{Re} = \frac{D_3 v_3 \rho}{\mu} = \frac{0.3353(2.523)(60.52)}{2.33 \times 10^{-4}} = 2.193 \times 10^5$$

Hence, the flow is turbulent. From Fig. 2.10-3, $\varepsilon = 4.6 \times 10^{-5}$ m $(1.5 \times 10^{-4}$ ft).

$$\frac{\varepsilon}{D_3} = \frac{0.00015}{0.3353} = 0.000448$$

Then, for $N_{Re} = 219\,300$, the Fanning friction factor $f = 0.0047$. Substituting to Eq. (2.10-6) for $\Delta L = 20.0$ ft of 4-in. pipe,

$$F_f = 4f \frac{\Delta L}{D} \frac{v^2}{2g_c} = 4(0.0047) \frac{20.0}{0.3353} \frac{(2.523)^2}{2(32.174)} = 0.111 \frac{\text{ft} \cdot \text{lb}_f}{\text{lb}_m}$$

3. *Friction in 4-in. elbow.* From Table 2.10-1, $K_f = 0.75$. Then, substituting into Eq. (2.10-17),

$$h_f = K_f \frac{v^2}{2g_c} = 0.75 \frac{(2.523)^2}{2(32.174)} = 0.074 \frac{\text{ft} \cdot \text{lb}_f}{\text{lb}_m}$$

4. *Contraction loss from 4- to 2-in. pipe.* Using Eq. (2.10-16) again for contraction from A_3 to A_4 cross-sectional area,

$$K_c = 0.55\left(1 - \frac{A_4}{A_3}\right) = 0.55\left(1 - \frac{0.02330}{0.0884}\right) = 0.405$$

$$h_c = K_c \frac{v_4^2}{2g_c} = 0.405 \frac{(9.57)^2}{2(32.174)} = 0.575 \frac{\text{ft} \cdot \text{lb}_f}{\text{lb}_m}$$

5. *Friction in the 2-in. pipe.* The Reynolds number is

$$N_{Re} = \frac{D_4 v_4 \rho}{\mu} = \frac{0.1722(9.57)(60.52)}{2.33 \times 10^{-4}} = 4.280 \times 10^5$$

$$\frac{\varepsilon}{D} = \frac{0.00015}{0.1722} = 0.00087$$

The Fanning friction factor from Fig. 2.10-3 is $f = 0.0048$. The total length $\Delta L = 125 + 10 + 50 = 185$ ft. Substituting into Eq. (2.10-6),

$$F_f = 4f \frac{\Delta L}{D} \frac{v^2}{2g_c} = 4(0.0048) \frac{185(9.57)^2}{(0.1722)(2)(32.174)} = 29.4 \frac{\text{ft} \cdot \text{lb}_f}{\text{lb}_m}$$

6. *Friction in the two 2-in. elbows.* For a $K_f = 0.75$ and two elbows,

$$h_f = 2K_f \frac{v^2}{2g_c} = \frac{2(0.75)(9.57)^2}{2(32.174)} = 2.136 \frac{\text{ft} \cdot \text{lb}_f}{\text{lb}_m}$$

The total frictional loss $\sum F$ is the sum of items (1) through (6).

$$\sum F = 0.054 + 0.111 + 0.074 + 0.575 + 29.4 + 2.136$$

$$= 32.35 \text{ ft} \cdot \text{lb}_f/\text{lb}_m$$

Using as a datum level z_2, $z_1 = H$ ft, $z_2 = 0$. Since turbulent flow exists, $\alpha = 1.0$. Also, $v_1 = 0$ and $v_2 = v_4 = 9.57$ ft/s. Since p_1 and p_2 are both at 1 atm abs pressure and $\rho_1 = \rho_2$,

$$\frac{p_1}{\rho} - \frac{p_2}{\rho} = 0$$

Also, since no pump is used, $W_S = 0$. Substituting these values into Eq. (2.10-20),

$$H\frac{g}{g_c} + 0 + 0 - 0 = 0 + \frac{1(9.57)^2}{2(32.174)} + 32.35$$

Solving, $H(g/g_c) = 33.77$ ft $\cdot$ lb$_f$/lb$_m$ (100.9 J/kg) and H is 33.77 ft (10.3 m) height of water level above the discharge outlet.

EXAMPLE 2.10-7. Friction Losses with Pump in Mechanical-Energy Balance

Water at 20°C is being pumped from a tank to an elevated tank at the rate of 5.0×10^{-3} m³/s. All of the piping in Fig. 2.10-5 is 4-in. schedule 40 pipe. The pump has an efficiency of 65%. Calculate the kW power needed for the pump.

Solution: The mechanical-energy-balance equation (2.7-28) is written between points 1 and 2, with point 1 being the reference plane.

$$\frac{1}{2\alpha}(v_{2\text{ av}}^2 - v_{1\text{ av}}^2) + g(z_2 - z_1) + \frac{p_2 - p_1}{\rho} + \sum F + W_S = 0 \quad \textbf{(2.7-28)}$$

From Appendix A.2 for water, $p = 998.2$ kg/m³, $\mu = 1.005 \times 10^{-3}$ Pa $\cdot$ s. For 4-in. pipe from Appendix A.5, $D = 0.1023$ m and $A = 8.219 \times 10^{-3}$ m². The velocity in the pipe is $v = 5.0 \times 10^{-3}/(8.219 \times 10^{-3}) = 0.6083$ m/s. The Reynolds number is

$$N_{\text{Re}} = \frac{Dv\rho}{\mu} = \frac{0.1023(0.6083)(998.2)}{1.005 \times 10^{-3}} = 6.181 \times 10^4$$

Hence, the flow is turbulent.

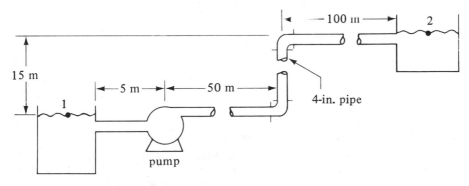

FIGURE 2.10-5. *Process flow diagram for Example 2.10-7.*

The $\sum F$ term for frictional losses includes the following: (1) contraction loss at tank exit, (2) friction in the straight pipe, (3) friction in the two elbows, and (4) expansion loss at the tank entrance.

1. *Contraction loss at tank exit.* From Eq. (2.10-16) for contraction from a large A_1 to a small A_2,

$$k_c = 0.55\left(1 - \frac{A_2}{A_1}\right) = 0.55(1 - 0) = 0.55$$

$$h_c = K_c \frac{v^2}{2\alpha} = (0.55)\frac{(0.6083)^2}{2(1.0)} = 0.102 \text{ J/kg}$$

2. *Friction in the straight pipe.* From Fig. 2.10-3, $\varepsilon = 4.6 \times 10^{-5}$ m and $\varepsilon/D = 4.6 \times 10^{-5}/0.1023 = 0.00045$. Then for $N_{Re} = 6.181 \times 10^4$, $f = 0.0051$. Substituting into Eq. (2.10-6) for $\Delta L = 5 + 50 + 15 + 100 = 170$ m,

$$F_f = 4f\frac{\Delta L}{D}\frac{v^2}{2} = 4(0.0051)\frac{170}{0.1023}\frac{(0.6083)^2}{2} = 6.272 \text{ J/kg}$$

3. *Friction in the two elbows.* From Table 2.10-1, $K_f = 0.75$. Then, substituting into Eq. (2.10-7) for two elbows,

$$h_f = 2K_f\frac{v^2}{2} = 2(0.75)\frac{(0.6083)^2}{2} = 0.278 \text{ J/kg}$$

4. *Expansion loss at the tank entrance.* Using Eq. (2.10-15),

$$K_{ex} = \left(1 - \frac{A_1}{A_2}\right)^2 = (1 - 0)^2 = 1.0$$

$$h_{ex} = K_{ex}\frac{v^2}{2} = 1.0\frac{(0.6083)^2}{2} = 0.185 \text{ J/kg}$$

The total frictional loss is $\sum F$.

$$\sum F = 0.102 + 6.272 + 0.278 + 0.185 = 6.837 \text{ J/kg}$$

Substituting into Eq. (2.7-28), where $(v_1^2 - v_2^2) = 0$ and $(p_2 - p_1) = 0$,

$$0 + 9.806(15.0 - 0) + 0 + 6.837 + W_S = 0$$

Solving, $W_S = -153.93$ J/kg. The mass flow rate is $m = 5.0 \times 10^{-3}(998.2) = 4.991$ kg/s. Using Eq. (2.7-30),

$$W_S = -\eta W_p$$

$$-153.93 = -0.65 W_p$$

Solving, $W_p = 236.8$ J/kg. The pump kW power is

$$\text{pump kW} = mW_p = \frac{4.991(236.8)}{1000} = 1.182 \text{ kW}$$

2.10G   Friction Loss in Noncircular Conduits

The friction loss in long straight channels or conduits of noncircular cross section can be estimated by using the same equations employed for circular pipes if the diameter in the Reynolds number and in the friction factor equation (2.10-6) is taken as the equivalent diameter. The equivalent diameter D is defined as four times the hydraulic radius r_H. The hydraulic radius is defined as the ratio of the cross-sectional area of the channel to the wetted perimeter of the channel for turbulent flow only. Hence,

$$D = 4r_H = 4 \frac{\text{cross-sectional area of channel}}{\text{wetted perimeter of channel}} \tag{2.10-21}$$

For example, for a circular tube,

$$D = \frac{4(\pi D^2/4)}{\pi D} = D$$

For an annular space with outside diameter D_1 and inside D_2,

$$D = \frac{4(\pi D_1^2/4 - \pi D_2^2/4)}{\pi D_1 + \pi D_2} = D_1 - D_2 \tag{2.10-22}$$

For a rectangular duct of sides a and b ft,

$$D = \frac{4(ab)}{2a + 2b} = \frac{2ab}{a + b} \tag{2.10-23}$$

For open channels and partly filled ducts in turbulent flow, the equivalent diameter and Eq. (2.10-6) are also used (P1). For a rectangle with depth of liquid y and width b,

$$D = \frac{4(by)}{b + 2y} \tag{2.10-24}$$

For a wide, shallow stream of depth y,

$$D = 4y \tag{2.10-25}$$

For laminar flow in ducts running full and in open channels with various cross-sectional shapes other than circular, equations are given elsewhere (P1).

2.10H Entrance Section of a Pipe

If the velocity profile at the entrance region of a tube is flat, a certain length of the tube is necessary for the velocity profile to be fully established. This length for the establishment of fully developed flow is called the transition length or entry length. This is shown in Fig. 2.10-6 for laminar flow. At the entrance the velocity profile is flat; i.e., the velocity is the same at all positions. As the fluid progresses down the tube, the boundary-layer thickness increases until finally they meet at the center of the pipe and the parabolic velocity profile is fully established.

The approximate entry length L_e of a pipe having a diameter of D for a fully

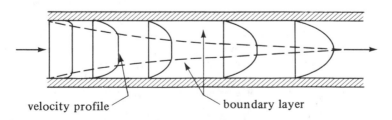

velocity profile   boundary layer

FIGURE 2.10-6. *Velocity profiles near a pipe entrance for laminar flow.*

developed velocity profile to be formed in laminar flow is ($L2$)

$$\frac{L_e}{D} = 0.0575 N_{Re} \qquad\qquad \textbf{(2.10-26)}$$

For turbulent flow, no relation is available to predict the entry length for a fully developed turbulent velocity profile to form. As an approximation, the entry length is nearly independent of the Reynolds number and is fully developed after 50 diameters downstream.

> *EXAMPLE 2.10-8. Entry Length for a Fluid in a Pipe*
> Water at 20°C is flowing through a tube having a diameter of 0.010 m at a velocity of 0.10 m/s.
>> (a) Calculate the entry length.
>> (b) Calculate the entry length for turbulent flow.

Solution: For part (a), from Appendix A.2, $\rho = 998.2$ kg/m^3, $\mu = 1.005 \times 10^{-3}$ Pa $\cdot$ s. The Reynolds number is

$$N_{Re} = \frac{D v \rho}{\mu} = \frac{0.010(0.10)(998.2)}{1.005 \times 10^{-3}} = 993.2$$

Using Eq. (2.10-26) for laminar flow,

$$\frac{L_e}{D} = \frac{L_e}{0.01} = 0.0575(993.2) = 57.1$$

Hence, $L_e = 0.571$ m.
For turbulent flow in part (b), $L_e = 50(0.01) = 0.50$ m.

The pressure drop or friction factor in the entry length is greater than in fully developed flow. For laminar flow the friction factor is highest at the entrance ($L2$) and then decreases smoothly to the fully developed flow value. For turbulent flow there will be some portion of the entrance over which the boundary layer is laminar and the friction factor profile is difficult to express. As an approximation the friction factor for the entry length can be taken as two to three times the value of the friction factor in fully developed flow.

2.10 I   Selection of Pipe Sizes

In large or complex process piping systems, the optimum size of pipe to use for a specific situation depends upon the relative costs of capital investment, power, maintenance, and so on. Charts are available for determining these optimum sizes (P1). However, for small installations approximations are usually sufficiently accurate. A table of representative values of ranges of velocity in pipes are shown in Table 2.10-3.

TABLE 2.10-3.   *Representative Ranges of Velocities in Steel Pipes*

		Velocity	
Type of Fluid	*Type of Flow*	*ft/s*	*m/s*
Nonviscous liquid	Inlet to pump	2–3	0.6–0.9
	Process line or pump discharge	5–8	1.5–2.5
Viscous liquid	Inlet to pump	0.2–0.8	0.06–0.25
	Process line or pump discharge	0.5–2	0.15–0.6
Gas		30–120	9–36
Steam		30–75	9–23

2.11 COMPRESSIBLE FLOW OF GASES

2.11A Introduction and Basic Equation for Flow in Pipes

When pressure changes in gases occur which are greater than about 10%, the friction-loss equations (2.10-9) and (2.10-10) may be in error since compressible flow is occurring. Then the solution of the energy balance is more complicated because of the variation of the density or specific volume with changes in pressure. The field of compressible flow is very large and covers a very wide range of variations in geometry, pressure, velocity, and temperature. In this section we restrict our discussion to isothermal and adiabatic flow in uniform, straight pipes and do not cover flow in nozzles, which is discussed in some detail in other references (M2, P1).

The general mechanical energy-balance equation (2.7-27) can be used as a starting point. Assuming turbulent flow, so that $\alpha = 1.0$; no shaft work, so that $W_S = 0$; and writing the equation for a differential length dL, Eq. (2.7-27) becomes

$$v \, dv + g \, dz + \frac{dp}{\rho} + dF = 0 \tag{2.11-1}$$

For a horizontal duct, $dz = 0$. Using only the wall shear frictional term for dF and writing Eq. (2.10-6) in differential form,

$$v \, dv + V \, dp + \frac{4 f v^2 \, dL}{2D} = 0 \tag{2.11-2}$$

where $V = 1/\rho$. Assuming steady-state flow and a uniform pipe diameter, G is constant and

$$G = v\rho = \frac{v}{V} \tag{2.11-3}$$

$$dv = G \, dV \tag{2.11-4}$$

Substituting Eqs. (2.11-3) and (2.11-4) into (2.11-2) and rearranging,

$$G^2 \frac{dV}{V} + \frac{dp}{V} + \frac{2f G^2}{D} dL = 0 \tag{2.11-5}$$

This is the basic differential equation that is to be integrated. To do this the relation between V and p must be known so that the integral of dp/V can be evaluated. This integral depends upon the nature of the flow and two important conditions used are isothermal and adiabatic flow in pipes.

2.11B Isothermal Compressible Flow

To integrate Eq. (2.11-5) for isothermal flow, an ideal gas will be assumed where

$$pV = \frac{1}{M} RT \tag{2.11-6}$$

Solving for V in Eq. (2.11-6) and substituting it into Eq. (2.11-5), and integrating assuming f is constant,

$$G^2 \int_1^2 \frac{dV}{V} + \frac{M}{RT} \int_1^2 p \, dp + 2f \frac{G^2}{D} \int_1^2 dL = 0 \tag{2.11-7}$$

$$G^2 \ln \frac{V_2}{V_1} + \frac{M}{2RT} (p_2^2 - p_1^2) + 2f \frac{G^2}{D} \Delta L = 0 \tag{2.11-8}$$

Substituting p_1/p_2 for V_2/V_1 and rearranging,

$$p_1^2 - p_2^2 = \frac{4f \Delta L G^2 RT}{DM} + \frac{2G^2 RT}{M} \ln \frac{p_1}{p_2} \qquad \text{(2.11-9)}$$

where M = molecular weight in kg mass/kg mol, $R = 8314.34$ N·m/kg mol·K, and T = temperature K. The quantity $RT/M = p_{av}/\rho_{av}$, where $p_{av} = (p_1 + p_2)/2$ and ρ_{av} is the average density at T and p_{av}. In English units, $R = 1545.3$ ft·lb$_f$/lb mol·°R and the right-hand terms are divided by g_c. Equation (2.11-9) then becomes

$$(p_1 - p_2)_f = \frac{4f \Delta L G^2}{2D\rho_{av}} + \frac{G^2}{\rho_{av}} \ln \frac{p_1}{p_2} \qquad \text{(2.11-10)}$$

The first term on the right of Eqs. (2.11-9) and (2.11-10) represents the frictional loss as given by Eqs. (2.10-9) and (2.10-10). The last term in both equations is generally negligible in ducts of appreciable lengths unless the pressure drop is very large.

EXAMPLE 2.11-1. Compressible Flow of a Gas in a Pipe Line

Natural gas, which is essentially methane, is being pumped through a 1.016-m-ID pipeline for a distance of 1.609×10^5 m (D1) at a rate of 2.077 kg mol/s. It can be assumed that the line is isothermal at 288.8 K. The pressure p_2 at the discharge end of the line is 170.3×10^3 Pa absolute. Calculate the pressure p_1 at the inlet of the line. The viscosity of methane at 288.8 K is 1.04×10^{-5} Pa·s.

Solution: $D = 1.016$ m, $A = \pi D^2/4 = \pi(1.016)^2/4 = 0.8107$ m^2. Then,

$$G = \left(2.077 \, \frac{\text{kg mol}}{\text{s}}\right)\left(16.0 \, \frac{\text{kg}}{\text{kg mol}}\right)\left(\frac{1}{0.8107 \text{ m}^2}\right) = 41.00 \, \frac{\text{kg}}{\text{s·m}^2}$$

$$N_{Re} = \frac{DG}{\mu} = \frac{1.016(41.00)}{1.04 \times 10^{-5}} = 4.005 \times 10^6$$

From Fig. 2.10-3, $\varepsilon = 4.6 \times 10^5$ m.

$$\frac{\varepsilon}{D} = \frac{4.6 \times 10^{-5}}{1.016} = 0.0000453$$

The friction factor $f = 0.0027$.

In order to solve for p_1 in Eq. (2.11-9), trial and error must be used. Estimating p_1 at 620.5×10^3 Pa, $R = 8314.34$ N·m/kg mol·K, and $\Delta L = 1.609 \times 10^5$ m. Substituting into Eq. (2.11-9),

$$p_1^2 - p_2^2 = \frac{4(0.0027)(1.609 \times 10^5)(41.00)^2(8314.34)(288.8)}{1.016(16.0)}$$

$$+ \frac{2(41.00)^2(8314.34)(288.8)}{(16.0)} \ln \frac{620.5 \times 10^3}{170.3 \times 10^3}$$

$$= 4.375 \times 10^{11} + 0.00652 \times 10^{11} = 4.382 \times 10^{11} \, (\text{Pa})^2$$

Now, $P_2 = 170.3 \times 10^3$ Pa. Substituting this into the above and solving for p_1, $p_1 = 683.5 \times 10^3$ Pa. Substituting this new value of p_1 into Eq. (2.11-9) again and solving for p_1, the final result is $p_1 = 683.5 \times 10^3$ Pa. Note that the last term in Eq. (2.11-9) in this case is almost negligible.

When the upstream pressure p_1 remains constant, the mass flow rate G changes as the downstream pressure p_2 is varied. From Eq. (2.11-9), when $p_1 = p_2$, $G = 0$ and when $p_2 = 0$, $G = 0$. This indicates that at some intermediate value of p_2, the flow G must be a maximum. This means that the flow is a maximum when $dG/dp_2 = 0$. Performing this differentiation on Eq. (2.11-9) for constant p_1 and f and solving for G,

$$G_{max} = \sqrt{\frac{Mp_2^2}{RT}} \qquad \textbf{(2.11-11)}$$

Using Eqs. (2.11-3) and (2.11-6),

$$v_{max} = \sqrt{\frac{RT}{M}} = \sqrt{p_2 V_2} \qquad \textbf{(2.11-12)}$$

This is the equation for the velocity of sound in the fluid at the conditions for isothermal flow. Thus, for isothermal compressible flow there is a maximum flow for a given upstream p_1 and further reduction of p_2 will not give any further increase in flow. Further details as to the length of pipe and the pressure at the maximum flow conditions are discussed elsewhere (D1, M2, P1).

EXAMPLE 2.11-2. *Maximum Flow for Compressible Flow of a Gas*
For the conditions of Example 2.11-1, calculate the maximum velocity that can be obtained and the velocity of sound at these conditions. Compare with Example 2.11-1.

Solution: Using Eq. (2.11-12) and the conditions in Example 2.11-1,

$$v_{max} = \sqrt{\frac{RT}{M}} = \sqrt{\frac{8314(288.8)}{16.0}} = 387.4 \text{ m/s}$$

This is the maximum velocity obtainable if p_2 is decreased. This is also the velocity of sound in the fluid at the conditions for isothermal flow. To compare with Example 2.11-1, the actual velocity at the exit pressure p_2 is obtained by combining Eqs. (2.11-3) and (2.11-6) to give

$$v_2 = \frac{RTG}{p_2 M} \qquad \textbf{(2.11-13)}$$

$$= \frac{8314.34(288.8)(41.00)}{(170.3 \times 10^3)16.0} = 36.13 \text{ m/s}$$

2.11C Adiabatic Compressible Flow

When heat transfer through the wall of the pipe is negligible, the flow of gas in compressible flow in a straight pipe of constant cross section is adiabatic. Equation (2.11-5) has been integrated for adiabatic flow and details are given elsewhere (D1, M1, P1). Convenient charts to solve this case are also available (P1). The results for adiabatic flow often deviate very little from isothermal flow, especially in long lines. For very short pipes and relatively large pressure drops, the adiabatic flow rate is greater than the isothermal, but the maximum possible difference is about 20% (D1). For pipes of length of about 1000 diameters or longer, the difference is generally less than 5%. Equation (2.11-8) can also be used when the temperature change over the conduit is small by using an arithmetic average temperature.

Using the same procedures for finding a maximum flow that were used in the isothermal case, the maximum flow occurs when the velocity at the downstream end of the pipe is the sonic velocity for adiabatic flow. This is

$$v_{max} = \sqrt{\gamma p_2 V_2} = \sqrt{\frac{\gamma RT}{M}} \qquad \textbf{(2.11-14)}$$

where, $\gamma = c_p/c_v$, the ratio of heat capacities. For air, $\gamma = 1.4$. Hence, the maximum velocity for adiabatic flow is about 20% greater than for isothermal flow. The rate of flow may not be limited by the flow conditions in the pipe, in practice, but by the development of sonic velocity in a fitting or valve in the pipe. Hence, care should be used in selection of fittings in such pipes for compressible flow. Further details as to the length of pipe and pressure at the maximum flow conditions are given elsewhere (D1, M2, P1).

A convenient parameter often used in compressible flow equations is the Mach number, N_{Ma}, which is defined as the ratio of v, the speed of the fluid in the conduit, to v_{max}, the speed of sound in the fluid at the actual flow conditions.

$$N_{Ma} = \frac{v}{v_{max}} \qquad (2.11\text{-}15)$$

At a Mach number of 1.0, the flow is sonic. At a value less than 1.0, the flow is subsonic, and supersonic at a number above 1.0.

PROBLEMS

2.2-1. Pressure in a Spherical Tank. Calculate the pressure in psia and kN/m² in a spherical tank at the bottom of the tank filled with oil having a diameter of 8.0 ft. The top of the tank is vented to the atmosphere having a pressure of 14.72 psia. The density of the oil is 0.922 g/cm³.

Ans. 17.92 lb$_f$/in.² (psia), 123.5 kN/m²

2.2-2. Pressure with Two Liquids, Hg and Water. An open test tube at 293 K is filled at the bottom with 12.1 cm of Hg and 5.6 cm of water is placed above the Hg. Calculate the pressure at the bottom of the test tube if the atmospheric pressure is 756 mm Hg. Use a density of 13.55 g/cm³ for Hg and 0.998 g/cm³ for water. Give the answer in terms of dyn/cm², psia, and kN/m². See Appendix A.1 for conversion factors.

Ans. 1.175×10^6 dyn/cm², 17.0 psia, 2.3 psig, 117.5 kN/m²

2.2-3. Head of a Fluid of Jet Fuel and Pressure. The pressure at the top of a tank of jet fuel is 180.6 kN/m². The depth of liquid in the tank is 6.4 m. The density of the fuel is 825 kg/m³. Calculate the head of the liquid in m which corresponds to the absolute pressure at the bottom of the tank.

2.2-4. Measurement of Pressure. An open U-tube manometer similar to Fig. 2.2-4a is being used to measure the absolute pressure p_a in a vessel containing air. The pressure p_b is atmospheric pressure, which is 754 mm Hg. The liquid in the manometer is water having a density of 1000 kg/m³. Assume that the density ρ_B is 1.30 kg/m³ and that the distance Z is very small. The reading R is 0.415 m. Calculate p_a in psia and kPa.

Ans. $p_a = 15.17$ psia, 104.6 kPa

2.2-5. Measurement of Small Pressure Differences. The two-fluid U-tube manometer is being used to measure the difference in pressure at two points in a line containing air at 1 atm abs pressure. The value of $R_0 = 0$ for equal pressures. The lighter fluid is a hydrocarbon with a density of 812 kg/m³ and the heavier water has a density of 998 kg/m³. The inside diameters of the U tube and reservoir are 3.2 mm and 54.2 mm, respectively. The reading R of the manometer is 117.2 mm. Calculate the pressure difference in mm Hg and pascal.

2.2-6. Pressure in a Sea Lab. A sea lab 5.0 m high is to be designed to withstand submersion to 150 m, measured from the sea level to the top of the sea lab. Calculate the pressure on top of the sea lab and also the pressure variation on the side of the container measured as the distance x in m from the top of the sea lab downward. The density of seawater is 1020 kg/m³.

Ans. $p = 10.00(150 + x)$ kN/m²

2.2-7. *Measurement of Pressure Difference in Vessels.* In Fig. 2.2-5b the differential manometer is used to measure the pressure difference between two vessels. Derive the equation for the pressure difference $p_A - p_B$ in terms of the liquid heights and densities.

2.2-8. *Design of Settler and Separator for Immiscible Liquids.* A vertical cylindrical settler-separator is to be designed for separating a mixture flowing at $20.0 \, m^3/h$ and containing equal volumes of a light petroleum liquid ($\rho_B = 875 \, kg/m^3$) and a dilute solution of wash water ($\rho_A = 1050 \, kg/m^3$). Laboratory experiments indicate a settling time of 15 min is needed to adequately separate the two phases. For design purposes use a 25-min settling time and calculate the size of the vessel needed, the liquid levels of the light and heavy liquids in the vessel, and the height h_{A2} of the heavy liquid overflow. Assume that the ends of the vessel are approximately flat, that the vessel diameter equals its height, and that one-third of the volume is vapor space vented to the atmosphere. Use the nomenclature given in Fig. 2.2-6.

Ans. $h_{A2} = 1.537 \, m$

2.3-1. *Molecular Transport of a Property with a Variable Diffusivity.* A property is being transported through a fluid at steady state through a constant cross-sectional area. At point 1 the concentration Γ_1 is 2.78×10^{-2} amount of property/m^3 and 1.50×10^{-2} at point 2 at a distance of 2.0 m away. The diffusivity depends on concentration Γ as follows.

$$\delta = A + B\Gamma = 0.150 + 1.65\Gamma$$

(a) Derive the integrated equation for the flux in terms of Γ_1 and Γ_2. Then, calculate the flux.
(b) Calculate Γ at $z = 1.0$ m and plot Γ versus z for the three points.

Ans. (a) $\psi_z = [A(\Gamma_1 - \Gamma_2) + (B/2)(\Gamma_1^2 - \Gamma_2^2)]/(z_2 - z_1)$

2.3-2. *Integration of General Property Equation for Steady State.* Integrate the general property equation (2.3-11) for steady state and no generation between the points Γ_1 at z_1 and Γ_2 at z_2. The final equation should relate Γ to z.

Ans. $\Gamma = (\Gamma_2 - \Gamma_1)(z - z_1)/(z_2 - z_1) + \Gamma_1$

2.4-1. *Shear Stress in Soybean Oil.* Using Fig. 2.4-1, the distance between the two parallel plates is 0.00914 m and the lower plate is being pulled at a relative velocity of 0.366 m/s greater than the top plate. The fluid used is soybean oil with viscosity of 4×10^{-2} Pa·s at 303 K (Appendix A.4).

(a) Calculate the shear stress τ and the shear rate using lb force, ft, and s units.
(b) Repeat, using SI units.
(c) If glycerol at 293 K having a viscosity of 1.069 kg/m·s is used instead of soybean oil, what relative velocity in m/s is needed using the same distance between plates so that the same shear stress is obtained as in part (a)? Also, what is the new shear rate?

Ans. (a) Shear stress = 3.34×10^{-2} lb$_f$/ft^2, shear rate = $40.0 \, s^{-1}$;
(b) 1.60 N/m^2; (c) relative velocity = 0.01369 m/s, shear rate = $1.50 \, s^{-1}$

2.4-2. *Shear Stress and Shear Rate in Fluids.* Using Fig. 2.4-1, the lower plate is being pulled at a relative velocity of 0.40 m/s greater than the top plate. The fluid used is water at 24°C.

(a) How far apart should the two plates be placed so that the shear stress τ is 0.30 N/m^2? Also, calculate the shear rate.
(b) If oil with a viscosity of 2.0×10^{-2} Pa·s is used instead at the same plate spacing and velocity as in part (a), what is the shear stress and the shear rate?

2.5-1. *Reynolds Number for Milk Flow.* Whole milk at 293 K having a density of 1030 kg/m^3 and viscosity of 2.12 cp is flowing at the rate of 0.605 kg/s in a glass pipe having a diameter of 63.5 mm.

(a) Calculate the Reynolds number. Is this turbulent flow?
(b) Calculate the flow rate needed in m³/s for a Reynolds number of 2100 and the velocity in m/s.

Ans. (a) $N_{Re} = 5723$, turbulent flow

2.5-2. Pipe Diameter and Reynolds Number. An oil is being pumped inside a 10.0-mm-diameter pipe at a Reynolds number of 2100. The oil density is $855 \, kg/m^3$ and the viscosity is $2.1 \times 10^{-2} \, Pa \cdot s$.
(a) What is the velocity in the pipe?
(b) It is desired to maintain the same Reynolds number of 2100 and the same velocity as in part (a) using a second fluid with a density of $925 \, kg/m^3$ and a viscosity of $1.5 \times 10^{-2} \, Pa \cdot s$. What pipe diameter should be used?

2.6-1. Average Velocity for Mass Balance in Flow Down Vertical Plate. For a layer of liquid flowing in laminar flow in the z direction down a vertical plate or surface, the velocity profile is

$$v_z = \frac{\rho g \delta^2}{2\mu}\left[1 - \left(\frac{x}{\delta}\right)^2\right]$$

where δ is the thickness of the layer, x is the distance from the free surface of the liquid toward the plate, and v_z is the velocity at a distance x from the free surface.
(a) What is the maximum velocity $v_{z\,max}$?
(b) Derive the expression for the average velocity $v_{z\,av}$ and also relate it to $v_{z\,max}$.

Ans. (a) $v_{z\,max} = \rho g \delta^2/2\mu$, (b) $v_{z\,av} = \frac{2}{3}v_{z\,max}$

2.6-2. Flow of Liquid in a Pipe and Mass Balance. A hydrocarbon liquid enters a simple flow system shown in Fig. 2.6-1 at an average velocity of 1.282 m/s, where $A_1 = 4.33 \times 10^{-3} \, m^2$ and $\rho_1 = 902 \, kg/m^3$. The liquid is heated in the process and the exit density is $875 \, kg/m^3$. The cross-sectional area at point 2 is $5.26 \times 10^{-3} \, m^2$. The process is steady state.
(a) Calculate the mass flow rate m at the entrance and exit.
(b) Calculate the average velocity v in 2 and the mass velocity G in 1.

Ans. (a) $m_1 = m_2 = 5.007 \, kg/s$, (b) $G_1 = 1156 \, kg/s \cdot m^2$

2.6-3. Average Velocity for Mass Balance in Turbulent Flow. For turbulent flow in a smooth circular tube with a radius of R, the velocity profile varies according to the following expression at a Reynolds number of about 10^5:

$$v = v_{max}\left(\frac{R - r}{R}\right)^{1/7}$$

where r is the radial distance from the center and v_{max} the maximum velocity at the center. Derive the equation relating the average velocity (bulk velocity) v_{av} to v_{max} for an incompressible fluid. (*Hint*: The integration can be simplified by substituting z for $R - r$.)

Ans. $v_{av} = \left(\frac{49}{60}\right)v_{max} = 0.817v_{max}$

2.6-4. Bulk Velocity for Flow Between Parallel Plates. A fluid flowing in laminar flow in the x direction between two parallel plates has a velocity profile given by the following.

$$v_x = v_{x\,max}\left[1 - \left(\frac{y}{y_0}\right)^2\right]$$

where $2y_0$ is the distance between the plates, y is the distance from the center line, and v_x is the velocity in the x direction at position y. Derive an equation relating $v_{x\,av}$ (bulk or average velocity) with $v_{x\,max}$.

2.6-5. ***Overall Mass Balance for Dilution Process.*** A well-stirred storage vessel contains 10 000 kg of solution of a dilute methanol solution ($w_A = 0.05$ mass fraction alcohol). A constant flow of 500 kg/min of pure water is suddenly introduced into the tank and a constant rate of withdrawal of 500 kg/min of solution is started. These two flows are continued and remain constant. Assuming that the densities of the solutions are the same and that the total contents of the tank remain constant at 10 000 kg of solution, calculate the time for the alcohol content to drop to 1.0 wt %.

Ans. 32.2 min

2.6-6. ***Overall Mass Balance for Unsteady-State Process.*** A storage vessel is well stirred and contains 500 kg of total solution with a concentration of 5.0 % salt. A constant flow rate of 900 kg/h of salt solution containing 16.67 % salt is suddenly introduced into the tank and a constant withdrawal rate of 600 kg/h is also started. These two flows remain constant thereafter. Derive an equation relating the outlet withdrawal concentration as a function of time. Also, calculate the concentration after 2.0 h.

2.6-7 ***Mass Balance for Flow of Sucrose Solution.*** A 20 wt % sucrose (sugar) solution having a density of 1074 kg/m^3 is flowing through the same piping system as Example 2.6-1 (Fig. 2.6-2). The flow rate entering pipe 1 is 1.892 m^3/h. The flow divides equally in each of pipes 3. Calculate the following:
(a) The velocity in m/s in pipes 2 and 3.
(b) The mass velocity G kg/m$^2 \cdot$ s in pipes 2 and 3.

2.7-1. ***Kinetic-Energy Velocity Correction Factor for Turbulent Flow.*** Derive the equation to determine the value of α, the kinetic-energy velocity correction factor, for turbulent flow. Use Eq. (2.7-20) to approximate the velocity profile and substitute this into Eq. (2.7-15) to obtain $(v^3)_{av}$. Then use Eqs. (2.7-20), (2.6-17), and (2.7-14) to obtain α.

Ans. $\alpha = 0.9448$

2.7-2. ***Flow Between Parallel Plates and Kinetic-Energy Correction Factor.*** The equation for the velocity profile for a fluid flowing in laminar flow between two parallel plates is given in Problem 2.6-4. Derive the equation to determine the value of the kinetic-energy velocity correction factor α. [*Hint*: First derive an equation relating v to v_{av}. Then derive the equation for $(v^3)_{av}$ and, finally, relate these results to α.]

2.7-3. ***Temperature Drop in Throttling Valve and Energy Balance.*** Steam is flowing through an adiabatic throttling valve (no heat loss or external work). Steam enters point 1 upstream of the valve at 689 kPa abs and 171.1°C and leaves the valve (point 2) at 359 kPa. Calculate the temperature t_2 at the outlet. [*Hint*: Use Eq. (2.7-21) for the energy balance and neglect the kinetic-energy and potential-energy terms as shown in Example 2.7-1. Obtain the enthalpy H_1 from Appendix A.2, steam tables. For H_2, linear interpolation of the values in the table will have to be done to obtain t_2.] Use SI units.

Ans. $t_2 = 160.6°C$

2.7-4. ***Energy Balance on a Heat Exchanger and a Pump.*** Water at 93.3°C is being pumped from a large storage tank at 1 atm abs at a rate of 0.189 m^3/min by a pump. The motor that drives the pump supplies energy to the pump at the rate of 1.49 kW. The water is pumped through a heat exchanger, where it gives up 704 kW of heat and is then delivered to a large open storage tank at an elevation of 15.24 m above the first tank. What is the final temperature of the water to the second tank? Also, what is the gain in enthalpy of the water due to the work input? (*Hint*: Be sure and use the steam tables for the enthalpy of the water. Neglect any kinetic-energy changes, but not potential-energy changes.)

Ans. $t_2 = 38.2°C$, work input gain = 0.491 kJ/kg

2.7-5. *Steam Boiler and Overall Energy Balance.* Liquid water under pressure at 150 kPa enters a boiler at 24°C through a pipe at an average velocity of 3.5 m/s in turbulent flow. The exit steam leaves at a height of 25 m above the liquid inlet at 150°C and 150 kPa absolute and the velocity in the outlet line is 12.5 m/s in turbulent flow. The process is steady state. How much heat must be added per kg of steam?

2.7-6. *Energy Balance on a Flow System with a Pump and Heat Exchanger.* Water stored in a large, well-insulated storage tank at 21.0°C and atmospheric pressure is being pumped at steady state from this tank by a pump at the rate of 40 m³/h. The motor driving the pump supplies energy at the rate of 8.5 kW. The water is used as a cooling medium and passes through a heat exchanger where 255 kW of heat is added to the water. The heated water then flows to a second, large vented tank, which is 25 m above the first tank. Determine the final temperature of the water delivered to the second tank.

2.7-7. *Mechanical-Energy Balance in Pumping Soybean Oil.* Soybean oil is being pumped through a uniform-diameter pipe at a steady mass-flow rate. A pump supplies 209.2 J/kg mass of fluid flowing. The entrance abs pressure in the inlet pipe to the pump is 103.4 kN/m². The exit section of the pipe downstream from the pump is 3.35 m above the entrance and the exit pressure is 172.4 kN/m². Exit and entrance pipes are the same diameter. The fluid is in turbulent flow. Calculate the friction loss in the system. See Appendix A.4 for the physical properties of soybean oil. The temperature is 303 K.

Ans. $\sum F = 101.3$ J/kg

2.7-8. *Pump Horsepower in Brine System.* A pump pumps 0.200 ft³/s of brine solution having a density of 1.15 g/cm³ from an open feed tank having a large cross-sectional area. The suction line has an inside diameter of 3.548 in. and the discharge line from the pump a diameter of 2.067 in. The discharge flow goes to an open overhead tank and the open end of this line is 75 ft above the liquid level in the feed tank. If the friction losses in the piping system are 18.0 ft · lb$_f$/lb$_m$, what pressure must the pump develop and what is the horsepower of the pump if the efficiency is 70%? The flow is turbulent.

2.7-9. *Pressure Measurements from Flows.* Water having a density of 998 kg/m³ is flowing at the rate of 1.676 m/s in a 3.068-in.-diameter horizontal pipe at a pressure p_1 of 68.9 kPa abs. It then passes to a pipe having an inside diameter of 2.067 in.
  (a) Calculate the new pressure p_2 in the 2.067-in. pipe. Assume no friction losses.
  (b) If the piping is vertical and the flow is upward, calculate the new pressure p_2. The pressure tap for p_2 is 0.457 m above the tap for p_1.

Ans. (a) $p_2 = 63.5$ kPa; (b) $p_2 = 59.1$ kPa

2.7-10. *Draining Cotton Seed Oil from a Tank.* A cylindrical tank 1.52 m in diameter and 7.62 m high contains cotton seed oil having a density of 917 kg/m³. The tank is open to the atmosphere. A discharge nozzle of inside diameter 15.8 mm and cross-sectional area A_2 is located near the bottom of the tank. The surface of the liquid is located at $H = 6.1$ m above the center line of the nozzle. The discharge nozzle is opened, draining the liquid level from $H = 6.1$ m to $H = 4.57$ m. Calculate the time in seconds to do this. [*Hint*: The velocity on the surface of the reservoir is small and can be neglected. The velocity v_2 m/s in the nozzle can be calculated for a given H by Eq. (2.7-36). However, H, and hence v_2, are varying. Set up an unsteady-state mass balance as follows. The volumetric flow rate in the tank is $(A_t \, dH)/dt$, where A_t is the tank cross section in m² and $A_t \, dH$ is the m^3 liquid flowing in dt s. This rate must equal the negative of the volumetric rate in the nozzle, or $-A_2 v_2$ m^3/s. The negative sign is present since dH is the negative of v_2. Rearrange this equation and integrate between $H = 6.1$ m at $t = 0$ and $H = 4.57$ m at $t = t_F$.]

Ans. $t_F = 1380$ s

2.7-11. Friction Loss in Turbine Water Power System. Water is stored in an elevated reservoir. To generate power, water flows from this reservoir down through a large conduit to a turbine and then through a similar-sized conduit. At a point in the conduit 89.5 m above the turbine, the pressure is 172.4 kPa and at a level 5 m below the turbine, the pressure is 89.6 kPa. The water flow rate is 0.800 m^3/s. The output of the shaft of the turbine is 658 kW. The water density is 1000 kg/m^3. If the efficiency of the turbine in converting the mechanical energy given up by the fluid to the turbine shaft is 89% ($\eta_t = 0.89$), calculate the friction loss in the turbine in J/kg. Note that in the mechanical-energy-balance equation, the W_s is equal to the output of the shaft of the turbine over η_t.

Ans. $\sum F = 85.3$ J/kg

2.7-12. Pipeline Pumping of Oil. A pipeline laid cross country carries oil at the rate of 795 m^3/d. The pressure of the oil is 1793 kPg gage leaving pumping station 1. The pressure is 862 kPa gage at the inlet to the next pumping station, 2. The second station is 17.4 m higher than the first station. Calculate the lost work ($\sum F$ friction loss) in J/kg mass oil. The oil density is 769 kg/m^3.

2.7-13. Test of Centrifugal Pump and Mechanical-Energy Balance. A centrifugal pump is being tested for performance and during the test the pressure reading in the 0.305-m-diameter suction line just adjacent to the pump casing is -20.7 kPa (vacuum below atmospheric pressure). In the discharge line with a diameter of 0.254 m at a point 2.53 m above the suction line, the pressure is 289.6 kPa gage. The flow of water from the pump is measured as 0.1133 m^3/s. (The density can be assumed as 1000 kg/m^3.) Calculate the kW input of the pump.

Ans. 38.11 kW

2.7-14. Friction Loss in Pump and Flow System. Water at 20°C is pumped from the bottom of a large storage tank where the pressure is 310.3 kPa gage to a nozzle which is 15.25 m above the tank bottom and discharges to the atmosphere with a velocity in the nozzle of 19.81 m/s. The water flow rate is 45.4 kg/s. The efficiency of the pump is 80% and 7.5 kW are furnished to the pump shaft. Calculate the following.
(a) The friction loss in the pump.
(b) The friction loss in the rest of the process.

2.7-15. Power for Pumping in Flow System. Water is being pumped from an open water reservoir at the rate of 2.0 kg/s at 10°C to an open storage tank 1500 m away. The pipe used is schedule 40 $3\frac{1}{2}$-in. pipe and the frictional losses in the system are 625 J/kg. The surface of the water reservoir is 20 m above the level of the storage tank. The pump has an efficiency of 75%.
(a) What is the kW power required for the pump?
(b) If the pump is not present in the system, will there be a flow?

Ans. (a) 1.143 kW

2.8-1. Momentum Balance in a Reducing Bend. Water is flowing at steady state through the reducing bend in Fig. 2.8-3. The angle $\alpha_2 = 90°$ (a right-angle bend). The pressure at point 2 is 1.0 atm abs. The flow rate is 0.020 m^3/s and the diameters at points 1 and 2 are 0.050 m and 0.030 m, respectively. Neglect frictional and gravitational forces. Calculate the resultant forces on the bend in newtons and lb force. Use $\rho = 1000$ kg/m^3.

Ans. $-R_x = +450.0$ N, $-R_y = -565.8$ N.

2.8-2. Forces on Reducing Bend. Water is flowing at steady state and 363 K at a rate of 0.0566 m^3/s through a 60° reducing bend ($\alpha_2 = 60°$) in Fig. 2.8-3. The inlet pipe diameter is 0.1016 m and the outlet 0.0762 m. The friction loss in the pipe bend can be estimated as $v_2^2/5$. Neglect gravity forces. The exit pressure $p_2 = 111.5 \, kN/m^2$ gage. Calculate the forces on the bend in newtons.

Ans. $-R_x = +1344 \, N$, $-R_y = -1026 \, N$

2.8-3. ***Force of Water Stream on a Wall.*** Water at 298 K discharges from a nozzle and travels horizontally hitting a flat vertical wall. The nozzle has a diameter of 12 mm and the water leaves the nozzle with a flat velocity profile at a velocity of 6.0 m/s. Neglecting frictional resistance of the air on the jet, calculate the force in newtons on the wall.

Ans. $-R_x = 4.059 \, N$

2.8-4. ***Flow Through an Expanding Bend.*** Water at a steady-state rate of 0.050 m³/s is flowing through an expanding bend that changes direction by 120°. The upstream diameter is 0.0762 m and downstream is 0.2112 m. The upstream pressure is 68.94 kPa gage. Neglect energy losses within the elbow and calculate the downstream pressure at 298 K. Also calculate R_x and R_y.

2.8-5. ***Force of Stream on a Wall.*** Repeat Problem 2.8-3 for the same conditions except that the wall is inclined 45° with the vertical. The flow is frictionless. Assume no loss in energy. The amount of fluid splitting in each direction along the plate can be determined by using the continuity equation and a momentum balance. Calculate this flow division and the force on the wall.

Ans. $m_2 = 0.5774 \, kg/s$, $m_3 = 0.09907 \, kg/s$, $-R_x = 2.030 \, N$, $-R_y = -2.030 \, N$ (force on wall).

2.8-6. ***Momentum Balance for Free Jet on a Curved, Fixed Vane.*** A free jet having a velocity of 30.5 m/s and a diameter of 5.08×10^{-2} m is deflected by a curved, fixed vane as in Fig. 2.8-5a. However, the vane is curved downward at an angle of 60° instead of upward. Calculate the force of the jet on the vane. The density is 1000 kg/m³.

Ans. $-R_x = 942.8 \, N$, $-R_y = 1633 \, N$

2.8-7. ***Momentum Balance for Free Jet on a U-Type, Fixed Vane.*** A free jet having a velocity of 30.5 m/s and a diameter of 1.0×10^{-2} m is deflected by a smooth, fixed vane as in Fig. 2.8-5a. However, the vane is in the form of a U so that the exit jet travels in a direction exactly opposite to the entering jet. Calculate the force of the jet on the vane. Use $\rho = 1000 \, kg/m^3$.

Ans. $-R_x = 146.1 \, N$, $-R_y = 0$

2.8-8. ***Momentum Balance on Reducing Elbow and Friction Losses.*** Water at 20°C is flowing through a reducing bend, where α_2 (see Fig. 2.8-3) is 120°. The inlet pipe diameter is 1.829 m, the outlet is 1.219 m, and the flow rate is 8.50 m³/s. The exit point z_2 is 3.05 m above the inlet and the inlet pressure is 276 kPa gage. Friction losses are estimated as $0.5v_2^2/2$ and the mass of water in the elbow is 8500 kg. Calculate the forces R_x and R_y and the resultant force on the control volume fluid.

2.8-9. ***Momentum Velocity Correction Factor β for Turbulent Flow.*** Determine the momentum velocity correction factor β for turbulent flow in a tube. Use Eq. (2.7-20) for the relationship between v and position.

2.9-1. ***Film of Water on Wetted-Wall Tower.*** Pure water at 20°C is flowing down a vertical wetted-wall column at a rate of 0.124 kg/s·m. Calculate the film thickness and the average velocity.

Ans. $\delta = 3.370 \times 10^{-4} \, m$, $v_{z \, av} = 0.3687 \, m/s$

2.9-2. ***Shell Momentum Balance for Flow Between Parallel Plates.*** A fluid of constant density is flowing in laminar flow at steady state in the horizontal x direction between two flat and parallel plates. The distance between the two plates in the vertical y direction is $2y_0$. Using a shell momentum balance, derive the equation for the velocity profile within this fluid and the maximum velocity for a distance

L m in the x direction. [*Hint :* See the method used in Section 2.9B to derive Eq. (2.9-9). One boundary condition used is $dv_x/dy = 0$ at $y = 0$.]

$$\text{Ans.} \quad v_x = \frac{p_0 - p_L}{2\mu L} y_0^2 \left[1 - \left(\frac{y}{y_0} \right)^2 \right]$$

2.9-3. Velocity Profile for Non-Newtonian Fluid. The stress rate of shear for a non-Newtonian fluid is given by

$$\tau_{rx} = K \left(-\frac{dx_x}{dr} \right)^n$$

where K and n are constants. Find the relation between velocity and radial position r for this incompressible fluid at steady state. [*Hint :* Combine the equation given here with Eq. (2.9-6). Then raise both sides of the resulting equation to the $1/n$ power and integrate.]

$$\text{Ans.} \quad v_x = \frac{n}{n+1} \left(\frac{p_0 - p_L}{2KL} \right)^{1/n} (R_0)^{(n+1)/n} \left[1 - \left(\frac{r}{R_0} \right)^{(n+1)/n} \right]$$

2.9-4. Shell Momentum Balance for Flow Down an Inclined Plane. Consider the case of a Newtonian fluid in steady-state laminar flow down an inclined plane surface that makes an angle θ with the horizontal. Using a shell momentum balance, find the equation for the velocity profile within the liquid layer having a thickness L and the maximum velocity of the free surface. (*Hint :* The convective momentum terms cancel for fully developed flow and the pressure-force terms also cancel, because of the presence of a free surface. Note that there is a gravity force on the fluid.)

$$\text{Ans.} \quad v_{x \, max} = \rho g L^2 \sin \theta / 2\mu$$

2.10-1. Viscosity Measurement of a Liquid. One use of the Hagen–Poiseuille equation (2.10-2) is in determining the viscosity of a liquid by measuring the pressure drop and velocity of the liquid in a capillary of known dimensions. The liquid used has a density of 912 kg/m³ and the capillary has a diameter of 2.222 mm and a length of 0.1585 m. The measured flow rate was 5.33×10^{-7} m³/s of liquid and the pressure drop 131 mm of water (density 996 kg/m³). Neglecting end effects, calculate the viscosity of the liquid in Pa·s.

$$\text{Ans.} \quad \mu = 9.06 \times 10^{-3} \text{ Pa·s}$$

2.10-2. Frictional Pressure Drop in Flow of Olive Oil. Calculate the frictional pressure drop in pascal for olive oil at 293 K flowing through a commercial pipe having an inside diameter of 0.0525 m and a length of 76.2 m. The velocity of the fluid is 1.22 m/s. Use the friction factor method. Is the flow laminar or turbulent? Use physical data from Appendix A.4.

2.10-3. Frictional Loss in Straight Pipe and Effect of Type of Pipe. A liquid having a density of 801 kg/m³ and a viscosity of 1.49×10^{-3} Pa·s is flowing through a horizontal straight pipe at a velocity of 4.57 m/s. The commercial steel pipe is $1\frac{1}{2}$-in. nominal pipe size, schedule 40. For a length of pipe of 61 m, do as follows.
(a) Calculate the friction loss F_f.
(b) For a smooth tube of the same inside diameter, calculate the friction loss. What is the percent reduction of the F_f for the smooth tube?

$$\text{Ans.} \quad \text{(a) 348.9 J/kg; (b) 274.2 J/kg(91.7 ft·lb}_f/\text{lb}_m\text{), 21.4\% reduction}$$

2.10-4. Trial-and-Error Solution for Hydraulic Drainage. In a hydraulic project a cast iron pipe having an inside diameter of 0.156 m and a 305-m length is used to drain wastewater at 293 K. The available head is 4.57 m of water. Neglecting any losses in fittings and joints in the pipe, calculate the flow rate in m^3/s. (*Hint :* Assume the physical properties of pure water. The solution is trial and error since the velocity appears in N_{Re}, which is needed to determine the friction factor. As a first trial, assume that $v = 1.7$ m/s.)

2.10-5. *Mechanical-Energy Balance and Friction Losses.* Hot water is being discharged from a storage tank at the rate of 0.223 ft^3/s. The process flow diagram and conditions are the same as given in Example 2.10-6, except for different nominal pipe sizes of schedule 40 steel pipe as follows. The 20-ft-long outlet pipe from the storage tank is $1\frac{1}{2}$-in. pipe instead of 4-in. pipe. The other piping, which was 2-in. pipe, is now 2.5-in. pipe. Note that now a sudden expansion occurs after the elbow in the $1\frac{1}{2}$-in. pipe to a $2\frac{1}{2}$-in pipe.

2.10-6. *Friction Losses and Pump Horsepower.* Hot water in an open storage tank at 82.2°C is being pumped at the rate of 0.379 m^3/min from this storage tank. The line from the storage tank to the pump suction is 6.1 m of 2-in. schedule 40 steel pipe and it contains three elbows. The discharge line after the pump is 61 m of 2-in. pipe and contains two elbows. The water discharges to the atmosphere at a height of 6.1 m above the water level in the storage tank.
(a) Calculate all frictional losses $\sum F$.
(b) Make a mechanical-energy balance and calculate W_S of the pump in J/kg.
(c) What is the kW power of the pump if its efficiency is 75%?

> **Ans.** (a) $\sum F = 122.8$ J/kg.
> (b) $W_S = -186.9$ J/kg, (c) 1.527 kW

2.10-7. *Pressure Drop of a Flowing Gas.* Nitrogen gas is flowing through a 4-in. schedule 40 commerical steel pipe at 298 K. The total flow rate is 7.40×10^{-2} kg/s and the flow can be assumed as isothermal. The pipe is 3000 m long and the inlet pressure is 200 kPa. Calculate the outlet pressure.

> **Ans.** $p_2 = 188.5$ kPa

2.10-8. *Entry Length for Flow in a Pipe.* Air at 10°C and 1.0 atm abs pressure is flowing at a velocity of 2.0 m/s inside a tube having a diameter of 0.012 m.
(a) Calculate the entry length.
(b) Calculate the entry length for water at 10°C and the same velocity.

2.10-9. *Friction Loss in Pumping Oil to Pressurized Tank.* An oil having a density of 833 kg/m^3 and a viscosity of 3.3×10^{-3} Pa·s is pumped from an open tank to a pressurized tank held at 345 kPa gage. The oil is pumped from an inlet at the side of the open tank through a line of commercial steel pipe having an inside diameter of 0.07792 m at the rate of 3.494×10^{-3} m^3/s. The length of straight pipe is 122 m and the pipe contains two elbows (90°) and a globe valve half open. The level of the liquid in the open tank is 20 m above the liquid level in the pressurized tank. The pump efficiency is 65%. Calculate the kW power of the pump.

2.10-10. *Flow in an Annulus and Pressure Drop.* Water flows in the annulus of a horizontal, concentric-pipe heat exchanger and is being heated from 40°C to 50°C in the exchanger which has a length of 30 m of equivalent straight pipe. The flow rate of the water is 2.90×10^{-3} m^3/s. The inner pipe is 1-in. schedule 40 and the outer is 2-in. schedule 40. What is the pressure drop? Use an average temperature of 45°C for bulk physical properties. Assume that the wall temperature is an average of 4°C higher than the average bulk temperature so that a correction can be made for the effect of heat transfer on the friction factor.

2.11-1. *Derivation of Maximum Velocity for Isothermal Compressible Flow.* Starting with Eq. (2.11-9), derive Eqs. (2.11-11) and (2.11-12) for the maximum velocity in isothermal compressible flow.

2.11-2. *Pressure Drop in Compressible Flow.* Methane gas is being pumped through a 305-m length of 52.5-mm-ID steel pipe at the rate of 41.0 kg/m^2·s. The inlet pressure is $p_1 = 345$ kPa abs. Assume isothermal flow at 288.8 K.

(a) Calculate the pressure p_2 at the end of the pipe. The viscosity is 1.04×10^{-5} Pa·s.

(b) Calculate the maximum velocity that can be attained at these conditions and compare with the velocity in part (a).

Ans. (a) $p_2 = 298.4$ kPa, (b) $v_{max} = 387.4$ m/s, $v_2 = 20.62$ m/s

2.11-3. *Pressure Drop in Isothermal Compressible Flow.* Air at 288 K and 275 kPa abs enters a pipe and is flowing in isothermal compressible flow in a commercial pipe having an ID of 0.080 m. The length of the pipe is 60 m. The mass velocity at the entrance to the pipe is 165.5 kg/m²·s. Assume 29 for the molecular weight of air. Calculate the pressure at the exit. Also, calculate the maximum allowable velocity that can be attained and compare with the actual.

REFERENCES

(B1) BENNETT, C. O., and MEYERS, J. E. *Momentum, Heat and Mass Transfer*, 3rd ed. New York: McGraw-Hill Book Company, 1982.

(D1) DODGE, B. F. *Chemical Engineering Thermodynamics.* New York: McGraw-Hill Book Company, 1944.

(E1) EARLE, R. L. *Unit Operations in Food Processing.* Oxford: Pergamon Press, Inc., 1966.

(K1) KITTRIDGE, C. P., and ROWLEY, D. S. *Trans. A.S.M.E.,* **79**, 1759 (1957).

(L1) LANGE, N. A. *Handbook of Chemistry*, 10th ed. New York: McGraw-Hill Book Company, 1967.

(L2) LANGHAAR, H. L. *Trans. A.S.M.E,* **64**, A-55 (1942).

(M1) MOODY, L. F. *Trans. A.S.M.E.,* **66**, 671 (1944); *Mech Eng.,* **69**, 1005 (1947).

(M2) MCCABE, W.L., SMITH, J. C., and HARRIOTT, P. *Unit Operations of Chemical Engineering*, 4th ed. New York: McGraw-Hill Book Company, 1985.

(N1) National Bureau of Standards. *Tables of Thermal Properties of Gases*, Circular 464 (1955).

(P1) PERRY, R. H., and GREEN, D. *Perry's Chemical Engineers' Handbook*, 6th ed. New York: McGraw-Hill Book Company, 1984.

(R1) REID, R. C., PRAUSNITZ, J. M., and SHERWOOD, T. K. *The Properties of Gases and Liquids,* 3rd ed. New York: McGraw-Hill Book Company, 1977.

(R2) *Reactor Handbook*, vol. 2, AECD-3646. Washington D.C.: Atomic Energy Commission, May 1955.

(S1) SWINDELLS, J. F., COE, J. R. Jr., and GODFREY, T. B. *J. Res. Nat. Bur. Standards,* **48**, 1 (1952).

(S2) SKELLAND, A. H. P. *Non-Newtonian Flow and Heat Transfer.* New York: John Wiley & Sons, Inc., 1967.

(S3) SIEDER, E. N., and TATE, G. E. *Ind. Eng. Chem.,* **28**, 1429 (1936).

(W1) WEAST, R. C. *Handbook of Chemistry and Physics*, 48th ed. Boca Raton, Fla.: Chemical Rubber Co., Inc., 1967–1968.

CHAPTER 3

Principles of
Momentum Transfer
and Applications

3.1 FLOW PAST IMMERSED OBJECTS AND PACKED AND FLUIDIZED BEDS

3.1A Definition of Drag Coefficient for Flow Past Immersed Objects

1. Introduction and types of drag. In Chapter 2 we were concerned primarily with the momentum transfer and the frictional losses for flow of fluids inside conduits or pipes. In this section we consider in some detail the flow of fluids around solid, immersed objects.

The flow of fluids outside immersed bodies appears in many chemical engineering applications and other processing applications. These occur, for example, in flow past spheres in settling, flow through packed beds in drying and filtration, flow past tubes in heat exchangers, and so on. It is useful to be able to predict the frictional losses and/or the force on the submerged objects in these various applications.

In the examples of fluid friction inside conduits that we considered in Chapter 2, the transfer of momentum perpendicular to the surface resulted in a tangential shear stress or drag on the smooth surface parallel to the direction of flow. This force exerted by the fluid on the solid in the direction of flow is called *skin* or *wall drag*. For any surface in contact with a flowing fluid, skin friction will exist. In addition to skin friction, if the fluid is not flowing parallel to the surface but must change directions to pass around a solid body such as a sphere, significant additional frictional losses will occur and this is called *form drag*.

In Fig. 3.1-1a the flow of fluid is parallel to the smooth surface of the flat, solid plate, and the force F in newtons on an element of area dA m^2 of the plate is the wall shear stress τ_w times the area dA or $\tau_w\,dA$. The total force is the sum of the integrals of these quantities evaluated over the entire area of the plate. Here the transfer of momentum to the surface results in a tangential stress or skin drag on the surface.

In many cases, however, the immersed body is a blunt-shaped solid which presents various angles to the direction of the fluid flow. As shown in Fig. 3.1-1b, the free-stream velocity is v_0 and is uniform on approaching the blunt-shaped body suspended in a very

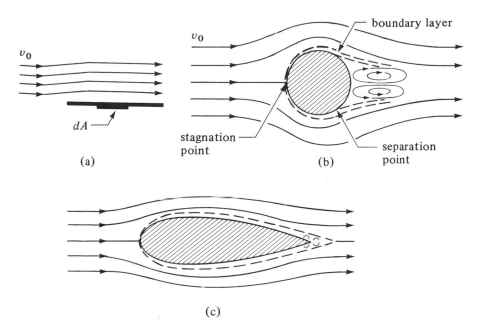

FIGURE 3.1-1. *Flow past immersed objects : (a) flat plate, (b) sphere, (c) streamlined object.*

large duct. Lines called *streamlines* represent the path of fluid elements around the suspended body. The thin boundary layer adjacent to the solid surface is shown as a dashed line and at the edge of this layer the velocity is essentially the same as the bulk fluid velocity adjacent to it. At the front center of the body, called the *stagnation point*, the fluid velocity will be zero and boundary-layer growth begins at this point and continues over the surface until it separates. The tangential stress on the body because of the velocity gradient in the boundary layer is the skin friction. Outside the boundary layer the fluid changes direction to pass around the solid and also accelerates near the front and then decelerates. Because of these effects, an additional force is exerted by the fluid on the body. This phenomenon, called *form drag*, is in addition to the skin drag in the boundary layer.

In Fig. 3.1-lb, as shown, separation of the boundary layer occurs and a wake, covering the entire rear of the object, occurs where large eddies are present and contribute to the form drag. The point of separation depends on the shape of the particle, Reynolds number, and so on, and is discussed in detail elsewhere (S3).

Form drag for bluff bodies can be minimized by streamlining the body (Fig. 3.1-1c), which forces the separation point toward the rear of the body, which greatly reduces the size of the wake. Additional discussion of turbulence and boundary layers is given in Section 3.10.

2. Drag coefficient. From the previous discussions it is evident that the geometry of the immersed solid is a main factor in determining the amount of total drag force exerted on the body. Correlations of the geometry and flow characteristics for solid objects suspended or held in a free stream (immersed objects) are similar in concept and form to the friction factor–Reynolds number correlation given for flow inside conduits. In flow through conduits, the friction factor was defined as the ratio of the drag force per unit area (shear stress) to the product of density times velocity head as given in Eq. (2.10-4).

In a similar manner for flow past immersed objects, the drag coefficient C_D is defined as the ratio of the total drag force per unit area to $\rho v_0^2/2$.

$$C_D = \frac{F_D/A_p}{\rho v_0^2/2} \quad \text{(SI)}$$

$$C_D = \frac{F_D/A_p}{\rho v_0^2/2g_c} \quad \text{(English)}$$

(3.1-1)

where F_D is the total drag force in N, A_p is an area in m^2, C_D is dimensionless, v_0 is free-stream velocity in m/s, and ρ is density of fluid in kg/m^3. In English units, F_D is in lb_f, v_0 is in ft/s, ρ is in lb_m/ft^3, and A_p in ft^2. The area A_p used is the area obtained by projecting the body on a plane perpendicular to the line of flow. For a sphere, $A_p = \pi D_p^2/4$, where D_p is sphere diameter; for a cylinder whose axis is perpendicular to the flow direction, $A_p = LD_p$, where L = cylinder length. Solving Eq. (3.1-1) for the total drag force,

$$F_D = C_D \frac{v_0^2}{2} \rho A_p$$

(3.1-2)

The Reynolds number for a given solid immersed in a flowing liquid is

$$N_{Re} = \frac{D_p v_0 \rho}{\mu} = \frac{D_p G_0}{\mu}$$

(3.1-3)

where $G_0 = v_0 \rho$.

3.1B Flow Past Sphere, Long Cylinder, and Disk

For each particular shape of object and orientation of the object with the direction of flow, a different relation of C_D versus N_{Re} exists. Correlations of drag coefficient versus Reynolds number are shown in Fig. 3.1-2 for spheres, long cylinders, and disks. The face of the disk and the axis of the cylinder are perpendicular to the direction of flow. These curves have been determined experimentally. However, in the laminar region for low Reynolds numbers less than about 1.0, the experimental drag force for a sphere is the same as the theoretical Stokes' law equation as follows.

$$F_D = 3\pi\mu D_p v_0$$

(3.1-4)

Combining Eqs. (3.1-2) and (3.1-4) and solving for C_D, the drag coefficient predicted by Stokes' law is

$$C_D = \frac{24}{D_p v_0 \rho/\mu} = \frac{24}{N_{Re}}$$

(3.1-5)

The variation of C_D with N_{Re} (Fig. 3.1-2) is quite complicated because of the interaction of the factors that control skin drag and form drag. For a sphere as the Reynolds number is increased beyond the Stokes' law range, separation occurs and a wake is formed. Further increases in N_{Re} cause shifts in the separation point. At about $N_{Re} = 3 \times 10^5$ the sudden drop in C_D is the result of the boundary layer becoming completely turbulent and the point of separation moving downstream. In the region of N_{Re} of about 1×10^3 to 2×10^5 the drag coefficient is approximately constant for each shape and $C_D = 0.44$ for a sphere. Above a N_{Re} of about 5×10^5 the drag coefficients are again approximately constant with C_D for a sphere being 0.13, 0.33 for a cylinder, and

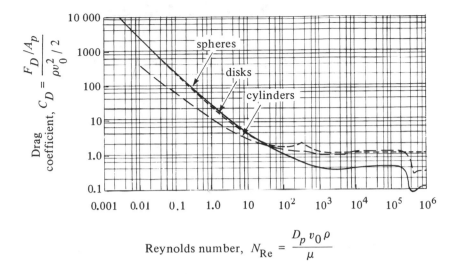

$$N_{Re} = \frac{D_p v_0 \rho}{\mu}$$

Reynolds number, $N_{Re} = \dfrac{D_p v_0 \rho}{\mu}$

FIGURE 3.1-2. *Drag coefficients for flow past immersed spheres, long cylinders, and disks. (Reprinted with permission from C. E. Lapple and C. B. Shepherd, Ind. Eng. Chem., 32, 606 (1940). Copyright by the American Chemical Society.)*

1.12 for a disk. Additional discussions and theory on flow past spheres are given in Section 3.9E.

For derivations of theory and detailed discussions of the drag force for flow parallel to a flat plate, Section 3.10 on boundary-layer flow and turbulence should be consulted. The flow of fluids normal to banks of cylinders or tubes occurs in heat exchangers and other processing applications. The banks of tubes can be arranged in a number of different geometries. Because of the many possible geometric tube configurations and spacings, it is not possible to have one correlation of the data on pressure drop and friction factors. Details of the many correlations available are given elsewhere (P1).

EXAMPLE 3.1-1. Force on Submerged Sphere

Air at 37.8°C and 101.3 kPa absolute pressure flows past a sphere having a diameter of 42 mm at a velocity of 23 m/s. What is the drag coefficient C_D and the force on the sphere?

Solution: From Appendix A.3 for air at 37.8°C, $\rho = 1.137$ kg/m^3, $\mu = 1.90 \times 10^{-5}$ Pa·s. Also, $D_p = 0.042$ m and $v_0 = 23.0$ m/s. Using Eq. (3.1-3),

$$N_{Re} = \frac{D_p v_0 \rho}{\mu} = \frac{0.042(23.0)(1.137)}{1.90 \times 10^{-5}} = 5.781 \times 10^4$$

From Fig. 3.1-2 for a sphere, $C_D = 0.47$. Substituting into Eq. (3.1-2), where $A_p = \pi D_p^2/4$ for a sphere,

$$F_D = C_D \frac{v_0^2}{2} \rho A_p = (0.47) \frac{(23.0)^2}{2} (1.137)(\pi) \frac{(0.042)^2}{4} = 0.1958 \text{ N}$$

EXAMPLE 3.1-2. Force on a Cylinder in a Tunnel

Water at 24°C is flowing past a long cylinder at a velocity of 1.0 m/s in a large tunnel. The axis of the cylinder is perpendicular to the direction of flow. The diameter of the cylinder is 0.090 m. What is the force per meter length on the cylinder?

Solution: From Appendix A.2 for water at 24°C, $\rho = 997.2$ kg/m^3, $\mu = 0.9142 \times 10^{-3}$ Pa·s. Also, $D_p = 0.090$ m, $L = 1.0$ m, and $v_0 = 1.0$ m/s. Using Eq. (3.1-3),

$$N_{Re} = \frac{D_p v_0 \rho}{\mu} = \frac{0.090(1.0)(997.2)}{0.9142 \times 10^{-3}} = 9.817 \times 10^4$$

From Fig. 3.1-2 for a long cylinder, $C_D = 1.4$. Substituting into Eq. (3.1-2), where $A_p = LD_p = 1.0(0.090) = 0.090$ m^2,

$$F_D = C_D \frac{v_0^2}{2} \rho A_p = (1.4) \frac{(1.0)^2}{2} (997.2)(0.09) = 62.82 \text{ N}$$

3.1C Flow in Packed Beds

1. Introduction. A system of considerable importance in chemical and other process engineering fields is the packed bed or packed column which is used for a fixed-bed catalytic reactor, adsorption of a solute, absorption, filter bed, and so on. The packing material in the bed may be spheres, irregular particles, cylinders, or various kinds of commercial packings. In the discussion to follow it is assumed that the packing is everywhere uniform and that little or no channeling occurs. The ratio of diameter of the tower to the packing diameter should be a minimum of 8 : 1 to 10 : 1 for wall effects to be small. In the theoretical approach used, the packed column is regarded as a bundle of crooked tubes of varying cross-sectional area. The theory developed in Chapter 2 for single straight tubes is used to develop the results for the bundle of crooked tubes.

2. Laminar flow in packed beds. Certain geometric relations for particles in packed beds are used in the derivations for flow. The void fraction ε in a packed bed is defined as

$$\varepsilon = \frac{\text{volume of voids in bed}}{\text{total volume of bed (voids plus solids)}} \tag{3.1-6}$$

The specific surface of a particle a_v in m^{-1} is defined as

$$a_v = \frac{S_p}{v_p} \tag{3.1-7}$$

where S_p is the surface area of a particle in m^2 and v_p the volume of a particle in m^3. For a spherical particle,

$$a_v = \frac{6}{D_p} \tag{3.1-8}$$

where D_p is diameter in m. For a packed bed of nonspherical particles, the effective particle diameter D_p is defined as

$$D_p = \frac{6}{a_v} \tag{3.1-9}$$

Since $(1 - \varepsilon)$ is the volume fraction of particles in the bed,

$$a = a_v(1 - \varepsilon) = \frac{6}{D_p}(1 - \varepsilon) \tag{3.1-10}$$

where a is the ratio of total surface area in the bed to total volume of bed (void volume plus particle volume) in m^{-1}.

EXAMPLE 3.1-3. *Surface Area in Packed Bed of Cylinders*
A packed bed is composed of cylinders having a diameter $D = 0.02$ m and a
length $h = D$. The bulk density of the overall packed bed is 962 kg/m³ and
the density of the solid cylinders is 1600 kg/m³.
  (a) Calculate the void fraction ε.
  (b) Calculate the effective diameter D_p of the particles.
  (c) Calculate the value of a in Eq. (3.1-10).

Solution: For part (a), taking 1.00 m³ of packed bed as a basis, the total
mass of the bed is (962 kg/m³) (1.00 m³) = 962 kg. This mass of 962 kg is
also the mass of the solid cylinders. Hence, volume of cylinders = 962 kg/
(1600 kg/m³) = 0.601 m³. Using Eq. (3.1-6),

$$\varepsilon = \frac{\text{volume of voids in bed}}{\text{total volume of bed}} = \frac{1.000 - 0.601}{1.000} = 0.399$$

For the effective particle diameter D_p in part (b), for a cylinder where $h = D$,
the surface area of a particle is

$$S_p = (2)\frac{\pi D^2}{4}\ (\text{ends}) + \pi D(D)\ (\text{sides}) = \tfrac{3}{2}\pi D^2$$

The volume v_p of a particle is

$$v_p = \frac{\pi}{4}D^2(D) = \frac{\pi D^3}{4}$$

Substituting into Eq. (3.1-7),

$$a_v = \frac{S_p}{v_p} = \frac{\tfrac{3}{2}\pi D^2}{\tfrac{1}{4}\pi D^3} = \frac{6}{D}$$

Finally, substituting into Eq. (3.1-9),

$$D_p = \frac{6}{a_v} = \frac{6}{6/D} = D = 0.02\text{ m}$$

Hence, the effective diameter to use is $D_p = D = 0.02$ m. For part (c), using
Eq. (3.1-10),

$$a = \frac{6}{D_p}(1 - \varepsilon) = \frac{6}{0.02}(1 - 0.399) = 180.3\text{ m}^{-1}$$

The average interstitial velocity in the bed is v m/s and it is related to the superficial
velocity v' based on the cross section of the empty container by

$$v' = \varepsilon v \tag{3.1-11}$$

The hydraulic radius r_H for flow defined in Eq. (2.10-21) is modified as follows (B2).

$$r_H = \frac{\left(\begin{array}{c}\text{cross-sectional area}\\ \text{available for flow}\end{array}\right)}{(\text{wetted perimeter})}$$

$$= \frac{\text{void volume available for flow}}{\text{total wetted surface of solids}}$$

$$= \frac{\text{volume of voids/volume of bed}}{\text{wetted surface/volume of bed}} = \frac{\varepsilon}{a} \tag{3.1-12}$$

Combining Eqs. (3.1-10) and (3.1-12),

$$r_H = \frac{\varepsilon}{6(1-\varepsilon)} D_p \tag{3.1-13}$$

Since the equivalent diameter D for a channel is $D = 4r_H$, the Reynolds number for a packed bed is as follows using Eq. (3.1-13) and $v' = \varepsilon v$.

$$N_{Re} = \frac{(4r_H)v\rho}{\mu} = \frac{4\varepsilon}{6(1-\varepsilon)} D_p \frac{v'}{\varepsilon} \frac{\rho}{\mu} = \frac{4}{6(1-\varepsilon)} \frac{D_p\, v'\rho}{\mu} \tag{3.1-14}$$

For packed beds Ergun (E1) defined the Reynolds number as above but without the 4/6 term.

$$N_{Re,\,p} = \frac{D_p v'\rho}{(1-\varepsilon)\mu} = \frac{D_p G'}{(1-\varepsilon)\mu} \tag{3.1-15}$$

where $G' = v'\rho$.

For laminar flow, the Hagen–Poiseuille equation (2.10-2) can be combined with Eq. (3.1-13) for r_H and Eq. (3.1-11) to give

$$\Delta p = \frac{32\mu v\,\Delta L}{D^2} = \frac{32\mu(v'/\varepsilon)\,\Delta L}{(4r_H)^2} = \frac{(72)\mu v'\,\Delta L(1-\varepsilon)^2}{\varepsilon^3 D_p^2} \tag{3.1-16}$$

The true ΔL is larger because of the tortuous path and use of the hydraulic radius predicts too large a v'. Experimental data show that the constant should be 150, which gives the Blake–Kozeny equation for laminar flow, void fractions less than 0.5, effective particle diameter D_p, and $N_{Re,\,p} < 10$.

$$\Delta p = \frac{150\mu v'\,\Delta L}{D_p^2} \frac{(1-\varepsilon)^2}{\varepsilon^3} \tag{3.1-17}$$

3. Turbulent flow in packed beds. For turbulent flow we use the same procedure by starting with Eq. (2.10-5) and substituting Eqs. (3.1-11) and (3.1-13) into this equation to obtain

$$\Delta p = \frac{3f\rho(v')^2\,\Delta L}{D_p} \frac{1-\varepsilon}{\varepsilon^3} \tag{3.1-18}$$

For highly turbulent flow the friction factor should approach a constant value. Also, it is assumed that all packed beds should have the same relative roughness. Experimental data indicated that $3f = 1.75$. Hence, the final equation for turbulent flow for $N_{Re,\,p} > 1000$, which is called the Burke–Plummer equation, becomes

$$\Delta p = \frac{1.75\rho(v')^2\,\Delta L}{D_p} \frac{1-\varepsilon}{\varepsilon^3} \tag{3.1-19}$$

Adding Eq. (3.1-17) for laminar flow and Eq. (3.1-19) for turbulent flow, Ergun (E1) proposed the following general equation for low, intermediate, and high Reynolds numbers which has been tested experimentally.

$$\Delta p = \frac{150\mu v'\,\Delta L}{D_p^2} \frac{(1-\varepsilon)^2}{\varepsilon^3} + \frac{1.75\rho(v')^2\,\Delta L}{D_p} \frac{1-\varepsilon}{\varepsilon^3} \tag{3.1-20}$$

Rewriting Eq. (3.1-20) in terms of dimensionless groups,

$$\frac{\Delta p\rho}{(G')^2} \frac{D_p}{\Delta L} \frac{\varepsilon^3}{1-\varepsilon} = \frac{150}{N_{Re,\,p}} + 1.75 \tag{3.1-21}$$

See also Eq. (3.1-33) for another form of Eq. (3.1-21). The Ergun equation (3.1-21) can be used for gases by using the density ρ of the gas as the arithmetic average of the inlet and outlet pressures. The velocity v' changes throughout the bed for a compressible fluid, but G' is a constant. At high values of $N_{Re, p}$, Eqs. (3.1-20) and (3.1-21) reduce to Eq. (3.1-19) and to Eq. (3.1-17) for low values. For large pressure drops with gases, Eq. (3.1-20) can be written in differential form (P1).

EXAMPLE 3.1-4. Pressure Drop and Flow of Gases in Packed Bed

Air at 311 K is flowing through a packed bed of spheres having a diameter of 12.7 mm. The void fraction ε of the bed is 0.38 and the bed has a diameter of 0.61 m and a height of 2.44 m. The air enters the bed at 1.10 atm abs at the rate of 0.358 kg/s. Calculate the pressure drop of the air in the packed bed. The average molecular weight of air is 28.97.

Solution: From Appendix A.3 for air at 311 K, $\mu = 1.90 \times 10^{-5}$ Pa·s. The cross-sectional area of the bed is $A = (\pi/4)D^2 = (\pi/4)(0.61)^2 = 0.2922$ m^2. Hence, $G' = 0.358/0.2922 = 1.225$ kg/m^2·s (based on empty cross section of container or bed). $D_p = 0.0127$ m, $\Delta L = 2.44$ m, inlet pressure $p_1 = 1.1(1.01325 \times 10^5) = 1.115 \times 10^5$ Pa.

From Eq. (3.1-15),

$$N_{Re, p} = \frac{D_p G'}{(1 - \varepsilon)\mu} = \frac{0.0127(1.225)}{(1 - 0.38)(1.90 \times 10^{-5})} = 1321$$

To use Eq. (3.1-21) for gases, the density ρ to use is the average at the inlet p_1 and outlet p_2 pressures or at $(p_1 + p_2)/2$. This is trial and error since p_2 is unknown. Assuming that $\Delta p = 0.05 \times 10^5$ Pa, $p_2 = 1.115 \times 10^5 - 0.05 \times 10^5 = 1.065 \times 10^5$ Pa. The average pressure is $p_{av} = (1.115 \times 10^5 + 1.065 \times 10^5)/2 = 1.090 \times 10^5$ Pa. The average density to use is

$$\rho_{av} = \frac{M}{RT} p_{av} \tag{3.1-22}$$

$$= \frac{28.97(1.090 \times 10^5)}{8314.34(311)} = 1.221 \text{ kg/m}^3$$

Substituting into Eq. (3.1-21) and solving for Δp,

$$\frac{\Delta p(1.221)}{(1.225)^2} \frac{0.0127}{2.44} \frac{(0.38)^3}{1 - 0.38} = \frac{150}{1321} + 1.75$$

Solving, $\Delta p = 0.0497 \times 10^5$ Pa. This is close enough to the assumed value, so a second trial is not needed.

4. **Shape factors.** Many particles in packed beds are often irregular in shape. The equivalent diameter of a particle is defined as the diameter of a sphere having the same volume as this particle. The sphericity shape factor ϕ_S of a particle is the ratio of the surface area of this sphere having the same volume as the particle to the actual surface area of the particle. For a sphere, the surface area $S_p = \pi D_p^2$ and the volume is $v_p = \pi D_p^3/6$. Hence, for any particle, $\phi_S = \pi D_p^2/S_p$, where S_p is the actual surface area of the particle and D_p is the diameter (equivalent diameter) of the sphere having the same volume as the particle. Then

$$\frac{S_p}{v_p} = \frac{\pi D_p^2/\phi_S}{\pi D_p^3/6} = \frac{6}{\phi_S D_p} \tag{3.1-23}$$

From Eq. (3.1–7),

$$a_v = \frac{S_p}{v_p} = \frac{6}{\phi_S D_p}$$ (3.1-24)

Then Eq. (3.1–10) becomes

$$a = \frac{6}{\phi_S D_p}(1 - \epsilon)$$ (3.1-25)

For a sphere, $\phi_S = 1.0$. For a cylinder where the diameter = length, ϕ_S is calculated to be 0.874 and for a cube, ϕ_S is calculated as 0.806. For granular materials it is difficult to measure the actual volume and surface area to obtain the equivalent diameter. Hence, D_p is usually taken to be the nominal size from a screen analysis or visual length measurements. The surface area is determined by adsorption measurements or measurement of the pressure drop in a bed of particles. Then Eq. (3.1–23) is used to calculate ϕ_S (Table 3.1–1). Typical values for many crushed materials are between 0.6 and 0.7. For convenience for the cylinder and the cube, the nominal diameter is sometimes used (instead of the equivalent diameter) which then gives a shape factor of 1.0.

5. Mixtures of particles. For mixtures of particles of various sizes we can define a mean specific surface a_{vm} as

$$a_{vm} = \sum x_i a_{vi}$$ (3.1-26)

where x_i is volume fraction. Combining Eqs. (3.1-24) and (3.1-26),

$$D_{pm} = \frac{6}{a_{vm}} = \frac{6}{\sum x_i (6/\phi_S D_{pi})} = \frac{1}{\sum x_i/(\phi_S D_{pi})}$$ (3.1-27)

where D_{pm} is the effective mean diameter for the mixture.

EXAMPLE 3.1-5. Mean Diameter for a Particle Mixture
A mixture contains three sizes of particles: 25% by volume of 25 mm size, 40% of 50 mm, and 35% of 75 mm. The sphericity is 0.68. Calculate the effective mean diameter.

TABLE 3.1-1. *Shape Factors (Sphericity) of Some Materials*

Material	Shape Factor, ϕ_S	Reference
Spheres	1.0	
Cubes	0.81	
Cylinders, $D_p = h$ (length)	0.87	
Berl saddles	0.3	(B4)
Raschig rings	0.3	(C2)
Coal dust, pulverized	0.73	(C2)
Sand, average	0.75	(C2)
Crushed glass	0.65	(C2)

Solution: The following data are given: $x_1 = 0.25$, $D_{p1} = 25$ mm; $x_2 = 0.40$, $D_{p2} = 50$; $x_3 = 0.35$, $D_{p3} = 75$; $\phi_S = 0.68$. Substituting into Eq. (3.1-27),

$$D_{pm} = \frac{1}{0.25/(0.68 \times 25) + 0.40/(0.68 \times 50) + 0.35/(0.68 \times 75)}$$

$$= 30.0 \text{ mm}$$

6. *Darcy's empirical law for laminar flow.* Equation (3.1-17) for laminar flow in packed beds shows that the flow rate is proportional to Δp and inversely proportional to the viscosity μ and length ΔL. This is the basis for Darcy's law as follows for purely viscous flow in consolidated porous media.

$$v' = \frac{q'}{A} = -\frac{k}{\mu}\frac{\Delta p}{\Delta L} \tag{3.1-28}$$

where v' is superficial velocity based on the empty cross section in cm/s, q' is flow rate cm^3/s, A is empty cross section in cm^2, μ is viscosity in cp, Δp is pressure drop in atm, ΔL is length in cm, and k is permeability in (cm^3 flow/s)·(cp)·(cm length)/(cm^2 area)·(atm pressure drop). The units used for k of cm^2·cp·atm are often given in darcy or in millidarcy (1/1000 darcy). Hence, if a porous medium has a permeability of 1 darcy, a fluid of 1-cp viscosity will flow at 1 cm^3/s per 1 cm^2 cross section with a Δp of 1 atm per cm length. This equation is often used in measuring permeabilities of underground oil reservoirs.

3.1D Flow in Fluidized Beds

1. Minimum velocity and porosity for fluidization. When a fluid flows upward through a packed bed of particles at low velocities, the particles remain stationary. As the fluid velocity is increased, the pressure drop increases according to the Ergun equation (3.1-20). Upon further increases in velocity, conditions finally occur where the force of the pressure drop times the cross-sectional area just equals the gravitational force on the mass of particles. Then the particles just begin to move, and this is the onset of fluidization or minimum fluidization. The fluid velocity at which fluidization begins is the minimum fluidization velocity v'_{mf} in m/s based on the empty cross section of the tower (superficial velocity).

The porosity of the bed when true fluidization occurs is the minimum porosity for fluidization and is ε_{mf}. Some typical values of ε_{mf} for various materials are given in Table 3.1-2. The bed expands to this voidage or porosity before particle motion appears. This minimum voidage can be experimentally determined by subjecting the bed to a rising gas stream and measuring the height of the bed L_{mf} in m. Generally, it appears best to use gas as the fluid rather than a liquid since liquids give somewhat higher values of ε_{mf}.

As stated earlier, the pressure drop increases as the gas velocity is increased until the onset of minimum fluidization. Then as the velocity is further increased, the pressure drop decreases very slightly and then remains practically unchanged as the bed continues to expand or increase in porosity with increases in velocity. The bed resembles a boiling liquid. As the bed expands with increase in velocity, the bed continues to retain its top horizontal surface. Eventually, as the velocity is increased much further, entrainment of particles from the actual fluidized bed becomes appreciable.

The relation between bed height L and porosity ε is as follows for a bed having a uniform cross-sectional area A. Since the volume $LA(1 - \varepsilon)$ is equal to the total volume

TABLE 3.1-2. *Void Fraction, ε_{mf}, at Minimum Fluidization Conditions (L2)*

	Particle Size, D_p (mm)			
Type of Particles	0.06	0.10	0.20	0.40
	Void fraction, ε_{mf}			
Sharp sand ($\phi_S = 0.67$)	0.60	0.58	0.53	0.49
Round sand ($\phi_S = 0.86$)	0.53	0.48	0.43	(0.42)
Anthracite coal ($\phi_S = 0.63$)	0.61	0.60	0.56	0.52

of solids if they existed as one piece,

$$L_1 A(1 - \varepsilon_1) = L_2 A(1 - \varepsilon_2) \tag{3.1-29}$$

$$\frac{L_1}{L_2} = \frac{1 - \varepsilon_2}{1 - \varepsilon_1} \tag{3.1-30}$$

where L_1 is height of bed with porosity ε_1 and L_2 is height with porosity ε_2.

2. Pressure drop and minimum fluidizing velocity. As a first approximation, the pressure drop at the start of fluidization can be determined as follows. The force obtained from the pressure drop times the cross-sectional area must equal the gravitational force exerted by the mass of the particles minus the buoyant force of the displaced fluid.

$$\Delta p A = L_{mf} A(1 - \varepsilon_{mf})(\rho_p - \rho)g \tag{3.1-31}$$

Hence,

$$\frac{\Delta p}{L_{mf}} = (1 - \varepsilon_{mf})(\rho_p - \rho)g \quad \text{(SI)}$$

$$\frac{\Delta p}{L_{mf}} = (1 - \varepsilon_{mf})(\rho_p - \rho)\frac{g}{g_c} \quad \text{(English)} \tag{3.1-32}$$

Often we have irregular-shaped particles in the bed, and it is more convenient to use the particle size and shape factor in the equations. First we substitute for the effective mean diameter D_p the term $\phi_S D_P$ where D_P now represents the particle size of a sphere having the same volume as the particle and ϕ_S the shape factor. Often, the value of D_P is approximated by using the nominal size from a sieve analysis. Then Eq. (3.1-20) for pressure drop in a packed bed becomes

$$\frac{\Delta p}{L} = \frac{150 \mu v'}{\phi_S^2 D_P^2} \frac{(1 - \varepsilon)^2}{\varepsilon^3} + \frac{1.75 \rho(v')^2}{\phi_S D_P} \frac{1 - \varepsilon}{\varepsilon^3} \tag{3.1-33}$$

where $\Delta L = L$, bed length in m.

Equation (3.1-33) can now be used by a small extrapolation for packed beds to calculate the minimum fluid velocity v'_{mf} at which fluidization begins by substituting v'_{mf} for v', ε_{mf} for ε, and L_{mf} for L and combining the result with Eq. (3.1-32) to give

$$\frac{1.75 D_P^2 (v'_{mf})^2 \rho^2}{\phi_S \varepsilon_{mf}^3 \mu^2} + \frac{150(1 - \varepsilon_{mf}) D_P v'_{mf} \rho}{\phi_S^2 \varepsilon_{mf}^3 \mu} - \frac{D_P^3 \rho(\rho_p - \rho)g}{\mu^2} = 0 \tag{3.1-34}$$

Defining a Reynolds number as

$$N_{Re, mf} = \frac{D_p v'_{mf} \rho}{\mu} \qquad (3.1\text{-}35)$$

Eq. (3.1-34) becomes

$$\frac{1.75(N_{Re, mf})^2}{\phi_s \varepsilon_{mf}^3} + \frac{150(1 - \varepsilon_{mf})(N_{Re, mf})}{\phi_s^2 \varepsilon_{mf}^3} - \frac{D_p^3 \rho(\rho_p - \rho)g}{\mu^2} = 0 \qquad (3.1\text{-}36)$$

When $N_{Re, mf} < 20$ (small particles), the first term of Eq. (3.1-36) can be dropped and when $N_{Re, mf} > 1000$ (large particles), the second term drops out.

If the terms ε_{mf} and/or ϕ_s are not known, Wen and Yu (W4) found for a variety of systems that

$$\phi_s \varepsilon_{mf}^3 \cong \frac{1}{14}, \qquad \frac{1 - \varepsilon_{mf}}{\phi_s^2 \varepsilon_{mf}^3} \cong 11 \qquad (3.1\text{-}37)$$

Substituting into Eq. (3.1-36), the following simplified equation is obtained.

$$N_{Re, mf} = \left[(33.7)^2 + 0.0408 \frac{D_p^3 \rho(\rho_p - \rho)g}{\mu^2} \right]^{1/2} - 33.7 \qquad (3.1\text{-}38)$$

This equation holds for a Reynolds number range of 0.001 to 4000 with an average deviation of $\pm 25\%$. Alternative equations are available in the literature (K1, W4).

EXAMPLE 3.1-6. *Minimum Velocity for Fluidization*

Solid particles having a size of 0.12 mm, a shape factor ϕ_s of 0.88, and a density of 1000 kg/m^3 are to be fluidized using air at 2.0 atm abs and 25°C. The voidage at minimum fluidizing conditions is 0.42.

(a) If the cross section of the empty bed is 0.30 m^2 and the bed contains 300 kg of solid, calculate the minimum height of the fluidized bed.
(b) Calculate the pressure drop at minimum fluidizing conditions.
(c) Calculate the minimum velocity for fluidization.
(d) Use Eq. (3.1-38) to calculate v'_{mf} assuming that data for ϕ_s and ε_{mf} are not available.

Solution: For part (a), the volume of solids = 300 kg/(1000 kg/m^3) = 0.300 m^3. The height the solids would occupy in the bed if $\varepsilon_1 = 0$ is $L_1 = 0.300$ m^3/(0.30 m^2 cross section) = 1.00 m. Using Eq. (3.1-30) and calling $L_{mf} = L_2$ and $\varepsilon_{mf} = \varepsilon_2$,

$$\frac{L_1}{L_{mf}} = \frac{1 - \varepsilon_{mf}}{1 - \varepsilon_1}$$

$$\frac{1.00}{L_{mf}} = \frac{1 - 0.42}{1 - 0}$$

Solving, $L_{mf} = 1.724$ m.

The physical properties of air at 2.0 atm and 25°C (Appendix A.3) are $\mu = 1.845 \times 10^{-5}$ Pa·s, $\rho = 1.187 \times 2 = 2.374$ kg/m^3, $p = 2.0265 \times 10^5$ Pa. For the particle, $D_p = 0.00012$ m, $\rho_p = 1000$ kg/m^3, $\phi_s = 0.88$, $\varepsilon_{mf} = 0.42$.

For part (b) using Eq. (3.1-32) to calculate Δp,

$$\Delta p = L_{mf}(1 - \varepsilon_{mf})(\rho_p - \rho)g$$

$$= 1.724(1 - 0.42)(1000 - 2.374)(9.80665) = 0.0978 \times 10^5 \text{ Pa}$$

To calculate v'_{mf} for part (c), Eq. (3.1-36) is used.

$$\frac{1.75(N_{Re,\,mf})^2}{(0.88)(0.42)^3} + \frac{150(1-0.42)(N_{Re,\,mf})}{(0.88)^2(0.42)^3}$$

$$- (0.00012)^3 \frac{2.374(1000 - 2.374)(9.80665)}{(1.845 \times 10^{-5})^2}$$

Solving,

$$N_{Re,\,mf} = 0.07764 = \frac{D_P v'_{mf}\,\rho}{\mu} = \frac{0.00012(v'_{mf})(2.374)}{1.845 \times 10^{-5}}$$

$$v'_{mf} = 0.005029 \text{ m/s}$$

Using the simplified Eq. (3.1-38) for part (d),

$$N_{Re,\,mf} = \left[(33.7)^2 + \frac{0.0408(0.00012)^3(2.374)(1000-2.374)(9.80665)}{(1.845 \times 10^{-5})^2} \right]^{1/2} - 33.7$$

$$= 0.07129$$

Solving, $v'_{mf} = 0.004618$ m/s.

3. *Expansion of fluidized beds.* For the case of small particles and where $N_{Re,\,f} = D_P v'\rho/\mu < 20$, we can estimate the variation of porosity or bed height L as follows. We assume that Eq. (3.1-36) applies over the whole range of fluid velocities with the first term being neglected. Then, solving for v',

$$v' = \frac{D_P^2(\rho_p - \rho)g\phi_S^2}{150\mu}\,\frac{\varepsilon^3}{1-\varepsilon} = K_1 \frac{\varepsilon^3}{1-\varepsilon} \qquad (3.1\text{-}39)$$

We find that all terms except ε are constant for the particular system and ε depends upon v'. This equation can be used with liquids to estimate ε with $\varepsilon < 0.80$. However, because of clumping and other factors, errors can occur when used for gases.

The flow rate in a fluidized bed is limited on one hand by the minimum v'_{mf} and on the other by entrainment of solids from the bed proper. This maximum allowable velocity is approximated as the terminal settling velocity v'_t of the particles. (See Section 13.3 for methods to calculate this settling velocity.) Approximate equations to calculate the operating range are as follows (P2). For fine solids and $N_{Re,\,f} < 0.4$,

$$\frac{v'_t}{v'_{mf}} \cong \frac{90}{1} \qquad (3.1\text{-}40)$$

For large solids and $N_{Re,\,f} > 1000$,

$$\frac{v'_t}{v'_{mf}} \cong \frac{9}{1} \qquad (3.1\text{-}41)$$

EXAMPLE 3.1-7. *Expansion of Fluidized Bed*
Using the data from Example 3.1-6, estimate the maximum allowable velocity v'_t. Using an operating velocity of 3.0 times the minimum, estimate the voidage of the bed.

Solution: From Example 3.1-6, $N_{Re,\,mf} = 0.07764$, $v'_{mf} = 0.005029$ m/s, $\varepsilon_{mf} = 0.42$. Using Eq. (3.1-40), the maximum allowable velocity is

$$v'_t \cong 90(v'_{mf}) = 90(0.005029) = 0.4526 \text{ m/s}$$

Using an operating velocity v' of 3.0 times the minimum,

$$v' = 3.0(v'_{mf}) = 3.0(0.005029) = 0.01509 \text{ m/s}$$

To determine the voidage at this new velocity, we substitute into Eq. (3.1-39) using the known values at minimum fluidizing conditions to determine K_1.

$$0.005029 = K_1 \frac{(0.42)^3}{1 - 0.42}$$

Solving, $K_1 = 0.03938$. Then using the operating velocity in Eq. (3.1-39),

$$0.01509 = (0.03938) \frac{\varepsilon^3}{1 - \varepsilon}$$

Solving, the voidage of the bed $\varepsilon = 0.555$ at the operating velocity.

3.2 MEASUREMENT OF FLOW OF FLUIDS

It is important to be able to measure and control the amount of material entering and leaving a chemical and other processing plants. Since many of the materials are in the form of fluids, they are flowing in pipes or conduits. Many different types of devices are used to measure the flow of fluids. The most simple are those that measure directly the volume of the fluids, such as ordinary gas and water meters and positive-displacement pumps. Current meters make use of an element, such as a propeller or cups on a rotating arm, which rotates at a speed determined by the velocity of the fluid passing through it. Very widely used for fluid metering are the pitot tube, venturi meter, orifice meter, and open-channel weirs.

3.2A Pitot Tube

The pitot tube is used to measure the local velocity at a given point in the flow stream and not the average velocity in the pipe or conduit. In Fig. 3.2-1a a sketch of this simple device is shown. One tube, the impact tube, has its opening normal to the direction of flow and the static tube has its opening parallel to the direction of flow.

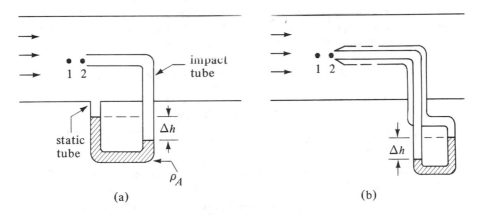

FIGURE 3.2-1. *Diagram of pitot tube : (a) simple tube, (b) tube with static pressure holes.*

The fluid flows into the opening at point 2, pressure builds up, and then remains stationary at this point, called the *stagnation point*. The difference in the stagnation pressure at this point 2 and the static pressure measured by the static tube represents the pressure rise associated with the deceleration of the fluid. The manometer measures this small pressure rise. If the fluid is incompressible, we can write the Bernoulli equation (2.7-32) between point 1, where the velocity v_1 is undisturbed before the fluid decelerates, and point 2, where the velocity v_2 is zero.

$$\frac{v_1^2}{2} - \frac{v_2^2}{2} + \frac{p_1 - p_2}{\rho} = 0 \tag{3.2-1}$$

Setting $v_2 = 0$ and solving for v_1,

$$v = C_p \sqrt{\frac{2(p_2 - p_1)}{\rho}} \tag{3.2-2}$$

where v is the velocity v_1 in the tube at point 1 in m/s, p_2 is the stagnation pressure, ρ is the density of the flowing fluid at the static pressure p_1, and C_p is a dimensionless coefficient to take into account deviations from Eq. (3.2-1) and generally varies between about 0.98 to 1.0. For accurate use, the coefficient should be determined by calibration of the pitot tube. This equation applies to incompressible fluids but can be used to approximate the flow of gases at moderate velocities and pressure changes of about 10% or less of the total pressure. For gases the pressure change is often quite low and, hence, accurate measurement of velocities is difficult.

The value of the pressure drop $p_2 - p_1$ or Δp in Pa is related to Δh, the reading on the manometer, by Eq. (2.2-14) as follows:

$$\Delta p = \Delta h(\rho_A - \rho)g \tag{3.2-3}$$

where ρ_A is the density of the fluid in the manometer in kg/m³ and Δh is the manometer reading in m. In Fig. 3.2-1b, a more compact design is shown with concentric tubes. In the outer tube, static pressure holes are parallel to the direction of flow. Further details are given elsewhere (P1).

Since the pitot tube measures velocity at one point only in the flow, several methods can be used to obtain the average velocity in the pipe. In the first method the velocity is measured at the exact center of the tube to obtain v_{max}. Then by using Fig. 2.10-2 the v_{av} can be obtained. Care should be taken to have the pitot tube at least 100 diameters downstream from any pipe obstruction. In the second method, readings are taken at several known positions in the pipe cross section and then using Eq. (2.6-17), a graphical or numerical integration is performed to obtain v_{av}.

EXAMPLE 3.2-1. Flow Measurement Using a Pitot Tube

A pitot tube similar to Fig. 3.2-1a is used to measure the airflow in a circular duct 600 mm in diameter. The flowing air temperature is 65.6°C. The pitot tube is placed at the center of the duct and the reading Δh on the manometer is 10.7 mm of water. A static pressure measurement obtained at the pitot tube position is 205 mm of water above atmospheric. The pitot tube coefficient $C_p = 0.98$.

(a) Calculate the velocity at the center and the average velocity.
(b) Calculate the volumetric flow rate of the flowing air in the duct.

Solution: For part (a), the properties of air at 65.6°C from Appendix A.3 are $\mu = 2.03 \times 10^{-5}$ Pa·s, $\rho = 1.043$ kg/m³ (at 101.325 kPa). To calculate the absolute static pressure, the manometer reading $\Delta h = 0.205$ m of water indicates the pressure above 1 atm abs. Using Eq. (2.2-14), the water density as 1000 kg/m³, and assuming 1.043 kg/m³ as the air density,

$$\Delta p = 0.205(1000 - 1.043)9.80665 = 2008 \text{ Pa}$$

Then the absolute static pressure $p_1 = 1.01325 \times 10^5 + 0.02008 \times 10^5 = 1.0333 \times 10^5$ Pa. The correct air density in the flowing air is $(1.0333 \times 10^5/1.01325 \times 10^5)(1.043) = 1.063$ kg/m³. This correct value when used instead of 1.043 would have a negligible effect on the recalculation of p_1.

To calculate the Δp for the pitot tube, Eq. (3.2-3) is used.

$$\Delta p = \Delta h(\rho_A - \rho)g = \frac{10.7}{1000}(1000 - 1.063)(9.80665) = 104.8 \text{ Pa}$$

Using Eq. (3.2-2), the maximum velocity at the center is

$$v = 0.98\sqrt{\frac{2(104.8)}{1.063}} = 13.76 \text{ m/s}$$

The Reynolds number using the maximum velocity is

$$N_{\text{Re}} = \frac{Dv_{\text{max}}\rho}{\mu} = \frac{0.600(13.76)(1.063)}{2.03 \times 10^{-5}} = 4.323 \times 10^5$$

From Fig. 2.10-2, $v_{\text{av}}/v_{\text{max}} = 0.85$. Then, $v_{\text{av}} = 0.85(13.76) = 11.70$ m/s.

To calculate the flow rate for part (b), the cross-sectional area of the duct, $A = (\pi/4)(0.600)^2 = 0.2827$ m². The flow rate $= 0.2827(11.70) = 3.308$ m³/s.

3.2B Venturi Meter

A *venturi meter* is shown in Fig. 3.2-2 and is usually inserted directly into a pipeline. A manometer or other device is connected to the two pressure taps shown and measures the pressure difference $p_1 - p_2$ between points 1 and 2. The average velocity at point 1 where the diameter is D_1 m is v_1 m/s, and at point 2 or the throat the velocity is v_2 and diameter D_2. Since the narrowing down from D_1 to D_2 and the expansion from D_2 back to D_1 is gradual, little frictional loss due to contraction and expansion is incurred.

To derive the equation for the venturi meter, friction is neglected and the pipe is assumed horizontal. Assuming turbulent flow and writing the mechanical-energy-balance equation (2.7-28) between points 1 and 2 for an incompressible fluid,

$$\frac{v_1^2}{2} + \frac{p_1}{\rho} = \frac{v_2^2}{2} + \frac{p_2}{\rho} \tag{3.2-4}$$

The continuity equation for constant p is

$$v_1\frac{\pi D_1^2}{4} = v_2\frac{\pi D_2^2}{4} \tag{3.2-5}$$

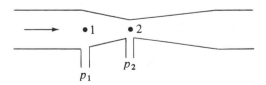

FIGURE 3.2-2. *Venturi flow meter.*

Combining Eqs. (3.2-4) and (3.2-5) and eliminating v_1,

$$v_2 = \frac{1}{\sqrt{1 - (D_2/D_1)^4}} \sqrt{\frac{2(p_1 - p_2)}{\rho}} \qquad (3.2\text{-}6)$$

To account for the small friction loss an experimental coefficient C_v is introduced to give

$$v_2 = \frac{C_v}{\sqrt{1 - (D_2/D_1)^4}} \sqrt{\frac{2(p_1 - p_2)}{\rho}} \qquad \text{(SI)}$$

$$v_2 = \frac{C_v}{\sqrt{1 - (D_2/D_1)^4}} \sqrt{\frac{2g_c(p_1 - p_2)}{\rho}} \qquad \text{(English)} \qquad (3.2\text{-}7)$$

For many meters and a Reynolds number $> 10^4$ at point 1, C_v is about 0.98 for pipe diameters below 0.2 m and 0.99 for larger sizes. However, these coefficients can vary and individual calibration is recommended if the manufacturer's calibration is not available.

To calculate the volumetric flow rate, the velocity v_2 is multiplied by the area A_2.

$$\text{flow rate} = v_2 \frac{\pi D_2^2}{4} \qquad \text{m}^3/\text{s} \qquad (3.2\text{-}8)$$

For the measurement of compressible flow of gases, the adiabatic expansion from p_1 to p_2 pressure must be allowed for in Eq. (3.2-7). A similar equation and the same coefficient C_v are used along with the dimensionless expansion correction factor Y (shown in Fig. 3.2-3 for air) as follows:

$$m = \frac{C_v A_2 Y}{\sqrt{1 - (D_2/D_1)^4}} \sqrt{2(p_1 - p_2)\rho_1} \qquad (3.2\text{-}9)$$

where m is flow rate in kg/s, ρ_1 is density of the fluid upstream at point 1 in kg/m^3, and A_2 is cross-sectional area at point 2 in m^2.

The pressure difference $p_1 - p_2$ occurs because the velocity is increased from v_1 to v_2. However, farther down the tube the velocity returns to its original value of v_1 for liquids. Because of some frictional losses, some of the difference $p_1 - p_2$ is not fully

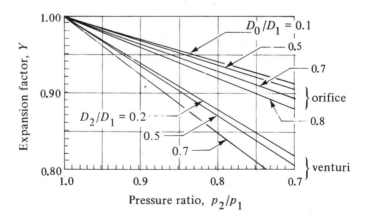

FIGURE 3.2-3. *Expansion factor for air in venturi and orifice. [Calculated from equations and data in references (A2, M2, S3).]*

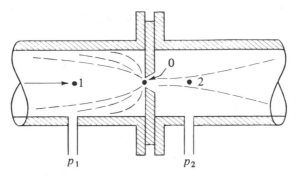

FIGURE 3.2-4. *Orifice flow meter.*

recovered. In a properly designed venturi meter, the permanent loss is about 10% of the differential $p_1 - p_2$, and this represents power loss. A venturi meter is often used to measure flows in large lines, such as city water systems.

3.2C Orifice Meter

For ordinary installations in a process plant the venturi meter has several disadvantages. It occupies considerable space and is expensive. Also, the throat diameter is fixed so that if the flow-rate range is changed considerably, inaccurate pressure differences may result. The *orifice meter* overcomes these objections but at the price of a much larger permanent head or power loss.

A typical sharp-edged orifice is shown in Fig. 3.2-4. A machined and drilled plate having a hole of diameter D_0 is mounted between two flanges in a pipe of diameter D_1. Pressure taps at point 1 upstream and 2 downstream measure $p_1 - p_2$. The exact positions of the two taps are somewhat arbitrary, and in one type of meter the taps are installed about 1 pipe diameter upstream and 0.3 to 0.8 pipe diameter downstream. The fluid stream, once past the orifice plate, forms a vena contracta or free-flowing jet.

The equation for the orifice is similar to Eq. (3.2-7).

$$v_0 = \frac{C_0}{\sqrt{1 - (D_0/D_1)^4}} \sqrt{\frac{2(p_1 - p_2)}{\rho}} \qquad (3.2\text{-}10)$$

where v_0 is the velocity in the orifice in m/s, D_0 is the orifice diameter in m, and C_0 is the dimensionless orifice coefficient. The orifice coefficient C_0 is always determined experimentally. If the N_{Re} at the orifice is above 20 000 and D_0/D_1 is less than about 0.5, the value of C_0 is approximately constant and it has the value of 0.61, which is adequate for design for liquids (M2, P1). Below 20 000 the coefficient rises sharply and then drops and a correlation for C_0 is given elsewhere (P1).

As in the case of the venturi, for the measurement of compressible flow of gases in an orifice, a correction factor Y given in Fig. 3.2-3 for air is used as follows.

$$m = \frac{C_0 A_0 Y}{\sqrt{1 - (D_0/D_1)^4}} \sqrt{2(p_1 - p_2)\rho_1} \qquad (3.2\text{-}11)$$

where m is flow rate in kg/s, ρ_1 is upstream density in kg/m^3, and A_0 is the cross-sectional area of the orifice.

The permanent pressure loss is much higher than for a venturi because of the eddies formed when the jet expands below the vena contracta. This loss depends on D_0/D_1 and is 73% of $p_1 - p_2$ for $D_0/D_1 = 0.5$, 56% for $D_0/D_1 = 0.65$, and 38% for $D_0/D_1 = 0.8$ (P1).

EXAMPLE 3.2-2. *Metering Oil Flow by an Orifice*

A sharp-edged orifice having a diameter of 0.0566 m is installed in a 0.1541-m pipe through which oil having a density of 878 kg/m³ and a viscosity of 4.1 cp is flowing. The measured pressure difference across the orifice is 93.2 kN/m². Calculate the volumetric flow rate in m³/s. Assume that $C_0 = 0.61$.

Solution:

$$p_1 - p_2 = 93.2 \text{ kN/m}^2 = 9.32 \times 10^4 \text{ N/m}^2$$

$$D_1 = 0.1541 \text{ m} \qquad D_0 = 0.0566 \text{ m} \qquad \frac{D_0}{D_1} = \frac{0.0566}{0.1541} = 0.368$$

Substituting into Eq. (3.2-10),

$$v_0 = \frac{C_0}{\sqrt{1 - (D_0/D_1)^4}} \sqrt{\frac{2(p_1 - p_2)}{\rho}}$$

$$= \frac{0.61}{\sqrt{1 - (0.368)^4}} \sqrt{\frac{2(9.32 \times 10^4)}{878}}$$

$$v_0 = 8.97 \text{ m/s}$$

$$\text{volumetric flow rate} = v_0 \frac{\pi D_0^2}{4} = (8.97) \frac{(\pi)(0.0566)^2}{4}$$

$$= 0.02257 \text{ m}^3/\text{s} \ (0.797 \text{ ft}^3/\text{s})$$

The N_{Re} is calculated to see if it is greater than 2×10^4 for $C_0 = 0.61$.

$$\mu = 4.1 \times 1 \times 10^{-3} = 4.1 \times 10^{-3} \text{ kg/m} \cdot \text{s} = 4.1 \times 10^{-3} \text{ Pa} \cdot \text{s}$$

$$N_{Re} = \frac{D_0 v_0 \rho}{\mu} = \frac{0.0566(8.97)(878)}{4.1 \times 10^{-3}} = 1.087 \times 10^5$$

Hence, the Reynolds number is above 2×10^4.

Other measuring devices for flow in closed conduits, such as rotameters, flow nozzles, and so on, are discussed elsewhere (P1).

3.2D Flow in Open Channels and Weirs

In many instances in process engineering and in agriculture, liquids are flowing in open channels rather than closed conduits. To measure the flow rates, weir devices are often used. A *weir* is a dam over which the liquid flows. The two main types of weirs are the rectangular weir and the triangular weir shown in Fig. 3.2-5. The liquid flows over the weir and the height h_0 (weir head) in m is measured above the flat base or the notch as shown. This head should be measured at a distance of about $3h_0$ m upstream of the weir by a level or float gage.

The equation for the volumetric flow rate q in m³/s for a rectangular weir is given by

$$q = 0.415(L - 0.2h_0)h_0^{1.5}\sqrt{2g} \tag{3.2-12}$$

where L = crest length in m, $g = 9.80665$ m/s², and h_0 = weir head in m. This is called the *modified Francis weir formula* and it agrees with experimental values within 3% if

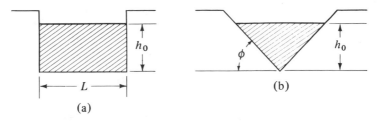

FIGURE 3.2-5. *Types of weirs : (a) rectangular, (b) triangular.*

$L > 2h_0$, velocity upstream is <0.6 m/s, $h_0 > 0.09$ m, and the height of the crest above the bottom of the channel is $>3h_0$. In English units L and h are in ft, q in ft^3/s, and $g = 32.174$ ft/s^2.

For the triangular notch weir,

$$q = \frac{0.31 h_0^{2.5}}{\tan \phi} \sqrt{2g} \tag{3.2-13}$$

Both Eqs. (3.2-12) and (3.2-13) apply only to water. For other liquids, see data given elsewhere (P1).

3.3 PUMPS AND GAS-MOVING EQUIPMENT

3.3A Introduction

In order to make a fluid flow from one point to another in a closed conduit or pipe, it is necessary to have a driving force. Sometimes this force is supplied by gravity, where differences in elevation occur. Usually, the energy or driving force is supplied by a mechanical device such as a pump or blower, which increases the mechanical energy of the fluid. This energy may be used to increase the velocity (move the fluid), the pressure, or the elevation of the fluid, as seen in the mechanical-energy-balance equation (2.7-28), which relates v, p, ρ, and work. The most common methods of adding energy are by positive displacement or centrifugal action.

Generally, the word "pump" designates a machine or device for moving an incompressible liquid. Fans, blowers, and compressors are devices for moving gas (usually air). Fans discharge large volumes of gas at low pressures of the order of several hundred mm of water. Blowers and compressors discharge gases at higher pressures. In pumps and fans the density of the fluid does not change appreciably, and incompressible flow can be assumed. Compressible flow theory is used for blowers and compressors.

3.3B Pumps

1. Power and work required.   Using the total mechanical-energy-balance equation (2.7-28) on a pump and piping system, the actual or theoretical mechanical energy W_S J/kg added to the fluid by the pump can be calculated. Example 2.7-5 shows such a case. If η is the fractional efficiency and W_p the shaft work delivered to the pump, Eq. (2.7-30) gives

$$W_p = - \frac{W_S}{\eta} \tag{3.3-1}$$

The actual or brake power of a pump is as follows.

$$\text{brake kW} = \frac{W_p m}{1000} = -\frac{W_S m}{\eta \times 1000} \quad \text{(SI)}$$

$$\text{brake hp} = \frac{W_p m}{550} = \frac{W_S m}{\eta \times 550} \quad \text{(English)} \tag{3.3-2}$$

where W_p is J/kg, m is the flow rate in kg/s, and 1000 is the conversion factor W/kW. In English units, W_S is in ft · lb_f/lb_m and m in lb_m/s. The theoretical or fluid power is

$$\text{theoretical power} = (\text{brake kW})(\eta) \tag{3.3-3}$$

The mechanical energy W_S in J/kg added to the fluid is often expressed as the developed head H of the pump in m of fluid being pumped, where

$$-W_S = Hg \quad \text{(SI)}$$

$$-W_S = H \frac{g}{g_c} \quad \text{(English)} \tag{3.3-4}$$

To calculate the power of a fan where the pressure difference is of the order of a few hundred mm of water, a linear average density of the gas between the inlet and outlet of the fan is used to calculate W_S and brake kW or horsepower.

Since most pumps are driven by electric motors, the efficiency of the electric motor must be taken into account to determine the total electric power input to the motor. Typical efficiencies η_e of electric motors are as follows: 75% for $\frac{1}{2}$-kW motors, 80% for 2 kW, 84% for 5 kW, 87% for 15 kW, and about 93% for over 150-kW motors. Hence, the total electric power input equals the brake power divided by the electric motor drive efficiency η_e.

$$\text{electric power input (kW)} = \frac{\text{brake kW}}{\eta_e} = \frac{-W_S m}{\eta \eta_e \cdot 1000} \tag{3.3-5}$$

2. Suction lift. The power calculated by Eq. (2.7-3) depends on the differences in pressures and not on the actual pressures being above or below atmospheric pressure. However, the lower limit of the absolute pressure in the suction (inlet) line to the pump is fixed by the vapor pressure of the liquid at the temperature of the liquid in the suction line. If the pressure on the liquid in the suction line drops to the vapor pressure, some of the liquid flashes into vapor (cavitation). Then no liquid can be drawn into the pump.

For the special case where the liquid is nonvolatile, the friction in the suction line to the pump is negligible, and the liquid is being pumped from an open reservoir, the maximum possible vertical suction lift which the pump can perform occurs. For cold water this would be about 10.4 m of water. Practically, however, because of friction, vapor pressure, dissolved gases, and the entrance loss, the actual value is much less. For details, see references elsewhere (P1, M2).

3. Centrifugal pumps. Process industries commonly use centrifugal pumps. They are available in sizes of about 0.004 to 380 m³/min (1 to 100 000 gal/min) and for discharge pressures from a few m of head to 5000 kPa or so. A centrifugal pump in its simplest form consists of an impeller rotating inside a casing. Figure 3.3-1 shows a schematic diagram of a simple centrifugal pump.

The liquid enters the pump axially at point 1 in the suction line and then enters the rotating eye of the impeller, where it spreads out radially. On spreading radially it enters

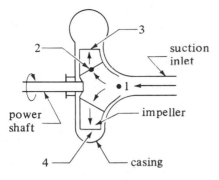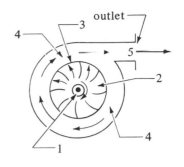

FIGURE 3.3-1. *Simple centrifugal pump.*

the channels between the vanes at point 2 and flows through these channels to point 3 at the periphery of the impeller. From here it is collected in the volute chamber 4 and flows out the pump discharge at 5. The rotation of the impeller imparts a high-velocity head to the fluid, which is changed to a pressure head as the liquid passes into the volute chamber and out the discharge. Some pumps are also made as two-stage or even multistage pumps.

Many complicating factors determine the actual efficiency and performance characteristics of a pump. Hence, the actual experimental performance of the pump is usually employed. The performance is usually expressed by the pump manufacturer by means of curves called *characteristic curves* and are usually for water. The head H in m produced will be the same for any liquid of the same viscosity. The pressure produced, which is $p = H\rho g$, will be in proportion to the density. Viscosities of less than 0.05 Pa·s (50 cp) have little effect on the head produced. The brake kW varies directly as the density.

As rough approximations, the following can be used for a given pump. The *capacity* q_1 in m^3/s is directly proportional to the rpm N_1, or

$$\frac{q_1}{q_2} = \frac{N_1}{N_2} \tag{3.3-6}$$

The *head* H_1 is proportional to q_1^2, or

$$\frac{H_1}{H_2} = \frac{q_1^2}{q_2^2} = \frac{N_1^2}{N_2^2} \tag{3.3-7}$$

The *power consumed* W_1 is proportional to the product of $H_1 q_1$, or

$$\frac{W_1}{W_2} = \frac{H_1 q_1}{H_2 q_2} = \frac{N_1^3}{N_2^3} \tag{3.3-8}$$

In most pumps, the speed is generally not varied. Characteristic curves for a typical single-stage centrifugal pump operating at a constant speed are given in Fig. 3.3-2. Most pumps are usually rated on the basis of head and capacity at the point of peak efficiency. The efficiency reaches a peak at about 50 gal/min flow rate. As the discharge rate in gal/min increases, the developed head drops. The brake hp increases, as expected, with flow rate.

EXAMPLE 3.3-1. *Calculation of Brake Horsepower of a Pump*
In order to see how the brake-hp curve is determined, calculate the brake hp at 40 gal/min flow rate for the pump in Fig. 3.3-2.

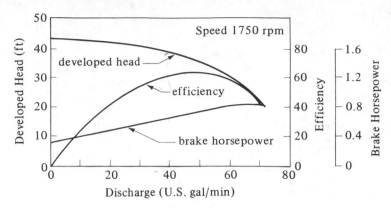

FIGURE 3.3-2. *Characteristic curves for a single-stage centrifugal pump with water. (From W. L. Badger and J. T. Banchero, Introduction to Chemical Engineering. New York: McGraw-Hill Book Company, 1955. With permission.)*

Solution: At 40 gal/min, the efficiency η from the curve is about 60% and the head H is 38.5 ft. A flow rate of 40 gal/min of water with a density of 62.4 lb mass/ft^3 is

$$m = \left(\frac{40 \text{ gal/min}}{60 \text{ s/min}}\right)\left(\frac{1 \text{ ft}^3}{7.481 \text{ gal}}\right)\left(62.4 \frac{\text{lb}_m}{\text{ft}^3}\right) = 5.56 \frac{\text{lb}_m}{\text{s}}$$

The work W_S is as follows, from Eq. (3.3-4):

$$W_S = -H\frac{g}{g_c} = -38.5 \frac{\text{ft} \cdot \text{lb}_f}{\text{lb}_m}$$

The brake hp from Eq. (3.3-2) is

$$\text{brake hp} = \frac{-W_S m}{\eta \cdot 550} = \frac{38.5(5.56)}{0.60(550)} = 0.65 \text{ hp } (0.48 \text{ kW})$$

This value checks the value on the curve in Fig. 3.3-2.

4. Positive-displacement pumps. In this class of pumps a definite volume of liquid is drawn into a chamber and then forced out of the chamber at a higher pressure. There are two main types of positive-displacement pumps. In the *reciprocating pump* the chamber is a stationary cylinder and liquid is drawn into the cylinder by withdrawal of a piston in the cylinder. Then the liquid is forced out by the piston on the return stroke. In the *rotary pump* the chamber moves from inlet to discharge and back again. In a gear rotary pump two intermeshing gears rotate and liquid is trapped in the spaces between the teeth and forced out the discharge.

Reciprocating and rotary pumps can be used to very high pressures, whereas centrifugal pumps are limited in their head and are used for lower pressures. Centrifugal pumps deliver liquid at uniform pressure without shocks or pulsations and can handle liquids with large amounts of suspended solids. In general, in chemical and biological processing plants, centrifugal pumps are primarily used.

Equations (3.3-1) through (3.3-5) hold for calculation of the power of positive displacement pumps. At a constant speed, the flow capacity will remain constant with different liquids. In general, the discharge rate will be directly dependent upon the speed. The power increases directly as the head, and the discharge rate remains nearly constant as the head increases.

3.3C Gas-Moving Machinery

Gas-moving machinery comprises mechanical devices used for compressing and moving gases. They are often classified or considered from the standpoint of the pressure heads produced and are fans for low pressures, blowers for intermediate pressures, and compressors for high pressures.

1. Fans. The commonest method for moving small volumes of gas at low pressures is by means of a *fan*. Large fans are usually centrifugal and their operating principle is similar to that of centrifugal pumps. The discharge heads are low, from about 0.1 m to 1.5 m H_2O. However, in some cases much of the added energy of the fan is converted to velocity energy and a small amount to pressure head.

In a centrifugal fan, the centrifugal force produced by the rotor produces a compression of the gas, called the *static pressure head*. Also, since the velocity of the gas is increased, a velocity head is produced. Both the static-pressure-head increase and velocity-head increase must be included in estimating efficiency and power. Operating efficiencies are in the range 40 to 70%. The operating pressure of a fan is generally given as inches of water gage and is the sum of the velocity head and static pressure of the gas leaving the fan. Incompressible flow theory can be used to calculate the power of fans.

> *EXAMPLE 3.3-2. Brake-kW Power of a Centrifugal Fan*
> It is desired to use 28.32 m³/min of air (metered at a pressure of 101.3 kPa and 294.1 K) in a process. This amount of air, which is at rest, enters the fan suction at a pressure of 741.7 mm Hg and a temperature of 366.3 K and is discharged at a pressure of 769.6 mm Hg and a velocity of 45.7 m/s. A centrifugal fan having a fan efficiency of 60% is to be used. Calculate the brake-kW power needed.
>
> *Solution:* Incompressible flow can be assumed, since the pressure drop is only (27.9/741.7)100, or 3.8% of the upstream pressure. The average density of the flowing gas can be used in the mechanical-energy-balance equation.
>   The density at the suction, point 1, is
>
> $$\rho_1 = \left(28.97 \frac{\text{kg air}}{\text{kg mol}}\right)\left(\frac{1 \text{ kg mol}}{22.414 \text{ m}^3}\right)\left(\frac{273.2}{366.3}\right)\left(\frac{741.7}{760}\right)$$
>
> $$= 0.940 \text{ kg/m}^3$$
>
> (The molecular weight of 28.97 for air, the volume of 22.414 m³/kg mol at 101.3 kPa, and 273.2 K were obtained from Appendix A.1.) The density at the discharge, point 2, is
>
> $$\rho_2 = (0.940)\frac{769.6}{741.7} = 0.975 \text{ kg/m}^3$$
>
> The average density of the gas is
>
> $$\rho_{av} = \frac{\rho_1 + \rho_2}{2} = \frac{0.940 + 0.975}{2} = 0.958 \text{ kg/m}^3$$

The mass flow rate of the gas is

$$m = \left(28.32 \, \frac{m^3}{\text{min}}\right)\left(\frac{1 \text{ min}}{60 \text{ s}}\right)\left(\frac{1 \text{ kg mol}}{22.414 \text{ m}^3}\right)\left(\frac{273.2}{294.1}\right)\left(28.97 \, \frac{\text{kg}}{\text{kg mol}}\right)$$

$$= 0.5663 \text{ kg/s}$$

The developed pressure head is

$$\frac{p_2 - p_1}{\rho_{av}} = \frac{(769.6 - 741.7) \text{ mm Hg}}{760 \text{ mm/atm}}\left(1.01325 \times 10^5 \, \frac{\text{N/m}^2}{\text{atm}}\right)\left(\frac{1}{0.958 \text{ kg/m}^3}\right)$$

$$= 3883 \text{ J/kg}$$

The developed velocity head for $v_1 = 0$ is

$$\frac{v_2^2}{2} = \frac{(45.7)^2}{2} = 1044 \text{ J/kg}$$

Writing the mechanical-energy-balance equation (2.7-28),

$$z_1 g + \frac{v_1^2}{2} + \frac{p_1}{\rho} - W_S = z_2 g + \frac{v_2^2}{2} + \frac{p_2}{\rho} + \sum F$$

Setting $z_1 = 0$, $z_2 = 0$, $v_1 = 0$, and $\sum F = 0$, and solving for W_S,

$$-W_S = \frac{p_2 - p_1}{\rho_{av}} + \frac{v_2^2}{2} = 3883 + 1044 = 4927 \text{ J/kg}$$

Substituting into Eq. (3.3-2),

$$\text{brake kW} = \frac{-W_S m}{\eta \cdot 1000} = \frac{4927(0.5663)}{0.60(1000)} = 4.65 \text{ kW (6.23 hp)}$$

2. *Blowers and compressors.* For handling gas volumes at higher pressure rises than fans, several distinct types of equipment are used. *Turboblowers* or *centrifugal compressors* are widely used to move large volumes of gas for pressure rises from about 5 kPa to several thousand kPa. The principles of operation of a turboblower are the same as those of a centrifugal pump. The turboblower resembles the centrifugal pump in appearance, the main difference being that the gas in the blower is compressible. The head of the turboblower, as in a centrifugal pump, is independent of the fluid handled. Multistage turboblowers are often used to go to the higher pressures.

Rotary blowers and compressors are machines of the positive-displacement type and are essentially constant-volume flow-rate machines with variable discharge pressure. Changing the speed will change the volume flow rate. Details of construction of the various types (P1) vary considerably and pressures up to about 1000 kPa can be obtained, depending on the type.

Reciprocating compressers which are of the positive displacement type using pistons are available for higher pressures. Multistage machines are also available for pressures up to 10 000 kPa or more.

3.3D Equations for Compression of Gases

In blowers and compressors pressure changes are large and compressible flow occurs. Since the density changes markedly, the mechanical-energy-balance equation must be written in differential form and then integrated to obtain the work of compression. In compression of gases the static-head terms, the velocity-head terms, and the friction terms are dropped and only the work term dW and the dp/ρ term remain in the differential form of the mechanical-energy equation; or,

$$dW = \frac{dp}{\rho} \qquad (3.3\text{-}9)$$

Integration between the suction pressure p_1 and discharge pressure p_2 gives the work of compression.

$$W = \int_{p_1}^{p_2} \frac{dp}{\rho} \qquad (3.3\text{-}10)$$

To integrate Eq. (3.3-10) for a perfect gas, either adiabatic or isothermal compression is assumed. For isothermal compression, where the gas is cooled on compression, p/ρ is a constant equal to RT/M, where $R = 8314.3$ J/kg mol · K in SI units and 1545.3 ft · lb$_f$/lb mol · °R in English units. Then,

$$\frac{p_1}{\rho_1} = \frac{p}{\rho} \qquad (3.3\text{-}11)$$

Solving for ρ in Eq. (3.3-11) and substituting it in Eq. (3.3-10), the work for isothermal compression is

$$-W_S = \frac{p_1}{\rho_1} \int_{p_1}^{p_2} \frac{dp}{p} = \frac{p_1}{\rho_1} \ln \frac{p_2}{p_1} = \frac{2.3026 RT_1}{M} \log \frac{p_2}{p_1} \qquad (3.3\text{-}12)$$

Also, $T_1 = T_2$, since the process is isothermal.

For adiabatic compression, the fluid follows an isentropic path and

$$\frac{p_1}{p} = \left(\frac{\rho_1}{\rho}\right)^{\gamma} \qquad (3.3\text{-}13)$$

where $\gamma = c_p/c_v$, the ratio of heat capacities. By combining Eqs. (3.3-10) and (3.3-13) and integrating,

$$-W_S = \frac{\gamma}{\gamma - 1} \frac{RT_1}{M} \left[\left(\frac{p_2}{p_1}\right)^{(\gamma - 1)/\gamma} - 1\right] \qquad (3.3\text{-}14)$$

The adiabatic temperatures are related by

$$\frac{T_2}{T_1} = \left(\frac{p_2}{p_1}\right)^{(\gamma - 1)/\gamma} \qquad (3.3\text{-}15)$$

To calculate the brake power when the efficiency is η,

$$\text{brake kW} = \frac{-W_S m}{(\eta)(1000)} \qquad (3.3\text{-}16)$$

where m = kg gas/s and W_S = J/kg.

The values of γ are approximately 1.40 for air, 1.31 for methane, 1.29 for SO_2, 1.20 for ethane, and 1.40 for N_2 (P1). For a given compression ratio, the work in isothermal compression in Eq. (3.3-12) is less than the work for adiabatic compression in Eq. (3.3-14). Hence, cooling is sometimes used in compressors.

EXAMPLE 3.3-3. Compression of Methane

A single-stage compressor is to compress 7.56×10^{-3} kg mol/s of methane gas at 26.7°C and 137.9 kPa abs to 551.6 kPa abs.

    (a) Calculate the power required if the mechanical efficiency is 80% and the compression is adiabatic.

    (b) Repeat, but for isothermal compression.

Solution: For part (a), $p_1 = 137.9$ kPa, $p_2 = 551.6$ kPa, $M = 16.0$ kg mass/kg mol, and $T_1 = 273.2 + 26.7 = 299.9$ K. The mass flow rate per sec is

$$m = (7.56 \times 10^{-3} \text{ kg mol/s})(16.0 \text{ kg/kg mol}) = 0.121 \frac{\text{kg}}{\text{s}}$$

Substituting into Eq. (3.3-14) for $\gamma = 1.31$ for methane and $p_2/p_1 = 551.6/137.9 = 4.0/1$,

$$-W_S = \frac{\gamma}{\gamma - 1} \frac{RT_1}{M} \left[\left(\frac{p_2}{p_1} \right)^{(\gamma - 1)/\gamma} - 1 \right]$$

$$= \left(\frac{1.31}{1.31 - 1} \right) \frac{8314.3(299.9)}{16.0} \left[\left(\frac{4}{1} \right)^{(1.31 - 1)/1.31} - 1 \right]$$

$$= 256\,300 \text{ J/kg}$$

Using Eq. (3.3-16),

$$\text{brake kW} = \frac{-W_S m}{\eta \cdot 1000} = \frac{(256\,300)0.121}{0.80(1000)} = 38.74 \text{ kW (52.0 hp)}$$

For part (b) using Eq. (3.3-12) for isothermal compression,

$$-W_S = \frac{2.3026 RT_1}{M} \log \frac{p_2}{p_1} = \frac{2.3026(8314.3)(299.9)}{16.0} \log \frac{4}{1}$$

$$= 216\,000 \text{ J/kg}$$

$$\text{brake kW} = \frac{-W_S m}{\eta \cdot 1000} = \frac{(216\,000)0.121}{0.80(1000)} = 32.67 \text{ kW (43.8 hp)}$$

Hence, isothermal compression uses 15.8% less power.

3.4 AGITATION AND MIXING OF FLUIDS AND POWER REQUIREMENTS

3.4A Purposes of Agitation

In the chemical and other processing industries, many operations are dependent to a great extent on effective agitation and mixing of fluids. Generally, *agitation* refers to forcing a fluid by mechanical means to flow in a circulatory or other pattern inside a vessel. *Mixing* usually implies the taking of two or more separate phases, such as a fluid and a powdered solid, or two fluids, and causing them to be randomly distributed through one another.

There are a number of purposes for agitating fluids and some of these are briefly summarized.

1. Blending of two miscible liquids, such as ethyl alcohol and water.
2. Dissolving solids in liquids, such as salt in water.
3. Dispersing a gas in a liquid as fine bubbles, such as oxygen from air in a suspension of microorganisms for fermentation or for the activated sludge process in waste treatment.
4. Suspending of fine solid particles in a liquid, such as in the catalytic hydrogenation of a liquid where solid catalyst particles and hydrogen bubbles are dispersed in the liquid.
5. Agitation of the fluid to increase heat transfer between the fluid and a coil or jacket in the vessel wall.

3.4B Equipment for Agitation

Generally, liquids are agitated in a cylindrical vessel which can be closed or open to the air. The height of liquid is approximately equal to the tank diameter. An impeller mounted on a shaft is driven by an electric motor. A typical agitator assembly is shown in Fig. 3.4-1.

1. Three-blade propeller agitator. There are several types of agitators commonly used. A common type, shown in Fig. 3.4-1, is a three-bladed marine-type propeller similar to the propeller blade used in driving boats. The propeller can be a side-entering type in a tank or be clamped on the side of an open vessel in an off-center position. These propellers turn at high speeds of 400 to 1750 rpm (revolutions per minute) and are used for liquids of low viscosity. The flow pattern in a baffled tank with a propeller positioned on the center of the tank is shown in Fig. 3.4-1. This type of flow pattern is called *axial flow* since the fluid flows axially down the center axis or propeller shaft and up on the sides of the tank as shown.

2. Paddle agitators. Various types of paddle agitators are often used at low speeds between about 20 and 200 rpm. Two-bladed and four-bladed flat paddles are often used, as shown in Fig. 3.4-2a. The total length of the paddle impeller is usually 60 to 80% of the tank diameter and the width of the blade $\frac{1}{6}$ to $\frac{1}{10}$ of its length. At low speeds mild

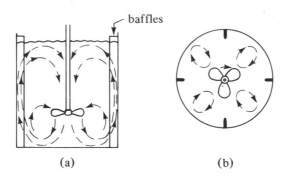

baffles

(a)   (b)

FIGURE 3.4-1. *Baffled tank and three-blade propeller agitator with axial-flow pattern: (a) side view, (b) bottom view.*

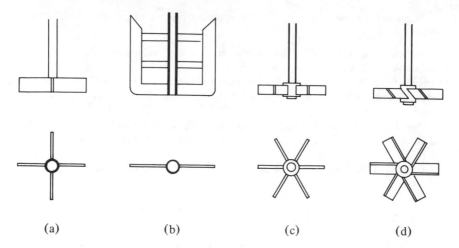

(a)   (b)   (c)   (d)

FIGURE 3.4-2.   *Various types of agitators: (a) four-blade paddle, (b) gate or anchor paddle, (c) six-blade open turbine, (d) pitched-blade (45°) turbine.*

agitation is obtained in an unbaffled vessel. At higher speeds baffles are used, since, without baffles, the liquid is simply swirled around with little actual mixing. The paddle agitator is ineffective for suspending solids since good radial flow is present but little vertical or axial flow. An anchor or gate paddle, shown in Fig. 3.4-2b, is often used. It sweeps or scrapes the tank walls and sometimes the tank bottom. It is used with viscous liquids where deposits on walls can occur and to improve heat transfer to the walls. However, it is a poor mixer. These are often used to process starch pastes, paints, adhesives, and cosmetics.

3. Turbine agitators.   Turbines that resemble multibladed paddle agitators with shorter blades are used at high speeds for liquids with a very wide range of viscosities. The diameter of a turbine is normally between 30 and 50% of the tank diameter. Normally, the turbines have four or six blades. Figure 3.4-3 shows a flat six-blade turbine agitator with disk. In Fig. 3.4-2c a flat, six-blade open turbine is shown. The turbines with flat blades give radial flow, as shown in Fig. 3.4-3. They are also useful for good gas dispersion where the gas is introduced just below the impeller at its axis and is drawn up to the blades and chopped into fine bubbles. In the pitched-blade turbine shown in Fig. 3.4-2d with the blades at 45°, some axial flow is imparted so that a combination of axial and radial flow is present. This type is useful in suspending solids since the currents flow downward and then sweep up the solids.

4. Helical-ribbon agitators.   This type of agitator is used in highly viscous solutions and operates at a low RPM in the laminar region. The ribbon is formed in a helical path and is attached to a central shaft. The liquid moves in a tortuous flow path down the center and up along the sides in a twisting motion. Similar types are the double helical ribbon and the helical ribbon with a screw.

5. Agitator selection and viscosity ranges.   The viscosity of the fluid is one of several factors affecting the selection of the type of agitator. Indications of the viscosity ranges of these agitators are as follows. Propellers are used for viscosities of the fluid below about

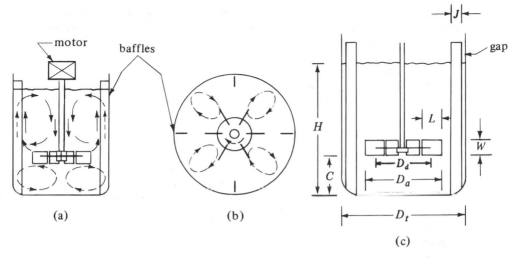

FIGURE 3.4-3. *Baffled tank with six-blade turbine agitator with disk showing flow patterns : (a) side view, (b) bottom view, (c) dimensions of turbine and tank.*

3 Pa·s (3000 cp); turbines can be used below about 100 Pa·s (100 000 cp); modified paddles such as anchor agitators can be used above 50 Pa·s to about 500 Pa·s (500 000 cp); helical and ribbon-type agitators are often used above this range to about 1000 Pa·s and have been used up to 25 000 Pa·s. For viscosities greater than about 2.5 to 5 Pa·s (5000 cp) and above, baffles are not needed since little swirling is present above these viscosities.

3.4C Flow Patterns in Agitation

The flow patterns in an agitated tank depend upon the fluid properties, the geometry of the tank, types of baffles in the tank, and the agitator itself. If a propeller or other agitator is mounted vertically in the center of a tank with no baffles, a swirling flow pattern usually develops. Generally, this is undesirable, because of excessive air entrainment, development of a large vortex, surging, and the like, especially at high speeds. To prevent this, an angular off-center position can be used with propellors with small horsepower. However, for vigorous agitation at higher power, unbalanced forces can become severe and limit the use of higher power.

For vigorous agitation with vertical agitators, baffles are generally used to reduce swirling and still promote good mixing. Baffles installed vertically on the walls of the tank are shown in Fig. 3.4-3. Usually four baffles are sufficient, with their width being about $\frac{1}{12}$ of the tank diameter for turbines and for propellers. The turbine impeller drives the liquid radially against the wall, where it divides, with one portion flowing upward near the surface and back to the impeller from above and the other flowing downward. Sometimes, in tanks with large liquid depths much greater than the tank diameter, two or three impellers are mounted on the same shaft, each acting as a separate mixer. The bottom impeller is about 1.0 impeller diameter above the tank bottom.

In an agitation system, the volume flow rate of fluid moved by the impeller, or circulation rate, is important to sweep out the whole volume of the mixer in a reasonable time. Also, turbulence in the moving stream is important for mixing, since it entrains the material from the bulk liquid in the tank into the flowing stream. Some agitation systems require high turbulence with low circulation rates, and others low turbulence and high circulation rates. This often depends on the types of fluids being mixed and on the amount of mixing needed.

3.4D Typical "Standard" Design of Turbine

The turbine agitator shown in Fig. 3.4-3 is the most commonly used agitator in the process industries. For design of an ordinary agitation system, this type of agitator is often used in the initial design. The geometric proportions of the agitation system which are considered as a typical "standard" design are given in Table 3.4-1. These relative proportions are the basis of the major correlations of agitator performance in numerous publications. (See Fig. 3.4-3c for nomenclature.)

In some cases $W/D_a = \frac{1}{8}$ for agitator correlations. The number of baffles is 4 in most uses. The clearance or gap between the baffles and the wall is usually 0.10 to 0.15 J to ensure that liquid does not form stagnant pockets next to the baffle and wall. In a few correlations the ratio of baffle to tank diameter is $J/D_t = \frac{1}{10}$ instead of $\frac{1}{12}$.

3.4E Power Used in Agitated Vessels

In the design of an agitated vessel, an important factor is the power required to drive the impeller. Since the power required for a given system cannot be predicted theoretically, empirical correlations have been developed to predict the power required. The presence or absence of turbulence can be correlated with the impeller Reynolds number N'_{Re}, defined as

$$N'_{Re} = \frac{D_a^2 N \rho}{\mu} \tag{3.4-1}$$

where D_a is the impeller (agitator) diameter in m, N is rotational speed in rev/s, ρ is fluid density in kg/m³, and μ is viscosity in kg/m·s. The flow is laminar in the tank for $N'_{Re} < 10$, turbulent for $N'_{Re} > 10^4$, and for a range between 10 and 10^4, the flow is transitional, being turbulent at the impeller and laminar in remote parts of the vessel.

Power consumption is related to fluid density ρ, fluid viscosity μ, rotational speed N,

TABLE 3.4-1. *Geometric Proportions for a "Standard" Agitation System*

$\dfrac{D_a}{D_t} = 0.3 \text{ to } 0.5$	$\dfrac{H}{D_t} = 1$	$\dfrac{C}{D_t} = \dfrac{1}{3}$
$\dfrac{W}{D_a} = \dfrac{1}{5}$ $\dfrac{D_d}{D_a} = \dfrac{2}{3}$	$\dfrac{L}{D_a} = \dfrac{1}{4}$	$\dfrac{J}{D_t} = \dfrac{1}{12}$

and impeller diameter D_a by plots of power number N_p versus N'_{Re}. The power number is

$$N_p = \frac{P}{\rho N^3 D_a^5} \quad \text{(SI)}$$

$$N_p = \frac{P g_c}{\rho N^3 D_a^5} \quad \text{(English)}$$

(3.4-2)

where P = power in J/s or W. In English units, $P = \text{ft} \cdot \text{lb}_f/\text{s}$.

Figure 3.4-4 is a correlation (B3, R1) for frequently used impellers with Newtonian liquids contained in baffled, cylindrical vessels. Dimensional measurements of baffle, tank, and impeller sizes are given in Fig. 3.4-3c. These curves may also be used for the same impellers in unbaffled tanks when N'_{Re} is 300 or less (B3, R1). When N'_{Re} is above 300, the power consumption for an unbaffled vessel is considerably less than for a baffled vessel. Curves for other impellers are also available (B3, R1).

EXAMPLE 3.4-1. Power Consumption in an Agitator

A flat-blade turbine agitator with disk having six blades is installed in a tank similar to Fig. 3.4-3. The tank diameter D_t is 1.83 m, the turbine diameter D_a

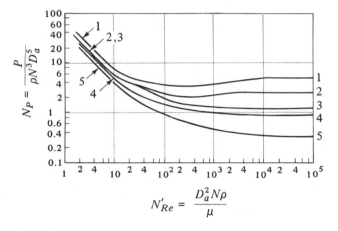

$$N'_{Re} = \frac{D_a^2 N \rho}{\mu}$$

FIGURE 3.4-4. *Power correlations for various impellers and baffles (see Fig. 3.4-3c for dimensions D_a, D_t, J, and W).*

Curve 1. Flat six-blade turbine with disk (like Fig. 3.4-3 but six blades); $D_a/W = 5$; four baffles each $D_t/J = 12$.

Curve 2. Flat six-blade open turbine (like Fig. 3.4-2c); $D_a/W = 8$; four baffles each $D_t/J = 12$.

Curve 3. Six-blade open turbine but blades at 45° (like Fig. 3.4-2d); $D_a/W = 8$; four baffles each $D_t/J = 12$.

Curve 4. Propeller (like Fig. 3.4-1); pitch = $2D_a$; four baffles each $D_t/J = 10$; also holds for same propeller in angular off-center position with no baffles.

Curve 5. Propeller; pitch = D_a; four baffles each $D_t/J = 10$; also holds for same propeller in angular off-center position with no baffles.

[Curves 1, 2, and 3 reprinted with permission from R. L. Bates, P. L. Fondy, and R. R. Corpstein, Ind. Eng. Chem. Proc. Des. Dev., 2, 310 (1963). Copyright by the American Chemical Society. Curves 4 and 5 from J. H. Rushton, E. W. Costich, and H. J. Everett, Chem. Eng. Progr., 46, 395, 467 (1950). With permission.]

is 0.61 m, $D_t = H$, and the width W is 0.122 m. The tank contains four baffles, each having a width J of 0.15 m. The turbine is operated at 90 rpm and the liquid in the tank has a viscosity of 10 cp and a density of 929 kg/m³.

(a) Calculate the required kW of the mixer.
(b) For the same conditions, except for the solution having a viscosity of 100 000 cp, calculate the required kW.

Solution: For part (a) the following data are given: $D_a = 0.61$ m, $W = 0.122$ m, $D_t = 1.83$ m, $J = 0.15$ m, $N = 90/60 = 1.50$ rev/s, $\rho = 929$ kg/m³, and

$$\mu = (10.0 \text{ cp})(1 \times 10^{-3}) = 0.01 \frac{\text{kg}}{\text{m} \cdot \text{s}} = 0.01 \text{ Pa} \cdot \text{s}$$

Using Eq. (3.4-1), the Reynolds number is

$$N'_{Re} = \frac{D_a^2 N \rho}{\mu} = \frac{(0.61)^2(1.50)929}{0.01} = 5.185 \times 10^4$$

Using curve 1 in Fig. 3.4-4 since $D_a/W = 5$ and $D_t/J = 12$, $N_P = 5$ for $N'_{Re} = 5.185 \times 10^4$. Solving for P in Eq. (3.4-2) and substituting known values,

$$P = N_P \rho N^3 D_a^5 = 5(929)(1.50)^3(0.61)^5$$

$$= 1324 \text{ J/s} = 1.324 \text{ kW (1.77 hp)}$$

For part (b),

$$\mu = 100\,000(1 \times 10^{-3}) = 100 \frac{\text{kg}}{\text{m} \cdot \text{s}}$$

$$N'_{Re} = \frac{(0.61)^2(1.50)929}{100} = 5.185.$$

This is in the laminar flow region. From Fig. 3.4-4, $N_P = 14$.

$$P = 14(929)(1.50)^3(0.61)^5 = 3707 \text{ J/s} = 3.71 \text{ kW (4.98 hp)}$$

Hence, a 10 000-fold increase in viscosity only increases the power from 1.324 to 3.71 kW.

Variations of various geometric ratios from the "standard" design can have different effects on the power number N_P in the turbulent region of the various turbine agitators as follows (B3).

1. For the flat six-blade open turbine, $N_P \propto (W/D_a)^{1.0}$.
2. For the flat, six-blade open turbine, varying D_a/D_t from 0.25 to 0.50 has practically no effect on N_P.
3. With two six-blade open turbines installed on the same shaft and the spacing between the two impellors (vertical distance between the bottom edges of the two turbines) being at least equal to D_a, the total power is 1.9 times a single flat-blade impeller. For two six-blade pitched-blade (45°) turbines, the power is also about 1.9 times that of a single pitched-blade impeller.
4. A baffled vertical square tank or a horizontal cylindrical tank has the same power number as a vertical cylindrical tank. However, marked changes in the flow patterns occur.

The power number for a plain anchor-type agitator similar to Fig. 3.4-2b but without the two horizontal cross bars is as follows for $N'_{Re} < 100$ (H2):

$$N_P = 215(N'_{Re})^{-0.955} \tag{3.4-3}$$

where $D_a/D_t = 0.90$, $W/D_t = 0.10$, and $C/D_t = 0.05$.

The power number for a helical-ribbon agitator for very viscous liquids for $N'_{Re} < 20$ is as follows (H2, P3).

$$N_p = 186(N'_{Re})^{-1} \quad \text{(agitator pitch/tank diameter} = 1.0) \tag{3.4-4}$$

$$N_p = 290(N'_{Re})^{-1} \quad \text{(agitator pitch/tank diameter} = 0.5) \tag{3.4-5}$$

The typical dimensional ratios used are $D_a/D_t = 0.95$, with some ratios as low as 0.75, and $W/D_t = 0.095$.

3.4F Agitator Scale-Up

1. Introduction. In the process industries experimental data are often available on a laboratory-size or pilot-unit-size agitation system and it is desired to scale-up the results to design a full-scale unit. Since there is much diversity in processes to be scaled up, no single method can handle all types of scale-up problems, and many approaches to scale-up exist. Geometric similarity is, of course, important and simplest to achieve. Kinematic similarity can be defined in terms of ratios of velocities or of times (R2). Dynamic similarity requires fixed ratios of viscous, inertial, or gravitational forces. Even if geometric similarity is achieved, dynamic and kinematic similarity cannot often be obtained at the same time. Hence, it is often up to the designer to rely on judgment and experience in the scale-up.

In many cases, the main objectives usually present in an agitation process are as follows: *equal liquid motion*, such as in liquid blending, where the liquid motion or corresponding velocities are approximately the same in both cases; *equal suspension of solids*, where the levels of suspension are the same; and *equal rates of mass transfer*, where mass transfer is occurring between a liquid and a solid phase, liquid–liquid phases, and so on, and the rates are the same.

2. Scale-up procedure. A suggested step-by-step procedure to follow in the scale-up is detailed as follows for scaling up from the initial conditions where the geometric sizes given in Table 3.4-1 are D_{a1}, D_{T1}, H_1, W_1, and so on, to the final conditions of D_{a2}, D_{T2}, and so on.

1. Calculate the scale-up ratio R. Assuming that the original vessel is a standard cylinder with $D_{T1} = H_1$, the volume V_1 is

$$V_1 = \left(\frac{\pi D_{T1}^2}{4}\right)(H_1) = \left(\frac{\pi D_{T1}^3}{4}\right) \tag{3.4-6}$$

Then the ratio of the volumes is

$$\frac{V_2}{V_1} = \frac{\pi D_{T2}^3/4}{\pi D_{T1}^3/4} = \frac{D_{T2}^3}{D_{T1}^3} \tag{3.4-7}$$

The scale-up ratio is then

$$R = \left(\frac{V_2}{V_1}\right)^{1/3} = \frac{D_{T2}}{D_{T1}} \tag{3.4-8}$$

2. Using this value of R, apply it to all of the dimensions in Table 3.4-1 to calculate the new dimensions. For example,

$$D_{a2} = RD_{a1}, \qquad J_2 = RJ_1, \cdots \qquad (3.4\text{-}9)$$

3. Then a scale-up rule must be selected and applied to determine the agitator speed N_2 to use to duplicate the small-scale results using N_1. This equation is as follows (R2):

$$N_2 = N_1 \left(\frac{1}{R}\right)^n = N_1 \left(\frac{D_{T1}}{D_{T2}}\right)^n \qquad (3.4\text{-}10)$$

where $n = 1$ for equal liquid motion, $n = \frac{3}{4}$ for equal suspension of solids, and $n = \frac{2}{3}$ for equal rates of mass transfer (which is equivalent to equal power per unit volume). This value of n is based on empirical and theoretical considerations.

4. Knowing N_2, the power required can be determined using Eq. (3.4-2) and Fig. 3.4-4.

EXAMPLE 3.4-2. Derivation of Scale-Up Rule Exponent
For the scale-up rule exponent n in Eq. (3.4-10), show the following for turbulent agitation.
   (a) That when $n = \frac{2}{3}$, the power per unit volume is constant in the scale-up.
   (b) That when $n = 1.0$, the tip speed is constant in the scale-up.

Solution: For part (a), from Fig. 3.4-4, N_P is constant for the turbulent region. Then, from Eq. (3.4-2),

$$P_1 = N_P \rho N_1^3 D_{a1}^5 \qquad (3.4\text{-}11)$$

Then for equal power per unit volume, $P_1/V_1 = P_2/V_2$, or using Eq. (3.4-6),

$$\frac{P_1}{V_1} = \frac{P_1}{\pi D_{T1}^3/4} = \frac{P_2}{V_2} = \frac{P_2}{\pi D_{T2}^3/4} \qquad (3.4\text{-}12)$$

Substituting P_1 from Eq. (3.4–11) and also a similar equation for P_2 into Eq. (3.4-12) and combining with Eq. (3.4-8),

$$N_2 = N_1 \left(\frac{1}{R}\right)^{2/3} \qquad (3.4\text{-}13)$$

For part (b), using Eq. (3.4-10) with $n = 1.0$, rearranging, and multiplying by π,

$$N_2 = N_1 \left(\frac{D_{T1}}{D_{T2}}\right)^{1.0} \qquad (3.4\text{-}14)$$

$$\pi D_{T2} N_2 = \pi D_{T1} N_1 \qquad (3.4\text{-}15)$$

where $\pi D_{T2} N_2$ is the tip speed in m/s.

To aid the designer of new agitation systems and to serve as a guide for evaluating existing systems, some approximate guidelines are given as follows for liquids of normal viscosities (M2): for mild agitation and blending, 0.1 to 0.2 kW/m^3 of fluid (0.0005 to 0.001 hp/gal); for vigorous agitation, 0.4 to 0.6 kW/m^3 (0.002 to 0.003 hp/gal); for intense agitation or where mass transfer is important, 0.8 to 2.0 kW/m^3 (0.004 to 0.010 hp/gal). This power in kW is the actual power delivered to the fluid as given in Fig. 3.4-4 and Eq. (3.4-2). This does not include the power used in the gear boxes and bearings. Typical efficiencies of electric motors are given in Section 3.3B. As an approximation the power lost in the gear boxes, bearings, and in inefficiency of the electric motor is about 30 to 40% of P, the actual power input to the fluid.

EXAMPLE 3.4-3. Scale-Up of Turbine Agitation System

An existing agitation system is the same as given in Example 3.4-1a for a flat-blade turbine with a disk and six blades. The given conditions and sizes are $D_{T1} = 1.83$ m, $D_{a1} = 0.61$ m, $W_1 = 0.122$ m, $J_1 = 0.15$ m, $N_1 = 90/60 = 1.50$ rev/s, $\rho = 929$ kg/m^3, and $\mu = 0.01$ Pa·s. It is desired to scale up these results for a vessel whose volume is 3.0 times as large. Do this for the following two process objectives.

(a) Where equal rate of mass transfer is desired.

(b) Where equal liquid motion is needed.

Solution: Since $H_1 = D_{T1} = 1.83$ m, the original tank volume $V_1 = (\pi D_{T1}^2/4)(H_1) = \pi(1.83)^3/4 = 4.813$ m^3. Volume $V_2 = 3.0(4.813) = 14.44$ m^3. Following the steps in the scale-up procedure, and using Eq. (3.4-8),

$$R = \left(\frac{V_2}{V_1}\right)^{1/3} = \left(\frac{14.44}{4.813}\right)^{1/3} = 1.442$$

The dimensions of the larger agitation system are as follows: $D_{T2} = RD_{T1} = 1.442(1.83) = 2.64$ m, $D_{a2} = 1.442(0.61) = 0.880$ m, $W_2 = 1.442(0.122) = 0.176$ m, and $J_2 = 1.442(0.15) = 0.216$ m.

For part (a) for equal mass transfer, $n = \frac{2}{3}$ in Eq. (3.4-10).

$$N_2 = N_1\left(\frac{1}{R}\right)^{2/3} = (1.50)\left(\frac{1}{1.442}\right)^{2/3} = 1.175 \text{ rev/s (70.5 rpm)}$$

Using Eq. (3.4-1),

$$N'_{Re} = \frac{D_a^2 N \rho}{\mu} = \frac{(0.880)^2(1.175)929}{0.01} = 8.453 \times 10^4$$

Using $N_P = 5.0$ in Eq. (3.4-2),

$$P_2 = N_P \rho N_2^3 D_{a2}^5 = 5.0(929)(1.175)^3(0.880)^5 = 3977 \text{ J/s} = 3.977 \text{ kW}$$

The power per unit volume is

$$\frac{P_1}{V_1} = \frac{1.324}{4.813} = 0.2752 \text{ kW/m}^3$$

$$\frac{P_2}{V_2} = \frac{3.977}{14.44} = 0.2752 \text{ kW/m}^3$$

The value of 0.2752 kW/m^3 is somewhat lower than the approximate guidelines of 0.8 to 2.0 for mass transfer.

For part (b) for equal liquid motion, $n = 1.0$.

$$N_2 = (1.50)\left(\frac{1}{1.442}\right)^{1.0} = 1.040 \text{ rev/s}$$

$$P_2 = 5.0(929)(1.040)^3(0.880)^5 = 2757 \text{ J/s} = 2.757 \text{ kW}$$

$$\frac{P_2}{V_2} = \frac{2.757}{14.44} = 0.1909 \text{ kW/m}^3$$

3.4G   Mixing Times of Miscible Liquids

In one method used to study the blending or mixing time of two miscible liquids, an amount of HCl acid is added to an equivalent of NaOH and the time required for the indicator to change color is noted. This is a measure of molecule–molecule mixing. Other experimental methods are also used. Rapid mixing takes place near the impeller, with slower mixing, which depends on the pumping circulation rate, in the outer zones.

In Fig. 3.4-5, a correlation of mixing time is given for a turbine agitator (B5, M5, N1). The dimensionless mixing factor f_t is defined as

$$f_t = t_T \frac{(ND_a^2)^{2/3} g^{1/6} D_a^{1/2}}{H^{1/2} D_t^{3/2}} \qquad (3.4\text{-}16)$$

where t_T is the mixing time in seconds. For $N'_{Re} > 1000$, since the f_t is approximately constant, then $t_T N^{2/3}$ is constant. For some other mixers it has been shown that $t_T N$ is approximately constant. For scaling up from vessel 1 to another size vessel 2 with similar geometry and with the same power/unit volume in the turbulent region, the mixing times are related by

$$\frac{t_{T_2}}{t_{T_1}} = \left(\frac{D_{a_2}}{D_{a_1}}\right)^{11/18} \qquad (3.4\text{-}17)$$

Hence, the mixing time increases for the larger vessel. For scaling up keeping the same mixing time, the power/unit volume P/V increases markedly.

$$\frac{(P_2/V_2)}{(P_1/V_1)} = \left(\frac{D_{a_2}}{D_{a_1}}\right)^{11/4} \qquad (3.4\text{-}18)$$

Usually, in scaling up to large-size vessels, a somewhat larger mixing time is used so that the power/unit volume does not increase markedly.

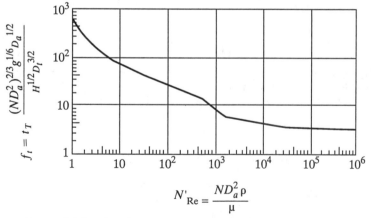

FIGURE 3.4-5. *Correlation of mixing time for miscible liquids using a turbine in a baffled tank (for a plain turbine, turbine with disk, and pitched-blade turbine). [From "Flow Patterns and Mixing Rates in Agitated Vessels" by K. W. Norwood and A. B. Metzner, A.I.Ch.E. J., 6, 432 (1960). Reproduced by permission of the American Institute of Chemical Engineers, 1960.]*

The mixing time for a helical-ribbon agitator is as follows for $N'_{Re} < 20$ (H2).

$$Nt_T = 126 \qquad \text{(agitator pitch/tank diameter} = 1.0) \qquad \text{(3.4-19)}$$

$$Nt_T = 90 \qquad \text{(agitator pitch/tank diameter} = 0.5) \qquad \text{(3.4-20)}$$

For very viscous liquids the helical-ribbon mixer gives a much smaller mixing time than a turbine for the same power/unit volume (M5). For nonviscous liquids, however, it gives longer times.

For a propellor agitator in a baffled tank, a mixing-time correlation is given by Biggs (B5) and that for an unbaffled tank by Fox and Gex (F1).

3.4H Flow Number and Circulation Rate in Agitation

An agitator acts like a centrifugal pump impeller without a casing and gives a flow at a certain pressure head. This circulation rate Q in m^3/s from the edge of the impeller is the flow rate perpendicular to the impeller discharge area. Fluid velocities have been measured in mixers and have been used to calculate the circulation rates. Data for baffled vessels have been correlated using the dimensionless flow number N_Q (U1).

$$N_Q = \frac{Q}{ND_a^3} \qquad \text{(3.4-21)}$$

$N_Q = 0.5 \qquad$ marine propeller (pitch = diameter)

$N_Q = 0.75 \qquad$ 6 blade turbine with disk $(W/D_a = \frac{1}{5})$

$N_Q = 0.5 \qquad$ 6 blade turbine with disk $(W/D_a = \frac{1}{8})$

$N_Q = 0.75 \qquad$ pitched-blade turbine $(W/D_a = \frac{1}{5})$

3.4I Special Agitation Systems

1. Suspension of solids. In some agitation systems a solid is suspended in the agitated liquid. Examples are where a finely dispersed solid is to be dissolved in the liquid, microorganisms are suspended in fermentation, a homogeneous liquid–solid mixture is to be produced for feed to a process, and a suspended solid is used as a catalyst to speed up a reaction. The suspension of solids is somewhat similar to a fluidized bed. In the agitated system circulation currents of the liquid keep the particles in suspension. The amount and type of agitation needed depends mainly on the terminal settling velocity of the particles, which can be calculated using the equations in Section 14.3. Empirical equations to predict the power required to suspend particles are given in references (M2, W1).

2. Dispersion of gases and liquids in liquids. In gas–liquid dispersion processes, the gas is introduced below the impeller, which chops the gas into very fine bubbles. The type and degree of agitation affects the size of the bubbles and the total interfacial area. Typical of such processes are aeration in sewage treatment plants, hydrogenation of liquids by hydrogen gas in the presence of a catalyst, absorption of a solute from the gas by the liquid, and fermentation. Correlations are available to predict the bubble size, holdup, and kW power needed (C3, L1, Z1). For liquids dispersed in inmiscible liquids, see reference (T1). The power required for the agitator in gas–liquid dispersion systems can be as much as 10 to 50% less than that required when no gas is present (C3, T2).

3. Motionless mixers. Mixing of two fluids can also be accomplished in motionless mixers with no moving parts. In such commercial devices stationary elements inside a pipe successively divide portions of the stream and then recombine these portions. In one type a short helical element divides the stream in two and rotates it 180°. The second element set at 90° to the first again divides the stream in two. For each element there are 2 divisions and recombinations, or 2^n for n elements in series. For 20 elements about 10^6 divisions occur.

Other types are available which consist of bars or flat sheets placed lengthwise in a pipe. Low-pressure drops are characteristic of all of these types of mixers. Mixing of even highly viscous materials is quite good in these mixers.

3.4J Mixing of Powders, Viscous Material, and Pastes

1. Powders. In mixing of solid particles or powders it is necessary to displace parts of the powder mixture with respect to other parts. The simplest class of devices suitable for gentle blending is the tumbler. However, it is not usually used for breaking up agglomerates. A common type of tumbler used is the *double-cone blender*, in which two cones are mounted with their open ends fastened together and rotated as shown in Fig. 3.4-6a. Baffles can also be used internally. If an internal rotating device is also used in the double cone, agglomerates can also be broken up. Other geometries used are a cylindrical drum with internal baffles or twin-shell V type. Tumblers used specifically for breaking up agglomerates are a rotating cylindrical or conical shell charged with metal or porcelain steel balls or rods.

Another class of devices for solids blending is the *stationary shell device*, in which the container is stationary and the material displacement is accomplished by single or multiple rotating inner devices. In the ribbon mixer in Fig. 3.4-6b, a shaft with two open helical screws numbers 1 and 2 attached to it rotates. One screw is left-handed and one right-handed. As the shaft rotates, sections of powder move in opposite directions and mixing occurs. Other types of internal rotating devices are available for special situations (P1). Also, in some devices both the shell and the internal device rotate.

2. Dough, pastes, and viscous materials. In the mixing of dough, pastes, and viscous materials, large amounts of power are required so that the material is divided, folded, or recombined, and also different parts of the material should be displaced relative to each other so that fresh surfaces recombine as often as possible. Some machines may require jacketed cooling to remove the heat generated.

The first class of device is somewhat similar to those for agitating fluids, with an impeller slowly rotating in a tank. The impeller can be a close-fitting anchor agitator as in Fig. 3.4-6b, where the outer sweep assembly may have scraper blades. A gate impeller can also be used which has horizontal and vertical bars that cut the paste at various levels and at the wall, which may have stationary bars. A modified gate mixer is the shear-bar mixer, which contains vertical rotating bars or paddles passing between vertical stationary fingers. Other modifications of these types are those where the can or container will rotate as well as the bars and scrapers. These are called *change-can mixers*.

The most commonly used mixer for heavy pastes and dough is the *double-arm*

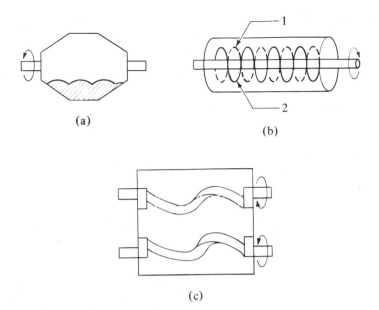

FIGURE 3.4-6. *Mixers for powders and pastes: (a) double-cone powder mixer, (b) ribbon powder mixer with two ribbons, (c) kneader mixer for pastes.*

kneader mixer. The mixing action is bulk movement, smearing, stretching, dividing, folding, and recombining. The most widely used design employs two contrarotating arms of sigmoid shape which may rotate at different speeds, as shown in Fig. 3.4-6c.

3.5 NON-NEWTONIAN FLUIDS

3.5A Types of Non-Newtonian Fluids

As discussed in Section 2.4, Newtonian fluids are those which follow Newton's law, Eq. (3.5-1).

$$\tau = -\mu \frac{dv}{dr} \qquad \text{(SI)}$$

$$\tau = -\frac{\mu}{g_c} \frac{dv}{dr} \qquad \text{(English)}$$

(3.5-1)

where μ is the viscosity and is a constant independent of shear rate. In Fig. 3.5-1 a plot is shown of shear stress τ versus shear rate $-dv/dr$. The line for a Newtonian fluid is straight, the slope being μ.

If a fluid does not follow Eq. (3.5-1), it is a non-Newtonian fluid. Then a plot of τ versus $-dv/dr$ is not linear through the origin for these fluids. Non-Newtonian fluids can be divided into two broad categories on the basis of their shear stress/shear rate behavior: those whose shear stress is independent of time or duration of shear (time-independent) and those whose shear stress is dependent on time or duration of shear (time-dependent). In addition to unusual shear-stress behavior, some non-Newtonian fluids also exhibit elastic (rubberlike) behavior which is a function of time and results in their being called *viscoelastic fluids*. These fluids exhibit normal stresses perpendicular to the direction of flow in addition to the usual tangential stresses. Most of the emphasis will be put on the time-independent class, which includes the majority of non-Newtonian fluids.

3.5B Time-Independent Fluids

1. Bingham plastic fluids.   These are the simplest because, as shown in Fig. 3.5-1, they differ from Newtonian only in that the linear relationship does not go through the origin. A finite shear stress τ_y (called *yield stress*) in N/m² is needed to initiate flow. Some fluids have a finite yield (shear) stress τ_y, but the plot of τ versus $-dv/dr$ is curved upward or downward. However, this departure from exact Bingham plasticity is often small. Examples of fluids with a yield stress are drilling muds, peat slurries, margarine, chocolate mixtures, greases, soap, grain-water suspensions, toothpaste, paper pulp, and sewage sludge.

2. Pseudoplastic fluids.   The majority of non-Newtonian fluids are in this category and include polymer solutions or melts, greases, starch suspensions, mayonnaise, biological fluids, detergent slurries, dispersion media in certain pharmaceuticals, and paints. The shape of the flow curve is shown in Fig. 3.5-1, and it generally can be represented by a power-law equation (sometimes called *Ostwald-deWaele equation*).

$$\tau = K\left(-\frac{dv}{dr}\right)^n \qquad (n < 1) \tag{3.5-2}$$

where K is the consistency index in $N \cdot s^n/m^2$ or $lb_f \cdot s^n/ft^2$, and n is the flow behavior index, dimensionless. The apparent viscosity, obtained from Eqs. (3.5-1) and (3.5-2), is $\mu_a = K(dv/dr)^{n-1}$ and decreases with increasing shear rate.

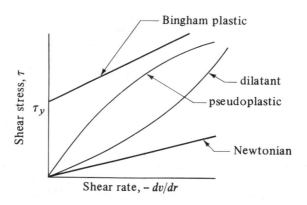

FIGURE 3.5-1.   *Shear diagram for Newtonian and time-independent non-Newtonian fluids.*

Chap. 3   Principles of Momentum Transfer and Applications

3. *Dilatant fluids.* These fluids are far less common than pseudoplastic fluids and their flow behavior in Fig. 3.5-1 shows an increase in apparent viscosity with increasing shear rate. The power law equation (3.5-2) is often applicable, but with $n > 1$.

$$\tau = K\left(-\frac{dv}{dr}\right)^n \qquad (n > 1) \tag{3.5-3}$$

For a Newtonian fluid, $n = 1$. Solutions showing dilatancy are some corn flour-sugar solutions, wet beach sand, starch in water, potassium silicate in water, and some solutions containing high concentrations of powder in water.

3.5C Time-Dependent Fluids

1. Thixotropic fluids. These fluids exhibit a reversible decrease in shear stress with time at a constant rate of shear. This shear stress approaches a limiting value that depends on the shear rate. Examples include some polymer solutions, shortening, some food materials, and paints. The theory for time-dependent fluids at present is still not completely developed.

2. Rheopectic fluids. These fluids are quite rare in occurrence and exhibit a reversible increase in shear stress with time at a constant rate of shear. Examples are bentonite clay suspensions, certain sols, and gypsum suspensions. In design procedures for thixotropic and rheopectic fluids for steady flow in pipes, the limiting flow-property values at a constant rate of shear are sometimes used (S2, W3).

3.5D Viscoelastic Fluids

Viscoelastic fluids exhibit elastic recovery from the deformations that occur during flow. They show both viscous and elastic properties. Part of the deformation is recovered upon removal of the stress. Examples are flour dough, napalm, polymer melts, and bitumens.

3.5E Laminar Flow of Time-Independent Non-Newtonian Fluids

1. Flow properties of a fluid. In determining the flow properties of a time-independent non-Newtonian fluid, a capillary-tube viscometer is often used. The pressure drop ΔP N/m^2 for a given flow rate q m^3/s is measured in a straight tube of length L m and diameter D m. This is repeated for different flow rates or average velocities V m/s. If the fluid is time-independent, these flow data can be used to predict the flow in any other pipe size.

A plot of $D \Delta p/4L$, which is τ_w, the shear stress at the wall in N/m^2, versus $8V/D$, which is proportional to the shear rate at the wall, is shown in Fig. 3.5-2 for a power-law fluid following Eq. (3.5-4).

$$\tau_w = \frac{D \Delta p}{4L} = K'\left(\frac{8V}{D}\right)^{n'} \tag{3.5-4}$$

where n' is the slope of the line when the data are plotted on logarithmic coordinates and K' has units of N $\cdot$ s$^{n'}$/m^2. For $n' = 1$, the fluid is Newtonian; for $n' < 1$, pseudoplastic, or Bingham plastic if the curve does not go through the origin; and for $n' > 1$, dilatant. The

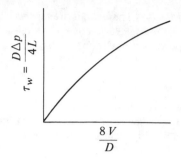

FIGURE 3.5-2. *General flow curve for a power-law fluid in laminar flow in a tube.*

K', the consistency index in Eq. (3.5-4), is the value of $D \, \Delta p / 4L$ for $8V/D = 1$. The shear rate at the wall, $(-dv/dr)_w$, is

$$\left(-\frac{dv}{dr}\right)_w = \left(\frac{3n' + 1}{4n'}\right)\left(\frac{8V}{D}\right) \tag{3.5-5}$$

Also, $K' = \mu$ for Newtonian fluids.

Equation (3.5-4) is simply another statement of the power-law model of Eq. (3.5-2) applied to flow in round tubes, and is more convenient to use for pipe-flow situations (D2). Hence, Eq. (3.5-4) defines the flow characteristics just as completely as Eq. (3.5-2). It has been found experimentally (M3) that for most fluids K' and n' are constant over wide ranges of $8V/D$ or $D \, \Delta p / 4L$. For some fluids this is not the case, and K' and n' vary. Then the particular values of K' and n' used must be valid for the actual $8V/D$ or $D \, \Delta P/4L$ with which one is dealing in a design problem. This method using flow in a pipe or tube is often used to determine the flow properties of a non-Newtonian fluid.

In many cases the flow properties of a fluid are determined using a rotational viscometer. The flow properties K and n in Eq. (3.5-2) are determined in this manner. A discussion of the rotational viscometer is given in Section 3.5I.

When the flow properties are constant over a range of shear stresses which occurs for many fluids, the following equations hold (M3):

$$n' = n \tag{3.5-6}$$

$$K' = K\left(\frac{3n' + 1}{4n'}\right)^{n'} \tag{3.5-7}$$

Often a generalized viscosity coefficient γ is defined as

$$\gamma = K'8^{n'-1} \quad \text{(SI)}$$

$$\gamma = g_c K'8^{n'-1} \quad \text{(English)} \tag{3.5-8}$$

where γ has units of $N \cdot s^{n'}/m^2$ or $lb_m/ft \cdot s^{2-n'}$

Typical flow-property constants (rheological constants) for some fluids are given in Table 3.5-1. Some data give γ values instead of K' values, but Eq. (3.5-8) can be used to convert these values if necessary. In some cases in the literature, K or K' values are given

TABLE 3.5-1. *Flow-Property Constants for Non-Newtonian Fluids*

Fluid	n'	$\gamma\left(\dfrac{N \cdot s^{n'}}{m^2}\right)$	$K\left(\dfrac{N \cdot s^{n'}}{m^2}\right)$	Ref.
1.5% carboxymethylcellulose in water	0.554	1.369		(S1)
3.0% CMC in water	0.566	4.17		(S1)
4% paper pulp in water	0.575	9.12		(A1)
14.3% clay in water	0.350	0.0512		(W2)
10% napalm in kerosene	0.520	1.756		(S1)
25% clay in water	0.185	0.3036		(W2)
Applesauce, brand A (297 K), density = 1.10 g/cm^3	0.645		0.500	(C1)
Banana purée, brand A (297 K), density = 0.977 g/cm^3	0.458		6.51	(C1)
Honey (297 K)	1.00		5.61	(C1)
Cream, 30% fat (276 K)	1.0		0.01379	(M4)
Tomato concentrate, 5.8% total solids (305 K)	0.59		0.2226	(H1)

as dyn $\cdot$ s$^{n'}$/cm^2 or lb$_f$ $\cdot$ s$^{n'}$/ft^2. From Appendix A.1, the conversion factors are

$$1 \text{ lb}_f \cdot s^{n'}/ft^2 = 47.880 \text{ N} \cdot s^{n'}/m^2$$

$$1 \text{ dyn} \cdot s^{n'}/cm^2 = 2.0886 \times 10^{-3} \text{ lb}_f \cdot s^{n'}/ft^2$$

2. Equations for flow in a tube. In order to predict the frictional pressure drop Δp in laminar flow in a tube, Eq. (3.5-4) is solved for Δp.

$$\Delta p = \frac{K'4L}{D}\left(\frac{8V}{D}\right)^{n'} \tag{3.5-9}$$

If the average velocity is desired, Eq. (3.5-4) can be rearranged to give

$$V = \frac{D}{8}\left(\frac{\Delta pD}{K'4L}\right)^{1/n'} \tag{3.5-10}$$

If the equations are desired in terms of K instead of K', Eqs. (3.5-6) and (3.5-7) can be substituted into (3.5-9) and (3.5-10). The flow must be laminar and the generalized Reynolds number has been defined as

$$N_{Re, gen} = \frac{D^{n'}V^{2-n'}\rho}{\gamma} = \frac{D^{n'}V^{2-n'}\rho}{K'8^{n'-1}} = \frac{D^{n}V^{2-n}\rho}{K8^{n-1}\left(\dfrac{3n+1}{4n}\right)^n} \quad \text{(SI)} \tag{3.5-11}$$

3. Friction factor method. Alternatively, using the Fanning friction factor method given in Eqs. (2.10-5) to (2.10-7) for Newtonian fluids, but using the generalized Reynolds numbers,

$$f = \frac{16}{N_{Re, gen}}$$ (3.5-12)

$$\Delta p = 4f\rho \frac{L}{D} \frac{V^2}{2} \quad \text{(SI)}$$

(3.5-13)

$$\Delta p = 4f\rho \frac{L}{D} \frac{V^2}{2g_c} \quad \text{(English)}$$

EXAMPLE 3.5-1. Pressure Drop of Power-Law Fluid in Laminar Flow

A power-law fluid having a density of 1041 kg/m³ is flowing through 14.9 m of a tubing having an inside diameter of 0.0524 m at an average velocity of 0.0728 m/s. The rheological or flow properties of the fluid are $K' = 15.23$ $N \cdot s^{n'}/m^2$ (0.318 $lb_f \cdot s^{n'}/ft^2$) and $n' = 0.40$.

    (a) Calculate the pressure drop and friction loss using Eq. (3.5-9) for laminar flow. Check the generalized Reynolds number to make sure that the flow is laminar.

    (b) Repeat part (a) but use the friction factor method.

Solution: The known data are as follows: $K' = 15.23$, $n' = 0.40$, $D = 0.0524$ m, $V = 0.0728$ m/s, $L = 14.9$ m, and $\rho = 1041$ kg/m³. For part (a), using Eq. (3.5-9),

$$\Delta p = \frac{K'4L}{D} \left(\frac{8V}{D} \right)^{n'} = \frac{15.23(4)(14.9)}{0.0524} \left(\frac{8 \times 0.0728}{0.0524} \right)^{0.4} = 45\,390 \text{ N/m}^2$$

Also, to calculate the friction loss,

$$F_f = \frac{\Delta p}{\rho} = \frac{45\,390}{1041} = 43.60 \text{ J/kg}$$

Using Eq. (3.5-11),

$$N_{Re, gen} = \frac{D^{n'}V^{2-n'}\rho}{K'8^{n'-1}} = \frac{(0.0524)^{0.40}(0.0728)^{1.60}(1041)}{15.23(8)^{-0.6}} = 1.106$$

Hence, the flow is laminar.

    For part (b), using Eq. (3.5-12),

$$f = \frac{16}{N_{Re, gen}} = \frac{16}{1.106} = 14.44$$

Substituting into Eq. (3.5-13),

$$\Delta p = 4f\rho \frac{L}{D} \frac{V^2}{2} = 4(14.44)(1041) \frac{14.9}{0.0524} \frac{(0.0728)^2}{2}$$

$$= 45.39 \frac{kN}{m^2} \left(946 \frac{lb_f}{ft^2} \right)$$

    To calculate the pressure drop for a Bingham plastic fluid with a yield stress, methods are available for laminar flow and are discussed in detail elsewhere (C1, P1, S2).

3.5F Friction Losses in Contractions, Expansions, and Fittings in Laminar Flow

Since non-Newtonian power-law fluids flowing in conduits are often in laminar flow because of their usually high effective viscosity, losses in sudden changes of velocity and fittings are important in laminar flow.

1. Kinetic energy in laminar flow. In application of the total mechanical-energy balance in Eq. (2.7-28), the average kinetic energy per unit mass of fluid is needed. For fluids, this is (S2)

$$\text{average kinetic energy/kg} = \frac{V^2}{2\alpha} \tag{3.5-14}$$

For Newtonian fluids, $\alpha = \frac{1}{2}$ for laminar flow. For power-law non-Newtonian fluids,

$$\alpha = \frac{(2n + 1)(5n + 3)}{3(3n + 1)^2} \tag{3.5-15}$$

For example, if $n = 0.50$, $\alpha = 0.585$. If $n = 1.00$, $\alpha = \frac{1}{2}$. For turbulent flow for Newtonian and non-Newtonian flow, $\alpha = 1.0$ (D1).

2. Losses in contractions and fittings. Skelland (S2) and Dodge and Metzner (D2) state that when a fluid leaves a tank and flows through a sudden contraction to a pipe of diameter D_2 or flows from a pipe of diameter D_1 through a sudden contraction to a pipe of D_2, a vena contracta is usually formed downstream from the contraction. General indications are that the frictional pressure losses for pseudoplastic and Bingham plastic fluids are very similar to those for Newtonian fluids at the same generalized Reynolds numbers in laminar and turbulent flow for contractions and also for fittings and valves.

For contraction losses, Eq. (2.10-16) can be used where $\alpha = 1.0$ for turbulent flow and for laminar flow Eq. (3.5-15) can be used to determine α, since n is not 1.00.

For fittings and valves, frictional losses should be determined using Eq. (2.10-17) and values from Table 2.10-1.

3. Losses in sudden expansion. For the frictional loss for a non-Newtonian fluid in laminar flow through a sudden expansion from D_1 to D_2 diameter, Skelland (S2) gives

$$h_{ex} = \frac{3n + 1}{2n + 1} V_1^2 \left[\frac{n + 3}{2(5n + 3)} \left(\frac{D_1}{D_2}\right)^4 - \left(\frac{D_1}{D_2}\right)^2 + \frac{3(3n + 1)}{2(5n + 3)} \right] \tag{3.5-16}$$

where h_{ex} is the frictional loss in J/kg. In English units Eq. (3.5-16) is divided by g_c and h_{ex} is in ft $\cdot$ lb$_f$/lb$_m$.

Equation (2.10-15) for laminar flow with $\alpha = \frac{1}{2}$ for a Newtonian fluid gives values reasonably close to those of Eq. (3.5-16) for $n = 1$ (Newtonian fluid). For turbulent flow the frictional loss can be approximated by Eq. (2.10-15), with $\alpha = 1.0$ for non-Newtonian fluids (S2).

3.5G Turbulent Flow and Generalized Friction Factors

In turbulent flow of time-independent fluids the Reynolds number at which turbulent flow occurs varies with the flow properties of the non-Newtonian fluid. Dodge and Metzner (D2) in a comprehensive study derived a theoretical equation for turbulent flow of non-Newtonian fluids through smooth round tubes. The final equation is plotted in Fig. 3.5-3, where the Fanning friction factor is plotted versus the generalized Reynolds

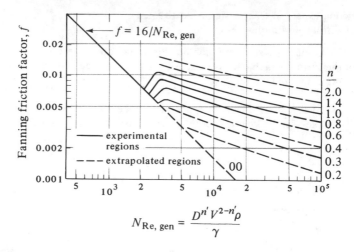

FIGURE 3.5-3. *Fanning friction factor versus generalized Reynolds number for time-independent non-Newtonian and Newtonian fluids flowing in smooth tubes. [From D. W. Dodge and A. B. Metzner, A.I.Ch.E. J., 5, 189 (1959). With permission.]*

number, $N_{Re, gen}$, given in Eq. (3.5-11). Power-law fluids with flow-behavior indexes n' between 0.36 and 1.0 were experimentally studied at Reynolds numbers up to 3.5×10^4 and confirmed the derivation.

The curves for different n' values break off from the laminar line at different Reynolds numbers to enter the transition region. For $n' = 1.0$ (Newtonian), the transition region starts at $N_{Re, gen} = 2100$. Since many non-Newtonian power-law fluids have high effective viscosities, they are often in laminar flow. The correlation for a smooth tube also holds for a rough pipe in laminar flow.

For rough commercial pipes with various values of roughness ε/D, Fig. 3.5-3 cannot be used for turbulent flow, since it is derived for smooth tubes. The functional dependence of the roughness values ε/D on n' requires experimental data which are not yet available. Metzner and Reed (M3, S3) recommend use of the existing relationship, Fig. 2.10-3, for Newtonian fluids in rough tubes using the generalized Reynolds number $N_{Re, gen}$. This is somewhat conservative since preliminary data indicate that friction factors for pseudoplastic fluids may be slightly smaller than for Newtonian fluids. This is also consistent with Fig. 3.5-3 for smooth tubes that indicate lower f values for fluids with n' below 1.0 (S2).

EXAMPLE 3.5-2. Turbulent Flow of Power-Law Fluid

A pseudoplastic fluid that follows the power law, having a density of 961 kg/m³, is flowing through a smooth circular tube having an inside diameter of 0.0508 m at an average velocity of 6.10 m/s. The flow properties of the fluid are $n' = 0.30$ and $K' = 2.744$ N·s$^{n'}$/m². Calculate the frictional pressure drop for a tubing 30.5 m long.

Solution: The data are as follows: $K' = 2.744$, $n' = 0.30$, $D = 0.0508$ m, $V = 6.10$ m/s, $\rho = 961$ kg/m³, and $L = 30.5$ m. Using the general Reynolds-number equation (3.5-11),

$$N_{Re, gen} = \frac{D^{n'} V^{2-n'} \rho}{K' 8^{n'-1}} = \frac{(0.0508)^{0.3}(6.10)^{1.7}(961)}{2.744(8^{-0.7})}$$

$$= 1.328 \times 10^4$$

Hence, the flow is turbulent. Using Fig. 3.5-3 for $N_{Re, gen} = 1.328 \times 10^4$ and $n' = 0.30$, $f = 0.0032$.

Substituting into Eq. (3.5-13),

$$\Delta p = 4 f \rho \frac{L}{D} \frac{V^2}{2} = 4(0.0032)(961)\left(\frac{30.5}{0.0508}\right)\frac{(6.10)^2}{2}$$

$$= 137.4 \text{ kN/m}^2 \ (2870 \text{ lb}_f/\text{ft}^2)$$

3.5H Velocity Profiles for Non-Newtonian Fluids

Starting with Eq. (3.5-2) written as

$$\tau_{rx} = K\left(-\frac{dv_x}{dr}\right)^n \tag{3.5-17}$$

the following equation can be derived relating the velocity v_x with the radial position r, which is the distance from the center. (See Problem 2.9-3 for this derivation.)

$$v_x = \frac{n}{n+1}\left(\frac{p_0 - p_L}{2KL}\right)^{1/n}(R_0)^{(n+1)/n}\left[1 - \left(\frac{r}{R_0}\right)^{(n+1)/n}\right] \tag{3.5-18}$$

At $r = 0$, $v_x = v_{x\,max}$ and Eq. (3.5-18) becomes

$$v_x = v_{x\,max}\left[1 - \left(\frac{r}{R_0}\right)^{(n+1)/n}\right] \tag{3.5-19}$$

The velocity profile can be calculated for laminar flow of a non-Newtonian fluid to show that the velocity profile for a Newtonian fluid given in Eq. (2.9-9) can differ greatly from that of a non-Newtonian fluid. For pseudoplastic fluids ($n < 1$), a relatively flat velocity profile is obtained compared to the parabolic profile for a Newtonian fluid. For $n = 0$, rodlike flow is obtained. For dilitant fluids ($n > 1$), a much sharper profile is obtained and for $n = \infty$, the velocity is a linear function of the radius.

3.5I Determination of Flow Properties of Non-Newtonian Fluids Using Rotational Viscometer

The flow-property or rheological constants of non-Newtonian fluids can be measured using pipe flow as discussed in Section 3.5E. Another, more important method for measuring flow properties is by use of a rotating concentric-cylinder viscometer. This was first described by Couette in 1890. In this device a concentric rotating cylinder (spindle) spins at a constant rotational speed inside another cylinder. Generally, there is a very small gap between the walls. This annulus is filled with the fluid. The torque needed to maintain this constant rotation rate of the inner spindle is measured by a torsion wire from which the spindle is suspended. A typical commercial instrument of this type is the Brookfield viscometer. Some types rotate the outer cylinder.

The shear stress at the wall of the bob or spindle is given by

$$\tau_w = \frac{T}{2\pi R_b^2 L}$$ (3.5-20)

where τ_w is the shear stress at the wall, N/m^2 or kg/s$^2 \cdot$m; T is the measured torque, kg$\cdot$m^2/s^2; R_b is the radius of the spindle, m; and L is the effective length of the spindle, m. Note that Eq. (3.5-20) holds for Newtonian and non-Newtonian fluids.

The shear rate at the surface of the spindle for non-Newtonian fluids is as follows (M6) for $0.5 < R_b/R_c < 0.99$:

$$\left(-\frac{dv}{dr}\right)_w = \frac{2\omega}{n[1 - (R_b/R_c)^{2/n}]}$$ (3.5-21)

where R_c is the radius of the outer cylinder or container, m; and ω is the angular velocity of the spindle, rad/s. Also, $\omega = 2\pi N/60$, when N is the RPM. Results calculated using Eq. (3.5-21) give values very close to those using the more complicated equation of Krieger and Maron (K2), also given in (P4, S2).

The power-law equation is given as

$$\tau = K\left(-\frac{dv}{dr}\right)^n$$ (3.5-2)

where $K = $ N$\cdot$s^n/m^2, kg$\cdot$s^{n-2}/m. Substituting Eqs. (3.5-20) and (3.5-21) into (3.5-2) gives

$$T = 2\pi R_b^2 LK\left[\frac{2}{n[1 - (R_b/R_c)^{2/n}]}\right]^n \omega^n$$ (3.5-22)

Or,

$$T = A\omega^n$$ (3.5-23)

where,

$$A = 2\pi R_b^2 LK\left[\frac{2}{n[1 - (R_b/R_c)^{2/n}]}\right]^n$$ (3.5-24)

Experimental data are obtained by measuring the torque T at different values of ω for a given fluid. The flow property constants may be evaluated by plotting log T versus log ω. The parameter, n, is the slope of the straight line and the intercept is log A. The consistency factor K is now easily evaluated from Eq. (3.5-24).

Various special cases can be derived for Eq. (3.5-21).
1. *Newtonian fluid.* $(n=1)$.

$$\left(-\frac{dv}{dr}\right)_w = \frac{2\omega}{1 - (R_b/R_c)^2}$$ (3.5-25)

2. *Very large gap* $(R_b/R_c < 0.1)$. This is the case of a spindle immersed in a large beaker of test fluid. Equation (3.5-21) becomes

$$\left(-\frac{dv}{dr}\right)_w = \frac{2\omega}{n}$$ (3.5-26)

Substituting Eqs. (3.5-20) and (3.5-26) into (3.5-2)

$$T = 2\pi R_b^2 LK\left(\frac{2}{n}\right)^n \omega^n \qquad (3.5\text{-}27)$$

Again, as before, the flow property constants can be evaluated by plotting log T versus log ω.

3. *Very narrow gap* $(R_b/R_c > 0.99)$. This is similar to flow between parallel plates. Taking the shear rate at radius $(R_b + R_c)/2$,

$$\left(-\frac{dv}{dr}\right)_w \cong \frac{\Delta v}{\Delta r} = \frac{2\omega}{1 - (R_b/R_c)^2} \qquad (3.5\text{-}28)$$

This equation, then, is the same as Eq. (3.5-25).

3.5J Power Requirements in Agitation and Mixing of Non-Newtonian Fluids

For correlating the power requirements in agitation and mixing of non-Newtonian fluids the power number N_P is defined by Eq. (3.4-2), which is also the same equation used for Newtonian fluids. However, the definition of the Reynolds number is much more complicated than for Newtonian fluids since the apparent viscosity is not constant for non-Newtonian fluids and varies with the shear rates or velocity gradients in the vessel. Several investigators (G1, M1) have used an average apparent viscosity μ_a, which is used in the Reynolds number as follows:

$$N'_{\text{Re}, n} = \frac{D_a^2 N \rho}{\mu_a} \qquad (3.5\text{-}29)$$

The average apparent viscosity can be related to the average shear rate or average velocity gradient by the following method. For a power-law fluid,

$$\tau = K\left(-\frac{dv}{dy}\right)_{\text{av}}^n \qquad (3.5\text{-}30)$$

For a Newtonian fluid,

$$\tau = \mu_a\left(-\frac{dv}{dy}\right)_{\text{av}} \qquad (3.5\text{-}31)$$

Combining Eqs. (3.5-30) and (3.5-31)

$$\mu_a = K\left(\frac{dv}{dy}\right)_{\text{av}}^{n-1} \qquad (3.5\text{-}32)$$

Metzner and others (G1, M1) found experimentally that the average shear rate $(dv/dy)_{\text{av}}$ for pseudoplastic liquids $(n < 1)$ varies approximately as follows with the rotational speed:

$$\left(\frac{dv}{dy}\right)_{\text{av}} = 11N \qquad (3.5\text{-}33)$$

Hence, combining Eqs. (3.5-32) and (3.5-33)

$$\mu_a = (11N)^{n-1}K \qquad (3.5\text{-}34)$$

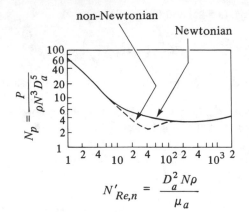

FIGURE 3.5-4. *Power correlation in agitation for a flat, six-blade turbine with disk in pseudoplastic non-Newtonian and Newtonian fluids (G1, M1, R1): $D_a/W = 5, L/W = 5/4, D_t/J = 10$.*

Substituting into Eq. (3.5-29),

$$N'_{\text{Re, }n} = \frac{D_a^2 N^{2-n}\rho}{11^{n-1}K} \tag{3.5-35}$$

Equation (3.5-35) has been used to correlate data for a flat six-blade turbine with disk in pseudoplastic liquids, and the dashed curve in Fig. 3.5-4 shows the correlation (M1). The solid curve applies to Newtonian fluids (R1). Both sets of data were obtained for four baffles with $D_t/J = 10$, $D_a/W = 5$, and $L/W = 5/4$. However. since it has been shown that the difference in results for $D_t/J = 10$ and $D_t/J = 12$ is very slight (R1), this Newtonian line can be considered the same as curve 1, Fig. 3.4-4. The curves in Fig. 3.5-4 show that the results are identical for the Reynolds number range 1 to 2000 except that they differ only in the Reynolds number range 10 to 100, where the pseudoplastic fluids use less power than the Newtonian fluids. The flow patterns for the pseudoplastic fluids show much greater velocity gradient changes than do the Newtonian fluids in the agitator. The fluid far from the impeller may be moving in slow laminar flow with a high apparent viscosity. Data for fan turbines and propellers are also available (M1).

3.6 DIFFERENTIAL EQUATIONS OF CONTINUITY

3.6A Introduction

In Sections 2.6, 2.7, and 2.8 overall mass, energy, and momentum balances allowed us to solve many elementary problems on fluid flow. These balances were done on an arbitrary finite volume sometimes called a *control volume*. In these total energy, mechanical energy, and momentum balances, we only needed to know the state of the inlet and outlet streams and the exchanges with the surroundings.

These overall balances were powerful tools in solving various flow problems because they did not require knowledge of what goes on inside the finite control volume. Also, in the simple shell momentum balances made in Section 2.9, expressions were obtained for the velocity distribution and pressure drop. However, to advance in our study of these flow systems, we must investigate in greater detail what goes on inside this finite control volume. To do this, we now use a differential element for a control volume. The differential balances will be somewhat similar to the overall and shell balances, but now we shall make the balance in a single phase and integrate to the phase boundary using the boundary conditions. In these balances done earlier, a balance was made for each new system studied. It is not necessary to formulate new balances for each new flow problem. It is often easier to start with the differential equations of the conservation of mass (equation of continuity) and the conservation of momentum in general form. Then these equations are simplified by discarding unneeded terms for each particular problem.

For nonisothermal systems a general differential equation of conservation of energy will be considered in Chapter 5. Also in Chapter 7 a general differential equation of continuity for a binary mixture will be derived. The differential-momentum-balance equation to be derived is based on Newton's second law and allows us to determine the way velocity varies with position and time and the pressure drop in laminar flow. The equation of momentum balance can be used for turbulent flow with certain modifications.

Often these conservation equations are called *equations of change*, since they describe the variations in the properties of the fluid with respect to position and time. Before we derive these equations, a brief review of the different types of derivatives with respect to time which occur in these equations and a brief description of vector notation will be given.

3.6B Types of Time Derivatives and Vector Notation

1. Partial time derivative. Various types of time derivatives are used in the derivations to follow. The most common type of derivative is the partial time derivative. For example, suppose that we are interested in the mass concentration or density ρ in kg/m^3 in a flowing stream as a function of position x, y, z and time t. The partial time derivative of ρ is $\partial \rho / \partial t$. This is the local change of density with time at a fixed point x, y, and z.

2. Total time derivative. Suppose that we want to measure the density in the stream while we are moving about in the stream with velocities in the x, y, and z directions of dx/dt, dy/dt and dz/dt, respectively. The total derivative $d\rho/dt$ is

$$\frac{d\rho}{dt} = \frac{\partial \rho}{\partial t} + \frac{\partial \rho}{\partial x}\frac{dx}{dt} + \frac{\partial \rho}{\partial y}\frac{dy}{dt} + \frac{\partial \rho}{\partial z}\frac{dz}{dt} \tag{3.6-1}$$

This means that the density is a function of t and of the velocity components dx/dt, dy/dt, and dz/dt at which the observer is moving.

3. Substantial time derivative. Another useful type of time derivative is obtained if the observer floats along with the velocity $\mathbf{v}$ of the flowing stream and notes the change in density with respect to time. This is called the derivative that follows the motion, or the *substantial time derivative*, $D\rho/Dt$.

$$\frac{D\rho}{Dt} = \frac{\partial \rho}{\partial t} + v_x \frac{\partial \rho}{\partial x} + v_y \frac{\partial \rho}{\partial y} + v_z \frac{\partial \rho}{\partial z} = \frac{\partial \rho}{\partial t} + (\mathbf{v} \cdot \nabla \rho) \tag{3.6-2}$$

where v_x, v_y, and v_z are the velocity components of the stream velocity $\mathbf{v}$, which is a vector. This substantial derivative is applied to both scalar and vector variables. The term $(\mathbf{v} \cdot \nabla \rho)$ will be discussed in part 6 of Section 3.6B.

4. Scalars. The physical properties encountered in momentum, heat, and mass transfer can be placed in several categories: scalars, vectors, and tensors. Scalars are quantities such as concentration, temperature, length, volume, time, and energy. They have magnitude but no direction and are considered to be zero-order tensors. The common mathematical algebraic laws hold for the algebra of scalars. For example, $bc = cd$, $b(cd) = (bc)d$, and so on.

5. Vectors. Velocity, force, momentum, and acceleration are considered vectors since they have magnitude and direction. They are regarded as first-order tensors and are written in boldface letters in this text, such as $\mathbf{v}$ for velocity. The addition of the two vectors $\mathbf{B} + \mathbf{C}$ by parallelogram construction and the subtraction of two vectors $\mathbf{B} - \mathbf{C}$ is shown in Fig. 3.6-1. The vector $\mathbf{B}$ is represented by its three projections B_x, B_y, and B_z on the x, y, and z axes and

$$\mathbf{B} = \mathbf{i}B_x + \mathbf{j}B_y + \mathbf{k}B_z \tag{3.6-3}$$

where $\mathbf{i}$, $\mathbf{j}$, and $\mathbf{k}$ are unit vectors along the axes x, y, and z, respectively.

In multiplying a scalar quantity r or s by a vector $\mathbf{B}$, the following hold.

$$r\mathbf{B} = \mathbf{B}r \tag{3.6-4}$$

$$(rs)\mathbf{B} = r(s\mathbf{B}) \tag{3.6-5}$$

$$r\mathbf{B} + s\mathbf{B} = (r + s)\mathbf{B} \tag{3.6-6}$$

The following also hold:

$$(\mathbf{B} \cdot \mathbf{C}) = (\mathbf{C} \cdot \mathbf{B}) \tag{3.6-7}$$

$$\mathbf{B} \cdot (\mathbf{C} + \mathbf{D}) = (\mathbf{B} \cdot \mathbf{C}) + (\mathbf{B} \cdot \mathbf{D}) \tag{3.6-8}$$

$$(\mathbf{B} \cdot \mathbf{C})\mathbf{D} \neq \mathbf{B}(\mathbf{C} \cdot \mathbf{D}) \tag{3.6-9}$$

$$(\mathbf{B} \cdot \mathbf{C}) = BC \cos \phi_{BC} \tag{3.6-10}$$

where ϕ_{BC} is the angle between two vectors and is $< 180°$.

Second-order tensors τ arise primarily in momentum transfer and have nine components. They are discussed elsewhere (B2).

6. Differential operations with scalars and vectors. The gradient or "grad" of a scalar field is

$$\nabla\rho = \mathbf{i}\frac{\partial\rho}{\partial x} + \mathbf{j}\frac{\partial\rho}{\partial y} + \mathbf{k}\frac{\partial\rho}{\partial z} \tag{3.6-11}$$

where ρ is a scalar such as density.

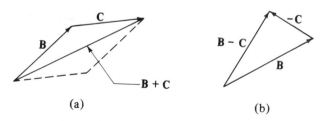

(a)   (b)

FIGURE 3.6-1.   *Addition and subtraction of vectors: (a) addition of vectors, $\mathbf{B} + \mathbf{C}$;*
(b) subtraction of vectors, $\mathbf{B} - \mathbf{C}$.

The divergence or "div" of a vector $\mathbf{v}$ is

$$(\nabla \cdot \mathbf{v}) = \frac{\partial v_x}{\partial x} + \frac{\partial v_y}{\partial y} + \frac{\partial v_z}{\partial z} \qquad (3.6\text{-}12)$$

where $\mathbf{v}$ is a function of v_x, v_y, and v_z.

The Laplacian of a scalar field is

$$\nabla^2 \rho = \frac{\partial^2 \rho}{\partial x^2} + \frac{\partial^2 \rho}{\partial y^2} + \frac{\partial^2 \rho}{\partial z^2} \qquad (3.6\text{-}13)$$

Other operations that may be useful are

$$\nabla rs = r\nabla s + s\nabla r \qquad (3.6\text{-}14)$$

$$(\nabla \cdot s\mathbf{v}) = (\nabla s \cdot \mathbf{v}) + s(\nabla \cdot \mathbf{v}) \qquad (3.6\text{-}15)$$

$$(\mathbf{v} \cdot \nabla s) = v_x \frac{\partial s}{\partial x} + v_y \frac{\partial s}{\partial y} + v_z \frac{\partial s}{\partial z} \qquad (3.6\text{-}16)$$

3.6C Differential Equation of Continuity

1. Derivation of equation of continuity. A mass balance will be made for a pure fluid flowing through a stationary volume element $\Delta x\, \Delta y\, \Delta z$ which is fixed in space as in Fig. 3.6-2. The mass balance for the fluid with a concentration of ρ kg/m^3 is

(rate of mass in) $-$ (rate of mass out) = (rate of mass accumulation) **(3.6-17)**

In the x direction the rate of mass entering the face at x having an area of $\Delta y\, \Delta z$ m^2 is $(\rho v_x)_x\, \Delta y\, \Delta z$ kg/s and that leaving at $x + \Delta x$ is $(\rho v_x)_{x+\Delta x}\, \Delta y\, \Delta z$. The term (ρv_x) is a mass flux in kg/s·m^2. Mass entering and that leaving in the y and the z directions are also shown in Fig. 3.6-2.

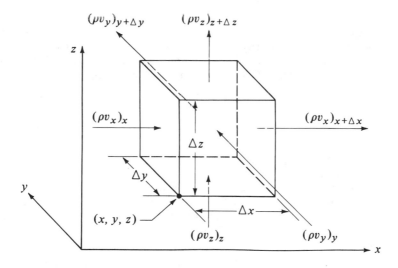

FIGURE 3.6-2. *Mass balance for a pure fluid flowing through a fixed volume $\Delta x\, \Delta y\, \Delta z$ in space.*

The rate of mass accumulation in the volume $\Delta x\, \Delta y\, \Delta z$ is

$$\text{rate of mass accumulation} = \Delta x\, \Delta y\, \Delta z\, \frac{\partial \rho}{\partial t} \tag{3.6-18}$$

Substituting all these expressions into Eq. (3.6-17) and dividing both sides by $\Delta x\, \Delta y\, \Delta z$,

$$\frac{[(\rho v_x)_x - (\rho v_x)_{x+\Delta x}]}{\Delta x} + \frac{[(\rho v_y)_y - (\rho v_y)_{y+\Delta y}]}{\Delta y} + \frac{[(\rho v_z)_z - (\rho v_z)_{z+\Delta z}]}{\Delta z} = \frac{\partial \rho}{\partial t} \tag{3.6-19}$$

Taking the limit as Δx, Δy, and Δz approach zero, we obtain the equation of continuity or conservation of mass for a pure fluid.

$$\frac{\partial \rho}{\partial t} = -\left[\frac{\partial(\rho v_x)}{\partial x} + \frac{\partial(\rho v_y)}{\partial y} + \frac{\partial(\rho v_z)}{\partial z}\right] = -(\nabla \cdot \rho \mathbf{v}) \tag{3.6-20}$$

The vector notation on the right side of Eq. (3.6-20) comes from the fact that $\mathbf{v}$ is a vector. Equation (3.6-20) tells us how density ρ changes with time at a fixed point resulting from the changes in the mass velocity vector $\rho \mathbf{v}$.

We can convert Eq. (3.6-20) into another form by carrying out the actual partial differentiation.

$$\frac{\partial \rho}{\partial t} = -\rho\left(\frac{\partial v_x}{\partial x} + \frac{\partial v_y}{\partial y} + \frac{\partial v_z}{\partial z}\right) - \left(v_x\frac{\partial \rho}{\partial x} + v_y\frac{\partial \rho}{\partial y} + v_z\frac{\partial \rho}{\partial z}\right) \tag{3.6-21}$$

Rearranging Eq. (3.6-21),

$$\frac{\partial \rho}{\partial t} + v_x\frac{\partial \rho}{\partial x} + v_y\frac{\partial \rho}{\partial y} + v_z\frac{\partial \rho}{\partial z} = -\rho\left(\frac{\partial v_x}{\partial x} + \frac{\partial v_y}{\partial y} + \frac{\partial v_z}{\partial z}\right) \tag{3.6-22}$$

The left-hand side of Eq. (3.6-22) is the same as the substantial derivative in Eq. (3.6-2). Hence, Eq. (3.6-22) becomes

$$\frac{D\rho}{Dt} = -\rho\left(\frac{\partial v_x}{\partial x} + \frac{\partial v_y}{\partial y} + \frac{\partial v_z}{\partial z}\right) = -\rho(\nabla \cdot \mathbf{v}) \tag{3.6-23}$$

2. Equation of continuity for constant density.   Often in engineering with liquids that are relatively incompressible, the density ρ is essentially constant. Then ρ remains constant for a fluid element as it moves along a path following the fluid motion, or $D\rho/Dt = 0$. Hence, Eq. (3.6-23) becomes for a fluid of constant density at steady or unsteady state,

$$(\nabla \cdot \mathbf{v}) = \frac{\partial v_x}{\partial x} + \frac{\partial v_y}{\partial y} + \frac{\partial v_z}{\partial z} = 0 \tag{3.6-24}$$

At steady state, $\partial \rho/\partial t = 0$ in Eq. (3.6-22).

EXAMPLE 3.6-1.   Flow over a Flat Plate
An incompressible fluid flows past one side of a flat plate. The flow in the x direction is parallel to the flat plate. At the leading edge of the plate the flow is uniform at the free stream velocity v_{x0}. There is no velocity in the z direction. The y direction is the perpendicular distance from the plate. Analyze this case using the equation of continuity.

Solution: For this case where ρ is constant, Eq. (3.6-24) holds.

$$\frac{\partial v_x}{\partial x} + \frac{\partial v_y}{\partial y} + \frac{\partial v_z}{\partial z} = 0 \tag{3.6-24}$$

Since there is no velocity in the z direction, we obtain

$$\frac{\partial v_x}{\partial x} = -\frac{\partial v_y}{\partial y} \qquad (3.6\text{-}25)$$

At a given small value of y close to the plate, the value of v_x must decrease from its free stream velocity v_{xo} as it passes the leading edge in the x direction because of fluid friction. Hence, $\partial v_x/\partial x$ is negative. Then from Eq. (3.6-25), $\partial v_y/\partial y$ is positive and there is a component of velocity away from the plate.

3. Continuity equation in cylindrical and spherical coordinates. It is often convenient to use cylindrical coordinates to solve the equation of continuity if fluid is flowing in a cylinder. The coordinate system as related to rectangular coordinates is shown in Fig. 3.6-3a. The relations between rectangular x, y, z and cylindrical r, θ, z coordinates are

$$x = r \cos \theta \qquad y = r \sin \theta \qquad z = z$$

$$\qquad (3.6\text{-}26)$$

$$r = +\sqrt{x^2 + y^2} \qquad \theta = \tan^{-1}\frac{y}{x}$$

Using the relations from Eq. (3.6-26) with Eq. (3.6-20), the equation of continuity in cylindrical coordinates is

$$\frac{\partial \rho}{\partial t} + \frac{1}{r}\frac{\partial(\rho r v_r)}{\partial r} + \frac{1}{r}\frac{\partial(\rho v_\theta)}{\partial \theta} + \frac{\partial(\rho v_z)}{\partial z} = 0 \qquad (3.6\text{-}27)$$

For spherical coordinates the variables r, θ, and ϕ are related to x, y, z by the following as shown in Fig. 3.6-3b.

$$x = r \sin \theta \cos \phi \qquad y = r \sin \theta \sin \phi \qquad z = r \cos \theta$$

$$\qquad (3.6\text{-}28)$$

$$r = +\sqrt{x^2 + y^2 + z^2} \qquad \theta = \tan^{-1}\frac{\sqrt{x^2 + y^2}}{z} \qquad \phi = \tan^{-1}\frac{y}{x}$$

The equation of continuity in spherical coordinates becomes

$$\frac{\partial \rho}{\partial t} + \frac{1}{r^2}\frac{\partial(\rho r^2 v_r)}{\partial r} + \frac{1}{r \sin \theta}\frac{\partial(\rho v_\theta \sin \theta)}{\partial \theta} + \frac{1}{r \sin \theta}\frac{\partial(\rho v_\phi)}{\partial \phi} = 0 \qquad (3.6\text{-}29)$$

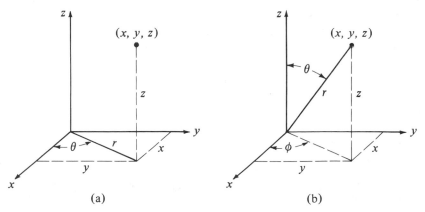

(a)   (b)

FIGURE 3.6-3. *Curvilinear coordinate systems : (a) cylindrical coordinates, (b) spherical coordinates.*

3.7   DIFFERENTIAL EQUATIONS OF MOMENTUM TRANSFER OR MOTION

3.7A   Derivation of Equations of Momentum Transfer

The *equation of motion* is really the equation for the conservation-of-momentum equation (2.8-3), which we can write as

$$\begin{pmatrix}\text{rate of}\\\text{momentum in}\end{pmatrix} - \begin{pmatrix}\text{rate of}\\\text{momentum out}\end{pmatrix}$$

$$+ \begin{pmatrix}\text{sum of forces}\\\text{acting on system}\end{pmatrix} = \begin{pmatrix}\text{rate of momentum}\\\text{accumulation}\end{pmatrix} \quad (3.7\text{-}1)$$

We will make a balance on an element as in Fig. 3.6-2. First we shall consider only the x component of each term in Eq. (3.6-30). The y and z components can be described in an analogous manner.

The rate at which the x component of momentum enters the face at x in the x direction by convection is $(\rho v_x v_x)_x \, \Delta y \, \Delta z$, and the rate at which it leaves at $x + \Delta x$ is $(\rho v_x v_x)_{x+\Delta x} \, \Delta y \, \Delta z$. The quantity (ρv_x) is the concentration in momentum/m³ or (kg · m/s)/m³, and it is multiplied by v_x to give the momentum flux as momentum/s · m².

The x component of momentum entering the face at y is $(\rho v_y v_x)_y \, \Delta x \, \Delta z$, and leaving at $y + \Delta y$ is $(\rho v_y v_x)_{y+\Delta y} \, \Delta x \, \Delta z$. For the face at z we have $(\rho v_z v_x)_z \, \Delta x \, \Delta y$ entering, and at $z + \Delta z$ we have $(\rho v_z v_x)_{z+\Delta z} \, \Delta x \, \Delta y$ leaving. Hence, the net convective x momentum flow into the volume element $\Delta x \, \Delta y \, \Delta z$ is

$$[(\rho v_x v_x)_x - (\rho v_x v_x)_{x+\Delta x}]\Delta y \, \Delta z + [(\rho v_y v_x)_y - (\rho v_y v_x)_{y+\Delta y}]\Delta x \, \Delta z$$

$$+ [(\rho v_z v_x)_z - (\rho v_z v_x)_{z+\Delta z}]\Delta x \, \Delta y \quad (3.7\text{-}2)$$

Momentum flows in and out of the volume element by the mechanisms of convection or bulk flow as given in Eq. (3.7-2) and also by molecular transfer (by virtue of the velocity gradients in laminar flow). The rate at which the x component of momentum enters the face at x by molecular transfer is $(\tau_{xx})_x \, \Delta y \, \Delta z$, and the rate at which it leaves the surface at $x + \Delta x$ is $(\tau_{xx})_{x+\Delta x} \, \Delta y \, \Delta z$. The rate at which it enters the face at y is $(\tau_{yx})_y \, \Delta x \, \Delta z$, and it leaves at $y + \Delta y$ at a rate of $(\tau_{yx})_{y+\Delta y} \, \Delta x \, \Delta z$. Note that τ_{yx} is the flux of x momentum through the face perpendicular to the y axis. Writing a similar equation for the remaining faces the net x component of momentum by molecular transfer is

$$[(\tau_{xx})_x - (\tau_{xx})_{x+\Delta x}]\Delta y \, \Delta z + [(\tau_{yx})_y - (\tau_{yx})_{y+\Delta y}]\Delta x \, \Delta z + [(\tau_{zx})_z - (\tau_{zx})_{z+\Delta z}]\Delta x \, \Delta y$$

$$(3.7\text{-}3)$$

These molecular fluxes of momentum may be considered as shear stresses and normal stresses. Hence, τ_{yx} is the x direction shear stress on the y face and τ_{zx} the shear stress on the z face. Also, τ_{xx} is the normal stress on the x face.

The net fluid pressure force acting on the element in the x direction is the difference between the force acting at x and $x + \Delta x$.

$$(p_x - p_{x+\Delta x})\Delta y \, \Delta z \quad (3.7\text{-}4)$$

The gravitational force g_x acting on a unit mass in the x direction is multiplied by the mass of the element to give

$$\rho g_x \, \Delta x \, \Delta y \, \Delta z \quad (3.7\text{-}5)$$

where g_x is the x component of the gravitational vector **g**.

The rate of accumulation of x momentum in the element is

$$\Delta x \, \Delta y \, \Delta z \, \frac{\partial(\rho v_x)}{\partial t} \qquad (3.7\text{-}6)$$

Substituting Eqs. (3.7-2)–(3.7-6) into (3.7-1), dividing by $\Delta x \, \Delta y \, \Delta z$, and taking the limit as Δx, Δy, and Δz approach zero, we obtain the x component of the differential equation of motion.

$$\frac{\partial(\rho v_x)}{\partial t} = \left[\frac{\partial(\rho v_x v_x)}{\partial x} + \frac{\partial(\rho v_y v_x)}{\partial y} + \frac{\partial(\rho v_z v_x)}{\partial z} \right]$$

$$- \left(\frac{\partial \tau_{xx}}{\partial x} + \frac{\partial \tau_{yx}}{\partial y} + \frac{\partial \tau_{zx}}{\partial z} \right) - \frac{\partial p}{\partial x} + \rho g_x \qquad (3.7\text{-}7)$$

The y and z components of the differential equation of motion are, respectively,

$$\frac{\partial(\rho v_y)}{\partial t} = -\left[\frac{\partial(\rho v_x v_y)}{\partial x} + \frac{\partial(\rho v_y v_y)}{\partial y} + \frac{\partial(\rho v_z v_y)}{\partial z} \right]$$

$$- \left(\frac{\partial \tau_{xy}}{\partial x} + \frac{\partial \tau_{yy}}{\partial y} + \frac{\partial \tau_{zy}}{\partial z} \right) - \frac{\partial p}{\partial y} + \rho g_y \qquad (3.7\text{-}8)$$

$$\frac{\partial(\rho v_z)}{\partial t} = -\left[\frac{\partial(\rho v_x v_z)}{\partial x} + \frac{\partial(\rho v_y v_z)}{\partial y} + \frac{\partial(\rho v_z v_z)}{\partial z} \right]$$

$$- \left(\frac{\partial \tau_{xz}}{\partial x} + \frac{\partial \tau_{yz}}{\partial y} + \frac{\partial \tau_{zz}}{\partial z} \right) - \frac{\partial p}{\partial y} + \rho g_z \qquad (3.7\text{-}9)$$

We can use Eq. (3.6-20), which is the continuity equation, and Eq. (3.7-7) and obtain an equation of motion for the x component and also do the same for the y and z components as follows:

$$\rho \left(\frac{\partial v_x}{\partial t} + v_x \frac{\partial v_x}{\partial x} + v_y \frac{\partial v_x}{\partial y} + v_z \frac{\partial v_x}{\partial z} \right) = -\left[\frac{\partial \tau_{xx}}{\partial x} + \frac{\partial \tau_{yx}}{\partial y} + \frac{\partial \tau_{zx}}{\partial z} \right] + \rho g_x - \frac{\partial p}{\partial x}$$

$$(3.7\text{-}10)$$

$$\rho \left(\frac{\partial v_y}{\partial t} + v_x \frac{\partial v_y}{\partial x} + v_y \frac{\partial v_y}{\partial y} + v_z \frac{\partial v_y}{\partial z} \right) = -\left[\frac{\partial \tau_{xy}}{\partial x} + \frac{\partial \tau_{yy}}{\partial y} + \frac{\partial \tau_{zy}}{\partial z} \right] + \rho g_y - \frac{\partial p}{\partial x}$$

$$(3.7\text{-}11)$$

$$\rho \left(\frac{\partial v_z}{\partial t} + v_x \frac{\partial v_z}{\partial x} + v_y \frac{\partial v_z}{\partial y} + v_z \frac{\partial v_z}{\partial z} \right) = -\left[\frac{\partial \tau_{xz}}{\partial x} + \frac{\partial \tau_{yz}}{\partial y} + \frac{\partial \tau_{zz}}{\partial z} \right] + \rho g_z - \frac{\partial p}{\partial x}$$

$$(3.7\text{-}12)$$

Adding vectorially, we obtain an equation of motion for a pure fluid.

$$\rho \frac{D\mathbf{v}}{Dt} = -(\nabla \cdot \boldsymbol{\tau}) - \nabla p + \rho \mathbf{g} \qquad (3.7\text{-}13)$$

We should note that Eqs. (3.7-7) to (3.7-13) are valid for any continuous medium.

3.7B Equations of Motion for Newtonian Fluids with Varying Density and Viscosity

In order to use Eqs. (3.7-7) to (3.7-13) to determine velocity distributions, expressions must be used for the various stresses in terms of velocity gradients and fluid properties. For Newtonian fluids the expressions for the stresses τ_{xx}, τ_{yx}, τ_{zx}, and so on, have been related to the velocity gradients and the fluid viscosity μ (B1, B2, D1) and are as follows.

1. Shear-stress components for Newtonian fluids in rectangular coordinates

$$\tau_{xx} = -2\mu \frac{\partial v_x}{\partial x} + \frac{2}{3} \mu (\nabla \cdot \mathbf{v}) \tag{3.7-14}$$

$$\tau_{yy} = -2\mu \frac{\partial v_y}{\partial y} + \frac{2}{3} \mu (\nabla \cdot \mathbf{v}) \tag{3.7-15}$$

$$\tau_{zz} = -2\mu \frac{\partial v_z}{\partial z} + \frac{2}{3} \mu (\nabla \cdot \mathbf{v}) \tag{3.7-16}$$

$$\tau_{xy} = \tau_{yx} = -\mu \left(\frac{\partial v_x}{\partial y} + \frac{\partial v_y}{\partial x} \right) \tag{3.7-17}$$

$$\tau_{yz} = \tau_{zy} = -\mu \left(\frac{\partial v_y}{\partial z} + \frac{\partial v_z}{\partial y} \right) \tag{3.7-18}$$

$$\tau_{zx} = \tau_{xz} = -\mu \left(\frac{\partial v_z}{\partial x} + \frac{\partial v_x}{\partial z} \right) \tag{3.7-19}$$

$$(\nabla \cdot \mathbf{v}) = \left(\frac{\partial v_x}{\partial x} + \frac{\partial v_y}{\partial y} + \frac{\partial v_z}{\partial z} \right) \tag{3.7-20}$$

2. Shear-stress components for Newtonian fluids in cylindrical coordinates

$$\tau_{rr} = -\mu \left[2 \frac{\partial v_r}{\partial r} - \frac{2}{3} (\nabla \cdot \mathbf{v}) \right] \tag{3.7-21}$$

$$\tau_{\theta\theta} = -\mu \left[2 \left(\frac{1}{r} \frac{\partial v_\theta}{\partial \theta} + \frac{v_r}{r} \right) - \frac{2}{3} (\nabla \cdot \mathbf{v}) \right] \tag{3.7-22}$$

$$\tau_{zz} = -\mu \left[2 \left(\frac{\partial v_z}{\partial z} \right) - \frac{2}{3} (\nabla \cdot \mathbf{v}) \right] \tag{3.7-23}$$

$$\tau_{r\theta} = \tau_{\theta r} = -\mu \left[r \frac{\partial (v_\theta / r)}{\partial r} + \frac{1}{r} \frac{\partial v_r}{\partial \theta} \right] \tag{3.7-24}$$

$$\tau_{\theta z} = \tau_{z\theta} = -\mu \left[\frac{\partial v_\theta}{\partial z} + \frac{1}{r} \frac{\partial v_z}{\partial \theta} \right] \tag{3.7-25}$$

$$\tau_{zr} = \tau_{rz} = -\mu \left[\frac{\partial v_\theta}{\partial r} + \frac{\partial v_r}{\partial z} \right] \tag{3.7-26}$$

$$(\nabla \cdot v) = \frac{1}{r} \frac{\partial (r v_r)}{\partial r} + \frac{1}{r} \frac{\partial v_\theta}{\partial \theta} + \frac{\partial v_z}{\partial z} \tag{3.7-27}$$

3. Shear-stress components for Newtonian fluids in spherical coordinates

$$\tau_{rr} = -\mu \left[2 \frac{\partial v_r}{\partial r} - \frac{2}{3} (\nabla \cdot v) \right] \tag{3.7-28}$$

$$\tau_{\theta\theta} = -\mu \left[2 \left(\frac{1}{r} \frac{\partial v_\theta}{\partial \theta} + \frac{v_r}{r} \right) - \frac{2}{3} (\nabla \cdot v) \right] \tag{3.7-29}$$

$$\tau_{\phi\phi} = -\mu \left[2 \left(\frac{1}{r \sin \theta} \frac{\partial v_\phi}{\partial \phi} + \frac{v_r}{r} + \frac{v_\theta \cot \theta}{r} \right) - \frac{2}{3} (\nabla \cdot v) \right] \tag{3.7-30}$$

$$\tau_{r\theta} = \tau_{\theta r} = -\mu \left[r \frac{\partial (v_\theta/r)}{\partial r} + \frac{1}{r} \frac{\partial v_r}{\partial \theta} \right] \tag{3.7-31}$$

$$\tau_{\theta\phi} = \tau_{\phi\theta} = -\mu \left[\frac{\sin \theta}{r} \frac{\partial (v_\phi/\sin \theta)}{\partial \theta} + \frac{1}{r \sin \theta} \frac{\partial v_\theta}{\partial \phi} \right] \tag{3.7-32}$$

$$\tau_{\phi r} = \tau_{r\phi} = -\mu \left[\frac{1}{r \sin \theta} \frac{\partial v_r}{\partial \phi} + r \frac{\partial (v_\phi/r)}{\partial r} \right] \tag{3.7-33}$$

$$(\nabla \cdot v) = \frac{1}{r^2} \frac{\partial (r^2 v_r)}{\partial r} + \left[\frac{1}{r \sin \theta} \frac{\partial (v_\theta \sin \theta)}{\partial \theta} + \frac{1}{r \sin \theta} \frac{\partial v_\phi}{\partial \phi} \right] \tag{3.7-34}$$

4. Equation of Motion for Newtonian fluids with varying density and viscosity After Eqs. (3.7-14)–(3.7-20) for shear-stress components are substituted into Eq. (3.7-10) for the *x*-component of momentum, we obtain the general equation of motion for a Newtonian fluid with varying density and viscosity.

$$\rho \frac{Dv_x}{Dt} = \frac{\partial}{\partial x} \left[2\mu \frac{\partial v_x}{\partial x} - \frac{2}{3} \mu (\nabla \cdot v) \right] + \frac{\partial}{\partial y} \left[\mu \left(\frac{\partial v_x}{\partial x} + \frac{\partial v_y}{\partial x} \right) \right]$$

$$+ \frac{\partial}{\partial z} \left[\mu \left(\frac{\partial v_z}{\partial x} + \frac{\partial v_x}{\partial z} \right) \right] - \frac{\partial p}{\partial x} + \rho g_x \tag{3.7-35}$$

Similar equations are obtained for the *y* and *z* components of momentum.

3.7C   Equations of Motion for Newtonian Fluids with Constant Density and Viscosity

The equations above are seldom used in their complete forms. When the density ρ and the viscosity μ are constant where $(\nabla \cdot v) = 0$, the equations are simplified and we obtain the equations of motion for Newtonian fluids. These equations are also called the *Navier–Stokes equations*.

1. Equation of motion in rectangular coordinates. For Newtonian fluids for constant ρ and μ for the x component, y component, and z component we obtain, respectively,

$$\rho\left(\frac{\partial v_x}{\partial t} + v_x\frac{\partial v_x}{\partial x} + v_y\frac{\partial v_x}{\partial y} + v_z\frac{\partial v_x}{\partial z}\right) = \mu\left(\frac{\partial^2 v_x}{\partial x^2} + \frac{\partial^2 v_x}{\partial y^2} + \frac{\partial^2 v_x}{\partial z^2}\right) - \frac{\partial p}{\partial x} + \rho g_x$$

(3.7-36)

$$\rho\left(\frac{\partial v_y}{\partial t} + v_x\frac{\partial v_y}{\partial x} + v_y\frac{\partial v_y}{\partial y} + v_z\frac{\partial v_y}{\partial z}\right) = \mu\left(\frac{\partial^2 v_y}{\partial x^2} + \frac{\partial^2 v_y}{\partial y^2} + \frac{\partial^2 v_y}{\partial z^2}\right) - \frac{\partial p}{\partial y} + \rho g_y$$

(3.7-37)

$$\rho\left(\frac{\partial v_z}{\partial t} + v_x\frac{\partial v_z}{\partial x} + v_y\frac{\partial v_z}{\partial y} + v_z\frac{\partial v_z}{\partial z}\right) = \mu\left(\frac{\partial^2 v_z}{\partial x^2} + \frac{\partial^2 v_z}{\partial y^2} + \frac{\partial^2 v_z}{\partial z^2}\right) - \frac{\partial p}{\partial z} + \rho g_z$$

(3.7-38)

Combining the three equations for the three components, we obtain

$$\rho\,\frac{D\mathbf{v}}{Dt} = -\nabla p + \rho\mathbf{g} + \mu\nabla^2\mathbf{v} \tag{3.7-39}$$

2. Equation of motion in cylindrical coordinates. These equations are as follows for Newtonian fluids for constant ρ and μ for the r, θ, and z components, respectively.

$$\rho\left(\frac{\partial v_r}{\partial t} + v_r\frac{\partial v_r}{\partial r} + \frac{v_\theta}{r}\frac{\partial v_r}{\partial \theta} - \frac{v_\theta^2}{r} + v_z\frac{\partial v_r}{\partial z}\right) = -\frac{\partial p}{\partial r}$$

$$+ \mu\left[\frac{\partial}{\partial r}\left(\frac{1}{r}\frac{\partial(rv_r)}{\partial r}\right) + \frac{1}{r^2}\frac{\partial^2 v_r}{\partial \theta^2} - \frac{2}{r^2}\frac{\partial v_\theta}{\partial \theta} + \frac{\partial^2 v_r}{\partial z^2}\right] + \rho g_r \qquad (3.7\text{-}40)$$

$$\rho\left(\frac{\partial v_\theta}{\partial t} + v_r\frac{\partial v_\theta}{\partial r} + \frac{v_\theta}{r}\frac{\partial v_\theta}{\partial \theta} + \frac{v_r v_\theta}{r} + v_z\frac{\partial v_\theta}{\partial z}\right) = -\frac{1}{r}\frac{\partial p}{\partial \theta}$$

$$+ \mu\left[\frac{\partial}{\partial r}\left(\frac{1}{r}\frac{\partial(rv_\theta)}{\partial r}\right) + \frac{1}{r^2}\frac{\partial^2 v_\theta}{\partial \theta^2} + \frac{2}{r^2}\frac{\partial v_r}{\partial \theta} + \frac{\partial^2 v_\theta}{\partial z^2}\right] + \rho g_\theta \qquad (3.7\text{-}41)$$

$$\rho\left(\frac{\partial v_z}{\partial t} + v_r\frac{\partial v_z}{\partial r} + \frac{v_\theta}{r}\frac{\partial v_z}{\partial \theta} + v_z\frac{\partial v_z}{\partial z}\right) = -\frac{\partial p}{\partial z}$$

$$+ \mu\left[\frac{1}{r}\frac{\partial}{\partial r}\left(r\frac{\partial v_z}{\partial r}\right) + \frac{1}{r^2}\frac{\partial^2 v_z}{\partial \theta^2} + \frac{\partial^2 v_z}{\partial z^2}\right] + \rho g_z \qquad (3.7\text{-}42)$$

3. Equation of motion in spherical coordinates. The equations for Newtonian fluids are given below for constant ρ and μ for the r, θ, and ϕ components, respectively.

$$\rho\left(\frac{\partial v_r}{\partial t} + v_r\frac{\partial v_r}{\partial r} + \frac{v_\theta}{r}\frac{\partial v_r}{\partial \theta} + \frac{v_\phi}{r\sin\theta}\frac{\partial v_r}{\partial \phi} - \frac{v_\theta^2 + v_\phi^2}{r}\right) = -\frac{\partial p}{\partial r}$$

$$+ \mu\left(\nabla^2 v_r - \frac{2}{r^2}v_r - \frac{2}{r^2}\frac{\partial v_\theta}{\partial \theta} - \frac{2}{r^2}v_\theta\cot\theta - \frac{2}{r^2\sin\theta}\frac{\partial v_\phi}{\partial \phi}\right) + \rho g_r \quad (3.7\text{-}43)$$

$$\rho\left(\frac{\partial v_\theta}{\partial t} + v_r \frac{\partial v_\theta}{\partial r} + \frac{v_\theta}{r}\frac{\partial v_\theta}{\partial \theta} + \frac{v_\phi}{r \sin \theta}\frac{\partial v_\theta}{\partial \phi} + \frac{v_r v_\theta}{r} - \frac{v_\phi^2 \cot \theta}{r}\right) = -\frac{1}{r}\frac{\partial p}{\partial \theta}$$

$$+ \mu\left(\nabla^2 v_\theta + \frac{2}{r^2}\frac{\partial v_r}{\partial \theta} - \frac{v_\theta}{r^2 \sin^2 \theta} - \frac{2}{r^2}\frac{\cos \theta}{\sin^2 \theta}\frac{\partial v_\phi}{\partial \phi}\right) + \rho g_\theta \quad (3.7\text{-}44)$$

$$\rho\left(\frac{\partial v_\phi}{\partial t} + v_r \frac{\partial v_\phi}{\partial r} + \frac{v_\theta}{r}\frac{\partial v_\phi}{\partial \theta} + \frac{v_\phi}{r \sin \theta}\frac{\partial v_\phi}{\partial \phi} + \frac{v_\phi v_r}{r} + \frac{v_\theta v_\phi}{r} \cot \theta\right) = -\frac{1}{r \sin \theta}\frac{\partial p}{\partial \phi}$$

$$+ \mu\left(\nabla^2 v_\phi - \frac{v_\phi}{r^2 \sin^2 \theta} + \frac{2}{r^2 \sin \theta}\frac{\partial v_r}{\partial \phi} + \frac{2}{r^2}\frac{\cos \theta}{\sin^2 \theta}\frac{\partial v_\theta}{\partial \phi}\right) + \rho g_\phi \quad (3.7\text{-}45)$$

where in the three equations above,

$$\nabla^2 = \frac{1}{r^2}\frac{\partial}{\partial r}\left(r^2 \frac{\partial}{\partial r}\right) + \frac{1}{r^2 \sin \theta}\frac{\partial}{\partial \theta}\left(\sin \theta \frac{\partial}{\partial \theta}\right) + \frac{1}{r^2 \sin^2 \theta}\left(\frac{\partial^2}{\partial \phi^2}\right) \quad (3.7\text{-}46)$$

Significant advantages and uses arise in the transformation from rectangular coordinates to cylindrical coordinates. For example, in Eq. (3.7-40) the term $\rho v_\theta^2/r$ is the centrifugal force. This gives the force in the r direction (radial) resulting from the motion of the fluid in the θ direction. Note that this term is obtained automatically from the transformation from rectangular to cylindrical coordinates. It does not have to be added to the equation on physical grounds.

The Coriolis force $\rho v_r v_\theta/r$ also arises automatically in the transformation of coordinates in Eq. (3.7-41). It is the effective force in the θ direction when there is flow in both the r and the θ directions, such as in the case of flow near a rotating disk.

3.8 USE OF DIFFERENTIAL EQUATIONS OF CONTINUITY AND MOTION

3.8A Introduction

The purpose and uses of the differential equations of motion and continuity, as mentioned previously, are to apply these equations to any viscous-flow problem. For a given specific problem, the terms that are zero or near zero are simply discarded and the remaining equations used in the solution to solve for the velocity, density, and pressure distributions. Of course, it is necessary to know the initial conditions and the boundary conditions to solve the equations. Several examples will be given to illustrate the general methods used.

We will consider cases for flow in specific geometries that can easily be described mathematically, such as for flow between parallel plates and in cylinders.

3.8B Differential Equations of Continuity and Motion for Flow between Parallel Plates

Two examples will be considered, one for horizontal plates and one for vertical plates.

EXAMPLE 3.8-1. Laminar Flow Between Horizontal Parallel Plates
Derive the equation giving the velocity distribution at steady state for laminar flow of a constant-density fluid with constant viscosity which is flowing between two flat and parallel plates. The velocity profile desired is

at a point far from the inlet or outlet of the channel. The two plates will be considered to be fixed and of infinite width, with the flow driven by the pressure gradient in the x direction.

Solution: Assuming that the channel is horizontal, Fig. 3.8-1 shows the axes selected with flow in the x direction and the width in the z direction. The velocities v_y and v_z are then zero. The plates are a distance $2y_0$ apart.

The continuity equation (3.6-24) for constant density is

$$\frac{\partial v_x}{\partial x} + \frac{\partial v_y}{\partial y} + \frac{\partial v_z}{\partial z} = 0 \qquad (3.6\text{-}24)$$

Since v_y and v_z are zero, Eq. (3.6-24) becomes

$$\frac{\partial v_x}{\partial x} = 0 \qquad (3.8\text{-}1)$$

The Navier–Stokes equation for the x component is

$$\rho\left(\frac{\partial v_x}{\partial t} + v_x \frac{\partial v_x}{\partial x} + v_y \frac{\partial v_x}{\partial y} + v_z \frac{\partial v_x}{\partial z}\right) = \mu\left(\frac{\partial^2 v_x}{\partial x^2} + \frac{\partial^2 v_x}{\partial y^2} + \frac{\partial^2 v_x}{\partial z^2}\right) - \frac{\partial p}{\partial x} + \rho g_x$$

$$(3.7\text{-}36)$$

Also, $\partial v_x/\partial t = 0$ for steady state, $v_y = 0$, $v_z = 0$, $\partial v_x/\partial x = 0$, $\partial^2 v_x/\partial x^2 = 0$. We can see that $\partial v_x/\partial z = 0$, since there is no change of v_x with z. Then $\partial^2 v_x/\partial z^2 = 0$. Making these substitutions into Eq. (3.7-36), we obtain

$$\frac{\partial p}{\partial x} - \rho g_x = \mu \frac{\partial^2 v_x}{\partial y^2} \qquad (3.8\text{-}2)$$

In fluid-flow problems we will be concerned with gravitational force only in the vertical direction for g_x, which is g, the gravitational force, in m/s². We shall combine the static pressure p and the gravitational force and call them simply p, as follows (note that $g_x = 0$ for the present case of a horizontal pipe but is not zero for the general case of a nonhorizontal pipe):

$$p = p + \rho g h \qquad (3.8\text{-}3)$$

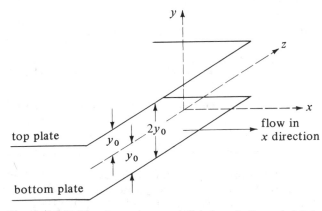

FIGURE 3.8-1. *Flow between two parallel plates in Example 3.8-1.*

where h is the distance upward from any chosen reference plane (h is in the direction opposed to gravity). Then Eq. (3.8-2) becomes

$$\frac{\partial p}{\partial x} = \mu \frac{\partial^2 v_x}{\partial y^2} \qquad (3.8\text{-}4)$$

We can see that p is not a function of z. Also, assuming that $2y_0$ is small, p is not a function of y. (Some references avoid this problem and simply use p as a dynamic pressure, which is rigorously correct since dynamic pressure gradients cause flow. In a fluid at rest the total pressure gradient is the hydrostatic pressure gradient and the dynamic pressure gradient is zero.) Also, $\partial p/\partial x$ is a constant in this problem since v_x is not a function of x. Then Eq. (3.8-4) becomes an ordinary differential equation.

$$\frac{d^2 v_x}{dy^2} = \frac{1}{\mu}\frac{dp}{dx} = \text{const} \qquad (3.8\text{-}5)$$

Integrating Eq. (3.8-5) once using the condition $dv_x/dy = 0$ at $y = 0$ for symmetry,

$$\frac{dv_x}{dy} = \left(\frac{1}{\mu}\frac{dp}{dx}\right) y \qquad (3.8\text{-}6)$$

Integrating again using $v_x = 0$ at $y = y_0$,

$$v_x = \frac{1}{2\mu}\frac{dp}{dx}(y^2 - y_0^2) \qquad (3.8\text{-}7)$$

The maximum velocity in Eq. (3.8-7) occurs when $y = 0$, giving

$$v_{x\,\text{max}} = \frac{1}{2\mu}\frac{dp}{dx}(-y_0^2) \qquad (3.8\text{-}8)$$

Combining Eqs. (3.8-7) and (3.8-8),

$$v_x = v_{x\,\text{max}}\left[1 - \left(\frac{y}{y_0}\right)^2\right] \qquad (3.8\text{-}9)$$

Hence, a parabolic velocity profile is obtained. This result was also obtained in Eq. (2.9-9) when using a shell momentum balance.

The results obtained in Example 3.8-1 could also have been obtained by making a force balance on a differential element of fluid and using the symmetry of the system to omit certain terms.

EXAMPLE 3.8-2. Laminar Flow Between Vertical Plates with One Plate Moving
A Newtonian fluid is confined between two parallel and vertical plates as shown in Fig. 3.8-2 (W6). The surface on the left is stationary and the other is moving vertically at a constant velocity v_0. Assuming that the flow is laminar, solve for the velocity profile.

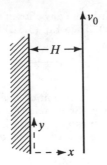

FIGURE 3.8-2. *Flow between vertical parallel plates in Example 3.8-2.*

Solution: The equation to use is the Navier-Stokes equation for the y coordinate, Eq. (3.7-37).

$$\rho\left(\frac{\partial v_y}{\partial t} + v_x \frac{\partial v_y}{\partial x} + v_y \frac{\partial v_y}{\partial y} + v_z \frac{\partial v_y}{\partial z}\right)$$

$$= \mu\left(\frac{\partial^2 v_y}{\partial x^2} + \frac{\partial^2 v_y}{\partial y^2} + \frac{\partial^2 v_y}{\partial z^2}\right) - \frac{\partial p}{\partial y} + \rho g_y \qquad (3.7\text{-}37)$$

At steady state, $\partial v_y/\partial t = 0$. The velocities v_x and $v_z = 0$. Also, $\partial v_y/\partial y = 0$ from the continuity equation, $\partial v_y/\partial z = 0$, and $\rho g_y = -\rho g$. The partial derivatives become derivatives and Eq. (3.7-37) becomes

$$\mu \frac{d^2 v_y}{d x^2} - \frac{dp}{dy} - \rho g = 0 \qquad (3.8\text{-}10)$$

This is similar to Eq. (3.8-2) in Example 3.8-1. The pressure gradient dp/dy is constant. Integrating Eq. (3.8-10) once yields

$$\frac{dv_y}{dx} - \frac{x}{\mu}\left(\frac{dp}{dy} + \rho g\right) = C_1 \qquad (3.8\text{-}11)$$

Integrating again gives

$$v_y - \frac{x^2}{2\mu}\left(\frac{dp}{dy} + \rho g\right) = C_1 x + C_2 \qquad (3.8\text{-}12)$$

The boundary conditions are at $x = 0$, $v_y = 0$ and at $x = H$, $v_y = v_0$. Solving for the constants, $C_1 = v_0/H - (H/2\mu)(dp/dy + \rho g)$ and $C_2 = 0$. Hence, Eq. (3.8-12) becomes

$$v_y = -\frac{1}{2\mu}\left(\frac{dp}{dy} + \rho g\right)(Hx - x^2) + v_0 \frac{x}{H} \qquad (3.8\text{-}13)$$

3.8C Differential Equations of Continuity and Motion for Flow in Stationary and Rotating Cylinders

Several examples will be given for flow in stationary and rotating cylinders.

EXAMPLE 3.8-3. *Laminar Flow in a Circular Tube*

Derive the equation for steady-state viscous flow in a horizontal tube of radius r_0, where the fluid is far from the tube inlet. The fluid is incompressible and μ is a constant. The flow is driven in one direction by a constant-pressure gradient.

Solution: The fluid will be assumed to flow in the z direction in the tube, as shown in Fig. 3.8-3. The y direction is vertical and the x direction horizontal. Since v_x and v_y are zero, the continuity equation becomes $\partial v_z/\partial z = 0$. For steady state $\partial v_z/\partial t = 0$. Then substituting into Eq. (3.7-38) for the z component, we obtain

$$\frac{dp}{dz} = \mu \left(\frac{\partial^2 v_z}{\partial x^2} + \frac{\partial^2 v_z}{\partial y^2} \right) \tag{3.8-14}$$

To solve Eq. (3.8-14) we can use cylindrical coordinates from Eq. (3.6-26), giving

$$z = z \qquad x = r \cos \theta \qquad y = r \sin \theta$$

$$r = +\sqrt{x^2 + y^2} \qquad \theta = \tan^{-1} \frac{y}{x} \tag{3.6-26}$$

Substituting these into Eq. (3.8-14),

$$\frac{1}{\mu} \frac{dp}{dz} = \frac{\partial^2 v_z}{\partial r^2} + \frac{1}{r} \frac{\partial v_z}{\partial r} + \frac{1}{r^2} \frac{\partial^2 v_z}{\partial \theta^2} \tag{3.8-15}$$

The flow is symmetrical about the z axis so $\partial^2 v_z/\partial \theta^2$ is zero in Eq. (3.8-15). As before, dp/dz is a constant, so Eq. (3.8-15) becomes

$$\frac{1}{\mu} \frac{dp}{dz} = \text{const} = \frac{d^2 v_z}{dr^2} + \frac{1}{r} \frac{dv_z}{dr} = \frac{1}{r} \frac{d}{dr} \left(r \frac{dv_z}{dr} \right) \tag{3.8-16}$$

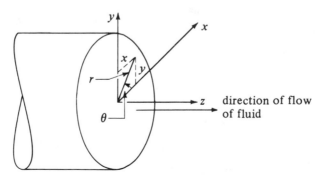

FIGURE 3.8-3. *Horizontal flow in a tube in Example 3.8-3.*

Alternatively, Eq. (3.7-42) in cylindrical coordinates can be used for the z component and the terms that are zero discarded.

$$\rho\left(\frac{\partial v_z}{\partial t} + v_r \frac{\partial v_z}{\partial r} + \frac{v_\theta}{r}\frac{\partial v_z}{\partial \theta} + v_z \frac{\partial v_z}{\partial z}\right) = -\frac{\partial p}{\partial z}$$

$$+ \mu\left[\frac{1}{r}\frac{\partial}{\partial r}\left(r\frac{\partial v_z}{\partial r}\right) + \frac{1}{r^2}\frac{\partial^2 v_z}{\partial \theta^2} + \frac{\partial^2 v_z}{\partial z^2}\right] + \rho g_z \quad \textbf{(3.7-42)}$$

As before, $\partial v_z/\partial t = 0$, $\partial^2 v_z/\partial \theta^2 = 0$, $v_r = 0$, $\partial v_z/\partial \theta = 0$, $\partial v_z/\partial z = 0$. Then Eq. (3.7-42) becomes identical to Eq. (3.8-16).

The boundary conditions for the first integration are $dv_z/dr = 0$ at $r = 0$. For the second integration, $v_z = 0$ at $r = r_0$ (tube radius). The result is

$$v_z = \frac{1}{4\mu}\frac{dp}{dz}(r^2 - r_0^2) \quad \textbf{(3.8-17)}$$

Converting to the maximum velocity as before,

$$v_z = v_{z\,max}\left[1 - \left(\frac{r}{r_0}\right)^2\right] \quad \textbf{(3.8-18)}$$

If Eq. (3.8-17) is integrated over the pipe cross section using Eq. (2.9-10) to give the average velocity $v_{z\,av}$,

$$v_{z\,av} = -\frac{r_0^2}{8\mu}\frac{dp}{dz} \quad \textbf{(3.8-19)}$$

Integrating to obtain the pressure drop from $z = 0$ for $p = p_1$ to $z = L$ for $p = p_2$, we obtain

$$p_1 - p_2 = \frac{8\mu v_{z\,av}L}{r_0^2} = \frac{32\mu v_{z\,av}L}{D^2} \quad \textbf{(3.8-20)}$$

where $D = 2r_0$. This is the Hagen-Poiseuille equation derived previously as Eq. (2.9-11).

EXAMPLE 3.8-4. *Laminar Flow in a Cylindrical Annulus*
Derive the equation for steady-state laminar flow inside the annulus between two concentric horizontal pipes. This type of flow occurs often in concentric pipe heat exchangers.

Solution: In this case Eq. (3.8-16) also still holds. However, the velocity in the annulus will reach a maximum at some radius $r = r_{max}$ which is between r_1 and r_2, as shown in Fig. 3.8-4. For the first integration of Eq. (3.8-16), the boundary conditions are $dv_z/dr = 0$ at $r = r_{max}$, which gives

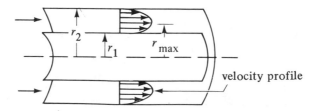

FIGURE 3.8-4. *Flow through a cylindrical annulus.*

$$\frac{r\,dv_z}{dr} = \left(\frac{1}{\mu}\frac{dp}{dz}\right)\left(\frac{r^2}{2} - \frac{r^2_{max}}{2}\right) \tag{3.8-21}$$

Also, for the second integration of Eq. (3.8-21), $v_z = 0$ at the inner wall where $r = r_1$, giving

$$v_z = \left(\frac{1}{2\mu}\frac{dp}{dz}\right)\left(\frac{r^2}{2} - \frac{r^2_1}{2} - r^2_{max}\ln\frac{r}{r_1}\right) \tag{3.8-22}$$

Repeating the second integration but for $v_z = 0$ at the outer wall where $r = r_2$, we obtain

$$v_z = \left(\frac{1}{2\mu}\frac{dp}{dz}\right)\left(\frac{r^2}{2} - \frac{r^2_2}{2} - r^2_{max}\ln\frac{r}{r_2}\right) \tag{3.8-23}$$

Combining Eqs. (3.8-22) and (3.8-23) and solving for r_{max}

$$r_{max} = \sqrt{\frac{1}{\ln{(r_2/r_1)}}(r^2_2 - r^2_1)/2} \tag{3.8-24}$$

In Fig. 3.8-4 the velocity profile predicted by Eq. (3.8-23) is plotted. For the case where $r_1 = 0$, r_{max} in Eq. (3.8-24) becomes zero and Eq. (3.8-23) reduces to Eq. (3.8-17) for a single circular pipe.

EXAMPLE 3.8-5. Velocity Distribution for Flow Between Coaxial Cylinders

Tangential laminar flow of a Newtonian fluid with constant density is occurring between two vertical coaxial cylinders in which the outer one is rotating (S4) with an angular velocity of ω as shown in Fig. 3.8-5. It can be assumed that end effects can be neglected. Determine the velocity and the shear stress distributions for this flow.

Solution: On physical grounds the fluid moves in a circular motion and the velocity v_r in the radial direction is zero and v_z in the axial direction is zero. Also, $\partial \rho/\partial t = 0$ at steady state. There is no pressure gradient in the θ direction. The equation of continuity in cylindrical coordinates as derived before is

$$\frac{\partial\rho}{\partial t} + \frac{1}{r}\frac{\partial(\rho r v_r)}{\partial r} + \frac{1}{r}\frac{\partial(\rho v_\theta)}{\partial\theta} + \frac{\partial(\rho v_z)}{\partial z} = 0 \tag{3.6-27}$$

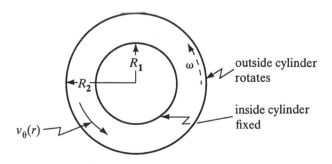

FIGURE 3.8-5. *Laminar flow in the region between two coaxial cylinders in Example 3.8-5.*

All terms in this equation are zero.

The equations of motion in cylindrical coordinates, Eqs. (3.7-40), (3.7-41), and (3.7-42) reduce to the following, respectively:

$$-\rho \frac{v_\theta^2}{r} = -\frac{\partial p}{\partial r} \qquad (r\text{-component}) \qquad (3.8\text{-}25)$$

$$0 = \frac{d}{dr}\left(\frac{1}{r}\frac{d(rv_\theta)}{dr}\right) \qquad (\theta\text{-component}) \qquad (3.8\text{-}26)$$

$$0 = -\frac{\partial p}{\partial z} + \rho g_z \qquad (z\text{-component}) \qquad (3.8\text{-}27)$$

Integrating Eq. (3.8-26),

$$v_\theta = C_1 r + \frac{C_2}{r} \qquad (3.8\text{-}28)$$

To solve for the integration constants C_1 and C_2, the following boundary conditions are used: at $r = R_1$, $v_\theta = 0$; at $r = R_2$, $v_\theta = \omega R_2$. The final equation is

$$v_\theta = \frac{\omega}{(R_1^2 - R_2^2)/(R_1 R_2^2)}\left[\frac{R_1}{r} - \frac{r}{R_1}\right] \qquad (3.8\text{-}29)$$

Using the shear-stress component for Newtonian fluids in cylindrical coordinates,

$$\tau_{r\theta} = -\mu\left[r\frac{\partial(v_\theta/r)}{\partial r} + \frac{1}{r}\frac{\partial v_r}{\partial \theta}\right] \qquad (3.7\text{-}31)$$

The last term in Eq. (3.7-31) is zero. Substituting Eq. (3.8-29) into (3.7-31) and differentiating gives

$$\tau_{r\theta} = -2\mu\omega R_2^2\left(\frac{1}{r^2}\right)\left[\frac{R_1^2/R_2^2}{1 - R_1^2/R_2^2}\right] \qquad (3.8\text{-}30)$$

The torque T that is necessary to rotate the outer cylinder is the product of the force times the lever arm.

$$T = (2\pi R_2 H)(-\tau_{r\theta})|_{r = R_2}(R_2)$$

$$= 4\pi\mu H\omega R_2^2\left[\frac{R_1^2/R_2^2}{1 - R_1^2/R_2^2}\right] \qquad (3.8\text{-}31)$$

where H is the length of the cylinder. This type of device has been used to measure fluid viscosities from observations of angular velocities and torque and also has been used as a model for some friction bearings.

EXAMPLE 3.8-6. *Rotating Liquid in a Cylindrical Container*
A Newtonian fluid of constant density is in a vertical cylinder of radius R (Fig. 3.8-6) with the cylinder rotating about its axis at angular velocity ω (B2). At steady state find the shape of the free surface.

Solution: The system can be described in cylindrical coordinates. As in Example 3.8-5, at steady state, $v_r = v_z = 0$ and $g_r = g_\theta = 0$. The final equations in cylindrical coordinates given below are the same as Eqs. (3.8-25) to (3.8-27) for Example 3.8-5 except that $g_z = -g$ in Eq. (3.8-27).

$$\rho \frac{v_\theta^2}{r} = \frac{\partial p}{\partial r} \tag{3.8-32}$$

$$0 = \mu \frac{\partial}{\partial r}\left(\frac{1}{r}\frac{\partial(rv_\theta)}{\partial r}\right) \tag{3.8-33}$$

$$\frac{\partial p}{\partial z} = -\rho g \tag{3.8-34}$$

Integration of Eq. (3.8-33) gives the same equation as in Example 3.8-5.

$$v_\theta = C_1 r + \frac{C_2}{r} \tag{3.8-28}$$

The constant C_2 must be zero since v_θ cannot be infinite at $r=0$. At $r=R$, the velocity $v_\theta = R\omega$. Hence, $C_1 = \omega$ and we obtain

$$v_\theta = \omega r \tag{3.8-35}$$

Combining Eqs. (3.8-35) and (3.8-32)

$$\frac{\partial p}{\partial r} = \rho \omega^2 r \tag{3.8-36}$$

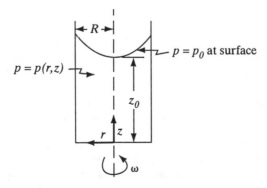

FIGURE 3.8-6. *Liquid being rotated in a container with a free surface in Example 3.8-6.*

Hence, we see that Eqs. (3.8-36) and (3.8-34) show that pressure depends upon r because of the centrifugal force and upon z because of the gravitational force.

$$\frac{\partial p}{\partial z} = -\rho g \qquad (3.8\text{-}34)$$

Since the term p is a function of position we can write the total differential of pressure as

$$dp = \frac{\partial p}{\partial r} \, dr + \frac{\partial p}{\partial z} \, dz \qquad (3.8\text{-}37)$$

Combining Eqs. (3.8-34) and (3.8-36) with (3.8-37) and integrating,

$$p = \frac{\rho \omega^2 r^2}{2} - \rho g z + C_3 \qquad (3.8\text{-}38)$$

The constant of integration can be determined since $p = p_o$ at $r = 0$ and $z = z_o$. The equation becomes

$$p - p_o = \frac{\rho \omega^2 r^2}{2} + \rho g(z_o - z) \qquad (3.8\text{-}39)$$

The free surface consists of all points on this surface at $p = p_o$. Hence,

$$z - z_o = \left(\frac{\omega^2}{2g}\right) r^2 \qquad (3.8\text{-}40)$$

This shows that the free surface is in the shape of a parabola.

3.9 OTHER METHODS FOR SOLUTION OF DIFFERENTIAL EQUATIONS OF MOTION

3.9A Introduction

In Section 3.8 we considered examples where the Navier–Stokes differential equations of motion could be solved analytically. These cases were used where there was only one nonvanishing component of the velocity. To solve these equations for flows in two and three directions is quite complex.

In this section we will consider some approximations that simplify the differential equations to allow us to obtain analytical solutions. Terms will be omitted which are quite small compared to the terms retained.

Three cases will be considered in this section: inviscid flow, potential flow, and creeping flow. The fourth case, for boundary-layer flow, will be considered in Section 3.10. The solution of these equations may be simplified by using a stream function $\psi(x, y)$ and/or a velocity potential $\phi(x, y)$ rather than the terms of the velocity components v_x, v_y, and v_z.

3.9B Stream Function

The stream function $\psi(x, y)$ is a convenient parameter by which we can represent two-dimensional, steady, incompressible flow. This stream function, ψ in m^2/s, is related to the velocity components v_x and v_y by

$$v_x = \frac{\partial \psi}{\partial y} \qquad v_y = -\frac{\partial \psi}{\partial x} \tag{3.9-1}$$

These definitions of v_x and v_y can then be used in the x and y components of the differential equation of motion, Eqs. (3.7-36) and (3.7-37), with $v_z = 0$ to obtain a differential equation for ψ that is equivalent to the Navier–Stokes equation. Details are given elsewhere (B2).

The stream function is very useful because its physical significance is that in steady flow lines defined by $\psi = constant$ are streamlines which are the actual curves traced out by the particles of fluid. A stream function exists for all two-dimensional, steady, incompressible flow whether viscous or inviscid and whether rotational or irrotational.

EXAMPLE 3.9-1. Stream Function and Streamlines

The stream function relationship is given as $\psi = xy$. Find the equations for the components of velocity. Also plot the streamlines for a constant $\psi = 4$ and $\psi = 1$.

Solution: Using Eq. (3.9-1),

$$v_x = \frac{\partial \psi}{\partial y} = \frac{\partial (xy)}{\partial y} = x$$

$$v_y = -\frac{\partial \psi}{\partial x} = -\frac{\partial (xy)}{\partial x} = -y$$

To determine the streamline for $\psi = constant = 1 = xy$, assume that $y = 0.5$ and solve for x.

$$\psi = 1 = xy = x(0.5)$$

Hence, $x = 2$. Repeating, for $y = 1$, $x = 1$; for $y = 2$, $x = 0.5$; for $y = 5$, $x = 0.2$, etc. Doing the same for $\psi = constant = 4$, the streamlines for $\psi = 1$ and $\psi = 4$ are plotted in Fig. 3.9-1. A possible flow model is flow around a corner.

3.9C Differential Equations of Motion for Ideal Fluids (Inviscid Flow)

Special equations for *ideal or inviscid fluids* can be obtained for a fluid having a constant density and zero viscosity. These are called the *Euler equations*. Equations (3.7-36)–(3.7-39) for the x, y, and z components of momentum become

$$\rho\left(\frac{\partial v_x}{\partial t} + v_x \frac{\partial v_x}{\partial x} + v_y \frac{\partial v_x}{\partial y} + v_z \frac{\partial v_x}{\partial z}\right) = -\frac{\partial p}{\partial x} + \rho g_x \tag{3.9-2}$$

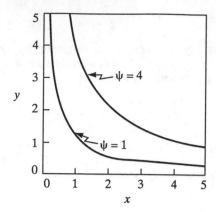

FIGURE 3.9-1. *Plot of streamlines for $\psi = xy$ for Example 3.9-1.*

$$\rho\left(\frac{\partial v_y}{\partial t} + v_x \frac{\partial v_y}{\partial x} + v_y \frac{\partial v_y}{\partial y} + v_z \frac{\partial v_y}{\partial z}\right) = -\frac{\partial p}{\partial y} + \rho g_y \qquad (3.9\text{-}3)$$

$$\rho\left(\frac{\partial v_z}{\partial t} + v_x \frac{\partial v_z}{\partial x} + v_y \frac{\partial v_z}{\partial y} + v_z \frac{\partial v_z}{\partial z}\right) = -\frac{\partial p}{\partial z} + \rho g_z \qquad (3.9\text{-}4)$$

At very high Reynolds numbers the viscous forces are quite small compared to the inertia forces and the viscosity can be assumed as zero. These equations are useful in calculating pressure distribution at the outer edge of the thin boundary layer in flow past immersed bodies. Away from the surface outside the boundary layer this assumption of an ideal fluid is often valid.

3.9D Potential Flow and Velocity Potential

The velocity potential or potential function $\phi(x,y)$ in m²/s is useful in inviscid flow problems and is defined as

$$v_x = \frac{\partial \phi}{\partial x} \qquad v_y = \frac{\partial \phi}{\partial y} \qquad v_z = \frac{\partial \phi}{\partial z} \qquad (3.9\text{-}5)$$

This potential exists only for a flow with zero angular velocity, or irrotationality. This type of flow of an ideal or inviscid fluid (ρ = constant, μ = 0) is called *potential flow*. Additionally, the velocity potential ϕ exists for three-dimensional flows, whereas the stream function does not.

The vorticity of a fluid is defined as follows:

$$\frac{\partial v_y}{\partial x} - \frac{\partial v_x}{\partial y} = 2\omega_z \qquad (3.9\text{-}6)$$

or,

$$\frac{\partial^2 \psi}{\partial x^2} + \frac{\partial^2 \psi}{\partial y^2} = -2\omega_z \qquad (3.9\text{-}7)$$

where $2\omega_z$ is the vorticity and ω_z in s^{-1} is angular velocity about the z axis. If $2\omega_z = 0$, the flow is irrotational and a potential function exists.

Using Eq. (3.6-24), the conservation of mass equation for flows in the x and the y direction is as follows for constant density:

$$\frac{\partial v_x}{\partial x} + \frac{\partial v_y}{\partial y} = 0 \tag{3.9-8}$$

Differentiating v_x in Eq. (3.9-5) with respect to x and v_y with respect to y and substituting into Eq. (3.9-8),

$$\frac{\partial^2 \phi}{\partial x^2} + \frac{\partial^2 \phi}{\partial y^2} = 0 \tag{3.9-9}$$

This is Laplace's equation in rectangular coordinates. If suitable boundary conditions exist or are known, Eq. (3.9-9) can be solved to give $\phi(x, y)$. Then the velocity at any point can be obtained using Eq. (3.9-5). Techniques for solving this equation include using numerical analysis, conformal mapping, and functions of a complex variable and are given elsewhere (B2, S3). Euler's equations can then be used to find the pressure distribution.

When the flow is inviscid and irrotational a similar type of Laplace equation is obtained from Eq. (3.9-7) for the stream function.

$$\frac{\partial^2 \psi}{\partial x^2} + \frac{\partial^2 \psi}{\partial y^2} = 0 \tag{3.9-10}$$

Lines of constant ϕ are called equal potential lines and for potential flow are everywhere perpendicular (orthogonal) to lines of constant ψ. This can be proved as follows. A line of constant ψ would be such that the change in ψ is zero.

$$d\psi = \frac{\partial \psi}{\partial x} dx + \frac{\partial \psi}{\partial y} dy = 0 \tag{3.9-11}$$

Then, substituting Eq. (3.9-1) into the above,

$$\left(\frac{dy}{dx}\right)_{\psi = \text{constant}} = \frac{v_y}{v_x} \tag{3.9-12}$$

Also, for lines of constant ϕ,

$$d\phi = \frac{\partial \phi}{\partial x} dx + \frac{\partial \phi}{\partial y} dy = 0 \tag{3.9-13}$$

$$\left(\frac{dy}{dx}\right)_{\phi = \text{constant}} = -\frac{v_y}{v_x} \tag{3.9-14}$$

Hence,

$$\left(\frac{dy}{dx}\right)_{\phi = \text{constant}} = -\frac{1}{(dy/dx)_{\psi = \text{constant}}} \tag{3.9-15}$$

An example of the use of the stream function is in obtaining the flow pattern for inviscid, irrotational flow past a cylinder of infinite length. The fluid approaching the cylinder has a steady and uniform velocity of v_∞ in the x direction. Laplace's Equation (3.9-10) can be converted to cylindrical coordinates to give

$$\frac{\partial^2 \psi}{\partial r^2} + \frac{1}{r}\frac{\partial \psi}{\partial r} + \frac{1}{r^2}\frac{\partial^2 \psi}{\partial \theta^2} = 0 \qquad (3.9\text{-}16)$$

where the velocity components are given by

$$v_r = \frac{1}{r}\frac{\partial \psi}{\partial \theta} \qquad v_\theta = -\frac{\partial \psi}{\partial r} \qquad (3.9\text{-}17)$$

Using four boundary conditions which are needed and the method of separation of variables, the stream function ψ is obtained. Converting to rectangular coordinates

$$\psi = v_\infty y\left(1 - \frac{R^2}{x^2 + y^2}\right) \qquad (3.9\text{-}18)$$

where R is the cylinder radius. The streamlines and the constant velocity potential lines are plotted in Fig. 3.9-2 as a flow net.

EXAMPLE 3.9-2. Stream Function for a Flow Field
The velocity components for a flow field are as follows:

$$v_x = a(x^2 - y^2) \qquad v_y = -2axy$$

Prove that it satisfies the conservation of mass and determine ψ.

Solution: First we determine $\partial v_x/\partial x = 2ax$ and $\partial v_y/\partial y = -2ax$. Substituting these values into Eq. (3.6-24), the conservation of mass for two-dimensional flow,

$$\frac{\partial v_x}{\partial x} + \frac{\partial v_y}{\partial y} = 0 \quad \text{or} \quad 2ax - 2ax = 0$$

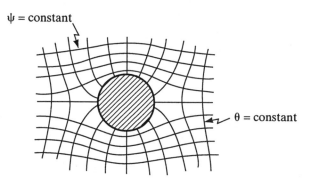

$\psi = \text{constant}$

$\theta = \text{constant}$

FIGURE 3.9-2. *Streamlines (ψ = constant) and constant velocity potential lines (ϕ = constant) for the steady and irrotational flow of an inviscid and incompressible fluid about an infinite circular cylinder.*

Then using Eq. (3.9-1),

$$v_x = \frac{\partial \psi}{\partial y} = ax^2 - ay^2 \qquad v_y = -\frac{\partial \psi}{\partial x} = -2axy \qquad \text{(3.9-19)}$$

Integrating Eq. (3.9-19) for v_x

$$\psi = ax^2y - \frac{ay^3}{3} + f(x) \qquad \text{(3.9-20)}$$

Differentiating Eq. (3.9-20) with respect to x and equating it to Eq. (3.9-19),

$$\frac{\partial \psi}{\partial x} = 2axy - 0 + f'(x) = +2axy \qquad \text{(3.9-21)}$$

Hence, $f'(x)=0$ and $f(x)=C$, a constant. Then Eq. (3.9-20) becomes

$$\psi = ax^2y - \frac{ay^3}{3} + C \qquad \text{(3.9-22)}$$

To plot the stream function, the constant C can be set equal to zero before plotting.

In potential flow, the stream function and the potential function are used to represent the flow in the main body of the fluid. These ideal fluid solutions do not satisfy the condition that $v_x = v_y = 0$ on the wall surface. Near the wall we have viscous drag and we use boundary-layer theory where we obtain approximate solutions for the velocity profiles in this thin boundary layer taking into account viscosity. This is discussed in Section 3.10. Then we splice this solution onto the ideal flow solution that describes flow outside the boundary layer.

3.9E Differential Equations of Motion for Creeping Flow

At very low Reynolds numbers below about 1, the term *creeping flow* is used to describe flow at very low velocities. This type of flow applies for the fall or settling of small particles through a fluid. Stokes' law is derived using this type of flow in problems of settling and sedimentation.

In flow around a sphere, for example, the fluid changes velocity and direction in a complex manner. If the inertia effects in this case were important, it would be necessary to keep all the terms in the three Navier–Stokes equations. Experiments show that at a Reynolds number below about 1, the inertia effects are small and can be omitted. Hence, the equations of motion, Eqs. (3.7-36)–(3.7-39) for creeping flow of an incompressible fluid, become

$$\frac{\partial p}{\partial x} = \mu \left(\frac{\partial^2 v_x}{\partial x^2} + \frac{\partial^2 v_x}{\partial y^2} + \frac{\partial^2 v_x}{\partial z^2} \right) \qquad \text{(3.9-23)}$$

$$\frac{\partial p}{\partial y} = \mu \left(\frac{\partial^2 v_y}{\partial x^2} + \frac{\partial^2 v_y}{\partial y^2} + \frac{\partial^2 v_y}{\partial z^2} \right) \qquad \text{(3.9-24)}$$

$$\frac{\partial p}{\partial z} = \mu \left(\frac{\partial^2 v_z}{\partial x^2} + \frac{\partial^2 v_z}{\partial y^2} + \frac{\partial^2 v_z}{\partial z^2} \right) \qquad \text{(3.9-25)}$$

For flow past a sphere the stream function ψ can be used in the Navier–Stokes equation in spherical coordinates to obtain the equation for the stream function and the velocity distribution and the pressure distribution over the sphere. Then by integration over the whole sphere, the form drag, caused by the pressure distribution, and the skin friction or viscous drag, caused by the shear stress at the surface, can be summed to give the total drag.

$$F_D = 3\pi\mu D_p v \quad \text{(SI)}$$

$$F_D = \frac{3\pi\mu D_p v}{g_c} \quad \text{(English)}$$

(3.9-26)

where F_D is total drag force in N, D_p is particle diameter in m, v is free stream velocity of fluid approaching the sphere in m/s, and μ is viscosity in kg/m · s. This is Stokes' equation for the drag force on a sphere.

Often Eq. (3.9-26) is rewritten as follows:

$$F_D = C_D \frac{v^2}{2} \rho A \quad \text{(SI)}$$

$$F_D = C_D \frac{v^2}{2g_c} \rho A \quad \text{(English)}$$

(3.9-27)

where C_D is a drag coefficient, which is equal to $24/N_{Re}$ for Stokes' law, and A is the projected area of the sphere, which is $\pi D_p^2/4$. This is discussed in more detail in Section 3.1 for flow past spheres.

3.10 BOUNDARY-LAYER FLOW AND TURBULENCE

3.10A Boundary-Layer Flow

In Sections 3.8 and 3.9 the Navier–Stokes equations were used to find relations that described laminar flow between flat plates and inside circular tubes, flow of ideal fluids, and creeping flow. In this section the flow of fluids around objects will be considered in more detail, with particular attention being given to the region close to the solid surface, called the *boundary layer*.

In the boundary-layer region near the solid, the fluid motion is greatly affected by this solid surface. In the bulk of the fluid away from the boundary layer the flow can often be adequately described by the theory of ideal fluids with zero viscosity. However, in the thin boundary layer, viscosity is important. Since the region is thin, simplified solutions can be obtained for the boundary-layer region. Prandtl originally suggested this division of the problem into two parts, which has been used extensively in fluid dynamics.

In order to help explain boundary layers, an example of boundary-layer formation in the steady-state flow of a fluid past a flat plate is given in Fig. 3.10-1. The velocity of the fluid upstream of the leading edge at $x = 0$ of the plate is uniform across the entire fluid stream and has the value v_∞. The velocity of the fluid at the interface is zero and the velocity v_x in the x direction increases as one goes farther from the plate. The velocity v_x approaches asymptotically the velocity v_∞ of the bulk of the stream.

The dashed line L is drawn so that the velocity at that point is 99% of the bulk velocity v_∞. The layer or zone between the plate and the dashed line constitutes the boundary layer. When the flow is laminar, the thickness δ of the boundary layer increases

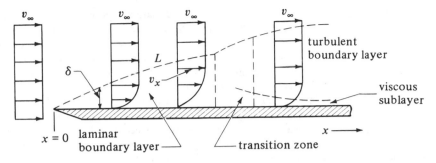

FIGURE 3.10-1. *Boundary layer for flow past a flat plate.*

with the $\sqrt{x}$ as we move in the x direction. The Reynolds number is defined as $N_{Re,x} = xv_\infty \rho/\mu$, where x is the distance downstream from the leading edge. When the Reynolds number is less than 2×10^5 the flow is laminar, as shown in Fig. 3.10-1.

The transition from laminar to turbulent flow on a smooth plate occurs in the Reynolds number range 2×10^5 to 3×10^6, as shown in Fig. 3.10-1. When the boundary layer is turbulent, a thin viscous sublayer persists next to the plate. The drag caused by the viscous shear in the boundary layers is called *skin friction* and it is the only drag present for flow past a flat plate.

The type of drag occurring when fluid flows by a bluff or blunt shape such as a sphere or cylinder, which is mostly caused by a pressure difference, is termed *form drag*. This drag predominates in flow past such objects at all except low values of the Reynolds numbers, and often a wake is present. Skin friction and form drag both occur in flow past a bluff shape, and the total drag is the sum of the skin friction and the form drag. (See also Section 3.1A).

3.10B Boundary-Layer Separation and Formation of Wakes

We discussed the growth of the boundary layer at the leading edge of a plate as shown in Fig. 3.10-2. However, some important phenomena also occur at the trailing edge of this plate and other objects. At the trailing edge or rear edge of the flat plate, the boundary layers are present at the top and bottom sides of the plate. On leaving the plate, the boundary layers gradually intermingle and disappear.

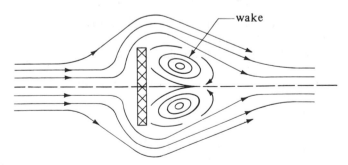

FIGURE 3.10-2. *Flow perpendicular to a flat plate and boundary-layer separation.*

If the direction of flow is at right angles to the plate as shown in Fig. 3.10-2, a boundary layer forms as before in the fluid that is flowing over the upstream face. Once at the edge of the plate, however, the momentum in the fluid prevents it from making the abrupt turn around the edge of the plate, and it separates from the plate. A zone of decelerated fluid is present behind the plate and large eddies (vortices), called the *wake*, are formed in this area. The eddies consume large amounts of mechanical energy. This separation of boundary layers occurs when the change in velocity of the fluid flowing by an object is too large in direction or magnitude for the fluid to adhere to the surface.

Since formation of a wake causes large losses in mechanical energy, it is often necessary to minimize or prevent boundary-layer separation by streamlining the objects or by other means. This is also discussed in Section 3.1A for flow past immersed objects.

3.10C   Laminar Flow and Boundary-Layer Theory

1. Boundary-layer equations. When laminar flow is occurring in a boundary layer, certain terms in the Navier-Stokes equations become negligible and can be neglected. The thickness of the boundary layer δ is arbitrarily taken as the distance away from the surface where the velocity reaches 99% of the free stream velocity. The concept of a relatively thin boundary layer leads to some important simplifications of the Navier-Stokes equations.

For two-dimensional laminar flow in the x and y directions of a fluid having a constant density, Eqs. (3.7-36) and (3.7-37) become as follows for flow at steady state as shown in Figure 3.10-1 when we neglect the body forces g_x and g_y.

$$v_x \frac{\partial v_x}{\partial x} + v_y \frac{\partial v_x}{\partial y} = -\frac{1}{\rho}\frac{\partial p}{\partial x} + \frac{\mu}{\rho}\left(\frac{\partial^2 v_x}{\partial x^2} + \frac{\partial^2 v_x}{\partial y^2}\right) \qquad (3.10\text{-}1)$$

$$v_x \frac{\partial v_y}{\partial x} + v_y \frac{\partial v_y}{\partial y} = -\frac{1}{\rho}\frac{\partial p}{\partial y} + \frac{\mu}{\rho}\left(\frac{\partial^2 v_y}{\partial x^2} + \frac{\partial^2 v_y}{\partial y^2}\right) \qquad (3.10\text{-}2)$$

The continuity equation for two-dimensional flow becomes

$$\frac{\partial v_x}{\partial x} + \frac{\partial v_y}{\partial y} = 0 \qquad (3.10\text{-}3)$$

In Eq. (3.10-1), the term $\mu/\rho(\partial^2 v_x/\partial x^2)$ is negligible in comparison with the other terms in the equation. Also, it can be shown that all the terms containing v_y and its derivatives are small. Hence, the final two boundary-layer equations to be solved are Eqs. (3.10-3) and (3.10-4).

$$v_x \frac{\partial v_x}{\partial x} + v_y \frac{\partial v_x}{\partial y} = -\frac{1}{\rho}\frac{dp}{dx} + \frac{\mu}{\rho}\frac{\partial^2 v_x}{\partial y^2} \qquad (3.10\text{-}4)$$

2. Solution for laminar boundary layer on a flat plate. An important case in which an analytical solution has been obtained for the boundary-layer equations is for the laminar boundary layer on a flat plate in steady flow, as shown in Fig. 3.10-1. A further simplification can be made in Eq. (3.10-4) in that dp/dx is zero since v_∞ is constant.

The final boundary-layer equations reduce to the equation of motion for the x direction and the continuity equation as follows:

$$v_x \frac{\partial v_x}{\partial x} + v_y \frac{\partial v_x}{\partial y} = \frac{\mu}{\rho}\frac{\partial^2 v_x}{\partial y^2} \qquad (3.10\text{-}5)$$

$$\frac{\partial v_x}{\partial x} + \frac{\partial v_y}{\partial y} = 0 \qquad (3.10\text{-}3)$$

The boundary conditions are $v_x = v_y = 0$ at $y = 0$ (y is distance from plate), and $v_x = v_\infty$ at $y = \infty$.

The solution of this problem for laminar flow over a flat plate giving v_x and v_y as a function of x and y was first obtained by Blasius and later elaborated by Howarth (B1, B2, S3). The mathematical details of the solution are quite tedious and complex and will not be given here. The general procedure will be outlined. Blasius reduced the two equations to a single ordinary differential equation which is nonlinear. The equation could not be solved to give a closed form but a series solution was obtained.

The results of the work by Blasius are given as follows. The boundary-layer thickness δ, where $v_x \cong 0.99 v_\infty$, is given approximately by

$$\delta = \frac{5.0x}{\sqrt{N_{Re, x}}} = 5.0 \sqrt{\frac{\mu x}{\rho v_\infty}} \qquad (3.10\text{-}6)$$

where $N_{Re, x} = x v_\infty \rho / \mu$. Hence, the thickness δ varies as $\sqrt{x}$.

The drag in flow past a flat plate consists only of skin friction and is calculated from the shear stress at the surface at $y = 0$ for any x as follows.

$$\tau_0 = \mu \left(\frac{\partial v_x}{\partial y}\right)_{y=0} \qquad (3.10\text{-}7)$$

From the relation of v_x as a function of x and y obtained from the series solution, Eq. (3.10-7) becomes

$$\tau_0 = 0.332 \mu v_\infty \sqrt{\frac{\rho v_\infty}{\mu x}} \qquad (3.10\text{-}8)$$

The total drag is given by the following for a plate of length L and width b:

$$F_D = b \int_0^L \tau_0 \, dx \qquad (3.10\text{-}9)$$

Substituting Eq. (3.10-8) into (3.10-9) and integrating,

$$F_D = 0.664 b \sqrt{\mu \rho v_\infty^3 L} \qquad (3.10\text{-}10)$$

The drag coefficient C_D related to the total drag on one side of the plate having an area $A = bL$ is defined as

$$F_D = C_D \frac{v_\infty^2}{2} \rho A \qquad (3.10\text{-}11)$$

Substituting the value for A and Eq. (3.10-10) into (3.10-11),

$$C_D = 1.328 \sqrt{\frac{\mu}{L v_\infty \rho}} = \frac{1.328}{N_{Re, L}^{1/2}} \qquad (3.10\text{-}12)$$

where $N_{Re, L} = L v_\infty \rho / \mu$. A form of Eq. (3.10-11) is used in Section 14.3 for particle movement through a fluid. The definition of C_D in Eq. (3.10-12) is similar to the Fanning friction factor f for pipes.

The equation derived for C_D applies only to the laminar boundary layer for $N_{Re, L}$ less than about 5×10^5. Also, the results are valid only for positions where x is sufficiently far from the leading edge so that x or L is much greater than δ. Experimental results on the drag coefficient to a flat plate confirm the validity of Eq. (3.10-12). Boundary-layer flow past many other shapes has been successfully analyzed using similar methods.

3.10D Nature and Intensity of Turbulence

1. Nature of turbulence. Since turbulent flow is important in many areas of engineering, the nature of turbulence has been extensively investigated. Measurements of the velocity fluctuations of the eddies in turbulent flow have helped explain turbulence.

For turbulent flow there are no exact solutions of flow problems as there are in laminar flow, since the approximate equations used depend on many assumptions. However, useful relations have been obtained by using a combination of experimental data and theory. Some of these relations will be discussed.

Turbulence can be generated by contact of two layers of fluid moving at different velocities or by a flowing stream in contact with a solid boundary, such as a wall or sphere. When a jet of fluid from an orifice flows into a mass of fluid, turbulence can arise. In turbulent flow at a given place and time large eddies are continually being formed which break down into smaller eddies and which finally disappear. Eddies are as small as about 0.1 or 1 mm or so and as large as the smallest dimension of the turbulent stream. Flow inside an eddy is laminar because of its large size.

In turbulent flow the velocity is fluctuating in all directions. In Fig. 3.10-3 a typical plot of the variation of the instantaneous velocity v_x in the x direction at a given point in turbulent flow is shown. The velocity v_x' is the deviation of the velocity from the mean velocity $\bar{v}_x$ in the x-direction of flow of the stream. Similar relations also hold for the y and z directions.

$$v_x = \bar{v}_x + v_x', \qquad v_y = \bar{v}_y + v_y', \; v_z = \bar{v}_z + v_z' \tag{3.10-13}$$

$$\bar{v}_x = \frac{1}{t} \int_0^t v_x \, dt \tag{3.10-14}$$

where the mean velocity $\bar{v}_x$ is the time-averaged velocity for time t, v_x the instantaneous total velocity in the x direction, and v_x' the instantaneous deviating or fluctuating velocity in the x direction. These fluctuations can also occur in the y and z directions. The value of v_x' fluctuates about zero as an average and, hence, the time-averaged values $\bar{v}_x' = 0$, $\bar{v}_y' = 0$, $\bar{v}_z' = 0$. However, the values of $v_x'^2$, $v_y'^2$, and $v_z'^2$ will not be zero. Similar expressions can also be written for pressure, which also fluctuates.

2. Intensity of turbulence. The time average of the fluctuating components vanishes over a time period of a few seconds. However, the time average of the mean square of the fluctuating components is a positive value. Since the fluctuations are random, the data have been analyzed by statistical methods. The level or intensity of turbulence can be

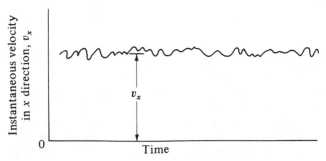

FIGURE 3.10-3. *Velocity fluctuations in turbulent flow.*

related to the square root of the sum of the mean squares of the fluctuating components. This intensity of turbulence is an important parameter in testing of models and theory of boundary layers.

The intensity of turbulence I can be defined mathematically as

$$I = \frac{\sqrt{\frac{1}{3}(\overline{v_x'^2} + \overline{v_y'^2} + \overline{v_z'^2})}}{\bar{v}_x} \tag{3.10-15}$$

This parameter I is quite important. Such factors as boundary-layer transition, separation, and heat- and mass-transfer coefficients depend upon the intensity of turbulence. Simulation of turbulent flows in testing of models requires that the Reynolds number and the intensity of turbulence be the same. One method used to measure intensity of turbulence is to utilize a hot-wire anemometer.

3.10E Turbulent Shear or Reynolds Stresses

In a fluid flowing in turbulent flow shear forces occur wherever there is a velocity gradient across a shear plane and these are much larger than those occurring in laminar flow. The velocity fluctuations in Eq. (3.10-13) give rise to turbulent shear stresses. The equations of motion and the continuity equation are still valid for turbulent flow. For an incompressible fluid having a constant density ρ and viscosity μ, the continuity equation (3.6-24) holds.

$$\frac{\partial v_x}{\partial x} + \frac{\partial v_y}{\partial y} + \frac{\partial v_z}{\partial z} = 0 \tag{3.6-24}$$

Also, the x component of the equation of motion, Eq. (3.7-36), can be written as follows if Eq. (3.6-24) holds:

$$\frac{\partial(\rho v_x)}{\partial t} + \frac{\partial(\rho v_x v_x)}{\partial_x} + \frac{\partial(\rho v_x v_y)}{\partial y} \frac{\partial(\rho v_x v_z)}{\partial z} = \mu\left(\frac{\partial^2 v_x}{\partial x^2} + \frac{\partial^2 v_x}{\partial y^2} + \frac{\partial^2 v_x}{\partial z^2}\right) - \frac{\partial p}{\partial x} + \rho g_x \tag{3.10-16}$$

We can rewrite the continuity equation (3.6-24) and Eq. (3.10-16) by replacing v_x by $\bar{v}_x + v_x'$, v_y by $\bar{v}_y + v_y'$, v_z by $\bar{v}_z + v_z'$, and p by $\bar{p} + p'$.

$$\frac{\partial(\bar{v}_x + v_x')}{\partial x} + \frac{\partial(\bar{v}_y + v_y')}{\partial y} + \frac{\partial(\bar{v}_z + v_z')}{\partial z} = 0 \tag{3.10-17}$$

$$\frac{\partial[\rho(\bar{v}_x + v_x')]}{\partial t} + \frac{\partial[\rho(\bar{v}_x + v_x')(\bar{v}_x + v_x')]}{\partial x} + \frac{\partial[\rho(\bar{v}_x + v_x')(\bar{v}_y + v_y')]}{\partial y}$$

$$+ \frac{\partial[\rho(\bar{v}_x + v_x')(\bar{v}_z + v_z')]}{\partial z} = \mu\nabla^2(\bar{v}_x + v_x') - \frac{\partial(\bar{p} + p')}{\partial x} + \rho g_x \tag{3.10-18}$$

Now we use the fact that the time-averaged value of the fluctuating velocities is zero ($\bar{v}_x'$, $\bar{v}_y'$, $\bar{v}_z'$ are zero), and that the time-averaged product $\overline{v_x' v_y'}$ is not zero. Then Eqs. (3.10-17) and (3.10-18) become

$$\frac{\partial\bar{v}_x}{\partial x} + \frac{\partial\bar{v}_y}{\partial y} + \frac{\partial\bar{v}_z}{\partial z} = 0 \tag{3.10-19}$$

$$\frac{\partial(\rho\bar{v}_x)}{\partial t} + \frac{\partial(\rho\bar{v}_x\bar{v}_x)}{\partial x} + \frac{\partial(\rho\bar{v}_x\bar{v}_y)}{\partial y} + \frac{\partial(\rho\bar{v}_x\bar{v}_z)}{\partial z}$$

$$+ \left[\frac{\partial(\rho\overline{v_x' v_x'})}{\partial x} + \frac{\partial(\rho\overline{v_x' v_y'})}{\partial y} + \frac{\partial(\rho\overline{v_x' v_z'})}{\partial z}\right] = \mu\nabla^2\bar{v}_x - \frac{\partial\bar{p}}{\partial x} + \rho g_x \tag{3.10-20}$$

By comparing these two time-smoothed equations with Eqs. (3.6-24) and (3.10-16) we see that the time-smoothed values everywhere replace the instantaneous values. However, in Eq. (3.10-20) new terms arise in the set of brackets which are related to turbulent velocity fluctuations. For convenience we use the notation

$$\bar{\tau}_{xx}^t = \rho\overline{v_x' v_x'}, \quad \bar{\tau}_{yx}^t = \rho\overline{v_x' v_y'}, \quad \bar{\tau}_{zx}^t = \rho\overline{v_x' v_z'} \tag{3.10-21}$$

These are the components of the turbulent momentum flux and are called *Reynolds stresses*.

3.10F   Prandtl Mixing Length

The equations derived for turbulent flow must be solved to obtain velocity profiles. To do this, more simplifications must be made before the expressions for the Reynolds stresses can be evaluated. A number of semiempirical equations have been used and the eddy diffusivity model of Boussinesq is one early attempt to evaluate these stresses. By analogy to the equation for shear stress in laminar flow, $\tau_{yx} = -\mu(dv_x/dy)$, the turbulent shear stress can be written as

$$\bar{\tau}_{yx}^t = -\eta_t \frac{d\bar{v}_x}{dy} \tag{3.10-22}$$

where η_t is a turbulent or eddy viscosity, which is a strong function of position and flow. This equation can also be written as follows:

$$\bar{\tau}_{yx}^t = -\rho\varepsilon_t \frac{d\bar{v}_x}{dy} \tag{3.10-23}$$

where $\varepsilon_t = \eta_t/\rho$ and ε_t is eddy diffusivity of momentum in m²/s by analogy to the momentum diffusivity μ/ρ for laminar flow.

Prandtl in his mixing-length model developed an expression to evaluate these stresses by assuming that eddies move in a fluid in a manner similar to the movement of molecules in a gas. The eddies move a distance called the mixing length L before they lose their identity.

Actually, the moving eddy or "lump" of fluid will gradually lose its identity. However, in the definition of the Prandtl mixing-length L, this small packet of fluid is assumed to retain its identity while traveling the entire length L and then to lose its identity or be absorbed in the host region.

Prandtl assumed that the velocity fluctuation v_x' is due to a "lump" of fluid moving a distance L in the y direction and retaining its mean velocity. At point L, the lump of fluid will differ in mean velocity from the adjacent fluid by $\bar{v}_{x|y+L} - \bar{v}_{x|y}$. Then, the value of $v_{x|y}'$ is

$$v_{x|y}' = \bar{v}_{x|y+L} - \bar{v}_{x|y} \tag{3.10-24}$$

The length L is small enough so that the velocity difference can be written as

$$v_{x|y}' = \bar{v}_{x|y+L} - \bar{v}_{x|y} = L\frac{d\bar{v}_x}{dy} \tag{3.10-25}$$

Hence,

$$v_x' = L\frac{d\bar{v}_x}{dy} \tag{3.10-26}$$

Prandtl also assumed $v'_x \simeq v'_y$. Then the time average, $\overline{v'_x x'_y}$, is

$$\overline{v'_x v'_y} = -L^2 \left| \frac{d\bar{v}_x}{dy} \right| \frac{d\bar{v}_x}{dy} \qquad \text{(3.10-27)}$$

The minus sign and the absolute value were used to make the quantity $\overline{v'_x v'_y}$ agree with experimental data. Substituting Eq. (3.10-27) into (3.10-21),

$$\bar{\tau}^t_{yx} = -\rho L^2 \left| \frac{d\bar{v}_x}{dy} \right| \frac{d\bar{v}_x}{dy} \qquad \text{(3.10-28)}$$

Comparing with Eq. (3.10-23),

$$\varepsilon_t = L^2 \left| \frac{d\bar{v}_x}{dy} \right| \qquad \text{(3.10-29)}$$

3.10G   Universal Velocity Distribution in Turbulent Flow

To determine the velocity distribution for turbulent flow at steady state inside a circular tube, we divide the fluid inside the pipe into two regions: a central core where the Reynolds stress approximately equals the shear stress; and a thin, viscous sublayer adjacent to the wall where the shear stress is due only to viscous shear and the turbulence effects are assumed negligible. Later we include a third region, the buffer zone, where both stresses are important.

Dropping the subscripts and superscripts on the shear stresses and velocity, and considering the thin, viscous sublayer, we can write

$$\tau_0 = -\mu \frac{dv}{dy} \qquad \text{(3.10-30)}$$

where τ_0 is assumed constant in this region. On integration,

$$\tau_0 y = \mu v \qquad \text{(3.10-31)}$$

Defining a friction velocity as follows and substituting into Eq. (3.10-31)

$$v^* = \sqrt{\frac{\tau_0}{\rho}} \qquad \text{(3.10-32)}$$

$$\frac{v}{v^*} = \frac{yv^*}{\mu/\rho} \qquad \text{(3.10-33)}$$

The dimensionless velocity ratio on the left can be written as

$$v^+ = v \sqrt{\frac{\rho}{\tau_0}} \qquad \text{(SI)}$$

$$v^+ = v \sqrt{\frac{\rho}{\tau_0 g_c}} \qquad \text{(English)} \qquad \text{(3.10-34)}$$

The dimensionless number on the right can be written as

$$y^+ = \frac{\sqrt{\tau_0 \rho}}{\mu} y \qquad \text{(SI)}$$

$$y^+ = \frac{\sqrt{\tau_0 g_c \rho}}{\mu} y \qquad \text{(English)} \qquad \text{(3.10-35)}$$

where y is the distance from the wall of the tube. For a tube of radius r_0, $y = r_0 - r$, where r is the distance from the center. Hence, for the viscous sublayer, the velocity distribution is

$$v^+ = y^+ \tag{3.10-36}$$

Next, considering the turbulent core where any viscous stresses are neglected, Eq. (3.10-28) becomes

$$\tau = \rho L^2 \left(\frac{dv}{dy}\right)^2 \tag{3.10-37}$$

where dv/dy is always positive and the absolute value sign is dropped. Prandtl assumed that the mixing length is proportional to the distance from the wall, or

$$L = Ky \tag{3.10-38}$$

and that $\tau = \tau_0 = $ constant. Equation (3.10-37) now becomes

$$\tau_0 = \rho K^2 y^2 \left(\frac{dv}{dy}\right)^2 \tag{3.10-39}$$

Hence,

$$v^* = Ky \frac{dv}{dy} \tag{3.10-40}$$

Upon integration,

$$v^* \ln y = Kv + K_1 \tag{3.10-41}$$

where K_1 is a constant. The constant K_1 can be found by assuming that v is zero at a small value of y, say y_0.

$$\frac{v}{v^*} = v^+ = \frac{1}{K} \ln \frac{y}{y_0} \tag{3.10-42}$$

Introducing the variable y^+ by multiplying the numerator and the denominator of the term y/y_0 by v^*/v, where $v = \mu/\rho$, we obtain

$$v^+ = \frac{1}{K} \left(\ln \frac{yv^*}{v} - \ln \frac{y_0 v^*}{v} \right) \tag{3.10-43}$$

$$v^+ = \frac{1}{K} \ln y^+ + C_1 \tag{3.10-44}$$

A large amount of velocity distribution data by Nikuradse and others for a range of Reynolds numbers of 4000 to 3.2×10^6 have been obtained and the data fit Eq. (3.10-36) in the region up to y^+ of 5 and also fit Eq. (3.10-44) above y^+ of 30 with K and C_1 being universal constants. For the region of y^+ from 5 to 30, which is defined as the buffer region, an empirical equation of the form of Eq. (3.10-44) fits the data. In Fig. 3.10-4 the following relations which are valid are plotted to give a universal velocity profile for fluids flowing in smooth circular tubes.

$$v^+ = y^+ \qquad (0 < y^+ < 5) \tag{3.10-45}$$

$$v^+ = 5.0 \ln y^+ - 3.05 \qquad (5 < y^+ < 30) \tag{3.10-46}$$

$$v^+ = 2.5 \ln y^+ + 5.5 \qquad (30 < y^+) \tag{3.10-47}$$

Chap. 3   Principles of Momentum Transfer and Applications

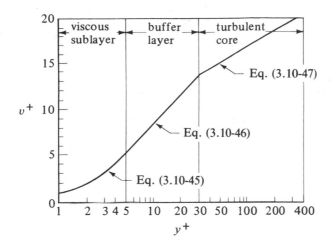

FIGURE 3.10-4. *Universal velocity profile for turbulent flow in smooth circular tubes.*

Three distinct regions are apparent in Fig. 3.10-4. The first region next to the wall is the *viscous sublayer* (historically called "laminar" sublayer), given by Eq. (3.10-45), where the velocity is proportional to the distance from the wall. The second region, called the *buffer layer*, is given by Eq. (3.10-46), which is a region of transition between the viscous sublayer with practically no eddy activity and the violent eddy activity in the *turbulent core region* given by Eq. (3.10-47). These equations can then be used and related to the Fanning friction factor discussed earlier in the chapter. They can also be used in solving turbulent boundary-layer problems.

3.10H Integral Momentum Balance for Boundary-Layer Analysis

1. Introduction and derivation of integral expression. In the solution for the laminar boundary layer on a flat plate, the Blasius solution is quite restrictive, since it is for laminar flow over a flat plate. Other more complex systems cannot be solved by this method. An approximate method developed by von Kármán can be used when the configuration is more complicated or the flow is turbulent. This is an approximate momentum integral analysis of the boundary layer using an empirical or assumed velocity distribution.

In order to derive the basic equation for a laminar or turbulent boundary layer, a small control volume in the boundary layer on a flat plate is used as shown in Fig. 3.10-5. The depth in the z direction is b. Flow is only through the surfaces A_1 and A_2 and also from the top curved surface at δ. An overall integral momentum balance using Eq. (2.8-8) and overall integral mass balance using Eq. (2.6-6) are applied to the control volume inside the boundary layer at steady state and the final integral expression by von Kármán is (B2, S3)

$$\frac{\tau_0}{\rho} = \frac{d}{dx} \int_0^{\delta} v_x(v_\infty - v_x) \, dy \tag{3.10-48}$$

where τ_0 is the shear stress at the surface $y = 0$ at point x along the plate. Also, δ and τ_0 are functions of x.

Equation (3.10-48) is an expression whose solution requires knowledge of the velocity

v_x as a function of the distance from the surface, y. The accuracy of the results will, of course, depend on how closely the assumed velocity profile approaches the actual profile.

2. Integral momentum balance for laminar boundary layer. Before we use Eq. (3.10-48) for the turbulent boundary layer, this equation will be applied to the laminar boundary layer over a flat plate so that the results can be compared with the exact Blasius solution in Eqs. (3.10-6)–(3.10-12).

In this analysis certain boundary conditions must be satisfied in the boundary layer.

$$v_x = 0 \qquad \text{at } y = 0$$
$$v_x \cong v_\infty \qquad \text{at } y = \delta \qquad\qquad \text{(3.10-49)}$$
$$\frac{dv_x}{dy} \cong 0 \qquad \text{at } y = \delta$$

The conditions above are fulfilled in the following simple, assumed velocity profile.

$$\frac{v_x}{v_\infty} = \frac{3}{2}\frac{y}{\delta} - \frac{1}{2}\left(\frac{y}{\delta}\right)^3 \qquad\qquad \text{(3.10-50)}$$

The shear stress τ_0 at a given x can be obtained from

$$\tau_0 = \mu\left(\frac{dv_x}{dy}\right)_{y=0} \qquad\qquad \text{(3.10-51)}$$

Differentiating Eq. (3.10-50) with respect to y and setting $y = 0$,

$$\left(\frac{dv_x}{dy}\right)_{y=0} = \frac{3v_\infty}{2\delta} \qquad\qquad \text{(3.10-52)}$$

Substituting Eq. (3.10-52) into (3.10-51),

$$\tau_0 = \frac{3\mu v_\infty}{2\delta} \qquad\qquad \text{(3.10-53)}$$

Substituting Eq. (3.10-50) into Eq. (3.10-48) and integrating between $y = 0$ and $y = \delta$, we obtain

$$\frac{d\delta}{dx} = \frac{280}{39}\frac{\tau_0}{v_\infty^2 \rho} \qquad\qquad \text{(3.10-54)}$$

Combining Eqs. (3.10-53) and (3.10-54) and integrating between $\delta = 0$ and $\delta = \delta$, and $x = 0$ and $x = L$,

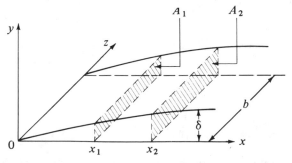

FIGURE 3.10-5. *Control volume for integral analysis of the boundary-layer flow.*

$$\delta = 4.64 \sqrt{\frac{\mu L}{\rho v_\infty}} \qquad (3.10\text{-}55)$$

where the length of plate is $x = L$. Proceeding in a manner similar to Eqs. (3.10-6)–(3.10-12), the drag coefficient is

$$C_D = 1.292 \sqrt{\frac{\mu}{L v_\infty \rho}} = \frac{1.292}{N_{\text{Re}, L}^{1/2}} \qquad (3.10\text{-}56)$$

A comparison of Eq. (3.10-6) with (3.10-55) and (3.10-12) with (3.10-56) shows the success of this method. Only the numerical constants differ slightly. This method can be used with reasonable accuracy for cases where an exact analysis is not feasible.

3. Integral momentum analysis for turbulent boundary layer. The procedures used for the integral momentum analysis for a laminar boundary layer can be applied to the turbulent boundary layer on a flat plate. A simple empirical velocity distribution for pipe flow which is valid up to a Reynolds number of 10^5 can be adapted for the boundary layer on a flat plate, to become

$$\frac{v_x}{v_\infty} = \left(\frac{y}{\delta}\right)^{1/7} \qquad (3.10\text{-}57)$$

This is the Blasius $\frac{1}{7}$-power law often used.

Equation (3.10-57) is substituted into the integral relation equation (3.10-48).

$$\frac{d}{dx} \int_0^\delta v_\infty^2 \left[\left(\frac{y}{\delta}\right)^{1/7} - \left(\frac{y}{\delta}\right)^{2/7} \right] dy = \frac{\tau_0}{\rho} \qquad (3.10\text{-}58)$$

The power-law equation does not hold, as y goes to zero at the wall. Another useful relation is the Blasius correlation for shear stress for pipe flow, which is consistent at the wall for the wall shear stress τ_0. For boundary-layer flow over a flat plate, it becomes

$$\frac{\tau_0}{\rho v_\infty^2} = 0.023 \left(\frac{\delta v_\infty \rho}{\mu}\right)^{-1/4} \qquad (3.10\text{-}59)$$

Integrating Eq. (3.10-58), combining the result with Eq. (3.10-59), and integrating between $\delta = 0$ and $\delta = \delta$, and $x = 0$ and $x = L$,

$$\delta = 0.376 \left(\frac{L v_\infty \rho}{\mu}\right)^{-1/5} L = \frac{0.376 L}{N_{\text{Re}, L}^{1/5}} \qquad (3.10\text{-}60)$$

Integration of the drag force as before gives

$$C_D = \frac{0.072}{N_{\text{Re}, L}^{1/5}} \qquad (3.10\text{-}61)$$

In this development the turbulent boundary layer was assumed to extend to $x = 0$. Actually, a certain length at the front has a laminar boundary layer. Experimental data check Eq. (3.10-61) reasonably well from a Reynolds number of 5×10^5 to 10^7. More accurate results at higher Reynolds numbers can be obtained by using a logarithmic velocity distribution, Eqs. (3.10-45)–(3.10-47).

3.11 DIMENSIONAL ANALYSIS IN MOMENTUM TRANSFER

3.11A Dimensional Analysis of Differential Equations

In this chapter we have derived several differential equations describing various flow situations. Dimensional homogeneity requires that each term in a given equation have the same units. Then, the ratio of one term in the equation to another term is dimensionless. Knowing the physical meaning of each term in the equation, we are then able to give a physical interpretation to each of the dimensionless parameters or numbers formed. These dimensionless numbers, such as the Reynolds number and others, are useful in correlating and predicting transport phenomena in laminar and turbulent flow.

Often it is not possible to integrate the differential equation describing a flow situation. However, we can use the equation to find out which dimensionless numbers can be used in correlating experimental data for this physical situation.

An important example of this involves the use of the Navier–Stokes equation, which often cannot be integrated for a given physical situation. To start, we use Eq. (3.7-36) for the x component of the Navier–Stokes equation. At steady state this becomes

$$v_x \frac{\partial v_x}{\partial x} + v_y \frac{\partial v_x}{\partial y} + v_z \frac{\partial v_x}{\partial z} = g_x - \frac{1}{\rho}\frac{\partial p}{\partial x} + \frac{\mu}{\rho}\left(\frac{\partial^2 v_x}{\partial x^2} + \frac{\partial^2 v_x}{\partial y^2} + \frac{\partial^2 v_x}{\partial z^2}\right) \quad (3.11\text{-}1)$$

Each term in this equation has the units length/time^2 or (L/t^2).

In this equation each term has a physical significance. First we use a single characteristic velocity v and a single characteristic length L for all terms. Then the expression of each term in Eq. (3.11-1) is as follows. The left-hand side can be expressed as v^2/L and the right-hand terms, respectively, as g, $p/\rho L$, and $\mu v/\rho L^2$. We then write

$$\left[\frac{v^2}{L}\right] = [g] - \left[\frac{p}{\rho L}\right] + \left[\frac{\mu v}{\rho L^2}\right] \quad (3.11\text{-}2)$$

This expresses a dimensional equality and not a numerical equality. Each term has dimensions L/t^2.

The left-hand term in Eq. (3.11-2) represents the inertia force and the terms on the right-hand side represent, respectively, the gravity force, pressure force, and viscous force. Dividing each of the terms in Eq. (3.11-2) by the inertia force $[v^2/L]$, the following dimensionless groups or their reciprocals are obtained.

$$\frac{[v^2/L]}{[g]} = \frac{\text{inertia force}}{\text{gravity force}} = \frac{v^2}{gL} = N_{\text{Fr}} \qquad \text{(Froude number)} \qquad (3.11\text{-}3)$$

$$\frac{[p/\rho L]}{[v^2/L]} = \frac{\text{pressure force}}{\text{inertia force}} = \frac{p}{\rho v^2} = N_{\text{Eu}} \qquad \text{(Euler number)} \qquad (3.11\text{-}4)$$

$$\frac{[v^2/L]}{[\mu v/\rho L^2]} = \frac{\text{inertia force}}{\text{viscous force}} = \frac{Lv\rho}{\mu} = N_{\text{Re}} \qquad \text{(Reynolds number)} \qquad (3.11\text{-}5)$$

Note that this method not only gives the various dimensionless groups for a differential equation but also gives physical meaning to these dimensionless groups. The length, velocity, etc., to be used in a given case will be that value which is most significant. For example, the length may be the diameter of a sphere, the length of a flat plate, and so on.

Systems that are geometrically similar are said to be *dynamically similar* if the parameters representing ratios of forces pertinent to the situation are equal. This means that the Reynolds, Euler, or Froude numbers must be equal between the two systems.

This dynamic similarity is an important requirement in obtaining experimental data on a small model and extending these data to scale up to the large prototype. Since experiments with full-scale prototypes would often be difficult and/or expensive, it is customary to study small models. This is done in the scaleup of chemical process equipment and in the design of ships and airplanes.

3.11B Dimensional Analysis Using Buckingham Method

The method of obtaining the important dimensionless numbers from the basic differential equations is generally the preferred method. In many cases, however, we are not able to formulate a differential equation which clearly applies. Then a more general procedure is required, which is known as the *Buckingham method*. In this method the listing of the important variables in the particular physical problem is done first. Then we determine the number of dimensionless parameters into which the variables may be combined by using the Buckingham pi theorem.

The *Buckingham theorem* states that the functional relationship among q quantities or variables whose units may be given in terms of u fundamental units or dimensions may be written as $(q - u)$ independent dimensionless groups, often called π's. [This quantity u is actually the maximum number of these variables which will not form a dimensionless group. However, only in a few cases is this u not equal to the number of fundamental units (B1).]

Let us consider the following example, to illustrate the use of this method. An incompressible fluid is flowing inside a circular tube of inside diameter D. The significant variables are pressure drop Δp, velocity v, diameter D, tube length L, viscosity μ, and density ρ. The total number of variables is $q = 6$.

The fundamental units or dimensions are $u = 3$ and are mass M, length L, and time t. The units of the variables are as follows: Δp in M/Lt^2, v in L/t, D in L, L in L, μ in M/Lt, and ρ in M/L^3. The number of dimensionless groups or π's is $q - u$, or $6 - 3 = 3$. Thus,

$$\pi_1 = f(\pi_2, \pi_3) \tag{3.11-6}$$

Next, we must select a core group of u (or 3) variables which will appear in each π group and among them contain all the fundamental dimensions. Also, no two of the variables selected for the core can have the same dimensions. In choosing the core, the variable whose effect one desires to isolate is often excluded (for example, Δp). This leaves us with the variables v, D, μ, and ρ to be used. (L and D have the same dimensions.)

We will select D, v, and ρ to be the core variables common to all three groups. Then the three dimensionless groups are

$$\pi_1 = D^a v^b \rho^c \, \Delta p^1 \tag{3.11-7}$$

$$\pi_2 = D^d v^e \rho^f L^1 \tag{3.11-8}$$

$$\pi_3 = D^g v^h \rho^i \mu^1 \tag{3.11-9}$$

To be dimensionless, the variables must be raised to certain exponents a, b, c, etc.

First we consider the π_1 group.

$$\pi_1 = D^a v^b \rho^c \, \Delta p^1 \tag{3.11-7}$$

To evaluate these exponents, we write Eq. (3.11-7) dimensionally by substituting the dimensions for each variable.

$$M^0 L^0 t^0 = 1 = L^a \left(\frac{L}{t}\right)^b \left(\frac{M}{L^3}\right)^c \frac{M}{Lt^2} \qquad (3.11\text{-}10)$$

Next we equate the exponents of L on both sides of this equation, of M, and finally of t.

$$(L) \quad 0 = a + b - 3c - 1$$

$$(M) \quad 0 = c + 1 \qquad (3.11\text{-}11)$$

$$(t) \quad 0 = -b - 2$$

Solving these equations, $a = 0$, $b = -2$, and $c = -1$.
    Substituting these values into Eq. (3.11-7),

$$\pi_1 = \frac{\Delta p}{v^2 \rho} = N_{\text{Eu}} \qquad (3.11\text{-}12)$$

Repeating this procedure for π_2 and π_3,

$$\pi_2 = \frac{L}{D} \qquad (3.11\text{-}13)$$

$$\pi_3 = \frac{Dv\rho}{\mu} = N_{\text{Re}} \qquad (3.11\text{-}14)$$

Finally, substituting π_1, π_2, and π_3 into Eq. (3.11-6),

$$\frac{\Delta p}{v^2 \rho} = f\left(\frac{L}{D}, \frac{Dv\rho}{\mu}\right) \qquad (3.11\text{-}15)$$

Combining Eq. (2.10-5) with the left-hand side of Eq. (3.11-15), the result obtained shows that the friction factor is a function of the Reynolds number (as was shown before in the empirical correlation of friction factor and Reynolds number) and of length/diameter ratio. In pipes with $L/D \gg 1$ or pipes with fully developed flow, the friction factor is found to be independent of L/D.

    This type of analysis is useful in empirical correlations of data. However, it does not tell us the importance of each dimensionless group, which must be determined by experimentation, nor does it select the variables to be used.

PROBLEMS

3.1-1. *Force on a Cylinder in a Wind Tunnel.* Air at 101.3 kPa absolute and 25°C is flowing at a velocity of 10 m/s in a wind tunnel. A long cylinder having a diameter of 90 mm is placed in the tunnel and the axis of the cylinder is held perpendicular to the air flow. What is the force on the cylinder per meter length?
    Ans. $C_D = 1.3$, $F_D = 6.94$ N

3.1-2. *Wind Force on a Steam Boiler Stack.* A cylindrical steam boiler stack has a diameter of 1.0 m and is 30.0 m high. It is exposed to a wind at 25°C having a velocity of 50 miles/h. Calculate the force exerted on the boiler stack.
    Ans. $C_D = 0.33$, $F_D = 2935$ N

3.1-3. *Effect of Velocity on Force on a Sphere and Stokes' Law.* A sphere is held in a small wind tunnel where air at 37.8°C and 1 atm abs and various velocities is forced by the sphere having a diameter of 0.042 m.

(a) Determine the drag coefficient and force on the sphere for a velocity of 2.30×10^{-4} m/s. Use Stokes' law here if it is applicable.

(b) Also determine the force for velocities of 2.30×10^{-3}, 2.30×10^{-2}, 2.30×10^{-1}, and 2.30 m/s. Make a plot of F_D versus velocity.

3.1-4. Drag Force on Bridge Pier in River. A cylindrical bridge pier 1.0 m in diameter is submerged to a depth of 10 m. Water in the river at 20°C is flowing past at a velocity of 1.2 m/s. Calculate the force on the pier.

3.1-5. Surface Area in a Packed Bed. A packed bed is composed of cubes 0.020 m on a side and the bulk density of the packed bed is 980 kg/m³. The density of the solid cubes is 1500 kg/m³.

(a) Calculate ε, effective diameter D_p, and a.

(b) Repeat for the same conditions but for cylinders having a diameter of $D = 0.02$ m and a length $h = 1.5D$.

        Ans. (a) $\varepsilon = 0.3467$, $D_p = 0.020$ m, $a = 196.0 \, \text{m}^{-1}$

3.1-6. Derivation for Number of Particles in a Bed of Cylinders. For a packed bed containing cylinders where the diameter D of the cylinders is equal to the length h, do as follows for a bed having a void fraction ε.

(a) Calculate the effective diameter.

(b) Calculate the number, n, of cylinders in 1 m³ of the bed.

        Ans. (a) $D_p = D$

3.1-7. Derivation of Dimensionless Equation for Packed Bed. Starting with Eq. (3.1-20), derive the dimensionless equation (3.1-21). Show all steps in the derivation.

3.1-8. Flow and Pressure Drop of Gases in Packed Bed. Air at 394.3 K flows through a packed bed of cylinders having a diameter of 0.0127 m and length the same as the diameter. The bed void fraction is 0.40 and the length of the packed bed is 3.66 m. The air enters the bed at 2.20 atm abs at the rate of 2.45 kg/m² · s based on the empty cross section of the bed. Calculate the pressure drop of air in the bed.

        Ans. $\Delta p = 0.1547 \times 10^5$ Pa

3.1-9. Flow of Water in a Filter Bed. Water at 24°C is flowing by gravity through a filter bed of small particles having an equivalent diameter of 0.0060 m. The void fraction of the bed is measured as 0.42. The packed bed has a depth of 1.50 m. The liquid level of water above the bed is held constant at 0.40 m. What is the water velocity v' based on the empty cross section of the bed?

3.1-10. Mean Diameter of Particles in Packed Bed. A mixture of particles in a packed bed contains the following volume percent of particles and sizes: 15%, 10 mm; 25%, 20 mm; 40%, 40 mm; 20%, 70 mm. Calculate the effective mean diameter, D_{pm}, if the shape factor is 0.74.

        Ans. $D_{pm} = 18.34$ mm

3.1-11. Permeability and Darcy's Law. A sample core of a porous rock obtained from an oil reservoir is 8 cm long and has a diameter of 2.0 cm. It is placed in a core holder. With a pressure drop of 1.0 atm, the water flow at 20.2°C through the core was measured as 2.60 cm³/s. What is the permeability in darcy?

3.1-12. Minimum Fluidization and Expansion of Fluid Bed. Particles having a size of 0.10 mm, a shape factor of 0.86, and a density of 1200 kg/m³ are to be fluidized using air at 25°C and 202.65 kPa abs pressure. The void fraction at minimum fluidizing conditions is 0.43. The bed diameter is 0.60 m and the bed contains 350 kg of solids.

(a) Calculate the minimum height of the fluidized bed.

(b) Calculate the pressure drop at minimum fluidizing conditions.

(c) Calculate the minimum velocity for fluidization.

(d) Using 4.0 times the minimum velocity, estimate the porosity of the bed.

Ans. (a) $L_{mf} = 1.810$ m, (b) $\Delta p = 0.1212 \times 10^5$ Pa,
(c) $v'_{mf} = 0.004374$ m/s, (d) $\varepsilon = 0.604$

3.1-13. *Minimum Fluidization Velocity Using a Liquid.* A tower having a diameter of
0.1524 m is being fluidized with water at 20.2°C. The uniform spherical beads in
the tower bed have a diameter of 4.42 mm and a density of 1603 kg/m³. Estimate
the minimum fluidizing velocity and compare with the experimental value of
0.02307 m/s of Wilhelm and Kwauk (W5).

3.1-14. *Fluidization of a Sand Bed Filter.* To clean a sand bed filter it is fluidized at
minimum conditions using water at 24°C. The round sand particles have a
density of 2550 kg/m³ and an average size of 0.40 mm. The sand has the
properties given in Table 3.1-2.
(a) The bed diameter is 0.40 m and the desired height of the bed at these
minimum fluidizing conditions is 1.75 m. Calculate the amount of solids
needed.
(b) Calculate the pressure drop at these conditions and the minimum velocity for
fluidization.
(c) Using 4.0 times the minimum velocity, estimate the porosity and height of the
expanded bed.

3.2-1. *Flow Measurement Using a Pitot Tube.* A pitot tube is used to measure the flow
rate of water at 20°C in the center of a pipe having an inside diameter of
102.3 mm. The manometer reading is 78 mm of carbon tetrachloride at 20°C.
The pitot tube coefficient is 0.98.
(a) Calculate the velocity at the center and the average velocity.
(b) Calculate the volumetric flow rate of the water.
Ans. (a) $v_{max} = 0.9372$ m/s, $v_{av} = 0.773$ m/s, (b) 6.35×10^{-3} m³/s

3.2-2. *Gas Flow Rate Using a Pitot Tube.* The flow rate of air at 37.8°C is being
measured at the center of a duct having a diameter of 800 mm by a pitot tube.
The pressure difference reading on the manometer is 12.4 mm of water. At the
pitot tube position, the static pressure reading is 275 mm of water above one
atmosphere absolute. The pitot tube coefficient is 0.97. Calculate the velocity at
the center and the volumetric flow rate of the air.

3.2-3. *Pitot-Tube Traverse for Flow Rate Measurement.* In a pitot tube traverse of a
pipe having an inside diameter of 155.4 mm in which water at 20°C is flowing, the
following data were obtained.

Distance from Wall (mm)	Reading in Manometer (mm of Carbon Tetrachloride)
26.9	122
52.3	142
77.7	157
103.1	137
128.5	112

The pitot tube coefficient is 0.98.
(a) Calculate the maximum velocity at the center.
(b) Calculate the average velocity. [*Hint :* Use Eq. (2.6-17) and do a graphical
integration.]

3.2-4. *Metering Flow by a Venturi.* A venturi meter having a throat diameter of
38.9 mm is installed in a line having an inside diameter of 102.3 mm. It meters
water having a density of 999 kg/m³. The measured pressure drop across the

venturi is 156.9 kPa. The venturi coefficient C_v is 0.98. Calculate the gal/min and m^3/s flow rate.

Ans.   330 gal/min, 0.0208 m^3/s

3.2-5. *Use of a Venturi to Meter Water Flow.* Water at 20°C is flowing in a 2-in. schedule 40 steel pipe. Its flow rate is measured by a venturi meter having a throat diameter of 20 mm. The manometer reading is 214 mm of mercury. The venturi coefficient is 0.98. Calculate the flow rate.

3.2-6. *Metering of Oil Flow by an Orifice.* A heavy oil at 20°C having a density of 900 kg/m^3 and a viscosity of 6 cp is flowing in a 4-in. schedule 40 steel pipe. When the flow rate is 0.0174 m^3/s it is desired to have a pressure drop reading across the manometer equivalent to 0.93 × 10^5 Pa. What size orifice should be used if the orifice coefficient is assumed as 0.61? What is the permanent pressure loss?

3.2-7. *Water Flow Rate in an Irrigation Ditch.* Water is flowing in an open channel in an irrigation ditch. A rectangular weir having a crest length $L = 1.75$ ft is used. The weir head is measured as $h_0 = 0.47$ ft. Calculate the flow rate in ft^3/s and m^3/s.

Ans.   Flow rate = 1.776 ft^3/s, 0.0503 m^3/s

3.3-1. *Brake Horsepower of Centrifugal Pump.* Using Fig. 3.3-2 and a flow rate of 60 gal/min, do as follows.
(a) Calculate the brake hp of the pump using water with a density of 62.4 lb_m/ft^3. Compare with the value from the curve.
(b) Do the same for a nonviscous liquid having a density of 0.85 g/cm^3.

Ans.   (b) 0.69 brake hp (0.51 kW)

3.3-2. *kW-Power of a Fan.* A centrifugal fan is to be used to take a flue gas at rest (zero velocity) and at a temperature of 352.6 K and a pressure of 749.3 mm Hg and to discharge this gas at a pressure of 800.1 mm Hg and a velocity of 38.1 m/s. The volume flow rate of gas is 56.6 std m^3/min of gas (at 294.3 K and 760 mm Hg). Calculate the brake kW of the fan if its efficiency is 65% and the gas has a molecular weight of 30.7. Assume incompressible flow.

3.3-3. *Adiabatic Compression of Air.* A compressor operating adiabatically is to compress 2.83 m^3/min of air at 29.4°C and 102.7 kN/m^2 to 311.6 kN/m^2. Calculate the power required if the efficiency of the compressor is 75%. Also, calculate the outlet temperature.

3.4-1. *Power for Liquid Agitation.* It is desired to agitate a liquid having a viscosity of 1.5 × 10^{-3} Pa·s and a density of 969 kg/m^3 in a tank having a diameter of 0.91 m. The agitator will be a six-blade open turbine having a diameter of 0.305 m operating at 180 rpm. The tank has four vertical baffles each with a width J of 0.076 m. Also, $W = 0.0381$ m. Calculate the required kW. Use curve 2, Fig. 3.4-4.

Ans. $N_P = 2.5$, power = 0.172 kW (0.231 hp)

3.4-2. *Power for Agitation and Scale-Up.* A turbine agitator having six flat blades and a disk has a diameter of 0.203 m and is used in a tank having a diameter of 0.61 m and height of 0.61 m. The width $W = 0.0405$ m. Four baffles are used having a width of 0.051 m. The turbine operates at 275 rpm in a liquid having a density of 909 kg/m^3 and viscosity of 0.020 Pa·s.
(a) Calculate the kW power of the turbine and kW/m^3 of volume.
(b) Scale up this system to a vessel having a volume of 100 times the original for the case of equal mass transfer rates.

Ans.   (a) $P = 0.1508$ kW, $P/V = 0.845$ kW/m^3,
(b) $P_2 = 15.06$ kW, $P_2/V_2 = 0.845$ kW/m^3

3.4-3. *Scale-Down of Process Agitation System.* An existing agitation process operates using the same agitation system and fluid as described in Example 3.4-1a. It is desired to design a small pilot unit with a vessel volume of 2.0 liters so that effects of various process variables on the system can be studied in the laboratory. The rates of mass transfer appear to be important in this system, so the scale-down should be on this basis. Design the new system specifying sizes, rpm, and kW power.

3.4-4. Anchor Agitation System. An anchor-type agitator similar to that described for Eq. (3.4-3) is to be used to agitate a fluid having a viscosity of 100 Pa·s and a density of 980 kg/m³. The vessel size is $D_t = 0.90$ m and $H = 0.90$ m. The rpm is 50. Calculate the power required.

3.4-5. Design of Agitation System. An agitation system is to be designed for a fluid having a density of 950 kg/m³ and viscosity of 0.005 Pa·s. The vessel volume is 1.50 m³ and a standard six-blade open turbine with blades at 45° (curve 3, Fig. 3.4-4) is to be used with $D_a/W = 8$ and $D_a/D_t = 0.35$. For the preliminary design a power of 0.5 kW/m³ volume is to be used. Calculate the dimensions of the agitation system, rpm, and kW power.

3.4-6. Scale-Up of Mixing Times for a Turbine. For scaling up a turbine-agitated system, do as follows:
(a) Derive Eq. (3.4-17) for the same power/unit volume.
(b) Derive Eq. (3.4-18) for the same mixing times.

3.4-7. Mixing Time in a Turbine-Agitated System. Do as follows:
(a) Predict the time of mixing for the turbine system in Example 3.4-1a.
(b) Using the same system as part (a) but with a tank having a volume of 10.0 m³ and the same power/unit volume, predict the new mixing time.
Ans. (a) $f_t = 4.1$, $t_T = 17.7$ s

3.5-1. Pressure Drop of Power-Law Fluid, Banana Purée. A power-law biological fluid, banana purée, is flowing at 23.9°C, with a velocity of 1.018 m/s, through a smooth tube 6.10 m long having an inside diameter of 0.01267 m. The flow properties of the fluid are $K = 6.00$ N·s$^{0.454}$/m² and $n = 0.454$. The density of the fluid is 976 kg/m³.
(a) Calculate the generalized Reynolds number and also the pressure drop using Eq. (3.5-9). Be sure to convert K to K' first.
(b) Repeat part (a), but use the friction factor method.
Ans. (a) $N_{Re, gen} = 63.6$, $\Delta p = 245.2$ kN/m² (5120 lb$_f$/ft²)

3.5-2. Pressure Drop of Pseudoplastic Fluid. A pseudoplastic power-law fluid having a density of 63.2 lb$_m$/ft³ is flowing through 100 ft of a pipe having an inside diameter of 2.067 in. at an average velocity of 0.500 ft/s. The flow properties of the fluid are $K = 0.280$ lb$_f$·s^n/ft² and $n = 0.50$. Calculate the generalized Reynolds number and also the pressure drop, using the friction factor method.

3.5-3. Turbulent Flow of Non-Newtonian Fluid, Applesauce. Applesauce having the flow properties given in Table 3.5-1 is flowing in a smooth tube having an inside diameter of 50.8 mm and a length of 3.05 m at a velocity of 4.57 m/s.
(a) Calculate the friction factor and the pressure drop in the smooth tube.
(b) Repeat, but for a commercial pipe having the same inside diameter with a roughness of $\varepsilon = 4.6 \times 10^{-5}$ m.
Ans. (a) $N_{Re, gen} = 4855$, $f = 0.0073$, (b) $f = 0.0100$

3.5-4. Agitation of a Non-Newtonian Liquid. A pseudoplastic liquid having the following properties of $n = 0.53$, $K = 26.49$ N·s^n/m², and $\rho = 975$ kg/m³ is being agitated in a system such as in Fig. 3.5-4 where $D_t = 0.304$ m, $D_a = 0.151$ m, and $N = 5$ rev/s. Calculate μ_a, $N'_{Re, n}$, and the kW power for this system.
Ans. $\mu_a = 4.028$ Pa·s, $N'_{Re, n} = 27.60$, $N_P = 3.1$, $P = 0.02966$ kW

3.5-5. *Flow Properties of a Non-Newtonian Fluid from Rotational Viscometer Data.* Following are data obtained on a fluid using a Brookfield rotational viscometer.

RPM	0.5	1	2.5	5	10	20	50
Torque (dyn-cm)	86.2	168.9	402.5	754	1365	2379	4636

The diameter of the inner concentric rotating spindle is 25.15 mm, the outer cylinder diameter is 27.62 mm, and the effective length is 92.39 mm. Determine the flow properties of this non-Newtonian fluid.

Ans. $n = 0.870$

3.6-1. *Equation of Continuity in a Cylinder.* Fluid having a constant density ρ is flowing in the z direction through a circular pipe with axial symmetry. The radial direction is designated by r.
(a) Using a cylindrical shell balance with dimensions dr and dz, derive the equation of continuity for this system.
(b) Use the equation of continuity in cylindrical coordinates to derive the equation.

3.6-2. *Change of Coordinates for Continuity Equation.* Using the general equation of continuity given in rectangular coordinates, convert it to Eq. (3.6-27), which is the equation of continuity in cylindrical coordinates. Use the relationships in Eq. (3.6-26) to do this.

3.7-1. *Combining Equations of Continuity and Motion.* Using the continuity equation and the equations of motion for the x, y, and z components, derive Eq. (3.7-13).

3.8-1. *Average Velocity in a Circular Tube.* Using Eq. (3.8-17) for the velocity in a circular tube as a function of radius r,

$$v_z = \frac{1}{4\mu}\frac{dp}{dz}(r^2 - r_0^2) \tag{3.8-17}$$

derive Eq. (3.8-19) for the average velocity.

$$v_{z\,av} = -\frac{r_0^2}{8\mu}\frac{dp}{dz} \tag{3.8-19}$$

3.8-2. *Laminar Flow in a Cylindrical Annulus.* Derive all the equations given in Example 3.8-4 showing all the steps. Also, derive the equation for the average velocity $v_{z\,av}$. Finally, integrate to obtain the pressure drop from $z = 0$ for $p = p_0$ to $z = L$ for $p = p_L$.

$$\text{Ans.} \quad v_{z\,av} = -\frac{1}{8\mu}\frac{dp}{dz}\left[r_2^2 + r_1^2 - \frac{r_2^2 - r_1^2}{\ln(r_2/r_1)}\right],$$

$$v_{z\,av} = \frac{p_0 - p_L}{8\mu L}\left[r_2^2 + r_1^2 - \frac{r_2^2 - r_1^2}{\ln(r_2/r_1)}\right]$$

3.8-3. *Velocity Profile in Wetted-Wall Tower.* In a vertical wetted-wall tower, the fluid flows down the inside as a thin film δ m thick in laminar flow in the vertical z direction. Derive the equation for the velocity profile v_z as a function of x, the distance from the liquid surface toward the wall. The fluid is at a large distance from the entrance. Also, derive expressions for $v_{z\,av}$ and $v_{z\,max}$. (*Hint:* At $x = \delta$, which is at the wall, $v_z = 0$. At $x = 0$, the surface of the flowing liquid, $v_z = v_{z\,max}$.) Show all steps.

Ans. $v_z = (\rho g \delta^2/2\mu)[1 - (x/\delta)^2]$, $v_{z\,av} = \rho g \delta^2/3\mu$, $v_{z\,max} = \rho g \delta^2/2\mu$

3.8-4. Velocity Profile in Falling Film and Differential Momentum Balance. A Newtonian liquid is flowing as a falling film on an inclined flat surface. The surface makes an angle of β with the vertical. Assume that in this case the section being considered is sufficiently far from both ends that there are no end effects on the velocity profile. The thickness of the film is δ. The apparatus is similar to Fig. 2.9-3 but is not vertical. Do as follows.
(a) Derive the equation for the velocity profile of v_z as a function of x in this film using the differential momentum balance equation.
(b) What are the maximum velocity and the average velocity?
(c) What is the equation for the momentum flux distribution of τ_{xz}? [Hint: Can Eq. (3.7-19) be used here?]

$$\text{Ans. (a) } v_z = (\rho g \delta^2 \cos \beta / 2\mu)[1 - (x/\delta)^2]$$
$$\text{(c) } \tau_{xz} = \rho g x \cos \beta$$

3.8-5. Velocity Profiles for Flow Between Parallel Plates. In Example 3.8-2 a fluid is flowing between vertical parallel plates with one plate moving. Do as follows.
(a) Determine the average velocity and the maximum velocity.
(b) Make a sketch of the velocity profile for three cases where the surface is moving upward, downward, and stationary.

3.8-6. Conversion of Shear Stresses in Terms of Fluid Motion. Starting with the x component of motion, Eq. (3.7-10), which is in terms of shear stresses, convert it to the equation of motion, Eq. (3.7-36), in terms of velocity gradients, for a Newtonian fluid with constant ρ and μ. Note that $(\nabla \cdot v) = 0$ in this case. Also, use of Eqs. (3.7-14) to (3.7-20) should be considered.

3.8-7. Derivation of Equation of Continuity in Cylindrical Coordinates. By means of a mass balance over a stationary element whose volume is $r \, \Delta r \, \Delta\theta \, \Delta z$, derive the equation of continuity in cylindrical coordinates.

3.8-8. Flow between Two Rotating Coaxial Cylinders. The geometry of two coaxial cylinders is the same as in Example 3.8-5. In this case, however, both cylinders are rotating with the inner rotating with an annular velocity of ω_1 and the outer at ω_2. Determine the velocity and the shear stress distributions using the differential equation of momentum.

$$\text{Ans. } v_\theta = \frac{R_2^2}{R_2^2 - R_1^2}\left[r\left(\omega_2 - \frac{\omega_1 R_1^2}{R_2^2}\right) - \frac{R_1^2}{r}(\omega_2 - \omega_1)\right]$$

3.9-1. Potential Function. The potential function ϕ for a given flow situation is $\phi = C(x^2 - y^2)$, where C is a constant. Check to see if it satisfies Laplace's equation. Determine the velocity components v_x and v_y.

$$\text{Ans. } v_x = 2Cx, v_y = -2Cy (C = \text{constant})$$

3.9-2. Determining the Velocities from the Potential Function. The potential function for flow is given as $\phi = Ax + By$, where A and B are constants. Determine the velocities v_x and v_y.

3.9-3. Stream Function and Velocity Vector. Flow of a fluid in two dimensions is given by the stream function $\psi = Bxy$, where $B = 50$ s^{-1} and the units of x and y are in cm. Determine the value of v_x, v_y, and the velocity vector at $x = 1$ cm and $y = 1$ cm.

$$\text{Ans. } v = 70.7 \text{ cm/s}$$

3.9-4. Stream Function and Potential Function. A liquid is flowing parallel to the x axis. The flow is uniform and is represented by $v_x = U$ and $v_y = 0$.
(a) Find the stream function ψ for this flow field and plot the streamlines.
(b) Find the potential function and plot the potential lines.

$$\text{Ans. (a) } \psi = Uy + C \ (C = \text{constant})$$

3.9-5. *Velocity Components and Stream Function.* A liquid is flowing in a uniform manner at an angle of β with respect to the x axis. Its velocity components are $v_x = U \cos \beta$ and $v_y = U \sin \beta$. Find the stream function and the potential function.

Ans. $\psi = Uy \cos \beta - Ux \sin \beta + C$ (C = constant)

3.9-6. *Flow Field with Concentric Streamlines.* The flow of a fluid that has concentric streamlines has a stream function represented by $\psi = 1/(x^2 + y^2)$. Find the components of velocity v_x and v_y. Also, determine if the flow is rotational, and if so, determine the vorticity, $2\omega_z$.

3.9-7. *Potential Function and Velocity Field.* In Example 3.9-2 the velocity components were given. Show if a velocity potential exists and, if so, also determine ϕ.

Ans. $\phi = ax^3/3 - axy^2 + C$ (C = constant)

3.9-8. *Euler's Equation of Motion for an Ideal Fluid.* Using the Euler equations (3.9-2)–(3.9-4) for ideal fluids with constant density and zero viscosity, obtain the following equation:

$$\rho \frac{D\mathbf{v}}{Dt} = -\nabla p + \rho \mathbf{g}$$

3.10-1. *Laminar Boundary Layer on Flat Plate.* Water at 20°C is flowing past a flat plate at 0.914 m/s. The plate is 0.305 m wide.
  (a) Calculate the Reynolds number 0.305 m from the leading edge to determine if the flow is laminar.
  (b) Calculate the boundary-layer thickness at $x = 0.152$ and $x = 0.305$ m from the leading edge.
  (c) Calculate the total drag on the 0.305-m-long plate.

Ans. (a) $N_{Re, L} = 2.77 \times 10^5$, (b) $\delta = 0.0029$ m at $x = 0.305$ m

3.10-2. *Air Flow Past a Plate.* Air at 294.3 K and 101.3 kPa is flowing past a flat plate at 6.1 m/s. Calculate the thickness of the boundary layer at a distance of 0.3 m from the leading edge and the total drag for a 0.3-m-wide plate.

3.10-3. *Boundary-Layer Flow Past a Plate.* Water at 293 K is flowing past a flat plate at 0.5 m/s. Do as follows.
  (a) Calculate the boundary-layer thickness in m at a point 0.1 m from the leading edge.
  (b) At the same point, calculate the point shear stress τ_0. Also calculate the total drag coefficient.

3.10-4. *Transition Point to Turbulent Boundary Layer.* Air at 101.3 kPa and 293 K is flowing past a smooth flat plate at 100 ft/s. The turbulence in the air stream is such that the transition from a laminar to a turbulent boundary layer occurs at $N_{Re, L} = 5 \times 10^5$.
  (a) Calculate the distance from the leading edge where the transition occurs.
  (b) Calculate the boundary-layer thickness δ at a distance of 0.5 ft and 3.0 ft from the leading edge. Also calculate the drag coefficient for both distances $L = 0.5$ and 3.0 ft.

3.11-1. *Dimensional Analysis for Flow Past a Body.* A fluid is flowing external to a solid body. The force F exerted on the body is a function of the fluid velocity v, fluid density ρ, fluid viscosity μ, and a dimension of the body L. By dimensional analysis, obtain the dimensionless groups formed from the variables given. (*Note:* Use the M, L, t system of units. The units of F are ML/t^2. Select v, ρ, and L as the core variables.)

Ans. $\pi_1 = (F/L^2)/\rho v^2$, $\pi_2 = \mu/Lv\rho$

3.11-2. *Dimensional Analysis for Bubble Formation.* Dimensional analysis is to be used to correlate data on bubble size with the properties of the liquid when gas bubbles are formed by a gas issuing from a small orifice below the liquid surface. Assume that the significant variables are bubble diameter D, orifice diameter d, liquid density ρ, surface tension σ in N/m, liquid viscosity μ, and g. Select d, ρ, and g as the core variables.

Ans. $\pi_1 = D/d, \pi_2 = \sigma/\rho d^2 g, \pi_3 = \mu^2/\rho^2 d^3 g$

REFERENCES

(A1)   Allis Chalmers Mfg. Co. *Bull. 1659.*

(A2)   American Gas Association, "Orifice Metering of Natural Gas," *Gas Measurement Rept. 3*, New York, 1955.

(B1)   BENNETT, C. O., and MYERS, J. E. *Momentum, Heat, and Mass Transfer*, 3rd ed. New York: McGraw-Hill Book Company, 1982.

(B2)   BIRD, R. B., STEWART, W. E., and LIGHTFOOT, E. N. *Transport Phenomena.* New York: John Wiley & Sons, Inc., 1960.

(B3)   BATES, R. L., FONDY, P. L., and CORPSTEIN, R. R. *I.E.C. Proc. Des. Dev.*, **2**, 310 (1963).

(B4)   BROWN, G. G., et al. *Unit Operations.* New York: John Wiley & Sons, Inc., 1950.

(B5)   BIGGS, R. D., *A.I.Ch.E.J.*, **9**, 636 (1963).

(C1)   CHARM, S. E. *The Fundamentals of Food Engineering.* 2nd ed. Westport, Conn.: Avi Publishing Co., Inc., 1971.

(C2)   CARMAN, P. C. *Trans. Inst. Chem. Eng. (London)*, **15**, 150 (1937).

(C3)   CALDERBANK, P. H. In *Mixing : Theory and Practice*, Vol. 2, V. W. Uhl and J. B. Gray (eds.). New York: Academic Press, Inc., 1967.

(D1)   DREW, T. B., and HOOPES, J. W., Jr. *Advances in Chemical Engineering.* New York: Academic Press, Inc., 1956.

(D2)   DODGE, D. W., and METZNER, A. B. *A.I.Ch.E.J.*, **5**, 189 (1959).

(E1)   ERGUN, S. *Chem. Eng. Progr.*, **48**, 89 (1952).

(F1)   FOX, E. A., and GEX, V. E. *A.I.Ch.E.J.*, **2**, 539 (1956).

(G1)   GODLESKI, E. S., and SMITH, J. C. *A.I.Ch.E. J.*, **8**, 617 (1962).

(H1)   HARPER, J. C., and EL SAHRIGI. *J. Food Sci.*, **30**, 470 (1965).

(H2)   HO, F. C., and KWONG, A. *Chem. Eng.*, July 23, 94 (1973).

(K1)   KUNII, D., and LEVENSPIEL, O. *Fluidization Engineering.* New York: John Wiley & Sons, Inc., 1969.

(K2)   KRIEGER, I. M., and MARON, S. H. *J. Appl. Phys.*, **25**, 72 (1954).

(L1)   LEAMY, G. H. *Chem. Eng.*, Oct. 15, 115 (1973).

(L2)   LEVA, M., WEINTRAUB, M., GRUMMER, M., POLLCHIK, M., and STORCH, H. H. *U.S. Bur. Mines Bull.*, 504 (1951).

(M1)   METZNER, A. B., FEEHS, R. H., RAMOS, H. L., OTTO, R. E., and TUTHILL, J. D. *A.I.Ch.E. J.*, **7**, 3 (1961).

(M2)   McCABE, W. L., SMITH, J. C., and HARRIOTT, P. *Unit Operations of Chemical Engineering, 4th ed. New York:* McGraw-Hill Book Company, 1985.

(M3)   METZNER, A. B., and REED, J. C. *A.I.Ch.E. J.*, **1**, 434 (1955).

(M4)   MOHSENIN, N. N. *Physical Properties of Plant and Animal Materials*, Vol. 1, Part II. New York: Gordon & Breach, Inc., 1970.

(M5)   MOO-YUNG, TICHAR, M. K., and DULLIEN, F. A. L. *A.I.Ch.E. J.*, **18**, 178 (1972).

(M6)   McKELVEY, J. M. *Polymer Processing*, New York: John Wiley & Sons, Inc., 1962.

(N1)   NORWOOD, K. W., and METZNER, A. B. *A.I.Ch.E. J.*, **6**, 432 (1960).

(P1)   PERRY, R. H., and GREEN, D. *Perry's Chemical Engineers' Handbook*, 6th ed. New York: McGraw-Hill Book Company, 1984.

(P2)   PINCHBECK, P. H., and POPPER, F. *Chem. Eng. Sci.*, **6**, 57 (1956).

(P3)   PATTERSON, W. I., CARREAU, P. J., and YAP, C. Y. *A.I.Ch.E. J.*, **25**, 208 (1979).

(P4)   PERRY, R. H., and CHILTON, C. H. *Chemical Engineers' Handbook*, 5th ed. New York: McGraw-Hill Book Company, 1973.

(R1)   RUSHTON, J. H., COSTICH, D. W., and EVERETT, H. J. *Chem. Eng. Progr.*, **46**, 395, 467 (1950).

(R2)   RAUTZEN, R. R., CORPSTEIN, R. R., and DICKEY, D. S. *Chem. Eng.*, Oct. 25, 119 (1976).

(S1)   STEVENS, W. E. Ph.D. thesis, University of Utah, 1953.

(S2)   SKELLAND, A. H. P. *Non-Newtonian Flow and Heat Transfer*. New York: John Wiley & Sons, Inc., 1967.

(S3)   STREETER, V. L. *Handbook of Fluid Dynamics*. New York: McGraw-Hill Book Company, 1961.

(S4)   SCHLICHTING, H. *Boundary Layer Theory*. New York: McGraw-Hill Book Company, 1955.

(T1)   TREYBAL, R. E. *Liquid Extraction*. 2nd ed. New York: McGraw-Hill Book Company, 1953.

(T2)   TREYBAL, R. E. *Mass Transfer Operations*, 3rd ed. New York: McGraw-Hill Book Company, 1980.

(U1)   UHL, V. W., and GRAY, J. B. (eds.), *Mixing: Theory and Practice*, Vol. I. New York: Academic, 1969.

(W1)   WEISMAN, J., and EFFERDING, L. E. *A.I.Ch.E. J.*, **6**, 419 (1960).

(W2)   WINNING, M. D. M.Sc. thesis, University of Alberta, 1948.

(W3)   WALTERS, K. *Rheometry*. London: Chapman & Hall Ltd., 1975.

(W4)   WEN, C. Y., and YU, Y. H. *A.I.Ch.E. J.*, **12**, 610 (1966).

(W5)   WILHELM, R. H., and KWAUK, M. *Chem. Eng. Progr.*, **44**, 201 (1948).

(W6)   WELTY, J. R., WICKS, C. E., and WILSON, R. E. *Fundamentals of Momentum, Heat, and Mass Transfer*, 3rd ed. New York: John Wiley & Sons, 1984.

(Z1)   ZLOKARNIK, M., and JUDAT, H. *Chem. Eng. Tech.*, **39**, 1163 (1967).

CHAPTER 4

Principles of
Steady-State
Heat Transfer

4.1 INTRODUCTION AND MECHANISMS OF HEAT TRANSFER

4.1A Introduction to Steady-State Heat Transfer

The transfer of energy in the form of heat occurs in many chemical and other types of processes. Heat transfer often occurs in combination with other unit operations, such as drying of lumber or foods, alcohol distillation, burning of fuel, and evaporation. The heat transfer occurs because of a temperature difference driving force and heat flows from the high- to the low-temperature region.

In Section 2.3 we derived an equation for a general property balance of momentum, thermal energy, or mass at unsteady state by writing Eq. (2.3-7). Writing a similar equation but specifically for heat transfer,

$$\begin{pmatrix} \text{rate of} \\ \text{heat in} \end{pmatrix} + \begin{pmatrix} \text{rate of gener-} \\ \text{ation of heat} \end{pmatrix} = \begin{pmatrix} \text{rate of} \\ \text{heat out} \end{pmatrix} + \begin{pmatrix} \text{rate of accumu-} \\ \text{lation of heat} \end{pmatrix} \tag{4.1-1}$$

Assuming the rate of transfer of heat occurs only by conduction, we can rewrite Eq. (2.3-14), which is *Fourier's law*, as

$$\frac{q_x}{A} = -k\frac{dT}{dx} \tag{4.1-2}$$

Making an unsteady-state heat balance for the x direction only on the element of volume or control volume in Fig. 4.1-1 by using Eqs. (4.1-1) and (4.1-2) with the cross-sectional area being A m^2,

$$q_{x|x} + \dot{q}(\Delta x \cdot A) = q_{x|x+\Delta x} + \rho c_p \frac{\partial T}{\partial t}(\Delta x \cdot A) \tag{4.1-3}$$

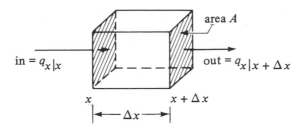

FIGURE 4.1-1. *Unsteady-state balance for heat transfer in control volume.*

where $\dot{q}$ is rate of heat generated per unit volume. Assuming no heat generation and also assuming steady-state heat transfer where the rate of accumulation is zero, Eq. (4.1-3) becomes

$$q_{x|x} = q_{x|x+\Delta x} \qquad (4.1\text{-}4)$$

This means the rate of heat input by conduction = the rate of heat output by conduction; or q_x is a constant with time for steady-state heat transfer.

In this chapter we are concerned with a control volume where the rate of accumulation of heat is zero and we have steady-state heat transfer. The rate of heat transfer is then constant with time, and the temperatures at various points in the system do not change with time. To solve problems in steady-state heat transfer, various mechanistic expressions in the form of differential equations for the different modes of heat transfer such as Fourier's law are integrated. Expressions for the temperature profile and heat flux are then obtained in this chapter.

In Chapter 5 the conservation-of-energy equations (2.7-2) and (4.1-3) will be used again when the rate of accumulation is not zero and unsteady-state heat transfer occurs. The mechanistic expression for Fourier's law in the form of a partial differential equation will be used where temperature at various points and the rate of heat transfer change with time. In Section 5.6 a general differential equation of energy change will be derived and integrated for various specific cases to determine the temperature profile and heat flux.

4.1B Basic Mechanisms of Heat Transfer

Heat transfer may occur by any one or more of the three basic mechanisms of heat transfer: conduction, convection, or radiation.

1. Conduction. In conduction, heat can be conduced through solids, liquids, and gases. The heat is conducted by the transfer of the energy of motion between adjacent molecules. In a gas the "hotter" molecules, which have greater energy and motions, impart energy to the adjacent molecules at lower energy levels. This type of transfer is present to some extent in all solids, gases, or liquids in which a temperature gradient exists. In conduction, energy can also be transferred by "free" electrons, which is quite important in metallic solids. Examples of heat transfer mainly by conduction are heat transfer through walls of exchangers or a refrigerator, heat treatment of steel forgings, freezing of the ground during the winter, and so on.

2. Convection. The transfer of heat by convection implies the transfer of heat by bulk transport and mixing of macroscopic elements of warmer portions with cooler portions of a gas or a liquid. It also often involves the energy exchange between a solid surface and a fluid. A distinction must be made between forced-convection heat transfer, where a fluid is forced to flow past a solid surface by a pump, fan, or other mechanical means, and natural or free convection, where warmer or cooler fluid next to the solid surface causes a circulation because of a density difference resulting from the temperature differences in the fluid. Examples of heat transfer by convection are loss of heat from a car radiator where the air is being circulated by a fan, cooking of foods in a vessel being stirred, cooling of a hot cup of coffee by blowing over the surface, and so on.

3. *Radiation.* Radiation differs from heat transfer by conduction and convection in that no physical medium is needed for its propagation. Radiation is the transfer of energy through space by means of electromagnetic waves in much the same way as electro-magnetic light waves transfer light. The same laws which govern the transfer of light govern the radiant transfer of heat. Solids and liquids tend to absorb the radiation being transferred through it, so that radiation is important primarily in transfer through space or gases. The most important example of radiation is the transport of heat to the earth from the sun. Other examples are cooking of food when passed below red-hot electric heaters, heating of fluids in coils of tubing inside a combustion furnace, and so on.

4.1C Fourier's Law of Heat Conduction

As discussed in Section 2.3 for the general molecular transport equation, all three main types of rate-transfer processes—momentum transfer, heat transfer, and mass transfer—are characterized by the same general type of equation. The transfer of electric current can also be included in this category. This basic equation is as follows:

$$\text{rate of a transfer process} = \frac{\text{driving force}}{\text{resistance}} \tag{2.3-1}$$

This equation states what we know intuitively: that in order to transfer a property such as heat or mass, we need a driving force to overcome a resistance.

The transfer of heat by conduction also follows this basic equation and is written as Fourier's law for heat conduction in fluids or solids.

$$\frac{q_x}{A} = -k \frac{dT}{dx} \tag{4.1-2}$$

where q_x is the heat-transfer rate in the x direction in watts (W), A is the cross-sectional area normal to the direction of flow of heat in m^2, T is temperature in K, x is distance in m, and k is the thermal conductivity in $W/m \cdot K$ in the SI system. The quantity q_x/A is called the heat flux in W/m^2. The quantity dT/dx is the temperature gradient in the x direction. The minus sign in Eq. (4.1-2) is required because if the heat flow is positive in a given direction, the temperature decreases in this direction.

The units in Eq. (4.1-2) may also be expressed in the cgs system with q_x in cal/s, A in cm^2, k in $cal/s \cdot °C \cdot cm$, T in °C, and x in cm. In the English system, q_x is in btu/h, A in ft^2, T in °F, x in ft, k in $btu/h \cdot °F \cdot ft$, and q_x/A in $btu/h \cdot ft^2$. From Appendix A.1, the conversion factors are, for thermal conductivity,

$$1 \text{ btu/h} \cdot ft \cdot °F = 4.1365 \times 10^{-3} \text{ cal/s} \cdot cm \cdot °C \tag{4.1-5}$$

$$1 \text{ btu/h} \cdot ft \cdot °F = 1.73073 \text{ W/m} \cdot K \tag{4.1-6}$$

For heat flux and power,

$$1 \text{ btu/h} \cdot \text{ft}^2 = 3.1546 \text{ W/m}^2 \qquad \text{(4.1-7)}$$

$$1 \text{ btu/h} = 0.29307 \text{ W} \qquad \text{(4.1-8)}$$

Fourier's law, Eq. (4.1-2), can be integrated for the case of steady-state heat transfer through a flat wall of constant cross-sectional area A, where the inside temperature at point 1 is T_1 and T_2 at point 2 a distance of $x_2 - x_1$ m away. Rearranging Eq. (4.1-2),

$$\frac{q_x}{A} \int_{x_1}^{x_2} dx = -k \int_{T_1}^{T_2} dT \qquad \text{(4.1-9)}$$

Integrating, assuming that k is constant and does not vary with temperature and dropping the subscript x on q_x for convenience,

$$\frac{q}{A} = \frac{k}{x_2 - x_1} (T_1 - T_2) \qquad \text{(4.1-10)}$$

EXAMPLE 4.1-1. Heat Loss Through an Insulating Wall
Calculate the heat loss per m^2 of surface area for an insulating wall composed of 25.4-mm-thick fiber insulating board, where the inside temperature is 352.7 K and the outside temperature is 297.1 K.

Solution: From Appendix A.3, the thermal conductivity of fiber insulating board is 0.048 W/m·K. The thickness $x_2 - x_1 = 0.0254$ m. Substituting into Eq. (4.1-10),

$$\frac{q}{A} = \frac{k}{x_2 - x_1}(T_1 - T_2) = \frac{0.048}{0.0254}(352.7 - 297.1)$$

$$= 105.1 \text{ W/m}^2$$

$$= (105.1 \text{ W/m}^2)\frac{1}{(3.1546 \text{ W/m}^2)/(\text{btu/h} \cdot \text{ft}^2)} = 33.30 \text{ btu/h} \cdot \text{ft}^2$$

4.1D Thermal Conductivity

The defining equation for thermal conductivity is given as Eq. (4.1-2), and with this definition, experimental measurements have been made to determine the thermal conductivity of different materials. In Table 4.1-1 thermal conductivities are given for a few materials for the purpose of comparison. More detailed data are given in Appendix A.3 for inorganic and organic materials and A.4 for food and biological materials. As seen in Table 4.1-1, gases have quite low values of thermal conductivity, liquids intermediate values, and solid metals very high values.

1. Gases. In gases the mechanism of thermal conduction is relatively simple. The molecules are in continuous random motion, colliding with one another and exchanging energy and momentum. If a molecule moves from a high-temperature region to a region of lower temperature, it transports kinetic energy to this region and gives up this energy through collisions with lower-energy molecules. Since smaller molecules move faster, gases such as hydrogen should have higher thermal conductivities, as shown in Table 4.1-1.

Theories to predict thermal conductivities of gases are reasonably accurate and are given elsewhere (R1). The thermal conductivity increases approximately as the square root of the absolute temperature and is independent of pressure up to a few atmospheres. At very low pressures (vacuum), however, the thermal conductivity approaches zero.

2. Liquids. The physical mechanism of conduction of energy in liquids is somewhat similar to that of gases, where higher-energy molecules collide with lower-energy molecules. However, the molecules are packed so closely together that molecular force fields exert a strong effect on the energy exchange. Since an adequate molecular theory of liquids is not available, most correlations to predict the thermal conductivities are empirical. Reid et al. (R1) discuss these in detail. The thermal conductivity of liquids varies moderately with temperature and often can be expressed as a linear variation,

$$k = a + bT \tag{4.1-11}$$

where a and b are empirical constants. Thermal conductivities of liquids are essentially independent of pressure.

Water has a high thermal conductivity compared to organic-type liquids such as benzene. As shown in Table 4.1-1, the thermal conductivities of most unfrozen foodstuffs, such as skim milk and applesauce, which contain large amounts of water have thermal conductivities near that of pure water.

3. Solids. The thermal conductivity of homogeneous solids varies quite widely, as may be seen for some typical values in Table 4.1-1. The metallic solids of copper and aluminum have very high thermal conductivities, and some insulating nonmetallic materials such as rock wool and corkboard have very low conductivities.

Heat or energy is conducted through solids by two mechanisms. In the first, which applies primarily to metallic solids, heat, like electricity, is conducted by free electrons which move through the metal lattice. In the second mechanism, present in all solids, heat is conducted by the transmission of energy of vibration between adjacent atoms.

TABLE 4.1-1. *Thermal Conductivities of Some Materials at 101.325 kPa (1 Atm) Pressure (k in W/m · K)*

Substance	Temp. (K)	k	Ref.	Substance	Temp. (K)	k	Ref.
Gases				Solids			
Air	273	0.0242	(K2)	Ice	273	2.25	(C1)
	373	0.0316		Fire claybrick	473	1.00	(P1)
H_2	273	0.167	(K2)	Paper	—	0.130	(M1)
n-Butane	273	0.0135	(P2)	Hard rubber	273	0.151	(M1)
Liquids				Cork board	303	0.043	(M1)
Water	273	0.569	(P1)	Asbestos	311	0.168	(M1)
	366	0.680		Rock wool	266	0.029	(K1)
Benzene	303	0.159	(P1)	Steel	291	45.3	(P1)
	333	0.151			373	45	
Biological materials				Copper	273	388	(P1)
and foods					373	377	
Olive oil	293	0.168	(P1)	Aluminum	273	202	(P1)
	373	0.164					
Lean beef	263	1.35	(C1)				
Skim milk	275	0.538	(C1)				
Applesauce	296	0.692	(C1)				
Salmon	277	0.502	(C1)				
	248	1.30					

Thermal conductivities of insulating materials such as rock wool approach that of air since the insulating materials contain large amounts of air trapped in void spaces. Superinsulations to insulate cryogenic materials such as liquid hydrogen are composed of multiple layers of highly reflective materials separated by evacuated insulating spacers. Values of thermal conductivity are considerably lower than for air alone.

Ice has a thermal conductivity much greater than water. Hence, the thermal conductivities of frozen foods such as lean beef and salmon given in Table 4.1-1 are much higher than for unfrozen foods.

4.1E Convective-Heat-Transfer Coefficient

It is well known that a hot piece of material will cool faster when air is blown or forced by the object. When the fluid outside the solid surface is in forced or natural convective motion, we express the rate of heat transfer from the solid to the fluid, or vice versa, by the following equation:

$$q = hA(T_w - T_f) \tag{4.1-12}$$

where q is the heat-transfer rate in W, A is the area in m^2, T_w is the temperature of the solid surface in K, T_f is the average or bulk temperature of the fluid flowing by in K, and h is the convective heat-transfer coefficient in $W/m^2 \cdot K$. In English units, h is in $btu/h \cdot ft^2 \cdot °F$.

The coefficient h is a function of the system geometry, fluid properties, flow velocity, and temperature difference. In many cases, empirical correlations are available to predict this coefficient, since it often cannot be predicted theoretically. Since we know that when a fluid flows by a surface there is a thin, almost stationary layer or film of fluid adjacent to the wall which presents most of the resistance to heat transfer, we often call the coefficient h a *film coefficient*.

In Table 4.1-2 some order-of-magnitude values of h for different convective mechanisms of free or natural convection, forced convection, boiling, and condensation are given. Water gives the highest values of the heat-transfer coefficients.

To convert the heat-transfer coefficient h from English to SI units,

$$1 \text{ btu/h} \cdot ft^2 \cdot °F = 5.6783 \text{ W/m}^2 \cdot K$$

TABLE 4.1-2. *Approximate Magnitude of Some Heat-Transfer Coefficients*

	Range of Values of h	
Mechanism	*btu/h·ft²·°F*	*W/m²·K*
Condensing steam	1000–5000	5700–28 000
Condensing organics	200–500	1100–2800
Boiling liquids	300–5000	1700–28 000
Moving water	50–3000	280–17 000
Moving hydrocarbons	10–300	55–1700
Still air	0.5–4	2.8–23
Moving air	2–10	11.3–55

4.2A Conduction Through a Flat Slab or Wall

In this section Fourier's equation (4.1-2) will be used to obtain equations for one-dimensional steady-state conduction of heat through some simple geometries. For a flat slab or wall where the cross-sectional area A and k in Eq. (4.1-2) are constant, we obtained Eq. (4.1-10), which we rewrite as

$$\frac{q}{A} = \frac{k}{x_2 - x_1}(T_1 - T_2) = \frac{k}{\Delta x}(T_1 - T_2) \tag{4.2-1}$$

This is shown in Fig. 4.2-1, where $\Delta x = x_2 - x_1$. Equation (4.2-1) indicates that if T is substituted for T_2 and x for x_2, the temperature varies linearly with distance as shown in Fig. 4.2-1b.

If the thermal conductivity is not constant but varies linearly with temperature, then substituting Eq. (4.1-11) into Eq. (4.1-2) and integrating,

$$\frac{q}{A} = \frac{a + b\dfrac{T_1 + T_2}{2}}{\Delta x}(T_1 - T_2) = \frac{k_m}{\Delta x}(T_1 - T_2) \tag{4.2-2}$$

where

$$k_m = a + b\frac{T_1 + T_2}{2} \tag{4.2-3}$$

This means that the mean value of k (i.e., k_m) to use in Eq. (4.2-2) is the value of k evaluated at the linear average of T_1 and T_2.

As stated in the introduction in Eq. (2.3-1), the rate of a transfer process equals the driving force over the resistance. Equation (4.2-1) can be rewritten in that form.

$$q = \frac{T_1 - T_2}{\Delta x/kA} = \frac{T_1 - T_2}{R} = \frac{\text{driving force}}{\text{resistance}} \tag{4.2-4}$$

where $R = \Delta x/kA$ and is the resistance in K/W or h · °F/btu.

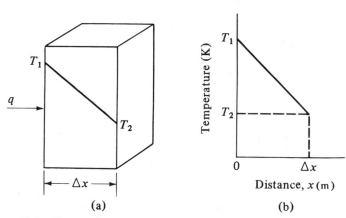

FIGURE 4.2-1. *Heat conduction in a flat wall: (a) geometry of wall, (b) temperature plot.*

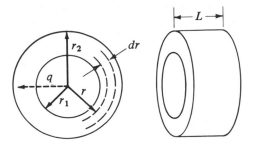

FIGURE 4.2-2. *Heat conduction in a cylinder.*

4.2B Conduction Through a Hollow Cylinder

In many instances in the process industries, heat is being transferred through the walls of a thick-walled cylinder as in a pipe that may or may not be insulated. Consider the hollow cylinder in Fig. 4.2-2 with an inside radius of r_1, where the temperature is T_1, an outside radius of r_2 having a temperature of T_2, and a length of L m. Heat is flowing radially from the inside surface to the outside. Rewriting Fourier's law, Eq. (4.1-2), with distance dr instead of dx,

$$\frac{q}{A} = -k \frac{dT}{dr} \qquad (4.2\text{-}5)$$

The cross-sectional area normal to the heat flow is

$$A = 2\pi r L \qquad (4.2\text{-}6)$$

Substituting Eq. (4.2-6) into (4.2-5), rearranging, and integrating,

$$\frac{q}{2\pi L} \int_{r_1}^{r_2} \frac{dr}{r} = -k \int_{T_1}^{T_2} dT \qquad (4.2\text{-}7)$$

$$q = k \frac{2\pi L}{\ln (r_2/r_1)} (T_1 - T_2) \qquad (4.2\text{-}8)$$

Multiplying numerator and denominator by $(r_2 - r_1)$,

$$q = k A_{\text{lm}} \frac{T_1 - T_2}{r_2 - r_1} = \frac{T_1 - T_2}{(r_2 - r_1)/(k A_{\text{lm}})} = \frac{T_1 - T_2}{R} \qquad (4.2\text{-}9)$$

where

$$A_{\text{lm}} = \frac{(2\pi L r_2) - (2\pi L r_1)}{\ln (2\pi L r_2/2\pi L r_1)} = \frac{A_2 - A_1}{\ln (A_2/A_1)} \qquad (4.2\text{-}10)$$

$$R = \frac{r_2 - r_1}{k A_{\text{lm}}} = \frac{\ln (r_2/r_1)}{2\pi k L} \qquad (4.2\text{-}11)$$

The log mean area is A_{lm}. In engineering practice, if $A_2/A_1 < 1.5/1$, the linear mean area of $(A_1 + A_2)/2$ is within 1.5% of the log mean area. From Eq. (4.2-8), if r is substituted for r_2 and T for T_2, the temperature is seen to be a linear function of $\ln r$ instead of r as in the case of a flat wall. If the thermal conductivity varies with temperature as in Eq. (4.1-10), it can be shown that the mean value to use in a cylinder is still k_m of Eq. (4.2-3).

EXAMPLE 4.2-1. Length of Tubing for Cooling Coil

A thick-walled cylindrical tubing of hard rubber having an inside radius of 5 mm and an outside radius of 20 mm is being used as a temporary cooling coil in a bath. Ice water is flowing rapidly inside and the inside wall temperature is 274.9 K. The outside surface temperature is 297.1 K. A total of 14.65 W must be removed from the bath by the cooling coil. How many m of tubing are needed?

Solution: From Appendix A.3, the thermal conductivity at 0°C (273 K) is $k = 0.151$ W/m·K. Since data at other temperatures are not available, this value will be used for the range of 274.9 to 297.1 K.

$$r_1 = \frac{5}{1000} = 0.005 \text{ m} \qquad r_2 = \frac{20}{1000} = 0.02 \text{ m}$$

The calculation will be done first for a length of 1.0 m of tubing. Solving for the areas A_1, A_2, and A_{lm} in Eq. (4.2-10),

$$A_1 = 2\pi L r_1 = 2\pi(1.0)(0.005) = 0.0314 \text{ m}^2 \qquad A_2 = 0.1257 \text{ m}^2$$

$$A_{lm} = \frac{A_2 - A_1}{\ln (A_2/A_1)} = \frac{0.1257 - 0.0314}{2.303 \log (0.1257/0.0314)} = 0.0680 \text{ m}^2$$

Substituting into Eq. (4.2-9) and solving,

$$q = kA_{lm} \frac{T_1 - T_2}{r_2 - r_1} = 0.151(0.0682)\left(\frac{274.9 - 297.1}{0.02 - 0.005}\right)$$

$$= -15.2 \text{ W (51.9 btu/h)}$$

The negative sign indicates that the heat flow is from r_2 on the outside to r_1 on the inside. Since 15.2 W is removed for a 1-m length, the needed length is

$$\text{length} = \frac{14.65 \text{ W}}{15.2 \text{ W/m}} = 0.964 \text{ m}$$

Note that the thermal conductivity of rubber is quite small. Generally, metal cooling coils are used, since the thermal conductivity of metals is quite high. The liquid film resistances in this case are quite small and are neglected.

4.2C Conduction Through a Hollow Sphere

Heat conduction through a hollow sphere is another case of one-dimensional conduction. Using Fourier's law for constant thermal conductivity with distance dr, where r is the radius of the sphere,

$$\frac{q}{A} = -k \frac{dT}{dr} \tag{4.2-5}$$

The cross-sectional area normal to the heat flow is

$$A = 4\pi r^2 \tag{4.2-12}$$

Substituting Eq. (4.2-12) into (4.2-5), rearranging, and integrating,

$$\frac{q}{4\pi} \int_{r_1}^{r_2} \frac{dr}{r^2} = -k \int_{T_1}^{T_2} dt \tag{4.2-13}$$

$$q = \frac{4\pi k(T_1 - T_2)}{1/r_1 - 1/r_2} = \frac{T_1 - T_2}{(1/r_1 - 1/r_2)/4\pi k} \tag{4.2-14}$$

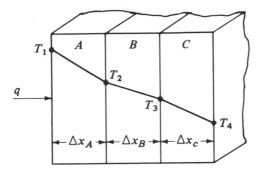

FIGURE 4.3-1.   *Heat flow through a multilayer wall.*

It can be easily shown that the temperature varies hyperbolically with the radius. (See Problem 4.2-5.)

4.3   CONDUCTION THROUGH SOLIDS IN SERIES

4.3A   Plane Walls in Series

In the case where there is a multilayer wall of more than one material present as shown in Fig. 4.3-1, we proceed as follows. The temperature profiles in the three materials A, B, and C are shown. Since the heat flow q must be the same in each layer, we can write Fourier's equation for each layer as

$$q = \frac{k_A A}{\Delta x_A}(T_1 - T_2) = \frac{k_B A}{\Delta x_B}(T_2 - T_3) = \frac{k_C A}{\Delta x_C}(T_3 - T_4) \qquad \textbf{(4.3-1)}$$

Solving each equation for ΔT,

$$T_1 - T_2 = q\frac{\Delta x_A}{k_A A} \qquad T_2 - T_3 = q\frac{\Delta x_B}{k_B A} \qquad T_3 - T_4 = q\frac{\Delta x_C}{k_C A} \qquad \textbf{(4.3-2)}$$

Adding the equations for $T_1 - T_2$, $T_2 - T_3$, and $T_3 - T_4$, the internal temperatures T_2 and T_3 drop out and the final rearranged equation is

$$q = \frac{T_1 - T_4}{\Delta x_A/(k_A A) + \Delta x_B/(k_B A) + \Delta x_C/(k_C A)} = \frac{T_1 - T_4}{R_A + R_B + R_C} \qquad \textbf{(4.3-3)}$$

where the resistance $R_A = \Delta x_A/k_A A$, and so on.

Hence, the final equation is in terms of the overall temperature drop $T_1 - T_4$ and the total resistance, $R_A + R_B + R_C$.

> ***EXAMPLE 4.3-1.   Heat Flow Through an Insulated Wall of a Cold Room***
> A cold-storage room is constructed of an inner layer of 12.7 mm of pine, a middle layer of 101.6 mm of cork board, and an outer layer of 76.2 mm of concrete. The wall surface temperature is 255.4 K inside the cold room and 297.1 K at the outside surface of the concrete. Use conductivities from Appendix A.3 for pine, 0.151; for cork board, 0.0433; and for concrete, 0.762 W/m · K. Calculate the heat loss in W for 1 m² and the temperature at the interface between the wood and cork board.

Solution: Calling $T_1 = 255.4$, $T_4 = 297.1$ K, pine as material A, cork as B, and concrete as C, a tabulation of the properties and dimensions is as follows:

$$k_A = 0.151 \qquad k_B = 0.0433 \qquad k_C = 0.762$$

$$\Delta x_A = 0.0127 \text{ m}$$

$$\Delta x_B = 0.1016 \text{ m}$$

$$\Delta x_C = 0.0762 \text{ m}$$

The resistances for each material are, from Eq. (4.3-3), for an area of 1 m^2,

$$R_A = \frac{\Delta x_A}{k_A A} = \frac{0.0127}{0.151(1)} = 0.0841 \text{ K/W}$$

$$R_B = \frac{\Delta x_B}{k_B A} = \frac{0.1016}{0.0433(1)} = 2.346$$

$$R_C = \frac{\Delta x_C}{k_C A} = \frac{0.0762}{0.762(1)} = 0.100$$

Substituting into Eq. (4.3-3),

$$q = \frac{T_1 - T_4}{R_A + R_B + R_C} = \frac{255.4 - 297.1}{0.0841 + 2.346 + 0.100}$$

$$= \frac{-41.7}{2.530} = -16.48 \text{ W} \ (-56.23 \text{ btu/h})$$

Since the answer is negative, heat flows in from the outside.

To calculate the temperature T_2 at the interface between the pine wood and cork,

$$q = \frac{T_1 - T_2}{R_A}$$

Substituting the known values and solving,

$$-16.48 = \frac{255.4 - T_2}{0.0841} \quad \text{and} \quad T_2 = 256.79 \text{ K at the interface}$$

An alternative procedure to use to calculate T_2 is to use the fact that the temperature drop is proportional to the resistance.

$$T_1 - T_2 = \frac{R_A}{R_A + R_B + R_C}(T_1 - T_4) \qquad \text{(4.3-4)}$$

Substituting,

$$255.4 - T_2 = \frac{0.0841(255.4 - 297.1)}{2.530} = -1.39 \text{ K}$$

Hence, $T_2 = 256.79$ K, as calculated before.

4.3B Multilayer Cylinders

In the process industries, heat transfer often occurs through multilayers of cylinders, as for example when heat is being transferred through the walls of an insulated pipe. Figure

FIGURE 4.3-2. *Radial heat flow through multi-ple cylinders in series.*

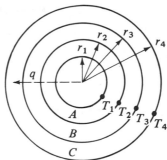

4.3-2 shows a pipe with two layers of insulation around it, i.e., a total of three concentric hollow cylinders. The temperature drop is $T_1 - T_2$ across material A, $T_2 - T_3$ across B, and $T_3 - T_4$ across C.

The heat-transfer rate q will, of course, be the same for each layer, since we are at steady state. Writing an equation similar to Eq. (4.2-9) for each concentric cylinder,

$$q = \frac{T_1 - T_2}{(r_2 - r_1)/(k_A A_{A\,\mathrm{lm}})} = \frac{T_2 - T_3}{(r_3 - r_2)/(k_B A_{B\,\mathrm{lm}})} = \frac{T_3 - T_4}{(r_4 - r_3)/(k_C A_{C\,\mathrm{lm}})} \tag{4.3-5}$$

where

$$A_{A\,\mathrm{lm}} = \frac{A_2 - A_1}{\ln (A_2/A_1)} \quad A_{B\,\mathrm{lm}} = \frac{A_3 - A_2}{\ln (A_3/A_2)} \quad A_{C\,\mathrm{lm}} = \frac{A_4 - A_3}{\ln (A_4/A_3)} \tag{4.3-6}$$

Using the same method to combine the equations to eliminate T_2 and T_3 as was done for the flat walls in series, the final equations are

$$q = \frac{T_1 - T_4}{(r_2 - r_1)/(k_A A_{A\,\mathrm{lm}}) + (r_3 - r_2)/(k_B A_{B\,\mathrm{lm}}) + (r_4 - r_3)/(k_C A_{C\,\mathrm{lm}})} \tag{4.3-7}$$

$$q = \frac{T_1 - T_4}{R_A + R_B + R_C} = \frac{T_1 - T_4}{\sum R} \tag{4.3-8}$$

Hence, the overall resistance is again the sum of the individual resistances in series.

EXAMPLE 4.3-2.   Heat Loss from an Insulated Pipe

A thick-walled tube of stainless steel (A) having a $k = 21.63$ W/m·K with dimensions of 0.0254 m ID and 0.0508 m OD is covered with a 0.0254-m layer of asbestos (B) insulation, $k - 0.2423$ W/m·K. The inside wall temper-ature of the pipe is 811 K and the outside surface of the insulation is at 310.8 K. For a 0.305-m length of pipe, calculate the heat loss and also the temperature at the interface between the metal and the insulation.

Solution: Calling $T_1 = 811$ K, T_2 the interface, and $T_3 = 310.8$ K, the dimensions are

$$r_1 = \frac{0.0254}{2} = 0.0127 \text{ m} \qquad r_2 = \frac{0.0508}{2} = 0.0254 \text{ m} \qquad r_3 = 0.0508 \text{ m}$$

The areas are as follows for $L = 0.305$ m.

$$A_1 = 2\pi L r_1 = 2\pi(0.305)(0.0127) = 0.0243 \text{ m}^2$$

$$A_2 = 2\pi L r_2 = 2\pi(0.305)(0.0254) = 0.0487 \text{ m}^2$$

$$A_3 = 2\pi L r_3 = 2\pi(0.305)(0.0508) = 0.0974 \text{ m}^2$$

From Eq. (4.3-6), the log mean areas for the stainless steel (A) and asbestos (B) are

$$A_{A\text{ lm}} = \frac{A_2 - A_1}{\ln (A_2/A_1)} = \frac{0.0487 - 0.0243}{\ln (0.0487/0.0243)} = 0.0351 \text{ m}^2$$

$$A_{B\text{ lm}} = \frac{A_3 - A_2}{\ln (A_3/A_2)} = \frac{0.0974 - 0.0487}{\ln (0.0974/0.0487)} = 0.0703 \text{ m}^2$$

From Eq. (4.3-7) the resistances are

$$R_A = \frac{r_2 - r_1}{k_A A_{A\text{ lm}}} = \frac{0.0127}{21.63(0.0351)} = 0.01673 \text{ K/W}$$

$$R_B = \frac{r_3 - r_2}{k_B A_{B\text{ lm}}} = \frac{0.0254}{0.2423(0.0703)} = 1.491 \text{ K/W}$$

Hence, the heat-transfer rate is

$$q = \frac{T_1 - T_3}{R_A + R_B} = \frac{811 - 310.8}{0.01673 + 1.491} = 331.7 \text{ W} \ (1132 \text{ btu/h})$$

To calculate the temperature T_2,

$$q = \frac{T_1 - T_2}{R_A} \quad \text{or} \quad 331.7 = \frac{811 - T_2}{0.01673}$$

Solving, $811 - T_2 = 5.5$ K and $T_2 = 805.5$ K. Only a small temperature drop occurs across the metal wall because of its high thermal conductivity.

4.3C   Conduction Through Materials in Parallel

Suppose that two plane solids A and B are placed side by side in parallel, and the direction of heat flow is perpendicular to the plane of the exposed surface of each solid. Then the total heat flow is the sum of the heat flow through solid A plus that through B. Writing Fourier's equation for each solid and summing,

$$q_T = q_A + q_B = \frac{k_A A_A}{\Delta x_A} (T_1 - T_2) + \frac{k_B A_B}{\Delta x_B} (T_3 - T_4) \qquad (4.3\text{-}9)$$

where q_T is total heat flow, T_1 and T_2 are the front and rear surface temperatures of solid A; T_3 and T_4, for solid B.

If we assume that $T_1 = T_3$ (front temperatures the same for A and B) and $T_2 = T_4$ (equal rear temperatures),

$$q_T = \frac{T_1 - T_2}{\Delta x_A/k_A A_A} + \frac{T_1 - T_2}{\Delta x_B/k_B A_B} = \left(\frac{1}{R_A} + \frac{1}{R_B}\right)(T_1 - T_2) \qquad (4.3\text{-}10)$$

An example would be an insulated wall (A) of a brick oven where steel reinforcing members (B) are in parallel and penetrate the wall. Even though the area A_B of the steel would be small compared to the insulated brick area A_A, the higher conductivity of the metal (which could be several hundred times larger than that of the brick) could allow a large portion of the heat lost to be conducted by the steel.

Another example is a method of increasing heat conduction to accelerate the freeze drying of meat. Spikes of metal in the frozen meat conduct heat more rapidly into the insides of the meat.

It should be mentioned that in some cases some two-dimensional heat flow can occur if the thermal conductivities of the materials in parallel differ markedly. Then the results using Eq. (4.3-10) would be affected somewhat.

4.3D   Combined Convection and Conduction and Overall Coefficients

In many practical situations the surface temperatures (or boundary conditions at the surface) are not known, but there is a fluid on both sides of the solid surfaces. Consider the plane wall in Fig. 4.3-3a with a hot fluid at temperature T_1 on the inside surface and a cold fluid at T_4 on the outside surface. The outside convective coefficient is h_o W/m^2 · K and h_i on the inside. (Methods to predict the convective h will be given later in Section 4.4 of this chapter.)

The heat-transfer rate using Eqs. (4.1-12) and (4.3-1) is given as

$$q = h_i A(T_1 - T_2) = \frac{k_A A}{\Delta x_A}(T_2 - T_3) = h_o A(T_3 - T_4) \qquad (4.3\text{-}11)$$

Expressing $1/h_i A$, $\Delta x_A/k_A A$, and $1/h_o A$ as resistances and combining the equations as before,

$$q = \frac{T_1 - T_4}{1/h_i A + \Delta x_A/k_A A + 1/h_o A} = \frac{T_1 - T_4}{\sum R} \qquad (4.3\text{-}12)$$

The overall heat transfer by combined conduction and convection is often expressed in terms of an overall heat-transfer coefficient U defined by

$$q = U A \Delta T_{\text{overall}} \qquad (4.3\text{-}13)$$

where $\Delta T_{\text{overall}} = T_1 - T_4$ and U is

$$U = \frac{1}{1/h_i + \Delta x_A/k_A + 1/h_o} \frac{W}{\text{m}^2 \cdot \text{K}} \left(\frac{\text{btu}}{\text{h} \cdot \text{ft}^2 \cdot {}^\circ\text{F}}\right) \qquad (4.3\text{-}14)$$

A more important application is heat transfer from a fluid outside a cylinder, through a metal wall, and to a fluid inside the tube, as often occurs in heat exchangers. In Fig. 4.3-3b, such a case is shown.

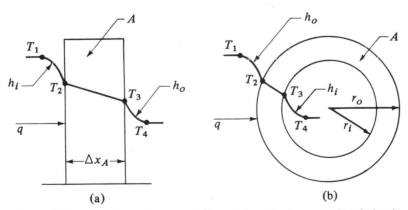

(a)   (b)

FIGURE 4.3-3.   *Heat flow with convective boundaries: (a) plane wall, (b) cylindrical wall.*

Using the same procedure as before, the overall heat-transfer rate through the cylinder is

$$q = \frac{T_1 - T_4}{1/h_i A_i + (r_o - r_i)/k_A A_{A \, lm} + 1/h_o A_o} = \frac{T_1 - T_4}{\sum R} \tag{4.3-15}$$

where A_i represents $2\pi L r_i$, the inside area of the metal tube; $A_{A \, lm}$ the log mean area of the metal tube; and A_o the outside area.

The overall heat-transfer coefficient U for the cylinder may be based on the inside area A_i or the outside area A_o of the tube. Hence,

$$q = U_i A_i (T_1 - T_4) = U_o A_o (T_1 - T_4) = \frac{T_1 - T_4}{\sum R} \tag{4.3-16}$$

$$U_i = \frac{1}{1/h_i + (r_o - r_i)A_i/k_A A_{A \, lm} + A_i/A_o h_o} \tag{4.3-17}$$

$$U_o = \frac{1}{A_o/A_i h_i + (r_o - r_i)A_o/k_A A_{A \, lm} + 1/h_o} \tag{4.3-18}$$

EXAMPLE 4.3-3. *Heat Loss by Convection and Conduction and Overall U*
Saturated steam at 267°F is flowing inside a $\frac{3}{4}$-in. steel pipe having an ID of 0.824 in. and an OD of 1.050 in. The pipe is insulated with 1.5 in. of insulation on the outside. The convective coefficient for the inside steam surface of the pipe is estimated as $h_i = 1000$ btu/h·ft²·°F, and the convective coefficient on the outside of the lagging is estimated as $h_o = 2$ btu/h·ft²·°F. The mean thermal conductivity of the metal is 45 W/m·K or 26 btu/h·ft·°F and 0.064 W/m·K or 0.037 btu/h·ft·°F for the insulation.
    (a) Calculate the heat loss for 1 ft of pipe using resistances if the surrounding air is at 80°F.
    (b) Repeat using the overall U_i based on the inside area A_i.

Solution: Calling r_i the inside radius of the steel pipe, r_1 the outside radius of the pipe, and r_o the outside radius of the lagging, then

$$r_i = \frac{0.412}{12} \text{ ft} \qquad r_i = \frac{0.525}{12} \text{ ft} \qquad r_o = \frac{2.025}{12} \text{ ft}$$

For 1 ft of pipe, the areas are as follows.

$$A_i = 2\pi L r_i = 2\pi(1)\left(\frac{0.412}{12}\right) = 0.2157 \text{ ft}^2$$

$$A_1 = 2\pi L r_1 = 2\pi(1)\left(\frac{0.525}{12}\right) = 0.2750 \text{ ft}^2$$

$$A_o = 2\pi L r_o = 2\pi(1)\left(\frac{2.025}{12}\right) = 1.060 \text{ ft}^2$$

From Eq. (4.3-6) the log mean areas for the steel (A) pipe and lagging (B) are

$$A_{A \, lm} = \frac{A_1 - A_i}{\ln (A_1/A_i)} = \frac{0.2750 - 0.2157}{\ln (0.2750/0.2157)} = 0.245$$

$$A_{B \, lm} = \frac{A_o - A_1}{\ln (A_o/A_1)} = \frac{1.060 - 0.2750}{\ln (1.060/0.2750)} = 0.583$$

From Eq. (4.3-15) the various resistances are

$$R_i = \frac{1}{h_i A_i} = \frac{1}{1000(0.2157)} = 0.00464$$

$$R_A = \frac{r_1 - r_i}{k_A A_{A \, lm}} = \frac{(0.525 - 0.412)/12}{26(0.245)} = 0.00148$$

$$R_B = \frac{r_o - r_1}{k_B A_{B \, lm}} = \frac{(2.025 - 0.525)/12}{0.037(0.583)} = 5.80$$

$$R_o = \frac{1}{h_o A_o} = \frac{1}{2(1.060)} = 0.472$$

Using an equation similar to Eq. (4.3-15),

$$q = \frac{T_i - T_o}{R_i + R_A + R_B + R_o} = \frac{267 - 80}{0.00464 + 0.00148 + 5.80 + 0.472} \qquad \textbf{(4.3-19)}$$

$$= \frac{267 - 80}{6.278} = 29.8 \text{ btu/h}$$

For part (b), the equation relating U_i to q is Eq. (4.3-16), which can be equated to Eq. (4.3-19).

$$q = U_i A_i (T_i - T_o) = \frac{T_i - T_o}{\sum R} \qquad \textbf{(4.3-20)}$$

Solving for U_i,

$$U_i = \frac{1}{A_i \sum R} \qquad \textbf{(4.3-21)}$$

Substituting known values,

$$U_i = \frac{1}{0.2157(6.278)} = 0.738 \; \frac{\text{btu}}{\text{h} \cdot \text{ft}^2 \cdot {}^\circ\text{F}}$$

Then to calculate q,

$$q = U_i A_i (T_i - T_o) = 0.738(0.2157)(267 - 80) = 29.8 \text{ btu/h} \; (8.73 \text{ W})$$

4.3E Conduction with Internal Heat Generation

In certain systems heat is generated inside the conducting medium; i.e., a uniformly distributed heat source is present. Examples of this are electric resistance heaters and nuclear fuel rods. Also, if a chemical reaction is occurring uniformly in a medium, a heat of reaction is given off. In the agricultural and sanitation fields, compost heaps and trash heaps in which biological activity is occurring will have heat given off.

Other important examples are in food processing, where the heat of respiration of fresh fruits and vegetables is present. These heats of generation can be as high as 0.3 to 0.6 W/kg or 0.5 to 1 btu/h · lb$_m$.

1. Heat generation in plane wall. In Fig. 4.3-4 a plane wall is shown with internal heat generation. Heat is conducted only in the one x direction. The other walls are assumed to be insulated. The temperature T_w in K at $x = L$ and $x = -L$ is held constant. The volumetric rate of heat generation is $\dot{q}$ W/m^3 and the thermal conductivity of the medium is k W/m · K.

To derive the equation for this case of heat generation at steady state, we start with Eq. (4.1-3) but drop the accumulation term.

$$q_{x|x} + \dot{q}(\Delta x \cdot A) = q_{x|x+\Delta x} + 0 \tag{4.3-22}$$

where A is the cross-sectional area of the plate. Rearranging, dividing by Δx, and letting Δx approach zero,

$$\frac{-dq_x}{dx} + \dot{q} \cdot A = 0 \tag{4.3-23}$$

Substituting Eq. (4.1-2) for q_x,

$$\frac{d^2 T}{dx^2} + \frac{\dot{q}}{k} = 0 \tag{4.3-24}$$

Integration gives the following for $\dot{q}$ constant:

$$T = -\frac{\dot{q}}{2k} x^2 + C_1 x + C_2 \tag{4.3-25}$$

where C_1, and C_2 are integration constants. The boundary condition's are at $x = L$ or $-L$, $T = T_w$, and at $x = 0$, $T = T_0$ (center temperature). Then, the temperature profile is

$$T = -\frac{\dot{q}}{2k} x^2 + T_0 \tag{4.3-26}$$

The center temperature is

$$T_0 = \frac{\dot{q}L^2}{2k} + T_w \tag{4.3-27}$$

The total heat lost from the two faces at steady state is equal to the total heat generated, $\dot{q}_T$, in W.

$$\dot{q}_T = \dot{q}(2LA) \tag{4.3-28}$$

where A is the cross-sectional area (surface area at T_w) of the plate.

2. *Heat generation in cylinder.* In a similar manner an equation can be derived for a cylinder of radius R with uniformly distributed heat sources and constant thermal

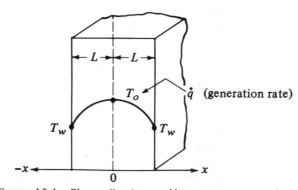

FIGURE 4.3-4. *Plane wall with internal heat generation at steady state.*

conductivity. The heat is assumed to flow only radially; i.e., the ends are neglected or insulated. The final equation for the temperature profile is

$$T = \frac{\dot{q}}{4k}(R^2 - r^2) + T_w \qquad (4.3\text{-}29)$$

where r is distance from the center. The center temperature T_0 is

$$T_0 = \frac{\dot{q}R^2}{4k} + T_w \qquad (4.3\text{-}30)$$

EXAMPLE 4.3-4.   Heat Generation in a Cylinder
An electric current of 200 A is passed through a stainless steel wire having a radius R of 0.001268 m. The wire is $L = 0.91$ m long and has a resistance R of 0.126 Ω. The outer surface temperature T_w is held at 422.1 K. The average thermal conductivity is $k = 22.5$ W/m · K. Calculate the center temperature.

Solution: First the value of $\dot{q}$ must be calculated. Since power $= I^2R$, where I is current in amps and R is resistance in ohms,

$$I^2R = \text{watts} = \dot{q}\pi R^2 L \qquad (4.3\text{-}31)$$

Substituting known values and solving,

$$(200)^2(0.126) = \dot{q}\pi(0.001268)^2(0.91)$$

$$\dot{q} = 1.096 \times 10^9 \text{ W/m}^3$$

Substituting into Eq. (4.3-30) and solving, $T_0 = 441.7$ K.

4.3F   Critical Thickness of Insulation for a Cylinder

In Fig. 4.3-5 a layer of insulation is installed around the outside of a cylinder whose radius r_1 is fixed with a length L. The cylinder has a high thermal conductivity and the inner temperature T_1 at point r_1 outside the cylinder is fixed. An example is the case where the cylinder is a metal pipe with saturated steam inside. The outer surface of the insulation at T_2 is exposed to an environment at T_0 where convective heat transfer occurs. It is not obvious if adding more insulation with a thermal conductivity of k will decrease the heat transfer rate.

At steady state the heat-transfer rate q through the cylinder and the insulation equals the rate of convection from the surface.

$$q = h_o A(T_2 - T_0) \qquad (4.3\text{-}32)$$

As insulation is added, the outside area, which is $A = 2\pi r_2 L$, increases but T_2 decreases.

FIGURE 4.3-5.   *Critical radius for insulation of cylinder or pipe.*

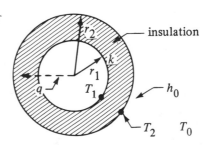

However, it is not apparent whether q increases or decreases. To determine this, an equation similar to Eq. (4.3-15) with the resistance of the insulation represented by Eq. (4.2-11) is written using the two resistances.

$$q = \frac{2\pi L(T_1 - T_0)}{\dfrac{\ln (r_2/r_1)}{k} + \dfrac{1}{r_2 h_o}} \qquad (4.3\text{-}33)$$

To determine the effect of the thickness of insulation on q, we take the derivative of q with respect to r_2, equate this result to zero, and obtain the following for maximum heat flow.

$$\frac{dq}{dr_2} = \frac{-2\pi L(T_1 - T_0)(1/r_2 k - 1/r_2^2 h_o)}{\left[\dfrac{\ln (r_2/r_1)}{k} + \dfrac{1}{r_2 h_o}\right]^2} = 0, \qquad (4.3\text{-}34)$$

Solving,

$$(r_2)_{cr} = \frac{k}{h_o} \qquad (4.3\text{-}35)$$

where $(r_2)_{cr}$ is the value of the critical radius when the heat-transfer rate is a maximum. Hence, if the outer radius r_2 is less than the critical value, adding more insulation will actually increase the heat-transfer rate q. Also, if the outer radius is greater than the critical, adding more insulation will decrease the heat-transfer rate. Using typical values of k and h_o often encountered, the critical radius is only a few mm. As a result, adding insulation on small electrical wires could increase the heat loss. Adding insulation to large pipes decreases the heat-transfer rate.

EXAMPLE 4.3-5. *Insulating an Electrical Wire and Critical Radius*
An electric wire having a diameter of 1.5 mm and covered with a plastic insulation (thickness = 2.5 mm) is exposed to air at 300 K and $h_o =$ 20 W/m² · K. The insulation has a k of 0.4 W/m · K. It is assumed that the wire surface temperature is constant at 400 K and is not affected by the covering.
    (a) Calculate the value of the critical radius.
    (b) Calculate the heat loss per m of wire length with no insulation.
    (c) Repeat (b) for the insulation present.

Solution: For part (a) using Eq. (4.3-35),

$$(r_2)_{cr} = \frac{k}{h_o} = \frac{0.4}{20} = 0.020 \text{ m} = 20 \text{ mm}$$

For part (b), $L = 1.0$ m, $r_2 = 1.5/(2 \times 1000) = 0.75 \times 10^{-3}$ m, $A = 2\pi r_2 L$. Substituting into Eq. (4.3-32),

$$q = h_o A(t_2 - T_0) = (20)(2\pi \times 0.75 \times 10^{-3} \times 1)(400 - 300) = 9.42 \text{ W}$$

For part (c) with insulation, $r_1 = 1.5/(2 \times 1000) = 0.75 \times 10^{-3}$ m, $r_2 = (2.5 + 1.5/2)/1000 = 3.25 \times 10^{-3}$ m. Substituting into Eq. (4.3-33),

$$q = \frac{2\pi(1.0)(400 - 300)}{\dfrac{\ln (3.25 \times 10^{-3}/0.75 \times 10^{-3})}{0.4} + \dfrac{1}{(3.25 \times 10^{-3})(20)}} = 32.98 \text{ W}$$

Hence, adding insulation greatly increases the heat loss.

4.3G Contact Resistance at an Interface

In the equations derived in this section for conduction through solids in series (see Fig. 4.3-1) it has been assumed that the adjacent touching surfaces are at the same temperature; i.e., completely perfect contact is made between the surfaces. For many engineering designs in industry, this assumption is reasonably accurate. However, in cases such as in nuclear power plants where very high heat fluxes are present, a significant drop in temperature may be present at the interface. This interface resistance, called *contact resistance*, occurs when the two solids do not fit tightly together and a thin layer of stagnant fluid is trapped between the two surfaces. At some points the solids touch at peaks in the surfaces and at other points the fluid occupies the open space.

This interface resistance is a complex function of the roughness of the two surfaces, the pressure applied to hold the surfaces in contact, the interface temperature, and the interface fluid. Heat transfer takes place by conduction, radiation, and convection across the trapped fluid and also by conduction through the points of contact of the solids. No completely reliable empirical correlations or theories are available to predict contact resistances for all types of materials. See references (C7, R2) for detailed discussions.

The equation for the contact resistance is often given as follows:

$$q = h_c \, A \, \Delta T = \frac{\Delta T}{1/h_c \, A} = \frac{\Delta T}{R_c} \qquad (4.3\text{-}36)$$

where h_c is the contact resistance coefficient in $W/m^2 \cdot K$, ΔT the temperature drop across the contact resistance in K, and R_c the contact resistance. The contact resistance R_c can be added with the other resistances in Eq. (4.3-3) to include this effect for solids in series. For contact between two ground metal surfaces h_c values of the order of magnitude of about 0.2×10^4 to 1×10^4 $W/m^2 \cdot K$ have been obtained.

An approximation of the maximum contact resistance can be obtained if the maximum gap Δx between the surfaces can be estimated. Then, assuming that the heat transfer across the gap is by conduction only through the stagnant fluid, h_c is estimated as

$$h_c = \frac{k}{\Delta x} \qquad (4.3\text{-}37)$$

If any actual convection, radiation, or point-to-point contact is present, this will reduce this assumed resistance.

4.4 STEADY-STATE CONDUCTION AND SHAPE FACTORS

4.4A Introduction and Graphical Method for Two-Dimensional Conduction

In previous sections of this chapter we discussed steady-state heat conduction in one direction. In many cases, however, steady-state heat conduction is occurring in two directions; i.e., two-dimensional conduction is occurring. The two-dimensional solutions are more involved and in most cases analytical solutions are not available. One important approximate method to solve such problems is to use a numerical method discussed in detail in Section 4.15. Another important approximate method is the graphical method, which is a simple method that can provide reasonably accurate answers for the heat-transfer rate. This method is particularly applicable to systems having isothermal boundaries.

In the graphical method we first note that for one-dimensional heat conduction through a flat slab (see Fig. 4.2-1) the direction of the heat flux or flux lines is always perpendicular to the isotherms. The graphical method for two-dimensional conduction is also based on the requirement that the heat flux lines and the isotherm lines intersect each other at right angles while forming a network of curvilinear squares. This means, as shown in Fig. 4.4-1, that we can sketch the isotherms and also the flux lines until they intersect at right angles (are perpendicular to each other). With care and experience we can obtain reasonably accurate results. General steps to use in this graphical method are as follows.

1. Draw a model to scale of the two-dimensional solid. Lable the isothermal boundaries. In Fig. 4.4-1, T_1 and T_2 are isothermal boundaries.
2. Select a number N that is the number of equal temperature subdivisions between the isothermal boundaries. In Fig. 4.4-1, $N = 4$ subdivisions between T_1 and T_2. Sketch in the isotherm lines and the heat flow or flux lines so that they are perpendicular to each other at the intersections. Note that isotherms are perpendicular to adiabatic (insulated) boundaries and also lines of symmetry.
3. Keep adjusting the isotherm and flux lines until for each curvilinear square the condition $\Delta x = \Delta y$ is satisfied.

In order to calculate the heat flux using the results of the graphical plot, we first assume unit depth of the material. The heat flow q' through the curvilinear section shown in Fig. 4.4-1 is given by Fourier's law.

$$q' = -kA \frac{dT}{dy} = k(\Delta x \cdot 1) \frac{\Delta T}{\Delta y}$$

(4.4-1)

This heat flow q' will be the same through each curvilinear square within this heat-flow lane. Since $\Delta x = \Delta y$, each temperature subdivision ΔT is equal. This temperature subdivision can be expressed in terms of the overall temperature difference $T_1 - T_2$ and N, which is the number of equal subdivisions.

$$\Delta T = \frac{T_1 - T_2}{N}$$

(4.4-2)

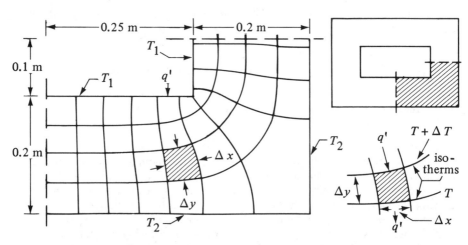

FIGURE 4.4-1. *Graphical curvilinear-square method for two-dimensional heat conduction in a rectangular flue.*

Also, the heat flow q' through each lane is the same since $\Delta x = \Delta y$ in the construction and in Eq. (4.4-1). Hence, the total heat transfer q through all of the lanes is

$$q = Mq' = Mk \, \Delta T \tag{4.4-3}$$

where M is the total number of heat-flow lanes as determined by the graphical procedure. Substituting Eq. (4.4-2) into (4.4-3),

$$q = \frac{M}{N} k(T_1 - T_2) \tag{4.4-4}$$

EXAMPLE 4.4-1. Two-Dimensional Conduction by Graphical Procedure
Determine the total heat transfer through the walls of the flue shown in Fig. 4.4-1 if $T_1 = 600$ K, $T_2 = 400$ K, $k = 0.90$ W/m · K, and L (length of flue) = 5 m.

Solution: In Fig. 4.4-1, $N = 4$ temperature subdivisions and $M = 9.25$. The total heat-transfer rate through the four identical sections with a depth or length L of 5 m is obtained by using Eq. (4.4-4).

$$q = 4\left[\frac{M}{N} kL(T_1 - T_2)\right] = 4\left[\frac{9.25}{4}(0.9)(5.0)(600-400)\right]$$

$$= 8325 \text{ W}$$

4.4B Shape Factors in Conduction

In Eq. (4.4-4) the factor M/N is called the conduction shape factor S, where

$$S = \frac{M}{N} \tag{4.4-5}$$

$$q = kS(T_1 - T_2) \tag{4.4-6}$$

This shape factor S has units of m and is used in two-dimensional heat conduction where only two temperatures are involved. The shape factors for a number of geometries have been obtained and some are given in Table 4.4-1.

For a three-dimensional geometry such as a furnace, separate shape factors are used to obtain the heat flow through the edge and corner sections. When each of the interior dimensions is greater than one-fifth of the wall thickness, the shape factors are as follows for a uniform wall thickness T_w:

$$S_{\text{wall}} = \frac{A}{T_w} \qquad S_{\text{edge}} = 0.54L \qquad S_{\text{corner}} = 0.15T_w \tag{4.4-7}$$

where A is the inside area of wall and L the length of inside edge. For a completely enclosed geometry, there are 6 wall sections, 12 edges, and 8 corners. Note that for a single flat wall, $q = kS_{\text{wall}}(T_1 - T_2) = k(A/T_w)(T_1 - T_2)$, which is the same as Eq. (4.2-1) for conduction through a single flat slab.

For a long hollow cylinder of length L such as that in Fig. 4.2-2,

$$S = \frac{2\pi L}{\ln (r_2/r_1)} \tag{4.4-8}$$

For a hollow sphere from Eq. (4.2-14),

$$S = \frac{4\pi r_2 r_1}{r_2 - r_1} \tag{4.4-9}$$

TABLE 4.4-1.   *Conduction Shape Factors for* $q = kS(T_1 - T_2)$*

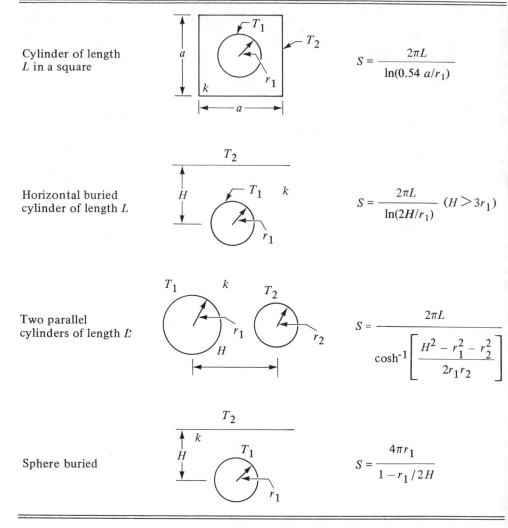

Cylinder of length L in a square	$S = \dfrac{2\pi L}{\ln(0.54\,a/r_1)}$
Horizontal buried cylinder of length L	$S = \dfrac{2\pi L}{\ln(2H/r_1)} \quad (H > 3r_1)$
Two parallel cylinders of length L	$S = \dfrac{2\pi L}{\cosh^{-1}\left[\dfrac{H^2 - r_1^2 - r_2^2}{2r_1 r_2}\right]}$
Sphere buried	$S = \dfrac{4\pi r_1}{1 - r_1/2H}$

* The thermal conductivity of the medium is k.

4.5   FORCED CONVECTION HEAT TRANSFER INSIDE PIPES

4.5A   Introduction and Dimensionless Numbers

In most situations involving a liquid or a gas in heat transfer, convective heat transfer usually occurs as well as conduction. In most industrial processes where heat transfer is occurring, heat is being transformed from one fluid through a solid wall to a second fluid. In Fig. 4.5-1 heat is being transferred from the hot flowing fluid to the cold flowing fluid. The temperature profile is shown.

The velocity gradient, when the fluid is in turbulent flow, is very steep next to the wall in the thin viscous sublayer where turbulence is absent. Here the heat transfer is mainly by conduction with a large temperature difference of $T_2 - T_3$ in the warm fluid.

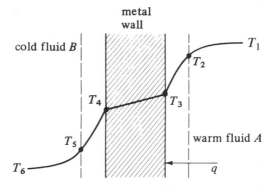

FIGURE 4.5-1. *Temperature profile for heat transfer by convection from one fluid to another.*

As we move farther away from the wall, we approach the turbulent region, where rapidly moving eddies tend to equalize the temperature. Hence, the temperature gradient is less and the difference $T_1 - T_2$ is small. The average temperature of fluid A is slightly less than the peak value T_1. A similar explanation can be given for the temperature profile in the cold fluid.

The convective coefficient for heat transfer through a fluid is given by

$$q = hA(T - T_w) \tag{4.5-1}$$

where h is the convective coefficient in $W/m^2 \cdot K$, A is the area in m^2, T is the bulk or average temperature of the fluid in K, T_w is the temperature of the wall in contact with the fluid in K, and q is the heat-transfer rate in W. In English units, q is in btu/h, h in btu/h $\cdot$ ft$^2 \cdot$ °F, A in ft^2, and T and T_w in °F.

The type of fluid flow, whether laminar or turbulent, of the individual fluid has a great effect on the heat-transfer coefficient h, which is often called a film coefficient, since most of the resistance to heat transfer is in a thin film close to the wall. The more turbulent the flow, the greater the heat-transfer coefficient.

There are two main classifications of convective heat transfer. The first is free or *natural convection*, where the motion of the fluid results from the density changes in heat transfer. The buoyant effect produces a natural circulation of the fluid, so it moves past the solid surface. In the second type, *forced convection*, the fluid is forced to flow by pressure differences, a pump, a fan, and so on.

Most of the correlations for predicting film coefficients h are semiempirical in nature and are affected by the physical properties of the fluid, the type and velocity of flow, the temperature difference, and by the geometry of the specific physical system. Some approximate values of convective coefficients were presented in Table 4.1-2. In the following correlations, SI or English units can be used since the equations are dimensionless.

To correlate these data for heat-transfer coefficients, dimensionless numbers such as the Reynolds and Prandtl numbers are used. The Prandtl number is the ratio of the shear component of diffusivity for momentum μ/ρ to the diffusivity for heat $k/\rho c_p$ and physically relates the relative thickness of the hydrodynamic layer and thermal boundary layer.

$$N_{Pr} = \frac{\mu/\rho}{k/\rho c_p} = \frac{c_p \mu}{k} \tag{4.5-2}$$

Values of the N_{Pr} for gases are given in Appendix A.3 and range from about 0.5 to 1.0. Values for liquids range from about 2 to well over 10^4. The dimensionless Nusselt number, N_{Nu}, is used to relate data for the heat-transfer coefficient h to the thermal conductivity k of the fluid and a characteristic dimension D.

$$N_{Nu} = \frac{hD}{k} \tag{4.5-3}$$

For example, for flow inside a pipe, D is the diameter.

4.5B Heat-Transfer Coefficient for Laminar Flow Inside a Pipe

Certainly, the most important convective heat-transfer process industrially is that of cooling or heating a fluid flowing inside a closed circular conduit or pipe. Different types of correlations for the convective coefficient are needed for laminar flow (N_{Re} below 2100), for fully turbulent flow (N_{Re} above 6000), and for the transition region (N_{Re} between 2100 and 6000).

For laminar flow of fluids inside horizontal tubes or pipes, the following equation of Sieder and Tate (S1) can be used for $N_{Re} < 2100$:

$$(N_{Nu})_a = \frac{h_a D}{k} = 1.86 \left(N_{Re} N_{Pr} \frac{D}{L} \right)^{1/3} \left(\frac{\mu_b}{\mu_w} \right)^{0.14} \tag{4.5-4}$$

where D = pipe diameter in m, L = pipe length before mixing occurs in the pipe in m, μ_b = fluid viscosity at bulk average temperature in Pa·s, μ_w = viscosity at the wall temperature, c_p = heat capacity in J/kg·K, k = thermal conductivity in W/m·K, h_a = average heat-transfer coefficient in W/m²·K, and N_{Nu} = dimensionless Nusselt number. All the physical properties are evaluated at the bulk fluid temperature except μ_w. The Reynolds number is

$$N_{Re} = \frac{D v \rho}{\mu} \tag{4.5-5}$$

and the Prandtl number,

$$N_{Pr} = \frac{c_p \mu}{k} \tag{4.5-6}$$

This equation holds for $(N_{Re} N_{Pr} D/L) > 100$. If used down to a$(N_{Re} N_{Pr} D/L) > 10$, it still holds to $\pm 20\%$ (B1). For $(N_{Re} N_{Pr} D/L) < 100$, another expression is available (P1).

In laminar flow the average coefficient h_a depends strongly on heated length. The average (arithmetic mean) temperature drop ΔT_a is used in the equation to calculate the heat-transfer rate q.

$$q = h_a A \, \Delta T_a = h_a A \frac{(T_w - T_{bi}) + (T_w - T_{bo})}{2} \tag{4.5-7}$$

where T_w is the wall temperature in K, T_{bi} the inlet bulk fluid temperature, and T_{bo} the outlet bulk fluid temperature.

For large pipe diameters and large temperature differences ΔT between pipe wall and bulk fluid, natural convection effects can increase h(P1). Equations are also available for laminar flow in vertical tubes.

4.5C Heat-Transfer Coefficient for Turbulent Flow Inside a Pipe

When the Reynolds number is above 2100, the flow is turbulent. Since the rate of heat transfer is greater in the turbulent region, many industrial heat-transfer processes are in the turbulent region.

The following equation has been found to hold for tubes but is also used for pipes. This holds for a $N_{Re} > 6000$, a N_{Pr} between 0.7 and 16 000, and $L/D > 60$.

$$N_{Nu} = \frac{h_L D}{k} = 0.027 N_{RE}^{0.8} N_{Pr}^{1/3} \left(\frac{\mu_b}{\mu_w} \right)^{0.14} \tag{4.5-8}$$

where h_L is the heat-transfer coefficient based on the log mean driving force ΔT_{lm} (see Section 4.5H). The fluid properties except for μ_w are evaluated at the mean bulk temperature. If the bulk fluid temperature varies from the inlet to the outlet of the pipe, the mean of the inlet and outlet temperatures is used. To correct for an $L/D < 60$, where the entry is an abrupt contraction, an approximate correction is to multiply the right-hand side of Eq. (4.5-8) by a correction factor given in Section 4.5F.

The use of Eq. (4.5-8) may be trial and error, since the value of h_L must be known to evaluate T_w, and hence μ_w, at the wall temperature. Also, if the mean bulk temperature increases or decreases in the tube length L because of heat transfer, the bulk temperature at length L must be estimated in order to have a mean bulk temperature of the entrance and exit to use.

The heat-transfer coefficient for turbulent flow is somewhat greater for a pipe than for a smooth tube. This effect is much less than in fluid friction, and it is usually neglected in calculations. Also, for liquid metals that have Prandtl numbers $\ll 1$, other correlations must be used to predict the heat-transfer coefficient. (See Section 4.5G.) For shapes of tubes other than circular, the equivalent diameter can be used as discussed in Section 4.5E.

For air at 1 atm total pressure, the following simplified equation holds for turbulent flow in pipes.

$$h_L = \frac{3.52 v^{0.8}}{D^{0.2}} \quad \text{(SI)}$$

$$ \tag{4.5-9}$$

$$h_L = \frac{0.5 v_s^{0.8}}{(D')^{0.2}} \quad \text{(English)}$$

where D is in m, v in m/s, and h_L in W/m$^2 \cdot$ K for SI units; and D' is in in., v_s in ft/s, and h_L in btu/h $\cdot$ ft$^2 \cdot$ °F in English units.

Water is often used in heat-transfer equipment. A simplified equation to use for a temperature range of $T = 4$ to 105°C (40 − 220°F) is

$$h_L = 1429(1 + 0.0146 T°C) \frac{v^{0.8}}{D^{0.2}} \quad \text{(SI)}$$

$$ \tag{4.5-10}$$

$$h_L = 150(1 + 0.011 T°F) \frac{v_s^{0.8}}{(D')^{0.2}} \quad \text{(English)}$$

A very simplified equation for organic liquids to use for approximations is as follows (P3):

$$h_L = 423 \frac{v^{0.8}}{D^{0.2}} \quad \text{(SI)}$$

$$ \tag{4.5-11}$$

$$h_L = 60 \frac{v_s^{0.8}}{(D')^{0.2}} \quad \text{(English)}$$

For flow inside helical coils and N_{Re} above 10^4, the predicted film coefficient for straight pipes should be increased by the factor $(1 + 3.5D/D_{coil})$.

EXAMPLE 4.5-1. *Heating of Air in Turbulent Flow*
Air at 206.8 kPa and an average of 477.6 K is being heated as it flows through a tube of 25.4 mm inside diameter at a velocity of 7.62 m/s. The heating medium is 488.7 K steam condensing on the outside of the tube. Since the heat-transfer coefficient of condensing steam is several thousand $W/m^2 \cdot K$ and the resistance of the metal wall is very small, it will be assumed that the surface wall temperature of the metal in contact with the air is 488.7 K. Calculate the heat-transfer coefficient for an $L/D > 60$ and also the heat-transfer flux q/A.

Solution: From Appendix A.3 for physical properties of air at 477.6 K (204.4°C), $\mu_b = 2.60 \times 10^{-5}$ Pa·s, $k = 0.03894$ W/m, $N_{Pr} = 0.686$. At 488.7 K (215.5°C), $\mu_w = 2.64 \times 10^{-5}$ Pa·s.

$$\mu_b = 2.60 \times 10^{-5} \text{ Pa} \cdot \text{s} = 2.60 \times 10^{-5} \text{ kg/m} \cdot \text{s}$$

$$\rho = (28.97)\left(\frac{1}{22.414}\right)\left(\frac{206.8}{101.33}\right)\left(\frac{273.2}{477.6}\right) = 1.509 \text{ kg/m}^3$$

The Reynolds number calculated at the bulk fluid temperature of 477.6 K is

$$N_{Re} = \frac{Dv\rho}{\mu} = \frac{0.0254(7.62)(1.509)}{2.6 \times 10^{-5}} = 1.122 \times 10^4$$

Hence, the flow is turbulent and Eq. (4.5-8) will be used. Substituting into Eq. (4.5-8),

$$N_{Nu} = \frac{h_L D}{k} = 0.027 N_{Re}^{0.8} N_{Pr}^{1/3} \left(\frac{\mu_b}{\mu_w}\right)^{0.14}$$

$$\frac{h_L(0.0254)}{0.03894} = 0.027(1.122 \times 10^4)^{0.8}(0.686)^{1/3}\left(\frac{0.0260}{0.0264}\right)^{0.14}$$

Solving, $h_L = 63.2$ $W/m^2 \cdot K$ (11.13 btu/h·ft²·°F).
To solve for the flux q/A,

$$\frac{q}{A} = h_L(T_w - T) = 63.2(488.7 - 477.6)$$

$$= 701.1 \text{ W/m}^2 \text{ (222.2 btu/h·ft}^2\text{)}$$

4.5D Heat-Transfer Coefficient for Transition Flow Inside a Pipe

In the transition region for a N_{Re} between 2100 and 6000, the empirical equations are not well defined just as in the case of fluid friction factors. No simple equation exists for accomplishing a smooth transition from heat transfer in laminar flow to turbulent flow, i.e., a transition from Eq. (4.5-4) at a $N_{Re} = 2100$ to Eq. (4.5-8) at a $N_{Re} = 6000$.

The plot in Fig. 4.5-2 represents an approximate relationship to use between the various heat-transfer parameters and the Reynolds number between 2100 and 6000. For below a N_{Re} of 2100, the curves represent Eq. (4.5-4) and above 10^4, Eq. (4.5-8). The mean ΔT_a of Eq. (4.5-7) should be used with the h_a in Fig. 4.5-2.

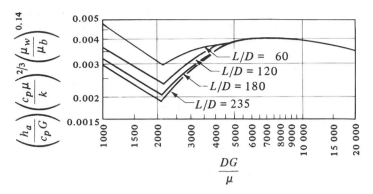

FIGURE 4.5-2. *Correlation of heat-transfer parameters for transition region for Reynolds numbers between 2100 and 6000. (From R. H. Perry and C. H. Chilton, Chemical Engineers' Handbook, 5th ed. New York: McGraw-Hill Book Company, 1973. With permission.)*

4.5E   Heat-Transfer Coefficient for Noncircular Conduits

A heat-transfer system often used is that in which fluids flow at different temperatures in concentric pipes. The heat-transfer coefficient of the fluid in the annular space can be predicted using the same equations as for circular pipes. However, the equivalent diameter defined in Section 2.10G must be used. For an annular space, D_{eq} is the ID of the outer pipe D_1 minus the OD of the inner pipe D_2. For other geometries, an equivalent diameter can also be used.

> *EXAMPLE 4.5-2.   Water Heated by Steam and Trial-and-Error Solution*
> Water is flowing in a horizontal 1-in. schedule 40 steel pipe at an average temperature of 65.6°C and a velocity of 2.44 m/s. It is being heated by condensing steam at 107.8°C on the outside of the pipe wall. The steam side coefficient has been estimated as $h_o = 10\,500$ W/m² · K.
> (a) Calculate the convective coefficient h_i for water inside the pipe.
> (b) Calculate the overall coefficient U_i based on the inside surface area.
> (c) Calculate the heat-transfer rate q for 0.305 m of pipe with the water at an average temperature of 65.6°C.

Solution: From Appendix A.5 the various dimensions are $D_i = 0.0266$ m and $D_o = 0.0334$ m. For water at a bulk average temperature of 65.6°C from   Appendix A.2, $N_{Pr} - 2.72$, $\rho = 0.980(1000) - 980$ kg/m³, $k - 0.633$ W/m · K, and $\mu = 4.32 \times 10^{-4}$ Pa · s $= 4.32 \times 10^{-4}$ kg/m · s.

The temperature of the inside metal wall is needed and will be assumed as about one third the way between 65.6 and 107.8 or 80°C $= T_w$ for the first trial. Hence, μ_w at 80°C $= 3.56 \times 10^{-4}$ Pa · s.

First, the Reynolds number of the water is calculated at the bulk average temperature.

$$N_{Re} = \frac{D_i v \rho}{\mu} = \frac{0.0266(2.44)(980)}{4.32 \times 10^{-4}} = 1.473 \times 10^5$$

Hence, the flow is turbulent. Using Eq. (4.5-8) and substituting known values,

$$\frac{h_L D}{k} = 0.027 N_{Re}^{0.8} N_{Pr}^{1/3} \left(\frac{\mu_b}{\mu_w}\right)^{0.14}$$

$$\frac{h_L(0.0266)}{0.663} = 0.027(1.473 \times 10^5)^{0.8}(2.72)^{1/3}\left(\frac{4.32 \times 10^{-4}}{3.56 \times 10^{-4}}\right)^{0.14}$$

Solving, $h_L = h_i = 13\ 324$ W/m$^2\cdot$K.

For part (b), the various areas are as follows for 0.305-m pipe.

$$A_i = \pi D_i L = \pi(0.0266)(0.305) = 0.0255 \text{ m}^2$$

$$A_{\text{lm}} = \pi\,\frac{(0.0266) + 0.0334)0.305}{2} = 0.0287 \text{ m}^2$$

$$A_o = \pi(0.0334)(0.305) = 0.0320 \text{ m}^2$$

The k for steel is 45.0 W/m $\cdot$ K. The resistances are

$$R_i = \frac{1}{h_i A_i} = \frac{1}{(13\ 324)0.0255} = 0.002943$$

$$R_m = \frac{r_o - r_i}{k A_{\text{lm}}} = \frac{0.0334 - 0.0266}{2}\,\frac{1}{45.0(0.0287)} = 0.002633$$

$$R_o = \frac{1}{h_o A_o} = \frac{1}{(10\ 500)(0.0320)} = 0.002976$$

$$\sum R = 0.002943 + 0.002633 + 0.002976 = 0.008552$$

The overall temperature difference is $(107.8 - 65.6)°C = 42.2°C = 42.2$ K. The temperature drop across the water film is

$$\text{temperature drop} = \frac{R_i}{\sum R}\,(42.2) = \left(\frac{0.002943}{0.008852}\right)(42.2) = 14.5 \text{ K} = 14.5°C$$

Hence, $T_w = 65.6 + 14.5 = 80.1°C$. This is quite close to the original estimate of 80°C. The only physical property changing in the second estimate would be μ_w. This will have a negligible effect on h_i and a second trial is not necessary.

For part (b), the overall coefficient is, by Eq. (4.3-16),

$$q = U_i A_i (T_o - T_i) = \frac{T_o - T_i}{\sum R}$$

$$U_i = \frac{1}{A_i \sum R} = \frac{1}{0.0255(0.008552)} = 4586 \text{ W/m}^2\cdot\text{K}.$$

For part (c), with the water at an average temperature of 65.6°C,

$$T_o - T_i = 107.8 - 65.6 = 42.2°C = 42.2 \text{ K}$$

$$q = U_i A_i (T_o - T_i) = 4586(0.0255)(42.2) = 4935 \text{ W}$$

4.5F Entrance-Region Effect on Heat-Transfer Coefficient

Near the entrance of a pipe where the fluid is being heated, the temperature profile is not fully developed and the local coefficient h is greater than the fully developed heat-transfer coefficient h_L for turbulent flow. At the entrance itself where no temperature gradient has been established, the value of h is infinite. The value of h drops rapidly and is approximately the same as h_L at $L/D \cong 60$, where L is the entrance length. These relations for turbulent flow inside a pipe are as follows where the entrance is an abrupt contraction.

$$\frac{h}{h_L} = 1 + \left(\frac{D}{L}\right)^{0.7} \qquad 2 < \frac{L}{D} < 20 \qquad\qquad \text{(4.5-12)}$$

$$\frac{h}{h_L} = 1 + 6\left(\frac{D}{L}\right) \qquad 20 < \frac{L}{D} < 60 \qquad\qquad \text{(4.5-13)}$$

where h is the average value for a tube of finite length L and h_L is the value for a very long tube.

4.5G Liquid-Metals Heat-Transfer Coefficient

Liquid metals are sometimes used as a heat-transfer fluid in cases where a fluid is needed over a wide temperature range at relatively low pressures. Liquid metals are often used in nuclear reactors and have high heat-transfer coefficients as well as a high heat capacity per unit volume. The high heat-transfer coefficients are due to the very high thermal conductivities and, hence, low Prandtl numbers. In liquid metals in pipes, the heat transfer by conduction is very important in the entire turbulent core because of the high thermal conductivity and is often more important than the convection effects.

For fully developed turbulent flow in tubes with uniform heat flux the following equation can be used (L1):

$$N_{Nu} = \frac{h_L D}{k} = 0.625 N_{Pe}^{0.4} \qquad\qquad \text{(4.5-14)}$$

where the Peclet number $N_{Pe} = N_{Re} N_{Pr}$. This holds for $L/D > 60$ and N_{Pe} between 100 and 10^4. For constant wall temperatures,

$$N_{Nu} = \frac{h_L D}{k} = 5.0 + 0.025 N_{Pe}^{0.8} \qquad\qquad \text{(4.5-15)}$$

for $L/D > 60$ and $N_{Pe} > 100$. All physical properties are evaluated at the average bulk temperature.

EXAMPLE 4.5-3. Liquid-Metal Heat Transfer Inside a Tube
A liquid metal flows at a rate of 4.00 kg/s through a tube having an inside diameter of 0.05 m. The liquid enters at 500 K and is heated to 505 K in the tube. The tube wall is maintained at a temperature of 30 K above the fluid bulk temperature and constant heat flux is maintained. Calculate the required tube length. The average physical properties are as follows: $\mu = 7.1 \times 10^{-4}$ Pa·s, $\rho = 7400$ kg/m³, $c_p = 120$ J/kg·K, $k = 13$ W/m·K.

Solution: The area is $A = \pi D^2/4 = \pi(0.05)^2/4 = 1.963 \times 10^{-3}$ m². Then $G = 4.0/1.963 \times 10^{-3} = 2.038 \times 10^3$ kg/m²·s. The Reynolds number is

$$N_{Re} = \frac{DG}{\mu} = \frac{0.05(2.038 \times 10^3)}{7.1 \times 10^{-4}} = 1.435 \times 10^5$$

$$N_{Pr} = \frac{c_p \mu}{k} = \frac{120(7.1 \times 10^{-4})}{13} = 0.00655$$

Using Eq. (4.5-14),

$$h_L = \frac{k}{D}(0.625)N_{Pe}^{0.4} = \frac{13}{0.05}(0.625)(1.435 \times 10^5 \times 0.00655)^{0.4}$$

$$= 2512 \text{ W/m}^2 \cdot \text{K}$$

Using a heat balance,

$$q = mc_p \, \Delta T$$

$$= 4.00(120)(505 - 500) = 2400 \text{ W} \tag{4.5-16}$$

Substituting into Eq. (4.5-1),

$$\frac{q}{A} = \frac{2400}{A} = h_L(T_w - T) = 2512(30) = 75\,360 \text{ W/m}^2$$

Hence, $A = 2400/75\,360 = 3.185 \times 10^{-2} \text{ m}^2$. Then,

$$A = 3.185 \times 10^{-2} = \pi D L = \pi(0.05)(L)$$

Solving, $L = 0.203$ m.

4.5H Log Mean Temperature Difference and Varying Temperature Drop

Equations (4.5-1) and (4.3-12) as written apply only when the temperature drop $(T_i - T_o)$ is constant for all parts of the heating surface. Hence, the equation

$$q = U_i A_i (T_i - T_o) = U_o A_o (T_i - T_o) = UA(\Delta T) \tag{4.5-17}$$

only holds at one point in the apparatus when the fluids are being heated or cooled. However, as the fluids travel through the heat exchanger, they become heated or cooled and both T_i and T_o or either T_i and T_o vary. Then $(T_i - T_o)$ or ΔT varies with position, and some mean ΔT_m must be used over the whole apparatus.

In a typical heat exchanger a hot fluid inside a pipe is cooled from T_1' to T_2' by a cold fluid which is flowing on the outside in a double pipe countercurrently (in the reverse direction) and is heated from T_2 to T_1 as shown in Fig. 4.5-3a. The ΔT shown is varying with distance. Hence, ΔT in Eq. (4.5-17) varies as the area A goes from 0 at the inlet to A at the outlet of the exchanger.

For countercurrent flow of the two fluids as in Fig. 4.5-3a, the heat-transfer rate is

$$q = UA \, \Delta T_m \tag{4.5-18}$$

where ΔT_m is a suitable mean temperature difference to be determined. For a dA area, a heat balance on the hot and the cold fluids gives

$$dq = -m'c_p' \, dT' = mc_p \, dT \tag{4.5-19}$$

where m is flow rate in kg/s. The values of m, m', c_p, c_p', and U are assumed constant.

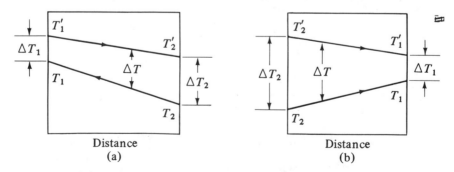

FIGURE 4.5-3. *Temperature profiles for one-pass double-pipe heat exchangers: (a) countercurrent flow; (b) cocurrent or parallel flow.*

Chap. 4   Principles of Steady-State Heat Transfer

Also,
$$dq = U(T' - T)\, dA \tag{4.5-20}$$

From Eq. (4.5-19), $dT' = -dq/m'c'_p$ and $dT = dq/mc_p$. Then,

$$dT' - dT = d(T' - T) = -dq\left(\frac{1}{m'c'_p} + \frac{1}{mc_p}\right) \tag{4.5-21}$$

Substituting Eq. (4.5-20) into (4.5-21),

$$\frac{d(T' - T)}{T' - T} = -U\left(\frac{1}{m'c'_p} + \frac{1}{mc_p}\right) dA \tag{4.5-22}$$

Integrating between points 1 and 2,

$$\ln\left(\frac{T'_2 - T_2}{T'_1 - T_1}\right) = -UA\left(\frac{1}{m'c'_p} + \frac{1}{mc_p}\right) \tag{4.5-23}$$

Making a heat balance between the inlet and outlet,

$$q = m'c'_p(T'_1 - T'_2) = mc_p(T_2 - T_1) \tag{4.5-24}$$

Solving for $m'c'_p$ and mc_p in Eq. (4.5-24) and substituting into Eq. (4.5-23),

$$q = \frac{UA[(T'_2 - T_2) - (T'_1 - T_1)]}{\ln\,[(T'_2 - T_2)/(T'_1 - T_1)]} \tag{4.5-25}$$

Comparing Eqs. (4.5-18) and (4.5-25), we see that ΔT_m is the log mean temperature difference ΔT_{lm}. Hence, in the case where the overall heat-transfer coefficient U is constant throughout the equipment and the heat capacity of each fluid is constant, the proper temperature driving force to use over the entire apparatus is the log mean driving force,

$$q = UA\Delta T_{lm} \tag{4.5-26}$$

where,

$$\Delta T_{lm} = \frac{\Delta T_2 - \Delta T_1}{\ln\,(\Delta T_2/\Delta T_1)} \tag{4.5-27}$$

It can be also shown that for parallel flow as pictured in Fig. 4.5-3b, the log mean temperature difference should be used. In some cases where steam is condensing, T'_1 and T'_2 may be the same. The equations still hold for this case. When U varies with distance or other complicating factors occur, other references should be consulted (B2, P3, W1).

EXAMPLE 4.5-4. Heat-Transfer Area and Log Mean Temperature Difference

A heavy hydrocarbon oil which has a $c_{pm} = 2.30$ kJ/kg·K is being cooled in a heat exchanger from 371.9 K to 349.7 K and flows inside the tube at a rate of 3630 kg/h. A flow of 1450 kg water/h enters at 288.6 K for cooling and flows outside the tube.

    (a) Calculate the water outlet temperature and heat-transfer area if the overall $U_i = 340$ W/m²·K and the streams are countercurrent.

    (b) Repeat for parallel flow.

Solution: Assume a $c_{pm} = 4.187$ kJ/kg·K for water. The water inlet $T_2 = 288.6$ K, outlet $= T_1$; oil inlet $T'_1 = 371.9$, outlet $T'_2 = 349.7$ K. Calculating the heat lost by the oil,

$$q = \left(3630\ \frac{\text{kg}}{\text{h}}\right)\left(2.30\ \frac{\text{kJ}}{\text{kg·K}}\right)(371.9 - 349.7)\text{K}$$

$$= 185\,400\ \text{kJ/h} \quad \text{or} \quad 51\,490\ \text{W}\ (175\,700\ \text{btu/h})$$

By a heat balance, the q must also equal the heat gained by the water.

$$q = 185\,400 \text{ kJ/h} = \left(1450\,\frac{\text{kg}}{\text{h}}\right)\left(4.187\,\frac{\text{kJ}}{\text{kg}\cdot\text{K}}\right)(T_1 - 288.6)\text{ K}$$

Solving, $T_1 = 319.1$ K.

To solve for the log mean temperature difference, $\Delta T_2 = T_2' - T_2 = 349.7 - 288.6 = 61.1$ K, $\Delta T_1 = T_1' - T_1 = 371.9 - 319.1 = 52.8$ K. Substituting into Eq. (4.5-27),

$$\Delta T_{lm} = \frac{\Delta T_2 - \Delta T_1}{\ln (\Delta T_2/\Delta T_1)} = \frac{61.1 - 52.8}{\ln (61.1/52.8)} = 56.9 \text{ K}$$

Using Eq. (4.5-26),

$$q = U_i A_i \Delta T_{lm}$$

$$51\,490 = 340(A_i)(56.9)$$

Solving, $A_i = 2.66 \text{ m}^2$.

For part (b), the water outlet is still $T_1 = 319.1$ K. Referring to Fig. 4.5-3b, $\Delta T_2 = 371.9 - 288.6 = 83.3$ K and $\Delta T_1 = 349.7 - 319.1 = 30.6$ K. Again, using Eq. (4.5-27) and solving, $\Delta T_{lm} = 52.7$ K. Substituting into Eq. (4.5-26), $A_i = 2.87 \text{ m}^2$. This is a larger area than for counterflow. This occurs because counterflow gives larger temperature driving forces and is usually preferred over parallel flow for this reason.

EXAMPLE 4.5-5. Laminar Heat Transfer and Trial and Error

A hydrocarbon oil at 150°F enters inside a pipe with an inside diameter of 0.0303 ft and a length of 15 ft with a flow rate of 80 lb_m/h. The inside pipe surface is assumed constant at 350°F since steam is condensing outside the pipe wall and has a very large heat-transfer coefficient. The properties of the oil are $c_{pm} = 0.50$ btu/$\text{lb}_m \cdot$ °F and $k_m = 0.083$ btu/h · ft · °F. The viscosity of the oil varies with temperature as follows: 150°F, 6.50 cp; 200°F, 5.05 cp; 250°F, 3.80 cp; 300°F, 2.82 cp; 350°F, 1.95 cp. Predict the heat-transfer coefficient and the oil outlet temperature, T_{bo}.

Solution: This is a trial-and-error solution since the outlet temperature of the oil T_{bo} is unknown. The value of $T_{bo} = 250$°F will be assumed and checked later. The bulk mean temperature of the oil to use for the physical properties is $(150 + 250)/2$ or 200°F. The viscosity at 200°F is

$$\mu_b = 5.05(2.4191) = 12.23\,\frac{\text{lb}_m}{\text{ft} \cdot \text{h}}$$

At the wall temperature of 350°F,

$$\mu_w = 1.95(2.4191) = 4.72\,\frac{\text{lb}_m}{\text{ft} \cdot \text{h}}$$

The cross-section area of the pipe A is

$$A = \frac{\pi D_i^2}{4} = \frac{\pi(0.0303)^2}{4} = 0.000722 \text{ ft}^2$$

$$G = \frac{m}{A} = \frac{80 \text{ lb}_m/\text{h}}{0.000722 \text{ ft}^2} = 111\,000\,\frac{\text{lb}_m}{\text{ft}^2 \cdot \text{h}}$$

The Reynolds number at the bulk mean temperature is

$$N_{Re} = \frac{D_i v \rho}{\mu} = \frac{D_i G}{\mu} = \frac{0.0303(111\,000)}{12.23} = 275.5$$

The Prandtl number is

$$N_{Pr} = \frac{c_p \mu}{k} = \frac{0.50(12.23)}{0.083} = 73.7$$

Since the N_{Re} is below 2100, the flow is in the laminar region and Eq. (4.5-4) will be used. Even at the outlet temperature of 250°F, the flow is still laminar. Substituting,

$$(N_{Nu})_a = \frac{h_a D}{k} = 1.86 \left(N_{Re} N_{Pr} \frac{D}{L} \right)^{1/3} \left(\frac{\mu_b}{\mu_w} \right)^{0.14}$$

$$\frac{h_a (0.0303)}{0.083} = 1.86 \left[275.5(73.7) \left(\frac{0.0303}{15.0} \right) \right]^{1/3} \left(\frac{12.23}{4.72} \right)^{0.14}$$

Solving, $h_a = 20.1$ btu/h · ft² · °F (114 W/m² · K).

Next, making a heat balance on the oil,

$$q = mc_{pm}(T_{bo} - T_{bi}) = 80.0(0.50)(T_{bo} - 150) \tag{4.5-28}$$

Using Eq. (4.5-7),

$$q = h_a A \Delta T_a \tag{4.5-7}$$

For ΔT_a,

$$\Delta T_a = \frac{(T_w - T_{bi}) + (T_w - T_{bo})}{2}$$

$$= \frac{(350 - 150) + (350 - T_{bo})}{2}$$

$$= 275 - 0.5 T_{bo}$$

Equating Eq. (4.5-28) to (4.5-7) and substituting,

$$80.0(0.50)(T_{bo} - 150) = hA \Delta T_a$$

$$= 20.1 [\pi(0.0303)(15)](275 - 0.5 T_{bo})$$

Solving, $T_{bo} = 255°F$.

This is higher than the assumed value of 250°F. For the second trial the mean bulk temperature of the oil would be $(150 + 255)/2$ or 202.5°F. The new viscosity is 5.0 cp compared with 5.05 for the first estimate. This only affects the $(\mu_b/\mu_w)^{0.14}$ factor in Eq. (4.5-4), since the viscosity effect in the $(N_{Re})(N_{Pr})$ factor cancels out. The heat-transfer coefficient will change by less than 0.2%, which is negligible. Hence, the outlet temperature of $T_1 = 255°F$ (123.9°C) is correct.

4.6 HEAT TRANSFER OUTSIDE VARIOUS GEOMETRIES IN FORCED CONVECTION

4.6A Introduction

In many cases a fluid is flowing over completely immersed bodies such as spheres, tubes, plates, and so on, and heat transfer is occurring between the fluid and the solid only. Many of these shapes are of practical interest in process engineering. The sphere, cylinder, and flat plate are perhaps of greatest importance with heat transfer between these surfaces and a moving fluid frequently encountered.

When heat transfer occurs during immersed flow, the flux is dependent on the geometry of the body, the position on the body (front, side, back, etc.), the proximity of other bodies, the flow rate, and the fluid properties. The heat-transfer coefficient varies over the body. The average heat-transfer coefficient is given in the empirical relationships to be discussed in the following sections.

In general, the average heat-transfer coefficient on immersed bodies is given by

$$N_{Nu} = CN_{Re}^m N_{Pr}^{1/3} \qquad (4.6-1)$$

where C and m are constants that depend on the various configurations. The fluid properties are evaluated at the film temperature $T_f = (T_w + T_b)/2$, where T_w is the surface or wall temperature and T_b the average bulk fluid temperature. The velocity in the N_{Re} is the undisturbed free stream velocity v of the fluid approaching the object.

4.6B Flow Parallel to Flat Plate

When the fluid is flowing parallel to a flat plate and heat transfer is occurring between the whole plate of length L m and the fluid, the N_{Nu} is as follows for a $N_{Re, L}$ below 3×10^5 in the laminar region and a $N_{Pr} > 0.7$.

$$N_{Nu} = 0.664 N_{Re, L}^{0.5} N_{Pr}^{1/3} \qquad (4.6-2)$$

where $N_{Re, L} = Lv\rho/\mu$.

For the completely turbulent region at a $N_{Re, L}$ above 3×10^5 (K1, K3) and $N_{Pr} > 0.7$,

$$N_{Nu} = 0.0366 N_{Re, L}^{0.8} N_{Pr}^{1/3} \qquad (4.6-3)$$

However, turbulence can start at a $N_{Re, L}$ below 3×10^5 if the plate is rough (K3) and then Eq. (4.6-3) will hold and give a N_{Nu} greater than by Eq. (4.6-2). Below about a $N_{Re, L}$ of 2×10^4, Eq. (4.6-2) gives the larger value of N_{Nu}.

> **EXAMPLE 4.6-1. Cooling a Copper Fin**
> A smooth, flat, thin fin of copper extending out from a tube is 51 mm by 51 mm square. Its temperature is approximately uniform at 82.2°C. Cooling air at 15.6°C and 1 atm abs flows parallel to the fin at a velocity of 12.2 m/s.
> (a) For laminar flow, calculate the heat-transfer coefficient, h.
> (b) If the leading edge of the fin is rough so that all of the boundary layer or film next to the fin is completely turbulent, calculate h.
>
> **Solution:** The fluid properties will be evaluated at the film temperature $T_f = (T_w + T_b)/2$.
> $$T_f = \frac{T_w + T_b}{2} = \frac{82.2 + 15.6}{2} = 48.9°C \ (322.1 \ K)$$
>
> The physical properties of air at 48.9°C from Appendix A.3 are $k = 0.0280$ W/m·K, $\rho = 1.097$ kg/m³, $\mu = 1.95 \times 10^{-5}$ Pa·s, $N_{Pr} = 0.704$. The Reynolds number is, for $L = 0.051$ m,
> $$N_{Re, L} = \frac{Lv\rho}{\mu} = \frac{(0.051)(12.2)(1.097)}{1.95 \times 10^{-5}} = 3.49 \times 10^4$$
>
> Substituting into Eq. (4.6-2),
> $$N_{Nu} = \frac{hL}{k} = 0.664 N_{Re, L}^{0.5} N_{Pr}^{1/3}$$
> $$\frac{h(0.051)}{(0.0280)} = 0.664(3.49 \times 10^4)^{0.5}(0.704)^{1/3}$$
>
> Solving, $h = 60.7$ W/m²·K (10.7 btu/h·ft²·°F).

For part (b), substituting into Eq. (4.6-3) and solving, $h = 77.2$ W/m^2 · K (13.6 btu/h · ft^2 · °F).

4.6C Cylinder with Axis Perpendicular to Flow

Often a cylinder containing a fluid inside is being heated or cooled by a fluid flowing perpendicular to its axis. The equation for predicting the average heat-transfer coefficient of the outside of the cylinder for gases and liquids is (K3, P3) Eq. (4.6-1) with C and m as given in Table 4.6-1. The $N_{Re} = Dv\rho/\mu$, where D is the outside tube diameter and all physical properties are evaluated at the film temperature T_f. The velocity is the undisturbed free stream velocity approaching the cylinder.

4.6D Flow Past Single Sphere

When a single sphere is being heated or cooled by a fluid flowing past it, the following equation can be used to predict the average heat-transfer coefficient for a $N_{Re} = Dv\rho/\mu$ of 1 to 70 000 and a N_{Pr} of 0.6 to 400.

$$N_{Nu} = 2.0 + 0.60N_{Re}^{0.5} N_{Pr}^{1/3} \qquad (4.6\text{-}4)$$

The fluid properties are evaluated at the film temperature T_f. A somewhat more accurate correlation is available for a N_{Re} range 1–17 000 by others (S2), which takes into account the effects of natural convection at these lower Reynolds numbers.

EXAMPLE 4.6-2. Cooling of a Sphere
Using the same conditions as Example 4.6-1, where air at 1 atm abs pressure and 15.6°C is flowing at a velocity of 12.2 m/s, predict the average heat-transfer coefficient for air flowing by a sphere having a diameter of 51 mm and an average surface temperature of 82.2°C. Compare this with the value of $h = 77.2$ W/m^2 · K for the flat plate in turbulent flow.

Solution: The physical properties at the average film temperature of 48.9°C are the same as for Example 4.6-1. The N_{Re} is

$$N_{Re} = \frac{Dv\rho}{\mu} = \frac{(0.051)(12.2)(1.097)}{1.95 \times 10^{-5}} = 3.49 \times 10^4$$

TABLE 4.6-1. *Constants for Use in Eq. (4.6-1) for Heat Transfer to Cylinders with Axis Perpendicular to Flow ($N_{Pr} > 0.6$)*

N_{Re}	m	C
1–4	0.330	0.989
4–40	0.385	0.911
40–4 × 10^3	0.466	0.683
4 × 10^3–4 × 10^4	0.618	0.193
4 × 10^4–2.5 × 10^5	0.805	0.0266

Substituting into Eq. (4.6-4) for a sphere,

$$N_{Nu} = \frac{hD}{k} = \frac{h(0.051)}{0.0280} = 2.0 + 0.60 N_{Re}^{0.5} N_{Pr}^{1/3}$$

$$= 2.0 + (0.60)(3.49 \times 10^4)^{0.5}(0.704)^{1/3}$$

Solving, $h = 56.1$ W/m$^2 \cdot$ K (9.88 btu/h $\cdot$ ft$^2 \cdot$ °F). This value is somewhat smaller than the value of $h = 77.2$ W/m$^2 \cdot$ K (13.6 btu/h $\cdot$ ft$^2 \cdot$ °F) for a flat plate.

4.6E Flow Past Banks of Tubes or Cylinders

Many types of commercial heat exchangers are constructed with multiple rows of tubes, where the fluid flows at right angles to the bank of tubes. An example is a gas heater in which a hot fluid inside the tubes heats a gas passing over the outside of the tubes. Another example is a cold liquid stream inside the tubes being heated by a hot fluid on the outside.

Figure 4.6-1 shows the arrangement for banks of tubes in-line and banks of tubes staggered where D is tube OD in m (ft), S_n is distance m (ft) between the centers of the tubes normal to the flow, and S_p parallel to the flow. The open area to flow for in-line tubes is $(S_n - D)$ and $(S_p - D)$, and for staggered tubes it is $(S_n - D)$ and $(S_p' - D)$. Values of C and m to be used in Eq. (4.6-1) for a Reynolds number range of 2000 to 40 000 for heat transfer to banks of tubes containing more than 10 transverse rows in the direction of flow are given in Table 4.6-2. For less than 10 rows, Table 4.6-3 gives correction factors.

For cases where S_n/D and S_p/D are not equal to each other, the reader should consult Grimison (G1) for more data. In baffled exchangers where there is normal leakage where all the fluid does not flow normal to the tubes, the average values of h obtained should be multiplied by about 0.6 (P3). The Reynolds number is calculated using the minimum area open to flow for the velocity. All physical properties are evaluated at T_f.

EXAMPLE 4.6-3. Heating Air by a Bank of Tubes
Air at 15.6°C and 1 atm abs flows across a bank of tubes containing four transverse rows in the direction of flow and 10 rows normal to the flow at a

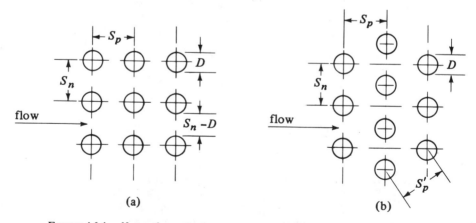

(a)   (b)

FIGURE 4.6-1. *Nomenclature for banks of tubes in Table 4.6-2: (a) in-line tube rows, (b) staggered tube rows.*

TABLE 4.6-2. *Values of C and m To Be Used in Eq. (4.6-1) for Heat Transfer to Banks of Tubes Containing More Than 10 Transverse Rows*

Arrangement	$\frac{S_n}{D} = \frac{S_p}{D} = 1.25$		$\frac{S_n}{D} = \frac{S_p}{D} = 1.50$		$\frac{S_n}{D} = \frac{S_p}{D} = 2.0$	
	C	m	C	m	C	m
In-line	0.386	0.592	0.278	0.620	0.254	0.632
Staggered	0.575	0.556	0.511	0.562	0.535	0.556

Source: E. D. Grimison, *Trans. ASME,* **59**, 583 (1937).

velocity of 7.62 m/s as the air approaches the bank of tubes. The tube surfaces are maintained at 57.2°C. The outside diameter of the tubes is 25.4 mm and the tubes are in-line to the flow. The spacing S_n of the tubes normal to the flow is 38.1 mm and also S_p is 38.1 mm parallel to the flow. For a 0.305 m length of the tube bank, calculate the heat-transfer rate.

Solution: Referring to Fig. 4.6-1a,

$$\frac{S_n}{D} = \frac{38.1}{25.4} = \frac{1.5}{1} \qquad \frac{S_p}{D} = \frac{38.1}{25.4} = \frac{1.5}{1}$$

Since the air is heated in passing through the four transverse rows, an outlet bulk temperature of 21.1°C will be assumed. The average bulk temperature is then

$$T_b = \frac{15.6 + 21.1}{2} = 18.3°C$$

The average film temperature is

$$T_j = \frac{T_w + T_b}{2} = \frac{57.2 + 18.3}{2} = 37.7°C$$

From Appendix A.3 for air at 37.7°C,

$$k = 0.02700 \text{ W/m} \cdot \text{K} \qquad N_{Pr} = 0.705$$
$$c_p = 1.0048 \text{ kJ/kg} \cdot \text{K} \qquad \rho = 1.137 \text{ kg/m}^3$$
$$\mu = 1.90 \times 10^{-5} \text{ Pa} \cdot \text{s}$$

TABLE 4.6-3. *Ratio of h for N Transverse Rows Deep to h for 10 Transverse Rows Deep (for Use with Table 4.6-2)*

N	1	2	3	4	5	6	7	8	9	10
Ratio for staggered tubes	0.68	0.75	0.83	0.89	0.92	0.95	0.97	0.98	0.99	1.00
Ratio for in-line tubes	0.64	0.80	0.87	0.90	0.92	0.94	0.96	0.98	0.99	1.00

Source: W. M. Kays and R. K. Lo, *Stanford Univ. Tech. Rept. 15,* Navy Contract N6-ONR-251 T.O.6, 1952.

The ratio of the minimum-flow area to the total frontal area is $(S_n - D)/S_n$. The maximum velocity in the tube banks is then

$$v_{max} = \frac{vS_n}{S_n - D} = \frac{7.62(0.0381)}{(0.0381 - 0.0254)} = 22.86 \text{ m/s}$$

$$N_{Re} = \frac{Dv_{max}\rho}{\mu} = \frac{0.0254(22.86)(1.137)}{1.90 \times 10^{-5}} = 3.47 \times 10^4$$

For $S_n/D = S_p/D = 1.5/1$, the values of C and m from Table 4.6-2 are 0.278 and 0.620, respectively. Substituting into Eq. (4.6-1) and solving for h,

$$h = \frac{k}{D} CN_{Re}^m N_{Pr}^{1/3} = \left(\frac{0.02700}{0.0254}\right)(0.278)(3.47 \times 10^4)^{0.62}(0.705)^{1/3}$$

$$= 171.8 \text{ W/m}^2 \cdot \text{K}$$

This h is for 10 rows. For only four rows in the transverse direction, the h must be multiplied by 0.90, as given in Table 4.6-3.

Since there are 10×4 or 40 tubes, the total heat-transfer area per 0.305 m length is

$$A = 40\pi DL = 40\pi(0.0254)(0.305) = 0.973 \text{ m}^2$$

The total heat-transfer rate q using an arithmetic average temperature difference between the wall and the bulk fluid is

$$q = hA(T_w - T_b) = (0.90 \times 171.8)(0.973)(57.2 - 18.3) = 5852 \text{ W}$$

Next, a heat balance on the air is made to calculate its temperature rise ΔT using the calculated q. First the mass flow rate of air m must be calculated. The total frontal area of the tube bank assembly of 10 rows of tubes each 0.305 m long is

$$A_t = 10S_n(1.0) = 10(0.0381)(0.305) = 0.1162 \text{ m}^2$$

The density of the entering air at 15.6°C is $\rho = 1.224 \text{ kg/m}^3$. The mass flow rate m is

$$m = v\rho A_t(3600) = 7.62(1.224)(0.1162) = 1.084 \text{ kg/s}$$

For the heat balance the mean c_p of air at 18.3°C is 1.0048 kJ/kg·K and then

$$q = 5852 = mc_p\Delta T = 1.084(1.0048 \times 10^3)\,\Delta T$$

Solving, $\Delta T = 5.37$°C.

Hence, the calculated outlet bulk gas temperature is $15.6 + 5.37 = 20.97$°C, which is close to the assumed value of 21.1°C. If a second trial were to be made, the new average T_b to use would be $(15.6 + 20.97)/2$ or 18.28°C.

4.6F   Heat Transfer for Flow in Packed Beds

Correlations for heat-transfer coefficients for packed beds are useful in designing fixed-bed systems such as catalytic reactors, dryers for solids, and pebble-bed heat exchangers. In Section 3.1C the pressure drop in packed beds was considered and discussions of the geometry factors in these beds were given. For determining the rate of heat transfer in packed beds for a differential length dz in m,

$$dq = h(a\ S\ dz)(T_1 - T_2) \tag{4.6-5}$$

where a is the solid particle surface area per unit volume of bed in m^{-1}, S the empty cross-sectional area of bed in m^2, T_1 the bulk gas temperature in K, and T_2 the solid surface temperature.

For the heat transfer of gases in beds of spheres (G2, G3) and a Reynolds number range of 10–10 000,

$$\varepsilon J_H = \varepsilon \frac{h}{c_p v' \rho} \left(\frac{c_p \mu}{k}\right)_f^{2/3} = \frac{2.876}{N_{Re}} + \frac{0.3023}{N_{Re}^{0.35}} \tag{4.6-6}$$

where v' is the superficial velocity based on the cross section of the empty container in m/s [see Eq. (3.1-11)], ε is the void fraction, $N_{Re} = D_p G'/\mu_f$, and $G' = v'\rho$ is the superficial mass velocity in $kg/m^2 \cdot s$. The subscript f indicates properties evaluated at the film temperature with others at the bulk temperature. This correlation can also be used for a fluidized bed. An alternate equation to use in place of Eq. (4.6-6) for fixed and fluidized beds is Eq. (7.3 36) for a Reynolds number range of 10–4000. The term J_H is called the Colburn J factor and is defined as in Eq. (4.6-6) in terms of h.

Equations for heat transfer to noncircular cylinders such as a hexagon, etc., are given elsewhere (H1, J1, P3).

4.7 NATURAL CONVECTION HEAT TRANSFER

4.7A Introduction

Natural convection heat transfer occurs when a solid surface is in contact with a gas or liquid which is at a different temperature from the surface. Density differences in the fluid arising from the heating process provide the buoyancy force required to move the fluid. Free or natural convection is observed as a result of the motion of the fluid. An example of heat transfer by natural convection is a hot radiator used for heating a room. Cold air encountering the radiator is heated and rises in natural convection because of buoyancy forces. The theoretical derivation of equations for natural convection heat-transfer coefficients requires the solution of motion and energy equations.

An important heat-transfer system occurring in process engineering is that in which heat is being transferred from a hot vertical plate to a gas or liquid adjacent to it by natural convection. The fluid is not moving by forced convection but only by natural or free convection. In Fig. 4.7-1 the vertical flat plate is heated and the free-convection boundary layer is formed. The velocity profile differs from that in a forced-convection system in that the velocity at the wall is zero and also is zero at the other edge of the boundary layer since the free-stream velocity is zero for natural convection. The boundary layer initially is laminar as shown, but at some distance from the leading edge it starts to become turbulent. The wall temperature is T_w K and the bulk temperature T_b.

The differential momentum balance equation is written for the x and y directions for the control volume $(dx\ dy \cdot 1)$. The driving force is the buoyancy force in the gravitational field and is due to the density difference of the fluid. The momentum balance becomes

$$\rho\left(v_x \frac{\partial v_x}{\partial x} + v_y \frac{\partial v_x}{\partial y}\right) = g(\rho_b - \rho) + \mu \frac{\partial^2 v_x}{\partial y^2} \tag{4.7-1}$$

where ρ_b is the density at the bulk temperature T_b and ρ at T. The density difference can be expressed in terms of the volumetric coefficient of expansion β and substituted back into Eq. (4.7-1).

$$\beta = \frac{\rho_b - \rho}{\rho(T - T_b)} \tag{4.7-2}$$

For gases, $\beta = 1/T$. The energy-balance equation can be expressed as follows:

$$\rho c_p\left(v_x \frac{\partial T}{\partial x} + v_y \frac{\partial T}{\partial y}\right) = k \frac{\partial^2 T}{\partial y^2} \qquad (4.7\text{-}3)$$

The solutions of these equations have been obtained by using integral methods of analysis discussed in Section 3.10. Results for a vertical plate have been obtained, which is the most simple case and serves to introduce the dimensionless Grashof number discussed below. However, in other physical geometries the relations are too complex and empirical correlations have been obtained. These are discussed in the following sections.

4.7B Natural Convection from Various Geometries

1. Natural convection from vertical planes and cylinders. For an isothermal vertical surface or plate with height L less than 1 m (P3), the average natural convection heat-transfer coefficient can be expressed by the following general equation:

$$N_{\text{Nu}} = \frac{hL}{k} = a\left(\frac{L^3\rho^2 g\beta\,\Delta T}{\mu^2}\,\frac{c_p\mu}{k}\right)^m = a(N_{\text{Gr}}\,N_{\text{Pr}})^m \qquad (4.7\text{-}4)$$

where a and m are constants from Table 4.7-1, N_{Gr} the Grashof number, ρ density in kg/m^3, μ viscosity in kg/m·s, ΔT the positive temperature difference between the wall and bulk fluid or vice versa in K, k the thermal conductivity in W/m·K, c_p the heat capacity in J/kg·K, β the volumetric coefficient of expansion of the fluid in 1/K [for gases β is $1/(T_f \text{ K})$], and g is 9.80665 m/s^2. All the physical properties are evaluated at the film temperature $T_f = (T_w + T_b)/2$. In general, for a vertical cylinder with length L m, the same equations can be used as for a vertical plate. In English units β is $1/(T_f°\text{F} + 460)$ in 1/°R and g is 32.174 × (3600)^2 ft/h^2.

The Grashof number can be interpreted physically as a dimensionless number that represents the ratio of the buoyancy forces to the viscous forces in free convection and plays a role similar to that of the Reynolds number in forced convection.

EXAMPLE 4.7-1. Natural Convection from Vertical Wall of an Oven
A heated vertical wall 1.0 ft (0.305 m) high of an oven for baking food with the surface at 450°F (505.4 K) is in contact with air at 100°F (311 K). Calculate the heat-transfer coefficient and the heat transfer/ft (0.305 m) width of wall. Note that heat transfer for radiation will not be considered. Use English and SI units.

FIGURE 4.7-1. *Boundary-layer velocity profile for natural convection heat transfer from a heated, vertical plate.*

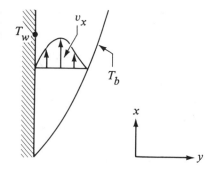

TABLE 4.7-1. *Constants for Use with Eq. (4.7-4) for Natural Convection*

Physical Geometry	$N_{Gr} N_{Pr}$	a	m	Ref.
Vertical planes and cylinders [vertical height $L < 1$ m (3 ft)]				
	$< 10^4$	1.36	$\frac{1}{5}$	(P3)
	10^4-10^9	0.59	$\frac{1}{4}$	(M1)
	$> 10^9$	0.13	$\frac{1}{3}$	(M1)
Horizontal cylinders [diameter D used for L and $D < 0.20$ m (0.66 ft)]				
	$< 10^{-5}$	0.49	0	(P3)
	$10^{-5}-10^{-3}$	0.71	$\frac{1}{25}$	(P3)
	$10^{-3}-1$	1.09	$\frac{1}{10}$	(P3)
	$1-10^4$	1.09	$\frac{1}{5}$	(P3)
	10^4-10^9	0.53	$\frac{1}{4}$	(M1)
	$> 10^9$	0.13	$\frac{1}{3}$	(P3)
Horizontal plates				
Upper surface of heated plates or lower surface of cooled plates	$10^5-2 \times 10^7$	0.54	$\frac{1}{4}$	(M1)
	$2 \times 10^7-3 \times 10^{10}$	0.14	$\frac{1}{3}$	(M1)
Lower surface of heated plates or upper surface of cooled plates	10^5-10^{11}	0.58	$\frac{1}{5}$	(F1)

Solution: The film temperature is

$$T_f = \frac{T_w + T_b}{2} = \frac{450 + 100}{2} = 275°F = \frac{505.4 + 311}{2} = 408.2 \text{ K}$$

The physical properties of air at 275°F are $k = 0.0198$ btu/h·ft·°F, 0.0343 W/m·K; $\rho = 0.0541$ lb$_m$/ft^3, 0.867 kg/m^3; $N_{Pr} = 0.690$; $\mu = (0.0232 \text{ cp}) \times (2.4191) = 0.0562$ lb$_m$/ft·h $= 2.32 \times 10^{-5}$ Pa·s; $\beta = 1/408.2 = 2.45 \times 10^{-3}$ K^{-1}, $\beta = 1/(460 + 275) = 1.36 \times 10^{-3}$ °R^{-1}; $\Delta T = T_w - T_b = 450 - 100 = 350°F$ (194.4 K). The Grashof number is, in English units,

$$N_{Gr} = \frac{L^3 \rho^2 g \beta \Delta T}{\mu^2} = \frac{(1.0)^3(0.0541)^2(32.174)(3600)^2(1.36 \times 10^{-3})(350)}{(0.0562)^2}$$

$$= 1.84 \times 10^8$$

In SI units,

$$N_{Gr} = \frac{(0.305)^3(0.867)^2(9.806)(2.45 \times 10^{-3})(194.4)}{(2.32 \times 10^{-5})^2} = 1.84 \times 10^8$$

The Grashof numbers calculated using English and SI units must, of course, be the same as shown.

$$N_{Gr} N_{Pr} = (1.84 \times 10^8)(0.690) = 1.270 \times 10^8$$

Hence, from Table 4.7-1, $a = 0.59$ and $m = \frac{1}{4}$ for use in Eq. (4.7-4).

Solving for h in Eq. (4.7-4) and substituting known values,

$$h = \frac{k}{L} a(N_{Gr} N_{Pr})^m = \left(\frac{0.0198}{1.0}\right)(0.59)(1.270 \times 10^8)^{1/4} = 1.24 \text{ btu/h} \cdot \text{ft}^2 \cdot {}^\circ\text{F}$$

$$h = \left(\frac{0.0343}{0.305}\right)(0.59)(1.27 \times 10^8)^{1/4} = 7.03 \text{ W/m}^2 \cdot \text{K}$$

For a 1-ft width of wall, $A = 1 \times 1 = 1.0 \text{ ft}^2$ ($0.305 \times 0.305 \text{ m}^2$). Then

$$q = hA(T_w - T_b) = (1.24)(1.0)(450 - 100) = 433 \text{ btu/h}$$

$$q = 7.03(0.305 \times 0.305)(194.4) = 127.1 \text{ W}$$

A considerable amount of heat will also be lost by radiation. This will be considered in Section 4.10.

Simplified equations for the natural convection heat transfer from air to vertical planes and cylinders at 1 atm abs pressure are given in Table 4.7-2. In SI units the equation for the range of $N_{Gr} N_{Pr}$ of 10^4 to 10^9 is the one usually encountered and this holds for ($L^3 \Delta T$) values below about $4.7 \text{ m}^3 \cdot \text{K}$ and film temperatures between 255 and 533 K. To correct the value of h to pressures other than 1 atm, the values of h in Table

TABLE 4.7-2. *Simplified Equations for Natural Convection from Various Surfaces*

Physical Geometry	$N_{Gr} N_{Pr}$	Equation		Ref.
		$h = btu/h \cdot ft^2 \cdot {}^\circ F$ $L = ft, \Delta T = {}^\circ F$ $D = ft$	$h = W/m^2 \cdot K$ $L = m, \Delta T = K$ $D = m$	
Air at 101.32 kPa (1 atm) abs pressure				
Vertical planes and	10^4-10^9	$h = 0.28(\Delta T/L)^{1/4}$	$h = 1.37(\Delta T/L)^{1/4}$	(P1)
cylinders	$>10^9$	$h = 0.18(\Delta T)^{1/3}$	$h = 1.24 \Delta T^{1/3}$	(P1)
Horizontal cylinders	10^3-10^9	$h = 0.27(\Delta T/D)^{1/4}$	$h = 1.32(\Delta T/D)^{1/4}$	(M1)
	$>10^9$	$h = 0.18(\Delta T)^{1/3}$	$h = 1.24 \Delta T^{1/3}$	(M1)
Horizontal plates				
Heated plate facing	$10^5-2 \times 10^7$	$h = 0.27(\Delta T/L)^{1/4}$	$h = 1.32(\Delta T/L)^{1/4}$	(M1)
upward or cooled plate facing downward	$2 \times 10^7-3 \times 10^{10}$	$h = 0.22(\Delta T)^{1/3}$	$h = 1.52 \Delta T^{1/3}$	(M1)
Heated plate facing downward or cooled plate facing upward	$3 \times 10^5-3 \times 10^{10}$	$h = 0.12(\Delta T/L)^{1/4}$	$h = 0.59(\Delta T/L)^{1/4}$	(M1)
Water at 70°F (294 K)				
Vertical planes and cylinders	10^4-10^9	$h = 26(\Delta T/L)^{1/4}$	$h = 127(\Delta T/L)^{1/4}$	(P1)
Organic liquids at 70°F (294 K)				
Vertical planes and cylinders	10^4-10^9	$h = 12(\Delta T/L)^{1/4}$	$h = 59(\Delta T/L)^{1/4}$	(P1)

4.7-2 can be multiplied by $(p/101.32)^{1/2}$ for $N_{Gr}N_{Pr}$ 10^4 to 10^9 and by $(p/101.32)^{2/3}$ for $N_{Gr}N_{Pr} > 10^9$, where p = pressure in kN/m^2. In English units the range of $N_{Gr}N_{Pr}$ of 10^4 to 10^9 is encountered when $(L^3 \, \Delta T)$ is less than about $300\,ft^3 \cdot {}^\circ F$. The value of h can be corrected to pressures other than 1.0 atm abs by multiplying the h at 1 atm by $p^{1/2}$ for $N_{Gr}N_{Pr}$ of 10^4 to 10^9 and by $p^{2/3}$ for $N_{Gr}N_{Pr}$ above 10^9, where p = atm abs pressure. Simplified equations are also given for water and organic liquids.

EXAMPLE 4.7-2. Natural Convections and Simplified Equation
Repeat Example 4.7-1 but use the simplified equation.

Solution: The film temperature of 408.2 K is in the range 255–533 K. Also,

$$L^3 \, \Delta T = (0.305)^3(194.4) = 5.5$$

This is slightly greater than the value of 4.7 given as the approximate maximum for use of the simplified equation. However, in Example 4.7-1 the value of $N_{Gr}N_{Pr}$ is below 10^9, so the simplified equation from Table 4.7-2 will be used.

$$h = 1.37\left(\frac{\Delta T}{L}\right)^{1/4} = 1.37\left(\frac{194.4}{0.305}\right)^{1/4}$$

$$= 6.88 \; W/m^2 \cdot K \; (1.21 \; btu/h \cdot ft^2 \cdot {}^\circ F)$$

The heat-transfer rate q is

$$q = hA(T_w - T_b) = 6.88(0.305 \times 0.305)(194.4) = 124.4 \; W \; (424 \; btu/h)$$

This value is reasonably close to the value of 127.1 W for Example 4.7-1.

2. Natural convection from horizontal cylinders. For a horizontal cylinder with an outside diameter of D m, Eq. (4.7-4) is used with the constants being given in Table 4.7-1. The diameter D is used for L in the equation. Simplified equations are given in Table 4.7-2. The usual case for pipes is for the $N_{Gr}N_{Pr}$ range 10^4 to 10^9 (M1).

3. Natural convection from horizontal plates. For horizontal flat plates Eq. (4.7-4) is also used with the constants given in Table 4.7-1 and simplified equations in Table 4.7-2. The dimension L to be used is the length of a side of a square plate, the linear mean of the two dimensions for a rectangle, and 0.9 times the diameter of a circular disk.

4. Natural convection in enclosed spaces. Free convection in enclosed spaces occurs in a number of processing applications. One example is in an enclosed double window in which two layers of glass are separated by a layer of air for energy conservation. The flow phenomena inside these enclosed spaces are complex since a number of different types of flow patterns can occur. At low Grashof numbers the heat transfer is mainly by conduction across the fluid layer. As the Grashof number is increased, different flow regimes are encountered.

The system for two vertical plates of height L m containing the fluid with a gap of δ m is shown in Fig. 4.7-2, where the plate surfaces are at T_1 and T_2 temperatures. The Grashof number is defined as

$$N_{Gr, \delta} = \frac{\delta^3 \rho^2 g \beta (T_1 - T_2)}{\mu^2} \qquad (4.7\text{-}5)$$

The Nusselt number is defined as

$$N_{Nu, \delta} = \frac{h\delta}{k} \qquad (4.7\text{-}6)$$

The heat flux is calculated from

$$\frac{q}{A} = h(T_1 - T_2) \qquad (4.7\text{-}7)$$

The physical properties are all evaluated at the mean temperature between the two plates.

For gases enclosed between vertical plates and $L/\delta > 3$ (H1, J1, K1, P1),

$$N_{\text{Nu}, \delta} = \frac{h\delta}{k} = 1.0 \qquad (N_{\text{Gr}, \delta} N_{\text{Pr}} < 2 \times 10^3) \qquad (4.7\text{-}8)$$

$$N_{\text{Nu}, \delta} = 0.20 \frac{(N_{\text{Gr}, \delta} N_{\text{Pr}})^{1/4}}{(L/\delta)^{1/9}} \qquad (6 \times 10^3 < N_{\text{Gr}, \delta} N_{\text{Pr}} < 2 \times 10^5) \qquad (4.7\text{-}9)$$

$$N_{\text{Nu}, \delta} = 0.073 \frac{(N_{\text{Gr}, \delta} N_{\text{Pr}})^{1/3}}{(L/\delta)^{1/9}} \qquad (2 \times 10^5 < N_{\text{Gr}, \delta} N_{\text{Pr}} < 2 \times 10^7) \qquad (4.7\text{-}10)$$

For liquids in vertical plates,

$$N_{\text{Nu}, \delta} = \frac{h\delta}{k} = 1.0 \qquad (N_{\text{Gr}, \delta} N_{\text{Pr}} < 1 \times 10^3) \qquad (4.7\text{-}11)$$

$$N_{\text{Nu}, \delta} = 0.28 \frac{(N_{\text{Gr}, \delta} N_{\text{Pr}})^{1/4}}{(L/\delta)^{1/4}} \qquad (1 \times 10^3 < N_{\text{Gr}, \delta} N_{\text{Pr}} < 1 \times 10^7) \qquad (4.7\text{-}12)$$

For gases or liquids in a vertical annulus, the same equations hold as for vertical plates.

For gases in horizontal plates with the lower plate hotter than the upper,

$$N_{\text{Nu}, \delta} = 0.21(N_{\text{Gr}, \delta} N_{\text{Pr}})^{1/4} \qquad (7 \times 10^3 < N_{\text{Gr}, \delta} N_{\text{Pr}} < 3 \times 10^5) \qquad (4.7\text{-}13)$$

$$N_{\text{Nu}, \delta} = 0.061(N_{\text{Gr}, \delta} N_{\text{Pr}})^{1/3} \qquad (N_{\text{Gr}, \delta} N_{\text{Pr}} > 3 \times 10^5) \qquad (4.7\text{-}14)$$

For liquids in horizontal plates with the lower plate hotter than the upper (G5),

$$N_{\text{Nu}, \delta} = 0.069(N_{\text{Gr}, \delta} N_{\text{Pr}})^{1/3} N_{\text{Pr}}^{0.074} \qquad (1.5 \times 10^5 < N_{\text{Gr}, \delta} N_{\text{Pr}} < 1 \times 10^9) \qquad (4.7\text{-}15)$$

EXAMPLE 4.7-3. *Natural Convection in Enclosed Vertical Space*
Air at 1 atm abs pressure is enclosed between two vertical plates where $L = 0.6$ m and $\delta = 30$ mm. The plates are 0.4 m wide. The plate temperatures are $T_1 = 394.3$ K and $T_2 = 366.5$ K. Calculate the heat-transfer rate across the air gap.

FIGURE 4.7-2.   *Natural convection in enclosed vertical space.*

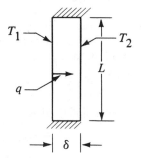

Solution: The mean temperature between the plates is used to evaluate the physical properties. $T_f = (T_1 + T_2)/2 = (394.3 + 366.5)/2 = 380.4$ K. Also, $\delta = 30/1000 = 0.030$ m. From Appendix A.3, $\rho = 0.9295$ kg/m^3, $\mu = 2.21 \times 10^{-5}$ Pa·s, $k = 0.03219$ W/m·K, $N_{Pr} = 0.693$, $\beta = 1/T_f = 1/380.4 = 2.629 \times 10^{-3}$K^{-1}.

$$N_{Gr,\,\delta} = \frac{(0.030)^3(0.9295)^2(9.806)(2.629 \times 10^{-3})(394.3 - 366.5)}{(2.21 \times 10^{-5})^2}$$

$$= 3.423 \times 10^4$$

Also, $N_{Gr,\,\delta}\,N_{Pr} = (3.423 \times 10^4)0.693 = 2.372 \times 10^4$. Using Eq. (4.7-9),

$$h = \frac{k}{\delta}\frac{(0.20)(N_{Gr,\,\delta}\,N_{Pr})^{1/4}}{(L/\delta)^{1/9}} = \frac{0.03219(0.20)(2.352 \times 10^4)^{1/4}}{0.030(0.6/0.030)^{1/9}}$$

$$= 1.909 \text{ W/m}^2 \cdot \text{K}$$

The area $A = (0.6 \times 0.4) = 0.24$ m^2. Substituting into Eq. (4.7-7),

$$q = hA(T_1 = T_2) = 1.909(0.24)(394.3 - 366.5) = 12.74 \text{ W}$$

5. Natural convection from other shapes. For spheres, blocks, and other types of enclosed air spaces, references elsewhere (H1, K1, M1, P1, P3) should be consulted. In some cases when a fluid is forced over a heated surface at low velocity in the laminar region, combined forced-convection plus natural-convection heat transfer occurs. For further discussion of this, see (H1, K1, M1).

4.8 BOILING AND CONDENSATION

4.8A Boiling

1. Mechanisms of boiling. Heat transfer to a boiling liquid is very important in evaporation and distillation and also in other kinds of chemical and biological processing, such as petroleum processing, control of the temperature of chemical reactions, evaporation of liquid foods, and so on. The boiling liquid is usually contained in a vessel with a heating surface of tubes or vertical or horizontal plates which supply the heat for boiling. The heating surfaces can be heated electrically or by a hot or condensing fluid on the other side of the heated surface.

In boiling the temperature of the liquid is the boiling point of this liquid at the pressure in the equipment. The heated surface is, of course, at a temperature above the boiling point. Bubbles of vapor are generated at the heated surface and rise through the mass of liquid. The vapor accumulates in a vapor space above the liquid level and is withdrawn.

Boiling is a complex phenomenon. Suppose we consider a small heated horizontal tube or wire immersed in a vessel containing water boiling at 373.2 K (100°C). The heat flux is q/A W/m^2, $\Delta T = T_w - 373.2$ K, where T_w is the tube or wire wall temperature and h is the heat-transfer coefficient in W/m^2·K. Starting with a low ΔT, the q/A and h values are measured. This is repeated at higher values of ΔT and the data obtained are shown in Fig. 4.8-1 plotted as q/A versus ΔT.

In the first region A of the plot in Fig. 4.8-1, at low temperature drops, the mechanism of boiling is essentially that of heat transfer to a liquid in natural convection.

The variation of h with $\Delta T^{0.25}$ is approximately the same as that for natural convection to horizontal plates or cylinders. The very few bubbles formed are released from the surface of the metal and rise and do not disturb appreciably the normal natural convection.

In the region B of nucleate boiling for a ΔT of about $5 - 25$ K ($9 - 45°$F), the rate of bubble production increases so that the velocity of circulation of the liquid increases. The heat-transfer coefficient h increases rapidly and is proportional to ΔT^2 to ΔT^3 in this region.

In the region C of transition boiling, many bubbles are formed so quickly that they tend to coalesce and form a layer of insulating vapor. Increasing the ΔT increases the thickness of this layer and the heat flux and h drop as ΔT is increased. In region D or film boiling, bubbles detach themselves regularly and rise upward. At higher ΔT values radiation through the vapor layer next to the surface helps increase the q/A and h.

The curve of h versus ΔT has approximately the same shape as Fig. 4.8-1. The values of h are quite large. At the beginning of region B in Fig. 4.8-1 for nucleate boiling, h has a value of about 5700–11 400 W/m$^2 \cdot$ K, or 1000–2000 btu/h $\cdot$ ft$^2 \cdot$ °F, and at the end of this region h has a peak value of almost 57 000 W/m$^2 \cdot$ K, or 10 000 btu/hr $\cdot$ ft$^2 \cdot$ °F. These values are quite high, and in most cases the percent resistance of the boiling film is only a few percent of the overall resistance to heat transfer.

The regions of commercial interest are the nucleate and film-boiling regions (P3). Nucleate boiling occurs in kettle-type and natural-circulation reboilers.

2. Nucleate boiling. In the nucleate boiling region the heat flux is affected by ΔT, pressure, nature and geometry of the surface and system, and physical properties of the vapor and liquid. Equations have been derived by Rohesenow et al. (P1). They apply to single tubes or flat surfaces and are quite complex.

Simplified empirical equations to estimate the boiling heat-transfer coefficients for water boiling on the outside of submerged surfaces at 1.0 atm abs pressure have been developed (J2).

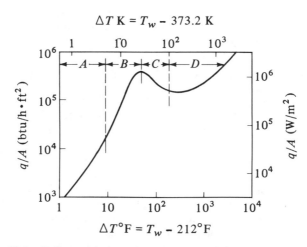

$$\Delta T \text{ K} = T_w - 373.2 \text{ K}$$

$$\Delta T °\text{F} = T_w - 212°\text{F}$$

FIGURE 4.8-1. *Boiling mechanisms for water at atmospheric pressure, heat flux vs. temperature drop: (A) natural convection, (B) nucleate boiling, (C) transition boiling, (D) film boiling.*

For a horizontal surface (SI and English units),

$$h, \text{btu/h} \cdot \text{ft}^2 \cdot {}^\circ\text{F} = 151(\Delta T{}^\circ\text{F})^{1/3} \qquad q/A, \text{btu/h} \cdot \text{ft}^2, < 5000$$
$$h, \text{W/m}^2 \cdot \text{K} = 1043(\Delta T\text{K})^{1/3} \qquad q/A, \text{kW/m}^2, < 16 \tag{4.8-1}$$

$$h, \text{btu/h} \cdot \text{ft}^2 \cdot {}^\circ\text{F} = 0.168(\Delta T{}^\circ\text{F})^3 \qquad 5000 < q/A, \text{btu/h} \cdot \text{ft}^2, < 75\,000$$
$$h, \text{W/m}^2 \cdot \text{K} = 5.56(\Delta T\text{K})^3 \qquad 16 < q/A, \text{kW/m}^2, < 240 \tag{4.8-2}$$

For a vertical surface,

$$h, \text{btu/h} \cdot \text{ft}^2 \cdot {}^\circ\text{F} = 87(\Delta T{}^\circ\text{F})^{1/7} \qquad q/A, \text{btu/h} \cdot \text{ft}^2, < 1000$$
$$h, \text{W/m}^2 \cdot \text{K} = 537(\Delta T\text{K})^{1/7} \qquad q/A, \text{kW/m}^2, < 3 \tag{4.8-3}$$

$$h, \text{btu/h} \cdot \text{ft}^2 \cdot {}^\circ\text{F} = 0.240(\Delta T{}^\circ\text{F})^3 \qquad 1000 < q/A, \text{btu/h} \cdot \text{ft}^2, < 20\,000$$
$$h, \text{W/m}^2 \cdot \text{K} = 7.95(\Delta T\text{K})^3 \qquad 3 < q/A, \text{kW/m}^2, < 63 \tag{4.8-4}$$

where $\Delta T = T_w - T_{sat}$ K or ${}^\circ$F.

If the pressure is p atm abs, the values of h at 1 atm given above are multiplied by $(p/1)^{0.4}$. Equations (4.8-1) and (4.8-3) are in the natural convection region.

For forced convection boiling inside tubes, the following simplified relation can be used (J3).

$$h = 2.55(\Delta T\text{K})^3 e^{p/1551} \text{ W/m}^2 \cdot \text{K} \qquad \text{(SI)}$$
$$h = 0.077(\Delta T{}^\circ\text{F})^3 e^{p/225} \text{ btu/h} \cdot \text{ft}^2 \cdot {}^\circ\text{F} \qquad \text{(English)} \tag{4.8-5}$$

where p in this case is in kPa (SI units) and psia (English units).

3. *Film boiling.* In the film-boiling region the heat-transfer rate is low in view of the large temperature drop used, which is not utilized effectively. Film boiling has been subjected to considerable theoretical analysis. Bromley (B3) gives the following equation to predict the heat-transfer coefficient in the film-boiling region on a horizontal tube.

$$h = 0.62 \left[\frac{k_v^3 \rho_v(\rho_l - \rho_v)g(h_{fg} + 0.4c_{pv}\,\Delta T)}{D\mu_v\,\Delta T} \right]^{1/4} \tag{4.8-6}$$

where k_v is the thermal conductivity of the vapor in W/m $\cdot$ K, ρ_v the density of the vapor in kg/m^3, ρ_l the density of the liquid in kg/m^3, h_{fg} the latent heat of vaporization in J/kg, $\Delta T = T_w - T_{sat}$, T_{sat} the temperature of saturated vapor in K, D the outside tube diameter in m, μ_v the viscosity of the vapor in Pa $\cdot$ s, and g the acceleration of gravity in m/s^2. The physical properties of the vapor are evaluated at the film temperature of $T_f = (T_w + T_{sat})/2$ and h_{fg} at the saturation temperature. If the temperature difference is quite high, some additional heat transfer occurs by radiation (H1).

EXAMPLE 4.8-1. Rate of Heat Transfer in a Jacketed Kettle
Water is being boiled at 1 atm abs pressure in a jacketed kettle with steam condensing in the jacket at 115.6°C. The inside diameter of the kettle is 0.656 m and the height is 0.984 m. The bottom is slightly curved but it will be assumed to be flat. Both the bottom and the sides up to a height of 0.656 m are jacketed. The kettle surface for heat transfer is 3.2-mm stainless steel with a k of 16.27 W/m $\cdot$ K. The condensing steam coefficient h_i inside the jacket has been estimated as 10 200 W/m$^2 \cdot$ K. Predict the boiling heat-transfer coefficient h_0 for the bottom surface of the kettle.

Solution: A diagram of the kettle is shown in Fig. 4.8-2. The simplified equations will be used for the boiling coefficient h_o. The solution is trial and error, since the inside metal surface temperature T_w is unknown. Assuming that $T_w = 110°C$,

$$\Delta T = T_w - T_{sat} = 110 - 100 = 10°C = 10 \text{ K}$$

Substituting into Eq. (4.8-2),

$$h_o = 5.56(\Delta T)^3 = 5.56(10)^3 = 5560 \text{ W/m}^2 \cdot \text{K}$$

$$\frac{q}{A} = h\Delta T = 5560(10) = 55\,600 \text{ W/m}^2$$

To check the assumed T_w, the resistances R_i of the condensing steam, R_w of the metal wall, and R_o of the boiling liquid must be calculated. Assuming equal areas of the resistances for $A = 1 \text{ m}^2$, then by Eq. (4.3-12),

$$R_i = \frac{1}{h_i A} = \frac{1}{10\,200(1)} = 9.80 \times 10^{-5}$$

$$R_w = \frac{\Delta x}{kA} = \frac{3.2/1000}{16.27(1.0)} = 19.66 \times 10^{-5}$$

$$R_o = \frac{1}{h_o A} = \frac{1}{5560(1)} = 17.98 \times 10^{-5}$$

$$\sum R = 9.80 \times 10^{-5} + 19.66 \times 10^{-5} + 17.98 \times 10^{-5} = 47.44 \times 10^{-5}$$

The temperature drop across the boiling film is then

$$\Delta T = \frac{R_o}{\sum R}(115.6 - 100) = \frac{17.98 \times 10^{-5}}{47.44 \times 10^{-5}}(15.6) = 5.9°C$$

Hence, $T_w = 100 + 5.9 = 105.9°C$. This is lower than the assumed value of $110°C$.

For the second trial $T_w = 108.3°C$ will be used. Then, $\Delta T = 108.3 - 100 = 8.3°C$ and, from Eq. (4.8-2), the new $h_0 = 3180$. Calculating the new $R_0 = 31.44 \times 10^{-5}$, and

$$\Delta T = \left(\frac{31.44 \times 10^{-5}}{60.90 \times 10^{-5}}\right)(115.6 - 100) = 8.1°C$$

and

$$T_w = 100 + 8.1 = 108.1°C$$

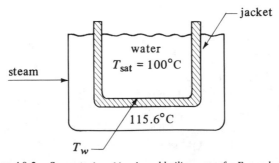

FIGURE 4.8-2. *Steam-jacketed kettle and boiling water for Example 4.8-1.*

This value is reasonably close to the assumed value of 108.3°C, so no further trials will be made.

4.8B Condensation

1. Mechanisms of condensation.   Condensation of a vapor to a liquid and vaporization of a liquid to a vapor both involve a change of phase of a fluid with large heat-transfer coefficients. Condensation occurs when a saturated vapor such as steam comes in contact with a solid whose surface temperature is below the saturation temperature, to form a liquid such as water.

Normally, when a vapor condenses on a surface such as a vertical or horizontal tube or other surfaces, a film of condensate is formed on the surface and flows over the surface by the action of gravity. It is this film of liquid between the surface and the vapor that forms the main resistance to heat transfer. This is called *film-type condensation.*

Another type of condensation, *dropwise condensation,* can occur where small drops are formed on the surface. These drops grow, coalesce, and the liquid flows from the surface. During this condensation, large areas of tube are devoid of any liquid and are exposed directly to the vapor. Very high rates of heat transfer occur on these bare areas. The average coefficient can be as high as $110\,000\ \text{W/m}^2 \cdot \text{K}$ ($20\,000\ \text{btu/h} \cdot \text{ft}^2 \cdot °\text{F}$), which is 5 to 10 times larger than film-type coefficients. Condensing film coefficients are normally much greater than those in forced convection and are of the order of magnitude of several thousand $\text{W/m}^2 \cdot \text{K}$ or more.

Dropwise condensation occurs on contaminated surfaces and when impurities are present. Film-type condensation is more dependable and more common. Hence, for normal design purposes, film-type condensation is assumed.

2. Film-condensation coefficients for vertical surfaces.   Film-type condensation on a vertical wall or tube can be analyzed analytically by assuming laminar flow of the condensate film down the wall. The film thickness is zero at the top of the wall or tube and increases in thickness as it flows downward because of condensation. Nusselt (H1, W1) assumed that the heat transfer from the condensing vapor at T_{sat} K, through this liquid film, and to the wall at T_w K was by conduction. Equating this heat transfer by conduction to that from condensation of the vapor, a final expression can be obtained for the average heat-transfer coefficient over the whole surface.

In Fig. 4.8-3a vapor at T_{sat} is condensing on a wall whose temperature is T_w K. The condensate is flowing downward in laminar flow. Assuming unit thickness, the mass of the element with liquid density ρ_l in Fig. 4.8-3b is $(\delta - y)(dx \cdot 1)\rho_l$. The downward force on this element is the gravitational force minus the buoyancy force or $(\delta - y)(dx) \times (\rho_l - \rho_v)g$ where ρ_v is the density of the saturated vapor. This force is balanced by the viscous-shear force at the plane y of $\mu_l(dv/dy)(dx \cdot 1)$. Equating these forces,

$$(\delta - y)(dx)(\rho_l - \rho_v)g = \mu_l\left(\frac{dv}{dy}\right)(dx) \tag{4.8-7}$$

Integrating and using the boundary condition that $v = 0$ at $y = 0$,

$$v = \frac{g(\rho_l - \rho_v)}{\mu_l}(\delta y - y^2/2) \tag{4.8-8}$$

The mass flow rate of film condensate at any point x for unit depth is

$$m = \int_0^\delta \rho_l v\,dy = \int_0^\delta \rho_l \frac{g(\rho_l - \rho_v)}{\mu_l}(\delta y - y^2/2)\,dy \tag{4.8-9}$$

Integrating,

$$m = \frac{\rho_l g(\rho_l - \rho_v)\delta^3}{3\mu_l} \tag{4.8-10}$$

At the wall for area $(dx \cdot 1)\,\mathrm{m}^2$, the rate of heat transfer is as follows if a linear temperature distribution is assumed in the liquid between the wall and the vapor:

$$q_x = -k_l(dx \cdot 1)\left.\frac{dT}{dy}\right)_{y=0} = k_l\, dx\, \frac{T_{\text{sat}} - T_w}{\delta} \tag{4.8-11}$$

In a dx distance, the rate of heat transfer is q_x. Also, in this dx distance, the increase in mass from condensation is dm. Using Eq. (4.8-10),

$$dm = d\left[\frac{\rho_l g(\rho_l - \rho_v)\delta^3}{3\mu_l}\right] = \frac{\rho_l g(\rho_l - \rho_v)\delta^2\, d\delta}{\mu_l} \tag{4.8-12}$$

Making a heat balance for dx distance, the mass flow rate dm times the latent heat h_{fg} must equal the q_x from Eq. (4.8-11).

$$h_{fg}\frac{\rho_l g(\rho_l - \rho_v)\delta^2\, d\delta}{\mu_l} = k_l\, dx\, \frac{T_{\text{sat}} - T_w}{\delta} \tag{4.8-13}$$

Integrating with $\delta = 0$ at $x = 0$ and $\delta = \delta$ at $x = x$,

$$\delta = \left[\frac{4\mu_l k_l x(T_{\text{sat}} - T_w)}{gh_{fg}\rho_l(\rho_l - \rho_v)}\right]^{1/4} \tag{4.8-14}$$

Using the local heat-transfer coefficient h_x at x, a heat balance gives

$$h_x(dx \cdot 1)(T_{\text{sat}} - T_w) = k_l(dx \cdot 1)\frac{T_{\text{sat}} - T_w}{\delta} \tag{4.8-15}$$

This gives

$$h_x = \frac{k_l}{\delta} \tag{4.8-16}$$

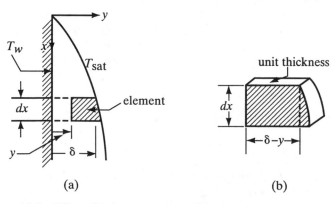

(a)   (b)

FIGURE 4.8-3.   *Film condensation on a vertical plate : (a) increase in film thickness with position, (b) balance on element of condensate.*

Chap. 4   Principles of Steady-State Heat Transfer

Combining Eqs. (4.8-14) and (4.8-16),

$$h_x = \left[\frac{\rho_l (\rho_l - \rho_v)gh_{fg} k_l^3}{4\mu_l \, x(T_{sat} - T_w)} \right]^{1/4}$$

(4.8-17)

By integrating over the total length L, the average value of h is obtained as follows.

$$h = \frac{1}{L} \int_0^L h_x \, dx = \tfrac{4}{3} h_{x=L}$$

(4.8-18)

$$h = 0.943 \left[\frac{\rho_l (\rho_l - \rho_v)gh_{fg} k_l^3}{\mu_l \, L(T_{sat} - T_w)} \right]^{1/4}$$

(4.8-19)

However, for laminar flow, experimental data show that the data are about 20% above Eq. (4.8-19).

Hence, the final recommended expression for vertical surfaces in laminar flow is (M1)

$$N_{Nu} = \frac{hL}{k_l} = 1.13 \left(\frac{\rho_l(\rho_l - \rho_v)gh_{fg} L^3}{\mu_l k_l \Delta T} \right)^{1/4}$$

(4.8-20)

where ρ_l is the density of liquid in kg/m^3 and ρ_v that of the vapor, g is 9.8066 m/s^2, L is the vertical height of the surface or tube in m, μ_l the viscosity of liquid in Pa·s, k_l the liquid thermal conductivity in W/m·K, $\Delta T = T_{sat} - T_w$ in K, and h_{fg} latent heat of condensation in J/kg at T_{sat}. All physical properties of the liquid except h_{fg} are evaluated at the film temperature $T_f = (T_{sat} + T_w)/2$. For long vertical surfaces the flow at the bottom can be turbulent. The Reynolds number is defined as

$$N_{Re} = \frac{4m}{\pi D \mu_l} = \frac{4\Gamma}{\mu_l} \qquad \text{(vertical tube, diameter } D)$$

(4.8-21)

$$N_{Re} = \frac{4m}{W \mu_l} = \frac{4\Gamma}{\mu_l} \qquad \text{(vertical plate, width } W)$$

(4.8-22)

where m is total kg mass/s of condensate at tube or plate bottom and $\Gamma = m/\pi D$ or m/W. The N_{Re} should be below about 1800 for Eq. (4.8-20) to hold. The reader should note that some references define N_{Re} as Γ/μ. Then this N_{Re} should be below 450.

For turbulent flow for $N_{Re} > 1800$ (M1),

$$N_{Nu} = \frac{hL}{k_l} = 0.0077 \left(\frac{g\rho_l^2 L^3}{\mu_l^2} \right)^{1/3} (N_{Re})^{0.4}$$

(4.8-23)

Solution of this equation is by trial and error since a value of N_{Re} must first be assumed to calculate h.

EXAMPLE 4.8-2. Condensation on a Vertical Tube
Steam saturated at 68.9 kPa (10 psia) is condensing on a vertical tube 0.305 m (1.0 ft) long having an OD of 0.0254 m (1.0 in.) and a surface temperature of 86.11°C (187°F). Calculate the average heat-transfer coefficient using English and SI units.

Solution: From Appendix A.2,

$$T_{sat} = 193°F \ (89.44°C) \qquad T_w = 187°F \ (86.11°C)$$

$$T_f = \frac{T_w + T_{sat}}{2} = \frac{187 + 193}{2} = 190°F \ (87.8°C)$$

latent heat $h_{fg} = 1143.3 - 161.0 = 982.3$ btu/lb$_m$

$$= 2657.8 - 374.6 = 2283.2 \text{ kJ/kg} = 2.283 \times 10^6 \text{ J/kg}$$

$$\rho_l = \frac{1}{0.01657} = 60.3 \text{ lb}_m/\text{ft}^3 = 60.3(16.018) = 966.7 \text{ kg/m}^3$$

$$\rho_v = \frac{1}{40.95} = 0.0244 \text{ lb}_m/\text{ft}^3 = 0.391 \text{ kg/m}^3$$

$$\mu_l = (0.324 \text{ cp})(2.4191) = 0.784 \text{ lb}_m/\text{ft} \cdot \text{h} = 3.24 \times 10^{-4} \text{ Pa} \cdot \text{s}$$

$$k_l = 0.390 \text{ btu/ft} \cdot \text{h} \cdot {}^\circ\text{F} = (0.390)(1.7307) = 0.675 \text{ W/m} \cdot \text{K}$$

$$L = 1 \text{ ft} = 0.305 \text{ m} \qquad \Delta T = T_{\text{sat}} - T_w = 193 - 187 = 6{}^\circ\text{F (3.33 K)}$$

Assuming a laminar film, using Eq. (4.8-20) in English and also SI units, and neglecting ρ_v as compared to ρ_l,

$$N_{\text{Nu}} = 1.13 \left(\frac{\rho_l^2 g h_{fg} L^3}{\mu_l k_l \Delta T} \right)^{1/4}$$

$$= 1.13 \left[\frac{(60.3)^2 (32.174)(3600)^2 (982.3)(1.0)^3}{(0.784)(0.390)(6)} \right]^{1/4} = 6040$$

$$N_{\text{Nu}} = 1.13 \left[\frac{(966.7)^2 (9.806)(2.283 \times 10^6)(0.305)^3}{(3.24 \times 10^{-4})(0.675)(3.33)} \right]^{1/4} = 6040$$

$$N_{\text{Nu}} = \frac{hL}{k_l} = \frac{h(1.0)}{0.390} = 6040 \qquad \text{SI units:} \frac{h(0.305)}{0.675} = 6040$$

Solving, $h = 2350$ btu/h $\cdot$ ft$^2 \cdot {}^\circ$F $= 13\,350$ W/m$^2 \cdot$ K.

Next, the N_{Re} will be calculated to see if laminar flow occurs as assumed. To calculate the total heat transferred for a tube of area

$$A = \pi DL = \pi (1/12)(1.0) = \pi/12 \text{ ft}^2, \qquad A = \pi (0.0254)(0.305) \text{ m}^2$$

$$q = hA \, \Delta T \tag{4.8-24}$$

However, this q must also equal that obtained by condensation of m lb$_m$/h or kg/s. Hence,

$$q = hA \, \Delta T = h_{fg} m \tag{4.8-25}$$

Substituting the values given and solving for m,

$$2350(\pi/12)(193 - 187) = 982.3(m) \qquad m = 3.77 \text{ lb}_m/\text{h}$$

$$13.350(\pi)(0.0254)(0.305)(3.33) = 2.284 \times 10^6 (m) \qquad m = 4.74 \times 10^{-4} \text{ kg/s}$$

Substituting into Eq. (4.8-21),

$$N_{\text{Re}} = \frac{4m}{\pi D \mu_l} = \frac{4(3.77)}{\pi (1/12)(0.784)} = 73.5 \quad N_{\text{Re}} = \frac{4(4.74 \times 10^{-4})}{\pi (0.0254)(3.24 \times 10^{-4})} = 73.5$$

Hence, the flow is laminar as assumed.

3. *Film-condensation coefficients outside horizontal cylinders.* The analysis of Nusselt can also be extended to the practical case of condensation outside a horizontal tube. For a single tube the film starts out with zero thickness at the top of the tube and increases in thickness as it flows around to the bottom and then drips off. If there is a

bank of horizontal tubes, the condensate from the top tube drips onto the one below; and so on.

For a vertical tier of N horizontal tubes placed one below the other with outside tube diameter D (M1),

$$N_{Nu} = \frac{hD}{k_l} = 0.725 \left(\frac{\rho_l(\rho_l - \rho_v)gh_{fg}D^3}{N\mu_l k_l \Delta T} \right)^{1/4} \tag{4.8-26}$$

In most practical applications, the flow is in the laminar region and Eq. (4.8-26) holds (C3, M1).

4.9 HEAT EXCHANGERS

4.9A Types of Exchangers

1. Introduction. In the process industries the transfer of heat between two fluids is generally done in heat exchangers. The most common type is one in which the hot and the cold fluid do not come into direct contact with each other but are separated by a tube wall or a flat or curved surface. The transfer of heat is accomplished from the hot fluid to the wall or tube surface by convection, through the tube wall or plate by conduction, and then by convection to the cold fluid. In the preceding sections of this chapter we have discussed the calculation procedures for these various steps. Now we will discuss some of the types of equipment used and overall thermal analyses of exchangers. Complete detailed design methods have been highly developed and will not be considered here.

2. Double-pipe heat exchanger. The simplest exchanger is the double-pipe or concentric-pipe exchanger. This is shown in Fig. 4.9-1, where the one fluid flows inside one pipe and the other fluid in the annular space between the two pipes. The fluids can be in cocurrent or countercurrent flow. The exchanger can be made from a pair of single lengths of pipe with fittings at the ends or from a number of pairs interconnected in series. This type of exchanger is useful mainly for small flow rates.

3. Shell-and-tube exchanger. If larger flows are involved, a shell and tube exchanger is used, which is the most important type of exchanger in use in the process industries. In these exchangers the flows are continuous. Many tubes in parallel are used where one fluid flows inside these tubes. The tubes, arranged in a bundle, are enclosed in a single shell and the other fluid flows outside the tubes in the shell side. The simplest shell and tube exchanger is shown in Fig. 4.9-2a for 1 shell pass and 1 tube pass, or a 1–1 counterflow exchanger. The cold fluid enters and flows inside through all the tubes in

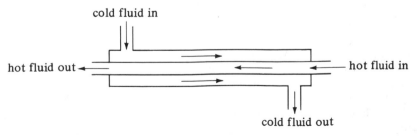

FIGURE 4.9-1. *Flow in a double-pipe heat exchanger.*

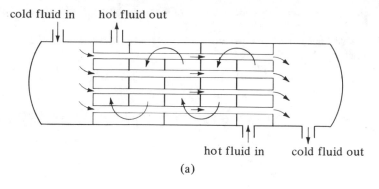

cold fluid in   hot fluid out

hot fluid in   cold fluid out

(a)

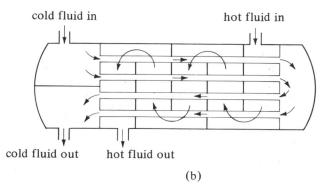

cold fluid in   hot fluid in

cold fluid out   hot fluid out

(b)

FIGURE 4.9-2.   *Shell-and-tube heat exchangers: (a) 1 shell pass and 1 tube pass (1–1 exchanger); (b) 1 shell pass and 2 tube passes (1–2 exchanger).*

parallel in one pass. The hot fluid enters at the other end and flows counterflow across the outside of the tubes. Cross baffles are used so that the fluid is forced to flow perpendicular across the tube bank rather than parallel with it. This added turbulence generated by this cross flow increases the shell-side heat-transfer coefficient.

In Fig. 4.9-2b a 1–2 parallel–counterflow exchanger is shown. The liquid on the tube side flows in two passes as shown and the shell-side liquid flows in one pass. In the first pass of the tube side the cold fluid is flowing counterflow to the hot shell-side fluid, and in the second pass of the tube side the cold fluid flows in parallel (cocurrent) with the hot fluid. Another type of exchanger has 2 shell-side passes and 4 tube passes. Other combinations of number of passes are also used sometimes, with the 1–2 and 2–4 types being the most common.

4. Cross-flow exchanger.   When a gas such as air is being heated or cooled, a common device used is the cross-flow heat exchanger shown in Fig. 4.9-3a. One of the fluids, which is a liquid, flows inside through the tubes and the exterior gas flows across the tube bundle by forced or sometimes natural convection. The fluid inside the tubes is considered to be unmixed since it is confined and cannot mix with any other stream. The gas flow outside the tubes is mixed since it can move about freely between the tubes and there will be a tendency for the gas temperature to equalize in the direction normal to the flow. For the unmixed fluid inside the tubes there will be a temperature gradient both parallel and normal to the direction of flow.

A second type of cross-flow heat exchanger shown in Fig. 4.9-3b is used typically in

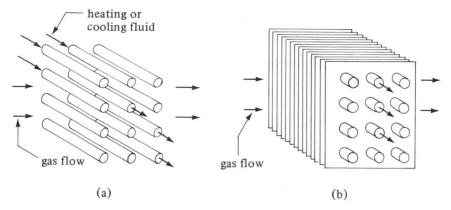

air-conditioning and space-heating applications. In this type the gas flows across a finned-tube bundle and is unmixed since it is confined in separate flow channels between the fins as it passes over the tubes. The fluid in the tubes is unmixed.

Discussions of other types of specialized heat-transfer equipment is deferred to Section 4.13. The remainder of this section deals primarily with a shell-and-tube and cross-flow heat exchangers.

4.9B Log Mean Temperature Difference Correction Factors

In Section 4.5H it was shown that when the hot and cold fluids in a heat exchanger are in true countercurrent flow or in cocurrent (parallel) flow, the log mean temperature difference should be used.

$$\Delta T_{\mathrm{lm}} = \frac{\Delta T_2 - \Delta T_1}{\ln (\Delta T_2/\Delta T_1)} \tag{4.9-1}$$

where ΔT_2 is the temperature difference at one end of the exchanger and ΔT_1 at the other end. This ΔT_{lm} holds for a double-pipe heat exchanger and a 1–1 exchanger with 1 shell pass and 1 tube pass in parallel or counterflow.

In the cases where a multiple-pass heat exchanger is involved, it is necessary to obtain a different expression for the mean temperature difference to use, depending on the arrangement of the shell and tube passes. Considering first the one-shell-pass, two-tube-pass exchanger in Fig. 4.9-2b, the cold fluid in the first tube pass is in counterflow with the hot fluid. In the second tube pass the cold fluid is in parallel flow with the hot fluid. Hence, the log mean temperature difference, which applies to either parallel or counterflow but not to a mixture of both types, as in a 1–2 exchanger, cannot be used to calculate the true mean temperature drop without a correction.

The mathematical derivation of the equation for the proper mean temperature to use is quite complex. The usual procedure is to use a correction factor F_T which is so defined that when it is multiplied by the ΔT_{lm}, the product is the correct mean temperature drop ΔT_m to use. In using the correction factors F_T it is immaterial whether the warmer fluid flows through the tubes or shell (K1). The factor F_T has been calculated

(B4) for a 1–2 exchanger and is shown in Fig. 4.9-4a. Two dimensionless ratios are used as follows:

$$Z = \frac{T_{hi} - T_{ho}}{T_{co} - T_{ci}}$$ (4.9-2)

$$Y = \frac{T_{co} - T_{ci}}{T_{hi} - T_{ci}}$$ (4.9-3)

where T_{hi} = inlet temperature of hot fluid in K (°F), T_{ho} = outlet of hot fluid, T_{ci} inlet of cold fluid, and T_{co} = outlet of cold fluid.

In Fig. 4.9-4b the factor F_T (B4) for a 2–4 exchanger is shown. In general, it is not recommended to use a heat exchanger for conditions under which $F_T < 0.75$. Another

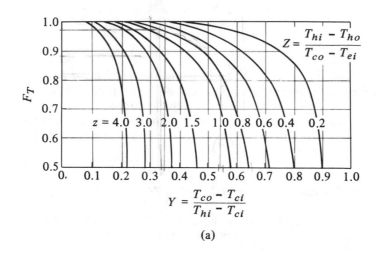

(a)

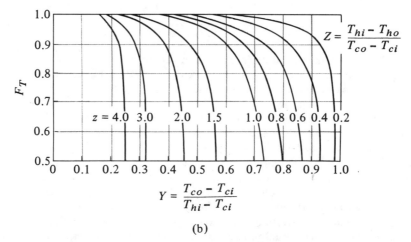

(b)

FIGURE 4.9-4. *Correction factor F_T to log mean temperature difference: (a) 1–2 exchangers, (b) 2–4 exchangers. [From R. A. Bowman, A. C. Mueller, and W. M. Nagle, Trans. A.S.M.E., 62, 284, 285 (1940). With permission.]*

Chap. 4   Principles of Steady-State Heat Transfer

shell and tube arrangement should be used. Correction factors for two types of cross-flow exchangers are given in Fig. 4.9-5. Other types are available elsewhere (B4, P1).

Using the nomenclature of Eqs. (4.9-2) and (4.9-3), the ΔT_{lm} of Eq. (4.9-1) can be written as

$$\Delta T_{lm} = \frac{(T_{hi} - T_{co}) - (T_{ho} - T_{ci})}{\ln \left[(T_{hi} - T_{co})/(T_{ho} - T_{ci}) \right]} \tag{4.9-4}$$

Then the equation for an exchanger is

$$q = U_i A_i \Delta T_m = U_o A_o \Delta T_m \tag{4.9-5}$$

where

$$\Delta T_m = F_T \Delta T_{lm} \tag{4.9-6}$$

EXAMPLE 4.9-1.   Temperature Correction Factor for a Heat Exchanger

A 1–2 heat exchanger containing one shell pass and two tube passes heats 2.52 kg/s of water from 21.1 to 54.4°C by using hot water under pressure entering at 115.6 and leaving at 48.9°C. The outside surface area of the tubes in the exchanger is $A_o = 9.30 \text{ m}^2$.

   (a) Calculate the mean temperature difference ΔT_m in the exchanger and the overall heat-transfer coefficient U_o.
   (b) For the same temperatures but using a 2–4 exchanger, what would be the ΔT_m?

Solution: The temperatures are as follows.

$$T_{hi} = 115.6°C \qquad T_{ho} = 48.9°C \qquad T_{ci} = 21.1°C \qquad T_{co} = 54.4°C$$

First making a heat balance on the cold water assuming a c_{pm} of water of 4187 J/kg·K and $T_{co} - T_{ci} = (54.4 - 21.1)°C = 33.3°C = 33.3$ K,

$$q = mc_{pm}(T_{co} - T_{ci}) = (2.52)(4187)(54.4 - 21.1) = 348\ 200 \text{ W}$$

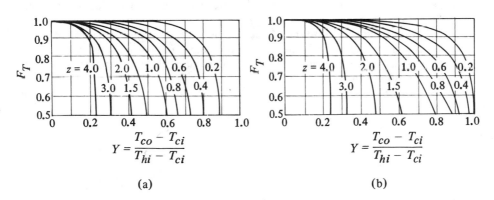

(a)   (b)

FIGURE 4.9-5.   *Correction factor F_T to log mean temperature difference for cross-flow exchangers $[Z = (T_{hi} - T_{ho})/(T_{co} - T_{ci})]$: (a) single pass, shell fluid mixed, other fluid unmixed, (b) single pass, both fluids unmixed. [From R. A. Bowman, A. C. Mueller, and W. M. Nagle, Trans. A.S.M.E., **62**, 288, 289 (1940). With permission.]*

The log mean temperature difference using Eq. (4.9-4) is

$$\Delta T_{lm} = \frac{(115.6 - 54.4) - (48.9 - 21.1)}{\ln \left[(115.6 - 54.4)/(48.9 - 21.1)\right]} = 42.3°C = 42.3 \text{ K}$$

Next substituting into Eqs. (4.9-2) and (4.9-3)

$$Z = \frac{T_{hi} - T_{ho}}{T_{co} - T_{ci}} = \frac{115.6 - 48.9}{54.4 - 21.1} = 2.00 \qquad \text{(4.9-2)}$$

$$Y = \frac{T_{co} - T_{ci}}{T_{hi} - T_{ci}} = \frac{54.4 - 21.1}{115.6 - 21.1} = 0.352 \qquad \text{(4.9-3)}$$

From Fig. 4.9-4a, $F_T = 0.74$. Then, by Eq. (4.9-6),

$$\Delta T_m = F_T \, \Delta T_{lm} = 0.74(42.3) = 31.3°C = 31.3 \text{ K} \qquad \text{(4.9-6)}$$

Rearranging Eq. (4.9-5) to solve for U_o and substituting the known values, we have

$$U_o = \frac{q}{A_o \, \Delta T_m} = \frac{348\,200}{(9.30)(31.3)} = 1196 \text{ W/m}^2 \cdot \text{K} \ (211 \text{ btu/h} \cdot \text{ft}^2 \cdot °\text{F})$$

For part (b), using a 2–4 exchanger and Fig. 4.9-4b, $F_T = 0.94$. Then,

$$\Delta T_m = F_T \, \Delta T_{lm} = 0.94(42.3) = 39.8°C = 39.8 \text{ K}$$

Hence, in this case the 2–4 exchanger utilizes more of the available temperature driving force.

4.9C Heat-Exchanger Effectiveness

1. Introduction. In the preceding section the log mean temperature difference was used in the equation $q = UA \, \Delta T_{lm}$ in the design of heat exchangers. This form is convenient when the inlet and outlet temperatures of the two fluids are known or can be determined by a heat balance. Then the surface area can be determined if U is known. However, when the temperatures of the fluids leaving the exchanger are not known and a given exchanger is to be used, a tedious trial-and-error procedure is necessary. To solve these cases, a method called the heat-exchanger effectiveness ε is used which does not involve any of the outlet temperatures.

The heat-exchanger effectiveness is defined as the ratio of the actual rate of heat transfer in a given exchanger to the maximum possible amount of heat transfer if an infinite heat-transfer area were available. The temperature profile for a counterflow heat exchanger is shown in Fig. 4.9-6.

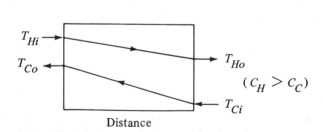

FIGURE 4.9-6. *Temperature profile for countercurrent heat exchanger.*

2. *Derivation of effectiveness equation.* The heat balance for the cold (C) and the hot (H) fluids is

$$q = (mc_p)_H(T_{Hi} - T_{Ho}) = (mc_p)_C(T_{Co} - T_{Ci}) \tag{4.9-7}$$

Calling $(mc_p)_H = C_H$ and $(mc_p)_C = C_C$, then in Fig. 4.9-6, $C_H > C_C$ and the cold fluid undergoes a greater temperature change than the hot fluid. Hence, we designate C_C as C_{min} or minimum heat capacity. Then, if there is an infinite area available for heat transfer, $T_{Co} = T_{Hi}$. Then the effectiveness ε is

$$\varepsilon = \frac{C_H(T_{Hi} - T_{Ho})}{C_C(T_{Hi} - T_{Ci})} = \frac{C_{max}(T_{Hi} - T_{Ho})}{C_{min}(T_{Hi} - T_{Ci})} \tag{4.9-8}$$

If the hot fluid is the minimum fluid, $T_{Ho} = T_{Ci}$, and

$$\varepsilon = \frac{C_C(T_{Co} - T_{Ci})}{C_H(T_{Hi} - T_{Ci})} = \frac{C_{max}(T_{Co} - T_{Ci})}{C_{min}(T_{Hi} - T_{Ci})} \tag{4.9-9}$$

In both equations the denominators are the same and the numerator gives the actual heat transfer.

$$q = \varepsilon C_{min}(T_{Hi} - T_{Ci}) \tag{4.9-10}$$

Note that Eq. (4.9-10) uses only inlet temperatures, which is an advantage when inlet temperatures are known and it is desired to predict the outlet temperatures for a given existing exchanger.

For the case of a single-pass, counterflow exchanger, combining Eqs. (4.9-8) and (4.9-9),

$$\varepsilon = \frac{C_H(T_{Hi} - T_{Ho})}{C_{min}(T_{Hi} - T_{Ci})} = \frac{C_C(T_{Co} - T_{Ci})}{C_{min}(T_{Hi} - T_{Ci})} \tag{4.9-11}$$

We consider first the case of the cold fluid to be the minimum fluid. Rewriting Eq. (4.5-25) using the present nomenclature,

$$q = C_C(T_{Co} - T_{Ci}) = UA \frac{(T_{Ho} - T_{Ci}) - (T_{Hi} - T_{Co})}{\ln[(T_{Ho} - T_{Ci})/(T_{Hi} - T_{Co})]} \tag{4.9-12}$$

Combining Eq. (4.9-7) with the left side of Eq. (4.9-11) and solving for T_{Hi},

$$T_{Hi} = T_{Ci} + \frac{1}{\varepsilon}(T_{Co} - T_{Ci}) \tag{4.9-13}$$

Subtracting T_{Co} from both sides,

$$T_{Hi} - T_{Co} = T_{Ci} - T_{Co} + \frac{1}{\varepsilon}(T_{Co} - T_{Ci}) = \left(\frac{1}{\varepsilon} - 1\right)(T_{Co} - T_{Ci}) \tag{4.9-14}$$

From Eq. (4.9-7) for $C_{min} = C_C$ and $C_{max} = C_H$,

$$T_{Ho} = T_{Hi} - \frac{C_{min}}{C_{max}}(T_{Co} - T_{Ci}) \tag{4.9-15}$$

This can be rearranged to give the following:

$$T_{Ho} - T_{Ci} = T_{Hi} - T_{Ci} - \frac{C_{min}}{C_{max}}(T_{Co} - T_{Ci}) \tag{4.9-16}$$

Substituting Eq. (4.9-13) into (4.9-16),

$$T_{Ho} - T_{Ci} = \frac{1}{\varepsilon}(T_{Co} - T_{Ci}) - \frac{C_{min}}{C_{max}}(T_{Co} - T_{Ci}) \tag{4.9-17}$$

Finally, substituting Eqs. (4.9-14) and (4.9-17) into (4.9-12), rearranging, taking the antilog of both sides, and solving for ε,

$$\varepsilon = \frac{1 - \exp\left[-\dfrac{UA}{C_{min}}\left(1 - \dfrac{C_{min}}{C_{max}}\right)\right]}{1 - \dfrac{C_{min}}{C_{max}}\exp\left[-\dfrac{UA}{C_{min}}\left(1 - \dfrac{C_{min}}{C_{max}}\right)\right]} \tag{4.9-18}$$

We define NTU as the number of transfer units as follows:

$$\text{NTU} = \frac{UA}{C_{min}} \tag{4.9-19}$$

The same result would have been obtained if $C_H = C_{min}$.

For parallel flow we obtain

$$\varepsilon = \frac{1 - \exp\left[\dfrac{-UA}{C_{min}}\left(1 + \dfrac{C_{min}}{C_{max}}\right)\right]}{1 + \dfrac{C_{min}}{C_{max}}} \tag{4.9-20}$$

In Fig. 4.9-7, Eqs. (4.9-18) and (4.9-20) have been plotted in convenient graphical form. Additional charts are available for different shell-and-tube and cross-flow arrangements (K1).

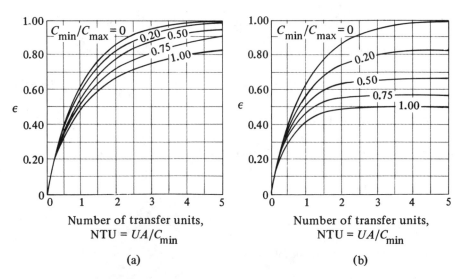

FIGURE 4.9-7. *Heat-exchanger effectiveness ε: (a) counterflow exchanger, (b) parallel flow exchanger.*

Chap. 4   *Principles of Steady-State Heat Transfer*

EXAMPLE 4.9-2. *Effectiveness of Heat Exchanger*
Water flowing at a rate of 0.667 kg/s enters a countercurrent heat exchanger at 308 K and is heated by an oil stream entering at 383 K at a rate of 2.85 kg/s ($c_p = 1.89$ kJ/kg·K). The overall $U = 300$ W/m²·K and the area $A = 15.0$ m². Calculate the heat-transfer rate and the exit water temperature.

Solution: Assuming that the exit water temperature is about 370 K, the c_p for water at an average temperature of $(308 + 370)/2 = 339$ K is 4.192 kJ/kg·K (Appendix A.2). Then, $(mc_p)_H = C_H = 2.85(1.89 \times 10^3) = 5387$ W/K and $(mc_p)_C = C_C = 0.667(4.192 \times 10^3) = 2796$ W/K $= C_{min}$. Since C_C is the minimum, $C_{min}/C_{max} = 2796/5387 = 0.519$.

Using Eq. (4.9-19), NTU $= UA/C_{min} = 300(15.0)/2796 = 1.607$. Using Fig. (4.9-7a) for a counterflow exchanger, $\varepsilon = 0.71$. Substituting into Eq. (4.9-10),

$$q = \varepsilon C_{min}(T_{Hi} - T_{Ci}) = 0.71(2796)(383 - 308) = 148\,900 \text{ W}$$

Using Eq. (4.9-7),

$$q = 148\,900 = 2796(T_{Co} - 308)$$

Solving, $T_{Co} = 361.3$ K.

4.9D Fouling Factors and Typical Overall U Values

In actual practice, heat-transfer surfaces do not remain clean. Dirt, soot, scale, and other deposits form on one or both sides of the tubes of an exchanger and on other heat-transfer surfaces. These deposits form additional resistances to the flow of heat and reduce the overall heat-transfer coefficient U. In petroleum processes coke and other substances can deposit. Silting and deposits of mud and other materials can occur. Corrosion products may form on the surfaces which could form a serious resistance to heat transfer. Biological growth such as algae can occur with cooling water and in the biological industries.

To avoid or lessen these fouling problems chemical inhibitors are often added to minimize corrosion, salt deposition, and algae growth. Water velocities above 1 m/s are generally used to help reduce fouling. Large temperature differences may cause excessive deposition of solids on surfaces and should be avoided if possible.

The effect of such deposits and fouling is usually taken care of in design by adding a term for the resistance of the fouling on the inside and the outside of the tube in Eq. (4.3-17) as follows.

$$U_i = \frac{1}{1/h_i + 1/h_{di} + (r_o - r_i)A_i/k_A A_{A\,lm} + A_i/A_o h_o + A_i/A_o h_{do}} \qquad (4.9\text{-}21)$$

where h_{di} is the fouling coefficient for the inside and h_{do} the fouling coefficient for the outside of the tube in W/m²·K. A similar expression can be written for U_o using Eq. (4.3-18).

Fouling coefficients recommended for use in designing heat-transfer equipment are available in many references (P3, N1). A short tabulation of some typical fouling coefficients is given in Table 4.9-1.

In order to do preliminary estimating of sizes of shell-and-tube heat exchangers, typical values of overall heat-transfer coefficients are given in Table 4.9-2. These values should be useful as a check on the results of the design methods described in this chapter.

TABLE 4.9-1. *Typical Fouling Coefficients (P3, N1)*

	h_d $(W/m^2 \cdot K)$	h_d $(btu/h \cdot ft^2 \cdot {}^\circ F)$
Distilled and seawater	11 350	2000
City water	5680	1000
Muddy water	1990–2840	350–500
Gases	2840	500
Vaporizing liquids	2840	500
Vegetable and gas oils	1990	350

4.10 INTRODUCTION TO RADIATION HEAT TRANSFER

4.10A Introduction and Basic Equation for Radiation

1. Nature of radiant heat transfer.   In the preceding sections of this chapter we have studied conduction and convection heat transfer. In conduction heat is transferred from one part of a body to another, and the intervening material is heated. In convection the heat is transferred by the actual mixing of materials and by conduction. In radiant heat transfer the medium through which the heat is transferred usually is not heated. Radiation heat transfer is the transfer of heat by electromagnetic radiation.

Thermal radiation is a form of electromagnetic radiation similar to x rays, light waves, gamma rays, and so on, differing only in wavelength. It obeys the same laws as light: travels in straight lines, can be transmitted through space and vacuum, and so on. It is an important mode of heat transfer and is especially important where large

TABLE 4.9-2. *Typical Values of Overall Heat-Transfer Coefficients in Shell-and-Tube Exchangers (H1, P3, W1)*

	U $(W/m^2 \cdot K)$	U $(btu/h \cdot ft^2 \cdot {}^\circ F)$
Water to water	1140–1700	200–300
Water to brine	570–1140	100–200
Water to organic liquids	570–1140	100–200
Water to condensing steam	1420–2270	250–400
Water to gasoline	340–570	60–100
Water to gas oil	140–340	25–60
Water to vegetable oil	110–285	20–50
Gas oil to gas oil	110–285	20–50
Steam to boiling water	1420–2270	250–400
Water to air (finned tube)	110–230	20–40
Light organics to light organics	230–425	40–75
Heavy organics to heavy organics	55–230	10–40

temperature differences occur, as, for example, in a furnace with boiler tubes, in radiant dryers, and in an oven baking food. Radiation often occurs in combination with conduction and convection. An elementary discussion of radiant heat transfer will be given here, with a more advanced and comprehensive discussion being given in Section 4.11.

In an elementary sense the mechanism of radiant heat transfer is composed of three distinct steps or phases:

1. The thermal energy of a hot source, such as the wall of a furnace at T_1, is converted into the energy of electromagnetic radiation waves.
2. These waves travel through the intervening space in straight lines and strike a cold object at T_2 such as a furnace tube containing water to be heated.
3. The electromagnetic waves that strike the body are absorbed by the body and converted back to thermal energy or heat.

2. Absorptivity and black bodies. When thermal radiation (like light waves) falls upon a body, part is absorbed by the body in the form of heat, part is reflected back into space, and part may be actually transmitted through the body. For most cases in process engineering, bodies are opaque to transmission, so this will be neglected. Hence, for opaque bodies,

$$\alpha + \rho = 1.0 \tag{4.10-1}$$

where α is absorptivity or fraction absorbed and ρ is reflectivity or fraction reflected.

A *black body* is defined as one that absorbs all radiant energy and reflects none. Hence, ρ is 0 and $\alpha = 1.0$ for a black body. Actually, in practice there are no perfect black bodies, but a close approximation to this is a small hole in a hollow body, as shown in Fig. 4.10-1. The inside surface of the hollow body is blackened by charcoal. The radiation enters the hole and impinges on the rear wall; part is absorbed there and part is reflected in all directions. The reflected rays impinge again, part is absorbed, and the process continues. Hence, essentially all of the energy entering is absorbed and the area of the hole acts as a perfect black body. The surface of the inside walls are "rough" and rays are scattered in all directions, unlike a mirror, where they are reflected at a definite angle.

As stated previously, a black body absorbs all radiant energy falling on it and reflects none. Such a black body also emits radiation, depending on its temperature, and does not reflect any. The ratio of the emissive power of a surface to that of a black body is called *emissivity* ε and it is 1.0 for a black body. Kirchhoff's law states that at the same temperature T_1, α_1 and ε_1 of a given surface are the same or

$$\alpha_1 = \varepsilon_1 \tag{4.10-2}$$

Equation (4.10-2) holds for any black or nonblack solid surface.

FIGURE 4.10-1. *Concept of a perfect black body.*

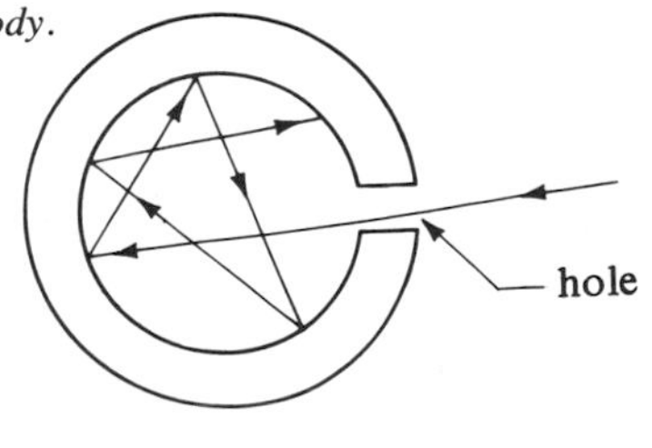

hole

3. *Radiation from a body and emissivity.* The basic equation for heat transfer by radiation from a perfect black body with an emissivity $\varepsilon = 1.0$ is

$$q = A\sigma T^4 \tag{4.10-3}$$

where q is heat flow in W, A is m^2 surface area of body, σ a constant 5.676×10^{-8} W/m$^2 \cdot$ K^4 (0.1714×10^{-8} btu/h $\cdot$ ft$^2 \cdot$ °R^4), and T is temperature of the black body in K (°R).

For a body that is not a black body and has an emissivity $\varepsilon < 1.0$, the emissive power is reduced by ε, or

$$q = A\varepsilon\sigma T^4 \tag{4.10-4}$$

Substances that have emissivities of less than 1.0 are called *gray bodies* when the emissivity is independent of the wavelength. All real materials have an emissivity $\varepsilon < 1$.

Since the emissivity ε and absorptivity α of a body are equal at the same temperature, the emissivity, like absorptivity, is low for polished metal surfaces and high for oxidized metal surfaces. Typical values are given in Table 4.10-1 but do vary some with temperature. Most nonmetallic substances have high values. Additional data are tabulated in Appendix A.3.

4.10B Radiation to a Small Object from Surroundings

When we have the case of a small gray object of area A_1 m^2 at temperature T_1 in a large enclosure at a higher temperature T_2, there is a net radiation to the small object. The small body emits an amount of radiation to the enclosure given by Eq. (4.10-4) of $A_1\varepsilon_1\sigma T_1^4$. The emissivity ε_1 of this body is taken at T_1. The small body also absorbs energy from the surroundings at T_2 given by $A_1\alpha_{12}\sigma T_2^4$. The α_{12} is the absorptivity of body 1 for radiation from the enclosure at T_2. The value of α_{12} is approximately the same as the emissivity of this body at T_2. The net heat of absorption is then, by the Stefan-Boltzmann equation,

$$q = A_1\varepsilon_1\sigma T_1^4 - A_1\alpha_{12}\sigma T_2^4 = A_1\sigma(\varepsilon_1 T_1^4 - \alpha_{12} T_2^4) \tag{4.10-5}$$

A further simplification of Eq. (4.10-5) is usually made for engineering purposes by using one emissivity of the small body at temperature T_2. Thus,

$$q = A_1\varepsilon\sigma(T_1^4 - T_2^4) \tag{4.10-6}$$

TABLE 4.10-1. *Total Emissivity, ε, of Various Surfaces*

Surface	T(K)	T(°F)	Emissivity, ε
Polished aluminum	500	440	0.039
	850	1070	0.057
Polished iron	450	350	0.052
Oxidized iron	373	212	0.74
Polished copper	353	176	0.018
Asbestos board	296	74	0.96
Oil paints, all colors	373	212	0.92–0.96
Water	273	32	0.95

EXAMPLE 4.10-1. *Radiation to a Metal Tube*

A small oxidized horizontal metal tube with an OD of 0.0254 m (1 in.) and being 0.61 m (2 ft) long with a surface temperature at 588 K (600°F) is in a very large furnace enclosure with fire-brick walls and the surrounding air at 1088 K (1500°F). The emissivity of the metal tube is 0.60 at 1088 K and 0.46 at 588 K. Calculate the heat transfer to the tube by radiation using SI and English units.

Solution: Since the large furnace surroundings are very large compared with the small enclosed tube, the surroundings, even if gray, when viewed from the position of the small body appear black and Eq. (4.10-6) is applicable. Substituting given values into Eq. (4.10-6) with an ε of 0.6 at 1088 K

$$A_1 = \pi DL = \pi(0.0254)(0.61) \text{ m}^2 = \pi(1/12)(2.0) \text{ ft}^2$$

$$q = A_1 \varepsilon \sigma(T_1^4 - T_2^4) = [\pi(0.0254)(0.61)](0.6)(5.676 \times 10^{-8})[(588)^4 - (1088)^4]$$

$$= -2130 \text{ W}$$

$$= [\pi(1/12)(2)](0.6)(0.1714 \times 10^{-8})[(1060)^4 - (1960)^4] = -7270 \text{ btu/h}$$

Other examples of small objects in large enclosures occurring in the process industries are a loaf of bread in an oven receiving radiation from the walls around it, a package of meat or food radiating heat to the walls of a freezing enclosure, a hot ingot of solid iron cooling and radiating heat in a large room, and a thermometer measuring the temperature in a large duct.

4.10C Combined Radiation and Convection Heat Transfer

When radiation heat transfer occurs from a surface it is usually accompanied by convective heat transfer unless the surface is in a vacuum. When the radiating surface is at a uniform temperature, we can calculate the heat transfer for natural or forced convection using the methods described in the previous sections of this chapter. The radiation heat transfer is calculated by the Stefan-Boltzmann equation (4.10-6). Then the total rate of heat transfer is the sum of convection plus radiation.

As discussed before, the heat-transfer rate by convection and the convective coefficient are given by

$$q_{conv} = h_c A_1(T_1 - T_2) \tag{4.10-7}$$

where q_{conv} is the heat-transfer rate by convection in W, h_c the natural or forced convection coefficient in W/m$^2 \cdot$ K, T_1 the temperature of the surface, and T_2 the temperature of the air and the enclosure. A radiation heat-transfer coefficient h_r in W/m$^2 \cdot$ K can be defined as

$$q_{rad} = h_r A_1(T_1 - T_2) \tag{4.10-8}$$

where q_{rad} is the heat-transfer rate by radiation in W. The total heat transfer is the sum of Eqs. (4.10-7) and (4.10-8),

$$q = q_{conv} + q_{rad} = (h_c + h_r)A_1(T_1 - T_2) \tag{4.10-9}$$

To obtain an expression for h_r, we equate Eq. (4.10-6) to (4.10-8) and solve for h_r.

$$h_r = \frac{\varepsilon\sigma(T_1^4 - T_2^4)}{T_1 - T_2} = \varepsilon(5.676)\frac{(T_1/100)^4 - (T_2/100)^4}{T_1 - T_2} \quad \text{(SI)}$$

$$\tag{4.10-10}$$

$$h_r = \varepsilon(0.1714)\frac{(T_1/100)^4 - (T_2/100)^4}{T_1 - T_2} \quad \text{(English)}$$

A convenient chart giving values of h_r in English units calculated from Eq. (4.10-10) with $\varepsilon = 1.0$ is given in Fig. 4.10-2. To use values from this figure, the value obtained from this figure should be multiplied by ε to give the value of h_r to use in Eq. (4.10-9). If the air temperature is not the same as T_2 of the enclosure, Eqs. (4.10-7) and (4.10-8) must be used separately and not combined together as in (4.10-9).

EXAMPLE 4.10-2. *Combined Convection Plus Radiation from a Tube*
Recalculate Example 4.10-1 for combined radiation plus natural convection to the horizontal 0.0254-m tube.

Solution: The area A of the tube $= \pi(0.0254)(0.61) = 0.0487$ m^2. For the natural convection coefficient to the 0.0254-m horizontal tube, the simplified equation from Table 4.7-2 will be used as an approximation even though the film temperature is quite high.

$$h_c = 1.32\left(\frac{\Delta T}{D}\right)^{1/4}$$

Substituting the known values,

$$h_c = 1.32\left(\frac{1088 - 588}{0.0254}\right)^{1/4} = 15.64 \text{ W/m}^2 \cdot \text{K}$$

Using Eq. (4.10-10) and $\varepsilon = 0.6$,

$$h_r = (0.60)(5.676)\frac{(1088/100)^4 - (588/100)^4}{1088 - 588} = 87.3 \text{ W/m}^2 \cdot \text{K}$$

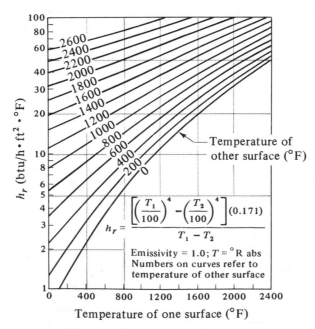

FIGURE 4.10-2. *Radiation heat-transfer coefficient as a function of temperature. (From R. H. Perry and C. H. Chilton, Chemical Engineers' Handbook, 5th ed. New York: McGraw-Hill Book Company, 1973. With permission.)*

Substituting into Eq. (4.10-9),

$$q = (h_c + h_r)A_1(T_1 - T_2) = (15.64 + 87.3)(0.0487)(588 - 1088)$$

$$= -2507 \text{ W}$$

Hence, the heat loss of -2130 W for radiation is increased to only -2507 W when natural convection is also considered. In this case, because of the large temperature difference, radiation is the most important factor.

Perry and Green (P3, p. 10-14) give a convenient table of natural convection plus radiation coefficients $(h_c + h_r)$ from single horizontal oxidized steel pipes as a function of the outside diameter and temperature difference. The coefficients for insulated pipes are about the same as those for a bare pipe (except that lower surface temperatures are involved for the insulated pipes), since the emissivity of cloth insulation wrapping is about that of oxidized steel, approximately 0.8. A more detailed discussion of radiation will be given in Section 4.11.

4.11 ADVANCED RADIATION HEAT-TRANSFER PRINCIPLES

4.11A Introduction and Radiation Spectrum

1. *Introduction.* This section will cover some basic principles and also some advanced topics on radiation that were not covered in Section 4.10. The exchange of radiation between two surfaces depends upon the size, shape, and relative orientation of these two surfaces and also upon their emissivities and absorptivities. In the cases to be considered the surfaces are separated by nonabsorbing media such as air. When gases such as CO_2 and H_2O vapor are present, some absorption by the gases occurs, which is not considered until later in this section.

2. *Radiation spectrum and thermal radiation.* Energy can be transported in the form of electromagnetic waves and these waves travel at the speed of light. Bodies may emit many forms of radiant energy, such as gamma rays, thermal energy, radio waves, and so on. In fact, there is a continuous spectrum of electromagnetic radiation. This electromagnetic spectrum is divided into a number of wavelength ranges such as cosmic rays $(\lambda < 10^{-13} \text{ m})$, gamma rays $(\lambda, 10^{-13} \text{ to } 10^{-10} \text{ m})$, thermal radiation $(\lambda, 10^{-7} \text{ to } 10^{-4} \text{ m})$, and so on. The electromagnetic radiation produced solely because of the temperature of the emitter is called thermal radiation and exists between the wavelengths of 10^{-7} and 10^{-4} m. This portion of the electromagnetic spectrum is of importance in radiant thermal heat transfer. Electromagnetic waves having wavelengths between 3.8×10^{-7} and 7.6×10^{-7} m, called *visible radiation*, can be detected by the human eye. This visible radiation lies within the thermal radiation range.

When different surfaces are heated to the same temperature, they do not all emit or absorb the same amount of thermal radiant energy. A body that absorbs and emits the maximum amount of energy at a given temperature is called a *black body*. A black body is a standard to which other bodies can be compared.

3. *Planck's law and emissive power.* When a black body is heated to a temperature T, photons are emitted from the surface which have a definite distribution of energy.

Planck's equation relates the monochromatic emissive power $E_{B\lambda}$ in W/m³ at a temperature T in K and a wavelength λ in m.

$$E_{B\lambda} = \frac{3.7418 \times 10^{-16}}{\lambda^5 [e^{1.4388 \times 10^{-2}/\lambda T} - 1]} \tag{4.11-1}$$

A plot of Eq. (4.11-1) is given in Fig. 4.11-1 and shows that the energy given off increases with T. Also, for a given T, the emissive power reaches a maximum value at a wavelength that decreases as the temperature T increases. At a given temperature the radiation emitted extends over a spectrum of wavelengths. The visible light spectrum occurs in the low λ region. The sun has a temperature of about 5800 K and the solar spectrum straddles this visible range.

For a given temperature, the wavelength at which the black-body emissive power is a maximum can be determined by differentiating Eq. (4.11-1) with respect to λ at constant T and setting the result equal to zero. The result is as follows and is known as *Wien's displacement law*:

$$\lambda_{max} T = 2.898 \times 10^{-3} \text{ m} \cdot \text{K} \tag{4.11-2}$$

The locus of the maximum values is shown in Fig. 4.11-1.

4. Stefan–Boltzmann law. The total emissive power is the total amount of radiation energy per unit area leaving a surface with temperature T over all wavelengths. For a black body, the total emissive power is given by the integral of Eq. (4.11-1) at a given T over all wavelengths or the area under the curve in Fig. 4.11-1.

$$E_B = \int_0^\infty E_{B\lambda} \, d\lambda \tag{4.11-3}$$

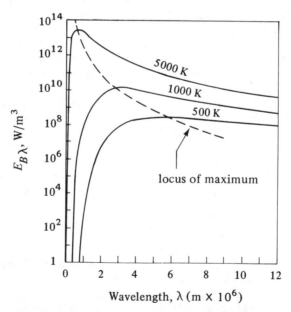

FIGURE 4.11-1. *Spectral distribution of total energy emitted by a black body at various temperatures of the black body.*

This gives

$$E_B = \sigma T^4 \tag{4.11-4}$$

The result is the Stefan–Boltzmann law with $\sigma = 5.676 \times 10^{-8}$ W/m$^2 \cdot$ K^4. The units of E_B are W/m^2.

5. *Emissivity and Kirchhoff's law.* An important property in radiation is the emissivity of a surface. The *emissivity* ε of a surface is defined as the total emitted energy of the surface divided by the total emitted energy of a black body at the same temperature.

$$\varepsilon = \frac{E}{E_B} = \frac{E}{\sigma T^4} \tag{4.11-5}$$

Since a black body emits the maximum amount of radiation, ε is always < 1.0.

We can derive a relationship between the absorptivity α_1 and emissitivy ε_1 of a material by placing this material in an isothermal enclosure and allowing the body and enclosure to reach the same temperature at thermal equilibrium. If G is the irradiation on the body, the energy absorbed must equal the energy emitted.

$$\alpha_1 G = E_1 \tag{4.11-6}$$

If this body is removed and replaced by a black body of equal size, then at equilibrium,

$$\alpha_2 G = E_B \tag{4.11-7}$$

Dividing Eq. (4.11-6) by (4.11-7),

$$\frac{\alpha_1}{\alpha_2} = \frac{E_1}{E_B} \tag{4.11-8}$$

But $\alpha_2 = 1.0$ for a black body. Hence, since $E_1/E_B = \varepsilon_1$,

$$\alpha_1 = \frac{E_1}{E_B} = \varepsilon_1 \tag{4.11-9}$$

This is Kirchhoff's law, which states that at thermal equilibrium $\alpha = \varepsilon$ of a body. When a body is not at equilibrium with its surroundings, the result is not valid.

6. *Concept of gray body.* A gray body is defined as a surface for which the monochromatic properties are constant over all wavelengths. For a gray surface,

$$\varepsilon_\lambda = \text{const.}, \qquad \alpha_\lambda = \text{const.} \tag{4.11-10}$$

Hence, the total absorptivity α and the monochromatic absorptivity α_λ of a gray surface are equal, as are ε and ε_λ.

$$\alpha = \alpha_\lambda, \qquad \varepsilon = \varepsilon_\lambda \tag{4.11-11}$$

Applying Kirchhoff's law to a gray body $\alpha_\lambda = \varepsilon_\lambda$ and

$$\alpha = \varepsilon \tag{4.11-12}$$

As a result, the total absorptivity and emissivity are equal for a gray body even if the body is not in thermal equilibrium with its surroundings.

Gray bodies do not exist in practice and the concept of a gray body is an idealized one. The absorptivity of a surface actually varies with the wavelength of the incident radiation. Engineering calculations can often be based on the assumption of a gray body with reasonable accuracy. The α is assumed constant even with a variation in λ of the incident radiation. Also, in actual systems, various surfaces may be at different temperatures. In these cases, α for a surface is evaluated by determining the emissivity not at the actual surface temperature but at the temperature of the source of the other radiating surface or emitter since this is the temperature the absorbing surface would reach if the absorber and emitter were at thermal equilibrium. The temperature of the absorber has only a slight effect on the absorptivity.

4.11B Derivation of View Factors in Radiation for Various Geometries

1. Introduction. The concepts and definitions presented in Section 4.11A form a sufficient foundation so that the net radiant exchange between surfaces can be determined. If two surfaces are arranged so that radiant energy can be exchanged, a net flow of energy will occur from the hotter surface to the colder surface. The size, shape, and orientation of two radiating surfaces or a system of surfaces are factors in determining the net heat-flow rate between them. To simplify the discussion we assume that the surfaces are separated by a nonabsorbing medium such as air. This assumption is adequate for many engineering applications. However, in cases such as a furnace, the presence of CO_2 and H_2O vapor make such a simplification impossible because of their high absorptivities.

The simplest geometrical configuration will be considered first, that of radiation exchange between parallel, infinite planes. This assumption implies that there are no edge effects in the case of finite surfaces. First, the simplest case will be treated in which the surfaces are black bodies and then more complicated geometries and gray bodies will be treated.

2. View factor for infinite parallel black planes. If two parallel and infinite black planes at T_1 and T_2 are radiating toward each other, plane 1 emits σT_1^4 radiation to plane 2, which is all absorbed. Also, plane 2 emits σT_2^4 radiation to plane 1, which is all absorbed. Then for plane 1, the net radiation is from plane 1 to 2,

$$q_{12} = A_1 \sigma (T_1^4 - T_2^4) \qquad (4.11\text{-}13)$$

In this case all the radiation from 1 to 2 is intercepted by 2; that is, the fraction of radiation leaving 1 that is intercepted by 2 is F_{12}, which is 1.0. The factor F_{12} is called the geometric view factor or view factor. Hence,

$$q_{12} = F_{12} A_1 \sigma (T_1^4 - T_2^4) \qquad (4.11\text{-}14)$$

where F_{12} is fraction of radiation leaving surface 1 in all directions which is intercepted by surface 2. Also,

$$q_{21} = F_{21} A_2 \sigma (T_1^4 - T_2^4) \qquad (4.11\text{-}15)$$

In the case for parallel plates $F_{12} = F_{21} = 1.0$ and the geometric factor is simply omitted.

3. View factor for infinite parallel gray planes. If both of the parallel plates A_1 and A_2 are gray with emissivities and absorptivities of $\varepsilon_1 = \alpha_1$ and $\varepsilon_2 = \alpha_2$, respectively, we can proceed as follows. Since each surface has an unobstructed view of each other, the view factor is 1.0. In unit time surface A_1 emits $\varepsilon_1 A_1 \sigma T_1^4$ radiation to A_2. Of this, the fraction ε_2 (where $\alpha_2 = \varepsilon_2$) is absorbed:

$$\text{absorbed by } A_2 = \varepsilon_2 (\varepsilon_1 A_1 \sigma T_1^4) \qquad (4.11\text{-}16)$$

Also, the fraction $(1 - \varepsilon_2)$ or the amount $(1 - \varepsilon_2)(\varepsilon_1 A_1 \sigma T_1^4)$ is reflected back to A_1. Of this amount A_1 reflects back to A_2 a fraction $(1 - \varepsilon_1)$ or an amount $(1 - \varepsilon_1)(1 - \varepsilon_2)(\varepsilon_1 A_1 \sigma T_1^4)$. The surface A_2 absorbs the fraction ε_2, or

$$\text{absorbed by } A_2 = \varepsilon_2(1 - \varepsilon_1)(1 - \varepsilon_2)(\varepsilon_1 A_1 \sigma T_1^4) \qquad (4.11\text{-}17)$$

The amount reflected back to A_1 from A_2 is $(1 - \varepsilon_2)(1 - \varepsilon_1)(1 - \varepsilon_2)(\varepsilon_1 A_1 \sigma T_1^4)$. Then A_1 absorbs ε_1 of this and reflects back to A_2 an amount $(1-\varepsilon_1)(1-\varepsilon_2)(1-\varepsilon_1)(1-\varepsilon_2) \times (\varepsilon_1 A_1 \sigma T_1^4)$. The surface A_2 then absorbs

$$\text{absorbed by } A_2 = \varepsilon_2(1 - \varepsilon_1)(1 - \varepsilon_2)(1 - \varepsilon_1)(1 - \varepsilon_2)(\varepsilon_1 A_1 \sigma T_1^4) \qquad (4.11\text{-}18)$$

This continues and the total amount absorbed at A_2 is the sum of Eqs. (4.11-16), (4.11-17), (4.11-18), and so on.

$$q_{1 \to 2} = A_1 \sigma T_1^4[\varepsilon_1\varepsilon_2 + \varepsilon_1\varepsilon_2(1 - \varepsilon_1)(1 - \varepsilon_2) + \varepsilon_1\varepsilon_2(1 - \varepsilon_1)^2(1 - \varepsilon_2)^2 + \cdots] \qquad (4.11\text{-}19)$$

The result is a geometric series (M1).

$$q_{1 \to 2} = A_1 \sigma T_1^4 \frac{\varepsilon_1\varepsilon_2}{1 - (1 - \varepsilon_1)(1 - \varepsilon_2)} = A_1 \sigma T_1^4 \frac{1}{1/\varepsilon_1 + 1/\varepsilon_2 - 1} \qquad (4.11\text{-}20)$$

Repeating the above for the amount absorbed at A_1, which comes from A_2,

$$q_{2 \to 1} = A_1 \sigma T_2^4 \frac{1}{1/\varepsilon_1 + 1/\varepsilon_2 - 1} \qquad (4.11\text{-}21)$$

The net radiation is the difference of Eqs. (4.11-20) and (4.11-21).

$$q_{12} = A_1 \sigma(T_1^4 - T_2^4) \frac{1}{1/\varepsilon_1 + 1/\varepsilon_2 - 1} \qquad (4.11\text{-}22)$$

If $\varepsilon_1 = \varepsilon_2 = 1.0$ for black bodies, Eq. (4.11-22) becomes Eq. (4.11-13).

EXAMPLE 4.11-1. Radiation Between Parallel Planes
Two parallel gray planes which are very large have emissivities of $\varepsilon_1 = 0.8$ and $\varepsilon_2 = 0.7$ and surface 1 is at 1100°F (866.5 K) and surface 2 at 600°F (588.8 K). Use English and SI units for the following.
  (a) What is the net radiation from 1 to 2?
  (b) If the surfaces are both black, what is the net radiation?

Solution: For part (a), using Eq. (4.11-22) and substituting the known values,

$$\frac{q_{12}}{A_1} = \frac{\sigma(T_1^4 - T_2^4)}{1/\varepsilon_1 + 1/\varepsilon_2 - 1} = (0.1714 \times 10^{-8}) \frac{(1100 + 460)^4 - (600 + 460)^4}{1/0.8 + 1/0.7 - 1}$$

$$= 4750 \text{ btu/h} \cdot \text{ft}^2$$

$$\frac{q_{12}}{A_1} = (5.676 \times 10^{-8}) \frac{(866.5)^4 - (588.8)^4}{1/0.8 + 1/0.7 - 1} = 15010 \text{ W/m}^2$$

For black surfaces in part (b), using Eq. (4.11-13),

$$q = 7960 \text{ btu/h} \cdot \text{ft}^2 \quad \text{or} \quad 25110 \text{ W/m}^2$$

Note the large reduction in radiation when surfaces with emissivities less than 1.0 are used.

Example 4.11-1 shows the large influence that emissivities less than 1.0 have on radiation. This fact is used to reduce radiation loss or gain from a surface by using planes as a radiation shield. For example, for two parallel surfaces of emissivity ε at T_1 and T_2, the interchange is, by Eq. (4.11-22),

$$\frac{(q_{12})_0}{A} = \frac{\sigma(T_1^4 - T_2^4)}{2/\varepsilon - 1} \qquad \textbf{(4.11-23)}$$

The subscript 0 indicates that there are no planes in between the two surfaces. Suppose that we now insert one or more radiation planes between the original surfaces. Then it can be shown that

$$\frac{(q_{12})_N}{A} = \frac{1}{N+1} \frac{\sigma(T_1^4 - T_2^4)}{2/\varepsilon - 1} \qquad \textbf{(4.11-24)}$$

where N is the number of radiation planes or shields between the original surfaces. Hence, a great reduction in radiation heat loss is obtained by using these shields.

4. Derivation of general equation for view factor between black bodies. Suppose that we consider radiation between two parallel black planes of finite size as in Fig. 4.11-2a. Since the planes are not infinite in size, some of the radiation from surface 1 does not strike surface 2, and vice versa. Hence, the net radiation interchange is less since some is lost to the surroundings. The fraction of radiation leaving surface 1 in all directions which is intercepted by surface 2 is called F_{12} and must be determined for each geometry by taking differential surface elements and integrating over the entire surfaces.

Before we can derive a general relationship for the view factor between two finite bodies we must consider and discuss two quantities, a solid angle and the intensity of radiation. A solid angle ω is a dimensionless quantity which is a measure of an angle in solid geometry. In Fig. 4.11-3a the differential solid angle $d\omega_1$ is equal to the normal projection of dA_2 divided by the square of the distance between the point P and area dA_2.

$$d\omega_1 = \frac{dA_2 \cos \theta_2}{r^2} \qquad \textbf{(4.11-25)}$$

The units of a solid angle are steradian or sr. For a hemisphere the number of sr subtended by this surface is 2π.

The intensity of radiation for a black body, I_B, is the rate of radiation emitted per unit area projected in a direction normal to the surface and per unit solid angle in a

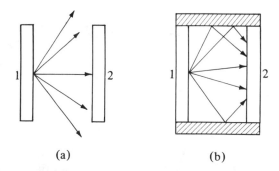

(a)   (b)

FIGURE 4.11-2.   *Radiation between two black surfaces: (a) two planes alone, (b) two planes connected by refractory reradiating walls.*

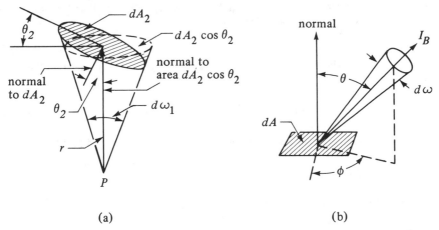

FIGURE 4.11-3.　*Geometry for a solid angle and intensity of radiation : (a) solid-angle geometry, (b) intensity of radiation from emitting area dA.*

specified direction as shown in Fig. 4.11-3b. The projection of dA on the line between centers is $dA \cos \theta$.

$$I_B = \frac{dq}{dA \cos d\omega} \tag{4.11-26}$$

where q is in W and I_B is in $W/m^2 \cdot sr$. We assume that the black body is a diffuse surface which emits with equal intensity in all directions, i.e., $I = $ constant. The emissive power E_B which leaves a black-body plane surface is determined by integrating Eq. (4.11-26) over all solid angles subtended by a hemisphere covering the surface. The final result is as follows. [See references (C3, H1, K1) for details.]

$$E_B = \pi I_B \tag{4.11-27}$$

where E_B is in W/m^2.

In order to determine the radiation heat-transfer rates between two black surfaces we must determine the general case for the fraction of the total radiant heat that leaves a surface and arrives on a second surface. Using only black surfaces, we consider the case shown in Fig. 4.11-4, in which radiant energy is exchanged between area elements dA_1 and dA_2. The line r is the distance between the areas and the angles between this line and the normals to the two surfaces are θ_1 and θ_2. The rate of radiant energy that leaves dA_1

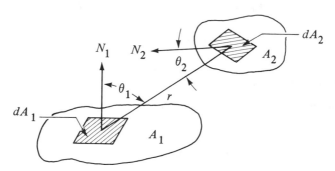

FIGURE 4.11-4.　*Area elements for radiation shape factor.*

in the direction given by the angle θ_1 is $I_{B1} \, dA \cos \theta_1$. The rate that leaves dA_1 and arrives on dA_2 is given by Eq. (4.11-28).

$$dq_{1 \to 2} = I_{B1} \, dA \cos \theta_1 \, d\omega_1 \qquad \text{(4.11-28)}$$

where $d\omega_1$ is the solid angle subtended by the area dA_2 as seen from dA_1. Combining Eqs. (4.11-25) and (4.11-28),

$$dq_{1 \to 2} = \frac{I_{B1} \, dA_1 \cos \theta_1 \cos \theta_2 \, dA_2}{r^2} \qquad \text{(4.11-29)}$$

From Eq. (4.11-27), $I_{B1} = E_{B1}/\pi$. Substituting E_{B1}/π for I_{B1} into Eq. (4.11-29),

$$dq_{1 \to 2} = \frac{E_{B1} \cos \theta_1 \cos \theta_2 \, dA_1 \, dA_2}{\pi r^2} \qquad \text{(4.11-30)}$$

The energy leaving dA_2 and arriving at dA_1 is

$$dq_{2 \to 1} = \frac{E_{B2} \cos \theta_2 \cos \theta_1 \, dA_2 \, dA_1}{\pi r^2} \qquad \text{(4.11-31)}$$

Substituting σT_1^4 for E_{B1} and σT_2^4 for E_{B2} from Eq. (4.11-4) and taking the difference of Eqs. (4.11-30) and (4.11-31) for the net heat flow,

$$dq_{12} = \sigma(T_1^4 - T_2^4) \frac{\cos \theta_1 \cos \theta_2 \, dA_1 \, dA_2}{\pi r^2} \qquad \text{(4.11-32)}$$

Performing the double integrations over surfaces A_1 and A_2 will yield the total net heat flow between the finite areas.

$$q_{12} = \sigma(T_1^4 - T_2^4) \int_{A2} \int_{A1} \frac{\cos \theta_1 \cos \theta_2 \, dA_1 \, dA_2}{\pi r^2} \qquad \text{(4.11-33)}$$

Equation (4.11-33) can also be written as

$$q_{12} = A_1 F_{12} \, \sigma(T_1^4 - T_2^4) = A_2 F_{21} \sigma(T_1^4 - T_2^4) \qquad \text{(4.11-34)}$$

where F_{12} is a geometric shape factor or view factor and designates the fraction of the total radiation leaving A_1 which strikes A_2 and F_{21} represents the fraction leaving A_2 which strikes A_1. Also, the following relation exists.

$$A_1 F_{12} = A_2 F_{21} \qquad \text{(4.11-35)}$$

which is valid for black surfaces and also nonblack surfaces. The view factor F_{12} is then

$$F_{12} = \frac{1}{A_1} \int_{A2} \int_{A1} \frac{\cos \theta_1 \cos \theta_2 \, dA_1 \, dA_2}{\pi r^2} \qquad \text{(4.11-36)}$$

Values of the view factor can be calculated for a number of geometrical arrangements.

5. *View factors between black bodies for various geometries.* A number of basic relationships between view factors are given below.

The reciprocity relationship given by Eq. (4.11-35) is

$$A_1 F_{12} = A_2 F_{21} \qquad \text{(4.11-35)}$$

This relationship can be applied to any two surfaces i and j.

$$A_i F_{ij} = A_j F_{ji} \qquad \text{(4.11-37)}$$

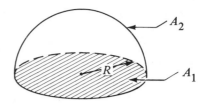

If surface A_1 can only see surface A_2, then $F_{12} = 1.0$.

If surface A_1 sees a number of surfaces A_2, A_3, ..., and all the surfaces form an enclosure then the enclosure relationship is

$$F_{11} + F_{12} + F_{13} + \cdots = 1.0 \qquad \text{(4.11-38)}$$

If the surface A_1 cannot see itself (surface is flat or convex), $F_{11} = 0$.

EXAMPLE 4.11-2. *View Factor from a Plane to a Hemisphere*

Determine the view factors between a plane A_1 covered by a hemisphere A_2 as shown in Fig. 4.11-5.

Solution: Since surface A_1 sees only A_2, the view factor $F_{12} = 1.0$. Using Eq. (4.11-35),

$$A_1 F_{12} = A_2 F_{21} \qquad \text{(4.11-35)}$$

The area $A_1 = \pi R^2$, $A_2 = 2\pi R^2$. Substituting into Eq. (4.11-35) and solving for F_{21},

$$F_{21} = F_{12} \frac{A_1}{A_2} = (1.0) \frac{\pi R^2}{2\pi R^2} = \frac{1}{2}$$

Using Eq. (4.11-38) for surface A_1, $F_{11} = 1.0 - F_{12} = 1.0 - 1.0 = 0$. Also, writing Eq. (4.11-38) for surface A_2,

$$F_{22} + F_{21} = 1.0 \qquad \text{(4.11-39)}$$

Solving for F_{22}, $F_{22} = 1.0 - F_{21} = 1.0 - \frac{1}{2} = \frac{1}{2}$.

EXAMPLE 4.11-3. *Radiation Between Parallel Disks*

In Fig. 4.11-6 a small disk of area A_1 is parallel to a large disk of area A_2 and A_1 is centered directly below A_2. The distance between the centers of the disks is R and the radius of A_2 is a. Determine the view factor for radiant heat transfer from A_1 to A_2.

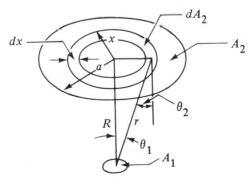

FIGURE 4.11-6. *View factor for radiation from a small element to a parallel disk for Example 4.11-3.*

Solution: The differential area for A_2 is taken as the circular ring of radius x so that $dA_2 = 2\pi x\, dx$. The angle $\theta_1 = \theta_2$. Using Eq. (4.11-36),

$$F_{12} = \frac{1}{A_1} \int_{A_2} \int_{A_1} \frac{\cos \theta_1 \cos \theta_1 \, dA_1 (2\pi x\, dx)}{\pi r^2}$$

In this case the area A_1 is very small compared to A_2, so dA_1 can be integrated to A_1 and the other terms inside the integral can be assumed constant. From the geometry shown, $r = (R^2 + x^2)^{1/2}$, $\cos \theta_1 = R/(R^2 + x^2)^{1/2}$. Making these substitutions into the equation for F_{12},

$$F_{12} = \int_0^a \frac{2R^2 x\, dx}{(R^2 + x^2)^2}$$

Integrating,

$$F_{12} = \frac{a^2}{R^2 + a^2}$$

The integration of Eq. (4.11-36) has been done for numerous geometrical configurations and values of F_{12} tabulated. Then,

$$q_{12} = F_{12} A_1 \sigma(T_1^4 - T_2^4) = F_{21} A_2 \sigma(T_1^4 - T_2^4) \qquad \textbf{(4.11-34)}$$

where F_{12} is the fraction of the radiation leaving A_1 which is intercepted by A_2 and F_{21} the fraction reaching A_1 from A_2. Since the flux from 1 to 2 must equal that from 2 to 1, Eq. (4.11-34) becomes Eq. (4.11-35) as given previously.

$$A_1 F_{12} = A_2 F_{21} \qquad \textbf{(4.11-35)}$$

Hence, one selects the surface whose view factor can be determined most easily. For example, the view factor F_{12} for a small surface A_1 completely enclosed by a large surface A_2 is 1.0, since all the radiation leaving A_1 is intercepted by A_2. In Fig. 4.11-7 the view factors F_{12} between parallel planes are given, and in Fig. 4.11-8 the view factors for

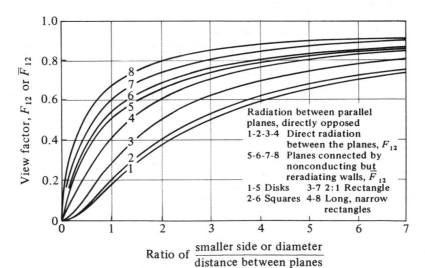

FIGURE 4.11-7. *View factor between parallel planes directly opposed. (From W. H. McAdams, Heat Transmission, 3rd ed. New York: McGraw-Hill Book Company, 1954. With permission.)*

Chap. 4   Principles of Steady-State Heat Transfer

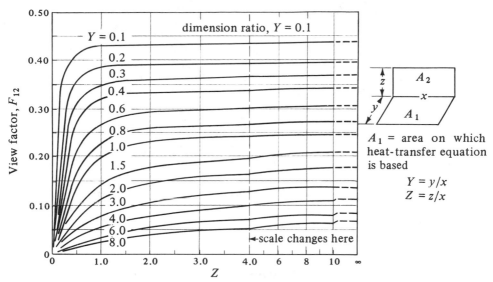

FIGURE 4.11-8. *View factor for adjacent perpendicular rectangles.* [*From H. C. Hottel, Mech. Eng.,* **52**, *699 (1930). With permission.*]

adjacent perpendicular rectangles. View factors for other geometries are given elsewhere (H1, K1, P3, W1).

4.11C View Factors When Surfaces Are Connected by Reradiating Walls

If the two black-body surfaces A_1 and A_2 are connected by nonconducting (refractory) but reradiating walls as in Fig. 4.11-2b, a larger fraction of the radiation from surface 1 is intercepted by 2. This view factor is called $\bar{F}_{12}$. The case of two surfaces connected by the walls of an enclosure such as a furnace is a common example of this. The general equation for this case assuming a uniform refractory temperature has been derived (M1, C3) for two radiant sources A_1 and A_2, which are not concave, so they do not see themselves.

$$\bar{F}_{12} = \frac{A_2 - A_1 F_{12}^2}{A_1 + A_2 - 2A_1 F_{12}} = \frac{1 - (A_1/A_2)F_{12}^2}{A_1/A_2 + 1 - 2(A_1/A_2)F_{12}} \tag{4.11-40}$$

Also, as before,

$$A_1 \bar{F}_{12} = A_2 \bar{F}_{21} \tag{4.11-41}$$

$$q_{12} = \bar{F}_{12} A_1 \sigma(T_1^4 - T_2^4) \tag{4.11-42}$$

The factor $\bar{F}_{12}$ for parallel planes is given in Fig. 4.11-7 and for other geometries can be calculated from Eq. (4.11-36). For view factors F_{12} and $\bar{F}_{12}$ for parallel tubes adjacent to a wall as in a furnace and also for variation in refractory wall temperature, see elsewhere (M1, P3). If there are no reradiating walls,

$$F_{12} = \bar{F}_{12} \tag{4.11-43}$$

4.11D View Factors and Gray Bodies

A general and more practical case, which is the same as for Eq. (4.11-40) but with the surfaces A_1 and A_2 being gray with emissivities ε_1 and ε_2, will be considered. Nonconducting reradiating walls are present as before. Since the two surfaces are now gray, there will be some reflection of radiation which will decrease the net radiant exchange between the surfaces below that for black surfaces. The final equations for this case are

$$q_{12} = \mathcal{F}_{12} A_1 \sigma (T_1^4 - T_2^4) \tag{4.11-44}$$

$$\mathcal{F}_{12} = \cfrac{1}{\cfrac{1}{\bar{F}_{12}} + \cfrac{A_1}{A_2}\left(\cfrac{1}{\varepsilon_2} - 1\right) + \left(\cfrac{1}{\varepsilon_1} - 1\right)} \tag{4.11-45}$$

where $\mathcal{F}_{12}$ is the new view factor for two gray surfaces A_1 and A_2 which cannot see themselves and are connected by reradiating walls. If no refractory walls are present, F_{12} is used in place of $\bar{F}_{12}$ in Eq. (4.11-41). Again,

$$A_1 \mathcal{F}_{12} = A_2 \mathcal{F}_{21} \tag{4.11-46}$$

EXAMPLE 4.11-4.   Radiation Between Infinite Parallel Gray Planes
Derive Eq. (4.11-22) by starting with the general equation for radiation between two gray bodies A_1 and A_2 which are infinite parallel planes having emissivities ε_1 and ε_2, respectively.

Solution: Since there are no reradiating walls, by Eq. (4.11-43), $\bar{F}_{12}$ becomes F_{12}. Also, since all the radiation from surface 1 is intercepted by surface 2, $F_{12} = 1.0$. Substituting into Eq. (4.11-45), noting that $A_1/A_2 = 1.0$,

$$\mathcal{F}_{12} = \cfrac{1}{\cfrac{1}{\bar{F}_{12}} + \cfrac{A_1}{A_2}\left(\cfrac{1}{\varepsilon_2} - 1\right) + \left(\cfrac{1}{\varepsilon_1} - 1\right)} = \cfrac{1}{\cfrac{1}{1} + 1\left(\cfrac{1}{\varepsilon_2} - 1\right) + \left(\cfrac{1}{\varepsilon_1} - 1\right)}$$

$$= \cfrac{1}{\cfrac{1}{\varepsilon_1} + \cfrac{1}{\varepsilon_2} - 1}$$

Then using Eq. (4.11-44),

$$q_{12} = \mathcal{F}_{12} A_1 \sigma (T_1^4 - T_2^4) = A_1 \sigma (T_1^4 - T_2^4)\cfrac{1}{\cfrac{1}{\varepsilon_1} + \cfrac{1}{\varepsilon_2} - 1}$$

This is identical to Eq. (4.11-22).

EXAMPLE 4.11-5.   Complex View Factor for Perpendicular Rectangles
Find the view factor F_{12} for the configuration shown in Fig. 4.11-9 of the rectangle with area A_2 displaced from the common edge of rectangle A_1 and perpendicular to A_1. The temperature of A_1 is T_1 and that of A_2 and A_3 is T_2.

Solution: The area A_3 is a fictitious area between areas A_2 and A_1. Call the area A_2 plus A_3 as $A_{(23)}$. The view factor $F_{1(23)}$ for areas A_1 and $A_{(23)}$ can be obtained from Fig. 4.11-8 for adjacent perpendicular rectangles. Also, F_{13} can be obtained from Fig. 4.11-8. The radiation interchange between A_1 and $A_{(23)}$ is equal to that intercepted by A_2 and by A_3.

$$A_1 F_{1(23)}\, \sigma (T_1^4 - T_2^4) = A_1 F_{12}\,\sigma (T_1^4 - T_2^4) + A_1 F_{13}(T_1^4 - T_2^4) \tag{4.11-47}$$

FIGURE 4.11-9. *Configuration for Example 4.11-5.*

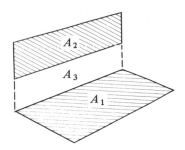

Hence,

$$A_1 F_{1(23)} = A_1 F_{12} + A_1 F_{13} \qquad \text{(4.11-48)}$$

Solving for F_{12},

$$F_{12} = F_{1(23)} - F_{13} \qquad \text{(4.11-49)}$$

Methods similar to those used in this example can be employed to find the shape factors for a general orientation of two rectangles in perpendicular planes or parallel rectangles (C3, H1, K1).

EXAMPLE 4.11-6. Radiation to a Small Package
A small cold package having an area A_1 and emissivity ε_1 is at temperature T_1. It is placed in a warm room with the walls at T_2 and an emissivity ε_2. Derive the view factor for this using Eq. (4.11-45), and the equation for the radiation heat transfer.

Solution: For the small surface A_1 completely enclosed by the enclosure A_2, $\bar{F}_{12} = F_{12}$ by Eq. (4.11-43), since there are no reradiating (refractory) walls. Also, $F_{12} = 1.0$, since all the radiation from A_1 is intercepted by the enclosure A_2 because A_1 does not have any concave surfaces and cannot "see" itself. Since A_2 is very large compared to A_1, $A_1/A_2 = 0$. Substituting into Eq. (4.11-45),

$$\mathscr{F}_{12} = \frac{1}{\dfrac{1}{\bar{F}_{12}} + \dfrac{A_1}{A_2}\left(\dfrac{1}{\varepsilon_2} - 1\right) + \left(\dfrac{1}{\varepsilon_1} - 1\right)} = \frac{1}{\dfrac{1}{1} + 0\left(\dfrac{1}{\varepsilon_2} - 1\right) + \dfrac{1}{\varepsilon_1} - 1} = \varepsilon_1$$

Substituting into Eq. (4.11-44),

$$q_{12} = \mathscr{F}_{12} A_1 \sigma (T_1^4 - T_2^4) = \varepsilon_1 A_1 \sigma (T_1^4 - T_2^4)$$

This is the same as Eq. (4.10-6) derived previously.

For methods to solve complicated radiation problems involving more than four or five heat-transfer surfaces, matrix methods to solve these problems have been developed and are discussed in detail elsewhere (H1, K1).

4.11E   Radiation in Absorbing Gases

1. Introduction to absorbing gases in radiation.   As discussed in this section, solids and liquids emit radiation over a continuous spectrum. However, most gases that are monoatomic or diatomic, such as He, Ar, H_2, O_2, and N_2, are virtually transparent to thermal radiation; i.e., they emit practically no radiation and also do not absorb radiation. Gases with a dipole moment and higher polyatomic gases emit significant amounts of radiation and also absorb radiant energy within the same bands in which they emit radiation. These gases include CO_2, H_2O, CO, SO_2, NH_3, and organic vapors.

For a particular gas, the width of the absorption or emission bands depends on the pressure and also the temperature. If an absorbing gas is heated, it radiates energy to the cooler surroundings. The net radiation heat-transfer rate between surfaces is decreased in these cases because the gas absorbs some of the radiant energy being transported between the surfaces.

2. Absorption of radiation by a gas. The absorption of radiation in a gas layer can be described analytically since the absorption by a given gas depends on the number of molecules in the path of radiation. Increasing the partial pressure of the absorbing gas or the path length increases the amount of absorption. We define $I_{\lambda 0}$ as the intensity of radiation at a particular wavelength before it enters the gas and $I_{\lambda L}$ as the intensity at the same wavelength after having traveled a distance of L in the gas. If the beam impinges on a gas layer of thickness dL, the decrease in intensity, dI_λ, is proportional to I_λ and dL.

$$dI_\lambda = -\alpha_\lambda I_\lambda \, dL \qquad (4.11\text{-}50)$$

where I_λ is in W/m^2. Integrating,

$$I_{\lambda L} = I_{\lambda 0} e^{-\alpha_\lambda L} \qquad (4.11\text{-}51)$$

The constant α_λ depends on the particular gas, its partial pressure, and the wavelength of radiation. This equation is called Beer's law. Gases frequently absorb only in narrow-wavelength bands.

3. Characteristic mean beam length of absorbing gas. The calculation methods for gas radiation are quite complicated. For the purpose of engineering calculations, Hottel (M1) has presented approximate methods to calculate radiaton and absorption when gases such as CO_2 and water vapor are present. Thick layers of a gas absorb more energy than do thin layers. Hence, in addition to specifying the pressure and temperature of a gas, we must specify a characteristic length (mean beam length) of a gas mass to determine the emissivity and absorptivity of a gas. The mean beam length L depends on the specific geometry.

For a black differential receiving surface area dA located in the center of the base of a hemisphere of radius L containing a radiating gas, the mean beam length is L. The mean beam length has been evaluated for various geometries and is given in Table 4.11-1. For other shapes, L can be approximated by

$$L = 3.6 \frac{V}{A} \qquad (4.11\text{-}52)$$

where V is volume of the gas in m^3, A the surface area of the enclosure in m^2, and L is in m.

4. Emissivity, absorptivity, and radiation of a gas. Gas emissivities have been correlated and Fig. 4.11-10 gives the gas emissivity ε_G of CO_2 at a total pressure of the system of 1.0 atm abs. The p_G is the partial pressure of CO_2 in atm and the mean beam length L in m. The emissivity ε_G is defined as the ratio of the rate of energy transfer from the hemispherical body of gas to a surface element at the midpoint divided by the rate of energy transfer from a black hemisphere surface of radius L and temperature T_G to the same element.

TABLE 4.11-1. *Mean Beam Length for Gas Radiation to*
Entire Enclosure Surface (M1, R2, P3)

Geometry of Enclosure	Mean Beam Length, L
Sphere, diameter D	$0.65D$
Infinite cylinder, diameter D	$0.95D$
Cylinder, length = diameter D	$0.60D$
Infinite parallel plates, separation distance D	$1.8D$
Hemisphere, radiation to element in base, radius R	R
Cube, radiation to any face, side D	$0.60D$
Volume surrounding bank of long tubes with centers on equilateral triangle, clearance = tube diameter D	$2.8D$

The rate of radiation emitted from the gas is $\sigma \varepsilon_G T_G^4$ in W/m^2 of receiving surface element, where ε_G is evaluated at T_G. If the surface element at the midpoint at T_1 is radiating heat back to the gas, the absorption rate of the gas will be $\sigma \alpha_G T_1^4$, where α_G is the absorptivity of the gas for blackbody radiation from the surface at T_1. The α_G of CO_2 is determined from Fig. 4.11-10 at T_1 but instead of using the parameter of $p_G L$, the parameter $p_G L(T_1/T_G)$ is used. The resulting value from the chart is then multiplied by

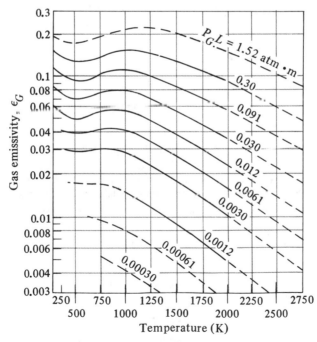

FIGURE 4.11-10.   *Total emissivity of the gas carbon dioxide at a total pressure of 1.0 atm. (From W. H. McAdams, Heat Transmission, 3rd ed. New York: McGraw-Hill Book Company, 1954. With permission.)*

$(T_G/T_1)^{0.65}$ to give α_G. The net rate of radiant transfer between a gas at T_G and a black surface of finite area A_1 at T_1 is then

$$q = \sigma A(\varepsilon_G T_G^4 - \alpha_G T_1^4) \qquad (4.11\text{-}53)$$

When the total pressure is not 1.0 atm, a correction chart is available to correct the emissivity of CO_2. Also, charts are available for water vapor (H1, K1, M1, P3). When both CO_2 and H_2O are present the total radiation is reduced somewhat, since each gas is somewhat opaque to radiation from the other gas. Charts for these interactions are also available (H1, K1, M1, P3).

EXAMPLE 4.11-7. Gas Radiation to a Furnace Enclosure

A furnace is in the form of a cube 0.3 m on a side inside, and these interior walls can be approximated as black surfaces. The gas inside at 1.0 atm total pressure and 1100 K contains 10 mole % CO_2 and the rest is O_2 and N_2. The small amount of water vapor present will be neglected. The walls of the furnace are maintained at 600 K by external cooling. Calculate the total heat transfer to the walls neglecting heat transfer by convection.

Solution: From Table 4.11-1, the mean beam length for radiation to a cube face is $L = 0.60D = 0.60(0.30) = 0.180$ m. The partial pressure of CO_2 is $p_G = 0.10(100) = 0.10$ atm. Then $p_G L = 0.10(0.180) = 0.0180$ atm · m. From Fig. 4.11-10, $\varepsilon_G = 0.064$ at $T_G = 1100$ K.

To obtain α_G, we evaluate α_G at $T_1 = 600$ K and $p_G L(T_1/T_G) = (0.0180)(600/1100) = 0.00982$ atm · m. From Fig. 4.11-10, the uncorrected value of $\alpha_G = 0.048$. Multiplying this by the correction factor $(T_G/T_1)^{0.65}$, the final correction value is

$$\alpha_G = 0.048(1100/600)^{0.65} = 0.0712$$

Substituting into Eq. (4.11-53),

$$\frac{q}{A} = \sigma(\varepsilon_G T_G^4 - \alpha_G T_1^4)$$

$$= (5.676 \times 10^{-8})[0.064(1100)^4 - 0.0712(600)^4]$$

$$= 4.795 \times 10^3 \text{ W/m}^2 = 4.795 \text{ kW/m}^2$$

For six sides, $A = 6(0.3 \times 0.3) = 0.540$ m². Then,

$$q = 4.795(0.540) = 2.589 \text{ kW}$$

For the case where the walls of the enclosure are not black, some of the radiation striking the walls is reflected back to the other walls and into the gas. As an approximation when the emissivity of the walls is greater than 0.7, an effective emissivity ε' can be used.

$$\varepsilon' = \frac{\varepsilon + 1.0}{2} \qquad (4.11\text{-}54)$$

where ε is the actual emissivity of the enclosure walls. Then Eq. (4.11-53) is modified to give the following (M1):

$$q = \sigma A\varepsilon'(\varepsilon_G T_G^4 - \alpha_G T_1^4) \qquad (4.11\text{-}55)$$

Other approximate methods are available for gases containing suspended luminous flames, clouds of nonblack particles, refractory walls and absorbing gases present, and so on (M1, P3).

4.12 HEAT TRANSFER OF NON-NEWTONIAN FLUIDS

4.12A Introduction

Most of the studies on heat transfer with fluids have been done with Newtonian fluids. However, a wide variety of non-Newtonian fluids are encountered in the industrial chemical, biological, and food processing industries. To design equipment to handle these fluids, the flow property constants (rheological constants) must be available or must be measured experimentally. Section 3.5 gave a detailed discussion of rheological constants for non-Newtonian fluids. Since many non-Newtonian fluids have high effective viscosities, they are often in laminar flow. Since the majority of non-Newtonian fluids are pseudoplastic fluids, which can usually be represented by the power law, Eq. (3.5-2), the discussion will be concerned with such fluids. For other fluids, the reader is referred to Skelland (S3).

4.12B Heat Transfer Inside Tubes

1. Laminar flow in tubes. A large portion of the experimental investigations have been concerned with heat transfer of non-Newtonian fluids in laminar flow through cylindrical tubes. The physical properties that are needed for heat transfer coefficients are density, heat capacity, thermal conductivity, and the rheological constants K' and n' or K and n.

In heat transfer in a fluid in laminar flow, the mechanism is one of primarily conduction. However, for low flow rates and low viscosities, natural convection effects can be present. Since many non-Newtonian fluids are quite "viscous," natural convection effects are reduced substantially. For laminar flow inside circular tubes of power-law fluids, the equation of Metzner and Gluck (M2) can be used with highly "viscous" non-Newtonian fluids with negligible natural convection for horizontal or vertical tubes for the Graetz number $N_{Gz} > 20$ and $n' > 0.10$.

$$(N_{Nu})_a = \frac{h_a D}{k} = 1.75 \delta^{1/3} (N_{Gz})^{1/3} \left(\frac{\gamma_b}{\gamma_w}\right)^{0.14} \qquad (4.12\text{-}1)$$

where,

$$\delta = \frac{3n' + 1}{4n'} \qquad (4.12\text{-}2)$$

$$N_{Gz} = \frac{mc_p}{kL} = \frac{\pi}{4} \frac{D v \rho}{\mu} \frac{c_p \mu}{k} \frac{D}{L} = \frac{\pi}{4} N_{Re} N_{Pr} \frac{D}{L} \qquad (4.12\text{-}3)$$

The viscosity coefficients γ_b at temperature T_b and γ_w at T_w are defined as

$$\frac{\gamma_b}{\gamma_w} = \frac{K'_b 8^{n'-1}}{K'_w 8^{n'-1}} = \frac{K'_b}{K'_w} = \frac{K_b}{K_w} \qquad (4.12\text{-}4)$$

The nomenclature is as follows: k in W/m · K, c_p in J/kg · K, ρ in kg/m³, flow rate m in kg/s, length of heated section of tube L in m, inside diameter D in m, the mean coefficient h_a in W/m² · K, and K and n' rheological constants (see Section 3.5). The physical properties and K_b are all evaluated at the mean bulk temperature T_b and K_w at the average wall temperature T_w.

The value of the rheological constant n' or n has been found to not vary appreciably over wide temperature ranges (S3). However, the rheological constant K' or K has been found to vary appreciably. A plot of log K' versus $1/T_{abs}$ (C1) or versus $T°C$ (S3) can often be approximated by a straight line. Often data for the temperature effect on K are not available. Since the ratio K_b/K_w is taken to the 0.14 power, this factor can sometimes be neglected without causing large errors. For a value of the ratio of 2:1, the error is only about 10%. A plot of log viscosity versus $1/T$ for Newtonian fluids is also often a straight line. The value of h_a obtained from Eq. (4.12-1) is the mean value to use over the tube length L with the arithmetic temperature difference ΔT_a.

$$\Delta T_a = \frac{(T_w - T_{bi}) + (T_w - T_{bo})}{2} \tag{4.12-5}$$

when T_w is the average wall temperature for the whole tube and T_{bi} is the inlet bulk temperature and T_{bo} the outlet bulk temperature. The heat flux q is

$$q = h_a A \, \Delta T_a = h_a (\pi D L) \, \Delta T_a \tag{4.12-6}$$

EXAMPLE 4.12-1. *Heating a Non-Newtonian Fluid in Laminar Flow*
A non-Newtonian fluid flowing at a rate of 7.56×10^{-2} kg/s inside a 25.4-mm-ID tube is being heated by steam condensing outside the tube. The fluid enters the heating section of the tube, which is 1.524 m long, at a temperature of 37.8°C. The inside wall temperature T_w is constant at 93.3°C. The mean physical properties of the fluid are $\rho = 1041$ kg/m³, $c_{pm} = 2.093$ kJ/kg·K, and $k = 1.212$ W/m·K. The fluid is a power-law fluid having the following flow property (rheological) constants: $n = n' = 0.40$, which is approximately constant over the temperature range encountered, and $K = 139.9$ N·s$^{n''}$/m² at 37.8°C and 62.5 at 93.3°C. For this fluid a plot of log K versus $T°C$ is approximately a straight line. Calculate the outlet bulk temperature of the fluid if it is in laminar flow.

Solution: The solution is trial and error since the outlet bulk temperature T_{bo} of the fluid must be known to calculate h_a from Eq. (4.12-2). Assuming $T_{bo} = 54.4°C$ for the first trial, the mean bulk temperature T_b is $(54.4 + 37.8)/2$, or 46.1°C.

Plotting the two values of K given at 37.8 and 93.3°C as log K versus $T°C$ and drawing a straight line through these two points, a value of K_b of 123.5 at $T_b = 46.1°C$ is read from the plot. At $T_w = 93.3°C$, $K_w = 62.5$.

Next, δ is calculated using Eq. (4.12-2).

$$\delta = \frac{3n' + 1}{4n'} = \frac{3(0.40) + 1}{4(0.40)} = 1.375$$

Substituting into Eq. (4.12-3),

$$N_{Gz} = \frac{mc_p}{kL} = \frac{(7.56 \times 10^{-2})(2.093 \times 10^3)}{1.212(1.524)} = 85.7$$

From Eq. (4.12-4),

$$\frac{\gamma_b}{\gamma_w} = \frac{K_b}{K_w} = \frac{123.5}{62.5}$$

Then substituting into Eq. (4.12-1),

$$\frac{h_a D}{k} = \frac{h_a(0.0254)}{1.212} = 1.75\delta^{1/3}(N_{Gz})^{1/3}\left(\frac{\gamma_b}{\gamma_w}\right)^{0.14}$$

$$= 1.75(1.375)^{1/3}(85.7)^{1/3}\left(\frac{123.5}{62.5}\right)^{0.14} \quad \textbf{(4.12-1)}$$

Solving, $h_a = 448.3 \text{ W/m}^2 \cdot \text{K}$.

By a heat balance, the value of q in W is as follows:

$$q = mc_{pm}(T_{bo} - T_{bi}) \quad \textbf{(4.12-7)}$$

This is equated to Eq. (4.12-6) to obtain

$$q = mc_{pm}(T_{bo} - T_{bi}) = h_a(\pi DL)\, \Delta T_a \quad \textbf{(4.12-8)}$$

The arithmetic mean temperature difference ΔT_a by Eq. (4.12-5) is

$$\Delta T_a = \frac{(T_w - T_{bi}) + (T_w - T_{bo})}{2}$$

$$= \frac{(93.3 - 37.8) + (93.3 - T_{bo})}{2} = 74.4 - 0.5T_{bo}$$

Substituting the known values in Eq. (4.12-8) and solving for T_{bo},

$$(7.56 \times 10^{-2})(2.093 \times 10^3)(T_{bo} - 37.8)$$

$$= 448.3(\pi \times 0.0254 \times 1.524)(74.4 - 0.5T_{bo})$$

$$T_{bo} = 54.1°\text{C}$$

This value of 54.1°C is close enough to the assumed value of 54.5°C so that a second trial is not needed. Only the value of K_b would be affected. Known values can be substituted into Eq. (3.5-11) for the Reynolds number to show that it is less than 2100 and is laminar flow.

For less "viscous" non-Newtonian power-law fluids in laminar flow, natural convection may affect the heat-transfer rates. Metzner and Gluck (M2) recommend use of an empirical correction to Eq. (4.12-1) for horizontal tubes.

2. *Turbulent flow in tubes.* For turbulent flow of power-law fluids through tubes, Clapp (C4) presents the following empirical equation for heat transfer:

$$N_{Nu} = \frac{h_L D}{k} = 0.0041(N_{Re,\, gen})^{0.99}\left[\frac{K'c_p}{k}\left(\frac{8V}{D}\right)^{n-1}\right]^{0.4} \quad \textbf{(4.12-9)}$$

where $N_{Re,\, gen}$ is defined by Eq. (3.5-11) and h_L is the heat-transfer coefficient based on the log mean temperature driving force. The fluid properties are evaluated at the bulk mean temperature. Metzner and Friend (M3) also give equations for turbulent heat transfer.

4.12C Natural Convection

Acrivos (A1, S3) gives relationships for natural convection heat transfer to power-law fluids from various geometries of surfaces such as spheres, cylinders, and plates.

4.13 SPECIAL HEAT-TRANSFER COEFFICIENTS

4.13A Heat Transfer in Agitated Vessels

1. Introduction. Many chemical and biological processes are often carried out in agitated vessels. As discussed in Section 3.4, the liquids are generally agitated in cylindrical vessels with an impeller mounted on a shaft and driven by an electric motor. Typical agitators and vessel assemblies have been shown in Figs. 3.4-1 and 3.4-3. Often it is necessary to cool or heat the contents of the vessel during the agitation. This is usually done by heat-transfer surfaces, which may be in the form of cooling or heating jackets in the wall of the vessel or coils of pipe immersed in the liquid.

2. Vessel with heating jacket. In Fig. 4.13-1a a vessel with a cooling or heating jacket is shown. When heating, the fluid entering is often steam, which condenses inside the jacket and leaves at the bottom. The vessel is equipped with an agitator and in most cases also with baffles (not shown).

Correlations for the heat-transfer coefficient from the agitated Newtonian liquid inside the vessel to the jacket walls of the vessel have the following form:

$$\frac{hD_t}{k} = a\left(\frac{D_a^2 N\rho}{\mu}\right)^b \left(\frac{c_p \mu}{k}\right)^{1/3} \left(\frac{\mu}{\mu_w}\right)^m \tag{4.13-1}$$

where h is the heat-transfer coefficient for the agitated liquid to the inner wall in $W/m^2 \cdot K$, D_t is the inside diameter of the tank in m, k is thermal conductivity in $W/m \cdot K$, D_a is diameter of agitator in m, N is rotational speed in revolutions per sec, ρ is fluid density in kg/m^3, and μ is liquid viscosity in $Pa \cdot s$. All the liquid physical properties are evaluated at the bulk liquid temperature except μ_w, which is evaluated at the wall temperature T_w. Below are listed some correlations available and the Reynolds number range $(N'_{Re} = D_a^2 N\rho/\mu)$.

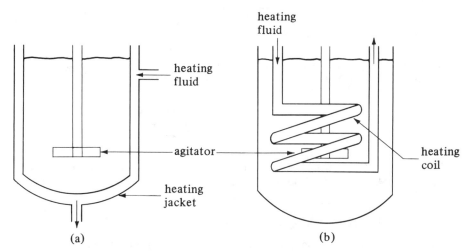

FIGURE 4.13-1. *Heat transfer in agitated vessels:* (a) *vessel with heating jacket,* (b) *vessel with heating coils.*

1. Paddle agitator with no baffles (C5, U1)

$$a = 0.36, \quad b = \tfrac{2}{3}, \quad m = 0.21, \quad N'_{Re} = 300 \text{ to } 3 \times 10^5$$

2. Flat-blade turbine agitator with no baffles (B4)

$$a = 0.54, \quad b = \tfrac{2}{3}, \quad m = 0.14, \quad N'_{Re} = 30 \text{ to } 3 \times 10^5$$

3. Flat-blade turbine agitator with baffles (B4, B5)

$$a = 0.74, \quad b = \tfrac{2}{3}, \quad m = 0.14. \quad N'_{Re} = 500 \text{ to } 3 \times 10^5$$

4. Anchor agitator with no baffles (U1)

$$a = 1.0, \quad b = \tfrac{1}{2}, \quad m = 0.18, \quad N'_{Re} = 10 \text{ to } 300$$

$$a = 0.36, \quad b = \tfrac{2}{3}, \quad m - 0.18, \quad N'_{Re} = 300 \text{ to } 4 \times 10^4$$

5. Helical ribbon agitator with no baffles (G4)

$$a = 0.633, \quad b = \tfrac{1}{2}, \quad m = 0.18, \quad N'_{Re} = 8 \text{ to } 10^5$$

Some typical overall U values for jacketed vessels for various process applications are tabulated in Table 4.13-1.

EXAMPLE 4.13-1. Heat-Transfer Coefficient in Agitated Vessel with Jacket

A jacketed 1.83-m-diameter agitated vessel with baffles is being used to heat a liquid which is at 300 K. The agitator is 0.61 m in diameter and is a flat-blade turbine rotating at 100 rpm. Hot water is in the heating jacket. The wall surface temperature is constant at 355.4 K. The liquid has the following bulk physical properties: $\rho = 961$ kg/m^3, $c_p = 2500$ J/kg $\cdot$ K, $k = 0.173$ W/m $\cdot$ K, and $\mu = 1.00$ Pa $\cdot$ s at 300 K and 0.084 Pa $\cdot$ s at 355.4 K. Calculate the heat-transfer coefficient to the wall of the jacket.

TABLE 4.13-1. *Typical Overall Heat-Transfer Coefficients in Jacketed Vessels*

Fluid in Jacket	Fluid in Vessel	Wall Material	Agitation	U $\dfrac{btu}{h \cdot ft^2 \cdot {}^\circ F}$	$\dfrac{W}{m^2 \cdot K}$	Ref.
Steam	Water	Copper	None	150	852	(P1)
			Simple stirring	250	1420	
Steam	Paste	Cast iron	Double scrapers	125	710	(P1)
Steam	Boiling water	Copper	None	250	1420	(P1)
Steam	Milk	Enameled cast iron	None	200	1135	(P1)
			Stirring	300	1700	
Hot water	Cold water	Enameled cast iron	None	70	398	(P1)
Steam	Tomato purée	Metal	Agitation	30	170	(C1)

Solution: The following are given:

$$D_t = 1.83 \text{ m} \qquad D_a = 0.61 \text{m} \qquad N = 100/60 \text{ rev/s}$$

$$\mu(300 \text{ K}) = 1.00 \text{ Pa} \cdot \text{s} = 1.00 \text{ kg/m} \cdot \text{s}$$

$$\mu_w(355.4 \text{ K}) = 0.084 \text{ Pa} \cdot \text{s} = 0.084 \text{ kg/m} \cdot \text{s}$$

First, calculating the Reynolds number at 300 K,

$$N'_{Re} = \frac{D_a^2 N \rho}{\mu} = \frac{(0.61)^2(100/60)(961)}{1.00} = 596$$

The Prandtl number is

$$N_{Pr} = \frac{c_p \mu}{k} = \frac{2500(1.00)}{0.173} = 14\,450$$

Using Eq. (4.13-1) with $a = 0.74$, $b = 2/3$, and $m = 0.14$

$$\frac{hD_t}{k} = 0.74(N'_{Re})^{2/3} (N_{Pr})^{1/3} \left(\frac{\mu}{\mu_w}\right)^{0.14} \qquad \textbf{(4.13-1)}$$

Substituting and solving for h,

$$\frac{h(1.83)}{0.173} = 0.74(596)^{2/3}(14\,450)^{1/3}\left(\frac{1000}{84}\right)^{0.14}$$

$$h = 170.6 \text{ W/m}^2 \cdot \text{K} \ (30.0 \text{ btu/h} \cdot \text{ft}^2 \cdot {}^\circ\text{F})$$

A correlation to predict the heat-transfer coefficient of a power-law non-Newtonian fluid in a jacketed vessel with a turbine agitator is also available elsewhere (C6).

3. *Vessel with heating coils.* In Fig. 4.13-1b an agitated vessel with a helical heating or cooling coil is shown. Correlations for the heat-transfer coefficient to the outside surface of the coils in agitated vessels are listed below for various types of agitators.

For a paddle agitator with no baffles (C5),

$$\frac{hD_t}{k} = 0.87\left(\frac{D_a^2 N \rho}{\mu}\right)^{0.62}\left(\frac{c_p \mu}{k}\right)^{1/3}\left(\frac{\mu}{\mu_w}\right)^{0.14} \qquad \textbf{(4.13-2)}$$

This holds for a Reynolds number range of 300 to 4×10^5.

For a flat-blade turbine agitator with baffles, see (O1).

When the heating or cooling coil is in the form of vertical tube baffles with a flat-blade turbine, the following correlation can be used (D1).

$$\frac{hD_o}{k} = 0.09\left(\frac{D_a^2 N \rho}{\mu}\right)^{0.65}\left(\frac{c_p \mu}{k}\right)^{1/3}\left(\frac{D_a}{D_t}\right)^{1/3}\left(\frac{2}{n_b}\right)^{0.2}\left(\frac{\mu}{\mu_f}\right)^{0.4} \qquad \textbf{(4.13-3)}$$

where D_o is the outside diameter of the coil tube in m, n_b is the number of vertical baffle tubes, and μ_f is the viscosity at the mean film temperature.

Perry and Green (P3) give typical values of overall heat-transfer coefficients U for coils immersed in various liquids in agitated and nonagitated vessels.

4.13B Scraped Surface Heat Exchangers

Liquid–solid suspensions, viscous aqueous and organic solutions, and numerous food products, such as margarine and orange juice concentrate, are often cooled or heated in a scraped-surface exchanger. This consists of a double-pipe heat exchanger with a jacketed

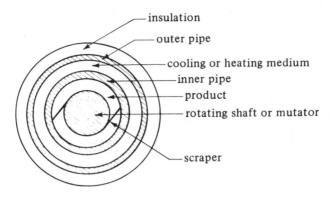

insulation

outer pipe

cooling or heating medium

inner pipe

product

rotating shaft or mutator

scraper

FIGURE 4.13-2. *Scraped surface heat exchanger.*

cylinder containing steam or cooling liquid and an internal shaft rotating and fitted with wiper blades, as shown in Fig. 4.13-2.

The viscous liquid product flows at low velocity through the central tube between the rotating shaft and the inner pipe. The rotating scrapers or wiper blades continually scrape the surface of liquid, preventing localized overheating and giving rapid heat transfer. This device in some cases is also called a *votator heat exchanger.*

Skelland et al. (S4) give the following equation to predict the inside heat-transfer coefficient for the votator.

$$\frac{hD}{k} = \alpha \left(\frac{c_p \mu}{k}\right)^\beta \left(\frac{(D - D_S)v\rho}{\mu}\right)^{1.0} \left(\frac{DN}{v}\right)^{0.62} \left(\frac{D_S}{D}\right)^{0.55} (n_B)^{0.53} \qquad \textbf{(4.13-4)}$$

$$\alpha = 0.014 \quad \beta = 0.96 \qquad \text{for viscous liquids}$$

$$\alpha = 0.039 \quad \beta = 0.70 \qquad \text{for nonviscous liquids}$$

where D = diameter of vessel in m, D_S = diameter of rotating shaft in m, v = axial flow velocity of liquid in m/s, N = agitator speed in rev/s, and n_B = number of blades on agitator. Data cover a region of axial flow velocities of 0.076 to 0.38 m/min and rotational speeds of 100 to 750 rpm.

Typical overall heat-transfer coefficients in food applications are $U = 1700$ W/m² · K (300 btu/h · ft² · °F) for cooling margarine with NH_3, 2270 (400) for heating applesauce with steam, 1420 (250) for chilling shortening with NH_3, and 2270 (400) for cooling cream with water (B6).

4.13C Extended Surface or Finned Exchangers

1. Introduction.   The use of fins or extended surfaces on the outside of a heat exchanger pipe wall to give relatively high heat-transfer coefficients in the exchanger is quite common. An automobile radiator is such a device, where hot water passes inside through a bank of tubes and loses heat to the air. On the outside of the tubes, extended surfaces receive heat from the tube walls and transmit it to the air by forced convection.

Two common types of fins attached to the outside of a tube wall are shown in Fig. 4.13-3. In Fig. 4.13-3a there are a number of longitudinal fins spaced around the tube wall and the direction of gas flow is parallel to the axis of the tube. In Fig. 4.13-3b the gas flows normal to the tubes containing many circular or transverse fins.

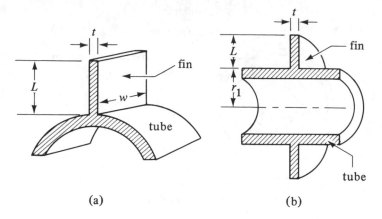

FIGURE 4.13-3. *Two common types of fins on a section of circular tube:*
(a) longitudinal fin, (b) circular or transverse fin.

The qualitative effect of using extended surfaces can be shown approximately in Eq. (4.13-5) for a fluid inside a tube having a heat-transfer coefficient of h_i and an outside coefficient of h_o.

$$\frac{1}{U_i A_i} = \sum R \simeq \frac{1}{h_i A_i} + R_{\text{metal}} + \frac{1}{h_o A_o} \tag{4.13-5}$$

The resistance R_{metal} of the wall can often be neglected. The presence of the fins on the outside increases A_o and hence reduces the resistance $1/h_o A_o$ of the fluid on the outside of the tube. For example, if we have h_i for condensing steam, which is very large, and h_o for air outside the tube, which is quite small, increasing A_o greatly reduces $1/h_o A_o$. This in turn greatly reduces the total resistance, which increases the heat-transfer rate. If the positions of the two fluids are reversed with air inside and steam outside, little increase in heat transfer could be obtained by using fins.

Equation (4.13-5) is only an approximation, since the temperature on the outside surface of the bare tube is not the same as that at the end of the fin because of the added resistance to heat flow by conduction from the fin tip to the base of the fin. Hence, a unit area of fin surface is not as efficient as a unit area of bare tube surface at the base of the fin. A fin efficiency η_f has been mathematically derived for various geometries of fins.

2. Derivation of equation for fin efficiency. We will consider a one-dimensional fin exposed to a surrounding fluid at temperature T_∞ as shown in Fig. 4.13-4. At the base of the fin the temperature is T_0 and at point x it is T. At steady state, the rate of heat conducted in to the element at x is $q_{x|x}$ and is equal to the rate of heat conducted out plus the rate of heat lost by convection.

$$q_{x|x} = q_{x|x+\Delta x} + q_c \tag{4.13-6}$$

Substituting Fourier's equation for conduction and the convection equation,

$$-kA \left.\frac{dT}{dx}\right|_x = -kA \left.\frac{dT}{dx}\right|_{x+\Delta x} + h(P\,\Delta x)(T - T_\infty) \tag{4.13-7}$$

where A is the cross-sectional area of the fin in m², P the perimeter of the fin in m, and $(P\,\Delta x)$ the area for convection. Rearranging Eq. (4.13-7), dividing by Δx, and letting Δx

Chap. 4   Principles of Steady-State Heat Transfer

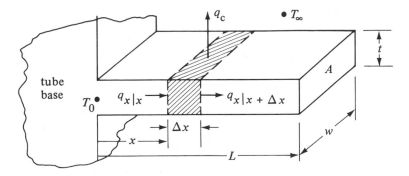

FIGURE 4.13-4. *Heat balance for one-dimensional conduction and convection in a rectangular fin with constant cross-sectional area.*

approach zero,

$$\frac{d^2T}{dx^2} - \frac{hP}{kA}(T - T_\infty) = 0 \tag{4.13-8}$$

Letting $\theta = T - T_\infty$, Eq. (4.13-8) becomes

$$\frac{d^2\theta}{dx^2} = \frac{hP}{kA}\theta = 0 \tag{4.13-9}$$

The first boundary condition is that $\theta = \theta_0 = T_0 - T_\infty$ at $x = 0$. For the second boundary condition needed to integrate Eq. (4.13-9), several cases can be considered, depending upon the physical conditions at $x = L$. In the first case, the end of the fin is insulated and $d\theta/dx = 0$ at $x = L$. In the second case the fin loses heat by convection from the tip surface so that $-k(dT/dx)_L = h(T_L - T_\infty)$. The solution using case 2 is quite involved and will not be considered here. Using the first case where the tip is insulated, integration of Eq. (4.13-9) gives

$$\frac{\theta}{\theta_0} = \frac{\cosh\,[m(L - x)]}{\cosh\,mL} \tag{4.13-10}$$

where $m = (hP/kA)^{1/2}$.
    The heat lost by the fin is expressed as

$$q = -kA\left.\frac{dT}{dx}\right|_{x=0} \tag{4.13-11}$$

Differentiating Eq. (4.13-10) with respect to x and combining it with Eq. (4.13-11),

$$q = (hPkA)^{1/2}(T_0 - T_\infty)\tanh\,mL \tag{4.13-12}$$

In the actual fin the temperature T in the fin decreases as the tip of the fin is approached. Hence, the rate of heat transfer per unit area decreases as the distance from the tube base is increased. To indicate this effectiveness of the fin to transfer heat, the fin efficiency η_f is defined as the ratio of the actual heat transferred from the fin to the heat transferred if the entire fin were at the base temperature T_0.

$$\eta_f = \frac{(hPkA)^{1/2}(T_0 - T_\infty)\tanh\,mL}{h(PL)(T_0 - T_\infty)} = \frac{\tanh\,mL}{mL} \tag{4.13-13}$$

where PL is the entire surface area of fin.

The expression for mL is

$$mL = \left(\frac{hP}{kA}\right)^{1/2} L = \left[\frac{h(2w + 2t)}{k(wt)}\right]^{1/2} L \qquad \text{(4.13-14)}$$

For fins which are thin, $2t$ is small compared to $2w$ and

$$mL = \left(\frac{2h}{kt}\right)^{1/2} L \qquad \text{(4.13-15)}$$

Equation (4.13-15) holds for a fin with an insulated tip. This equation can be modified to hold for the case where the fin loses heat from its tip. This can be done by extending the length of the fin by $t/2$, where the corrected length L_c to use in Eqs. (4.13-13) to (4.13-15) is

$$L_c = L + \frac{t}{2} \qquad \text{(4.13-16)}$$

The fin efficiency calculated from Eq. (4.13-13) for a longitudinal fin is shown in Fig. 4.13-5a. In Fig. 4.13-5b the fin efficiency for a circular fin is presented. Note that the abscissa on the curves is $L_c(h/kt)^{1/2}$ and not $L_c(2h/kt)^{1/2}$ as in Eq. (4.13-15).

EXAMPLE 4.13.-2.   Fin Efficiency and Heat Loss from Fin
A circular aluminum fin as shown in Fig. 4.13-3b ($k = 222$ W/m·K) is attached to a copper tube having an outside radius of 0.04 m. The length of the fin is 0.04 m and the thickness is 2 mm. The outside wall or tube base is at 523.2 K and the external surrounding air at 343.2 K has a convective coefficient of 30 W/m²·K. Calculate the fin efficiency and the rate of heat loss from the fin.

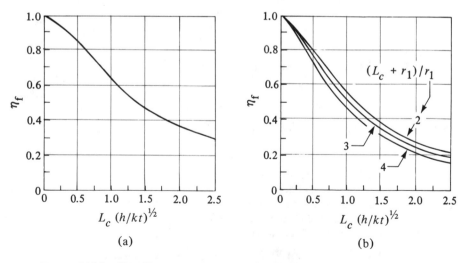

FIGURE 4.13-5.   *Fin efficiency η_f for various fins: (a) longitudinal or straight fins, (b) circular or transverse fins. (See Fig. 4.13-3 for the dimensions of the fins.)*

Chap. 4   Principles of Steady-State Heat Transfer

Solution: The given data are $T_0 = 523.2$ K, $T_\infty = 343.2$ K, $L = 0.04$ m, $r_1 = 0.04$ m, $t = 0.002$ m, $k = 222$ W/m · K, $h = 30$ W/m² · K. By Eq. (4.13-16), $L_c = L + t/2 = 0.040 + 0.002/2 = 0.041$ m. Then,

$$L_c\left(\frac{h}{kt}\right)^{1/2} = (0.041)\left[\frac{30}{222(0.002)}\right]^{1/2} = 0.337$$

Also, $(L_c + r_1)/r_1 = (0.041 + 0.040)/0.040 = 2.025$. Using Fig. 4.13-5b, $\eta_f = 0.89$. The heat transfer from the fin itself is

$$q_f = \eta_f h A_f (T_0 - T_\infty) \tag{4.13-17}$$

where A_f is the outside surface area (annulus) of the fin and is given by the following for both sides of the fin:

$$A_f = 2\pi[(L_c + r_1)^2 - (r_1)^2] \quad \text{(circular fin)}$$

$$A_f = 2\pi(L_c\,xw) \quad \text{(longitudinal fin)} \tag{4.13-18}$$

Hence,

$$A_f = 2\pi[(0.041 + 0.040)^2 - (0.040)^2] = 3.118 \times 10^{-2}\ \text{m}^2$$

Substituting into Eq. (4.13-17),

$$q_f = 0.89(30)(3.118 \times 10^{-2})(523.2 - 343.2) = 149.9\ \text{W}$$

3. *Overall heat-transfer coefficient for finned tubes.* We consider here the general case similar to Fig. 4.3-3b, where heat transfer occurs from a fluid inside a cylinder or tube, through the cylinder metal wall A of thickness Δx_A, and then to the fluid outside the tube, where the tube has fins on the outside. The heat is transferred through a series of resistances. The total heat q leaving the outside of the tube is the sum of heat loss by convection from the base of the bare tube q_t and the loss by convection from the fins, q_f.

$$q = q_t + q_f = h_o\,A_t(T_0 - T_\infty) + h_o\,A_f\,\eta_f(T_0 - T_\infty) \tag{4.13-19}$$

This can be written as follows as a resistance since the paths are in parallel.

$$q = (h_o\,A_t + h_o\,A_f\,\eta_f)(T_1 - T_\infty) = \frac{T_0 - T_\infty}{\dfrac{1}{h_o(A_t + A_f\,\eta_f)}} = \frac{T_0 - T_\infty}{R} \tag{4.13-20}$$

where A_t is the area of the bare tube between the fins, A_f the area of the fins, and h_o the outside convective coefficient. The resistance in Eq. (4.3-20) can be substituted for the resistance $(1/h_o A_o)$ in Eq. (4.3-15) for a bare tube to give the overall equation for a finned tube exchanger.

$$q = \frac{T_4 - T_1}{1/h_i A_i + \Delta x_A/k_A A_{A\,\text{lm}} + 1/h_o(A_t + A_f\,\eta_f)} = \frac{T_4 - T_1}{\sum R} \tag{4.13-21}$$

where T_4 is the temperature of the fluid inside the tube and T_1 the outside fluid temperature. Writing Eq. (4.13-21) in the form of an overall heat-transfer coefficient U_i based on the inside area A_i, $q = U_i A_i(T_4 - T_1)$ and

$$U_i = \frac{1}{1/h_i + \Delta x_A A_i/k_A A_{A\,\text{lm}} + A_i/h_o(A_t + A_f\,\eta_f)} \tag{4.13-22}$$

The presence of fins on the outside of the tube changes the characteristics of the fluid flowing by the tube (either flowing parallel to the longitudinal finned tube or transverse to the circular finned tube). Hence, the correlations for fluid flow parallel to or transverse to bare tubes cannot be used to predict the outside convective coefficient h_o. Correlations are available in the literature (K4, M1, P1, P3) for heat transfer to various types of fins.

4.14 DIMENSIONAL ANALYSIS IN HEAT TRANSFER

4.14A Introduction

As seen in many of the correlations for fluid flow and heat transfer, many dimensionless groups, such as the Reynolds number and Prandtl number, occur in these correlations. Dimensional analysis is often used to group the variables in a given physical situation into dimensionless parameters or numbers which can be useful in experimentation and correlating data.

An important way of obtaining these dimensionless groups is to use dimensional analysis of differential equations described in Section 3.11. Another useful method is the Buckingham method, in which the listing of the significant variables in the particular physical problem is done first. Then we determine the number of dimensionless parameters into which the variables may be combined.

4.14B Buckingham Method

1. Heat transfer inside a pipe. The Buckingham theorem, given in Section 3.11 states that the function relationship among q quantities or variables whose units may be given in terms of u fundamental units or dimensions may be written as $(q - u)$ dimensionless groups.

As an additional example to illustrate the use of this method, let us consider a fluid flowing in turbulent flow at velocity v inside a pipe of diameter D and undergoing heat transfer to the wall. We wish to predict the dimensionless groups relating the heat-transfer coefficient h to the variables D, ρ, μ, c_p, k, and v. The total number of variables is $q = 7$.

The fundamental units or dimensions are $u = 4$ and are mass M, length L, time t, and temperature T. The units of the variables in terms of these fundamental units are as follows:

$$h = \frac{M}{t^3 T} \qquad D = L \qquad \rho = \frac{M}{L^3} \qquad \mu = \frac{M}{Lt} \qquad c_p = \frac{L^2}{t^2 T} \qquad k = \frac{ML}{t^3 T} \qquad v = \frac{T}{t}$$

Hence, the number of dimensionless groups or π's can be assumed to be $7 - 4$, or 3. Then

$$\pi_1 = f(\pi_2, \pi_3) \tag{4.14-1}$$

We will choose the four variables D, k, μ, and v to be common to all the dimensionless groups. Then the three dimensionless groups are

$$\pi_1 = D^a k^b \mu^c v^d \rho \tag{4.14-2}$$

$$\pi_2 = D^e k^f \mu^g v^h c_p \tag{4.14-3}$$

$$\pi_3 = D^i k^j \mu^k v^l h \tag{4.14-4}$$

For π_1, substituting the actual dimensions,

$$M^0 L^0 t^0 T^0 = 1 = L^a \left(\frac{ML}{t^3 T}\right)^b \left(\frac{M}{LT}\right)^c \left(\frac{L}{T}\right)^d \left(\frac{M}{L^3}\right) \qquad (4.14\text{-}5)$$

Summing for each exponent,

$$
\begin{aligned}
(L) \quad & 0 = a + b - c + d - 3 \\
(M) \quad & 0 = b + c + 1 \\
(t) \quad & 0 = -3b - c - d \\
(T) \quad & 0 = -b
\end{aligned}
\qquad (4.14\text{-}6)
$$

Solving these equations simultaneously, $a = 1, b = 0, c = -1, d = 1$.
    Substituting these values into Eq. (4.14-2),

$$\pi_1 = \frac{Dv\rho}{\mu} = N_{Re} \qquad (4.14\text{-}7)$$

Repeating for π_2 and π_3 and substituting the actual dimensions,

$$\pi_2 = \frac{c_p \mu}{k} = N_{Pr} \qquad (4.14\text{-}8)$$

$$\pi_3 = \frac{hD}{k} = N_{Nu} \qquad (4.14\text{-}9)$$

Substituting π_1, π_2, and π_3 into Eq. (4.14-1) and rearranging,

$$\frac{hD}{k} = f\left(\frac{Dv\rho}{\mu}, \frac{c_p \mu}{k}\right) \qquad (4.14\text{-}10)$$

This is in the form of the familiar equation for heat transfer inside pipes, Eq. (4.5-8).
    This type of analysis is useful in empirical correlations of heat-transfer data. The importance of each dimensionless group, however, must be determined by experimentation (B1, M1).

2. *Natural convection heat transfer outside a vertical plane.* In the case of natural-convection heat transfer from a vertical plane wall of length L to an adjacent fluid, different dimensionless groups should be expected when compared to forced convection inside a pipe since velocity is not a variable. The buoyant force due to the difference in density between the cold and heated fluid should be a factor. As seen in Eqs. (4.7-1) and (4.7-2), the buoyant force depends upon the variables β, g, ρ, and ΔT. Hence, the list of variables to be considered and their fundamental units are as follows:

$$L = L \qquad \rho = \frac{M}{L^3} \qquad \mu = \frac{M}{Lt} \qquad c_p = \frac{L^2}{t^2 T} \qquad \beta = \frac{1}{T}$$

$$g = \frac{L}{t^2} \qquad \Delta T = T \qquad h = \frac{M}{t^3 T} \qquad k = \frac{ML}{t^3 T}$$

The number of variables is $q = 9$. Since $u = 4$, the number of dimensionless groups or π's is $9 - 4$, or 5. Then $\pi_1 = f(\pi_2, \pi_3, \pi_4, \pi_5)$.

We will choose the four variables, L, μ, k, and g to be common to all the dimensionless groups.

$$\pi_1 = L^a \mu^b k^c g^d \rho \qquad \pi_2 = L^e \mu^f k^g g^h c_p \qquad \pi_3 = L^i \mu^j k^k g^l \beta$$

$$\pi_4 = L^m \mu^n k^o g^p \Delta T \qquad \pi_5 = L^q \mu^r k^s g^t h$$

For π_1, substituting the dimensions,

$$1 = L^a \left(\frac{M}{Lt}\right)^b \left(\frac{ML}{t^3 T}\right)^c \left(\frac{L}{t^2}\right)^d \left(\frac{M}{L^3}\right) \tag{4.14-11}$$

Solving for the exponents as before, $a = \frac{3}{2}, b = -1, c = 0, d = \frac{1}{2}$. Then π_1 becomes

$$\pi_1 = \frac{L^{3/2} \rho g^{1/2}}{\mu} \tag{4.14-12}$$

Taking the square of both sides to eliminate fractional exponents,

$$\pi_1 = \frac{L^3 \rho^2 g}{\mu^2} \tag{4.14-13}$$

Repeating for the other π equations,

$$\pi_1 = \frac{L^3 \rho^2 g}{\mu^2} \qquad \pi_2 = \frac{c_p \mu}{k} = N_{Pr} \qquad \pi_3 = \frac{L \mu g \beta}{k}$$

$$\pi_4 = \frac{k \Delta T}{L \mu g} \qquad \pi_5 = \frac{hL}{k} = N_{Nu}$$

Combining the dimensionless groups π_1, π_3, and π_4 as follows,

$$\pi_1 \pi_3 \pi_4 = \frac{L^3 \rho^2 g}{\mu^2} \frac{L \mu g \beta}{k} \frac{k \Delta T}{L \mu g} = \frac{L^3 \rho^2 g \beta \Delta T}{\mu^2} = N_{Gr} \tag{4.14-14}$$

Equation (4.14-14) is the Grashof group given in Eq. (4.7-4). Hence,

$$N_{Nu} = f(N_{Gr}, N_{Pr}) \tag{4.14-15}$$

4.15 NUMERICAL METHODS FOR STEADY-STATE CONDUCTION IN TWO DIMENSIONS

4.15A Analytical Equation for Conduction

In Section 4.4 we discussed methods for solving two-dimensional heat-conduction problems using graphical procedures and shape factors. In this section we consider analytical and numerical methods.

The equation for conduction in the x direction is as follows:

$$q_x = -kA \frac{\partial T}{\partial x} \tag{4.15-1}$$

Now we shall derive an equation for steady-state conduction in two directions x and y. Referring to Fig. 4.15-1, a rectangular block Δx by Δy by L is shown. The total heat input to the block is equal to the output.

$$q_{x|x} + q_{y|y} = q_{x|x + \Delta x} + q_{y|y + \Delta y} \tag{4.15-2}$$

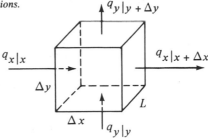

FIGURE 4.15-1. *Steady-state conduction in two directions.*

Now, from Eq. (4.15-1),

$$q_{x|x} = -k(\Delta yL) \left.\frac{\partial T}{\partial x}\right|_x \qquad (4.15\text{-}3)$$

Writing similar equations for the other three terms and substituting into Eq. (4.15-2),

$$-k(\Delta yL) \left.\frac{\partial T}{\partial x}\right|_x - k(\Delta xL) \left.\frac{\partial T}{\partial y}\right|_y = -k(\Delta yL) \left.\frac{\partial T}{\partial x}\right|_{x+\Delta x} - k(\Delta xL) \left.\frac{\partial T}{\partial y}\right|_{y+\Delta y} \qquad (4.15\text{-}4)$$

Dividing through by $\Delta x \, \Delta yL$ and letting Δx and Δy approach zero, we obtain the final equation for steady-state conduction in two directions.

$$\frac{\partial^2 T}{\partial x^2} + \frac{\partial^2 T}{\partial y^2} = 0 \qquad (4.15\text{-}5)$$

This is called the *Laplace equation*. There are a number of analytical methods to solve this equation. In the method of separation of variables, the final solution is expressed as an infinite Fourier series (H1, G2, K1). We consider the case shown in Fig. 4.15-2. The solid is called a semi-infinite solid since one of its dimensions is ∞. The two edges or boundaries at $x = 0$ and $x = L$ are held constant at T_1 K. The edge at $y = 0$ is held at T_2. And at $y = \infty$, $T = T_1$. The solution relating T to position y and x is

$$\frac{T - T_1}{T_2 - T_1} = \frac{4}{\pi} \left[\frac{1}{1} e^{-(\pi/L)y} \sin \frac{1\pi x}{L} + \frac{1}{3} e^{-(3\pi/L)y} \sin \frac{3\pi x}{L} + \cdots \right] \qquad (4.15\text{-}6)$$

Other analytical methods are available and are discussed in many texts (C2, H1, G2, K1). A large number of such analytical solutions have been given in the literature. However, there are many practical situations where the geometry or boundary conditions are too complex for analytical solutions, so that finite-difference numerical methods are used. These are discussed in the next section.

FIGURE 4.15-2. *Steady-state heat conduction in two directions in a semi-infinite plate.*

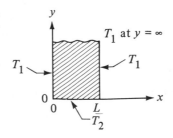

4.15B Finite-Difference Numerical Methods

1. Derivation of the method. Since the advent of the fast digital computers, solutions to many complex two-dimensional heat-conduction problems by numerical methods are readily possible. In deriving the equations we can start with the partial differential equation (4.15-5). Setting up the finite difference of $\partial^2 T/\partial x^2$,

$$\frac{\partial^2 T}{\partial x^2} = \frac{\partial(\partial T/\partial x)}{\partial x} = \frac{\dfrac{T_{n+1,\,m} - T_{n,\,m}}{\Delta x} - \dfrac{T_{n,\,m} - T_{n-1,\,m}}{\Delta x}}{\Delta x}$$

$$= \frac{T_{n+1,\,m} - 2T_{n,\,m} + T_{n-1,\,m}}{(\Delta x)^2} \tag{4.15-7}$$

where the index m stands for a given value of y, $m + 1$ stands for $y + 1\ \Delta y$, and n is the index indicating the position of T on the x scale. This is shown in Fig. 4.15-3. The two-dimensional solid is divided into squares. The solid inside a square is imagined to be concentrated at the center of the square, and this concentrated mass is a "node." Each node is imagined to be connected to the adjacent nodes by a small conducting rod as shown.

The finite difference of $\partial^2 T/\partial y^2$ is written in a similar manner.

$$\frac{\partial^2 T}{\partial y^2} = \frac{T_{n,\,m+1} - 2T_{n,\,m} + T_{n,\,m-1}}{(\Delta y)^2} \tag{4.15-8}$$

Substituting Eqs. (4.15-7) and (4.15-8) into Eq. (4.15-5) and setting $\Delta x = \Delta y$,

$$T_{n,\,m+1} + T_{n,\,m-1} + T_{n+1,\,m} + T_{n-1,\,m} - 4T_{n,\,m} = 0 \tag{4.15-9}$$

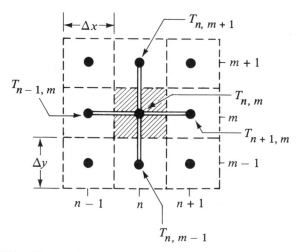

FIGURE 4.15-3. *Temperatures and arrangement of nodes for two-dimensional steady-state heat conduction.*

This equation states that the net heat flow into any point or node is zero at steady state. The shaded area in Fig. 4.15-3 represents the area on which the heat balance was made. Alternatively, Eq. (4.15-9) can be derived by making a heat balance on this shaded area. The total heat in for unit thickness is

$$\frac{k\,\Delta y}{\Delta x}(T_{n-1,\,m} - T_{n,\,m}) + \frac{k\,\Delta y}{\Delta x}(T_{n+1,\,m} - T_{n,\,m})$$

$$+ \frac{k\,\Delta x}{\Delta y}(T_{n,\,m+1} - T_{n,\,m}) + \frac{k\,\Delta x}{\Delta y}(T_{n,\,m-1} - T_{n,\,m}) = 0 \quad \textbf{(4.15-10)}$$

Rearranging, this becomes Eq. (4.15-9). In Fig. 4.15-3 the rods connecting the nodes act as fictitious heat conducting rods.

To use the numerical method, Eq. (4.15-9) is written for each node or point. Hence, for N unknown nodes, N linear algebraic equations must be written and the system of equations solved for the various node temperatures. For a hand calculation using a modest number of nodes, the iteration method can be used to solve the system of equations.

2. Iteration method of solution. In using the iteration method, the right-hand side of Eq. (4.15-9) is set equal to a residual $\bar{q}_{n,\,m}$.

$$\bar{q}_{n,\,m} = T_{n-1,\,m} + T_{n+1,\,m} + T_{n,\,m+1} + T_{n,\,m-1} - 4T_{n,\,m} \quad \textbf{(4.15-11)}$$

Since $\bar{q}_{n,\,m} = 0$ at steady state, solving for $T_{n,\,m}$ in Eq. (4.15-11) or (4.15-9),

$$T_{n,\,m} = \frac{T_{n-1,\,m} + T_{n+1,\,m} + T_{n,\,m+1} + T_{n,\,m-1}}{4} \quad \textbf{(4.15-12)}$$

Equations (4.15-11) and (4.15-12) are the final equations to be used. Their use is illustrated in the following example.

EXAMPLE 4.15-1. *Steady-State Heat Conduction in Two Directions*
Figure 4.15-4 shows a cross section of a hollow rectangular chamber with the inside dimensions 4×2 m and outside dimensions 8×8 m. The chamber is 20 m long. The inside walls are held at 600 K and the outside at

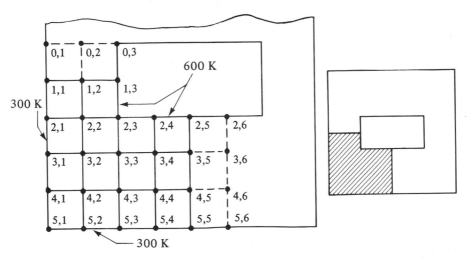

FIGURE 4.15-4. *Square grid pattern for Example 4.15-1.*

300 K. The k is 1.5 W/m · K. For steady-state conditions find the heat loss per unit chamber length. Use grids 1×1 m.

Solution: Since the chamber is symmetrical, one-fourth of the chamber (shaded part) will be used. Preliminary estimates will be made for the first approximation. These are for node $T_{1,2} = 450$ K, $T_{2,2} = 400$, $T_{3,2} = 400$, $T_{3,3} = 400$, $T_{3,4} = 450$, $T_{3,5} = 500$, $T_{4,2} = 325$, $T_{4,3} = 350$, $T_{4,4} = 375$, and $T_{4,5} = 400$. Note that $T_{0,2} = T_{2,2}$, $T_{3,6} = T_{3,4}$, and $T_{4,6} = T_{4,4}$ by symmetry.

To start the calculation, one can select any interior point, but it is usually better to start near a boundary. Using $T_{1,2}$, we calculate the residual $\bar{q}_{1,2}$ by Eq. (4.15-11).

$$\bar{q}_{1,2} = T_{1,1} + T_{1,3} + T_{0,2} + T_{2,2} - 4T_{1,2}$$

$$= 300 + 600 + 400 + 400 - 4(450) = -100$$

Hence, $T_{1,2}$ is not at steady state. Next, we set $\bar{q}_{1,2}$ to 0 and calculate a new value of $T_{1,2}$ by Eq. (4.15-12).

$$T_{1,2} = \frac{T_{1,1} + T_{1,3} + T_{0,2} + T_{2,2}}{4} = \frac{300 + 600 + 400 + 400}{4} = 425$$

This new value of $T_{1,2}$ of 425 K will replace the old one of 450 and be used to calculate the other nodes. Next,

$$\bar{q}_{2,2} = T_{2,1} + T_{2,3} + T_{1,2} + T_{3,2} - 4T_{2,2}$$

$$= 300 + 600 + 425 + 400 - 4(400) = 125$$

Setting $\bar{q}_{2,2}$ to zero and using Eq. (4.15-12),

$$T_{2,2} = \frac{T_{2,1} + T_{2,3} + T_{1,2} + T_{3,2}}{4} = \frac{300 + 600 + 425 + 400}{4} = 431$$

Continuing for all the rest of the interior nodes,

$$\bar{q}_{3,2} = 300 + 400 + 431 + 325 - 4(400) = -144$$

Using Eq. (4.15-12), $T_{3,2} = 364$,

$$\bar{q}_{3,3} = 364 + 450 + 600 + 350 - 4(400) = 164$$

$$T_{3,3} = 441$$

$$\bar{q}_{3,4} = 441 + 500 + 600 + 375 - 4(450) = 116$$

$$T_{3,4} = 479$$

$$\bar{q}_{3,5} = 479 + 479 + 600 + 400 - 4(500) = -42$$

$$T_{3,5} = 489$$

$$\bar{q}_{4,2} = 300 + 350 + 364 + 300 - 4(325) = 14$$

$$T_{4,2} = 329$$

$$\bar{q}_{4,3} = 329 + 375 + 441 + 300 - 4(350) = 45$$

$$T_{4,3} = 361$$

$$\bar{q}_{4,4} = 361 + 400 + 479 + 300 - 4(375) = 40$$

$$T_{4,4} = 385$$

$$\bar{q}_{4,5} = 385 + 385 + 489 + 399 - 4(400) = -41$$

$$T_{4,5} = 390$$

Having completed one sweep across the grid map, we can start a second approximation, using, of course, the new values calculated. We can start again with $T_{1,2}$ or we can select the node with the largest residual. Starting with $T_{1,2}$ again,

$$\bar{q}_{1,2} = 300 + 600 + 431 + 431 - 4(425) = 62$$

$$T_{1,2} = 440$$

$$\bar{q}_{2,2} = 300 + 600 + 440 + 364 - 4(431) = -20$$

$$T_{2,2} = 426$$

This is continued until the residuals are as small as desired. The final values arc as follows:

$$T_{1,2} = 441, \quad T_{2,2} = 432, \quad T_{3,2} = 384, \quad T_{3,3} = 461, \quad T_{3,4} = 485,$$

$$T_{3,5} = 490, \quad T_{4,2} = 340, \quad T_{4,3} = 372, \quad T_{4,4} = 387, \quad T_{4,5} = 391$$

To calculate the total heat loss from the chamber per unit chamber length, we use Fig. 4.15-5. For node $T_{2,4}$ to $T_{3,4}$ with $\Delta x = \Delta y$ and 1 m deep,

$$q = \frac{kA\ \Delta T}{\Delta x} = \frac{k[\Delta x(1)]}{\Delta x}(T_{2,4} - T_{3,4}) = k(T_{2,4} - T_{3,4}) \quad \textbf{(4.15-13)}$$

The heat flux for node $T_{2,5}$ to $T_{3,5}$ and for $T_{1,3}$ to $T_{1,2}$ should be multiplied by $\frac{1}{2}$ because of symmetry. The total heat conducted is the sum of the five paths for $\frac{1}{4}$ of the solid. For four duplicate parts,

$$q_I = 4k[\tfrac{1}{2}(T_{1,3} - T_{1,2}) + (T_{2,3} - T_{2,2}) + (T_{2,3} - T_{3,3})$$

$$+ (T_{2,4} - T_{3,4}) + \tfrac{1}{2}(T_{2,5} - T_{3,5})] \quad \textbf{(4.15-14)}$$

$$= 4(1.5)[\tfrac{1}{2}(600 - 441) + (600 - 432) + (600 - 461)$$

$$+ (600 - 485) + \tfrac{1}{2}(600 - 490)]$$

$$= 3340 \text{ W per } 1.0 \text{ m deep}$$

Also, the total heat conducted can be calculated using the nodes at the outside, as shown in Fig. 4.15-5. This gives $q_{II} = 3430$ W. The average value is

$$q_{av} = \frac{3340 + 3430}{2} = 3385 \text{ W per } 1.0 \text{ m deep}$$

FIGURE 4.15-5. *Drawing for calculation of total heat conduction.*

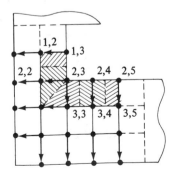

If a larger number of nodes, i.e., a smaller grid size, is used, a more accurate solution can be obtained. Using a grid size of 0.5 m instead of 1.0 m for Example 4.15-1, a q_{av} of 3250 W is obtained. If a very fine grid is used, more accuracy can be obtained but a digital computer would be needed for the large number of calculations. Matrix methods are also available for solving a set of simultaneous equations on a computer. The iteration method used here is often called the Gauss–Seidel method. Conte (C7) gives an actual subroutine to solve such a system of equations. Most computers have standard subroutines for solving these equations (G2, K1).

3. Equations for other boundary conditions.   In Example 4.15-1 the conditions at the boundaries were such that the node points were known and constant. For the case where there is convection at the boundary to a constant temperature T_∞, a heat balance on the node n,m in Fig. 4.15-6a is as follows, where heat in = heat out (K1):

$$\frac{k\,\Delta y}{\Delta x}(T_{n-1,\,m} - T_{n,m}) + \frac{k\,\Delta x}{2\,\Delta y}(T_{n,\,m+1} - T_{n,m})$$

$$+ \frac{k\,\Delta x}{2\,\Delta y}(T_{n,\,m-1} - T_{n,m}) = h\,\Delta y(T_{n,m} - T_\infty) \quad \textbf{(4.15-15)}$$

Setting $\Delta x = \Delta y$, rearranging, and setting the resultant equation $= \bar{q}_{n,m}$ residual, the following results.

(a) For convection at a boundary,

$$\frac{h\,\Delta x}{k}T_\infty + \tfrac{1}{2}(2T_{n-1,\,m} + T_{n,\,m+1} + T_{n,\,m-1}) - T_{n,m}\!\left(\frac{h\,\Delta x}{k} + 2\right) = \bar{q}_{n,m} \quad \textbf{(4.15-16)}$$

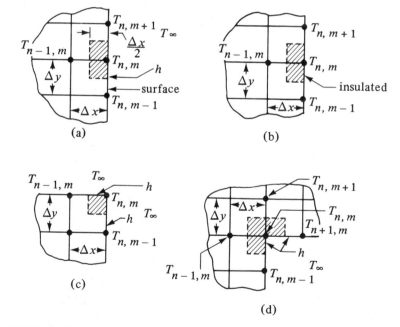

FIGURE 4.15-6.   *Other types of boundary conditions : (a) convection at a boundary,*
(b) insulated boundary, (c) exterior corner with convective boundary,
(d) interior corner with convective boundary.

In a similar manner for the cases in Fig. 4.15-6:

(b) For an insulated boundary,

$$\tfrac{1}{2}(T_{n,m+1} + T_{n,m-1}) + T_{n-1,m} - 2T_{n,m} = \bar{q}_{n,m} \qquad \text{(4.15-17)}$$

(c) For an exterior corner with convection at the boundary,

$$\frac{h\,\Delta x}{k}\,T_\infty + \tfrac{1}{2}(T_{n-1,m} + T_{n,m-1}) - \left(\frac{h\,\Delta x}{k} + 1\right)T_{n,m} = \bar{q}_{n,m} \qquad \text{(4.15-18)}$$

(d) For an interior corner with convection at the boundary,

$$\frac{h\,\Delta x}{k}\,T_\infty + T_{n-1,m} + T_{n,m+1} + \tfrac{1}{2}(T_{n+1,m} + T_{n,m-1}) - \left(3 + \frac{h\,\Delta x}{k}\right)T_{n,m} = \bar{q}_{n,m} \qquad \text{(4.15-19)}$$

For curved boundaries and other types of boundaries, see (C3, K1). To use Eqs. (4.15-16)–(4.15-19), the residual $\bar{q}_{n,m}$ is first obtained using the proper equation. Then $\bar{q}_{n,m}$ is set equal to zero and $T_{n,m}$ solved for in the resultant equation.

PROBLEMS

4.1-1. Insulation in a Cold Room. Calculate the heat loss per m² of surface area for a temporary insulating wall of a food cold storage room where the outside temperature is 299.9 K and the inside temperature 276.5 K. The wall is composed of 25.4 mm of corkboard having a k of 0.0433 W/m · K.

Ans.   39.9 W/m²

4.1-2. Determination of Thermal Conductivity. In determining the thermal conductivity of an insulating material, the temperatures were measured on both sides of a flat slab of 25 mm of the material and were 318.4 and 303.2 K. The heat flux was measured as 35.1 W/m². Calculate the thermal conductivity in btu/h · ft · °F and in W/m · K.

4.2-1. Mean Thermal Conductivity in a Cylinder. Prove that if the thermal conductivity varies linearly with temperature as in Eq. (4.1-11), the proper mean value k_m to use in the cylindrical equation is given by Eq. (4.2-3) as in a slab.

4.2-2. Heat Removal of a Cooling Coil. A cooling coil of 1.0 ft of 304 stainless steel tubing having an inside diameter of 0.25 in. and an outside diameter of 0.40 in. is being used to remove heat from a bath. The temperature at the inside surface of the tube is 40°F and 80°F on the outside. The thermal conductivity of 304 stainless steel is a function of temperature.

$$k = 7.75 + 7.78 \times 10^{-3}\,T$$

where k is in btu/h · ft · °F and T is in °F. Calculate the heat removal in btu/s and watts.

Ans.   1.225 btu/s, 1292 W

4.2-3. Removal of Heat from a Bath. Repeat Problem 4.2-2 but for a cooling coil made of 308 stainless steel having an average thermal conductivity of 15.23 W/m · K.

4.2-4. Variation of Thermal Conductivity. A flat plane of thickness Δx has one surface maintained at T_1 and the other at T_2. If the thermal conductivity varies according to temperature as

$$k = A + bT + cT^3$$

where a, b, and c are constants, derive an expression for the one-dimensional heat flux q/A.

4.2-5. Temperature Distribution in a Hollow Sphere. Derive Eq. (4.2-14) for the steady-state conduction of heat in a hollow sphere. Also, derive an equation which shows that the temperature varies hyperbolically with the radius r.

$$\text{Ans.} \quad \frac{T - T_1}{T_2 - T_1} = \frac{r_2}{r_2 - r_1}\left(1 - \frac{r_1}{r}\right)$$

4.3-1. Insulation Needed for Food Cold Storage Room. A food cold storage room is to be constructed of an inner layer of 19.1 mm of pine wood, a middle layer of cork board, and an outer layer of 50.8 mm of concrete. The inside wall surface temperature is $-17.8°C$ and the outside surface temperature is $29.4°C$ at the outer concrete surface. The mean conductivities are for pine, 0.151; cork, 0.0433; and concrete, 0.762 W/m · K. The total inside surface area of the room to use in the calculation is approximately 39 m² (neglecting corner and end effects). What thickness of cork board is needed to keep the heat loss to 586 W?

$$\text{Ans.} \quad 0.128 \text{ m thickness}$$

4.3-2. Insulation of a Furnace. A wall of a furnace 0.244 m thick is constructed of material having a thermal conductivity of 1.30 W/m · K. The wall will be insulated on the outside with material having an average k of 0.346 W/m · K, so the heat loss from the furnace will be equal to or less than 1830 W/m². The inner surface temperature is 1588 K and the outer 299 K. Calculate the thickness of insulation required.

$$\text{Ans.} \quad 0.179 \text{ m}$$

4.3-3. Heat Loss Through Thermopane Double Window. A double window called thermopane is one in which two layers of glass are used separated by a layer of dry stagnant air. In a given window, each of the glass layers is 6.35 mm thick separated by a 6.35 mm space of stagnant air. The thermal conductivity of the glass is 0.869 W/m · K and that of air is 0.026 over the temperature range used. For a temperature drop of 27.8 K over the system, calculate the heat loss for a window 0.914 m × 1.83 m. (*Note:* This calculation neglects the effect of the convective coefficient on the one outside surface of one side of the window, the convective coefficient on the other outside surface, and convection inside the window.)

4.3-4. Heat Loss from Steam Pipeline. A steel pipeline, 2-in. schedule 40 pipe, contains saturated steam at 121.1°C. The line is insulated with 25.4 mm of asbestos. Assuming that the inside surface temperature of the metal wall is at 121.1°C and the outer surface of the insulation is at 26.7°C, calculate the heat loss for 30.5 m of pipe. Also, calculate the kg of steam condensed per hour in the pipe due to the heat loss. The average k for steel from Appendix A.3 is 45 W/m · K and by linear interpolation for an average temperature of $(121.1 + 26.7)/2$ or 73.9°C, the k for asbestos is 0.182.

$$\text{Ans.} \quad 5384 \text{ W}, 8.81 \text{ kg steam/h}$$

4.3-5. Heat Loss with Trial-and-Error Solution. The exhaust duct from a heater has an inside diameter of 114.3 mm with ceramic walls 6.4 mm thick. The average $k = 1.52$ W/m · K. Outside this wall, an insulation of rock wool 102 mm thick is installed. The thermal conductivity of the rock wool is $k = 0.046 + 1.56 \times 10^{-4}T°C$(W/m · K). The inside surface temperature of the ceramic is $T_1 = 588.7$ K, and the outside surface temperature of the insulation is $T_3 = 311$ K. Calculate the heat loss for 1.5 m of duct and the interface temperature T_2 between the ceramic and the insulation. [*Hint:* The correct value of k_m for the insulation is that evaluated at the mean temperature of $(T_2 + T_3)/2$. Hence, for the first trial assume a mean temperature of, say, 448 K. Then calculate the heat loss and T_2. Using this new T_2, calculate a new mean temperature and proceed as before.]

4.3-6. Heat Loss by Convection and Conduction. A glass window with an area of 0.557 m² is installed in the wooden outside wall of a room. The wall dimensions are 2.44 × 3.05 m. The wood has a k of 0.1505 W/m·K and is 25.4 mm thick. The glass is 3.18 mm thick and has a k of 0.692. The inside room temperature is 299.9 K (26.7°C) and the outside air temperature is 266.5 K. The convection coefficient h_i on the inside wall of the glass and of the wood is estimated as 8.5 W/m²·K and the outside h_o also as 8.5 for both surfaces. Calculate the heat loss through the wooden wall, through the glass, and the total.

> **Ans.** 569.2 W (wood) (1942 btu/h); 77.6 W (glass) (265 btu/h); 646.8 W (total) (2207 btu/h)

4.3-7. Convection, Conduction, and Overall U. A gas at 450 K is flowing inside a 2-in. steel pipe, schedule 40. The pipe is insulated with 51 mm of lagging having a mean $k = 0.0623$ W/m·K. The convective heat-transfer coefficient of the gas inside the pipe is 30.7 W/m²·K and the convective coefficient on the outside of the lagging is 10.8. The air is at a temperature of 300 K.
(a) Calculate the heat loss per unit length of 1 m of pipe using resistances.
(b) Repeat using the overall U_o based on the outside area A_o.

4.3-8. Heat Transfer in Steam Heater. Water at an average of 70°F is flowing in a 2-in. steel pipe, schedule 40. Steam at 220°F is condensing on the outside of the pipe. The convective coefficient for the water inside the pipe is $h = 500$ btu/h·ft²·°F and the condensing steam coefficient on the outside is $h = 1500$.
(a) Calculate the heat loss per unit length of 1 ft of pipe using resistances.
(b) Repeat using the overall U_i based on the inside area A_i.
(c) Repeat using U_o.

> **Ans.** (a) $q = 26\,710$ btu/h (7.828 kW),
> (b) $U_i = 329.1$ btu/h·ft²·°F (1869 W/m²·K),
> (c) $U_o = 286.4$ btu/h·ft²·°F (1626 W/m²·K)

4.3-9. Heat Loss from Temperature Measurements. A steel pipe carrying steam has an outside diameter of 89 mm. It is lagged with 76 mm of insulation having an average $k = 0.043$ W/m·K. Two thermocouples, one located at the interface between the pipe wall and the insulation and the other at the outer surface of the insulation, give temperatures of 115°C and 32°C, respectively. Calculate the heat loss in W per m of pipe.

4.3-10. Effect of Convective Coefficients on Heat Loss in Double Window. Repeat Problem 4.3-3 for heat loss in the double window. However, include a convective coefficient of $h = 11.35$ W/m²·K on the one outside surface of one side of the window and an h of 11.35 on the other outside surface. Also calculate the overall U.

> **Ans.** $q = 106.7$ W, $U = 2.29$ W/m²·K

4.3-11. Uniform Chemical Heat Generation. Heat is being generated uniformly by a chemical reaction in a long cylinder of radius 91.4 mm. The generation rate is constant at 46.6 W/m³. The walls of the cylinder are cooled so that the wall temperature is held at 311.0 K. The thermal conductivity is 0.865 W/m·K. Calculate the center-line temperature at steady state.

> **Ans.** $T_o = 311.112$ K

4.3-12. Heat of Respiration of a Food Product. A fresh food product is held in cold storage at 278.0 K. It is packed in a container in the shape of a flat slab with all faces insulated except for the top flat surface, which is exposed to the air at 278.0 K. For estimation purposes the surface temperature will be assumed to be 278 K. The slab is 152.4 mm thick and the exposed surface area is 0.186 m². The density of the foodstuff is 641 kg/m³. The heat of respiration is 0.070 kJ/kg·h and the thermal conductivity is 0.346 W/m·K. Calculate the maximum temperature in the food product at steady state and the total heat given off in W. (Note: It is assumed in this problem that there is no air circulation inside the foodstuff. Hence, the results will be conservative, since circulation during respiration will reduce the temperature.)

Ans. 278.42 K, 0.353 W (1.22 btu/h)

4.3-13. *Temperature Rise in Heating Wire*. A current of 250 A is passing through a stainless steel wire having a diameter of 5.08 mm. The wire is 2.44 m long and has a resistance of 0.0843 Ω. The outer surface is held constant at 427.6 K. The thermal conductivity is $k = 22.5$ W/m · K. Calculate the center-line temperature at steady state.

4.3-14. *Critical Radius for Insulation*. A metal steam pipe having an outside diameter of 30 mm has a surface temperature of 400 K and is to be insulated with an insulation having a thickness of 20 mm and a k of 0.08 W/m · K. The pipe is exposed to air at 300 K and a convection coefficient of 30 W/m² · K.
(a) Calculate the critical radius and the heat loss per m of length for the bare pipe.
(b) Calculate the heat loss for the insulated pipe assuming that the surface temperature of the pipe remains constant.

Ans. (b) $q = 54.4$ W

4.4-1. *Curvilinear-Squares Graphical Method*. Repeat Example 4.4-1 but with the following changes.
(a) Select the number of equal temperature subdivisions between the isothermal boundaries to be five instead of four. Draw in the curvilinear squares and determine the total heat flux. Also calculate the shape factor S. Label each isotherm with the actual temperature.
(b) Repeat part (a), but in this case the thermal conductivity is not constant but $k = 0.85 (1 + 0.00040T)$, where T is temperature in K. [*Note:* To calculate the overall q, the mean value of k at the mean temperature is used. The spacing of the isotherms is independent of how k varies with T (M1). However, the temperatures corresponding to the individual isotherms are a function of how the value of k depends upon T. Write the equation for q' for a given curvilinear section using the mean value of k over the temperature interval. Equate this to the overall value of q divided by M or q/M. Then solve for the isotherm temperature.]

4.4-2. *Heat Loss from a Furnace*. A rectangular furnace with inside dimensions of $1.0 \times 1.0 \times 2.0$ m has a wall thickness of 0.20 m. The k of the walls is 0.95 W/m · K. The inside of the furnace is held at 800 K and the outside at 350 K. Calculate the total heat loss from the furnace.

Ans. $q = 25081$ W

4.4-3. *Heat Loss from a Buried Pipe*. A water pipe whose wall temperature is 300 K has a diameter of 150 mm and a length of 10 m. It is buried horizontally in the ground at a depth of 0.40 m measured to the center line of the pipe. The ground surface temperature is 280 K and $k = 0.85$ W/m · K. Calculate the loss of heat from the pipe.

Ans. $q = 451.2$ W

4.5-1. *Heating Air by Condensing Steam*. Air is flowing through a tube having an inside diameter of 38.1 mm at a velocity of 6.71 m/s, average temperature of 449.9 K, and pressure of 138 kPa. The inside wall temperature is held constant at 204.4°C (477.6 K) by steam condensing outside the tube wall. Calculate the heat-transfer coefficient for a long tube and the heat-transfer flux.

Ans. $h = 39.27$ W/m² · K (6.91 btu/h · ft² · °F)

4.5-2. *Trial-and-Error Solution for Heating Water*. Water is flowing inside a horizontal $1\frac{1}{4}$-in. schedule 40 steel pipe at 37.8°C and a velocity of 1.52 m/s. Steam at 108.3°C is condensing on the outside of the pipe wall and the steam coefficient is assumed constant at 9100 W/m² · K.

(a) Calculate the convective coefficient h_i for the water. (Note that this is trial and error. A wall temperature on the inside must be assumed first.)

(b) Calculate the overall coefficient U_i based on the inside area and the heat transfer flux q/A_i in W/m^2.

4.5-3. Heat-Transfer Area and Use of Log Mean Temperature Difference. A reaction mixture having a $c_{pm} = 2.85$ kJ/kg·K is flowing at a rate of 7260 kg/h and is to be cooled from 377.6 K to 344.3 K. Cooling water at 288.8 K is available and the flow rate is 4536 kg/h. The overall U_o is 653 W/m^2·K.

(a) For counterflow, calculate the outlet water temperature and the area A_o of the exchanger.

(b) Repeat for cocurrent flow.

Ans. (a) $T_1 = 325.2$ K, $A_o = 5.43$ m^2, (b) $A_o = 6.46$ m^2

4.5-4. Heating Water with Hot Gases and Heat-Transfer Area. A water flow rate of 13.85 kg/s is to be heated from 54.5 to 87.8°C in a heat exchanger by 54 430 kg/h of hot gas flowing counterflow and entering at 427°C ($c_{pm} = 1.005$ kJ/kg·K). The overall $U_o = 69.1$ W/m^2·K. Calculate the exit-gas temperature and the heat-transfer area.

Ans. $T = 299.5$°C

4.5-5. Cooling Oil and Overall U. Oil flowing at the rate of 7258 kg/h with a $c_{pm} = 2.01$ kJ/kg·K is cooled from 394.3 K to 338.9 K in a counterflow heat exchanger by water entering at 294.3 K and leaving at 305.4 K. Calculate the flow rate of the water and the overall U_i if the A_i is 5.11 m^2.

Ans. 17 420 kg/h, $U_i = 686$ W/m^2·K

4.5-6. Laminar Flow and Heating of Oil. A hydrocarbon oil having the same physical properties as the oil in Example 4.5-5 enters at 175°F inside a pipe having an inside diameter of 0.0303 ft and a length of 15 ft. The inside pipe surface temperature is constant at 325°F. The oil is to be heated to 250°F in the pipe. How many lb$_m$/h oil can be heated? (*Hint*: This solution is trial and error. One method is to assume a flow rate of say $m = 75$ lb mass/h. Calculate the N_{Re} and the value of h_a. Then make a heat balance to solve for q in terms of m. Equate this q to the q from the equation $q = h_a A \Delta T_a$. Solve for m. This is the new m to use for the second trial.)

Ans. $m = 84.2$ lb$_m$/h (38.2 kg/h)

4.5-7. Heating Air by Condensing Steam. Air at a pressure of 101.3 kPa and 288.8 K enters inside a tube having an inside diameter of 12.7 mm and a length of 1.52 m with a velocity of 24.4 m/s. Condensing steam on the outside of the tube maintains the inside wall temperature at 372.1 K. Calculate the convection coefficient of the air. (*Note*: This solution is trial and error. First assume an outlet temperature of the air.)

4.5-8. Heat Transfer with a Liquid Metal. The liquid metal bismuth at a flow rate of 2.00 kg/s enters a tube having an inside diameter of 35 mm at 425°C and is heated to 430°C in the tube. The tube wall is maintained at a temperature of 25°C above the liquid bulk temperature. Calculate the tube length required. The physical properties are as follows (H1): $k = 15.6$ W/m·K, $c_p = 149$ J/kg·K, $\mu = 1.34 \times 10^{-3}$ Pa·s.

4.6-1. Heat Transfer from a Flat Plate. Air at a pressure of 101.3 kPa and a temperature of 288.8 K is flowing over a thin, smooth flat plate at 3.05 m/s. The plate length in the direction of flow is 0.305 m and is at 333.2 K. Calculate the heat-transfer coefficient assuming laminar flow.

Ans. $h = 12.35$ W/m^2·K (2.18 btu/h·ft^2·°F)

4.6-2. Chilling Frozen Meat. Cold air at -28.9°C and 1 atm is recirculated at a velocity of 0.61 m/s over the exposed top flat surface of a piece of frozen meat. The sides and bottom of this rectangular slab of meat are insulated and the top surface is 254 mm by 254 mm square. If the surface of the meat is at -6.7°C,

predict the average heat-transfer coefficient to the surface. As an approximation, assume that either Eq. (4.6-2) or (4.6-3) can be used, depending on the $N_{Re\,L}$.
Ans. $h = 6.05\ \text{W/m}^2 \cdot \text{K}$

4.6-3. *Heat Transfer to an Apple.* It is desired to predict the heat-transfer coefficient for air being blown by an apple lying on a screen with large openings. The air velocity is 0.61 m/s at 101.32 kPa pressure and 316.5 K. The surface of the apple is at 277.6 K and its average diameter is 114 mm. Assume that it is a sphere.

4.6-4. *Heating Air by a Steam Heater.* A total of 13 610 kg/h of air at 1 atm abs pressure and 15.6°C is to be heated by passing over a bank of tubes in which steam at 100°C is condensing. The tubes are 12.7 mm OD, 0.61 m long, and arranged in-line in a square pattern with $S_p = S_n = 19.05$ mm. The bank of tubes contains 6 transverse rows in the direction of flow and 19 rows normal to the flow. Assume that the tube surface temperature is constant at 93.33°C. Calculate the outlet air temperature.

4.7-1. *Natural Convection from an Oven Wall.* The oven wall in Example 4.7-1 is insulated so that the surface temperature is 366.5 K instead of 505.4 K. Calculate the natural convection heat-transfer coefficient and the heat-transfer rate per m of width. Use both Eq. (4.7-4) and the simplified equation. (*Note:* Radiation is being neglected in this calculation.) Use both SI and English units.

4.7-2. *Losses by Natural Convection from a Cylinder.* A vertical cylinder 76.2 mm in diameter and 121.9 mm high is maintained at 397.1 K at its surface. It loses heat by natural convection to air at 294.3 K. Heat is lost from the cylindrical side and the flat circular end at the top. Calculate the heat loss neglecting radiation losses. Use the simplified equations of Table 4.7-2 and those equations for the lowest range of $N_{Gr} N_{Pr}$. The equivalent L to use for the top flat surface is 0.9 times the diameter.
Ans. $q = 26.0\ \text{W}$

4.7-3. *Heat Loss from a Horizontal Tube.* A horizontal tube carrying hot water has a surface temperature of 355.4 K and an outside diameter of 25.4 mm. The tube is exposed to room air at 294.3 K. What is the natural convection heat loss for a 1-m length of pipe?

4.7-4. *Natural Convection Cooling of an Orange.* An orange 102 mm in diameter having a surface temperature of 21.1°C is placed on an open shelf in a refrigerator held at 4.4°C. Calculate the heat loss by natural convection, neglecting radiation. As an approximation, the simplified equation for vertical planes can be used with L replaced by the radius of the sphere (M1). For a more accurate correlation, see (S2).

4.7-5. *Natural Convection in Enclosed Horizontal Space.* Repeat Example 4.7-3 but for the case where the two plates are horizontal and the bottom plate is hotter than the upper plate. Compare the results.
Ans. $q = 12.54\ \text{W}$

4.7-6. *Natural Convection Heat Loss in Double Window.* A vertical double plate-glass window has an enclosed air-gap space of 10 mm. The window is 2.0 m high by 1.2 m wide. One window surface is at 25°C and the other at 10°C. Calculate the free-convection heat-transfer rate through the air gap.

4.7-7. *Natural Convection Heat Loss for Water in Vertical Plates.* Two vertical square metal plates having dimensions of 0.40 × 0.40 m are separated by a gap of 12 mm and this enclosed space is filled with water. The average surface temperature of one plate is 65.6°C and the other plate is at 37.8°C. Calculate the heat-transfer rate through this gap.

4.7-8. *Heat Loss from a Furnace.* Two horizontal metal plates having dimensions of 0.8 × 1.0 m comprise the top of a furnace and are separated by a distance of

15 mm. The lower plate is at 400°C and the upper at 100°C and air at 1 atm abs is enclosed in the gap. Calculate the heat-transfer rate between the plates.

4.8-1. *Boiling Coefficient in a Jacketed Kettle.* Predict the boiling heat-transfer coefficient for the vertical jacketed sides of the kettle given in Example 4.8-1. Then, using this coefficient for the sides and the coefficient from Example 4.8-1 for the bottom, predict the total heat transfer.

Ans. $T_w = 107.65°C, \Delta T = 7.65$ K, and h(vertical) $= 3560$ W/m² · K.

4.8-2. *Boiling Coefficient on a Horizontal Tube.* Predict the boiling heat-transfer coefficient for water under pressure boiling at 250°F for a horizontal surface of $\frac{1}{16}$-in.-thick stainless steel having a k of 9.4 btu/h · ft · °F. The heating medium on the other side of this surface is a hot fluid at 290°F having an h of 275 btu/h · ft² · °F. Use the simplified equations. Be sure and correct this h value for the effect of pressure.

4.8-3. *Condensation on a Vertical Tube.* Repeat Example 4.8-2 but for a vertical tube 1.22 m (4.0 ft) high instead of 0.305 m (1.0 ft) high. Use SI and English units.

Ans. $h = 9438$ W/m² · K, 1663 btu/h · ft² · °F; $N_{Re} = 207.2$ (laminar flow)

4.8-4. *Condensation of Steam on Vertical Tubes.* Steam at 1 atm pressure abs and 100°C is condensing on a bank of five vertical tubes each 0.305 m high and having an OD of 25.4 mm. The tubes are arranged in a bundle spaced far enough apart so that they do not interfere with each other. The surface temperature of the tubes is 97.78°C. Calculate the average heat-transfer coefficient and the total kg condensate per hour.

Ans. $h = 15\,240$ W/m² · K

4.8-5. *Condensation on a Bank of Horizontal Tubes.* Steam at 1 atm abs pressure and 100°C is condensing on a horizontal tube bank with five layers of tubes ($N = 5$) placed one below the other. Each layer has four tubes (total tubes $= 4 \times 5 = 20$) and the OD of each tube is 19.1 mm. The tubes are each 0.61 m long and the tube surface temperature is 97.78°C. Calculate the average heat-transfer coefficient and the kg condensate per second for the whole condenser. Make a sketch of the tube bank.

4.9-1. *Mean Temperature Difference in an Exchanger.* A 1–2 exchanger with one shell pass and two tube passes is used to heat a cold fluid from 37.8°C to 121.1°C by using a hot fluid entering at 315.6°C and leaving at 148.9°C. Calculate the ΔT_{lm} and the mean temperature difference ΔT_m in K.

Ans. $\Delta T_{lm} = 148.9$ K; $\Delta T_m = 131.8$ K

4.9-2. *Cooling Oil by Water in an Exchanger.* Oil flowing at the rate of 5.04 kg/s ($c_{pm} = 2.09$ kJ/kg · K) is cooled in a 1–2 heat exchanger from 366.5 K to 344.3 K by 2.02 kg/s of water entering at 283.2 K. The overall heat-transfer coefficient U_o is 340 W/m² · K. Calculate the area required. (*Hint :* A heat balance must first be made to determine the outlet water temperature.)

4.9-3. *Heat Exchange Between Oil and Water.* Water is flowing at the rate of 1.13 kg/s in a 1–2 shell-and-tube heat exchanger and is heated from 45°C to 85°C by an oil having a heat capacity of 1.95 kJ/kg · K. The oil enters at 120°C and leaves at 85°C. Calculate the area of the exchanger if the overall heat-transfer coefficient is 300 W/m² · K.

4.9-4. *Outlet Temperature and Effectiveness of an Exchanger.* Hot oil at a flow rate of 3.00 kg/s ($c_p = 1.92$ kJ/kg · K) enters an existing counterflow exchanger at 400 K and is cooled by water entering at 325 K (under pressure) and flowing at a rate of 0.70 kg/s. The overall $U = 350$ W/m² · K and $A = 12.9$ m². Calculate the heat-transfer rate and the exit oil temperature.

4.10-1. *Radiation to a Tube from a Large Enclosure.* Repeat Example 4.10-1 but use the slightly more accurate Eq. (4.10-5) with two different emissivities.

Ans. $q = -2171$ W (-7410 btu/h)

4.10-2. Baking a Loaf of Bread in an Oven. A loaf of bread having a surface temperature of 373 K is being baked in an oven whose walls and the air are at 477.4 K. The bread moves continuously through the large oven on an open chain belt conveyor. The emissivity of the bread is estimated as 0.85 and the loaf can be assumed a rectangular solid 114.3 mm high × 114.3 mm wide × 330 mm long. Calculate the radiation heat-transfer rate to the bread, assuming that it is small compared to the oven and neglecting natural convection heat transfer.

Ans. $q = 278.4$ W (950 btu/h)

4.10-3. Radiation and Convection from a Steam Pipe. A horizontal oxidized steel pipe carrying steam and having an OD of 0.1683 m has a surface temperature of 374.9 K and is exposed to air at 297.1 K in a large enclosure. Calculate the heat loss for 0.305 m of pipe from natural convection plus radiation. For the steel pipe, use an ε of 0.79.

Ans. $q = 163.3$ W (557 btu/h)

4.10-4. Radiation and Convection to a Loaf of Bread. Calculate the total heat-transfer rate to the loaf of bread in Problem 4.10-2, including the radiation plus natural convection heat transfer. For radiation first calculate a value of h_r. For natural convection, use the simplified equations for the lower $N_{Gr}N_{Pr}$ range. For the four vertical sides, the equation for vertical planes can be used with an L of 114.3 mm. For the top surface, use the equation for a cooled plate facing upward and for the bottom, a cooled plate facing downward. The characteristic L for a horizontal rectangular plate is the linear mean of the two dimensions.

4.10-5. Heat Loss from a Pipe. A bare stainless steel tube having an outside diameter of 76.2 mm and an ε of 0.55 is placed horizontally in air at 294.2 K. The pipe surface temperature is 366.4 K. Calculate the value of $h_c + h_r$ for convection plus radiation and the heat loss for 3 m of pipe.

4.11-1. Radiation Shielding. Two very large and parallel planes each have an emissivity of 0.7. Surface 1 is at 866.5 K and surface 2 is at 588.8 K. Use SI and English units.

(a) What is the net radiation loss of surface 1?

(b) To reduce this loss, two additional radiation shields also having an emissivity of 0.7 are placed between the original surfaces. What is the new radiation loss?

Ans. (a) 13 565 W/m², 4300 btu/h · ft²; (b) 4521 W/m², 1433 btu/h · ft²

4.11-2. Radiation from a Craft in Space. A space satellite in the shape of a sphere is traveling in outer space, where its surface temperature is held at 283.2 K. The sphere "sees" only outer space, which can be considered as a black body with a temperature of 0 K. The polished surface of the sphere has an emissivity of 0.1. Calculate the heat loss per m² by radiation.

Ans. $q_{12}/A_1 = 36.5$ W/m²

4.11-3. Radiation and Complex View Factor. Find the view factor F_{12} for the configuration shown in Fig. P4.11-3. The areas A_4 and A_3 are fictitious areas (C3).

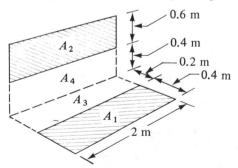

FIGURE P4.11-3. *Geometric configuration for Problem 4.11-3.*

The area $A_2 + A_4$ is called $A_{(24)}$ and $A_1 + A_3$ is called $A_{(13)}$. Areas $A_{(24)}$ and $A_{(13)}$ are perpendicular to each other. [*Hint:* Follow the methods in Example 4.11-5. First, write an equation similar to Eq. (4.11-48) which relates the interchange between A_3 and $A_{(24)}$. Then relate the interchange between $A_{(13)}$ and $A_{(24)}$. Finally, relate $A_{(13)}$ and A_4.]

Ans. $A_1 F_{12} = A_{(13)} F_{(13)(24)} + A_3 F_{34} - A_3 F_{3(24)} - A_{(13)} F_{(13)4}$

4.11-4. *Radiation Between Parallel Surfaces.* Two parallel surfaces each 1.83×1.83 m square are spaced 0.91 m apart. The surface temperature of A_1 is 811 K and that of A_2 is 533 K. Both are black surfaces.
   (a) Calculate the radiant heat transfer between the two surfaces.
   (b) Do the same as for part (a), but for the case where the two surfaces are connected by nonconducting reradiating walls.
   (c) Repeat part (b), but A_1 has an emissivity of 0.8 and A_2 an emissivity of 0.7.

4.11-5. *Radiation Between Adjacent Perpendicular Plates.* Two adjacent rectangles are perpendicular to each other. The first rectangle is 1.52×2.44 m and the second is 1.83×2.44 m with the 2.44-m side common to both. The temperature of the first surface is 699 K and that of the second is 478 K. Both surfaces are black. Calculate the radiant heat transfer between the two surfaces.

4.11-6. *View Factor for Complex Geometry.* Using the dimensions given in Fig. P4.11-3, calculate the individual view factors and also F_{12}.

4.11-7. *Radiation from a Surface to the Sky.* A plane surface having an area of 1.0 m^2 is insulated on the bottom side and is placed on the ground exposed to the atmosphere at night. The upper surface is exposed to air at 290 K and the convective heat-transfer coefficient from the air to the plane is 12 W/m$^2 \cdot$ K. The plane radiates to the clear sky. The effective radiation temperature of the sky can be assumed as 80 K. If the plane is a black body, calculate the temperature of the plane at equilibrium.

Ans. $T = 266.5$ K $= -6.7°$C

4.11-8. *Radiation and Heating of Planes.* Two plane disks each 1.25 m in diameter are parallel and directly opposed to each other. They are separated by a distance of 0.5 m. Disk 1 is heated by electrical resistance to 833.3 K. Both disks are insulated on all faces except the two faces directly opposed to each other. Assume that the surroundings emit no radiation and that the disks are in space. Calculate the temperature of disk 2 at steady state and also the electrical energy input to disk 1. (*Hint:* The fraction of heat lost from area number 1 to space is $1 - F_{12}$.)

Ans. $F_{12} = 0.45$, $T_2 = 682.5$ K

4.11-9. *Radiation by Disks to Each Other and to Surroundings.* Two disks each 2.0 m in diameter are parallel and directly opposite each other and separated by a distance of 2.0 m. Disk 1 is held at 1000 K by electric heating and disk 2 at 400 K by cooling water in a jacket at the rear of the disk. The disks radiate only to each other and to the surrounding space at 300 K. Calculate the electric heat input and also the heat removed by the cooling water.

4.11-10. *View Factor by Integration.* A small black disk is vertical having an area of 0.002 m^2 and radiates to a vertical black plane surface that is 0.03 m wide and 2.0 m high and is opposite and parallel to the small disk. The disk source is 2.0 m away from the vertical plane and placed opposite the bottom of the plane. Determine F_{12} by integration of the view-factor equation.

Ans. $F_{12} = 0.00307$

4.11-11. *Gas Radiation to Gray Enclosure.* Repeat Example 4.11-7 but with the following changes
   (a) The interior walls are not black surfaces but gray surfaces with an emissivity of 0.75.

(b) The same conditions as part (a) with gray walls, but in addition heat is transferred by natural convection to the interior walls. Assume an average convective coefficient of 8.0 W/m$^2 \cdot$ K.

Ans. (b) q(convection + radiation) = 4.426 W

4.11-12. *Gas Radiation and Convection to a Stack.* A furnace discharges hot flue gas at 1000 K and 1 atm abs pressure containing 5% CO_2 into a stack having an inside diameter of 0.50 m. The inside walls of the refractory lining are at 900 K and the emissivity of the lining is 0.75. The convective heat-transfer coefficient of the gas has been estimated as 10 W/m$^2 \cdot$ K. Calculate the rate of heat transfer q/A from the gas by radiation plus convection.

4.12-1. *Laminar Heat Transfer of a Power-Law Fluid.* A non-Newtonian power-law fluid banana purée flowing at a rate of 300 lb$_m$/h inside a 1.0-in.-ID tube is being heated by a hot fluid flowing outside the tube. The banana purée enters the heating section of the tube, which is 5 ft long, at a temperature of 60°F. The inside wall temperature is constant at 180°F. The fluid properties as given by Charm (C1) are $\rho = 69.9$ lb$_m$/ft^3, $c_p = 0.875$ btu/lb$_m \cdot$ °F, and $k = 0.320$ btu/h $\cdot$ ft $\cdot$ °F. The fluid has the following rheological constants: $n = n' = 0.458$, which can be assumed constant and $K = 0.146$ lb$_f \cdot$ s$^n \cdot$ ft^{-2} at 70°F and 0.0417 at 190°F. A plot of log K versus T°F can be assumed to be a straight line. Calculate the outlet bulk temperature of the fluid in laminar flow.

4.12-2. *Heating a Power-Law Fluid in Laminar Flow.* A non-Newtonian power-law fluid having the same physical properties and rheological constants as the fluid in Example 4.12-1 is flowing in laminar flow at a rate of 6.30×10^{-2} kg/s inside a 25.4 mm-ID tube. It is being heated by a hot fluid outside the tube. The fluid enters the heating section of the tube at 26.7°C and leaves the heating section at an outlet bulk temperature of 46.1°C. The inside wall temperature is constant at 82.2°C. Calculate the length of tube needed in m. (*Note:* In this case the unknown tube length L appears in the equation for h_a and in the heat-balance equation.)

Ans. $L = 1.722$ m

4.13-1. *Heat Transfer in a Jacketed Vessel with a Paddle Agitator.* A vessel with a paddle agitator and no baffles is used to heat a liquid at 37.8°C in this vessel. A steam-heated jacket furnishes the heat. The vessel inside diameter is 1.22 m and the agitator diameter is 0.406 m and is rotating at 150 rpm. The wall surface temperature is 93.3°C. The physical properties of the liquid are $\rho = 977$ kg/m^3, $c_p = 2.72$ kJ/kg$\cdot$K, $k = 0.346$ W/m$\cdot$K, and $\mu = 0.100$ kg/m$\cdot$s at 37.8°C and 7.5×10^{-3} at 93.3°C. Calculate the heat-transfer coefficient to the wall of the jacket.

4.13-2. *Heat Loss from Circular Fins.* Use the same data and conditions from Example 4.13-2 and calculate the fin efficiency and rate of heat loss from the following different fin materials.
(a) Carbon steel ($k = 44$ W/m $\cdot$ K).
(b) Stainless steel ($k = 17.9$ W/m $\cdot$ K).

Ans. (a) $\eta_f = 0.66, q = 111.1$ W

4.13-3. *Heat Loss from Longitudinal Fin.* A longitudinal aluminum fin as shown in Fig. 4.13-3a ($k = 230$ W/m$\cdot$K) is attached to a copper tube having an outside radius of 0.04 m. The length of the fin is 0.080 m and the thickness is 3 mm. The tube base is held at 450 K and the external surrounding air at 300 K has a convective coefficient of 25 W/m$^2 \cdot$ K. Calculate the fin efficiency and the heat loss from the fin per 1.0 m of length.

4.13-4. *Heat Transfer is Finned Tube Exchanger.* Air at an average temperature of 50°C is being heated by flowing outside a steel tube ($k = 45.1$ W/m$\cdot$K) having an inside diameter of 35 mm and a wall thickness of 3 mm. The outside of the tube is covered with 16 longitudinal steel fins with a length $L = 13$ mm and a thickness of $t = 1.0$ mm. Condensing steam inside at 120°C has a coefficient of 7000 W/m² · K. The outside coefficient of the air has been estimated as 30 W/m² · K. Neglecting fouling factors and using a tube 1.0 m long, calculate the overall heat-transfer coefficient U_i based on the inside area A_i.

4.14-1. *Dimensional Analysis for Natural Convection.* Repeat the dimensional analysis for natural convection heat transfer to a vertical plate as given in Section 4.14. However, do as follows.
(a) Carry out all the detailed steps solving for all the exponents in the π's.
(b) Repeat, but in this case select the four variables L, μ, c_p, and g to be common to all the dimensionless groups.

4.14-2. *Dimensional Analysis for Unsteady-State Conduction.* For unsteady-state conduction in a solid the following variables are involved: ρ, c_p, L (dimension of solid), t, k, and z (location in solid). Determine the dimensionless groups relating the variables.

$$\text{Ans.} \quad \pi_1 = \frac{kt}{\rho c_p L^2}, \quad \pi_2 = \frac{z}{L}$$

4.15-1. *Temperatures in a Semi-Infinite Plate.* A semi-infinite plate is similar to that in Fig. 4.15-2. At the surfaces $x = 0$ and $x = L$, the temperature is held constant at 200 K. At the surface $y = 0$, the temperature is held at 400 K. If $L = 1.0$ m, calculate the temperature at the point $y = 0.5$ m and $x = 0.5$ m at steady state.

4.15-2. *Heat Conduction in a Two-Dimensional Solid.* For two-dimensional heat conduction as given in Example 4.15-1, derive the equation to calculate the total heat loss from the chamber per unit length using the nodes at the outside. There should be eight paths for one-fourth of the chamber. Substitute the actual temperatures into the equation and obtain the heat loss.

$$\text{Ans.} \quad q = 3426 \text{ W}$$

4.15-3. *Steady-State Heat Loss from a Rectangular Duct.* A chamber that is in the shape of a long hollow rectangular duct has outside dimensions of 3×4 m and inside dimensions of 1×2 m. The walls are 1 m thick. The inside surface temperature is constant at 800 K and the outside constant at 200 K. The $k = 1.4$ W/m · K. Calculate the steady-state heat loss per unit m of length of duct. Use a grid size of $\Delta x = \Delta y = 0.5$ m. Also, use the outside nodes to calculate the total heat conduction.

$$\text{Ans.} \quad q = 7428 \text{ W}$$

4.15-4. *Two-Dimensional Heat Conduction and Different Boundary Conditions.* A very long solid piece of material 1 by 1 m square has its top face maintained at a constant temperature of 1000 K and its left face at 200 K. The bottom face and right face are exposed to an environment at 200 K and have a convection coefficient of $h = 10$ W/m² · K. The $k = 10$ W/m · K. Use a grid size of $\Delta x = \Delta y = \frac{1}{3}$ m and calculate the steady-state temperatures of the various nodes.

4.15-5. *Nodal Point at Exterior Corner Between Insulated Surfaces.* Derive the finite-difference equation for the case of the nodal point $T_{n,m}$ at an exterior corner between insulated surfaces. The diagram is similar to Fig. 4.15-6c except that the two boundaries are insulated.

$$\text{Ans.} \quad \bar{q}_{n,m} = \tfrac{1}{2}(T_{n-1,m} + T_{n,m-1}) - T_{n,m}$$

REFERENCES

(A1) ACRIVOS, A. *A.I.Ch.E. J.*, **6**, 584 (1960).

(B1) BIRD, R. B., STEWART, W. E., and LIGHTFOOT, E. N. *Transport Phenomena.* New York: John Wiley & Sons, Inc., 1960.

(B2)   BADGER, W. L., and BANCHERO, J. T. *Introduction to Chemical Engineering.* New York: McGraw-Hill Book Company, 1955.

(B3)   BROMLEY, L. A. *Chem. Eng. Progr.*, **46**, 221 (1950).

(B4)   BOWMAN, R. A., MUELLER, A. C., and NAGLE, W. M. *Trans. A.S.M.E.*, **62**, 283 (1940).

(B5)   BROOKS, G., and SU, G. *Chem. Eng. Progr.*, **55**, 54 (1959).

(B6)   BOLANOWSKI, J. P., and LINEBERRY, D. D. *Ind. Eng. Chem.*, **44**, 657 (1952).

(C1)   CHARM, S. E. *The Fundamentals of Food Engineering*, 2nd ed. Westport, Conn.: Avi Publishing Co., Inc., 1971.

(C2)   CARSLAW, H. S., and JAEGER, J. E. *Conduction of Heat in Solids*, 2nd ed. New York: Oxford University Press, Inc., 1959.

(C3)   CHAPMAN, A. J. *Heat Transfer.* New York: Macmillan Publishing Co., Inc., 1960.

(C4)   CLAPP, R. M. *International Developments in Heat Transfer*, Part III. New York: American Society of Mechanical Engineers, 1961.

(C5)   CHILTON, T. H., DREW, T. B., and JEBENS, R. H. *Ind. Eng. Chem.*, **36**, 510 (1944).

(C6)   CARREAU, P., CHAREST, G., and CORNEILLE, J. L. *Can. J. Chem. Eng.*, **44**, 3 (1966).

(C7)   CLAUSING, A. M. *Int. J. Heat Mass Transfer*, **9**, 791 (1966).

(D1)   DUNLAP, I. R., and RUSHTON, J. H. *Chem. Eng. Progr. Symp.*, **49**(5), 137 (1953).

(F1)   FUJII, T., and IMURA, H. *Int. J. Heat Mass Transfer*, **15**, 755 (1972).

(G1)   GRIMISON, E. D. *Trans. A.S.M.E.*, **59**, 583 (1937).

(G2)   GEANKOPLIS, C. J. *Mass Transport Phenomena.* Columbus, Ohio: Ohio State University Bookstores, 1972.

(G3)   GUPTA, A. S., CHAUBE, R. B., and UPADHYAY, S. N. *Chem. Eng. Sci.*, **29**, 839 (1974).

(G4)   GLUZ, M. D., and PAVLUSHENKO, L. S. *J.Appl. Chem., U.S.S.R.*, **39**, 2323 (1966).

(G5)   GLOBE, S., and DROPKIN, D. *J. Heat Transfer*, **81**, 24 (1959).

(H1)   HOLMAN, J. P. *Heat Transfer*, 4th ed. New York: McGraw-Hill Book Company, 1976.

(J1)   JACOB, M. *Heat Transfer*, Vol. 1. New York: John Wiley & Sons, Inc., 1949.

(J2)   JACOB, M., and HAWKINS, G. *Elements of Heat Transfer*, 3rd ed. New York: John Wiley & Sons, Inc., 1957.

(J3)   JACOB, M. *Heat Transfer*, Vol. 2. New York: John Wiley & Sons, Inc., 1957.

(K1)   KREITH, F., and BLACK, W. Z. *Basic Heat Transfer.* New York: Harper & Row, Publishers, 1980.

(K2)   KEYES, F. G. *Trans. A.S.M.E.*, **73**, 590 (1951); **74**, 1303 (1952).

(K3)   KNUDSEN, J. G., and KATZ, D. L. *Fluid Dynamics and Heat Transfer.* New York: McGraw-Hill Book Company, 1958.

(K4)   KERN, D. Q. *Process Heat Transfer.* New York: McGraw-Hill Book Company, 1950.

(L1)   LUBARSKY, B., and KAUFMAN, S. J. *NACA Tech. Note No. 3336* (1955).

(M1)   MCADAMS, W. H. *Heat Transmission*, 3rd ed. New York: McGraw-Hill Book Company, 1954.

(M2)   METZNER, A. B., and GLUCK, D. F. *Chem. Eng. Sci.*, **12**, 185 (1960).

(M3)   METZNER, A. B., and FRIEND, P. S. *Ind. Eng. Chem.*, **51**, 879 (1959).

(N1)   NELSON, W. L. *Petroleum Refinery Engineering*, 4th ed. New York: McGraw-Hill Book Company, 1949.

(O1)   OLDSHUE, J. Y., and GRETTON, A. I. *Chem. Eng. Progr.*, **50**, 615 (1954).

(P1)   PERRY, R. H., and CHILTON, C. H. *Chemical Engineers' Handbook*, 5th ed. New York: McGraw-Hill Book Company, 1973.

(P2) PERRY, J. H. *Chemical Engineers' Handbook*, 4th ed. New York: McGraw-Hill Book Company, 1963.

(P3) PERRY, R. H., and GREEN, D. *Perry's Chemical Engineers' Handbook*, 6th ed. New York: McGraw-Hill Book Company, 1984.

(R1) REID, R. C., PRAUSNITZ, J. M., and SHERWOOD, T. K. *The Properties of Gases and Liquids*, 3rd ed. New York: McGraw-Hill Book Company, 1977.

(R2) ROHESENOW, W. M., and HARTNETT, J. P., eds. *Handbook of Heat Transfer*. New York: McGraw-Hill Book Company, 1973.

(S1) SIEDER, E. N., and TATE, G. E. *Ind. Eng. Chem.*, **28**, 1429 (1936).

(S2) STEINBERGER, R. L., and TREYBAL, R. E. *A.I.Ch.E. J.*, **6**, 227 (1960).

(S3) SKELLAND, A. H. P. *Non-Newtonian Flow and Heat Transfer*. New York: John Wiley & Sons, Inc., 1967.

(S4) SKELLAND, A. H. P., OLIVER, D. R., and TOOKE, S. *Brit. Chem. Eng.*, **7**(5), 346 (1962).

(U1) UHL, V. W. *Chem. Eng. Progr. Symp.*, **51**(17), 93 (1955).

(W1) WELTY, J. R., WICKS, C. E., and WILSON, R. E. *Fundamentals of Momentum, Heat and Mass Transfer,* 3rd ed. New York: John Wiley & Sons, Inc., 1984.

CHAPTER 5

*Principles of
Unsteady-State
Heat Transfer*

5.1 DERIVATION OF BASIC EQUATION

5.1A Introduction

In Chapter 4 we considered various heat-transfer systems in which the temperature at any given point and the heat flux were always constant with time, i.e., in steady state. In the present chapter we will study processes in which the temperature at any given point in the system changes with time, i.e., heat transfer is unsteady state or transient.

Before steady-state conditions can be reached in a process, some time must elapse after the heat-transfer process is initiated to allow the unsteady-state conditions to disappear. For example, in Section 4.2A we determined the heat flux through a wall at steady state. We did not consider the period during which the one side of the wall was being heated up and the temperatures were increasing.

Unsteady-state heat transfer is important because of the large number of heating and cooling problems occurring industrially. In metallurgical processes it is necessary to predict cooling and heating rates of various geometries of metals in order to predict the time required to reach certain temperatures. In food processing, such as in the canning industry, perishable canned foods are heated by immersion in steam baths or chilled by immersion in cold water. In the paper industry wood logs are immersed in steam baths before processing. In most of these processes the material is suddenly immersed into a fluid of higher or lower temperature.

5.1B Derivation of Unsteady-State Conduction Equation

To derive the equation for unsteady-state condition in one direction in a solid, we refer to Fig. 5.1-1. Heat is being conducted in the x direction in the cube Δx, Δy, Δz in size. For conduction in the x direction, we write

$$q_x = -kA \frac{\partial T}{\partial x} \tag{5.1-1}$$

The term $\partial T/\partial x$ means the partial or derivative of T with respect to x with the other variables, y, z, and time t, being held constant. Next, making a heat balance on the cube, we can write

rate of heat input + rate of heat generation = rate of heat output

+ rate of heat accumulation   (5.1-2)

The rate of heat input to the cube is

$$\text{rate of heat input} = q_{x|x} = -k(\Delta y\ \Delta z) \left.\frac{\partial T}{\partial x}\right|_x \qquad (5.1\text{-}3)$$

Also,

$$\text{rate of heat output} = q_{x|x+\Delta x} = -k(\Delta y\ \Delta z) \left.\frac{\partial T}{\partial x}\right|_{x+\Delta x} \qquad (5.1\text{-}4)$$

The rate of accumulation of heat in the volume $\Delta x\ \Delta y\ \Delta z$ in time ∂t is

$$\text{rate of heat accumulation} = (\Delta x\ \Delta y\ \Delta z)\rho c_p \frac{\partial T}{\partial t} \qquad (5.1\text{-}5)$$

The rate of heat generation in volume $\Delta x\ \Delta y\ \Delta z$ is

$$\text{rate of heat generation} = (\Delta x\ \Delta y\ \Delta z)\dot{q} \qquad (5.1\text{-}6)$$

Substituting Eqs. (5.1-3)–(5.1-6) into (5.1-2) and dividing by $\Delta x\ \Delta y\ \Delta z$,

$$\dot{q} + \frac{-k\left(\left.\dfrac{\partial T}{\partial x}\right|_x - \left.\dfrac{\partial T}{\partial x}\right|_{x+\Delta x}\right)}{\Delta x} = \rho c_p \frac{\partial T}{\partial t} \qquad (5.1\text{-}7)$$

Letting Δx approach zero, we have the second partial of T with respect to x or $\partial^2 T/\partial x^2$ on the left side. Then, rearranging,

$$\frac{\partial T}{\partial t} = \frac{k}{\rho c_p} \frac{\partial^2 T}{\partial x^2} + \frac{\dot{q}}{\rho c_p} = \alpha \frac{\partial^2 T}{\partial x^2} + \frac{\dot{q}}{\rho c_p} \qquad (5.1\text{-}8)$$

where α is $k/\rho c_p$, thermal diffusivity. This derivation assumes constant k, ρ, and c_p. In SI units, $\alpha = \text{m}^2/\text{s}$, $T = \text{K}$, $t = \text{s}$, $k = \text{W/m} \cdot \text{K}$, $\rho = \text{kg/m}^3$, $\dot{q} = \text{W/m}^3$, and $c_p = \text{J/kg} \cdot \text{K}$. In English units, $\alpha = \text{ft}^2/\text{h}$, $T = {}^\circ\text{F}$, $t = \text{h}$, $k = \text{btu/h} \cdot \text{ft} \cdot {}^\circ\text{F}$, $\rho = \text{lb}_m/\text{ft}^3$, $\dot{q} = \text{btu/h} \cdot \text{ft}^3$, and $c_p = \text{btu/lb}_m \cdot {}^\circ\text{F}$.

For conduction in three dimensions, a similar derivation gives

$$\frac{\partial T}{\partial t} = \alpha\left(\frac{\partial^2 T}{\partial x^2} + \frac{\partial^2 T}{\partial y^2} + \frac{\partial^2 T}{\partial z^2}\right) + \frac{\dot{q}}{\rho c_p} \qquad (5.1\text{-}9)$$

FIGURE 5.1-1.   *Unsteady-state conduction in one direction.*

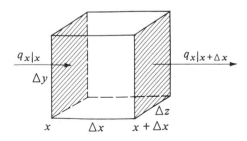

In many cases, unsteady-state heat conduction is occurring but the rate of heat generation is zero. Then Eqs. (5.1-8) and (5.1-9) become

$$\frac{\partial T}{\partial t} = \alpha \frac{\partial^2 T}{\partial x^2} \tag{5.1-10}$$

$$\frac{\partial T}{\partial t} = \alpha \left(\frac{\partial^2 T}{\partial x^2} + \frac{\partial^2 T}{\partial y^2} + \frac{\partial^2 T}{\partial z^2} \right) \tag{5.1-11}$$

Equations (5.1-10) and (5.1-11) relate the temperature T with position x, y, and z and time t. The solutions of Eqs. (5.1-10) and (5.1-11) for certain specific cases as well as for the more general cases are considered in much of the remainder of this chapter.

5.2 SIMPLIFIED CASE FOR SYSTEMS WITH NEGLIGIBLE INTERNAL RESISTANCE

5.2A Basic Equation

We begin our treatment of transient heat conduction by analyzing a simplified case. In this situation we consider a solid which has a very high thermal conductivity or very low internal conductive resistance compared to the external surface resistance, where convection occurs from the external fluid to the surface of the solid. Since the internal resistance is very small, the temperature within the solid is essentially uniform at any given time.

An example would be a small, hot cube of steel at T_0 K at time $t = 0$, suddenly immersed into a large bath of cold water at T_∞ which is held constant with time. Assume that the heat-transfer coefficient h in W/m$^2 \cdot$ K is constant with time. Making a heat balance on the solid object for a small time interval of time dt s, the heat transfer from the bath to the object must equal the change in internal energy of the object.

$$hA(T_\infty - T)\, dt = c_p \rho V\, dT \tag{5.2-1}$$

where A is the surface area of the object in m^2, T the average temperature of the object at time t in s, ρ the density of the object in kg/m^3, and V the volume in m^3. Rearranging the equation and integrating between the limits of $T = T_0$ when $t = 0$ and $T = T$ when $t = t$,

$$\int_{T=T_0}^{T=T} \frac{dT}{T_\infty - T} = \frac{hA}{c_p \rho V} \int_{t=0}^{t=t} dt \tag{5.2-2}$$

$$\frac{T - T_\infty}{T_0 - T_\infty} = e^{-(hA/c_p \rho V)t} \tag{5.2-3}$$

This equation describes the time–temperature history of the solid object. The term $c_p \rho V$ is often called the *lumped thermal capacitance* of the system. This type of analysis is often called the *lumped capacity method* or *Newtonian heating or cooling method*.

5.2B Equation for Different Geometries

In using Eq. (5.2-3) the surface/volume ratio of the object must be known. The basic assumption of negligible internal resistance was made in the derivation. This assumption is reasonably accurate when

$$N_{Bi} = \frac{hx_1}{k} < 0.1 \tag{5.2-4}$$

where hx_1/k is called the Biot number N_{Bi}, which is dimensionless, and x_1 is a characteristic dimension of the body obtained from $x_1 = V/A$. The Biot number compares the relative values of internal conduction resistance and surface convective resistance to heat transfer.

For a sphere,

$$x_1 = \frac{V}{A} = \frac{4\pi r^3/3}{4\pi r^2} = \frac{r}{3} \tag{5.2-5}$$

For a long cylinder,

$$x_1 = \frac{V}{A} = \frac{\pi D^2 L/4}{\pi D L} = \frac{D}{4} = \frac{r}{2} \tag{5.2-6}$$

For a long square rod,

$$x_1 = \frac{V}{A} = \frac{(2x)^2 L}{4(2x)L} = \frac{x}{2} \qquad (x = \tfrac{1}{2} \text{ thickness}) \tag{5.2-7}$$

EXAMPLE 5.2-1. Cooling of a Steel Ball

A steel ball having a radius of 1.0 in. (25.4 mm) is at a uniform temperature of 800°F (699.9 K). It is suddenly plunged into a medium whose temperature is held constant at 250°F (394.3 K). Assuming a convective coefficient of $h = 2.0$ btu/h · ft² · °F (11.36 W/m² · K), calculate the temperature of the ball after 1 h (3600 s). The average physical properties are $k = 25$ btu/h · ft · °F (43.3 W/m · K), $\rho = 490$ lb$_m$/ft³ (7849 kg/m³), and $c_p = 0.11$ btu/lb$_m$ · °F (0.4606 kJ/kg · K). Use SI and English units.

Solution: For a sphere from Eq. (5.2-5),

$$x_1 = \frac{V}{A} = \frac{r}{3} = \frac{\frac{1}{12}}{3} = \tfrac{1}{36} \text{ ft}$$

$$= \frac{25.4}{1000 \times 3} = 8.47 \times 10^{-3} \text{ m}$$

From Eq. (5.2-4) for the Biot number,

$$N_{Bi} = \frac{hx_1}{k} = \frac{2(\frac{1}{36})}{25} = 0.00222$$

$$N_{Bi} = \frac{11.36(8.47 \times 10^{-3})}{43.3} = 0.00222$$

This value is < 0.1; hence, the lumped capacity method can be used. Then,

$$\frac{hA}{c_p \rho V} = \frac{2}{0.11(490)(\frac{1}{36})} = 1.335 \text{ h}^{-1}$$

$$\frac{hA}{c_p \rho V} = \frac{11.36}{(0.4606 \times 1000)(7849)(8.47 \times 10^{-3})} = 3.71 \times 10^{-4} \text{ s}^{-1} \;(1.335 \text{ h}^{-1})$$

Substituting into Eq. (5.2-3) for $t = 1.0$ h and solving for T,

$$\frac{T - T_\infty}{T_0 - T_\infty} = \frac{T - 250°F}{800 - 250} = e^{-(hA/c_p\rho V)t} = e^{-(1.335)(1.0)} \qquad T = 395°F$$

$$\frac{T - 394.3 \text{ K}}{699.9 - 394.3} = e^{-(3.71 \times 10^{-4})(3600)} \qquad T = 474.9 \text{ K}$$

5.2C Total Amount of Heat Transferred

The temperature of the solid at any time t can be calculated from Eq. (5.2-3). At any time t, the instantaneous rate of heat transfer $q(t)$ in W from the solid of negligible internal resistance can be calculated from

$$q(t) = hA(T - T_\infty) \tag{5.2-8}$$

Substituting the instantaneous temperature T from Eq. (5.2-3) into Eq. (5.2-8),

$$q(t) = hA(T_0 - T_\infty)e^{-(hA/c_p\rho V)t} \tag{5.2-9}$$

To determine the total amount of heat Q in W · s or J transferred from the solid from time $t = 0$ to $t = t$, we can integrate Eq. (5.2-9).

$$Q = \int_{t=0}^{t=t} q(t)dt = \int_{t=0}^{t=t} hA(t_0 - T_\infty)e^{-(hA/c_p\rho V)t}\, dt \tag{5.2-10}$$

$$Q = c_p\rho V(T_0 - T_\infty)[1 - e^{-(hA/c_p\rho V)t}] \tag{5.2-11}$$

EXAMPLE 5.2-2. Total Amount of Heat in Cooling
For the conditions in Example 5.2-1, calculate the total amount of heat removed up to time $t = 3600$ s.

Solution: From Example 5.2-1, $hA/c_p\rho V = 3.71 \times 10^{-4}$ s^{-1}. Also, $V = 4\pi r^3/3 = 4(\pi)(0.0254)^3/3 = 6.864 \times 10^{-5}$ m^3. Substituting into Eq. (5.2-11),

$$Q = (0.4606 \times 1000)(7849)(6.864 \times 10^{-5})(699.9 - 394.3)$$

$$\cdot [1 - e^{-(3.71 \times 10^{-4})(3600)}]$$

$$= 5.589 \times 10^4 \text{ J}$$

5.3 UNSTEADY-STATE HEAT CONDUCTION IN VARIOUS GEOMETRIES

5.3A Introduction and Analytical Methods

In Section 5.2 we considered a simplified case of negligible internal resistance where the object has a very high thermal conductivity. Now we will consider the more general situation where the internal resistance is not small, and hence the temperature is not constant in the solid. The first case that we shall consider is one where the surface convective resistance is negligible compared to the internal resistance. This could occur because of a very large heat-transfer coefficient at the surface or because of a relatively large conductive resistance in the object.

To illustrate an analytical method of solving this first case, we will derive the equation for unsteady-state conduction in the x direction only in a flat plate of thickness $2H$ as shown in Fig. 5.3-1. The initial profile of the temperature in the plate at $t = 0$ is uniform at $T = T_0$. At time $t = 0$, the ambient temperature is suddenly changed to T_1 and held there. Since there is no convection resistance, the temperature of the surface is also held constant at T_1. Since this is conduction in the x direction, Eq. (5.1-10) holds.

$$\frac{\partial T}{\partial t} = \alpha \frac{\partial^2 T}{\partial x^2} \tag{5.1-10}$$

The initial and boundary conditions are

$$T = T_0, \quad t = 0, \quad x = x$$
$$T = T_1, \quad t = t, \quad x = 0 \tag{5.3-1}$$
$$T = T_1, \quad t = t, \quad x = 2H$$

Generally, it is convenient to define a dimensionless temperature Y so that it varies between 0 and 1. Hence,

$$Y = \frac{T_1 - T}{T_1 - T_0} \tag{5.3-2}$$

Substituting Eq. (5.3-2) into (5.1-10),

$$\frac{\partial Y}{\partial t} = \alpha \frac{\partial^2 Y}{\partial x^2} \tag{5.3-3}$$

Redefining the boundary and initial conditions,

$$Y = \frac{T_1 - T_0}{T_1 - T_0} = 1, \quad t = 0, \quad x = x$$
$$Y = \frac{T_1 - T_1}{T_1 - T_0} = 0, \quad t = t, \quad x = 0 \tag{5.3-4}$$
$$Y = \frac{T_1 - T_1}{T_1 - T_0} = 0, \quad t = t, \quad x = 2H$$

A convenient procedure to use to solve Eq. (5.3-3) is the method of separation of variables, which leads to a product solution

$$Y = e^{-a^2 \alpha t}(A \cos ax + B \sin ax) \tag{5.3-5}$$

where A and B are constants and a is a parameter. Applying the boundary and initial conditions of Eq. (5.3-4) to solve for these constants in Eq. (5.3-5), the final solution is an infinite Fourier series (G1).

FIGURE 5.3-1. *Unsteady-state conduction in a flat plate with negligible surface resistance.*

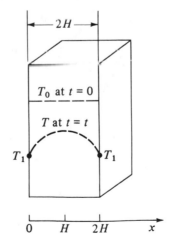

$$\frac{T_1 - T}{T_1 - T_0} = \frac{4}{\pi}\left(\frac{1}{1}\exp\frac{-1^2\pi^2\alpha t}{4H^2}\sin\frac{1\pi x}{2H} + \frac{1}{3}\exp\frac{-3^2\pi^2\alpha t}{4H^2}\sin\frac{3\pi x}{2H}\right.$$

$$\left. + \frac{1}{5}\exp\frac{-5^2\pi^2\alpha t}{4H^2}\sin\frac{5\pi x}{2H} + \cdots\right) \tag{5.3-6}$$

Hence from Eq. (5.3-6), the temperature T at any position x and time t can be determined. However, these types of equations are very time consuming to use, and convenient charts have been prepared which are discussed in Sections 5.3B, 5.3C, 5.3D, and 5.3E, where a surface resistance is present.

5.3B   Unsteady-State Conduction in a Semiinfinite Solid

In Fig. 5.3-2 a semiinfinite solid is shown that extends to ∞ in the $+x$ direction. Heat conduction occurs only in the x direction. Originally, the temperature in the solid is uniform at T_0. At time $t = 0$, the solid is suddenly exposed to or immersed in a large mass of ambient fluid at temperature T_1, which is constant. The convection coefficient h in W/m² · K or btu/h · ft² · °F is present and is constant; i.e., a surface resistance is present. Hence, the temperature T_S at the surface is not the same as T_1.

The solution of Eq. (5.1-10) for these conditions has been obtained (S1) and is

$$\frac{T - T_0}{T_1 - T_0} = 1 - Y = \text{erfc}\frac{x}{2\sqrt{\alpha t}}$$

$$- \exp\left[\frac{h\sqrt{\alpha t}}{k}\left(\frac{x}{\sqrt{\alpha t}} + \frac{h\sqrt{\alpha t}}{k}\right)\right]\text{erfc}\left(\frac{x}{2\sqrt{\alpha t}} + \frac{h}{k}\sqrt{\alpha t}\right) \tag{5.3-7}$$

where x is the distance into the solid from the surface in SI units in m, $t = $ time in s, $\alpha = k/\rho c_p$ in m²/s. In English units, $x = $ ft, $t = $ h, and $\alpha = $ ft²/h. The function erfc is $(1 - \text{erf})$, where erf is the error function and numerical values are tabulated in standard tables and texts (G1, P1, S1), Y is fraction of unaccomplished change $(T_1 - T)/(T_1 - T_0)$, and $1 - Y$ is fraction of change.

Figure 5.3-3, calculated using Eq. (5.3-7), is a convenient plot used for unsteady-state heat conduction into a semiinfinite solid with surface convection. If conduction into the solid is slow enough or h is very large, the top line with $h\sqrt{\alpha t}/k = \infty$ is used.

EXAMPLE 5.3-1.   Freezing Temperature in the Ground
The depth in the soil of the earth at which freezing temperatures penetrate is often of importance in agriculture and construction. During a certain fall day, the temperature in the earth is constant at 15.6°C (60°F) to a depth of several meters. A cold wave suddenly reduces the air temperature from 15.6 to -17.8°C (0°F). The convective coefficient above the soil is 11.36 W/m² · K (2 btu/h · ft² · °F). The soil properties can be assumed as $\alpha = 4.65$

FIGURE 5.3-2.   *Unsteady-state conduction in a semiinfinite solid.*

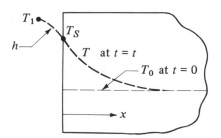

Chap. 5   Principles of Unsteady-State Heat Transfer

$\times\ 10^{-7}\ \mathrm{m^2/s}$ (0.018 ft^2/h) and $k = 0.865\ \mathrm{W/m \cdot K}$ (0.5 btu/h·ft·°F). Neglect any latent heat effects. Use SI and English units.

(a) What is the surface temperature after 5 h?

(b) To what depth in the soil will the freezing temperature of 0°C (32°F) penetrate in 5 h?

Solution: This is a case of unsteady-state conduction in a semiinfinite solid. For part (a), the value of x which is the distance from the surface is $x = 0$ m. Then the value of $x/2\sqrt{\alpha t}$ is calculated as follows for $t = 5$ h, $\alpha = 4.65 \times 10^{-7}\ \mathrm{m^2/s}$, $k = 0.865\ \mathrm{W/m \cdot °C}$, and $h = 11.36\ \mathrm{W/m^2 \cdot °C}$. Using SI and English units,

$$\frac{x}{2\sqrt{\alpha t}} = \frac{0}{2\sqrt{(4.65 \times 10^{-7})(5 \times 3600)}} \qquad \frac{x}{2\sqrt{\alpha t}} = \frac{0}{2\sqrt{0.018(5)}}$$

Also,

$$\frac{h\sqrt{\alpha t}}{k} = \frac{11.36\sqrt{(4.65 \times 10^{-7})(5 \times 3600)}}{0.865} \qquad \frac{h\sqrt{\alpha t}}{k} = \frac{2\sqrt{0.018(5)}}{0.5}$$

$$= 1.2 \qquad\qquad\qquad\qquad\qquad\qquad = 1.2$$

Using Fig. 5.3-3, for $x/2\sqrt{\alpha t} = 0$ and $h\sqrt{\alpha t}/k = 1.2$, the value of $1 - Y = 0.63$ is read off the curve. Converting temperatures to K, $T_0 = 15.6°C + 273.2 = 288.8$ K (60°F) and $T_1 = -17.8°C + 273.2 = 255.4$ K (0°F). Then

$$1 - Y = \frac{T - T_0}{T_1 - T_0} = 0.63 = \frac{T - 288.8}{255.4 - 288.8}$$

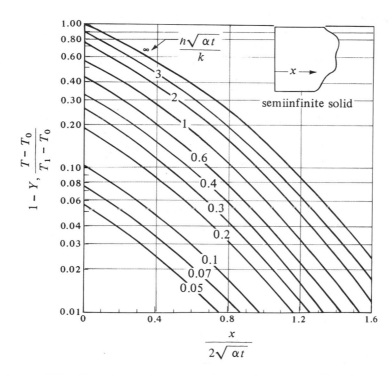

FIGURE 5.3-3. *Unsteady-state heat conducted in a semiinfinite solid with surface convection. Calculated from Eq. (5.3-7)(SI).*

Solving for T at the surface after 5 h,

$$T = 267.76 \; K \quad \text{or} \quad -5.44°C \; (22.2°F)$$

For part (b), $T = 273.2$ K or 0°C, and the distance x is unknown. Substituting the known values,

$$\frac{T - T_0}{T_1 - T_0} = \frac{273.2 - 288.8}{255.4 - 288.8} = 0.467$$

From Fig. 5.3-3 for $(T - T_0)/(T_1 - T_0) = 0.467$ and $h\sqrt{\alpha t}/k = 1.2$, a value of 0.16 is read off the curve for $x/2\sqrt{\alpha t}$. Hence,

$$\frac{x}{2\sqrt{\alpha t}} = \frac{x}{2\sqrt{(4.65 \times 10^{-7})(5 \times 3600)}} = 0.16 \qquad \frac{x}{2\sqrt{\alpha t}} = \frac{x}{2\sqrt{0.018(5)}} = 0.16$$

Solving for x, the distance the freezing temperature penetrates in 5 h,

$$x = 0.0293 \text{ m } (0.096 \text{ ft})$$

5.3C Unsteady-State Conduction in a Large Flat Plate

A geometry that often occurs in heat-conduction problems is a flat plate of thickness $2x_1$ in the x direction and having large or infinite dimensions in the y and z directions, as shown in Fig. 5.3-4. Heat is being conducted only from the two flat and parallel surfaces in the x direction. The original uniform temperature of the plate is T_0, and at time $t = 0$, the solid is exposed to an environment at temperature T_1 and unsteady-state conduction occurs. A surface resistance is present.

The numerical results of this case are presented graphically in Figs. 5.3-5 and 5.3-6. Figure 5.3-5 by Gurney and Lurie (G2) is a convenient chart for determining the temperatures at any position in the plate and at any time t. The dimensionless parameters used in these and subsequent unsteady-state charts in this section are given in Table 5.3-1 (x is the distance from the center of the flat plate, cylinder, or sphere. x_1 is one half the thickness of the flat plate radius of cylinder, or radius of sphere. $x =$ distance from the surface for a semiinfinite solid.)

When $n = 0$, the position is at the center of the plate in Fig. 5.3-5. Often the temperature history at the center of the plate is quite important. A more accurate chart for determining only the center temperature is given in Fig. 5.3-6 in the Heisler (H1) chart. Heisler (H1) has also prepared multiple charts for determining the temperatures at other positions.

EXAMPLE 5.3-2. *Heat Conduction in a Slab of Butter*
A rectangular slab of butter which is 46.2 mm thick at a temperature of 277.6 K (4.4°C) in a cooler is removed and placed in an environment at 297.1 K (23.9°C). The sides and bottom of the butter container can be

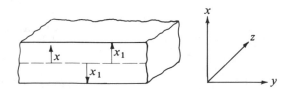

FIGURE 5.3-4. *Unsteady-state conduction in a large flat plate.*

considered to be insulated by the container side walls. The flat top surface of the butter is exposed to the environment. The convective coefficient is constant and is 8.52 W/m² · K. Calculate the temperature in the butter at the surface, at 25.4 mm below the surface, and at 46.2 mm below the surface at the insulated bottom after 5 h of exposure.

Solution: The butter can be considered as a large flat plate with conduction vertically in the x direction. Since heat is entering only at the top face and the bottom face is insulated, the 46.2 mm of butter is equivalent to a half plate with thickness $x_1 = 46.2$ mm. In a plate with two exposed surfaces as in Fig. 5.3-4, the center at $x = 0$ acts as an insulated surface and both halves are mirror images of each other.

The physical properties of butter from Appendix A.4 are $k = 0.197$ W/m · K, $c_p = 2.30$ kJ/kg · K = 2300 J/kg · K, and $\rho = 998$ kg/m³. The thermal diffusivity is

$$\alpha = \frac{k}{\rho c_p} = \frac{0.197}{998(2300)} = 8.58 \times 10^{-8} \text{ m}^2/\text{s}$$

Also, $x_1 = 46.2/1000 = 0.0462$ m.

The parameters needed for use in Fig. 5.3-5 are

$$m = \frac{k}{hx_1} = \frac{0.197}{8.52(0.0462)} = 0.50$$

$$X = \frac{\alpha t}{x_1^2} = \frac{(8.58 \times 10^{-8})(5 \times 3600)}{(0.0462)^2} = 0.72$$

For the top surface where $x = x_1 = 0.0462$ m,

$$n = \frac{x}{x_1} = \frac{0.0462}{0.0462} = 1.0$$

Then using Fig. 5.3-5,

$$Y = 0.25 = \frac{T_1 - T}{T_1 - T_0} = \frac{297.1 - T}{297.1 - 277.6}$$

Solving, $T = 292.2$ K (19.0°C).

TABLE 5.3-1. *Dimensionless Parameters for Use in Unsteady-State Conduction Charts*

$$Y = \frac{T_1 - T}{T_1 - T_0} \qquad m = \frac{k}{hx_1}$$

$$1 - Y = \frac{T - T_0}{T_1 - T_0} \qquad n = \frac{x}{x_1}$$

$$X = \frac{\alpha t}{x_1^2}$$

SI units: $\alpha = \text{m}^2/\text{s}$, $T = \text{K}$, $t = \text{s}$, $x = \text{m}$, $x_1 = \text{m}$, $k = \text{W/m} \cdot \text{K}$, $h = \text{W/m}^2 \cdot \text{K}$
English units: $\alpha = \text{ft}^2/\text{h}$, $T = °\text{F}$, $t = \text{h}$, $x = \text{ft}$, $x_1 = \text{ft}$, $k = \text{btu/h} \cdot \text{ft} \cdot °\text{F}$,
   $h = \text{btu/h} \cdot \text{ft}^2 \cdot °\text{F}$
Cgs units: $\alpha = \text{cm}^2/\text{s}$, $T = °\text{C}$, $t = \text{s}$, $x = \text{cm}$, $x_1 = \text{cm}$, $k = \text{cal/s} \cdot \text{cm} \cdot °\text{C}$,
   $h = \text{cal/s} \cdot \text{cm}^2 \cdot °\text{C}$

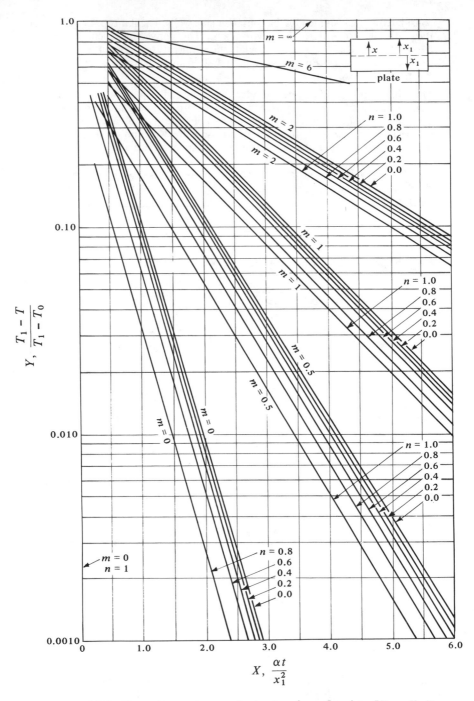

FIGURE 5.3-5. *Unsteady-state heat conduction in a large flat plate.* [*From H. P. Gurney and J. Lurie, Ind. Eng. Chem., 15, 1170 (1923).*]

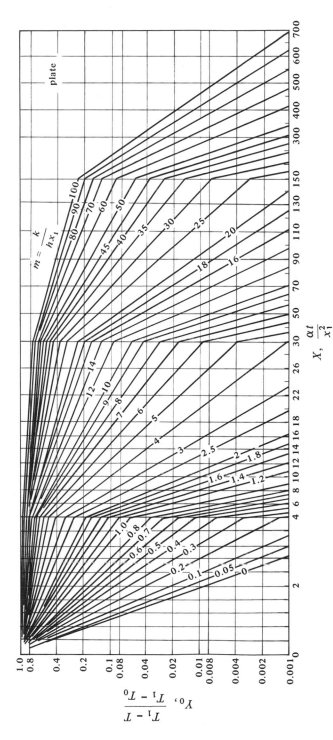

FIGURE 5.3-6.   Chart for determining temperature at the center of a large flat plate for unsteady-state heat conduction. [From H. P. Heisler, Trans. A.S.M.E., **69**, 227 (1947). With permission.]

At the point 25.4 mm from the top surface or 20.8 mm from the center, $x = 0.0208$ m, and

$$n = \frac{x}{x_1} = \frac{0.0208}{0.0462} = 0.45$$

From Fig. 5.3-5,

$$Y = 0.45 = \frac{T_1 - T}{T_1 - T_0} = \frac{297.1 - T}{297.1 - 277.6}$$

Solving, $T = 288.3$ K $(15.1°C)$.

For the bottom point or 0.0462 m from the top, $x = 0$ and

$$n = \frac{x}{x_1} = \frac{0}{x_1} = 0$$

Then, from Fig. 5.3-5,

$$Y = 0.50 = \frac{T_1 - T}{T_1 - T_0} = \frac{297.1 - T}{297.1 - 277.6}$$

Solving, $T = 287.4$ K $(14.2°C)$. Alternatively, using Fig. 5.3-6, which is only for the center point, $Y = 0.53$ and $T = 286.8$ K $(13.6°C)$.

5.3D Unsteady-State Conduction in a Long Cylinder

Here we consider unsteady-state conduction in a long cylinder where conduction occurs only in the radial direction. The cylinder is long so that conduction at the ends can be neglected or the ends are insulated. Charts for this case are presented in Fig. 5.3-7 for determining the temperatures at any position and Fig. 5.3-8 for the center temperature only.

EXAMPLE 5.3-3. *Transient Heat Conduction in a Can of Pea Purée*
A cylindrical can of pea purée (C2) has a diameter of 68.1 mm and a height of 101.6 mm and is initially at a uniform temperature of 29.4°C. The cans are stacked vertically in a retort and steam at 115.6°C is admitted. For a heating time of 0.75 h at 115.6°C, calculate the temperature at the center of the can. Assume that the can is in the center of a vertical stack of cans and that it is insulated on its two ends by the other cans. The heat capacity of the metal wall of the can will be neglected. The heat-transfer coefficient of the steam is estimated as 4540 W/m² · K. Physical properties of purée are $k = 0.830$ W/m · K and $\alpha = 2.007 \times 10^{-7}$ m²/s.

Solution: Since the can is insulated at the two ends we can consider it as a long cylinder. The radius is $x_1 = 0.0681/2 = 0.03405$ m. For the center with $x = 0$,

$$n = \frac{x}{x_1} = \frac{0}{x_1} = 0$$

Also,

$$m = \frac{k}{hx_1} = \frac{0.830}{4540(0.03405)} = 0.00537$$

$$X = \frac{\alpha t}{x_1^2} = \frac{(2.007 \times 10^{-7})(0.75 \times 3600)}{(0.03405)^2} = 0.468$$

Using Fig. 5.3-8 by Heisler for the center temperature,

$$Y = 0.13 = \frac{T_1 - T}{T_1 - T_0} = \frac{115.6 - T}{115.6 - 29.4}$$

Solving, $T = 104.4°C$.

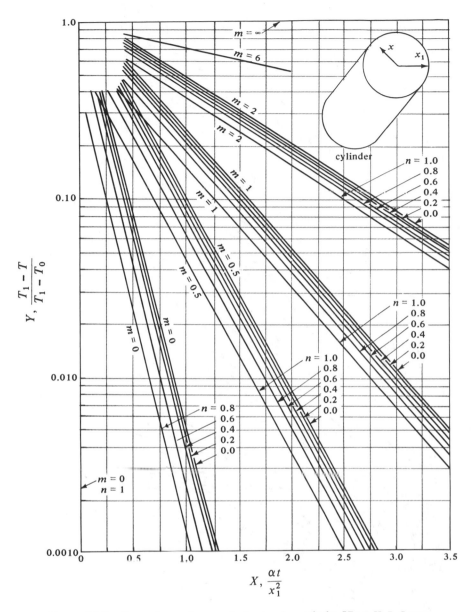

$$Y, \frac{T_1 - T}{T_1 - T_0}$$

$$X, \frac{\alpha t}{x_1^2}$$

FIGURE 5.3-7.   *Unsteady-state heat conduction in a long cylinder. [From H. P. Gurney and J. Lurie, Ind. Eng. Chem., 15, 1170 (1923).]*

5.3E   Unsteady-State Conduction in a Sphere

In Fig. 5.3-9 a chart is given by Gurney and Lurie for determining the temperatures at any position in a sphere. In Fig. 5.3-10 a chart by Heisler is given for determining the center temperature only in a sphere.

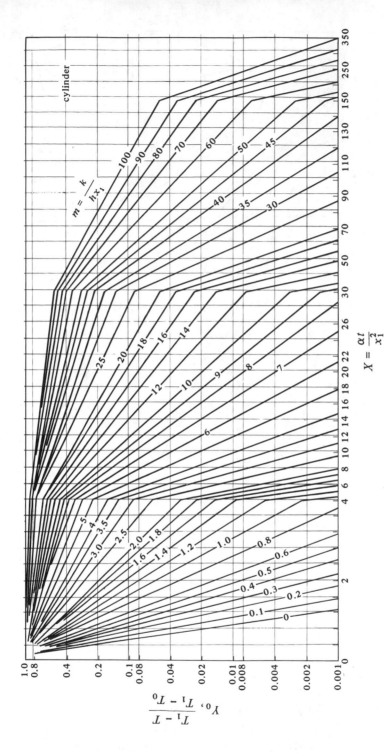

FIGURE 5.3-8. *Chart for determining temperature at the center of a long cylinder for unsteady-state heat conduction.* [*From H. P. Heisler, Trans. A.S.M.E.,* **69**, *227 (1947). With permission.*]

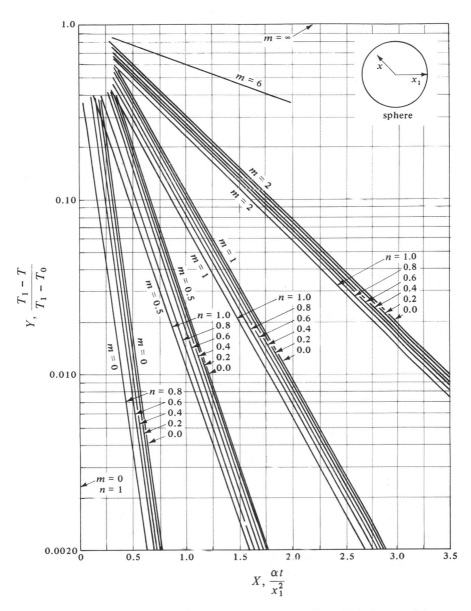

FIGURE 5.3-9. *Unsteady-state heat conduction in a sphere.* [*From H. P. Gurney and J. Lurie, Ind. Eng. Chem.,* **15**, *1170 (1923).*]

5.3F Unsteady-State Conduction in Two- and Three-Dimensional Systems

The heat-conduction problems considered so far have been limited to one dimension. However, many practical problems are involved with simultaneous unsteady-state conduction in two and three directions. We shall illustrate how to combine one-dimensional solutions to yield solutions for several-dimensional systems.

Newman (N1) used the principle of superposition and showed mathematically how

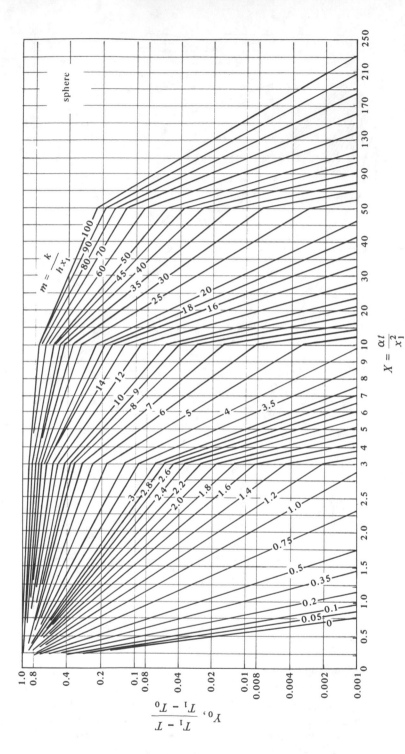

FIGURE 5.3-10. *Chart for determining the temperature at the center of a sphere for unsteady-state heat conduction. [From H. P. Heisler, Trans. A.S.M.E., **69**, 227 (1947). With permission.]*

FIGURE 5.3-11. *Unsteady-state conduction in three directions in a rectangular block.*

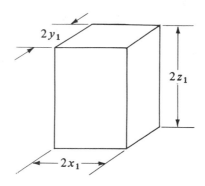

to combine the solutions for one-dimensional heat conduction in the x, the y, and the z direction into an overall solution for simultaneous conduction in all three directions. For example, a rectangular block with dimensions $2x_1$, $2y_1$, and $2z_1$ is shown in Fig. 5.3-11. For the Y value in the x direction, as before,

$$Y_x = \frac{T_1 - T_x}{T_1 - T_0} \tag{5.3-8}$$

where T_x is the temperature at time t and position x distance from the center line, as before. Also, $n = x/x_1$, $m = k/hx_1$, and $X_x = \alpha t/x_1^2$, as before. Then for the y direction,

$$Y_y = \frac{T_1 - T_y}{T_1 - T_0} \tag{5.3-9}$$

and $n = y/y_1$, $m = k/hy_1$ and $X_y = \alpha t/y_1^2$. Similarly, for the z direction,

$$Y_z = \frac{T_1 - T_z}{T_1 - T_0} \tag{5.3-10}$$

Then, for the simultaneous transfer in all three directions,

$$Y_{x, y, z} = (Y_x)(Y_y)(Y_z) = \frac{T_1 - T_{x, y, z}}{T_1 - T_0} \tag{5.3-11}$$

where $T_{x, y, z}$ is the temperature at the point x, y, z from the center of the rectangular block. The value of Y_x for the two parallel faces is obtained from Figs. 5.3-5 and 5.3-6 for conduction in a flat plate. The values of Y_y and Y_z are similarly obtained from the same charts.

For a short cylinder with radius x_1 and length $2y_1$, the following procedure is followed. First Y_x for the radial conduction is obtained from the figures for a long cylinder. Then Y_y for conduction between two parallel planes is obtained from Fig. 5.3-5 or 5.3-6 for conduction in a flat plate. Then,

$$Y_{x, y} = (Y_x)(Y_y) = \frac{T_1 - T_{x, y}}{T_1 - T_0} \tag{5.3-12}$$

EXAMPLE 5.3-4. Two-Dimensional Conduction in a Short Cylinder
Repeat Example 5.3-3 for transient conduction in a can of pea purée but assume that conduction also occurs from the two flat ends.

Solution: The can, which has a diameter of 68.1 mm and a height of 101.6 mm, is shown in Fig. 5.3-12. The given values from Example 5.3-3 are

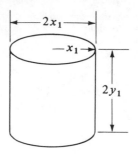

FIGURE 5.3-12. *Two dimensional conduction in a short cylinder in Example 5.3-4.*

$x_1 = 0.03405$ m, $y_1 = 0.1016/2 = 0.0508$ m, $k = 0.830$ W/m·K, $\alpha = 2.007 \times 10^{-7}$ m²/s, $h = 4540$ W/m²·K and $t = 0.75(3600) = 2700$ s.

For conduction in the x (radial) direction as calculated previously,

$$n = \frac{x}{x_1} = \frac{0}{x_1} = 0, \qquad m = \frac{k}{hx_1} = \frac{0.830}{4540(0.03405)} = 0.00537$$

$$X = \frac{\alpha t}{x_1^2} = \frac{(2.007 \times 10^{-7})2700}{(0.03405)^2} = 0.468$$

From Fig. 5.3-8 for the center temperature,

$$Y_x = 0.13$$

For conduction in the y (axial) direction for the center temperature,

$$n = \frac{y}{y_1} = \frac{0}{0.0508} = 0$$

$$m = \frac{k}{hy_1} = \frac{0.830}{4540(0.0508)} = 0.00360$$

$$X = \frac{\alpha t}{y_1^2} = \frac{(2.007 \times 10^{-7})2700}{(0.0508)^2} = 0.210$$

Using Fig. 5.3-6 for the center of a large plate (two parallel opposed planes),

$$Y_y = 0.80$$

Substituting into Eq. (5.3-12),

$$Y_{x, y} = (Y_x)(Y_y) = 0.13(0.80) = 0.104$$

Then,

$$\frac{T_1 - T_{x, y}}{T_1 - T_0} = \frac{115.6 - T_{x, y}}{115.6 - 29.4} = 0.104$$

$$T_{x, y} = 106.6°C$$

This compares with 104.4°C obtained in Example 5.3-3 for only radial conduction.

5.3G Charts for Average Temperature in a Plate, Cylinder, and Sphere with Negligible Surface Resistance

If the surface resistance is negligible, the curves given in Fig. 5.3-13 will give the total fraction of unaccomplished change, E, for slabs, cylinders, or spheres for unsteady-state conduction. The value of E is

$$E = \frac{T_1 - T_{av}}{T_1 - T_0} \tag{5.3-13}$$

where T_0 is the original uniform temperature, T_1 is the environment temperature to which the solid is suddenly subjected, and T_{av} is the average temperature of the solid after t hours.

The values of E_a, E_b, and E_c are each used for conduction between a pair of parallel faces as in a plate. For example, for conduction in the a and b directions in a rectangular bar,

$$E = F_a E_b \tag{5.3-14}$$

For conduction from all three sets of faces,

$$E = E_a E_b E_c \tag{5.3-15}$$

For conduction in a short cylinder $2c$ long and radius a,

$$E = E_c E_r \tag{5.3-16}$$

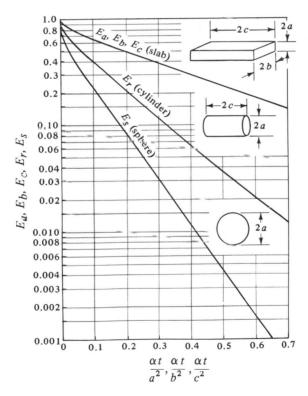

FIGURE 5.3-13. *Unsteady-state conduction and average temperatures for negligible surface resistance. (From R. E. Treybal, Mass Transfer Operations, 2nd ed. New York: McGraw-Hill Book Company, 1968. With permission.)*

5.4 NUMERICAL FINITE-DIFFERENCE METHODS FOR UNSTEADY-STATE CONDUCTION

5.4A Unsteady-State Conduction in a Slab

1. Introduction. As discussed in previous sections of this chapter, the partial differential equations for unsteady-state conduction in various simple geometries can be solved analytically if the boundary conditions are constant at $T = T_1$ with time. Also, in the solutions the initial profile of the temperature at $t = 0$ is uniform at $T = T_0$. The unsteady-state charts used also have these same boundary conditions and initial condition. However, when the boundary conditions are not constant with time and/or the initial conditions are not constant with position, numerical methods must be used.

Numerical calculation methods for unsteady-state heat conduction are similar to numerical methods for steady state discussed in Section 4.15. The solid is subdivided into sections or slabs of equal length and a fictitious node is placed at the center of each section. Then a heat balance is made for each node. This method differs from the steady-state method in that we have heat accumulation in a node for unsteady-state conduction.

2. Equations for a slab. The unsteady-state equation for conduction in the x direction in a slab is

$$\frac{\partial T}{\partial t} = \alpha \frac{\partial^2 T}{\partial x^2} \tag{5.1-10}$$

This can be set up for a numerical solution by expressing each partial derivative as an actual finite difference in ΔT, Δt, and Δx. However, an alternative method will be used to derive the final result by making a heat balance. Figure 5.4-1 shows a slab centered at position n, represented by the shaded area. The slab has a width of Δx m and a cross-sectional area of A m^2. The node at position n having a temperature of T_n is placed at the center of the shaded section and this node represents the total mass and heat capacity of the section or slab. Each node is imagined to be connected to the adjacent node by a fictitious, small conducting rod. (See Fig. 4.15-3 for an example.)

The figure shows the temperature profile at a given instant of time t s. Making a heat balance on this node or slab, the rate of heat in − the rate of heat out = the rate of heat accumulation in Δt s.

$$\frac{kA}{\Delta x}(_tT_{n-1} - {_tT_n}) - \frac{kA}{\Delta x}(_tT_n - {_tT_{n+1}}) = \frac{(A\,\Delta x)\rho c_p}{\Delta t}(_{t+\Delta t}T_n - {_tT_n}) \tag{5.4-1}$$

where $_tT_n$ is the temperature at point n at time t and $_{t+\Delta t}T_n$ is the temperature at point n at time $t + 1$ Δt later. Rearranging and solving for $_{t+\Delta t}T_n$,

$$_{t+\Delta t}T_n = \frac{1}{M}\left[_tT_{n+1} + (M - 2)_tT_n + {_tT_{n-1}}\right] \tag{5.4-2}$$

where

$$M = \frac{(\Delta x)^2}{\alpha\,\Delta t} \tag{5.4-3}$$

Note that in Eq. (5.4-2) the temperature $_{t+\Delta t}T_n$ at position or node n and at a new time $t + \Delta t$ is calculated from the three points which are known at time t, the starting time. This is called the *explicit method*, because the temperature at a new time can be calculated explicitly from the temperatures at the previous time. In this method the calculation proceeds directly from one time increment to the next until the final temperature distribution is calculated at the desired final time. Of course, the temperature distribution at the initial time and the boundary conditions must be known.

Once the value of Δx has been selected, then from Eq. (5.4-3) a value of M or the time increment Δt may be picked. For a given value of M, smaller values of Δx mean smaller values of Δt. The value of M must be as follows:

$$M \geq 2 \tag{5.4-4}$$

If M is less than 2, the second law of thermodynamics is violated. It also can be shown that for stability and convergence of the finite-difference solution M must be ≥ 2.

Stability means the errors in the solution do not grow exponentially as the solution proceeds but damp out. Convergence means that the solution of the difference equation approaches the exact solution of the partial differential equation as Δt and Δx go to zero with M fixed. Using smaller sizes of Δt and Δx increases the accuracy in general but greatly increases the number of calculations required. Hence, a digital computer is often ideally suited for this type of calculation.

3. Simplified Schmidt method for a slab. If the value of $M = 2$, then a great simplification of Eq. (5.4-2) occurs, giving the Schmidt method.

$$_{t+\Delta t}T_n = \frac{_tT_{n-1} + _tT_{n+1}}{2} \tag{5.4-5}$$

This means that when a time 1 Δt has elapsed, the new temperature at a given point n at $t + \Delta t$ is the arithmetic average of the temperatures at the two adjacent nodes $n + 1$ and $n - 1$ at the original time t.

5.4B Boundary Conditions for Numerical Method for a Slab

1. Convection at the boundary. For the case where there is a finite convective resistance at the boundary and the temperature of the environment or fluid outside is suddenly changed to T_a, we can derive the following for a slab. Referring to Fig. 5.4-1, we make a

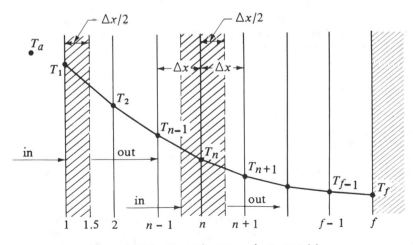

FIGURE 5.4-1. *Unsteady-state conduction in a slab.*

heat balance on the outside $\frac{1}{2}$ element. The rate of heat in by convection − the rate of heat out by conduction − the rate of heat accumulations in Δt s.

$$hA(_tT_a - _tT_1) - \frac{kA}{\Delta x}(_tT_1 - _tT_2) = \frac{(A\,\Delta x/2)\rho c_p}{\Delta t}(_{t+\Delta t}T_{1.25} - _tT_{1.25}) \qquad (5.4-6)$$

where $_tT_{1.25}$ is the temperature at the midpoint of the 0.5 Δx outside slab. As an approximation, the temperature T_1 at the surface can be used to replace that of $T_{1.25}$. Rearranging,

$$_{t+\Delta t}T_1 = \frac{1}{M}[2N_tT_a + [M - (2N + 2)]_tT_1 + 2_tT_2] \qquad (5.4-7)$$

where

$$N = \frac{h\,\Delta x}{k} \qquad (5.4-8)$$

Note that the value of M must be such that

$$M \geq 2N + 2 \qquad (5.4-9)$$

2. *Insulated boundary condition.* In the case for the boundary condition where the rear face is insulated, a heat balance is made on the rear $\frac{1}{2}$ Δx slab just as on the front $\frac{1}{2}$ Δx slab in Fig. 5.4-1. The resulting equation is the same as Eqs. (5.4-6) and (5.4-7), but $h = 0$ or $N = 0$ and $_tT_{f-1} = _tT_{f+1}$ because of symmetry.

$$_{t+\Delta t}T_f = \frac{1}{M}[(M - 2)_tT_f + 2_tT_{f-1}] \qquad (5.4-10)$$

3. *Alternative convective condition.* To use the equations above for a given problem, the same values of M, Δx, and Δt must be used. If N gets too large, so that M may be inconveniently too large, another form of Eq. (5.4-7) can be derived. By neglecting the heat accumulation in the front half-slab in Eq. (5.4-6),

$$_{t+\Delta t}T_1 = \frac{N}{N+1}{}_{t+\Delta t}T_a + \frac{1}{N+1}{}_{t+\Delta t}T_2 \qquad (5.4-11)$$

Here the value of M is not restricted by the N value. This approximation works fairly well when a large number of increments in Δx are used so that the amount of heat neglected is small compared to the total.

4. *Procedures for use of initial boundary temperature.* When the temperature of the environment outside is suddenly changed to T_a, the following procedures should be used.

1. When $M = 2$ and a hand calculation of a limited number of increments is used, a special procedure should be used in Eqs. (5.4-5) and (5.4-7) or (5.4-11). For the first time increment one should use an average value for $_1T_a$ of $(T_a + _0T_1)/2$, where $_0T_1$ is the initial temperature at point 1. For all succeeding Δt values, the value of T_a should be used (D1, K1). This special procedure for the value of T_a to use for the first time increment increases the accuracy of the numerical method, especially after a few time intervals. If T_a varies with time t, a new value can be used for each Δt interval.
2. When $M = 2$ and many time increments are used with a digital computer, this special procedure is not needed and the one value of T_a is used for all time increments.
3. When $M = 3$ or more and a hand calculation of a limited number of increments or a

digital computer calculation of many increments is used, only the one value of T_a is used for all time increments. Note that when $M = 3$ or more, many more calculations are needed compared to the case for $M = 2$. The most accurate results are obtained when $M = 4$, which is the preferred method, with slightly less accurate results for $M = 3$ (D1, K1, K2).

EXAMPLE 5.4-1. Unsteady-State Conduction and the Schmidt Numerical Method

A slab of material 1.00 m thick is at a uniform temperature of 100°C. The front surface is suddenly exposed to a constant environmental temperature of 0°C. The convective resistance is zero ($h = \infty$). The back surface of the slab is insulated. The thermal diffusivity $\alpha = 2.00 \times 10^{-5}$ m²/s. Using five slices each 0.20 m thick and the Schmidt numerical method with $M = 2.0$, calculate the temperature profile at $t = 6000$ s. Use the special procedure for the first time increment.

Solution: Figure 5.4-2 shows the temperature profile at $t = 0$ and the environmental temperature of $T_a = 0°C$ with five slices used. For the Schmidt method, $M = 2$. Substituting into Eq. (5.4-3) with $\alpha = 2.00 \times 10^{-5}$ and $\Delta x = 0.20$ and solving for Δt,

$$M = 2 = \frac{(\Delta x)^2}{\alpha \, \Delta t} = \frac{(0.20)^2}{(2.00 \times 10^{-5}) \, \Delta t} \qquad \Delta t = 1000 \text{ s} \qquad \textbf{(5.4-3)}$$

This means that $(6000 \text{ s})/(1000 \text{ s})/$increment, or six time increments must be used to reach 6000 s.

For the front surface where $n = 1$, the temperature $_1T_a$ to use for the first Δt time increment, as stated previously, is

$$_1T_a = \frac{T_a + {_0}T_1}{2} = {_1}T_1 \qquad n = 1 \qquad \textbf{(5.4-12)}$$

where $_0T_1$ is the initial temperature at point 1. For the remaining time

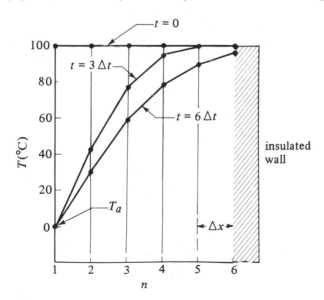

FIGURE 5.4-2. *Temperature for numerical method, Example 5.4-1.*

increments,

$$T_1 = T_a \qquad n = 1 \tag{5.4-13}$$

To calculate the temperatures for all time increments for the slabs $n = 2$ to 5, using Eq. (5.4-5),

$$_{t+\Delta t}T_n = \frac{_tT_{n-1} + {}_tT_{n+1}}{2} \qquad n = 2, 3, 4, 5 \tag{5.4-14}$$

For the insulated end for all time increments at $n = 6$, substituting $M = 2$ and $f = 6$ into Eq. (5.4-12),

$$_{t+\Delta t}T_6 = \frac{(2-2)_tT_6 + 2_tT_5}{2} = {}_tT_5 \tag{5.4-15}$$

For the first time increment of $t + \Delta t$ and calculating the temperature at $n = 1$ by Eq. (5.4-12),

$$_{t+\Delta t}T_1 = \frac{T_a + {}_0T_1}{2} = \frac{0 + 100}{2} = 50°C = {}_1T_a$$

For $n = 2$, using Eq. (5.4-14),

$$_{t+\Delta t}T_2 = \frac{_tT_1 + {}_tT_3}{2} = \frac{50 + 100}{2} = 75$$

Continuing for $n = 3, 4, 5$, we have

$$_{t+\Delta t}T_3 = \frac{_tT_2 + {}_tT_4}{2} = \frac{100 + 100}{2} = 100$$

$$_{t+\Delta t}T_4 = \frac{_tT_3 + {}_tT_5}{2} = \frac{100 + 100}{2} = 100$$

$$_{t+\Delta t}T_5 = \frac{_tT_4 + {}_tT_6}{2} = \frac{100 + 100}{2} = 100$$

For $n = 6$, using Eq. (5.4-15),

$$_{t+\Delta t}T_6 = {}_tT_5 = 100$$

For $2\,\Delta t$ using Eq. (5.4-13) for $n = 1$ and continuing for $n = 2$ to 6, using Eqs. (5.4-14) and (5.4-15),

$$_{t+2\,\Delta t}T_1 = T_a = 0$$

$$_{t+2\,\Delta t}T_2 = \frac{_{t+\Delta t}T_1 + {}_{t+\Delta t}T_3}{2} = \frac{0 + 100}{2} = 50$$

$$_{t+2\,\Delta t}T_3 = \frac{_{t+\Delta t}T_2 + {}_{t+\Delta t}T_4}{2} = \frac{75 + 100}{2} = 87.5$$

$$_{t+2\,\Delta t}T_4 = \frac{_{t+\Delta t}T_3 + {}_{t+\Delta t}T_5}{2} = \frac{100 + 100}{2} = 100$$

$$_{t+2\,\Delta t}T_5 = \frac{_{t+\Delta t}T_4 + {}_{t+\Delta t}T_6}{2} = \frac{100 + 100}{2} = 100$$

$$_{t+2\,\Delta t}T_6 = {}_{t+\Delta t}T_5 = 100$$

For 3 Δt,

$$_{t+3\,\Delta t}T_1 = 0$$

$$_{t+3\,\Delta t}T_2 = \frac{0 + 87.5}{2} = 43.75$$

$$_{t+3\,\Delta t}T_3 = \frac{50 + 100}{2} = 75$$

$$_{t+3\,\Delta t}T_4 = \frac{87.5 + 100}{2} = 93.75$$

$$_{t+3\,\Delta t}T_5 = \frac{100 + 100}{2} = 100$$

$$_{t+3\,\Delta t}T_6 = 100$$

For 4 Δt,

$$_{t+4\,\Delta t}T_1 = 0$$

$$_{t+4\,\Delta t}T_2 = \frac{0 + 75}{2} = 37.5$$

$$_{t+4\,\Delta t}T_3 = \frac{43.75 + 93.75}{2} = 68.75$$

$$_{t+4\,\Delta t}T_4 = \frac{75 + 100}{2} = 87.5$$

$$_{t+4\,\Delta t}T_5 = \frac{93.75 + 100}{2} = 96.88$$

$$_{t+4\,\Delta t}T_6 = 100$$

For 5 Δt,

$$_{t+5\,\Delta t}T_1 = 0$$

$$_{t+5\,\Delta t}T_2 = \frac{0 + 68.75}{2} = 34.38$$

$$_{t+5\,\Delta t}T_3 = \frac{37.5 + 87.5}{2} = 62.50$$

$$_{t+5\,\Delta t}T_4 = \frac{68.75 + 96.88}{2} = 82.81$$

$$_{t+5\,\Delta t}T_5 = \frac{87.5 + 100}{2} = 93.75$$

$$_{t+5\,\Delta t}T_6 = 96.88$$

For 6 Δt (final time),

$$_{t+6\,\Delta t}T_1 = 0$$

$$_{t+6\,\Delta t}T_2 = \frac{0 + 62.5}{2} = 31.25$$

$$_{t+6\,\Delta t}T_3 = \frac{34.38 + 82.81}{2} = 58.59$$

$$_{t+6\,\Delta t}T_4 = \frac{62.50 + 93.75}{2} = 78.13$$

$$_{t+6\,\Delta t}T_5 = \frac{82.81 + 96.88}{2} = 89.84$$

$$_{t+6\,\Delta t}T_6 = 93.75$$

The temperature profiles for 3 Δt increments and the final time of 6 Δt increments are plotted in Fig. 5.4-2. This example shows how a hand calculation can be done. To increase the accuracy, more slab increments and more time increments are required. This, then, is ideally suited for computation using the digital computer.

EXAMPLE 5.4-2. Unsteady-State Conduction Using the Digital Computer
Repeat Example 5.4-1 using the digital computer. Use a $\Delta x = 0.05$ m. Write the computer program and compare the final temperatures with Example 5.4-1. Use the explicit method of Schmidt for $M = 2$. Although not needed for many time increments using the digital computer, use the special procedure for the value of $_1T_a$ for the first time increment. Hence, a direct comparison of the effect of the number of increments on the results can be made with Example 5.4-1.

Solution: The number of slabs to use is 1.00 m/(0.05 m/slab) or 20 slabs. Substituting into Eq. (5.4-3) with $\alpha = 2.00 \times 10^{-5}$ m²/s, $\Delta x = 0.05$ m, and $M = 2$, and solving for Δt,

$$M = 2 = \frac{(\Delta x)^2}{\alpha\,\Delta t} = \frac{(0.05)^2}{(2.00 \times 10^{-5})(\Delta t)}$$

$$\Delta t = 62.5 \text{ s}$$

Hence, (6000/62.5) = 96 time increments to be used. The value of n goes from $n = 1$ to 21.
The equations to use to calculate the temperatures are again Eqs. (5.4-12)–(5.4-15). However, the only differences are that in Eq. (5.4-14) n goes from 2 to 20, and in Eq. (5.4-15) $n = 21$, so that $_{t+\Delta t}T_{21} = {}_tT_{20}$.
The computer program for these equations is easily written and is left up to the reader. The results are tabulated in Table 5.4-1 for comparison with Example 5.4-1, where only 5 slices were used. The table shows that the

TABLE 5.4-1. *Comparison of Results for Examples 5.4-1 and 5.4-2*

Distance from Front Face	Results Using $\Delta x = 0.20$ m		Results Using $\Delta x = 0.05$ m	
		Temperature		Temperature
m	*n*	°C	*n*	°C
0	1	0.0	1	0.0
0.20	2	31.25	5	31.65
0.40	3	58.59	9	58.47
0.60	4	78.13	13	77.55
0.80	5	89.84	17	88.41
1.00	6	93.75	21	91.87

results for 5 slices are reasonably close to those for 20 slices with values from both cases deviating by 2% or less from each other.

As a rule-of-thumb guide for hand calculations, using a minimum of five slices and at least 8 to 10 time increments should give sufficient accuracy for most purposes. Only when very high accuracy is desired or several cases are to be solved is it desirable to solve the problem using a digital computer. In some cases the physical properties are not known with sufficient accuracy to justify a computer solution.

EXAMPLE 5.4-3. Unsteady-State Conduction with Convective Boundary Condition

Use the same conditions as Example 5.4-1, but a convective coefficient of $h = 25.0 \ W/m^2 \cdot K$ is now present at the surface. The thermal conductivity $k = 10.0 \ W/m \cdot K$.

Solution: Equations (5.4-7) and (5.4-8) can be used for convection at the surface. From Eq. (5.4-8), $N = h \, \Delta x/k = 25.0(0.20)/10.0 = 0.50$. Then $2N + 2 = 2(0.50) + 2 = 3.0$. However, by Eq. (5.4-9), the value of M must be equal to or greater than $2N + 2$. This means that a value of $M = 2$ cannot be used. We will select the preferred method where $M = 4.0$. [Another less accurate alternative is to use Eq. (5.4-11) for convection and then the value of M is not restricted by the N value.]
Substituting into Eq. (5.4-3) and solving for Δt,

$$M = 4 = \frac{(\Delta x)^2}{\alpha \, \Delta t} = \frac{(0.20)^2}{(2.00 \times 10^{-5})(\Delta t)} \qquad \Delta t = 500 \ s$$

Hence, $6000/500 = 12$ time increments must be used.
For the first Δt time increment and for all time increments, the value of the environmental temperature T_a to use is $T_a = 0°C$ since $M > 3$. For convection at the node or point $n = 1$ we use Eq. (5.4-7), where $M = 4$ and $N = 0.50$.

$$_{t+\Delta t}T_1 = \tfrac{1}{4}[2(0.5)T_a + [4 - (2 \times 0.5 + 2)]_t T_1 + 2_t T_2]$$

$$= 0.25 T_a + 0.25_t T_1 + 0.50_t T_2 \qquad n = 1 \qquad \textbf{(5.4-16)}$$

For $n = 2, 3, 4, 5$, we use Eq. (5.4-2),

$$_{t+\Delta t}T_n = \tfrac{1}{4}[_t T_{n+1} + (4 - 2)_n T_t + _t T_{n-1}]$$

$$= 0.25_t T_{n+1} + 0.50_t T_n + 0.25_t T_{n-1} \qquad n = 2, 3, 4, 5 \ \textbf{(5.4-17)}$$

For $n = 6$ (insulated boundary), we use Eq. (5.4-10) and $f = 6$.

$$_{t+\Delta t}T_6 = \tfrac{1}{4}[(4 - 2)_t T_6 + 2_t T_5]$$

$$= 0.50_t T_6 + 0.50_t T_5 \qquad n = 6 \qquad \textbf{(5.4-18)}$$

For 1 Δt, for the first time increment of $t + \Delta t$, $T_a = 0$. Using Eq. (5.4-16) to calculate the temperature at node 1,

$$_{t+\Delta t}T_1 = 0.25(0) + 0.25(100) + 0.50(100) = 75.0$$

For $n = 2, 3, 4, 5$ using Eq. (5.4-17),

$$_{t+\Delta t}T_2 = 0.25_t T_3 + 0.50_t T_2 + 0.25_t T_1$$

$$= 0.25(100) + 0.50(100) + 0.25(100) = 100.0$$

Also, in a similar calculation, T_3, T_4, and $T_5 = 100.0$. For $n = 6$, using Eq. (5.4-18), $T_6 = 100.0$.

For 2 Δt, $T_a = 0$. Using Eq. (5.4-16),

$$_{t+2\,\Delta t}T_1 = 0.25(0) + 0.25(75.0) + 0.50(100) = 68.75$$

Using Eq. (5.4-17) for $n = 2, 3, 4, 5$,

$$_{t+2\,\Delta t}T_2 = 0.25(100) + 0.50(100) + 0.25(75.00) = 93.75$$

$$_{t+2\,\Delta t}T_3 = 0.25(100) + 0.50(100) + 0.25(100) = 100.0$$

Also, T_4 and $T_5 = 100.0$.
For $n = 6$, using Eq. (5.4-18), $T_6 = 100.0$.
For 3 Δt, $T_a = 0$. Using Eq. (5.4-16),

$$_{t+3\,\Delta t}T_1 = 0.25(0) + 0.25(68.75) + 0.50(93.75) = 64.07$$

Using Eq. (5.4-17) for $n = 2, 3, 4, 5$,

$$_{t+3\,\Delta t}T_2 = 0.25(100) + 0.50(93.75) + 0.25(68.75) = 89.07$$

$$_{t+3\,\Delta t}T_3 = 0.25(100) + 0.50(100) + 0.25(93.75) = 98.44$$

$$_{t+3\,\Delta t}T_4 = 0.25(100) + 0.50(100) + 0.25(100) = 100.0$$

Also, $T_5 = 100.0$ and $T_6 = 100.0$.
    In a similar manner the calculations can be continued for the remaining time until a total of 12 Δt increments have been used.

5.4C. Other Numerical Methods for Unsteady-State Conduction

1. Unsteady-state conduction in a cylinder. In deriving the numerical equations for unsteady-state conduction in a flat slab, the cross-sectional area was constant throughout. In a cylinder it changes radially. To derive the equation for a cylinder, Fig. 5.4-3 is used where the cylinder is divided into concentric hollow cylinders whose walls are Δx m thick. Assuming a cylinder 1 m long and making a heat balance on the slab at point n, the rate of heat in − rate of heat out = rate of heat accumulation.

$$\frac{k[2\pi(n + 1/2)\,\Delta x]}{\Delta x}(_tT_{n+1} - {}_tT_n) - k\frac{[2\pi(n - 1/2)\,\Delta x]}{\Delta x}(_tT_n - {}_tT_{n-1})$$

$$= \frac{2\pi n(\Delta x)^2 \rho c_p}{\Delta t}(_{t+\Delta t}T_n - {}_tT_n) \qquad (5.4\text{-}19)$$

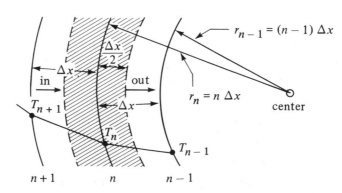

FIGURE 5.4-3. *Unsteady-state conduction in a cylinder.*

Chap. 5 *Principles of Unsteady-State Heat Transfer*

Rearranging, the final equation is

$$_{t+\Delta t}T_n = \frac{1}{M}\left[\frac{2n+1}{2n}\,_tT_{n+1} + (M-2)_tT_n + \frac{2n-1}{2n}\,_tT_{n-1}\right] \qquad (5.4\text{-}20)$$

where $M = (\Delta x)^2/(\alpha\,\Delta t)$ as before. Also, at the center where $n = 0$,

$$_{t+\Delta t}T_0 = \frac{4}{M}\,_tT_1 + \frac{M-4}{M}\,_tT_0 \qquad (5.4\text{-}21)$$

To use Equations (5.4-20) and (5.4-21),

$$M \geq 4 \qquad (5.4\text{-}22)$$

Equations for convection at the outer surface of the cylinder have been derived (D1). If the heat capacity of the outer $\frac{1}{2}$ slab is neglected,

$$_{t+\Delta t}T_n = \frac{nN}{\dfrac{2n-1}{2} + nN}\,_{t+\Delta t}T_a + \frac{(2n-1)/2}{\dfrac{2n-1}{2} + nN}\,_{t+\Delta t}T_{n-1} \qquad (5.4\text{-}23)$$

where T_n is the temperature at the surface and T_{n-1} at a position in the solid 1 Δx below the surface.

Equations for numerical methods for two-dimensional unsteady-state conduction have been derived and are available in a number of references (D1, K2).

2. Unsteady-state conduction and implicit numerical method. In some practical problems the restrictions imposed on the value $M \geq 2$ by stability requirements may prove inconvenient. Also, to minimize the stability problems, implicit methods using different finite-difference formulas have been developed. An important one of these formulas is the Crank–Nicolson method, which will be considered here.

In deriving Eqs. (5.4-1) and (5.4-2), the rate at which heat entered the slab in Fig. 5.4-1 was taken to be the rate at time t.

$$\text{Rate of heat in at } t = \frac{kA}{\Delta x}(_tT_{n-1} - _tT_n) \qquad (5.4\text{-}24)$$

It was then assumed that this rate could be used during the whole interval from t to $t + \Delta t$. However, this is an approximation since the rate changes during this Δt interval. A better value would be the average value of the rate at t and at $t + \Delta t$, or

$$\text{average rate of heat in} = \frac{kA}{\Delta x}\left[\frac{(_tT_{n-1} - _tT_n) + (_{t+\Delta t}T_{n-1} - _{t+\Delta t}T_n)}{2}\right] \qquad (5.4\text{-}25)$$

Also, for the heat leaving, a similar type of average is used. The final equation is

$$_{t+\Delta t}T_{n+1} - (2M+2)_{t+\Delta t}T_n + _{t+\Delta t}T_{n-1} = -_tT_{n+1} + (2-2M)_tT_n - _tT_{n-1} \qquad (5.4\text{-}26)$$

This means that now a new value of $_{t+\Delta t}T_n$ cannot be calculated only from values at time t as in Eq. (5.4-2) but that all the new values of T at $t + \Delta t$ at all points must be calculated simultaneously. To do this an equation is written similar to Eq. (5.4-26) for each of the internal points. Each of these equations and the boundary equations are linear algebraic equations. These then can be solved simultaneously by the standard methods used, such as the Gauss–Seidel iteration technique, matrix inversion technique, and so on (G1, K1).

An important advantage of Eq. (5.4-26) is that the stability and convergence criteria are satisfied for all positive values of M. This means that M can have values less than 2.0. A disadvantage of the implicit method is the larger number of calculations needed for each time step. Explicit methods are simpler to use but because of stability considerations, especially in complex situations, implicit methods are often used.

5.5 CHILLING AND FREEZING OF FOOD AND BIOLOGICAL MATERIALS

5.5A Introduction

Unlike many inorganic and organic materials which are relatively stable, food and other biological materials decay and deteriorate more or less rapidly with time at room temperature. This spoilage is due to a number of factors. Tissues of foods such as fruits and vegetables, after harvesting, continue to undergo metabolic respiration and ripen and eventually spoil. Enzymes of the dead tissues of meats and fish remain active and induce oxidation and other deteriorating effects. Microorganisms attack all types of foods by decomposing the foods so that spoilage occurs; also, chemical reactions occur, such as the oxidation of fats.

At low temperatures the growth rate of microorganisms will be slowed if the temperature is below that which is optimum for growth. Also, enzyme activity and chemical reaction rates are reduced at low temperatures. The rates of most chemical and biological reactions in storage of chilled or frozen foods and biological materials are reduced by factors of $\frac{1}{2}$ to $\frac{1}{3}$ for each 10 K (10°C) drop in temperature.

Water plays an important part in these rates of deterioration, and it is present to a substantial percentage in most biological materials. To reach a low enough temperature for most of these rates to approximately cease, most of the water must be frozen. Materials such as food do not freeze at 0°C (32°F) as pure water does but at a range of temperatures below 0°C. However, because of some of the physical effects of ice crystals and other effects, such as concentrating of solutions, chilling of biological materials is often used instead of freezing for preservation.

Chilling of materials involves removing the sensible heat and heat of metabolism and reducing the temperature usually to a range of 4.4°C (40°F) to just above freezing. Essentially no latent heat of freezing is involved. The materials can be stored for a week or so up to a few months, depending on the product stored and the gaseous atmosphere. Each material has its optimum chill storage temperature.

In the freezing of food and biological materials, the temperature is reduced so that most of the water is frozen to ice. Depending on the final storage temperatures of down to −30°C, the materials can be stored for up to 1 year or so. Often in the production of many frozen foods, they are first treated by blanching or scalding to destroy enzymes.

5.5B Chilling of Food and Biological Materials

In the chilling of food and biological materials, the temperature of the materials is reduced to the desired chill storage temperature, which can be about −1.1°C (30°F) to 4.4°C (40°F). For example, after slaughter, beef has a temperature of 37.8°C (100°F) to 40°C (104°F), and it is often cooled to about 4.4°C (40°F). Milk from cows must be chilled quickly to temperatures just above freezing. Some fish fillets at the time of packing are at a temperature of 7.2°C (45°F) to 10°C (50°F) and are chilled to close to 0°C.

These rates of chilling or cooling are governed by the laws of unsteady-state heat conduction discussed in Sections 5.1 to 5.4. The heat is removed by convection at the surface of the material and by unsteady-state conduction in the material. The fluid outside the foodstuff or biological materials is used to remove this heat, and in many cases it is air. The air has previously been cooled by refrigeration to $-1.1°C$ to $+4.4°C$, depending on the material and other conditions. The convective heat-transfer coefficients, which usually include radiation effects, can also be predicted by the methods in Chapter 4, and for air the coefficient varies from about 8.5 to 40 $W/m^2 \cdot K$ (1.5 to 7 btu/h $\cdot$ ft$^2 \cdot$ °F), depending primarily on air velocity.

In some cases the fluid used for chilling is a liquid flowing over the surface and the values of h can vary from about 280 to 1700 $W/m^2 \cdot K$ (50–300 btu/h $\cdot$ ft$^2 \cdot$ °F). Also, in other cases, a contact or plate cooler is used where chilled plates are in direct contact with the material. Then the temperature of the surface of the material is usually assumed to be equal to or close to that of the contact plates. Contact freezers are used for freezing biological materials.

Where the food is packaged in boxes or where the material is tightly covered by a film of plastic, this additional resistance must be considered. One method to do this is to add the resistance of the package covering to that of the convective film:

$$R_T = R_P + R_C \tag{5.5-1}$$

where R_P is the resistance of covering, R_C the resistance of the outside convective film, and R_T the total resistance. Then, for each resistance,

$$R_C = \frac{1}{h_c A} \tag{5.5-2}$$

$$R_P = \frac{\Delta x}{k A} \tag{5.5-3}$$

$$R_T = \frac{1}{h A} \tag{5.5-4}$$

where h_c is the convective gas or liquid coefficient, A is the area, Δx is the thickness of the covering, k is the thermal conductivity of the covering, and h is the overall coefficient. The overall coefficient h is the one to use in the unsteady-state charts. This assumes a negligible heat capacity of the covering, which is usually the case. Also, it assumes that the covering closely touches the food material so there is no resistance between the covering and the food.

The major sources of error in using the unsteady-state charts are the inadequate data on the density, heat capacity, and thermal conductivity of the foods and the prediction of the convective coefficient. Food materials are irregular anisotropic substances, and the physical properties are often difficult to evaluate. Also, if evaporation of water occurs on chilling, latent heat losses can affect the accuracy of the results.

EXAMPLE 5.5-1. Chilling Dressed Beef
Hodgson (H2) gives physical properties of beef carcasses during chilling of $\rho = 1073$ kg/m^3, $c_p = 3.48$ kJ/kg $\cdot$ K, and $k = 0.498$ W/m $\cdot$ K. A large slab of beef 0.203 m thick and initially at a uniform temperature of 37.8°C is to be cooled so that the center temperature is 10°C. Chilled air at 1.7°C (assumed constant) and having an $h = 39.7$ W/m$^2 \cdot$ K is used. Calculate the time needed.

Solution: The thermal diffusivity α is

$$\alpha = \frac{k}{\rho c_p} = \frac{0.498}{(1073)(3.48 \times 1000)} = 1.334 \times 10^{-7} \text{ m}^2/\text{s}$$

Then for the half-thickness x_1 of the slab,

$$x_1 = \frac{0.203}{2} = 0.1015 \text{ m}$$

For the center of the slab,

$$n = \frac{x}{x_1} = \frac{0}{x_1} = 0$$

Also,

$$m = \frac{k}{hx_1} = \frac{0.498}{(39.7)(0.1015)} = 0.123$$

$$T_1 = 1.7°C + 273.2 = 274.9 \text{ K} \qquad T_0 = 37.8 + 273.2 = 311.0 \text{ K}$$

$$T = 10 + 273.2 = 283.2 \text{ K}$$

$$Y = \frac{T_1 - T}{T_1 - T_0} = \frac{274.9 - 283.2}{274.9 - 311.0} = 0.230$$

Using Fig. 5.3-6 for the center of a large flat plate,

$$X = 0.90 = \frac{\alpha t}{x_1^2} = \frac{(1.334 \times 10^{-7})(t)}{(0.1015)^2}$$

Solving, $t = 6.95 \times 10^4 \text{ s}$ (19.3 h).

5.5C Freezing of Food and Biological Materials

1. Introduction.   In the freezing of food and other biological materials, the removal of sensible heat in chilling first occurs and then the removal of the latent heat of freezing. The latent heat of freezing water of 335 kJ/kg (144 btu/lb$_m$) is a substantial portion of the total heat removed on freezing. Other slight effects, such as the heats of solution of salts, and so on, may be present but are quite small. Actually, when materials such as meats are frozen to $-29°C$, only about 90% of the water is frozen to ice, with the rest thought to be bound water (B1).

Riedel (R1) gives enthalpy–temperature–composition charts for the freezing of many different foods. These charts show that freezing does not occur at a given temperature but extends over a range of several degrees. As a consequence, there is no one freezing point with a single latent heat of freezing.

Since the latent heat of freezing is present in the unsteady-state process of freezing, the standard unsteady-state conduction equations and charts given in this chapter cannot be used for prediction of freezing times. A full analytical solution of the rate of freezing of food and biological materials is very difficult because of the variation of physical properties with temperature, the amount of freezing varying with temperature, and other factors. An approximate solution by Plank is often used.

2. Approximate solution of Plank for freezing.   Plank (P2) has derived an approximate solution for the time of freezing which is often sufficient for engineering purposes. The assumptions in the derivation are as follows. Initially, all the food is at the freezing temperature but is unfrozen. The thermal conductivity of the frozen part is constant. All the material freezes at the freezing point, with a constant latent heat. The heat transfer by conduction in the frozen layer occurs slowly enough so that it is under pseudo-steady-state conditions.

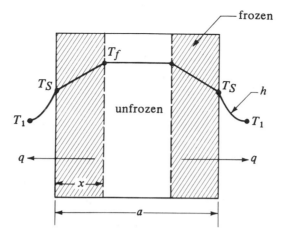

FIGURE 5.5-1. *Temperature profile during freezing.*

In Fig. 5.5-1 a slab of thickness a m is cooled from both sides by convection. At a given time t s, a thickness of x m of frozen layer has formed on both sides. The temperature of the environment is constant at T_1 K and the freezing temperature is constant at T_f. An unfrozen layer in the center at T_f is present.

The heat leaving at time t is q W. Since we are at pseudo-steady state, at time t, the heat leaving by convection on the outside is

$$q = hA(T_S - T_1) \tag{5.5-5}$$

where A is the surface area. Also, the heat being conducted through the frozen layer of x thickness at steady state is

$$q = \frac{kA}{x}(T_f - T_S) \tag{5.5-6}$$

where k is the thermal conductivity of the frozen material. In a given time dt s, a layer dx thick of material freezes. Then multiplying A times dx times ρ gives the kg mass frozen. Multiplying this by the latent heat λ in J/kg and dividing by dt,

$$q = \frac{A\, dx\, \rho\lambda}{dt} = A\rho\lambda\,\frac{dx}{dt} \tag{5.5-7}$$

where ρ is the density of the unfrozen material.

Next, to eliminate T_S from Eqs. (5.5-5) and (5.5-6), Eq. (5.5-5) is solved for T_S and substituted into Eq. (5.5-6), giving

$$q = \frac{(T_f - T_1)A}{x/k + 1/h} \tag{5.5-8}$$

Equating Eq. (5.5-8) to (5.5-7),

$$\frac{(T_f - T_1)A}{x/k + 1/h} = A\rho\lambda\,\frac{dx}{dt} \tag{5.5-9}$$

Rearranging and integrating between $t = 0$ and $x = 0$, to $t = t$ and $x = a/2$,

$$(T_f - T_1)\int_0^t dt = \lambda\rho \int_0^{a/2} \left(\frac{x}{k} + \frac{1}{h}\right) dx \tag{5.5-10}$$

Integrating and solving for t,

$$t = \frac{\lambda\rho}{T_f - T_1}\left(\frac{a}{2h} + \frac{a^2}{8k}\right)$$ (5.5-11)

To generalize the equation for other shapes,

$$t = \frac{\lambda\rho}{T_f - T_1}\left(\frac{Pa}{h} + \frac{Ra^2}{k}\right)$$ (5.5-12)

where a is the thickness of an infinite slab (as in Fig. 5.5-1), diameter of a sphere, diameter of a long cylinder, or the smallest dimension of a rectangular block or brick. Also,

$P = \frac{1}{2}$ for infinite slab, $\frac{1}{6}$ for sphere, $\frac{1}{4}$ for infinite cylinder

$R = \frac{1}{8}$ for infinite slab, $\frac{1}{24}$ for sphere, $\frac{1}{16}$ for infinite cylinder

For a rectangular brick having dimensions a by $\beta_1 a$ by $\beta_2 a$, where a is the shortest side, Ede (B1) has prepared a chart to determine the values of P and R to be used to calculate t in Eq. (5.5-12). Equation (5.5-11) can also be used for calculation of thawing times by replacing the k of the frozen material by the k of the thawed material.

EXAMPLE 5.5-2. Freezing of Meat

Slabs of meat 0.0635 m thick are to be frozen in an air-blast freezer at 244.3 K ($-28.9°C$). The meat is initially at the freezing temperature of 270.4 K ($-2.8°C$). The meat contains 75% moisture. The heat-transfer coefficient is $h = 17.0$ W/m$^2 \cdot$ K. The physical properties are $\rho = 1057$ kg/m^3 for the unfrozen meat and $k = 1.038$ W/m $\cdot$ K for the frozen meat. Calculate the freezing time.

Solution: Since the latent heat of fusion of water to ice is 335 kJ/kg (144 btu/lb$_m$), for meat with 75% water,

$$\lambda = 0.75(335) = 251.2 \text{ kJ/kg}$$

The other given variables are $a = 0.0635$ m, $T_f = 270.4$ K, $T_1 = 244.3$ K, $\rho = 1057$ kg/m^3, $h = 17.0$ W/m$^2 \cdot$ K, $k = 1.038$ W/m $\cdot$ K. Substituting into Eq. (5.5-11),

$$t = \frac{\lambda\rho}{T_f - T_1}\left(\frac{a}{2h} + \frac{a^2}{8k}\right) = \frac{(2.512 \times 10^5)1057}{270.4 - 244.3}\left[\frac{0.0635}{2(17.0)} + \frac{(0.0635)^2}{8(1.038)}\right]$$
$$= 2.395 \times 10^4 \text{ s } (6.65 \text{ h})$$

3. *Other methods to calculate freezing times.* Neumann (C1, C2) has derived a complicated equation for freezing in a slab. He assumes the following conditions. The surface temperature is the same as the environment, i.e., no surface resistance. The temperature of freezing is constant. This method suffers from this limitation that a convection coefficient cannot be used at the surface, since it assumes no surface resistance. However, the method does include the effect of cooling from an original temperature, which may be above the freezing point.

Plank's equation does not make provision for an original temperature, which may be above the freezing point. An approximate method to calculate the additional time necessary to cool from temperature T_0 down to the freezing point T_f is as follows. Calculate by means of the unsteady-state charts the time for the average temperature in the material to reach T_f assuming that no freezing occurs using the physical properties of the unfrozen material. If there is no surface resistance, Fig. 5.3-13 can be used directly for this. If a resistance is present, the temperature at several points in the material will have to be obtained from the unsteady-state charts and the average temperature calculated from these point temperatures. This may be partial trial and error since the time is unknown, which must be assumed. If the average temperature calculated is not at the freezing point, a new time must be assumed. This is an approximate method since some material will actually freeze.

5.6 DIFFERENTIAL EQUATION OF ENERGY CHANGE

5.6A Introduction

In Sections 3.6 and 3.7 we derived a differential equation of continuity and a differential equation of momentum transfer for a pure fluid. These equations were derived because overall mass, energy, and momentum balances made on a finite volume in the earlier parts of Chapter 2 did not tell us what goes on inside a control volume. In the overall balances performed, a new balance was made for each new system studied. However, it is often easier to start with the differential equations of continuity and momentum transfer in general form and then to simplify the equations by discarding unneeded terms for each specific problem.

In Chapter 4 on steady-state heat transfer and Chapter 5 on unsteady-state heat transfer new overall energy balances were made on a finite control volume for each new situation. To advance further in our study of heat or energy transfer in flow and nonflow systems we must use a differential volume to investigate in greater detail what goes on inside this volume. The balance will be made on a single phase and the boundary conditions at the phase boundary will be used for integration.

In the next section we derive a general differential equation of energy change: the conservation-of-energy equation. Then this equation is modified for certain special cases that occur frequently. Finally, applications of the uses of these equations are given. Cases for both steady-state and unsteady-state energy transfer are studied using this conservation-of-energy equation, which is perfectly general and holds for steady- or unsteady-state conditions.

5.6B Derivation of Differential Equation of Energy Change

As in the derivation of the differential equation of momentum transfer, we write a balance on an element of volume of size Δx, Δy, and Δz which is stationary. We then write the law of conservation of energy, which is really the first law of thermodynamics for the fluid in this volume element at any time. The following is the same as Eq. (2.7-7) for a control volume given in Section 2.7.

$$\begin{pmatrix} \text{rate of} \\ \text{energy in} \end{pmatrix} - \begin{pmatrix} \text{rate of} \\ \text{energy out} \end{pmatrix} - \begin{pmatrix} \text{rate of} \\ \text{external work done by} \\ \text{system on surroundings} \end{pmatrix} = \begin{pmatrix} \text{rate of} \\ \text{accumulation of} \\ \text{energy} \end{pmatrix} \quad \text{(5.6-1)}$$

As in momentum transfer, the transfer of energy into and out of the volume element is by convection and molecular transport or conduction. There are two kinds of energy being transferred. The first is internal energy U in J/kg (btu/lb$_m$) or any other set of units.

This is the energy associated with random translational and internal motions of the molecules plus molecular interactions. The second is kinetic energy $\rho v^2/2$, which is the energy associated with the bulk fluid motion, where v is the local fluid velocity, m/s (ft/s). Hence, the total energy per unit volume is $(\rho U + \rho v^2/2)$. The rate of accumulation of energy in the volume element in m^3 (ft^3) is then

$$\Delta x \, \Delta y \, \Delta z \frac{\partial}{\partial t}\left(\rho U + \frac{\rho v^2}{2}\right) \tag{5.6-2}$$

The total energy coming in by convection in the x direction at x minus that leaving at $x + \Delta x$ is

$$\Delta y \, \Delta z \left[v_x\left(\rho U + \frac{\rho v^2}{2}\right)\right]_x - \Delta y \, \Delta z \left[v_x\left(\rho U + \frac{\rho v^2}{2}\right)\right]_{x+\Delta x} \tag{5.6-3}$$

Similar equations can be written for the y and z directions using velocities v_y and v_z, respectively.

The net rate of energy into the element by conduction in the x direction is

$$\Delta y \, \Delta z[(q_x)_x - (q_x)_{x+\Delta x}] \tag{5.6-4}$$

Similar equations can be written for the y and z directions where q_x, q_y, and q_z are the components of the heat flux vector $\mathbf{q}$, which is in W/m^2 ($btu/s \cdot ft^2$) or any other convenient set of units.

The net work done by the system on its surroundings is the sum of the following three parts for the x direction. For the net work done against the gravitational force,

$$-\rho \, \Delta x \, \Delta y \, \Delta z(v_x g_x) \tag{5.6-5}$$

where g_x is gravitational force. The net work done against the static pressure p is

$$\Delta y \, \Delta z[(pv_x)_{x+\Delta x} - (pv_x)_x] \tag{5.6-6}$$

where p is N/m^2 (lb_f/ft^2) or any other convenient set of units. For the net work against the viscous forces,

$$(\Delta y \, \Delta z)[(\tau_{xx}v_x + \tau_{xy}v_y + \tau_{xz}v_z)_{x+\Delta x} - (\tau_{xx}v_x + \tau_{xy}v_y + \tau_{xz}v_z)_x] \tag{5.6-7}$$

In Section 3.7 these viscous forces are discussed in more detail.

Writing equations similar to (5.6-3)–(5.6-7) in all three directions; substituting these equations and Eq. (5.6-2) into (5.6-1); dividing by Δx, Δy, and Δz; and letting Δx, Δy, and Δz approach zero, we obtain

$$\frac{\partial}{\partial t}\left(\rho U + \frac{\rho v^2}{2}\right) = -\left[\frac{\partial}{\partial x}v_x\left(\rho U + \frac{\rho v^2}{2}\right) + \frac{\partial}{\partial y}v_y\left(\rho U + \frac{\rho v^2}{2}\right) + \frac{\partial}{\partial z}v_z\left(\rho U + \frac{\rho v^2}{2}\right)\right]$$

$$-\left(\frac{\partial q_x}{\partial x} + \frac{\partial q_y}{\partial y} + \frac{\partial q_z}{\partial z}\right) + \rho(v_x g_x + v_y g_y + v_z g_z)$$

$$-\left[\frac{\partial}{\partial x}(pv_x) + \frac{\partial}{\partial y}(pv_y) + \frac{\partial}{\partial z}(pv_z)\right]$$

$$-\left[\frac{\partial}{\partial x}(\tau_{xx}v_x + \tau_{xy}v_y + \tau_{xz}v_z) + \frac{\partial}{\partial y}(\tau_{yx}v_x + \tau_{yy}v_y + \tau_{yz}v_z)\right.$$

$$\left. + \frac{\partial}{\partial z}(\tau_{zx}v_x + \tau_{zy}v_y + \tau_{zz}v_z)\right] \tag{5.6-8}$$

For further details of this derivation, see (B2).

Equation (5.6-8) is the final equation of energy change relative to a stationary point. However, it is not in a convenient form. We first combine Eq. (5.6-8) with the equation of continuity, Eq. (3.6-23), with the equation of motion, Eq. (3.7-13), and express the internal energy in terms of fluid temperature T and heat capacity. Then writing the resultant equation for a Newtonian fluid with constant thermal conductivity k, we obtain

$$\rho c_v \frac{DT}{Dt} = k\nabla^2 T - T\left(\frac{\partial p}{\partial T}\right)_\rho (\nabla \cdot \mathbf{v}) + \mu\phi \tag{5.6-9}$$

This equation utilizes Fourier's second law in three directions, where

$$k\nabla^2 T = k\left(\frac{\partial^2 T}{\partial x^2} + \frac{\partial^2 T}{\partial y^2} + \frac{\partial^2 T}{\partial z^2}\right) \tag{5.6-10}$$

The viscous dissipation term $\mu\phi$ is generally negligible except where extremely large velocity gradients exist. It will be omitted in the discussions to follow. Equation (5.6-9) is the equation of energy change for a Newtonian fluid with constant k in terms of the fluid temperature T.

5.6C Special Cases of the Equation of Energy Change

The following special forms of Eq. (5.6-9) for a Newtonian fluid with constant thermal conductivity are commonly encountered. First, Eq. (5.6-9) will be written in rectangular coordinates without the $\mu\phi$ term.

$$\rho c_v\left(\frac{\partial T}{\partial t} + v_x \frac{\partial T}{\partial x} + v_y \frac{\partial T}{\partial y} + v_z \frac{\partial T}{\partial z}\right)$$

$$= k\left(\frac{\partial^2 T}{\partial x^2} + \frac{\partial^2 T}{\partial y^2} + \frac{\partial^2 T}{\partial z^2}\right) - T\left(\frac{\partial p}{\partial T}\right)_\rho \left(\frac{\partial v_x}{\partial x} + \frac{\partial v_y}{\partial y} + \frac{\partial v_z}{\partial z}\right) \tag{5.6-11}$$

1. Fluid at constant pressure. The equations below can be used for constant-density fluids as well as for constant pressure.

$$\rho c_p \frac{DT}{Dt} = k\nabla^2 T \tag{5.6-12}$$

In rectangular coordinates,

$$\rho c_p\left(\frac{\partial T}{\partial t} + v_x \frac{\partial T}{\partial x} + v_y \frac{\partial T}{\partial y} + v_z \frac{\partial T}{\partial z}\right) = k\left(\frac{\partial^2 T}{\partial x^2} + \frac{\partial^2 T}{\partial y^2} + \frac{\partial^2 T}{\partial z^2}\right) \tag{5.6-13}$$

In cylindrical coordinates,

$$\rho c_p\left(\frac{\partial T}{\partial t} + v_r \frac{\partial T}{\partial r} + \frac{v_\theta}{r} \frac{\partial T}{\partial \theta} + v_z \frac{\partial T}{\partial z}\right) = k\left(\frac{\partial^2 T}{\partial r^2} + \frac{1}{r} \frac{\partial T}{\partial r} + \frac{1}{r^2} \frac{\partial^2 T}{\partial \theta^2} + \frac{\partial^2 T}{\partial z^2}\right) \tag{5.6-14}$$

In spherical coordinates,

$$\rho c_p\left(\frac{\partial T}{\partial t} + v_r \frac{\partial T}{\partial r} + \frac{v_\theta}{r} \frac{\partial T}{\partial \theta} + \frac{v_\phi}{r \sin \theta} \frac{\partial T}{\partial \phi}\right)$$

$$= k\left[\frac{1}{r^2} \frac{\partial}{\partial r}\left(r^2 \frac{\partial T}{\partial r}\right) + \frac{1}{r^2 \sin \theta} \frac{\partial}{\partial \theta}\left(\sin \theta \frac{\partial T}{\partial \theta}\right) + \frac{1}{r^2 \sin^2 \theta} \frac{\partial^2 T}{\partial \phi^2}\right] \tag{5.6-15}$$

For definitions of cylindrical and spherical coordinates, see Section 3.6. If the velocity v is zero, DT/Dt becomes $\partial T/\partial t$.

2. Fluid at constant density

$$\rho c_p \frac{DT}{Dt} = k\nabla^2 T \qquad (5.6\text{-}16)$$

Note that this is identical to Eq. (5.6-12) for constant pressure.

3. Solid. Here we consider ρ is constant and $v = 0$.

$$\rho c_p \frac{\partial T}{\partial t} = k\nabla^2 T \qquad (5.6\text{-}17)$$

This is often referred to as *Fourier's second law* of heat conduction. This also holds for a fluid with zero velocity at constant pressure.

4. Heat generation. If there is heat generation in the fluid by electrical or chemical means, then $\dot q$ can be added to the right side of Eq. (5.6-17).

$$\rho c_p \frac{\partial T}{\partial t} = k\nabla^2 T + \dot q \qquad (5.6\text{-}18)$$

where $\dot q$ is the rate of heat generation in W/m^3 ($btu/h \cdot ft^3$) or other suitable units. Viscous dissipation is also a heat source, but its inclusion greatly complicates problem solving because the equations for energy and motion are then coupled.

5. Other coordinate systems. Fourier's second law of unsteady-state heat conduction can be written as follows.

For rectangular coordinates,

$$\frac{\partial T}{\partial t} = \frac{k}{\rho c_p}\nabla^2 T = \alpha\left(\frac{\partial^2 T}{\partial x^2} + \frac{\partial^2 T}{\partial y^2} + \frac{\partial^2 T}{\partial z^2}\right) \qquad (5.6\text{-}19)$$

where α is $k/\rho c_p$ and is thermal diffusivity in m^2/s (ft^2/h).

For cylindrical coordinates,

$$\frac{\partial T}{\partial t} = \alpha\left(\frac{\partial^2 T}{\partial r^2} + \frac{1}{r}\frac{\partial T}{\partial r} + \frac{1}{r^2}\frac{\partial^2 T}{\partial \theta^2} + \frac{\partial^2 T}{\partial z^2}\right) \qquad (5.6\text{-}20)$$

For spherical coordinates,

$$\frac{\partial T}{\partial t} = \alpha\left[\frac{1}{r^2}\frac{\partial}{\partial r}\left(r^2\frac{\partial T}{\partial r}\right) + \frac{1}{r^2 \sin\theta}\frac{\partial}{\partial\theta}\left(\sin\theta\frac{\partial T}{\partial\theta}\right) + \frac{1}{r^2 \sin^2\theta}\frac{\partial^2 T}{\partial\phi^2}\right] \qquad (5.6\text{-}21)$$

5.6D Uses of Equation of Energy Change

In Section 3.8 we used the differential equations of continuity and of motion to set up fluid flow problems. We did this by discarding the terms that are zero or near zero and using the remaining equations to solve for the velocity and pressure distributions. This was done instead of making new mass and momentum balances for each new situation. In a similar manner, to solve problems of heat transfer, the differential equations of continuity, motion, and energy will be used with the unneeded terms being discarded. Several examples will be given to illustrate the general methods used.

EXAMPLE 5.6-1. Temperature Profile with Heat Generation
A solid cylinder in which heat generation is occurring uniformly as $\dot{q}$ W/m^3 is insulated on the ends. The temperature of the surface of the cylinder is held constant at T_w K. The radius of the cylinder is $r = R$ m. Heat flows only in the radial direction. Derive the equation for the temperature profile at steady state if the solid has a constant thermal conductivity.

Solution: Equation (5.6-20) will be used for cylindrical coordinates. Also the term $\dot{q}/\rho c_p$ for generation will be added to the right side, giving

$$\frac{\partial T}{\partial t} = \frac{k}{\rho c_p}\left(\frac{\partial^2 T}{\partial r^2} + \frac{1}{r}\frac{\partial T}{\partial r} + \frac{1}{r^2}\frac{\partial^2 T}{\partial \theta^2} + \frac{\partial^2 T}{\partial z^2}\right) + \frac{\dot{q}}{\rho c_p} \tag{5.6-22}$$

For steady state $\partial T/\partial t = 0$. Also, for conduction only in the radial direction $\partial^2 T/\partial z^2 = 0$ and $\partial^2 T/\partial \theta^2 = 0$. This gives the following differential equation:

$$\frac{d^2 T}{dr^2} + \frac{1}{r}\frac{dT}{dr} = -\frac{\dot{q}}{k} \tag{5.6-23}$$

This can be rewritten as

$$r\frac{d^2 T}{dr^2} + \frac{dT}{dr} = -\frac{\dot{q}r}{k} \tag{5.6-24}$$

Note that Eq. (5.6-24) can be rewritten as follows:

$$\frac{d}{dr}\left(r\frac{dT}{dr}\right) = -\frac{\dot{q}r}{k} \tag{5.6-25}$$

Integrating Eq. (5.6-25) once,

$$r\frac{dT}{dr} = -\frac{\dot{q}r^2}{2k} + K_1 \tag{5.6-26}$$

where K_1 is a constant. Integrating again,

$$T = -\frac{\dot{q}r^2}{4k} + K_1 \ln r + K_2 \tag{5.6-27}$$

where K_2 is a constant. The boundary conditions are when $r = 0, dT/dr = 0$ (by symmetry); and when $r = R, T = T_w$. The final equation is

$$T = \frac{\dot{q}(R^2 - r^2)}{4k} + T_w \tag{5.6-28}$$

This is the same as Eq. (4.3-29), which was obtained by another method.

EXAMPLE 5.6-2. Laminar Flow and Heat Transfer Using Equation of Energy Change
Using the differential equation of energy change, derive the partial differential equation and boundary conditions needed for the case of laminar flow of a constant density fluid in a horizontal tube which is being heated. The fluid is flowing at a constant velocity v_z. At the wall of the pipe where the radius $r = r_0$, the heat flux is constant at q_0. The process is at steady state and it is assumed at $z = 0$ at the inlet that the velocity profile is established. Constant physical properties will be assumed.

Solution: From Example 3.8-3, the equation of continuity gives $\partial v_z/\partial z = 0$. Solution of the equation of motion for steady state using cylindrical coordinates gives the parabolic velocity profile.

$$v_z = v_{z\,max}\left[1 - \left(\frac{r}{r_0}\right)^2\right] \tag{5.6-29}$$

Since the fluid has a constant density, Eq. (5.6-14) in cylindrical coordinates will be used for the equation of energy change. For this case $v_r = 0$ and $v_\theta = 0$. Since this will be symmetrical $\partial T/\partial \theta$ and $\partial^2 T/\partial \theta^2$ will be zero. For steady state, $\partial T/\partial t = 0$. Hence, Eq. (5.6-14) reduces to

$$v_z \frac{\partial T}{\partial z} = \frac{k}{\rho c_p}\left(\frac{\partial^2 T}{\partial r^2} + \frac{1}{r}\frac{\partial T}{\partial r} + \frac{\partial^2 T}{\partial z^2}\right) \tag{5.6-30}$$

Usually conduction in the z direction ($\partial^2 T/\partial z^2$ term) is small compared to the convective term $v_z\,\partial T/\partial z$ and can be dropped. Finally, substituting Eq. (5.6-29) into (5.6-30), we obtain

$$v_{z\,max}\left[1 - \left(\frac{r}{r_0}\right)^2\right]\frac{\partial T}{\partial z} = \frac{k}{\rho c_p}\left(\frac{\partial^2 T}{\partial r^2} + \frac{1}{r}\frac{\partial T}{\partial r}\right) \tag{5.6-31}$$

The boundary conditions are

At $z = 0$,　　$T = T_0$　　(all r)

At $r = 0$,　　$T = $ finite

At $r = r_0$,　　$q_0 = -k\dfrac{\partial T}{\partial r}$　　(constant)

For details on the actual solution of this equation, see Siegel et al. (S2).

5.7 BOUNDARY-LAYER FLOW AND TURBULENCE IN HEAT TRANSFER

5.7A Laminar Flow and Boundary-Layer Theory in Heat Transfer

In section 3.10C an exact solution was obtained for the velocity profile for isothermal laminar flow past a flat plate. The solution of Blasius can be extended to include the convective heat-transfer problem for the same geometry and laminar flow. In Fig. 5.7-1 the thermal boundary layer is shown. The temperature of the fluid approaching the plate is T_∞ and that of the plate is T_S at the surface.

We start by writing the differential energy balance, Eq. (5.6-13).

$$\frac{\partial T}{\partial t} + v_x \frac{\partial T}{\partial x} + v_y \frac{\partial T}{\partial y} + v_z \frac{\partial T}{\partial z} = \frac{k}{\rho c_p}\left(\frac{\partial^2 T}{\partial x^2} + \frac{\partial^2 T}{\partial y^2} + \frac{\partial^2 T}{\partial z^2}\right) \tag{5.7-1}$$

If the flow is in the x and y directions, $v_z = 0$. At steady state, $\partial T/\partial t = 0$. Conduction is neglected in the x and z directions, so $\partial^2 T/\partial x^2 = \partial^2 T/\partial z^2 = 0$. Conduction occurs in the y direction. The result is

$$v_x \frac{\partial T}{\partial x} + v_y \frac{\partial T}{\partial y} = \frac{k}{\rho c_p}\frac{\partial^2 T}{\partial y^2} \tag{5.7-2}$$

The simplified momentum balance equation used in the velocity boundary-layer derivation is very similar and is

$$v_x \frac{\partial v_x}{\partial x} + v_y \frac{\partial v_x}{\partial y} = \frac{\mu}{\rho} \frac{\partial^2 v_x}{\partial y^2} \tag{3.10-5}$$

The continuity equation used previously is

$$\frac{\partial v_x}{\partial x} + \frac{\partial v_y}{\partial y} = 0 \tag{3.10-3}$$

Equations (3.10-5) and (3.10-3) were used by Blasius for solving the case for laminar boundary-layer flow. The boundary conditions used were

$$\frac{v_x}{v_\infty} = \frac{v_y}{v_\infty} = 0 \qquad \text{at } y = 0$$

$$\frac{v_x}{v_\infty} = 1 \qquad \text{at } y = \infty \tag{5.7-3}$$

$$\frac{v_x}{v_\infty} = 1 \qquad \text{at } x = 0$$

The similarity between Eqs. (3.10-5) and (5.7-2) is obvious. Hence, the Blasius solution can be applied if $k/\rho c_p = \mu/\rho$. This means the Prandtl number $c_p \mu/k = 1$. Also, the boundary conditions must be the same. This is done by replacing the temperature T in Eq. (5.7-2) by the dimensionless variable $(T - T_S)/(T_\infty - T_S)$. The boundary conditions become

$$\frac{v_x}{v_\infty} = \frac{v_y}{v_\infty} = \frac{T - T_S}{T_\infty - T_S} = 0 \qquad \text{at } y = 0$$

$$\frac{v_x}{v_\infty} = \frac{T - T_S}{T_\infty - T_S} = 1 \qquad \text{at } y = \infty \tag{5.7-4}$$

$$\frac{v_x}{v_\infty} = \frac{T - T_S}{T_\infty - T_S} = 1 \qquad \text{at } x = 0$$

We see that the equations and boundary conditions are identical for the temperature profile and the velocity profile. Hence, for any point x, y in the flow system, the dimensionless velocity variables v_x/v_∞ and $(T - T_S)/(T_\infty - T_S)$ are equal. The velocity-profile solution is the same as the temperature-profile solution.

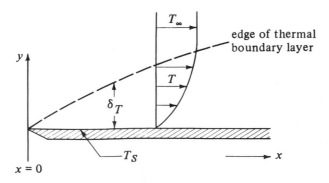

FIGURE 5.7-1.   *Laminar flow of fluid past a flat plate and thermal boundary layer.*

This means that the transfer of momentum and heat are directly analogous and the boundary-layer thickness δ for the velocity profile (hydrodynamic boundary layer) and the thermal boundary-layer thickness δ_T are equal. This is important for gases, where the Prandtl numbers are close to 1.

By combining Eqs. (3.10-7) and (3.10-8), the velocity gradient at the surface is

$$\left(\frac{\partial v_x}{\partial y}\right)_{y=0} = 0.332 \frac{v_\infty}{x} N_{Re,x}^{1/2} \tag{5.7-5}$$

where $N_{Re,x} = xv_\infty \rho/\mu$. Also,

$$\frac{v_x}{v_\infty} = \frac{T - T_S}{T_\infty - T_S} \tag{5.7-6}$$

Combining Eqs. (5.7-5) and (5.7-6),

$$\left(\frac{\partial T}{\partial y}\right)_{y=0} = (T_\infty - T_S)\left(\frac{0.332}{x} N_{Re,x}^{1/2}\right) \tag{5.7-7}$$

The convective equation can be related to the Fourier equation by the following, where q_y is in J/s or W (btu/h).

$$\frac{q_y}{A} = h_x(T_S - T_\infty) = -k\left(\frac{\partial T}{\partial y}\right)_{y=0} \tag{5.7-8}$$

Combining Eqs. (5.7-7) and (5.7-8),

$$\frac{h_x x}{k} = N_{Nu,x} = 0.332 N_{Re,x}^{1/2} \tag{5.7-9}$$

where $N_{Nu,x}$ is the dimensionless Nusselt number and h_x is the local heat-transfer coefficient at point x on the plate.

Pohlhausen (K1) was able to show that the relation between the hydrodynamic and thermal boundary layers for fluids with Prandtl number >0.6 gives approximately

$$\frac{\delta}{\delta_T} = N_{Pr}^{1/3} \tag{5.7-10}$$

As a result, the equation for the local heat-transfer coefficient is

$$h_x = 0.332 \frac{k}{x} N_{Re,x}^{1/2} N_{Pr}^{1/3} \tag{5.7-11}$$

Also,

$$\frac{h_x x}{k} = N_{Nu,x} = 0.332 N_{Re,x}^{1/2} N_{Pr}^{1/3} \tag{5.7-12}$$

The equation for the mean heat-transfer coefficient h from $x = 0$ to $x = L$ is for a plate of width b and area bL,

$$h = \frac{b}{A}\int_0^L h_x \, dx$$

$$= \frac{1}{L} 0.332k \left(\frac{\rho v_\infty}{\mu}\right)^{1/2} N_{Pr}^{1/3} \int_0^L \frac{dx}{x^{1/2}} \tag{5.7-13}$$

Integrating,

$$h = 0.644 \frac{k}{L} N_{Re, L}^{1/2} N_{Pr}^{1/3} \tag{5.7-14}$$

$$\frac{hL}{k} = N_{Nu} = 0.644 N_{Re, L}^{1/2} N_{Pr}^{1/3} \tag{5.7-15}$$

As pointed out previously, this laminar boundary layer on smooth plates holds up to a Reynolds number of about 5×10^5. In using the results above, the fluid properties are usually evaluated at the film temperature $T_f = (T_S + T_\infty)/2$.

5.7B Approximate Integral Analysis of the Thermal Boundary Layer

As discussed in the analysis of the hydrodynamic boundary layer, the Blasius solution is accurate but limited in its scope. Other more complex systems cannot be solved by this method. The approximate integral analysis was used by von Kármán to calculate the hydrodynamic boundary layer and was covered in Section 3.10. This approach can be used to analyze the thermal boundary layer.

This method will be outlined briefly. First, a control volume, as previously given in Fig. 3.10-5, is used to derive the final energy integral expression.

$$\frac{k}{\rho c_p} \left(\frac{\partial T}{\partial y} \right)_{y=0} = \frac{d}{dx} \int_0^{\delta_T} v_x (T_\infty - T) \, dy \tag{5.7-16}$$

This equation is analogous to Eq. (3.10-48) combined with Eq. (3.10-51) for the momentum analysis, giving

$$\frac{\mu}{\rho} \left(\frac{\partial v_x}{\partial y} \right)_{y=0} = \frac{d}{dx} \int_0^{\delta} v_x (v_\infty - v_x) \, dy \tag{5.7-17}$$

Equation (5.7-16) can be solved if both a velocity profile and temperature profile are known. The assumed velocity profile used is Eq. (3.10-50).

$$\frac{v_x}{v_\infty} = \frac{3}{2} \frac{y}{\delta} - \frac{1}{2} \left(\frac{y}{\delta} \right)^3 \tag{3.10-50}$$

The same form of temperature profile is assumed.

$$\frac{T - T_S}{T_\infty - T_S} = \frac{3}{2} \frac{y}{\delta_T} - \frac{1}{2} \left(\frac{y}{\delta_T} \right)^3 \tag{5.7-18}$$

Substituting Eqs. (3.10-50) and (5.7-18) into the integral expression and solving,

$$N_{Nu, x} = 0.36 N_{Re, x}^{1/2} N_{Pr}^{1/3} \tag{5.7-19}$$

This is only about 8% greater than the exact result in Eq. (5.7-11), which indicates that this approximate integral method can be used with confidence in cases where exact solutions cannot be obtained.

In a similar fashion, the integral momentum analysis method used for the turbulent hydrodynamic boundary layer in Section 3.10 can be used for the thermal boundary layer in turbulent flow. Again, the Blasius $\frac{1}{7}$-power law is used for the temperature distribution. These give results that are quite similar to the experimental equations as given in Section 4.6.

5.7C Prandtl Mixing Length and Eddy Thermal Diffusivity

1. Eddy momentum diffusivity in turbulent flow. In Section 3.10F the total shear stress $\bar{\tau}^t_{yx}$ for turbulent flow was written as follows when the molecular and turbulent contributions are summed together:

$$\bar{\tau}^t_{yx} = -\rho\left(\frac{\mu}{\rho} + \varepsilon_t\right)\frac{d\bar{v}_x}{dy} \tag{5.7-20}$$

The molecular momentum diffusivity μ/ρ in m^2/s is a function only of the fluid molecular properties. However, the turbulent momentum eddy diffusivity ε_t depends on the fluid motion. In Eq. (3.10-29) we related ε_t to the Prandtl mixing length L as follows:

$$\varepsilon_t = L^2\left|\frac{d\bar{v}_x}{dy}\right| \tag{3.10-29}$$

2. Prandtl mixing length and eddy thermal diffusivity. We can derive in a similar manner the eddy thermal diffusivity α_t for turbulent heat transfer as follows. Eddies or clumps of fluid are transported a distance L in the y direction. At this point L the clump of fluid differs in mean velocity from the adjacent fluid by the velocity v'_x, which is the fluctuating velocity component discussed in Section 3.10F. Energy is also transported the distance L with a velocity v'_y in the y direction together with the mass being transported. The instantaneous temperature of the fluid is $T = T' + \bar{T}$, where $\bar{T}$ is the mean value and T' the deviation from the mean value. This fluctuating T' is similar to the fluctuating velocity v'_x. The mixing length is small enough so that the temperature difference can be written as

$$T' = L\frac{d\bar{T}}{dy} \tag{5.7-21}$$

The rate of energy transported per unit area is q_y/A and is equal to the mass flux in the y direction times the heat capacity times the temperature difference.

$$\frac{q_y}{A} = \frac{-v'_y\,\rho c_p L\,d\bar{T}}{dy} \tag{5.7-22}$$

In Section 3.10F we assumed $v'_x \cong v'_y$ and that

$$v'_x = v'_y = L\left|\frac{d\bar{v}_x}{dy}\right| \tag{5.7-23}$$

Substituting Eq. (5.7-23) into (5.7-22),

$$\frac{q_y}{A} = -\rho c_p L^2\left|\frac{d\bar{v}_x}{d_y}\right|\frac{d\bar{T}}{dy} \tag{5.7-24}$$

The term $L^2|d\bar{v}_x/dy|$ by Eq. (3.10-29) is the momentum eddy diffusivity ε_t. When this term is in the turbulent heat-transfer equation (5.7-24), it is called α_t, eddy thermal diffusivity. Then Eq. (5.7-24) becomes

$$\frac{q_y}{A} = -\rho c_p \alpha_t \frac{d\bar{T}}{dy} \tag{5.7-25}$$

Combining this with the Fourier equation written in terms of the molecular thermal diffusivity α,

$$\frac{q_y}{A} = -\rho c_p(\alpha + \alpha_t)\frac{d\bar{T}}{dy} \tag{5.7-26}$$

3. *Similarities among momentum, heat, and mass transport.* Equation (5.7-26) is similar to Eq. (5.7-20) for total momentum transport. The eddy thermal diffusivity α_t and the eddy momentum diffusivity ε_t have been assumed equal in the derivations. Experimental data show that this equality is only approximate. An eddy mass diffusivity for mass transfer has also been defined in a similar manner using the Prandtl mixing length theory and is assumed equal to α_t and ε_t.

PROBLEMS

5.2-1. *Temperature Response in Cooling a Wire.* A small copper wire with a diameter of 0.792 mm and initially at 366.5 K is suddenly immersed into a liquid held constant at 311 K. The convection coefficient $h = 85.2$ W/m^2 · K. The physical properties can be assumed constant and are $k = 374$ W/m · K, $c_p = 0.389$ kJ/kg · K, and $\rho = 8890$ kg/m^3.
   (a) Determine the time in seconds for the average temperature of the wire to drop to 338.8 K (one half the initial temperature difference).
   (b) Do the same but for $h = 11.36$ W/m^2 · K.
   (c) For part (b), calculate the total amount of heat removed for a wire 1.0 m long.
   Ans. (a) $t = 5.66$ s

5.2-2. *Quenching Lead Shot in a Bath.* Lead shot having an average diameter of 5.1 mm is at an initial temperature of 204.4°C. To quench the shot it is added to a quenching oil bath held at 32.2°C and falls to the bottom. The time of fall is 15 s. Assuming an average convection coefficient of $h = 199$ W/m^2 · K, what will be the temperature of the shot after the fall? For lead, $\rho = 11\,370$ kg/m^3 and $c_p = 0.138$ kJ/kg · K.

5.2-3. *Unsteady-State Heating of a Stirred Tank.* A vessel is filled with 0.0283 m^3 of water initially at 288.8 K. The vessel, which is well stirred, is suddenly immersed into a steam bath held at 377.6 K. The overall heat-transfer coefficient U between the steam and water is 1136 W/m^2 · K and the area is 0.372 m^2. Neglecting the heat capacity of the walls and agitator, calculate the time in hours to heat the water to 338.7 K. [*Hint :* Since the water is well stirred, its temperature is uniform. Show that Eq. (5.2-3) holds by starting with Eq. (5.2-1).]

5.3-1. *Temperature in a Refractory Lining.* A combustion chamber has a 2-in.-thick refractory lining to protect the outer shell. To predict the thermal stresses at startup, the temperature 0.2 in. below the surface is needed 1 min after startup. The following data are available. The initial temperature $T_0 = 100°F$, the hot gas temperature $T_1 = 3000°F$, $h = 40$ btu/h · ft^2 · °F, $k = 0.6$ btu/h · ft · °F, and $\alpha = 0.020$ ft^2/h. Calculate the temperature at a 0.2 in. depth and at a 0.6 in. depth. Use Fig. 5.3-3 and justify its use by seeing if the lining acts as a semiinfinite solid during this 1-min period.
   Ans. For $x = 0.2$ in., $(T - T_0)/(T_1 - T_0) = 0.28$ and $T = 912°F$ (489°C); for $x = 0.6$ in., $(T - T_0)/(T_1 - T_0) = 0.02$ and $T = 158°F (70°C)$

5.3-2. *Freezing Temperature in the Soil.* The average temperature of the soil to a considerable depth is approximately 277.6 K (40°F) during a winter day. If the outside air temperature suddenly drops to 255.4 K (0°F) and stays there, how long will it take for a pipe 3.05 m (10 ft) below the surface to reach 273.2 K (32°F)? The convective coefficient is $h = 8.52$ W/m^2 · K (1.5 btu/h · ft^2 · °F). The soil physical properties can be taken as 5.16×10^{-7} m^2/s (0.02 ft^2/h) for the thermal diffusivity and 1.384 W/m · K (0.8 btu/h · ft · °F) for the thermal conductivity. (*Note*: The solution is trial and error, since the unknown time appears twice in the graph for a semiinfinite solid.)

5.3-3. *Cooling a Slab of Aluminum.* A large piece of aluminum that can be considered a semiinfinite solid initially has a uniform temperature of 505.4 K. The surface is suddenly exposed to an environment at 338.8 K with a surface convection coefficient of 455 W/m² · K. Calculate the time in hours for the temperature to reach 388.8 K at a depth of 25.4 mm. The average physical properties are $\alpha = 0.340$ m²/h and $k = 208$ W/m · K.

5.3-4. *Transient Heating of a Concrete Wall.* A wall made of concrete 0.305 m thick is insulated on the rear side. The wall at a uniform temperature of 10°C (283.2 K) is exposed on the front side to a gas at 843°C (1116.2 K). The convection coefficient is 28.4 W/m² · K, the thermal diffusivity is 1.74×10^{-3} m²/h, and the thermal conductivity is 0.935 W/m · K.
   (a) Calculate the time for the temperature at the insulated face to reach 232°C (505.2 K).
   (b) Calculate the temperature at a point 0.152 m below the surface at this same time.

   Ans. (a) $\alpha t/x_1^2 = 0.25$, $t = 13.4$ h

5.3-5. *Cooking a Slab of Meat.* A slab of meat 25.4 mm thick originally at a uniform temperature of 10°C is to be cooked from both sides until the center reaches 121°C in an oven at 177°C. The convection coefficient can be assumed constant at 25.6 W/m² · K. Neglect any latent heat changes and calculate the time required. The thermal conductivity is 0.69 W/m · K and the thermal diffusivity 5.85×10^{-4} m²/h. Use the Heisler chart.

   Ans. 0.80 h (2880 s)

5.3-6. *Unsteady-State Conduction in a Brick Wall.* A flat brick wall 1.0 ft thick is the lining on one side of a furnace. If the wall is at a uniform temperature of 100°F and the one side is suddenly exposed to a gas at 1100°F, calculate the time for the furnace wall at a point 0.5 ft from the surface to reach 500°F. The rear side of the wall is insulated. The convection coefficient is 2.6 btu/h · ft² · °F and the physical properties of the brick are $k = 0.65$ btu/h · ft · °F and $\alpha = 0.02$ ft²/h.

5.3-7. *Cooling a Steel Rod.* A long steel rod 0.305 m in diameter is initially at a temperature of 588 K. It is immersed in an oil bath maintained at 311 K. The surface convective coefficient is 125 W/m² · K. Calculate the temperature at the center of the rod after 1 h. The average physical properties of the steel are $k = 38$ W/m · K and $\alpha = 0.0381$ m²/h.

   Ans. $T = 391$ K

5.3-8. *Effect of Size on Heat-Processing Meat.* An autoclave held at 121.1°C is being used to process sausage meat 101.6 mm in diameter and 0.61 m long which is originally at 21.1°C. After 2 h the temperature at the center is 98.9°C. If the diameter is increased to 139.7 mm, how long will it take for the center to reach 98.9°C? The heat transfer coefficient to the surface is $h = 1100$ W/m² · K, which is very large so that the surface resistance can be considered negligible. (Show this.) Neglect the heat transfer from the ends of the cylinder. The thermal conductivity $k = 0.485$ W/m · K.

   Ans. 3.78 h

5.3-9. *Temperature of Oranges on Trees During Freezing Weather.* In orange-growing areas, the freezing of the oranges on the trees during cold nights is economically important. If the oranges are initially at a temperature of 21.1°C, calculate the center temperature of the orange if exposed to air at -3.9°C for 6 h. The oranges are 102 mm in diameter and the convective coefficient is estimated as 11.4 W/m² · K. The thermal conductivity k is 0.431 W/m · K and α is 4.65×10^{-4} m²/h. Neglect any latent heat effects.

   Ans. $(T_1 - T)/(T_1 - T_0) = 0.05$, $T = -2.65$°C

5.3-10. *Hardening a Steel Sphere.* To harden a steel sphere having a diameter of 50.8 mm, it is heated to 1033 K and then dunked into a large water bath at 300 K. Determine the time for the center of the sphere to reach 366.5 K. The surface coefficient can be assumed as 710 W/m² · K, $k = 45$ W/m · K, and $\alpha = 0.0325$ m²/h.

5.3-11. *Unsteady-State Conduction in a Short Cylinder.* An aluminum cylinder is initially heated so it is at a uniform temperature of 204.4°C. Then it is plunged into a large bath held at 93.3°C, where $h = 568$ W/m² · K. The cylinder has a diameter of 50.8 mm and is 101.6 mm long. Calculate the center temperature after 60 s. The physical properties are $\alpha = 9.44 \times 10^{-5}$ m²/s and $k = 207.7$ W/m · K.

5.3-12. *Conduction in Three Dimensions in a Rectangular Block.* A rectangular steel block 0.305 m by 0.457 m by 0.61 m is initially at 315.6°C. It is suddenly immersed into an environment at 93.3°C. Determine the temperature at the center of the block after 1 h. The surface convection coefficient is 34 W/m² · K. The physical properties are $k = 38$ W/m · K and $\alpha = 0.0379$ m²/h.

5.4-1. *Schmidt Numerical Method for Unsteady-State Conduction.* A material in the form of an infinite plate 0.762 m thick is at an initial uniform temperature of 366.53 K. The rear face of the plate is insulated. The front face is suddenly exposed to a temperature of 533.2 K. The convective resistance at this face can be assumed as zero. Calculate the temperature profile after 0.875 h using the Schmidt numerical method with $M = 2$ and slabs 0.1524 m thick. The thermal diffusivity is 0.0929 m²/h.

Ans. $\Delta t = 0.125$ h, seven time increments needed

5.4-2. *Unsteady-State Conduction with Nonuniform Initial Temperature Profile.* Use the same conditions as Problem 5.4-1 but with the following change. The initial temperature profile is not uniform but is 366.53 K at the front face and 422.1 K at the rear face with a linear variation between the two faces.

5.4-3. *Unsteady-State Conduction Using the Digital Computer.* Repeat Problem 5.4-2 but use the digital computer and write the Fortran program. Use slabs 0.03048 m thick and $M = 2.0$. Calculate the temperature profile after 0.875 h.

5.4-4. *Chilling Meat Using Numerical Methods.* A slab of beef 45.7 mm thick and initially at a uniform temperature of 283 K is being chilled by a surface contact cooler at 274.7 K on the front face. The rear face of the meat is insulated. Assume that the convection resistance at the front surface is zero. Using five slices and $M = 2$, calculate the temperature profile after 0.54 h. The thermal diffusivity is 4.64×10^{-4} m²/h.

Ans. $\Delta t = 0.090$ h, six time increments

5.4-5. *Cooling Beef with Convective Resistance.* A large slab of beef is 45.7 mm thick and is at an initial uniform temperature of 37.78°C. It is being chilled at the front surface in a chilled air blast at −1.11°C with a convective heat-transfer coefficient of $h = 38.0$ W/m² · K. The rear face of the meat is insulated. The thermal conductivity of the beef is $k = 0.498$ W/m · K and $\alpha = 4.64 \times 10^{-4}$ m²/h. Using a numerical method with five slices and $M = 4.0$, calculate the temperature profile after 0.27 h. [*Hint:* Since there is a convective resistance, the value of N must be calculated. Also, Eq. (5.4-7) should be used to calculate the surface temperature $_{t+\Delta t}T_1$.]

5.4-6. *Cooling Beef Using the Digital Computer.* Repeat Problem 5.4-5 using the digital computer. Use 20 slices and $M = 4.0$. Write the Fortran program.

5.4-7. *Convection and Unsteady-State Conduction.* For the conditions of Example 5.4-3, continue the calculations for a total of 12 time increments. Plot the temperature profile.

5.4-8. *Alternative Convective Boundary Condition for Numerical Method.* Repeat Example 5.4-3 but instead use the alternative boundary condition, Eq. (5.4-11). Also, use $M = 4$. Calculate the profile for the full 12 time increments.

5.4-9. Numerical Method for Semi-infinite Solid and Convection. A semiinfinite solid initially at a uniform temperature of 200°C is cooled at its surface by convection. The cooling fluid at a constant temperature of 100°C has a convective coefficient of $h = 250$ W/m$^2 \cdot$ K. The physical properties of the solid are $k = 20$ W/m $\cdot$ K and $\alpha = 4 \times 10^{-5}$ m^2/s. Using a numerical method with $\Delta x = 0.040$ m and $M = 4.0$, calculate the temperature profile after 50 s total time.

Ans. $T_1 = 157.72$, $T_2 = 181.84$, $T_3 = 194.44$, $T_4 = 198.93$, $T_5 = 199.90$°C

5.5-1. Chilling Slab of Beef. Repeat Example 5.5-1, where the slab of beef is cooled to 10°C at the center but use air of 0°C at a lower value of $h = 22.7$ W/m$^2 \cdot$ K.

Ans. $(T_1 - T)/(T_1 - T_0) = 0.265$, $X = 0.92$, $t = 19.74$ h

5.5-2. Chilling Fish Fillets. Cod fish fillets originally at 10°C are packed to a thickness of 102 mm. Ice is packed on both sides of the fillets and wet-strength paper separates the ice and fillets. The surface temperature of the fish can be assumed as essentially 0°C. Calculate the time for the center of the filets to reach 2.22°C and the temperature at this time at a distance of 25.4 mm from the surface. Also, plot temperature versus position for the slab. The physical properties are (B1) $k = 0.571$ W/m $\cdot$ K, $\rho = 1052$ kg/m^3, and $c_p = 4.02$ kJ/kg $\cdot$ K.

5.5-3. Average Temperature in Chilling Fish. Fish fillets having the same physical properties given in Problem 5.5-2 are originally at 10°C. They are packed to a thickness of 102 mm with ice on each side. Assuming that the surface temperature of the fillets is 0°C, calculate the time for the average temperature to reach 1.39°C. (*Note:* This is a case where the surface resistance is zero. Can Fig. 5.3-13 be used for this case?)

5.5-4. Time to Freeze a Slab of Meat. Repeat Example 5.5-2 using the same conditions except that a plate or contact freezer is used where the surface coefficient can be assumed as $h = 142$ W/m$^2 \cdot$ K.

Ans. $t = 2.00$ h

5.5-5. Freezing a Cylinder of Meat. A package of meat containing 75% moisture and in the form of a long cylinder 5 in. in diameter is to be frozen in an air-blast freezer at -25°F. The meat is initially at the freezing temperature of 27°F. The heat-transfer coefficient is $h = 3.5$ btu/h $\cdot$ ft$^2 \cdot$ °F. The physical properties are $\rho = 64$ lb$_m$/ft^3 for the unfrozen meat and $k = 0.60$ btu/h $\cdot$ ft $\cdot$ °F for the frozen meat. Calculate the freezing time.

5.6-1. Heat Generation Using Equation of Energy Change. A plane wall with uniform internal heat generation of $\dot{q}$ W/m^3 is insulated at four surfaces with heat conduction only in the x direction. The wall has a thickness of $2L$ m. The temperature at the one wall at $x = +L$ and at the other wall at $x = -L$ is held constant at T_w K. Using the differential equation of energy change, Eq. (5.6-18), derive the equation for the final temperature profile.

Ans. $T = \dfrac{\dot{q}(L^2 - x^2)}{2k} + T_w$

5.6-2. Heat Transfer in a Solid Using Equation of Energy Change. A solid of thickness L is at a uniform temperature of T_0 K. Suddenly the front surface temperature of the solid at $z = 0$ m is raised to T_1 at $t = 0$ and held there and at $z = L$ at the rear to T_2 and held constant. Heat transfer occurs only in the z direction. For constant physical properties and using the differential equation of energy change, do as follows.

(a) Derive the partial differential equation and the boundary conditions for unsteady-state energy transfer.

(b) Do the same for steady state and integrate the final equation.

Ans. (a) $\partial T/\partial t = \alpha \, \partial^2 T/\partial z^2$; B.C.(1): $t = 0$, $z = z$, $T = T_0$; B.C.(2): $t = t$, $z = 0$, $T = T_1$; B.C.(3): $t = t$, $z = L$, $T = T_2$; (b) $T = (T_2 - T_1)z/L + T_1$

5.6-3. Radial Temperature Profile Using the Equation of Energy Change. Radial heat transfer is occurring by conduction through a long hollow cylinder of length L with the ends insulated.

(a) What is the final differential equation for steady-state conduction? Start with Fourier's second law in cylindrical coordinates, Eq. (5.6-20).

(b) Solve the equation for the temperature profile from part (a) for the boundary conditions given as follows: $T = T_i$ for $r = r_i$, $T = T_o$ for $r = r_o$.

(c) Using part (b), derive an expression for the heat flow q in W.

Ans. (b) $T = T_i - \dfrac{T_i - T_o}{\ln (r_o/r_i)} \ln \dfrac{r}{r_i}$

5.6-4. Heat Conduction in a Sphere. Radial energy flow is occurring in a hollow sphere with an inside radius of r_i and an outside radius of r_o. At steady state the inside surface temperature is constant at T_i and constant at T_o on the outside surface.

(a) Using the differential equation of energy change, solve the equation for the temperature profile.

(b) Using part (a), derive an expression for the heat flow in W.

5.6-5. Variable Heat Generation and Equation of Energy Change. A plane wall is insulated so that conduction occurs only in the x direction. The boundary conditions which apply at steady state are $T = T_0$ at $x = 0$ and $T = T_L$ at $x = L$. Internal heat generation per unit volume is occurring and varies as $\dot{q} = \dot{q}_0 e^{-\beta x/L}$, where $\dot{q}_0$, and β are constants. Solve the general differential equation of energy change for the temperature profile.

5.7-1. Thermal and Hydrodynamic Boundary Layer Thicknesses. Air at 294.3 K and 101.3 kPa with a free stream velocity of 12.2 m/s is flowing parallel to a smooth flat plate held at a surface temperature of 383 K. Do the following.

(a) At the critical $N_{Re, L} = 5 \times 10^5$, calculate the critical length $x = L$ of the plate, the thickness δ of the hydrodynamic boundary layer, and the thickness δ_T of the thermal boundary layer. Note that the Prandtl number is not 1.0.

(b) Calculate the average heat-transfer coefficient over the plate covered by the laminar boundary layer.

5.7-2. Boundary-Layer Thicknesses and Heat Transfer. Air at 37.8°C and 1 atm abs flows at a velocity of 3.05 m/s parallel to a flat plate held at 93.3°C. The plate is 1 m wide. Calculate the following at a position 0.61 m from the leading edge.

(a) The thermal boundary-layer thickness δ_T and the hydrodynamic boundary-layer thickness δ.

(b) Total heat transfer from the plate.

REFERENCES

(B1) BLAKEBROUGH, N. *Biochemical and Biological Engineering Science*, Vol. 2. New York: Academic Press, Inc., 1968.

(B2) BIRD, R. B., STEWART, W. E., and LIGHTFOOT, E. N. *Transport Phenomena*. New York: John Wiley & Sons, Inc., 1960.

(C1) CARSLAW, H. S., and JAEGER, J. C. *Conduction of Heat in Solids*. Oxford: Clarendon Press, 1959.

(C2) CHARM, S. E. *The Fundamentals of Food Engineering*, 2nd ed. Westport, Conn.: Avi Publishing Co., Inc., 1971.

(D1) DUSINBERRE, G. M. *Heat Transfer Calculations by Finite Differences*. Scranton, Pa.: International Textbook Co., Inc., 1961.

(G1) GEANKOPLIS, C. J. *Mass Transport Phenomena.* Columbus, Ohio: Ohio State University Bookstores, 1972.

(G2) GURNEY, H. P., and LURIE, J. *Ind. Eng. Chem.*, **15**, 1170 (1923).

(H1) HEISLER, H. P. *Trans. A.S.M.E.*, **69**, 227 (1947).

(H2) HODGSON, T. *Fd. Inds. S. Afr.*, **16**, 41 (1964); *Int. Inst. Refrig. Annexe*, **1966**, 633 (1966).

(K1) KREITH, F. *Principles of Heat Transfer*, 2nd ed. Scranton, Pa.: International Textbook Company, 1965.

(K2) KREITH, F., and BLACK, W. Z. *Basic Heat Transfer.* New York: Harper & Row, Publishers, 1980.

(N1) NEWMAN, A. H. *Ind. Eng. Chem.*, **28**, 545 (1936).

(P1) PERRY, R. H., and CHILTON, C. H. *Chemical Engineers' Handbook*, 5th ed. New York: McGraw-Hill Book Company, 1973.

(P2) PLANK, R. *Z. Ges. Kalteind.*, **20**, 109 (1913); *Z. Ges. Kalteind. Bieh. Reih*, **10** (3), 1 (1941).

(R1) RIEDEL, L. *Kaltetchnik*, **8**, 374 (1956); **9**, 38 (1957); **11**, 41 (1959); **12**, 4 (1960).

(S1) SCHNEIDER, P. J. *Conduction Heat Transfer.* Reading, Mass.: Addison-Wesley Publishing Company, Inc., 1955.

(S2) SIEGEL, R., SPARROW, E. M., and HALLMAN, T. M. *Appl. Sci. Res.*, **A7**, 386 (1958).

CHAPTER 6

Principles of
Mass Transfer

6.1 INTRODUCTION TO MASS TRANSFER AND DIFFUSION

6.1A Similarity of Mass, Heat, and Momentum Transfer Processes

1. Introduction. In Chapter 1 we noted that the various unit operations could be classified into three fundamental transfer (or "transport") processes: momentum transfer, heat transfer, and mass transfer. The fundamental process of *momentum transfer* occurs in such unit operations as fluid flow, mixing, sedimentation, and filtration. *Heat transfer* occurs in conductive and convective transfer of heat, evaporation, distillation, and drying.

The third fundamental transfer process, *mass transfer*, occurs in distillation, absorption, drying, liquid–liquid extraction, adsorption, and membrane processes. When mass is being transferred from one distinct phase to another or through a single phase, the basic mechanisms are the same whether the phase is a gas, liquid, or solid. This was also shown in heat transfer, where the transfer of heat by conduction followed Fourier's law in a gas, solid, or liquid.

2. General molecular transport equation. All three of the molecular transport processes of momentum, heat, and mass are characterized by the same general type of equation given in Section 2.3A.

$$\text{rate of a transfer process} = \frac{\text{driving force}}{\text{resistance}} \qquad \textbf{(2.3-1)}$$

This can be written as follows for molecular diffusion of the property momentum, heat, and mass:

$$\psi_z = -\delta \frac{d\Gamma}{dz} \qquad \textbf{(2.3-2)}$$

3. Molecular diffusion equations for momentum, heat, and mass transfer. Newton's equation for momentum transfer for constant density can be written as follows in a manner similar to Eq. (2.3-2):

$$\tau_{zx} = -\frac{\mu}{\rho} \frac{d(v_x \rho)}{dz} \qquad \textbf{(6.1-1)}$$

where τ_{zx} is momentum transferred/s · m², μ/ρ is kinematic viscosity in m²/s, z is distance in m, and $v_x \rho$ is momentum/m³, where the momentum has units of kg · m/s.

Fourier's law for heat conduction can be written as follows for constant ρ and c_p:

$$\frac{q_z}{A} = -\alpha \frac{d(\rho c_p T)}{dz} \qquad (6.1\text{-}2)$$

where q_z/A is heat flux in W/m², α is the thermal diffusivity in m²/s, and $\rho c_p T$ is J/m³.

The equation for molecular diffusion of mass is Fick's law and is also similar to Eq. (2.3-2). It is written as follows for constant total concentration in a fluid:

$$J_{Az}^* = -D_{AB} \frac{dc_A}{dz} \qquad (6.1\text{-}3)$$

where J_{Az}^* is the molar flux of component A in the z direction due to molecular diffusion in kg mol A/s · m², D_{AB} the molecular diffusivity of the molecule A in B in m²/s, c_A the concentration of A in kg mol/m³, and z the distance of diffusion in m. In cgs units J_{Az}^* is g mol A/s · cm², D_{AB} is cm²/s, and c_A is g mol A/cm³. In English units, J_{Az}^* is lb mol/h · ft², D_{AB} is ft²/h, and c_A is lb mol/ft³.

The similarity of Eqs. (6.1-1), (6.1-2), and (6.1-3) for momentum, heat, and mass transfer should be obvious. All the fluxes on the left-hand side of the three equations have as units transfer of a quantity of momentum, heat, or mass per unit time per unit area. The transport properties μ/ρ, α, and D_{AB} all have units of m²/s, and the concentrations are represented as momentum/m³, J/m³, or kg mol/m³.

4. Turbulent diffusion equations for momentum, heat, and mass transfer. In Section 5.7C equations were given discussing the similarities among momentum, heat, and mass transfer in turbulent transfer. For turbulent momentum transfer and constant density,

$$\tau_{zx} = -\left(\frac{\mu}{\rho} + \varepsilon_t\right) \frac{d(v_x \rho)}{dz} \qquad (6.1\text{-}4)$$

For turbulent heat transfer for constant ρ and c_p,

$$\frac{q_z}{A} = -(\alpha + \alpha_t) \frac{d(\rho c_p T)}{dz} \qquad (6.1\text{-}5)$$

For turbulent mass transfer for constant c,

$$J_{Az}^* = -(D_{AB} + \varepsilon_M) \frac{dc_A}{dz} \qquad (6.1\text{-}6)$$

In these equations ε_t is the turbulent or eddy momentum diffusivity in m²/s, α_t the turbulent or eddy thermal diffusivity in m²/s, and ε_M the turbulent or eddy mass diffusivity in m²/s. Again, these equations are quite similar to each other. Many of the theoretical equations and empirical correlations for turbulent transport to various geometries are also quite similar.

6.1B Examples of Mass-Transfer Processes

Mass transfer is important in many areas of science and engineering. Mass transfer occurs when a component in a mixture migrates in the same phase or from phase to phase because of a difference in concentration between two points. Many familiar phenomena involve mass transfer. Liquid in an open pail of water evaporates into still air because of the difference in concentration of water vapor at the water surface and the surrounding air. There is a "driving force" from the surface to the air. A piece of sugar added to a cup of coffee eventually dissolves by itself and diffuses to the surrounding solution. When newly cut and moist green timber is exposed to the atmosphere, the wood will dry partially when water in the timber diffuses through the wood, to the surface, and then to the atmosphere. In a fermentation process nutrients and oxygen dissolved in the solution diffuse to the microorganisms. In a catalytic reaction the reactants diffuse from the surrounding medium to the catalyst surface, where reaction occurs.

Many purification processes involve mass transfer. In uranium processing, a uranium salt in solution is extracted by an organic solvent. Distillation to separate alcohol from water involves mass transfer. Removal of SO_2 from flue gas is done by absorption in a basic liquid solution.

We can treat mass transfer in a manner somewhat similar to that used in heat transfer with Fourier's law of conduction. However, an important difference is that in molecular mass transfer one or more of the components of the medium is moving. In heat transfer by conduction the medium is usually stationary and only energy in the form of heat is being transported. This introduces some differences between heat and mass transfer that will be discussed in this chapter.

6.1C Fick's Law for Molecular Diffusion

Molecular diffusion or molecular transport can be defined as the transfer or movement of individual molecules through a fluid by means of the random, individual movements of the molecules. We can imagine the molecules traveling only in straight lines and changing direction by bouncing off other molecules after collisions. Since the molecules travel in a random path, molecular diffusion is often called a *random-walk process*.

In Fig. 6.1-1 the *molecular diffusion process* is shown schematically. A random path that molecule A might take in diffusing through B molecules from point (1) to (2) is shown. If there are a greater number of A molecules near point (1) than at (2), then, since molecules diffuse randomly in both directions, more A molecules will diffuse from (1) to (2) than from (2) to (1). The net diffusion of A is from high- to low-concentration regions.

As another example, a drop of blue liquid dye is added to a cup of water. The dye molecules will diffuse slowly by molecular diffusion to all parts of the water. To increase this rate of mixing of the dye, the liquid can be mechanically agitated by a spoon and *convective mass transfer* will occur. The two modes of heat transfer, conduction and convective heat transfer, are analogous to molecular diffusion and convective mass transfer.

First, we will consider the diffusion of molecules when the whole bulk fluid is not moving but is stationary. Diffusion of the molecules is due to a concentration gradient.

FIGURE 6.1-1.   *Schematic diagram of molecular diffusion process.*

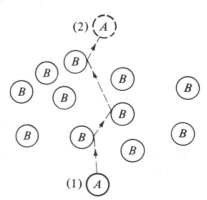

The general Fick's law equation can be written as follows for a binary mixture of A and B:

$$J_{Az}^* = -cD_{AB}\frac{dx_A}{dz} \tag{6.1-7}$$

where c is total concentration of A and B in kg mol $A + B/m^3$, and x_A is the mole fraction of A in the mixture of A and B. If c is constant, then since $c_A = cx_A$,

$$c\,dx_A = d(cx_A) = dc_A \tag{6.1-8}$$

Substituting into Eq. (6.1-7) we obtain Eq. (6.1-3) for constant total concentration.

$$J_{Az}^* = -D_{AB}\frac{dc_A}{dz} \tag{6.1-3}$$

This equation is the more commonly used one in many molecular diffusion processes. If c varies some, an average value is often used with Eq. (6.1-3).

EXAMPLE 6.1-1. Molecular Diffusion of Helium in Nitrogen
A mixture of He and N_2 gas is contained in a pipe at 298 K and 1 atm total pressure which is constant throughout. At one end of the pipe at point 1 the partial pressure p_{A1} of He is 0.60 atm and at the other end 0.2 m (20 cm) $p_{A2} = 0.20$ atm. Calculate the flux of He at steady state if D_{AB} of the He–N_2 mixture is 0.687×10^{-4} m²/s (0.687 cm²/s). Use SI and cgs units.

Solution: Since total pressure P is constant, then c is constant, where c is as follows for a gas from the perfect gas law.

$$PV = nRT \tag{6.1-9}$$

$$\frac{n}{V} = \frac{P}{RT} = c \tag{6.1-10}$$

where n is kg mol A plus B, V is volume in m³, T is temperature in K, R is 8314.3 m³·Pa/kg mol·K or R is 82.057×10^{-3} m³·atm/kg mol·K, and c is kg mol A plus B/m^3. In cgs units, R is 82.057 cm³·atm/g mol·K.

For steady state the flux J_{Az}^* in Eq. (6.1-3) is constant. Also, D_{AB} for a gas is constant. Rearranging Eq. (6.1-3) and integrating,

$$J_{Az}^* \int_{z_1}^{z_2} dz = -D_{AB} \int_{c_{A1}}^{c_{A2}} dc_A$$

$$J_{Az}^* = \frac{D_{AB}(c_{A1} - c_{A2})}{z_2 - z_1} \tag{6.1-11}$$

Also, from the perfect gas law, $p_A V = n_A RT$, and

$$c_{A1} = \frac{p_{A1}}{RT} = \frac{n_A}{V} \tag{6.1-12}$$

Substituting Eq. (6.1-12) into (6.1-11),

$$J_{Az}^* = \frac{D_{AB}(p_{A1} - p_{A2})}{RT(z_2 - z_1)} \tag{6.1-13}$$

This is the final equation to use, which is in the form easily used for gases. Partial pressures are $p_{A1} = 0.6$ atm $= 0.6 \times 1.01325 \times 10^5 = 6.08 \times 10^4$ Pa and $p_{A2} = 0.2$ atm $= 0.2 \times 1.01325 \times 10^5 = 2.027 \times 10^4$ Pa. Then, using SI units,

$$J^*_{Az} = \frac{(0.687 \times 10^{-4})(6.08 \times 10^4 - 2.027 \times 10^4)}{8314(298)(0.20 - 0)}$$

$$= 5.63 \times 10^{-6} \text{ kg mol } A/\text{s} \cdot \text{m}^2$$

If pressures in atm are used with SI units,

$$J^*_{Az} = \frac{(0.687 \times 10^{-4})(0.60 - 0.20)}{(82.06 \times 10^{-3})(298)(0.20 - 0)} = 5.63 \times 10^{-6} \text{ kg mol } A/\text{s} \cdot \text{m}^2$$

For cgs units, substituting into Eq. (6.1-13),

$$J^*_{Az} = \frac{0.687(0.60 - 0.20)}{82.06(298)(20 - 0)} = 5.63 \times 10^{-7} \text{ g mol } A/\text{s} \cdot \text{cm}^2$$

Other driving forces (besides concentration differences) for diffusion also occur because of temperature, pressure, electrical potential, and other gradients. Details are given elsewhere (B3).

6.1D Convective Mass-Transfer Coefficient

When a fluid is flowing outside a solid surface in forced convection motion, we can express the rate of convective mass transfer from the surface to the fluid, or vice versa, by the following equation:

$$N_A = k_c(c_{L1} - c_{Li}) \tag{6.1-14}$$

where k_c is a mass-transfer coefficient in m/s, c_{L1} the bulk fluid concentration in kg mol A/m^3, and c_{Li} the concentration in the fluid next to the surface of the solid. This mass-transfer coefficient is very similar to the heat-transfer coefficient h and is a function of the system geometry, fluid properties, and flow velocity. In Chapter 7 we consider convective mass transfer in detail.

6.2 MOLECULAR DIFFUSION IN GASES

6.2A Equimolar Counterdiffusion in Gases

In Fig. 6.2-1 a diagram is given of two gases A and B at constant total pressure P in two large chambers connected by a tube where molecular diffusion at steady state is oc-

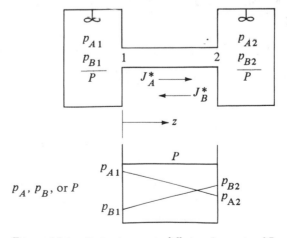

FIGURE 6.2-1. *Equimolar counterdiffusion of gases A and B.*

curring. Stirring in each chamber keeps the concentrations in each chamber uniform. The partial pressure $p_{A1} > p_{A2}$ and $p_{B2} > p_{B1}$. Molecules of A diffuse to the right and B to the left. Since the total pressure P is constant throughout, the net moles of A diffusing to the right must equal the net moles of B to the left. If this is not so, the total pressure would not remain constant. This means that

$$J^*_{Az} = -J^*_{Bz} \tag{6.2-1}$$

The subscript z is often dropped when the direction is obvious. Writing Fick's law for B for constant c,

$$J^*_B = -D_{BA} \frac{dc_B}{dz} \tag{6.2-2}$$

Now since $P = p_A + p_B = $ constant, then

$$c = c_A + c_B \tag{6.2-3}$$

Differentiating both sides,

$$dc_A = -dc_B \tag{6.2-4}$$

Equating Eq. (6.1-3) to (6.2-2),

$$J^*_A = -D_{AB} \frac{dc_A}{dz} = -J^*_B = -(-)D_{BA} \frac{dc_B}{dz} \tag{6.2-5}$$

Substituting Eq. (6.2-4) into (6.2-5) and canceling like terms,

$$D_{AB} = D_{BA} \tag{6.2-6}$$

This shows that for a binary gas mixture of A and B the diffusivity coefficient D_{AB} for A diffusing in B is the same as D_{BA} for B diffusing into A.

EXAMPLE 6.2-1. Equimolar Counterdiffusion

Ammonia gas (A) is diffusing through a uniform tube 0.10 m long containing N_2 gas (B) at 1.0132×10^5 Pa press and 298 K. The diagram is similar to Fig. 6.2-1. At point 1, $p_{A1} = 1.013 \times 10^4$ Pa and at point 2, $p_{A2} = 0.507 \times 10^4$ Pa. The diffusivity $D_{AB} = 0.230 \times 10^{-4}$ m^2/s.
  (a) Calculate the flux J^*_A at steady state.
  (b) Repeat for J^*_B.

Solution: Equation (6.1-13) can be used where $P = 1.0132 \times 10^5$ Pa, $z_2 - z_1 = 0.10$ m, and $T = 298$ K. Substituting into Eq. (6.1-13) for part (a),

$$J^*_A = \frac{D_{AB}(p_{A1} - p_{A2})}{RT(z_2 - z_1)} = \frac{(0.23 \times 10^{-4})(1.013 \times 10^4 - 0.507 \times 10^4)}{8314(298)(0.10 - 0)}$$

$$= 4.70 \times 10^{-7} \text{ kg mol } A/\text{s} \cdot \text{m}^2$$

Rewriting Eq. (6.1-13) for component B for part (b) and noting that $p_{B1} = P - p_{A1} = 1.0132 \times 10^5 - 1.013 \times 10^4 = 9.119 \times 10^4$ Pa   and $p_{B2} = P - p_{A2} = 1.0132 \times 10^5 - 0.507 \times 10^4 = 9.625 \times 10^4$ Pa,

$$J^*_B = \frac{D_{AB}(p_{B1} - p_{B2})}{RT(z_2 - z_1)} = \frac{(0.23 \times 10^{-4})(9.119 \times 10^4 - 9.625 \times 10^4)}{8314(298)(0.10 - 0)}$$

$$= -4.70 \times 10^{-7} \text{ kg mol } B/\text{s} \cdot \text{m}^2$$

The negative value for J^*_B means the flux goes from point 2 to 1.

6.2B General Case for Diffusion of Gases A and B Plus Convection

Up to now we have considered Fick's law for diffusion in a stationary fluid; i.e., there has been no net movement or convective flow of the entire phase of the binary mixture A and B. The diffusion flux J_A^* occurred because of the concentration gradient. The rate at which moles of A passed a fixed point to the right, which will be taken as a positive flux, is J_A^* kg mol $A/s \cdot m^2$. This flux can be converted to a velocity of diffusion of A to the right by

$$J_A^* \text{ (kg mol } A/s \cdot m^2) = v_{Ad} c_A \left(\frac{m}{s} \frac{\text{kg mol } A}{m^3} \right) \tag{6.2-7}$$

where v_{Ad} is the diffusion velocity of A in m/s.

Now let us consider what happens when the whole fluid is moving in bulk or convective flow to the right. The molar average velocity of the whole fluid relative to a stationary point is v_M m/s. Component A is still diffusing to the right, but now its diffusion velocity v_{Ad} is measured relative to the moving fluid. To a stationary observer A is moving faster than the bulk of the phase, since its diffusion velocity v_{Ad} is added to that of the bulk phase v_M. Expressed mathematically, the velocity of A relative to the stationary point is the sum of the diffusion velocity and the average or convective velocity.

$$v_A = v_{Ad} + v_M \tag{6.2-8}$$

where v_A is the velocity of A relative to a stationary point. Expressed pictorially,

Multiplying Eq. (6.2-8) by c_A,

$$c_A v_A = c_A v_{Ad} + c_A v_M \tag{6.2-9}$$

Each of the three terms represents a flux. The first term, $c_A v_A$, can be represented by the flux N_A kg mol $A/s \cdot m^2$. This is the total flux of A relative to the stationary point. The second term is J_A^*, the diffusion flux relative to the moving fluid. The third term is the convective flux of A relative to the stationary point. Hence, Eq. (6.2-9) becomes

$$N_A = J_A^* + c_A v_M \tag{6.2-10}$$

Let N be the total convective flux of the whole stream relative to the stationary point. Then,

$$N = c v_M = N_A + N_B \tag{6.2-11}$$

Or, solving for v_M,

$$v_M = \frac{N_A + N_B}{c} \tag{6.2-12}$$

Substituting Eq. (6.2-12) into (6.2-10),

$$N_A = J_A^* + \frac{c_A}{c} (N_A + N_B) \tag{6.2-13}$$

Since J_A^* is Fick's law, Eq. (6.1-7),

$$N_A = -cD_{AB}\frac{dx_A}{dz} + \frac{c_A}{c}(N_A + N_B) \tag{6.2-14}$$

Equation (6.2-14) is the final general equation for diffusion plus convection to use when the flux N_A is used, which is relative to a stationary point. A similar equation can be written for N_B.

$$N_B = -cD_{BA}\frac{dx_B}{dz} + \frac{c_B}{c}(N_A + N_B) \tag{6.2-15}$$

To solve Eq. (6.2-14) or (6.2-15), the relation between the flux N_A and N_B must be known. Equations (6.2-14) and (6.2-15) hold for diffusion in a gas, liquid, or solid.

For equimolar counterdiffusion, $N_A = -N_B$ and the convective term in Eq. (6.2-14) becomes zero. Then, $N_A = J_A^* = -N_B = -J_B^*$.

6.2C Special Case for A Diffusing Through Stagnant, Nondiffusing B

The case of diffusion of A through stagnant or nondiffusing B at steady state often occurs. In this case one boundary at the end of the diffusion path is impermeable to component B, so it cannot pass through. One example shown in Fig. 6.2-2a is in evaporation of a pure liquid such as benzene (A) at the bottom of a narrow tube, where a large amount of inert or nondiffusing air (B) is passed over the top. The benzene vapor (A) diffuses through the air (B) in the tube. The boundary at the liquid surface at point 1 is impermeable to air, since air is insoluble in benzene liquid. Hence, air (B) cannot diffuse into or away from the surface. At point 2 the partial pressure $p_{A2} = 0$, since a large volume of air is passing by.

Another example shown in Fig. 6.2-2b occurs in the absorption of NH_3 (A) vapor which is in air (B) by water. The water surface is impermeable to the air, since air is only very slightly soluble in water. Thus, since B cannot diffuse, $N_B = 0$.

To derive the case for A diffusing in stagnant, nondiffusing B, $N_B = 0$ is substituted into the general Eq. (6.2-14).

$$N_A = -cD_{AB}\frac{dx_A}{dz} + \frac{c_A}{c}(N_A + 0) \tag{6.2-16}$$

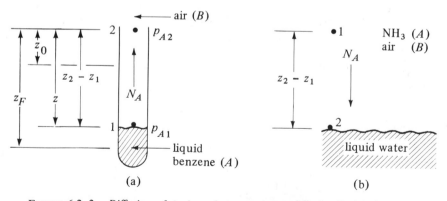

FIGURE 6.2.-2. *Diffusion of A through stagnant, nondiffusing B: (a) benzene evaporating into air, (b) ammonia in air being absorbed into water.*

Keeping the total pressure P constant, substituting $c = P/RT$, $p_A = x_A P$, and $c_A/c = p_A/P$ into Eq. (6.2-16),

$$N_A = -\frac{D_{AB}}{RT}\frac{dp_A}{dz} + \frac{p_A}{P}N_A \qquad (6.2\text{-}17)$$

Rearranging and integrating,

$$N_A\left(1 - \frac{p_A}{P}\right) = -\frac{D_{AB}}{RT}\frac{dp_A}{dz} \qquad (6.2\text{-}18)$$

$$N_A\int_{z_1}^{z_2} dz = -\frac{D_{AB}}{RT}\int_{p_{A1}}^{p_{A2}}\frac{dp_A}{1 - p_A/P} \qquad (6.2\text{-}19)$$

$$N_A = \frac{D_{AB}P}{RT(z_2 - z_1)}\ln\frac{P - p_{A2}}{P - p_{A1}} \qquad (6.2\text{-}20)$$

Equation (6.2-20) is the final equation to be used to calculate the flux of A. However, it is often written in another form as follows. A log mean value of the inert B is defined as follows. Since $P = p_{A1} + p_{B1} = p_{A2} + p_{B2}$, $p_{B1} = P - p_{A1}$, and $p_{B2} = P - p_{A2}$,

$$p_{BM} = \frac{p_{B2} - p_{B1}}{\ln(p_{B2}/p_{B1})} = \frac{p_{A1} - p_{A2}}{\ln[(P - p_{A2})/(P - p_{A1})]} \qquad (6.2\text{-}21)$$

Substituting Eq. (6.2-21) into (6.2-20),

$$N_A = \frac{D_{AB}P}{RT(z_2 - z_1)p_{BM}}(p_{A1} - p_{A2}) \qquad (6.2\text{-}22)$$

EXAMPLE 6.2-2. *Diffusion of Water Through Stagnant, Nondiffusing Air*
Water in the bottom of a narrow metal tube is held at a constant temperature of 293 K. The total pressure of air (assumed dry) is 1.01325×10^5 Pa (1.0 atm) and the temperature is 293 K (20°C). Water evaporates and diffuses through the air in the tube and the diffusion path $z_2 - z_1$ is 0.1524 m (0.5 ft) long. The diagram is similar to Fig. 6.2-2a. Calculate the rate of evaporation at steady state in lb mol/h·ft² and kg mol/s·m² The diffusivity of water vapor at 293 K and 1 atm pressure is 0.250×10^{-4} m²/s. Assume that the system is isothermal. Use SI and English units.

Solution: The diffusivity is converted to ft²/h by using the conversion factor from Appendix A.1.

$$D_{AB} = 0.250 \times 10^{-4}\,(3.875 \times 10^4) = 0.969\ \text{ft}^2/\text{h}$$

From Appendix A.2 the vapor pressure of water at 20°C is 17.54 mm or $p_{A1} = 17.54/760 = 0.0231$ atm $= 0.0231(1.01325 \times 10^5) = 2.341 \times 10^3$ Pa, $p_{A2} = 0$ (pure air). Since the temperature is 20°C (68°F), $T = 460 + 68 = 528°R = 293$ K. From Appendix A.1, $R = 0.730$ ft³· atm/lb mol·°R. To calculate the value of p_{BM} from Eq. (6.2-21),

$$p_{B1} = P - p_{A1} = 1.00 - 0.0231 = 0.9769\ \text{atm}$$

$$p_{B2} = P - p_{A2} = 1.00 - 0 = 1.00\ \text{atm}$$

$$p_{BM} = \frac{p_{B2} - p_{B1}}{\ln(p_{B2}/p_{B1})} = \frac{1.00 - 0.9769}{\ln(1.00/0.9769)} = 0.988\ \text{atm} = 1.001 \times 10^5\ \text{Pa}$$

Since p_{B1} is close to p_{B2}, the linear mean $(p_{B1} + p_{B2})/2$ could be used and would be very close to p_{BM}.

Substituting into Eq. (6.2-22) with $z_2 - z_1 = 0.5$ ft (0.1524 m),

$$N_A = \frac{D_{AB}P}{RT(z_2 - z_1)p_{BM}} (p_{A1} - p_{A2}) = \frac{0.969(1.0)(0.0231 - 0)}{0.730(528)(0.5)(0.988)}$$

$$= 1.175 \times 10^{-4} \text{ lb mol/h} \cdot \text{ft}^2$$

$$N_A = \frac{(0.250 \times 10^{-4})(1.01325 \times 10^5)(2.341 \times 10^3 - 0)}{8314(293)(0.1524)(1.001 \times 10^5)}$$

$$= 1.595 \times 10^{-7} \text{ kg mol/s} \cdot \text{m}^2$$

EXAMPLE 6.2-3. Diffusion in a Tube with Change in Path Length

Diffusion of water vapor in a narrow tube is occurring as in Example 6.2-2 under the same conditions. However, as shown in Fig. 6.2-2a, at a given time t, the level is z m from the top. As diffusion proceeds, the level drops slowly. Derive the equation for the time t_F for the level to drop from a starting point of z_0 m at $t = 0$ to z_F at $t = t_F$ s as shown.

Solution: We assume a pseudo-steady-state condition since the level drops very slowly. As time progresses, the path length z increases. At any time t, Eq. (6.2-22) holds; but the path length is z and Eq. (6.2-22) becomes as follows where N_A and z are now variables.

$$N_A = \frac{D_{AB}P}{RTzp_{BM}} (p_{A1} - p_{A2}) \tag{6.2-23}$$

Assuming a cross-sectional area of 1 m^2, the level drops dz m in dt s and $\rho_A (dz \cdot 1)/M_A$ is the kg mol of A that have left and diffused. Then,

$$N_A \cdot 1 = \frac{\rho_A (dz \cdot 1)}{M_A \, dt} \tag{6.2-24}$$

Equating Eq. (6.2-24) to (6.2-23), rearranging, and integrating between the limits of $z = z_0$ when $t = 0$ and $z = z_F$ when $t = t_F$,

$$\frac{\rho_A}{M_A} \int_{z_0}^{z_F} z \, dz = \frac{D_{AB}P(p_{A1} - p_{A2})}{RTp_{BM}} \int_0^{t_F} dt \tag{6.2-25}$$

Solving for t_F,

$$t_F = \frac{\rho_A (z_F^2 - z_0^2) RTp_{BM}}{2M_A D_{AB} P(p_{A1} - p_{A2})} \tag{6.2-26}$$

The method shown in Example 6.2-3 has been used to experimentally determine the diffusivity D_{AB}. In this experiment the starting path length z_0 is measured at $t = 0$ and also the final z_F at t_F. Then Eq. (6.2-26) is used to calculate D_{AB}.

6.2D   Diffusion Through a Varying Cross-Sectional Area

In the cases so far at steady state we have considered N_A and J_A^* as constants in the integrations. In these cases the cross-sectional area A m^2 through which the diffusion occurs has been constant with varying distance z. In some situations the area A may vary. Then it is convenient to define N_A as

$$N_A = \frac{\bar{N}_A}{A} \tag{6.2-27}$$

where $\bar{N}_A$ is kg moles of A diffusing per second or kg mol/s. At steady state, $\bar{N}_A$ will be constant but not A for a varying area.

1. Diffusion from a sphere. To illustrate the use of Eq. (6.2-27), the important case of diffusion to or from a sphere in a gas will be considered. This situation appears often in such cases as the evaporation of a drop of liquid, the evaporation of a ball of naphthalene, and the diffusion of nutrients to a spherical-like microorganism in a liquid. In Fig. 6.2-3a is shown a sphere of fixed radius r_1 m in an infinite gas medium. Component (A) at partial pressure p_{A1} at the surface is diffusing into the surrounding stagnant medium (B), where $p_{A2} = 0$ at some large distance away. Steady-state diffusion will be assumed.

The flux N_A can be represented by Eq. (6.2-27), where A is the cross-sectional area $4\pi r^2$ at point r distance from the center of the sphere. Also, $\bar{N}_A$ is a constant at steady state.

$$N_A = \frac{\bar{N}_A}{4\pi r^2} \qquad \text{(6.2-28)}$$

Since this is a case of A diffusing through stagnant, nondiffusing B, Eq. (6.2-18) will be used in its differential form and N_A will be equated to Eq. (6.2-28), giving

$$N_A = \frac{\bar{N}_A}{4\pi r^2} = -\frac{D_{AB}}{RT} \frac{dp_A}{(1 - p_A/P)\, dr} \qquad \text{(6.2-29)}$$

Note that dr was substituted for dz. Rearranging and integrating between r_1 and some point r_2 a large distance away,

$$\frac{\bar{N}_A}{4\pi} \int_{r_1}^{r_2} \frac{dr}{r^2} = -\frac{D_{AB}}{RT} \int_{p_{A1}}^{p_{A2}} \frac{dp_A}{(1 - p_A/P)} \qquad \text{(6.2-30)}$$

$$\frac{\bar{N}_A}{4\pi} \left(\frac{1}{r_1} - \frac{1}{r_2} \right) = \frac{D_{AB} P}{RT} \ln \frac{P - p_{A2}}{P - p_{A1}} \qquad \text{(6.2-31)}$$

Since $r_2 \gg r_1$, $1/r_2 \cong 0$. Substituting p_{BM} from Eq. (6.2-21) into Eq. (6.2-31),

$$\frac{\bar{N}_A}{4\pi_1^2} = N_{A1} = \frac{D_{AB} P}{RTr_1} \frac{p_{A1} - p_{A2}}{p_{BM}} \qquad \text{(6.2-32)}$$

This equation can be simplified further. If p_{A1} is small compared to P (a dilute gas

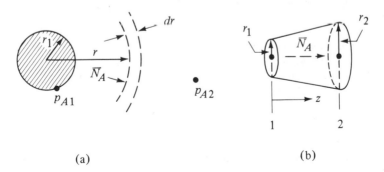

(a)   (b)

FIGURE 6.2-3.   *Diffusion through a varying cross-sectional area: (a) from a sphere to a surrounding medium, (b) through a circular conduit that is tapered uniformly.*

phase), $p_{BM} \cong P$. Also, setting $2r_1 = D_1$, diameter, and $c_{A1} = p_{A1}/RT$, we obtain

$$N_{A1} = \frac{2D_{AB}}{D_1}(c_{A1} - c_{A2}) \tag{6.2-33}$$

This equation can also be used for liquids, where D_{AB} is the diffusivity of A in the liquid.

EXAMPLE 6.2-4. Evaporation of a Naphthalene Sphere

A sphere of naphthalene having a radius of 2.0 mm is suspended in a large volume of still air at 318 K and 1.01325×10^5 Pa (1 atm). The surface temperature of the naphthalene can be assumed to be at 318 K and its vapor pressure at 318 K is 0.555 mm Hg. The D_{AB} of naphthalene in air at 318 K is 6.92×10^{-6} m²/s. Calculate the rate of evaporation of naphthalene from the surface.

Solution: The flow diagram is similar to Fig. 6.2-3a. $D_{AB} = 6.92 \times 10^{-6}$ m²/s, $p_{A1} = (0.555/760)(1.01325 \times 10^5) = 74.0$ Pa, $p_{A2} = 0$, $r_1 = 2/1000$ m, $R = 8314$ m³·Pa/kg mol·K, $p_{B1} = P - p_{A1} = 1.01325 \times 10^5 - 74.0 = 1.01251 \times 10^5$ Pa, $p_{B2} = P - p_{A2} = 1.01325 \times 10^5 - 0$. Since the values of p_{B1} and p_{B2} are close to each other,

$$p_{BM} = \frac{p_{B1} + p_{B2}}{2} = \frac{(1.0125 + 1.01325) \times 10^5}{2} = 1.0129 \times 10^5 \text{ Pa}$$

Substituting into Eq. (6.2-32),

$$N_{A1} = \frac{D_{AB}P(p_{A1} - p_{A2})}{RTr_1 p_{BM}} = \frac{(6.92 \times 10^{-6})(1.01325 \times 10^5)(74.0 - 0)}{8314(318)(2/1000)(1.0129 \times 10^5)}$$

$$= 9.68 \times 10^{-8} \text{ kg mol } A/\text{s}\cdot\text{m}^2$$

If the sphere in Fig. 6.2-3a is evaporating, the radius r of the sphere decreases slowly with time. The equation for the time for the sphere to evaporate completely can be derived by assuming pseudo-steady state and by equating the diffusion flux equation (6.2-32), where r is now a variable, to the moles of solid A evaporated per dt time and per unit area as calculated from a material balance. (See Problem 6.2-9 for this case.) The material-balance method is similar to Example 6.2-3. The final equation is

$$t_F = \frac{\rho_A r_1^2 RT p_{BM}}{2M_A D_{AB} P(p_{A1} - p_{A2})} \tag{6.2-34}$$

where r_1 is the original sphere radius, ρ_A the density of the sphere, and M_A the molecular weight.

2. Diffusion through a conduit of nonuniform cross-sectional area.

In Fig. 6.2-3b component A is diffusion at steady state through a circular conduit which is tapered uniformly as shown. At point 1 the radius is r_1 and at point 2 it is r_2. At position z in the conduit, for A diffusing through stagnant, nondiffusing B,

$$N_A = \frac{\bar{N}_A}{\pi r^2} = -\frac{D_{AB}}{RT}\frac{dp_A}{(1 - p_A/P)\,dz} \tag{6.2-35}$$

Using the geometry shown, the variable radius r can be related to position z in the path as follows:

$$r = \left(\frac{r_2 - r_1}{z_2 - z_1}\right)z + r_1 \tag{6.2-36}$$

This value of r is then substituted into Eq. (6.2-35) to eliminate r and the equation integrated.

$$\frac{\bar{N}_A}{\pi} \int_{z_1}^{z_2} \frac{dz}{\left[\left(\dfrac{r_2 - r_1}{z_2 - z_1}\right)z + r_1\right]^2} = -\frac{D_{AB}}{RT} \int_{p_{A1}}^{p_{A2}} \frac{dp_A}{1 - p_A/P} \tag{6.2-37}$$

A case similar to this is given in Problem 6.2-10.

6.2E Diffusion Coefficients for Gases

1. Experimental determination of diffusion coefficients. A number of different experimental methods have been used to determine the molecular diffusivity for binary gas mixtures. Several of the important methods are as follows. One method is to evaporate a pure liquid in a narrow tube with a gas passed over the top as shown in Fig. 6.2-2a. The fall in liquid level is measured with time and the diffusivity calculated from Eq. (6.2-26).

In another method, two pure gases having equal pressures are placed in separate sections of a long tube separated by a partition. The partition is slowly removed and diffusion proceeds. After a given time the partition is reinserted and the gas in each section analyzed. The diffusivities of the vapor of solids such as naphthalene, iodine, and benzoic acid in a gas have been obtained by measuring the rate of evaporation of a sphere. Equation (6.2-32) can be used. See Problem 6.2-9 for an example of this.

A useful method often used is the two-bulb method (N1). The apparatus consists of two glass bulbs with volumes V_1 and V_2 m³ connected by a capillary of cross-sectional area A m² and length L whose volume is small compared to V_1 and V_2, as shown in Fig. 6.2-4. Pure gas A is added to V_1 and pure B to V_2 at the same pressures. The valve is opened, diffusion proceeds for a given time, and then the valve is closed and the mixed contents of each chamber are sampled separately.

The equations can be derived by neglecting the capillary volume and assuming each bulb is always of a uniform concentration. Assuming quasi-steady-state diffusion in the capillary,

$$J_A^* = -D_{AB} \frac{dc}{dz} = -\frac{D_{AB}(c_2 - c_1)}{L} \tag{6.2-38}$$

where c_2 is the concentration of A in V_2 at time t and c_1 in V_1. The rate of diffusion of A going to V_2 is equal to the rate of accumulation in V_2.

$$AJ_A^* = -\frac{D_{AB}(c_2 - c_1)A}{L} = V_2 \frac{dc_2}{dt} \tag{6.2-39}$$

The average value c_{av} at equilibrium can be calculated by a material balance from the starting compositions c_1^0 and c_2^0 at $t = 0$.

$$(V_1 + V_2)c_{av} = V_1 c_1^0 + V_2 c_2^0 \tag{6.2-40}$$

FIGURE 6.2-4. *Diffusivity measurement of gases by the two-bulb method.*

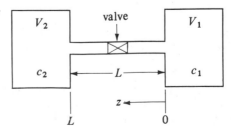

A similar balance at time t gives

$$(V_1 + V_2)c_{av} = V_1 c_1 + V_2 c_2 \tag{6.2-41}$$

Substituting c_1 from Eq. (6.2-41) into (6.2-39), rearranging, and integrating between $t = 0$ and $t = t$, the final equation is

$$\frac{c_{av} - c_2}{c_{av} - c_2^0} = \exp\left[-\frac{D_{AB}(V_1 + V_2)}{(L/A)(V_2 V_1)} t\right] \tag{6.2-42}$$

If c_2 is obtained by sampling at t, D_{AB} can be calculated.

2. Experimental diffusivity data. Some typical data are given in Table 6.2-1. Other data are tabulated in Perry and Green (P1) and Reid et al. (R1). The values range from about 0.05×10^{-4} m²/s, where a large molecule is present, to about 1.0×10^{-4} m²/s, where H_2 is present at room temperatures. The relation between diffusivity in m²/s and ft²/h is 1 m²/s = 3.875×10^4 ft²/h.

3. Prediction of diffusivity for gases. The diffusivity of a binary gas mixture in the dilute gas region, i.e., at low pressures near atmospheric, can be predicted using the kinetic theory of gases. The gas is assumed to consist of rigid spherical particles that are completely elastic on collision with another molecule, which implies that momentum is conserved.

In a simplified treatment it is assumed that there are no attractive or repulsive forces between the molecules. The derivation uses the mean free path λ, which is the average distance that a molecule has traveled between collisions. The final equation is

$$D_{AB} = \tfrac{1}{3}\bar{u}\lambda \tag{6.2-43}$$

where $\bar{u}$ is the average velocity of the molecules. The final equation obtained after substituting expressions for $\bar{u}$ and λ into Eq. (6.2-43) is approximately correct, since it correctly predicts D_{AB} proportional to 1/pressure and approximately predicts the temperature effect.

A more accurate and rigorous treatment must consider the intermolecular forces of attraction and repulsion between molecules and also the different sizes of molecules A and B. Chapman and Enskog (H3) solved the Boltzmann equation, which does not utilize the mean free path λ but uses a distribution function. To solve the equation, a relation between the attractive and repulsive forces between a given pair of molecules must be used. For a pair of nonpolar molecules a reasonable approximation to the forces is the Lennard–Jones function.

The final relation to predict the diffusivity of a binary gas pair of A and B molecules is

$$D_{AB} = \frac{1.8583 \times 10^{-7} T^{3/2}}{P\sigma_{AB}^2 \Omega_{D,\,AB}} \left(\frac{1}{M_A} + \frac{1}{M_B}\right)^{1/2} \tag{6.2-44}$$

where D_{AB} is the diffusivity in m²/s, T temperature in K, M_A molecular weight of A in kg mass/kg mol, M_B molecular weight of B, and P absolute pressure in atm. The term σ_{AB} is an "average collision diameter" and $\Omega_{D,\,AB}$ is a collision integral based on the Lennard–Jones potential. Values of σ_A and σ_B and $\Omega_{D,\,AB}$ can be obtained from a number of sources (B3, G2, H3, R1).

The collision integral $\Omega_{D,\,AB}$ is a ratio giving the deviation of a gas with interactions compared to a gas of rigid, elastic spheres. This value would be 1.0 for a gas with no interactions. Equation (6.2-44) predicts diffusivities with an average deviation of about

TABLE 6.2-1. *Diffusion Coefficients of Gases at 101.32 kPa Pressure*

System	Temperature		Diffusivity $[(m^2/s)10^4$ or $cm^2/s]$	Ref.
	°C	K		
Air–NH_3	0	273	0.198	(W1)
Air–H_2O	0	273	0.220	(N2)
	25	298	0.260	(L1)
	42	315	0.288	(M1)
Air–CO_2	3	276	0.142	(H1)
	44	317	0.177	
Air–H_2	0	273	0.611	(N2)
Air–C_2H_5OH	25	298	0.135	(M1)
	42	315	0.145	
Air–CH_3COOH	0	273	0.106	(N2)
Air–n-hexane	21	294	0.080	(C1)
Air–benzene	25	298	0.0962	(L1)
Air–toluene	25.9	298.9	0.086	(G1)
Air–n-butanol	0	273	0.0703	(N2)
	25.9	298.9	0.087	
H_2–CH_4	25	298	0.726	(C2)
H_2–N_2	25	298	0.784	(B1)
	85	358	1.052	
H_2–benzene	38.1	311.1	0.404	(H2)
H_2–Ar	22.4	295.4	0.83	(W2)
H_2–NH_3	25	298	0.783	(B1)
H_2–SO_2	50	323	0.61	(S1)
H_2–C_2H_5OH	67	340	0.586	(T1)
He–Ar	25	298	0.729	(S2)
He–n-butanol	150	423	0.587	(S2)
He–air	44	317	0.765	(H1)
He–CH_4	25	298	0.675	(C2)
He–N_2	25	298	0.687	(S2)
He–O_2	25	298	0.729	(S2)
Ar–CH_4	25	298	0.202	(C2)
CO_2–N_2	25	298	0.167	(W3)
CO_2–O_2	20	293	0.153	(W4)
N_2–n-butane	25	298	0.0960	(B2)
H_2O–CO_2	34.3	307.3	0.202	(S3)
CO–N_2	100	373	0.318	(A1)
CH_3Cl–SO_2	30	303	0.0693	(C3)
$(C_2H_5)_2O$–NH_3	26.5	299.5	0.1078	(S4)

8% up to about 1000 K (R1). For a polar–nonpolar gas mixture Eq. (6.2-44) can be used if the correct force constant is used for the polar gas (M1, M2). For polar–polar gas pairs the potential-energy function commonly used is the *Stockmayer potential* (M2).

The effect of concentration of A in B in Eq. (6.2-44) is not included. However, for real gases with interactions the maximum effect of concentration on diffusivity is about 4% (G2). In most cases the effect is considerably less, and hence it is usually neglected.

Equation (6.2-44) is relatively complicated to use and often some of the constants such as σ_{AB} are not available or difficult to estimate. Hence, the semiempirical method of Fuller et al. (F1), which is much more convenient to use, is often utilized. The equation was obtained by correlating many recent data and uses atomic volumes from Table 6.2-2, which are summed for each gas molecule. The equation is

$$D_{AB} = \frac{1.00 \times 10^{-7} T^{1.75} (1/M_A + 1/M_B)^{1/2}}{P[(\sum v_A)^{1/3} + (\sum v_B)^{1/3}]^2} \tag{6.2-45}$$

where $\sum v_A$ = sum of structural volume increments, Table 6.2-2, and $D_{AB} = m^2/s$. This method can be used for mixtures of nonpolar gases or for a polar–nonpolar mixture. Its accuracy is not quite as good as that of Eq. (6.2-44).

TABLE 6.2-2. *Atomic Diffusion Volumes for Use with the Fuller, Schettler, and Giddings Method**

Atomic and structural diffusion volume increments, v

C	16.5	(Cl)	19.5
H	1.98	(S)	17.0
O	5.48	Aromatic ring	−20.2
(N)	5.69	Heterocyclic ring	−20.2

Diffusion volumes for simple molecules, $\sum v$

H_2	7.07	CO	18.9
D_2	6.70	CO_2	26.9
He	2.88	N_2O	35.9
N_2	17.9	NH_3	14.9
O_2	16.6	H_2O	12.7
Air	20.1	(CCl_2F_2)	114.8
Ar	16.1	(SF_6)	69.7
Kr	22.8	(Cl_2)	37.7
(Xe)	37.9	(Br_2)	67.2
Ne	5.59	(SO_2)	41.1

* Parentheses indicate that the value listed is based on only a few data points.
Source: Reprinted with permission from E. N. Fuller, P. D. Schettler, and J. C. Giddings, *Ind. Eng. Chem.*, **58**, 19(1966). Copyright by the American Chemical Society.

The equation shows that D_{AB} is proportional to $1/P$ and to $T^{1.75}$. If an experimental value of D_{AB} is available at a given T and P and it is desired to have a value of D_{AB} at another T and P, one should correct the experimental value to the new T and P by the relationship $D_{AB} \propto T^{1.75}/P$.

4. Schmidt number of gases. The *Schmidt number* of a gas mixture of dilute A in B is dimensionless and is defined as

$$N_{Sc} = \frac{\mu}{\rho D_{AB}} \tag{6.2-46}$$

where μ is viscosity of the gas mixture, which is viscosity of B for a dilute mixture in Pa $\cdot$ s or kg/m $\cdot$ s, D_{AB} is diffusivity in m^2/s, and ρ is the density of the mixture in kg/m^3. For a gas the Schmidt number can be assumed independent of temperature over moderate ranges and independent of pressure up to about 10 atm or 10×10^5 Pa.

The Schmidt number is the dimensionless ratio of the molecular momentum diffusivity μ/ρ to the molecular mass diffusivity D_{AB}. Values of the Schmidt number for gases range from about 0.5 to 2. For liquids Schmidt numbers range from about 100 to over 10 000 for viscous liquids.

EXAMPLE 6.2-5. Estimation of Diffusivity of a Gas Mixture
Normal butanol (A) is diffusing through air (B) at 1 atm abs. Using the Fuller et al. method, estimate the diffusivity D_{AB} for the following temperatures and compare with the experimental data.
   (a) For 0°C.
   (b) For 25.9°C.
   (c) For 0°C and 2.0 atm abs.

Solution: For part (a), $P = 1.00$ atm, $T = 273 + 0 = 273$ K, M_A (butanol) $= 74.1$, M_B (air) $= 29$. From Table 6.2-2,

$$\sum v_A = 4(16.5) + 10(1.98) + 1(5.48) = 91.28 \text{ (butanol)}$$

$$\sum v_B = 20.1 \text{ (air)}$$

Substituting into Eq. (6.2-45),

$$D_{AB} = \frac{1.0 \times 10^{-7}(273)^{1.75}(1/74.1 + 1/29)^{1/2}}{1.0[(91.28)^{1/3} + (20.1)^{1/3}]^2}$$

$$= 7.73 \times 10^{-6} \text{ m}^2/\text{s}$$

This value deviates by $+10\%$ from the experimental value of 7.03×10^{-6} m^2/s from Table 6.2-1.

For part (b), $T = 273 + 25.9 = 298.9$. Substituting into Eq. (6.2-45), $D_{AB} = 9.05 \times 10^{-6}$ m^2/s. This value deviates by $+4\%$ from the experimental of 8.70×10^{-6} m^2/s.

For part (c), the total pressure $P = 2.0$ atm. Using the value predicted in part (a) and correcting for pressure,

$$D_{AB} = 7.73 \times 10^{-6}(1.0/2.0) = 3.865 \times 10^{-6} \text{ m}^2/\text{s}$$

6.3 MOLECULAR DIFFUSION IN LIQUIDS

6.3A Introduction

Diffusion of solutes in liquids is very important in many industrial processes, especially in such separation operations as liquid–liquid extraction or solvent extraction, gas absorption, and distillation. Diffusion in liquids also occurs in many situations in nature, such as oxygenation of rivers and lakes by the air and diffusion of salts in blood.

It should be apparent that the rate of molecular diffusion in liquids is considerably slower than in gases. The molecules in a liquid are very close together compared to a gas. Hence, the molecules of the diffusing solute A will collide with molecules of liquid B more often and diffuse more slowly than in gases. In general, the diffusion coefficient in a gas will be of the order of magnitude of about 10^5 times greater than in a liquid. However, the flux in a gas is not that much greater, being only about 100 times faster, since the concentrations in liquids are considerably higher than in gases.

6.3B Equations for Diffusion in Liquids

Since the molecules in a liquid are packed together much more closely than in gases, the density and the resistance to diffusion in a liquid are much greater. Also, because of this closer spacing of the molecules, the attractive forces between molecules play an important role in diffusion. Since the kinetic theory of liquids is only partially developed, we write the equations for diffusion in liquids similar to those for gases.

In diffusion in liquids an important difference from diffusion in gases is that the diffusivities are often quite dependent on the concentration of the diffusing components.

1. Equimolar counterdiffusion. Starting with the general equation (6.2-14), we can obtain for equimolal counterdiffusion where $N_A = -N_B$, an equation similar to Eq. (6.1-11) for gases at steady state.

$$N_A = \frac{D_{AB}(c_{A1} - c_{A2})}{z_2 - z_1} = \frac{D_{AB}\, c_{av}(x_{A1} - x_{A2})}{z_2 - z_1} \tag{6.3-1}$$

where N_A is the flux of A in kg mol $A/s \cdot m^2$, D_{AB} the diffusivity of A in B in m^2/s, c_{A1} the concentration of A in kg mol A/m^3 at point 1, x_{A1} the mole fraction of A at point 1, and c_{av} defined by

$$c_{av} = \left(\frac{\rho}{M}\right)_{av} = \left(\frac{\rho_1}{M_1} + \frac{\rho_2}{M_2}\right)\Big/ 2 \tag{6.3-2}$$

where c_{av} is the average total concentration of $A + B$ in kg mol/m^3, M_1 the average molecular weight of the solution at point 1 in kg mass/kg mol, and ρ_1 the average density of the solution in kg/m^3 at point 1.

Equation (6.3-1) uses the average value of D_{AB}, which may vary some with concentration and the average value of c, which also may vary with concentration. Usually, the linear average of c is used as in Eq. (6.3-2). The case of equimolar counterdiffusion in Eq. (6.3-1) occurs only very infrequently in liquids.

2. Diffusion of A through nondiffusing B. The most important case of diffusion in liquids is that where solute A is diffusing and solvent B is stagnant or nondiffusing. An example is a dilute solution of propionic acid (A) in a water (B) solution being contacted with toluene. Only the propionic acid (A) diffuses through the water phase, to the boundary, and then into the toluene phase. The toluene–water interface is a barrier to diffusion of B and $N_B = 0$. Such cases often occur in industry (T2). If Eq. (6.2-22) is rewritten in terms of concentrations by substituting $c_{av} = P/RT$, $c_{A1} = p_{A1}/RT$, and $x_{BM} = p_{BM}/P$, we obtain the equation for liquids at steady state.

$$N_A = \frac{D_{AB}\, c_{av}}{(z_2 - z_1)x_{BM}} (x_{A1} - x_{A2}) \tag{6.3-3}$$

where

$$x_{BM} = \frac{x_{B2} - x_{B1}}{\ln (x_{B2}/x_{B1})} \tag{6.3-4}$$

Note that $x_{A1} + x_{B1} = x_{A2} + x_{B2} = 1.0$. For dilute solutions x_{BM} is close to 1.0 and c is essentially constant. Then Eq. (6.3-3) simplifies to

$$N_A = \frac{D_{AB}(c_{A1} - c_{A2})}{z_2 - z_1} \tag{6.3-5}$$

EXAMPLE 6.3-1. Diffusion of Ethanol (A) Through Water (B)

An ethanol (A)–water (B) solution in the form of a stagnant film 2.0 mm thick at 293 K is in contact at one surface with an organic solvent in which ethanol is soluble and water is insoluble. Hence, $N_B = 0$. At point 1 the concentration of ethanol is 16.8 wt % and the solution density is $\rho_1 = 972.8$ kg/m³. At point 2 the concentration of ethanol is 6.8 wt % and $\rho_2 = 988.1$ kg/m³ (P1). The diffusivity of ethanol is 0.740×10^{-9} m²/s (T2). Calculate the steady-state flux N_A.

Solution: The diffusivity is $D_{AB} = 0.740 \times 10^{-9}$ m²/s. The molecular weights of A and B are $M_A = 46.05$ and $M_B = 18.02$. For a wt % of 6.8, the mole fraction of ethanol (A) is as follows when using 100 kg of solution.

$$x_{A2} = \frac{6.8/46.05}{6.8/46.05 + 93.2/18.02} = \frac{0.1477}{0.1477 + 5.17} = 0.0277$$

Then $x_{B2} = 1 - 0.0277 = 0.9723$. Calculating x_{A1} in a similar manner, $x_{A1} = 0.0732$ and $x_{B1} = 1 - 0.0732 = 0.9268$. To calculate the molecular weight M_2 at point 2.

$$M_2 = \frac{100 \text{ kg}}{(0.1477 + 5.17) \text{ kg mol}} = 18.75 \text{ kg/kg mol}$$

Similarly, $M_1 = 20.07$. From Eq. (6.3-2),

$$c_{av} = \frac{\rho_1/M_1 + \rho_2/M_2}{2} = \frac{972.8/20.07 + 988.1/18.75}{2} = 50.6 \text{ kg mol/m}^3$$

To calculate x_{BM} from Eq. (6.3-4), we can use the linear mean since x_{B1} and x_{B2} are close to each other.

$$x_{BM} = \frac{x_{B1} + x_{B2}}{2} = \frac{0.9268 + 0.9723}{2} = 0.949$$

Substituting into Eq. (6.3-3) and solving,

$$N_A = \frac{D_{AB} c_{av}}{(z_2 - z_1) x_{BM}} (x_{A1} - x_{A2}) = \frac{(0.740 \times 10^{-9})(50.6)(0.0732 - 0.0277)}{(2/1000)0.949}$$

$$= 8.99 \times 10^{-7} \text{ kg mol/s} \cdot \text{m}^2$$

6.3C Diffusion Coefficients for Liquids

1. Experimental determination of diffusivities. Several different methods are used to determine diffusion coefficients experimentally in liquids. In one method unsteady-state diffusion in a long capillary tube is carried out and the diffusivity determined from the concentration profile. If the solute A is diffusing in B, the diffusion coefficient determined is D_{AB}. Also, the value of diffusivity is often very dependent upon the concentration of the diffusing solute A. Unlike gases, the diffusivity D_{AB} does not equal D_{BA} for liquids.

In a relatively common method a relatively dilute solution and a slightly more concentrated solution are placed in chambers on opposite sides of a porous membrane of sintered glass as shown in Fig. 6.3-1. Molecular diffusion takes place through the narrow passageways of the pores in the sintered glass while the two compartments are stirred. The effective diffusion length is $K_1 \delta$, where $K_1 > 1$ is a constant and corrects for the fact the path is actually greater than δ cm. In this method, discussed by Bidstrup and Geankoplis (B4), the effective diffusion length is obtained by calibrating with a solute such as KCl having a known diffusivity.

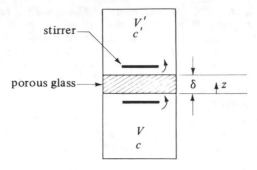

FIGURE 6.3-1. *Diffusion cell for determination of diffusivity in a liquid.*

To derive the equation, quasi-steady-state diffusion in the membrane is assumed.

$$N_A = \varepsilon D_{AB} \frac{c - c'}{K_1 \delta} \tag{6.3-6}$$

where c is the concentration in the lower chamber at a time t, c' is the concentration in the upper, and ε is the fraction of area of the glass open to diffusion. Making a balance on solute A in the upper chamber, where the rate in = rate out + rate of accumulation, making a similar balance on the lower chamber, using volume $V = V'$, and combining and integrating the final equation is

$$\ln \frac{c_0 - c'_0}{c - c'} = \frac{2\varepsilon A}{K_1 \delta V} D_{AB} t \tag{6.3-7}$$

where $2\varepsilon A/K_1 \delta V$ is a cell constant that can be determined using a solute of known diffusivity, such as KCl. The values c_0 and c'_0 are initial concentrations and c and c' final concentrations.

2. Experimental liquid diffusivity data.   Experimental diffusivity data for binary mixtures in the liquid phase are given in Table 6.3-1. All the data are for dilute solutions of the diffusing solute in the solvent. In liquids the diffusivities often vary quite markedly with concentration. Hence, the values in Table 6.3-1 should be used with some caution when outside the dilute range. Additional data are given in (P1). Values for biological solutes in solution are given in the next section. As noted in the table, the diffusivity values are quite small and in the range of about 0.5×10^{-9} to 5×10^{-9} m²/s for relatively nonviscous liquids. Diffusivities in gases are larger by a factor of 10^4 to 10^5.

6.3D   Prediction of Diffusivities in Liquids

The equations for predicting diffusivities of dilute solutes in liquids are by necessity semiempirical, since the theory for diffusion in liquids is not well established as yet. The Stokes–Einstein equation, one of the first theories, was derived for a very large spherical molecule (A) diffusing in a liquid solvent (B) of small molecules. Stokes' law was used to describe the drag on the moving solute molecule. Then the equation was modified by assuming that all molecules are alike and arranged in a cubic lattice and by expressing the molecular radius in terms of the molar volume (W5),

$$D_{AB} = \frac{9.96 \times 10^{-16} T}{\mu V_A^{1/3}} \tag{6.3-8}$$

TABLE 6.3-1.   *Diffusion Coefficients for Dilute Liquid Solutions*

Solute	Solvent	Temperature °C	Temperature K	Diffusivity $[(m^2/s)10^9$ or $(cm^2/s)10^5]$	Ref.
NH_3	Water	12	285	1.64	(N2)
		15	288	1.77	
O_2	Water	18	291	1.98	(N2)
		25	298	2.41	(V1)
CO_2	Water	25	298	2.00	(V1)
H_2	Water	25	298	4.8	(V1)
Methyl alcohol	Water	15	288	1.26	(J1)
Ethyl alcohol	Water	10	283	0.84	(J1)
		25	298	1.24	(J1)
n-Propyl alcohol	Water	15	288	0.87	(J1)
Formic acid	Water	25	298	1.52	(B4)
Acetic acid	Water	9.7	282.7	0.769	(B4)
		25	298	1.26	(B4)
Propionic acid	Water	25	298	1.01	(B4)
HCl (9 g mol/liter)	Water	10	283	3.3	(N2)
(2.5 g mol/liter)		10	283	2.5	(N2)
Benzoic acid	Water	25	298	1.21	(C4)
Acetone	Water	25	298	1.28	(A2)
Acetic acid	Benzene	25	298	2.09	(C5)
Urea	Ethanol	12	285	0.54	(N2)
Water	Ethanol	25	298	1.13	(H4)
KCl	Water	25	298	1.870	(P2)
KCl	Ethylene glycol	25	298	0.119	(P2)

where D_{AB} is diffusivity in m^2/s, T is temperature in K, μ is viscosity of solution in Pa·s or kg/m·s, and V_A is the solute molar volume at its normal boiling point in m^3/kg mol. This equation applies very well to very large unhydrated and spherical-like solute molecules of about 1000 molecular weight or greater (R1), or where the V_A is above about 0.500 m^3/kg mol (W5) in aqueous solution.

For smaller solute molar volumes, Eq. (6.3-8) does not hold. Several other theoretical derivations have been attempted, but the equations do not predict diffusivities very accurately. Hence, a number of semitheoretical expressions have been developed (R1). The Wilke–Chang (T3, W5) correlation can be used for most general purposes where the solute (A) is dilute in the solvent (B).

$$D_{AB} = 1.173 \times 10^{-16}(\varphi M_B)^{1/2} \frac{T}{\mu_B V_A^{0.6}} \qquad (6.3\text{-}9)$$

where M_B is the molecular weight of solvent B, μ_B is the viscosity of B in Pa·s or kg/m·s, V_A is the solute molar volume at the boiling point (L2) that can be obtained from Table 6.3-2, and φ is an "association parameter" of the solvent, where φ is 2.6 for water, 1.9 methanol, 1.5 ethanol, 1.0 benzene, 1.0 ether, 1.0 heptane, and 1.0 other unassociated solvents. When values of V_A are above 0.500 m^3/kg mol (500 cm^3/g mol), Eq. (6.3-8) should be used.

TABLE 6.3-2. *Atomic and Molar Volumes at the Normal Boiling Point*

Material	Atomic Volume $(m^3/kg\ mol)\ 10^3$	Material	Atomic Volume $(m^3/kg\ mol)\ 10^3$
C	14.8	Ring, 3-membered	−6
H	3.7	as in ethylene	
O (except as below)	7.4	oxide	
Doubly bound as	7.4	4-membered	−8.5
carbonyl		5-membered	−11.5
Coupled to two		6-membered	−15
other elements		Naphthalene ring	−30
In aldehydes, ketones	7.4	Anthracene ring	−47.5
In methyl esters	9.1		
In methyl ethers	9.9		
In ethyl esters	9.9		*Molecular Volume*
In ethyl ethers	9.9		$(m^3/kg\ mol)\ 10^3$
In higher esters	11.0		
In higher ethers	11.0	Air	29.9
In acids (—OH)	12.0	O_2	25.6
Joined to S, P, N	8.3	N_2	31.2
N		Br_2	53.2
Doubly bonded	15.6	Cl_2	48.4
In primary amines	10.5	CO	30.7
In secondary amines	12.0	CO_2	34.0
Br	27.0	H_2	14.3
Cl in RCHClR′	24.6	H_2O	18.8
Cl in RCl (terminal)	21.6	H_2S	32.9
F	8.7	NH_3	25.8
I	37.0	NO	23.6
S	25.6	N_2O	36.4
P	27.0	SO_2	44.8

Source : G. Le Bas, *The Molecular Volumes of Liquid Chemical Compounds.* New York: David McKay Co., Inc., 1915.

When water is the solute, values from Eq. (6.3-9) should be multiplied by a factor of 1/2.3 (R1). Equation (6.3-9) predicts diffusivities with a mean deviation of 10–15% for aqueous solutions and about 25% in nonaqueous solutions. Outside the range 278–313 K, the equation should be used with caution. For water as the diffusing solute, an equation by Reddy and Doraiswamy is preferred (R2). Skelland (S5) summarizes the correlations available for binary systems. Geankoplis (G2) discusses and gives an equation to predict diffusion in a ternary system, where a dilute solute A is diffusing in a mixture of B and C solvents. This case is often approximated in industrial processes.

EXAMPLE 6.3-2. *Prediction of Liquid Diffusivity*

Predict the diffusion coefficient of acetone (CH_3COCH_3) in water at 25° and 50°C using the Wilke–Chang equation. The experimental value is $1.28 \times 10^{-9}\ m^2/s$ at 25°C (298 K).

Solution: From Appendix A.2 the viscosity of water at 25.0°C is $\mu_B = 0.8937 \times 10^{-3}\ Pa \cdot s$ and at 50°C, 0.5494×10^{-3}. From Table 6.3-2 for

CH_3COCH_3 with 3 carbons + 6 hydrogens + 1 oxygen,

$$V_A = 3(0.0148) + 6(0.0037) + 1(0.0074) = 0.0740 \text{ m}^3/\text{kg mol}$$

For water the association parameter $\varphi = 2.6$ and $M_B = 18.02$ kg mass/kg mol. For 25°C, $T = 298$ K. Substituting into Eq. (6.3-9),

$$D_{AB} = (1.173 \times 10^{-16})(\varphi M_B)^{1/2} \frac{T}{\mu_B V_A^{0.6}}$$

$$= \frac{(1.173 \times 10^{-16})(2.6 \times 18.02)^{1/2}(298)}{(0.8937 \times 10^{-3})(0.0740)^{0.6}}$$

$$= 1.277 \times 10^{-9} \text{ m}^2/\text{s}$$

For 50°C or $T = 323$ K,

$$D_{AB} = \frac{(1.173 \times 10^{-16})(2.6 \times 18.02)^{1/2}(323)}{(0.5494 \times 10^{-3})(0.0740)^{0.6}}$$

$$= 2.251 \times 10^{-9} \text{ m}^2/\text{s}$$

Electrolytes in solution such as KCl dissociate into cations and anions and diffuse more rapidly than the undissociated molecule because of their small size. Diffusion coefficients can be estimated using ionic conductance at infinite dilution in water. Equations and data are given elsewhere (S5, T2). Both the negatively and positively charged ions diffuse at the same rate so that electrical neutrality is preserved.

6.4 MOLECULAR DIFFUSION IN BIOLOGICAL SOLUTIONS AND GELS

6.4A Diffusion of Biological Solutes in Liquids

1. Introduction. The diffusion of small solute molecules and especially macromolecules (e.g., proteins) in aqueous solutions are important in the processing and storing of biological systems and in the life processes of microorganisms, animals, and plants. Food processing is an important area where diffusion plays an important role. In the drying of liquid solutions of fruit juice, coffee, and tea, water and often volatile flavor or aroma constituents are removed. These constituents diffuse through the liquid during evaporation.

In fermentation processes, nutrients, sugars, oxygen, and so on, diffuse to the microorganisms and waste products and at times enzymes diffuse away. In the artificial kidney machine various waste products diffuse through the blood solution to a membrane and then through the membrane to an aqueous solution.

Macromolecules in solution having molecular weights of tens of thousands or more were often called *colloids*, but now we know they generally form true solutions. The diffusion behavior of protein macromolecules in solution is affected by their large sizes and shapes, which can be random coils, rodlike, or globular (spheres or ellipsoids). Also, interactions of the large molecules with the small solvent and/or solute molecules affect the diffusion of the macromolecules and also of the small solute molecules.

Besides the Fickian diffusion to be discussed here, mediated transport often occurs in biological systems where chemical interactions occur. This latter type of transport will not be discussed here.

2. Interaction and "binding" in diffusion. Protein macromolecules are very large compared to small solute molecules such as urea, KCl, and sodium caprylate, and often have a number of sites for interaction or "binding" of the solute or ligand molecules. An example is the binding of oxygen to hemoglobin in the blood. Human serum albumin protein binds most of the free fatty acids in the blood and increases their apparent solubility. Bovine serum albumin, which is in milk, binds 23 mol sodium caprylate/mol albumin when the albumin concentration is $30 \, kg/m^3$ solution and the sodium caprylate is about 0.05 molar (G6). Hence, Fickian-type diffusion of macromolecules and small solute molecules can be greatly affected by the presence together of both types of molecules, even in dilute solutions.

3. Experimental methods to determine diffusivity. Methods to determine the diffusivity of biological solutes are similar to those discussed previously in Section 6.3 with some modifications. In the diaphragm diffusion cell shown in Fig. 6.3-1, the chamber is made of Lucite or Teflon instead of glass, since protein molecules bind to glass. Also, the porous membrane through which the molecular diffusion occurs is composed of cellulose acetate or other polymers (G5, G6, K1).

4. Experimental data for biological solutes. Most of the experimental data in the literature on protein diffusivities have been extrapolated to zero concentration since the diffusivity is often a function of concentration. A tabulation of diffusivities of a few proteins and also of small solutes often present in biological systems is given in Table 6.4-1.

   The diffusion coefficients for the large protein molecules are of the order of magnitude of $5 \times 10^{-11} \, m^2/s$ compared to the values of about $1 \times 10^{-9} \, m^2/s$ for the small solutes in Table 6.4-1. This means macromolecules diffuse at a rate about 20 times as slow as small solute molecules for the same concentration differences.

TABLE 6.4-1.   *Diffusion Coefficients for Dilute Biological Solutes in Aqueous Solution*

	Temperature		Diffusivity (m^2/s)	Molecular Weight	Ref.
Solute	°C	K			
Urea	20	293	1.20×10^{-9}	60.1	(N2)
	25	298	1.378×10^{-9}		(G5)
Glycerol	20	293	0.825×10^{-9}	92.1	(G3)
Glycine	25	298	1.055×10^{-9}	75.1	(L3)
Sodium caprylate	25	298	8.78×10^{-10}	166.2	(G6)
Bovine serum albumin	25	298	6.81×10^{-11}	67 500	(C6)
Urease	25	298	4.01×10^{-11}	482 700	(C7)
	20	293	3.46×10^{-11}		(S6)
Soybean protein	20	293	2.91×10^{-11}	361 800	(S6)
Lipoxidase	20	293	5.59×10^{-11}	97 440	(S6)
Fibrinogen, human	20	293	1.98×10^{-11}	339 700	(S6)
Human serum albumin	20	293	5.93×10^{-11}	72 300	(S6)
γ-Globulin, human	20	293	4.00×10^{-11}	153 100	(S6)
Creatinine	37	310	1.08×10^{-9}	113.1	(C8)
Sucrose	37	310	0.697×10^{-9}	342.3	(C8)
	20	293	0.460×10^{-9}		(P3)

When the concentration of macromolecules such as proteins increases, the diffusion coefficient would be expected to decrease, since the diffusivity of small solute molecules decreases with increasing concentration. However, experimental data (G4, C7) show that the diffusivity of macromolecules such as proteins decreases in some cases and increases in other cases as protein concentration increases. Surface charges on the molecules appear to play a role in these phenomena.

When small solutes such as urea, KCl, and sodium caprylate, which are often present with protein macromolecules in solution diffuse through these protein solutions, the diffusivity decreases with increasing polymer concentration (C7, G5, G6, N3). Experimental data for the diffusivity of the solute sodium caprylate (A) diffusing through bovine serum albumin (P) solution show that the diffusivity D_{AP} of A through P is markedly reduced as the protein (P) concentration is increased (G5, G6). A large part of the reduction is due to the binding of A to P so that there is less free A to diffuse. The rest is due to blockage by the large molecules.

5. Prediction of diffusivities for biological solutes.   For predicting the diffusivity of small solutes alone in aqueous solution with molecular weights less than about 1000 or solute molar volumes less than about 0.500 m^3/kg mol, Eq. (6.3-9) should be used. For larger solutes the equations to be used are not as accurate. As an approximation the Stokes–Einstein equation (6.3-8) can be used.

$$D_{AB} = \frac{9.96 \times 10^{-16}T}{\mu V_A^{1/3}} \qquad (6.3\text{-}8)$$

Probably a better approximate equation to use is the semiempirical equation of Polson (P3), which is recommended for a molecular weight above 1000. A modification of his equation to take into account different temperatures is as follows for dilute aqueous solutions.

$$D_{AB} = \frac{9.40 \times 10^{-15}T}{\mu (M_A)^{1/3}} \qquad (6.4\text{-}1)$$

where M_A is the molecular weight of the large molecule A. When the shape of the molecule deviates greatly from a sphere, the equation should be used with caution.

EXAMPLE 6.4-1.   Prediction of Diffusivity of Albumin
Predict the diffusivity of bovine serum albumin at 298 K in water as a dilute solution using the modified Polson equation (6.4-1) and compare with the experimental value in Table 6.4-1.

Solution: The molecular weight of bovine serum albumin (A) from Table 6.4-1 is $M_A = 67\,500$ kg/kg mol. The viscosity of water at 25°C is 0.8937×10^{-3} Pa·s and $T = 298$ K. Substituting into Eq. (6.4-1),

$$D_{AB} = \frac{9.40 \times 10^{-15}T}{\mu (M_A)^{1/3}} = \frac{(9.40 \times 10^{-15})298}{(0.8937 \times 10^{-3})(67\,500)^{1/3}}$$

$$= 7.70 \times 10^{-11} \text{ m}^2/\text{s}$$

This value is 11% higher than the experimental value of 6.81×10^{-11} m^2/s.

6. Prediction of diffusivity of small solutes in protein solution.   When a small solute (A) diffuses through a macromolecule (P) protein solution, Eq. (6.3-9) cannot be used for prediction for the small solute because of blockage to diffusion by the large molecules. The data needed to predict these effects are the diffusivity D_{AB} of solute A in water alone, the water of hydration on the protein, and an obstruction factor. A semi-theoretical equation that can be used to approximate the diffusivity D_{AP} of A in globular-type protein P solutions is as follows, where only the blockage effect is considered (C8, G5, G6) and no binding is present:

$$D_{AP} = D_{AB}(1 - 1.81 \times 10^{-3}c_p) \tag{6.4-2}$$

where $c_p = $ kg P/m^3. Then the diffusion equation is

$$N_A = \frac{D_{AP}(c_{A1} - c_{A2})}{z_2 - z_1} \tag{6.4-3}$$

where c_{A1} is concentration of A in kg mol A/m^3.

7. *Prediction of diffusivity with binding present.* When A is in a protein solution P and binds to P, the diffusion flux of A is equal to the flux of unbound solute A in the solution plus the flux of the protein–solute complex. Methods to predict this flux are available (G5, G6) when binding data have been experimentally obtained. The equation used is

$$D_{AP} = \left[D_{AB}(1 - 1.81 \times 10^{-3}c_p)\left(\frac{\% \text{ free } A}{100}\right) + D_P\left(\frac{\% \text{ bound } A}{100}\right) \right] \tag{6.4-4}$$

where D_P is the diffusivity of the protein alone in the solution, m^2/s; and free A is that A not bound to the protein which is determined from the experimental binding coefficient. Then Eq. (6.4-3) is used to calculate the flux where c_A is the total concentration of A in the solution.

6.4B Diffusion in Biological Gels

Gels can be looked upon as semisolid materials which are "porous." They are composed of macromolecules which are usually in dilute aqueous solution with the gel comprising a few wt % of the water solution. The "pores" or open spaces in the gel structure are filled with water. The rates of diffusion of small solutes in the gels are somewhat less than in aqueous solution. The main effect of the gel structure is to increase the path length for diffusion, assuming no electrical-type effects (S7).

Recent studies by electron microscopy (L4) have shown that the macromolecules of the gel agarose (a major constituent of agar) exist as long and relatively straight threads. This suggests a gel structure of loosely interwoven, extensively hydrogen-bonded polysaccharide macromolecules.

Some typical gels are agarose, agar, and gelatin. A number of organic polymers exist as gels in various types of solutions. To measure the diffusivity of solutes in gels, unsteady-state methods are used. In one method the gel is melted and poured into a narrow tube open at one end. After solidification, the tube is placed in an agitated bath containing the solute for diffusion. The solute leaves the solution at the gel boundary and diffuses through the gel itself. After a period of time the amount diffusing in the gel is determined to give the diffusion coefficient of the solute in the gel.

A few typical values of diffusivities of some solutes in various gels are given in Table 6.4-2. In some cases the diffusivity of the solute in pure water is given so that the decrease in diffusivity due to the gel can be seen. For example, from Table 6.4-2 at 278 K, urea in water has a diffusivity of 0.880×10^{-9} m^2/s and in 2.9 wt % gelatin, has a value of 0.640×10^{-9} m^2/s, a decrease of 27%.

TABLE 6.4-2. *Typical Diffusivities of Solutes in Dilute Biological Gels in Aqueous Solution*

Solute	Gel	Wt % Gel in Solution	Temperature K	°C	Diffusivity (m^2/s)	Ref.
Sucrose	Gelatin	0	278	5	0.285×10^{-9}	(F2)
		3.8	278	5	0.209×10^{-9}	(F2)
		10.35	278	5	0.107×10^{-9}	(F2)
		5.1	293	20	0.252×10^{-9}	(F3)
Urea	Gelatin	0	278	5	0.880×10^{-9}	(F2)
		2.9	278	5	0.644×10^{-9}	(F2)
		5.1	278	5	0.609×10^{-9}	(F3)
		10.0	278	5	0.542×10^{-9}	(F2)
		5.1	293	20	0.859×10^{-9}	(F3)
Methanol	Gelatin	3.8	278	5	0.626×10^{-9}	(F3)
Urea	Agar	1.05	278	5	0.727×10^{-9}	(F3)
		3.16	278	5	0.591×10^{-9}	(F3)
		5.15	278	5	0.472×10^{-9}	(F3)
Glycerin	Agar	2.06	278	5	0.297×10^{-9}	(F3)
		6.02	278	5	0.199×10^{-9}	(F3)
Dextrose	Agar	0.79	278	5	0.327×10^{-9}	(F3)
Sucrose	Agar	0.79	278	5	0.247×10^{-9}	(F3)
Ethanol	Agar	5.15	278	5	0.393×10^{-9}	(F3)
NaCl (0.05 M)	Agarose	0	298	25	1.511×10^{-9}	(S7)
		2	298	25	1.398×10^{-9}	(S7)

In both agar and gelatin the diffusivity of a given solute decreases approximately linearly with an increase in wt % gel. However, extrapolation to 0% gel gives a value smaller than that shown for pure water. It should be noted that in different preparations or batches of the same type of gel, the diffusivities can vary by as much as 10 to 20%.

EXAMPLE 6.4-2. Diffusion of Urea in Agar

A tube or bridge of a gel solution of 1.05 wt % agar in water at 278 K is 0.04 m long and connects two agitated solutions of urea in water. The urea concentration in the first solution is 0.2 g mol urea per liter solution and 0 in the other. Calculate the flux of urea in kg mol/s · m² at steady state.

Solution: From Table 6.4-2 for the solute urea at 278 K, $D_{AB} = 0.727 \times 10^{-9}$ m²/s. For urea diffusing through stagnant water in the gel, Eq. (6.3-3) can be used. However, since the value of x_{A1} is less than about 0.01, the solution is quite dilute and $x_{BM} \cong 1.00$. Hence, Eq. (6.3-5) can be used. The concentrations are $c_{A1} = 0.20/1000 = 0.0002$ g mol/cm³ = 0.20 kg mol/m³ and $c_{A2} = 0$. Substituting into Eq. (6.3-5),

$$N_A = \frac{D_{AB}(c_{A1} - c_{A2})}{z_2 - z_1} = \frac{0.727 \times 10^{-9}(0.20 - 0)}{0.04 - 0}$$

$$= 3.63 \times 10^{-9} \text{ kg mol/s} \cdot \text{m}^2$$

6.5 MOLECULAR DIFFUSION IN SOLIDS

6.5A Introduction and Types of Diffusion in Solids

Even though rates of diffusion of gases, liquids, and solids in solids are generally slower than rates in liquids and gases, mass transfer in solids is quite important in chemical and biological processing. Some examples are leaching of foods, such as soybeans, and of metal ores; drying of timber, salts, and foods; diffusion and catalytic reaction in solid catalysts; separation of fluids by membranes; diffusion of gases through polymer films used in packaging; and treating of metals at high temperatures by gases.

We can broadly classify transport in solids into two types of diffusion: diffusion that can be considered to follow Fick's law and does not depend primarily on the actual structure of the solid, and diffusion in porous solids where the actual structure and void channels are important. These two broad types of diffusion will be considered.

6.5B Diffusion in Solids Following Fick's Law

1. Derivation of equations.   This type of diffusion in solids does not depend on the actual structure of the solid. The diffusion occurs when the fluid or solute diffusing is actually dissolved in the solid to form a more or less homogeneous solution—for example, in leaching, where the solid contains a large amount of water and a solute is diffusing through this solution, or in the diffusion of zinc through copper, where solid solutions are present. Also, the diffusion of nitrogen or hydrogen through rubber, or in some cases diffusion of water in foodstuffs can be classified here, since equations of similar type can be used.

Generally, simplified equations are used. Using the general Eq. (6.2-14) for binary diffusion,

$$N_A = -cD_{AB}\frac{dx_A}{dz} + \frac{c_A}{c}(N_A + N_B) \qquad (6.2\text{-}14)$$

the bulk flow term, $(c_A/c)(N_A + N_B)$, even if present, is usually small, since c_A/c or x_A is quite small. Hence, it is neglected. Also, c is assumed constant giving for diffusion in solids,

$$N_A = -\frac{D_{AB}\,dc_A}{dz} \qquad (6.5\text{-}1)$$

where D_{AB} is diffusivity in m²/s of A through B and usually is assumed constant independent of pressure for solids. Note that $D_{AB} \neq D_{BA}$ in solids.

Integration of Eq. (6.5-1) for a solid slab at steady state gives

$$N_A = \frac{D_{AB}(c_{A1} - c_{A2})}{z_2 - z_1} \qquad (6.5\text{-}2)$$

For the case of diffusion radially through a cylinder wall of inner radius r_1 and outer r_2 and length L,

$$\frac{\bar{N}_A}{2\pi rL} = -D_{AB}\frac{dc_A}{dr} \qquad (6.5\text{-}3)$$

$$\bar{N}_A = D_{AB}(c_{A1} - c_{A2})\frac{2\pi L}{\ln(r_2/r_1)} \qquad (6.5\text{-}4)$$

This case is similar to conduction heat transfer radially through a hollow cylinder in Fig. 4.3-2.

The diffusion coefficient D_{AB} in the solid as stated above is not dependent upon the pressure of the gas or liquid on the outside of the solid. For example, if CO_2 gas is outside a slab of rubber and is diffusing through the rubber, D_{AB} would be independent of p_A, the partial pressure of CO_2 at the surface. The solubility of CO_2 in the solid, however, is directly proportional to p_A. This is similar to the case of the solubility of O_2 in water being directly proportional to the partial pressure of O_2 in the air by Henry's law.

The solubility of a solute gas (A) in a solid is usually expressed as S in m^3 solute (at STP of 0°C and 1 atm) per m^3 solid per atm partial pressure of (A). Also, $S = cm^3$ (STP)/atm · cm^3 solid in the cgs system. To convert this to c_A concentration in the solid in kg mol A/m^3 using SI units,

$$c_A = \frac{S \text{ m}^3\text{(STP)/m}^3 \text{ solid} \cdot \text{atm}}{22.414 \text{ m}^3\text{(STP)/kg mol } A} p_A \text{ atm} = \frac{Sp_A}{22.414} \frac{\text{kg mol } A}{\text{m}^3 \text{ solid}} \tag{6.5-5}$$

Using cgs units,

$$c_A = \frac{Sp_A}{22\,414} \frac{\text{g mol } A}{\text{cm}^3 \text{ solid}} \tag{6.5-6}$$

EXAMPLE 6.5-1. Diffusion of H_2 Through Neoprene Membrane

The gas hydrogen at 17°C and 0.010 atm partial pressure is diffusing through a membrane of vulcanized neoprene rubber 0.5 mm thick. The pressure of H_2 on the other side of the neoprene is zero. Calculate the steady-state flux, assuming that the only resistance to diffusion is in the membrane. The solubility S of H_2 gas in neoprene at 17°C is 0.051 m^3 (at STP of 0°C and 1 atm)/m^3 solid · atm and the diffusivity D_{AB} is 1.03×10^{-10} m^2/s at 17°C.

Solution: A sketch showing the concentration is shown in Fig. 6.5-1. The equilibrium concentration c_{A1} at the inside surface of the rubber is, from Eq. (6.5-5),

$$c_{A1} = \frac{S}{22.414} p_{A1} = \frac{0.051(0.010)}{22.414} = 2.28 \times 10^{-5} \text{ kg mol } H_2/\text{m}^3 \text{ solid}$$

Since p_{A2} at the other side is 0, $c_{A2} = 0$. Substituting into Eq. (6.5-2) and solving,

$$N_A = \frac{D_{AB}(c_{A1} - c_{A2})}{z_2 - z_1} = \frac{(1.03 \times 10^{-10})(2.28 \times 10^{-5} - 0)}{(0.5 - 0)/1000}$$

$$= 4.69 \times 10^{-12} \text{ kg mol } H_2/\text{s} \cdot \text{m}^2$$

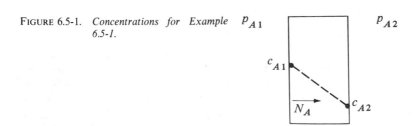

FIGURE 6.5-1. Concentrations for Example 6.5-1.

2. *Permeability equations for diffusion in solids.* In many cases the experimental data for diffusion of gases in solids are not given as diffusivities and solubilities but as permeabilities, P_M, in m^3 of solute gas A at STP (0°C and 1 atm press) diffusing per second per m^2 cross-sectional area through a solid 1 m thick under a pressure difference of 1 atm pressure. This can be related to Fick's equation (6.5-2) as follows.

$$N_A = \frac{D_{AB}(c_{A1} - c_{A2})}{z_2 - z_1} \tag{6.5-2}$$

From Eq. (6.5-5),

$$c_{A1} = \frac{Sp_{A1}}{22.414} \qquad c_{A2} = \frac{Sp_{A2}}{22.414} \tag{6.5-7}$$

Substituting Eq. (6.5-7) into (6.5-2),

$$N_A = \frac{D_{AB}S(p_{A1} - p_{A2})}{22.414(z_2 - z_1)} = \frac{P_M(p_{A1} - p_{A2})}{22.414(z_2 - z_1)} \text{ kg mol/s} \cdot m^2 \tag{6.5-8}$$

where the permeability P_M is

$$P_M = D_{AB}S \frac{m^3(STP)}{s \cdot m^2 C.S. \cdot atm/m} \tag{6.5-9}$$

Permeability is also given in the literature in several other ways. For the cgs system, the permeability is given as P'_M, cc(STP)/(s · cm²C.S. · atm/cm). This is related to P_M by

$$P_M = 10^{-4}P'_M \tag{6.5-10}$$

In some cases in the literature the permeability is given as P''_M, cc(STP)/(s · cm²C.S. · cm Hg/mm thickness). This is related to P_M by

$$P_M = 7.60 \times 10^{-4}P''_M \tag{6.5-11}$$

When there are several solids 1, 2, 3, ..., in series and L_1, L_2, ..., represent the thickness of each, then Eq. (6.5-8) becomes

$$N_A = \frac{p_{A1} - p_{A2}}{22.414} \frac{1}{L_1/P_{M1} + L_2/P_{M2} + \cdots} \tag{6.5-12}$$

where $p_{A1} - p_{A2}$ is the overall partial pressure difference.

3. *Experimental diffusivities, solubilities, and permeabilities.* Accurate prediction of diffusivities in solids is generally not possible because of the lack of knowledge of the theory of the solid state. Hence, experimental values are needed. Some experimental data for diffusivities, solubilities, and permeabilities are given in Table 6.5-1 for gases diffusing in solids and solids diffusing in solids.

For the simple gases such as He, H_2, O_2, N_2, and CO_2, with gas pressures up to 1 or 2 atm, the solubility in solids such as polymers and glasses generally follows Henry's law and Eq. (6.5-5) holds. Also, for these gases the diffusivity and permeability are independent of concentration, and hence pressure. For the effect of temperature T in K, the ln P_M is approximately a linear function of $1/T$. Also, the diffusion of one gas, say H_2, is approximately independent of the other gases present, such as O_2 and N_2.

For metals such as Ni, Cd, and Pt, where gases such as H_2 and O_2 are diffusing, it is found experimentally that the flux is approximately proportional to $(\sqrt{p_{A1}} - \sqrt{p_{A2}})$, so Eq. (6.5-8) does not hold (B5). When water is diffusing through polymers, unlike the

TABLE 6.5-1. *Diffusivities and Permeabilities in Solids*

Solute (A)	Solid (B)	T (K)	D_{AB}, Diffusion Coefficient $[m^2/s]$	Solubility, S $\left[\dfrac{m^3\ solute(STP)}{m^3\ solid \cdot atm}\right]$	Permeability, P_M $\left[\dfrac{m^3\ solute(STP)}{s \cdot m^2 \cdot atm/m}\right]$	Ref.
H_2	Vulcanized rubber	298	$0.85(10^{-9})$	0.040	$0.342(10^{-10})$	(B5)
O_2		298	$0.21(10^{-9})$	0.070	$0.152(10^{-10})$	(B5)
N_2		298	$0.15(10^{-9})$	0.035	$0.054(10^{-10})$	(B5)
CO_2		298	$0.11(10^{-9})$	0.90	$1.01(10^{-10})$	(B5)
H_2	Vulcanized neoprene	290	$0.103(10^{-9})$	0.051		(B5)
		300	$0.180(10^{-9})$	0.053		(B5)
H_2	Polyethylene	298			$6.53(10^{-12})$	(R3)
O_2		303			$4.17(10^{-12})$	(R3)
N_2		303			$1.52(10^{-12})$	(R3)
O_2	Nylon	303			$0.029(10^{-12})$	(R3)
N_2		303			$0.0152(10^{-12})$	(R3)
Air	English leather	298			0.15–0.68×10^{-4}	(B5)
H_2O	Wax	306			$0.16(10^{-10})$	(B5)
H_2O	Cellophane	311			0.91–$1.82(10^{-10})$	(B5)
He	Pyrex glass	293			$4.86(10^{-15})$	(B5)
		373			$20.1(10^{-15})$	(B5)
He	SiO_2	293	2.4–$5.5(10^{-14})$	0.01		(B5)
H_2	Fe	293	$2.59(10^{-13})$			(B5)
Al	Cu	293	$1.3(10^{-34})$			(B5)

simple gases, P_M may depend somewhat on the relative pressure difference (C9, B5). Further data are available in monographs by Crank and Park (C9) and Barrer (B5).

EXAMPLE 6.5-2. Diffusion Through a Packaging Film Using Permeability

A polyethylene film 0.00015 m (0.15 mm) thick is being considered for use in packaging a pharmaceutical product at 30°C. If the partial pressure of O_2 outside is 0.21 atm and inside the package it is 0.01 atm, calculate the diffusion flux of O_2 at steady state. Use permeability data from Table 6.5-1. Assume that the resistances to diffusion outside the film and inside are negligible compared to the resistance of the film.

Solution: From Table 6.5-1 $P_M = 4.17(10^{-12})$ m³ solute(STP)/(s·m²·atm/m). Substituting into Eq. (6.5-8),

$$N_A = \frac{P_M\,(p_{A1} - p_{A2})}{22.414(z_2 - z_1)} = \frac{4.17(10^{-12})(0.21 - 0.01)}{22.414(0.00015 - 0)}$$

$$= 2.480 \times 10^{-10} \text{ kg mol/s} \cdot \text{m}^2$$

Note that a film made of nylon has a much smaller value of permeability P_M for O_2 and would make a more suitable barrier.

4. *Membrane separation processes.* In Chapter 13 a detailed discussion is given of the various membrane separation processes of gas separation by membranes, dialysis, reverse osmosis, and ultrafiltration.

6.5C Diffusion in Porous Solids That Depends on Structure

1. Diffusion of liquids in porous solids. In Section 6.5B we used Fick's law and treated the solid as a uniform homogeneous-like material with an experimental diffusivity D_{AB}. In this section we are concerned with porous solids that have pores or interconnected voids in the solid which affect the diffusion. A cross section of such a typical porous solid is shown in Fig. 6.5-2.

For the situation where the voids are filled completely with liquid water, the concentration of salt in water at boundary 1 is c_{A1} and at point 2 is c_{A2}. The salt in diffusing through the water in the void volume takes a tortuous path which is unknown and greater than $(z_2 - z_1)$ by a factor τ, called *tortuosity*. Diffusion does not occur in the inert solid. For a dilute solution using Eq. (6.3-5) for diffusion of salt in water at steady state,

$$N_A = \frac{\varepsilon D_{AB}(c_{A1} - c_{A2})}{\tau(z_2 - z_1)} \tag{6.5-13}$$

where ε is the open void fraction, D_{AB} is the diffusivity of salt in water, and τ is a factor which corrects for the path longer than $(z_2 - z_1)$. For inert-type solids τ can vary from about 1.5 to 5. Often the terms are combined into an effective diffusivity.

$$D_{A \text{ eff}} = \frac{\varepsilon}{\tau} D_{AB} \quad m^2/s \tag{6.5-14}$$

EXAMPLE 6.5-3. Diffusion of KCl in Porous Silica
A sintered solid of silica 2.0 mm thick is porous with a void fraction ε of 0.30 and a tortuosity τ of 4.0. The pores are filled with water at 298 K. At one face the concentration of KCl is held at 0.10 g mol/liter, and fresh water flows rapidly by the other face. Neglecting any other resistances but that in the porous solid, calculate the diffusion of KCl at steady state.

Solution: The diffusivity of KCl in water from Table 6.3-1 is $D_{AB} = 1.87 \times 10^{-9}$ m²/s. Also, $c_{A1} = 0.10/1000 = 1.0 \times 10^{-4}$ g mol/cm³ $= 0.10$ kg mol/m³, and $c_{A2} = 0$. Substituting into Eq. (6.5-13),

$$N_A = \frac{\varepsilon D_{AB}(c_{A1} - c_{A2})}{\tau(z_2 - z_1)} = \frac{0.30(1.870 \times 10^{-9})(0.10 - 0)}{4.0(0.002 - 0)}$$

$$= 7.01 \times 10^{-9} \text{ kg mol KCl/s} \cdot m^2$$

FIGURE 6.5-2. *Sketch of a typical porous solid.*

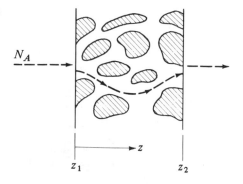

N_A

z

z_1 z_2

2. *Diffusion of gases in porous solids.* If the voids shown in Fig. 6.5-2 are filled with gases, then a somewhat similar situation exists. If the pores are very large so that diffusion occurs only by Fickian-type diffusion, then Eq. (6.5-13) becomes, for gases,

$$N_A = \frac{\varepsilon D_{AB}(c_{A1} - c_{A2})}{\tau(z_2 - z_1)} = \frac{\varepsilon D_{AB}(p_{A1} - p_{A2})}{\tau RT(z_2 - z_1)} \tag{6.5-15}$$

Again the value of the tortuosity must be determined experimentally. Diffusion is assumed to occur only through the voids or pores and not through the actual solid particles.

A correlation of tortuosity versus the void fraction of various unconsolidated porous media of beds of glass spheres, sand, salt, talc, and so on (S8), gives the following approximate values of τ for different values of ε: $\varepsilon = 0.2$, $\tau = 2.0$; $\varepsilon = 0.4$, $\tau = 1.75$; $\varepsilon = 0.6$, $\tau = 1.65$.

When the pores are quite small in size and of the order of magnitude of the mean free path of the gas, other types of diffusion occur, which are discussed in Section 7.6.

6.6 NUMERICAL METHODS FOR STEADY-STATE MOLECULAR DIFFUSION IN TWO DIMENSIONS

6.6A Derivation of Equations for Numerical Method

1. Derivation of method for steady state. In Fig. 6.6-1 a two-dimensional solid shown with unit thickness is divided into squares. The numerical methods for steady-state molecular diffusion are very similar to those for steady-state heat conduction discussed in Section 4.15. Hence, only a brief summary will be given here. The solid inside of a square is imagined to be concentrated at the center of the square at $c_{n,m}$ and is called a "node," which is connected to the adjacent nodes by connecting rods through which the mass diffuses.

A total mass balance is made at steady state by stating that the sum of the molecular diffusion to the shaded area for unit thickness must equal zero.

$$\frac{D_{AB}\,\Delta y}{\Delta x}(c_{n-1,m} - c_{n,m}) + \frac{D_{AB}\,\Delta y}{\Delta x}(c_{n+1,m} - c_{n,m})$$

$$+ \frac{D_{AB}\,\Delta x}{\Delta y}(c_{n,m+1} - c_{n,m}) + \frac{D_{AB}\,\Delta x}{\Delta y}(c_{n,m-1} - c_{n,m}) = 0 \tag{6.6-1}$$

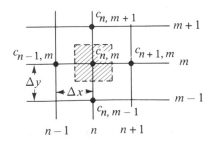

FIGURE 6.6-1. *Concentrations and spacing of nodes for two-dimensional steady-state molecular diffusion.*

where $c_{n, m}$ is concentration of A at node n, m in kg mol A/m^3. Setting $\Delta x = \Delta y$, and rearranging,

$$c_{n, m+1} + c_{n, m-1} + c_{n+1, m} + c_{n-1, m} - 4c_{n, m} = 0 \tag{6.6-2}$$

2. *Iteration method of numerical solution.* In order to solve Eq. (6.6-2), a separate equation is written for each unknown point giving N linear algebraic equations for N unknown points. For a hand calculation using a modest number of nodes, the iteration method can be used to solve the equations, where the right-hand side of Eq. (6.6-2) is set equal to a residual $\bar{N}_{n, m}$.

$$c_{n, m+1} + c_{n, m-1} + c_{n+1, m} + c_{n-1, m} - 4c_{n, m} = \bar{N}_{n, m} \tag{6.6-3}$$

Setting the equation equal to zero, $\bar{N}_{n, m} = 0$ for steady state and $c_{n, m}$ is calculated by

$$c_{n, m} = \frac{c_{n-1, m} + c_{n+1, m} + c_{n, m+1} + c_{n, m-1}}{4} \tag{6.6-4}$$

Equations (6.6-3) and (6.6-4) are the final equations to be used to calculate all the concentrations at steady state.

Example 4.15-1 for steady-state heat conduction illustrates the detailed steps for the iteration method, which are identical to those for steady-state diffusion.

Once the concentrations have been calculated, the flux can be calculated for each element as follows. Referring to Fig. 6.6-1, the flux for the node or element $c_{n, m}$ to $c_{n, m-1}$ is $(\Delta x = \Delta y)$

$$N = \frac{AD_{AB}}{\Delta y}(c_{n, m} - c_{n, m-1}) = \frac{[(\Delta x)(1)]D_{AB}}{\Delta y}(c_{n, m} - c_{n, m-1})$$

$$= D_{AB}(c_{n, m} - c_{n, m-1}) \tag{6.6-5}$$

where the area A is Δx times 1 m deep and N is kg mol A/s. Equations are written for the other appropriate elements and the sum of the fluxes calculated.

6.6B Equations for Special Boundary Conditions for Numerical Method

1. *Equations for boundary conditions.* When one of the nodal points $c_{n, m}$ is at a boundary where convective mass transfer is occurring to a constant concentration c_∞ in the bulk fluid shown in Fig. 6.6-2a a different equation must be derived. Making a mass balance on the node n, m, where mass in = mass out at steady state,

$$\frac{D_{AB} \Delta y}{\Delta x}(c_{n-1, m} - c_{n, m}) + \frac{D_{AB} \Delta x}{2 \Delta y}(c_{n, m+1} - c_{n, m})$$

$$+ \frac{D_{AB} \Delta x}{2 \Delta y}(c_{n, m-1} - c_{n, m}) = k_c \Delta y(c_{n, m} - c_\infty) \tag{6.6-6}$$

where k_c is the convective mass-transfer coefficient in m/s defined by Eq. (6.1-14).

Setting $\Delta x = \Delta y$, rearranging, and setting the resultant equation = $\bar{N}_{n, m}$, the residual, the following results.

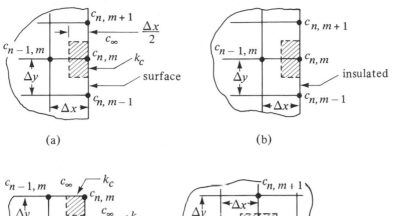

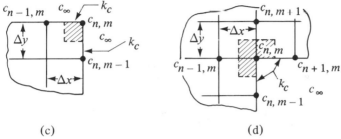

FIGURE 6.6-2. *Different boundary conditions for steady-state diffusion:* (a) *convection at a boundary,* (b) *insulated boundary,* (c) *exterior corner with convective boundary,* (d) *interior corner with convective boundary.*

1. For convection at a boundary (Fig. 6.6-2a),

$$\frac{k_c \, \Delta x}{D_{AB}} c_\infty + \tfrac{1}{2}(2c_{n-1, m} + c_{n, m+1} + c_{n, m-1}) - c_{n, m}\left(\frac{k_c \, \Delta x}{D_{AB}} + 2\right) = \bar{N}_{n, m} \quad (6.6\text{-}7)$$

This equation is similar to Eq. (4.15-16) for heat conduction and convection with $k_c \, \Delta x/D_{AB}$ being used in place of $h \, \Delta x/k$. Similarly, Eqs. (6.6-8)–(6.6-10) have been derived for the other boundary conditions shown in Fig. 6.6-2.

2. For an insulated boundary (Fig. 6.6-2b),

$$\tfrac{1}{2}(c_{n, m+1} + c_{n, m-1}) + c_{n-1, m} - 2c_{n, m} = \bar{N}_{n, m} \quad (6.6\text{-}8)$$

3. For an exterior corner with convection at the boundary (Fig. 6.6-2c),

$$\frac{k_c \, \Delta x}{D_{AB}} c_\infty + \tfrac{1}{2}(c_{n-1, m} + c_{n, m-1}) - \left(\frac{k_c \, \Delta x}{D_{AB}} + 1\right)c_{n, m} = \bar{N}_{n, m} \quad (6.6\text{-}9)$$

4. For an interior corner with convection at the boundary (Fig. 6.6-2d),

$$\frac{k_c \, \Delta x}{D_{AB}} c_\infty + c_{n-1, m} + c_{n, m+1} + \tfrac{1}{2}(c_{n+1, m} + c_{n, m-1}) - \left(3 + \frac{k_c \, \Delta x}{D_{AB}}\right)c_{n, m} = \bar{N}_{n, m} \quad (6.6\text{-}10)$$

2. Boundary conditions with distribution coefficient. When Eq. (6.6-7) was derived, the distribution coefficient K between the liquid and the solid at the surface interface was 1.0. The distribution coefficient as shown in Fig. 6.6-3 is defined as

$$K = \frac{c_{n, mL}}{c_{n, m}} \quad (6.6\text{-}11)$$

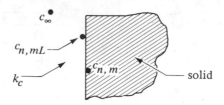

$c_{n, mL}$

c_∞

k_c

$c_{n, m}$

solid

FIGURE 6.6-3. *Interface concentrations for convective mass transfer at a solid surface and an equilibrium distribution coefficient* $K = c_{n, mL}/c_{n, m}$.

where $c_{n, mL}$ is the concentration in the liquid adjacent to the surface and $c_{n, m}$ is the concentration in the solid adjacent to the surface. Then in deriving Eq. (6.6-6), the right-hand side $k_c \, \Delta y (c_{n, m} - c_\infty)$ becomes

$$k_c \, \Delta y (c_{n, mL} - c_\infty) \tag{6.6-12}$$

where c_∞ is the concentration in the bulk fluid. Substituting $K c_{n, m}$ for $c_{n, mL}$ from Eq. (6.6-11) into (6.6-12) and multiplying and dividing by K,

$$K k_c \, \Delta y \left(\frac{K c_{n, m}}{K} - \frac{c_\infty}{K} \right) = K k_c \, \Delta y \left(c_{n, m} - \frac{c_\infty}{K} \right) \tag{6.6-13}$$

Hence, whenever k_c appears as in Eq. (6.6-7), $K k_c$ should be substituted and when c_∞ appears, c_∞/K should be used. Then Eq. (6.6-7) becomes as follows.

1. For convection at a boundary (Fig. 6.6-2a),

$$\left(\frac{K k_c \, \Delta x}{D_{AB}} \right) \frac{c_\infty}{K} + \tfrac{1}{2}(2c_{n-1, m} + c_{n, m+1} + c_{n, m-1})$$

$$- c_{n, m} \left(\frac{K k_c \, \Delta x}{D_{AB}} + 2 \right) = \bar{N}_{n, m} \quad \text{(6.6-14)}$$

Equations (6.6-9) and (6.6-10) can be rewritten in a similar manner as follows,

2. For an exterior corner with convection at the boundary (Fig. 6.6-2c),

$$\left(\frac{K k_c \, \Delta x}{D_{AB}} \right) \frac{c_\infty}{K} + \tfrac{1}{2}(c_{n-1, m} + c_{n, m-1}) - \left(\frac{K k_c \, \Delta x}{D_{AB}} + 1 \right) c_{n, m} = \bar{N}_{n, m} \quad \text{(6.6-15)}$$

3. For an interior corner with convection at the boundary (Fig. 6.6-2d),

$$\left(\frac{K k_c \, \Delta x}{D_{AB}} \right) \frac{c_\infty}{K} + c_{n-1, m} + c_{n, m+1}$$

$$+ \tfrac{1}{2}(c_{n+1, m} + c_{n, m-1}) - \left(3 + \frac{K k_c \, \Delta x}{D_{AB}} \right) c_{n, m} = \bar{N}_{n, m} \quad \text{(6.6-16)}$$

EXAMPLE 6.6-1. *Numerical Method for Convection and Steady-State Diffusion*

For the two-dimensional hollow solid chamber shown in Fig. 6.6-4, determine the concentrations at the nodes as shown at steady state. At the inside surfaces the concentrations remain constant at 6.00×10^{-3} kg mol/m³. At the outside surfaces the convection coefficient $k_c = 2 \times 10^{-7}$ m/s and $c_\infty = 2.00 \times 10^{-3}$ kg mol/m³. The diffusivity in the solid is $D_{AB} = 1.0 \times 10^{-9}$

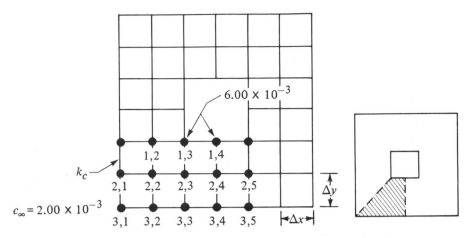

FIGURE 6.6-4. *Concentrations for hollow chamber for Example 6.6-1.*

m^2/s. The grid size is $\Delta x = \Delta y = 0.005$ m. Also, determine the diffusion rates per 1.0-m depth. The distribution coefficient $K = 1.0$.

Solution: To simplify the calculations, all the concentrations will be multiplied by 10^3. Since the chamber is symmetrical, we do the calculations on the $\frac{1}{8}$ shaded portion shown. The fixed known values are $c_{1,3} = 6.00$, $c_{1,4} = 6.00$, $c_\infty = 2.00$. Because of symmetry, $c_{1,2} = c_{2,3}$, $c_{2,5} = c_{2,3}$, $c_{2,1} = c_{3,2}$, $c_{3,3} = c_{3,5}$. To speed up the calculations, we will make estimates of the unknown concentrations as follows: $c_{2,2} = 3.80$, $c_{2,3} = 4.20$, $c_{2,4} = 4.40$, $c_{3,1} = 2.50$, $c_{3,2} = 2.70$, $c_{3,3} = 3.00$, $c_{3,4} = 3.20$.

For the interior points $c_{2,2}$, $c_{2,3}$ and $c_{2,4}$ we use Eqs. (6.6-3) and (6.6-4); for the corner convection point $c_{3,1}$, Eq. (6.6-9); for the other convection points $c_{3,2}$, $c_{3,3}$, $c_{3,4}$, Eq. (6.6-7). The term $k_c\,\Delta x/D_{AB} = (2 \times 10^{-7})(0.005)/(1.0 \times 10^{-9}) = 1.00$.

First approximation. Starting with $c_{2,2}$ we use Eq. (6.6-3) and calculate the residual $\bar{N}_{2,2}$.

$$c_{1,2} + C_{3,2} + c_{2,1} + c_{2,3} - 4c_{2,2} = \bar{N}_{2,2}$$
$$4.20 + 2.70 + 2.70 + 4.20 - 4(3.80) - -1.40$$

Hence, $c_{2,2}$ is not at steady state. Next we set $\bar{N}_{2,2}$ to zero and calculate a new value of $c_{2,2}$ from Eq. (6.6-4).

$$c_{2,2} = \frac{c_{1,2} + c_{3,2} + c_{2,1} + c_{2,3}}{4} = \frac{4.20 + 2.70 + 2.70 + 4.20}{4} = 3.45$$

This new value of $c_{2,2}$ replaces the old value.
For $c_{2,3}$,

$$c_{2,2} + c_{2,4} + c_{1,3} + c_{3,3} - 4c_{2,3} = \bar{N}_{2,3}$$
$$3.45 + 4.40 + 6.00 + 3.00 - 4(4.20) = 0.05$$

$$c_{2,3} = \frac{c_{2,2} + c_{2,4} + c_{1,3} + c_{3,3}}{4} = \frac{3.45 + 4.40 + 6.00 + 3.00}{4} = 4.21$$

For $c_{2,4}$,

$$c_{2,3} + c_{2,5} + c_{1,4} + C_{3,4} - 4c_{2,4} = \bar{N}_{2,4}$$

$$4.21 + 4.21 + 6.00 + 3.20 - 4(4.40) = 0.02$$

$$c_{2,4} = \frac{c_{2,3} + c_{2,5} + c_{1,4} + c_{3,4}}{4} = \frac{4.21 + 4.21 + 6.00 + 3.20}{4} = 4.41$$

For $c_{3,1}$, we use Eq. (6.6-9):

$$(1.0)c_\infty + \tfrac{1}{2}(c_{2,1} + c_{3,2}) - (1.0 + 1)c_{3,1} = \bar{N}_{3,1}$$

$$(1.0)2.00 + \tfrac{1}{2}(2.70 + 2.70) - 2.0(2.50) = -0.30$$

Setting Eq. (6.6-9) to zero and solving for $c_{3,1}$,

$$(1.0)2.00 + \tfrac{1}{2}(2.70 + 2.70) - (2.0)c_{3,1} = 0 \qquad c_{3,1} = 2.35$$

For $c_{3,2}$, we use Eq. (6.6-7).

$$(1.0)c_\infty + \tfrac{1}{2}(2 \times c_{2,2} + c_{3,1} + c_{3,3}) - (1.0 + 2)c_{3,2} = \bar{N}_{3,2}$$

$$(1.0)2.00 + \tfrac{1}{2}(2 \times 3.45 + 2.35 + 3.00) - (3.0)2.70 = 0.03$$

$$(1.0)2.00 + \tfrac{1}{2}(2 \times 3.45 + 2.35 + 3.00) - (3.0)c_{3,2} = 0 \qquad c_{3,2} = 2.71$$

For $c_{3,3}$,

$$(1.0)c_\infty + \tfrac{1}{2}(2 \times c_{2,3} + c_{3,2} + c_{3,4}) - (3.0)c_{3,3} = \bar{N}_{3,3}$$

$$(1.0)2.00 + \tfrac{1}{2}(2 \times 4.21 + 2.71 + 3.20) - (3.0)3.00 = 0.17$$

$$(1.0)2.00 + \tfrac{1}{2}(2 \times 4.21 + 2.71 + 3.20) - (3.0)c_{3,3} = 0 \qquad c_{3,3} = 3.06$$

For $c_{3,4}$,

$$(1.0)c_\infty + \tfrac{1}{2}(2 \times c_{2,4} + c_{3,3} + c_{3,5}) - (3.0)c_{3,4} = \bar{N}_{3,4}$$

$$(1.0)2.00 + \tfrac{1}{2}(2 \times 4.41 + 3.06 + 3.06) - (3.0)3.20 = -0.13$$

$$(1.0)2.00 + \tfrac{1}{2}(2 \times 4.41 + 3.06 + 3.06) - (3.0)c_{3,4} = 0 \qquad c_{3,4} = 3.16$$

Having completed one sweep across the grid map, we can start a second approximation using the new values calculated and starting with $c_{2,2}$ or any other node. This is continued until all the residuals are as small as desired. The final values after three approximations are $c_{2,2} = 3.47$, $c_{2,3} = 4.24$, $c_{2,4} = 4.41, c_{3,1} = 2.36, c_{3,2} = 2.72, c_{3,3} = 3.06, c_{3,4} = 3.16$.

To calculate the diffusion rates we first calculate the total convective diffusion rate, leaving the bottom surface at nodes $c_{3,1}, c_{3,2}, c_{3,3}$, and $c_{3,4}$ for 1.0-m depth.

$$N = k_c(\Delta x \cdot 1)\left[\frac{c_{3,1} - c_\infty}{2} + (c_{3,2} - c_\infty) + (c_{3,3} - c_\infty) + \frac{c_{3,4} - c_\infty}{2}\right]$$

$$= (2 \times 10^{-7})(0.005 \times 1)\left[\frac{2.36 - 2.00}{2} + (2.72\text{-}2.00)\right.$$

$$\left. + (3.06 - 2.00) + \frac{3.16 - 2.00}{2}\right] \times 10^{-3}$$

$$= 2.540 \times 10^{-12} \text{ kg mol/s}$$

Note that the first and fourth paths include only $\tfrac{1}{2}$ of a surface. Next we calculate the total diffusion rate in the solid entering the top surface inside, using an equation similar to Eq. (6.6-5).

$$N = \frac{D_{AB}(\Delta x \cdot 1)}{\Delta y}\left[(c_{1,3} - c_{2,3}) + \frac{c_{1,4} - c_{2,4}}{2}\right]$$

$$= \frac{(1.0 \times 10^{-9})(0.005 \times 1)}{0.005}\left[(6.00 - 4.24) + \frac{6.00 - 4.41}{2}\right] \times 10^{-3}$$

$$= 2.555 \times 10^{-12} \text{ kg mol/s}$$

At steady state the diffusion rate leaving by convection should equal that entering by diffusion. These results indicate a reasonable check. Using smaller grids would give even more accuracy. Note that the results for the diffusion rate should be multiplied by 8.0 for the whole chamber.

PROBLEMS

6.1-1. Diffusion of Methane Through Helium. A gas of CH_4 and He is contained in a tube at 101.32 kPa pressure and 298 K. At one point the partial pressure of methane is $p_{A1} = 60.79$ kPa and at a point 0.02 m distance away, $p_{A2} = 20.26$ kPa. If the total pressure is constant throughout the tube, calculate the flux of CH_4 (methane) at steady-state for equimolar counterdiffusion.

Ans. $J_{Az}^* = 5.52 \times 10^{-5}$ kg mol $A/s \cdot m^2$ (5.52×10^{-6} g mol $A/s \cdot cm^2$)

6.1-2. Diffusion of CO_2 in a Binary Gas Mixture. The gas CO_2 is diffusing at steady state through a tube 0.20 m long having a diameter of 0.01 m and containing N_2 at 298 K. The total pressure is constant at 101.32 kPa. The partial pressure of CO_2 at one end is 456 mm Hg and 76 mm Hg at the other end. The diffusivity D_{AB} is 1.67×10^{-5} m^2/s at 298 K. Calculate the flux of CO_2 in cgs and SI units for equimolar counterdiffusion.

6.2-1. Equimolar Counterdiffusion of a Binary Gas Mixture. Helium and nitrogen gas are contained in a conduit 5 mm in diameter and 0.1 m long at 298 K and a uniform constant pressure of 1.0 atm abs. The partial pressure of He at one end of the tube is 0.060 atm and 0.020 atm at the other end. The diffusivity can be obtained from Table 6.2-1. Calculate the following for steady-state equimolar counterdiffusion.
(a) Flux of He in kg mol/s · m^2 and g mol/s · cm^2.
(b) Flux of N_2.
(c) Partial Pressure of He at a point 0.05 m from either end.

6.2-2. Equimolar Counterdiffusion of NH_3 and N_2 at Steady State. Ammonia gas (A) and nitrogen gas (B) are diffusing in counterdiffusion through a straight glass tube 2.0 ft (0.610 m) long with an inside diameter of 0.080 ft (24.4 mm) at 298 K and 101.32 kPa. Both ends of the tube are connected to large mixed chambers at 101.32 kPa. The partial pressure of NH_3 in one chamber is constant at 20.0 kPa and 6.666 kPa in the other chamber. The diffusivity at 298 K and 101.32 kPa is 2.30×10^{-5} m^2/s.
(a) Calculate the diffusion of NH_3 in lb mol/h and kg mol/s.
(b) Calculate the diffusion of N_2.
(c) Calculate the partial pressures at a point 1.0 ft (0.305 m) in the tube and plot p_A, p_B, and P versus distance z.

Ans. (a) Diffusion of $NH_3 = 7.52 \times 10^{-7}$ lb mol A/h, 9.48×10^{-11} kg mol A/s; (c) $p_A = 1.333 \times 10^4$ Pa

6.2-3. Diffusion of A Through Stagnant B and Effect of Type of Boundary on Flux. Ammonia gas is diffusing through N_2 under steady-state conditions with

N_2 nondiffusing since it is insoluble in one boundary. The total pressure is 1.013×10^5 Pa and the temperature is 298 K. The partial pressure of NH_3 at one point is 1.333×10^4 Pa and at the other point 20 mm away it is 6.666×10^3 Pa. The D_{AB} for the mixture at 1.013×10^5 Pa and 298 K is 2.30×10^{-5} m^2/s.
(a) Calculate the flux of NH_3 in kg mol/s · m^2.
(b) Do the same as (a) but assume that N_2 also diffuses; i.e., both boundaries are permeable to both gases and the flux is equimolar counterdiffusion. In which case is the flux greater?

Ans. (a) $N_A = 3.44 \times 10^{-6}$ kg mol/s · m^2

6.2-4. Diffusion of Methane Through Nondiffusing Helium. Methane gas is diffusing in a straight tube 0.1 m long containing helium at 298 K and a total pressure of 1.01325×10^5 Pa. The partial pressure of CH_4 at one end is 1.400×10^4 Pa and 1.333×10^3 Pa at the other end. Helium is insoluble in one boundary, and hence is nondiffusing or stagnant. The diffusivity is given in Table 6.2-1. Calculate the flux of methane in kg mol/s · m^2 at steady state.

6.2-5. Mass Transfer from a Naphthalene Sphere to Air. Mass transfer is occurring from a sphere of naphthalene having a radius of 10 mm. The sphere is in a large volume of still air at 52.6°C and 1 atm abs pressure. The vapor pressure of naphthalene at 52.6°C is 1.0 mm Hg. The diffusivity of naphthalene in air at 0°C is 5.16×10^{-6} m^2/s. Calculate the rate of evaporation of naphthalene from the surface in kg mol/s · m^2. [*Note :* The diffusivity can be corrected for temperature using the temperature-correction factor of the Fuller et al. Eq. (6.2-45).]

6.2-6. Estimation of Diffusivity of a Binary Gas. For a mixture of ethanol (CH_3CH_2OH) vapor and methane (CH_4), predict the diffusivity using the method of Fuller et al.
(a) At 1.0132×10^5 Pa and 298 and 373 K.
(b) At 2.0265×10^5 Pa and 298 K.

Ans. (a) $D_{AB} = 1.43 \times 10^{-5}$ m^2/s (298 K)

6.2-7. Diffusion Flux and Effect of Temperature and Pressure. Equimolar counterdiffusion is occurring at steady state in a tube 0.11 m long containing N_2 and CO gases at a total pressure of 1.0 atm abs. The partial pressure of N_2 is 80 mm Hg at one end and 10 mm at the other end. Predict the D_{AB} by the method of Fuller et al.
(a) Calculate the flux in kg mol/s · m^2 at 298 K for N_2.
(b) Repeat at 473 K. Does the flux increase?
(c) Repeat at 298 K but for a total pressure of 3.0 atm abs. The partial pressure of N_2 remains at 80 and 10 mm Hg, as in part (a). Does the flux change?

Ans. (a) $D_{AB} = 2.05 \times 10^{-5}$ m^2/s, $N_A = 7.02 \times 10^{-7}$ kg mol/s · m^2;
(b) $N_A = 9.92 \times 10^{-7}$ kg mol/s · m^2;
(c) $N_A = 2.34 \times 10^{-7}$ kg mol/s · m^2

6.2-8. Evaporation Losses of Water in Irrigation Ditch. Water at 25°C is flowing in a covered irrigation ditch below ground. Every 100 ft there is a vent line 1.0 in. inside diameter and 1.0 ft long to the outside atmosphere at 25°C. There are 10 vents in the 1000-ft ditch. The outside air can be assumed to be dry. Calculate the total evaporation loss of water in lb$_m$/d. Assume that the partial pressure of water vapor at the surface of the water is the vapor pressure, 23.76 mm Hg at 25°C. Use the diffusivity from Table 6.2-1.

6.2-9. Time to Completely Evaporate a Sphere. A drop of liquid toluene is kept at a uniform temperature of 25.9°C and is suspended in air by a fine wire. The initial radius $r_1 = 2.00$ mm. The vapor pressure of toluene at 25.9°C is $P_{A1} = 3.84$ kPa and the density of liquid toluene is 866 kg/m^3.
(a) Derive Eq. (6.2-34) to predict the time t_F for the drop to evaporate completely in a large volume of still air. Show all steps.
(b) Calculate the time in seconds for complete evaporation.

Ans. (b) $t_F = 1388$ s

6.2-10. Diffusion in a Nonuniform Cross-Sectional Area. The gas ammonia (A) is diffusing at steady state through $N_2(B)$ by equimolar counterdiffusion in a conduit 1.22 m long at 25°C and a total pressure of 101.32 kPa abs. The partial pressure of ammonia at the left end is 25.33 kPa and 5.066 kPa at the other end. The cross section of the conduit is in the shape of an equilateral triangle, the length of each side of the triangle being 0.0610 m at the left end and tapering uniformly to 0.0305 m at the right end. Calculate the molar flux of ammonia. The diffusivity is $D_{AB} = 0.230 \times 10^{-4}$ m²/s.

6.3-1. Diffusion of A Through Stagnant B in a Liquid. The solute HCl (A) is diffusing through a thin film of water (B) 2.0 mm thick at 283 K. The concentration of HCl at point 1 at one boundary of the film is 12.0 wt % HCl (density $\rho_1 = 1060.7$ kg/m³), and at the other boundary at point 2 it is 6.0 wt % HCl ($\rho_2 = 1030.3$ kg/m³). The diffusion coefficient of HCl in water is 2.5×10^{-9} m²/s. Assuming steady state and one boundary impermeable to water, calculate the flux of HCl in kg mol/s·m².

$$\text{Ans.} \quad N_A = 2.372 \times 10^{-6} \text{ kg mol/s} \cdot \text{m}^2$$

6.3-2. Diffusion of Ammonia in an Aqueous Solution. An ammonia (A)–water (B) solution at 278 K and 4.0 mm thick is in contact at one surface with an organic liquid at this interface. The concentration of ammonia in the organic phase is held constant and is such that the equilibrium concentration of ammonia in the water at this surface is 2.0 wt % ammonia (density of aqueous solution is 991.7 kg/m³) and the concentration of ammonia in water at the other end of the film 4.0 mm away is 10 wt % (density of 961.7 kg/m³). Water and the organic are insoluble in each other. The diffusion coefficient of NH_3 in water is 1.24×10^{-9} m²/s.
(a) At steady state, calculate the flux N_A in kg mol/s·m².
(b) Calculate the flux N_B. Explain.

6.3-3. Estimation of Liquid Diffusivity. It is desired to predict the diffusion coefficient of dilute acetic acid (CH_3COOH) in water at 282.9 K and at 298 K using the Wilke–Chang method. Compare the predicted values with the experimental values in Table 6.3-1.

$$\text{Ans.} \quad D_{AB} = 0.897 \times 10^{-9} \text{ m}^2/\text{s (282.9 K)}; D_{AB} = 1.396 \times 10^{-9} \text{ m}^2/\text{s (298 K)}$$

6.3-4. Estimation of Diffusivity of Methanol in H_2O. The diffusivity of dilute methanol in water has been determined experimentally to be 1.26×10^{-9} m²/s at 288 K.
(a) Estimate the diffusivity at 293 K using the Wilke–Chang equation.
(b) Estimate the diffusivity at 293 K by correcting the experimental value at 288 K to 293 K. (*Hint*: Do this by using the relationship $D_{AB} \propto T/\mu_B$.)

6.4-1. Prediction of Diffusivity of Enzyme Urease in Solution. Predict the diffusivity of the enzyme urease in a dilute solution in water at 298 K using the modified Polson equation and compare the result with the experimental value in Table 6.4-1.

$$\text{Ans.} \quad \text{Predicted } D_{AB} = 3.995 \times 10^{-11} \text{ m}^2/\text{s}$$

6.4-2. Diffusion of Sucrose in Gelatin. A layer of gelatin in water 5 mm thick containing 5.1 wt % gelatin at 293 K separates two solutions of sucrose. The concentration of sucrose in the solution at one surface of the gelatin is constant at 2.0 g sucrose/100 mL solution and 0.2 g/100 mL at the other surface. Calculate the flux of sucrose in kg sucrose/s·m² through the gel at steady state.

6.4-3. Diffusivity of Oxygen in Protein Solution. Oxygen is diffusing through a solution of bovine serum albumin (BSA) at 298 K. Oxygen has been shown to not bind to BSA. Predict the diffusivity D_{AP} of oxygen in a protein solution containing 11 g protein/100 mL solution. (*Note*: See Table 6.3-1 for the diffusivity of O_2 in water.)

$$\text{Ans.} \quad D_{AP} = 1.930 \times 10^{-9} \text{ m}^2/\text{s}$$

6.4-4. Diffusion of Uric Acid in Protein Solution and Binding. Uric acid (A) at 37°C is diffusing in an aqueous solution of proteins (P) containing 8.2 g protein/100 mL solution. Uric acid binds to the proteins and over the range of concentrations present, 1.0 g mol of acid binds to the proteins for every 3.0 g mol of total acid present in the solution. The diffusivity D_{AB} of uric acid in water is 1.21×10^{-5} cm^2/s and $D_P = 0.091 \times 10^{-5}$ cm^2/s.

(a) Assuming no binding, predict the ratio D_{AP}/D_{AB} due only to blockage effects.

(b) Assuming blockage plus binding effects, predict the ratio D_{AP}/D_{AB}. Compare this with the experimental value for D_{AP}/D_{AB} of 0.616 (C8).

(c) Predict the flux in g uric acid/s · cm^2 for a concentration of acid of 0.05 g/L at point (1) and 0 g/L at point (2) a distance 1.5 μm away.

Ans. (c) $N_A = 2.392 \times 10^{-6}$ g/s · cm^2

6.5-1. Diffusion of CO_2 Through Rubber. A flat plug 30 mm thick having an area of 4.0×10^{-4} m^2 and made of vulcanized rubber is used for closing an opening in a container. The gas CO_2 at 25°C and 2.0 atm pressure is inside the container. Calculate the total leakage or diffusion of CO_2 through the plug to the outside in kg mol CO_2/s at steady state. Assume that the partial pressure of CO_2 outside is zero. From Barrer (B5) the solubility of the CO_2 gas is 0.90 m^3 gas (at STP of 0°C and 1 atm) per m^3 rubber per atm pressure of CO_2. The diffusivity is 0.11×10^{-9} m^2/s.

Ans. 1.178×10^{-13} kg mol CO_2/s

6.5-2. Leakage of Hydrogen Through Neoprene Rubber. Pure hydrogen gas at 2.0 atm abs pressure and 27°C is flowing past a vulcanized neoprene rubber slab 5 mm thick. Using the data from Table 6.5-1, calculate the diffusion flux in kg mol/s · m^2 at steady state. Assume no resistance to diffusion outside the slab and zero partial pressure of H_2 on the outside.

6.5-3. Relation Between Diffusivity and Permeability. The gas hydrogen is diffusing through a sheet of vulcanized rubber 20 mm thick at 25°C. The partial pressure of H_2 inside is 1.5 atm and 0 outside. Using the data from Table 6.5-1, calculate the following.

(a) The diffusivity D_{AB} from the permeability P_M and solubility S and compare with the value in Table 6.5-1.

(b) The flux N_A of H_2 at steady state.

Ans. (b) $N_A = 1.144 \times 10^{-10}$ kg mol/s · m^2

6.5-4. Loss from a Tube of Neoprene. Hydrogen gas at 2.0 atm and 27°C is flowing in a neoprene tube 3.0 mm inside diameter and 11 mm outside diameter. Calculate the leakage of H_2 through a tube 1.0 m long in kg mol H_2/s at steady state.

6.5-5. Diffusion Through Membranes in Series. Nitrogen gas at 2.0 atm and 30°C is diffusing through a membrane of nylon 1.0 mm thick and polyethylene 8.0 mm thick in series. The partial pressure at the other side of the two films is 0 atm. Assuming no other resistances, calculate the flux N_A at steady state.

6.5-6. Diffusion of CO_2 in a Packed Bed of Sand. It is desired to calculate the rate of diffusion of CO_2 gas in air at steady state through a loosely packed bed of sand at 276 K and a total pressure of 1.013×10^5 Pa. The bed depth is 1.25 m and the void fraction ε is 0.30. The partial pressure of CO_2 at the top of the bed is 2.026×10^3 Pa and 0 Pa at the bottom. Use a τ of 1.87.

Ans. $N_A = 1.609 \times 10^{-9}$ kg mol CO_2/s · m^2

6.5-7. Packaging to Keep Food Moist. Cellophane is being used to keep food moist at 38°C. Calculate the loss of water vapor in g/d at steady state for a wrapping 0.10 mm thick and an area of 0.200 m^2 when the vapor pressure of water vapor inside is 10 mm Hg and the air outside contains water vapor at a pressure of 5 mm Hg. Use the larger permeability in Table 6.5-1.

Ans. 0.1667 g H_2O/day.

6.5-8. *Loss of Helium and Permeability.* A window of SiO_2 2.0 mm thick and 1.0×10^{-4} m² area is used to view the contents in a metal vessel at 20°C. Helium gas at 202.6 kPa is contained in the vessel. To be conservative use $D_{AB} = 5.5 \times 10^{-14}$ m²/s from Table 6.5-1.

    (a) Calculate the loss of He in kg mol/h at steady state.

    (b) Calculate the permeability P_M and P_M''.

                  Ans. (a) Loss = 8.833×10^{-15} kg mol He/h

6.6-1. *Numerical Method for Steady-State Diffusion.* Using the results from Example 6.6-1, calculate the total diffusion rate in the solid using the bottom nodes and paths of $c_{2,2}$ to $c_{3,2}$, $c_{2,3}$ to $c_{3,3}$, and so on. Compare with the other diffusion rates in Example 6.6-1.

                  Ans. $N = 2.555 \times 10^{-12}$ kg mol/s

6.6-2. *Numerical Method for Steady-State Diffusion with Distribution Coefficient.* Use the conditions given in Example 6.6-1 except that the distribution coefficient defined by Eq. (6.6-11) between the concentration in the liquid adjacent to the external surface and the concentration in the solid adjacent to the external surface is $K = 1.2$. Calculate the steady-state concentrations and the diffusion rates.

6.6-3. *Digital Solution for Steady-State Diffusion.* Use the conditions given in Example 6.6-1 but instead of using $\Delta x = 0.005$ m, use $\Delta x = \Delta y = 0.001$ m. The overall dimensions of the hollow chamber remain as in Example 6.6-1. The only difference is that many more nodes will be used. Write the computer program and solve for the steady-state concentrations using the numerical method. Also, calculate the diffusion rates and compare with Example 6.6-1.

6.6-4. *Numerical Method with Fixed Surface Concentrations.* Steady-state diffusion is occurring in a two-dimensional solid as shown in Fig. 6.6-4. The grid $\Delta x = \Delta y = 0.010$ m. The diffusivity $D_{AB} = 2.00 \times 10^{-9}$ m²/s. At the inside of the chamber the surface concentration is held constant at 2.00×10^{-3} kg mol/m³. At the outside surfaces, the concentration is constant at 8.00×10^{-3}. Calculate the steady-state concentrations and the diffusion rates per m of depth.

REFERENCES

(A1) AMDUR, I., and SHULER, L. M. *J. Chem. Phys.*, **38**, 188 (1963).

(A2) ANDERSON, D. K., HALL. J. R., and BABB, A. L. *J. Phys. Chem.*, **62**, 404 (1958).

(B1) BUNDE, R. E. *Univ. Wisconsin Naval Res. Lab. Rept. No. CM-850*, August 1955.

(B2) BOYD, C. A., STEIN, N., STEINGRIMISSON, V., and RUMPEL, W. F. *J. Chem. Phys.*, **19**, 548 (1951).

(B3) BIRD, R. B., STEWART, W. E., and LIGHTFOOT, E. N. *Transport Phenomena*. New York: John Wiley & Sons, Inc., 1960.

(B4) BIDSTRUP, D. E., and GEANKOPLIS, C. J. *J. Chem. Eng. Data*, **8**, 170 (1963).

(B5) BARRER, R. M. *Diffusion in and Through Solids*. London: Cambridge University Press, 1941.

(C1) CARMICHAEL, L. T., SAGE, B. H., and LACEY, W. N. *A.I.Ch.E.J.*, **1**, 385 (1955).

(C2) CARSWELL, A. J., and STRYLAND, J. C. *Can. J. Phys.*, **41**, 708 (1963).

(C3) CHAKRABORTI, P. K., and GRAY, P. *Trans. Faraday Soc.*, **62**, 3331 (1961).

(C4) CHANG, S. Y. M. S. thesis, Massachusetts Institute of Technology, 1959.

(C5) CHANG, PIN, and WILKE, C. R. *J. Phys. Chem.*, **59**, 592 (1955).

(C6) CHARLWOOD, P. A. *J. Phys. Chem.*, **57**, 125 (1953).

(C7) CAMERON, J. R. M. S. thesis, Ohio State University, 1973.

(C8) COLTON, C. K., SMITH, K. A., MERRILL, E. W., and REECE, J. M. *Chem. Eng. Progr. Symp.*, **66**(99), 85(1970).

(C9) CRANK, J., and PARK, G. S. *Diffusion in Polymers*. New York: Academic Press, Inc., 1968.

(F1) FULLER, E. N., SCHETTLER, P. D., and GIDDINGS, J. C. *Ind. Eng. Chem.*, **58**, 19 (1966)

(F2) FRIEDMAN, L., and KRAMER, E. O. *J. Am. Chem. Soc.*, **52**, 1298 (1930).

(F3) FRIEDMAN, L. *J. Am. Chem. Soc.*, **52**, 1305, 1311 (1930).

(G1) GILLILAND, E. R. *Ind. Eng. Chem.*, **26**, 681 (1934)

(G2) GEANKOPLIS, C. J. *Mass Transport Phenomena*. Columbus, Ohio: Ohio State University Bookstores, 1972.

(G3) GARNER, G. H., and MARCHANT, P. J. M. *Trans. Inst. Chem. Eng. (London)*, **39**, 397 (1961).

(G4) GOSTING, L. S. *Advances in Protein Chemistry*, Vol. 1. New York: Academic Press, Inc., 1956.

(G5) GEANKOPLIS, C. J., OKOS, M. R., and GRULKE, E. A. *J. Chem. Eng. Data*, **23**, 40 (1978).

(G6) GEANKOPLIS, C. J., GRULKE, E. A., and OKOS, M. R. *Ind. Eng. Chem. Fund.*, **18**, 233 (1979).

(H1) HOLSEN, J. N., and STRUNK, M. R. *Ind. Eng. Chem. Fund.*, **3**, 143 (1964).

(H2) HUDSON, G. H., McCOUBREY, J. C., and UBBELOHDE, A. R. *Trans. Faraday Soc.*, **56**, 1144 (1960).

(H3) HIRSCHFELDER, J. O., CURTISS, C. F., and BIRD, R. B. *Molecular Theory of Gases and Liquids*. New York: John Wiley & Sons, Inc., 1954.

(H4) HAMMOND, B. R., and STOKES, R. H. *Trans. Faraday Soc.*, **49**, 890 (1953).

(J1) JOHNSON, P. A., and BABB, A. L. *Chem. Revs.*, **56**, 387 (1956).

(K1) KELLER, K. H., and FRIEDLANDER, S. K. *J. Gen. Physiol.*, **49**, 68 (1966).

(L1) LEE, C. Y., and WILKE, C. R. *Ind. Eng. Chem.*, **46**, 2381 (1954).

(L2) LE BAS, G. *The Molecular Volumes of Liquid Chemical Compounds*. New York: David McKay Co., Inc., 1915.

(L3) LONGSWORTH, L. G. *J. Phys. Chem.*, **58**, 770 (1954).

(L4) LANGDON, A. G., and THOMAS, H. C. *J. Phys. Chem.*, **75**, 1821 (1971).

(M1) MASON, E. A., and MONCHICK, L. *J. Chem. Phys.*, **36**, 2746 (1962).

(M2) MONCHICK, L., and MASON, E. A. *J. Chem. Phys.*, **35**, 1676 (1961).

(N1) NEY, E. P., and ARMISTEAD, F. C. *Phys. Rev.*, **71**, 14 (1947).

(N2) National Research Council, *International Critical Tables*, Vol. V. New York: McGraw-Hill Book Company, 1929.

(N3) NARVARI, R. M., GAINER, J. L., and HALL, K. R. *A.I.Ch.E.J.*, **17**, 1028 (1971).

(P1) PERRY, R. H., and GREEN, D. *Perry's Chemical Engineers' Handbook*, 6th ed. New York: McGraw-Hill Book Company, 1984.

(P2) PERKINS, L. R., and GEANKOPLIS, C. J. *Chem. Eng. Sci.*, **24**, 1035 (1969).

(P3) POLSON, A. *J. Phys. Colloid Chem.*, **54**, 649 (1950).

(R1) REID, R. C., PRAUSNITZ, J. M., and SHERWOOD, T. K. *The Properties of Gases and Liquids*, 3rd ed. New York: McGraw-Hill Book Company, 1977.

(R2) REDDY, K. A., and DORAISWAMY, L. K. *Ind. Eng. Chem. Fund.*, **6**, 77 (1967).

(R3) ROGERS, C. E. *Engineering Design for Plastics*. New York: Reinhold Publishing Co., Inc., 1964.

(S1) SCHAFER, K. L. *Z. Elecktrochem.*, **63**, 111 (1959).

(S2) SEAGER, S. L., GEERTSON, L. R., and GIDDINGS, J. C. *J. Chem. Eng. Data*, **8**, 168 (1963).

(S3) SCHWERTZ, F. A., and BROW, J. E. *J. Chem. Phys.*, **19**, 640 (1951).

(S4)   SRIVASTAVA, B. N., and SRIVASTAVA, I. B. *J. Chem. Phys.*, **38**, 1183 (1963).

(S5)   SKELLAND, A. H. P. *Diffusional Mass Transfer*. New York: McGraw-Hill Book Company, 1974.

(S6)   SORBER, H. A. *Handbook of Biochemistry, Selected Data for Molecular Biology*. Cleveland: Chemical Rubber Co., Inc., 1968.

(S7)   SPALDING, G. E. *J. Phys. Chem.*, **73**, 3380 (1969).

(S8)   SATTERFIELD, C. N. *Mass Transfer in Heterogeneous Catalysis*. Cambridge, Mass: The MIT Press, 1978.

(T1)   TRAUTZ, M., and MULLER, W. *Ann. Physik*, **22**, 333 (1935).

(T2)   TREYBAL, R. E. *Liquid Extraction*, 2nd ed. New York: McGraw-Hill Book Company, 1963.

(T3)   TREYBAL, R. E. *Mass Transfer Operations*, 3rd ed. New York: McGraw-Hill Book Company, 1980.

(V1)   VIVIAN, J. E., and KING, C. J. *A.I.Ch.E.J.*, **10**, 220 (1964).

(W1)   WINTERGERST, V. E. *Ann. Physik*, **4**, 323, (1930).

(W2)   WESTENBERG, A. A., and FRAZIER, G. *J. Chem. Phys.*, **36**, 3499 (1962).

(W3)   WALKER, R. E., and WESTENBERG, A. A. *J. Chem. Phys.*, **29**, 1139 (1958).

(W4)   WALKER, R. E., and WESTENBERG, A. A. *J. Chem. Phys.*, **32**, 436 (1960).

(W5)   WILKE, C. R., and CHANG, PIN. *A.I.Ch.E.J.*, **1**, 264 (1955).

CHAPTER 7

Principles of
Unsteady-State and
Convective Mass Transfer

7.1 UNSTEADY-STATE DIFFUSION

7.1A Derivation of Basic Equation

In Chapter 6 we considered various mass-transfer systems where the concentration or partial pressure at any point and the diffusion flux were constant with time, hence at steady state. Before steady state can be reached, time must elapse after the mass-transfer process is initiated for the unsteady-state conditions to disappear.

In Section 2.3 a general property balance for unsteady-state molecular diffusion was made for the properties momentum, heat, and mass. For no generation present, this was

$$\frac{\partial \Gamma}{\partial t} = \delta \frac{\partial^2 \Gamma}{\partial z^2} \tag{2.13-12}$$

In Section 5.1 an unsteady-state equation for heat conduction was derived,

$$\frac{\partial T}{\partial t} = \alpha \frac{\partial^2 T}{\partial x^2} \tag{5.1-10}$$

The derivation of the unsteady-state diffusion equation in one direction for mass transfer is similar to that done for heat transfer in obtaining Eq. (5.1-10). We refer to Fig. 7.1-1 where mass is diffusing in the x direction in a cube composed of a solid, stagnant gas, or stagnant liquid and having dimensions Δx, Δy, and Δz. For diffusion in the x direction we write

$$N_{Ax} = -D_{AB} \frac{\partial c_A}{\partial x} \tag{7.1-1}$$

The term $\partial c_A / \partial x$ means the partial of c_A with respect to x or the rate of change of c_A with x when the other variable time t is kept constant.

Next we make a mass balance on component A in terms of moles for no generation.

$$\text{rate of input} = \text{rate of output} + \text{rate of accumulation} \tag{7.1-2}$$

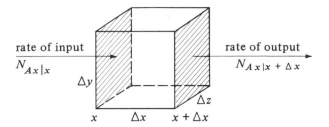

FIGURE 7.1-1.   *Unsteady-state diffusion in one direction.*

The rate of input and rate of output in kg mol A/s are

$$\text{rate of input} = N_{Ax|x} = -D_{AB} \left. \frac{\partial c_A}{\partial x} \right|_x \qquad (7.1\text{-}3)$$

$$\text{rate of output} = N_{Ax|x+\Delta x} = -D_{AB} \left. \frac{\partial c_A}{\partial x} \right|_{x+\Delta x} \qquad (7.1\text{-}4)$$

The rate of accumulation is as follows for the volume $\Delta x\, \Delta y\, \Delta z\, \text{m}^3$:

$$\text{rate of accumulation} = (\Delta x\, \Delta y\, \Delta z)\, \frac{\partial c_A}{\partial t} \qquad (7.1\text{-}5)$$

Substituting Eqs. (7.1-3), (7.1-4), and (7.1-5) into (7.1-2) and dividing by $\Delta x\, \Delta y\, \Delta z$,

$$-D_{AB} \frac{\left. \dfrac{\partial c_A}{\partial x} \right|_x - \left. \dfrac{\partial c_A}{\partial x} \right|_{x+\Delta x}}{\Delta x} = \frac{\partial c_A}{\partial t} \qquad (7.1\text{-}6)$$

Letting Δx approach zero,

$$\frac{\partial c_A}{\partial t} = D_{AB} \frac{\partial^2 c_A}{\partial x^2} \qquad (7.1\text{-}7)$$

The above holds for a constant diffusivity D_{AB}. If D_{AB} is a variable,

$$\frac{\partial c_A}{\partial t} = \frac{\partial (D_{AB}\, \partial c_A/\partial x)}{\partial x} \qquad (7.1\text{-}8)$$

Equation (7.1-7) relates the concentration c_A with position x and time t. For diffusion in all three directions a similar derivation gives

$$\frac{\partial c_A}{\partial t} = D_{AB}\left(\frac{\partial^2 c_A}{\partial x^2} + \frac{\partial^2 c_A}{\partial y^2} + \frac{\partial^2 c_A}{\partial z^2} \right) \qquad (7.1\text{-}9)$$

In the remainder of this section, the solutions of Eqs. (7.1-7) and (7.1-9) will be considered. Note the mathematical similarity between the equation for heat conduction,

$$\frac{\partial T}{\partial t} = \alpha \frac{\partial^2 T}{\partial x^2} \qquad (5.1\text{-}6)$$

and Eq. (7.1-7) for diffusion. Because of this similarity, the mathematical methods used for solution of the unsteady-state heat-conduction equation can be used for unsteady-state mass transfer. This is discussed more fully in Sections 7.1B, 7.1C, and 7.7.

7.1B Diffusion in a Flat Plate with Negligible Surface Resistance

To illustrate an analytical method of solving Eq. (7.1-7), we will derive the solution for unsteady-state diffusion in the x direction for a plate of thickness $2x_1$, as shown in Fig. 7.1-2. For diffusion in one direction,

$$\frac{\partial c_A}{\partial t} = D_{AB} \frac{\partial^2 c_A}{\partial x^2} \tag{7.1-7}$$

Dropping the subscripts A and B for convenience,

$$\frac{\partial c}{\partial t} = D \frac{\partial^2 c}{\partial x^2} \tag{7.1-10}$$

The initial profile of the concentration in the plate at $t = 0$ is uniform at $c = c_0$ at all x values, as shown in Fig. 7.1-2. At time $t = 0$ the concentration of the fluid in the environment outside is suddenly changed to c_1. For a very high mass-transfer coefficient outside the surface resistance will be negligible and the concentration at the surface will be equal to that in the fluid, which is c_1.

The initial and boundary conditions are

$$c = c_0, \qquad t = 0, \qquad x = x, \qquad Y = \frac{c_1 - c_0}{c_1 - c_0} = 1$$

$$c = c_1, \qquad t = t, \qquad x = 0, \qquad Y = \frac{c_1 - c_1}{c_1 - c_0} = 0 \tag{7.1-11}$$

$$c = c_1, \qquad t = t, \qquad x = 2x_1, \qquad Y = \frac{c_1 - c_1}{c_1 - c_0} = 0$$

Redefining the concentration so it goes between 0 and 1,

$$Y = \frac{c_1 - c}{c_1 - c_0} \tag{7.1-12}$$

$$\frac{\partial Y}{\partial t} = D \frac{\partial^2 Y}{\partial x^2} \tag{7.1-13}$$

The solution of Eq. (7.1-13) is an infinite Fourier series and is identical to the solution of Eq. (5.1-6) for heat transfer.

FIGURE 7.1-2. *Unsteady-state diffusion in a flat plate with negligible surface resistance.*

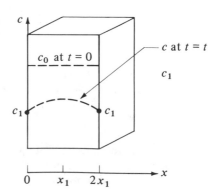

c_0 at $t = 0$

c at $t = t$

c_1

c_1 c_1

0 x_1 $2x_1$

$$Y = \frac{c_1 - c}{c_1 - c_0} = \frac{4}{\pi} \left[\frac{1}{1} \exp\left(-\frac{1^2 \pi^2 X}{4} \right) \sin \frac{1\pi x}{2x_1} \right.$$

$$\left. + \frac{1}{3} \exp\left(-\frac{3^2 \pi^2 X}{4} \right) \sin \frac{3\pi x}{2x_1} + \frac{1}{5} \exp\left(-\frac{5^2 \pi^2 X}{4} \right) \sin \frac{5\pi x}{2x_1} + \cdots \right] \quad \textbf{(7.1-14)}$$

where,

$X = Dt/x_1^2$, dimensionless

c = concentration at point x and time t in slab

$Y = (c_1 - c)/(c_1 - c_0)$ = fraction of unaccomplished change, dimensionless

$1 - Y = (c - c_0)/(c_1 - c_0)$ = fraction of change

Solution of equations similar to Eq. (7.1-14) are time consuming; convenient charts for various geometries are available and will be discussed in the next section.

7.1C   Unsteady-State Diffusion in Various Geometries

1. Convection and boundary conditions at the surface.   In Fig. 7.1-2 there was no convective resistance at the surface. However, in many cases when a fluid is outside the solid, convective mass transfer is occurring at the surface. A convective mass-transfer coefficient k_c, similar to a convective heat-transfer coefficient, is defined as follows:

$$N_A = k_c(c_{L1} - c_{Li}) \quad \textbf{(7.1-15)}$$

where k_c is a mass-transfer coefficient in m/s, c_{L1} is the bulk fluid concentration in kg mol A/m^3, and c_{Li} is the concentration in the fluid just adjacent to the surface of the solid. The coefficient k_c is an empirical coefficient and will be discussed more fully in Section 7.2.

In Fig. 7.1-3a the case for a mass-transfer coefficient being present at the boundary is shown. The concentration drop across the fluid is $c_{L1} - c_{Li}$. The concentration in the solid c_i at the surface is in equilibrium with c_{Li}.

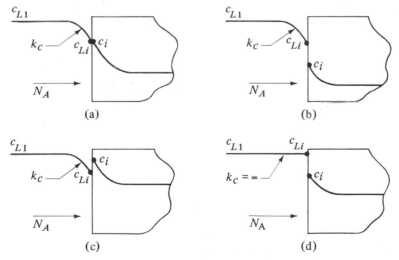

FIGURE 7.1-3.   *Interface conditions for convective mass transfer and an equilibrium distribution coefficient $K = c_{Li}/c_i$: (a) $K = 1$, (b) $K > 1$, (c) $K < 1$, (d) $K > 1$ and $k_c = \infty$.*

In Fig. 7.1-3a the concentration c_{Li} in the liquid adjacent to the solid and c_i in the solid at the surface are in equilibrium and also equal. However, unlike heat transfer, where the temperatures are equal, the concentrations are in equilibrium and are related by

$$K = \frac{c_{Li}}{c_i} \tag{7.1-16}$$

where K is the equilibrium distribution coefficient (similar to Henry's law coefficient for a gas and liquid). The value of K in Fig. 7.1-3a is 1.0.

In Fig. 7.1-3b the distribution coefficient K is > 1 and $c_{Li} > c_i$, even though they are in equilibrium. Other cases are shown in Fig. 7.1-3c and d. This was also discussed in Section 6.6B.

2. *Relation between mass- and heat-transfer parameters.* In order to use the unsteady-state heat-conduction charts in Chapter 5 for solving unsteady-state diffusion problems, the dimensionless variables or parameters for heat transfer must be related to those for mass transfer. In Table 7.1-1 the relations between these variables are tabulated. For $K \neq 1.0$, whenever k_c appears, it is given as Kk_c, and whenever c_1 appears, it is given as c_1/K.

TABLE 7.1-1. *Relation Between Mass- and Heat-Transfer Parameters for Unsteady-State Diffusion**

	Heat Transfer	Mass Transfer	
		$K = c_L/c = 1.0$	$K = c_L/c \neq 1.0$
$Y,$	$\dfrac{T_1 - T}{T_1 - T_0}$	$\dfrac{c_1 - c}{c_1 - c_0}$	$\dfrac{c_1/K - c}{c_1/K - c_0}$
$1 - Y,$	$\dfrac{T - T_0}{T_1 - T_0}$	$\dfrac{c - c_0}{c_1 - c_0}$	$\dfrac{c - c_0}{c_1/K - c_0}$
$X,$	$\dfrac{\alpha t}{x_1^2}$	$\dfrac{D_{AB} t}{x_1^2}$	$\dfrac{D_{AB} t}{x_1^2}$
	$\dfrac{x}{2\sqrt{\alpha t}}$	$\dfrac{x}{2\sqrt{D_{AB} t}}$	$\dfrac{x}{2\sqrt{D_{AB} t}}$
$m,$	$\dfrac{k}{hx_1}$	$\dfrac{D_{AB}}{k_c x_1}$	$\dfrac{D_{AB}}{Kk_c x_1}$
	$\dfrac{h}{k}\sqrt{\alpha t}$	$\dfrac{k_c}{D_{AB}}\sqrt{D_{AB} t}$	$\dfrac{Kk_c}{D_{AB}}\sqrt{D_{AB} t}$
$n,$	$\dfrac{x}{x_1}$	$\dfrac{x}{x_1}$	$\dfrac{x}{x_1}$

* x is the distance from the center of the slab, cylinder, or sphere; for a semiinfinite slab, x is the distance from the surface. c_0 is the original uniform concentration in the solid, c_1 the concentration in the fluid outside the slab, and c the concentration in the solid at position x and time t.

3. *Charts for diffusion in various geometries.* The various heat-transfer charts for unsteady-state conduction can be used for unsteady-state diffusion and are as follows.

1. Semiinfinite solid, Fig. 5.3-3.
2. Flat plate, Figs. 5.3-5 and 5.3-6.
3. Long cylinder, Figs. 5.3-7 and 5.3-8.
4. Sphere, Figs. 5.3-9 and 5.3-10.
5. Average concentrations, zero convective resistance, Fig. 5.3-13.

EXAMPLE 7.1-1.   Unsteady-State Diffusion in a Slab of Agar Gel

A solid slab of 5.15 wt % agar gel at 278 K is 10.16 mm thick and contains a uniform concentration of urea of 0.1 kg mol/m^3. Diffusion is only in the x direction through two parallel flat surfaces 10.16 mm apart. The slab is suddenly immersed in pure turbulent water so that the surface resistance can be assumed to be negligible; i.e., the convective coefficient k_c is very large. The diffusivity of urea in the agar from Table 6.4-2 is 4.72×10^{-10} m^2/s.

(a) Calculate the concentration at the midpoint of the slab (5.08 mm from the surface) and 2.54 mm from the surface after 10 h.
(b) If the thickness of the slab is halved, what would be the midpoint concentration in 10 h?

Solution: For part (a), $c_0 = 0.10$ kg mol/m^3, $c_1 = 0$ for pure water, and c = concentration at distance x from center line and time t s. The equilibrium distribution coefficient K in Eq, (7.1-16) can be assumed to be 1.0, since the water in the aqueous solution in the gel and outside should be very similar in properties. From Table 7.1-1,

$$Y = \frac{c_1/K - c}{c_1/K - c_0} = \frac{0/1.0 - c}{0/1.0 - 0.10}$$

Also, $x_1 = 10.16/(1000 \times 2) = 5.08 \times 10^{-3}$ m (half-slab thickness), $x = 0$ (center), $X = D_{AB}t/x_1^2 = (4.72 \times 10^{-10})(10 \times 3600)/(5.08 \times 10^{-3})^2 = 0.658$. The relative position $n = x/x_1 = 0/5.08 \times 10^{-3} = 0$ and relative resistance $m = D_{AB}/Kk_c x_1 = 0$, since k_c is very large (zero resistance).
From Fig. 5.3-5 for $X = 0.658$, $m = 0$, and $n = 0$,

$$Y = 0.275 = \frac{0 - c}{0 - 0.10}$$

Solving, $c = 0.0275$ kg mol/m^3 for $x = 0$.
For the point 2.54 mm from the surface or 2.54 mm from center, $x = 2.54/1000 = 2.54 \times 10^{-3}$ m, $X = 0.658$, $m = 0$, $n = x/x_1 = 2.54 \times 10^{-3}/5.08 \times 10^{-3} = 0.5$. Then from Fig. 5.3-5, $Y = 0.172$. Solving, $c = 0.0172$ kg mol/m^3.
For part (b) and half the thickness, $X = 0.658/(0.5)^2 = 2.632$, $n = 0$, and $m = 0$. Hence, $Y = 0.0020$ and $c = 2.0 \times 10^{-4}$ kg mol/m^3.

EXAMPLE 7.1-2.   Unsteady-State Diffusion in a Semiinfinite Slab

A very thick slab has a uniform concentration of solute A of $c_0 = 1.0 \times 10^{-2}$ kg mol A/m^3. Suddenly, the front face of the slab is exposed to a flowing fluid having a concentration $c_1 = 0.10$ kg mol A/m^3 and a convective coefficient $k_c = 2 \times 10^{-7}$ m/s. The equilibrium distribution coefficient $K = c_{Li}/c_i = 2.0$. Assuming that the slab is a semiinfinite solid, calculate the concentration in the solid at the surface ($x = 0$) and $x = 0.01$ m from the surface after $t = 3 \times 10^4$ s. The diffusivity in the solid is $D_{AB} = 4 \times 10^{-9}$ m^2/s.

Solution: To use Fig. 5.3-3,

$$\frac{Kk_c}{D_{AB}}\sqrt{D_{AB}t} = \frac{2.0(2 \times 10^{-7})}{4 \times 10^{-9}}\sqrt{(4 \times 10^{-9})(3 \times 10^4)} = 1.095$$

For $x = 0.01$ m from the surface in the solid,

$$\frac{x}{2\sqrt{D_{AB}t}} = \frac{0.01}{2\sqrt{(4 \times 10^{-9})(3 \times 10^4)}} = 0.457$$

From the chart, $1 - Y = 0.26$. Then, substituting into the equation for $(1 - Y)$ from Table 7.1-1 and solving,

$$1 - Y = \frac{c - c_0}{c_1/K - c_0} = \frac{c - 1 \times 10^{-2}}{(10 \times 10^{-2})/2 - (1 \times 10^{-2})} = 0.26$$

$$c = 2.04 \times 10^{-2} \text{ kg mol/m}^3 \qquad \text{(for } x = 0.01 \text{ m)}$$

For $x = 0$ m (i.e., at the surface of the solid),

$$\frac{x}{2\sqrt{D_{AB}t}} = 0$$

From the chart, $1 - Y = 0.62$. Solving, $c = 3.48 \times 10^{-2}$. This value is the same as c_i, as shown in Fig. 7.1-3b. To calculate the concentration c_{Li} in the liquid at the interface,

$$C_{Li} = Kc_i = 2.0(3.48 \times 10^{-2}) = 6.96 \times 10^{-2} \text{ kg mol/m}^3$$

A plot of these values will be similar to Fig. 7.1-3b.

4. Unsteady-state diffusion in more than one direction. In Section 5.3F a method was given for unsteady-state heat conduction to combine the one-dimensional solutions to yield solutions for several-dimensional systems. The same method can be used for unsteady-state diffusion in more than one direction. Rewriting Eq. (5.3-11) for diffusion in a rectangular block in the x, y, and z directions,

$$Y_{x,\,y,\,z} = (Y_x)(Y_y)(Y_z) = \frac{c_1/K - c_{x,\,y,\,z}}{c_1/K - c_0} \tag{7.1-17}$$

where $c_{x,\,y,\,z}$ is the concentration at the point x, y, z from the center of the block. The value of Y_x for the two parallel faces is obtained from Fig. 5.3-5 or 5.3-6 for a flat plate in the x direction. The values of Y_y and Y_z are similarly obtained from the same charts. For a short cylinder, an equation similar to Eq. (5.3-12) is used, and for average concentrations, ones similar to Eqs. (5.3-14), (5.3-15), and (5.3-16) are used.

7.2 CONVECTIVE MASS-TRANSFER COEFFICIENTS

7.2A Introduction to Convective Mass Transfer

In the previous sections of this chapter and Chapter 6 we have emphasized molecular diffusion in stagnant fluids or fluids in laminar flow. In many cases the rate of diffusion is slow, and more rapid transfer is desired. To do this, the fluid velocity is increased until turbulent mass transfer occurs.

 To have a fluid in convective flow usually requires the fluid to be flowing by another immiscible fluid or by a solid surface. An example is a fluid flowing in a pipe, where part of the pipe wall is made by a slightly dissolving solid material such as benzoic acid. The benzoic acid dissolves and is transported perpendicular to the main stream from the wall. When a fluid is in turbulent flow and is flowing past a surface, the actual velocity of small particles of fluid cannot be described clearly as in laminar flow. In laminar flow the fluid flows in streamlines and its behavior can usually be described mathematically. However, in turbulent motion there are no streamlines, but there are large eddies or "chunks" of fluid moving rapidly in seemingly random fashion.

When a solute A is dissolving from a solid surface there is a high concentration of this solute in the fluid at the surface, and its concentration, in general, decreases as the distance from the wall increases. However, minute samples of fluid adjacent to each other do not always have concentrations close to each other. This occurs because eddies having solute in them move rapidly from one part of the fluid to another, transferring relatively large amounts of solute. This turbulent diffusion or eddy transfer is quite fast in comparison to molecular transfer.

Three regions of mass transfer can be visualized. In the first, which is adjacent to the surface, a thin viscous sublayer film is present. Most of the mass transfer occurs by molecular diffusion, since few or no eddies are present. A large concentration drop occurs across this film as a result of the slow diffusion rate.

The transition or buffer region is adjacent to the first region. Some eddies are present and the mass transfer is the sum of turbulent and molecular diffusion. There is a gradual transition in this region from the transfer by mainly molecular diffusion at the one end to mainly turbulent at the other end.

In the *turbulent region* adjacent to the buffer region, most of the transfer is by turbulent diffusion, with a small amount by molecular diffusion. The concentration decrease is very small here since the eddies tend to keep the fluid concentration uniform. A more detailed discussion of these three regions is given in Section 3.10G.

A typical plot for the mass transfer of a dissolving solid from a surface to a turbulent fluid in a conduit is given in Fig. 7.2-1. The concentration drop from c_{A1} adjacent to the surface is very abrupt close to the surface and then levels off. This curve is very similar to the shapes found for heat and momentum transfer. The average or mixed concentration $\bar{c}_A$ is shown and is slightly greater than the minimum c_{A2}.

7.2B Types of Mass-Transfer Coefficients

1. Definition of mass-transfer coefficient.   Since our understanding of turbulent flow is incomplete, we attempt to write the equations for turbulent diffusion in a manner similar to that for molecular diffusion. For turbulent mass transfer for constant c, Eq. (6.1-6) is

FIGURE 7.2-1.   *Concentration profile in turbulent mass transfer from a surface to a fluid.*

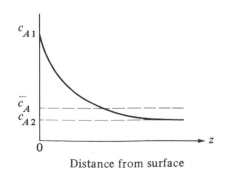

Distance from surface

written as

$$J_A^* = -(D_{AB} + \varepsilon_M) \frac{dc_A}{dz} \qquad (7.2\text{-}1)$$

where D_{AB} is the molecular diffusivity in m^2/s and ε_M is the mass eddy diffusivity in m^2/s. The value of ε_M is a variable and is near zero at the interface or surface and increases as the distance from the wall increases. We then use an average value $\bar{\varepsilon}_M$ since the variation of ε_M is not generally known. Integrating Eq. (7.2-1) between points 1 and 2,

$$J_{A1}^* = \frac{D_{AB} + \bar{\varepsilon}_M}{z_2 - z_1}(c_{A1} - c_{A2}) \qquad (7.2\text{-}2)$$

The flux J_{A1}^* is based on the surface area A_1 since the cross-sectional area may vary. The value of $z_2 - z_1$, the distance of the path, is often not known. Hence, Eq. (7.2-2) is simplified and is written using a convective mass-transfer coefficient k_c'.

$$J_{A1}^* = k_c'(c_{A1} - c_{A2}) \qquad (7.2\text{-}3)$$

where J_{A1}^* is the flux of A from the surface A_1 relative to the whole bulk phase, k_c' is $(D_{AB} + \bar{\varepsilon}_M)/(z_2 - z_1)$ an experimental mass-transfer coefficient in kg mol/s $\cdot$ $m^2 \cdot$ (kg mol/m^3) or simplified as m/s, and c_{A2} is the concentration at point 2 in kg mol A/m^3 or more usually the average bulk concentration $\bar{c}_{A2}$. This defining of a convective mass-transfer coefficient k_c' is quite similar to the convective heat-transfer coefficient h.

2. Mass-transfer coefficient for equimolar counterdiffusion. Generally, we are interested in N_A, the flux of A relative to stationary coordinates. We can start with the following, which is similar to that for molecular diffusion but the term ε_M is added.

$$N_A = -c(D_{AB} + \varepsilon_M) \frac{dx_A}{dz} + x_A(N_A + N_B) \qquad (7.2\text{-}4)$$

For the case of equimolar counterdiffusion, where $N_A = -N_B$, and integrating at steady state, calling $k_c' = (D_{AB} + \bar{\varepsilon}_M)/(z_2 - z_1)$,

$$N_A = k_c'(c_{A1} - c_{A2}) \qquad (7.2\text{-}5)$$

Equation (7.2-5) is the defining equation for the mass-transfer coefficient. Often, however, we define the concentration in terms of mole fraction if a liquid or gas or in terms of partial pressure if a gas. Hence, we can define the mass-transfer coefficient in several ways. If y_A is mole fraction in a gas phase and x_A in a liquid phase, then Eq. (7.2-5) can be written as follows for equimolar counterdiffusion:

Gases: $\qquad N_A = k_c'(c_{A1} - c_{A2}) = k_G'(p_{A1} - p_{A2}) = k_y'(y_{A1} - y_{A2}) \qquad (7.2\text{-}6)$

Liquids: $\qquad N_A = k_c'(c_{A1} - c_{A2}) = k_L'(c_{A1} - c_{A2}) = k_x'(x_{A1} - x_{A2}) \qquad (7.2\text{-}7)$

All of these mass-transfer coefficients can be related to each other. For example, using Eq. (7.2-6) and substituting $y_{A1} = c_{A1}/c$ and $y_{A2} = c_{A2}/c$ into the equation,

$$N_A = k_c'(c_{A1} - c_{A2}) = k_y'(y_{A1} - y_{A2}) = k_y'\left(\frac{c_{A1}}{c} - \frac{c_{A2}}{c}\right) = \frac{k_y'}{c}(c_{A1} - c_{A2}) \qquad (7.2\text{-}8)$$

Hence,

$$k_c' = \frac{k_y'}{c} \qquad (7.2\text{-}9)$$

These relations among mass-transfer coefficients, and the various flux equations, are given in Table 7.2-1.

TABLE 7.2-1. Flux Equations and Mass-Transfer Coefficients

Flux equations for equimolar counterdiffusion

Gases: $N_A = k'_c(c_{A1} - c_{A2}) = k'_G(p_{A1} - p_{A2}) = k'_y(y_{A1} - y_{A2})$

Liquids: $N_A = k'_c(c_{A1} - c_{A2}) = k'_L(c_{A1} - c_{A2}) = k'_x(x_{A1} - x_{A2})$

Flux equations for A diffusing through stagnant, nondiffusing B

Gases: $N_A = k_c(c_{A1} - c_{A2}) = k_G(p_{A1} - p_{A2}) = k_y(y_{A1} - y_{A2})$

Liquids: $N_A = k_c(c_{A1} - c_{A2}) = k_L(c_{A1} - c_{A2}) = k_x(x_{A1} - x_{A2})$

Conversions between mass-transfer coefficients

Gases:

$$k'_c c = k'_c \frac{P}{RT} = k_c \frac{p_{BM}}{RT} = k'_G P = k_G p_{BM} = k_y y_{BM} = k'_y = k_c y_{BM}\, c = k_G y_{BM}\, P$$

Liquids:

$$k'_c c = k'_L c = k_L x_{BM} c = k'_L \rho/M = k'_x = k_x x_{BM}$$

(where ρ is density of liquid and M is molecular weight)

Units of mass-transfer coefficients

	SI Units	Cgs Units	English Units
k_c, k_L, k'_c, k'_L	m/s	cm/s	ft/h
k_x, k_y, k'_x, k'_y	$\dfrac{\text{kg mol}}{\text{s} \cdot \text{m}^2 \cdot \text{mol frac}}$	$\dfrac{\text{g mol}}{\text{s} \cdot \text{cm}^2 \cdot \text{mol frac}}$	$\dfrac{\text{lb mol}}{\text{h} \cdot \text{ft}^2 \cdot \text{mol frac}}$
k_G, k'_G	$\dfrac{\text{kg mol}}{\text{s} \cdot \text{m}^2 \cdot \text{Pa}}$ $\dfrac{\text{kg mol}}{\text{s} \cdot \text{m}^2 \cdot \text{atm}}$ (preferred)	$\dfrac{\text{g mol}}{\text{s} \cdot \text{cm}^2 \cdot \text{atm}}$	$\dfrac{\text{lb mol}}{\text{h} \cdot \text{ft}^2 \cdot \text{atm}}$

3. Mass-transfer coefficient for A diffusing through stagnant, nondiffusing B. For A diffusing through stagnant, nondiffusing B where $N_B = 0$, Eq. (7.2-4) gives for steady state

$$N_A - \frac{k'_c}{x_{BM}}(c_{A1} - c_{A2}) = k_c(c_{A1} - c_{A2}) \qquad (7.2\text{-}10)$$

where the x_{BM} and its counterpart y_{BM} are similar to Eq. (6.2-21) and k_c is the mass-transfer coefficient for A diffusing through stagnant B. Also,

$$x_{BM} = \frac{x_{B2} - x_{B1}}{\ln (x_{B2}/x_{B1})} \qquad y_{BM} = \frac{y_{B2} - y_{B1}}{\ln (y_{B2}/y_{B1})} \qquad (7.2\text{-}11)$$

Rewriting Eq. (7.2-10) using other units,

(Gases): $N_A = k_c(c_{A1} - c_{A2}) = k_G(p_{A1} - p_{A2}) = k_y(y_{A1} - y_{A2})$ (7.2-12)

(Liquids): $N_A = k_c(c_{A1} - c_{A2}) = k_L(c_{A1} - c_{A2}) = k_x(x_{A1} - x_{A2})$ (7.2-13)

Again all the mass-transfer coefficients can be related to each other and are given in

Table 7.2-1. For example, setting Eq. (7.2-10) equal to (7.2-13),

$$N_A = \frac{k'_c}{x_{BM}}(c_{A1} - c_{A2}) = k_x(x_{A1} - x_{A2}) = k_x\left(\frac{c_{A1}}{c} - \frac{c_{A2}}{c}\right) \qquad (7.2\text{-}14)$$

Hence,

$$\frac{k'_c}{x_{BM}} = \frac{k_x}{c} \qquad (7.2\text{-}15)$$

EXAMPLE 7.2-1. *Vaporizing A and Convective Mass Transfer*
A large volume of pure gas B at 2 atm pressure is flowing over a surface from which pure A is vaporizing. The liquid A completely wets the surface, which is a blotting paper. Hence, the partial pressure of A at the surface is the vapor pressure of A at 298 K, which is 0.20 atm. The k'_y has been estimated to be 6.78×10^{-5} kg mol/s·m²·mol frac. Calculate N_A, the vaporization rate, and also the value of k_y and k_G.

Solution: This is the case of A diffusing through B, where the flux of B normal to the surface is zero, since B is not soluble in liquid A. $p_{A1} = 0.20$ atm and $p_{A2} = 0$ in the pure gas B. Also, $y_{A1} = p_{A1}/P = 0.20/2.0 = 0.10$ and $y_{A2} = 0$. We can use Eq. (7.2-12) with mole fractions.

$$N_A = k_y(y_{A1} - y_{A2}) \qquad (7.2\text{-}12)$$

However, we have a value for k'_y which is related to k_y from Table 7.2-1 by

$$k_y y_{BM} = k'_y \qquad (7.2\text{-}16)$$

The term y_{BM} is similar to x_{BM} and is, from Eq. (7.2-11),

$$y_{BM} = \frac{y_{B2} - y_{B1}}{\ln(y_{B2}/y_{B1})} \qquad (7.2\text{-}11)$$

$$y_{B1} = 1 - y_{A1} = 1 - 0.10 = 0.90 \qquad y_{B2} = 1 - y_{A1} = 1 - 0 = 1.0$$

Substituting into Eq. (7.2-11),

$$y_{BM} = \frac{1.0 - 0.90}{\ln(1.0/0.90)} = 0.95$$

Then, from Eq. (7.2-16),

$$k_y = \frac{k'_y}{y_{BM}} = \frac{6.78 \times 10^{-5}}{0.95} = 7.138 \times 10^{-5} \text{ kg mol/s·m²·mol frac}$$

Also, from Table 7.2-1,

$$k_G y_{BM} P = k_y y_{BM} \qquad (7.2\text{-}17)$$

Hence, solving for k_G and substituting knowns,

$$k_G = \frac{k_y}{P} = \frac{7.138 \times 10^{-5}}{2 \times 1.01325 \times 10^5 \text{ Pa}} = 3.522 \times 10^{-10} \text{ kg mol/s·m²·Pa}$$

$$k_G = \frac{k_y}{P} = \frac{7.138 \times 10^{-5}}{2.0 \text{ atm}} = 3.569 \times 10^{-5} \text{ kg mol/s·m²·atm}$$

For the flux using Eq. (7.2-12),

$$N_A = k_y(y_{A1} - y_{A2}) = 7.138 \times 10^{-5}(0.10 - 0) = 7.138 \times 10^{-6} \text{ kg mol/s·m²}$$

Also,

$$p_{A1} = 0.20 \text{ atm} = 0.20(1.01325 \times 10^5) = 2.026 \times 10^4 \text{ Pa}$$

Using Eq. (7.2-12) again,

$$N_A = k_G(p_{A1} - p_{A2}) = 3.522 \times 10^{-10}(2.026 \times 10^4 - 0)$$
$$= 7.138 \times 10^{-6} \text{ kg mol/s} \cdot \text{m}^2$$
$$N_A = k_G(p_{A1} - p_{A2}) = 3.569 \times 10^{-5}(0.20 - 0)$$
$$= 7.138 \times 10^{-6} \text{ kg mol/s} \cdot \text{m}^2$$

Note that in this case, since the concentrations were dilute, y_{BM} is close to 1.0 and k_y and k'_y differ very little.

7.2C Methods to Determine Mass-Transfer Coefficients

Many different experimental methods have been employed to experimentally obtain mass-transfer coefficients. In determining the mass-transfer coefficient to a sphere, Steele and Geankoplis (S3) used a solid sphere of benzoic acid held rigidly by a rear support in a pipe. Before the run the sphere was weighed. After flow of the fluid for a timed interval, the sphere was removed, dried, and weighed again to give the amount of mass transferred, which was small compared to the weight of the sphere. From the mass transferred and the area of the sphere, the flux N_A was calculated. Then the driving force $(c_{AS} - 0)$ was used to calculate k_L, where c_{AS} is the solubility and the water contained no benzoic acid.

Another method used is to flow gases over various geometries wet with evaporating liquids. For mass transfer from a flat plate, a porous blotter wet with the liquid serves as the plate.

7.3 MASS-TRANSFER COEFFICIENTS FOR VARIOUS GEOMETRIES

7.3A Dimensionless Numbers Used to Correlate Data

The experimental data for mass-transfer coefficients obtained using various kinds of fluids, different velocities, and different geometries are correlated using dimensionless numbers similar to those of heat and momentum transfer. Methods of dimensional analysis are discussed in Sections 3.11, 4.14, and 7.8.

The most important dimensionless number is the Reynolds number N_{Re}, which indicates degree of turbulence.

$$N_{Re} = \frac{Lv\rho}{\mu} \tag{7.3-1}$$

where L is diameter D_p for a sphere, diameter D for a pipe, or length L for a flat plate. The velocity v is the mass average velocity if in a pipe. In a packed bed the superficial velocity v' in the empty cross section is often used or sometimes $v = v'/\varepsilon$ is used, where v is interstitial velocity and ε void fraction of bed.

The *Schmidt number* is

$$N_{Sc} = \frac{\mu}{\rho D_{AB}} \tag{7.3-2}$$

The viscosity μ and density ρ used are the actual flowing mixture of solute A and fluid B. If the mixture is dilute, properties of the pure fluid B can be used. The Prandtl number $c_p \mu / k$ for heat transfer is analogous to the Schmidt number for mass transfer. The Schmidt number is the ratio of the shear component for diffusivity μ/ρ to the diffusivity for mass transfer D_{AB}, and it physically relates the relative thickness of the hydrodynamic layer and mass-transfer boundary layer.

The *Sherwood number*, which is dimensionless, is

$$N_{Sh} = k_c' \frac{L}{D_{AB}} = k_c \, y_{BM} \frac{L}{D_{AB}} = \frac{k_x'}{c} \frac{L}{D_{AB}} = \cdots \tag{7.3-3}$$

Other substitutions from Table 7.2-1 can be made for k_c' in Eq. (7.3-3).

The *Stanton number* occurs often and is

$$N_{St} = \frac{k_c'}{v} = \frac{k_y'}{G_M} = \frac{k_G' P}{G_M} = \cdots \tag{7.3-4}$$

Again, substitution for k_c' can be made. $G_M = v\rho / M_{av} = vc$.

Often the mass-transfer coefficient is correlated as a dimensionless J_D factor which is related to k_c' and N_{Sh} as follows.

$$J_D = \frac{k_c'}{v} (N_{Sc})^{2/3} = \frac{k_G' P}{G_M} (N_{Sc})^{2/3} = \cdots = N_{Sh}/(N_{Re} N_{Sc}^{1/3}) \tag{7.3-5}$$

For heat transfer a dimensionless J_H factor is as follows:

$$J_H = \frac{h}{c_p G} (N_{Pr})^{2/3} \tag{7.3-6}$$

7.3B Analogies Among Mass, Heat, and Momentum Transfer

1. Introduction. In molecular transport of momentum, heat, or mass there are many similarities, which were pointed out in Chapters 2 to 6. The molecular diffusion equations of Newton for momentum, Fourier for heat, and Fick for mass are very similar and we can say that we have analogies among these three molecular transport processes. There are also similarities in turbulent transport, as discussed in Sections 5.7C and 6.1A, where the flux equations were written using the turbulent eddy momentum diffusivity ε_t, the turbulent eddy thermal diffusivity α_t, and the turbulent eddy mass diffusivity ε_M. However, these similarities are not as well defined mathematically or physically and are more difficult to relate to each other.

A great deal of effort has been devoted in the literature to developing analogies among these three transport processes for turbulent transfer so as to allow prediction of one from any of the others. We discuss several next.

2. Reynolds analogy. Reynolds was the first to note similarities in transport processes and relate turbulent momentum and heat transfer. Since then, mass transfer has also been related to momentum and heat transfer. We derive this analogy from Eqs. (6.1-4)–(6.1-6) for turbulent transport. For fluid flow in a pipe for heat transfer from the fluid to the wall, Eq. (6.1-5) becomes as follows, where z is distance from the wall:

$$\frac{q}{A} = -\rho c_p (\alpha + \alpha_t) \frac{dT}{dz} \tag{7.3-7}$$

For momentum transfer, Eq. (6.1-4) becomes

$$\tau = -\rho\left(\frac{\mu}{\rho} + \varepsilon_t\right)\frac{dv}{dz} \tag{7.3-8}$$

Next we assume α and μ/ρ are negligible and that $\alpha_t = \varepsilon_t$. Then dividing Eq. (7.3-7) by (7.3-8),

$$\left(\frac{\tau}{q/A}\right)c_p\,dT = dv \tag{7.3-9}$$

If we assume that heat flux q/A in a turbulent system is analogous to momentum flux τ, the ratio $\tau/(q/A)$ must be constant for all radial positions. We now integrate between conditions at the wall where $T = T_i$ and $v = 0$ to some point in the fluid where T is the same as the bulk T and assume that the velocity at this point is the same as v_{av}, the bulk velocity. Also, q/A is understood to be the flux at the wall, as is the shear at the wall, written as τ_s. Hence,

$$\frac{\tau_s}{q/A}\,c_p(T - T_i) = v_{av} - 0 \tag{7.3-10}$$

Also, substituting $q/A = h(T - T_i)$ and $\tau_s = fv_{av}^2\,\rho/2$ from Eq. (2.10-4) into Eq. (7.3-10),

$$\frac{f}{2} = \frac{h}{c_p v_{av}\,\rho} = \frac{h}{c_p G} \tag{7.3-11}$$

In a similar manner using Eq. (6.1-6) for J_A^* and also $J_A^* = k_c'(c_A - c_{Ai})$, we can relate this to Eq. (7.3-8) for momentum transfer. Then, the complete Reynolds analogy is

$$\frac{f}{2} = \frac{h}{c_p G} = \frac{k_c'}{v_{av}} \tag{7.3-12}$$

Experimental data for gas streams agree approximately with Eq. (7.3-12) if the Schmidt and Prandtl numbers are near 1.0 and only skin friction is present in flow past a flat plate or inside a pipe. When liquids are present and/or form drag is present, the analogy is not valid.

3. *Other analogies.* The Reynolds analogy assumes that the turbulent diffusivities ε_t, α_t, and ε_M are all equal and that the molecular diffusivities μ/ρ, α, and D_{AB} are negligible compared to the turbulent diffusivities. When the Prandtl number $(\mu/\rho)/\alpha$ is 1.0, then $\mu/\rho = \alpha$; also, for $N_{Sc} = 1.0$, $\mu/\rho = D_{AB}$. Then, $(\mu/\rho + \varepsilon_t) = (\alpha + \alpha_t) = (D_{AB} + \varepsilon_M)$ and the Reynolds analogy can be obtained with the molecular terms present. However, the analogy breaks down when the viscous sublayer becomes important since the eddy diffusivities diminish to zero and the molecular diffusivities become important.

Prandtl modified the Reynolds analogy by writing the regular molecular diffusion equation for the viscous sublayer and a Reynolds-analogy equation for the turbulent core region. Then since these processes are in series, these equations were combined to produce an overall equation (G1). The results also are poor for fluids where the Prandtl and Schmidt numbers differ from 1.0.

Von Kármán further modified the Prandtl analogy by considering the buffer region in addition to the viscous sublayer and the turbulent core. These three regions are shown in the universal velocity profile in Fig. 3.10-4. Again, an equation is written for molecular diffusion in the viscous sublayer using only the molecular diffusivity and a Reynolds analogy equation for the turbulent core. Both the molecular and eddy diffusivity are used in an equation for the buffer layer, where the velocity in this layer is used to obtain an equation for the eddy diffusivity. These three equations are then combined to give the von Kármán analogy. Since then, numerous other analogies have appeared (P1, S4).

4. *Chilton and Colburn J-factor analogy.* The most successful and most widely used analogy is the Chilton and Colburn *J*-factor analogy (C2). This analogy is based on experimental data for gases and liquids in both the laminar and turbulent flow regions and is written as follows:

$$\frac{f}{2} = J_H = \frac{h}{c_p G} (N_{Pr})^{2/3} = J_D = \frac{k_c'}{v_{av}} (N_{Sc})^{2/3} \tag{7.3-13}$$

Although this is an equation based on experimental data for both laminar and turbulent flow, it can be shown to satisfy the exact solution derived from laminar flow over a flat plate in Sections 3.10 and 5.7.

Equation (7.3-13) has been shown to be quite useful in correlating momentum, heat, and mass transfer data. It permits the prediction of an unknown transfer coefficient when one of the other coefficients is known. In momentum transfer the friction factor is obtained for the total drag or friction loss, which includes form drag or momentum losses due to blunt objects and also skin friction. For flow past a flat plate or in a pipe where no form drag is present, $f/2 = J_H = J_D$. When form drag is present, such as in flow in packed beds or past other blunt objects, $f/2$ is greater than J_H or J_D and $J_H \cong J_D$.

7.3C Derivation of Mass-Transfer Coefficients in Laminar Flow

1. Introduction. When a fluid is flowing in laminar flow and mass transfer by molecular diffusion is occurring, the equations are very similar to those for heat transfer by conduction in laminar flow. The phenomena of heat and mass transfer are not always completely analogous since in mass transfer several components may be diffusing. Also, the flux of mass perpendicular to the direction of the flow must be small so as not to distort the laminar velocity profile.

In theory it is not necessary to have experimental mass-transfer coefficients for laminar flow, since the equations for momentum transfer and for diffusion can be solved. However, in many actual cases it is difficult to describe mathematically the laminar flow for geometries, such as flow past a cylinder or in a packed bed. Hence, experimental mass-transfer coefficients are often obtained and correlated. A simplified theoretical derivation will be given for two cases in laminar flow.

2. Mass transfer in laminar flow in a tube. We consider the case of mass transfer from a tube wall to a fluid inside in laminar flow, where, for example, the wall is made of solid benzoic acid which is dissolving in water. This is similar to heat transfer from a wall to the flowing fluid where natural convection is negligible. For fully developed flow, the parabolic velocity derived as Eqs. (2.6-18) and (2.6-20) is

$$v_x = v_{max}\left[1 - \left(\frac{r}{R}\right)^2\right] = 2v_{av}\left[1 - \left(\frac{r}{R}\right)^2\right] \tag{7.3-14}$$

where v_x is the velocity in the x direction at the distance r from the center. For steady-state diffusion in a cylinder, a mass balance can be made on a differential element where the rate in by convection plus diffusion equals the rate out radially by diffusion to give

$$v_x \frac{\partial c_A}{\partial x} = D_{AB}\left(\frac{1}{r}\frac{\partial c_A}{\partial r} + \frac{\partial^2 c_A}{\partial r^2} + \frac{\partial^2 c_A}{\partial x^2}\right) \tag{7.3-15}$$

Then, $\partial^2 c_A/\partial x^2 = 0$ if the diffusion in the x direction is negligible compared to that by convection. Combining Eqs. (7.3-14) and (7.3-15), the final solution (S1) is a complex series similar to the Graetz solution for heat transfer and a parabolic velocity profile.

If it is assumed that the velocity profile is flat as in rodlike flow, the solution is more easily obtained (S1). A third solution, called the approximate *Leveque solution*, has been obtained, where there is a linear velocity profile near the wall and the solute diffuses only a short distance from the wall into the fluid. This is similar to the parabolic velocity profile solution at high flow rates. Experimental design equations are presented in Section 7.3D for this case.

3. Diffusion in a laminar falling film. In Section 2.9C we derived the equation for the velocity profile in a falling film shown in Fig. 7.3-1a. We will consider mass transfer of solute A into a laminar falling film, which is important in wetted-wall columns, in developing theories to explain mass transfer in stagnant pockets of fluids, and in turbulent mass transfer. The solute A in the gas is absorbed at the interface and then diffuses a distance into the liquid so that it has not penetrated the whole distance $x = \delta$ at the wall. At steady state the inlet concentration $c_A = 0$. At a point z distance from the inlet the concentration profile of c_A is shown in Fig. 7.3-1a.

A mass balance will be made on the element shown in Fig. 7.3-1b. For steady state, rate of input = rate of output.

$$N_{Ax|x}(1\ \Delta z) + N_{Az|z}(1\ \Delta x) = N_{Ax|x+\Delta x}(1\ \Delta z) + N_{Az|z+\Delta z}(1\ \Delta x) \qquad (7.3\text{-}16)$$

For a dilute solution the diffusion equation for A in the x direction is

$$N_{Ax} = -D_{AB}\frac{\partial c_A}{\partial x} + \text{zero convection} \qquad (7.3\text{-}17)$$

For the z direction the diffusion is negligible.

$$N_{Az} = 0 + c_A v_z \qquad (7.3\text{-}18)$$

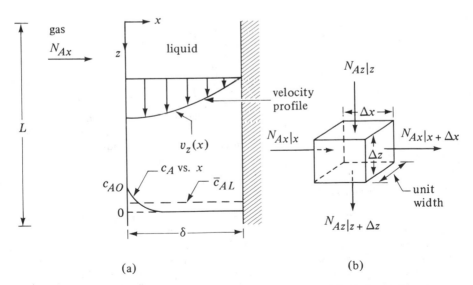

(a)   (b)

FIGURE 7.3-1.   *Diffusion of solute A in a laminar falling film: (a) velocity profile and concentration profile, (b) small element for mass balance.*

Dividing Eq. (7.3-16) by $\Delta x\,\Delta z$, letting Δx and Δz approach zero, and substituting Eqs. (7.3-17) and (7.3-18) into the result, we obtain

$$v_z \frac{\partial c_A}{\partial z} = D_{AB} \frac{\partial^2 c_A}{\partial x^2} \tag{7.3-19}$$

From Eqs. (2.9-24) and (2.9-25), the velocity profile is parabolic and is $v_z = v_{z\,max}\,[1 - (x/\delta)^2]$. Also, $v_{z\,max} = (3/2)v_{z\,av}$. If the solute has penetrated only a short distance into the fluid, i.e., short contact times of t seconds equals z/v_{max}, then the A that has diffused has been carried along at the velocity $v_{z\,max}$ or v_{max} if the subscript z is dropped. Then Eq. (7.3-19) becomes

$$\frac{\partial c_A}{\partial(z/v_{max})} = D_{AB} \frac{\partial^2 c_A}{\partial x^2} \tag{7.3-20}$$

Using the boundary conditions of $c_A = 0$ at $z = 0$, $c_A = c_{A0}$ at $x = 0$, and $c_A = 0$ at $x = \infty$, we can integrate Eq. (7.3-20) to obtain

$$\frac{c_A}{c_{A0}} = \operatorname{erfc}\left(\frac{x}{\sqrt{4D_{AB}\,z/v_{max}}}\right) \tag{7.3-21}$$

where erf y is the error function and erfc $y = 1 - \operatorname{erf} y$. Values of erf y are standard tabulated functions.

To determine the local molar flux at the surface $x = 0$ at position z from the top entrance, we write (B1)

$$N_{Ax}(z)\bigg|_{x=0} = -D_{AB} \frac{\partial c_A}{\partial x}\bigg|_{x=0} = c_{A0}\sqrt{\frac{D_{AB}\,v_{max}}{\pi z}} \tag{7.3-22}$$

The total moles of A transferred per second to the liquid over the entire length $z = 0$ to $z = L$, where the vertical surface is unit width is

$$N_A(L \cdot 1) = (1) \int_0^L (N_{Ax|x=0})\,dz$$

$$= (1) \int_0^L c_{A0} \left(\frac{D_{AB}\,v_{max}}{\pi}\right)^{1/2} \frac{1}{z^{1/2}}\,dz$$

$$= (L \cdot 1)c_{A0} \sqrt{\frac{4D_{AB}\,v_{max}}{\pi L}} \tag{7.3-23}$$

The term L/v_{max} is t_L, time of exposure of the liquid to the solute A in the gas. This means the rate of mass transfer is proportional to $D_{AB}^{0.5}$ and $1/t_L^{0.5}$. This is the basis for the penetration theory in turbulent mass transfer where pockets of liquid are exposed to unsteady-state diffusion (penetration) for short contact times.

7.3D Mass Transfer for Flow Inside Pipes

1. Mass transfer for laminar flow inside pipes. When a liquid or a gas is flowing inside a pipe and the Reynolds number $Dv\rho/\mu$ is below 2100, laminar flow occurs. Experimental data obtained for mass transfer from the walls for gases (G2, L1) are plotted in Fig. 7.3-2 for values of $W/D_{AB}\rho L$ less than about 70. The ordinate is $(c_A - c_{A0})/(c_{Ai} - c_{A0})$, where c_A is the exit concentration, c_{A0} inlet concentration, and c_{Ai} concentration at the interface between the wall and the gas. The dimensionless abscissa is $W/D_{AB}\rho L$ or $N_{Re}N_{Sc}(D/L)(\pi/4)$, where W is flow in kg/s and L is length of mass-transfer section in m.

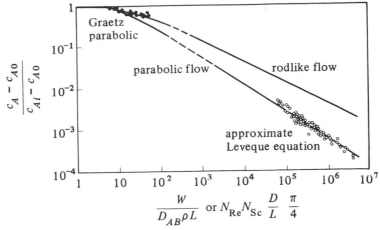

$$\frac{W}{D_{AB}\rho L} \text{ or } N_{Re}N_{Sc}\frac{D}{L}\frac{\pi}{4}$$

FIGURE 7.3-2. *Data for diffusion in a fluid in streamline flow inside a pipe: filled circles, vaporization data of Gilliland and Sherwood; open circles, dissolving-solids data of Linton and Sherwood. [From W. H. Linton and T. K. Sherwood, Chem. Eng. Progr., 46, 258 (1950). With permission.]*

Since the experimental data follow the rodlike plot, that line should be used. The velocity profile is assumed fully developed to parabolic form at the entrance.

For liquids that have small values of D_{AB}, data follow the parabolic flow line, which is as follows for $W/D_{AB}\rho L$ over 400.

$$\frac{c_A - c_{A0}}{c_{Ai} - c_{A0}} = 5.5\left(\frac{W}{D_{AB}\rho L}\right)^{-2/3} \tag{7.3-24}$$

2. Mass transfer for turbulent flow inside pipes. For turbulent flow for $Dv\rho/\mu$ above 2100 for gases or liquids flowing inside a pipe,

$$N_{Sh} = k_c'\frac{D}{D_{AB}} = \frac{k_c p_{BM}}{P}\frac{D}{D_{AB}} = 0.023\left(\frac{Dv\rho}{\mu}\right)^{0.83}\left(\frac{\mu}{\rho D_{AB}}\right)^{0.33} \tag{7.3-25}$$

The equation holds for N_{Sc} of 0.6 to 3000 (G2, L1). Note that the N_{Sc} for gases is in the range 0.5–3 and for liquids is above 100 in general. Equation (7.3-25) for mass transfer and Eq. (4.5-8) for heat transfer inside a pipe are similar to each other.

3. Mass transfer for flow inside wetted-wall towers. When a gas is flowing inside the core of a wetted-wall tower the same correlations that are used for mass transfer of a gas in laminar or turbulent flow in a pipe are applicable. This means that Eqs. (7.3-24) and (7.3-25) can be used to predict mass transfer for the gas. For the mass transfer in the liquid film flowing down the wetted-wall tower, Eqs. (7.3-22) and (7.3-23) can be used for Reynolds numbers of $4\Gamma/\mu$ as defined by Eq. (2.9-29) up to about 1200, and the theoretically predicted values should be multiplied by about 1.5 because of ripples and other factors. These equations hold for short contact times or Reynolds numbers above about 100 (S1).

EXAMPLE 7.3-1. *Mass Transfer Inside a Tube*
A tube is coated on the inside with naphthalene and has an inside diameter of 20 mm and a length of 1.10 m. Air at 318 K and an average pressure of

101.3 kPa flows through this pipe at a velocity of 0.80 m/s. Assuming that the absolute pressure remains essentially constant, calculate the concentration of naphthalene in the exit air. Use the physical properties given in Example 6.2-4.

Solution: From Example 6.2-4, $D_{AB} = 6.92 \times 10^{-6}$ m²/s and the vapor pressure $p_{Ai} = 74.0$ Pa or $c_{Ai} = p_{Ai}/RT = 74.0/(8314.3 \times 318) = 2.799 \times 10^{-5}$ kg mol/m³. For air from Appendix A.3, $\mu = 1.932 \times 10^{-5}$ Pa·s, $\rho = 1.114$ kg/m³. The Schmidt number is

$$N_{Sc} = \frac{\mu}{\rho D_{AB}} = \frac{1.932 \times 10^{-5}}{1.114 \times 6.92 \times 10^{-6}} = 2.506$$

The Reynolds number is

$$N_{Re} = \frac{Dv\rho}{\mu} = \frac{0.020(0.80)(1.114)}{1.932 \times 10^{-5}} = 922.6$$

Hence, the flow is laminar. Then,

$$N_{Re} N_{Sc} \frac{D}{L} \frac{\pi}{4} = 922.6(2.506) \frac{0.020}{1.10} \frac{\pi}{4} = 33.02$$

Using Fig. 7.3-2 and the rodlike flow line, $(c_A - c_{A0})/(c_{Ai} - c_{A0}) = 0.55$. Also, $c_{A0}(\text{inlet}) = 0$. Then, $(c_A - 0)/(2.799 \times 10^{-5} - 0) = 0.55$. Solving, $c_A(\text{exit concentration}) = 1.539 \times 10^{-5}$ kg mol/m³.

7.3E Mass Transfer for Flow Outside Solid Surfaces

1. Mass transfer in flow parallel to flat plates. The mass transfer and vaporization of liquids from a plate or flat surface to a flowing stream is of interest in the drying of inorganic and biological materials, in evaporation of solvents from paints, for plates in wind tunnels, and in flow channels in chemical process equipment.

When the fluid flows past a plate in a free stream in an open space the boundary layer is not fully developed. For gases or evaporation of liquids in the gas phase and for the laminar region of $N_{Re, L} = Lv\rho/\mu$ less than 15 000, the data can be represented within ±25% by the equation (S4)

$$J_D = 0.664 N_{Re, L}^{-0.5} \tag{7.3-26}$$

Writing Eq. (7.3-26) in terms of the Sherwood number N_{Sh},

$$\frac{k_c' L}{D_{AB}} = N_{Sh} = 0.664 N_{Re, L}^{0.5} N_{Sc}^{1/3} \tag{7.3-27}$$

where L is the length of plate in the direction of flow. Also, $J_D = J_H = f/2$ for this geometry. For gases and $N_{Re, L}$ of 15 000–300 000, the data are represented within ±30% by $J_D = J_H = f/2$ as

$$J_D = 0.036 N_{Re, L}^{-0.2} \tag{7.3-28}$$

Experimental data for liquids are correlated within about ±40% by the following for a $N_{Re, L}$ of 600–50 000 (L2):

$$J_D = 0.99 N_{Re, L}^{-0.5} \tag{7.3-29}$$

EXAMPLE 7.3-2. Mass Transfer from a Flat Plate

A large volume of pure water at 26.1°C is flowing parallel to a flat plate of solid benzoic acid, where $L = 0.244$ m in the direction of flow. The water velocity is 0.061 m/s. The solubility of benzoic acid in water is 0.02948 kg mol/m³. The diffusivity of benzoic acid is 1.245×10^{-9} m²/s. Calculate the mass-transfer coefficient k_L and the flux N_A.

Solution: Since the solution is quite dilute, the physical properties of water at 26.1°C from Appendix A.2 can be used.

$$\mu = 8.71 \times 10^{-4} \text{ Pa·s}$$

$$\rho = 996 \text{ kg/m}^3$$

$$D_{AB} = 1.245 \times 10^{-9} \text{ m}^2/\text{s}$$

The Schmidt number is

$$N_{Sc} = \frac{8.71 \times 10^{-4}}{996(1.245 \times 10^{-9})} = 702$$

The Reynolds number is

$$N_{Re, L} = \frac{Lv\rho}{\mu} = \frac{0.244(0.0610)(996)}{8.71 \times 10^{-4}} = 1.700 \times 10^4$$

Using Eq. (7.3-29),

$$J_D = 0.99N_{Re, L}^{-0.5} = 0.99(1.700 \times 10^4)^{-0.5} = 0.00758$$

The definition of J_D from Eq. (7.3-5) is

$$J_D = \frac{k_c'}{v} (N_{Sc})^{2/3} \qquad (7.3-5)$$

Solving for k_c', $k_c' = J_D v (N_{Sc})^{-2/3}$. Substituting known values and solving,

$$k_c' = 0.00758(0.0610)(702)^{-2/3} = 5.85 \times 10^{-6} \text{ m/s}$$

In this case, diffusion is for A through nondiffusing B, so k_c in Eq. (7.2-10) should be used.

$$N_A = \frac{k_c'}{x_{BM}} (c_{A1} - c_{A2}) = k_c(c_{A1} - c_{A2}) \qquad (7.2-10)$$

Since the solution is very dilute, $x_{BM} \cong 1.0$ and $k_c' \cong k_c$. Also, $c_{A1} = 2.948 \times 10^{-2}$ kg mol/m³ (solubility) and $c_{A2} = 0$ (large volume of fresh water). Substituting into Eq. (7.2-10),

$$N_A = (5.85 \times 10^{-6})(0.02948 - 0) = 1.726 \times 10^{-7} \text{ kg mol/s·m}^2$$

2. *Mass transfer for flow past single spheres.* For flow past single spheres and for very low $N_{Re} = D_p v\rho/\mu$, where v is the average velocity in the empty test section before the sphere, the Sherwood number, which is $k_c' D_p/D_{AB}$, should approach a value of 2.0. This can be shown from Eq. (6.2-33), which was derived for a stagnant medium. Rewriting Eq. (6.2-33) as follows, where D_p is the sphere diameter,

$$N_A = \frac{2D_{AB}}{D_p} (c_{A1} - c_{A2}) = k_c(c_{A1} - c_{A2}) \qquad (7.3-30)$$

The mass-transfer coefficient k_c, which is k_c' for a dilute solution, is then

$$k_c' = \frac{2D_{AB}}{D_p} \qquad (7.3-31)$$

Rearranging,

$$\frac{k_c' D_p}{D_{AB}} = N_{Sh} = 2.0 \qquad (7.3\text{-}32)$$

Of course, natural convection effects could increase k_c'.

For gases for a Schmidt number range of 0.6–2.7 and a Reynolds number range of 1–48 000, a modified equation (G1) can be used.

$$N_{Sh} = 2 + 0.552 N_{Re}^{0.53} N_{Sc}^{1/3} \qquad (7.3\text{-}33)$$

This equation also holds for heat transfer where the Prandtl number replaces the Schmidt number and the Nusselt number hD_p/k replaces the Sherwood number.

For liquids (G3) and a Reynolds number range of 2 to about 2000, the following can be used.

$$N_{Sh} = 2 + 0.95 N_{Re}^{0.50} N_{Sc}^{1/3} \qquad (7.3\text{-}34)$$

For liquids and a Reynolds number of 2000–17 000, the following can be used (S5).

$$N_{Sh} = 0.347 N_{Re}^{0.62} N_{Sc}^{1/3} \qquad (7.3\text{-}35)$$

EXAMPLE 7.3-3. Mass Transfer from a Sphere
Calculate the value of the mass-transfer coefficient and the flux for mass transfer from a sphere of napthalene to air at 45°C and 1 atm abs flowing at a velocity of 0.305 m/s. The diameter of the sphere is 25.4 mm. The diffusivity of naphthalene in air at 45°C is 6.92×10^{-6} m²/s and the vapor pressure of solid naphthalene is 0.555 mm Hg. Use English and SI units.

Solution: In English units $D_{AB} = 6.92 \times 10^{-6}(3.875 \times 10^4) = 0.2682$ ft²/h. The diameter $D_p = 0.0254$ m $= 0.0254(3.2808) = 0.0833$ ft. From Appendix A.3 the physical properties of air will be used since the concentration of naphthalene is low.

$$\mu = 1.93 \times 10^{-5} \text{ Pa} \cdot \text{s} = 1.93 \times 10^{-5}(2.4191 \times 10^3) = 0.0467 \text{ lb}_m/\text{ft} \cdot \text{h}$$

$$\rho = 1.113 \text{ kg/m}^3 = \frac{1.113}{16.0185} = 0.0695 \text{ lb}_m/\text{ft}^3$$

$$v = 0.305 \text{ m/s} = 0.305(3600 \times 3.2808) = 3600 \text{ ft/h}$$

The Schmidt number is

$$N_{Sc} = \frac{\mu}{\rho D_{AB}} = \frac{0.0467}{0.0695(0.2682)} = 2.505$$

$$N_{Sc} = \frac{1.93 \times 10^{-5}}{1.113(6.92 \times 10^{-6})} = 2.505$$

The Reynolds number is

$$N_{Re} = \frac{D_p v \rho}{\mu} = \frac{0.0833(3600)(0.0695)}{0.0467} = 446$$

$$N_{Re} = \frac{0.0254(0.3048)(1.113)}{1.93 \times 10^{-5}} = 446$$

Equation (7.3-33) for gases will be used.

$$N_{Sh} = 2 + 0.552(N_{Re})^{0.53}(N_{Sc})^{1/3} = 2 + 0.552(446)^{0.53}(2.505)^{1/3} = 21.0$$

From Eq. (7.3-3),

$$N_{Sh} = k'_c \frac{L}{D_{AB}} = k'_c \frac{D_p}{D_{AB}}$$

Substituting the knowns and solving,

$$21.0 = \frac{k'_c(0.0833)}{0.2682} \qquad k'_c = 67.6 \text{ ft/h}$$

$$21.0 = \frac{k'_c(0.0254)}{6.92 \times 10^{-6}} \qquad k'_c = 5.72 \times 10^{-3} \text{ m/s}$$

From Table 7.2-1,

$$k'_c c = k'_c \frac{P}{RT} = k'_G P$$

Hence, for $T = 45 + 273 = 318 \text{ K} = 318(1.8) = 574°\text{R}$,

$$k'_G = \frac{k'_c}{RT} = \frac{67.6}{(0.730)(573)} = 0.1616 \text{ lb mol/h} \cdot \text{ft}^2 \cdot \text{atm}$$

$$k'_G = \frac{5.72 \times 10^{-3}}{8314(318)} = 2.163 \times 10^{-9} \text{ kg mol/s} \cdot \text{m}^2 \cdot \text{Pa}$$

Since the gas is very dilute, $y_{BM} \cong 1.0$ and $k'_G \cong k_G$. Substituting into Eq. (7.2-12) for A diffusing through stagnant B and noting that $p_{A1} = 0.555/760 = 7.303 \times 10^{-4} \text{ atm} = 74.0 \text{ Pa}$ and $p_{A2} = 0$ (pure air),

$$N_A = k_G(p_{A1} - p_{A2}) = 0.1616(7.303 \times 10^{-4} - 0)$$

$$= 1.180 \times 10^{-4} \text{ lb mol/h} \cdot \text{ft}^2$$

$$= 2.163 \times 10^{-9}(74.0 - 0) = 1.599 \times 10^{-7} \text{ kg mol/s} \cdot \text{m}^2$$

The area of the sphere is

$$A = \pi D_p^2 = \pi(0.0833)^2 = 2.18 \times 10^{-2} \text{ ft}^2$$

$$= (2.18 \times 10^{-2})\left(\frac{1}{3.2808}\right)^2 = 2.025 \times 10^{-3} \text{ m}^2$$

Total amount evaporated $= N_A A = (1.18 \times 10^{-4})(2.18 \times 10^{-2}) = 2.572 \times 10^{-6}$ lb mol/h $= (1.599 \times 10^{-7})(2.025 \times 10^{-3}) = 3.238 \times 10^{-10}$ kg mol/s

3. *Mass transfer to packed beds.* Mass transfer to and from packed beds occurs often in processing operations, including drying operations, adsorption or desorption of gases or liquids by solid particles such as charcoal, and mass transfer of gases and liquids to catalyst particles. Using a packed bed a large amount of mass-transfer area can be contained in a relatively small volume.

The void fraction in a bed is ε, m³ volume void space divided by the m³ total volume of void space plus solid. The values range from 0.3 to 0.5 in general. Because of flow channeling, nonuniform packing, etc., accurate experimental data are difficult to obtain and data from different investigators can deviate considerably.

For a Reynolds number range of 10–10 000 for gases in a packed bed of spheres (D4), the recommended correlation with an average deviation of about $\pm20\%$ and a maximum of about $\pm50\%$ is

$$J_D = J_H = \frac{0.4548}{\varepsilon} N_{Re}^{-0.4069} \tag{7.3-36}$$

It has been shown (G4, G5) that J_D and J_H are approximately equal. The Reynolds number is defined as $N_{Re} = D_p v' \rho / \mu$, where D_p is diameter of the spheres and v' is the superficial mass average velocity in the empty tube without packing. For Eqs. (7.3-36)–(7.3-39) and Eqs. (7.3-5)–(7.3-6), v' is used.

For mass transfer of liquids in packed beds, the correlations of Wilson and Geankoplis (W1) should be used. For a Reynolds number $D_p v' \rho / \mu$ range of 0.0016–55 and a Schmidt number range of 165–70 600, the equation to use is

$$J_D = \frac{1.09}{\varepsilon} N_{Re}^{-2/3} \qquad (7.3\text{-}37)$$

For liquids and a Reynolds number range of 55–1500 and a Schmidt number range of 165–10 690,

$$J_D = \frac{0.250}{\varepsilon} N_{Re}^{-031} \qquad (7.3\text{-}38)$$

Or, as an alternate, Eq. (7.3-36) can be used for liquids for a Reynolds number range of 10–1500.

For fluidized beds of spheres, Eq. (7.3-36) can be used for gases and liquids and a Reynolds number range of 10–4000. For liquids in a fluidized bed and a Reynolds number range of 1–10 (D4),

$$\varepsilon J_D = 1.1068 N_{Re}^{-0.72} \qquad (7.3\text{-}39)$$

If packed beds of solids other than spheres are used, approximate correction factors can be used with Eqs. (7.3-36)–(7.3-38) for spheres. This is done, for example, for a given nonspherical particle as follows. The particle diameter to use in the equations to predict J_D is the diameter of a sphere with the same surface area as the given solid particle. The flux to these particles in the bed is then calculated using the area of the given particles. An alternative approximate procedure to use is given elsewhere (G6).

4. *Calculation method for packed beds.* To calculate the total flux in a packed bed, J_D is first obtained and then k_c in m/s from the J_D. Then knowing the total volume V_b m^3 of the bed (void plus solids), the total external surface area A m^2 of the solids for mass transfer is calculated using Eqs. (7.3-40) and (7.3-41).

$$a = \frac{6(1 - \varepsilon)}{D_p} \qquad (7.3\text{-}40)$$

where a is the m^2 surface area/m^3 total volume of bed when the solids are spheres.

$$A = aV_b \qquad (7.3\text{-}41)$$

To calculate the mass-transfer rate the log mean driving force at the inlet and outlet of the bed should be used.

$$N_A A = A k_c \frac{(c_{Ai} - c_{A1}) - (c_{Ai} - c_{A2})}{\ln \dfrac{c_{Ai} - c_{A1}}{c_{Ai} - c_{A2}}} \qquad (7.3\text{-}42)$$

where the final term is the log mean driving force: c_{Ai} is the concentration at the surface of the solid, in kg mol/m^3; c_{A1} is the inlet bulk fluid concentration; and c_{A2} is the outlet. The material-balance equation on the bulk stream is

$$N_A A = V(c_{A2} - c_{A1}) \qquad (7.3\text{-}43)$$

where V is volumetric flow rate of fluid entering in m^3/s. Equations (7.3-42) and (7.3-43) must both be satisfied. The use of these two equations is similar to the use of the log mean temperature difference and heat balance in heat exchangers. These two equations can also be used for a fluid flowing in a pipe or past a flat plate, where A is the pipe wall area or plate area.

EXAMPLE 7.3-4. Mass Transfer of a Liquid in a Packed Bed
Pure water at $26.1°C$ flows at the rate of 5.514×10^{-7} m^3/s through a packed bed of benzoic acid spheres having a diameter of 6.375 mm. The total surface area of the spheres in the bed is 0.01198 m^2 and the void fraction is 0.436. The tower diameter is 0.0667 m. The solubility of benzoic acid in water is 2.948×10^{-2} kg mol$/m^3$.
   (a) Predict the mass-transfer coefficient k_c. Compare with the experimental value of 4.665×10^{-6} m/s by Wilson and Geankoplis (W1).
   (b) Using the experimental value of k_c, predict the outlet concentration of benzoic acid in the water.

Solution: Since the solution is dilute, the physical properties of water will be used at $26.1°C$ from Appendix A.2. At $26.1°C$, $\mu = 0.8718 \times 10^{-3}$ Pa·s, $\rho = 996.7$ kg$/m^3$. At $25.0°C$, $\mu = 0.8940 \times 10^{-3}$ Pa·s and from Table 6.3-1, $D_{AB} = 1.21 \times 10^{-9}$ m^2/s. To correct D_{AB} to $26.1°C$ using Eq. (6.3-9), $D_{AB} \propto T/\mu$. Hence,

$$D_{AB}(26.1°C) = (1.21 \times 10^{-9})\left(\frac{299.1}{298}\right)\left(\frac{0.8940 \times 10^{-3}}{0.8718 \times 10^{-3}}\right)$$

$$= 1.254 \times 10^{-9} \ m^2/s$$

The tower cross-sectional area $= (\pi/4)(0.0667)^2 = 3.494 \times 10^{-3}$ m^2. Then $v' = (5.514 \times 10^{-7})/(3.494 \times 10^{-3}) = 1.578 \times 10^{-4}$ m/s. Then,

$$N_{Sc} = \frac{\mu}{\rho D_{AB}} = \frac{0.8718 \times 10^{-3}}{996.7(1.245 \times 10^{-9})} = 702.6$$

The Reynolds number is

$$N_{Re} = \frac{Dv'\rho}{\mu} = \frac{0.006375(1.578 \times 10^{-4})(996.7)}{0.8718 \times 10^{-3}} = 1.150$$

Using Eq. (7.3-37) and assuming $k_c = k'_c$ for dilute solutions,

$$J_D = \frac{1.09}{\varepsilon}(N_{Re})^{-2/3} = \frac{1.09}{0.436}(1.150)^{-2/3} = 2.277$$

Then, using Eq. (7.3-5) and solving,

$$J_D = \frac{k'_c}{v'}(N_{Sc})^{2/3} \qquad 2.277 = \frac{k'_c}{1.578 \times 10^{-4}}(702.6)^{2/3}$$

The predicted $k'_c = 4.447 \times 10^{-6}$ m/s. This compares with the experimental value of 4.665×10^{-6} m/s.
   For part (b), using Eqs. (7.3-42) and (7.3-43),

$$Ak_c \frac{(c_{Ai} - c_{A1}) - (c_{Ai} - c_{A2})}{\ln\dfrac{c_{Ai} - c_{A1}}{c_{Ai} - c_{A2}}} = V(c_{A2} - c_{A1}) \qquad \textbf{(7.3-44)}$$

The values to substitute into Eq. (7.3-44) are $c_{Ai} = 2.948 \times 10^{-2}$, $c_{A1} = 0$, $A = 0.01198$, $V = 5.514 \times 10^{-7}$.

$$\frac{0.01198(4.665 \times 10^{-6})(c_{A2} - 0)}{\ln \dfrac{2.948 \times 10^{-2} - 0}{2.948 \times 10^{-2} - c_{A2}}} = (5.514 \times 10^{-7})(c_{A2} - 0)$$

Solving, $c_{A2} = 2.842 \times 10^{-3}$ kg mol/m^3.

5. Mass transfer for flow past single cylinders. Experimental data have been obtained for mass transfer from single cylinders when the flow is perpendicular to the cylinder. The cylinders are long and mass transfer to the ends of the cylinder is not considered. For the Schmidt number range of 0.6 to 2.6 for gases and 1000 to 3000 for liquids and a Reynolds number range of 50 to 50 000, data of many references (B3, L1, M1, S4, V1) have been plotted and the correlation to use is as follows:

$$J_D = 0.600(N_{Re})^{-0.487} \tag{7.3-45}$$

The data scatter considerably by up to $\pm 30\%$. This correlation can also be used for heat transfer with $J_D = J_H$.

6. Liquid metals mass transfer. In recent years several correlations for mass-transfer coefficients of liquid metals have appeared in the literature. It has been found (G1) that with moderate safety factors, the correlations for nonliquid metals mass transfer may be used for liquid metals mass transfer. Care must be taken to ensure that the solid surface is wetted. Also, if the solid is an alloy, there may exist a resistance to diffusion in the solid phase.

7.4 MASS TRANSFER TO SUSPENSIONS OF SMALL PARTICLES

7.4A Introduction

Mass transfer from or to small suspended particles in an agitated solution occurs in a number of process applications. In liquid-phase hydrogenation, hydrogen diffuses from gas bubbles, through an organic liquid, and then to small suspended catalyst particles. In fermentations, oxygen diffuses from small gas bubbles, through the aqueous medium, and then to small suspended microorganisms.

For a liquid–solid dispersion, increased agitation over and above that necessary to freely suspend very small particles has very little effect on the mass-transfer coefficient k_L to the particle (B2). When the particles in a mixing vessel are just completely suspended, turbulence forces balance those due to gravity, and the mass-transfer rates are the same as for particles freely moving under gravity. With very small particles of say a few μm or so, which is the size of many microorganisms in fermentations and some catalyst particles, their size is smaller than eddies, which are about 100 μm or so in size. Hence, increased agitation will have little effect on mass transfer except at very high agitation.

For a gas–liquid–solid dispersion, such as in fermentation, the same principles hold. However, increased agitation increases the number of gas bubbles and hence the interfacial area. The mass-transfer coefficients from the gas bubble to the liquid and from the liquid to the solid are relatively unaffected.

7.4B Equations for Mass Transfer to Small Particles

1. Mass transfer to small particles <0.6 mm. Equations to predict mass transfer to small particles in suspension have been developed which cover three size ranges of particles. The equation for particles <0.6 mm (600 μm) is discussed first.

The following equation has been shown to hold to predict mass-transfer coefficients from small gas bubbles such as oxygen or air to the liquid phase or from the liquid phase to the surface of small catalyst particles, microorganisms, other solids, or liquid drops (B2, C3).

$$k'_L = \frac{2D_{AB}}{D_p} + 0.31N_{Sc}^{-2/3}\left(\frac{\Delta\rho\mu_c g}{\rho_c^2}\right)^{1/3} \tag{7.4-1}$$

where D_{AB} is the diffusivity of the solute A in solution in m^2/s, D_p is the diameter of the gas bubble or the solid particle in m, μ_c is the viscosity of the solution in kg/m·s, $g = 9.80665$ m/s^2, $\Delta\rho = (\rho_c - \rho_p)$ or $(\rho_p - \rho_c)$, ρ_c is the density of the continuous phase in kg/m^3, and ρ_p is the density of the gas or solid particle. The value of $\Delta\rho$ is always positive.

The first term on the right in Eq. (7.4-1) is the molecular diffusion term, and the second term is that due to free fall or rise of the sphere by gravitational forces. This equation has been experimentally checked for dispersions of low-density solids in agitated dispersions and for small gas bubbles in agitated systems.

EXAMPLE 7.4-1. *Mass Transfer from Air Bubbles in Fermentation*
Calculate the maximum rate of absorption of O_2 in a fermenter from air bubbles at 1 atm abs pressure having diameters of 100 μm at 37°C into water having a zero concentration of dissolved O_2. The solubility of O_2 from air in water at 37°C is 2.26×10^{-7} g mol O_2/cm^3 liquid or 2.26×10^{-4} kg mol O_2/m^3. The diffusivity of O_2 in water at 37°C is 3.25×10^{-9} m^2/s. Agitation is used to produce the air bubbles.

Solution: The mass-transfer resistance inside the gas bubble to the outside interface of the bubble can be neglected since it is negligible (B2). Hence, the mass-transfer coefficient k'_L outside the bubble is needed. The given data are

$$D_p = 100 \ \mu m = 1 \times 10^{-4} \text{ m} \qquad D_{AB} = 3.25 \times 10^{-9} \text{ m}^2/\text{s}$$

At 37°C,

$$\mu_c(\text{water}) = 6.947 \times 10^{-4} \text{ Pa·s} = 6.947 \times 10^{-4} \text{ kg/m·s}$$

$$\rho_c(\text{water}) = 994 \text{ kg/m}^3 \qquad \rho_p(\text{air}) = 1.13 \text{ kg/m}^3$$

$$N_{Sc} = \frac{\mu_c}{\rho_c D_{AB}} = \frac{6.947 \times 10^{-4}}{(994)(3.25 \times 10^{-9})} = 215$$

$$N_{Sc}^{2/3} = (215)^{2/3} = 35.9 \qquad \Delta\rho = \rho_c - \rho_p = 994 - 1.13 = 993 \text{ kg/m}^3$$

Substituting into Eq. (7.4-1),

$$k'_L = \frac{2D_{AB}}{D_p} + 0.31N_{Sc}^{-2/3}\left(\frac{\Delta\rho\mu_c g}{\rho_c^2}\right)^{1/3}$$

$$= \frac{2(3.25 \times 10^{-9})}{1 \times 10^{-4}} + \frac{0.31}{35.9}\left[\frac{993 \times 6.947 \times 10^{-4} \times 9.806}{(994)^2}\right]^{1/3}$$

$$= 6.50 \times 10^{-5} + 16.40 \times 10^{-5} = 2.290 \times 10^{-4} \text{ m/s}$$

The flux is as follows assuming $k_L = k'_L$ for dilute solutions.

$$N_A = k_L(c_{A1} - c_{A2}) = 2.290 \times 10^{-4} (2.26 \times 10^{-4} - 0)$$

$$= 5.18 \times 10^{-8} \text{ kg mol } O_2/\text{s·m}^2$$

Knowing the total number of bubbles and their area, the maximum possible rate of transfer of O_2 to the fermentation liquid can be calculated.

In Example 7.4-1, k_L was small. For mass transfer of O_2 in a solution to a microorganism with $D_p \cong 1$ μm, the term $2D_{AB}/D_p$ would be 100 times larger. Note that at large diameters the second term in Eq. (7.4-1) becomes small and the mass-transfer coefficient k_L becomes essentially independent of size D_p. In agitated vessels with gas introduced below the agitator in aqueous solutions, or when liquids are aerated with sintered plates, the gas bubbles are often in the size range covered by Eq. (7.4-1) (B2, C3, T1).

In aerated mixing vessels the mass-transfer coefficients are essentially independent of the power input. However, as the power is increased, the bubble size decreases and the mass transfer coefficient continues to follow Eq. (7.4-1). The dispersions include those in which the solid particles are just completely suspended in mixing vessels. Increase in agitation intensity above the level needed for complete suspension of these small particles results in only a small increase in k_L (C3).

Equation (7.4-1) has also been shown to apply to heat transfer and can be written as follows (B2, C3):

$$N_{Nu} = \frac{hD_p}{k} = 2.0 + 0.31 N_{Pr}^{1/3} \left(\frac{D_p^3 \rho_c \, \Delta\rho g}{\mu_c^2} \right)^{1/3} \tag{7.4-2}$$

2. *Mass transfer to large gas bubbles > 2.5 mm.* For large gas bubbles or liquid drops > 2.5 mm, the mass-transfer coefficient can be predicted by

$$k_L' = 0.42 N_{Sc}^{-0.5} \left(\frac{\Delta\rho \mu_c g}{\rho_c^2} \right)^{1/3} \tag{7.4-3}$$

Large gas bubbles are produced when pure liquids are aerated in mixing vessels and sieve-plate columns (C1). In this case the mass-transfer coefficient k_L' or k_L is independent of the bubble size and is constant for a given set of physical properties. For the same physical properties the large bubble Eq. (7.4-3) gives values of k_L about three to four times larger than Eq. (7.4-1) for small particles. Again, Eq. (7.4-3) shows that the k_L is essentially independent of agitation intensity in an agitated vessel and gas velocity in a sieve-tray tower.

3. *Mass transfer to particles in transition region.* In mass transfer in the transition region between small and large bubbles in the size range 0.6 to 2.5 mm, the mass-transfer coefficient can be approximated by assuming that it increases linearly with bubble diameter (B2, C3).

4. *Mass transfer to particles in highly turbulent mixers.* In the preceding three regions, the density difference between phases is sufficiently large to cause the force of gravity to primarily determine the mass-transfer coefficient. This also includes solids just completely suspended in mixing vessels. When agitation power is increased beyond that needed for suspension of solid or liquid particles and the turbulence forces become larger than the gravitational forces, Eq. (7.4-1) is not followed and Eq. (7.4-4) should be used where small increases in k_L' are observed (B2, C3).

$$k_L' N_{Sc}^{2/3} = 0.13 \left(\frac{(P/V)\mu_c}{\rho_c^2} \right)^{1/4} \tag{7.4-4}$$

where P/V is power input per unit volume defined in Section 3.4. The data deviate substantially by up to 60% from this correlation. In the case of gas–liquid dispersions it is quite impractical to exceed gravitational forces by agitation systems.

The experimental data are complicated by the fact that very small particles are easily suspended and if their size is of the order of the smallest eddies, the mass-transfer coefficient will remain constant until a large increase in power input is added above that required for suspension.

7.5 MOLECULAR DIFFUSION PLUS CONVECTION AND CHEMICAL REACTION

7.5A Different Types of Fluxes and Fick's Law

In Section 6.2B the flux J_A^* was defined as the molar flux of A in kg mol $A/s \cdot m^2$ relative to the molar average velocity v_M of the whole or bulk stream. Also, N_A was defined as the molar flux of A relative to stationary coordinates. Fluxes and velocities can also be defined in other ways. Table 7.5-1 lists the different types of fluxes and velocities often used in binary systems.

TABLE 7.5-1. *Different Types of Fluxes and Velocities in Binary Systems*

	Mass Flux (kg $A/s \cdot m^2$)	Molar Flux (kg mol $A/s \cdot m^2$)
Relative to fixed coordinates	$n_A = \rho_A v_A$	$N_A = c_A v_A$
Relative to molar average velocity v_M	$j_A^* = \rho_A(v_A - v_M)$	$J_A^* = c_A(v_A - v_M)$
Relative to mass average velocity v	$j_A = \rho_A(v_A - v)$	$J_A = c_A(v_A - v)$

Relations Between Fluxes Above

$$N_A + N_B = c v_M \qquad N_A = n_A/M_A \qquad N_A = J_A + c_A v$$

$$J_A^* + J_B^* = 0 \qquad J_A = j_A/M_A \qquad n_A + n_B = \rho v$$

$$j_A + j_B = 0 \qquad N_A = J_A^* + c_A v_M \qquad n_A = j_A + \rho_A v$$

Different Forms of Fick's Law for Diffusion Flux

$$J_A^* = -c D_{AB}\, dx_A/dz \qquad j_A = -\rho D_{AB}\, dw_A/dz$$

The velocity v is the mass average velocity of the stream relative to stationary coordinates and can be obtained by actually weighing the flow for a timed increment. It is related to the velocity v_A and v_B by

$$v = w_A v_A + w_B v_B = \frac{\rho_A}{\rho} v_A + \frac{\rho_B}{\rho} v_B \tag{7.5-1}$$

where w_A is ρ_A/ρ, the weight fraction of A; w_B the weight fraction of B; and v_A is the velocity of A relative to stationary coordinates in m/s. The molar average velocity v_M in m/s is relative to stationary coordinates.

$$v_M = x_A v_A + x_B v_B = \frac{c_A}{c} v_A + \frac{c_B}{c} v_B \tag{7.5-2}$$

The molar diffusion flux relative to the molar average velocity v_M defined previously is

$$J_A^* = c_A(v_A - v_M) \tag{7.5-3}$$

The molar diffusion flux J_A relative to the mass average velocity v is

$$J_A = c_A(v_A - v) \tag{7.5-4}$$

Fick's law from Table 7.5-1 as given previously is relative to v_M and is

$$J_A^* = -cD_{AB} \frac{dx_A}{dz} \tag{7.5-5}$$

Fick's law can also be defined in terms of a mass flux relative to v as given in Table 7.5-1.

$$j_A = -\rho D_{AB} \frac{dw_A}{dz} \tag{7.5-6}$$

EXAMPLE 7.5-1. *Proof of Mass Flux Equation*
Table 7.5-1 gives the following relation:

$$j_A + j_B = 0 \tag{7.5-7}$$

Prove this relationship using the definitions of the fluxes in terms of velocities.

Solution: From Table 7.5-1 substituting $\rho_A(v_A - v)$ for j_A and $\rho_B(v_B - v)$ for j_B, and rearranging,

$$\rho_A v_A - \rho_A v + \rho_B v_B - \rho_B v = 0 \tag{7.5-8}$$

$$\rho_A v_A + \rho_B v_B - v(\rho_A + \rho_B) = 0 \tag{7.5-9}$$

Substituting Eq. (7.5-1) for v and ρ for $\rho_A + \rho_B$, the identity is proved.

7.5B Equation of Continuity for a Binary Mixture

A general equation can be derived for a binary mixture of A and B for diffusion and convection that also includes the terms for unsteady-state diffusion and chemical reaction. We shall make a mass balance on component A on an element $\Delta x \, \Delta y \, \Delta z$ fixed in space as shown in Fig. 7.5-1. The general mass balance on A is

$$\begin{pmatrix} \text{rate of} \\ \text{mass } A \text{ in} \end{pmatrix} - \begin{pmatrix} \text{rate of} \\ \text{mass } A \text{ out} \end{pmatrix}$$

$$+ \begin{pmatrix} \text{rate of} \\ \text{generation of mass } A \end{pmatrix} = \begin{pmatrix} \text{rate of} \\ \text{accumulation of mass } A \end{pmatrix} \tag{7.5-10}$$

The rate of mass A entering in the direction relative to stationary coordinates is $(n_{Ax}|_x)\Delta y \, \Delta z$ kg A/s and leaving is $(n_{Ax}|_{x+\Delta x})\Delta y \, \Delta z$. Similar terms can be written for the y and z directions. The rate of chemical production of A is r_A kg A generated/s $\cdot$ m^3 volume and the total rate generated is $r_A(\Delta x \, \Delta y \, \Delta z)$ kg A/s. The rate of accumulation of A is $(\partial \rho_A/\partial t)\Delta x \, \Delta y \, \Delta z$. Substituting into Eq. (7.5-10) and letting Δx, Δy, and Δz approach zero,

$$\frac{\partial \rho_A}{\partial t} + \left(\frac{\partial n_{Ax}}{\partial x} + \frac{\partial n_{Ay}}{\partial y} + \frac{\partial n_{Az}}{\partial z} \right) = r_A \tag{7.5-11}$$

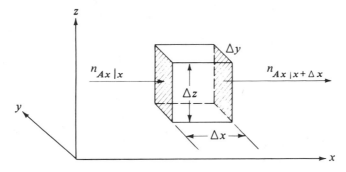

FIGURE 7.5-1. *Mass balance for A in a binary mixture.*

In vector notation,

$$\frac{\partial \rho_A}{\partial t} + (\nabla \cdot \mathbf{n}_A) = r_A \tag{7.5-12}$$

Dividing both sides of Eq. (7.5-11) by M_A,

$$\frac{\partial c_A}{\partial t} + \left(\frac{\partial N_{Ax}}{\partial x} + \frac{\partial N_{Ay}}{\partial y} + \frac{\partial N_{Az}}{\partial z} \right) = R_A \tag{7.5-13}$$

where R_A is kg mol A generated/s · m³. Substituting N_A and Fick's law from Table 7.5-1,

$$N_A = -cD_{AB}\frac{dx_A}{dz} + c_A v_M \tag{7.5-14}$$

and writing the equation for all three directions, Eq. (7.5-13) becomes

$$\frac{\partial c_A}{\partial t} + (\nabla \cdot c_A v_M) - (\nabla \cdot cD_{AB}\nabla x_A) = R_A \tag{7.5-15}$$

This is the final general equation

7.5C Special Cases of the Equation of Continuity

1. Equation for constant c and D_{AB}. In diffusion with gases the total pressure P is often constant. Then, since $c = P/RT$, c is constant for constant temperature T. Starting with the general equation (7.5-15) and substituting $\nabla x_A = \nabla c_A/c$, we obtain

$$\frac{\partial c_A}{\partial t} + c_A(\nabla \cdot v_M) + (v_M \cdot \nabla c_A) - D_{AB}\nabla^2 c_A = R_A \tag{7.5-16}$$

2. Equimolar counterdiffusion for gases. For the special case of equimolar counterdiffusion of gases at constant pressure and no reaction, $c = $ constant, $v_M = 0$, $D_{AB} = $ constant, $R_A = 0$, and Eq. (7.5-15) becomes

$$\frac{\partial c_A}{\partial t} = D_{AB}\left(\frac{\partial^2 c_A}{\partial x^2} + \frac{\partial^2 c_A}{\partial y^2} + \frac{\partial^2 c_A}{\partial z^2} \right) \tag{7.1-9}$$

This equation is Eq. (7.1-9) derived previously and this equation is also used for unsteady-state diffusion of a dilute solute A in a solid or a liquid when D_{AB} is constant.

3. *Equation for constant ρ and D_{AB} (liquids).* In dilute liquid solutions the mass density ρ and D_{AB} can often be considered constant. Starting with Eq. (7.5-12) we substitute $n_A = -\rho D_{AB} \nabla w_A + \rho_A v$ from Table 7.5-1 into this equation. Then using the fact that for constant ρ, $\nabla w_A = \nabla \rho_A / \rho$ and also that $(\nabla \cdot v) = 0$, substituting these into the resulting equation, and dividing both sides by M_A, we obtain

$$\frac{\partial c_A}{\partial t} + (v \cdot \nabla c_A) - D_{AB} \nabla^2 c_A = R_A \qquad (7.5\text{-}17)$$

7.5D Special Cases of the General Diffusion Equation at Steady State

1. Introduction and physical conditions at the boundaries. The general equation for diffusion and convection of a binary mixture in one direction with no chemical reaction has been given previously.

$$N_A = -cD_{AB} \frac{dx_A}{dz} + \frac{c_A}{c}(N_A + N_B) \qquad (6.2\text{-}14)$$

To integrate this equation at steady state it is necessary to specify the boundary conditions at z_1 and at z_2. Often in many mass-transfer problems the molar ratio N_A/N_B is determined by the physical conditions occurring at the two boundaries.

As an example, one boundary of the diffusion path may be impermeable to species B because B is insoluble in the phase at this boundary. Diffusion of ammonia (A) and nitrogen (B) through a gas phase to a water phase at the boundary is such a case since nitrogen is essentially insoluble in water. Hence, $N_B = 0$, since at steady state N_B must have the same value at all points in the path of $z_2 - z_1$. In some cases, a heat balance in the adjacent phase at the boundary can determine the flux ratios. For example, if component A condenses at a boundary and releases its latent heat to component B, which vaporizes and diffuses back, the ratios of the latent heats determine the flux ratio.

In another example, the boundary concentration can be fixed by having a large volume of a phase flowing rapidly by with a given concentration x_{A1}. In some cases the concentration x_{A1} may be set by an equilibrium condition, whereby x_{A1} is in equilibrium with some fixed composition at the boundary. Chemical reactions can also influence the rates of diffusion and the boundary conditions.

2. Equimolar counterdiffusion. For the special case of equimolar counterdiffusion where $N_A = -N_B$, Eq. (6.2-14) becomes, as shown previously, for steady state and constant c,

$$N_A = J_A^* = -cD_{AB} \frac{dx_A}{dz} = \frac{D_{AB}(c_{A1} - c_{A2})}{z_2 - z_1} \qquad (7.5\text{-}18)$$

3. Diffusion of A through stagnant, nondiffusing B. For gas A diffusing through stagnant nondiffusing gas B, $N_B = 0$, and integration of Eq. (6.2-14) gives Eq. (6.2-22).

$$N_A = \frac{D_{AB} P}{RT(z_2 - z_1) p_{BM}}(p_{A1} - p_{A2}) \qquad (6.2\text{-}22)$$

Several other more complicated cases of integration of Eq. (6.2-14) are considered next.

4. Diffusion and chemical reaction at a boundary. Often in catalytic reactions where A and B are diffusing to and from a catalyst surface, the relation between the fluxes N_A and

N_B at steady state is controlled by the stoichiometry of a reaction at a boundary. An example is gas A diffusing from the bulk gas phase to the catalyst surface, where it reacts instantaneously and irreversibly in a heterogeneous reaction as follows:

$$A \rightarrow 2B \qquad (7.5\text{-}19)$$

Gas B then diffuses back, as is shown in Fig. 7.5-2.

At steady state 1 mol of A diffuses to the catalyst for every 2 mol of B diffusing away, or $N_B = -2N_A$. The negative sign indicates that the fluxes are in opposite directions. Rewriting Eq. (6.2-14) in terms of mole fractions,

$$N_A = -cD_{AB} \frac{dx_A}{dz} + x_A(N_A + N_B) \qquad (7.5\text{-}20)$$

Next, substituting $N_B = -2N_A$ into Eq. (7.5-20),

$$N_A = -cD_{AB} \frac{dx_A}{dz} + x_A(N_A - 2N_A) \qquad (7.5\text{-}21)$$

Rearranging and integrating with constant c ($P =$ constant), we obtain the following:

1. *Instantaneous surface reaction :*

$$N_A \int_{z_1 = 0}^{z_2 = \delta} dz = -cD_{AB} \int_{x_{A1}}^{x_{A2}} \frac{dx_A}{1 + x_A} \qquad (7.5\text{-}22)$$

$$N_A = \frac{cD_{AB}}{\delta} \ln \frac{1 + x_{A1}}{1 + x_{A2}} \qquad (7.5\text{-}23)$$

Since the reaction is instantaneous, $x_{A2} = 0$, because no A can exist next to the catalyst surface. Equation (7.5-23) describes the overall rate of the process of diffusion plus instantaneous chemical reaction.

2. *Slow surface reaction.* If the heterogeneous reaction at the surface is not instantaneous but slow for the reaction $A \rightarrow 2B$, and the reaction is first order,

$$N_{Az = \delta} = k_1' c_A = k_1' c x_A \qquad (7.5\text{-}24)$$

where k_1' is the first-order heterogeneous reaction velocity constant in m/s. Equation (7.5-23) still holds for this case, but the boundary condition x_{A2} at $z = \delta$ is obtained by

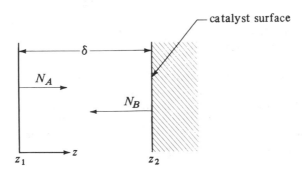

FIGURE 7.5-2. *Diffusion of A and heterogeneous reaction at a surface.*

solving for x_A in Eq. (7.5-24),

$$x_A = x_{A2} = \frac{N_{Az=\delta}}{k_1'c} = \frac{N_A}{k_1'c} \tag{7.5-25}$$

For steady state, $N_{Az=\delta} = N_A$. Substituting Eq. (7.5-25) into (7.5-23),

$$N_A = \frac{cD_{AB}}{\delta} \ln \frac{1 + x_{A1}}{1 + N_A/k_1'c} \tag{7.5-26}$$

The rate in Eq. (7.5-26) is less than in Eq. (7.5-23), since the denominator in the latter equation is $1 + x_{A2} = 1 + 0$ and in the former is $1 + N_A/k_1'c$.

EXAMPLE 7.5-2. Diffusion and Chemical Reaction at a Boundary

Pure gas A diffuses from point 1 at a partial pressure of 101.32 kPa to point 2 a distance 2.00 mm away. At point 2 it undergoes a chemical reaction at the catalyst surface and $A \rightarrow 2B$. Component B diffuses back at steady state. The total pressure is $P = 101.32$ kPa. The temperature is 300 K and $D_{AB} = 0.15 \times 10^{-4}$ m²·s.
  (a) For instantaneous rate of reaction, calculate x_{A2} and N_A.
  (b) For a slow reaction where $k_1' = 5.63 \times 10^{-3}$ m/s, calculate x_{A2} and N_A.

Solution: For part (a), $p_{A2} = x_{A2} = 0$ since no A can exist next to the catalyst surface. Since $N_B = -2N_A$, Eq. (7.5-23) will be used as follows: $\delta = 2.00 \times 10^{-3}$ m, $T = 300$ K, $c = P/RT = 101.32 \times 10^3/(8314 \times 300) = 4.062 \times 10^{-2}$ kg mol/m³, $x_{A1} = p_{A1}/P = 101.32 \times 10^3/101.32 \times 10^3 = 1.00$.

$$N_A = \frac{cD_{AB}}{\delta} \ln \frac{1 + x_{A1}}{1 + x_{A2}} = \frac{(4.062 \times 10^{-2})(0.15 \times 10^{-4})}{2.00 \times 10^{-3}} \ln \frac{1 + 1.00}{1 + 0}$$

$$= 2.112 \times 10^{-4} \text{ kg mol } A/\text{s} \cdot \text{m}^2$$

For part (b), from Eq. (7.5-25), $x_{A2} = N_A/k_1'c = N_A/(5.63 \times 10^{-3} \times 4.062 \times 10^{-2})$. Substituting into Eq. (7.5-26),

$$N_A = \frac{(4.062 \times 10^{-2})(0.15 \times 10^{-4})}{2.00 \times 10^{-3}} \ln$$

$$\frac{1 + 1.00}{1 + N_A/(5.63 \times 10^{-3} \times 4.062 \times 10^{-2})}$$

Solving by trial and error, $N_A = 1.004 \times 10^{-4}$ kg mol $A/\text{s} \cdot \text{m}^2$. Then, $x_{A2} = (1.004 \times 10^{-4})/(5.63 \times 10^{-3} \times 4.062 \times 10^{-2}) = 0.4390$.

Even though in part (a) of Example 7.5-2 the rate of reaction is instantaneous, the flux N_A is diffusion controlled. As the reaction rate slows, the flux N_A is decreased also.

5. Diffusion and homogeneous reaction in a phase. Equation (7.5-23) was derived for the case of chemical reaction of A at the boundary on a catalyst surface. In some cases component A undergoes an irreversible chemical reaction in the homogeneous phase B while diffusing as follows, $A \rightarrow C$. Assume that component A is very dilute in phase B, which can be a gas or a liquid. Then at steady state the equation for diffusion of A is as follows where the bulk-flow term is dropped.

$$N_{Az} = -D_{AB} \frac{dc_A}{dz} + 0 \tag{7.5-27}$$

FIGURE 7.5-3. *Homogeneous chemical reaction and diffusion in a fluid.*

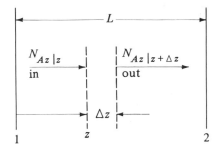

Writing a material balance on A shown in Fig. 7.5-3 for the Δz element for steady state,

$$\begin{pmatrix}\text{rate of} \\ A \text{ in}\end{pmatrix} + \begin{pmatrix}\text{rate of} \\ \text{generation of } A\end{pmatrix} = \begin{pmatrix}\text{rate of} \\ A \text{ out}\end{pmatrix} + \begin{pmatrix}\text{rate of} \\ \text{accumulation of } A\end{pmatrix} \quad (7.5\text{-}28)$$

The first-order reaction rate of A per m³ volume is

$$\text{rate of generation} = -k'c_A \quad (7.5\text{-}29)$$

where k' is the reaction velocity constant in s⁻¹. Substituting into Eq. (7.5-28) for a cross-sectional area of 1 m² with the rate of accumulation being 0 at steady state,

$$N_{Az|z}(1) - k'c_A(1)(\Delta z) = N_{Az|z+\Delta z}(1) + 0 \quad (7.5\text{-}30)$$

Next we divide through by Δz and let Δz approach zero.

$$-\frac{dN_{Az}}{dz} = k'c_A \quad (7.5\text{-}31)$$

Substituting Eq. (7.5-27) into (7.5-31),

$$\frac{d^2c_A}{dz^2} = \frac{k'}{D_{AB}} c_A \quad (7.5\text{-}32)$$

The boundary conditions are $c_A = c_{A1}$ for $z = 0$ and $c_A = c_{A2}$ for $z = L$. Solving,

$$c_A = \frac{c_{A2} \sinh\left(\sqrt{\dfrac{k'}{D_{AB}}}\, z\right) + c_{A1} \sinh\left[\sqrt{\dfrac{k'}{D_{AB}}}\,(L - z)\right]}{\sinh\left(\sqrt{\dfrac{k'}{D_{AB}}}\, L\right)} \quad (7.5\text{-}33)$$

This equation can be used at steady state to calculate c_A at any z and can be used for reaction in gases, liquids, or even solids, where the solute A is dilute.

As an alternative derivation of Eq. (7.5-32), we can use Eq. (7.5-17) for constant ρ and D_{AB}.

$$\frac{\partial c_A}{\partial t} + (\mathbf{v} \cdot \nabla c_A) - D_{AB}\nabla^2 c_A = R_A \quad (7.5\text{-}17)$$

We set the first term $\partial c_A/\partial t = 0$ for steady state. Since we are assuming dilute solutions and neglecting the bulk flow term, $\mathbf{v} = 0$, making the second term in Eq. (7.5-17) zero. For a first-order reaction of A where A disappears, $R_A = -k'c_A$ kg mol A generated/s · m³.

Writing the diffusion term $-D_{AB}\nabla^2 c_A$ for only the z direction, we obtain

$$D_{AB}\frac{d^2c_A}{dz^2} = k'c_A \tag{7.5-34}$$

which is, of course, identical to Eq. (7.5-32).

7.5E Unsteady-State Diffusion and Reaction in a Semiinfinite Medium

Here we consider a case where dilute A is absorbed at the surface of a solid or stagnant fluid phase and then unsteady-state diffusion and reaction occur in the phase. The fluid or solid phase B is considered semiinfinite. At the surface where $z = 0$, the concentration of c_A is kept constant at c_{A0}. The dilute solute A reacts by a first-order mechanism

$$A + B \rightarrow C \tag{7.5-35}$$

and the rate of generation is $-k'c_A$. The same diagram as in Fig. 7.5-3 holds. Using Eq. (7.5-30) but substituting $(\partial c_A/\partial t)(\Delta z)(1)$ for the rate of accumulation,

$$N_{Az|z}(1) - k'c_A(1)(\Delta z) = N_{Az|z+\Delta z}(1) + \left(\frac{\partial c_A}{\partial t}\right)(\Delta z)(1) \tag{7.5-36}$$

This becomes

$$\frac{\partial c_A}{\partial t} = D_{AB}\frac{\partial^2 c_A}{\partial z^2} - k'c_A \tag{7.5-37}$$

The initial and boundary conditions are

$$
\begin{array}{llll}
t = 0, & c_A = 0 & \text{for } z > 0 & \\
z = 0, & c_A = c_{A0} & \text{for } t > 0 & \tag{7.5-38} \\
z = \infty, & c_A = 0 & \text{for } t > 0 &
\end{array}
$$

The solution by Danckwerts (D1) is

$$\frac{c_A}{c_{A0}} = \tfrac{1}{2}\exp(-z\sqrt{k'/D_{AB}}) \cdot \text{erfc}\left(\frac{z}{2\sqrt{tD_{AB}}} - \sqrt{k't}\right)$$

$$+ \tfrac{1}{2}\exp(z\sqrt{k'/D_{AB}}) \cdot \text{erfc}\left(\frac{z}{2\sqrt{tD_{AB}}} + \sqrt{k't}\right) \tag{7.5-39}$$

The total amount Q of A absorbed up to time t is

$$Q = c_{A0}\sqrt{D_{AB}/k'}[(k't + \tfrac{1}{2})\text{erf}\sqrt{k't} + \sqrt{k't/\pi}e^{-k't}] \tag{7.5-40}$$

where Q is kg mol A absorbed/m^2. Many actual cases are approximated by this case. The equation is useful where absorption occurs at the surface of a stagnant fluid or a solid and unsteady-state diffusion and reaction occurs in the solid or fluid. The results can be used to measure the diffusivity of a gas in a solution, to determine reaction rate constants k' of dissolved gases, and to determine solubilities of gas in liquids with which they react. Details are given elsewhere (D3).

EXAMPLE 7.5-3. Reaction and Unsteady-State Diffusion
Pure CO_2 gas at 101.32 kPa pressure is absorbed into a dilute alkaline buffer solution containing a catalyst. The dilute, absorbed solute CO_2 undergoes a first-order reaction with $k' = 35$ s^{-1} and $D_{AB} = 1.5 \times 10^{-9}$

m^2/s. The solubility of CO_2 is 2.961×10^{-7} kg mol/m$^3 \cdot$ Pa (D3). The surface is exposed to the gas for 0.010 s. Calculate the kg mol CO_2 absorbed/m^2 surface.

Solution: For use in Eq. (7.5-40), $k't = 35(0.01) = 0.350$. Also, $c_{A0} = 2.961 \times 10^{-7}$ (kg mol/m$^3 \cdot$ Pa)(101.32 $\times 10^3$ Pa) $= 3.00 \times 10^{-2}$ kg mol SO_2/m^3.

$$Q = (3.00 \times 10^{-2})\sqrt{1.5 \times 10^{-9}/35}[(0.35 + \tfrac{1}{2})\text{erf}\sqrt{0.35} + \sqrt{0.35/\pi}e^{-0.35}]$$

$$= 1.458 \times 10^{-7} \text{ kg mol } CO_2/m^2$$

7.5F Multicomponent Diffusion of Gases

The equations derived in this chapter have been for a binary system of A and B, which is probably the most important and most useful one. However, multicomponent diffusion sometimes occurs where three or more components A, B, C, ..., are present. The simplest case is for diffusion of A in a gas through a stagnant nondiffusing mixture of B, C, D, ..., at constant total pressure. Hence, $N_B = 0$, $N_c = 0$, ... The final equation derived using the Stefan–Maxwell method (G1) for steady-state diffusion is

$$N_A = \frac{D_{Am}P}{RT(z_2 - z_1)p_{iM}}(p_{A1} - p_{A2}) \qquad (7.5\text{-}41)$$

where p_{iM} is the log mean of $p_{i1} = P - p_{A1}$ and $p_{i2} = P - p_{A2}$. Also,

$$D_{Am} = \frac{1}{x'_B/D_{AB} + x'_C/D_{AC} + \cdots} \qquad (7.5\text{-}42)$$

where x'_B = mol B/mol inerts $= x_B/(1 - x_A)$, $x'_C = x_C/(1 - x_A)$, ...

EXAMPLE 7.5-4. Diffusion of A Through Nondiffusing B and C
At 298 K and 1 atm total pressure, methane (A) is diffusing at steady state through nondiffusing argon (B) and helium (C). At $z_1 = 0$, the partial pressures in atm are $p_{A1} = 0.4$, $p_{B1} = 0.4$, $p_{C1} = 0.2$ and at $z_2 = 0.005$ m, $p_{A2} = 0.1$, $p_{B2} = 0.6$, and $p_{C2} = 0.3$. The binary diffusivities from Table 6.2-1 are $D_{AB} = 2.02 \times 10^{-5}$ m^2/s, $D_{AC} = 6.75 \times 10^{-5}$ m^2/s, and $D_{BC} = 7.29 \times 10^{-5}$ m^2/s. Calculate N_A.

Solution: At point 1, $x'_B = x_B/(1 - x_A) = 0.4/(1 - 0.4) = 0.667$. At point 2, $x'_B = 0.6/(1 - 0.1) = 0.667$. The value of x'_B is constant throughout the path. Also, $x'_C = x_C/(1 - x_A) = 0.2/(1 - 0.4) = 0.333$.
Substituting into Eq. (7.5-42),

$$D_{Am} = \frac{1}{x'_B/D_{AB} + x'_C/D_{AC}} = \frac{1}{0.667/2.02 \times 10^{-5} + 0.333/6.75 \times 10^{-5}}$$

$$= 2.635 \times 10^{-5} \text{ m}^2/\text{s}$$

For calculating p_{iM}, $p_{i1} = P - p_{A1} = 1.0 - 0.4 = 0.6$ atm, $p_{i2} = P - p_{A2} = 1.0 - 0.1 = 0.90$. Then,

$$p_{iM} = \frac{p_{i2} - p_{i1}}{\ln(p_{i2}/p_{i1})} = \frac{0.90 - 0.60}{\ln(0.90/0.60)} = 0.740 \text{ atm} = 7.496 \times 10^4 \text{ Pa}$$

$$p_{A1} = 0.4(1.01325 \times 10^5) = 4.053 \times 10^4 \text{ Pa}$$

$$p_{A2} = 0.1(1.01325 \times 10^5) = 1.013 \times 10^4 \text{ Pa}$$

Substituting into Eq. (7.5-41),

$$N_A = \frac{D_{Am} P}{RT(z_2 - z_1) p_{iM}} (p_{A1} - p_{A2})$$

$$= \frac{(2.635 \times 10^{-5})(1.01325 \times 10^5)(4.053 - 1.013)(10^4)}{(8314)(298)(0.005 - 0)(7.496 \times 10^4)}$$

$$= 8.74 \times 10^{-5} \text{ kg mol } A/s \cdot m^2$$

Using atm pressure units,

$$N_A = \frac{D_{Am} P}{RT(z_2 - z_1) p_{iM}} (p_{A1} - p_{A2}) = \frac{(2.635 \times 10^{-5})(1.0)(0.4 - 0.1)}{(82.06 \times 10^{-3})(298)(0.005 - 0)(0.740)}$$

$$= 8.74 \times 10^{-5} \text{ kg mol } A/s \cdot m^2$$

A number of analytical solutions have been obtained for other cases such as for equimolar diffusion of three components, diffusion of components A and B through stagnant C, and the general case of two or more components diffusing in a multicomponent mixture. These are discussed in detail with examples by Geankoplis (G1) and the reader is referred there for further details.

7.6 DIFFUSION OF GASES IN POROUS SOLIDS AND CAPILLARIES

7.6A Introduction

In Section 6.5C diffusion in porous solids that depends on structure was discussed for liquids and for gases. For gases it was assumed that the pores were very large and Fickian-type diffusion occured. However, often the pores are small in diameter and the mechanism of diffusion is basically changed.

Diffusion of gases in small pores occurs often in heterogeneous catalysis where gases diffuse through very small pores to react on the surface of the catalyst. In freeze drying of foods such as turkey meat, gaseous H_2O diffuses through very fine pores of the porous structure.

Since the pores or capillaries of porous solids are often small, the diffusion of gases may depend upon the diameter of the pores. We first define a mean free path λ, which is the average distance a gas molecule travels before it collides with another gas molecule.

$$\lambda = \frac{3.2\mu}{P} \sqrt{\frac{RT}{2\pi M}} \tag{7.6-1}$$

where λ is in m, μ is viscosity in Pa·s, P is pressure in N/m^2, T is temperature in K, M = molecular weight in kg/kg mol, and R = 8.3143×10^3 N·m/kg mol·K. Note that low pressures give large values of λ. For liquids, since λ is so small, diffusion follows Fick's law.

In the next sections we shall consider what happens to the basic mechanisms of diffusion in gases as the relative value of the mean free path compared to the pore diameter varies. The total pressure P in the system will be constant, but partial pressures of A and B may be different.

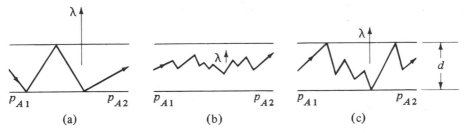

FIGURE 7.6-1. Types of diffusion of gases in small capillary tubes: (a) Knudsen gas diffusion, (b) molecular or Fick's gas diffusion, (c) transition gas diffusion.

7.6B Knudsen Diffusion of Gases

In Fig. 7.6-1a a gas molecule A at partial pressure p_{A1} at the entrance to a capillary is diffusing through the capillary having a diameter of d m. The total pressure P is constant throughout. The mean free path λ is large compared to the diameter d. As a result, the molecule collides with the wall and molecule–wall collisions are important. This type is called *Knudsen diffusion*.

The Knudsen diffusivity is independent of pressure P and is calculated from

$$D_{KA} = \tfrac{2}{3}\bar{r}\bar{v}_A \tag{7.6-2}$$

where D_{KA} is diffusivity in m^2/s, $\bar{r}$ is average pore radius in m, and $\bar{v}_A$ is the average molecular velocity for component A in m/s. Using the kinetic theory of gases to evaluate $\bar{v}_A$ the final equation for D_{KA} is

$$D_{KA} = 97.0\bar{r}\left(\frac{T}{M_A}\right)^{1/2} \tag{7.6-3}$$

where M_A is molecular weight of A in kg/kg mol and T is temperature in K.

EXAMPLE 7.6-1. Knudsen Diffusivity of Hydrogen

A $H_2(A)$–$C_2H_6(B)$ gas mixture is diffusing in a pore of a nickel catalyst used for hydrogenation at 1.01325×10^5 Pa pressure and 373 K. The pore radius is 60 Å (angstrom). Calculate the Knudsen diffusivity D_{KA} of H_2.

Solution: Substituting into Eq. (7.6-3) for $\bar{r} = 6.0 \times 10^{-9}$ m, $M_A = 2.016$, and $T = 373$ K,

$$D_{KA} = 97.0\bar{r}\left(\frac{T}{M_A}\right)^{1/2} = 97.0(6.0 \times 10^{-9})\left(\frac{373}{2.016}\right)^{1/2}$$

$$= 7.92 \times 10^{-6} \text{ m}^2/s$$

The flux equation for Knudsen diffusion in a pore is

$$N_A = -D_{KA}\frac{dc_A}{dz} = -\frac{D_{KA}}{RT}\frac{dp_A}{dz} \tag{7.6-4}$$

Integrating between $z_1 = 0, p_A = p_{A1}$ and $z_2 = L, p_A = p_{A2}$,

$$N_A = \frac{D_{KA}P}{RTL}(x_{A1} - x_{A2}) = \frac{D_{KA}}{RTL}(p_{A1} - p_{A2}) \tag{7.6-5}$$

The diffusion of A for Knudsen diffusion is completely independent of B, since A collides with the walls of the pore and not with B. A similar equation can be written for component B.

When the Knudsen number N_{Kn} defined as

$$N_{Kn} = \frac{\lambda}{2\bar{r}} \tag{7.6-6}$$

is $\geq 10/1$, the diffusion is primarily Knudsen and Eq. (7.6-5) predicts the flux to within about a 10% error. As N_{Kn} gets larger, this error decreases, since the diffusion approaches the Knudsen type.

7.6C Molecular Diffusion of Gases

As shown in Fig. 7.6-1b when the mean free path λ is small compared to the pore diameter d or where $N_{Kn} \leq 1/100$, molecule–molecule collisions predominate and molecule–wall collisions are few. Ordinary molecular or Fickian diffusion holds and Fick's law predicts the diffusion to within about 10%. The error diminishes as N_{Kn} gets smaller since the diffusion approaches more closely the Fickian type.

The equation for molecular diffusion given in previous sections is

$$N_A = -\frac{D_{AB}P}{RT}\frac{dx_A}{dz} + x_A(N_A + N_B) \tag{7.6-7}$$

A flux ratio factor α can be defined as

$$\alpha = 1 + \frac{N_B}{N_A} \tag{7.6-8}$$

Combining Eqs. (7.6-7) and (7.6-8) and integrating for a path length of L cm,

$$N_A = \frac{D_{AB}P}{\alpha RTL} \ln \frac{1 - \alpha x_{A2}}{1 - \alpha x_{A1}} \tag{7.6-9}$$

If the diffusion is equimolar, $N_A = -N_B$ and Eq. (7.6-7) becomes Fick's law. The molecular diffusivity D_{AB} is inversely proportional to the total pressure P.

7.6D Transition-Region Diffusion of Gases

As shown in Fig. 7.6-1c, when the mean free path λ and pore diameter are intermediate in size between the two limits given for Knudsen and molecular diffusion, transition-type diffusion occurs where molecule–molecule and molecule–wall collisions are important in diffusion.

The transition-region diffusion equation can be derived by adding the momentum loss due to molecule–wall collisions in Eq. (7.6-4) and that due to molecule–molecule collisions in Eq. (7.6-7) on a slice of capillary. No chemical reactions are occurring. The final differential equation is (G1)

$$N_A = -\frac{D_{NA}P}{RT}\frac{dx_A}{dz} \tag{7.6-10}$$

where

$$D_{NA} = \frac{1}{(1 - \alpha x_A)/D_{AB} + 1/D_{KA}} \tag{7.6-11}$$

This transition region diffusivity D_{NA} depends slightly on concentration x_A.

Integrating Eq. (7.6-10),

$$N_A = \frac{D_{AB}\,P}{\alpha RTL}\ln\frac{1 - \alpha x_{A2} + D_{AB}/D_{KA}}{1 - \alpha x_{A1} + D_{AB}/D_{KA}} \qquad (7.6\text{-}12)$$

This equation has been shown experimentally to be valid over the entire transition region (R1). It reduces to the Knudsen equation at low pressures and to the molecular diffusion equation at high pressures. An equation similar to Eq. (7.6-12) can also be written for component B.

The term D_{AB}/D_{KA} is proportional to $1/P$. Hence, as the total pressure P increases, the term D_{AB}/D_{KA} becomes very small and N_A in Eq. (7.6-12) becomes independent of total pressure since $D_{AB}P$ is independent of P. At low total pressures Eq. (7.6-12) becomes the Knudsen diffusion equation (7.6-5) and the flux N_A becomes directly proportional to P for constant x_{A1} and x_{A2}.

This is illustrated in Fig. 7.6-2 for a fixed capillary diameter where the flux increases as total pressure increases and then levels off at high pressure. The relative position of the curve depends, of course, on the capillary diameter and the molecular and Knudsen diffusivities. Using only a smaller diameter, D_{KA} would be smaller, and the Knudsen flux line would be parallel to the existing line at low pressures. At high pressures the flux line would asymptotically approach the existing horizontal line since molecular diffusion is independent of capillary diameter.

If A is diffusing in a catalytic pore and reacts at the surface at the end of the pore so that $A \to B$, then at steady state, equimolar counterdiffusion occurs or $N_A = -N_B$. Then from Eq. (7.6-8), $\alpha = 1 - 1 = 0$. The effective diffusivity D_{NA} from Eq. (7.6-11) becomes

$$D'_{NA} = \frac{1}{1/D_{AB} + 1/D_{KA}} \qquad (7.6\text{-}13)$$

The diffusivity is then independent of concentration and is constant. Integration of

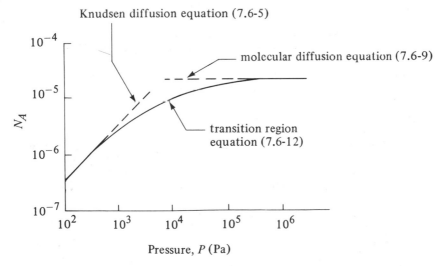

FIGURE 7.6-2. *Effect of total pressure P on the diffusion flux N_A in the transition region.*

(7.6-10) then gives

$$N_A = \frac{D'_{NA}P}{RTL}(x_{A1} - x_{A2}) = \frac{D'_{NA}}{RTL}(p_{A1} - p_{A2}) \tag{7.6-14}$$

This simplified diffusivity D'_{NA} is often used in diffusion in porous catalysts even when equimolar counterdiffusion is not occurring. This greatly simplifies the equations for diffusion and reaction by using this simplified diffusivity.

An alternative simplified diffusivity to use is to use an average value of x_A in Eq. (7.6-11), to give

$$D''_{NA} = \frac{1}{(1 - \alpha x_{A\,av})/D_{AB} + 1/D_{KA}} \tag{7.6-15}$$

where $x_{A\,av} = (x_{A1} + x_{A2})/2$. This diffusivity is more accurate than D'_{NA}. Integration of Eq. (7.6-10) gives

$$N_A = \frac{D''_{NA}P}{RTL}(x_{A1} - x_{A2}) = \frac{D''_{NA}}{RTL}(p_{A1} - p_{A2}) \tag{7.6-16}$$

7.6E   Flux Ratios for Diffusion of Gases in Capillaries

1. Diffusion in open system.   If diffusion in porous solids or channels with no chemical reaction is occurring where the total pressure P remains constant, then for an open binary counterdiffusing system, the ratio of N_A/N_B is constant in all of the three diffusion regimes and is (G1)

$$\frac{N_B}{N_A} = -\sqrt{\frac{M_A}{M_B}} \tag{7.6-17}$$

Hence,

$$\alpha = 1 - \sqrt{\frac{M_A}{M_B}} \tag{7.6-18}$$

In this case, gas flows by the two open ends of the system. However, when chemical reaction occurs, stoichiometry determines the ratio N_B/N_A and not Eq. (7.6-17).

2. Diffusion in closed system.   When molecular diffusion is occurring in a closed system shown in Fig. 6.2-1 at constant total pressure P, equimolar counterdiffusion occurs.

> **EXAMPLE 7.6-2.   Transition-Region Diffusion of He and N_2**
> A gas mixture at a total pressure of 0.10 atm abs and 298 K is composed of N_2 (A) and He (B). The mixture is diffusing through an open capillary 0.010 m long having a diameter of 5×10^{-6} m. The mole fraction of N_2 at one end is $x_{A1} = 0.8$ and at the other end is $x_{A2} = 0.2$. The molecular diffusivity D_{AB} is 6.98×10^{-5} m^2/s at 1 atm, which is an average value by several investigators.
> (a) Calculate the flux N_A at steady state.
> (b) Use the approximate equations (7.6-14) and (7.6-16), for this case.
>
> **Solution:**   The given values are $T = 273 + 25 = 298$ K, $\bar{r} = 5 \times 10^{-6}/2 = 2.5 \times 10^{-6}$ m, $L = 0.01$ m, $P = 0.1(1.01325 \times 10^5) = 1.013 \times 10^4$ Pa, $x_{A1} = 0.8, x_{A2} = 0.2, D_{AB} = 6.98 \times 10^{-5}$ m^2/s at 1 atm. Other values needed are $M_A = 28.02$ kg/kg mol, $M_B = 4.003$.

The molecular diffusivity at 0.1 atm is $D_{AB} = 6.98 \times 10^{-5}/0.1 = 6.98 \times 10^{-4}$ m²/s. Substituting into Eq. (7.6-3) for the Knudsen diffusivity,

$$D_{KA} = 97.0(2.5 \times 10^{-6})\sqrt{298/28.02} = 7.91 \times 10^{-4} \text{ m}^2/\text{s}$$

From Eq. (7.6-17),

$$\frac{N_B}{N_A} = -\sqrt{\frac{M_A}{M_B}} = -\sqrt{\frac{28.02}{4.003}} = -2.645$$

From Eq. (7.6-8),

$$\alpha = 1 + \frac{N_B}{N_A} = 1 - 2.645 = -1.645$$

Substituting into Eq. (7.6-12) for part (a),

$$N_A = \frac{(6.98 \times 10^{-4})(1.013 \times 10^4)}{(-1.645)(8314)(298)(0.01)} \ln \frac{1 + 1.645(0.2) + 6.98/7.91}{1 + 1.645(0.8) + 6.98/7.91}$$

$$= 6.40 \times 10^{-5} \text{ kg mol/s} \cdot \text{m}^2$$

For part (b), the approximate equation (7.6-13) is used.

$$D'_{NA} = \frac{1}{1/D_{AB} + 1/D_{KA}} = \frac{1}{1/6.98 \times 10^{-4} + 1/7.91 \times 10^{-4}}$$

$$= 3.708 \times 10^{-4} \text{ m}^2/\text{s}$$

Substituting into Eq. (7.6-14), the approximate flux is

$$N_A = \frac{D'_{NA} P}{RTL}(x_{A1} - x_{A2}) = \frac{(3.708 \times 10^{-4})(1.013 \times 10^4)}{8314(298)(0.01)}(0.8 - 0.2)$$

$$= 9.10 \times 10^{-5} \text{ kg mol/s} \cdot \text{m}^2$$

Hence, the calculated flux is approximately 40% high when using the approximation of equimolar counterdiffusion ($\alpha = 0$).

The more accurate approximate equation (7.6-15) is used next. The average concentration is $x_{A \text{ av}} = (x_{A1} + x_{A2})/2 = (0.8 + 0.2)/2 = 0.50$.

$$D''_{NA} = \frac{1}{(1 - \alpha x_{A \text{ av}})/D_{AB} + 1/D_{KA}}$$

$$= \frac{1}{(1 + 1.645 \times 0.5)/(6.98 \times 10^{-4}) + 1/(7.91 \times 10^{-4})}$$

$$= 2.581 \times 10^{-4} \text{ m}^2/\text{s}$$

Substituting into Eq. (7.6-16),

$$N_A = \frac{D''_{NA}}{RTL}(x_{A1} - x_{A2})$$

$$= \frac{(2.581 \times 10^{-4})(1.013 \times 10^4)}{8314(298)(0.01)}(0.8 - 0.2)$$

$$= 6.33 \times 10^{-5} \text{ kg mol/s} \cdot \text{m}^2$$

In this case the flux is only -1.1% low.

7.6F Diffusion of Gases in Porous Solids

In actual diffusion in porous solids the pores are not straight and cylindrical but are irregular. Hence, the equations for diffusion in pores must be modified somewhat for actual porous solids. The problem is further complicated by the fact that the pore diameters vary and the Knudsen diffusivity is a function of pore diameter.

As a result of these complications, investigators often measure effective diffusivities $D_{A\,\text{eff}}$ in porous media, where

$$N_A = \frac{D_{A\,\text{eff}}\,P}{RTL}(x_{A1} - x_{A2}) \tag{7.6-19}$$

If a tortuosity factor τ is used to correct the length L in Eq. (7.6-16), and the right-hand side is multiplied by the void fraction ε, Eq. (7.6-16) becomes

$$N_A = \frac{\varepsilon D''_{NA}}{\tau}\frac{P}{RTL}(x_{A1} - x_{A2}) \tag{7.6-20}$$

Comparing Eqs. (7.6-19) and (7.6-20),

$$D_{A\,\text{eff}} = \frac{\varepsilon D''_{NA}}{\tau} \tag{7.6-21}$$

In some cases investigators measure $D_{A\,\text{eff}}$ but use D'_{NA} instead of the more accurate D''_{NA} in Eq. (7.6-21).

Experimental data (C4, S2, S6) show that τ varies from about 1.5 to over 10. A reasonable range for many commercial porous solids is about 2–6 (S2). If the porous solid consists of a bidispersed system of micropores and macropores instead of a monodispersed pore system, the approach above should be modified (C4, S6).

Discussions and references for diffusion in porous inorganic-type solids, organic solids, and freeze-dried foods such as meat and fruit are given elsewhere (S2, S6).

Another type of diffusion that may occur is surface diffusion. When a molecular layer of absorption occurs on the solid, the molecules can migrate on the surface. Details are given elsewhere (S2, S6).

7.7 NUMERICAL METHODS FOR UNSTEADY-STATE MOLECULAR DIFFUSION

7.7A Introduction

Unsteady-state diffusion often occurs in inorganic, organic, and biological solid materials. If the boundary conditions are constant with time, if they are the same on all sides or surfaces of the solid, and if the initial concentration profile is uniform throughout the solid, the methods described in Section 7.1 can be used. However, these conditions are not always fulfilled. Hence, numerical methods must be used.

7.7B Unsteady-State Numerical Methods for Diffusion

1. Derivation for unsteady state for a slab. For unsteady-state diffusion in one direction, Eq. (7.1-9) becomes

$$\frac{\partial c_A}{\partial t} = D_{AB}\frac{\partial^2 c_A}{\partial x^2} \tag{7.7-1}$$

Since this equation is identical mathematically to the unsteady-state heat-conduction Eq. (5.10-10),

$$\frac{\partial T}{\partial t} = \alpha \frac{\partial^2 T}{\partial x^2} \qquad (5.10\text{-}10)$$

identical mathematical methods can be used for solving both diffusion and conduction numerically.

Figure 7.7-1 shows a slab with width Δx centered at point n represented by the shaded area. Making a mole balance of A on this slab at the time t when the rate in − rate out = rate of accumulation in Δt s,

$$\frac{D_{AB} A}{\Delta x}(_t c_{n-1} - _t c_n) - \frac{D_{AB} A}{\Delta x}(_t c_n - _t c_{n+1}) = \frac{(A\,\Delta x)}{\Delta t}(_{t+\Delta t} c_n - _t c_n) \qquad (7.7\text{-}2)$$

where A is cross-sectional area and $_{t+\Delta t} c_n$ is concentration at point n one Δt later. Rearranging,

$$_{t+\Delta t} c_n = \frac{1}{M}\left[_t c_{n+1} + (M-2)_t c_n + _t c_{n-1}\right] \qquad (7.7\text{-}3)$$

where M is a constant.

$$M = \frac{(\Delta x)^2}{D_{AB}\,\Delta t} \qquad (7.7\text{-}4)$$

As in heat conduction, $M \geq 2$.

In using Eq. (7.7-3), the concentration $_{t+\Delta t} c_n$ at position n and the new time $t + \Delta t$ is calculated explicitly from the known three points at t. In this calculation method, starting with the known concentrations at $t = 0$, the calculations proceed directly from one time increment to the next until the final time is reached.

2. *Simplified Schmidt method for a slab.* If the value of $M = 2$, a simplification of Eq. (7.7-3) occurs, giving the Schmidt method.

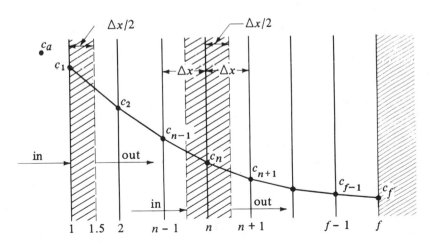

FIGURE 7.7-1. *Unsteady-state diffusion in a slab.*

$$t + \Delta t c_n = \frac{t c_{n-1} + t c_{n+1}}{2} \tag{7.7-5}$$

7.7C Boundary Conditions for Numerical Method for a Slab

1. Convection at a boundary. For the case where convection occurs outside in the fluid and the concentration of the fluid outside is suddenly changed to c_a, we can make a mass balance on the outside $\frac{1}{2}$ slab in Fig. 7.7-1. Following the methods used for heat transfer to derive Eq. (5.4-7), we write rate of mass entering by convection − rate of mass leaving by diffusion = rate of mass accumulation in Δt hours.

$$k_c A({}_t c_a - {}_t c_1) - \frac{D_{AB} A}{\Delta x}({}_t c_1 - {}_t c_2) = \frac{(A \, \Delta x/2)}{\Delta t}({}_{t+\Delta t} c_{1.25} - {}_t c_{1.25}) \tag{7.7-6}$$

where ${}_t c_{1.25}$ is the concentration at the midpoint of the 0.5 Δx outside slab. As an approximation using ${}_t c_1$ for ${}_t c_{1.25}$ and rearranging Eq. (7.7-6),

$$_{t+\Delta t} c_1 = \frac{1}{M} [2N {}_t c_a + [M - (2N + 2)] {}_t c_1 + 2 {}_t c_2] \tag{7.7-7}$$

$$N = \frac{k_c \, \Delta x}{D_{AB}} \tag{7.7-8}$$

where k_c is the convective mass-transfer coefficient in m/s. Again, note that $M \geq (2N + 2)$.

2. Insulated boundary condition. For the insulated boundary at f in Fig. 7.7-1, setting $k_c = 0 \, (N = 0)$ in Eq. (7.7-7), we obtain

$$_{t+\Delta t} c_f = \frac{1}{M} [(M - 2) {}_t c_f + 2 {}_t c_{f-1}] \tag{7.7-9}$$

3. Alternative convective equation at the boundary. Another form of Eq. (7.7-7) to use if N gets too large can be obtained by neglecting the accumulation in the front half-slab of Eq. (7.7-6) to give

$$_{t+\Delta t} c_1 = \frac{N}{N + 1} {}_{t+\Delta t} c_a + \frac{1}{N + 1} {}_{t+\Delta t} c_2 \tag{7.7-10}$$

The value of M is not restricted by the N value in this equation. When a large number of increments in Δx are used, the amount of mass neglected is small compared to the total.

4. Procedure for use of initial boundary concentration. For the first time increment we should use an average value for ${}_t c_a$ of $(c_a + {}_0 c_1)/2$, where ${}_0 c_1$ is the initial concentration at point 1. For succeeding times, the full value of c_a should be used. This special procedure for the value of c_a increases the accuracy of the numerical method, especially after a few time intervals.

In Section 5.4B for heat-transfer numerical methods, a detailed discussion is given on the best value of M to use in Eq. (7.7-3). The most accurate results are obtained for $M = 4$.

5. Boundary conditions with distribution coefficient. Equations for boundary conditions in Eqs. (7.7-7) and (7.7-10) were derived for the distribution coefficient K given in Eq.

(7.7-7) being 1.0. When K is not 1.0, as in the boundary conditions for steady state, Kk_c should be substituted for k_c in Eq. (7.7-8) to become as follows. (See also Sections 6.6B and 7.1C.)

$$N = \frac{Kk_c \, \Delta x}{D_{AB}} \tag{7.7-11}$$

Also, in Eqs. (7.7-7) and (7.7-10), the term c_a/K should be substituted for c_a.

Other cases such as for diffusion between dissimilar slabs in series, resistance between slabs in series, and so on, are covered in detail elsewhere (G1), with actual numerical examples being given. Also, in reference (G1) the implicit numerical method is discussed.

EXAMPLE 7.7-1. Numerical Solution for Unsteady-State Diffusion with a Distribution Coefficient

A slab of material 0.004 m thick has an initial concentration profile of solute A as follows, where x is distance in m from the exposed surface:

$x(m)$	Concentration (kg mol A/m^3)	Position, n
0	1.0×10^{-3}	1 (exposed)
0.001	1.25×10^{-3}	2
0.002	1.5×10^{-3}	3
0.003	1.75×10^{-3}	4
0.004	2.0×10^{-3}	5 (insulated)

The diffusivity $D_{AB} = 1.0 \times 10^{-9}$ m²/s. Suddenly, the top surface is exposed to a fluid having a constant concentration $c_a = 6 \times 10^{-3}$ kg mol A/m^3. The distribution coefficient $K = c_a/c_n = 1.50$. The rear surface is insulated and unsteady-state diffusion is occurring only in the x direction. Calculate the concentration profile after 2500 s. The convective mass-transfer coefficient k_c can be assumed as infinite. Use $\Delta x = 0.001$ m and $M = 2.0$.

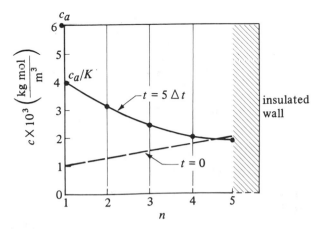

FIGURE 7.7-2. *Concentrations for numerical method for unsteady-state diffusion, Example 7.7-1.*

Solution: Figure 7.7-2 shows the initial concentration profile for four slices and $c_a = 6 \times 10^{-3}$. Since $M = 2$, substituting into Eq. (7.7-4) with $\Delta x = 0.001$ m, and solving for Δt,

$$M = 2 = \frac{(\Delta x)^2}{D_{AB} \, \Delta t} = \frac{(0.001)^2}{(1 \times 10^{-9})(\Delta t)}$$

$$\Delta t = 500 \text{ s}$$

Hence, 2500 s/(500 s/increment) or five time increments are needed.

For the front surface where $n = 1$, the concentration to use for the first time increment, as stated previously, is

$$_1c_a = \frac{c_a/K + _0c_1}{2} = {}_1c_1 \qquad (n = 1) \tag{7.7-12}$$

where $_0c_1$ is the initial concentration at $n = 1$. For the remaining time increments,

$$c_1 = \frac{c_a}{K} \qquad (n = 1) \tag{7.7-13}$$

To calculate the concentrations for all time increments for slabs $n = 2, 3, 4$, using Eq. (7.7-5) for $M = 2$,

$$_{t+\Delta t}c_n = \frac{_tc_{n-1} + {}_tc_{n+1}}{2} \qquad (n = 2, 3, 4) \tag{7.7-14}$$

For the insulated end at $n = 5$, substituting $M = 2$ and $f = n = 5$ into Eq. (7.7-9),

$$_{t+\Delta t}c_5 = \frac{(2 - 2)_tc_5 + 2_tc_4}{2} = {}_tc_4 \qquad (n = 5) \tag{7.7-15}$$

For 1 Δt or $t + 1$ Δt, the first time increment, calculating the concentration for $n = 1$ by Eq. (7.7-12),

$$_{t+\Delta t}c_1 = \frac{c_a/K + {}_0c_1}{2} = \frac{6 \times 10^{-3}/1.5 + 1 \times 10^{-3}}{2} = 2.5 \times 10^{-3}$$

For $n = 2, 3$, and 4, using Eq. (7.7-14),

$$_{t+\Delta t}c_2 = \frac{_tc_{n-1} + {}_tc_{n+1}}{2} = \frac{_tc_1 + {}_tc_3}{2} = \frac{2.5 \times 10^{-3} + 1.5 \times 10^{-3}}{2} = 2 \times 10^{-3}$$

$$_{t+\Delta t}c_3 = \frac{_tc_2 + {}_tc_4}{2} = \frac{1.25 \times 10^{-3} + 1.75 \times 10^{-3}}{2} = 1.5 \times 10^{-3}$$

$$_{t+\Delta t}c_4 = \frac{_tc_3 + {}_tc_5}{2} = \frac{1.5 \times 10^{-3} + 2 \times 10^{-3}}{2} = 1.75 \times 10^{-3}$$

For $n = 5$, using Eq. (7.7-15),

$$_{t+\Delta t}c_5 = {}_tc_4 = 1.75 \times 10^{-3}$$

For 2 Δt using Eq. (7.7-13) for $n = 1$, using Eq. (7.7-14) for $n = 2$–4, and using Eq. (7.7-15) for $n = 5$,

$$_{t+2 \, \Delta t}c_1 = \frac{c_a}{K} = \frac{6 \times 10^{-3}}{1.5} = 4.0 \times 10^{-3} \qquad \text{(constant for rest of time)}$$

$$_{t+2\,\Delta t}c_2 = \frac{_{t+\Delta t}c_1 + {}_{t+\Delta t}c_3}{2} = \frac{4 \times 10^{-3} + 1.5 \times 10^{-3}}{2} = 2.75 \times 10^{-3}$$

$$_{t+2\,\Delta t}c_3 = \frac{_{t+\Delta t}c_2 + {}_{t+\Delta t}c_4}{2} = \frac{2 \times 10^{-3} + 1.75 \times 10^{-3}}{2} = 1.875 \times 10^{-3}$$

$$_{t+2\,\Delta t}c_4 = \frac{_{t+\Delta t}c_3 + {}_{t+\Delta t}c_5}{2} = \frac{1.5 \times 10^{-3} + 1.75 \times 10^{-3}}{2} = 1.625 \times 10^{-3}$$

$$_{t+2\,\Delta t}c_5 = {}_{t+\Delta t}c_4 = 1.75 \times 10^{-3}$$

For $3\,\Delta t$,

$$_{t+3\,\Delta t}c_1 = 4 \times 10^{-3}$$

$$_{t+3\,\Delta t}c_2 = \frac{4 \times 10^{-3} + 1.875 \times 10^{-3}}{2} = 2.938 \times 10^{-3}$$

$$_{t+3\,\Delta t}c_3 = \frac{2.75 \times 10^{-3} + 1.625 \times 10^{-3}}{2} = 2.188 \times 10^{-3}$$

$$_{t+3\,\Delta t}c_4 = \frac{1.875 \times 10^{-3} + 1.75 \times 10^{-3}}{2} = 1.813 \times 10^{-3}$$

$$_{t+3\,\Delta t}c_5 = 1.625 \times 10^{-3}$$

For $4\,\Delta t$,

$$_{t+4\,\Delta t}c_1 = 4 \times 10^{-3}$$

$$_{t+4\,\Delta t}c_2 = \frac{4 \times 10^{-3} + 2.188 \times 10^{-3}}{2} = 3.094 \times 10^{-3}$$

$$_{t+4\,\Delta t}c_3 = \frac{2.938 \times 10^{-3} + 1.813 \times 10^{-3}}{2} = 2.376 \times 10^{-3}$$

$$_{t+4\,\Delta t}c_4 = \frac{2.188 \times 10^{-3} + 1.625 \times 10^{-3}}{2} = 1.906 \times 10^{-3}$$

$$_{t+4\,\Delta t}c_5 = 1.813 \times 10^{-3}$$

For $5\,\Delta t$,

$$_{t+5\,\Delta t}c_1 = 4 \times 10^{-3}$$

$$_{t+5\,\Delta t}c_2 = \frac{4 \times 10^{-3} + 2.376 \times 10^{-3}}{2} = 3.188 \times 10^{-3}$$

$$_{t+5\,\Delta t}c_3 = \frac{3.094 \times 10^{-3} + 1.906 \times 10^{-3}}{2} = 2.500 \times 10^{-3}$$

$$_{t+5\,\Delta t}c_4 = \frac{2.376 \times 10^{-3} + 1.813 \times 10^{-3}}{2} = 2.095 \times 10^{-3}$$

$$_{t+5\,\Delta t}c_5 = 1.906 \times 10^{-3}$$

The final concentration profile is plotted in Fig. 7.7-2. To increase the accuracy, more slab increments and more time increments are needed. This type of calculation is suitable for a digital computer.

7.8 DIMENSIONAL ANALYSIS IN MASS TRANSFER

7.8A Introduction

The use of dimensional analysis enables us to predict the various dimensional groups which are very helpful in correlating experimental mass-transfer data. As we saw in fluid flow and in heat transfer, the Reynolds number, the Prandtl number, the Grashof number, and the Nusselt number were often used in correlating experimental data. The Buckingham theorem discussed in Section 3.11 and Section 4.14 states that the functional relationship among q quantities or variables whose units may be given in terms of u fundamental units or dimensions may be written as $(q - u)$ dimensionless groups.

7.8B Dimensional Analysis for Convective Mass Transfer

We consider a case of convective mass transfer where a fluid is flowing by forced convection in a pipe and mass transfer is occurring from the wall to the fluid. The fluid flows at a velocity v inside a pipe of diameter D and we wish to relate the mass-transfer coefficient k_c' to the variables D, ρ, μ, v, and D_{AB}. The total number of variables is $q = 6$. The fundamental units or dimensions are $u = 3$ and are mass M, length L, and time t. The units of the variables are

$$k_c' = \frac{L}{t} \qquad \rho = \frac{M}{L^3} \qquad \mu = \frac{M}{Lt} \qquad v = \frac{L}{t} \qquad D_{AB} = \frac{L^2}{t} \qquad D = L$$

The number of dimensionless groups or π' are then $6 - 3$, or 3. Then,

$$\pi_1 = f(\pi_2, \pi_3) \tag{7.8-1}$$

We choose the variables D_{AB}, ρ, and D to be the variables common to all the dimensionless groups, which are

$$\pi_1 = D_{AB}^a \rho^b D^c k_c' \tag{7.8-2}$$

$$\pi_2 = D_{AB}^d \rho^e D^f v \tag{7.8-3}$$

$$\pi_3 = D_{AB}^g \rho^h D^i \mu \tag{7.8-4}$$

For π_1 we substitute the actual dimensions as follows:

$$1 = \left(\frac{L^2}{t}\right)^a \left(\frac{M}{L^3}\right)^b (L)^c \left(\frac{L}{t}\right) \tag{7.8-5}$$

Summing for each exponent,

$$\begin{align}
\text{(L)} \quad & 0 = 2a - 3b + c + 1 \notag \\
\text{(M)} \quad & 0 = b \tag{7.8-6} \\
\text{(t)} \quad & 0 = a - 1 \notag
\end{align}$$

Solving these equation simultaneously, $a = -1$, $b = 0$, $c = 1$. Substituting these values into Eq. (7.8-2),

$$\pi_1 = \frac{k_c' D}{D_{AB}} = N_{Sh} \tag{7.8-7}$$

Repeating for π_2 and π_3,

$$\pi_2 = \frac{vD}{D_{AB}} \tag{7.8-8}$$

$$\pi_3 = \frac{\mu}{\rho D_{AB}} = N_{Sc} \tag{7.8-9}$$

If we divide π_2 by π_3 we obtain the Reynolds number.

$$\frac{\pi_2}{\pi_3} = \frac{vD}{D_{AB}} \Bigg/ \left(\frac{\mu}{\rho D_{AB}}\right) = \frac{Dv\rho}{\mu} = N_{Re} \tag{7.8-10}$$

Hence, substituting into Eq. (7.8-1),

$$N_{Sh} = f(N_{Re}, N_{Sc}) \tag{7.8-11}$$

7.9 BOUNDARY-LAYER FLOW AND TURBULENCE IN MASS TRANSFER

7.9A Laminar Flow and Boundary-Layer Theory in Mass Transfer

In Section 3.10C an exact solution was obtained for the hydrodynamic boundary layer for isothermal laminar flow past a plate and in Section 5.7A an extension of the Blasius solution was also used to derive an expression for convective heat transfer. In an analogous manner we use the Blasius solution for convective mass transfer for the same geometry and laminar flow. In Fig. 7.9-1 the concentration boundary layer is shown where the concentration of the fluid approaching the plate is $c_{A\infty}$ and c_{AS} in the fluid adjacent to the surface.

We start by using the differential mass balance, Eq. (7.5-17), and simplifying it for steady state where $\partial c_A/\partial t = 0$, $R_A = 0$, flow only in the x and y directions, so $v_z = 0$, and neglecting diffusion in the x and z directions to give

$$v_x \frac{\partial c_A}{\partial x} + v_y \frac{\partial c_A}{\partial y} = D_{AB} \frac{\partial^2 c_A}{\partial y^2} \tag{7.9-1}$$

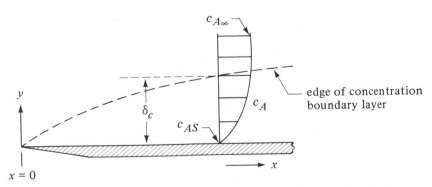

FIGURE 7.9-1. *Laminar flow of fluid past a flat plate and concentration boundary layer.*

The momentum boundary-layer equation is very similar.

$$v_x \frac{\partial v_x}{\partial x} + v_y \frac{\partial v_x}{\partial y} = \frac{\mu}{\rho} \frac{\partial^2 v_x}{\partial y^2} \tag{3.10-5}$$

The thermal boundary-layer equation is also similar.

$$v_x \frac{\partial T}{\partial x} + v_y \frac{\partial T}{\partial y} = \frac{k}{\rho c_p} \frac{\partial^2 T}{\partial y^2} \tag{5.7-2}$$

The continuity equation used previously is

$$\frac{\partial v_x}{\partial x} + \frac{\partial v_y}{\partial y} = 0 \tag{3.10-3}$$

The dimensionless concentration boundary conditions are

$$\frac{v_x}{v_\infty} = \frac{T - T_S}{T_\infty - T_S} = \frac{c_A - c_{AS}}{c_{A\infty} - c_{AS}} = 0 \quad \text{at } y = 0$$

$$\frac{v_x}{v_\infty} = \frac{T - T_S}{T_\infty - T_S} = \frac{c_A - c_{AS}}{c_{A\infty} - c_{AS}} = 1 \quad \text{at } y = \infty \tag{7.9-2}$$

The similarity between the three differential equations (7.9-1), (3.10-5), and (5.7-2) is obvious, as is the similarity among the three sets of boundary conditions in Eq. (7.9-2). In Section 5.7A the Blasius solution was applied to convective heat transfer when $(\mu/\rho)/\alpha = N_{Pr} = 1.0$. We use the same type of solution for laminar convective mass transfer when $(\mu/\rho)/D_{AB} = N_{Sc} = 1.0$.

The velocity gradient at the surface was derived previously.

$$\left(\frac{\partial v_x}{\partial y} \right)_{y=0} = 0.332 \frac{v_\infty}{x} N_{Re, x}^{1/2} \tag{5.7-5}$$

where $N_{Re, x} = x v_\infty \rho / \mu$. Also, from Eq. (7.9-2),

$$\frac{v_x}{v_\infty} = \frac{c_A - c_{AS}}{c_{A\infty} - c_{AS}} \tag{7.9-3}$$

Differentiating Eq. (7.9-3) and combining the result with Eq. (5.7-5),

$$\left(\frac{\partial c_A}{\partial y} \right)_{y=0} = (c_{A\infty} - c_{AS}) \left(\frac{0.332}{x} N_{Re, x}^{1/2} \right) \tag{7.9-4}$$

The convective mass-transfer equation can be written as follows and also related with Fick's equation for dilute solutions:

$$N_{Ay} = k_c'(c_{AS} - c_{A\infty}) = -D_{AB} \left(\frac{\partial c_A}{\partial y} \right)_{y=0} \tag{7.9-5}$$

Combining Eqs. (7.9-4) and (7.9-5),

$$\frac{k_c' x}{D_{AB}} = N_{Sh, x} = 0.332 N_{Re, x}^{1/2} \tag{7.9-6}$$

This relationship is restricted to gases with a $N_{Sc} = 1.0$.

The relationship between the thickness δ of the hydrodynamic and δ_c of the concentration boundary layers where the Schmidt number is not 1.0 is

$$\frac{\delta}{\delta_c} = N_{Sc}^{1/3} \qquad (7.9\text{-}7)$$

As a result, the equation for the local convective mass-transfer coefficient is

$$\frac{k_c'x}{D_{AB}} = N_{Sh,\,x} = 0.332 N_{Re,\,x}^{1/2} N_{Sc}^{1/3} \qquad (7.9\text{-}8)$$

We can obtain the equation for the mean mass-transfer coefficient k_c' from $x = 0$ to $x = L$ for a plate of width b by integrating as follows:

$$k_c' = \frac{b}{bL} \int_0^L k_c' \, dx \qquad (7.9\text{-}9)$$

The result is

$$\frac{k_c'L}{D_{AB}} = N_{Sh} = 0.664 N_{Re,\,L}^{1/2} N_{Sc}^{1/3} \qquad (7.9\text{-}10)$$

This is similar to the heat-transfer equation for a flat plate, Eq. (5.7-15), and also checks the experimental mass-transfer equation (7.3-27) for a flat plate.

In Section 3.10 an approximate integral analysis was made for the laminar hydrodynamic and also for the turbulent hydrodynamic boundary layer. This was also done in Section 5.7 for the thermal boundary layer. This approximate integral analysis can also be done in exactly the same manner for the laminar and turbulent concentration boundary layers.

7.9B Prandtl Mixing Length and Turbulent Eddy Mass Diffusivity

In many applications the flow in mass transfer is turbulent and not laminar. The turbulent flow of a fluid is quite complex and the fluid undergoes a series of random eddy movements throughout the turbulent core. When mass transfer is occurring, we refer to this as eddy mass diffusion. In Sections 3.10 and 5.7 we derived equations for turbulent eddy thermal diffusivity and momentum diffusivity using the Prandtl mixing length theory.

In a similar manner we can derive a relation for the turbulent eddy mass diffusivity, ε_M. Eddies are transported a distance L, called the Prandtl mixing length, in the y direction. At this point L the fluid eddy differs in velocity from the adjacent fluid by the velocity v_x', which is the fluctuating velocity component given in Section 3.10F. The instantaneous rate of mass transfer of A at a velocity v_y' for a distance L in the y direction is

$$J_{Ay}^* = c_A' v_y' \qquad (7.9\text{-}11)$$

where c_A' is the instantaneous fluctuating concentration. The instantaneous concentration of the fluid is $c_A = c_A' + \bar{c}_A$ where $\bar{c}_A$ is the mean value and c_A' the deviation from the mean value. The mixing length L is small enough so the concentration difference is

$$c_A' = L \frac{d\bar{c}_A}{dy} \qquad (7.9\text{-}12)$$

The rate of mass transported per unit area is J_{Ay}^*. Combining Eqs. (7.9-11) and (7.9-12),

$$J_{Ay}^* = -v_y' L \frac{d\bar{c}_A}{dy} \tag{7.9-13}$$

From Eq. (5.7-23),

$$v_y' = v_x' = L \left| \frac{d\bar{v}_x}{dy} \right| \tag{7.9-14}$$

Substituting Eq. (7.9-14) into (7.9-13),

$$J_{Ay}^* = -L^2 \left| \frac{d\bar{v}_x}{dy} \right| \frac{d\bar{c}_A}{dy} \tag{7.9-15}$$

The term $L^2 |d\bar{v}_x/dy|$ is called the turbulent eddy mass diffusivity ε_M. Combining Eq. (7.9-15) with the diffusion equation in terms of D_{AB}, the total flux is

$$J_{Ay}^* = -(D_{AB} + \varepsilon_M) \frac{d\bar{c}_A}{dy} \tag{7.9-16}$$

The similarities between Eq. (7.9-16) for mass transfer and heat and momentum transfer have been pointed out in detail in Section 6.1A.

7.9C Models for Mass-Transfer Coefficients

1. *Introduction.* For many years mass-transfer coefficients, which were based primarily on empirical correlations, have been used in the design of process equipment. A better understanding of the mechanisms of turbulence is needed before we can give a theoretical explanation of convective-mass-transfer coefficients. Some theories of convective mass transfer, such as the eddy diffusivity theory, have been presented in this chapter. In the following sections we present briefly some of these theories and also discuss how they can be used to extend empirical correlations.

2. *Film mass-transfer theory.* The film theory, which is the simplest and most elementary theory, assumes the presence of a fictitious laminar film next to the boundary. This film, where only molecular diffusion is assumed to be occurring, has the same resistance to mass transfer as actually exists in the viscous, transition, and turbulent core regions. Then the actual mass transfer coefficient k_c' is related to this film thickness δ_f by

$$J_A^* = k_c'(c_{A1} - c_{A2}) = \frac{D_{AB}}{\delta_f}(c_{A1} - c_{A2}) \tag{7.9-17}$$

$$k_c' = \frac{D_{AB}}{\delta_f} \tag{7.9-18}$$

The mass-transfer coefficient is proportional to $D_{AB}^{1.0}$. However, since we have shown that in Eq. (7.3-13), J_D is proportional to $(\mu/\rho D_{AB})^{2/3}$, then $k_c' \propto D_{AB}^{2/3}$. Hence, the film theory is not correct. The great advantage of the film theory is its simplicity where it can be used in complex situations such as simultaneous diffusion and chemical reaction.

3. *Penetration theory.* The penetration theory derived by Higbie and modified by Danckwerts (D3) was derived for diffusion or penetration into a laminar falling film for short contact times in Eq. (7.3-23) and is as follows:

$$k_c' = \sqrt{\frac{4 D_{AB}}{\pi t_L}} \tag{7.9-19}$$

where t_L is the time of penetration of the solute in seconds. This was extended by Danckwerts. He modified this for turbulent mass transfer and postulated that a fluid eddy has a uniform concentration in the turbulent core and is swept to the surface and undergoes unsteady-state diffusion. Then the eddy is swept away to the eddy core and other eddies are swept to the surface and stay for a random amount of time. A mean surface renewal factor s in s^{-1} is defined as follows:

$$k'_c = \sqrt{D_{AB}s} \qquad (7.9\text{-}20)$$

The mass-transfer coefficient k'_c is proportional to $D_{AB}^{0.5}$. In some systems, such as where liquid flows over packing and semistagnant pockets occur where the surface is being renewed, the results approximately follow Eq. (7.9-20). The value of s must be obtained experimentally. Others (D3, T2) have derived more complex combination film-surface renewal theories predicting a gradual change of the exponent on D_{AB} from 0.5 to 1.0 depending on turbulence and other factors. Penetration theories have been used in cases where diffusion and chemical reaction are occurring (D3).

4. *Boundary-layer theory.* The boundary-layer theory has been discussed in detail in Section 7.9 and is useful in predicting and correlating data for fluids flowing past solid surfaces. For laminar flow and turbulent flow the mass-transfer coefficient $k'_c \propto D_{AB}^{2/3}$. This has been experimentally verified for many cases.

PROBLEMS

7.1-1. *Unsteady-State Diffusion in a Thick Slab.* Repeat Example 7.1-2 but use a distribution coefficient $K = 0.50$ instead of 2.0. Plot the data.
  Ans. $c = c_i = 5.75 \times 10^{-2}$ ($x = 0$), $c = 2.78 \times 10^{-2}$ ($x = 0.01$ m), $c_{Li} = 2.87 \times 10^{-2}$ kg mol/m^3

7.1-2. *Plot of Concentration Profile in Unsteady-State Diffusion.* Using the same conditions as in Example 7.1-2, calculate the concentration at the points $x = 0$, 0.005, 0.01, 0.015, and 0.02 m from the surface. Also calculate c_{Li} in the liquid at the interface. Plot the concentrations in a manner similar to Fig. 7.1-3b, showing interface concentrations.

7.1-3. *Unsteady-State Diffusion in Several Directions.* Use the same conditions as in Example 7.1-1 except that the solid is a rectangular block 10.16 mm thick in the x direction, 7.62 mm thick in the y direction, and 10.16 mm thick in the z direction, and diffusion occurs at all six faces. Calculate the concentration at the midpoint after 10 h.
  Ans. $c = 6.20 \times 10^{-4}$ kg mol/m^3

7.1-4 *Drying of Moist Clay.* A very thick slab of clay has an initial moisture content of $c_0 = 14$ wt %. Air is passed over the top surface to dry the clay. Assume a relative resistance of the gas at the surface of zero. The equilibrium moisture content at the surface is constant at $c_1 = 3.0$ wt %. The diffusion of the moisture in the clay can be approximated by a diffusivity of $D_{AB} = 1.29 \times 10^{-8}$ m^2/s. After 1.0 h of drying, calculate the concentration of water at points 0.005, 0.01, and 0.02 m below the surface. Assume that the clay is a semiinfinite solid and that the Y value can be represented using concentrations of wt % rather than kg mol/m^3. Plot the values versus x.

7.1-5. *Unsteady-State Diffusion in a Cylinder of Agar Gel.* A wet cylinder of agar gel at 278 K containing a uniform concentration of urea of 0.1 kg mol/m^3 has a diameter of 30.48 mm and is 38.1 mm long with flat parallel ends. The diffusivity is 4.72×10^{-10} m^2/s. Calculate the concentration at the midpoint of the cylinder after 100 h for the following cases if the cylinder is suddenly immersed in turbulent pure water.

(a) For radial diffusion only.

(b) Diffusion occurs radially and axially.

7.1-6. Drying of Wood. A flat slab of Douglas fir wood 50.8 mm thick containing 30 wt % moisture is being dried from both sides (neglecting ends and edges). The equilibrium moisture content at the surface of the wood due to the drying air blown over it is held at 5 wt % moisture. The drying can be assumed to be represented by a diffusivity of 3.72×10^{-6} m^2/h. Calculate the time for the center to reach 10% moisture.

7.2-1. Flux and Conversion of Mass-Transfer Coefficient. A value of k_G was experimentally determined to be 1.08 lb mol/h·ft^2·atm for A diffusing through stagnant B. For the same flow and concentrations it is desired to predict k'_G and the flux of A for equimolar counterdiffusion. The partial pressures are $p_{A1} = 0.20$ atm, $p_{A2} = 0.05$ atm, and $P = 1.0$ atm abs total. Use English and SI units.

Ans. $k'_G = 0.942$ lb mol/h·ft^2·atm, 1.261×10^{-8} kg mol/s·m^2·Pa, $N_A = 0.141$ lb mol A/h·ft^2, 1.912×10^{-4} kg mol/s·m^2

7.2-2. Conversion of Mass-Transfer Coefficients. Prove or show the following relationships starting with the flux equations.

(a) Convert k'_c to k_y and k_G.

(b) Convert k_L to k_x and k'_x.

(c) Convert k_G to k_y and k_c.

7.2-3. Absorption of H$_2$S by Water. In a wetted-wall tower an air-H$_2$S mixture is flowing by a film of water which is flowing as a thin film down a vertical plate. The H$_2$S is being absorbed from the air to the water at a total pressure of 1.50 atm abs and 30°C. The value of k'_c of 9.567×10^{-4} m/s has been predicted for the gas-phase mass-transfer coefficient. At a given point the mole fraction of H$_2$S in the liquid at the liquid–gas interface is $2.0(10^{-5})$ and p_A of H$_2$S in the gas is 0.05 atm. The Henry's law equilibrium relation is p_A (atm) $= 609x_A$ (mole fraction in liquid). Calculate the rate of absorption of H$_2$S. (*Hint:* Call point 1 the interface and point 2 the gas phase. Then calculate p_{A1} from Henry's law and the given x_A. The value of p_{A2} is 0.05 atm.)

Ans. $N_A = -1.485 \times 10^{-6}$ kg mol/s·m^2

7.3-1. Mass Transfer from a Flat Plate to a Liquid. Using the data and physical properties of Example 7.3-2 calculate the flux for a water velocity of 0.152 m/s and a plate length of $L = 0.137$ m. Do not assume that $x_{BM} = 1.0$, but actually calculate its value.

7.3-2. Mass Transfer from a Pipe Wall. Pure water at 26.1°C is flowing at a velocity of 0.0305 m/s in a tube having an inside diameter of 6.35 mm. The tube is 1.829 m long with the last 1.22 m having the walls coated with benzoic acid. Assuming that the velocity profile is fully developed, calculate the average concentration of benzoic acid at the outlet. Use the physical property data of Example 7.3-2. [*Hint:* First calculate the Reynolds number $Dv\rho/\mu$. Then calculate $N_{Re}N_{Sc}$ $(D/L)(\pi/4)$, which is the same as $W/D_{AB}\rho L$.]

Ans. $(c_A - c_{A0})/(c_{Ai} - c_{A0}) = 0.0744$, $c_A = 2.193 \times 10^{-3}$ kg mol/m^3

7.3-3. Mass-Transfer Coefficient for Various Geometries. It is desired to estimate the mass-transfer coefficient k_G in kg mol/s·m^2·Pa for water vapor in air at 338.6 K and 101.32 kPa flowing in a large duct past different geometry solids. The velocity in the duct is 3.66 m/s. The water vapor concentration in the air is small, so the physical properties of air can be used. Water vapor is being transferred to the solids. Do this for the following geometries.

(a) A single 25.4-mm-diameter sphere.

(b) A packed bed of 25.4-mm spheres with $\varepsilon = 0.35$.

Ans. (a) $k_G = 1.98 \times 10^{-8}$ kg mol/s·m^2·Pa (1.48 lb mol/h·ft^2·atm)

7.3-4. Mass Transfer to Definite Shapes. Estimate the value of the mass-transfer coefficient in a stream of air at 325.6 K flowing in a duct by the following shapes made of solid naphthalene. The velocity of the air is 1.524 m/s at 325.6 K and 202.6 kPa. The D_{AB} of naphthalene in air is 5.16×10^{-6} m²/s at 273 K and 101.3 kPa.
(a) For air flowing parallel to a flat plate 0.152 m in length.
(b) For air flowing by a single sphere 12.7 mm in diameter.

7.3-5. Mass Transfer to Packed Bed and Driving Force. Pure water at 26.1°C is flowing at a rate of 0.0701 ft³/h through a packed bed of 0.251-in. benzoic acid spheres having a total surface area of 0.129 ft². The solubility of benzoic acid in water is 0.00184 lb mol benzoic acid/ft³ solution. The outlet concentration c_{A2} is 1.80×10^{-4} lb mol/ft³. Calculate the mass-transfer coefficient k_c.

7.3-6. Mass Transfer in Liquid Metals. Mercury at 26.5°C is flowing through a packed bed of lead spheres having a diameter of 2.096 mm with a void fraction of 0.499. The superficial velocity is 0.02198 m/s. The solubility of lead in mercury is 1.721 wt %, the Schmidt number is 124.1, the viscosity of the solution is 1.577×10^{-3} Pa·s, and the density is 13 530 kg/m³.
(a) Predict the value of J_D. Use Eq. (7.3-38) if applicable. Compare with the experimental of $J_D = 0.076$ (D2).
(b) Predict the value of k_c for the case of A diffusing through nondiffusing B.
$\qquad$ **Ans.** (a) $J_D = 0.0784$, (b) $k_c = 6.986 \times 10^{-5}$ m/s

7.3-7. Mass Transfer from a Pipe and Log Mean Driving Force. Use the same physical conditions as Problem 7.3-2, but the velocity in the pipe is now 3.05 m/s. Do as follows.
(a) Predict the mass-transfer coefficient k_c'. (Is this turbulent flow?)
(b) Calculate the average benzoic acid concentration at the outlet. [*Note:* In this case, Eqs. (7.3-42) and (7.3-43) must be used with the log mean driving force, where A is the surface area of the pipe.]
(c) Calculate the total kg mol of benzoic acid dissolved per second.

7.3-8. Derivation of Relation Between J_D and N_{Sh}. Equation (7.3-3) defines the Sherwood number and Eq. (7.3-5) defines the J_D factor. Derive the relation between N_{Sh} and J_D in terms of N_{Re} and N_{Sc}.
$\qquad$ **Ans.** $N_{Sh} = J_D N_{Re} N_{Sc}^{1/3}$

7.3-9. Driving Force to Use in Mass Transfer. Derive Eq. (7.3-42) for the log mean driving force to use for a fluid flowing in a packed bed or in a tube. (*Hint :* Start by making a mass balance and a diffusion rate balance over a differential area dA as follows:

$$N_A \, dA = k_c(c_{Ai} - c_A) \, dA = V \, dc_A$$

where $V = m^3/s$ flow rate. Assume dilute solutions.)

7.4-1. Maximum Oxygen Uptake of a Microorganism. Calculate the maximum possible rate of oxygen uptake at 37°C of microorganisms having a diameter of $\frac{2}{3}$ μm suspended in an agitated aqueous solution. It is assumed that the surrounding liquid is saturated with O_2 from air at 1 atm abs pressure. It will be assumed that the microorganism can utilize the oxygen much faster than it can diffuse to it. The microorganism has a density very close to that of water. Use physical property data from Example 7.4-1. (*Hint :* Since the oxygen is consumed faster than it is supplied, the concentration c_{A2} at the surface is zero. The concentration c_{A1} in the solution is at saturation.)
$\qquad$ **Ans.** $k_c = 9.75 \times 10^{-3}$ m/s, $N_A = 2.20 \times 10^{-6}$ kg mol O_2/s·m²

7.4-2. Mass Transfer of O_2 in Fermentation Process. A total of 5.0 g of wet microorganisms having a density of 1100 kg/m³ and a diameter of 0.667 μm are added to 0.100 L of aqueous solution at 37°C in a shaker flask for a fermentation. Air can enter through a porous stopper. Use physical property data from Example 7.4-1.

(a) Calculate the maximum rate possible of mass transfer of oxygen in kg mol O_2/s to the surface of the microorganisms assuming that the solution is saturated with air at 101.32 kPa abs pressure.

(b) By material balances on other nutrients, the actual utilization of O_2 by the microorganisms is 6.30×10^{-6} kg mol O_2/s. What would be the actual concentration of O_2 in the solution as percent saturation during the fermentation?

Ans. (a) $k'_L = 9.82 \times 10^{-3}$ m/s, $N_A A = 9.07 \times 10^{-5}$ kg mol O_2/s, (b) 6.95% saturation

7.5-1. Sum of Molar Fluxes. Prove the following equation using the definitions in Table 7.5-1.

$$N_A + N_B = c v_M$$

7.5-2. Proof of Derived Relation. Using the definitions from Table 7.5-1, prove the following:

$$j_A = n_A - w_A(n_A + n_B)$$

7.5-3. Different Forms of Fick's Law. Using Eq. (1), prove Eq. (2).

$$j_A = -\rho D_{AB} \frac{dw_A}{dz} \tag{1}$$

$$j_A = -\frac{c^2}{\rho} M_A M_B D_{AB} \frac{dx_A}{dz} \tag{2}$$

(*Hint:* First relate w_A to x_A. Then differentiate this equation to relate dw_A and dx_A. Finally, use $M = x_A M_A + x_B M_B$ to simplify.)

7.5-4. Other Form of Fick's Law. Show that the following form of Fick's law is valid:

$$c(v_A - v_B) = -\frac{c D_{AB}}{x_A x_B} \frac{dx_A}{dz}$$

(*Hint:* Start with $N_A = J_A^* + c_A v_M$. Substitute the expression for J_A^* from Table 7.5-1 and simplify.)

7.5-5. Different Form of Equation of Continuity. Starting with Eq. (7.5-12),

$$\frac{\partial \rho_A}{\partial t} + (\nabla \cdot \mathbf{n}_A) = r_A \tag{7.5-12}$$

convert this to the following for constant ρ:

$$\frac{\partial \rho_A}{\partial t} + (\mathbf{v} \cdot \nabla \rho_A) - (\nabla \cdot D_{AB} \nabla \rho_A) = r_A \tag{1}$$

[*Hint:* From Table 7.5-1, substitute $\mathbf{n}_A = \mathbf{j}_A + \rho_A \mathbf{v}$ into Eq. (7.5-12). Note that $(\nabla \cdot \mathbf{v}) = 0$ for constant ρ. Then substitute Fick's law in terms of $\mathbf{j}_A$.]

7.5-6. Diffusion and Reaction at a Surface. Gas A is diffusing from a gas stream at point 1 to a catalyst surface at point 2 and reacts instantaneously and irreversibly as follows:

$$2A \rightarrow B$$

Gas B diffuses back to the gas stream. Derive the final equation for N_A at constant pressure P and steady state in terms of partial pressures.

Ans. $N_A = \dfrac{2 D_{AB} P}{RT(z_2 - z_1)} \ln \dfrac{1 - p_{A2}/2P}{1 - p_{A1}/2P}$

7.5-7. Unsteady-State Diffusion and Reaction. Solute A is diffusing at unsteady state into a semiinfinite medium of pure B and undergoes a first order reaction with B. Solute A is dilute. Calculate the concentration c_A at points $z = 0$, 4, and 10 mm from the surface for $t = 1 \times 10^5$ s. Physical property data are $D_{AB} = 1 \times 10^{-9}$ m²/s, $k' = 1 \times 10^{-4}$ s⁻¹, $c_{A0} = 1.0$ kg mol/m³. Also calculate the kg mol absorbed/m².

7.5-8. Multicomponent Diffusion. At a total pressure of 202.6 kPa and 358 K, ammonia gas (A) is diffusing at steady state through an inert nondiffusing mixture of nitrogen (B) and hydrogen (C). The mole fractions at $z_1 = 0$ are $x_{A1} = 0.8$, $x_{B1} = 0.15$, and $x_{C1} = 0.05$; and at $z_2 = 4.0$ mm, $x_{A2} = 0.2$, $x_{B2} = 0.6$, and $x_{C2} = 0.2$. The diffusivities at 358 K and 101.3 kPa are $D_{AB} = 3.28 \times 10^{-5}$ m²/s and $D_{AC} = 1.093 \times 10^{-4}$ m²/s. Calculate the flux of ammonia.

Ans. $N_A = 4.69 \times 10^{-4}$ kg mol $A/s \cdot m^2$

7.5-9. Diffusion in Liquid Metals and Variable Diffusivity. The diffusion of tin (A) in liquid lead (B) at 510°C was carried out by using a 10.0-mm-long capillary tube and maintaining the mole fraction of tin at x_{A1} at the left end and x_{A2} at the right end of the tube. In the range of concentrations of $0.2 \leq x_A \leq 0.4$ the diffusivity of tin in lead has been found to be a linear function of x_A (S7).

$$D_{AB} = A + Bx_A$$

where A and B are constants and D_{AB} is in m²/s.

(a) Assuming the molar density to be constant at $c = c_A + c_B = c_{av}$, derive the final integrated equation for the flux N_A assuming steady state and that A diffuses through stagnant B.

(b) For this experiment, $A = 4.8 \times 10^{-9}$, $B = -6.5 \times 10^{-9}$, $c_{av} = 50$ kg mol/m³, $x_{A1} = 0.4$, $x_{A2} = 0.2$. Calculate N_A.

Ans. (b) $N_A = 4.055 \times 10^{-6}$ kg mol $A/s \cdot m^2$

7.5-10. Diffusion and Chemical Reaction of Molten Iron in Process Metallurgy. In a steelmaking process using molten pig iron containing carbon, a spray of molten iron particles containing 4.0 wt % carbon fall through a pure oxygen atmosphere. The carbon diffuses through the molten iron to the surface of the drop, where it is assumed that it reacts instantly at the surface, because of the high temperature, as follows by a first-order reaction:

$$C + \tfrac{1}{2} O_2(g) \rightarrow CO(g)$$

Calculate the maximum drop size allowable so that the final drop after a 2.0-s fall contains on an average 0.1 wt % carbon. Assume that the mass transfer rate of gases at the surface is very great, so there is no outside resistance. Assume no internal circulation of the liquid. Hence, the decarburization rate is controlled by the rate of diffusion of carbon to the surface of the droplet. The diffusivity of carbon in iron is 7.5×10^{-9} m²/s (S7). (*Hint*: Can Fig. 5.3-13 be used for this case?)

Ans. radius = 0.217 mm

7.5-11. Effect of Slow Reaction Rate on Diffusion. Gas A diffuses from point 1 to a catalyst surface at point 2, where it reacts as follows: $2A \rightarrow B$. Gas B diffuses back a distance δ to point 1.

(a) Derive the equation for N_A for a very fact reaction using mole fraction units x_{A1}, and so on.

(b) For $D_{AB} = 0.2 \times 10^{-4}$ m²/s, $x_{A1} = 0.97$, $P = 101.32$ kPa, $\delta = 1.30$ mm, and $T = 298$ K, solve for N_A.

(c) Do the same as part (a) but for a slow first-order reaction where k'_1 is the reaction velocity constant.

(d) Calculate N_A and x_{A2} for part (c) where $k'_1 = 0.53 \times 10^{-2}$ m/s.

Ans. (b) $N_A = 8.35 \times 10^{-4}$ kg mol/s $\cdot m^2$

7.5-12. Diffusion and Heterogeneous Reaction on Surface. In a tube of radius R m filled with a liquid dilute component A is diffusing in the nonflowing liquid phase represented by

$$N_A = -D_{AB}\frac{dc_A}{dz}$$

where z is distance along the tube axis. The inside wall of the tube exerts a catalytic effect and decomposes A so that the heterogeneous rate of decomposition on the wall in kg mol A/s is equal to $kc_A A_w$, where k is a first-order constant and A_w is the wall area in m^2. Neglect any radial gradients (this means a uniform radial concentration).

Derive the differential equation for unsteady state for diffusion and reaction for this system. [*Hint :* First make a mass balance for A for a Δz length of tube as follows: rate of input (diffusion) + rate of generation (heterogeneous) = rate of output (diffusion) + rate of accumulation.]

Ans. $\dfrac{\partial c_A}{\partial t} = D_{AB}\dfrac{\partial^2 c_A}{\partial z^2} - \dfrac{2k}{R}c_A$

7.6-1. Knudsen Diffusivities. A mixture of He(A) and Ar(B) is diffusing at 1.013×10^5 Pa total pressure and 298 K through a capillary having a radius of 100 Å.
(a) Calculate the Knudsen diffusivity of He (A).
(b) Calculate the Knudsen diffusivity of Ar (B).
(c) Compare with the molecular diffusivity D_{AB}.

Ans. (a) $D_{KA} = 8.37 \times 10^{-6}$ m^2/s; (c) $D_{AB} = 7.29 \times 10^{-5}$ m^2/s

7.6-2. Transition-Region Diffusion. A mixture of He (A) and Ar (B) at 298 K is diffusing through an open capillary 15 mm long with a radius of 1000 Å. The total pressure is 1.013×10^5 Pa. The molecular diffusivity D_{AB} at 1.013 $\times 10^5$ Pa is 7.29×10^{-5} m^2/s.
(a) Calculate the Knudsen diffusivity of He (A).
(b) Predict the flux N_A using Eq. (7.6-18) and Eq. (7.6-12) if $x_{A1} = 0.8$ and $x_{A2} = 0.2$. Assume steady state.
(c) Predict the flux N_A using the approximate Eqs. (7.6-14) and (7.6-16).

7.6-3. Diffusion in a Pore in the Transition Region. Pure H$_2$ gas (A) at one end of a noncatalytic pore of radius 50 Å and length 1.0 mm ($x_{A1} = 1.0$) is diffusing through this pore with pure C$_2$H$_6$ gas (B) at the other end at $x_{A2} = 0$. The total pressure is constant at 1013.2 kPa. The predicted molecular diffusivity of H$_2$–C$_2$H$_6$ is 8.60×10^{-5} m^2/s at 101.32 kPa and 373 K. Calculate the Knudsen diffusivity of H$_2$ and flux N_A of H$_2$ in the mixture at 373 K and steady state.

Ans. $D_{KA} = 6.60 \times 10^{-6}$ m^2/s, $N_A = 1.472 \times 10^{-3}$ kg mol A/s·m^2

7.6-4. Transition Region Diffusion in Capillary. A mixture of nitrogen gas (A) and helium (B) at 298 K is diffusing through a capillary 0.10 m long in an open system with a diameter of 10 μm. The mole fractions are constant at $x_{A1} = 1.0$ and $x_{A2} = 0$. See Example 7.6-2 for physical properties.
(a) Calculate the Knudsen diffusivity D_{KA} and D_{KB} at the total pressures of 0.001, 0.1, and 10.0 atm.
(b) Calculate the flux N_A at steady state at these pressures.
(c) Plot N_A versus P on log–log paper. What are the limiting lines at lower pressures and very high pressures? Calculate and plot these lines.

7.7-1. Numerical Method for Unsteady-State Diffusion. A solid slab 0.01 m thick has an initial uniform concentration of solute A of 1.00 kg mol/m^3. The diffusivity of A in the solid is $D_{AB} = 1.0 \times 10^{-10}$ m^2/s. All surfaces of the slab are insulated except the top surface. The surface concentration is suddenly dropped to zero concentration and held there. Unsteady-state diffusion occurs in the one x direction with the rear surface insulated. Using a numerical method, determine the concentrations after 12×10^4 s. Use $\Delta x = 0.002$ m and $M = 2.0$. The value of K is 1.0.

Ans. $c_1 = 0$(front surface, $x = 0$ m),
$c_2 = 0.3125$ kg mol/m^3 ($x = 0.002$ m)
$c_3 = 0.5859$ ($x = 0.004$ m),
$c_4 = 0.7813$ ($x = 0.006$ m)
$c_5 = 0.8984$ ($x = 0.008$ m),
$c_6 = 0.9375$ (insulated surface, $x = 0.01$ m)

7.7-2. *Digital Computer and Unsteady-State Diffusion.* Using the conditions of Problem 7.7-1, solve that problem by the digital computer. Use $\Delta x = 0.0005$ m. Write the computer program and plot the final concentrations. Use the explicit method, $M = 2$.

7.7-3. *Numerical Method and Different Boundary Condition.* Use the same conditions as in Example 7.7-1, but in this new case the rear surface is not insulated. At time $t = 0$ the concentration at the rear surface is also suddenly changed to $c_5 = 0$ and held there. Calculate the concentration profile after 2500 s. Plot the initial and final concentration profiles and compare with the final profile of Example 7.7-1.

7.8-1. *Dimensional Analysis in Mass Transfer.* A fluid is flowing in a vertical pipe and mass transfer is occurring from the pipe wall to the fluid. Relate the convective mass-transfer coefficient k_c' to the variables D, ρ, μ, v, D_{AB}, g, and $\Delta \rho$, where D is pipe diameter, L is pipe length, and $\Delta \rho$ is the density difference.

Ans. $$\frac{k_c' D}{D_{AB}} = f\left(\frac{gL^3 \rho \, \Delta \rho}{\mu^2}, \frac{Dv\rho}{\mu}, \frac{\mu}{\rho D_{AB}}\right)$$

7.9-1. *Mass Transfer and Turbulence Models.* Pure water at a velocity of 0.11 m/s is flowing at 26.1°C past a flat plate of solid benzoic acid where $L = 0.40$ m. Do as follows.
(a) Assuming dilute solutions, calculate the mass-transfer coefficient k_c. Use physical property data from Example 7.3-2.
(b) Using the film model, calculate the equivalent film thickness.
(c) Using the penetration model, calculate the time of penetration.
(d) Calculate the mean surface renewal factor using the modified penetration model.

Ans. (b) $\delta_f = 0.2031$ mm, (d) $s = 3.019 \times 10^{-2}$ s^{-1}

REFERENCES

(B1) BIRD, R B., STEWART, W. E., and LIGHTFOOT, E. N. *Transport Phenomena.* New York: John Wiley & Sons, Inc., 1960.

(B2) BLAKEBROUGH, N. *Biochemical and Biological Engineering Science*, Vol. 1. New York: Academic Press, Inc., 1967.

(B3) BEDDINGTON, C. H., Jr., and DREW, T. B. *Ind. Eng. Chem.*, **42**, 1164 (1950).

(C1) CARMICHAEL, L. T., SAGE, B. H., and LACEY, W. N. *A.I.Ch.E. J.*, **1**, 385 (1955).

(C2) CHILTON, T. H., and COLBURN, A. P. *Ind. Eng. Chem.*, **26**, 1183 (1934).

(C3) CALDERBANK, P. H., and MOO-YOUNG, M. B. *Chem. Eng. Sci.*, **16**, 39 (1961).

(C4) CUNNINGHAM, R. S., and GEANKOPLIS, C. J. *Ind. Eng. Chem. Fund.*, **7**, 535 (1968).

(D1) DANCKWERTS, P. V. *Trans. Faraday Soc.*, **46**, 300 (1950).

(D2) DUNN, W. E., BONILLA, C. F., FERSTENBERG, C., and GROSS, B. *A.I.Ch.E. J.*, **2**, 184 (1956).

(D3) DANCKWERTS, P. V. *Gas–Liquid Reactions*. New York: McGraw-Hill Book Company, 1970.

(D4) DWIVEDI, P. N., and UPADHYAY, S. N. *Ind. Eng. Chem., Proc. Des. Dev.*, **16**, 157 (1977).

(G1) GEANKOPLIS, C. J. *Mass Transport Phenomena*. Columbus, Ohio: Ohio State University Bookstores, 1972.

(G2) GILLILAND, E. R., and SHERWOOD, T. K. *Ind. Eng. Chem.*, **26**, 516 (1934).

(G3) GARNER, F. H., and SUCKLING, R. D. *A.I.Ch.E. J.*, **4**, 114 (1958).

(G4) GUPTA, A. S., and THODOS, G. *Ind. Chem. Eng. Fund.*, **3**, 218 (1964).

(G5) GUPTA, A. S., and THODOS, G. *A.I.Ch.E. J.*, **8**, 609 (1962).

(G6) GUPTA, A. S., and THODOS, G. *Chem. Eng. Progr.*, **58** (7), 58 (1962).

(L1) LINTON, W. H., Jr., and SHERWOOD, T. K. *Chem. Eng. Progr.*, **46**, 258 (1950).

(L2) LITT, M., and FRIEDLANDER, S. K. *A.I.Ch.E. J.*, **5**, 483 (1959).

(M1) MCADAMS, W. H. *Heat Transmission*, 3rd ed. New York: McGraw-Hill Book Company, 1954.

(P1) PERRY, R. H., and GREEN, D. *Perry's Chemical Engineers' Handbook*, 6th ed. New York: McGraw-Hill Book Company, 1984.

(R1) REMICK, R. R., and GEANKOPLIS, C. J. *Ind. Eng. Chem. Fund.*, **12**, 214 (1973).

(S1) SEAGER, S. L., GEERTSON, L. R., and GIDDINGS, J. C. *J. Chem. Eng. Data*, **8**, 168 (1963).

(S2) SATTERFIELD, C. N. *Mass Transfer in Heterogeneous Catalysis*. Cambridge, Mass.: The MIT Press, 1970.

(S3) STEELE, L. R., and GEANKOPLIS, C. J. *A.I.Ch.E. J.*, **5**, 178 (1959).

(S4) SHERWOOD, T. K., PIGFORD, R. L., and WILKE, C. R. *Mass Transfer*. New York: McGraw-Hill Book Company, 1975.

(S5) STEINBERGER, W. L., and TREYBAL, R. E. *A.I.Ch.E. J.*, **6**, 227 (1960).

(S6) SMITH, J. M. *Chemical Engineering Kinetics*, 2nd ed. New York: McGraw-Hill Book Company, 1970.

(S7) SZEKELY, J., and THEMELIS, N. *Rate Phenomena in Process Metallurgy*. New York: Wiley-Interscience, 1971.

(T1) TREYBAL, R. E. *Mass Transfer Operations*, 3rd ed. New York: McGraw-Hill Book Company, 1980.

(T2) TOOR, H. L., and MARCHELLO, J. M. *A.I.Ch.E. J.*, **1**, 97 (1958).

(V1) VOGTLANDER, P. H., and BAKKER, C. A. P. *Chem. Eng. Sci.*, **18**, 583 (1963).

(W1) WILSON, E. J., and GEANKOPLIS, C. J. *Ind. Eng. Chem. Fund.*, **5**, 9 (1966).

PART 2

Unit Operations

CHAPTER 8

Evaporation

8.1 INTRODUCTION

8.1A Purpose

In Section 4.8 we discussed the case of heat transfer to a boiling liquid. An important instance of this type of heat transfer occurs quite often in the process industries and is given the general name *evaporation*. In evaporation the vapor from a boiling liquid solution is removed and a more concentrated solution remains. In the majority of cases the unit operation evaporation refers to the removal of water from an aqueous solution.

Typical examples of evaporation are concentration of aqueous solutions of sugar, sodium chloride, sodium hydroxide, glycerol, glue, milk, and orange juice. In these cases the concentrated solution is the desired product and the evaporated water is normally discarded. In a few cases, water, which has a small amount of minerals, is evaporated to give a solids-free water which is used as boiler feed, for special chemical processes, or for other purposes. Evaporation processes to evaporate seawater to provide drinking water have been developed and used. In some cases, the primary purpose of evaporation is to concentrate the solution so that upon cooling, salt crystals will form and be separated. This special evaporation process, termed *crystallization*, is discussed in Chapter 12.

8.1B Processing Factors

The physical and chemical properties of the solution being concentrated and of the vapor being removed have a great effect on the type of evaporator used and on the pressure and temperature of the process. Some of these properties which affect the processing methods are discussed next.

1. Concentration in the liquid. Usually, the liquid feed to an evaporator is relatively dilute, so its viscosity is low, similar to water, and relatively high heat-transfer coefficients are obtained. As evaporation proceeds, the solution may become very concentrated and quite viscous, causing the heat-transfer coefficient to drop markedly. Adequate circulation and/or turbulence must be present to keep the coefficient from becoming too low.

2. Solubility.   As solutions are heated and concentration of the solute or salt increases, the solubility limit of the material in solution may be exceeded and crystals may form. This may limit the maximum concentration in solution which can be obtained by evaporation. In Fig. 8.1-1 some solubilities of typical salts in water are shown as a function of temperature. In most cases the solubility of the salt increases with temperature. This means that when a hot concentrated solution from an evaporator is cooled to room temperature, crystallization may occur.

3. Temperature sensitivity of materials.   Many products, especially food and other biological materials, may be temperature-sensitive and degrade at higher temperatures or after prolonged heating. Such products are pharmaceutical products; food products such as milk, orange juice, and vegetable extracts; and fine organic chemicals. The amount of degradation is a function of the temperature and the length of time.

4. Foaming or frothing.   In some cases materials composed of caustic solutions, food solutions such as skim milk, and some fatty acid solutions form a foam or froth during boiling. This foam accompanies the vapor coming out of the evaporator and entrainment losses occur.

5. Pressure and temperature.   The boiling point of the solution is related to the pressure of the system. The higher the operating pressure of the evaporator, the higher the temperature at boiling. Also, as the concentration of the dissolved material in solution increases by evaporation, the temperature of boiling may rise. This phenomenon is called *boiling-point rise* or *elevation* and is discussed in Section 8.4. To keep the temperatures low in heat-sensitive materials, it is often necessary to operate under 1 atm pressure, i.e., under vacuum.

6. Scale deposition and materials of construction.   Some solutions deposit solid materials called *scale* on the heating surfaces. These could be formed by decomposition products or solubility decreases. The result is that the overall heat-transfer coefficient decreases and the evaporator must eventually be cleaned. The materials of construction of the evaporator are important to minimize corrosion.

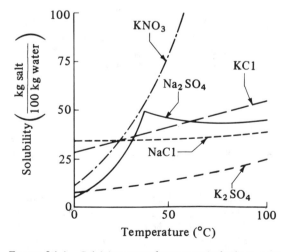

FIGURE 8.1-1.   *Solubility curves for some typical salts in water.*

8.2 TYPES OF EVAPORATION EQUIPMENT AND OPERATION METHODS

8.2A General Types of Evaporators

In evaporation, heat is added to a solution to vaporize the solvent, which is usually water. The heat is generally provided by the condensation of a vapor such as steam on one side of a metal surface with the evaporating liquid on the other side. The type of equipment used depends primarily on the configuration of the heat-transfer surface and on the means employed to provide agitation or circulation of the liquid. These general types are discussed below.

1. Open kettle or pan. The simplest form of evaporator consists of an open pan or kettle in which the liquid is boiled. The heat is supplied by condensation of steam in a jacket or in coils immersed in the liquid. In some cases the kettle is direct-fired. These evaporators are inexpensive and simple to operate, but the heat economy is poor. In some cases, paddles or scrapers for agitation are used.

2. Horizontal-tube natural circulation evaporator. The horizontal-tube natural circulation evaporator is shown in Fig. 8.2-1a. The horizontal bundle of heating tubes is similar to the bundle of tubes in a heat exchanger. The steam enters into the tubes, where it condenses. The steam condensate leaves at the other end of the tubes. The boiling-liquid solution covers the tubes. The vapor leaves the liquid surface, often goes through some deentraining device such as a baffle to prevent carryover of liquid droplets, and leaves out the top. This type is relatively cheap and is used for nonviscous liquids having high heat-transfer coefficients and liquids that do not deposit scale. Since liquid circulation is poor, they are unsuitable for viscous liquids. In almost all cases, this evaporator and the types discussed below are operated continuously, where the feed enters at a constant rate and the concentrate leaves at a constant rate.

3. Vertical-type natural circulation evaporator. In this type of evaporator, vertical rather than horizontal tubes are used, and the liquid is inside the tubes and the steam condenses outside the tubes. Because of boiling and decreases in density, the liquid rises in the tubes by natural circulation as shown in Fig. 8.2-1b and flows downward through a large central open space or downcomer. This natural circulation increases the heat-transfer coefficient. It is not used with viscous liquids. This type is often called the *short-tube evaporator.* A variation of this is the basket type, where vertical tubes are used, but the heating element is held suspended in the body so there is an annular open space as the downcomer. The basket type differs from the vertical natural circulation evapor ator, which has a central instead of annular open space as the downcomer. This type is widely used in the sugar, salt, and caustic soda industries.

4. Long-tube vertical-type evaporator. Since the heat-transfer coefficient on the steam side is very high compared to that on the evaporating liquid side, high liquid velocities are desirable. In a long-tube vertical-type evaporator shown in Fig. 8.2-1c, the liquid is inside the tubes. The tubes are 3 to 10 m long and the formation of vapor bubbles inside the tubes causes a pumping action giving quite high liquid velocities. Generally, the liquid passes through the tubes only once and is not recirculated. Contact times can be quite low in this type. In some cases, as when the ratio of feed to evaporation rate is low, natural recirculation of the product through the evaporator is done by adding a large pipe connection between the outlet concentrate line and the feed line. This is widely used for producing condensed milk.

5. *Falling-film-type evaporator.* A variation of the long-tube type is the falling-film evaporator, wherein the liquid is fed to the top of the tubes and flows down the walls as a thin film. Vapor–liquid separation usually takes place at the bottom. This type is widely used for concentrating heat-sensitive materials such as orange juice and other fruit juices, because the holdup time is very small (5 to 10 s or more) and the heat-transfer coefficients are high.

6. *Forced-circulation-type evaporator.* The liquid-film heat-transfer coefficient can be increased by pumping to cause forced circulation of the liquid inside the tubes. This could be done in the long-tube vertical type in Fig. 8.2-1c by adding a pipe connection with a pump between the outlet concentrate line and the feed line. However, usually in a forced-circulation type, the vertical tubes are shorter than in the long-tube type, and this

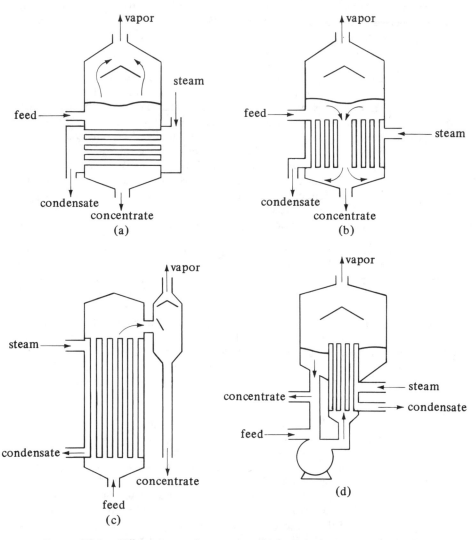

FIGURE 8.2-1. *Different types of evaporators: (a) horizontal-tube type, (b) vertical-tube type, (c) long-tube vertical type, (d) forced-circulation type.*

type is shown in Fig. 8.2-1d. Also, in other cases a separate and external horizontal heat exchanger is used. This type is very useful for viscous liquids.

7. Agitated-film evaporator. In an evaporator the main resistance to heat transfer is on the liquid side. One way to increase turbulence in this film, and hence the heat-transfer coefficient, is by actual mechanical agitation of this liquid film. This is done in a modified falling-film evaporator with only a single large jacketed tube containing an internal agitator. Liquid enters at the top of the tube and as it flows downward, it is spread out into a turbulent film by the vertical agitator blades. The concentrated solution leaves at the bottom and vapor leaves through a separator and out the top. This type of evaporator is very useful with highly viscous materials, since the heat-transfer coefficient is greater than in forced-circulation evaporators. It is used with heat-sensitive viscous materials such as rubber latex, gelatin, antibiotics, and fruit juices. However, it has a high cost and small capacity. For interested readers, Perry and Green (P2) give more detailed discussions and descriptions of evaporation equipment.

8. Open-pan solar evaporator. A very old but yet still used process is solar evaporation in open pans. Salt water is put in shallow open pans or troughs and allowed to evaporate slowly in the sun to crystallize the salt.

8.2B Methods of Operation of Evaporators

1. Single-effect evaporators. A simplified diagram of a single-stage or single-effect evaporator is given in Fig. 8.2-2. The feed enters at T_F K and saturated steam at T_S enters the heat-exchange section. Condensed steam leaves as condensate or drips. Since the solution in the evaporator is assumed to be completely mixed, the concentrated product and the solution in the evaporator have the same composition and temperature T_1, which is the boiling point of the solution. The temperature of the vapor is also T_1, since it is in equilibrium with the boiling solution. The pressure is P_1, which is the vapor pressure of the solution at T_1.

If the solution to be evaporated is assumed to be dilute and like water, then 1 kg of steam condensing will evaporate approximately 1 kg of vapor. This will hold if the feed entering has a temperature T_F near the boiling point.

The concept of an overall heat-transfer coefficient is used in the calculation of the rate of heat transfer in an evaporator. The general equation can be written

$$q = UA \ \Delta T = UA(T_S - T_1) \tag{8.2-1}$$

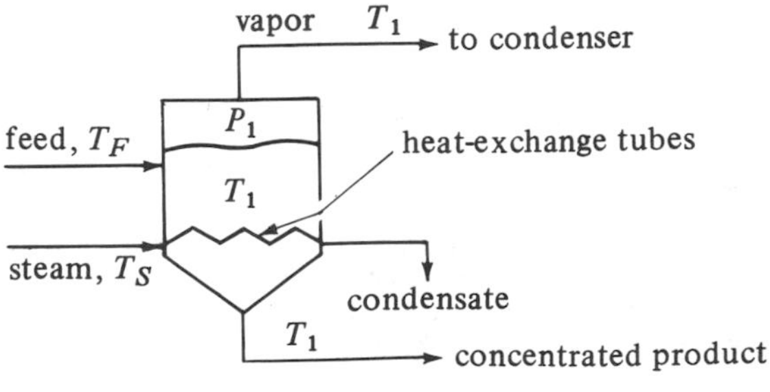

FIGURE 8.2-2. *Simplified diagram of single-effect evaporator.*

where q is the rate of heat transfer in W (btu/h), U is the overall heat-transfer coefficient in $W/m^2 \cdot K$ (btu/h $\cdot$ ft$^2 \cdot$ °F), A is the heat-transfer area in m^2 (ft^2), T_S is the temperature of the condensing steam in K (°F), and T_1 is the boiling point of the liquid in K (°F).

Single-effect evaporators are often used when the required capacity of operation is relatively small and/or the cost of steam is relatively cheap compared to the evaporator cost. However, for large-capacity operation, using more than one effect will markedly reduce steam costs.

2. Forward-feed multiple-effect evaporators.　A single-effect evaporator as shown in Fig. 8.2-2 is wasteful of energy since the latent heat of the vapor leaving is not used but is discarded. However, much of this latent heat can be recovered and reused by employing multiple-effect evaporators. A simplified diagram of a forward-feed triple-effect evaporation system is shown in Fig. 8.2-3.

If the feed to the first effect is near the boiling point at the pressure in the first effect, 1 kg of steam will evaporate almost 1 kg of water. The first effect operates at a high-enough temperature so that the evaporated water serves as the heating medium to the second effect. Here, again, almost another kg of water is evaporated, which can be used as the heating medium to the third effect. As a very rough approximation, almost 3 kg of water will be evaporated for 1 kg of steam for a three-effect evaporator. Hence, the *steam economy*, which is kg vapor evaporated/kg steam used, is increased. This also approximately holds for a number of effects over three. However, this increased steam economy of a multiple-effect evaporator is gained at the expense of the original first cost of these evaporators.

In forward-feed operation as shown in Fig. 8.2-3, the fresh feed is added to the first effect and flows to the next in the same direction as the vapor flow. This method of operation is used when the feed is hot or when the final concentrated product might be damaged at high temperatures. The boiling temperatures decrease from effect to effect. This means that if the first effect is at $P_1 = 1$ atm abs pressure, the last effect will be under vacuum at a pressure P_3.

3. Backward-feed multiple-effect evaporators.　In the backward-feed operation shown in Fig. 8.2-4 for a triple-effect evaporator, the fresh feed enters the last and coldest effect and continues on until the concentrated product leaves the first effect. This method of reverse feed is advantageous when the fresh feed is cold, since a smaller amount of liquid must be heated to the higher temperatures in the second and first effects. However, liquid pumps

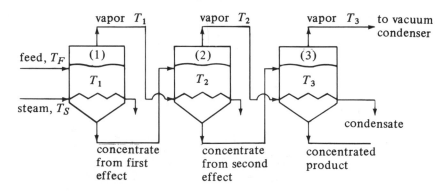

FIGURE 8.2-3.　*Simplified diagram of forward-feed triple-effect evaporator.*

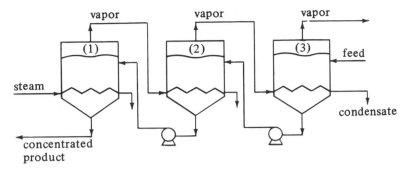

FIGURE 8.2-4. *Simplified diagram of backward-feed triple-effect evaporator.*

are used in each effect, since the flow is from low to high pressure. This method is also used when the concentrated product is highly viscous. The high temperatures in the early effects reduce the viscosity and give reasonable heat-transfer coefficients.

4. Parallel-feed multiple-effect evaporators. Parallel feed in multiple-effect evaporators involves the adding of fresh feed and the withdrawal of concentrated product from each effect. The vapor from each effect is still used to heat the next effect. This method of operation is mainly used when the feed is almost saturated and solid crystals are the product, as in the evaporation of brine to make salt.

8.3 OVERALL HEAT-TRANSFER COEFFICIENTS IN EVAPORATORS

The overall heat-transfer coefficient U in an evaporator is composed of the steam-side condensing coefficient, which has a value of about 5700 W/m$^2 \cdot$ K (1000 btu/h $\cdot$ ft$^2 \cdot$ °F); the metal wall, which has a high thermal conductivity and usually has a negligible resistance; the resistance of the scale on the liquid side; and the liquid film coefficient, which is usually inside the tubes.

The steam-side condensing coefficient outside the tubes can be estimated using Eqs. (4.8-20)–(4.8-26). The resistance due to scale formation usually cannot be predicted. Increasing the velocity of the liquid in the tubes greatly decreases the rate of scale formation. This is one important advantage of forced-circulation evaporators. The scale can be salts, such as calcium sulfate and sodium sulfate, which decrease in solubility with an increase in temperature and hence tend to deposit on the hot tubes.

For forced-circulation evaporators the coefficient h inside the tubes can be predicted if there is little or no vaporization inside the tube. The liquid hydrostatic head in the tubes prevents most boiling in the tubes. The standard equations for predicting the h value of liquids inside tubes can be used. Velocities used often range from 2 to 5 m/s (7 to 15 ft/s). The heat-transfer coefficient can be predicted from Eq. (4.5-8), but using a constant of 0.028 instead of 0.027 (B1). If there is some or appreciable boiling in part or all of the tubes, the use of the equation assuming no boiling will give conservative safe results (P1).

For long-tube vertical natural circulation evaporators the heat-transfer coefficient is more difficult to predict, since there is a nonboiling zone in the bottom of the tubes and a boiling zone in the top. The length of the nonboiling zone depends on the heat transfer in the two zones and the pressure drop in the boiling two-phase zone. The film heat-transfer

Type of Evaporator	Overall U	
	$W/m^2 \cdot K$	$Btu/h \cdot ft^2 \cdot °F$
Short-tube vertical, natural circulation	1100–2800	200–500
Horizontal-tube, natural circulation	1100–2800	200–500
Long-tube vertical, natural circulation	1100–4000	200–700
Long-tube vertical, forced circulation	2300–11 000	400–2000
Agitated film	680–2300	120–400

* Generally, nonviscous liquids have the higher coefficients and viscous liquids the lower coefficients in the ranges given.

coefficient in the nonboiling zone can be estimated using Eq. (4.5-8) with a constant of 0.028. For the boiling two-phase zone, a number of equations are given by Perry and Green (P2).

For short-tube vertical evaporators the heat-transfer coefficients can be estimated by using the same methods as for the long-tube vertical natural circulation evaporators. Horizontal-tube evaporators have heat-transfer coefficients of the same order of magnitude as the short-tube vertical evaporators.

For the agitated-film evaporator, the heat-transfer coefficient may be estimated using Eq. (4.13-4) for a scraped surface heat exchanger.

The methods given above are useful for actual evaporator design and/or for evaluating the effects of changes in operating conditions on the coefficients. In making preliminary designs or cost estimates, it is helpful to have available overall heat-transfer coefficients usually encountered in commercial practice. Some preliminary values and ranges of values for various types of evaporators are given in Table 8.3-1.

8.4   CALCULATION METHODS FOR SINGLE-EFFECT EVAPORATORS

8.4A   Heat and Material Balances for Evaporators

The basic equation for solving for the capacity of a single-effect evaporator is Eq. (8.2-1), which can be written as

$$q = UA \, \Delta T \qquad (8.4-1)$$

where ΔT K (°F) is the difference in temperature between the condensing steam and the boiling liquid in the evaporator. In order to solve Eq. (8.4-1) the value of q in W (btu/h) must be determined by making a heat and material balance on the evaporator shown in Fig. 8.4-1. The feed to the evaporator is F kg/h (lb$_m$/h) having a solids content of x_F mass fraction, temperature T_F, and enthalpy h_F J/kg (btu/lb$_m$). Coming out as a liquid is the concentrated liquid L kg/h (lb$_m$/h) having a solids content of x_L, temperature T_1, and enthalpy h_L. The vapor V kg/h (lb$_m$/h) is given off as pure solvent having a solids content of $y_V = 0$, temperature T_1, and enthalpy H_V. Saturated steam entering is S kg/h (lb$_m$/h) and has a temperature of T_S and enthalpy of H_S. The condensed steam leaving of S kg/h is assumed usually to be at T_S, the saturation temperature, with an enthalpy of h_S. This means that the steam gives off only its latent heat, which is λ, where

$$\lambda = H_S - h_S \tag{8.4-2}$$

Since the vapor V is in equilibrium with the liquid L, the temperatures of vapor and liquid are the same. Also, the pressure P_1 is the saturation vapor pressure of the liquid of composition x_L at its boiling point T_1. (This assumes no boiling-point rise.)

For the material balance since we are at steady state, the rate of mass in = rate of mass out. Then, for a total balance,

$$F = L + V \tag{8.4-3}$$

For a balance on the solute (solids) alone,

$$Fx_F = Lx_L \tag{8.4-4}$$

For the heat balance, since the total heat entering = total heat leaving,

heat in feed + heat in steam

= heat in concentrated liquid + heat in vapor + heat in condensed steam   (8.4-5)

This assumes no heat lost by radiation or convection. Substituting into Eq. (8.4-5),

$$Fh_F + SH_S = Lh_L + VH_V + Sh_S \tag{8.4-6}$$

Substituting Eq. (8.4-2) into (8.4-6),

$$Fh_F + S\lambda = Lh_L + VH_V \tag{8.4-7}$$

The heat q transferred in the evaporator is then

$$q = S(H_S - h_S) = S\lambda \tag{8.4-8}$$

In Eq. (8.4-7) the latent heat λ of steam at the saturation temperature T_S can be obtained from the steam tables in Appendix A.2. However, the enthalpies of the feed and products are often not available. These enthalpy–concentration data are available for only a few substances in solution. Hence, some approximations are made in order to make a heat balance. These are as follows.

1. It can be demonstrated as an approximation that the latent heat of evaporation of 1 kg mass of the water from an aqueous solution can be obtained from the steam tables using the temperature of the boiling solution T_1 (exposed surface temperature) rather than the equilibrium temperature for pure water at P_1.
2. If the heat capacities c_{pF} of the liquid feed and c_{pL} of the product are known, they can be used to calculate the enthalpies. (This neglects heats of dilution, which in most cases are not known.)

FIGURE 8.4-1. *Heat and mass balance for single-effect evaporator.*

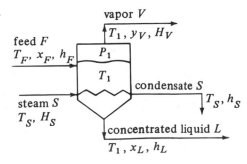

EXAMPLE 8.4-1. Heat-Transfer Area in Single-Effect Evaporator

A continuous single-effect evaporator concentrates 9072 kg/h of a 1.0 wt % salt solution entering at 311.0 K (37.8°C) to a final concentration of 1.5 wt %. The vapor space of the evaporator is at 101.325 kPa (1.0 atm abs) and the steam supplied is saturated at 143.3 kPa. The overall coefficient $U = 1704$ W/m² · K. Calculate the amounts of vapor and liquid product and the heat-transfer area required. Assume that, since it is dilute, the solution has the same boiling point as water.

Solution: The flow diagram is the same as that in Fig. 8.4-1. For the material balance, substituting into Eq. (8.4-3),

$$F = L + V \tag{8.4-3}$$

$$9072 = L + V$$

Substituting into Eq. (8.4-4) and solving,

$$Fx_F = Lx_L \tag{8.4-4}$$

$$9072(0.01) = L(0.015)$$

$$L = 6048 \text{ kg/h of liquid}$$

Substituting into Eq. (8.4-3) and solving,

$$V = 3024 \text{ kg/h of vapor}$$

The heat capacity of the feed is assumed to be $c_{pF} = 4.14$ kJ/kg · K. (Often, for feeds of inorganic salts in water, the c_p can be assumed approximately as that of water alone.) To make a heat balance using Eq. (8.4-7), it is convenient to select the boiling point of the dilute solution in the evaporator, which is assumed to be that of water at 101.32 kPa, $T_1 = 373.2$ K (100°C), as the datum temperature. Then H_V is simply the latent heat of water at 373.2 K, which from the steam tables in Appendix A.2 is 2257 kJ/kg (970.3 btu/lb$_m$). The latent heat λ of the steam at 143.3 kPa [saturation temperature $T_S = 383.2$ K (230°F)] is 2230 kJ/kg (958.8 btu/lb$_m$).

The enthalpy of the feed can be calculated from

$$h_F = c_{pF}(T_F - T_1) \tag{8.4-9}$$

Substituting into Eq. (8.4-7) with $h_L = 0$, since it is at the datum of 373.2 K,

$$9072(4.14)(311.0 - 373.2) + S(2230) = 6048(0) + 3024(2257)$$

$$S = 4108 \text{ kg steam/h}$$

The heat q transferred through the heating surface area A is, from Eq. (8.4-8),

$$q = S(\lambda) \tag{8.4-8}$$

$$q = 4108(2230)(1000/3600) = 2\,544\,000 \text{ W}$$

Substituting into Eq. (8.4-1), where $\Delta T = T_S - T_1$,

$$q = 2\,544\,000 = UA\,\Delta T = 1704(A)(383.2 - 373.2)$$

Solving, $A = 149.3$ m².

8.4B Effects of Processing Variables on Evaporator Operation

1. Effect of feed temperature. The inlet temperature of the feed has a large effect on the operation of the evaporator. In Example 8.4-1 the feed entering was at a cold temperature of 311.0 K compared to the boiling temperature of 373.2 K. About $\frac{1}{4}$ of the steam used for heating was used to heat the cold feed to the boiling point. Hence, only about $\frac{3}{4}$ of the steam was left for vaporization of the feed. If the feed is under pressure and enters the evaporator at a temperature above the boiling point in the evaporator, additional vaporization is obtained by the flashing of part of the entering hot feed. Preheating the feed can reduce the size of evaporator heat-transfer area needed.

2. Effect of pressure. In Example 8.4-1 the pressure of 101.32 kPa abs was used in the vapor space of the evaporator. This set the boiling point of the solution at 373.2 K and gave a ΔT for use in Eq. (8.4-1) of $383.2 - 373.2$, or 10 K. In many cases a larger ΔT is desirable, since, as the ΔT increases, the heating-surface area A and cost of evaporator decrease. To reduce the pressure below 101.32 kPa, i.e., to be under vacuum, a condenser and vacuum pump can be used. For example, if the pressure were reduced to 41.4 kPa, the boiling point of water would be 349.9 K and the new ΔT would be $383.2 - 349.9$, or 33.3 K. A large decrease in heating-surface area would be obtained.

3. Effect of steam pressure. Using higher pressure, saturated steam increases the ΔT, which decreases the size and cost of the evaporator. However, high-pressure steam is more costly and also is often more valuable as a source of power elsewhere. Hence, overall economic balances are really needed to determine the optimum steam pressures.

8.4C Boiling-Point Rise of Solutions

In the majority of cases in evaporation the solutions are not such dilute solutions as those considered in Example 8.4-1. In most cases, the thermal properties of the solution being evaporated may differ considerably from those of water. The concentrations of the solutions are high enough so that the heat capacity and boiling point are quite different from that of water.

For strong solutions of dissolved solutes the boiling-point rise due to the solutes in the solution usually cannot be predicted. However, a useful empirical law known as *Dühring's rule* can be used. In this rule a straight line is obtained if the boiling point of a solution in °C or °F is plotted against the boiling point of pure water at the same pressure for a given concentration at different pressures. A different straight line is obtained for each given concentration. In Fig. 8.4-2 such a Dühring line chart is given for solutions of sodium hydroxide in water. It is necessary to know the boiling point of a given solution at only two pressures to determine a line.

EXAMPLE 8.4-2. Use of Dühring Chart for Boiling-Point Rise
As an example of use of the chart, the pressure in an evaporator is given as 25.6 kPa (3.72 psia) and a solution of 30% NaOH is being boiled. Determine the boiling temperature of the NaOH solution and the boiling-point rise BPR of the solution over that of water at the same pressure.

Solution: From the steam tables in Appendix A.2, the boiling point of water at 25.6 kPa is 65.6°C. From Fig. 8.4-2 for 65.6°C (150°F) and 30% NaOH, the boiling point of the NaOH solution is 79.5°C (175°F). The boiling-point rise is $79.5 - 65.6 = 13.9$°C (25°F).

In Perry and Green (P2) a chart is given to estimate the boiling-point rise of a large number of common aqueous solutions used in chemical and biological processes. In addition to the common salts and solutes, such as $NaNO_3$, $NaOH$, $NaCl$, and H_2SO_4, the biological solutes sucrose, citric acid, kraft solution, and glycerol are given. These

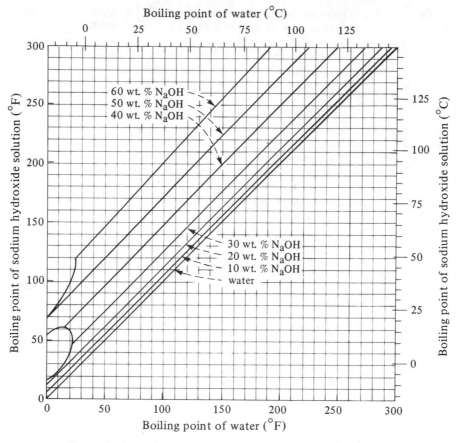

FIGURE 8.4-2. *Dühring lines for aqueous solutions of sodium hydroxide.*

biological solutes have quite small boiling-point-rise values compared to those of common salts.

8.4D Enthalpy–Concentration Charts of Solutions

If the heat of solution of the aqueous solution being concentrated in the evaporator is large, neglecting it could cause errors in the heat balances. This heat-of-solution phenomenon can be explained as follows. If pellets of NaOH are dissolved in a given amount of water, it is found that a considerable temperature rise occurs; i.e., heat is evolved, called *heat of solution*. The amount of heat evolved depends on the type of substance and on the amount of water used. Also, if a strong solution of NaOH is diluted to a lower concentration, heat is liberated. Conversely, if a solution is concentrated from a low to a high concentration, heat must be added.

In Fig. 8.4-3 an enthalpy–concentration chart for NaOH is given (M1) where the enthalpy is in kJ/kg (btu/lb$_m$) solution, temperature in °C (°F), and concentration in weight fraction NaOH in solution. Such enthalpy–concentration charts are usually not made for solutions having negligible heats of solution, since the heat capacities can be easily used to calculate enthalpies. Also, such charts are available for only a few solutions.

The enthalpy of the liquid water in Fig. 8.4-3 is referred to the same datum or reference state as in the steam tables, i.e., liquid water at 0°C (273 K). This means that enthalpies from the figure can be used with those in the steam tables. In Eq. (8.4-7) values for h_F and h_L can be taken from Fig. 8.4-3 and values for λ and H_V from the steam tables. The uses of Fig. 8.4-3 will be best understood in the following example.

EXAMPLE 8.4-3. Evaporation of an NaOH Solution

An evaporator is used to concentrate 4536 kg/h (10 000 lb_m/h) of a 20% solution of NaOH in water entering at 60°C (140°F) to a product of 50% solids. The pressure of the saturated steam used is 172.4 kPa (25 psia) and the pressure in the vapor space of the evaporator is 11.7 kPa (1.7 psia). The overall heat-transfer coefficient is 1560 W/m$^2 \cdot$ K (275 btu/h $\cdot$ ft$^2 \cdot$ °F). Calculate the steam used, the steam economy in kg vaporized/kg steam used, and the heating surface area in m^2.

Solution: The process flow diagram and nomenclature are the same as given in Fig. 8.4-1. The given variables are F = 4536 kg/h, x_F = 0.20 wt fraction, T_F = 60°C, P_1 = 11.7 kPa, steam pressure = 172.4 kPa, and x_L = 0.50 wt fraction. For the overall material balance, substituting into Eq. (8.4-3),

$$F = 4536 = L + V \qquad\qquad (8.4-3)$$

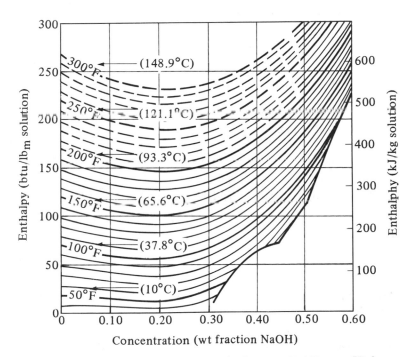

FIGURE 8.4-3. *Enthalpy–concentration chart for the system NaOH–water. [Reference state liquid water at 0°C (273 K) or 32°F.] [From W. L. McCabe, Trans. A.I.Ch.E., 31, 129 (1935). With permission.]*

Substituting into Eq. (8.4-4) and solving (8.4-3) and (8.4-4) simultaneously,

$$Fx_F = Lx_L \qquad (8.4\text{-}4)$$

$$4536(0.20) = L(0.50)$$

$$L = 1814 \text{ kg/h} \qquad V = 2722 \text{ kg/h}$$

To determine the boiling point T_1 of the 50% concentrated solution, we first obtain the boiling point of pure water at 11.7 kPa from the steam tables, Appendix A.2, as 48.9°C (120°F). From the Dühring chart, Fig. 8.4-2, for a boiling point of water of 48.9°C and 50% NaOH, the boiling point of the solution is $T_1 = 89.5$°C (193°F). Hence,

$$\text{boiling-point rise} = T_1 - 48.9 = 89.5 - 48.9 = 40.6\text{°C (73°F)}$$

From the enthalpy–concentration chart (Fig. 8.4-3), for 20% NaOH at 60°C (140°F), $h_f = 214$ kJ/kg (92 btu/lb$_m$). For 50% NaOH at 89.5°C (193°F), $h_L = 505$ kJ/kg (217 btu/lb$_m$).

For the superheated vapor V at 89.5°C (193°F) and 11.7 kPa [superheated 40.6°C (73°F) since the boiling point of water is 48.9°C (120°F) at 11.7 kPa], from the steam tables, $H_V = 2667$ kJ/kg (1147 btu/lb$_m$). An alternative method to use to calculate the H_V is to first obtain the enthalpy of saturated vapor at 48.9°C (120°F) and 11.7 kPa of 2590 kJ/kg (1113.5 btu/lb$_m$). Then using a heat capacity of 1.884 kJ/kg·K for superheated steam with the superheat of $(89.5 - 48.9)$°C $= (89.5 - 48.9)$ K,

$$H_V = 2590 + 1.884(89.5 - 48.9) = 2667 \text{ kJ/kg}$$

For the saturated steam at 172.4 kPa, the saturation temperature from the steam tables is 115.6°C (240°F) and the latent heat is $\lambda = 2214$ kJ/kg (952 btu/lb$_m$).

Substituting into Eq. (8.4-7) and solving for S,

$$Fh_f + S\lambda = Lh_L + VH_V \qquad (8.4\text{-}7)$$

$$4535(214) + S(2214) = 1814(505) + 2722(2667)$$

$$S = 3255 \text{ kg steam/h}$$

Substituting into Eq. (8.4-8),

$$q = S\lambda = 3255(2214)\left(\frac{1}{3600}\right) = 2002 \text{ kW}$$

Substituting into Eq. (8.4-1) and solving,

$$2002(1000) = 1560(A)(115.6 - 89.5)$$

Hence, $A = 49.2$ m^2. Also, steam economy $= 2722/3255 = 0.836$.

8.5 CALCULATION METHODS FOR MULTIPLE-EFFECT EVAPORATORS

8.5A Introduction

In evaporation of solutions in a single-effect evaporator, a major cost is the cost of the steam used to evaporate the water. A single-effect evaporator is wasteful of steam costs, since the latent heat of the vapor leaving the evaporator is usually not used. However, to reduce this cost, multiple-effect evaporators are used which recover the latent heat of the vapor leaving and reuse it.

A three-effect evaporator, discussed briefly in Section 8.2B, is shown in Fig. 8.2-3. In this system each effect in itself acts as a single-effect evaporator. In the first effect raw steam is used as the heating medium to this first effect, which is boiling at temperature T_1 and pressure P_1. The vapor removed from the first effect is used as the heating medium, condensing in the second effect and vaporizing water at temperature T_2 and pressure P_2 in this effect. To transfer heat from the condensing vapor to the boiling liquid in this second effect, the boiling temperature T_2 must be less than the condensing temperature. This means that the pressure P_2 in the second effect is lower than P_1 in the first effect. In a similar manner, vapor from the second effect is condensed in heating the third effect. Hence, pressure P_3 is less than P_2. If the first effect is operating at 1 atm abs pressure, the second and third effects will be under vacuum.

In the first effect, raw dilute feed is added and it is partly concentrated. Then this partly concentrated liquid (Fig. 8.2-3) flows to the second evaporator in series, where it is further concentrated. This liquid from the second effect flows to the third effect for final concentration.

When a multiple-effect evaporator is at steady-state operation, the flow rates and rate of evaporation in each effect are constant. The pressures, temperatures, and internal flow rates are automatically kept constant by the steady-state operation of the process itself. To change the concentration in the final effect, the feed rate to the first effect must be changed. The overall material balance made over the whole system and over each evaporator itself must be satisfied. If the final solution is too concentrated, the feed rate is increased, and vice versa. Then the final solution will reach a new steady state at the desired concentration.

8.5B Temperature Drops and Capacity of Multiple-Effect Evaporators

1. *Temperature drops in multiple-effect evaporators.* The amount of heat transferred per hour in the first effect of a triple-effect evaporator with forward feed as in Fig. 8.2-3 will be

$$q_1 = U_1 A_1 \, \Delta T_1 \tag{8.5-1}$$

where ΔT_1 is the difference between the condensing steam and the boiling point of the liquid, $T_s - T_1$. Assuming that the solutions have no boiling-point rise and no heat of solution and neglecting the sensible heat necessary to heat the feed to the boiling point, approximately all the latent heat of the condensing steam appears as latent heat in the vapor. This vapor then condenses in the second effect, giving up approximately the same amount of heat.

$$q_2 = U_2 A_2 \, \Delta T_2 \tag{8.5-2}$$

This same reasoning holds for q_3. Then since $q_1 = q_2 = q_3$, then, approximately,

$$U_1 A_1 \Delta T_1 = U_2 A_2 \, \Delta T_2 = U_3 A_3 \, \Delta T_3 \tag{8.5-3}$$

Usually, in commercial practice the areas in all effects are equal and

$$\frac{q}{A} = U_1 \, \Delta T_1 = U_2 \, \Delta T_2 = U_3 \, \Delta T_3 \tag{8.5-4}$$

Hence, the temperature drops ΔT in a multiple-effect evaporator are approximately inversely proportional to the values of U. Calling $\sum \Delta T$ as follows for no boiling-point rise,

$$\sum \Delta T = \Delta T_1 + \Delta T_2 + \Delta T_3 = T_S - T_3 \qquad (8.5\text{-}5)$$

Note that ΔT_1 °C $= \Delta T_1$ K, ΔT_2 °C $= \Delta T_2$ K, and so on. Since ΔT_1 is proportional to $1/U_1$, then

$$\Delta T_1 = \sum \Delta T \, \frac{1/U_1}{1/U_1 + 1/U_2 + 1/U_3} \qquad (8.5\text{-}6)$$

Similar equations can be written for ΔT_2 and ΔT_3.

2. Capacity of multiple-effect evaporators. A rough estimate of the capacity of a three-effect evaporator compared to a single effect can be obtained by adding the value of q for each evaporator.

$$q = q_1 + q_2 + q_3 = U_1 A_1 \, \Delta T_1 + U_2 A_2 \, \Delta T_2 + U_3 A_3 \, \Delta T_3 \qquad (8.5\text{-}7)$$

If we make the assumption that the value of U is the same in each effect and the values of A are equal, Eq. (8.5-7) becomes

$$q = U A (\Delta T_1 + \Delta T_2 + \Delta T_3) = U A \, \Delta T \qquad (8.5\text{-}8)$$

where $\Delta T = \sum \Delta T = \Delta T_1 + \Delta T_2 + \Delta T_3 = T_S - T_3$.

If a single-effect evaporator is used with the same area A, the same value of U, and the same total temperature drop ΔT, then

$$q = U A \, \Delta T \qquad (8.5\text{-}9)$$

This, of course, gives the same capacity as for the multiple-effect evaporators. Hence, the increase in steam economy obtained by using multiple-effect evaporators is obtained at the expense of reduced capacity.

8.5C Calculations for Multiple-Effect Evaporators

In doing calculations for a multiple-effect evaporator system, the values to be obtained are usually the area of the heating surface in each effect, the kg of steam per hour to be supplied, and the amount of vapor leaving each effect, especially the last effect. The given or known values are usually as follows: (1) steam pressure to first effect, (2) final pressure in vapor space of the last effect, (3) feed conditions and flow to first effect, (4) the final concentration in the liquid leaving the last effect, (5) physical properties such as enthalpies and/or heat capacities of the liquid and vapors, and (6) the overall heat-transfer coefficients in each effect. Usually, the areas of each effect are assumed equal.

The calculations are done using material balances, heat balances, and the capacity equations $q = U A \, \Delta T$ for each effect. A convenient way to solve these equations is by trial and error. The basic steps to follow are given as follows for a triple-effect evaporator.

8.5D Step-by-Step Calculation Methods
for Triple-Effect Evaporators

1. From the known outlet concentration and pressure in the last effect, determine the boiling point in the last effect. (If a boiling-point rise is present, this can be determined from a Dühring line plot.)

2. Determine the total amount of vapor evaporated by an overall material balance. For this first trial apportion this total amount of vapor among the three effects. (Usually, equal vapor produced in each effect, so that $V_1 = V_2 = V_3$ is assumed for the first trial.) Make a total material balance on effects 1, 2, and 3 to obtain L_1, L_2, and L_3. Then calculate the solids concentration in each effect by a solids balance on each effect.

3. Using Eq. (8.5-6), estimate the temperature drops ΔT_1, ΔT_2, and ΔT_3 in the three effects. Any effect that has an extra heating load, such as a cold feed, requires a proportionately larger ΔT. Then calculate the boiling point in each effect.

   [If a boiling-point rise (BPR) in °C is present, estimate the pressure in effects 1 and 2 and determine the BPR in each of the three effects. Only a crude pressure estimate is needed since BPR is almost independent of pressure. Then the $\sum \Delta T$ available for heat transfer without the superheat is obtained by subtracting the sum of all three BPRs from the overall ΔT of $T_S - T_3$ (saturation). Using Eq. (8.5-6), estimate $\Delta T_1, \Delta T_2$, and ΔT_3. Then calculate the boiling point in each effect.]

4. Using heat and material balances in each effect, calculate the amount vaporized and the flows of liquid in each effect. If the amounts vaporized differ appreciably from those assumed in step 2, then steps 2, 3, and 4 can be repeated using the amounts of evaporation just calculated. (In step 2 only the solids balance is repeated.)

5. Calculate the value of q transferred in each effect. Using the rate equation $q = UA \, \Delta T$ for each effect, calculate the areas A_1, A_2, and A_3. Then calculate the average value A_m by

$$A_m = \frac{A_1 + A_2 + A_3}{3} \tag{8.5-10}$$

If these areas are reasonably close to each other, the calculations are complete and a second trial is not needed. If these areas are not nearly equal, a second trial should be performed as follows.

6. To start trial 2, use the new values of L_1, L_2, L_3, V_1, V_2, and V_3 calculated by the heat balances in step 4 and calculate the new solids concentrations in each effect by a solids balance on each effect.

7. Obtain new values of $\Delta T'_1, \Delta T'_2$, and $\Delta T'_3$ from

$$\Delta T'_1 = \frac{\Delta T_1 A_1}{A_m} \qquad \Delta T'_2 = \frac{\Delta T_2 A_2}{A_m} \qquad \Delta T'_3 = \frac{\Delta T_3 A_3}{A_m} \tag{8.5-11}$$

The sum $\Delta T'_1 + \Delta T'_2 + \Delta T'_3$ must equal the original $\sum \Delta T$. If not, proportionately readjust all $\Delta T'$ values so that this is so. Then calculate the boiling point in each effect.

   [If a boiling point rise is present, then using the new concentrations from step 6, determine the new BPRs in the three effects. This gives a new value of $\sum \Delta T$ available for heat transfer by subtracting the sum of all three BPRs from the overall ΔT. Calculate the new values of $\Delta T'$ by Eq. (8.5-11). The sum of the $\Delta T'$ values just calculated must be readjusted to this new $\sum \Delta T$ value. Then calculate the boiling point in each effect.] Step 7 is essentially a repeat of step 3 but using Eq. (8.5-11) to obtain a better estimate of the $\Delta T'$ values.

8. Using the new $\Delta T'$ values from step 7, repeat the calculations starting with step 4. Two trials are usually sufficient so that the areas are reasonably close to being equal.

EXAMPLE 8.5-1. Evaporation of Sugar Solution in a Triple-Effect Evaporator

A triple-effect forward-feed evaporator is being used to evaporate a sugar solution containing 10 wt % solids to a concentrated solution of 50%. The boiling-point rise of the solutions (independent of pressure) can be estimated from $BPR°C = 1.78x + 6.22x^2$ ($BPR°F = 3.2x + 11.2x^2$), where x is wt fraction of sugar in solution (K1). Saturated steam at 205.5 kPa (29.8 psia) [121.1°C (250°F) saturation temperature] is being used. The pressure in the vapor space of the third effect is 13.4 kPa (1.94 psia). The feed rate is 22 680 kg/h (50 000 lb_m/h) at 26.7°C (80°F). The heat capacity of the liquid solutions is (K1) $c_p = 4.19 - 2.35x$ kJ/kg·K ($1.0 - 0.56x$ btu/lb_m·°F). The heat of solution is considered to be negligible. The coefficients of heat transfer have been estimated as $U_1 = 3123$, $U_2 = 1987$, and $U_3 = 1136$ W/m²·K or 550, 350, and 200 btu/h·ft²·°F. If each effect has the same surface area, calculate the area, the steam rate used, and the steam economy.

Solution: The process flow diagram is given in Fig. 8.5-1. Following the eight steps outlined, the calculations are as follows.

Step 1. For 13.4 kPa (1.94 psia), the saturation temperature is 51.67°C (125°F) from the steam tables. Using the equation for BPR for evaporator number 3 with $x = 0.5$,

$$BPR_3 = 1.78x + 6.22x^2 = 1.78(0.5) + 6.22(0.5)^2 = 2.45°C (4.4°F)$$

$$T_3 = 51.67 + 2.45 = 54.12°C (129.4°F)$$

Step 2. Making an overall and a solids balance to calculate the total amount vaporized $(V_1 + V_2 + V_3)$ and L_3,

$$F = 22\,680 = L_3 + (V_1 + V_2 + V_3)$$

$$Fx_F = 22\,680(0.1) = L_3(0.5) + (V_1 + V_2 + V_3)(0)$$

$$L_3 = 4536 \text{ kg/h } (10\,000 \text{ lb}_m/\text{h})$$

$$\text{total vaporized} = (V_1 + V_2 + V_3) = 18\,144 \text{ kg/h } (40\,000 \text{ lb}_m/\text{h})$$

Assuming equal amount vaporized in each effect, $V_1 = V_2 = V_3 = 6048$ kg/h (13 333 lb_m/h). Making a total material balance on effects 1, 2, and 3 and solving,

(1) $F = 22\,680 = V_1 + L_1 = 6048 + L_1$, $L_1 = 16\,632$ kg/h (33 667 lb_m/h)

(2) $L_1 = 16\,632 = V_2 + L_2 = 6048 + L_2$, $L_2 = 10\,584(23\,334)$

(3) $L_2 = 10\,584 = V_3 + L_3 = 6048 + L_3$, $L_3 = 4536(10\,000)$

Making a solids balance on effects 1, 2, and 3 and solving for x,

(1) $22\,680(0.1) = L_1 x_1 = 16\,632(x_1)$, $x_1 = 0.136$

(2) $16\,632(0.136) = L_2 x_2 = 10\,584(x_2)$, $x_2 = 0.214$

(3) $10\,584(0.214) = L_3 x_3 = 4536(x_3)$, $x_3 = 0.500$ (check balance)

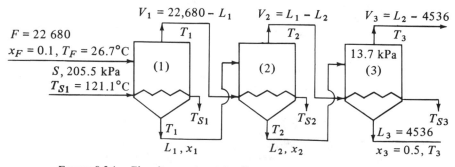

FIGURE 8.5-1.   *Flow diagram for triple-effect evaporation for Example 8.5-1.*

Step 3. The BPR in each effect is calculated as follows:

(1) $\text{BPR}_1 = 1.78x_1 + 6.22x_1^2 = 1.78(0.136) + 6.22(0.136)^2 = 0.36°\text{C} \ (0.7°\text{F})$

(2) $\text{BPR}_2 = 1.78(0.214) + 6.22(0.214)^2 = 0.65°\text{C} \ (1.2°\text{F})$

(3) $\text{BPR}_3 = 1.78(0.5) + 6.22(0.5)^2 = 2.45°\text{C} \ (4.4°\text{F})$

$$\sum \Delta T \text{ available} = T_{S1} - T_3 \text{ (saturation)} - (\text{BPR}_1 + \text{BPR}_2 + \text{BPR}_3)$$

$$= 121.1 - 51.67 - (0.36 + 0.65 + 2.45) = 65.97°\text{C} \ (118.7°\text{F})$$

Using Eq. (8.5-6) for ΔT_1 and similar equations for ΔT_2 and ΔT_3,

$$\Delta T_1 = \sum \Delta T \frac{1/U_1}{1/U_1 + 1/U_2 + 1/U_3} = \frac{(65.97)(1/3123)}{(1/3123) + (1/1987) + (1/1136)}$$

$$\Delta T_1 = 12.40°\text{C} \qquad \Delta T_2 = 19.50°\text{C} \qquad \Delta T_3 = 34.07°\text{C}$$

However, since a cold feed enters effect number 1, this effect requires more heat. Increasing ΔT_1 and lowering ΔT_2 and ΔT_3 proportionately as a first estimate,

$$\Delta T_1 = 15.56°\text{C} = 15.56 \text{ K} \qquad \Delta T_2 = 18.34°\text{C} = 18.34 \text{ K}$$

$$\Delta T_3 = 32.07°\text{C} = 32.07 \text{ K}$$

To calculate the actual boiling point of the solution in each effect,

(1) $T_1 = T_{S1} - \Delta T_1$

$$= 121.1 - 15.56 = 105.54°\text{C}$$

$T_{S1} = 121.1°\text{C}$ (condensing temperature of saturated steam to effect 1)

(2) $T_2 = T_1 - \text{BPR}_1 - \Delta T_2$

$$= 105.54 - 0.36 - 18.34 = 86.84°\text{C}$$

$T_{S2} = T_1 - \text{BPR}_1 = 105.54 - 0.36$

$$= 105.18°\text{C} \quad \text{(condensing temperature of steam to effect 2)}$$

(3) $T_3 = T_2 - \text{BPR}_2 - \Delta T_3$

$$= 86.84 - 0.65 - 32.07 = 54.12°\text{C}$$

$T_{S3} = T_2 - \text{BPR}_2$

$$= 86.84 - 0.65 = 86.19°\text{C} \quad \text{(condensing temperature of steam to effect 3)}$$

The temperatures in the three effects are as follows:

Effect 1	Effect 2	Effect 3	Condenser
$T_{S1} = 121.1°\text{C}$	$T_{S2} = 105.18$	$T_{S3} = 86.19$	$T_{S4} = 51.67$
$T_1 = 105.54$	$T_2 = 86.84$	$T_3 = 54.12$	

Step 4. The heat capacity of the liquid in each effect is calculated from the equation $c_p = 4.19 - 2.35x$.

$$F: \quad c_p = 4.19 - 2.35(0.1) = 3.955 \text{ kJ/kg} \cdot \text{K}$$

$$L_1: \quad c_p = 4.19 - 2.35(0.136) = 3.869$$

$$L_2: \quad c_p = 4.19 - 2.35(0.214) = 3.684$$

$$L_3: \quad c_p = 4.19 - 2.35(0.5) = 3.015$$

The values of the enthalpy H of the various vapor streams relative to water at 0°C as a datum are obtained from the steam table as follows:

Effect 1:

$$T_1 = 105.54°C, \ T_{S2} = 105.18 \ (221.3°F), \ \text{BPR}_1 = 0.36, \ T_{S1} = 121.1 \ (250°F)$$

$$H_1 = H_{S2} \ \text{(saturation enthalpy at } T_{S2}) + 1.884 \ (0.36°C \ \text{superheat})$$

$$= 2684 + 1.884(0.36) = 2685 \ \text{kJ/kg}$$

$$\lambda_{S1} = H_{S1} \ \text{(vapor saturation enthalpy)} - h_{S1} \ \text{(liquid enthalpy at } T_{S1})$$

$$= (2708 - 508) = 2200 \ \text{kJ/kg latent heat of condensation}$$

Effect 2:

$$T_2 = 86.84°C, \quad T_{S3} = 86.19, \quad \text{BPR}_2 = 0.65$$

$$H_2 = H_{S3} + 1.884(0.65) = 2654 + 1.884(0.65) = 2655 \ \text{kJ/kg}$$

$$\lambda_{S2} = H_1 - h_{S2} = 2685 - 441 = 2244 \ \text{kJ/kg}$$

Effect 3:

$$T_3 = 54.12°C, \quad T_{S4} = 51.67, \quad \text{BPR}_3 = 2.45$$

$$H_3 = H_{S4} + 1.884(2.45) = 2595 + 1.884(2.45) = 2600 \ \text{kJ/kg}$$

$$\lambda_{S3} = H_2 - h_{S3} = 2655 - 361 = 2294 \ \text{kJ/kg}$$

(Note that the superheat corrections in this example are small and could possibly have been neglected. However, the corrections were used to demonstrate the method of calculation.) Flow relations to be used in heat balances are

$$V_1 = 22\,680 - L_1, \quad V_2 = L_1 - L_2, \quad V_3 = L_2 - 4536, \quad L_3 = 4536$$

Write a heat balance on each effect. Use 0°C as a datum since the values of H of the vapors are relative to 0°C (32°F) and note that $(T_F - 0)°C = (T_F - 0) \ \text{K}$ and $(T_1 - 0)°C = (T_1 - 0) \ \text{K}$,

(1) $$Fc_p(T_F - 0) + S\lambda_{S1} = L_1 c_p(T_1 - 0) + V_1 H_1$$

Substituting the known values,

$$22\,680(3.955)(26.7 - 0) + S(2200) = L_1(3.869)(105.54 - 0)$$

$$+ (22\,680 - L_1)(2685)$$

(2) $$L_1 c_p(T_1 - 0) + V_1 \lambda_{S2} = L_2 c_p(T_2 - 0) + V_2 H_2$$

$$L_1(3.869)(105.54 - 0) + (22\,680 - L_1)(2244) = L_2(3.684)(86.84 - 0)$$

$$+ (L_1 - L_2)(2655)$$

(3) $$L_2 c_p(T_2 - 0) + V_2 \lambda_{S3} = L_3 c_p(T_3 - 0) + V_3 H_3$$

$$L_2(3.684)(86.84 - 0) + (L_1 - L_2)(2294) = 4536(3.015)(54.12 - 0)$$

$$+ (L_2 - 4536)(2600)$$

Solving the last two equations simultaneously for L_1 and L_2 and substituting into the first equation,

$$L_1 = 17\,078 \ \text{kg/h} \qquad L_2 = 11\,068 \qquad L_3 = 4536$$

$$S = 8936 \qquad V_1 = 5602 \qquad V_2 = 6010 \qquad V_3 = 6532$$

The calculated values of V_1, V_2, and V_3 are close enough to the assumed values so that steps 2, 3, and 4 do not need to be repeated. If the calculation were repeated, the calculated values of V_1, V_2, and V_3 would be used starting with step 2 and a solids balance in each effect would be made.

Step 5. Solving for the values of q in each effect and area,

$$q_1 = S\lambda_{S1} = \left(\frac{8936}{3600}\right)(2200 \times 1000) = 5.460 \times 10^6 \text{ W}$$

$$q_2 = V_1\lambda_{S2} = \left(\frac{5602}{3600}\right)(2244 \times 1000) = 3.492 \times 10^6 \text{ W}$$

$$q_3 = V_2\lambda_{S3} = \left(\frac{6010}{3600}\right)(2294 \times 1000) - 3.830 \times 10^6 \text{ W}$$

$$A_1 = \frac{q_1}{U_1 \, \Delta T_1} = \frac{5.460 \times 10^6}{3123(15.56)} = 112.4 \text{ m}^2$$

$$A_2 = \frac{q_2}{U_2 \, \Delta T_2} = \frac{3.492 \times 10^6}{1987(18.34)} = 95.8 \text{ m}^2$$

$$A_3 = \frac{q_3}{U_3 \, \Delta T_3} = \frac{3.830 \times 10^6}{1136(32.07)} = 105.1 \text{ m}^2$$

The average area $A_m = 104.4 \text{ m}^2$. The areas differ from the average value by less than 10% and a second trial is not really necessary. However, a second trial will be made starting with step 6 to indicate the calculation methods used.

Step 6. Making a new solids balance on effects 1, 2, and 3 using the new $L_1 = 17\,078$, $L_2 = 11\,068$, and $L_3 = 4536$ and solving for x,

(1) $\qquad 22\,680(0.1) = 17\,078(x_1), \qquad x_1 = 0.133$

(2) $\qquad 17\,078(0.133) = 11\,068(x_2), \qquad x_2 = 0.205$

(3) $\qquad 11\,068(0.205) = 4536(x_3), \qquad x_3 = 0.500 \text{ (check balance)}$

Step 7. The new BPR in each effect is then

(1) $\text{BPR}_1 = 1.78x_1 + 6.22x_1^2 = 1.78(0.133) + 6.22(0.133)^2 = 0.35°C$

(2) $\text{BPR}_2 = 1.78(0.205) + 6.22(0.205)^2 = 0.63°C$

(3) $\text{BPR}_3 = 1.78(0.5) + 6.22(0.5)^2 = 2.45°C$

$\qquad \sum \Delta T \text{ available} = 121.1 - 51.67 - (0.35 + 0.63 + 2.45) = 66.00°C$

The new values for ΔT are obtained using Eq. (8.5-11).

$$\Delta T'_1 = \frac{\Delta T_1 A_1}{A_m} = \frac{15.56(112.4)}{104.4} = 16.77 \text{ K} = 16.77°C$$

$$\Delta T'_2 = \frac{\Delta T_2 A_2}{A_m} = \frac{18.34(95.8)}{104.4} = 16.86°C$$

$$\Delta T'_3 = \frac{\Delta T_3 A_3}{A_m} = \frac{32.07(105.1)}{104.4} = 32.34°C$$

$$\sum \Delta T = 16.77 + 16.86 + 32.34 = 65.97°C$$

These $\Delta T'$ values are readjusted so that $\Delta T'_1 = 16.77$, $\Delta T'_2 = 16.87$, $\Delta T'_3 = 32.36$, and $\sum \Delta T = 16.77 + 16.87 + 32.36 = 66.00°C$. To calculate the actual boiling point of the solution in each effect,

(1) $T_1 = T_{S1} - \Delta T'_1 = 121.1 - 16.77 = 104.33°C,$ $T_{S1} = 121.1°C$

(2) $T_2 = T_1 - BPR_1 - \Delta T'_2 = 104.33 - 0.35 - 16.87 = 87.11°C$

  $T_{S2} = T_1 - BPR_1 = 104.33 - 0.35 = 103.98°C$

(3) $T_3 = T_2 - BPR_2 - \Delta T'_3 = 87.11 - 0.63 - 32.36 = 54.12°C$

  $T_{S3} = T_2 - BPR_2 = 87.11 - 0.63 = 86.48°C$

Step 8. Following step 4, the heat capacity of the liquid is $c_p = 4.19 - 2.35x$.

$$F: \quad c_p = 3.955 \text{ kJ/kg} \cdot \text{K}$$

$$L_1: \quad c_p = 4.19 - 2.35(0.133) = 3.877$$

$$L_2: \quad c_p = 4.19 - 2.35(0.205) = 3.708$$

$$L_3: \quad c_p = 3.015$$

The new values of the enthalpy H are as follows in each effect.

(1) $H_1 = H_{S2} + 1.884(°C \text{ superheat}) = 2682 + 1.884(0.35) = 2683 \text{ kJ/kg}$

  $\lambda_{S1} = H_{S1} - h_{S1} = 2708 - 508 = 2200 \text{ kJ/kg}$

(2) $H_2 = H_{S3} + 1.884(0.63) = 2654 + 1.884(0.63) = 2655 \text{ kJ/kg}$

  $\lambda_{S2} = H_1 - h_{S2} = 2683 - 440 = 2243 \text{ kJ/kg}$

(3) $H_3 = H_{S4} + 1.884(2.45) = 2595 + 1.884(2.45) = 2600 \text{ kJ/kg}$

  $\lambda_{S3} = H_2 - h_{S3} = 2655 - 362 = 2293 \text{ kJ/kg}$

Writing a heat balance on each effect,

(1) $22\,680(3.955)(26.7 - 0) + S(2200) = L_1(3.877)(104.33 - 0)$

  $+ (22\,680 - L_1)(2683)$

(2) $L_1(3.877)(104.33 - 0) + (22\,680 - L_1)(2243) = L_2(3.708)(87.11 - 0)$

  $+ (L_1 - L_2)(2655)$

(3) $L_2(3.708)(87.11 - 0) + (L_1 - L_2)(2293) = 4536(3.015)(54.12 - 0)$

  $+ (L_2 - 4536)(2600)$

Solving,

$$L_1 = 17\,005 \text{ kg/h} \quad L_2 = 10\,952 \quad L_3 = 4536 \quad S = 8960 \text{ (steam used)}$$
$$V_1 = 5675 \quad\quad V_2 = 6053 \quad\quad V_3 = 6416$$

Note that these values from trial 2 differ very little from the trial 1 results. Following step 5, and solving for q in each effect and A,

$$q_1 = S\lambda_{S1} = \frac{8960}{3600}(2200 \times 1000) = 5.476 \times 10^6 \text{ W}$$

$$q_2 = V_1\lambda_{S2} = \frac{5675}{3600}(2243 \times 1000) = 3.539 \times 10^6 \text{ W}$$

$$q_3 = V_2 \lambda_{s3} = \frac{6053}{3600}(2293 \times 1000) = 3.855 \times 10^6 \text{ W}$$

$$A_1 = \frac{q_1}{U_1 \, \Delta T_1'} = \frac{5.476 \times 10^6}{3123(16.77)} = 104.6 \text{ m}^2$$

$$A_2 = \frac{q_2}{U_2 \, \Delta T_2'} = \frac{3.539 \times 10^6}{1987(16.87)} = 105.6 \text{ m}^2$$

$$A_3 = \frac{q_3}{U_3 \, \Delta T_3'} = \frac{3.855 \times 10^6}{1136(32.36)} = 104.9 \text{ m}^2$$

The average area $A_m = 105.0 \text{ m}^2$ to use in each effect. Note that this value of 105.0 m^2 is quite close to the average value of 104.4 m^2 from the first trial.

$$\text{steam economy} = \frac{V_1 + V_2 + V_3}{S} = \frac{5675 + 6053 + 6416}{8960} = 2.025$$

8.6 CONDENSERS FOR EVAPORATORS

8.6A Introduction

In multiple-effect evaporators the vapors from the last effect are usually leaving under vacuum, i.e., at less than atmospheric pressure. These vapors must be condensed and discharged as a liquid at atmospheric pressure. This is done by condensing the vapors using cooling water. The condenser can be a surface condenser, where the vapor to be condensed and the cooling liquid are separated by a metal wall, or a direct contact condenser, where the vapor and cooling liquid are mixed directly.

8.6B Surface Condensers

Surface condensers are employed where actual mixing of the condensate with condenser cooling water is not desired. In general, they are shell and tube condensers with the vapor on the shell side and cooling water in multipass flow on the tube side. Noncondensable gases are usually present in the vapor stream. These can be air, CO_2, N_2, or another gas which may have entered as dissolved gases in the liquid feed or occur because of decomposition in the solutions. These noncondensable gases may be vented from any well-cooled point in the condenser. If the vapor being condensed is below atmospheric pressure, the condensed liquid leaving the surface condenser can be removed by pumping and the noncondensable gases by a vacuum pump. Surface condensers are much more expensive and use more cooling water, so they are usually not used in cases where a direct-contact condenser is suitable.

8.6C Direct-Contact Condensers

In *direct-contact condensers* cooling water directly contacts the vapors and condenses the vapors. One of the most common types of direct-contact condensers is the countercurrent barometric condenser shown in Fig. 8.6-1. The vapor enters the condenser and is condensed by rising upward against a shower of cooling water droplets. The condenser is located on top of a long discharge tail pipe. The condenser is high enough above the discharge point in the tail pipe so that the water column established in the pipe more than compensates for the difference in pressure between the low absolute pressure in the

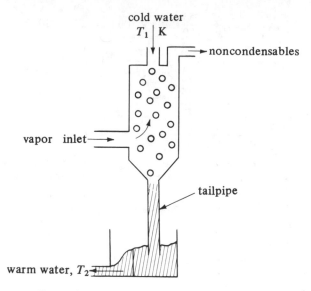

cold water
T_1 K

noncondensables

vapor inlet →

tailpipe

warm water, T_2

FIGURE 8.6-1. *Schematic of barometric condenser.*

condenser and the atmosphere. The water can then discharge by gravity through a seal pot at the bottom. A height of about 10.4 m (34 ft) is used.

The barometric condenser is inexpensive and economical of water consumption. It can maintain a vacuum corresponding to a saturated vapor temperature within about 2.8 K (5°F) of the water temperature leaving the condenser. For example, if the discharge water is at 316.5 K (110°F), the pressure corresponding to 316.5 + 2.8 or 319.3 K is 10.1 kPa (1.47 psia).

The water consumption can be estimated by a simple heat balance for a barometric condenser. If the vapor flow to the condenser is V kg/h at temperature T_S and the water flow is W kg/h at an entering temperature of T_1 and a leaving temperature of T_2, the derivation is as follows.

$$VH_S + Wc_p(T_1 - 273.2) = (V + W)c_p(T_2 - 273.2) \qquad \text{(8.6-1)}$$

where H_S is the enthalpy from the steam tables of the vapor at T_S K and the pressure in the vapor stream. Solving,

$$\frac{W}{V} = \frac{\text{kg water}}{\text{kg vapor}} = \frac{H_S - c_p(T_2 - 273.2)}{c_p(T_2 - T_1)} \qquad \text{(8.6-2)}$$

The noncondensable gases can be removed from the condenser by a vacuum pump. The vacuum pump used can be a mechanical pump or a steam-jet ejector. In the ejector high-pressure steam enters a nozzle at high speed and entrains the noncondensable gases from the space under vacuum.

Another type of direct-contact condenser is the jet barometric condenser. High-velocity jets of water act both as a vapor condenser and as an entrainer of the non-condensables out of the tail pipe. Jet condensers usually require more water than the more common barometric condensers and are more difficult to throttle at low vapor rates.

8.7 EVAPORATION OF BIOLOGICAL MATERIALS

8.7A Introduction and Properties of Biological Materials

The evaporation of many biological materials often differs from evaporation of inorganic materials such as NaCl and NaOH and organic materials such as ethanol and acetic acid. Biological materials such as pharmaceuticals, milk, citrus juices, and vegetable extracts are usually quite heat-sensitive and often contain fine particles of suspended matter in the solution. Also, because of problems due to bacteria growth, the equipment must be designed for easy cleaning. Many biological materials in solution exhibit only a small boiling-point rise when concentrated. This is due to the fact that suspended solids in a fine dispersed form and dissolved solutes of large molecular weight contribute little to this rise.

The amount of degradation of biological materials on evaporation is a function of the temperature and the length of time. To keep the temperature low, the evaporation must be done under vacuum, which reduces the boiling point of the solution. To keep the time of contact low, the equipment must provide for a low holdup time (contact time) of the material being evaporated. Typical types of equipment used and some biological materials processed are given below. Detailed discussions of the equipment are given in Section 8.2.

1. Long-tube vertical evaporator. Condensed milk.
2. Falling-film evaporator. Fruit juices.
3. Agitated-film (wiped-film) evaporator. Rubber latex, gelatin, antibiotics, fruit juices.
4. Heat-pump cycle evaporator. Fruit juices, milk, pharmaceuticals.

8.7B Fruit Juices

In evaporation of fruit juices such as orange juice the problems are quite different from evaporation of a typical salt such as NaCl. The fruit juices are heat-sensitive and the viscosity increases greatly as concentration increases. Also, solid suspended matter in fruit juices has a tendency to cling to the heating surface and thus causes overheating, leading to burning and spoilage of the matter (B2).

To reduce this tendency to stick and to reduce residence time, high rates of circulation over the heat-transfer surface are necessary. Since the material is heat-sensitive, low-temperature operation is also necessary. Hence, a fruit juice concentration plant usually employs a single and not a multiple evaporation unit. Vacuum is used to reduce the temperature of evaporation.

A typical fruit juice evaporation system using the heat pump cycle is shown (P1, C1), which uses low-temperature ammonia as the heating fluid. A frozen concentrated citrus juice process is described by Charm (C1). The process uses a multistage falling-film evaporator. A major fault of concentrated orange juice is a flat flavor due to the loss of volatile constituents during evaporation. To overcome this, a portion of the fresh pulpy juice bypasses the evaporation cycle and is blended with the evaporated concentrate.

8.7C Sugar Solutions

Sugar (sucrose) is obtained primarily from the sugar cane and sugar beet. Sugar tends to caramelize if kept at high temperatures for long periods (B2). The general tendency is to use short-tube evaporators of the natural circulation type. In the evaporation process of sugar solutions, the clear solution of sugar having a concentration of 10–13° Brix (10–13 wt %) is evaporated to 40–60° Brix (K1, S1).

The feed is first preheated by exhaust steam and then typically enters a six-effect forward-feed evaporator system. The first effect operates at a pressure in the vapor space of the evaporator of about 207 kPa (30 psia) [121.1°C (250°F) saturation temperature] and the last effect under vacuum at about 24 kPa (63.9°C saturation). Examples of the relatively small boiling-point rise of sugar solutions and the heat capacity are given in Example 8.5-1.

8.7D Paper-Pulp Waste Liquors

In the making of paper pulp in the sulfate process, wood chips are digested or cooked and spent black liquor is obtained after washing the pulp. This solution contains primarily sodium carbonate and organic sulfide compounds. This solution is concentrated by evaporation in a six-effect system (K1, S1).

8.8. EVAPORATION USING VAPOR RECOMPRESSION

8.8A Introduction

In the single-effect evaporator the vapor from the unit is generally condensed and discarded. In the multiple-effect evaporator, the pressure in each succeeding effect is lowered so that the boiling point of the liquid is lowered in each effect. Hence, there is a temperature difference created for the vapor from one effect to condense in the next effect and boil the liquid to form vapor.

In a single-effect vapor recompression (sometimes called vapor compression) evaporator the vapor is compressed so that its condensing or saturation temperature is increased. This compressed vapor is returned back to the heater of steam chest and condenses so that vapor is formed in the evaporator (B5, W1, Z1). In this manner the latent heat of the vapor is used and not discarded. The two types of vapor recompression units are the mechanical and the thermal type.

8.8B Mechanical Vapor Recompression Evaporator.

In a mechanical vapor recompression evaporator, a conventional single effect evaporator similar to that in Fig. 8.2-2 is used and is shown in Fig. 8.8-1. The cold feed is preheated by exchange with the hot outlet liquid product and then flows to the unit. The vapor coming overhead does not go to a condenser but is sent to a centrifugal or positive displacement compressor driven by an electric motor or steam. This compressed vapor or steam is sent back to the heat exchanger or steam chest. The compressed vapor condenses at a higher temperature than the boiling point of the hot liquid in the effect and a temperature difference is set up. Vapor is again generated and the cycle repeated.

Sometimes it is necessary to add a small amount of makeup steam to the vapor line before the compressor (B5, K2). Also, a small amount of condensate may be added to the compressed vapor to remove any superheat, if present.

Vapor recompression units generally operate at low optimum temperature differences of 5 to 10°C. Hence, large heat transfer areas are needed. These units usually have higher capital costs than multiple-effect units because of the larger area and the

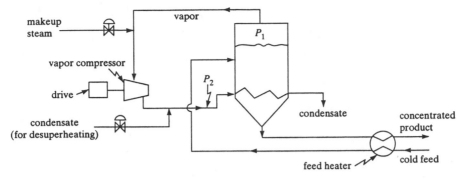

FIGURE 8.8-1. *Simplified process flow for mechanical vapor recompression evaporator.*

costs of the relatively expensive compressor and drive unit. The main advantage of vapor recompression units is in the lower energy costs. Using the steam equivalent of the power to drive the compressor, the steam economy is equivalent to a multiple-effect evaporator of up to 10 or more units (Z1).

Some typical applications of mechanical vapor recompression units are evaporation of sea water to give distilled water, evaporation of kraft black liquor in the paper industry (L2), evaporation of heat-sensitive materials such as fruit juices, and crystallizing of salts having inverse solubility curves where the solubility decreases with increasing temperature (K2, M3).

Falling-film evaporators are well suited for vapor recompression systems (W1) because they operate at low-temperature-difference values and have very little entrained liquid which can cause problems in the compressor. Vapor recompression has been used in distillation towers where the overhead vapor is recompressed and used in the reboiler as the heating medium (M2).

8.8C Thermal Vapor Recompression Evaporator

A steam jet can also be used to compress the vapor in a thermal vapor recompression unit. The main disadvantages are the low efficiency of the steam jet, necessitating the removal of this excess heat, and the lack of flexibility to changes in process variables (M3). Steam jets are cheaper and more durable than mechanical compressors and can more easily handle large volumes of low-pressure vapors.

PROBLEMS

8.4-1. Heat-Transfer Coefficient in Single-Effect Evaporator. A feed of 4535 kg/h of a 2.0 wt % salt solution at 311 K enters continuously a single-effect evaporator and is being concentrated to 3.0%. The evaporation is at atmospheric pressure and the area of the evaporator is 69.7 m². Saturated steam at 383.2 K is supplied for heating. Since the solution is dilute, it can be assumed to have the same boiling point as water. The heat capacity of the feed can be taken as $c_p = 4.10$ kJ/kg·K. Calculate the amounts of vapor and liquid product and the overall heat-transfer coefficient U.

> **Ans.** $U = 1823$ W/m²·K

8.4-2. Effects of Increased Feed Rate in Evaporator. Using the same area, value of U, steam pressure, evaporator pressure, and feed temperature as in Problem 8.4-1, calculate the amounts of liquid and vapor leaving and the liquid outlet concentration if the feed rate is increased to 6804 kg/h.

Ans. $V = 1256$ kg/h, $L = 5548$ kg/h, $x_L = 2.45\%$

8.4-3. Effect of Evaporator Pressure on Capacity and Product Composition. Recalculate Example 8.4-1 but use an evaporator pressure of 41.4 kPa instead of 101.32 kPa abs. Use the same steam pressure, area A, and heat-transfer coefficient U in the calculations.

(a) Do this to obtain the new capacity or feed rate under these new conditions. The composition of the liquid product will be the same as before.

(b) Do this to obtain the new product composition if the feed rate is increased to 18 144 kg/h.

8.4-4. Production of Distilled Water. An evaporator having an area of 83.6 m^2 and a $U = 2270$ W/m$^2 \cdot$ K is used to produce distilled water for a boiler feed. Tap water having 400 ppm of dissolved solids at 15.6°C is fed to the evaporator operating at 1 atm pressure abs. Saturated steam at 115.6°C is available for use. Calculate the amount of distilled water produced per hour if the outlet liquid contains 800 ppm solids.

8.4-5. Boiling-Point Rise of NaOH Solutions. Determine the boiling temperature of the solution and the boiling-point rise for the following cases.

(a) A 30% NaOH solution boiling in an evaporator at a pressure of 172.4 kPa (25 psia).

(b) A 60% NaOH solution boiling in an evaporator at a pressure of 3.45 kPa (0.50 psia).

Ans. (a) Boiling point = 130.6°C, boiling point rise = 15°C

8.4-6. Boiling-Point Rise of Biological Solutes in Solution. Determine the boiling-point rise for the following solutions of biological solutes in water. Use the figure in (P1), p. 11–31.

(a) A 30 wt % solution of citric acid in water boiling at 220°F (104.4°C).

(b) A 40 wt % solution of sucrose in water boiling at 220°F (104.4°C).

Ans. (a) Boiling-point rise = 2.2°F (1.22°C)

8.4-7. Effect of Feed Temperature on Evaporating an NaOH Solution. A single-effect evaporator is concentrating a feed of 9072 kg/h of a 10 wt % solution of NaOH in water to a product of 50% solids. The pressure of the saturated steam used is 42 kPa (gage) and the pressure in the vapor space of the evaporator is 20 kPa (abs). The overall heat-transfer coefficient is 1988 W/m$^2 \cdot$ K. Calculate the steam used, the steam economy in kg vaporized/kg steam, and the area for the following feed conditions.

(a) Feed temperature of 288.8 K (15.6°C).

(b) Feed temperature of 322.1 K (48.9°C).

Ans. (a) $S = 8959$ kg/h of steam, $A = 295.4$ m^2

8.4-8. Heat-Transfer Coefficient to Evaporate NaOH. In order to concentrate 4536 kg/h of an NaOH solution containing 10 wt % NaOH to a 20 wt % solution, a single-effect evaporator is being used with an area of 37.6 m^2. The feed enters at 21.1°C (294.3 K). Saturated steam at 110°C (383.2 K) is used for heating and the pressure in the vapor space of the evaporator is 51.7 kPa. Calculate the kg/h of steam used and the overall heat-transfer coefficient.

8.4-9. Throughput of a Single-Effect Evaporator. An evaporator is concentrating F kg/h at 311 K of a 20 wt % solution of NaOH to 50%. The saturated steam used for heating is at 399.3 K. The pressure in the vapor space of the evaporator is 13.3 kPa abs. The overall coefficient is 1420 W/m$^2 \cdot$ K and the area is 86.4 m^2. Calculate the feed rate F of the evaporator.

Ans. $F = 9072$ kg/h

8.4-10. Surface Area and Steam Consumption of an Evaporator. A single-effect evaporator is concentrating a feed solution of organic colloids from 5 to 50 wt %. The solution has a negligible boiling-point elevation. The heat capacity of the feed is $c_p = 4.06$ kJ/kg·K (0.97 btu/lb$_m$·°F) and the feed enters at 15.6°C (60°F). Saturated steam at 101.32 kPa is available for heating, and the pressure in the vapor space of the evaporator is 15.3 kPa. A total of 4536 kg/h (10 000 lb$_m$/h) of water is to be evaporated. The overall heat-transfer coefficient is 1988 W/m²·K (350 btu/h·ft²·°F). What is the required surface area in m² and the steam consumption?

8.4-11. *Evaporation of Tomato Juice Under Vacuum.* Tomato juice having a concentration of 12 wt % solids is being concentrated to 25% solids in a film-type evaporator. The maximum allowable temperature for the tomato juice is 135°F, which will be the temperature of the product. The feed enters at 100°F. Saturated steam at 25 psia is used for heating. The overall heat-transfer coefficient U is 600 btu/h·ft²·°F and the area A is 50 ft². The heat capacity of the feed c_p is estimated as 0.95 btu/lb$_m$·°F. Neglect any boiling-point rise if present. Calculate the feed rate of tomato juice to the evaporator.

8.4-12. *Concentration of Cane Sugar Solution.* A single-effect evaporator is being used to concentrate a feed of 10 000 lb$_m$/h of a cane sugar solution at 80°F and containing a sugar content of 15° Brix (degrees Brix is wt % sugar) to 30° Brix for use in a food process. Saturated steam at 240°F is available for heating. The vapor space in the evaporator will be at 1 atm abs pressure. The overall $U = 350$ btu/h·ft²·°F and the heat capacity of the feed is $c_p = 0.91$ btu/lb$_m$·°F. The boiling-point rise can be estimated from Example 8.5-1. The heat of solution can be considered negligible and neglected. Calculate the area required for the evaporator and the amount of steam used per hour.

Ans. Boiling-point rise = 2.0°F (1.1°C), $A = 667$ ft² (62.0 m²)

8.5-1. *Boiling Points in a Triple-Effect Evaporator.* A solution with a negligible boiling-point rise is being evaporated in a triple-effect evaporator using saturated steam at 121.1°C (394.3 K). The pressure in the vapor of the last effect is 25.6 kPa abs. The heat-transfer coefficients are $U_1 = 2840$, $U_2 = 1988$, and $U_3 = 1420$ W/m²·K and the areas are equal. Estimate the boiling point in each of the evaporators.

Ans. $T_1 = 108.6$°C (381.8 K)

8.5-2. *Evaporation of Sugar Solution in a Multiple-Effect Evaporator.* A triple-effect evaporator with forward feed is evaporating a sugar solution with negligible boiling-point rise (less than 1.0 K, which will be neglected) and containing 5 wt % solids to 25% solids. Saturated steam at 205 kPa abs is being used. The pressure in the vapor space of the third effect is 13.65 kPa. The feed rate is 22 680 kg/h and the temperature 299.9 K. The liquid heat capacity is $c_p = 4.19 - 2.35 x$, where c_p is in kJ/kg·K and x in wt fraction (K1). The heat-transfer coefficients are $U_1 = 3123$, $U_2 = 1987$, and $U_3 = 1136$ W/m²·K. Calculate the surface area of each effect if each effect has the same area, and the steam rate.

Ans. Area $A = 99.1$ m², steam rate $S = 8972$ kg/h

8.5-3. *Evaporation in Double-Effect Reverse-Feed Evaporators.* A feed containing 2 wt % dissolved organic solids in water is fed to a double-effect evaporator with reverse feed. The feed enters at 100°F and is concentrated to 25% solids. The boiling-point rise can be considered negligible as well as the heat of solution. Each evaporator has a 1000-ft² surface area and the heat-transfer coefficients are $U_1 = 500$ and $U_2 = 700$ btu/h·ft²·°F. The feed enters evaporator number 2 and steam at 100 psia is fed to number 1. The pressure in the vapor space of evaporator number 2 is 0.98 psia. Assume that the heat capacity of all liquid solutions is that of liquid water. Calculate the feed rate F and the product rate L_1 of a solution containing 25% solids. (*Hint:* Assume a feed rate of, say, $F = 1000$ lb$_m$/h. Calculate the area. Then calculate the actual feed rate by multiplying 1000 by 1000/calculated area.)

Ans. $F = 133 800$ lb$_m$/h (60 691 kg/h), $L_1 = 10 700$ lb$_m$/h (4853 kg/h)

8.5-4. *Concentration of NaOH Solution in Triple-Effect Evaporator.* A forced-circulation triple-effect evaporator using forward feed is to be used to concentrate a 10 wt % NaOH solution entering at 37.8°C to 50%. The steam used enters at 58.6 kPa gage. The absolute pressure in the vapor space of the third effect is 6.76 kPa. The feed rate is 13 608 kg/h. The heat-transfer coefficients are $U_1 = 6246$, $U_2 = 3407$, and $U_3 = 2271$ W/m$^2 \cdot$ K. All effects have the same area. Calculate the surface area and steam consumption.

Ans. $A = 97.3$ m^2, $S = 5284$ kg steam/h

8.5-5. *Triple-Effect Evaporator with Reverse Feed.* A feed rate of 20 410 kg/h of 10 wt % NaOH solution at 48.9°C is being concentrated in a triple-effect reverse-feed evaporator to produce a 50% solution. Saturated steam at 178.3°C is fed to the first evaporator and the pressure in the third effect is 10.34 kPa abs. The heat-transfer coefficient for each effect is assumed to be 2840 W/m$^2 \cdot$ K. Calculate the heat-transfer area and the steam consumption rate.

8.5-6. *Evaporation of Sugar Solution in Double Effect Evaporator.* A double-effect evaporator with reverse feed is used to concentrate 4536 kg/h of a 10 wt % sugar solution to 50%. The feed enters the second effect at 37.8°C. Saturated steam at 115.6°C enters the first effect and the vapor from this effect is used to heat the second effect. The absolute pressure in the second effect is 13.65 kPa abs. The overall coefficients are $U_1 = 2270$ and $U_2 = 1705$ W/m$^2 \cdot$ K. The heating areas for both effects are equal. Use boiling-point-rise and heat-capacity data from Example 8.5-1. Calculate the area and steam consumption.

8.6-1. *Water Consumption and Pressure in Barometric Condenser.* The concentration of NaOH solution leaving the third effect of a triple-effect evaporator is 50 wt %. The vapor flow rate leaving is 5670 kg/h and this vapor goes to a barometric condenser. The discharge water from the condenser leaves at 40.5°C. Assuming that the condenser can maintain a vacuum in the third effect corresponding to a saturated vapor pressure of 2.78°C above 40.5°C, calculate the pressure in the third effect and the cooling water flow to the condenser. The cooling water enters at 29.5°C. (*Note*: The vapor leaving the evaporator will be superheated because of the boiling-point rise.)

Ans. Pressure $= 8.80$ kPa abs, $W = 306\,200$ kg water/h

REFERENCES

(B1) BADGER, W. L., and BANCHERO, J. T. *Introduction to Chemical Engineering.* New York: McGraw-Hill Book Company, 1955.

(B2) BLAKEBROUGH, N. *Biochemical and Biological Engineering Science*, Vol. 2. New York: Academic Press, Inc., 1968.

(B3) BADGER, W. L., and McCABE, W. L. *Elements of Chemical Engineering*, 2nd ed. New York: McGraw-Hill Book Company, 1936.

(B4) BROWN, G. G., et al. *Unit Operations.* New York: John Wiley & Sons, Inc., 1950.

(B5) BEESLEY, A. H., and RHINESMITH, R. D. *Chem. Eng. Progr.*, **76**(8), 37 (1980).

(C1) CHARM, S. E. *The Fundamentals of Food Engineering*, 2nd ed. Westport, Conn.: Avi Publishing Co., Inc., 1971.

(K1) KERN, D. Q. *Process Heat Transfer.* New York: McGraw-Hill Book Company, 1950.

(K2) KING, R. J. *Chem. Eng. Progr.*, **80**(7), 63 (1984).

(L1) LINDSEY, E. *Chem. Eng.*, **60** (4), 227 (1953).

(L2) LOGSDON, J. D. *Chem. Eng. Progr.*, **79**(9), 36 (1983).

(M1) McCABE, W. L. *Trans. A.I.Ch.E.*, **31**, 129 (1935).

(M2) MEILI, A., and STUECHELI, A. *Chem. Eng.,* **94** (Feb. 16), 133 (1987).

(M3) MEHRA, D. K. *Chem. Eng.,* **93**(Feb. 3), 56 (1986).

(P1) PERRY, R. H., and CHILTON, C. H. *Chemical Engineers' Handbook*, 5th ed. New York: McGraw-Hill Book Company, 1973.

(P2) PERRY, R. H., and GREEN, D. *Perry's Chemical Engineers' Handbook*, 6th ed. New York: McGraw-Hill Book Company, 1984.

(S1) SHREVE, R. N., and BRINK, J. A., JR. *Chemical Process Industries*, 4th ed. New York: McGraw-Hill Book Company, 1977.

(W1) WEIMER, L. D., DOLF, H. R., and AUSTIN, D. A. *Chem. Eng. Progr.*, **76** (11), 70 (1980).

(Z1) ZIMMER, A. *Chem. Eng. Progr.*, **76**(9), 37 (1980).

CHAPTER 9

Drying of
Process Materials

9.1 INTRODUCTION AND METHODS OF DRYING

9.1A   Purposes of Drying

The discussions of drying in this chapter are concerned with the removal of water from process materials and other substances. The term *drying* is also used to refer to removal of other organic liquids, such as benzene or organic solvents, from solids. Many of the types of equipment and calculation methods discussed for removal of water can also be used for removal of organic liquids.

Drying, in general, usually means removal of relatively small amounts of water from material. Evaporation refers to removal of relatively large amounts of water from material. In evaporation the water is removed as vapor at its boiling point. In drying the water is usually removed as a vapor by air.

In some cases water may be removed mechanically from solid materials by presses, centrifuging, and other methods. This is cheaper than drying by thermal means for removal of water, which will be discussed here. The moisture content of the final dried product varies depending upon the type of product. Dried salt contains about 0.5% water, coal about 4%, and many food products about 5%. Drying is usually the final processing step before packaging and makes many materials, such as soap powders and dyestuffs, more suitable for handling.

Drying or dehydration of biological materials, especially foods, is used as a preservation technique. Microorganisms that cause food spoilage and decay cannot grow and multiply in the absence of water. Also, many enzymes that cause chemical changes in food and other biological materials cannot function without water. When the water content is reduced below about 10 wt %, the microorganisms are not active. However, it is usually necessary to lower the moisture content below 5 wt % in foods to preserve flavor and nutrition. Dried foods can be stored for extended periods of time.

Some biological materials and pharmaceuticals, which may not be heated for ordinary drying, may be freeze-dried as discussed in Section 9.11. Also, in Section 9.12, sterilization of foods and other biological materials is discussed, which is another method often employed to preserve such materials.

9.1B General Methods of Drying

Drying methods and processes can be classified in several different ways. Drying processes can be classified as *batch*, where the material is inserted into the drying equipment and drying proceeds for a given period of time, or as *continuous*, where the material is continuously added to the dryer and dried material continuously removed.

Drying processes can also be categorized according to the physical conditions used to add heat and remove water vapor: (1) in the first category, heat is added by direct contact with heated air at atmospheric pressure, and the water vapor formed is removed by the air; (2) in vacuum drying, the evaporation of water proceeds more rapidly at low pressures, and the heat is added indirectly by contact with a metal wall or by radiation (low temperatures can also be used under vacuum for certain materials that may discolor or decompose at higher temperatures); and (3) in freeze drying, water is sublimed from the frozen material.

9.2 EQUIPMENT FOR DRYING

9.2A Tray Dryer

In *tray dryers*, which are also called shelf, cabinet, or compartment dryers, the material, which may be a lumpy solid or a pasty solid, is spread uniformly on a metal tray to a depth of 10 to 100 mm. Such a typical tray dryer, shown in Fig. 9.2-1, contains removable trays loaded in a cabinet.

Steam-heated air is recirculated by a fan over and parallel to the surface of the trays. Electrical heat is also used, especially for low heating loads. About 10 to 20% of the air passing over the trays is fresh air, the remainder being recirculated air.

After drying, the cabinet is opened and the trays are replaced with a new batch of trays. A modification of this type is the tray–truck type, where trays are loaded on trucks which are pushed into the dryer. This saves considerable time, since the trucks can be loaded and unloaded outside the dryer.

In the case of granular materials, the material can be loaded on screens which are the bottom of each tray. Then in this through-circulation dryer, heated air passes through the permeable bed, giving shorter drying times because of the greater surface area exposed to the air.

9.2B Vacuum-Shelf Indirect Dryers

Vacuum-shelf dryers are indirectly heated batch dryers similar to tray dryers. Such a dryer consists of a cabinet made of cast-iron or steel plates fitted with tightly fitted doors

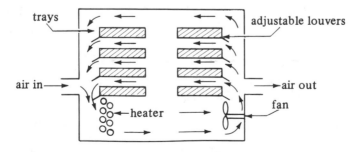

FIGURE 9.2-1. *Tray or shelf dryer.*

so that it can be operated under vacuum. Hollow shelves of steel are fastened permanently inside the chamber and are connected in parallel to inlet and outlet steam headers. The trays containing the solids to be dried rest upon the hollow shelves. The heat is conducted through the metal walls and added by radiation from the shelf above. For low-temperature operation, circulating warm water is used instead of steam for furnishing the heat to vaporize the moisture. The vapors usually pass to a condenser.

These dryers are used to dry expensive, or temperature-sensitive, or easily oxidizable materials. They are useful for handling materials with toxic or valuable solvents.

9.2C Continuous Tunnel Dryers

Continuous tunnel dryers are often batch truck or tray compartments operated in series, as shown in Fig. 9.2-2a. The solids are placed on trays or on trucks which move continuously through a tunnel with hot gases passing over the surface of each tray. The hot air flow can be countercurrent, cocurrent, or a combination. Many foods are dried in this way.

When granular particles of solids are to be dried, perforated or screen-belt continuous conveyors are often used, as in Fig. 9.2-2b. The wet granular solids are conveyed as a layer 25 to about 150 mm deep on a screen or perforated apron while heated air is blown upward through the bed, or downward. The dryer consists of several sections in series, each with a fan and heating coils. A portion of the air is exhausted to the atmosphere by a fan. In some cases pasty materials can be preformed into cylinders and placed on the bed to be dried.

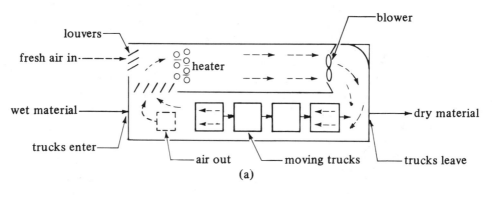

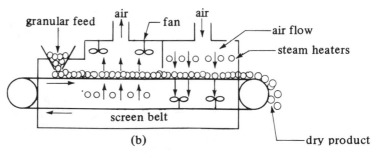

FIGURE 9.2-2. *Continuous tunnel dryers: (a) tunnel dryer trucks with countercurrent air flow, (b) through-circulation screen conveyor dryer.*

Chap. 9 *Drying of Process Materials*

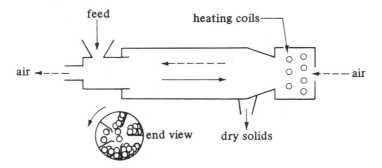

FIGURE 9.2-3. *Schematic drawing of a direct-heat rotary dryer.*

9.2D Rotary Dryers

A *rotary dryer* consists of a hollow cylinder which is rotated and usually slightly inclined toward the outlet. The wet granular solids are fed at the high end as shown in Fig. 9.2-3 and move through the shell as it rotates. The heating shown is by direct contact with hot gases in countercurrent flow. In some cases the heating is by indirect contact through the heated wall of the cylinder.

The granular particles move forward slowly a short distance before they are showered downward through the hot gases as shown. Many other variations of this rotary dryer are available, and these are discussed elsewhere (P1).

9.2E Drum Dryers

A *drum dryer* consists of a heated metal roll shown in Fig. 9.2-4, on the outside of which a thin layer of liquid or slurry is evaporated to dryness. The final dry solid is scraped off the roll, which is revolving slowly.

Drum dryers are suitable for handling slurries or pastes of solids in fine suspension and for solutions. The drum functions partly as an evaporator and also as a dryer. Other variations of the single-drum type are twin rotating drums with dip feeding or with top feeding to the two drums. Potato slurry is dried using drum dryers, to give potato flakes.

9.2F Spray Dryers

In a *spray dryer* a liquid or slurry solution is sprayed into a hot gas stream in the form of a mist of fine droplets. The water is rapidly vaporized from the droplets, leaving particles

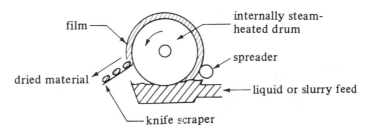

FIGURE 9.2-4. *Rotary-drum dryer.*

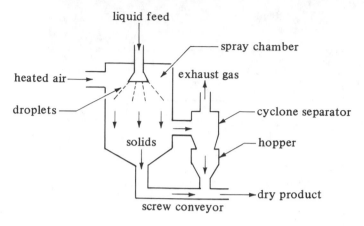

FIGURE 9.2-5. *Process flow diagram of spray-drying apparatus.*

of dry solid which are separated from the gas stream. The flow of gas and liquid in the spray chamber may be countercurrent, cocurrent, or a combination.

The fine droplets are formed from the liquid feed by spray nozzles or high-speed rotating spray disks inside a cylindrical chamber, as in Fig. 9.2-5. It is necessary to ensure that the droplets or wet particles of solid do not strike and stick to solid surfaces before drying has taken place. Hence, large chambers are used. The dried solids leave at the bottom of the chamber through a screw conveyor. The exhaust gases flow through a cyclone separator to remove any fines. The particles produced are usually light and quite porous. Dried milk powder is made from spray-drying milk.

9.2G Drying of Crops and Grains

In the drying of grain from a harvest, the grain contains about 30 to 35% moisture and for safe storage for about 1 year should be dried to about 13 wt % moisture (H1). A typical continuous-flow dryer is shown in Fig. 9.2-6. In the drying bin the thickness of the layer of grain is 0.5 m or less, through which the hot air passes. Unheated air in the bottom section cools the dry grain before it leaves. Other types of crop dryers and storage bins are described by Hall (H1).

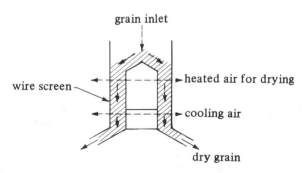

FIGURE 9.2-6. *Vertical continuous-flow grain dryer.*

9.3 VAPOR PRESSURE OF WATER AND HUMIDITY

9.3A Vapor Pressure of Water

1. Introduction. In a number of the unit operations and transport processes it is necessary to make calculations involving the properties of mixtures of water vapor and air. These calculations involve knowledge of the concentration of water vapor in air under various conditions of temperature and pressure, the thermal properties of these mixtures, and the changes occurring when this mixture is brought into contact with water or with wet solids in drying.

Humidification involves the transfer of water from the liquid phase into a gaseous mixture of air and water vapor. *Dehumidification* involves the reverse transfer, whereby water vapor is transferred from the vapor state to the liquid state. Humidification and dehumidification can also refer to vapor mixtures of materials such as benzene, but most practical applications occur with water. To better understand humidity, it is first necessary to discuss the vapor pressure of water.

2. Vapor pressure of water and physical states. Pure water can exist in three different physical states: solid ice, liquid, and vapor. The physical state in which it exists depends on the pressure and temperature.

Figure 9.3-1 illustrates the various physical states of water and the pressure–temperature relationships at equilibrium. In Fig. 9.3-1 the regions of the solid, liquid, and vapor states are shown. Along the line *AB*, the phases liquid and vapor coexist. Along line *AC*, the phases ice and liquid coexist. Along line *AD*, ice and vapor coexist. If ice at point (1) is heated at constant pressure, the temperature rises and the physical condition is shown moving horizontally. As the line crosses *AC*, the solid melts, and on crossing *AB* the liquid vaporizes. Moving from point (3) to (4), ice sublimes (vaporizes) to a vapor without becoming a liquid.

Liquid and vapor coexist in equilibrium along the line *AB*, which is the vapor-pressure line of water. Boiling occurs when the vapor pressure of the water is equal to the total pressure above the water surface. For example, at 100°C (212°F) the vapor pressure of water is 101.3 kPa (1.0 atm), and hence it will boil at 1 atm pressure. At 65.6°C (150°F), from the steam tables in Appendix A.2, the vapor pressure of water is 25.7 kPa (3.72 psia). Hence, at 25.7 kPa and 65.6°C, water will boil.

If a pan of water is held at 65.6°C in a room at 101.3 kPa abs pressure, the vapor pressure of water will again be 25.7 kPa. This illustrates an important property of the

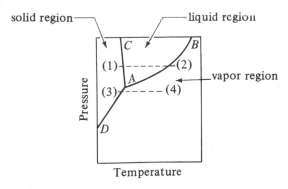

FIGURE 9.3-1. *Phase diagram for water.*

vapor pressure of water, which is not influenced by the presence of an inert gas such as air; i.e., the vapor pressure of water is essentially independent of the total pressure of the system.

9.3B Humidity and Humidity Chart

1. Definition of humidity. The humidity H of an air–water vapor mixture is defined as the kg of water vapor contained in 1 kg of dry air. The humidity so defined depends only on the partial pressure p_A of water vapor in the air and on the total pressure P (assumed throughout this chapter to be 101.325 kPa, 1.0 atm abs, or 760 mm Hg). Using the molecular weight of water (A) as 18.02 and of air as 28.97, the humdiity H in kg H_2O/kg dry air or in English units as lb H_2O/lb dry air is as follows:

$$H \frac{\text{kg } H_2O}{\text{kg dry air}} = \frac{p_A}{P - p_A} \frac{\text{kg mol } H_2O}{\text{kg mol air}} \times \frac{18.02 \text{ kg } H_2O}{\text{kg mol } H_2O} \times \frac{1}{28.97 \text{ kg air/kg mol air}}$$

$$H = \frac{18.02}{28.97} \frac{p_A}{P - p_A} \tag{9.3-1}$$

Saturated air is air in which the water vapor is in equilibrium with liquid water at the given conditions of pressure and temperature. In this mixture the partial pressure of the water vapor in the air–water mixture is equal to the vapor pressure p_{AS} of pure water at the given temperature. Hence, the saturation humidity H_S is

$$H_S = \frac{18.02}{28.97} \frac{p_{AS}}{P - p_{AS}} \tag{9.3-2}$$

2. Percentage humidity. The percentage humidity H_P is defined as 100 times the actual humidity H of the air divided by the humidity H_S if the air were saturated at the same temperature and pressure.

$$H_P = 100 \frac{H}{H_S} \tag{9.3-3}$$

3. Percentage relative humidity. The amount of saturation of an air–water vapor mixture is also given as percentage relative humidity H_R using partial pressures.

$$H_R = 100 \frac{p_A}{p_{AS}} \tag{9.3-4}$$

Note that $H_R \neq H_P$, since H_P expressed in partial pressures by combining Eqs. (9.3-1), (9.3-2), and (9.3-3) is

$$H_P = 100 \frac{H}{H_S} = (100) \frac{18.02}{28.97} \frac{p_A}{P - p_A} \bigg/ \frac{18.02}{28.97} \frac{p_{AS}}{P - p_{AS}} = \frac{p_A}{p_{AS}} \frac{P - p_{AS}}{P - p_A} (100) \tag{9.3-5}$$

This, of course, is not the same as Eq. (9.3-4).

EXAMPLE 9.3-1. Humidity from Vapor-Pressure Data
The air in a room is at 26.7°C (80°F) and a pressure of 101.325 kPa and contains water vapor with a partial pressure $p_A = 2.76$ kPa. Calculate the following.
(a) Humidity, H.
(b) Saturation humidity, H_S, and percentage humidity, H_P.
(c) Percentage relative humidity, H_R.

Solution: From the steam tables at 26.7°C, the vapor pressure of water is $p_{AS} = 3.50$ kPa (0.507 psia). Also, $p_A = 2.76$ kPa and $P = 101.3$ kPa (14.7 psia). For part (a), using Eq. (9.3-1),

$$H = \frac{18.02}{28.97} \frac{p_A}{P - p_A} = \frac{18.02(2.76)}{28.97(101.3 - 2.76)} = 0.01742 \text{ kg H}_2\text{O/kg air}$$

For part (b), using Eq. (9.3-2), the saturation humidity is

$$H_S = \frac{18.02}{28.97} \frac{p_{AS}}{P - p_{AS}} = \frac{18.02(3.50)}{28.97(101.3 - 3.50)} = 0.02226 \text{ kg H}_2\text{O/kg air}$$

The percentage humidity, from Eq. (9.3-3), is

$$H_P = 100 \frac{H}{H_S} = \frac{100(0.01742)}{0.02226} = 78.3\%$$

For part (c), from Eq. (9.3-4), the percentage relative humidity is

$$H_R = 100 \frac{p_A}{p_{AS}} = \frac{100(2.76)}{3.50} = 78.9\%$$

4. *Dew point of an air–water vapor mixture.* The temperature at which a given mixture of air and water vapor would be saturated is called the *dew-point temperature* or simply the *dew point*. For example, at 26.7°C (80°F), the saturation vapor pressure of water is $p_{AS} = 3.50$ kPa (0.507 psia). Hence, the dew point of a mixture containing water vapor having a partial pressure of 3.50 kPa is 26.7°C. If an air–water vapor mixture is at 37.8°C (often called the dry bulb temperature, since this is the actual temperature a dry thermometer bulb would indicate in this mixture) and contains water vapor of $p_A = 3.50$ kPa, the mixture would not be saturated. On cooling to 26.7°C, the air would be saturated, i.e., at the dew point. On further cooling, some water vapor would condense, since the partial pressure cannot be greater than the saturation vapor pressure.

5. *Humid heat of an air–water vapor mixture.* The humid heat c_S is the amount of heat in J (or kJ) required to raise the temperature of 1 kg of dry air plus the water vapor present by 1 K or 1°C. The heat capacity of air and water vapor can be assumed constant over the temperature ranges usually encountered at 1.005 kJ/kg dry air · K and 1.88 kJ/kg water vapor · K, respectively. Hence, for SI and English units,

$$c_S \text{ kJ/kg dry air} \cdot K = 1.005 + 1.88H \qquad \text{(SI)}$$
$$c_S \text{ btu/lb}_m \text{ dry air} \cdot °F = 0.24 + 0.45H \qquad \text{(English)}$$

(9.3-6)

[In some cases c_S will be given as $(1.005 + 1.88H)10^3$ J/kg · K.]

6. *Humid volume of an air–water vapor mixture.* The humid volume v_H is the total volume in m³ of 1 kg of dry air plus the vapor it contains at 101.325 kPa (1.0 atm) abs pressure and the given gas temperature. Using the ideal gas law,

$$v_H \text{ m}^3\text{/kg dry air} = \frac{22.41}{273} T \text{ K} \left(\frac{1}{28.97} + \frac{1}{18.02} H \right)$$
$$= (2.83 \times 10^{-3} + 4.56 \times 10^{-3} H)T \text{ K}$$

(9.3-7)

$$v_H \text{ ft}^3\text{/lb}_m \text{ dry air} = \frac{359}{492} T°\text{R} \left(\frac{1}{28.97} + \frac{1}{18.02} H \right)$$
$$= (0.0252 + 0.0405H)T°\text{R}$$

For a saturated air–water vapor mixture, $H = H_S$, and v_H is the saturated volume.

7. *Total enthalpy of an air–water vapor mixture.* The total enthalpy of 1 kg of air plus its water vapor is H_y J/kg or kJ/kg dry air. If T_0 is the datum temperature chosen for both components, the total enthalpy is the sensible heat of the air–water vapor mixture plus the latent heat λ_0 in J/kg or kJ/kg water vapor of the water vapor at T_0. Note that $(T - T_0)°C = (T - T_0)$ K and that this enthalpy is referred to liquid water.

$$H_y \text{ kJ/kg dry air} = c_S(T - T_0) + H\lambda_0 = (1.005 + 1.88H)(T - T_0 \, °C) + H\lambda_0$$

$$H_y \text{ btu/lb}_m \text{ dry air} = (0.24 + 0.45H)(T - T_0 \, °F) + H\lambda_0$$

(9.3-8)

If the total enthalpy is referred to a base temperature T_0 of 0°C (32°F), the equation for H_y becomes

$$H_y \text{ kJ/kg dry air} = (1.005 + 1.88H)(T°C - 0) + 2501.4H \qquad \text{(SI)}$$

$$H_y \text{ btu/lb}_m \text{ dry air} = (0.24 + 0.45H)(T°F - 32) + 1075.4H \qquad \text{(English)}$$

(9.3-9)

8. *Humidity chart of air–water vapor mixtures.* A convenient chart of the properties of air–water vapor mixtures at 1.0 atm abs pressure is the humidity chart in Fig. 9.3-2. In this figure the humidity H is plotted versus the actual temperature of the air–water vapor mixture (dry bulb temperature).

The curve marked 100% running upward to the right gives the saturation humidity H_S as a function of temperature. In Example 9.3-1, for 26.7°C H_S was calculated as 0.02226 kg H_2O/kg air. Plotting this point of 26.7°C (80°F) and $H_S = 0.02226$ on Fig. 9.3-2, it falls on the 100% saturated line.

Any point below the saturation line represents unsaturated air–water vapor mixtures. The curved lines below the 100% saturation line and running upward to the right represent unsaturated mixtures of definite percentage humidity H_P. Going downward vertically from the saturation line at a given temperature, the line between 100% saturation and zero humidity H (the bottom horizontal line) is divided evenly into 10 increments of 10% each.

All the percentage humidity lines H_P mentioned and the saturation humidity line H_S can be calculated from the data of vapor pressure of water.

EXAMPLE 9.3-2. Use of Humidity Chart

Air entering a dryer has a temperature (dry bulb temperature) of 60°C (140°F) and a dew point of 26.7°C (80°F). Using the humidity chart, determine the actual humidity H, percentage humidity H_P, humid heat c_S, and the humid volume v_H in SI and English units.

Solution: The dew point of 26.7°C is the temperature when the given mixture is at 100% saturation. Starting at 26.7°C, Fig. 9.3-2, and drawing a vertical line until it intersects the line for 100% humidity, a humidity of $H = 0.0225$ kg H_2O/kg dry air is read off the plot. This is the actual humidity of the air at 60°C. Stated in another way, if air at 60°C and having a humidity $H = 0.0225$ is cooled, its dew point will be 26.7°C. In English units, $H = 0.0225$ lb H_2O/lb dry air.

Locating this point of $H = 0.0225$ and $t = 60°C$ on the chart, the percentage humidity H_P is found to be 14%, by linear interpolation vertically between the 10 and 20% lines. The humid heat for $H = 0.0225$ is, from

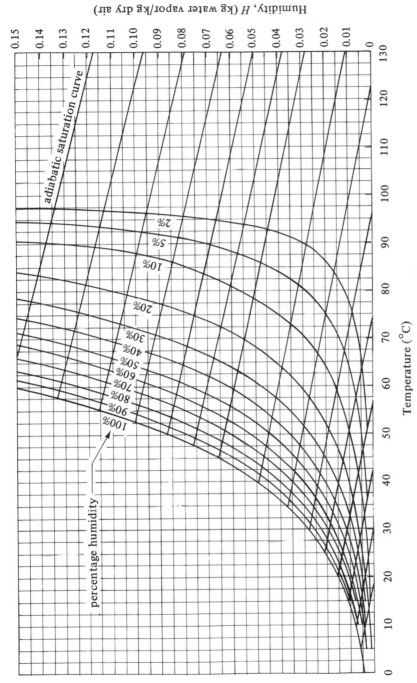

FIGURE 9.3-2. *Humidity chart for mixtures of air and water vapor at a total pressure of 101.325 kPa (760 mm Hg). (From R. E. Treybal, Mass-Transfer Operations, 3rd ed. New York: McGraw-Hill Book Company, 1980. With permission.)*

Eq. (9.3-6),

$$c_s = 1.005 + 1.88(0.0225)$$

$$= 1.047 \text{ kJ/kg dry air} \cdot \text{K} \quad \text{or} \quad 1.047 \times 10^3 \text{ J/kg} \cdot \text{K}$$

$$c_s = 0.24 + 0.45(0.0225)$$

$$= 0.250 \text{ btu/lb}_m \text{ dry air} \cdot °\text{F} \quad \text{(English)}$$

The humid volume at 60°C (140°F), from Eq. (9.3-7), is

$$v_H = (2.83 \times 10^{-3} + 4.56 \times 10^{-3} \times 0.0225)(60 + 273)$$

$$= 0.977 \text{ m}^3/\text{kg dry air}$$

In English units,

$$v_H = (0.0252 + 0.0405 \times 0.0225)(460 + 140) = 15.67 \text{ ft}^3/\text{lb}_m \text{ dry air}$$

9.3C Adiabatic Saturation Temperatures

Consider the process shown in Fig. 9.3-3, where the entering gas of air–water vapor mixture is contacted with a spray of liquid water. The gas leaves having a different humidity and temperature and the process is adiabatic. The water is recirculated, with some makeup water added.

The temperature of the water being recirculated reaches a steady-state temperature called the *adiabatic saturation temperature*, T_S. If the entering gas at temperature T having a humidity of H is not saturated, T_S will be lower than T. If the contact between the entering gas and the spray of droplets is enough to bring the gas and liquid to equilibrium, the leaving air is saturated at T_S, having a humidity H_S.

Writing an enthalpy balance (heat balance) over the process, a datum of T_S is used. The enthalpy of the makeup H_2O is then zero. This means that the total enthalpy of the entering gas mixture = enthalpy of the leaving gas mixture, or, using Eq. (9.3-8),

$$c_S(T - T_S) + H\lambda_S = c_S(T_S - T_S) + H_S \lambda_S \tag{9.3-10}$$

Or, rearranging, and using Eq. (9.3-6) for c_S,

$$\frac{H - H_S}{T - T_S} = -\frac{c_S}{\lambda_S} = \frac{1.005 + 1.88H}{\lambda_S} \quad \text{(SI)}$$

$$\frac{H - H_S}{T - T_S} = \frac{0.24 + 0.45H}{\lambda_S} \quad \text{(English)} \tag{9.3-11}$$

Equation (9.3-11) is the equation of an adiabatic humidification curve when plotted

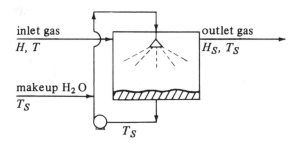

FIGURE 9.3-3. *Adiabatic air–water vapor saturator.*

on Fig. 9.3-2, which passes through the point H_S and T_S on the 100% saturation curve and other points of H and T. These series of lines, running upward to the left, are called *adiabatic humidification lines* or *adiabatic saturation lines*. Since c_S contains the term H, the adiabatic lines are not quite straight when plotted on the humidity chart.

If a given gas mixture at T_1 and H_1 is contacted for a sufficiently long time in an adiabatic saturator, it will leave saturated at H_{S1} and T_{S1}. The values of H_{S1} and T_{S1} are determined by following the adiabatic saturation line going through point T_1, H_1 until it intersects the 100% saturation line. If contact is not sufficient, the leaving mixture will be at a percentage saturation less than 100 but on the same line.

EXAMPLE 9.3-3. Adiabatic Saturation of Air
An air stream at 87.8°C having a humidity $H = 0.030$ kg H_2O/kg dry air is contacted in an adiabatic saturator with water. It is cooled and humidified to 90% saturation.
    (a) What are the final values of H and T?
    (b) For 100% saturation, what would be the values of H and T?

Solution: For part (a), the point $H = 0.030$ and $T = 87.8$°C is located on the humidity chart. The adiabatic saturation curve through this point is followed upward to the left until it intersects the 90% line at 42.5°C and $H = 0.0500$ kg H_2O/kg dry air.
    For part (b), the same line is followed to 100% saturation, where $T = 40.5$°C and $H = 0.0505$ kg H_2O/kg dry air.

9.3D Wet Bulb Temperature

The adiabatic saturation temperature is the steady-state temperature attained when a large amount of water is contacted with the entering gas. The *wet bulb temperature* is the steady-state nonequilibrium temperature reached when a small amount of water is contacted under adiabatic conditions by a continuous stream of gas. Since the amount of liquid is small, the temperature and humidity of the gas are not changed, contrary to the adiabatic saturation case, where the temperature and humidity of the gas are changed.

The method used to measure the wet bulb temperature is illustrated in Fig. 9.3-4, where a thermometer is covered by a wick or cloth. The wick is kept wet by water and is immersed in a flowing stream of air–water vapor having a temperature of T (dry bulb temperature) and humidity H. At steady state, water is evaporating to the gas stream. The wick and water are cooled to T_W and stay at this constant temperature. The latent heat of evaporation is exactly balanced by the convective heat flowing from the gas stream at T to the wick at a lower temperature T_W.

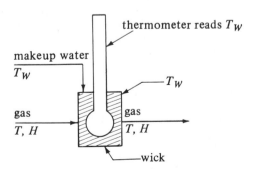

FIGURE 9.3-4. *Measurement of wet bulb temperature.*

A heat balance on the wick can be made. The datum temperature is taken at T_W. The amount of heat lost by vaporization, neglecting the small sensible heat change of the vaporized liquid and radiation, is

$$q = M_A N_A \lambda_W A \tag{9.3-12}$$

where q is kW(kJ/s), M_A is molecular weight of water, N_A is kg mol H_2O evaporating/ $s \cdot m^2$, A is surface area m^2, and λ_W is the latent heat of vaporization at T_W in kJ/kg H_2O. In English units, q is btu/h, N_A is lb mol/h $\cdot$ ft^2, and λ_W is btu/lb$_m$ H_2O. The flux N_A is

$$N_A = \frac{k_y'}{x_{BM}} (y_W - y) = k_y(y_W - y) \tag{9.3-13}$$

where k_y' is the mass-transfer coefficient in kg mol/s $\cdot$ m^2 $\cdot$ mol frac, x_{BM} is the log mean inert mole fraction of the air, y_W is the mole fraction of water vapor in the gas at the surface, and y is the mole fraction in the gas. For a dilute mixture $x_{BM} \cong 1.0$ and $k_y' \cong k_y$. The relation between H and y is.

$$y = \frac{H/M_A}{1/M_B + H/M_A} \tag{9.3-14}$$

where M_B is the molecular weight of air and M_A the molecular weight of H_2O. Since H is small, as an approximation,

$$y \cong \frac{H M_B}{M_A} \tag{9.3-15}$$

Substituting Eq. (9.3-15) into (9.3-13) and then substituting the resultant into Eq. (9.3-12),

$$q = M_B k_y \lambda_W (H_W - H)A \tag{9.3-16}$$

The rate of convective heat transfer from the gas stream at T to the wick at T_W is

$$q = h(T - T_W)A \tag{9.3-17}$$

where h is the heat-transfer coefficient in $kW/m^2 \cdot K$ (btu/h $\cdot$ ft^2 $\cdot$ °F).
Equating Eq. (9.3-16) to (9.3-17) and rearranging,

$$\frac{H - H_W}{T - T_W} = -\frac{h/M_B k_y}{\lambda_W} \tag{9.3-18}$$

Experimental data on the value of $h/M_B k_y$, called the *psychrometric ratio*, show that for water vapor–air mixtures, the value is approximately 0.96–1.005. Since this value is close to the value of c_S in Eq. (9.3-11), approximately 1.005, Eqs. (9.3-18) and (9.3-11) are almost the same. This means that the adiabatic saturation lines can also be used for wet bulb lines with reasonable accuracy. (Note that this is only true for water vapor and not for others, such as benzene.) Hence, the wet bulb determination is often used to determine the humidity of an air–water vapor mixture.

EXAMPLE 9.3-4. Wet Bulb Temperature and Humidity
A water vapor–air mixture having a dry bulb temperature of $T = 60°C$ is passed over a wet bulb as shown in Fig. 9.3-4, and the wet bulb temperature obtained is $T_W = 29.5°C$. What is the humidity of the mixture?

Solution: The wet bulb temperature of 29.5°C can be assumed to be the same as the adiabatic saturation temperature T_S, as discussed. Following the adiabatic saturation curve of 29.5°C until it reaches the dry bulb temperature of 60°C, the humidity is $H = 0.0135$ kg H_2O/kg dry air.

9.4 EQUILIBRIUM MOISTURE CONTENT OF MATERIALS

9.4A Introduction

As in other transfer processes, such as mass transfer, the process of drying of materials must be approached from the viewpoint of the equilibrium relationships and also the rate relationships. In most of the drying apparatus discussed in Section 9.2, material is dried in contact with an air–water vapor mixture. The equilibrium relationships between the air–water vapor and the solid material will be discussed in this section.

An important variable in the drying of materials is the humidity of the air in contact with a solid of given moisture content. Suppose that a wet solid containing moisture is brought into contact with a stream of air having a constant humidity H and temperature. A large excess of air is used, so its conditions remain constant. Eventually, after exposure of the solid sufficiently long for equilibrium to be reached, the solid will attain a definite moisture content. This is known as the equilibrium moisture content of the material under the specified humidity and temperature of the air. The moisture content is usually expressed on a dry basis as kg of water per kg of moisture-free (bone-dry) solid or kg $H_2O/100$ kg dry solid; in English units as lb $H_2O/100$ lb dry solid.

For some solids the value of the equilibrium moisture content depends on the direction from which equilibrium is approached. A different value of the equilibrium moisture content is obtained according to whether a wet sample is allowed to dry by desorption or whether a dry sample adsorbs moisture by adsorption. For drying calculations it is the desorption equilibrium that is the larger value and is of particular interest.

9.4B Experimental Data of Equilibrium Moisture Content for Inorganic and Biological Materials

1. Typical data for various materials. If the material contains more moisture than its equilibrium value in contact with a gas of a given humidity and temperature, it will dry until it reaches its equilibrium value. If the material contains less moisture than its equilibrium value, it will adsorb water until it reaches its equilibrium value. For air having 0% humidity, the equilibrium moisture value of all materials is zero.

The equilibrium moisture content varies greatly with the type of material for any given percent relative humidity, as shown in Fig. 9.4-1 for some typical materials at room temperature. Nonporous insoluble solids tend to have equilibrium moisture contents which are quite low, as shown for glass wool and kaolin. Certain spongy, cellular materials of organic and biological origin generally show large equilibrium moisture contents. Examples of this in Fig. 9.4-1 are wool, leather, and wood.

2. Typical food materials. In Fig. 9.4-2 the equilibrium moisture contents of some typical food materials are plotted versus percent relative humidity. These biological materials also show large values of equilibrium moisture contents. Data in this figure and in Fig. 9.4-1 for biological materials show that at high percent relative humidities of about 60 to 80%, the equilibrium moisture content increases very rapidly with increases of relative humidity.

In general, at low relative humidities the equilibrium moisture content is greatest for food materials high in protein, starch, or other high-molecular-weight polymers and lower for food materials high in soluble solids. Crystalline salts and sugars and also fats generally adsorb small amounts of water.

3. Effect of temperature. The equilibrium moisture content of a solid decreases some-

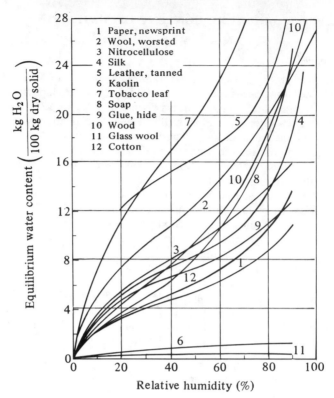

1	Paper, newsprint
2	Wool, worsted
3	Nitrocellulose
4	Silk
5	Leather, tanned
6	Kaolin
7	Tobacco leaf
8	Soap
9	Glue, hide
10	Wood
11	Glass wool
12	Cotton

FIGURE 9.4-1. *Typical equilibrium moisture contents of some solids at approximately 298 K (25°C). [From National Research Council, International Critical Tables, Vol. II. New York: McGraw-Hill Book Company, 1929. Reproduced with permission of the National Academy of Sciences.]*

what with an increase in temperature. For example, for raw cotton at a relative humidity of 50%, the equilibrium moisture content decreased from 7.3 kg H_2O/100 kg dry solid at 37.8°C (311 K) to about 5.3 at 93.3°C (366.5 K), a decrease of about 25%. Often for moderate temperature ranges, the equilibrium moisture content will be assumed constant when experimental data are not available at different temperatures.

At present, theoretical understanding of the structure of solids and surface phenomena does not enable us to predict the variation of equilibrium moisture content of various materials from first principles. However, by using models such as those used for adsorption isotherms of multilayers of molecules and others, attempts have been made to correlate experimental data. Henderson (H2) gives an empirical relationship between equilibrium moisture content and percent relative humidity for some agricultural materials. In general, empirical relationships are not available for most materials, and equilibrium moisture contents must be determined experimentally. Also, equilibrium moisture relationships often vary from sample to sample of the same kind of material.

9.4C Bound and Unbound Water in Solids

In Fig. 9.4-1, if the equilibrium moisture content of a given material is continued to its intersection with the 100% humidity line, the moisture is called *bound water*. This water

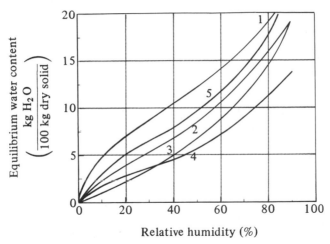

FIGURE 9.4-2. *Typical equilibrium moisture contents of some food materials at approximately 298 K (25°C): (1) macaroni, (2) flour, (3) bread, (4) crackers, (5) egg albumin. [Curve (5) from ref. (E1). Curves (1) to (4) from National Research Council, International Critical Tables, Vol. II. New York: McGraw-Hill Book Company, 1929. Reproduced with permission of the National Academy of Sciences.]*

in the solid exerts a vapor pressure less than that of liquid water at the same temperature. If such a material contains more water than indicated by intersection with the 100% humidity line, it can still exert only a vapor pressure as high as that of ordinary water at the same temperature. This excess moisture content is called *unbound water*, and it is held primarily in the voids of the solid. Substances containing bound water are often called *hygroscopic materials*.

As an example, consider curve 10 for wood in Fig. 9.4-1. This intersects the curve for 100% humidity at about 30 kg H_2O/100 kg dry solid. Any sample of wood containing less than 30 kg H_2O/100 kg dry solid contains only bound water. If the wood sample contained 34 kg H_2O/100 kg dry solid, 4 kg H_2O would be unbound and 30 kg H_2O bound per 100 kg dry solid.

The bound water in a substance may exist under several different conditions. Moisture in cell or fiber walls may have solids dissolved in it and have a lower vapor pressure. Liquid water in capillaries of very small diameter will exert a lowered vapor pressure because of the concave curvature of the surface. Water in natural organic materials is in chemical and physical–chemical combination.

9.4D Free and Equilibrium Moisture of a Substance

Free moisture content in a sample is the moisture above the equilibrium moisture content. This free moisture is the moisture that can be removed by drying under the given percent relative humidity. For example, in Fig. 9.4-1 silk has an equilibrium moisture content of 8.5 kg H_2O/100 kg dry material in contact with air of 50% relative humidity and 25°C. If a sample contains 10 kg H_2O/100 kg dry material, only 10.0 − 8.5, or 1.5, kg H_2O/100 kg dry material is removable by drying, and this is the free moisture of the sample under these drying conditions.

In many texts and references, the moisture content is given as percent moisture on a dry basis. This is exactly the same as the kg H_2O/100 kg dry material multiplied by 100.

9.5 RATE OF DRYING CURVES

9.5A Introduction and Experimental Methods

1. Introduction. In the drying of various types of process materials from one moisture content to another, it is usually desired to estimate the size of dryer needed, the various operating conditions of humidity and temperature for the air used, and the time needed to perform the amount of drying required. As discussed in Section 9.4, equilibrium moisture contents of various materials cannot be predicted and must be determined experimentally. Similarly, since our knowledge of the basic mechanisms of rates of drying is quite incomplete, it is necessary in most cases to obtain some experimental measurements of drying rates.

2. Experimental determination of rate of drying. To experimentally determine the rate of drying for a given material, a sample is usually placed on a tray. If it is a solid material it should fill the tray so that only the top surface is exposed to the drying air stream. By suspending the tray from a balance in a cabinet or duct through which the air is flowing, the loss in weight of moisture during drying can be determined at different intervals without interrupting the operation.

In doing batch-drying experiments, certain precautions should be observed to obtain usable data under conditions that closely resemble those to be used in the large-scale operations. The sample should not be too small in weight and should be supported in a tray or frame similar to the large-scale one. The ratio of drying to nondrying surface (insulated surface) and the bed depth should be similar. The velocity, humidity, temperature, and direction of the air should be the same and constant to simulate drying under constant drying conditions.

9.5B Rate of Drying Curves for Constant-Drying Conditions

1. Conversion of data to rate-of-drying curve. Data obtained from a batch-drying experiment are usually obtained as W total weight of the wet solid (dry solid plus moisture) at different times t hours in the drying period. These data can be converted to rate-of-drying data in the following ways. First, the data are recalculated. If W is the weight of the wet solid in kg total water plus dry solid and W_S is the weight of the dry solid in kg,

$$X_t = \frac{W - W_S}{W_S} \frac{\text{kg total water}}{\text{kg dry solid}} \left(\frac{\text{lb total water}}{\text{lb dry solid}} \right) \qquad (9.5\text{-}1)$$

For the given constant drying conditions, the equilibrium moisture content X^* kg equilibrium moisture/kg dry solid is determined. Then the free moisture content X in kg free water/kg dry solid is calculated for each value of X_t.

$$X = X_t - X^* \qquad (9.5\text{-}2)$$

Using the data calculated from Eq. (9.5-2), a plot of free moisture content X versus time t in h is made as in Fig. 9.5-1a. To obtain the rate-of-drying curve from this plot, the slopes of the tangents drawn to the curve in Fig. 9.5-1a can be measured, which give values of dX/dt at given values of t. The rate R is calculated for each point by

$$R = -\frac{L_S}{A} \frac{dX}{dt} \qquad (9.5\text{-}3)$$

where R is drying rate in kg $H_2O/h \cdot m^2$, L_S kg of dry solid used, and A exposed surface area for drying in m^2. In English units, R is lb_m $H_2O/h \cdot ft^2$, L_S is lb_m dry solid, and A is ft^2. For obtaining R from Fig. 9.5-1a, a value of L_S/A of 21.5 kg/m^2 was used. The drying-rate curve is then obtained by plotting R versus the moisture content, as in Fig. 9.5-1b.

Another method to obtain the rate-of-drying curve is to first calculate the weight loss ΔX for a Δt time. For example, if $X_1 = 0.350$ at a time $t_1 = 1.68$ h and $x_2 = 0.325$ at a time $t_2 = 2.04$ h, $\Delta X/\Delta t = (0.350 - 0.325)/(2.04 - 1.68)$. Then, using Eq. (9.5-4) and

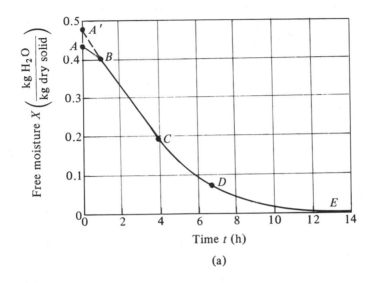

(a)

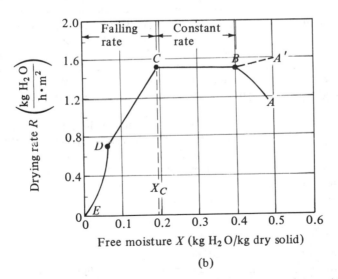

(b)

FIGURE 9.5-1. *Typical drying-rate curve for constant drying conditions: (a) plot of data as free moisture versus time, (b) rate of drying curve as rate versus free moisture content.*

$$L_S/A = 21.5,$$

$$R = -\frac{L_S}{A}\frac{\Delta X}{\Delta t} = 21.5\left(\frac{0.350 - 0.325}{2.04 - 1.68}\right) = 1.493$$

This rate R is the average over the period 1.68 to 2.04 h and should be plotted at the average concentration $X = (0.350 + 0.325)/2 = 0.338$.

2. Plot of rate-of-drying curve. In Fig. 9.5-1b the rate-of-drying curve for constant-drying conditions is shown. At zero time the initial free moisture content is shown at point A. In the beginning the solid is usually at a colder temperature than its ultimate temperature, and the evaporation rate will increase. Eventually at point B the surface temperature rises to its equilibrium value. Alternatively, if the solid is quite hot to start with, the rate may start at point A'. This initial unsteady-state adjustment period is usually quite short and it is often ignored in the analysis of times of drying.

From point B to C in Fig. 9.5-1a the line is straight, and hence the slope and rate are constant during this period. This *constant-rate-of-drying period* is shown as line BC in Fig. 9.5-1b.

At point C on both plots, the drying rate starts to decrease in the *falling-rate period* until it reaches point D. In this first falling-rate period, the rate shown as line CD in Fig. 9.5-1b is often linear.

At point D the rate of drying falls even more rapidly, until it reaches point E, where the equilibrium moisture content is X^* and $X = X^* - X^* = 0$. In some materials being dried, the region CD may be missing completely or it may constitute all of the falling-rate period.

9.5C Drying in the Constant-Rate Period

Drying of different solids under different constant conditions of drying will often give curves of different shapes in the falling-rate period, but in general the two major portions of the drying-rate curve—constant-rate period and falling-rate period—are present.

In the constant-rate drying period, the surface of the solid is initially very wet and a continuous film of water exists on the drying surface. This water is entirely unbound water and the water acts as if the solid were not present. The rate of evaporation under the given air conditions is independent of the solid and is essentially the same as the rate from a free liquid surface. Increased roughness of the solid surface, however, may lead to higher rates than from a flat surface.

If the solid is porous, most of the water evaporated in the constant-rate period is supplied from the interior of the solid. This period continues only as long as the water is supplied to the surface as fast as it is evaporated. Evaporation during this period is similar to that in determining the wet bulb temperature, and in the absence of heat transfer by radiation or conduction, the surface temperature is approximately that of the wet bulb temperature.

9.5D Drying in the Falling-Rate Period

Point C in Fig. 9.5-1b is at the *critical free moisture content* X_C. At this point there is insufficient water on the surface to maintain a continuous film of water. The entire surface is no longer wetted, and the wetted area continually decreases in this first falling-rate period until the surface is completely dry at point D in Fig. 9.5-1b.

The second falling-rate period begins at point D when the surface is completely dry. The plane of evaporation slowly recedes from the surface. Heat for the evaporation is transferred through the solid to the zone of vaporization. Vaporized water moves through the solid into the air stream.

In some cases no sharp discontinuity occurs at point D, and the change from partially wetted to completely dry conditions at the surface is so gradual that no sharp change is detectable.

The amount of moisture removed in the falling-rate period may be relatively small but the time required may be long. This can be seen in Fig. 9.5-1. The period BC for constant-rate drying lasts for about 3.0 h and reduces X from 0.40 to about 0.19, a reduction of 0.21 kg H_2O/kg dry solid. The falling-rate period CE lasts about 9.0 h and reduces X only from 0.19 to 0.

9.5E Moisture Movements in Solids During Drying in the Falling-Rate Period

When drying occurs by evaporation of moisture from the exposed surface of a solid, moisture must move from the depths of the solid to the surface. The mechanisms of the movement affect the drying during the constant-rate and falling-rate periods. Some of the theories advanced to explain the various types of falling-rate curves will be briefly reviewed.

1. Liquid diffusion theory. In this theory diffusion of liquid moisture occurs when there is a concentration difference between the depths of the solid and the surface. This method of transport of moisture is usually found in nonporous solids where single-phase solutions are formed with the moisture, such as in paste, soap, gelatin, and glue. This is also found in drying the last portions of moisture from clay, flour, wood, leather, paper, starches, and textiles. In drying many food materials, the movement of water in the falling-rate period occurs by diffusion.

The shapes of the moisture distribution curves in the solid at given times are qualitatively consistent with use of the unsteady-state diffusion equations given in Chapter 7. The moisture diffusivity D_{AB} usually decreases with decreased moisture content, so that the diffusivities are usually average values over the range of concentrations used. Materials drying in this way are usually said to be drying by diffusion, although the actual mechanisms may be quite complicated. Since the rate of evaporation from the surface is quite fast, i.e., the resistance is quite low, compared to the diffusion rate through the solid in the falling-rate period, the moisture content at the surface is at the equilibrium value.

The shape of a diffusion-controlled curve in the falling-rate period is similar to Fig. 9.5-2a. If the initial constant-rate drying is quite high, the first falling-rate period of

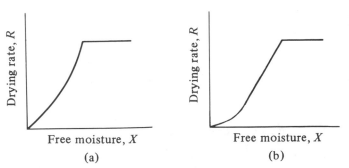

FIGURE 9.5-2. *Typical drying-rate curves: (a) diffusion-controlled falling-rate period, (b) capillary-controlled falling-rate period in a fine porous solid.*

unsaturated surface evaporation may not appear. If the constant-rate drying is quite low, the period of unsaturated surface evaporation is usually present in region CD in Fig. 9.5-1b and the diffusion-controlled curve is in region DE. Equations for calculating drying in this period where diffusion controls are given in Section 9.9. Also, Problem 7.1-4 for the drying of clay and Problem 7.1-6 for the drying of wood using diffusion theory are given in the Chapter 7 Problems.

2. Capillary movement in porous solids. When granular and porous solids such as clays, sand, soil, paint pigments, and minerals are being dried, unbound or free moisture moves through the capillaries and voids of the solids by *capillary action*, not by diffusion. This mechanism, involving surface tension, is similar to the movement of oil in a lamp wick.

A porous solid contains interconnecting pores and channels of varying pore sizes. As water is evaporated, a meniscus of liquid water is formed across each pore in the depths of the solid. This sets up capillary forces by the interfacial tension between the water and solid. These capillary forces provide the driving force for moving water through the pores to the surface. Small pores develop greater forces than those developed by large pores.

At the beginning of the falling-rate period at point C in Fig. 9.5-1b, the water is being brought to the surface by capillary action, but the surface layer of water starts to recede below the surface. Air rushes in to fill the voids. As the water is continuously removed, a point is reached where there is insufficient water left to maintain continuous films across the pores, and the rate of drying suddenly decreases at the start of the second falling-rate period at point D. Then the rate of diffusion of water vapor in the pores and rate of conduction of heat in the solid may be the main factors in drying.

In fine pores in solids, the rate-of-drying curve in the second falling-rate period may conform to the diffusion law and the curve is concave upward, as shown in Fig. 9.5-2b. For very porous solids, such as a bed of sand, where the pores are large, the rate-of-drying curve in the second falling-rate period is often straight, and hence the diffusion equations do not apply.

3. Effect of shrinkage. A factor often greatly affecting the drying rate is the shrinkage of the solid as moisture is removed. Rigid solids do not shrink appreciably, but colloidal and fibrous materials such as vegetables and other foodstuffs do undergo shrinkage. The most serious effect is that there may be developed a hard layer on the surface which is impervious to the flow of liquid or vapor moisture and slows the drying rate; examples are clay and soap. In many foodstuffs, if drying occurs at too high a temperature, a layer of closely packed shrunken cells, which are sealed together, forms at the surface. This presents a barrier to moisture migration and is known as *case hardening.* Another effect of shrinkage is to cause the material to warp and change its structure. This can happen in drying wood.

Sometimes to decrease these effects of shrinkage, it is desirable to dry with moist air. This decreases the rate of drying so that the effects of shrinkage on warping or hardening at the surface are greatly reduced.

9.6 CALCULATION METHODS FOR CONSTANT-RATE DRYING PERIOD

9.6A Method Using Experimental Drying Curve

1. Introduction. Probably the most important factor in drying calculations is the length of time required to dry a material from a given initial free moisture content X_1 to a final moisture content X_2. For drying in the constant-rate period, we can estimate the time needed by using experimental batch drying curves or by using predicted mass- and heat-transfer coefficients.

2. *Method using drying curve.* To estimate the time of drying for a given batch of material, the best method is based on actual experimental data obtained under conditions where the feed material, relative exposed surface area, gas velocity, temperature, and humidity are essentially the same as in the final drier. Then the time required for the constant-rate period can be determined directly from the drying curve of free moisture content versus time.

> **EXAMPLE 9.6-1.   Time of Drying from Drying Curve**
> A solid whose drying curve is represented by Fig. 9.5-1a is to be dried from a free moisture content $X_1 = 0.38$ kg H_2O/kg dry solid to $X_2 = 0.25$ kg H_2O/kg dry solid. Estimate the time required.
>
> *Solution:* From Fig. 9.5-1a for $X_1 = 0.38$, t_1 is read off as 1.28 h. For $X_2 = 0.25$, $t_2 = 3.08$ h. Hence, the time required is
>
> $$t = t_2 - t_1 = 3.08 - 1.28 = 1.80 \text{ h.}$$

3. *Method using rate-of-drying curve for constant-rate period.* Instead of using the drying curve, the rate-of-drying curve can be used. The drying rate R is defined by Eq. (9.5-3) as

$$R = -\frac{L_S}{A}\frac{dX}{dt} \tag{9.5-3}$$

This can be rearranged and integrated over the time interval to dry from X_1 at $t_1 = 0$ to X_2 at $t_2 = t$.

$$t = \int_{t_1=0}^{t_2=t} dt = \frac{L_S}{A}\int_{X_2}^{X_1}\frac{dX}{R} \tag{9.6-1}$$

If the drying takes place within the constant-rate period so that both X_1 and X_2 are greater than the critical moisture content X_C, then $R = $ constant $= R_C$. Integrating Eq. (9.6-1) for the constant-rate period,

$$t = \frac{L_S}{AR_C}(X_1 - X_2) \tag{9.6-2}$$

> **EXAMPLE 9.6-2.   Drying Time from Rate-of-Drying Curve**
> Repeat Example 9.6-1 but use Eq. (9.6-2) and Fig. 9.5-1b.
>
> *Solution:* As given previously, a value of 21.5 for L_S/A was used to prepare Fig. 9.5-1b from 9.5-1a. From Fig. 9.5-1b, $R_C = 1.51$ kg H_2O/h·m². Substituting into Eq. (9.6-2),
>
> $$t = \frac{L_S}{AR_C}(X_1 - X_2) = \frac{21.5}{1.51}(0.38 - 0.25) = 1.85 \text{ h}$$
>
> This is close to the value of 1.80 h of Example 9.6-1.

9.6B   Method Using Predicted Transfer Coefficients for Constant-Rate Period

1. *Introduction.* In the constant-rate period of drying, the surfaces of the grains of solid in contact with the drying air flow remain completely wetted. As stated previously, the

rate of evaporation of moisture under a given set of air conditions is independent of the type of solid and is essentially the same as the rate of evaporation from a free liquid surface under the same conditions. However, surface roughness may increase the rate of evaporation.

During this constant-rate period, the solid is so wet that the water acts as if the solid were not there. The water evaporated from the surface is supplied from the interior of the solid. The rate of evaporation from a porous material occurs by the same mechanism as that occurring at a wet bulb thermometer, which is essentially constant-rate drying.

2. Equations for predicting constant-rate drying. Drying of a material occurs by mass transfer of water vapor from the saturated surface of the material through an air film to the bulk gas phase or environment. The rate of moisture movement within the solid is sufficient to keep the surface saturated. The rate of removal of the water vapor (drying) is controlled by the rate of heat transfer to the evaporating surface, which furnishes the latent heat of evaporation for the liquid. At steady state, the rate of mass transfer balances the rate of heat transfer.

To derive the equation for drying, we neglect heat transfer by radiation to the solid surface and also assume no heat transfer by conduction from metal pans or surfaces. In Section 9.8, convection and radiation will also be considered. Assuming only heat transfer to the solid surface by convection from the hot gas to the surface of the solid and mass transfer from the surface to the hot gas (Fig. 9.6-1), we can write equations which are the same as those for deriving the wet bulb temperature T_W in Eq. (9.3-18).

The rate of convective heat transfer q in W (J/s, btu/h) from the gas at $T°C$ (°F) to the surface of the solid at T_W °C, where $(T - T_W)°C = (T - T_W)$ K is

$$q = h(T - T_W)A \qquad (9.6\text{-}3)$$

where h is the heat-transfer coefficient in $W/m^2 \cdot K$ (btu/h·ft²·°F) and A is the exposed drying area in m^2 (ft²). The equation of the flux of water vapor from the surface is the same as Eq. (9.3-13) and is

$$N_A = k_y(y_W - y) \qquad (9.6\text{-}4)$$

Using the approximation from Eq. (9.3-15) and substituting into Eq. (9.6-4),

$$N_A = k_y \frac{M_B}{M_A}(H_W - H) \qquad (9.6\text{-}5)$$

The amount of heat needed to vaporize N_A kg mol/s·m² (lb mol/h·ft²) water, neglecting the small sensible heat changes, is the same as Eq. (9.3-12).

$$q = M_A N_A \lambda_W A \qquad (9.6\text{-}6)$$

where λ_W is the latent heat at T_W in J/kg (btu/lb$_m$).

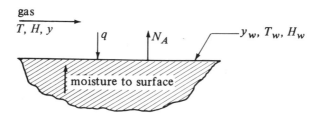

FIGURE 9.6-1. *Heat and mass transfer in constant-rate drying.*

Equating Eqs. (9.6-3) and (9.6-6) and substituting Eq. (9.6-5) for N_A,

$$R_C = \frac{q}{A\lambda_W} = \frac{h(T - T_W)}{\lambda_W} = k_y M_B(H_W - H) \tag{9.6-7}$$

Equation (9.6-7) is identical to Eq. (9.3-18) for the wet bulb temperature. Hence, in the absence of heat transfer by conduction and radiation, the temperature of the solid is at the wet bulb temperature of the air during the constant-rate drying period. Hence, the rate of drying R_C can be calculated using the heat-transfer equation $h(T - T_W)/\lambda_W$ or the mass-transfer equation $k_y M_B(H_W - H)$. However, it has been found more reliable to use the heat-transfer equation (9.6-8), since an error in determining the interface temperature T_W at the surface affects the driving force $(T - T_W)$ much less than it affects $(H_W - H)$.

$$R_C \text{ kg H}_2\text{O/h} \cdot \text{m}^2 = \frac{h}{\lambda_W} (T - T_W \,^\circ\text{C})(3600) \quad \text{(SI)}$$

$$\tag{9.6-8}$$

$$R_C \text{ lb}_m \text{ H}_2\text{O/h} \cdot \text{ft}^2 = \frac{h}{\lambda_W} (T - T_W \,^\circ\text{F}) \quad \text{(English)}$$

To predict R_C in Eq. (9.6-8), the heat-transfer coefficient must be known. For the case where the air is flowing parallel to the drying surface, Eq. (4.6-3) can be used for air. However, because the shape of the leading edge of the drying surface causes more turbulence, the following can be used for an air temperature of 45–150°C and a mass velocity G of 2450–29 300 kg/h · m² (500–6000 lb$_m$/h · ft²) or a velocity of 0.61–7.6 m/s (2–25 ft/s).

$$h = 0.0204 G^{0.8} \quad \text{(SI)}$$

$$\tag{9.6-9}$$

$$h = 0.0128 G^{0.8} \quad \text{(English)}$$

where in SI units G is $v\rho$ kg/h · m² and h is W/m² · K. In English units, G is in lb$_m$/h · ft² and h in btu/h · ft² · °F. When air flows perpendicular to the surface for a G of 3900–19 500 kg/h · m² or a velocity of 0.9–4.6 m/s (3–15 ft/s),

$$h = 1.17 G^{0.37} \quad \text{(SI)}$$

$$\tag{9.6-10}$$

$$h = 0.37 G^{0.37} \quad \text{(English)}$$

Equations (9.6-8) to (9.6-10) can be used to estimate the rate of drying during the constant-rate period. However, when possible, experimental measurements of the drying rate are preferred.

To estimate the time of drying during the constant-rate period, substituting Eq. (9.6-7) into (9.6-2),

$$t = \frac{L_S \lambda_W (X_1 - X_2)}{Ah(T - T_W)} = \frac{L_S(X_1 - X_2)}{Ak_y M_B(H_W - H)} \tag{9.6-11}$$

EXAMPLE 9.6-3. Prediction of Constant-Rate Drying
An insoluble wet granular material is dried in a pan 0.457 × 0.457 m (1.5 × 1.5 ft) and 25.4 mm deep. The material is 25.4 mm deep in the pan, and the sides and bottom can be considered to be insulated. Heat transfer is by convection from an air stream flowing parallel to the surface at a velocity of 6.1 m/s (20 ft/s). The air is at 65.6°C (150°F) and has a humidity of 0.010 kg H₂O/kg dry air. Estimate the rate of drying for the constant-rate period using SI and English units.

Solution: For a humidity $H = 0.010$ and dry bulb temperature of 65.6°C and using the humidity chart, Fig. 9.3-2, the wet bulb temperature T_W is found as 28.9°C (84°F) and $H_W = 0.026$ by following the adiabatic saturation line (the same as the wet bulb line) to the saturated humidity. Using Eq. (9.3-7) to calculate the humid volume,

$$v_H = (2.83 \times 10^{-3} + 4.56 \times 10^{-3}H)T$$
$$= (2.83 \times 10^{-3} + 4.56 \times 10^{-3} \times 0.01)(273 + 65.6)$$
$$= 0.974 \text{ m}^3/\text{kg dry air}$$

The density for 1.0 kg dry air + 0.010 kg H_2O is

$$\rho = \frac{1.0 + 0.010}{0.974} = 1.037 \text{ kg/m}^3 \ (0.0647 \text{ lb}_m/\text{ft}^3)$$

The mass velocity G is

$$G = v\rho = 6.1(3600)(1.037) = 22\,770 \text{ kg/h} \cdot \text{m}^2$$
$$G = v\rho = 20(3600)(0.0647) = 4660 \text{ lb}_m/\text{h} \cdot \text{ft}^2$$

Using Eq. (9.6-9),

$$h = 0.0204G^{0.8} = 0.0204(22\,770)^{0.8} = 62.45 \text{ W/m}^2 \cdot \text{K}$$
$$h = 0.0128G^{0.8} = 0.0128(4660)^{0.8} = 11.01 \text{ btu/h} \cdot \text{ft}^2 \cdot °\text{F}$$

At $T_W = 28.9°C$ (84°F), $\lambda_W = 2433$ kJ/kg (1046 btu/lb$_m$) from steam tables. Substituting into Eq. (9.6-8), noting that $(65.6 - 28.9)°C = (65.6 - 28.9)$ K,

$$R_C = \frac{h}{\lambda_W}(T - T_W)(3600) = \frac{62.45}{2433 \times 1000}(65.6 - 28.9)(3600)$$
$$= 3.39 \text{ kg/h} \cdot \text{m}^2$$

$$R_C = \frac{11.01}{1046}(150 - 84) = 0.695 \text{ lb}_m/\text{h} \cdot \text{ft}^2$$

The total evaporation rate for a surface area of $0.457 \times 0.457 \text{ m}^2$ is

$$\text{total rate} = R_C A = 3.39(0.457 \times 0.457) = 0.708 \text{ kg } H_2O/\text{h}$$
$$= 0.695(1.5 \times 1.5) = 1.564 \text{ lb}_m \ H_2O/\text{h}$$

9.6C Effect of Process Variables on Constant-Rate Period

As stated previously, experimental measurements of the drying rate are usually preferred over using the equations for prediction. However, these equations are quite helpful to predict the effect of changing the drying process variables when limited experimental data are available.

1. *Effect of air velocity.* When conduction and radiation heat transfer are not present, the rate R_C of drying in the constant-rate region is proportional to h and hence to $G^{0.8}$ as given by Eq. (9.6-9) for air flow parallel to the surface. The effect of gas velocity is less important when radiation and conduction are present.

2. *Effect of gas humidity.* If the gas humidity H is decreased for a given T of the gas, then from the humidity chart the wet bulb temperature T_W will decrease. Then using Eq. (9.6-7), R_C will increase. For example, if the original conditions are R_{C1}, T_1, T_{W1}, H_1, and

H_{W1}, then if H_1 is changed to H_2 and H_{W1} is changed to H_{W2}, R_{C2} becomes

$$R_{C2} = R_{C1} \frac{T - T_{W2}}{T - T_{W1}} \frac{\lambda_{W1}}{\lambda_{W2}} = R_{C1} \frac{H_{W2} - H_2}{H_{W1} - H_1} \qquad (9.6\text{-}12)$$

However, since $\lambda_{W1} \cong \lambda_{W2}$,

$$R_{C2} = R_{C1} \frac{T - T_{W2}}{T - T_{W1}} = R_{C1} \frac{H_{W2} - H_2}{H_{W1} - H_1} \qquad (9.6\text{-}13)$$

3. *Effect of gas temperature.* If the gas temperature T is increased, T_W is also increased some, but not as much as the increase in T. Hence, R_C increases as follows:

$$R_{C2} = R_{C1} \frac{T_2 - T_{W2}}{T_1 - T_{W1}} = R_{C1} \frac{H_{W2} - H_2}{H_{W1} - H_1} \qquad (9.6\text{-}14)$$

4. *Effect of thickness of solid being dried.* For heat transfer by convection only, the rate R_C is independent of the thickness x_1 of the solid. However, the time t for drying between fixed moisture contents X_1 and X_2 will be directly proportional to the thickness x_1. This is shown by Eq. (9.6-2), where increasing the thickness with a constant A will directly increase the amount of L_S kg dry solid.

5. *Experimental effect of process variables.* Experimental data tend to bear out the conclusions reached on the effects of material thickness, humidity, air velocity, and $T - T_W$.

9.7 CALCULATION METHODS FOR FALLING-RATE DRYING PERIOD

9.7A Method Using Graphical Integration

In the falling-rate drying period as shown in Fig. 9.5-1b, the rate of drying R is not constant but decreases when drying proceeds past the critical free moisture content X_C. When the free moisture content X is zero, the rate drops to zero.

The time for drying for any region between X_1 and X_2 has been given by Eq. (9.6-1).

$$t = \frac{L_S}{A} \int_{X_2}^{X_1} \frac{dX}{R} \qquad (9.6\text{-}1)$$

If the rate is constant, Eq. (9.6-1) can be integrated to give Eq. (9.6-2). However, in the falling-rate period, R varies. For any shape of falling-rate drying curve, Eq. (9.6-1) can be graphically integrated by plotting $1/R$ versus X and determining the area under the curve.

EXAMPLE 9.7-1. Graphical Integration in Falling-Rate Drying Period
A batch of wet solid whose drying-rate curve is represented by Fig. 9.5-1b is to be dried from a free moisture content of $X_1 = 0.38$ kg H_2O/kg dry solid to $X_2 = 0.04$ kg H_2O/kg dry solid. The weight of the dry solid is $L_S = 399$ kg dry solid and $A = 18.58$ m^2 of top drying surface. Calculate the time for drying. Note that $L_S/A = 399/18.58 = 21.5$ kg/m^2.

Solution: From Fig. 9.5-1b, the critical free moisture content is $X_C = 0.195$ kg H_2O/kg dry solid. Hence, the drying is in the constant-rate and falling-rate periods.

For the constant-rate period, $X_1 = 0.38$ and $X_2 = X_C = 0.195$. From Fig. 9.5-1b, $R_C = 1.51 \text{ kg H}_2\text{O/h} \cdot \text{m}^2$. Substituting into Eq. (9.6-2),

$$t = \frac{L_S}{AR_C}(X_1 - X_2) = \frac{399(0.38 - 0.195)}{(18.58)(1.51)} = 2.63 \text{ h}$$

For the falling-rate period, reading values of R for various values of X from Fig. 9.5-1b, the following table is prepared:

X	R	$1/R$	X	R	$1/R$
0.195	1.51	0.663	0.065	0.71	1.41
0.150	1.21	0.826	0.050	0.37	2.70
0.100	0.90	1.11	0.040	0.27	3.70

In Fig. 9.7-1 a plot of $1/R$ versus X is made and the area under the curve from $X_1 = 0.195$ (point C) to $X_2 = 0.040$ is determined:

$$\text{area} = A_1 + A_2 + A_3 = (2.5 \times 0.024) + (1.18 \times 0.056) + (0.84 + 0.075)$$

$$= 0.189$$

Substituting into Eq. (9.6-1),

$$t = \frac{L_S}{A}\int_{X_2}^{X_1} \frac{dX}{R} = \frac{399}{18.58}(0.189) = 4.06 \text{ h}$$

The total time is $2.63 + 4.06 = 6.69$ h.

9.7B Calculation Methods for Special Cases in Falling-Rate Region

In certain special cases in the falling-rate region, the equation for the time for drying, Eq. (9.6-1), can be integrated analytically.

1. Rate is a linear function of X. If both X_1 and X_2 are less than X_C and the rate R is linear in X over this region,

$$R = aX + b \tag{9.7-1}$$

where a is the slope of the line and b is a constant. Differentiating Eq. (9.7-1) gives $dR = a\,dX$. Substituting this into Eq. (9.6-1),

$$t = \frac{L_S}{aA}\int_{R_2}^{R_1} \frac{dR}{R} = \frac{L_S}{aA}\ln\frac{R_1}{R_2} \tag{9.7-2}$$

Since $R_1 = aX_1 + b$ and $R_2 = aX_2 + b$,

$$a = \frac{R_1 - R_2}{X_1 - X_2} \tag{9.7-3}$$

Substituting Eq. (9.7-3) into (9.7-2),

$$t = \frac{L_S(X_1 - X_2)}{A(R_1 - R_2)}\ln\frac{R_1}{R_2} \tag{9.7-4}$$

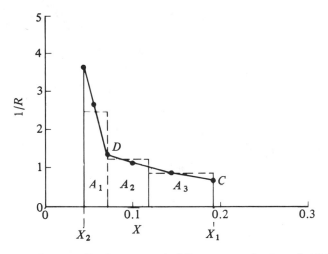

FIGURE 9.7-1.   *Graphical integration for falling-rate period in Example 9.7-1.*

2. Rate is a linear function through origin.   In some cases a straight line from the critical moisture content passing through the origin adequately represents the whole falling-rate period. In Fig. 9.5-1b this would be a straight line from C to E at the origin. Often for lack of more detailed data, this assumption is made. Then, for a straight line through the origin, where the rate of drying is directly proportional to the free moisture content,

$$R = aX \tag{9.7-5}$$

Differentiating, $dX = dR/a$. Substituting into Eq. (9.6-1),

$$t = \frac{L_S}{aA} \int_{R_2}^{R_1} \frac{dR}{R} = \frac{L_S}{aA} \ln \frac{R_1}{R_2} \tag{9.7-6}$$

The slope a of the line is R_C/X_C, and for $X_1 = X_C$ at $R_1 = R_C$,

$$t = \frac{L_S X_C}{AR_C} \ln \frac{R_C}{R_2} \tag{9.7-7}$$

Noting also that $R_C/R_2 = X_C/X_2$,

$$t = \frac{L_S X_C}{AR_C} \ln \frac{X_C}{X_2} \tag{9.7-8}$$

or

$$R = R_C \frac{X}{X_C} \tag{9.7-9}$$

EXAMPLE 9.7-2.   *Approximation of Straight Line for Falling-Rate Period*
Repeat Example 9.7-1, but as an approximation assume a straight line of the rate R versus X through the origin from point X_C to $X = 0$ for the falling-rate period.

Solution: $R_C = 1.51$ kg $H_2O/h \cdot m^2$ and $X_C = 0.195$. Drying in the falling-rate region is from X_C to $X_2 = 0.040$. Substituting into Eq. (9.7-8),

$$t = \frac{L_S X_C}{A R_C} \ln \frac{X_C}{X_2} = \frac{399(0.195)}{18.58(1.51)} \ln \frac{0.195}{0.040}$$

$$= 4.39 \text{ h}$$

This value of 4.39 h compares with the value of 4.06 h obtained in Example 9.7-1 by graphical integration.

9.8 COMBINED CONVECTION, RADIATION, AND CONDUCTION HEAT TRANSFER IN CONSTANT-RATE PERIOD

9.8A Introduction

In Section 9.6B an equation was derived for predicting the rate of drying in the constant-rate period. Equation (9.6-7) was derived assuming heat transfer to the solid by convection only from the surrounding air to the drying surface. Often the drying is done in an enclosure, where the enclosure surface radiates heat to the drying solid. Also, in some cases the solid may be resting on a metal tray, and heat transfer by conduction through the metal to the bottom of the solid may occur.

9.8B Derivation of Equation for Convection, Conduction, and Radiation

In Fig. 9.8-1 a solid material being dried by a stream of air is shown. The total rate of heat transfer to the drying surface is

$$q = q_C + q_R + q_K \tag{9.8-1}$$

where q_C is the convective heat transfer from the gas at $T°C$ to the solid surface at $T_S°C$ in W (J/s), q_R is the radiant heat transfer from the surface at T_R to T_S in W (J/s), and q_K is the rate of heat transfer by conduction from the bottom in W. The rate of convective

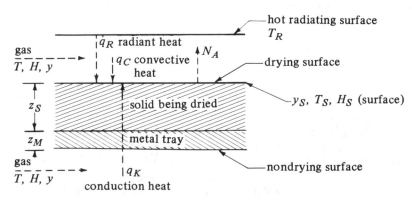

FIGURE 9.8-1. *Heat and mass transfer in drying a solid from the top surface.*

heat transfer is similar to Eq. (9.6-3) and is as follows, where $(T - T_S)°C = (T - T_S)$ K,

$$q_C = h_C(T - T_S)A \qquad (9.8-2)$$

where A is the exposed surface area in m². The radiant heat transfer is

$$q_R = h_R(T_R - T_S)A \qquad (9.8-3)$$

where h_R is the radiant-heat-transfer coefficient defined by Eq. (4.10-10).

$$h_R = \varepsilon(5.676) \frac{\left(\dfrac{T_R}{100}\right)^4 - \left(\dfrac{T_S}{100}\right)^4}{T_R - T_S} \qquad (4.10-10)$$

Note that in Eq. (4.10-10) T_R and T_S are in K. For the heat transfer by conduction from the bottom, the heat transfer is first by convection from the gas to the metal plate, then by conduction through the metal, and finally by conduction through the solid. Radiation to the bottom of the tray is often small if the tray is placed above another tray, and it will be neglected here. Also, if the gas temperatures are not too high, radiation from the top surface to the tray will be small. Hence, the heat by radiation should not be overemphasized. The heat by conduction is

$$q_K = U_K(T - T_S)A \qquad (9.8-4)$$

$$U_K = \frac{1}{1/h_C + z_M/k_M + z_S/k_S} \qquad (9.8-5)$$

where z_M is the metal thickness in m, k_M the metal thermal conductivity in W/m · K, z_S the solid thickness in m, and k_S the solid thermal conductivity. The value of h_C in Eq. (9.8-4) is assumed to be the same as in Eq. (9.8-2).

The equation for the rate of mass transfer is similar to Eq. (9.6-5) and is

$$N_A = k_y \frac{M_B}{M_A}(H_S - H) \qquad (9.8-6)$$

Also, rewriting Eq. (9.6-6),

$$q = M_A N_A \lambda_S A \qquad (9.8-7)$$

Combining Eqs. (9.8-1), (9.8-2), (9.8-3), (9.8-4), (9.8-6), and (9.8-7),

$$R_C = \frac{q}{A\lambda_S} = \frac{(h_C + U_K)(T - T_S) + h_R(T_R - T_S)}{\lambda_S} = k_y M_B(H_S - H) \qquad (9.8-8)$$

This equation can be compared to Eq. (9.6-7), which gives the wet bulb temperature T_W when radiation and conduction are absent. Equation (9.8-8) gives surface temperature T_S greater than the wet bulb temperature T_W. Equation (9.8-8) must also intersect the saturated humidity line at T_S and H_S, and $T_S > T_W$ and $H_S > H_W$. The equation must be solved by trial and error.

To facilitate solution of Eq. (9.8-8), it can be rearranged (T1) to the following:

$$\frac{(H_S - H)\lambda_S}{h_C/k_y M_B} = \left(1 + \frac{U_K}{h_C}\right)(T - T_S) + \frac{h_R}{h_C}(T_R - T_S) \qquad (9.8-9)$$

The ratio $h_C/k_y M_B$ was shown in the wet bulb derivation of Eq. (9.3-18) to be approximately c_S in Eq. (9.3-6).

$$c_S = (1.005 + 1.88H)10^3 \text{ J/kg} \cdot \text{K} \qquad (9.3-6)$$

EXAMPLE 9.8-1. Constant-Rate Drying When Radiation and Convection Are Present

An insoluble granular material wet with water is being dried in a pan 0.457×0.457 m and 25.4 mm deep. The material is 25.4 mm deep in the metal pan, which has a metal bottom with thickness $z_M = 0.610$ mm having a thermal conductivity $k_M = 43.3$ W/m·K. The thermal conductivity of the solid can be assumed as $k_S = 0.865$ W/m·K. Heat transfer is by convection from an air stream flowing parallel to the top drying surface and the bottom metal surface at a velocity of 6.1 m/s and having a temperature of 65.6°C and humidity $H = 0.010$ kg H_2O/kg dry air. The top surface also receives direct radiation from steam-heated pipes whose surface temperature $T_R = 93.3$°C. The emissivity of the solid is $\varepsilon = 0.92$. Estimate the rate of drying for the constant-rate period.

Solution: Some of the given values are as follows:

$$T = 65.6°C, \qquad z_S = 0.0254 \text{ m}, \qquad k_M = 43.3, \qquad k_S = 0.865$$

$$z_M = 0.00061 \text{ m} \qquad \varepsilon = 0.92, \qquad H = 0.010$$

The velocity, temperature, and humidity of air are the same as Example 9.6-3 and the convective coefficient was predicted as $h_C = 62.45$ W/m²·K.

The solution of Eq. (9.8-9) is by trial and error. The temperature T_S will be above the wet bulb temperature of $T_W = 28.9°C$ and will be estimated as $T_S = 32.2°C$. Then $\lambda_S = 2424$ kJ/kg from the steam tables. To predict h_R from Eq. (4.10-10) for $\varepsilon = 0.92$, $T_1 = 93.3 + 273.2 = 366.5$ K, and $T_2 = 32.2 + 273.2 = 305.4$ K,

$$h_R = (0.92)(5.676) \frac{(366.5/100)^4 - (305.4/100)^4}{366.5 - 305.4} = 7.96 \text{ W/m}^2 \cdot \text{K}$$

Using Eq. (9.8-5),

$$U_K = \frac{1}{1/h_C + z_M/k_M + z_S/k_S} = \frac{1}{1/62.45 + 0.00061/43.3 + 0.0254/0.865}$$

$$= 22.04 \text{ W/m}^2 \cdot \text{K}$$

From Eq. (9.3-6),

$$c_S = (1.005 + 1.88H)10^3 = (1.005 + 1.88 \times 0.010)10^3$$

$$= 1.024 \times 10^3 \text{ J/kg} \cdot \text{K}$$

This can be substituted for $(h_C/k_y M_B)$ into Eq. (9.8-9). Also, substituting other knowns,

$$\frac{(H_S - 0.01)\lambda_S}{1.024 \times 10^3} = (1 + 22.04/62.45)(65.6 - T_S)$$

$$+ (7.96/62.45)(93.3 - T_S)$$

$$= 1.353(65.6 - T_S) + 0.1275(93.3 - T_S) \qquad \textbf{(9.8-10)}$$

For T_S assumed as 32.2°C, $\lambda_S = 2424 \times 10^3$ J/kg. From the humidity chart for $T_S = 32.2°C$, the saturation humidity $H_S = 0.031$. Substituting into Eq. (9.8-10) and solving for T_S,

$$\frac{(0.031 - 0.010)(2424 \times 10^3)}{1.024 \times 10^3} = 1.353(65.6 - T_S) + 0.1275(93.3 - T_S)$$

$$T_S = 34.4°C$$

For the second trial, assuming that $T_S = 32.5°C$, $\lambda_S = 2423 \times 10^3$ and H_S from the humidity chart at saturation is 0.032. Substituting into Eq. (9.8-10) while assuming that h_R does not change appreciably, a value of $T_S = 32.8°C$ is obtained. Hence, the final value is 32.8°C. This is 3.9°C greater than the wet bulb temperature of 28.9°C in Example 9.6-3, where radiation and conduction were absent.

Using Eq. (9.8-8),

$$R_C = \frac{(h_C + U_K)(T - T_S) + h_R(T_R - T_S)}{\lambda_S} (3600)$$

$$= \frac{(62.45 + 22.04)(65.6 - 32.8) + 7.96(93.3 - 32.8)}{2423 \times 10^3} (3600)$$

$$= 4.83 \text{ kg/h} \cdot \text{m}^2$$

This compares with 3.39 kg/h·m² for Example 9.6-3 for no radiation or conduction.

9.9 DRYING IN FALLING-RATE PERIOD BY DIFFUSION AND CAPILLARY FLOW

9.9A Introduction

In the falling-rate period, the surface of the solid being dried is no longer completely wetted, and the rate of drying steadily falls with time. In Section 9.7 empirical methods were used to predict the time of drying. In one method the actual rate of drying curve was graphically integrated to determine the time of drying.

In another method an approximate straight line from the critical free moisture content to the origin at zero free moisture was assumed. Here the rate of drying was assumed to be a linear function of the free moisture content. The rate of drying R is defined by Eq. (9.5-3).

$$R = -\frac{L_S}{A}\frac{dX}{dt} \tag{9.5-3}$$

When R is a linear function of X in the falling-rate period,

$$R = aX \tag{9.7-5}$$

where a is a constant. Equating Eq. (9.7-5) to Eq. (9.5-3),

$$R = -\frac{L_S}{A}\frac{dX}{dt} = aX \tag{9.9-1}$$

Rearranging,

$$\frac{dX}{dt} = -\frac{aA}{L_S}X \tag{9.9-2}$$

In many instances, however, as mentioned briefly in Section 9.5E, the rate of moisture movement in the falling-rate period is governed by the rate of internal movement of the liquid by liquid diffusion or by capillary movement. These two methods of moisture movement will be considered in more detail and the theories related to experimental data in the falling-rate region.

9.9B Liquid Diffusion of Moisture in Drying

When liquid diffusion of moisture controls the rate of drying in the falling-rate period, the equations for diffusion described in Chapter 7 can be used. Using the concentrations as X kg free moisture/kg dry solid instead of concentrations kg mol moisture/m^3, Fick's second law for unsteady-state diffusion, Eq. (7.10-10), can be written as

$$\frac{\partial X}{\partial t} = D_L \frac{\partial^2 X}{\partial x^2} \tag{9.9-3}$$

where D_L is the liquid diffusion coefficient in m^2/h and x is distance in the solid in m.

This type of diffusion is often characteristic of relatively slow drying in nongranular materials such as soap, gelatin, and glue, and in the later stages of drying of bound water in clay, wood, textiles, leather, paper, foods, starches, and other hydrophilic solids.

A major difficulty in analyzing diffusion drying data is that the initial moisture distribution is not uniform throughout the solid at the start if a drying period at constant rate precedes this falling-rate period. During diffusion-type drying, the resistance to mass transfer of water vapor from the surface is usually very small, and the diffusion in the solid controls the rate of drying. Then the moisture content at the surface is at the equilibrium value X^*. This means that the free moisture content X at the surface is essentially zero.

Assuming that the initial moisture distribution is uniform at $t = 0$, Eq. (9.9-3) may be integrated by the methods in Chapter 7 to give the following:

$$\frac{X_t - X^*}{X_{t1} - X^*} = \frac{X}{X_1} = \frac{8}{\pi^2} \left[e^{-D_L t (\pi/2x_1)^2} + \frac{1}{9} e^{-9D_L t (\pi/2x_1)^2} + \frac{1}{25} e^{-25D_L t (\pi/2x_1)^2} + \cdots \right] \tag{9.9-4}$$

where X = average free moisture content at time t h, X_1 = initial free moisture content at $t = 0$, X^* = equilibrium free moisture content, $x_1 = \frac{1}{2}$ the thickness of the slab when drying occurs from the top and the bottom parallel faces, and x_1 = total thickness of slab if drying only from the top face.

Equation (9.9-4) assumes that D_L is constant, but D_L is rarely constant; it varies with moisture content, temperature, and humidity. For long drying times, only the first term in Eq. (9.9-4) is significant, and the equation becomes

$$\frac{X}{X_1} = \frac{8}{\pi^2} e^{-D_L t (\pi/2x_1)^2} \tag{9.9-5}$$

Solving for the time of drying,

$$t = \frac{4x_1^2}{\pi^2 D_L} \ln \frac{8X_1}{\pi^2 X} \tag{9.9-6}$$

In this equation if the diffusion mechanism starts at $X = X_C$, then $X_1 = X_C$. Differentiating Eq. (9.9-6) with respect to time and rearranging,

$$\frac{dX}{dt} = -\frac{\pi^2 D_L X}{4x_1^2} \tag{9.9-7}$$

Multiplying both sides by $-L_S/A$,

$$R = -\frac{L_S}{A} \frac{dX}{dt} = \frac{\pi^2 L_S D_L}{4x_1^2 A} X \tag{9.9-8}$$

Hence, Eqs. (9.9-7) and (9.9-8) state that when internal diffusion controls for long times, the rate of drying is directly proportional to the free moisture X and the liquid diffusivity and that the rate of drying is inversely proportional to the thickness squared.

Or, stated as the time of drying between fixed moisture limits, the time varies directly as the square of the thickness. The drying rate should be independent of gas velocity and humidity.

EXAMPLE 9.9-1. Drying Slabs of Wood When Diffusion of Moisture Controls

The experimental average diffusion coefficient of moisture in a given wood is 2.97×10^{-6} m^2/h (3.20×10^{-5} ft^2/h). Large planks of wood 25.4 mm thick are dried from both sides by air having a humidity such that the equilibrium moisture content in the wood is $X^* = 0.04$ kg H$_2$O/kg dry wood. The wood is to be dried from a total average moisture content of $X_{t1} = 0.29$ to $X_t = 0.09$. Calculate the time needed.

Solution: The free moisture content $X_1 = X_{t1} - X^* = 0.29 - 0.04 = 0.25$, $X = X_t - X^* = 0.09 - 0.04 = 0.05$. The half-slab thickness $x_1 = 25.4/(2 \times 1000) = 0.0127$ m. Substituting into Eq. (9.9-6),

$$t = \frac{4x_1^2}{\pi^2 D_L} \ln \frac{8 X_1}{\pi^2 X} = \frac{4(0.0127)^2}{\pi^2 (2.97 \times 10^{-6})} \ln \frac{8 \times 0.25}{\pi^2 \times 0.05}$$

$$= 30.8 \text{ h}$$

Alternatively, Fig. 5.3-13 for the average concentration in a slab can be used. The ordinate $E_a = X/X_1 = 0.05/0.25 = 0.20$. Reading off the plot a value of $0.56 = D_L t/x_1^2$, substituting, and solving for t,

$$t = \frac{x_1^2(0.56)}{D_L} = \frac{(0.0127)^2(0.56)}{2.97 \times 10^{-6}} = 30.4 \text{ h}$$

9.9C Capillary Movement of Moisture in Drying

Water can flow from regions of high concentrations to those of low concentrations as a result of capillary action rather than by diffusion if the pore sizes of granular materials are suitable.

The capillary theory (P1) assumes that a packed bed of nonporous spheres contains a void space between the spheres called *pores*. As water is evaporated, capillary forces are set up by the interfacial tension between the water and solid. These forces provide the driving force for moving the water through the pores to the drying surface.

A modified form of Poiseuille's equation for laminar flow can be used in conjunction with the capillary-force equation to derive an equation for the rate of drying when flow is by capillary movement. If the moisture movement follows the capillary-flow equations, the rate of drying R will vary linearly with X. Since the mechanism of evaporation during this period is the same as in the constant-rate period, the effects of the variables of the drying gas of gas velocity, temperature of the gas, humidity of the gas, and so on, will be the same as for the constant-rate drying period.

The defining equation for the rate of drying is

$$R = -\frac{L_S}{A} \frac{dX}{dt} \tag{9.5-3}$$

For the rate R varying linearly with X given previously,

$$R = R_C \frac{X}{X_C} \tag{9.7-9}$$

$$t = \frac{L_S X_C}{A R_C} \ln \frac{X_C}{X_2} \tag{9.7-8}$$

We define t as the time when $X = X_2$ and

$$L_S = x_1 A \rho_S \tag{9.9-9}$$

where $\rho_S =$ solid density kg dry solid/m^3. Substituting Eq. (9.9-9) and $X = X_2$ into Eq. (9.7-8),

$$t = \frac{x_1 \rho_S X_C}{R_C} \ln \frac{X_C}{X} \tag{9.9-10}$$

Substituting Eq. (9.6-7) for R_C,

$$t = \frac{x_1 \rho_S \lambda_W X_C}{h(T - T_W)} \ln \frac{X_C}{X} \tag{9.9-11}$$

Hence, Eqs. (9.9-10) and (9.9-11) state that when capillary flow controls in the falling-rate period, the rate of drying is inversely proportional to the thickness. The time of drying between fixed moisture limits varies directly as the thickness and depends upon the gas velocity, temperature, and humidity.

9.9D Comparison of Liquid Diffusion and Capillary Flow

To determine the mechanism of drying in the falling-rate period, the experimental data obtained of moisture content at various times using constant drying conditions are often analyzed as follows. The *unaccomplished moisture change*, defined as the ratio of free moisture present in the solid after drying for t hours to the total free moisture content present at the start of the falling-rate period, X/X_C, is plotted versus time on semilog paper. If a straight line is obtained, such as curve B in Fig. 9.9-1 using the upper scale for the abscissa, then either Eqs. (9.9-4)–(9.9-6) for diffusion are applicable or Eqs. (9.9-10) and (9.9-11) for capillary flow are applicable.

If the relation for capillary flow applies, the slope of the falling-rate drying line B in Fig. 9.9-1 is related to Eq. (9.9-10), which contains the constant drying rate R_C. The value of R_C is calculated from the measured slope of the line, which is $-R_C/x_1 \rho_S X_C$, and if it agrees with the experimental value of R_C in the constant drying period or the predicted value of R_C, the moisture movement is by capillary flow.

If the values of R_C do not agree, the moisture movement is by diffusion and the slope of line B in Fig. 9.9-1 from Eq. (9.9-6) should equal $-\pi^2 D_L/4x_1^2$. In actual practice, however, the diffusivity D_L is usually less at small moisture contents than at large moisture contents, and an average value of D_L is usually determined experimentally over the moisture range of interest. A plot of Eq. (9.9-4) is shown as line A, where $\ln(X/X_1)$ or $\ln(X/X_C)$ is plotted versus $D_L t/x_1^2$. This is the same plot as Fig. 5.3-13 for a slab and shows a curvature in the line for values of X/X_C between 1.0 and 0.6 and a straight line for $X/X_C < 0.6$.

When the experimental data show that the movement of moisture follows the diffusion law, the average experimental diffusivities can be calculated as follows for different concentration ranges. A value of X/X_C is chosen at 0.4, for example. From an experimental plot similar to curve B, Fig. 9.9-1, the experimental value of t is obtained. From curve A at $X/X_C = 0.4$, the theoretical value of $(D_L t/x_1^2)_{\text{theor}}$ is obtained. Then, by substituting the known values of t and x_1 into Eq. (9.9-12), the experimental average value of D_L over the range $X/X_C = 1.0$–0.4 is obtained.

$$D_L = \left(\frac{D_L t}{x_1^2}\right)_{\text{theor}} \frac{x_1^2}{t} \tag{9.9-12}$$

This is repeated for various values of X/X_C. Values of D_L obtained for $X/X_C > 0.6$ are in error because of the curvature of line A.

EXAMPLE 9.9-2. Diffusion Coefficient in the Tapioca Root

Tapioca flour is obtained from drying and then milling the tapioca root. Experimental data on drying thin slices of the tapioca root 3 mm thick on both sides in the falling-rate period under constant drying conditions are tabulated below. The time $t = 0$ is the start of the falling-rate period.

X/X_C	t (h)	X/X_C	t (h)	X/X_C	t (h)
1.0	0	0.55	0.40	0.23	0.94
0.80	0.15	0.40	0.60	0.18	1.07
0.63	0.27	0.30	0.80		

It has been determined that the data do not follow the capillary-flow equation but appear to follow the diffusion equation. Plot the data as X/X_C versus t on semilog coordinates and determine the average diffusivity of the moisture up to a value of $X/X_C = 0.20$.

Solution: In Fig. 9.9-2 the data are plotted as X/X_C on the log scale versus t on a linear scale and a smooth curve drawn through the data. At $X/X_C =$

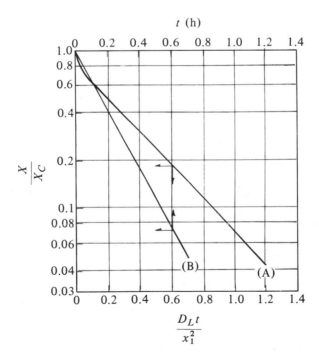

FIGURE 9.9-1. *Plot of equations for falling-rate period: (A) Eq. (9.9-4) for moisture movement by diffusion, (B) Eq. (9.9-10) for moisture movement by capillary flow. (From R. H. Perry and C. H. Chilton, Chemical Engineers Handbook, 5th ed. New York: McGraw-Hill Book Company, 1973. With permission.)*

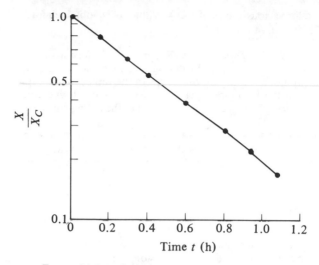

FIGURE 9.9-2. *Plot of drying data for Example 9.9-2.*

0.20, a value of $t = 1.02$ h is read off the plot. The value of $x_1 = 3$ mm/2 = 1.5 mm for drying from both sides. From Fig. 9.9-1, line A, for $X/X_C = 0.20$, $(D_L t/x_1^2)_{theor} = 0.56$. Then substituting into Eq. (9.9-12),

$$D_L = (D_L t/x_1^2)_{theor} \frac{x_1^2}{t} = \frac{0.56(1.5/1000)^2}{1.02 \times 3600} = 3.44 \times 10^{-10} \text{ m}^2/\text{s}$$

9.10 EQUATIONS FOR VARIOUS TYPES OF DRYERS

9.10A Through Circulation Drying in Packed Beds

For through circulation drying where the drying gas passes upward or downward through a bed of wet granular solids, both a constant-rate period and a falling-rate period of drying may result. Often the granular solids are arranged on a screen so that the gas passes through the screen and through the open spaces or voids between the solid particles.

1. Derivation of equations.   To derive the equations for this case, no heat losses will be assumed, so the system is adiabatic. The drying will be for unbound moisture in the wet granular solids. We shall consider a bed of uniform cross-sectional area A m², where a gas flow of G kg dry gas/h·m² cross section enters with a humidity of H_1. By a material balance on the gas at a given time, the gas leaves the bed with a humidity H_2. The amount of water removed from the bed by the gas is equal to the rate of drying at this time.

$$R = G(H_2 - H_1) \tag{9.10-1}$$

where $R = $ kg H$_2$O/h·m² cross section and $G = $ kg dry air/h·m² cross section.

In Fig. 9.10-1 the gas enters at T_1 and H_1 and leaves at T_2 and H_2. Hence, the temperature T and humidity H both vary through the bed. Making a heat balance over the short section dz m of the bed,

$$dq = -Gc_S A \, dT \qquad (9.10\text{-}2)$$

where $A = $ m^2 cross-sectional area, q is the heat-transfer rate in W (J/s), and c_S is the humid heat of the air–water vapor mixture in Eq. (9.3-6). Note that G in this equation is in kg/s · m^2. The heat-transfer equation gives

$$dq = haA \, dz(T - T_W) \qquad (9.10\text{-}3)$$

where $T_W = $ wet bulb temperature of solid, h is the heat-transfer coefficient in W/m^2 · K, and a is m^2 surface area of solids/m^3 bed volume. Equating Eq. (9.10-2) to (9.10-3), rearranging, and integrating,

$$\frac{ha}{Gc_S} \int_0^z dz = -\int_{T_1}^{T_2} \frac{d'T}{T - T_W} \qquad (9.10\text{-}4)$$

$$\frac{haz}{Gc_S} = \ln \frac{T_1 - T_W}{T_2 - T_W} \qquad (9.10\text{-}5)$$

where $z = $ bed thickness $= x_1$ m.

For the constant-rate period of drying by air flowing parallel to a surface, Eq. (9.6-11) was derived.

$$t = \frac{L_S \lambda_W (X_1 - X_2)}{Ah(T - T_W)} = \frac{L_S(X_1 - X_2)}{Ak_y M_B(H_W - H)} \qquad (9.6\text{-}11)$$

Using Eq. (9.9-9) and the definition of a, we obtain

$$\frac{L_S}{A} = \frac{\rho_S}{a} \qquad (9.10\text{-}6)$$

Substituting Eq. (9.10-6) into (9.6-11) and setting $X_2 = X_C$ for drying to X_C, we obtain the equation for through circulation drying in the constant-rate period.

$$t = \frac{\rho_S \lambda_W (X_1 - X_C)}{ah(T - T_W)} = \frac{\rho_S(X_1 - X_C)}{ak_y M_B(H_W - H)} \qquad (9.10\text{-}7)$$

In a similar manner, Eq. (9.7-8) for the falling-rate period, which assumes that R is proportional to X, becomes, for through circulation drying,

$$t = \frac{\rho_S \lambda_W X_C \ln (X_C/X)}{ah(T - T_W)} = \frac{\rho_S X_C \ln (X_C/X)}{ak_y M_B(H_W - H)} \qquad (9.10\text{-}8)$$

FIGURE 9.10-1. *Heat and material balances in a through circulation dryer in a packed bed.*

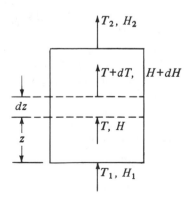

Both Eqs. (9.10-7) and (9.10-8), however, hold only for one point in the bed in Fig. 9.10-1, since the temperature T of the gas varies throughout the bed. Hence, in a manner similar to the derivation in heat transfer, a log mean temperature difference can be used as an approximation for the whole bed in place of $T - T_W$ in Eqs. (9.10-7) and (9.10-8).

$$(T - T_W)_{LM} = \frac{(T_1 - T_W) - (T_2 - T_W)}{\ln [(T_1 - T_W)/(T_2 - T_W)]} = \frac{T_1 - T_2}{\ln [(T_1 - T_W)/(T_2 - T_W)]} \quad (9.10\text{-}9)$$

Substituting Eq. (9.10-5) for the denominator of Eq. (9.10-9) and also substituting the value of T_2 from Eq. (9.10-5) into (9.10-9),

$$(T - T_W)_{LM} = \frac{(T_1 - T_W)(1 - e^{-haz/Gc_S})}{haz/Gc_S} \quad (9.10\text{-}10)$$

Substituting Eq. (9.10-10) into (9.10-7) for the constant-rate period and setting $x_1 = z$,

$$t = \frac{\rho_S \lambda_W x_1 (X_1 - X_C)}{Gc_S (T_1 - T_W)(1 - e^{-hax_1/Gc_S})} \quad (9.10\text{-}11)$$

Similarly, for the falling-rate period an approximate equation is obtained.

$$t = \frac{\rho_S \lambda_W x_1 X_C \ln (X_C/X)}{Gc_S (T_1 - T_W)(1 - e^{-hax_1/Gc_S})} \quad (9.10\text{-}12)$$

A major difficulty with the use of Eq. (9.10-12) is that the critical moisture content is not easily estimated. Different forms of Eqs. (9.10-11) and (9.10-12) can also be derived, using humidity instead of temperature (T1).

2. *Heat-transfer coefficients.* For through circulation drying, where the gases pass through a bed of wet granular solids, the following equations for estimating h for adiabatic evaporation of water can be used (G1, W1).

$$\left.\begin{array}{ll} h = 0.151 \dfrac{G_t^{0.59}}{D_p^{0.41}} & \text{(SI)} \\[3mm] h = 0.11 \dfrac{G_t^{0.59}}{D_p^{0.41}} & \text{(English)} \end{array}\right\} \quad \dfrac{D_p G_t}{\mu} > 350 \quad (9.10\text{-}13)$$

$$\left.\begin{array}{ll} h = 0.214 \dfrac{G_t^{0.49}}{D_p^{0.51}} & \text{(SI)} \\[3mm] h = 0.15 \dfrac{G_t^{0.49}}{D_p^{0.51}} & \text{(English)} \end{array}\right\} \quad \dfrac{D_p G_t}{\mu} < 350 \quad (9.10\text{-}14)$$

where h is in W/m$^2 \cdot$ K, D_p is diameter in m of a sphere having the same surface area as the particle in the bed, G_t is the total mass velocity entering the bed in kg/h $\cdot$ m^2, and μ is viscosity in kg/m $\cdot$ h. In English units, h is btu/h $\cdot$ ft$^2 \cdot$ °F, D_p is ft, G_t is lb$_m$/h $\cdot$ ft^2, and μ is lb$_m$/ft $\cdot$ h.

3. *Geometry factors in a bed.* To determine the value of a, m^2 surface area/m^3 of bed, in a packed bed for spherical particles having a diameter D_p m,

$$a = \frac{6(1 - \varepsilon)}{D_p} \quad (9.10\text{-}15)$$

where ε is the void fraction in the bed. For cylindrical particles,

$$a = \frac{4(1 - \varepsilon)(h + 0.5D_c)}{D_c h} \qquad (9.10\text{-}16)$$

where D_c is diameter of cylinder in m and h is length of cylinder in m. The value of D_p to use in Eqs. (9.10-13) and (9.10-14) for a cylinder is the diameter of a sphere having the same surface area as the cylinder, as follows:

$$D_p = (D_c h + 0.5D_c^2)^{1/2} \qquad (9.10\text{-}17)$$

4. *Equations for very fine particles.* The equations derived for the constant- and falling-rate periods in packed beds hold for particles of about 3–19 mm in diameter in shallow beds about 10–65 mm thick (T1, M1). For very fine particles of 10–200 mesh (1.66–0.079 mm) and bed depth greater than 11 mm, the interfacial area a varies with the moisture content. Empirical expressions are available to estimate a and the mass-transfer coefficient (T1, A1).

EXAMPLE 9.10-1. Through Circulation Drying in a Bed

A granular paste material is extruded into cylinders with a diameter of 6.35 mm and length of 25.4 mm. The initial total moisture content $X_{t1} = 1.0$ kg H_2O/kg dry solid and the equilibrium moisture is $X^* = 0.01$. The density of the dry solid is 1602 kg/m³ (100 lb$_m$/ft³). The cylinders are packed on a screen to a depth of $x_1 = 50.8$ mm. The bulk density of the dry solid in the bed is $\rho_S = 641$ kg/m³. The inlet air has a humidity $H_1 = 0.04$ kg H_2O/kg dry air and a temperature $T_1 = 121.1°C$. The gas superficial velocity is 0.811 m/s and the gas passes through the bed. The total critical moisture content is $X_{tC} = 0.50$. Calculate the total time to dry the solids to $X_t = 0.10$ kg H_2O/kg dry solid.

Solution: For the solid,

$$X_1 = X_{t1} - X^* = 1.00 - 0.01 = 0.99 \text{ kg } H_2O/\text{kg dry solid}$$

$$X_C = X_{tC} - X^* = 0.50 - 0.01 = 0.49$$

$$X - X_t - X^* = 0.10 - 0.01 = 0.09$$

For the gas, $T_1 = 121.1°C$ and $H_1 = 0.04$ kg H_2O/kg dry air. The wet bulb temperature $T_W = 47.2°C$ and $H_W = 0.074$. The solid temperature is at T_W if radiation and conduction are neglected. The density of the entering air at 121.1°C and 1 atm is as follows.

$$v_H = (2.83 \times 10^{-3} + 4.56 \times 10^{-3} \times 0.04)(273 + 121.1) \quad (9.3\text{-}7)$$

$$= 1.187 \text{ m}^3/\text{kg dry air}$$

$$\rho = \frac{1.00 + 0.04}{1.187} = 0.876 \text{ kg dry air} + H_2O/\text{m}^3$$

The mass velocity of the dry air is

$$G = v\rho\left(\frac{1.0}{1.0 + 0.04}\right) = 0.811(3600)(0.876)\left(\frac{1}{1.04}\right) = 2459 \text{ kg dry air/h} \cdot \text{m}^2$$

Since the inlet $H_1 = 0.040$ and the outlet will be less than 0.074, an approximate average H of 0.05 will be used to calculate the total average mass velocity. The approximate average G_t is

$$G_t = 2459 + 2459(0.05) = 2582 \text{ kg air} + H_2O/\text{h} \cdot \text{m}^2$$

For the packed bed, the void fraction ε is calculated as follows for 1 m³ of bed containing solids plus voids. A total of 641 kg dry solid is present. The density of the dry solid is 1602 kg dry solid/m³ solid. The volume of the solids in 1 m³ of bed is then 641/1602, or 0.40 m³ solid. Hence, $\varepsilon = 1 - 0.40 = 0.60$. The solid cylinder length $h = 0.0254$ m. The diameter $D_c = 0.00635$ m. Substituting into Eq. (9.10-16),

$$a = \frac{4(1 - \varepsilon)(h + 0.5D_c)}{D_c h} = \frac{4(1 - 0.6)[0.0254 + 0.5(0.00635)]}{0.00635(0.0254)}$$

$$= 283.5 \text{ m}^2 \text{ surface area/m}^3 \text{ bed volume}$$

To calculate the diameter D_p of a sphere with the same area as the cylinder using Eq. (9.10-17),

$$D_p = (D_c h + 0.5D_c^2)^{1/2} = [0.00635 \times 0.0254 + 0.5(0.00635)^2]^{1/2}$$

$$= 0.0135 \text{ m}$$

The bed thickness $x_1 = 50.8$ mm $= 0.0508$ m.

To calculate the heat-transfer coefficient, the Reynolds number is first calculated. Assuming an approximate average air temperature of 93.3°C, the viscosity of air is $\mu = 2.15 \times 10^{-5}$ kg/m · s $= 2.15 \times 10^{-5}$ (3600) $= 7.74 \times 10^{-2}$ kg/m · h. The Reynolds number is

$$N_{Re} = \frac{D_p G_t}{\mu} = \frac{0.0135(2582)}{7.74 \times 10^{-2}} = 450$$

Using Eq. (9.10-13),

$$h = 0.151 \frac{G_t^{0.59}}{D_p^{0.41}} = \frac{0.151(2582)^{0.51}}{(0.0135)^{0.41}} = 90.9 \text{ W/m}^2 \cdot \text{K}$$

For $T_W = 47.2$°C, $\lambda_W = 2389$ kJ/kg, or 2.389×10^6 J/kg (1027 btu/lb$_m$), from steam tables. The average humid heat, from Eq. (9.3-6), is

$$c_S = 1.005 + 1.88H = 1.005 + 1.88(0.05) = 1.099 \text{ kJ/kg dry air} \cdot \text{K}$$

$$= 1.099 \times 10^3 \text{ J/kg} \cdot \text{K}$$

To calculate the time of drying for the constant-rate period using Eq. (9.10-11) and $G = 2459/3600 = 0.6831$ kg/s · m²,

$$t = \frac{\rho_S \lambda_W x_1 (X_1 - X_C)}{G c_S (T_1 - T_W)(1 - e^{-hax_1/Gc_S})}$$

$$= \frac{641(2.389 \times 10^6)(0.0508)(0.99 - 0.49)}{(0.683)(1.099 \times 10^3)(121.1 - 47.2)[1 - e^{-(90.9 \times 283.5 \times 0.0508)/(0.683 \times 1.099 \times 10^3)}]}$$

$$= 850 \text{ s} = 0.236 \text{ h}$$

For the time of drying for the falling-rate period, using Eq. (9.10-12),

$$t = \frac{\rho_S \lambda_W x_1 X_C \ln (X_C/X)}{G c_S (T_1 - T_W)(1 - e^{-hax_1/Gc_S})}$$

$$= \frac{641(2.389 \times 10^6)(0.0508)(0.49) \ln (0.49/0.09)}{(0.6831)(1.099 \times 10^3)(121.1 - 47.2)[1 - e^{-(90.9 \times 283.5 \times 0.0508)/(0.683 \times 1.099 \times 10^3)}]}$$

$$= 1412 \text{ s} = 0.392 \text{ h}$$

$$\text{total time } t = 0.236 + 0.392 = 0.628 \text{ h}$$

9.10B Tray Drying with Varying Air Conditions

For drying in a compartment or tray dryer where the air passes in parallel flow over the surface of the tray, the air conditions do not remain constant. Heat and material balances similar to those for through circulation must be made to determine the exit-gas temperature and humidity.

In Fig. 9.10-2 air is shown passing over a tray. It enters having a temperature of T_1 and humidity H_1 and leaves at T_2 and H_2. The spacing between the trays is b m and dry air flow is G kg dry air/s·m² cross-sectional area. Writing a heat balance over a length dL_t of tray for a section 1 m wide,

$$dq = Gc_S(1 \times b) \, dT \qquad\qquad (9.10\text{-}18)$$

The heat-transfer equation is

$$dq = h(1 \times dL_t)(T - T_W) \qquad\qquad (9.10\text{-}19)$$

Rearranging and integrating,

$$\frac{hL_t}{Gc_S b} = \ln \frac{T_1 - T_W}{T_2 - T_W} \qquad\qquad (9.10\text{-}20)$$

Defining a log mean temperature difference similar to Eq. (9.10-10) and substituting into Eqs. (9.6-11) and (9.7-8), we obtain the following. For the constant-rate period,

$$t = \frac{x_1 \rho_S L_t \lambda_W (X_1 - X_C)}{Gc_S b(T_1 - T_W)(1 - e^{-hL_t/Gc_S b})} \qquad\qquad (9.10\text{-}21)$$

For the falling-rate period, an approximate equation is obtained.

$$t = \frac{x_1 \rho_S L_t \lambda_W X_C \ln (X_C/X)}{Gc_S b(T_1 - T_W)(1 - e^{-hL_t/Gc_S b})} \qquad\qquad (9.10\text{-}22)$$

9.10C Material and Heat Balances for Continuous Dryers

1. Simple heat and material balances. In Fig. 9.10-3 a flow diagram is given for a continuous-type dryer where the drying gas flows countercurrently to the solids flow. The solid enters at a rate of L_S kg dry solid/h, having a free moisture content X_1 and a

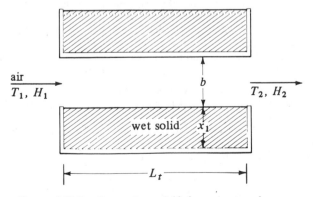

FIGURE 9.10-2. *Heat and material balances in a tray dryer.*

FIGURE 9.10-3. *Process flow for a counter-current continuous dryer.*

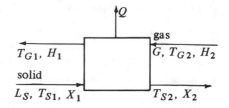

temperature T_{S1}. It leaves at X_2 and T_{S2}. The gas enters at a rate G kg dry air/h, having a humidity H_2 kg H_2O/kg dry air and a temperature of T_{G2}. The gas leaves at T_{G1} and H_1.

For a material balance on the moisture,

$$GH_2 + L_S X_1 = GH_1 + L_S X_2 \qquad (9.10\text{-}23)$$

For a heat balance a datum of $T_0\,°C$ is selected. A convenient temperature is $0°C$ ($32°F$). The enthalpy of the wet solid is composed of the enthalpy of the dry solid plus that of the liquid as free moisture. The heat of wetting is usually neglected. The enthalpy of the gas H_G' in kJ/kg dry air is

$$H_G' = c_S(T_G - T_0) + H\lambda_0 \qquad (9.10\text{-}24)$$

where λ_0 is the latent heat of water at $T_0\,°C$, 2501 kJ/kg (1075.4 btu/lb$_m$) at $0°C$, and c_S is the humid heat, given as kJ/kg dry air $\cdot$ K.

$$c_S = 1.005 + 1.88H \qquad (9.3\text{-}6)$$

The enthalpy of the wet solid H_S' in kJ/kg dry solid, where $(T_S - T_0)°C = (T_S - T_0)$ K, is

$$H_S' = c_{pS}(T_S - T_0) + Xc_{pA}(T_S - T_0) \qquad (9.10\text{-}25)$$

where c_{pS} is the heat capacity of the dry solid in kJ/kg dry solid $\cdot$ K and c_{pA} is the heat capacity of liquid moisture in kJ/kg $H_2O \cdot$ K. The heat of wetting or adsorption is neglected.

A heat balance on the dryer is

$$GH_{G2}' + L_S H_{S1}' = GH_{G1}' + L_S H_{S2}' + Q \qquad (9.10\text{-}26)$$

where Q is the heat loss in the dryer in kJ/h. For an adiabatic process $Q = 0$, and if heat is added, Q is negative.

EXAMPLE 9.10-2. Heat Balance on a Dryer

A continuous countercurrent dryer is being used to dry 453.6 kg dry solid/h containing 0.04 kg total moisture/kg dry solid to a value of 0.002 kg total moisture/kg dry solid. The granular solid enters at 26.7°C and is to be discharged at 62.8°C. The dry solid has a heat capacity of 1.465 kJ/kg $\cdot$ K, which is assumed constant. Heating air enters at 93.3°C, having a humidity of 0.010 kg H_2O/kg dry air, and is to leave at 37.8°C. Calculate the air flow rate and the outlet humidity, assuming no heat losses in the dryer.

Solution: The flow diagram is given in Fig. 9.10-3. For the solid, $L_S = 453.6$ kg/h dry solid, $c_{pS} = 1.465$ kJ/kg dry solid $\cdot$ K, $X_1 = 0.040$ kg H_2O/kg dry solid, $c_{pA} = 4.187$ kJ/kg $H_2O \cdot$ K, $T_{S1} = 26.7°C$, $T_{S2} = 62.8°C$, $X_2 = 0.002$. (Note that X values used are X_t values.) For the gas, $T_{G2} = 93.3°C$, $H_2 = 0.010$ kg H_2O/kg dry air, and $T_{G1} = 37.8°C$.

Making a material balance on the moisture using Eq. (9.10-23),

$$GH_2 + L_S X_1 = GH_1 + L_S X_2$$

$$G(0.010) + 453.6(0.040) = GH_1 + 453.6(0.002) \qquad (9.10\text{-}27)$$

For the heat balance, the enthalpy of the entering gas at 93.3°C using 0°C as a datum is, by Eq. (9.10-24), $\Delta T°C = \Delta T$ K, and $\lambda_0 = 2501$ kJ/kg, from the steam tables,

$$H'_{G2} = c_S(T_{G2} - T_0) + H_2\lambda_0$$

$$= [1.005 + 1.88(0.010)](93.3 - 0) + 0.010(2501)$$

$$= 120.5 \text{ kJ/kg dry air}$$

For the exit gas,

$$H'_{G1} = c_S(T_{G1} - T_0) + H_1\lambda_0$$

$$= (1.005 + 1.88H_1)(37.8 - 0) + H_1(2501) = 37.99 + 2572H_1$$

For the entering solid using Eq. (9.10-25),

$$H'_{S1} = c_{pS}(T_{S1} - T_0) + X_1c_{pA}(T_{S1} - T_0)$$

$$= 1.465(26.7 - 0) + 0.040(4.187)(26.7 - 0) = 43.59 \text{ kJ/kg dry solid}$$

$$H'_{S2} = c_{pS}(T_{S2} - T_0) + X_2c_{pA}(T_{S2} - T_0)$$

$$= 1.465(62.8 - 0) + 0.002(4.187)(62.8 - 0) = 92.53 \text{ kJ/kg}$$

Substituting into Eq. (9.10-26) for the heat balance with $Q = 0$ for no heat loss,

$$G(120.5) + 453.6(43.59) = G(37.99 + 2572H_1) + 453.6(92.53) + 0$$

$$\text{(9.10-28)}$$

Solving Eqs. (9.10-27) and (9.10-28) simultaneously,

$$G = 1166 \text{ kg dry air/h} \qquad H_1 = 0.0248 \text{ kg H}_2\text{O/kg dry air}$$

2. *Air recirculation in dryers.* In many dryers it is desired to control the wet bulb temperature at which the drying of the solid occurs. Also, since steam costs are often important in heating the drying air, recirculation of the drying air is sometimes used to reduce costs and control humidity. Part of the moist hot air leaving the dryer is recirculated (recycled) and combined with the fresh air. This is shown in Fig. 9.10-4. Fresh air having a temperature T_{G1} and humidity H_1 is mixed with recirculated air at T_{G2} and H_2 to give air at T_{G3} and H_3. This mixture is heated to T_{G4} with $H_4 = H_3$. After drying, the air leaves at a lower temperature T_{G2} and a higher humidity H_2.

The following material balances on the water can be made. For a water balance on

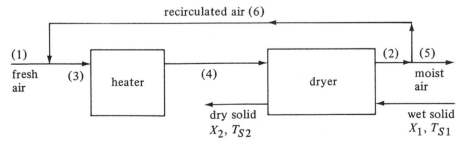

FIGURE 9.10-4. *Process flow for air recirculation in drying.*

the heater, noting that $H_6 = H_5 = H_2$,

$$G_1 H_1 + G_6 H_2 = (G_1 + G_6)H_4 \tag{9.10-29}$$

Making a water balance on the dryer,

$$(G_1 + G_6)H_4 + L_S X_1 = (G_1 + G_6)H_2 + L_S C_2 \tag{9.10-30}$$

In a similar manner heat balances can be made on the heater and dryer and on the overall system.

9.10D Continuous Countercurrent Drying

1. Introduction and temperature profiles. Drying continuously offers a number of advantages over batch drying. Smaller sizes of equipment can often be used and the product has a more uniform moisture content. In a continuous dryer the solid is moved through the dryer while in contact with a moving gas stream that may flow parallel or countercurrent to the solid. In countercurrent adiabatic operation, the entering hot gas contacts the leaving solid, which has been dried. In parallel adiabatic operation, the entering hot gas contacts the entering wet solid.

In Fig. 9.10-5 typical temperature profiles of the gas T_G and the solid T_S are shown for a continuous countercurrent dryer. In the preheat zone, the solid is heated up to the wet bulb or adiabatic saturation temperature. Little evaporation occurs here, and for low-temperature drying this zone is usually ignored. In the constant-rate zone, I, unbound and surface moisture are evaporated and the temperature of the solid remains essentially constant at the adiabatic saturation temperature if heat is transferred by convection. The rate of drying would be constant here but the gas temperature is changing and also the humidity. The moisture content falls to the critical value X_C at the end of this period.

In zone II, unsaturated surface and bound moisture are evaporated and the solid is dried to its final value X_2. The humidity of the entering gas entering zone II is H_2 and it rises to H_C. The material-balance equation (9.10-23) may be used to calculate H_C as follows.

$$L_S(X_C - X_2) = G(H_C - H_2) \tag{9.10-31}$$

where L_S is kg dry solid/h and G is kg dry gas/h.

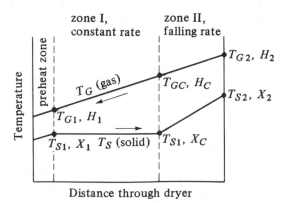

FIGURE 9.10-5. *Temperature profiles for a continuous countercurrent dryer.*

2. Equation for constant-rate period. The rate of drying in the constant-rate region in zone I would be constant if it were not for the varying gas conditions. The rate of drying in this section is given by an equation similar to Eq. (9.6-7).

$$R = k_y M_B (H_W - H) = \frac{h}{\lambda_W} (T_G - T_W) \tag{9.10-32}$$

The time for drying is given by Eq. (9.6-1) using limits between X_1 and X_C.

$$t = \left(\frac{L_S}{A}\right) \int_{X_C}^{X_1} \frac{dX}{R} \tag{9.10-33}$$

where A/L_S is the exposed drying surface m²/kg dry solid. Substituting Eq. (9.10-32) into (9.10-33) and $(G/L_S)\, dH$ for dX,

$$t = \frac{G}{L_S} \left(\frac{L_S}{A}\right) \frac{1}{k_y M_B} \int_{H_C}^{H_1} \frac{dH}{H_W - H} \tag{9.10-34}$$

where G = kg dry air/h, L_S = kg dry solid/h, and A/L_S = m²/kg dry solid. This can be integrated graphically.

For the case where T_W or H_W is constant for adiabatic drying, Eq. (9.10-34) can be integrated.

$$t = \frac{G}{L_S} \left(\frac{L_S}{A}\right) \frac{1}{k_y M_B} \ln \frac{H_W - H_C}{H_W - H_1} \tag{9.10-35}$$

The above can be modified by use of a log mean humidity difference.

$$\Delta H_{LM} = \frac{(H_W - H_C) - (H_W - H_1)}{\ln\left[(H_W - H_C)/(H_W - H_1)\right]} = \frac{H_1 - H_C}{\ln\left[(H_W - H_C)/(H_W - H_1)\right]} \tag{9.10-36}$$

Substituting Eq. (9.10-36) into (9.10-35), an alternative equation is obtained.

$$t = \frac{G}{L_S} \left(\frac{L_S}{A}\right) \frac{1}{k_y M_B} \frac{H_1 - H_C}{\Delta H_{LM}} \tag{9.10-37}$$

From Eq. (9.10-31), H_C can be calculated as follows.

$$H_C = H_2 + \frac{L_S}{G} (X_C - X_2) \tag{9.10-38}$$

3. Equation for falling-rate period. For the situation where unsaturated surface drying occurs, H_W is constant for adiabatic drying, the rate of drying is directly dependent upon X as in Eq. (9.7-9), and Eq. (9.10-32) applies.

$$R = R_C \frac{X}{X_C} = k_y M_R (H_W - H) \frac{X}{X_C} \tag{9.10-39}$$

Substituting Eq. (9.10-39) into (9.6-1),

$$t = \left(\frac{L_S}{A}\right) \frac{X_C}{k_y M_B} \int_{X_2}^{X_C} \frac{dX}{(H_W - H)X} \tag{9.10-40}$$

Substituting $G\, dH/L_S$ for dX and $(H - H_2)G/L_S + X_2$ for X,

$$t = \frac{G}{L_S} \left(\frac{L_S}{A}\right) \frac{X_C}{k_y M_B} \int_{H_2}^{H_C} \frac{dH}{(H_W - H)[(H - H_2)G/L_S + X_2]} \tag{9.10-41}$$

$$t = \frac{G}{L_S} \left(\frac{L_S}{A}\right) \frac{X_C}{k_y M_B} \frac{1}{(H_W - H_2)G/L_S + X_2} \ln \frac{X_C(H_W - H_2)}{X_2(H_W - H_C)} \tag{9.10-42}$$

Again, to calculate H_C, Eq. (9.10-38) can be used.

These equations for the two periods can also be derived using the last part of Eq. (9.10-32) and temperatures instead of humidities.

9.11 FREEZE DRYING OF BIOLOGICAL MATERIALS

9.11A Introduction

Certain foodstuffs, pharmaceuticals, and biological materials, which may not be heated even to moderate temperatures in ordinary drying, may be freeze-dried. The substance to be dried is usually frozen by exposure to very cold air. In freeze drying the water is removed as a vapor by sublimation from the frozen material in a vacuum chamber. After the moisture sublimes to a vapor, it is removed by mechanical vacuum pumps or steam jet ejectors.

As a rule, freeze drying produces the highest-quality food product obtainable by any drying method. A prominent factor is the structural rigidity afforded by the frozen substance when sublimation occurs. This prevents collapse of the remaining porous structure after drying. When water is added later, the rehydrated product retains much of its original structural form. Freeze drying of biological and food materials also has the advantage of little loss of flavor and aroma. The low temperatures involved minimize the degradative reactions which normally occur in ordinary drying processes. However, freeze drying is an expensive form of dehydration for foods because of the slow drying rate and the use of vacuum.

Since the vapor pressure of ice is very small, freeze drying requires very low pressures or high vacuum. If the water were in a pure state, freeze drying at or near 0°C (273 K) at a pressure of 4580 μm (4.58 mm Hg abs) could be performed. (See Appendix A.2 for the properties of ice.) However, since the water usually exists in a solution or a combined state, the material must be cooled below 0°C to keep the water in the solid phase. Most freeze drying is done at −10°C (263 K) or lower at pressures of about 2000 μm or less.

9.11B Derivation of Equations for Freeze Drying

In the freeze-drying process the original material is composed of a frozen core of material. As the ice sublimes, the plane of sublimation, which started at the outside surface, recedes and a porous shell of material already dried remains. The heat for the latent heat of sublimation of 2838 kJ/kg (1220 btu/lb$_m$) ice is usually conducted inward through the layer of dried material. In some cases it is also conducted through the frozen layer from the rear. The vaporized water vapor is transferred through the layer of dried material. Hence, heat and mass transfer are occurring simultaneously.

In Fig. 9.11-1 a material being freeze-dried is pictured. Heat by conduction, convection, and/or radiation from the gas phase reaches the dried surface and is then transferred by conduction to the ice layer. In some cases heat may also be conducted through the frozen material to reach the sublimation front or plane. The total drying time must be long enough so that the final moisture content is below about 5 wt % to prevent degradation of the final material on storage. The maximum temperatures reached in the dried food and reached in the frozen food must be low enough to keep degradation to a minimum.

The most widely used freeze-drying process is based upon the heat of sublimation

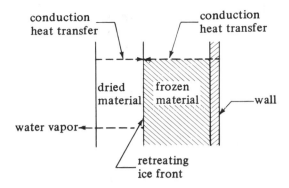

FIGURE 9.11-1. *Heat and mass transfer in freeze drying.*

being supplied from the surrounding gases to the sample surface. Then the heat is transferred by conduction through the dried material to the ice surface. A simplified model by Sandall et al. (S1) is shown in Fig. 9.11-2.

The heat flux to the surface of the material in Fig. 9.11-2 occurs by convection and in the dry solid by conduction to the sublimation surface. The heat flux to the surface is equal to that conducted through the dry solid, assuming pseudo steady state.

$$q = h(T_e - T_s) = \frac{k}{\Delta L}(T_s - T_f) \tag{9.11-1}$$

where q is heat flux in W (J/s), h is external heat-transfer coefficient in $W/m^2 \cdot K$, T_e is external temperature of the gas in °C, T_s is surface temperature of the dry solid in °C, T_f is temperature of the sublimation front or ice layer in °C, k is thermal conductivity of the dry solid in $W/m \cdot K$, and ΔL is the thickness of the dry layer in m. Note that $(T_s - T_f)°C = (T_s - T_f) K$.

In a similar manner, the mass flux of the water vapor from the sublimation front is

$$N_A = \frac{D'}{RT\,\Delta L}(p_{fw} - p_{sw}) = k_g(p_{sw} - p_{ew}) \tag{9.11-2}$$

where N_A is flux of water vapor in $kg\,mol/s \cdot m^2$, k_g is external mass-transfer coefficient in

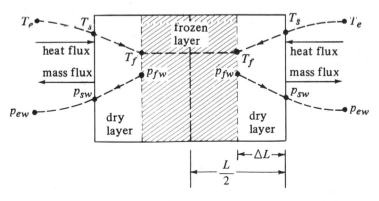

FIGURE 9.11-2. *Model for uniformly retreating ice front in freeze drying.*

kg mol/s · m^2 · atm, p_{sw} is partial pressure of water vapor at the surface in atm, p_{ew} is partial pressure of water vapor in the external bulk gas phase in atm, T is the average temperature in the dry layer, D' is an average effective diffusivity in the dry layer in m^2/s, and p_{fw} is the partial pressure of water vapor in equilibrium with the sublimation ice front in atm.

Equation (9.11-1) can be rearranged to give

$$q = \frac{1}{1/h + \Delta L/k} (T_e - T_f) \tag{9.11-3}$$

Also, Eq. (9.11-2) can be rearranged to give

$$N_A = \frac{1}{1/k_g + RT \, \Delta L/D'} (p_{fw} - p_{ew}) \tag{9.11-4}$$

The coefficients h and k_g are determined by the gas velocities and characteristics of the dryer and hence are constant. The values of T_e and p_{ew} are set by the external operating conditions. The values of k and D' are determined by the nature of the dried material.

The heat flux and mass flux at pseudo steady state are related by

$$q = \Delta H_s N_A \tag{9.11-5}$$

where ΔH_s is the latent heat of sublimation of ice in J/kg mol. Also, p_{fw} is uniquely determined by T_f, since it is the equilibrium vapor pressure of ice at that temperature; or

$$p_{fw} = f(T_f) \tag{9.11-6}$$

Substituting Eqs. (9.11-3) and (9.11-4) into (9.11-5),

$$\frac{1}{1/h + \Delta L/k} (T_e - T_f) = \Delta H_s \frac{1}{1/k_g + RT \, \Delta L/D'} (p_{fw} - p_{ew}) \tag{9.11-7}$$

Also, substituting Eqs. (9.11-1) and (9.11-4) into (9.11-5),

$$\frac{1}{\Delta L/k} (T_s - T_f) = \Delta H_s \frac{1}{1/k_g + RT \, \Delta L/D'} (p_{fw} - p_{ew}) \tag{9.11-8}$$

As T_e and, hence, T_s are raised to increase the rate of drying, two limits may possibly be reached. First, the outer surface temperature T_s cannot go too high because of thermal damage. Second, the temperature T_f must be kept well below the melting point. For the situation where $k/\Delta L$ is small compared to k_g and $D'/RT \, \Delta L$, the outer-surface temperature limit will be encountered first as T_s is raised. To further increase the drying rate, k must be raised. Hence, the process is considered to be heat-transfer-controlled. Most commercial freeze-drying processes are heat-transfer-controlled (K1).

In order to solve the given equations, ΔL is related to x, the fraction of the original free moisture remaining.

$$\Delta L = (1 - x) \frac{L}{2} \tag{9.11-9}$$

The rate of freeze drying can be related to N_A by

$$N_A = \frac{L}{2} \frac{1}{M_A V_S} \left(-\frac{dx}{dt} \right) \tag{9.11-10}$$

where M_A is molecular weight of water, V_S is the volume of solid material occupied by a unit kg of water initially ($V_S = 1/X_0 \rho_S$), X_0 is initial free moisture content in $kg\,H_2O/kg$ dry solid, and ρ_S is bulk density of dry solid in kg/m^3.

Combining Eqs. (9.11-3), (9.11-5), (9.11-9), and (9.11-10), we obtain, for heat transfer,

$$\frac{L}{2}\frac{\Delta H_s}{M_A V_S}\left(-\frac{dx}{dt}\right) = \frac{1}{1/h + (1-x)L/2k}(T_e - T_f) \qquad (9.11\text{-}11)$$

Similarly for mass transfer,

$$\frac{L}{2}\frac{1}{M_A V_S}\left(-\frac{dx}{dt}\right) = \frac{1}{1/k_g + RT(1-x)L/2D'}(p_{fw} - p_{ew}) \qquad (9.11\text{-}12)$$

Integrating Eq. (9.11-11) between the limits of $t = 0$ at $x_1 = 1.0$ and $t = t$ at $x_2 = x_2$, the equation for the time of drying to x_2 is as follows for h being very large (negligible external resistance):

$$t = \frac{L^2}{4kV_S}\frac{\Delta H_s}{M_A}\frac{1}{T_e - T_f}\left(x_1 - x_2 - \frac{x_1^2}{2} + \frac{x_2^2}{2}\right) \qquad (9.11\text{-}13)$$

where $\Delta H_s/M_A$ is heat of sublimation in $J/kg\,H_2O$. For $x_2 = 0$, the slab is completely dry.

Assuming that the physical properties and mass- and heat-transfer coefficients are known, Eq. (9.11-8) can be used to calculate the ice sublimation temperature T_f when the environment temperature T_e and the environment partial pressure p_{ew} are set. Since h is very large, $T_e \cong T_s$. Then Eq. (9.11-8) can be solved for T_f since T_f and p_{fw} are related by the equilibrium-vapor-pressure relation, Eq. (9.11-6). In Eq. (9.11-8) the value to use for T can be approximated by $(T_f + T_s)/2$.

The uniformly retreating ice-front model was tested by Sandall et al. (S1) against actual freeze-drying data. The model satisfactorily predicted the drying times for removal of 65–90% of the total initial water (S1, K1). The temperature T_f of the sublimation interface did remain essentially constant as assumed in the derivation. However, during removal of the last 10–35% of the water, the drying rate slowed markedly and the actual time was considerably greater than the predicted for this period.

The effective thermal conductivity k in the dried material has been found to vary significantly with the total pressure and with the type of gas present. Also, the type of material affects the value of k (S1, K1). The effective diffusivity D' of the dried material is a function of the structure of the material, Knudsen diffusivity, and molecular diffusivity (K1).

9.12 UNSTEADY-STATE THERMAL PROCESSING AND STERILIZATION OF BIOLOGICAL MATERIALS

9.12A Introduction

Materials of biological origin are usually not as stable as most inorganic and some organic materials. Hence, it is necessary to use certain processing methods to preserve biological materials, especially foods. Physical and chemical processing methods for preservation can be used such as drying, smoking, salting, chilling, freezing, and heating. Freezing and chilling of foods were discussed in Section 5.5 as methods of slowing the spoilage of biological materials. Also, in Section 9.11, freeze drying of biological materials was discussed.

An important method is heat or thermal processing, whereby contaminating microorganisms that occur primarily on the outer surface of foods and cause spoilage and health problems are destroyed. This leads to longer storage times of the food and other biological materials. A common method for preservation is to heat seal cans of food. Also, thermal processing is used to sterilize aqueous fermentation media to be used in fermentation processes so that organisms which do not survive are unable to compete with the organism that is to be cultured.

The sterilization of food materials by heating destroys bacteria, yeast, molds, and so on, which cause the spoilage and also destroys pathogenic (disease-producing) organisms, which may produce deadly toxins if not destroyed. The rate of destruction of microorganisms varies with the amount of heating and the type of organism. Some bacteria can exist in a vegetative growing form and in a dormant or spore form. The spore forms are much more resistant to heat. This mechanism of heat resistance is not clear.

For foods it is desired to kill essentially all the spores of *Clostridium botulinum*, which produces a toxin that is a deadly poison. Complete sterility with respect to this spore is the purpose of thermal processing. Since *Cl. botulinum* is so dangerous and often difficult to use, other spores, such as *Bacillus stearothermophilus*, which is a nonpathogenic organism of similar heat resistance, are often used for testing the heat-treating processes (A2, C1).

Temperature has a great effect on the growth rate of microorganisms, which have no temperature-regulating mechanism. Each organism has a certain optimal temperature range in which it grows best. If any microorganism is heated to a sufficiently high temperature for a sufficient time, it will be rendered sterile or killed.

The exact mechanism of thermal death of vegetative bacteria and spores is still somewhat uncertain. It is thought, however, to be due to the breakdown of the enzymes, which are essential to the functioning of the living cell (B1).

9.12B Thermal Death-Rate Kinetics of Microorganisms

The destruction of microorganisms by heating means loss of viability and not destruction in the physical sense. If it is assumed that inactivation of a single enzyme in a cell will inactivate the cell, then in a suspension of organisms of a single species at a constant temperature, the death rate can be expressed as a first-order kinetic equation (A2). The rate of destruction (number dying per unit time) is proportional to the number of organisms.

$$\frac{dN}{dt} = -kN \tag{9.12-1}$$

where N is the number of viable organisms at a given time, t is time in min, and k is a reaction velocity constant in $\min^{-1}$. The reaction velocity constant is a function of temperature and the type of microorganism.

After rearranging, Eq. (9.12-1) can be integrated as follows:

$$\int_{N_0}^{N} \frac{dN}{N} = -\int_{t=0}^{t} k \, dt \tag{9.12-2}$$

$$\ln \frac{N_0}{N} = kt \tag{9.12-3}$$

where N_0 is the original number at $t = 0$ and N is the number at time t. Often N_0 is called the *contamination level* (original number of contaminating microbes before sterilization) and N the sterility level. Also, Eq. (9.12-3) can be written as

$$N = N_0 e^{-kt} \tag{9.12-4}$$

Sometimes microbiologists use the term *decimal reduction time D*, which is the time in min during which the original number of viable microbes is reduced by 1/10. Substituting into Eq. (9.12-4),

$$\frac{N}{N_0} = \frac{1}{10} = e^{-kD} \tag{9.12-5}$$

Taking the $\log_{10}$ of both sides and solving for D,

$$D = \frac{2.303}{k} \tag{9.12-6}$$

Combining Eqs. (9.12-3) and (9.12-6),

$$t = D \log_{10} \frac{N_0}{N} \tag{9.12-7}$$

If the $\log_{10}(N/N_0)$ is plotted versus t, a straight line should result from Eq. (9.12-3). Experimental data bear this out for vegetative cells and approximately for spores. Data for the vegetative cell *E. coli* (A1) at constant temperature follow this logarithmic death-rate curve. Bacterial spore plots sometimes deviate somewhat from the logarithmic rate of death, particularly during a short period immediately following exposure to heat. However, for thermal-processing purposes for use with spores such as *Cl. botulinum*, a logarithmic-type curve is used.

To experimentally measure the microbial death rate, the spore or cell suspension in a solution is usually scaled in a capillary or test tube. A number of these tubes are then suddenly dipped into a hot bath for a given time. Then they are removed and immediately chilled. The number of viable organisms before and after exposure to the high temperature is then usually determined biologically with a plate count.

The effect of temperature on the reaction-rate constant k may be expressed by an Arrhenius-type equation.

$$k = ae^{-E/RT} \tag{9.12-8}$$

where a = an empirical constant, R is the gas constant in kJ/g mol $\cdot$ K (cal/g $\cdot$ mol $\cdot$ K), T is absolute temperature in K, and E is the activation energy in kJ/g mol (cal/g mol). The value of E is in the range 210 to about 418 kJ/g mol (50–100 kcal/g mol) for vegetative cells and spores (A2) and much less for enzymes and vitamins.

Substituting Eq. (9.12-8) into (9.12-2) and integrating,

$$\ln \frac{N_0}{N} = a \int_{t=0}^{t} e^{-E/RT} \, dt \tag{9.12-9}$$

At constant temperature T, Eq. (9.12-9) becomes (9.12-3). Since k is a function of temperature, the decimal reduction time D, which is related to k by Eq. (9.12-6), is also a function of temperature. Hence, D is often written as D_T to show that it is temperature-dependent.

9.12C Determination of Thermal Process Time for Sterilization

For canned foods, *Cl. botulinum* is the primary organism to be reduced in number (S2). It has been established that the minimum heating process should reduce the number of the spores by a factor of 10^{-12}. This means that since D is the time to reduce the original number by 10^{-1}, substituting $N/N_0 = 10^{-12}$ into Eq. (9.12-4) and solving for t,

$$t = 12\frac{2.303}{k} = 12D \tag{9.12-10}$$

This means that the time t is equal to $12D$ (often called the $12D$ concept). This time in Eq. (9.12-10) to reduce the number by 10^{-12} is called the *thermal death time*. Usually, the sterility level N is a number much less than one organism. These times do not represent complete sterilization but a mathematical concept which has been found empirically to give effective sterilization.

Experimental data of thermal death rates of *Cl. botulinum*, when plotted as the decimal reduction time D_T at a given T versus the temperature T in °F on a semilog plot, give essentially straight lines over the range of temperatures used in food sterilization (S2). A typical thermal destruction curve is shown in Fig. 9.12-1. Actually, by combining Eqs. (9.12-6) and (9.12-8), it can be shown that the plot of $\log_{10} D_T$ versus $1/T$ (T in degrees absolute) is a straight line, but over small ranges of temperature a straight line is obtained when $\log_{10} D_T$ is plotted versus T°F or °C.

In Fig. 9.12-1 the term z represents the temperature range in °F for a 10 : 1 change in D_T. Since the plot is a straight line, the equation can be represented as

$$\log_{10} D_{T_2} - \log_{10} D_{T_1} = \frac{1}{z}(T_1 - T_2) \tag{9.12-11}$$

Letting $T_1 = 250$°F (121.1°C), which is the standard temperature against which thermal processes are compared, and calling $T_2 = T$, Eq. (9.12-11) becomes

$$D_T = D_{250} \cdot 10^{(250 - T)/z} \tag{9.12-12}$$

For the organism *Cl. botulinum* the experimental value of $z = 18$°F. This means that each increase in temperature of 18°F (10°C) will increase the death rate by a factor of 10. This compares with the factor of 2 for many chemical reactions for an 18°F increase in temperature.

Using Eq. (9.12-7),

$$t = D_T \log_{10}\frac{N_0}{N} \tag{9.12-7}$$

Substituting $T = 250$°F (121.1°C) as the standard temperature into this equation and

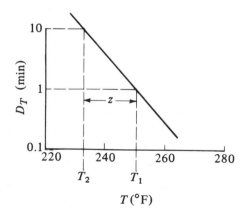

FIGURE 9.12-1. *Thermal destruction curve: plot of decimal reduction time versus temperature.*

substituting F_0 for t,

$$F_0 = D_{250} \log_{10} \frac{N_0}{N} \qquad\qquad \textbf{(9.12-13)}$$

where the F_0 value of a process is the time t in min at 250°F that will produce the same degree of sterilization as the given process at its temperature T. Combining Eqs. (9.12-7), (9.12-12), and (9.12-13), the F_0 of the given process at temperature T is

$$F_0 = t \cdot 10^{(T°C - 121.1)/(z°C)} \qquad (SI)$$

$$F_0 = t \cdot 10^{(T°F - 250)/(z°F)} \qquad \text{(English)} \qquad \textbf{(9.12-14)}$$

This is the F_0 value in min for the given thermal process at a given constant temperature $T°F$ and a given time t in min. Values for F_0 and z for adequate sterilization with *Cl. botulinum* vary somewhat with the type of food. Data are tabulated by Stumbo (S2) and Charm (C2) for various foods and microorganisms.

The effects of different but successive sterilization processes in a given material are additive. Hence, for several different temperature stages T_1, T_2, and so on, each having different times $t_1, t_2, \ldots$, the F_0 values for each stage are added to give the total F_0.

$$F_0 = t_1 \cdot 10^{(T_1 - 250)/z} + t_2 \cdot 10^{(T_2 - 250)/z} + \cdots \qquad \text{(English)} \qquad \textbf{(9.12-15)}$$

EXAMPLE 9.12-1. Sterilization of Cans of Food

Cans of a given food were heated in a retort for sterilization. The F_0 for *Cl. botulinum* in this type of food is 2.50 min and $z = 18°F$. The temperatures in the center of a can (the slowest-heating region) were measured and were approximately as follows, where the average temperature during each time period is listed: t_1 (0–20 min), $T_1 = 160°F$; t_2 (20–40 min), $T_2 = 210°F$; t_3 (40–73 min), $T_3 = 230°F$. Determine if this sterilization process is adequate. Use English and SI units.

Solution: For the three time periods the data are as follows:

$$t_1 = 20 - 0 = 20 \text{ min}, \quad T_1 = 160°F \text{ (71.1°C)}, \quad z = 18°F \text{ (10°C)}$$

$$t_2 = 40 - 20 = 20 \text{ min}, \quad T_2 = 210°F \text{ (98.9°C)}$$

$$t_3 = 73 - 40 = 33 \text{ min}, \quad T_3 = 230°F \text{ (110°C)}$$

Substituting into Eq. (9.12-15) and solving using English and SI units,

$$F_0 = t_1 \cdot 10^{(T_1 - 250)/z} + t_2 \cdot 10^{(T_2 - 250)/z} + t_3 \cdot 10^{(T_3 - 250)/z} \qquad \textbf{(9.12-15)}$$

$$= (20)10^{(160 - 250)/18} + (20)10^{(210 - 250)/18} + (33)10^{(230 - 250)/18}$$

$$= 0.0020 + 0.1199 + 2.555 = 2.68 \text{ min} \qquad \text{(English)}$$

$$= (20)10^{(71.1 - 121.1)/10} + (20)10^{(98.9 - 121.1)/10} + (33)10^{(110 - 121.1)/10}$$

$$= 2.68 \text{ min} \qquad (SI)$$

Hence, this thermal processing is adequate since only 2.50 min is needed for complete sterilization. Note that the time period at 160°F (71.1°C) contributes an insignificant amount to the final F_0. The major contribution is at 230°F (110°C), which is the highest temperature.

In the general case when cans of food are being sterilized in a retort, the temperature is not constant for a given time period but varies continuously with time. Hence, Eq. (9.12-15) can be modified and written for a continuously varying temperature T by taking small time increments of dt min for each value of T and summing. The final equation is

$$F_0 = \int_{t=0}^{t=t} 10^{(T°F - 250)/(z°F)} \; dt \quad \text{(English)}$$

$$F_0 = \int_{t=0}^{t=t} 10^{(T°C - 121.1)/(z°C)} \; dt \quad \text{(SI)}$$

(9.12-16)

This equation can be used as follows. Suppose that the temperature of a process is varying continuously and a graph or a table of values of T versus t is known or is calculated by the unsteady-state methods given in Chapter 5. The Eq. (9.12-16) can be graphically integrated by plotting values of $10^{(T-250)/z}$ versus t and taking the area under the curve.

In many cases the temperature of a process that is varying continuously with time is determined experimentally by measuring the temperature in the slowest-heating region. In cans this is the center of the can. Methods given in Chapter 5 for unsteady-state heating of short, fat cylinders by conduction can be used to predict the center temperature of the can as a function of time. However, these predictions can be somewhat in error, since physical and thermal properties of foods are difficult to measure accurately and often can vary. Also, trapped air in the container and unknown convection effects can affect the accuracy of predictions.

EXAMPLE 9.12-2. Thermal Process Evaluation by Graphical Integration
In the sterilization of a canned purée, the temperature in the slowest-heating region (center) of the can was measured giving the following time–temperature data for the heating and holding time. The cooling time data will be neglected as a safety factor.

t (min)	T (°F)	t (min)	T (°F)
0	80 (26.7°C)	40	225 (107.2°C)
15	165 (73.9)	50	230.5 (110.3)
25	201 (93.9)	64	235 (112.8)
30	212.5 (100.3)		

The F_0 value of *Cl. botulinum* is 2.45 min and z is 18°F. Calculate the F_0 value of the process above and determine if the sterilization is adequate.

Solution: In order to use Eq. (9.12-16), the values of $10^{(T-250)/z}$ must be calculated for each time. For $t = 0$ min, $T = 80°F$, and $z = 18°F$,

$$10^{(T-250)/z} = 10^{(80-250)/18} = 3.6 \times 10^{-10}$$

For $t = 15$ min, $T = 165°F$

$$10^{(165-250)/18} = 0.0000189$$

For $t = 25$ min, $T = 201°F$

$$10^{(201-250)/18} = 0.00189$$

For $t = 30$ min,

$$10^{(212.5-250)/18} = 0.00825$$

For $t = 40$ min,

$$10^{(225 - 250)/18} = 0.0408$$

For $t = 50$ min,

$$10^{(230.5 - 250)/18} = 0.0825$$

For $t = 64$ min,

$$10^{(235 - 250)/18} = 0.1465$$

These values are plotted versus t in Fig. 9.12-2. The areas of the various rectangles shown are

$$F_0 = A_1 + A_2 + A_3 + A_4$$
$$= 10(0.0026) + 10(0.0233) + 10(0.0620) + 14(0.1160)$$
$$= 0.026 + 0.233 + 0.620 + 1.621 = 2.50 \text{ min}$$

The process value of 2.50 min is greater than the required 2.45 min and the sterilization is adequate.

9.12D Sterilization Methods Using Other Design Criteria

In types of thermal processing which are not necessarily involved with sterilization of foods, other types of design criteria are used. In foods the minimum heat process should reduce the number of spores by a factor of 10^{-12}, i.e., $N/N_0 = 10^{-12}$. However, in other batch sterilization processes, such as in the sterilization of fermentation media, other criteria are often used. Often the equation for k, the reaction velocity constant for the specific organism to be used, is available.

$$k = ae^{-E/RT} \tag{9.12-8}$$

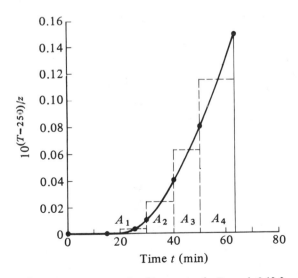

FIGURE 9.12-2. *Graphical integration for Example 9.12-2.*

Then Eq. (9.12-9) is written as

$$\nabla = \ln \frac{N_0}{N} = a \int_{t=0}^{t} e^{-E/RT} \, dt = \int_{t=0}^{t} k \, dt \qquad (9.12\text{-}17)$$

where ∇ is the design criterion. Usually, the contamination level N_0 is available and either the sterility level N is the unknown or the time of sterilization at a given temperature is the unknown. In either case a graphical integration is usually done, where $ae^{-E/RT}$ is plotted versus t and the area under the curve is obtained.

In sterilization of food in a container, the time required to render the material safe is calculated at the slowest-heating region of the container (usually the center). Other regions of the container are usually heated to higher temperatures and are overtreated. Hence, another method used is based on the probability of survival in the whole container. These details are given by others (C2, S2). In still another processing method, a short-time, continuous-flow process is used instead of a batch process in a container (B2).

9.12E Pasteurization

The term *pasteurization* is used today to apply to mild heat treatment of foods that is less drastic than sterilization. It is used to kill organisms that are relatively low in thermal resistance compared to those for which the more drastic sterilization processes are designed to eliminate. Pasteurization usually involves killing vegetative microorganisms and not heat-resistant spores.

The most common process is the pasteurization of milk to kill *Mycobacterium tuberculosis*, which is a non-spore-forming bacterium. This pasteurization does not sterilize the milk but kills the *M. tuberculosis* and reduces the other bacterial count sufficiently so that the milk can be stored if refrigerated.

For the pasteurization of such foods as milk, fruit juices, and beer, the same mathematical and graphical procedures covered for sterilization processes in this section are used to accomplish the degree of sterilization desired in pasteurization (B1, S2). The times involved are much shorter and the temperatures used in pasteurization are much lower. Generally, the F_0 value is given at 150°F (65.6°C) or a similar temperature rather than at 250°F as in sterilization. Also, the concept of the z value is employed, in which a rise in temperature of z°F will increase the death rate by a factor of 10. An F_0 value written as F_{150}^9 means the F value at 150°F with a z value of 9°F (S2).

In pasteurizing milk, batch and continuous processes are used. The U.S. health regulations specify two equivalent sets of conditions, where in one the milk is held at 145°F (62.8°C) for 30 min and in the other at 161°F (71.7°C) for 15 s.

The general equations used for pasteurization are similar to sterilization and can be written as follows. Rewriting Eq. (9.12-13),

$$F_{T_1}^z = D_{T_1} \log_{10} \frac{N_0}{N} \qquad (9.12\text{-}18)$$

Rewriting Eq. (9.12-14),

$$F_{T_1}^z = t \cdot 10^{(T - T_1)/z} \qquad (9.12\text{-}19)$$

where T_1 is the standard temperature being used such as 150°F, z is the value of z in °F for a tenfold increase in death rate, and T is the temperature of the actual process.

EXAMPLE 9.12-3. Pasteurization of Milk
A typical F value given for the thermal processing of milk in a tubular heat exchanger is $F_{150}^9 = 9.0$ min and $D_{150} = 0.6$ min. Calculate the reduction in the number of viable cells for these conditions.

Solution: The z value is 9°F (5°C) and the temperature of the process is 150°F (65.6°C). Substituting into Eq. (9.12-18) and solving,

$$F^9_{150} = 9.0 = 0.6 \log_{10} \frac{N_0}{N}$$

$$\frac{N_0}{N} = \frac{10^{15}}{1}$$

This gives a reduction in the number of viable cells of 10^{15}.

9.12F Effects of Thermal Processing on Food Constituents

Thermal processing is used to cause the death of various undesirable microorganisms, but it also causes undesirable effects, such as the reduction of certain nutritional values. Ascorbic acid (vitamin C) and thiamin and riboflavin (vitamins B_1 and B_2) are partially destroyed by thermal processing. The reduction of these desirable constituents can also be given kinetic parameters such as F_0 and z values in the same way as for sterilization and pasteurization. Examples and data are given by Charm (C2).

These same kinetic methods of thermal death rates can also be applied to predict the time for detecting a flavor change in a food product. Dietrich et al. (D1) determined a curve for the number of days to detect a flavor change of frozen spinach versus temperature of storage. The data followed Eq. (9.12-8) and a first-order kinetic relation.

PROBLEMS

9.3-1. *Humidity from Vapor Pressure.* The air in a room is at 37.8°C and a total pressure of 101.3 kPa abs containing water vapor with a partial pressure $p_A = 3.59$ kPa. Calculate:
(a) Humidity.
(b) Saturation humidity and percentage humidity.
(c) Percentage relative humidity.

9.3-2. *Percentage and Relative Humidity.* The air in a room has a humidity H of 0.021 kg H_2O/kg dry air at 32.2°C and 101.3 kPa abs pressure. Calculate:
(a) Percentage humidity H_P.
(b) Percentage relative humidity H_R.

        Ans. (a) $H_P = 67.5\%$; (b) $H_R = 68.6\%$

9.3-3. *Use of the Humidity Chart.* The air entering a dryer has a temperature of 65.6°C (150°F) and dew point of 15.6°C (60°F). Using the humidity chart, determine the actual humidity and percentage humidity. Calculate the humid volume of this mixture and also calculate c_S using SI and English units.

        Ans. $H = 0.0113$ kg H_2O/kg dry air, $H_P = 5.3\%$,
        $c_S = 1.026$ kJ/kg·K (0.245 btu/lb$_m$·°F),
        $v_H = 0.976$ m³ air + water vapor/kg dry air

9.3-4. *Properties of Air to a Dryer.* An air–water vapor mixture going to a drying process has a dry bulb temperature of 57.2°C and a humidity of 0.030 kg H_2O/kg dry air. Using the humidity chart and appropriate equations, determine the percentage humidity, saturation humidity at 57.2°C, dew point, humid heat, and humid volume.

9.3-5. *Adiabatic Saturation Temperature.* Air at 82.2°C and having a humidity $H = 0.0655$ kg H_2O/kg dry air is contacted in an adiabatic saturator with water. It leaves at 80% saturation.
(a) What are the final values of H and $T°C$?
(b) For 100% saturation, what would be the values of H and T?

        Ans. (a) $H = 0.079$ kg H_2O/kg dry air, $T = 52.8°C$

9.3-6. *Adiabatic Saturation of Air.* Air enters an adiabatic saturator having a temperature of 76.7°C and a dew-point temperature of 40.6°C. It leaves the saturator 90% saturated. What are the final values of H and $T°C$?

9.3-7. *Humidity from Wet and Dry Bulb Temperatures.* An air–water vapor mixture has a dry bulb temperature of 65.6°C and a wet bulb temperature of 32.2°C. What is the humidity of the mixture?

$$\text{Ans.} \quad H = 0.0175 \text{ kg H}_2\text{O/kg dry air}$$

9.3-8. *Humidity and Wet Bulb Temperature.* The humidity of an air–water vapor mixture is $H = 0.030$ kg H_2O/kg dry air. The dry bulb temperature of the mixture is 60°C. What is the wet bulb temperature?

9.3-9. *Dehumidification of Air.* Air having a dry bulb temperature of 37.8°C and a wet bulb of 26.7°C is to be dried by first cooling to 15.6°C to condense water vapor and then heating to 23.9°C.
(a) Calculate the initial humidity and percentage humidity.
(b) Calculate the final humidity and percentage humidity. [*Hint*: Locate the initial point on the humidity chart. Then go horizontally (cooling) to the 100% saturation line. Follow this line to 15.6°C. Then go horizontally to the right to 23.9°C.]

$$\text{Ans.} \quad \text{(b) } H = 0.0115 \text{ kg H}_2\text{O/kg dry air, } H_P = 60\%$$

9.3-10. *Cooling and Dehumidifying Air.* Air entering an adiabatic cooling chamber has a temperature of 32.2°C and a percentage humidity of 65%. It is cooled by a cold water spray and saturated with water vapor in the chamber. After leaving, it is heated to 23.9°C. The final air has a percentage humidity of 40%.
(a) What is the initial humidity of the air?
(b) What is the final humidity after heating?

9.6-1. *Time for Drying in Constant-Rate Period.* A batch of wet solid was dried on a tray dryer using constant drying conditions and a thickness of material on the tray of 25.4 mm. Only the top surface was exposed. The drying rate during the constant-rate period was $R = 2.05$ kg $H_2O/h \cdot m^2$ (0.42 lb_m $H_2O/h \cdot ft^2$). The ratio L_S/A used was 24.4 kg dry solid/m^2 exposed surface (5.0 lb_m dry solid/ft^2). The initial free moisture was $X_1 = 0.55$ and the critical moisture content $X_C = 0.22$ kg free moisture/kg dry solid.
   Calculate the time to dry a batch of this material from $X_1 = 0.45$ to $X_2 = 0.30$ using the same drying conditions but a thickness of 50.8 mm, with drying from the top and bottom surfaces. (*Hint*: First calculate L_S/A for this new case.)

$$\text{Ans.} \quad t = 1.785 \text{ h}$$

9.6-2. *Prediction of Effect of Process Variables on Drying Rate.* Using the conditions in Example 9.6-3 for the constant-rate drying period, do as follows.
(a) Predict the effect on R_C if the air velocity is only 3.05 m/s.
(b) Predict the effect if the gas temperature is raised to 76.7°C and H remains the same.
(c) Predict the effect on the time t for drying between moisture contents X_1 to X_2 if the thickness of material dried is 38.1 mm instead of 25.4 mm and the drying is still in the constant-rate period.

$$\text{Ans.} \quad \text{(a) } R_C = 1.947 \text{ kg H}_2\text{O/h} \cdot m^2 \text{ (0.399 lb}_m \text{ H}_2\text{O/h} \cdot ft^2)$$
$$\text{(b) } R_C = 4.21 \text{ kg H}_2\text{O/h} \cdot m^2$$

9.6-3. *Prediction in Constant-Rate Drying Region.* A granular insoluble solid material wet with water is being dried in the constant-rate period in a pan 0.61 m × 0.61 m and the depth of material is 25.4 mm. The sides and bottom are insulated. Air flows parallel to the top drying surface at a velocity of 3.05 m/s and has a dry bulb temperature of 60°C and wet bulb temperature of 29.4°C. The pan contains 11.34 kg of dry solid having a free moisture content of 0.35 kg H_2O/kg dry solid and the material is to be dried in the constant-rate period to 0.22 kg H_2O/kg dry solid.

(a) Predict the drying rate and the time in hours needed.

(b) Predict the time needed if the depth of material is increased to 44.5 mm.

9.6-4. Drying a Filter Cake in the Constant-Rate Region. A wet filter cake in a pan 1 ft × 1 ft square and 1 in. thick is dried on the top surface with air at a wet bulb temperature of 80°F and a dry bulb of 120°F flowing parallel to the surface at a velocity of 2.5 ft/s. The dry density of the cake is 120 lb_m/ft^3 and the critical free moisture content is 0.09 lb H_2O/lb dry solid. How long will it take to dry the material from a free moisture content of 0.20 lb H_2O/lb dry material to the critical moisture content?

> **Ans.** $t = 13.3$ h

9.7-1. Graphical Integration for Drying in Falling-Rate Region. A wet solid is to be dried in a tray dryer under steady-state conditions from a free moisture content of $X_1 = 0.40$ kg H_2O/kg dry solid to $X_2 = 0.02$ kg H_2O/kg dry solid. The dry solid weight is 99.8 kg dry solid and the top surface area for drying is 4.645 m^2. The drying-rate curve can be represented by Fig. 9.5-1b.

(a) Calculate the time for drying using graphical integration in the falling-rate period.

(b) Repeat but use a straight line through the origin for the drying rate in the falling-rate period.

> **Ans.** (a) t(constant rate) = 2.91 h, t(falling rate) = 6.36 h, t(total) = 9.27 h

9.7-2. Drying Tests with a Foodstuff. In order to test the feasibility of drying a certain foodstuff, drying data were obtained in a tray dryer with air flow over the top exposed surface having an area of 0.186 m^2. The bone-dry sample weight was 3.765 kg dry solid. At equilibrium after a long period, the wet sample weight was 3.955 kg H_2O + solid. Hence, 3.955 − 3.765, or 0.190, kg of equilibrium moisture was present. The following sample weights versus time were obtained in the drying test.

Time (h)	Weight (kg)	Time (h)	Weight (kg)	Time (h)	Weight (kg)
0	4.944	2.2	4.554	7.0	4.019
0.4	4.885	3.0	4.404	9.0	3.978
0.8	4.808	4.2	4.241	12.0	3.955
1.4	4.699	5.0	4.150		

(a) Calculate the free moisture content X kg H_2O/kg dry solid for each data point and plot X versus time. (*Hint:* For 0 h, 4.944 − 0.190 − 3.765 = 0.989 kg free moisture in 3.765 kg dry solid. Hence, $X = 0.989/3.765$.)

(b) Measure the slopes, calculate the drying rates R in kg H_2O/h · m^2, and plot R versus X.

(c) Using this drying-rate curve, predict the total time to dry the sample from $X = 0.20$ to $X = 0.04$. Use graphical integration for the falling-rate period. What is the drying rate R_C in the constant-rate period and X_C?

> **Ans.** (c) $R_C = 0.996$ kg H_2O/h · m^2
> $X_C = 0.12$, $t = 4.1$ h (total)

9.7-3. Prediction of Drying Time. A material was dried in a tray-type batch dryer using constant drying conditions. When the initial free moisture content was 0.28 kg free moisture/kg dry solid, 6.0 h was required to dry the material to a free moisture content of 0.08 kg free moisture/kg dry solid. The critical free moisture content is 0.14. Assuming a drying rate in the falling-rate region where the rate is a straight line from the critical point to the origin, predict the time to dry a sample from a free moisture content of 0.33 to 0.04 kg free moisture/kg dry solid. (*Hint*: First use the analytical equations for the constant-rate and the linear falling-rate periods with the known total time of 6.0 h. Then use the same equations for the new conditions.)

9.8-1. Drying of Biological Material in Tray Dryer. A granular biological material wet with water is being dried in a pan 0.305 × 0.305 m and 38.1 mm deep. The material is 38.1 mm deep in the pan, which is insulated on the sides and the bottom. Heat transfer is by convection from an air stream flowing parallel to the top surface at a velocity of 3.05 m/s, having a temperature of 65.6°C and humidity $H = 0.010$ kg H_2O/kg dry air. The top surface receives radiation from steam-heated pipes whose surface temperature $T_R = 93.3°C$. The emissivity of the solid is $\varepsilon = 0.95$. It is desired to keep the surface temperature of the solid below 32.2°C so that decomposition will be kept low. Calculate the surface temperature and the rate of drying for the constant-rate period.

Ans. $T_S = 31.3°C$, $R_C = 2.583$ kg H_2O/h · m^2

9.8-2. Drying When Radiation, Conduction, and Convection Are Present. A material is granular and wet with water and is being dried in a layer 25.4 mm deep in a batch-tray dryer pan. The pan has a metal bottom having a thermal conductivity of $k_M = 43.3$ W/m · K and a thickness of 1.59 mm. The thermal conductivity of the solid is $k_S = 1.125$ W/m · K. The air flows parallel to the top exposed surface and the bottom metal at a velocity of 3.05 m/s and a temperature of 60°C and humidity $H = 0.010$ kg H_2O/kg dry solid. Direct radiation heat from steam pipes having a surface temperature of 104.4°C falls on the exposed top surface, whose emissivity is 0.94. Estimate the surface temperature and the drying rate for the constant-rate period.

9.9-1. Diffusion Drying in Wood. Repeat Example 9.9-1 using the physical properties given but with the following changes.
(a) Calculate the time needed to dry the wood from a total moisture of 0.22 to 0.13. Use Fig. 5.3-13.
(b) Calculate the time needed to dry planks of wood 12.7 mm thick from $X_{t1} = 0.29$ to $X_t = 0.09$. Compare with the time needed for 25.4 mm thickness.

Ans. (b) $t = 7.60$ h (12.7 mm thick)

9.9-2. Diffusivity in Drying Tapioca Root. Using the data given in Example 9.9-2, determine the average diffusivity of the moisture up to a value of $X/X_C = 0.50$.

9.9-3. Diffusion Coefficient. Experimental drying data of a typical nonporous biological material obtained under constant drying conditions in the falling-rate region are tabulated below.

X/X_C	$t(h)$	X/X_C	$t(h)$
1.00	0	0.17	11.4
0.65	2.50	0.10	14.0
0.32	7.00	0.06	16.0

Drying from one side occurs with the material having a thickness of 10.1 mm. The data appear to follow the diffusion equation. Determine the average diffusivity over the range $X/X_C = 1.0 - 0.10$.

9.10-1. Drying a Bed of Solids by Through Circulation. Repeat Example 9.10-1 for drying of a packed bed of wet cylinders by through circulation of the drying air. Use the same conditions except that the air velocity is 0.381 m/s.

9.10-2. *Derivation of Equation for Through Circulation Drying.* Different forms of Eqs. (9.10-11) and (9.10-12) can be derived using humidity and mass-transfer equations rather than temperature and heat-transfer equations. This can be done by writing a mass-balance equation similar to Eq. (9.10-2) for a heat balance and a mass-transfer equation similar to Eq. (9.10-3).

    (a) Derive the final equation for the time of drying in the constant-rate period using humidity and mass-transfer equations.

    (b) Repeat for the falling-rate period.

    Ans. (a) $t = \dfrac{\rho_S x_1 (X_1 - X_C)}{G(H_W - H_1)(1 - e^{-k_y M_B a x_1/G})}$

9.10-3. *Through Circulation Drying in the Constant-Rate Period.* Spherical wet catalyst pellets having a diameter of 12.7 mm are being dried in a through circulation dryer. The pellets are in a bed 63.5 mm thick on a screen. The solids are being dried by air entering with a superficial velocity of 0.914 m/s at 82.2°C and having a humidity $H = 0.01$ kg H_2O/kg dry air. The dry solid density is determined as 1522 kg/m^3, and the void fraction in the bed is 0.35. The initial free moisture content is 0.90 kg H_2O/kg solid and the solids are to be dried to a free moisture content of 0.45, which is above the critical free moisture content. Calculate the time for drying in this constant-rate period.

9.10-4. *Material and Heat Balances on a Continuous Dryer.* Repeat Example 9.10-2, making heat and material balances, but with the following changes. The solid enters at 15.6°C and leaves at 60°C. The gas enters at 87.8°C and leaves at 32.2°C. Heat losses from the dryer are estimated as 2931 W.

9.10-5. *Drying in a Continuous Tunnel Dryer.* A rate of feed of 700 lb$_m$ dry solid/h containing a free moisture content of $X_1 = 0.4133$ lb H_2O/lb dry solid is to be dried to $X_2 = 0.0374$ lb H_2O/lb dry solid in a continuous counterflow tunnel dryer. A flow of 13 280 lb$_m$ dry air/h enters at 203°F with an $H_2 = 0.0562$ lb H_2O/lb dry air. The stock enters at the wet bulb temperature of 119°F and remains essentially constant in temperature in the dryer. The saturation humidity at 119°F from the humidity chart is $H_W = 0.0786$ lb H_2O/lb dry air. The surface area available for drying is $(A/L_S) = 0.30$ ft^2/lb$_m$ dry solid.

    A small batch experiment was performed using constant drying conditions, air velocity, and temperature of the solid approximately the same as in the continuous dryer. The equilibrium critical moisture content was found to be $X_C = 0.0959$ lb H_2O/lb dry solid, and the experimental value of $k_y M_B$ was found as 30.15 lb$_m$ air/h·ft^2. In the falling-rate period, the drying rate was directly proportional to X.

    For the continuous dryer, calculate the time in the dryer in the constant-rate zone and in the falling-rate zone.

    Ans. $H_C = 0.0593$ lb H_2O/lb dry air, $H_1 = 0.0760$ lb H_2O/lb dry air, $t = 4.24$ h in the constant-rate zone, $t = 0.47$ h in the falling-rate zone

9.10-6. *Air Recirculation in a Continuous Dryer.* The wet feed material to a continuous dryer contains 50 wt % water on a wet basis and is dried to 27 wt % by countercurrent air flow. The dried product leaves at the rate of 907.2 kg/h. Fresh air to the system is at 25.6°C and has a humidity of $H = 0.007$ kg H_2O/kg dry air. The moist air leaves the dryer at 37.8°C and $H = 0.020$ and part of it is recirculated and mixed with the fresh air before entering a heater. The heated mixed air enters the dryer at 65.6°C and $H = 0.010$. The solid enters at 26.7°C and leaves at 26.7°C. Calculate the fresh-air flow, the percent air leaving the dryer that is recycled, the heat added in the heater, and the heat loss from the dryer.

    Ans. 32 094 kg fresh dry air/h, 23.08% recycled, 440.6 kW in heater

9.12-1. *Sterilizing Canned Foods.* In a sterilizing retort, cans of a given food were heated and the average temperature in the center of a can is approximately 98.9°C for the first 30 min. The average temperature for the next period is 110°C. If the F_0 for the spore organism is 2.50 min and $z = 10°C$, calculate the time of heating at 110°C to make the process safe.

Ans. 29.9 min

9.12-2. *Temperature Effect on Decimal Reduction Time.* Prove by combining Eqs. (9.12-6) and (9.12-8) that a plot of $\log_{10} D_T$ versus $1/T$ (T in degrees absolute) is a straight line.

9.12-3. *Thermal Process Time for Pea Purée.* For cans of pea purée, the $F_0 = 2.45$ min and $z = 9.94°C$ (C2). Neglecting heat-up time, determine the process time for adequate sterilization at 112.8°C at the center of the can.

Ans. $t = 16.76$ min

9.12-4. *Process Time for Adequate Sterilization.* The F_0 value for a given canned food is 2.80 min and z is 18°F (10°C). The center temperatures of a can of this food when heated in a retort were as follows for the time periods given: t_1 (0–10 min), $T_1 = 140°F$; t_2 (10–30 min), $T_2 = 185°F$; t_3 (30–50 min), $T_3 = 220°F$; t_4 (50–80 min), $T_4 = 230°F$; t_5 (80–100 min), $T_5 = 190°F$. Determine if adequate sterilization is obtained.

9.12-5. *Process Time and Graphical Integration.* The following time–temperature data were obtained for the heating, holding, and cooling of a canned food product in a retort, the temperature being measured in the center of the can.

$t(min)$	$T(°F)$	$t(min)$	$T(°F)$
0	110 (43.3°C)	80	232 (111.1)
20	165 (73.9)	90	225 (107.2)
40	205 (96.1)	100	160 (71.1)
60	228 (108.9)		

The F_0 value used is 2.60 min and z is 18°F (10°C). Calculate the F_0 value for this process and determine if the thermal processing is adequate. Use SI and English units.

9.12-6. *Sterility Level of Fermentation Medium.* The aqueous medium in a fermentor is being sterilized and the time–temperature data obtained are as follows.

Time (min)	0	10	20	25	30	35
Temperature (°C)	100	110	120	120	110	100

The reaction velocity constant k in min^{-1} for the contaminating bacterial spores can be represented as (A1)

$$k = 7.94 \times 10^{38} e^{-(68.7 \times 10^3)/1.987T}$$

where $T = K$. The contamination level $N_0 = 1 \times 10^{12}$ spores. Calculate the sterility level N at the end and ∇.

9.12-7. *Time for Pasteurization of Milk.* Calculate the time in min at 62.8°C for pasteurization of milk. The F_0 value to be used at 65.6°C is 9.0 min. The z value is 5°C.

Ans. $t = 32.7$ min

582

9.12-8. *Reduction in Number of Viable Cells in Pasteurization.* In a given pasteurization process the reduction in the number of viable cells used is 10^{15} and the F_0 value used is 9.0 min. If the reduction is to be increased to 10^{16} because of increased contamination, what would be the new F_0 value?

REFERENCES

(A1)   ALLERTON, J., BROWNELL, L. E., and KATZ, D. L. *Chem. Eng. Progr.*, **45**, 619 (1949).

(A2)   AIBA, S., HUMPHREY, A. E., and MILLIS, N. F. *Biochemical Engineering*, 2nd ed. New York: Academic Press, Inc., 1973.

(B1)   BLAKEBROUGH, N. *Biochemical and Biological Engineering Science*, Vol. 2. New York: Academic Press, Inc., 1968.

(B2)   BATESON, R. N. *Chem. Eng. Progr. Symp. Ser.*, **67** (108), 44 (1971).

(C1)   CRUESS, W. V. *Commercial Fruit and Vegetable Products*, 4th ed. New York: McGraw-Hill Book Company, 1958.

(C2)   CHARM, S. E. *The Fundamentals of Food Engineering*, 2nd ed. Westport, Conn.: Avi Publishing Co., Inc., 1971.

(D1)   DIETRICH, W. C., BOGGS, M. M., NUTTING, M. D., and WEINSTEIN, N. E. *Food Technol.*, **14**, 522 (1960).

(E1)   EARLE, R. L. *Unit Operations in Food Processing*. Oxford: Pergamon Press, Inc., 1966.

(G1)   GAMSON, B. W., THODOS, G., and HOUGEN, O. A. *Trans. A.I.Ch.E.*, **39**, 1 (1943).

(H1)   HALL, C. W. *Processing Equipment for Agricultural Products*. Westport, Conn.: Avi Publishing Co., Inc., 1972.

(H2)   HENDERSON, S. M. *Agr. Eng.*, **33** (1), 29 (1952).

(K1)   KING, C. J. *Freeze Drying of Foods*. Boca Raton, Fla.: Chemical Rubber Co., Inc., 1971.

(M1)   MARSHALL, W. R., JR., and HOUGEN, O. A. *Trans. A.I.Ch.E.*, **38**, 91 (1942).

(P1)   PERRY, R. H., and GREEN, D. *Perry's Chemical Engineers' Handbook*, 6th ed. New York: McGraw-Hill Book Company, 1984.

(S1)   SANDALL, O. C., KING, C. J., and WILKE, C. R. *A.I.Ch.E. J.*, **13**, 428 (1967); *Chem. Eng. Progr.*, **64**(86), 43 (1968).

(S2)   STUMBO, C. R. *Thermobacteriology in Food Processing*, 2nd ed. New York: Academic Press, Inc., 1973.

(T1)   TREYBAL, R. E. *Mass Transfer Operations*, 3rd ed. New York: McGraw-Hill Book Company, 1980.

(W1)   WILKE, C. R., and HOUGEN, O. A. *Trans. A.I.Ch.E.*, **41**, 441 (1945).

CHAPTER 10

Stage and Continuous Gas–Liquid Separation Processes

10.1 TYPES OF SEPARATION PROCESSES AND METHODS

10.1A Introduction

Many chemical process materials and biological substances occur as mixtures of different components in the gas, liquid, or solid phase. In order to separate or remove one or more of the components from its original mixture, it must be contacted with another phase. The two phases are brought into more or less intimate contact with each other so that a solute or solutes can diffuse from one to the other. The two bulk phases are usually only somewhat miscible in each other. The two-phase pair can be gas–liquid, gas–solid, liquid–liquid, or liquid–solid. During the contact of the two phases the components of the original mixture redistribute themselves between the two phases. The phases are then separated by simple physical methods. By choosing the proper conditions and phases, one phase is enriched while the other is depleted in one or more components.

10.1B Types of Separation Processes

1. Absorption. When the two contacting phases are a gas and a liquid, the unit operation is called *absorption*. A solute *A* or several solutes are absorbed from the gas phase into a liquid phase in absorption. This process involves molecular and turbulent diffusion or mass transfer of solute *A* through a stagnant nondiffusing gas *B* into a stagnant liquid *C*. An example is absorption of ammonia *A* from air *B* by the liquid water *C*. Usually, the exit ammonia–water solution is distilled to recover relatively pure ammonia.

Another example is absorbing SO_2 from the flue gases by absorption in alkaline solutions. In the hydrogenation of edible oils in the food industry, hydrogen gas is bubbled into oil and absorbed. The hydrogen in solution then reacts with the oil in the presence of a catalyst. The reverse of absorption is called *stripping* or *desorption*, and the same theories and basic principles hold. An example is the steam stripping of nonvolatile

584

oils, in which the steam contacts the oil and small amounts of volatile components of the oil pass out with the steam.

When the gas is pure air and the liquid is pure water, the process is called *humidification*. *Dehumidification* involves removal of water vapor from air.

2. Distillation. In the *distillation* process, a volatile vapor phase and a liquid phase that vaporizes are involved. An example is distillation of an ethanol–water solution, where the vapor contains a concentration of ethanol greater than in the liquid. Another example is distillation of an ammonia–water solution to produce a vapor richer in ammonia. In the distillation of crude petroleum, various fractions, such as gasoline, kerosene, and heating oils, are distilled off.

3. Liquid–liquid extraction. When the two phases are liquids, where a solute or solutes are removed from one liquid phase to another liquid phase, the process is called *liquid–liquid extraction*. One example is extraction of acetic acid from a water solution by isopropyl ether. In the pharmaceutical industry, antibiotics in an aqueous fermentation solution are sometimes removed by extraction with an organic solvent.

4. Leaching. If a fluid is being used to extract a solute from a solid, the process is called *leaching*. Sometimes this process is also called *extraction*. Examples are leaching copper from solid ores by sulfuric acid and leaching vegetable oils from solid soybeans by organic solvents such as hexane. Vegetable oils are also leached from other biological products, such as peanuts, rape seeds, and sunflower seeds. Soluble sucrose is leached by water extraction from sugar cane and beets.

5. Membrane processing. Separation of molecules by the use of membranes is a relatively new unit operation and is becoming more important. The relatively thin, solid membrane controls the rate of movement of molecules between two phases. It is used to remove salt from water, purify gases, in food processing, and so on.

6. Crystallization. Solute components soluble in a solution can be removed from the solution by adjusting the conditions, such as temperature or concentration, so that the solubility of one or more solute components is exceeded and they *crystallize* out as a solid phase. Examples of this separation process are crystallization of sugar from solution and crystallization of metal salts in the processing of metal ore solutions.

7. Adsorption. In an adsorption process one or more components of a liquid or gas stream are adsorbed on the surface or in the pores of a solid adsorbent and a separation is obtained. Examples include removal of organic compounds from polluted water, separation of paraffins from aromatics, and removal of solvents from air.

10.1C Processing Methods

Several methods of processing are used in the separations discussed above. The two phases, such as gas and liquid, or liquid and liquid, can be mixed together in a vessel and then separated. This is a *single-stage process*. Often the phases are mixed in one stage, separated, and then contacted again in a *multiple-stage process*. These two methods can be carried out batchwise or continuously. In still another general method, the two phases can be contacted continuously in a packed tower.

In this chapter, humidification and absorption will be considered; in Chapter 11, distillation; in Chapter 12, adsorption, liquid–liquid extraction, leaching, and crystallization, and in Chapter 13, membrane processes. In all of these processes the equilibrium relations between the two phases being considered must be known. This is discussed for gas–liquid systems in Section 10.2 and for the other systems in Chapters 11, 12, and 13.

10.2 EQUILIBRIUM RELATIONS BETWEEN PHASES

10.2A Phase Rule and Equilibrium

In order to predict the concentration of a solute in each of two phases in equilibrium, experimental equilibrium data must be available. Also, if the two phases are not at equilibrium, the rate of mass transfer is proportional to the driving force, which is the departure from equilibrium. In all cases involving equilibria, two phases are involved, such as gas–liquid or liquid–liquid. The important variables affecting the equilibrium of a solute are temperature, pressure, and concentration.

The equilibrium between two phases in a given situation is restricted by the phase rule:

$$F = C - P + 2 \qquad (10.2\text{-}1)$$

where P is the number of phases at equilibrium, C the number of total components in the two phases when no chemical reactions are occurring, and F the number of variants or degrees of freedom of the system. For example, for the gas–liquid system of CO_2–air–water, there are two phases and three components (considering air as one inert component). Then, by Eq. (10.2-1),

$$F = C - P + 2 = 3 - 2 + 2 = 3$$

This means that there are 3 degrees of freedom. If the total pressure and the temperature are set, only one variable is left that can be arbitrarily set. If the mole fraction composition x_A of CO_2 (A) in the liquid phase is set, the mole fraction composition y_A or pressure p_A in the gas phase is automatically determined.

The phase rule does not tell us the partial pressure p_A in equilibrium with the selected x_A. The value of p_A must be determined experimentally. The two phases can, of course, be gas–liquid, liquid–solid, and so on. For example, the equilibrium distribution of acetic acid between a water phase and an isopropyl ether phase has been determined experimentally for various conditions.

10.2B Gas–Liquid Equilibrium

1. Gas–liquid equilibrium data. To illustrate the obtaining of experimental gas–liquid equilibrium data, the system SO_2–air–water will be considered. An amount of gaseous SO_2, air, and water are put in a closed container and shaken repeatedly at a given temperature until equilibrium is reached. Samples of the gas and liquid are analyzed to give the partial pressure p_A in atm of SO_2 (A) in the gas and mole fraction x_A in the liquid. Figure 10.2-1 shows a plot of data from Appendix A.3 of the partial pressure p_A of SO_2 in the vapor in equilibrium with the mole fraction x_A of SO_2 in the liquid at 293 K (20°C).

2. Henry's law. Often the equilibrium relation between p_A in the gas phase and x_A can be expressed by a straight-line Henry's law equation at low concentrations.

$$p_A = Hx_A \qquad (10.2\text{-}2)$$

where H is the Henry's law constant in atm/mole fraction for the given system. If both sides of Eq. (10.2-2) are divided by total pressure P in atm,

$$y_A = H'x_A \tag{10.2-3}$$

where H' is the Henry's law constant in mole frac gas/mole frac liquid and is equal to H/P. Note that H' depends on total pressure, whereas H does not.

In Fig. 10.2-1 the data follow Henry's law up to a concentration x_A of about 0.005, where $H = 29.6$ atm/mol frac. In general, up to a total pressure of about 5×10^5 Pa (5 atm) the value of H is independent of P. Data for some common gases with water are given in Appendix A.3.

EXAMPLE 10.2-1.   Dissolved Oxygen Concentration in Water
What will be the concentration of oxygen dissolved in water at 298 K when the solution is in equilibrium with air at 1 atm total pressure? The Henry's law constant is 4.38×10^4 atm/mol fraction.

Solution: The partial pressure p_A of oxygen (A) in air is 0.21 atm. Using Eq. (10.2-2),

$$0.21 = Hx_A = 4.38 \times 10^4 x_A$$

Solving, $x_A = 4.80 \times 10^{-6}$ mol fraction. This means that 4.80×10^{-6} mol O_2 is dissolved in 1.0 mol water plus oxygen or 0.000853 part O_2/100 parts water.

10.3   SINGLE AND MULTIPLE EQUILIBRIUM CONTACT STAGES

10.3A   Single-Stage Equilibrium Contact

In many operations of the chemical and other process industries, the transfer of mass from one phase to another occurs, usually accompanied by a separation of the components of the mixture, since one component will be transferred to a larger extent than will another component.

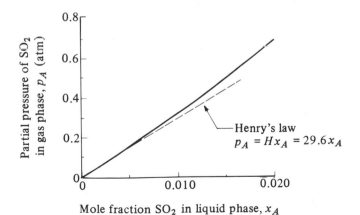

FIGURE 10.2-1.   *Equilibrium plot for SO_2–water system at 293 K ($20°C$).*

FIGURE 10.3-1. *Single-stage equilibrium process.*

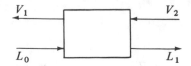

A single-stage process can be defined as one in which two different phases are brought into intimate contact with each other and then are separated. During the time of contact, intimate mixing occurs and the various components diffuse and redistribute themselves between the two phases. If mixing time is long enough, the components are essentially at equilibrium in the two phases after separation and the process is considered a single equilibrium stage.

A single equilibrium stage can be represented as in Fig. 10.3-1. The two entering phases, L_0 and V_2, of known amounts and compositions, enter the stage, mixing and equilibration occur, and the two exit streams L_1 and V_1 leave in equilibrium with each other. Making a total mass balance,

$$L_0 + V_2 = L_1 + V_1 = M \tag{10.3-1}$$

where L is kg (lb$_m$), V is kg, and M is total kg.

Assuming that three components, A, B, and C, are present in the streams and making a balance on A and C,

$$L_0 x_{A0} + V_2 y_{A2} = L_1 x_{A1} + V_1 y_{A1} = M x_{AM} \tag{10.3-2}$$

$$L_0 x_{C0} + V_2 y_{C2} = L_1 x_{C1} + V_1 y_{C1} = M x_{CM} \tag{10.3-3}$$

An equation for B is not needed since $x_A + x_B + x_C = 1.0$. The mass fraction of A in the L stream is x_A and y_A in the V stream. The mass fraction of A in the M stream is x_{AM}.

To solve the three equations, the equilibrium relations between the components must be known. In Section 10.3B, this will be done for a gas–liquid system and in Chapter 11 for a vapor–liquid system. Note that Eqs. (10.3-1)–(10.3-3) can also be written using mole units, with L and V having units of moles and x_A and y_A units of mole fraction.

10.3B Single-Stage Equilibrium Contact for Gas–Liquid System

In the usual gas–liquid system the solute A is in the gas phase V, along with inert air B and in the liquid phase L along with inert water C. Assuming that air is essentially insoluble in the water phase and that water does not vaporize to the gas phase, the gas phase is a binary A–B and the liquid phase is a binary A–C. Using moles and mole fraction units, Eq. (10.3-1) holds for a single-stage process for the total material balance. Since component A is the only component that redistributes between the two phases, a balance on A can be written as follows.

$$L'\left(\frac{x_{A0}}{1 - x_{A0}}\right) + V'\left(\frac{y_{A2}}{1 - y_{A2}}\right) = L'\left(\frac{x_{A1}}{1 - x_{A1}}\right) + V'\left(\frac{y_{A1}}{1 - y_{A1}}\right) \tag{10.3-4}$$

where L' is moles inert water C and V' is moles inert air B. Both L' and V' are constant and usually known.

To solve Eq. (10.3-4), the relation between y_{A1} and x_{A1} in equilibrium is given by Henry's law.

$$y_{A1} = H'x_{A1} \qquad (10.3\text{-}5)$$

If the solution is not dilute, equilibrium data in the form of a plot of p_A or y_A versus x_A must be available, such as in Fig. 10.2-1.

EXAMPLE 10.3-1.   Equilibrium Stage Contact for CO_2–Air–Water
A gas mixture at 1.0 atm pressure abs containing air and CO_2 is contacted in a single-stage mixer continuously with pure water at 293 K. The two exit gas and liquid streams reach equilibrium. The inlet gas flow rate is 100 kg mol/h, with a mole fraction of CO_2 of $y_{A2} = 0.20$. The liquid flow rate entering is 300 kg mol water/h. Calculate the amounts and compositions of the two outlet phases. Assume that water does not vaporize to the gas phase.

Solution: The flow diagram is the same as given in Fig. 10.3-1. The inert water flow is $L' = L_0 = 300$ kg mol/h. The inert air flow V' is obtained from Eq. (10.3-6).
$$V' = V(1 - y_A) \qquad (10.3\text{-}6)$$

Hence, the inert air flow is $V' = V_2(1 - y_{A2}) = 100(1 - 0.20) = 80$ kg mol/h. Substituting into Eq. (10.3-4) to make a balance on CO_2 (A),

$$300\left(\frac{0}{1-0}\right) + 80\left(\frac{0.20}{1-0.20}\right) = 300\left(\frac{x_{A1}}{1-x_{A1}}\right) + 80\left(\frac{y_{A1}}{1-y_{A1}}\right) \quad (10.3\text{-}7)$$

At 293 K, the Henry's law constant from Appendix A.3 is $H = 0.142 \times 10^4$ atm/mol frac. Then $H' = H/P = 0.142 \times 10^4/1.0 = 0.142 \times 10^4$ mol frac gas/mol frac liquid. Substituting into Eq. (10.3-5),

$$y_{A1} = 0.142 \times 10^4 x_{A1} \qquad (10.3\text{-}8)$$

Substituting Eq. (10.3-8) into (10.3-7) and solving, $x_{A1} = 1.41 \times 10^{-4}$ and $y_{A1} = 0.20$. To calculate the total flow rates leaving,

$$L_1 = \frac{L'}{1 - x_{A1}} = \frac{300}{1 - 1.41 \times 10^{-4}} = 300 \text{ kg mol/h}$$

$$V_1 = \frac{V'}{1 - y_{A1}} = \frac{80}{1 - 0.20} = 100 \text{ kg mol/h}$$

In this case, since the liquid solution is so dilute, $L_0 \cong L_1$.

10.3C   Countercurrent Multiple-Contact Stages

1. Derivation of general equation.   In Section 10.3A we used single-stage contact to transfer the solute A between the V and L phases. In order to transfer more solute from, say, the V_1 stream, the single-stage contact can be repeated by again contacting the V_1 stream leaving the first stage with fresh L_0. This can be repeated using multiple stages. However, this is wasteful of the L_0 stream and gives a dilute product in the outlet L_1 streams. To conserve use of the L_0 stream and to get a more concentrated product, countercurrent multiple-stage contacting is generally used. This is somewhat similar to countercurrent heat transfer in a heat exchanger, where the outlet heated stream approaches more closely the temperature of the inlet hot stream.

The process flow diagram for a countercurrent stage process is shown in Fig. 10.3-2. The inlet L stream is L_0 and the inlet V stream is V_{N+1} instead of V_2 as for a single-stage in Fig. 10.3-1. The outlet product streams are V_1 and L_N and the total number of stages is N. The component A is being exchanged between the V and L streams. The V stream is

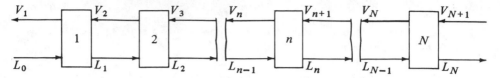

FIGURE 10.3-2. *Countercurrent multiple-stage process.*

composed mainly of component B and the L stream of component C. Components B and C may or may not be somewhat miscible in each other. The two-phase system can be gas–liquid, vapor–liquid, liquid–liquid, or other.

Making a total overall balance on all stages,

$$L_0 + V_{N+1} = L_N + V_1 = M \tag{10.3-9}$$

where V_{N+1} is mol/h entering, L_N is mol/h leaving the process, and M is the total flow. Note in Fig. 10.3-2 that any two streams leaving a stage are in equilibrium with each other. For example, in stage n, V_n and L_n are in equilibrium. For an overall component balance on A, B, or C,

$$L_0 x_0 + V_{N+1} y_{N+1} = L_N x_N + V_1 y_1 = M x_M \tag{10.3-10}$$

where x and y are mole fractions. Flows in kg/h (lb$_m$/h) and mass fraction can also be used in these equations.

Making a total balance over the first n stages,

$$L_0 + V_{n+1} = L_n + V_1 \tag{10.3-11}$$

Making a component balance over the first n stages,

$$L_0 x_0 + V_{n+1} y_{n+1} = L_n x_n + V_1 y_1 \tag{10.3-12}$$

Solving for y_{n+1} in Eq. (10.3-12).

$$y_{n+1} = \frac{L_n x_n}{V_{n+1}} + \frac{V_1 y_1 - L_0 x_0}{V_{n+1}} \tag{10.3-13}$$

This is an important material-balance equation, often called an *operating line*. It relates the concentration y_{n+1} in the V stream with x_n in the L stream passing it. The terms V_1, y_1, L_0, and x_0 are constant and usually known or can be determined from Eqs. (10.3-9)–(10.3-12).

2. Countercurrent contact with immiscible streams. An important case where the solute A is being transferred occurs when the solvent stream V contains components A and B with no C and the solvent stream L contains A and C with no B. The two streams L and V are immiscible in each other with only A being transferred. When Eq. (10.3-13) is plotted on an xy plot (x_A and y_A of component A) as in Fig. 10.3-3, it is often curved, since the slope L_n/V_{n+1} of the operating line varies if the L and V streams vary from stage to stage.

In Fig. 10.3-3 is plotted the equilibrium line that relates the compositions of two streams leaving a stage in equilibrium with each other. To determine the number of ideal stages required to bring about a given separation or reduction of the concentration of A from y_{N+1} to y_1, the calculation is often done graphically. Starting at stage 1, y_1 and x_0 are on the operating line, Eq. (10.3-13), plotted in the figure. The vapor y_1 leaving is in equilibrium with the leaving x_1 and both compositions are on the equilibrium line. Then y_2 and x_1 are on the operating line and y_2 is in equilibrium with x_2, and so on. Each stage is represented by a step drawn on Fig. 10.3-3. The steps are continued on the graph until y_{N+1} is reached. Alternatively, we can start at y_{N+1} and draw the steps going to y_1.

If the streams L and V are dilute in component A, the streams are approximately constant and the slope L_n/V_{n+1} of Eq. (10.3-13) is nearly constant. Hence, the operating line is essentially a straight line on an xy plot. In distillation, where only components A and B are present, Eq. (10.3-13) also holds for the operating line, and this will be covered in Chapter 11. Cases where A, B, and C are appreciably soluble in each other often occur in liquid–liquid extraction and will be discussed in Chapter 12.

EXAMPLE 10.3-2. Absorption of Acetone in a Countercurrent Stage Tower

It is desired to absorb 90% of the acetone in a gas containing 1.0 mol % acetone in air in a countercurrent stage tower. The total inlet gas flow to the tower is 30.0 kg mol/h, and the total inlet pure water flow to be used to absorb the acetone is 90 kg mol H_2O/h. The process is to operate isothermally at 300 K and a total pressure of 101.3 kPa. The equilibrium relation for the acetone (A) in the gas–liquid is $y_A = 2.53x_A$. Determine the number of theoretical stages required for this separation.

Solution: The process flow diagram is similar to Fig. 10.3-3. Given values are $y_{AN+1} = 0.01$, $x_{A0} = 0$, $V_{N+1} = 30.0$ kg mol/h, and $L_0 = 90.0$ kg mol/h. Making an acetone material balance,

$$\text{amount of entering acetone} = y_{AN+1}V_{N+1} = 0.01(30.0) = 0.30 \text{ kg mpl/h}$$

$$\text{entering air} = (1 - y_{AN+1})V_{N+1} = (1 - 0.01)(30.0)$$

$$= 29.7 \text{ kg mol air/h}$$

$$\text{acetone leaving in } V_1 = 0.10(0.30) = 0.030 \text{ kg mol/h}$$

$$\text{acetone leaving in } L_N = 0.90(0.30) = 0.27 \text{ kg mol/h}$$

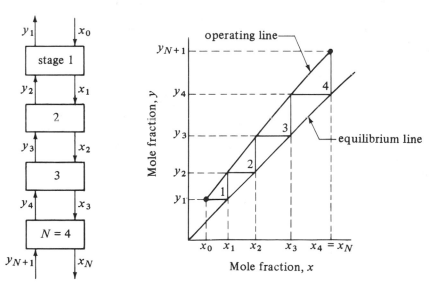

FIGURE 10.3-3. *Number of stages in a countercurrent multiple-stage contact process.*

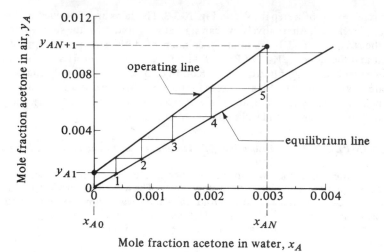

FIGURE 10.3-4. *Theoretical stages for countercurrent absorption in Example 10.3-2.*

$$V_1 = 29.7 + 0.03 = 29.73 \text{ kg mol air + acetone/h}$$

$$y_{A1} = \frac{0.030}{29.73} = 0.00101$$

$$L_N = 90.0 + 0.27 = 90.27 \text{ kg mol water + acetone/h}$$

$$x_{AN} = \frac{0.27}{90.27} = 0.00300$$

Since the flow of liquid varies only slightly from $L_0 = 90.0$ at the inlet to $L_N = 90.27$ at the outlet and V from 30.0 to 29.73, the slope L_n/V_{n+1} of the operating line in Eq. (10.3-13) is essentially constant. This line is plotted in Fig. 10.3-4 and the equilibrium relation $y_A = 2.53x_A$ is also plotted. Starting at point y_{A1}, x_{A0}, the stages are drawn as shown. About 5.2 theoretical stages are required.

10.3D Analytical Equations for Countercurrent Stage Contact

When the flow rates V and L in a countercurrent process are essentially constant, the operating-line equation (10.3-13) becomes straight. If the equilibrium line is also a straight line over the concentration range, simplified analytical expressions can be derived for the number of equilibrium stages in a countercurrent stage process.

Referring again to Fig. 10.3-2, Eq. (10.3-14) is an overall component balance on component A.

$$L_0 x_0 + V_{N+1} y_{N+1} = L_N x_N + V_1 y_1 \qquad (10.3\text{-}14)$$

Rearranging,

$$L_N x_N - V_{N+1} y_{N+1} = L_0 x_0 - V_1 y_1 \qquad (10.3\text{-}15)$$

Making a component balance for A on the first n stages,

$$L_0 x_0 + V_{n+1} y_{n+1} = L_n x_n + V_1 y_1 \qquad (10.3\text{-}16)$$

Rearranging,

$$L_0 x_0 - V_1 y_1 = L_n x_n - V_{n+1} y_{n+1} \qquad (10.3\text{-}17)$$

Equating Eq. (10.3-15) to (10.3-17),

$$L_n x_n - V_{n+1} y_{n+1} = L_N x_N - V_{N+1} y_{N+1} \tag{10.3-18}$$

Since the molar flows are constant, $L_n = L_N = \text{constant} = L$ and $V_{n+1} = V_{N+1} = \text{constant} = V$. Then Eq. (10.3-18) becomes

$$L(x_n - x_N) = V(y_{n+1} - y_{N+1}) \tag{10.3-19}$$

Since y_{n+1} and x_{n+1} are in equilibrium, and the equilibrium line is straight, $y_{n+1} = m x_{n+1}$. Also $y_{N+1} = m x_{N+1}$. Substituting $m x_{n+1}$ for y_{n+1} and calling $A = L/mV$, Eq. (10.3-19) becomes

$$x_{n+1} - A x_n = \frac{y_{N+1}}{m} - A x_N \tag{10.3-20}$$

where A is an absorption factor and is constant.

All factors on the right-hand side of Eq. (10.3-20) are constant. This equation is a linear first-order difference equation and can be solved by the calculus of finite-difference methods (G1, M1). The final derived equations are as follows.

For transfer of solute A from phase L to V (stripping),

$$\frac{x_0 - x_N}{x_0 - (y_{N+1}/m)} = \frac{(1/A)^{N+1} - (1/A)}{(1/A)^{N+1} - 1} \tag{10.3-21}$$

$$N = \frac{\log\left[\dfrac{x_0 - (y_{N+1}/m)}{x_N - (y_{N+1}/m)}(1 - A) + A\right]}{\log(1/A)} \tag{10.3-22}$$

When $A = 1$,

$$N = \frac{x_0 - x_N}{x_N - (y_{N+1}/m)} \tag{10.3-23}$$

For transfer of solute A from phase V to L (absorption),

$$\frac{y_{N+1} - y_1}{y_{N+1} - m x_0} = \frac{A^{N+1} - A}{A^{N+1} - 1} \tag{10.3-24}$$

$$N = \frac{\log\left[\dfrac{y_{N+1} - m x_0}{y_1 - m x_0}\left(1 - \dfrac{1}{A}\right) + \dfrac{1}{A}\right]}{\log A} \tag{10.3-25}$$

When $A = 1$,

$$N = \frac{y_{N+1} - y_1}{y_1 - m x_0} \tag{10.3-26}$$

Often the term A is called the *absorption factor* and S the *stripping factor*, where $S = 1/A$. These equations can be used with any consistent set of units such as mass flow and mass fraction or molar flow and mole fraction. Such series of equations are often called *Kremser equations* and are convenient to use. If A varies slightly from the inlet to the outlet, the geometric average of the two values can be used, with the value of m at the dilute end being used for both values of A.

EXAMPLE 10.3-3. Number of Stages by Analytical Equation.
Repeat Example 10.3-2 but use the Kremser analytical equations for countercurrent stage processes.

Solution: At one end of the process at stage 1, $V_1 = 29.73$ kg mol/h, $y_{A1} = 0.00101$, $L_0 = 90.0$, and $x_{A0} = 0$. Also, the equilibrium relation is $y_A = 2.53x_A$ where $m = 2.53$. Then,

$$A_1 = \frac{L}{mV} = \frac{L_0}{mV_1} = \frac{90.0}{2.53 \times 29.73} = 1.20$$

At stage N, $V_{N+1} = 30.0$, $y_{AN+1} = 0.01$, $L_N = 90.27$, and $x_{AN} = 0.00300$.

$$A_N = \frac{L_N}{mV_{N+1}} = \frac{90.27}{2.53 \times 30.0} = 1.19$$

The geometric average $A = \sqrt{A_1 A_N} = \sqrt{1.20 \times 1.19} = 1.195$.

The acetone solute is transferred from the V to the L phase (absorption). Substituting into Eq. (10.3-25),

$$N = \frac{\log\left[\dfrac{0.01 - 2.53(0)}{0.00101 - 2.53(0)} \left(1 - \dfrac{1}{1.195}\right) + \dfrac{1}{1.195} \right]}{\log (1.195)} = 5.04 \text{ stages}$$

This compares closely with 5.2 stages obtained using the graphical method.

10.4 MASS TRANSFER BETWEEN PHASES

10.4A Introduction and Equilibrium Relations

1. Introduction to interphase mass transfer. In Chapter 7 we considered mass transfer from a fluid phase to another phase, which was primarily a solid phase. The solute A was usually transferred from the fluid phase by convective mass transfer and through the solid by diffusion. In the present section we shall be concerned with the mass transfer of solute A from one fluid phase by convection and then through a second fluid phase by convection. For example, the solute may diffuse through a gas phase and then diffuse through and be absorbed in an adjacent and immiscible liquid phase. This occurs in the case of absorption of ammonia from air by water.

The two phases are in direct contact with each other, such as in a packed, tray, or spray-type tower, and the interfacial area between the phases is usually not well defined. In two-phase mass transfer, a concentration gradient will exist in each phase, causing mass transfer to occur. At the interface between the two fluid phases, equilibrium exists in most cases.

2. Equilibrium relations. Even when mass transfer is occurring equilibrium relations are important to determine concentration profiles for predicting rates of mass transfer. In Section 10.2 the equilibrium relation in a gas–liquid system and Henry's law were discussed. In Section 7.1C a discussion covered equilibrium distribution coefficients between two phases. These equilibrium relations will be used in discussion of mass transfer between phases in this section.

10.4B Concentration Profiles in Interphase Mass Transfer

In the majority of mass-transfer systems, two phases, which are essentially immiscible in each other, are present and also an interface between these two phases. Assuming the solute A is diffusing from the bulk gas phase G to the liquid phase L, it must pass through phase G, through the interface, and then into phase L in series. A concentration gradient

FIGURE 10.4-1. *Concentration profile of solute A diffusing through two phases.*

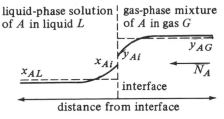

liquid-phase solution of A in liquid L | gas-phase mixture of A in gas G

y_{AG}

x_{Ai} | y_{Ai}

x_{AL}

N_A

interface

distance from interface

must exist to cause this mass transfer through the resistances in each phase, as shown in Fig. 10.4-1. The average or bulk concentration of A in the gas phase in mole fraction units is y_{AG}, where $y_{AG} = p_A/P$, and x_{AL} in the bulk liquid phase in mole fraction units.

The concentration in the bulk gas phase y_{AG} decreases to y_{Ai} at the interface. The liquid concentration starts at x_{Ai} at the interface and falls to x_{AL}. At the interface, since there would be no resistance to transfer across this interface, y_{Ai} and x_{Ai} are in equilibrium and are related by the equilibrium distribution relation

$$y_{Ai} = f(x_{Ai}) \tag{10.4-1}$$

where y_{Ai} is a function of x_{Ai}. They are related by an equilibrium plot such as Fig. 10.1-1. If the system follows Henry's law, $y_A P$ or p_A and x_A are related by Eq. (10.2-2) at the interface.

Experimentally, the resistance at the interface has been shown to be negligible for most cases of mass transfer where chemical reactions do not occur, such as absorption of common gases from air to water and extraction of organic solutes from one phase to another. However, there are some exceptions. Certain surface-active compounds may concentrate at the interface and cause an "interfacial resistance" that slows down the diffusion of solute molecules. Theories to predict when interfacial resistance may occur are still obscure and unreliable.

10.4C Mass Transfer Using Film Mass-Transfer Coefficients and Interface Concentrations

1. Equimolar counterdiffusion. For equimolar counterdiffusion the concentrations of Fig. 10.4-1 can be plotted on an xy diagram in Fig. 10.4-2. Point P represents the bulk phase compositions x_{AG} and x_{AL} of the two phases and point M the interface concentrations y_{Ai} and x_{Ai}. For A diffusing from the gas to liquid and B in equimolar counterdiffusion from liquid to gas,

$$N_A = k'_y(y_{AG} - y_{Ai}) = k'_x(x_{Ai} - x_{AL}) \tag{10.4-2}$$

where k'_y is the gas-phase mass-transfer coefficient in kg mol/s · m² · mol frac (g mol/s · cm² · mol frac, lb mol/h · ft² · mol frac) and k'_x the liquid-phase mass-transfer coefficient in kg mol/s · m² · mol frac (g mol/s · cm² · mol frac, lb mol/h · ft² · mol frac). Rearranging Eq. (10.4-2),

$$-\frac{k'_x}{k'_y} = \frac{y_{AG} - y_{Ai}}{x_{AL} - x_{Ai}} \tag{10.4-3}$$

The driving force in the gas phase is $(y_{AG} - y_{Ai})$ and in the liquid phase it is $(x_{Ai} - x_{AL})$. The slope of the line PM is $-k'_x/k'_y$. This means if the two film coefficients k'_x and k'_y are known, the interface compositions can be determined by drawing line PM with a slope $-k'_x/k'_y$ intersecting the equilibrium line.

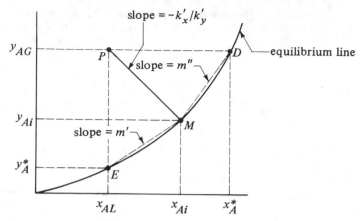

FIGURE 10.4-2. *Concentration driving forces and interface concentrations in inter-phase mass transfer (equimolar counterdiffusion).*

The bulk-phase concentrations y_{AG} and x_{AL} can be determined by simply sampling the mixed bulk gas phase and sampling the mixed bulk liquid phase. The interface concentrations are determined by Eq. (10.4-3).

2. Diffusion of A through stagnant or nondiffusing B. For the common case of A diffusing through a stagnant gas phase and then through a stagnant liquid phase, the concentrations are shown in Fig. 10.4-3, where P again represents bulk-phase compositions and M interface compositions. The equations for A diffusing through a stagnant gas and then through a stagnant liquid are

$$N_A = k_y(y_{AG} - y_{Ai}) = k_x(x_{Ai} - x_{AL}) \tag{10.4-4}$$

Now,

$$k_y = \frac{k_y'}{(1 - y_A)_{iM}} \qquad k_x = \frac{k_x'}{(1 - x_A)_{iM}} \tag{10.4-5}$$

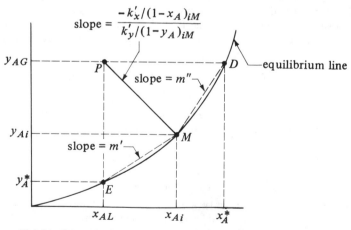

FIGURE 10.4-3. *Concentration driving forces and interface concentrations in inter-phase mass transfer (A diffusing through stagnant B).*

where

$$(1 - y_A)_{iM} = \frac{(1 - y_{Ai}) - (1 - y_{AG})}{\ln \left[(1 - y_{Ai})/(1 - y_{AG})\right]} \qquad (10.4\text{-}6)$$

$$(1 - x_A)_{iM} = \frac{(1 - x_{AL}) - (1 - x_{Ai})}{\ln \left[(1 - x_{AL})/(1 - x_{Ai})\right]} \qquad (10.4\text{-}7)$$

Then,

$$N_A = \frac{k_y'}{(1 - y_A)_{iM}} (y_{AG} - y_{Ai}) = \frac{k_x'}{(1 - x_A)_{iM}} (x_{Ai} - x_{AL}) \qquad (10.4\text{-}8)$$

Note that $(1 - y_A)_{iM}$ is the same as y_{BM} of Eq. (7.2-11) but is written for the interface, and $(1 - x_A)_{iM}$ is the same as x_{BM} of Eq. (7.2-11). Using Eq. (10.4-8) and rearranging,

$$\frac{- k_x'/(1 - x_A)_{iM}}{k_y'/(1 - y_A)_{iM}} = \frac{y_{AG} - y_{Ai}}{x_{AL} - x_{Ai}} \qquad (10.4\text{-}9)$$

The slope of the line PM in Fig. 10.4-3 to obtain the interface compositions is given by the left-hand side of Eq. (10.4-9). This differs from the slope of Eq. (10.4-3) for equimolar counterdiffusion by the terms $(1 - y_A)_{iM}$ and $(1 - x_A)_{iM}$. When A is diffusing through stagnant B and the solutions are dilute, $(1 - y_A)_{iM}$ and $(1 - x_A)_{iM}$ are close to 1.

A trial-and-error method is needed to use Eq. (10.4-9) to get the slope, since the left-hand side contains y_{Ai} and x_{Ai} that are being sought. For the first trial $(1 - y_A)_{iM}$ and $(1 - x_A)_{iM}$ are assumed to be 1.0 and Eq. (10.4-9) is used to get the slope and y_{Ai} and x_{Ai} values. Then for the second trial, these values of y_{Ai} and x_{Ai} are used to calculate a new slope to get new values of y_{Ai} and x_{Ai}. This is repeated until the interface compositions do not change. Three trials are usually sufficient.

EXAMPLE 10.4-1. *Interface Compositions in Interphase Mass Transfer*
The solute A is being absorbed from a gas mixture of A and B in a wetted-wall tower with the liquid flowing as a film downward along the wall. At a certain point in the tower the bulk gas concentration $y_{AG} = 0.380$ mol fraction and the bulk liquid concentration is $x_{AL} = 0.100$. The tower is operating at 298 K and 1.013×10^5 Pa and the equilibrium data are as follows:

x_A	y_A	x_A	y_A
0	0	0.20	0.131
0.05	0.022	0.25	0.187
0.10	0.052	0.30	0.265
0.15	0.087	0.35	0.385

The solute A diffuses through stagnant B in the gas phase and then through a nondiffusing liquid.

Using correlations for dilute solutions in wetted-wall towers, the film mass-transfer coefficient for A in the gas phase is predicted as $k_y = 1.465 \times 10^{-3}$ kg mol A/s · m² · mol frac (1.08 lb mol/h · ft² · mol frac) and for the liquid phase as $k_x = 1.967 \times 10^{-3}$ kg mol A/s · m² · mol frac (1.45 lb mol/h · ft² · mol frac). Calculate the interface concentrations y_{Ai} and x_{Ai} and the flux N_A.

Solution: Since the correlations are for dilute solutions, $(1 - y_A)_{iM}$ and

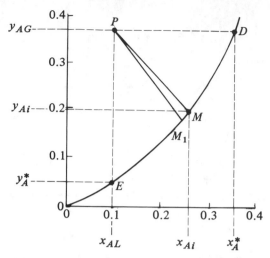

FIGURE 10.4-4. *Location of interface concentrations for Example 10.4-1.*

$(1 - x_A)_{iM}$ are approximately 1.0 and the coefficients are the same as k'_y and k'_x. The equilibrium data are plotted in Fig. 10.4-4. Point P is plotted at $y_{AG} = 0.380$ and $x_{AL} = 0.100$. For the first trial $(1 - y_A)_{iM}$ and $(1 - x_A)_{iM}$ are assumed as 1.0 and the slope of line PM is, from Eq. (10.4-9),

$$-\frac{k'_x/(1 - x_A)_{iM}}{k'_y/(1 - y_A)_{iM}} = -\frac{1.967 \times 10^{-3}/1.0}{1.465 \times 10^{-3}/1.0} = -1.342$$

A line through point P with a slope of -1.342 is plotted in Fig. 10.4-4 intersecting the equilibrium line at M_1, where $y_{Ai} = 0.183$ and $x_{Ai} = 0.247$.

For the second trial we use y_{Ai} and x_{Ai} from the first trial to calculate the new slope. Substituting into Eqs. (10.4-6) and (10.4-7),

$$(1 - y_A)_{iM} = \frac{(1 - y_{Ai}) - (1 - y_{AG})}{\ln\left[(1 - y_{Ai})/(1 - y_{AG})\right]}$$

$$= \frac{(1 - 0.183) - (1 - 0.380)}{\ln\left[(1 - 0.183)/(1 - 0.380)\right]} = 0.715$$

$$(1 - x_A)_{iM} = \frac{(1 - x_{AL}) - (1 - x_{Ai})}{\ln\left[(1 - x_{AL})/(1 - x_{Ai})\right]}$$

$$= \frac{(1 - 0.100) - (1 - 0.247)}{\ln\left[(1 - 0.100)/(1 - 0.247)\right]} = 0.825$$

Substituting into Eq. (10.4-9) to obtain the new slope,

$$-\frac{k'_x/(1 - x_A)_{iM}}{k'_y/(1 - y_A)_{iM}} = -\frac{1.967 \times 10^{-3}/0.825}{1.465 \times 10^{-3}/0.715} = -1.163$$

A line through point P with a slope of -1.163 is plotted and intersects the equilibrium line at M, where $y_{Ai} = 0.197$ and $x_{Ai} = 0.257$.

Using these new values for the third trial, the following values are calculated:

$$(1 - y_A)_{iM} = \frac{(1 - 0.197) - (1 - 0.380)}{\ln\left[(1 - 0.197)/(1 - 0.380)\right]} = 0.709$$

$$(1 - x_A)_{iM} = \frac{(1 - 0.100) - (1 - 0.257)}{\ln\,[(1 - 0.100)/(1 - 0.257)]} = 0.820$$

$$-\frac{k_x'/(1 - x_A)_{iM}}{k_y'/(1 - y_A)_{iM}} = -\frac{1.967 \times 10^{-3}/0.820}{1.465 \times 10^{-3}/0.709} = -1.160$$

This slope of -1.160 is essentially the same as the slope of -1.163 for the second trial. Hence, the final values are $y_{Ai} = 0.197$ and $x_{Ai} = 0.257$ and are shown as point M.

To calculate the flux, Eq. (10.4-8) is used.

$$N_A = \frac{k_y'}{(1 - y_A)_{iM}} (y_{AG} - y_{Ai}) = \frac{1.465 \times 10^{-3}}{0.709} (0.380 - 0.197)$$

$$= 3.78 \times 10^{-4} \text{ kg mol/s} \cdot \text{m}^2$$

$$N_A = \frac{1.08}{0.709} (0.380 - 0.197) = 0.2785 \text{ lb mol/h} \cdot \text{ft}^2$$

$$N_A = \frac{k_x'}{(1 - x_A)_{iM}} (x_{Ai} - x_{AL}) = \frac{1.967 \times 10^{-3}}{0.820} (0.257 - 0.100)$$

$$= 3.78 \times 10^{-4} \text{ kg mol/s} \cdot \text{m}^2$$

Note that the flux N_A through each phase is the same as in the other phase, which should be the case at steady state.

10.4D Overall Mass-Transfer Coefficients and Driving Forces

1. Introduction. Film or single-phase mass-transfer coefficients k_y' and k_x' or k_y and k_x are often difficult to measure experimentally, except in certain experiments designed so that the concentration difference across one phase is small and can be neglected. As a result, overall mass-transfer coefficients K_y' and K_x' are measured based on the gas phase or liquid phase. This method is used in heat transfer, where overall heat-transfer coefficients are measured based on inside or outside areas instead of film coefficients.

The overall mass transfer K_y' is defined as

$$N_A = K_y'(y_{AG} - y_A^*) \tag{10.4-10}$$

where K_y' is based on the overall gas-phase driving force in kg mol/s $\cdot$ m^2 $\cdot$ mol frac, and y_A^* is the value that would be in equilibrium with x_{AL}, as shown in Fig. 10.4-2. Also, K_x' is defined as

$$N_A = K_x'(x_A^* - x_{AL}) \tag{10.4-11}$$

where K_x' is based on the overall liquid-phase driving force in kg mol/s $\cdot$ m^2 $\cdot$ mol frac and x_A^* is the value that would be in equilibrium with y_{AG}.

2. Equimolar counterdiffusion and/or diffusion in dilute solutions. Equation (10.4-2) holds for equimolar counterdiffusion, or when the solutions are dilute, Eqs. (10.4-8) and (10.4-2) are identical.

$$N_A = k_y'(y_{AG} - y_{Ai}) = k_x'(x_{Ai} - x_{AL}) \tag{10.4-2}$$

From Fig. 10.4-2,

$$y_{AG} - y_A^* = (y_{AG} - y_{Ai}) + (y_{Ai} - y_A^*) \tag{10.4-12}$$

Between the points E and M the slope m' can be given as

$$m' = \frac{y_{Ai} - y_A^*}{x_{Ai} - x_{AL}} \tag{10.4-13}$$

Solving Eq. (10.4-13) for $(y_{Ai} - y_A^*)$ and substituting into Eq. (10.4-12),

$$y_{AG} - y_A^* = (y_{AG} - y_{Ai}) + m'(x_{Ai} - x_{AL}) \tag{10.4-14}$$

Then on substituting Eqs. (10.4-10) and (10.4-2) into (10.4-14) and canceling out N_A,

$$\frac{1}{K_y'} = \frac{1}{k_y'} + \frac{m'}{k_x'} \tag{10.4-15}$$

The left-hand side of Eq. (10.4-15) is the total resistance based on the overall gas driving force and equals the gas film resistance $1/k_y'$ plus the liquid film resistance m'/k_x'.

In a similar manner from Fig. 10.4-2,

$$x_A^* - x_{AL} = (x_A^* - x_{Ai}) + (x_{Ai} - x_{AL}) \tag{10.4-16}$$

$$m'' = \frac{y_{AG} - y_{Ai}}{x_A^* - x_{Ai}} \tag{10.4-17}$$

Proceeding as before,

$$\frac{1}{K_x'} = \frac{1}{m''k_y'} + \frac{1}{k_x'} \tag{10.4-18}$$

Several special cases of Eqs. (10.4-15) and (10.4-18) will now be discussed. The numerical values of k_x' and k_y' are very roughly similar. The values of the slopes m' or m'' are very important. If m' is quite small, so that the equilibrium curve in Fig. 10.4-2 is almost horizontal, a small value of y_A in the gas will give a large value of x_A in equilibrium in the liquid. The gas solute A is then very soluble in the liquid phase, and hence the term m'/k_x' in Eq. (10.4-15) is very small. Then,

$$\frac{1}{K_y'} \cong \frac{1}{k_y'} \tag{10.4-19}$$

and the major resistance is in the gas phase, or the "gas phase is controlling." The point M has moved down very close to E, so that

$$y_{AG} - y_A^* \cong y_{AG} - y_{Ai} \tag{10.4-20}$$

Similarly, when m'' is very large, the solute A is very insoluble in the liquid, $1/(m''k_y')$ becomes small, and

$$\frac{1}{K_x'} \cong \frac{1}{k_x'} \tag{10.4-21}$$

The "liquid phase is controlling" and $x_{Ai} \cong x_A^*$. Systems for absorption of oxygen or CO_2 from air by water are similar to Eq. (10.4-21).

3. *Diffusion of A through stagnant or nondiffusing B.* For the case of A diffusing through nondiffusing B, Eqs. (10.4-8) and (10.4-14) hold and Fig. 10.4-3 is used.

$$N_A = \frac{k_y'}{(1 - y_A)_{iM}}(y_{AG} - y_{Ai}) = \frac{k_x'}{(1 - x_A)_{iM}}(x_{Ai} - x_{AL}) \tag{10.4-8}$$

$$y_{AG} - y_A^* = (y_{AG} - y_{Ai}) + m'(x_{Ai} - x_{AL}) \tag{10.4-14}$$

We must, however, define the equations for the flux using overall coefficients as follows:

$$N_A = \left[\frac{K'_y}{(1 - y_A)_{*M}} \right] (y_{AG} - y^*_A) = \left[\frac{K'_x}{(1 - x_A)_{*M}} \right] (x^*_A - x_{AL}) \qquad \text{(10.4-22)}$$

The bracketed terms are often written as follows:

$$K_y = \frac{K'_y}{(1 - y_A)_{*M}} \qquad K_x = \frac{K'_x}{(1 - x_A)_{*M}} \qquad \text{(10.4-23)}$$

where K_y is the overall gas mass-transfer coefficient for A diffusing through stagnant B and K_x the overall liquid mass-transfer coefficient. These two coefficients are concentration-dependent. Substituting Eqs. (10.4-8) and (10.4-22) into (10.4-14), we obtain

$$\frac{1}{K'_y/(1 - y_A)_{*M}} = \frac{1}{k'_y/(1 - y_A)_{iM}} + \frac{m'}{k'_x/(1 - x_A)_{iM}} \qquad \text{(10.4-24)}$$

where

$$(1 - y_A)_{*M} = \frac{(1 - y^*_A) - (1 - y_{AG})}{\ln [(1 - y^*_A)/(1 - y_{AG})]} \qquad \text{(10.4-25)}$$

Similarly, for K'_x,

$$\frac{1}{K'_x/(1 - x_A)_{*M}} = \frac{1}{m'' k'_y/(1 - y_A)_{iM}} + \frac{1}{k'_x/(1 - x_A)_{iM}} \qquad \text{(10.4-26)}$$

where

$$(1 - x_A)_{*M} = \frac{(1 - x_{AL}) - (1 - x^*_A)}{\ln [(1 - x_{AL})/(1 - x^*_A)]} \qquad \text{(10.4-27)}$$

It should be noted that the relations derived here also hold for any two-phase system where y stands for the one phase and x for the other phase. For example, for the extraction of the solute acetic acid (A) from water (y phase) by isopropyl ether (x phase), the same relations will hold.

EXAMPLE 10.4-2. Overall Mass-Transfer Coefficients from Film Coefficients

Using the same data as in Example 10.4-1, calculate the overall mass-transfer coefficient K'_y, the flux, and the percent resistance in the gas and liquid films. Do this for the case of A diffusing through stagnant B.

Solution: From Fig. 10.4-4, $y^*_A = 0.052$, which is in equilibrium with the bulk liquid $x_{AL} = 0.10$. Also, $y_{AG} = 0.380$. The slope of chord m' between E and M from Eq. (10.4-13) is, for $y_{Ai} = 0.197$ and $x_{Ai} = 0.257$,

$$m' = \frac{y_{Ai} - y^*_A}{x_{Ai} - x_{AL}} = \frac{0.197 - 0.052}{0.257 - 0.100} = 0.923$$

From Example 10.4-1,

$$\frac{k'_y}{(1 - y_A)_{iM}} = \frac{1.465 \times 10^{-3}}{0.709} \qquad \frac{k'_x}{(1 - x_A)_{iM}} = \frac{1.967 \times 10^{-3}}{0.820}$$

Using Eq. (10.4-25),

$$(1 - y_A)_{*M} = \frac{(1 - y^*_A) - (1 - y_{AG})}{\ln [(1 - y^*_A)/(1 - y_{AG})]}$$

$$= \frac{(1 - 0.052) - (1 - 0.380)}{\ln [(1 - 0.052)/(1 - 0.380)]} = 0.773$$

Then, using Eq. (10.4-24),

$$\frac{1}{K_y'/0.773} = \frac{1}{1.465 \times 10^{-3}/0.709} + \frac{0.923}{1.967 \times 10^{-3}/0.820}$$

$$= 484.0 + 384.8 = 868.8$$

Solving, $K_y' = 8.90 \times 19^{-4}$. The percent resistance in the gas film is $(484.0/868.8)100 = 55.7\%$ and 44.3% in the liquid film. The flux is as follows, using Eq. (10.4-22):

$$N_A = \frac{K_y'}{(1 - y_A)_{*M}} (y_{AG} - y_A^*) = \frac{8.90 \times 10^{-4}}{0.773} (0.380 - 0.052)$$

$$= 3.78 \times 10^{-4} \text{ kg mol/s} \cdot \text{m}^2$$

This, of course, is the same flux value as was calculated in Example 10.4-1 using the film equations.

4. Discussion of overall coefficients. If the two-phase system is such that the major resistance is in the gas phase as in Eq. (10.4-19), then to increase the overall rate of mass transfer, efforts should be centered on increasing the gas-phase turbulence, not the liquid-phase turbulence. For a two-phase system where the liquid film resistance is controlling, turbulence should be increased in this phase to increase rates of mass transfer.

To design mass-transfer equipment, the overall mass-transfer coefficient is synthesized from the individual film coefficients, as discussed in this section.

10.5 CONTINUOUS HUMIDIFICATION PROCESSES

10.5A Introduction and Types of Equipment for Humidification

1. Introduction to gas–liquid contactors. When a relatively warm liquid is directly contacted with gas that is unsaturated, some of the liquid is vaporized. The liquid temperature will drop mainly because of the latent heat of evaporation. This direct contact of a gas with a pure liquid occurs most often in contacting air with water. This is done for the following purposes: humidifying air for control of the moisture content of air in drying or air conditioning; dehumidifying air, where cold water condenses some water vapor from warm air; and water cooling, where evaporation of water to the air cools warm water.

In Chapter 9 the fundamentals of humidity and adiabatic humidification were discussed. In this section the performance and design of continuous air–water contactors is considered. The emphasis is on cooling of water, since this is the most important type of process in the process industries. There are many cases in industry in which warm water is discharged from heat exchangers and condensers when it would be more economical to cool and reuse it than to discard it.

2. Towers for water cooling. In a typical water-cooling tower, warm water flows countercurrently to an air stream. Typically, the warm water enters the top of a packed tower and cascades down through the packing, leaving at the bottom. Air enters at the bottom of the tower and flows upward through the descending water. The tower packing often consists of slats of wood or plastic or of a packed bed. The water is distributed by troughs and overflows to cascade over slat gratings or packing that provide large interfacial areas of contact between the water and air in the form of droplets and films of water. The flow of air upward through the tower can be induced by the buoyancy of the warm air in the tower (natural draft) or by the action of a fan. Detailed descriptions of towers are given in other texts (B1, T1).

The water cannot be cooled below the wet bulb temperature. The driving force for the evaporation of the water is approximately the vapor pressure of the water less the vapor pressure it would have at the wet bulb temperature. The water can be cooled only to the wet bulb temperature, and in practice it is cooled to about 3 K or more above this. Only a small amount of water is lost by evaporation in cooling water. Since the latent heat of vaporization of water is about 2300 kJ/kg, a typical change of about 8 K in water temperature corresponds to an evaporation loss of about 1.5%. Hence, the total flow of water is usually assumed to be constant in calculations of tower size.

In humidification and dehumidification, intimate contact between the gas phase and liquid phase is needed for large rates of mass transfer and heat transfer. The gas-phase resistance controls the rate of transfer. Spray or packed towers are used to give large interfacial areas and to promote turbulence in the gas phase.

10.5B Theory and Calculation of Water-Cooling Towers

1. Temperature and concentration profiles at interface. In Fig. 10.5-1 the temperature profile and the concentration profile in terms of humidity are shown at the water–gas interface. Water vapor diffuses from the interface to the bulk gas phase with a driving force in the gas phase of $(H_i - H_G)$ kg H_2O/kg dry air. There is no driving force for mass transfer in the liquid phase, since water is a pure liquid. The temperature driving force is $T_L - T_i$ in the liquid phase and $T_i - T_G$ K or °C in the gas phase. Sensible heat flows from the bulk liquid to the interface in the liquid. Sensible heat also flows from the interface to the gas phase. Latent heat also leaves the interface in the water vapor, diffusing to the gas phase. The sensible heat flow from the liquid to the interface equals the sensible heat flow in the gas plus the latent heat flow in the gas.

The conditions in Fig. 10.5-1 occur at the upper part of the cooling tower. In the lower part of the cooling tower the temperature of the bulk water is higher than the wet bulb temperature of the air but may be below the dry bulb temperature. Then the direction of the sensible heat flow in Fig. 10.5-1 is reversed.

2. Rate equations for heat and mass transfer. We shall consider a packed water-cooling

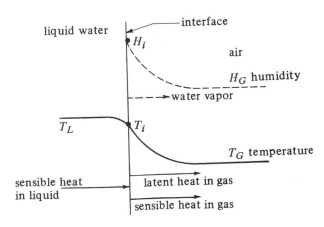

FIGURE 10.5-1. *Temperature and concentration profiles in upper part of cooling tower.*

tower with air flowing upward and water countercurrently downward in the tower. The total interfacial area between the air and water phases is unknown, since the surface area of the packing is not equal to the interfacial area between the water droplets and the air. Hence, we define a quantity a, defined as m^2 of interfacial area per m^3 volume of packed section, or m^2/m^3. This is combined with the gas-phase mass-transfer coefficient k_G in kg mol/s $\cdot m^2 \cdot$ Pa or kg mol/s $\cdot m^2 \cdot$ atm to give a volumetric coefficient $k_G a$ in kg mol/s $\cdot m^3$ volume $\cdot$ Pa or kg mol/s $\cdot m^3 \cdot$ atm (lb mol/h $\cdot$ ft$^3 \cdot$ atm).

The process is carried out adiabatically and the various streams and conditions are shown in Fig. 10.5-2, where

$\quad L$ = water flow, kg water/s $\cdot m^2$ (lb$_m$/h $\cdot$ ft^2)

$\quad T_L$ = temperature of water, °C or K (°F)

$\quad G$ = dry air flow, kg/s $\cdot m^2$ (lb$_m$/h $\cdot$ ft^2)

$\quad T_G$ = temperature of air, °C or K (°F)

$\quad H$ = humidity of air, kg water/kg dry air (lb water/lb dry air)

$\quad H_y$ = enthalpy of air–water vapor mixture, J/kg dry air (btu/lb$_m$ dry air)

The enthalpy H_y as given in Eq. (9.3-8) is

$$H_y = c_S(T - T_0) + H\lambda_0 = (1.005 + 1.88H)10^3(T - 0) + 2.501 \times 10^6 H \quad \text{(SI)}$$

$$H_y = c_S(T - T_0) + H\lambda_0 = (0.24 + 0.45H)(T - 32) + 1075.4H \quad\quad\quad \text{(English)}$$

$$(9.3\text{-}8)$$

The base temperature selected is 0°C or 273 K (32°F). Note that $(T - T_0)$°C $= (T - T_0)$ K.

Making a total heat balance for the dashed-line box shown in Fig. 10.5-2, an operating line is obtained.

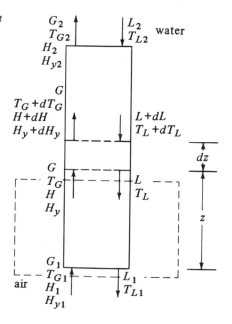

FIGURE 10.5-2. *Continuous countercurrent adiabatic water cooling.*

$$G(H_y - H_{y1}) = Lc_L(T_L - T_{L1}) \tag{10.5-1}$$

This assumes that L is essentially constant, since only a small amount is evaporated. The heat capacity c_L of the liquid is assumed constant at 4.187×10^3 J/kg·K (1.00 btu/lb$_m$· °F). When plotted on a chart of H_y versus T_L, this Eq. (10.5-1) is a straight line with a slope of Lc_L/G. Making an overall heat balance over both ends of the tower,

$$G(H_{y2} - H_{y1}) = Lc_L(T_{L2} - T_{L1}) \tag{10.5-2}$$

Again making a heat balance for the dz column height and neglecting sensible heat terms compared to the latent heat,

$$Lc_L \, dT_L = G \, dH_y \tag{10.5-3}$$

The total sensible heat transfer from the bulk liquid to the interface is (refer to Fig. 10.5-1)

$$Lc_L \, dT_L = G \, dH_y = h_L a \, dz(T_L - T_i) \tag{10.5-4}$$

where $h_L a$ is the liquid-phase volumetric heat-transfer coefficient in W/m^3·K (btu/h· ft^3·°F) and T_i is the interface temperature.

For adiabatic mass transfer the rate of heat transfer due to the latent heat in the water vapor being transferred can be obtained from Eq. (9.3-16) by rearranging and using a volumetric basis.

$$\frac{q_\lambda}{A} = M_B k_G aP\lambda_0 (H_i - H_G) \, dz \tag{10.5-5}$$

where q_λ/A is in W/m^2 (btu/h·ft^2), M_B = molecular weight of air, $k_G a$ is a volumetric mass-transfer coefficient in the gas in kg mol/s·m^3·Pa, P = atm pressure in Pa, λ_0 is the latent heat of water in J/kg water, H_i is the humidity of the gas at the interface in kg water/kg dry air, and H_G is the humidity of the gas in the bulk gas phase in kg water/kg dry air. The rate of sensible heat transfer in the gas is

$$\frac{q_S}{A} = h_G a(T_i - T_G) \, dz \tag{10.5-6}$$

where q_S/A is in W/m^2 and $h_G a$ is a volumetric heat-transfer coefficient in the gas in W/m^3·K.

Now from Fig. 10.5-1, Eq. (10.5-4) must equal the sum of Eqs. (10.5-5) and (10.5-6).

$$GdH_y = M_B k_G aP\lambda_0 (H_i - H_G) \, dz + h_G a(T_i - T_G) \, dz \tag{10.5-7}$$

Equation (9.3-18) states that

$$\frac{h_G a}{M_B k_y a} \cong c_S \tag{10.5-8}$$

Substituting $Pk_G a$ for $k_y a$,

$$\frac{h_G a}{M_B Pk_G a} \cong c_S \tag{10.5-9}$$

Substituting Eq. (10.5-9) into Eq. (10.5-7) and rearranging,

$$G \, dH_y = M_B k_G aP \, dz \, [(c_S T_i + \lambda_0 H_i) - (c_S T_G + \lambda_0 H_G)] \tag{10.5-10}$$

Adding and subtracting $c_S T_0$ inside the brackets,

$$G \, dH_y = M_B k_G aP \, dz\{c_S(T_i - T_0) + H_i \lambda_0 - [c_S(T_G - T_0) + H_G \lambda_0]\} \tag{10.5-11}$$

Sec. 10.5   *Continuous Humidification Processes*

The terms inside the braces are $(H_{yi} - H_y)$, and Eq. (10.5-11) becomes

$$G \, dH_y = M_B k_G aP \, dz(H_{yi} - H_y) \tag{10.5-12}$$

Integrating, the final equation to use to calculate the tower height is

$$\int_0^z dz = z = \frac{G}{M_B k_G aP} \int_{H_{y1}}^{H_{y2}} \frac{dH_y}{H_{yi} - H_y} \tag{10.5-13}$$

If Eq. (10.5-4) is equated to Eq. (10.5-12) and the result rearranged,

$$-\frac{h_L a}{k_G a M_B P} = \frac{H_{yi} - H_y}{T_i - T_L} \tag{10.5-14}$$

10.5C   Design of Water-Cooling Tower Using Film Mass-Transfer Coefficients

The tower design is done using the following steps.

1. The enthalpy of saturated air H_{yi} is plotted versus T_i on an H versus T plot as shown in Fig. 10.5-3. This enthalpy is calculated with Eq. (9.3-8) using the saturation humidity from the humidity chart for a given temperature, with 0°C (273 K) as a base temperature. Calculated values are tabulated in Table 10.5-1.
2. Knowing the entering air conditions T_{G1} and H_1, the enthalpy of this air H_{y1} is calculated from Eq. (9.3-8). The point H_{y1} and T_{L1} (desired leaving water temperature) is plotted in Fig. 10.5-3 as one point on the operating line. The operating line is plotted with a slope Lc_L/G and ends at point T_{L2}, which is the entering water temperature. This gives H_{y2}. Alternatively, H_{y2} can be calculated from Eq. (10.5-2).
3. Knowing $h_L a$ and $k_G a$, lines with a slope of $-h_L a/k_G a M_B P$ are plotted as shown in Fig. 10.5-3. From Eq. (10.5-14) point P represents H_y and T_L on the operating line, and point M represents H_{yi} and T_i, the interface conditions. Hence, line MS or $H_{yi} - H_y$ represents the driving force in Eq. (10.5-13).

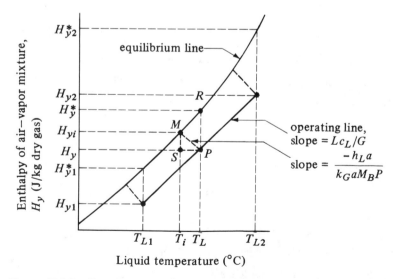

FIGURE 10.5-3.   *Temperature enthalpy diagram and operating line for water-cooling tower.*

TABLE 10.5-1. *Enthalpies of Saturated Air–Water Vapor Mixtures (0°C Base Temperature)*

T_L		H_y		T_L		H_y	
		btu	J			btu	J
°F	°C	lb_m dry air	kg dry air	°F	°C	lb_m dry air	kg dry air
60	15.6	18.78	43.68×10^3	100	37.8	63.7	148.2×10^3
80	26.7	36.1	84.0×10^3	105	40.6	74.0	172.1×10^3
85	29.4	41.8	97.2×10^3	110	43.3	84.8	197.2×10^3
90	32.2	48.2	112.1×10^3	115	46.1	96.5	224.5×10^3
95	35.0	55.4	128.9×10^3	140	60.0	198.4	461.5×10^3

4. The driving force $H_{yi} - H_y$ is computed for various values of T_L between T_{L1} and T_{L2}. Then by plotting $1/(H_{yi} - H_y)$ versus H_y from H_{y1} to H_{y2}, a graphical integration is performed to obtain the value of the integral in Eq. (10.5-13). Finally, the height z is calculated from Eq. (10.5-13).

10.5D Design of Water-Cooling Tower Using Overall Mass-Transfer Coefficients

Often, only an overall mass-transfer coefficient $K_G a$ in kg mol/s·m³·Pa or kg mol/s·m³·atm is available, and Eq. (10.5-13) becomes

$$z = \frac{G}{M_B K_G a P} \int_{H_{y1}}^{H_{y2}} \frac{dH_y}{H_y^* - H_y} \qquad (10.5\text{-}15)$$

The value of H_y^* is determined by going vertically from the value of H_y at point P up to the equilibrium line to give H_y^* at point R, as shown in Fig. 10.5-3. In many cases the experimental film coefficients $k_G a$ and $h_L a$ are not available. The few experimental data available indicate that $h_L a$ is quite large and the slope of the lines $-h_L a/(k_G a M_B P)$ in Eq. (10.5-14) would be very large and the value of H_{yi} would approach that of H_y^* in Fig. 10.5-3.

The tower design using the overall mass-transfer coefficient is done using the following steps.

1. The enthalpy–temperature data from Table 10.5-1 are plotted as shown in Fig. 10.5-3.
2. The operating line is calculated as in steps 1 and 2 for the film coefficients and plotted in Fig. 10.5-3.
3. In Fig. 10.5-3 point P represents H_y and T_L on the operating line and point R represents H_y^* on the equilibrium line. Hence, the vertical line RP or $H_y^* - H_y$ represents the driving force in Eq. (10.5-15).
4. The driving force $H_y^* - H_y$ is computed for various values of T_L between T_{L1} and T_{L2}. Then by plotting $1/(H_y^* - H_y)$ versus H_y from H_{y1} to H_{y2}, a graphical integration is performed to obtain the value of the integral in Eq. (10.5-15). Finally, the height z is obtained from Eq. (10.5-15).

If experimental cooling data in an actual run in a cooling tower with known height z are available, then using Eq. (10.5-15), the experimental value of $K_G a$ can be obtained.

EXAMPLE 10.5-1. *Design of Water-Cooling Tower Using Film Coefficients*
A packed countercurrent water-cooling tower using a gas flow rate of
$G = 1.356$ kg dry air/s $\cdot$ m^2 and a water flow rate of $L = 1.356$ kg
water/s $\cdot$ m^2 is to cool the water from $T_{L2} = 43.3°$C (110°F) to $T_{L1} = 29.4°$C
(85°F). The entering air at 29.4°C has a wet bulb temperature of 23.9°C. The
mass-transfer coefficient $k_G a$ is estimated as 1.207×10^{-7} kg mol/s $\cdot$ m$^3 \cdot$ Pa
and $h_L a/k_G a M_B P$ as 4.187×10^4 J/kg $\cdot$ K (10.0 btu/lb$_m \cdot$ °F). Calculate the
height of packed tower z. The tower operates at a pressure of 1.013×10^5 Pa.

Solution: Following the steps outlined, the enthalpies from the saturated
air–water vapor mixtures from Table 10.5-1 are plotted in Fig. 10.5-4. The
inlet air at $T_{G1} = 29.4°$C has a wet bulb temperature of 23.9°C. The humidity from the humidity chart is $H_1 = 0.0165$ kg H$_2$O/kg dry air. Substituting
into Eq. (9.3-8), noting that $(29.4 - 0)°$C $= (29.4 - 0)$ K,

$$H_{y1} = (1.005 + 1.88 \times 0.0165)10^3(29.4 - 0) + 2.501 \times 10^6(0.0165)$$

$$= 71.7 \times 10^3 \text{ J/kg}$$

The point $H_{y1} = 71.7 \times 10^3$ and $T_{L1} = 29.4°$C is plotted. Then substituting
into Eq. (10.5-2) and solving,

$$1.356(H_{y2} - 71.7 \times 10^3) = 1.356(4.187 \times 10^3)(43.3 - 29.4)$$

$H_{y2} = 129.9 \times 10^3$ J/kg dry air (55.8 btu/lb$_m$). The point $H_{y2} = 129.9 \times 10^3$
and $T_{L2} = 43.3°$C is also plotted, giving the operating line. Lines with slope
$-h_L a/k_G a M_B P = -41.87 \times 10^3$ J/kg $\cdot$ K are plotted giving H_{yi} and H_y
values, which are tabulated in Table 10.5-2 along with derived values as
shown. Values of $1/(H_{yi} - H_y)$ are plotted versus H_y and the area under the
curve from $H_{y1} = 71.7 \times 10^3$ to $H_{y2} = 129.9 \times 10^3$ is

$$\int_{H_{y1}}^{H_{y2}} \frac{dH_y}{H_{yi} - H_y} = 1.82$$

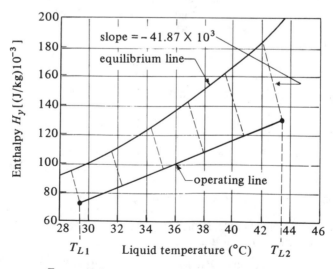

FIGURE 10.5-4. *Graphical solution of Example 10.5-1.*

     Chap. 10   *Stage and Continuous Gas–Liquid Separation Processes*

TABLE 10.5-2. *Enthalpy Values for Solution to Example 10.5-1 (enthalpy in J/kg dry air)*

H_{yi}	H_y	$H_{yi} - H_y$	$1/(H_{yi} - H_y)$
94.4×10^3	71.7×10^3	22.7×10^3	4.41×10^{-5}
108.4×10^3	83.5×10^3	24.9×10^3	4.02×10^{-5}
124.4×10^3	94.9×10^3	29.5×10^3	3.39×10^{-5}
141.8×10^3	106.5×10^3	35.3×10^3	2.83×10^{-5}
162.1×10^3	118.4×10^3	43.7×10^3	2.29×10^{-5}
184.7×10^3	129.9×10^3	54.8×10^3	1.82×10^{-5}

Substituting into Eq. (10.5-13),

$$z = \frac{G}{M_B k_G aP} \int \frac{dH_y}{H_{yi} - H_y} = \frac{1.356}{29(1.207 \times 10^{-7})(1.013 \times 10^5)} \quad (1.82)$$

$$= 6.98 \text{ m (22.9 ft)}$$

10.5E Minimum Value of Air Flow

Often the air flow G is not fixed but must be set for the design of the cooling tower. As shown in Fig. 10.5-5 for a minimum value of G, the operating line MN is drawn through the point H_{y1} and T_{L1} with a slope that touches the equilibrium line at T_{L2}, point N. If the equilibrium line is quite curved, line MN could become tangent to the equilibrium line at a point farther down the equilibrium line than point N. For the actual tower, a value of G greater than G_{min} must be used. Often, a value of G equal to 1.3 to 1.5 times G_{min} is used.

10.5F Design of Water-Cooling Tower Using Height of a Transfer Unit

Often another form of the film mass-transfer coefficient is used in Eq. (10.5-13):

$$z = H_G \int_{H_{y1}}^{H_{y2}} \frac{dH_y}{H_{yi} - H_y} \quad (10.5\text{-}16)$$

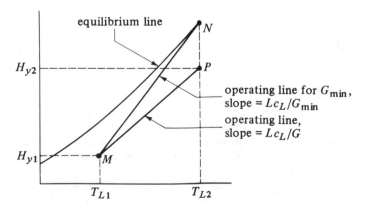

FIGURE 10.5-5. *Operating-line construction for minimum gas flow.*

$$H_G = \frac{G}{M_B k_G a P} \qquad (10.5\text{-}17)$$

where H_G is the height of a gas enthalpy transfer unit in m and the integral term is called the number of transfer units. The term H_G is often used since it is less dependent upon flow rates than $k_G a$.

Often another form of the overall mass-transfer coefficient $K_G a$ in kg mol/s · m^3 · Pa or kg mol/s · m^3 · atm is used and Eq. (10.5-15) becomes

$$z = \frac{G}{M_B K_G a P} \int_{H_{y1}}^{H_{y2}} \frac{dH_y}{H_y^* - H_y} = H_{OG} \int_{H_{y1}}^{H_{y2}} \frac{dH_y}{H_y^* - H_y} \qquad (10.5\text{-}18)$$

where H_{OG} is the height of an overall gas enthalpy transfer unit in m. The value of H_y^* is determined by going vertically from the value of H_y up to the equilibrium line as shown in Fig. 10.5-3. This method should be used only when the equilibrium line is almost straight over the range used. However, the H_{OG} is often used even if the equilibrium line is somewhat curved because of the lack of film mass-transfer coefficient data.

10.5G Temperature and Humidity of Air Stream in Tower

The procedures outlined above do not yield any information on the changes in temperature and humidity of the air–water vapor stream through the tower. If this information is of interest, a graphical method by Mickley (M2) is available. The equation used for the graphical method is derived by first setting Eq. (10.5-6) equal to $G c_S \, dT_G$ and combining it with Eqs. (10.5-12) and (10.5-9) to yield Eq. (10.5-19).

$$\frac{dH_y}{dT_G} = \frac{H_{yi} - H_y}{T_i - T_G} \qquad (10.5\text{-}19)$$

10.5H Dehumidification Tower

For the cooling or humidification tower discussed, the operating line lies below the equilibrium line and water is cooled and air humidified. In a dehumidification tower cool water is used to reduce the humidity and temperature of the air that enters. In this case the operating line is above the equilibrium line. Similar calculation methods are used (T1).

10.6 ABSORPTION IN PLATE AND PACKED TOWERS

10.6A Equipment for Absorption and Distillation

1. Introduction to absorption. As discussed briefly in Section 10.1B, absorption is a mass-transfer process in which a vapor solute A in a gas mixture is absorbed by means of a liquid in which the solute is more or less soluble. The gas mixture consists mainly of an inert gas and the solute. The liquid also is primarily immiscible in the gas phase; i.e., its vaporization into the gas phase is relatively slight. A typical example is absorption of the solute ammonia from an air–ammonia mixture by water. Subsequently, the solute is recovered from the solution by distillation. In the reverse process of desorption or stripping, the same principles and equations hold.

Equilibrium relations for gas–liquid systems in absorption were discussed in Section 10.2, and such data are needed for design of absorption towers. Some data are tabulated in Appendix A.3. Other more extensive data are available in Perry and Green (P1).

2. Various types of tray (plate) towers for absorption and distillation. In order to efficiently contact the vapor and liquid in absorption and distillation, tray (plate) towers are often used. A very common type of tray contacting device is the sieve tray, which is shown schematically in Fig. 10.6-1a and in Section 11.4A for distillation.

1. *Sieve tray.* Essentially, the same type of sieve tray is used in gas absorption and in distillation. In the sieve tray, vapor bubbles up through simple holes in the tray through the flowing liquid. Hole sizes range from 3 to 12 mm in diameter, with 5 mm a common size. The vapor area of the holes varies between 5 to 15% of the tray area. The liquid is maintained on the tray surface and prevented from flowing down through the holes by the kinetic energy of the gas or vapor. The depth of liquid on the tray is maintained by an overflow, outlet weir. The overflow liquid flows into the downspout to the next tray below.

2. *Valve tray.* A modification of the sieve tray is the valve tray, which consists of openings in the tray and a lift-valve cover for each opening, providing a variable open area which is varied by the vapor flow inhibiting leakage of liquid down the opening at low vapor rates. Hence, this type of tray can operate at a greater range of flow rates than the sieve tray, with a cost of only about 20% higher than a sieve tray. The valve tray is being increasingly used today.

3. *Bubble-cap tray.* Bubble-cap trays, shown in Fig. 10.6-1b, have been used for over 100 years, but since 1950 they have been generally superseded by sieve-type or valve trays because of their cost, which is almost double that of sieve-type trays. In the bubble tray, the vapor or gas rises through the opening in the tray into the bubble caps. Then the gas flows through slots in the periphery of each cap and bubbles upward through the flowing liquid. Details and design procedures for many of these and other types of trays are given elsewhere (B2, P1, T1). The different types of tray efficiencies are discussed in Section 11.5.

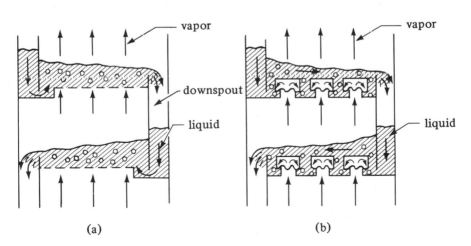

FIGURE 10.6-1. *Tray contacting devices: (a) detail of sieve-tray tower, (b) detail of bubble-cap tower tray.*

3. Packed towers for absorption and distillation. Packed towers are used for continuous countercurrent contacting of gas and liquid in absorption and also for vapor–liquid contacting in distillation. The tower in Fig. 10.6-2 consists of a cylindrical column containing a gas inlet and distributing space at the bottom, a liquid inlet and distributing device at the top, a gas outlet at the top, a liquid outlet at the bottom, and a packing or filling in the tower. The entering gas enters the distributing space below the packed section and rises upward through the openings or interstices in the packing and contacts the descending liquid flowing through the same openings. A large area of intimate contact between the liquid and gas is provided by the packing.

Many different types of tower packing have been developed and a number are used quite commonly. Common types of packing which are dumped at random in the tower are shown in Fig. 10.6-3. Such packings and other commercial packings are available in sizes of 3 mm to about 75 mm. Most of the tower packings are made of inert and cheap materials such as clay, porcelain, graphite, or plastic. High void spaces of 60 to 90% are characteristic of good packings. The packings permit relatively large volumes of liquid to pass countercurrently to the gas flow through the openings with relatively low pressure drops for the gas. These same types of packing are also used in vapor–liquid separation processes of distillation.

Stacked packing having sizes of 75 mm or so and larger is also used. The packing is stacked vertically, with open channels running uninterruptedly through the bed. The advantage of the lower pressure drop of the gas is offset in part by the poorer gas–liquid contact in stacked packings. Typical stacked packings are wood grids, drip-point grids, spiral partition rings, and others.

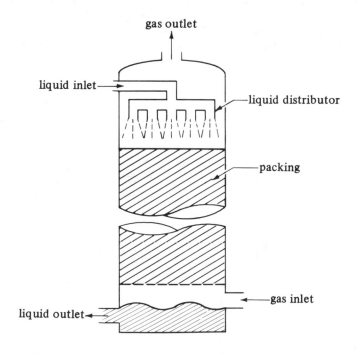

FIGURE 10.6-2. *Packed tower flows and characteristics for absorption.*

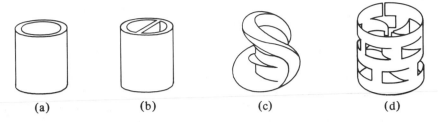

FIGURE 10.6-3. *Typical tower packings: (a) Raschig ring, (b) Lessing ring, (c) Berl saddle, (d) Pall ring.*

In a given packed tower with a given type and size of packing and with a definite flow of liquid, there is an upper limit to the rate of gas flow, called the *flooding velocity*. Above this gas velocity the tower cannot operate. At low gas velocities the liquid flows downward through the packing essentially uninfluenced by the upward gas flow. As the gas flow rate is increased at low gas velocities, the pressure drop is proportional to the flow rate to the 1.8 power. At a gas flow rate called the *loading point*, the gas starts to hinder the liquid downflow and local accumulations or pools of liquid start to appear in the packing. The pressure drop of the gas starts to rise at a faster rate. As the gas flow rate is increased, the liquid holdup or accumulation increases. At the flooding point, the liquid can no longer flow down through the packing and is blown out with the gas.

In an actual operating tower the gas velocity is well below flooding. The optimum economic gas velocity is about one half or so of the flooding velocity. It depends upon an economic balance between the cost of power and the fixed charges on the equipment cost (S1). Detailed design methods for predicting the pressure drop in various types of packing are given elsewhere (P1, L1, T1).

10.6B Design of Plate Absorption Towers

1. Operating-line derivation. A plate (tray) absorption tower has the same process flow diagram as the countercurrent multiple-stage process in Fig. 10.3-2 and is shown as a vertical tray tower in Fig. 10.6-4. In the case of solute A diffusing through a stagnant gas (B) and then into a stagnant fluid, as in the absorption of acetone (A) from air (B) by water, the moles of inert or stagnant air and inert water remain constant throughout the entire tower. If the rates are V' kg mol inert air/s and L' kg mol inert solvent water/s or in kg mol inert/s $\cdot$ m^2 units (lb mol inert/h $\cdot$ ft^2), an overall material balance on component A in Fig. 10.6-4 is

$$L'\left(\frac{x_0}{1 - x_0}\right) + V'\left(\frac{y_{N+1}}{1 - y_{N+1}}\right) = L'\left(\frac{x_N}{1 - x_N}\right) + V'\left(\frac{y_1}{1 - y_1}\right) \tag{10.6-1}$$

A balance around the dashed-line box gives

$$L'\left(\frac{x_0}{1 - x_0}\right) + V'\left(\frac{y_{n+1}}{1 - y_{n+1}}\right) = L'\left(\frac{x_n}{1 - x_n}\right) + V'\left(\frac{y_1}{1 - y_1}\right) \tag{10.6-2}$$

where x is the mole fraction A in the liquid, y the mole fraction of A in the gas, L_n the total moles liquid/s, and V_{n+1} the total moles gas/s. The total flows/s of liquid and of gas vary throughout the tower.

Equation (10.6-2) is the material balance or operating line for the absorption tower and is similar to Eq. (10.3-13) for a countercurrent-stage process, except that the inert streams L' and V' are used instead of the total flow rates L and V. Equation (10.6-2) relates the concentration y_{n+1} in the gas stream with x_n in the liquid stream passing it. The terms V', L', x_0, and y_1 are constant and usually known or can be determined.

2. *Graphical determination of the number of trays.* A plot of the operating-line equation (10.6-2) as y versus x will give a curved line. If x and y are very dilute, the denominators $1 - x$ and $1 - y$ will be close to 1.0, and the line will be approximately straight, with a slope $\cong L/V'$. The number of theoretical trays are determined by simply stepping off the number of trays, as done in Fig. 10.3-3 for a countercurrent multiple-stage process.

EXAMPLE 10.6-1.   *Absorption of SO_2 in a Tray Tower*

A tray tower is to be designed to absorb SO_2 from an air stream by using pure water at 293 K (68°F). The entering gas contains 20 mol % SO_2 and that leaving 2 mol % at a total pressure of 101.3 kPa. The inert air flow rate is 150 kg air/h·m², and the entering water flow rate is 6000 kg water/h·m². Assuming an overall tray efficiency of 25%, how many theoretical trays and actual trays are needed? Assume that the tower operates at 293 K (20°C).

Solution: First calculating the molar flow rates,

$$V' = \frac{150}{29} = 5.18 \text{ kg mol inert air/h} \cdot m^2$$

$$L' = \frac{6000}{18.0} = 333 \text{ kg mol inert water/h} \cdot m^2$$

Referring to Fig. 10.6-4, $y_{N+1} = 0.20$, $y_1 = 0.02$, and $x_0 = 0$. Substituting into Eq. (10.6-1) and solving for x_N,

$$333\left(\frac{0}{1-0}\right) + 5.18\left(\frac{0.20}{1-0.20}\right) = 333\left(\frac{x_N}{1-x_N}\right) + 5.18\left(\frac{0.02}{1-0.02}\right)$$

$$x_N = 0.00355$$

Substituting into Eq. (10.6-2), using V' and L' as kg mol/h·m² instead of

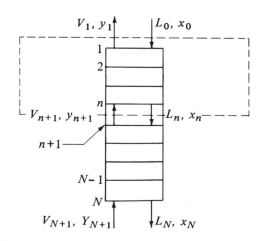

FIGURE 10.6-4.   *Material balance in an absorption tray tower.*

kg mol/s · m²,

$$333\left(\frac{0}{1-0}\right) + 5.18\left(\frac{y_{n+1}}{1-y_{n+1}}\right) = 333\left(\frac{x_n}{1-x_n}\right) + 5.18\left(\frac{0.02}{1-0.02}\right)$$

In order to plot the operating line, several intermediate points will be calculated. Setting $y_{n+1} = 0.07$ and substituting into the operating equation,

$$0 + 5.18\left(\frac{0.07}{1-0.07}\right) = 333\left(\frac{x_n}{1-x_n}\right) + 5.18\left(\frac{0.02}{1-0.02}\right)$$

Hence, $x_n = 0.000855$. To calculate another intermediate point, we set $y_{n+1} = 0.13$, and x_n is calculated as 0.00201. The two end points and the two intermediate points on the operating line are plotted in Fig. 10.6-5, as are the equilibrium data from Appendix A.3. The operating line is somewhat curved.

The number of theoretical trays is determined by stepping off the steps to give 2.4 theoretical trays. The actual number of trays is $2.4/0.25 = 9.6$ trays.

10.6C Design of Packed Towers for Absorption

1. Operating-line derivation. For the case of solute A diffusing through a stagnant gas and then into a stagnant fluid, an overall material balance on component A in Fig. 10.6-6 for a packed absorption tower is

$$L'\left(\frac{x_2}{1-x_2}\right) + V'\left(\frac{y_1}{1-y_1}\right) = L'\left(\frac{x_1}{1-x_1}\right) + V'\left(\frac{y_2}{1-y_2}\right) \qquad (10.6\text{-}3)$$

where L' is kg mol inert liquid/s or kg mol inert liquid/s · m², V' is kg mol inert gas/s or kg mol inert gas/s · m², and y_1 and x_1 are mole fractions A in gas and liquid, respectively.

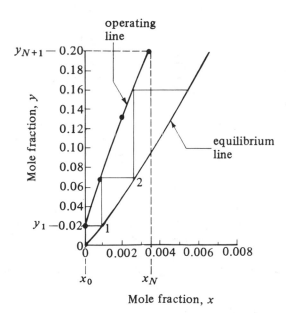

FIGURE 10.6-5. *Theoretical number of trays for absorption of SO_2 in Example 10.6-1.*

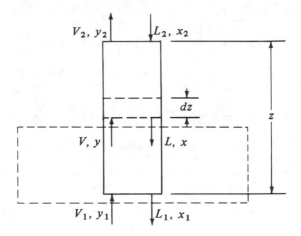

FIGURE 10.6-6.   *Material balance for a countercurrent packed absorption tower.*

The flows L' and V' are constant throughout the tower, but the total flows L and V are not constant.

A balance around the dashed-line box in Fig. 10.6-6 gives the operating-line equation.

$$L'\left(\frac{x}{1-x}\right) + V'\left(\frac{y_1}{1-y_1}\right) = L'\left(\frac{x_1}{1-x_1}\right) + V'\left(\frac{y}{1-y}\right) \qquad (10.6\text{-}4)$$

This equation, when plotted on yx coordinates, will give a curved line, as shown in Fig. 10.6-7a. Equation (10.6-4) can also be written in terms of partial pressure p_1 of A, where $y_1/(1 - y_1) = p_1/(P - p_1)$, and so on. If x and y are very dilute, $(1 - x)$ and $(1 - y)$ can be taken as 1.0 and Eq. (10.6-4) becomes

$$L'x + V'y_1 \cong L'x_1 + V'y \qquad (10.6\text{-}5)$$

This has a slope L'/V' and the operating line is essentially straight.

When the solute is being transferred from the L to the V stream, the process is called *stripping*. The operating line is below the equilibrium line, as shown in Fig. 10.6-7b.

2. Limiting and optimum L'/V' ratios. In the absorption process, the inlet gas flow V_1 (Fig. 10.6-6) and its composition y_1 are generally set. The exit concentration y_2 is also usually set by the designer and the concentration x_2 of the entering liquid is often fixed by process requirements. Hence, the amount of the entering liquid flow L_2 or L' is open to choice.

In Fig. 10.6-8 the flow V_1 and the concentrations y_2, x_2, and y_1 are set. When the operating line has a minimum slope and touches the equilibrium line at point P, the liquid flow L' is a minimum at L'_{min}. The value of x_1 is a maximum at $x_{1\,max}$ when L' is a minimum. At point P the driving forces $y - y^*$, $y - y_i$, $x^* - x$, or $x_i - x$ are all zero. To solve for L'_{min}, the values y_1 and $x_{1\,max}$ are substituted into the operating-line equation. In some cases if the equilibrium line is curved concavely downward, the minimum value of L is reached by the operating line becoming tangent to the equilibrium line instead of intersecting it.

The choice of the optimum L'/V' ratio to use in the design depends on an economic balance. In absorption, too high a value requires a large liquid flow, and hence a

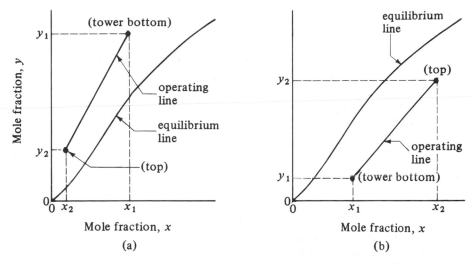

FIGURE 10.6-7. *Location of operating lines : (a) for absorption of A from V to L stream, (b) for stripping of A from L to V stream.*

large-diameter tower. The cost of recovering the solute from the liquid by distillation will be high. A small liquid flow results in a high tower, which is costly. As an approximation, the optimum liquid flow is obtained by using a value of about 1.5 for the ratio of the average slope of the operating line to that of the equilibrium line for absorption. This factor can vary depending on the value of the solute and tower type.

3. Film and overall mass-transfer coefficients in packed towers. As discussed in Section 10.5, it is very difficult to measure experimentally the interfacial area A m^2 between phases L and V. Also, it is difficult to measure the film coefficients k'_x and k'_y and the overall coefficients K'_x and K'_y. Usually, experimental measurements in a packed tower yield a volumetric mass-transfer coefficient that combines the interfacial area and mass-transfer coefficient.

Defining a as interfacial area in m^2 per m^3 volume of packed section, the volume of

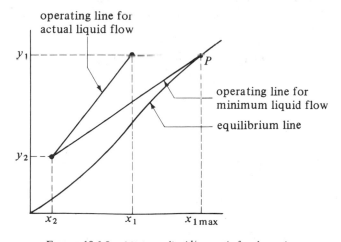

FIGURE 10.6-8. *Minimum liquid/gas ratio for absorption.*

packing in a height dz m (Fig. 10.6-6) is $S\,dz$ and

$$dA = aS\,dz \tag{10.6-6}$$

where S is m^2 cross-sectional area of tower. The volumetric film and overall mass-transfer coefficients are then defined as

$$k'_y a = \frac{\text{kg mol}}{\text{s}\cdot\text{m}^3 \text{ packing}\cdot\text{mol frac}} \qquad k'_x a = \frac{\text{kg mol}}{\text{s}\cdot\text{m}^3 \text{ packing}\cdot\text{mol frac}} \quad \text{(SI)}$$

$$K'_y a = \frac{\text{kg mol}}{\text{s}\cdot\text{m}^3 \text{ packing}\cdot\text{mol frac}} \qquad K'_x a = \frac{\text{kg mol}}{\text{s}\cdot\text{m}^3 \text{ packing}\cdot\text{mol frac}} \quad \text{(SI)}$$

$$k'_y a = \frac{\text{lb mol}}{\text{h}\cdot\text{ft}^3 \text{ packing}\cdot\text{mol frac}} \qquad k'_x a = \frac{\text{lb mol}}{\text{h}\cdot\text{ft}^3 \text{ packing}\cdot\text{mol frac}} \quad \text{(English)}$$

4. Design method for packed towers. For absorption of A from stagnant B, the operating-line equation (10.6-4) holds. For the differential height of tower dz in Fig. (10.6-6), the moles of A leaving V equal the moles entering L.

$$d(Vy) = d(Lx) \tag{10.6-7}$$

where V = kg mol total gas/s, L = kg mol total liquid/s, and $d(Vy) = d(Lx)$ = kg mol A transferred/s in height dz m. The kg mol A transferred/s from Eq. (10.6-7) must equal the kg mol A transferred/s from the mass-transfer equation for N_A. Equation (10.4-8) gives the flux N_A using the gas-film and liquid-film coefficients.

$$N_A = \frac{k'_y}{(1 - y_A)_{iM}}(y_{AG} - y_{Ai}) = \frac{k'_x}{(1 - x_A)_{iM}}(x_{Ai} - x_{AL}) \tag{10.4-8}$$

where $(1 - y_A)_{iM}$ and $(1 - x_A)_{iM}$ are defined by Eqs. (10.4-6) and (10.4-7). Multiplying the left-hand side of Eq. (10.4-8) by dA and the two right-side terms by $aS\,dz$ from Eq. (10.6-6),

$$N_A\,dA = \frac{k'_y a}{(1 - y_A)_{iM}}(y_{AG} - y_{Ai})S\,dz = \frac{k'_x a}{(1 - x_A)_{iM}}(x_{Ai} - x_{AL})S\,dz \tag{10.6-8}$$

where $N_A\,dA$ = kg mol A transferred/s in height dz m (lb mol/h).

Equating Eq. (10.6-7) to (10.6-8) and using y_{AG} for the bulk gas phase and x_{AL} for the bulk liquid phase,

$$d(Vy_{AG}) = \frac{k'_y a}{(1 - y_A)_{iM}}(y_{AG} - y_{Ai})S\,dz \tag{10.6-9}$$

$$d(Lx_{AL}) = \frac{k'_x a}{(1 - x_A)_{iM}}(x_{Ai} - x_{AL})S\,dz \tag{10.6-10}$$

Since $V' = V(1 - y_{AG})$ or $V = V'/(1 - y_{AG})$,

$$d(Vy_{AG}) = d\left(\frac{V'}{(1 - y_{AG})}y_{AG}\right) = V'd\left(\frac{y_{AG}}{1 - y_{AG}}\right) = \frac{V'\,dy_{AG}}{(1 - y_{AG})^2} \tag{10.6-11}$$

Substituting V for $V'/(1 - y_{AG})$ in Eq. (10.6-11) and then equating Eq. (10.6-11) to (10.6-9),

$$\frac{V\,dy_{AG}}{1 - y_{AG}} = \frac{k'_y a}{(1 - y_A)_{iM}}(y_{AG} - y_{Ai})S\,dz \tag{10.6-12}$$

Repeating for Eq. (10.6-10) since $L = L'/(1 - x_{AL})$,

$$\frac{L \, dx_{AL}}{1 - x_{AL}} = \frac{k'_x a}{(1 - x_A)_{iM}} (x_{Ai} - x_{AL}) S \, dz \tag{10.6-13}$$

Dropping the subscripts A, G, and L and integrating, the final equations are as follows using film coefficients:

$$\int_0^z dz = z = \int_{y_2}^{y_1} \frac{V \, dy}{\dfrac{k'_y aS}{(1 - y)_{iM}} (1 - y)(y - y_i)} \tag{10.6-14}$$

$$\int_0^z dz = z = \int_{x_2}^{x_1} \frac{L \, dx}{\dfrac{k'_x aS}{(1 - x)_{iM}} (1 - x)(x_i - x)} \tag{10.6-15}$$

In a similar manner, the final equations can be derived using overall coefficients.

$$z = \int_{y_2}^{y_1} \frac{V \, dy}{\dfrac{K'_y aS}{(1 - y)_{*M}} (1 - y)(y - y^*)} \tag{10.6-16}$$

$$z = \int_{x_2}^{x_1} \frac{L \, dx}{\dfrac{K'_x aS}{(1 - x)_{*M}} (1 - x)(x^* - x)} \tag{10.6-17}$$

In the general case, the equilibrium and the operating lines are usually curved and $k'_x a$, $k'_y a$, $K'_y a$, and $K'_x a$ vary somewhat with total gas and liquid flows. Then Eqs. (10.6-14)–(10.6-17) must be integrated graphically. The methods to do this for concentrated mixtures will be discussed in Section 10.7. Methods for dilute gases will be considered below.

10.6D Simplified Design Methods for Absorption of Dilute Gas Mixtures in Packed Towers

Since a considerable percentage of the absorption processes include absorption of a dilute gas A, these cases will be considered using a simplified design procedure.

The concentrations can be considered dilute for engineering design purposes when the mole fractions y and x in the gas and liquid streams are less than about 0.10, i.e., 10%. The flows will vary by less than 10% and the mass-transfer coefficients by considerably less than this. As a result, the average values of the flows V and L and the mass-transfer coefficients at the top and bottom of the tower can be taken outside the integral. Likewise, the terms $(1 - y)_{iM}/(1 - y)$, $(1 - y)_{*M}/(1 - y)$, $(1 - x)_{iM}/(1 - x)$, and $(1 - x)_{*M}/(1 - x)$ can be taken outside and average values of the values at the top and bottom of the tower used. (Often these terms are close to 1.0 and can be dropped out entirely.) Then Eqs. (10.6-14)–(10.6-17) become

$$z = \left[\frac{V}{k'_y aS} \frac{(1 - y)_{iM}}{1 - y} \right]_{av} \int_{y_2}^{y_1} \frac{dy}{y - y_i} \tag{10.6-18}$$

$$z = \left[\frac{L}{k'_x aS} \frac{(1 - x)_{iM}}{1 - x} \right]_{av} \int_{x_2}^{x_1} \frac{dx}{x_i - x} \tag{10.6-19}$$

$$z = \left[\frac{V}{K_y' aS}\frac{(1-y)_{*M}}{1-y}\right]_{av} \int_{y_2}^{y_1} \frac{dy}{y-y^*} \tag{10.6-20}$$

$$z = \left[\frac{L}{K_x' aS}\frac{(1-x)_{*M}}{1-x}\right]_{av} \int_{x_2}^{x_1} \frac{dx}{x^*-x} \tag{10.6-21}$$

Since the solutions are dilute, the operating line will be essentially straight. Assuming the equilibrium line is approximately straight over the range of concentrations used, $(y - y_i)$ varies linearly with y and also with x.

$$y - y_i = ky + b \tag{10.6-22}$$

where k and b are constants. Therefore, the integral of Eq. (10.6-18) can be integrated to give the following.

$$\int_{y_2}^{y_1} \frac{dy}{y-y_i} = \frac{y_1-y_2}{(y-y_i)_M} \tag{10.6-23}$$

where $(y - y_i)_M$ is the log mean driving force.

$$(y-y_i)_M = \frac{(y_1-y_{i1})-(y_2-y_{i2})}{\ln\left[(y_1-y_{i1})/(y_2-y_{i2})\right]} \tag{10.6-24}$$

$$(y-y^*)_M = \frac{(y_1-y_1^*)-(y_2-y_2^*)}{\ln\left[(y_1-y_1^*)/(y_2-y_2^*)\right]} \tag{10.6-25}$$

If the term $(1 - y)_{iM}/(1 - y)$ is considered 1.0, then substituting Eq. (10.6-23) into (10.6-18) and doing the same for Eqs. (10.6-19)–(10.6-21), the final results are as follows:

$$\frac{V}{S}(y_1-y_2) = k_y' az(y-y_i)_M \tag{10.6-26}$$

$$\frac{L}{S}(x_1-x_2) = k_x' az(x_i-x)_M \tag{10.6-27}$$

$$\frac{V}{S}(y_1-y_2) = K_y' az(y-y^*)_M \tag{10.6-28}$$

$$\frac{L}{S}(x_1-x_2) = K_x' az(x^*-x)_M \tag{10.6-29}$$

where the left side is the kg mol absorbed/s $\cdot$ m^2 (lb mol/h $\cdot$ ft^2) by material balance and the right-hand side is the rate equation for mass transfer. The value of V is the average $(V_1 + V_2)/2$ and of L is $(L_1 + L_2)/2$.

Equations (10.6-26) to (10.6-29) can be used in slightly different ways. The general steps to follow are discussed below and shown in Fig. 10.6-9.

1. The operating-line equation (10.6-4) is plotted as in Fig. 10.6-9 as a straight line. Calculate V_1, V_2, and $V_{av} = (V_1 + V_2)/2$; also calculate L_1, L_2, and $L_{av} = (L_1 + L_2)/2$.
2. Average experimental values of the film coefficients $k_y' a$ and $k_x' a$ are available or are obtained from empirical correlations. The interface compositions y_{i1} and x_{i1} at point y_1, x_1 in the tower are determined by plotting line $P_1 M_1$ whose slope is calculated by Eq. (10.6-30):

$$\text{slope} = -\frac{k_x' a/(1-x)_{iM}}{k_y' a/(1-y)_{iM}} = -\frac{k_x a}{k_y a} \tag{10.6-30}$$

$$\text{slope} \cong -\frac{k'_x a/(1-x_1)}{k'_y a/(1-y_1)} \qquad \text{(10.6-31)}$$

If terms $(1-x)_{iM}$ and $(1-y)_{iM}$ are used, the procedure is trial and error, as in Example 10.4-1. However, since the solutions are dilute, the terms $(1-x_1)$ and $(1-y_1)$ can be used in Eq. (10.6-31) without trial and error and with a small error in the slope. If the coefficients $k_y a$ and $k_x a$ are available for the approximate concentration range, they can be used, since they include the terms $(1-x)_{iM}$ and $(1-y)_{iM}$. For line $P_2 M_2$ at the other end of the tower, values of y_{i2} and x_{i2} are determined using Eq. (10.6-30) or (10.6-31) and y_2 and x_2.

3. If the overall coefficient $K'_y a$ is being used, y^*_1 and y^*_2 are determined as shown in Fig. 10.6-9. If $K'_x a$ is used, x^*_1 and x^*_2 are obtained.

4. Calculate the log mean driving force $(y-y_i)_M$ by Eq. (10.6-24) if $k'_y a$ is used. For $K'_y a$, $(y-y^*)_M$ is calculated by Eq. (10.6-25). Using the liquid coefficients, the appropriate driving forces are calculated.

5. Calculate the column height z m by substituting into the appropriate form of Eqs. (10.6-26)–(10.6-29).

EXAMPLE 10.6-2. Absorption of Acetone in a Packed Tower

Acetone is being absorbed by water in a packed tower having a cross-sectional area of 0.186 m^2 at 293 K and 101.32 kPa (1 atm). The inlet air contains 2.6 mol % acetone and outlet 0.5%. The gas flow is 13.65 kg mol inert air/h (30.1 lb mol/h). The pure water inlet flow is 45.36 kg mol water/h (100 lb mol/h). Film coefficients for the given flows in the tower are $k'_y a = 3.78 \times 10^{-2}$ kg mol/s · m^3 · mol frac (8.50 lb mol/h · ft^3 · mol frac) and $k'_x a = 6.16 \times 10^{-2}$ kg mol/s · m^3 · mol frac (13.85 lb mol/h · ft^3 · mol frac). Equilibrium data are given in Appendix A.3.

    (a) Calculate the tower height using $k'_y a$.
    (b) Repeat using $k'_x a$.
    (c) Calculate $K'_y a$ and the tower height.

Solution: From Appendix A.3 for acetone–water and $x_A = 0.0333$ mol frac, $p_A = 30/760 = 0.0395$ atm or $y_A = 0.0395$ mol frac. Hence, the equilibrium line is $y_A = mx_A$ or $0.0395 = m (0.0333)$. Then, $y = 1.186x$. This equi-

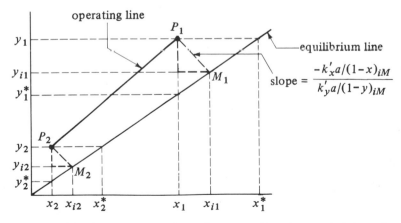

FIGURE 10.6-9. Operating-line and interface compositions in a packed tower for absorption of dilute gases.

librium line is plotted in Fig. 10.6-10. The given data are $L' = 45.36$ kg mol/h, $V' = 13.65$ kg mol/h, $y_1 = 0.026$, $y_2 = 0.005$, and $x_2 = 0$.

Substituting into Eq. (10.6-3) for an overall material balance using flow rates as kg mol/h instead of kg mol/s,

$$45.36\left(\frac{0}{1-0}\right) + 13.65\left(\frac{0.026}{1-0.026}\right) = 45.36\left(\frac{x_1}{1-x_1}\right) + 13.65\left(\frac{0.005}{1-0.005}\right)$$

$$x_1 = 0.00648$$

The points y_1, x_1 and y_2, x_2 are plotted in Fig. 10.6-10 and a straight line is drawn for the operating line.

Using Eq. (10.6-31) the approximate slope at y_1, x_1 is

$$\text{slope} \cong -\frac{k'_x a/(1-x_1)}{k'_y a/(1-y_1)} = -\frac{6.16 \times 10^{-2}/(1-0.00648)}{3.78 \times 10^{-2}/(1-0.026)} = -1.60$$

Plotting this line through y_1, x_1, the line intersects the equilibrium line at $y_{i1} = 0.0154$ and $x_{i1} = 0.0130$. Also, $y_1^* = 0.0077$. Using Eq. (10.6-30) to calculate a more accurate slope, the preliminary values of y_{i1} and x_{i1} will be used in the trial-and-error solution. Substituting into Eq. (10.4-6),

$$(1-y)_{iM} = \frac{(1-y_{i1})-(1-y_1)}{\ln{[(1-y_{i1})/(1-y_1)]}}$$

$$= \frac{(1-0.0154)-(1-0.026)}{\ln{[(1-0.0154)/(1-0.020)]}} = 0.979$$

Using Eq. (10.4-7),

$$(1-x)_{iM} = \frac{(1-x_1)-(1-x_{i1})}{\ln{[(1-x_1)/(1-x_{i1})]}}$$

$$= \frac{(1-0.00648)-(1-0.0130)}{\ln{[(1-0.00648)/(1-0.0130)]}} = 0.993$$

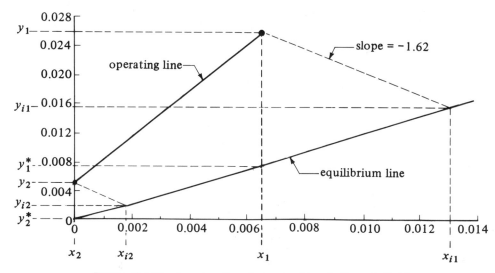

FIGURE 10.6-10. *Location of interface compositions for Example 10.6-2.*

Substituting into Eq. (10.6-30),

$$\text{slope} = -\frac{k_x'a/(1-x)_{iM}}{k_y'a/(1-y)_{iM}} = -\frac{6.16 \times 10^{-2}/0.993}{3.78 \times 10^{-2}/0.929} = -1.61$$

Hence, the approximate slope and interface values are accurate enough.
For the slope at point y_2, x_2,

$$\text{slope} \cong -\frac{k_x'a/(1-x_2)}{k_y'a/(1-y_2)} = -\frac{6.16 \times 10^{-2}/(1-0)}{3.78 \times 10^{-2}/(1-0.005)} = -1.62$$

The slope changes little in the tower. Plotting this line, $y_{i2} = 0.0020$, $x_{i2} = 0.0018$, and $y_2^* = 0$.
Substituting into Eq. (10.6-24),

$$(y - y_i)_M = \frac{(y_1 - y_{i1}) - (y_2 - y_{i2})}{\ln \left[(y_1 - y_{i1})/(y_2 - y_{i2}) \right]}$$

$$= \frac{(0.026 - 0.0154) - (0.005 - 0.0020)}{\ln \left[(0.026 - 0.0154)/(0.005 - 0.0020) \right]} = 0.00602$$

To calculate the total molar flow rates in kg mol/s,

$$V_1 = \frac{V'}{1-y_1} = \frac{13.65/3600}{1-0.026} = 3.893 \times 10^{-3} \text{ kg mol/s}$$

$$V_2 = \frac{V'}{1-y_2} = \frac{13.65/3600}{1-0.005} = 3.811 \times 10^{-3} \text{ kg mol/s}$$

$$V_{av} = \frac{V_1 + V_2}{2} = \frac{3.893 \times 10^{-3} + 3.811 \times 10^{-3}}{2} = 3.852 \times 10^{-3} \text{ kg mol/s}$$

$$L' \cong L_1 \cong L_2 \cong L_{av} = \frac{45.36}{3600} = 1.260 \times 10^{-2} \text{ kg mol/s}$$

For part (a), substituting into Eq. (10.6-26) and solving,

$$\frac{V_{av}}{S}(y_1 - y_2) = k_y'az(y - y_i)_M$$

$$\frac{3.852 \times 10^{-3}}{0.186}(0.0260 - 0.005) = (3.78 \times 10^{-2})z(0.00602)$$

$$z = 1.911 \text{ m } (6.27 \text{ ft})$$

For part (b), using an equation similar to Eq. (10.6-24),

$$(x_i - x)_M = \frac{(x_{i1} - x_1) - (x_{i2} - x_2)}{\ln \left[(x_{i1} - x_1)/(x_{i2} - x_2) \right]}$$

$$= \frac{(0.0130 - 0.00648) - (0.0018 - 0)}{\ln \left[(0.0130 - 0.00648)/(0.0018 - 0) \right]} = 0.00368$$

Substituting into Eq. (10.6-27) and solving,

$$\frac{1.260 \times 10^{-2}}{0.186}(0.00648 - 0) = (6.16 \times 10^{-2})z(0.00368)$$

$$z = 1.936 \text{ m}$$

This checks part (a) quite closely.

For part (c), substituting into Eq. (10.4-25) for point y_1, x_1,

$$(1 - y)_{*M} = \frac{(1 - y_1^*) - (1 - y_1)}{\ln\left[(1 - y_1^*)/(1 - y_1)\right]} = \frac{(1 - 0.0077) - (1 - 0.026)}{\ln\left[(1 - 0.0077)/(1 - 0.026)\right]} = 0.983$$

The overall mass-transfer coefficient $K_y' a$ at point y_1, x_1 is calculated by substituting into Eq. (10.4-24).

$$\frac{1}{K_y' a/(1 - y)_{*M}} = \frac{1}{k_y' a/(1 - y)_{iM}} + \frac{m'}{k_x' a/(1 - x)_{iM}}$$

$$\frac{1}{K_y' a/0.983} = \frac{1}{3.78 \times 10^{-2}/0.979} + \frac{1.186}{6.16 \times 10^{-2}/0.993}$$

$$K_y' a = 2.183 \times 10^{-2} \text{ kg mol/s} \cdot \text{m}^3 \cdot \text{mol frac}$$

Substituting into Eq. (10.6-25),

$$(y - y^*)_M = \frac{(y_1 - y_1^*) - (y_2 - y_2^*)}{\ln\left[(y_1 - y_1^*)/(y_2 - y_2^*)\right]}$$

$$= \frac{(0.0260 - 0.0077) - (0.0050 - 0)}{\ln\left[(0.0260 - 0.0077)/(0.0050 - 0)\right]} = 0.01025$$

Finally substituting into Eq. (10.6-28),

$$\frac{3.852 \times 10^{-3}}{0.186}(0.0260 - 0.0050) = (2.183 \times 10^{-2})z(0.01025)$$

$$z = 1.944 \text{ m}$$

This checks parts (a) and (b).

10.6E Design of Packed Towers Using Transfer Units

Another and in some ways a more useful design method of packed towers is the use of the transfer unit concept. For the most common case of A diffusing through stagnant and nondiffusing B, Eqs. (10.6-14)–(10.6-17) can be rewritten as

$$z = H_G \int_{y_2}^{y_1} \frac{(1 - y)_{iM} \, dy}{(1 - y)(y - y_i)} \tag{10.6-32}$$

$$z = H_L \int_{x_2}^{x_1} \frac{(1 - x)_{iM} \, dx}{(1 - x)(x_i - x)} \tag{10.6-33}$$

$$z = H_{OG} \int_{y_2}^{y_1} \frac{(1 - y)_{*M} \, dy}{(1 - y)(y - y^*)} \tag{10.6-34}$$

$$z = H_{OL} \int_{x_2}^{x_1} \frac{(1 - x)_{*M} \, dx}{(1 - x)(x^* - x)} \tag{10.6-35}$$

where

$$H_G = \frac{V}{k_y' aS} = \frac{V}{k_y a(1 - y)_{iM} S} \tag{10.6-36}$$

$$H_L = \frac{L}{k_x' aS} = \frac{L}{k_x a(1 - x)_{iM} S} \tag{10.6-37}$$

$$H_{OG} = \frac{V}{K'_y\, aS} = \frac{V}{K_y a(1-y)_{*M}\, S} \qquad (10.6\text{-}38)$$

$$H_{OL} = \frac{L}{K'_x\, aS} = \frac{L}{K_x a(1-x)_{*M}\, S} \qquad (10.6\text{-}39)$$

The units of H are in m (ft). The H_G is the height of a transfer unit based on the gas film. The values of the heights of transfer units are more constant than the mass-transfer coefficients. For example, $k'_y a$ is often proportional to $V^{0.7}$, then $H_G \propto V^{1.0}/\bar{V}^{0.7} \propto V^{0.3}$. The average values of the mass-transfer coefficients, $(1-y)_{iM}$, $(1-y)_{*M}$, $(1-x)_{iM}$, and $(1-x)_{*M}$, must be used in Eqs. (10.6-36)–(10.6-39).

The integrals on the right side of Eqs. (10.6-32)–(10.6-35) are the number of transfer units N_G, N_L, N_{OG}, and N_{OL}, respectively. The height of the packed tower is then

$$z = H_G N_G = H_L N_L = H_{OG} N_{OG} = H_{OL} N_{OL} \qquad (10.6\text{-}40)$$

These equations are basically no different than those using mass-transfer coefficients. One still needs $k'_y a$ and $k'_x a$ to determine interface concentrations. Disregarding $(1-y)_{iM}/(1-y)$, which is near 1.0 in Eq. (10.6-32), the greater the amount of absorption $(y_1 - y_2)$ or the smaller the driving force $(y - y_i)$, the larger the number of transfer units N_G and the taller the tower.

When the solutions are dilute with concentrations below 10%, the terms $(1-y)_{iM}/(1-y)$, $(1-x)_{iM}/(1-x)$, $(1-y)_{*M}/(1-y)$, and $(1-x)_{*M}/(1-x)$ can be taken outside the integral and average values used. Often they are quite close to 1 and can be dropped out. The equations become

$$z = H_G N_G = H_G \left[\frac{(1-y)_{iM}}{1-y}\right]_{av} \int_{y_2}^{y_1} \frac{dy}{y-y_i} \qquad (10.6\text{-}41)$$

$$z = H_L N_L = H_L \left[\frac{(1-x)_{iM}}{1-x}\right]_{av} \int_{x_2}^{x_1} \frac{dx}{x_i-x} \qquad (10.6\text{-}42)$$

$$z = H_{OG} N_{OG} = H_{OG} \left[\frac{(1-y)_{*M}}{1-y}\right]_{av} \int_{y_2}^{y_1} \frac{dy}{y-y^*} \qquad (10.6\text{-}43)$$

$$z = H_{OL} N_{OL} = H_{OL} \left[\frac{(1-x)_{*M}}{1-x}\right]_{av} \int_{x_2}^{x_1} \frac{dx}{x^*-x} \qquad (10.6\text{-}44)$$

If the operating and equilibrium lines are both straight and the solutions dilute, the integral shown in Eq. (10.6-23) is valid.

$$\int_{y_2}^{y_1} \frac{dy}{y-y_i} = \frac{y_1 - y_2}{(y-y_i)_M} \qquad (10.6\text{-}23)$$

This then can be substituted into Eq. (10.6-41) and similar expressions into Eqs. (10.6-42)–(10.6-44).

EXAMPLE 10.6-3. Use of Transfer Units for Packed Tower

Repeat Example 10.6-2 using transfer units and height of a transfer unit as follows.

(a) Use H_G and N_G to calculate tower height.

(b) Use H_{OG} and N_{OG} to calculate tower height.

Solution: For part (a), $k'_y a = 3.78 \times 10^{-2}$ kg mol/s·m³·mol frac from

Example 10.6-2. From Eq. (10.6-36),

$$H_G = \frac{V}{k_y' aS} \qquad (10.6\text{-}36)$$

The average V is 3.852×10^{-3} kg mol/s and $S = 0.186 \text{ m}^2$. Substituting and solving,

$$H_G = \frac{3.852 \times 10^{-3}}{(3.78 \times 10^{-2})(0.186)} = 0.548 \text{ m}$$

Since the solution is dilute, the number of transfer units from Eq. (10.6-41) is

$$N_G = \left[\frac{(1 - y)_{iM}}{1 - y}\right]_{av} \int_{y_1}^{y_2} \frac{dy}{y - y_i} \qquad (10.6\text{-}45)$$

The term in the brackets will be evaluated at point 1 and point 2. At point $y_1 = 0.026$, $y_{i1} = 0.0154$ from Example 10.6-2. Also, from Eq. (10.4-6) in Example 10.6-2, $(1 - y)_{iM} = 0.979$. Also, $1 - y = 1 - 0.026 = 0.974$. Then at point 1,

$$\frac{(1 - y)_{iM}}{1 - y} = \frac{0.979}{0.974} = 1.005$$

At point $y_2 = 0.005$, $y_{i2} = 0.002$. Substituting into Eq. (10.4-6), $(1 - y)_{iM} = 0.997$ and $1 - y = 1 - 0.005 = 0.995$. Then at point 2,

$$\frac{(1 - y)_{iM}}{1 - y} = \frac{0.997}{0.995} = 1.002$$

Hence, the average value of the term in the brackets in Eq. (10.6-45) is

$$\left[\frac{(1 - y)_{iM}}{1 - y}\right]_{av} = \frac{1.005 + 1.002}{2} = 1.003$$

Using Eq. (10.6-23), the integral is

$$\int_{y_2}^{y_1} \frac{dy}{y - y_i} = \frac{y_1 - y_2}{(y - y_i)_M} \qquad (10.6\text{-}23)$$

From Example 10.6-2, $(y - y_i)_M = 0.00602$. Substituting Eq. (10.6-23) into Eq. (10.6-45),

$$N_G = \left[\frac{(1 - y)_{iM}}{1 - y}\right]_{av} \frac{y_1 - y_2}{(y - y_i)_M} \qquad (10.6\text{-}46)$$

Substituting into Eq. (10.6-46),

$$N_G = (1.003)\left(\frac{0.026 - 0.005}{0.00602}\right) = 3.50 \text{ transfer units}$$

Finally, substituting into Eq. (10.6-40),

$$z = H_G N_G = (0.548)(3.50) = 1.918 \text{ m}$$

For part (b), using $K_y' a = 2.183 \times 10^{-2}$ kg mol/s·m³·mol frac from Example 10.6-2 and substituting into Eq. (10.6-38),

$$H_{OG} = \frac{V}{K_y' aS} = \frac{3.852 \times 10^{-3}}{(2.183 \times 10^{-2})(0.186)} = 0.949 \text{ m}$$

The number of transfer units in Eq. (10.6-43) becomes as follows when the integration similar to Eq. (10.6-23) is carried out.

$$N_{OG} = \left[\frac{(1-y)_{*M}}{1-y} \right]_{av} \frac{y_1 - y_2}{(y - y^*)_M} \qquad (10.6\text{-}47)$$

Substituting $(y - y^*)_M = 0.01025$ from Example 10.6-2 and the other knowns into Eq. (10.6-47) and calling the bracketed term 1.0,

$$N_{OG} = (1.0) \left(\frac{0.026 - 0.005}{0.01025} \right) = 2.05 \text{ transfer units}$$

Finally by Eq. (10.6-40),

$$z = H_{OG} N_{OG} = 0.949(2.05) = 1.945 \text{ m}$$

Note that the number of transfer units N_{OG} of 2.05 is not the same as N_G of 3.50.

10.7 ABSORPTION OF CONCENTRATED MIXTURES IN PACKED TOWERS

In Section 10.6D simplified design methods were given for absorption of dilute gases in packed towers when the mole fractions in the gas and liquid streams were less than about 10%. Straight operating lines and approximately straight equilibrium lines are obtained. In concentrated gas mixtures the operating line and usually the equilibrium line will be substantially curved and $k_x' a$ and $k_y' a$ may vary with total flows. Then the design equations (10.6-14)–(10.6-17) must be integrated graphically or numerically.

$$\int_0^z dz = z = \int_{y_2}^{y_1} \frac{V \, dy}{\dfrac{k_y' aS}{(1-y)_{iM}} (1-y)(y-y_i)} \qquad (10.6\text{-}14)$$

$$\int_0^z dz = z = \int_{x_2}^{x_1} \frac{L \, dx}{\dfrac{k_x' aS}{(1-x)_{iM}} (1-x)(x_i - x)} \qquad (10.6\text{-}15)$$

$$\int_0^z dz = z = \int_{y_2}^{y_1} \frac{V \, dy}{\dfrac{K_y' aS}{(1-y)_{*M}} (1-y)(y-y^*)} \qquad (10.6\text{-}16)$$

$$\int_0^z dz = z = \int_{x_2}^{x_1} \frac{L \, dx}{\dfrac{K_x' aS}{(1-x)_{*M}} (1-x)(x^* - x)} \qquad (10.6\text{-}17)$$

The detailed general steps to follow are as follows:

1. The operating-line equation (10.6-4) and the equilibrium line are plotted.
2. The values of the film coefficients $k_y' a$ and $k_x' a$ are obtained from empirical equations. These two film coefficients are functions of G_y^n, kg total gas/s·m², and G_x^m, kg total liquids/s·m², where n and m are in the range 0.2–0.8. Using the operating-line-equation values, total V and L are calculated for different values of y and x in the tower and converted to G_y and G_x. Then values of $k_y' a$ and $k_x' a$ are calculated. If the variation between $k_y' a$ or $k_x' a$ at the top and bottom of the tower is small, an average value can be used.

3. Starting with the tower bottom at point $P_1(y_1, x_1)$, the interface compositions y_{i1}, x_{i1} are determined by plotting a line $P_1 M_1$ with a slope calculated by Eq. (10.6-30). This line intersects the equilibrium line at the interface concentrations at point M_1:

$$\text{slope} = -\frac{k'_x a/(1-x)_{iM}}{k'_y a/(1-y)_{iM}} = -\frac{k_x a}{k_y a} \qquad (10.6\text{-}30)$$

where $(1-y)_{iM}$ and $(1-x)_{iM}$ are determined from Eqs. (10.4-6) and (10.4-7), respectively. This is trial and error. As a first trial, $(1-x_1)$ can be used for $(1-x)_{iM}$ and $(1-y_1)$ for $(1-y)_{iM}$. The values of y_{i1} and x_{i1} determined in the first trial are used in Eq. (10.6-30) for the second trial.
4. At point $P_2(y_2, x_2)$ determine a new slope using Eq. (10.6-30) repeating step 3. Do this for several intermediate points in the tower. This slope may vary throughout the tower.
5. Using the values of y_i and x_i determined, graphically integrate Eq. (10.6-14) to obtain the tower height by plotting $f(y)$, where $f(y)$ is as follows:

$$f(y) = \frac{V}{\dfrac{k'_y aS}{(1-y)_{iM}}(1-y)(y-y_i)} \qquad (10.7\text{-}1)$$

versus y between y_2 and y_1. Then determine the area under the curve to give the tower height. If $k'_x a$ or other coefficients are used the appropriate functions indicated in Eqs. (10.6-15)–(10.6-17) are plotted. If a stream is quite dilute, $(1-y)_{iM}$ or $(1-x)_{iM}$ can be assumed to be 1.0.

EXAMPLE 10.7-1. Design of an Absorption Tower with a Concentrated Gas Mixture

A tower packed with 25.4-mm ceramic rings is to be designed to absorb SO_2 from air by using pure water at 293 K and 1.013×10^5 Pa abs pressure. The entering gas contains 20 mol % SO_2 and that leaving 2 mol %. The inert air flow is 6.53×10^{-4} kg mol air/s and the inert water flow is 4.20×10^{-2} kg mol water/s. The tower cross-sectional area is 0.0929 m². For dilute SO_2, the film mass-transfer coefficients at 293 K are for 25.4-mm (1-in.) rings (W1),

$$k'_y a = 0.0594 G_y^{0.7} G_x^{0.25} \qquad k'_x a = 0.152 G_x^{0.82}$$

where $k'_y a$ is kg mol/s $\cdot$ m³ $\cdot$ mol frac, $k'_x a$ is kg mol/s $\cdot$ m³ $\cdot$ mol frac, and G_x and G_y are kg total liquid or gas, respectively, per sec per m² tower cross section. Calculate the tower height.

Solution: The given data are $V' = 6.53 \times 10^{-4}$ kg mol air/s (5.18 lb mol/h), $L' = 4.20 \times 10^{-2}$ kg mol/s (333 lb mol/h), $y_1 = 0.20$, $y_2 = 0.02$, and $x_2 = 0$.

Substituting into the overall material-balance equation (10.6-3),

$$L'\left(\frac{x_2}{1-x_2}\right) + V'\left(\frac{y_1}{1-y_1}\right) = L'\left(\frac{x_1}{1-x_1}\right) + V'\left(\frac{y_2}{1-y_2}\right)$$

$$4.20 \times 10^{-2}\left(\frac{0}{1-0}\right) + 6.53 \times 10^{-4}\left(\frac{0.2}{1-0.2}\right) = 4.20 \times 10^{-2}\left(\frac{x_1}{1-x_1}\right)$$

$$+ 6.53 \times 10^{-4}\left(\frac{0.02}{1-0.02}\right)$$

Solving, $x_1 = 0.00355$. The operating line Eq. (10.6-4) is

$$4.20 \times 10^{-2}\left(\frac{x}{1-x}\right) + 6.53 \times 10^{-4}\left(\frac{0.2}{1-0.2}\right) = 4.20 \times 10^{-2}\left(\frac{0.00355}{1-0.00355}\right)$$
$$+ 6.53 \times 10^{-4}\left(\frac{y}{1-y}\right)$$

Setting $y = 0.04$ in the operating-line equation above and solving for x, $x = 0.000332$. Selecting other values of y and solving for x, points on the operating line were calculated as shown in Table 10.7-1 and plotted in Fig. 10.7-1 together with the equilibrium data from Appendix A.3.

The total molar flow V is calculated from $V = V'/(1-y)$. At $y = 0.20$, $V = 6.53 \times 10^{-4}/(1-0.2) = 8.16 \times 10^{-4}$. Other values are calculated and tabulated in Table 10.7-1. The total mass flow G_y in kg/s·m² is equal to the mass flow of air plus SO_2 divided by the cross-sectional area.

$$G_y = \frac{6.53 \times 10^{-4}(29) \text{ kg air/s} + 6.53 \times 10^{-4}\left(\frac{y}{1-y}\right)(64.1) \text{ kg } SO_2/s}{0.0929 \text{ m}^2}$$

Setting $y = 0.20$,

$$G_y = \frac{6.53 \times 10^{-4}(29) + 6.53 \times 10^{-4}\left(\frac{0.2}{1-0.2}\right)64.1}{0.0929} = 0.3164 \text{ kg/s} \cdot \text{m}^2$$

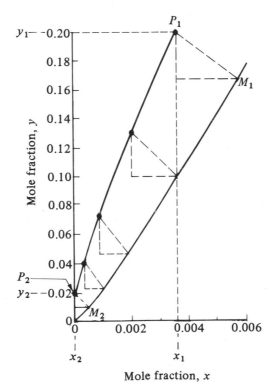

FIGURE 10.7-1. *Operating line and interface compositions for Example 10.7-1.*

TABLE 10.7-1. *Calculated Data for Example 10.7-1*

y	x	V	L	G_y	G_x	$k'_y a$	$k'_x a$	x_i	y_i	$1-y$	$(1-y)_{lM}$	$y-y_i$	$\dfrac{V}{\dfrac{k'_y aS}{(1-y)_{lM}}(1-y)(y-y_i)}$
0.02	0	6.66×10^{-4}	0.04200	0.2130	8.138	0.03398	0.848	0.00046	0.0090	0.980	0.985	0.0110	19.25
0.04	0.000332	6.80×10^{-4}	0.04201	0.2226	8.147	0.03504	0.849	0.00103	0.0235	0.960	0.968	0.0165	12.77
0.07	0.000855	7.02×10^{-4}	0.04203	0.2378	8.162	0.03673	0.850	0.00185	0.0476	0.930	0.941	0.0224	9.29
0.13	0.00201	7.51×10^{-4}	0.04208	0.2712	8.196	0.04032	0.853	0.00355	0.1015	0.870	0.885	0.0285	7.16
0.20	0.00355	8.16×10^{-4}	0.04215	0.3164	8.241	0.04496	0.857	0.00565	0.1685	0.800	0.816	0.0315	6.33

Similarly, G_y is calculated for all points and tabulated. For the liquid flow, $L = L'/(1 - x)$. Also, for the total liquid mass flow rate,

$$G_x = \frac{4.20 \times 10^{-2}(18) + 4.20 \times 10^{-2}\left(\dfrac{x}{1-x}\right)64.1}{0.0929}$$

Calculated values of L and G_x for various values of x are tabulated in Table 10.7-1.

To calculate values of $k'_x a$, for $x = 0$, $G_x = 8.138$ and

$$k'_x a = 0.152 G_x^{0.82} = 0.152(8.138)^{0.82} = 0.848 \text{ kg mol/s} \cdot \text{m}^3 \cdot \text{mol frac}$$

The rest of these values are calculated and given in Table 10.7-1. For the value of $k'_y a$, for $y = 0.02$, $G_y - 0.2130$, $G_x - 8.138$, and

$$k'_y a = 0.0594 G_y^{0.7} G_x^{0.25} = 0.0594(0.2130)^{0.7}(8.138)^{0.25}$$

$$= 0.03398 \text{ kg mol/s} \cdot \text{m}^3 \cdot \text{mol frac}$$

This and other calculated values of k'_y are tabulated. Note that these values vary considerably in the tower.

Next the interface compositions y_i and x_i must be determined for the given y and x values on the operating line. For the point $y_1 = 0.20$ and $x_1 = 0.00355$ we make a preliminary estimate of $(1 - y)_{iM} \cong 1 - y \cong 1 - 0.20 \cong 0.80$. Also, the estimate of $(1 - x)_{iM} \cong 1 - x \cong 1 - 0.00355 \cong 0.996$. The slope of the line $P_1 M_1$ by Eq. (10.6-30) is approximately

$$\text{slope} = -\frac{k'_x a/(1 - x)_{iM}}{k'_y a/(1 - y)_{iM}} = -\frac{0.857/(0.996)}{0.04496/(0.80)} = -15.3$$

Plotting this on Fig. 10.7-1, $y_i = 0.1688$ and $x_i = 0.00566$. Using these values for the second trial in Eqs. (10.4-6) and (10.4-7),

$$(1 - y)_{iM} = \frac{(1 - 0.1688) - (1 - 0.20)}{\ln{[(1 - 0.1688)/(1 - 0.20)]}} = 0.816$$

$$(1 - x)_{iM} = \frac{(1 - 0.00355) - (1 - 0.00566)}{\ln{[(1 - 0.00355)/(1 - 0.00566)]}} = 0.995$$

The new slope by Eq. (10.6-30) is $(-0.857/0.995)/(0.04496/0.816) = -15.6$. Plotting, $y_i = 0.1685$ and $x_i = 0.00565$. This is shown as point M_1. This calculation is repeated until point y_2, x_2 is reached. The slope of Eq. (10.6-30) increases markedly in going up the tower, being -24.6 at the top of the tower. The values of y_i and x_i are given in Table 10.7-1.

In order to integrate Eq. (10.6-14), values of $(1 - y)$, $(1 - y)_{iM}$, and $(y - y_i)$ are needed and are tabulated in Table 10.7-1. Then for $y = 0.20$, $f(y)$ is calculated from Eq. (10.7-1).

$$f(y) = \frac{V}{\dfrac{k'_y aS}{(1 - y)_{iM}}(1 - y)(y - y_i)} = \frac{8.16 \times 10^{-4}}{\dfrac{0.04496(0.0929)}{0.816}(0.800)(0.0315)} = 6.33$$

This is repeated for other values of y. Then the function $f(y)$ is plotted versus y. The total area is then the sum of four rectangles:

$$\text{total area} = 0.312 + 0.418 + 0.318 + 0.540 = 1.588$$

Hence, the tower height is equal to the area by Eq. (10.6-14) and $z = 1.588$ m.

10.8 ESTIMATION OF MASS-TRANSFER COEFFICIENTS FOR PACKED TOWERS

10.8A Experimental Determination of Film Coefficients

The individual film mass-transfer coefficients $k_y'a$ and $k_x'a$ depend generally upon Schmidt number, Reynolds number, and the size and shape of the packing. The interactions among these factors are quite complex. Hence, the correlations for mass-transfer coefficients are highly empirical. The reliability of these correlations is not too satisfactory. Deviations of up to 25% are not uncommon. A main difficulty arises because an overall coefficient or resistance is measured experimentally that represents the two film resistances in series. To obtain the single-phase film coefficient, the experiment is so arranged that the other film resistance is negligible or can be approximately calculated.

To measure the liquid film mass-transfer coefficient $k_x'a$, a system for absorption or desorption of very insoluble gases such as O_2 or CO_2 in water is used. The experiment gives $K_x'a$, which equals $k_x'a$, since the gas-phase resistance is negligible.

To measure the gas-phase film coefficient $k_y'a$, we desire to use a system such that the solute is very soluble in the liquid and the liquid-phase resistance is negligible. Most such systems as NH_3–air–water have a liquid-phase resistance of about 10%. By subtracting this known liquid phase resistance (obtained by correcting $k_x'a$ data for absorption of CO_2 or O_2 to NH_3 data for $k_x'a$) from the overall resistance in Eq. (10.4-24), we obtain the coefficient $k_y'a$. Details of these are discussed elsewhere (G1, S1, S2).

10.8B Correlations for Film Coefficients

The experimental data for the gas film coefficient in dilute mixtures have been correlated in terms of H_G, where

$$H_G = \frac{V}{k_y'aS} \tag{10.6-36}$$

The empirical equation is as follows:

$$H_G = \alpha G_y^\beta G_x^\gamma N_{Sc}^{0.5} \tag{10.8-1}$$

where G_y = kg total gas/s · m²; G_x = kg total liquid/s · m²; and α, β, and γ are constants for a packing as given in Table 10.8-1 (T2). The temperature effect, which is small, is included in the Schmidt number $\mu/\rho D$, where μ is the viscosity of the gas mixture in kg/m · s, ρ the density in kg/m³, and D the diffusivity of solute A in the gas in m²/s. The coefficients $k_y'a$ and H_G can be shown to be independent of pressure.

Equation (10.8-1) can be used to correct existing data for absorption of solute A in a gas on a specific packing to absorption of solute E in the same system and the same mass-flow rates. This is done by Eq. (10.8-2).

$$H_{G(E)} = H_{G(A)}[N_{Sc(E)}/N_{Sc(A)}]^{0.5} \tag{10.8-2}$$

The correlations for liquid film coefficients in dilute mixtures show that H_L is independent of gas rate until loading occurs, as given by the following:

$$H_L = \theta \left(\frac{G_x}{\mu_L}\right)^n N_{Sc}^{0.5} \tag{10.8-3}$$

where H_L is in m, μ_L is liquid viscosity in kg/m · s, N_{Sc} is Schmidt number $\mu_L/\rho D$, ρ is

TABLE 10.8-1. *Gas Film Height of a Transfer Unit H_G in Meters**

Packing Type	α	β	γ	Range of Values	
				G_y	G_x
Raschig rings					
9.5 mm ($\frac{3}{8}$ in.)	0.620	0.45	−0.47	0.271–0.678	0.678–2.034
25.4 mm (1 in.)	0.557	0.32	−0.51	0.271–0.814	0.678–6.10
38.1 mm (1.5 in.)	0.830	0.38	−0.66	0.271–0.950	0.678–2.034
38.1 mm (1.5 in.)	0.689	0.38	−0.40	0.271–0.950	2.034–6.10
50.8 mm (2 in.)	0.894	0.41	−0.45	0.271–1.085	0.678–6.10
Berl saddles					
12.7 mm (0.5 in.)	0.541	0.30	−0.74	0.271–0.950	0.678–2.034
12.7 mm (0.5 in.)	0.367	0.30	−0.24	0.271–0.950	2.034–6.10
25.4 mm (1 in.)	0.461	0.36	−0.40	0.271–1.085	0.542–6.10
38.1 mm (1.5 in.)	0.652	0.32	−0.45	0.271–1.356	0.542–6.10

* $H_G = \alpha G_y^\beta G_x^\gamma N_{Sc}^{0.5}$, where G_y = kg total gas/s · m^2, G_x = kg total liquid/s · m^2, and $N_{Sc} = \mu/\rho D$.
Source: Data from Fellinger (P2) as given by R. E. Treybal, *Mass Transfer Operations*. New York: McGraw-Hill Book Company, 1955, p. 239. With permission.

liquid density in kg/m^3, and D is diffusivity of solute A in the liquid in m^2/s. Data are given in Table 10.8-2 for different packings. Equation (10.8-3) can be used to correct existing data on a given packing and solute to another solute.

EXAMPLE 10.8-1. *Prediction of Film Coefficients for Ammonia Absorption*

Predict H_G, H_L, and $K_y'a$ for absorption of NH$_3$ from water in a dilute solution in a packed tower with 25.4-mm Raschig rings at 303 K (86°F) and

TABLE 10.8-2. *Liquid Film Height of a Transfer Unit H_L in Meters**

Packing	θ	η	Range of G_x
Raschig rings			
9.5 mm ($\frac{3}{8}$ in.)	3.21×10^{-4}	0.46	0.542–20.34
12.7 mm (0.5 in.)	7.18×10^{-4}	0.35	0.542–20.34
25.4 mm (1 in.)	2.35×10^{-3}	0.22	0.542–20.34
38.1 mm (1.5 in.)	2.61×10^{-3}	0.22	0.542–20.34
50.8 mm (2 in.)	2.93×10^{-3}	0.22	0.542–20.34
Berl saddles			
12.7 mm (0.5 in.)	1.456×10^{-3}	0.28	0.542–20.34
25.4 mm (1 in.)	1.285×10^{-3}	0.28	0.542–20.34
38.1 mm (1.5 in.)	1.366×10^{-3}	0.28	0.542–20.34

* $H_L = \theta(G_x/\mu_L)^\eta N_{Sc}^{0.5}$, where G_x = kg total liquid/s · m^2, μ_L = viscosity of liquid in kg/m · s, and $N_{Sc} = \mu_L/\rho D$. G_y is less than loading.
Source: Based on data by Sherwood and Holloway (S3) as given by R. E. Treybal, *Mass Transfer Operations*. New York: McGraw-Hill Book Company, 1955, p. 237. With permission.

101.32 kPa pressure. The flow rates are $G_x = 2.543$ kg/s·m² and $G_y = 0.339$ kg/s·m².

Solution: From Appendix A.3 the equilibrium relation in a dilute solution is $0.0151 = m(0.0126)$ or $y = 1.20x$. Also, from Appendix A.3 for air, $\mu = 1.86 \times 10^{-5}$ kg/m·s. Density $\rho = 1.168$ kg/m³. The diffusivity of NH_3 in air at 273 K from Table 6.2-1 is 1.98×10^{-5} m²/s. Correcting to 303 K by Eq. (6.2-45), $D_{AB} = 2.379 \times 10^{-5}$ m²/s. Hence,

$$N_{Sc} = \frac{\mu}{\rho D} = \frac{1.86 \times 10^{-5}}{(1.168)(2.379 \times 10^{-5})} = 0.669$$

Substituting into Eq. (10.8-1) using data from Table 10.8-1,

$$H_G = \alpha G_y^\beta G_x^\gamma N_{Sc}^{0.5} = 0.557(0.339)^{0.32}(2.543)^{-0.51}(0.669)^{0.5} = 0.200 \text{ m}$$

The viscosity of water $= 1.1404 \times 10^{-3}$ kg/m·s at 15°C and 0.8007×10^{-3} at 30°C from Appendix A.2. The D_{AB} of NH_3 in water at 288 K (15°C) from Table 6.3-1 is 1.77×10^{-9} m²/s. Correcting this to 303 K (30°C), using Eq. (6.3-9),

$$D_{AB} = \left(\frac{1.1404 \times 10^{-3}}{0.8007 \times 10^{-3}}\right)\left(\frac{303}{288}\right)(1.77 \times 10^{-9}) = 2.652 \times 10^{-9} \text{ m}^2/\text{s}$$

Then, using $\rho = 996$ kg/m³ for water,

$$N_{Sc} = \frac{\mu}{\rho D} = \frac{0.8007 \times 10^{-3}}{(996)(2.652 \times 10^{-9})} = 303.1$$

Substituting into Eq. (10.8-3), using data from Table 10.8-2,

$$H_L = \theta\left(\frac{G_x}{\mu_L}\right)^\eta N_{Sc}^{0.5} = (2.35 \times 10^{-3})\left(\frac{2.543}{0.8007 \times 10^{-3}}\right)^{0.22}(303.1)^{0.5}$$
$$= 0.2412 \text{ m}$$

Converting to $k_y'a$ using Eq. (10.6-36),

$$k_y'a = \frac{V}{H_G S} = \frac{0.339/29}{0.200} = 0.0584 \text{ kg mol/s·m}^3 \cdot \text{mol frac}$$

For $k_x'a$ using Eq. (10.6-37),

$$k_x'a = \frac{L}{H_L S} = \frac{2.543/18}{0.2412} = 0.586 \text{ kg mol/s·m}^3 \cdot \text{mol frac}$$

Substituting into Eq. (10.4-24) for dilute solutions,

$$\frac{1}{K_y'a} = \frac{1}{k_y'a} + \frac{m'}{k_x'a} = \frac{1}{0.0584} + \frac{1.20}{0.586} = 17.12 + 2.048 = 19.168$$

$$K_y'a = 0.0522 \text{ kg mol/s·m}^3 \cdot \text{mol frac}$$

Note that the percent resistance in the gas film is $(17.12/19.168)(100) = 89.3\%$.

PROBLEMS

10.2-1. Equilibrium and Henry's Law Constant. The partial pressure of CO_2 in air is 1.333×10^4 Pa and the total pressure is 1.133×10^5 Pa. The gas phase is in equilibrium with a water solution at 303 K. What is the value of x_A of CO_2 in equilibrium in the solution? See Appendix A.3 for the Henry's law constant.

Ans. $x_A = 7.07 \times 10^{-5}$ mol frac CO_2

10.2-2. *Gas Solubility in Aqueous Solution.* At 303 K the concentration of CO_2 in water is 0.90×10^{-4} kg CO_2/kg water. Using the Henry's law constant from Appendix A.3, what partial pressure of CO_2 must be kept in the gas to keep the CO_2 from vaporizing from the aqueous solution?

Ans. $p_A = 6.93 \times 10^3$ Pa (0.0684 atm)

10.2-3. *Phase Rule for a Gas–Liquid System.* For the system SO_2–air–water, the total pressure is set at 1 atm abs and the partial pressure of SO_2 in the vapor is set at 0.20 atm. Calculate the number of degrees of freedom, F. What variables are unspecified that can be arbitrarily set?

10.3-1. *Equilibrium Stage Contact for Gas–Liquid System.* A gas mixture at 2.026 $\times 10^5$ Pa total pressure containing air and SO_2 is contacted in a single-stage equilibrium mixer with pure water at 293 K. The partial pressure of SO_2 in the original gas is 1.52×10^4 Pa. The inlet gas contains 5.70 total kg mol and the inlet water 2.20 total kg mol. The exit gas and liquid leaving are in equilibrium. Calculate the amounts and compositions of the outlet phases. Use equilibrium data from Fig. 10.2-1.

Ans. $x_{A1} = 0.00495$, $y_{A1} = 0.0733$, $L_1 = 2.211$ kg mol, $V_1 = 5.69$ kg mol

10.3-2. *Absorption in a Countercurrent Stage Tower.* Repeat Example 10.3-2 using the same conditions but with the following change. Use a pure water flow to the tower of 108 kg mol H_2O/h, that is, 20% above the 90 used in Example 10.3-2. Determine the number of stages required graphically. Repeat using the analytical Kremser equation.

10.3-3. *Stripping Taint from Cream by Steam.* Countercurrent stage stripping is to be used to remove a taint from cream. The taint is present in the original cream to the stripper at a concentration of 20 parts per million (ppm). For every 100 kg of cream entering per unit time, 50 kg of steam will be used for stripping. It is desired to reduce the concentration of the taint in the cream to 1 ppm. The equilibrium relation between the taint in the steam vapor and the liquid cream is $y_A = 10x_A$, where y_A is ppm of taint in the steam and x_A ppm in the cream (E1). Determine the number of theoretical stages needed. [*Hint :* In this case, for stripping from the liquid (L) stream to the vapor (V) stream the operating line will be below the equilibrium line on the $y_A - x_A$ diagram. It is assumed that none of the steam condenses in the stripping. Use ppm in the material balances.]

Ans. Number stages = 1.85 (stepping down starting from the concentrated end)

10.4-1. *Overall Mass-Transfer Coefficient from Film Coefficients.* Using the same data as in Example 10.4-1, calculate the overall mass-transfer coefficients K'_x and K_x, the flux, and the percent resistance in the gas film.

Ans. $K'_x - 1.173 \times 10^{-3}$ kg mol/s $\cdot$ m^2 mol frac, $K_x - 1.519 \times 10^{-3}$, $N_A = 3.78 \times 10^{-4}$ kg mol/s $\cdot$ m^2, 36.7% resistance

10.4-2. *Interface Concentrations and Overall Mass-Transfer Coefficients.* Use the same equilibrium data and film coefficients k'_y and k'_x as in Example 10.4-1. However, use bulk concentrations of $y_{AG} = 0.25$ and $x_{AL} = 0.05$. Calculate the following.
(a) Interface concentrations y_{Ai} and x_{Ai} and flux N_A.
(b) Overall mass-transfer coefficients K'_y and K_y and flux N_A.
(c) Overall mass-transfer coefficient K'_x and flux N_A.

10.5-1. *Countercurrent Water-Cooling Tower.* A forced-draft countercurrent water-cooling tower is to cool water from 43.3 to 26.7°C. The air enters the bottom of the tower at 23.9°C with a wet bulb temperature of 21.1°C. The value of H_G for the flow conditions is $H_G = 0.533$ m. The heat-transfer resistance in the liquid phase will be neglected; i.e., h_L is very large. Hence, values of H_y^* should be used. Calculate the tower height needed if 1.5 times the minimum air rate is used.

10.5-2. Minimum Gas Rate and Height of Water-Cooling Tower. It is planned to cool water from 110°F to 85°F in a packed countercurrent water-cooling tower using entering air at 85°F with a wet bulb temperature of 75°F. The water flow is 2000 $lb_m/h \cdot ft^2$ and the air flow is 1400 lb_m air/h · ft². The overall mass-transfer coefficient is $K_G a = 6.90$ lb mol/h · ft³ · atm.

(a) Calculate the minimum air rate that can be used.

(b) Calculate the tower height needed if the air flow of 1400 lb_m air/h · ft² is used.

Ans. (a) $G_{min} = 935$ lb_m air/h · ft² (4241 kg air/h · m²); (b) $z = 21.8$ ft (6.64 m)

10.5-3. Design of Water-Cooling Tower. Recalculate Example 10.5-1, but calculate the minimum air rate and use 1.75 times the minimum air rate.

10.5-4. Effect of Changing Air Conditions on Cooling Tower. For the cooling tower in Example 10.5-1, to what temperature would the water be cooled if the entering air enters at 29.4°C but the wet bulb temperature is 26.7°C? The same gas and liquid flow rates are used. The water enters at 43.3°C, as before. (*Hint :* In this case T_{L1} is the unknown. The tower height is the same as in Example 10.5-1. The slope of the operating line is as before. The solution is trial and error. Assume a value of T_{L1} that is greater than 29.4°C. Do the graphical integration to see if the same height z is obtained.)

10.6-1. Amount of Absorption in a Tray Tower. An existing tower contains the equivalent of 3.0 theoretical trays and is being used to absorb SO_2 from air by pure water at 293 K and 1.013×10^5 Pa. The entering gas contains 20 mol % SO_2 and the inlet air flow rate is 150 kg inert air/h · m². The entering water rate is 6000 kg/h · m². Calculate the outlet composition of the gas. (*Hint :* This is a trial-and-error solution. Assume an outlet gas composition of, say, $y_1 = 0.01$. Plot the operating line and determine the number of theoretical trays needed. If this number is not 3.0 trays, assume another value of y_1, and so on.)

Ans. $y_1 = 0.011$

10.6-2. Analytical Method for Number of Trays in Absorption. Use the analytical equations in Section 10.3 for countercurrent stage contact to calculate the number of theoretical trays needed for Example 10.6-1.

10.6-3. Absorption of Ammonia in a Tray Tower. A tray tower is to be used to remove 99% of the ammonia from an entering air stream containing 6 mol % ammonia at 293 K and 1.013×10^5 Pa. The entering pure water flow rate is 188 kg $H_2O/h \cdot m^2$ and the inert air flow is 128 kg air/h · m². Calculate the number of theoretical trays needed. Use equilibrium data from Appendix A.3. For the dilute end of the tower, plot an expanded diagram to step off the number of trays more accurately.

Ans. $y_1 = 0.000639$ (exit), $x_N = 0.0260$ (exit), 3.8 theoretical trays

10.6-4. Minimum Liquid Flow in a Packed Tower. The gas stream from a chemical reactor contains 25 mol % ammonia and the rest inert gases. The total flow is 181.4 kg mol/h to an absorption tower at 303 K and 1.013×10^5 Pa pressure, where water containing 0.005 mol frac ammonia is the scrubbing liquid. The outlet gas concentration is to be 2.0 mol % ammonia. What is the minimum flow L'_{min}? Using 1.5 times the minimum plot the equilibrium and operating lines.

Ans. $L'_{min} = 262.6$ kg mol/h

10.6-5. Steam Stripping and Number of Trays. A relatively nonvolatile hydrocarbon oil contains 4.0 mol % propane and is being stripped by direct superheated steam in a stripping tray tower to reduce the propane content to 0.2%. The temperature is held constant at 422 K by internal heating in the tower at 2.026×10^5

Pa pressure. A total of 11.42 kg mol of direct steam is used for 300 kg mol of total entering liquid. The vapor–liquid equilibria can be represented by $y = 25x$, where y is mole fraction propane in the steam and x is mole fraction propane in the oil. Steam can be considered as an inert gas and will not condense. Plot the operating and equilibrium lines and determine the number of theoretical trays needed.

Ans. 5.6 theoretical trays (stepping down from the tower top)

10.6-6. *Absorption of Ammonia in Packed Tower.* A gas stream contains 4.0 mol % NH_3 and its ammonia content is reduced to 0.5 mol % in a packed absorption tower at 293 K and 1.013×10^5 Pa. The inlet pure water flow is 68.0 kg mol/h and the total inlet gas flow is 57.8 kg mol/h. The tower diameter is 0.747 m. The film mass-transfer coefficients are $k'_y a = 0.0739$ kg mol/s·m³·mol frac and $k'_x a = 0.169$ kg mol/s·m³·mol frac. Using the design methods for dilute gas mixtures, do as follows.
(a) Calculate the tower height using $k'_y a$.
(b) Calculate the tower height using $K'_y a$.

Ans. (a) $z = 2.362$ m (7.75 ft)

10.6-7. *Tower Height Using Overall Mass-Transfer Coefficient.* Repeat Example 10.6-2, using the overall liquid mass-transfer coefficient $K'_x a$ to calculate the tower height.

10.6-8. *Experimental Overall Mass-Transfer Coefficient.* In a tower 0.254 m in diameter absorbing acetone from air at 293 K and 101.32 kPa using pure water, the following experimental data were obtained. Height of 25.4-mm Raschig rings = 4.88 m, $V' = 3.30$ kg mol air/h, $y_1 = 0.01053$ mol frac acetone, $y_2 = 0.00072$, $L = 9.03$ kg mol water/h, $x_1 = 0.00363$ mol frac acetone. Calculate the experimental value of $K_y a$.

10.6-9. *Conversion to Transfer Unit Coefficients from Mass-Transfer Coefficients.* Experimental data on absorption of dilute acetone in air by water at 80°F and 1 atm abs pressure in a packed tower with 25.4-mm Raschig rings were obtained. The inert gas flow was 95 lb_m air/h·ft² and the pure water flow was 987 lb_m/h·ft². The experimental coefficients are $k_G a = 4.03$ lb mol/h·ft³· atm and $k_L a = 16.6$ lb mol/h·ft³·lb mol/ft³. The equilibrium data can be expressed by $c_A = 1.37 p_A$, where $c_A =$ lb mol/ft³ and $p_A =$ atm partial pressure of acetone.
(a) Calculate the film height of transfer units H_G and H_L.
(b) Calculate H_{OG}.

Ans. (b) $H_{OG} = 0.957$ ft (0.292 m)

10.6-10. *Height of Tower Using Transfer Units.* Repeat Example 10.6-2 but use transfer units and calculate H_L, N_L, and tower height.

10.6-11. *Experimental Value of H_{OG}.* Using the experimental data given in Problem 10.6-8, calculate the number of transfer units N_{OG} and the experimental value of H_{OG}.

Ans. $H_{OG} = 1.265$ m

10.7-1. *Liquid Film Coefficients and Design of SO_2 Tower.* Using the data of Example 10.7-1, calculate the height of the tower using Eq. (10.6-15), which is based on the liquid film mass-transfer coefficient $k'_x a$. [*Note:* The interface values x_i have already been obtained. Use a graphical integration of Eq. (10.6-15).]

Ans. $z = 1.586$ m

10.7-2. *Design of SO_2 Tower Using Overall Coefficients.* Using the data of Example 10.7-1, calculate the tower height using the overall mass-transfer coefficient $K'_y a$. [*Hint:* Calculate $K'_y a$ at the top of the tower and at the bottom of the tower from the film coefficients. Then use a linear average of the two values for the design. Obtain the values of y^* from the operating and equilibrium line plot. Graphically integrate Eq. (10.6-16), keeping $K'_y a$ outside the integral.]

10.7-3. *Height of Packed Tower Using Transfer Units.* For Example 10.7-1 calculate the tower height using the H_G and the number of transfer units N_G. [*Hint:* Calculate H_G at the tower top using Eq. (10.6-36) and at the tower bottom. Use the linear average value for H_G. Calculate the number of transfer units N_G by graphical integration of the integral of Eq. (10.6-32). Then calculate the tower height.]

Ans. $H_G = 0.2036$ m (average value)

10.7-4. *Design of Absorption Tower Using Transfer Units.* The gas SO_2 is being scrubbed from a gas mixture by pure water at 303 K and 1.013×10^5 Pa. The inlet gas contains 6.00 mol % SO_2 and the outlet 0.3 mol % SO_2. The tower cross-sectional area of packing is 0.426 m^2. The inlet gas flow is 13.65 kg mol inert air/h and the inlet water flow is 984 kg mol inert water/h. The mass-transfer coefficients are $H_L = 0.436$ m and $k_G a = 6.06 \times 10^{-7}$ kg mol/s·m^3 · Pa and are to be assumed constant in the tower for the given concentration range. Use equilibrium data from Appendix A.3. By graphical integration, determine N_G. Calculate the tower height. (*Note:* The equilibrium line is markedly curved, so graphical integration is necessary even for this dilute mixture.)

Ans. $N_G = 8.47$ transfer units, $z = 1.311$ m

10.8-1. *Prediction of Mass-Transfer Coefficients.* Predict the mass-transfer coefficients H_G, H_L, $k'_x a$, $k'_y a$, and $K'_x a$ for absorption of CO_2 from air by water in the same packing and using 1.6 times the flow rates in Example 10.8-1 at 303 K. (*Hint:* Use equilibrium data from Appendix A.3 for Henry's law constants for CO_2 in water. Use diffusivity data for CO_2 in water from Table 6.3-1 and for CO_2 in air from Table 6.2-1. Correct data to 303 K.)

Ans. $H_G = 0.2186$ m, $H_L = 0.2890$ m

10.8-2. *Correction of Film Mass-Transfer Coefficients.* For absorption of dilute NH_3 from air by water at 20°C and 101.3 kPa abs pressure, experimental values are $H_G = 0.1372$ m and $H_L = 0.2103$ m for heights of transfer units. The flow rates are $G_x = 13\,770$ kg/h·m^2 and $G_y = 1343$ kg/h·m^2. For absorption of acetone from air by water under the same conditions in the same packing, predict $k'_y a$, $k'_x a$, and $K'_y a$. Use equilibrium data from Appendix A.3. Use the diffusivity for acetone in water from Table 6.3-1. The diffusivity of acetone in air at 0°C is 0.109×10^{-4} m^2/s at 101.3 kPa abs pressure.

REFERENCES

(B1)   BADGER, W. L., and BANCHERO, J. T. *Introduction to Chemical Engineering.* New York: McGraw-Hill Book Company, 1955.

(B2)   *Bubble Tray Design Manual.* American Institute of Chemical Engineers, New York, 1958.

(E1)   EARLE, R. L. *Unit Operations in Food Processing.* Oxford: Pergamon Press, Inc., 1966.

(G1)   GEANKOPLIS, C. J. *Mass Transport Phenomena.* Columbus, Ohio: Ohio State University Bookstores, 1972.

(L1)   LEVA, M. *Tower Packings and Packed Tower Design,* 2nd ed. Akron, Ohio: U.S. Stoneware, Inc., 1953.

(M1)   MICKLEY, H. S., SHERWOOD, T. K., and REED, C. E. *Applied Mathematics in Chemical Engineering,* 2nd ed. New York: McGraw-Hill Book Company, 1957.

(M2)   MICKLEY, H. S. *Chem. Eng. Progr.,* **45,** 739 (1949).

(P1) PERRY, R. H., and GREEN, D. *Perry's Chemical Engineers' Handbook*, 6th ed. New York: McGraw-Hill Book Company, 1984.

(S1) SHERWOOD, T. K., PIGFORD, R. L., and WILKE, C. R. *Mass Transfer*. New York: McGraw-Hill Book Company, 1975.

(S2) SHULMAN, H. L., and coworkers. *A.I.Ch.E. J.*, **1**, 274 (1955); **9**, 479 (1963).

(S3) SHERWOOD, T. K., and HOLLOWAY, F. A. L. *Trans. A.I.Ch.E.*, **30**, 39 (1940).

(T1) TREYBAL, R. E. *Mass Transfer Operations*, 3rd ed. New York: McGraw-Hill Book Company, 1980.

(T2) TREYBAL, R. E. *Mass Transfer Operations*. New York: McGraw-Hill Book Company, 1955.

(W1) WHITNEY, R. P., and VIVIAN, J. E. *Chem. Eng. Progr.*, **45**, 323 (1949).

CHAPTER 11

Vapor–Liquid
Separation Processes

11.1 VAPOR–LIQUID EQUILIBRIUM RELATIONS

11.1A Phase Rule and Raoult's Law

As in the gas–liquid systems, the equilibrium in vapor–liquid systems is restricted by the phase rule, Eq. (10.2-1). As an example we shall use the ammonia–water, vapor–liquid system. For two components and two phases, F from Eq. (10.2-1) is 2 degrees of freedom. The four variables are temperature, pressure, and the composition y_A of NH_3 in the vapor phase and x_A in the liquid phase. The composition of water (B) is fixed if y_A or x_A is specified, since $y_A + y_B = 1.0$ and $x_A + x_B = 1.0$. If the pressure is fixed, only one more variable can be set. If we set the liquid composition, the temperature and vapor composition are automatically set.

An ideal law, *Raoult's law*, can be defined for vapor–liquid phases in equilibrium.

$$p_A = P_A x_A \tag{11.1-1}$$

where p_A is the partial pressure of component A in the vapor in Pa (atm), P_A is the vapor pressure of pure A in Pa (atm), and x_A is the mole fraction of A in the liquid. This law holds only for ideal solutions, such as benzene–toluene, hexane–heptane, and methyl alcohol–ethyl alcohol, which are usually substances very similar to each other. Many systems that are ideal or nonideal solutions follow Henry's law in dilute solutions.

11.1B Boiling-Point Diagrams and xy Plots

Often the vapor–liquid equilibrium relations for a binary mixture of A and B are given as a boiling-point diagram shown in Fig. 11.1-1 for the system benzene (A)–toluene (B) at a total pressure of 101.32 kPa. The upper line is the saturated vapor line (the *dew-point line*) and the lower line is the saturated liquid line (the *bubble-point line*). The two-phase region is in the region between these two lines.

In Fig. 11.1-1, if we start with a cold liquid mixture of $x_{A1} = 0.318$ and heat the mixture, it will start to boil at 98°C (371.2 K) and the composition of the first vapor in

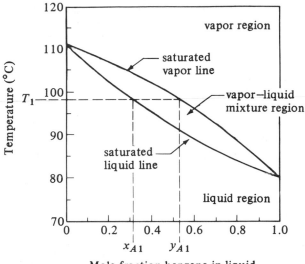

FIGURE 11.1-1. *Boiling point diagram for benzene (A)–toluene (B) at 101.325 kPa (1 atm) total pressure.*

equilibrium is $y_{A1} = 0.532$. As we continue boiling, the composition x_A will move to the left since y_A is richer in A.

The system benzene–toluene follows Raoult's law, so the boiling-point diagram can be calculated from the pure vapor-pressure data in Table 11.1-1 and the following equations:

$$p_A + p_B = P \tag{11.1-2}$$

$$P_A x_A + P_B(1 - x_A) = P \tag{11.1-3}$$

$$y_A = \frac{p_A}{P} = \frac{P_A x_A}{P} \tag{11.1-4}$$

EXAMPLE 11.1-1. *Use of Raoult's Law for Boiling-Point Diagram*
Calculate the vapor and liquid compositions in equilibrium at 95°C (368.2 K) for benzene–toluene using the vapor pressure from Table 11.1-1 at 101.32 kPa.

Solution: At 95°C from Table 11.1-1 for benzene, $P_A = 155.7$ kPa and $P_B = 63.3$ kPa. Substituting into Eq. (11.1-3) and solving,

$$155.7(x_A) + 63.3(1 - x_A) = 101.32 \text{ kPa (760 mm Hg)}$$

Hence, $x_A = 0.411$ and $x_B = 1 - x_A = 1 - 0.411 = 0.589$. Substituting into Eq. (11.1-4),

$$y_A = \frac{P_A x_A}{P} = \frac{155.7(0.411)}{101.32} = 0.632$$

A common method of plotting the equilibrium data is shown in Fig. 11.1-2, where y_A

TABLE 11.1-1. *Vapor-Pressure and Equilibrium-Mole-Fraction Data for Benzene–Toluene System*

		Vapor Pressure					
Temperature		*Benzene*		*Toluene*		*Mole Fraction Benzene at 101.325 kPa*	
K	*°C*	*kPa*	*mm Hg*	*kPa*	*mm Hg*	x_A	y_A
353.3	80.1	101.32	760			1.000	1.000
358.2	85	116.9	877	46.0	345	0.780	0.900
363.2	90	135.5	1016	54.0	405	0.581	0.777
368.2	95	155.7	1168	63.3	475	0.411	0.632
373.2	100	179.2	1344	74.3	557	0.258	0.456
378.2	105	204.2	1532	86.0	645	0.130	0.261
383.8	110.6	240.0	1800	101.32	760	0	0

is plotted versus x_A for the benzene–toluene system. The 45° line is given to show that y_A is richer in component A than is x_A.

The boiling-point diagram in Fig. 11.1-1 is typical of an ideal system following Raoult's law. Nonideal systems differ considerably. In Fig. 11.1-3a the boiling-point diagram is shown for a maximum-boiling azeotrope. The maximum temperature T_{max} corresponds to a concentration x_{Az} and $x_{Az} = y_{Az}$ at this point. The plot of y_A versus x_A would show the curve crossing the 45° line at this point. Acetone–chloroform is an example of such a system. In Fig. 11.1-3b a minimum-boiling azeotrope is shown with $y_{Az} = x_{Az}$ at T_{min}. Ethanol–water is such a system.

11.2 SINGLE-STAGE EQUILIBRIUM CONTACT FOR VAPOR–LIQUID SYSTEM

If a vapor–liquid system is being considered, where the stream V_2 is a vapor and L_0 is a liquid, and the two streams are contacted in a single equilibrium stage which is quite similar to Fig. 10.3-1, the boiling point or the xy equilibrium diagram must be used because an equilibrium relation similar to Henry's law is not available. Since we are considering only two components A and B, only Eqs. (10.3-1) and (10.3-2) are used for the material balances. If sensible heat effects are small and the latent heats of both compounds are the same, then when 1 mol of A condenses, 1 mol of B must vaporize. Hence, the total moles of vapor V_2 entering will equal V_1 leaving. Also, moles $L_0 = L_1$. This case is called one of *constant molal overflow*. An example is the benzene–toluene system.

EXAMPLE 11.2-1. Equilibrium Contact of Vapor–Liquid Mixture
A vapor at the dew point and 101.32 kPa containing a mole fraction of 0.40 benzene (A) and 0.60 toluene (B) and 100 kg mol total is contacted with 110 kg mol of a liquid at the boiling point containing a mole fraction of 0.30 benzene and 0.70 toluene. The two streams are contacted in a single stage, and the outlet streams leave in equilibrium with each other. Assume constant molal overflow. Calculate the amounts and compositions of the exit streams.

Solution: The process flow diagram is the same as in Fig. 10.3-1. The given

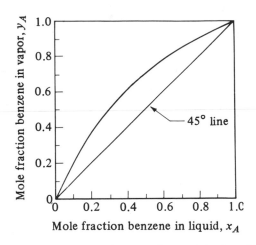

FIGURE 11.1-2 *Equilibrium diagram for system benzene (A)-toluene (B) at 101.32 kPa (1 atm).*

values are $V_2 = 100$ kg mol, $y_{A2} = 0.40$, $L_0 = 110$ kg mol, and $x_{A0} = 0.30$. For constant molal overflow, $V_2 = V_1$ and $L_0 = L_1$. Substituting into Eq. (10.3-2) to make a material balance on component A,

$$L_0 x_{A0} + V_2 y_{A2} = L_1 x_{A1} + V_1 y_{A1} \qquad \text{(10.3-2)}$$

$$110(0.30) + 100(0.40) = 110 x_{A1} + 100 y_{A1} \qquad \text{(11.2-1)}$$

To solve Eq. (11.2-1), the equilibrium relation between y_{A1} and x_{A1} in Fig. 11.1-1 must be used. This is by trial and error since an analytical expression is not available that relates y_A and x_A.

First, we assume that $x_{A1} = 0.20$ and substitute into Eq. (11.2-1) to solve for y_{A1}.

$$110(0.30) + 100(0.40) = 110(0.20) + 100 y_{A1}$$

Solving, $y_{A1} = 0.51$. The equilibrium relations for benzene–toluene are plotted in Fig. 11.2-1. It is evident that $y_{A1} = 0.51$ and $x_{A1} = 0.20$ do not fall on

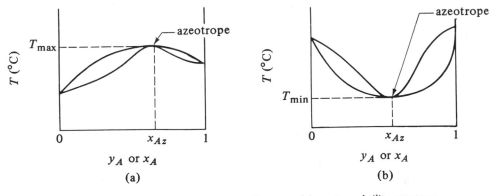

FIGURE 11.1-3 *Equilibrium boiling-point diagrams : (a) maximum-boiling azeotrope, (b) minimum-boiling azeotrope.*

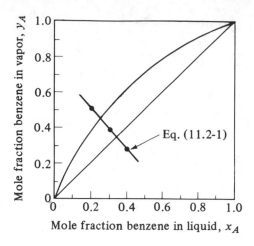

FIGURE 11.2-1. *Solution to Example 11.2-1.*

the curve. This point is plotted on the graph. Next, assuming that $x_{A1} = 0.40$ and solving, $y_{A1} = 0.29$. This point is also plotted in Fig. 11.2-1. Assuming that $x_{A1} = 0.30$, $y_{A1} = 0.40$. A straight line is drawn between these three points which represents Eq. (11.2-1). At the intersection of this line with the equilibrium curve, $y_{A1} = 0.455$ and $x_{A1} = 0.25$, which check Eq. (11.2-1).

11.3 SIMPLE DISTILLATION METHODS

11.3A Introduction

The unit operation distillation is a method used to separate the components of a liquid solution, which depends upon the distribution of these various components between a vapor and a liquid phase. All components are present in both phases. The vapor phase is created from the liquid phase by vaporization at the boiling point.

The basic requirement for the separation of the components by distillation is that the composition of the vapor be different from the composition of the liquid with which it is in equilibrium at the boiling point of the liquid. Distillation is concerned with solutions where all components are appreciably volatile, such as in ammonia–water or ethanol–water solutions, where both components will be in the vapor phase. In evaporation, however, of a solution of salt and water, the water is vaporized but the salt is not. The process of absorption differs from distillation in that one of the components in absorption is essentially insoluble in the liquid phase. An example is absorption of ammonia from air by water, where air is insoluble in the water–ammonia solution.

11.3B Relative Volatility of Vapor–Liquid Systems

In Fig. 11.1-2 for the equilibrium diagram for a binary mixture of A and B, the greater the distance between the equilibrium line and the 45° line, the greater the difference between the vapor composition y_A and liquid composition x_A. Hence, the separation is more easily made. A numerical measure of this separation is the relative volatility α_{AB}. This is defined as the ratio of the concentration of A in the vapor over the concentration of A in

Chap. 11   Vapor–Liquid Separation Processes

the liquid divided by the ratio of the concentration of B in the vapor over the concentration of B in the liquid.

$$\alpha_{AB} = \frac{y_A/x_A}{y_B/x_B} = \frac{y_A/x_A}{(1 - y_A)/(1 - x_A)} \tag{11.3-1}$$

where α_{AB} is the relative volatility of A with respect to B in the binary system.

If the system obeys Raoult's law, such as the benzene–toluene system,

$$y_A = \frac{P_A x_A}{P} \qquad y_B = \frac{P_B x_B}{P} \tag{11.3-2}$$

Substituting Eq. (11.3-2) into (11.3-1) for an ideal system,

$$\alpha_{AB} = \frac{P_A}{P_B} \tag{11.3-3}$$

Equation (11.3-1) can be rearranged to give

$$y_A = \frac{\alpha x_A}{1 + (\alpha - 1)x_A} \tag{11.3-4}$$

where $\alpha = \alpha_{AB}$. When the value of α is above 1.0, a separation is possible. The value of α may change as concentration changes. When binary systems follow Raoult's law, the relative volatility often varies only slightly over a large concentration range at constant total pressure.

EXAMPLE 11.3-1. *Relative Volatility for Benzene–Toluene System*
Using the data from Table 11.1-1, calculate the relative volatility for the benzene–toluene system at 85°C (358.2 K) and 105°C (378.2 K).

Solution: At 85°C, substituting into Eq. (11.3-3) for a system following Raoult's law,

$$\alpha = \frac{P_A}{P_B} = \frac{116.9}{46.0} = 2.54$$

Similarly at 105°C,

$$\alpha = \frac{204.2}{86.0} = 2.38$$

The variation in α is about 7%.

11.3C   Equilibrium or Flash Distillation

1. Introduction to distillation methods.   Distillation can be carried out by either of two main methods in practice. The first method of distillation involves the production of a vapor by boiling the liquid mixture to be separated in a single stage and recovering and condensing the vapors. No liquid is allowed to return to the single-stage still to contact the rising vapors. The second method of distillation involves the returning of a portion of the condensate to the still. The vapors rise through a series of stages or trays, and part of the condensate flows downward through the series of stages or trays countercurrently to the vapors. This second method is called *fractional distillation, distillation with reflux,* or *rectification.*

There are three important types of distillation that occur in a single stage or still and that do not involve rectification. The first of these is equilibrium or flash distillation, the second is simple batch or differential distillation, and the third is simple steam distillation.

2. Equilibrium or flash distillation. In *equilibrium* or *flash distillation*, which occurs in a single stage, a liquid mixture is partially vaporized. The vapor is allowed to come to equilibrium with the liquid, and the vapor and liquid phases are then separated. This can be done batchwise or continuously.

In Fig. 11.3-1 a binary mixture of components A and B flowing at the rate of F mol/h into a heater is partially vaporized. Then the mixture reaches equilibrium and is separated. The composition of F is x_F mole fraction of A. A total material balance on component A is as follows:

$$Fx_F = Vy + Lx \tag{11.3-5}$$

Since $L = F - V$, Eq. (11.3-5) becomes

$$Fx_F = Vy + (F - V)x \tag{11.3-6}$$

Usually, the moles per hour of feed F, moles per hour of vapor V, and moles per hour of L are known or set. Hence, there are two unknowns x and y in Eq. (11.3-6). The other relationship needed to solve Eq. (11.3-6) is the equilibrium line. A convenient method to use is to plot Eq. (11.3-6) on the xy equilibrium diagram. The intersection of the equation and the equilibrium line is the desired solution. This is similar to Example 11.2-1 and shown in Fig. 11.2-1.

11.3D Simple Batch or Differential Distillation

In *simple batch* or *differential distillation*, liquid is first charged to a heated kettle. The liquid charge is boiled slowly and the vapors are withdrawn as rapidly as they form to a condenser, where the condensed vapor (distillate) is collected. The first portion of vapor condensed will be richest in the more volatile component A. As vaporization proceeds, the vaporized product becomes leaner in A.

In Fig. 11.3-2 a simple still is shown. Originally, a charge of L_1 moles of components A and B with a composition of x_1 mole fraction of A is placed in the still. At any given time, there are L moles of liquid left in the still with composition x and the composition of the vapor leaving in equilibrium is y. A differential amount of dL is vaporized.

The composition in the still pot changes with time. For deriving the equation for this process, we assume that a small amount of dL is vaporized. The composition of the liquid

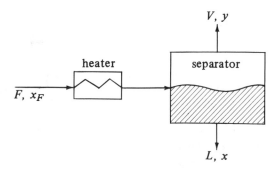

FIGURE 11.3-1. *Equilibrium or flash distillation.*

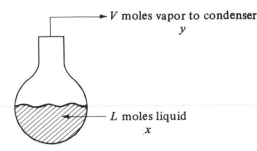

FIGURE 11.3-2   *Simple batch or differential distillation.*

changes from x to $x - dx$ and the amount of liquid from L to $L - dL$. A material balance on A can be made where the original amount = the amount left in the liquid + the amount of vapor.

$$xL = (x - dx)(L - dL) + y\,dL \qquad (11.3-7)$$

Multiplying out the right side,

$$xL = xL - x\,dL - L\,dx + dx\,dL + y\,dL \qquad (11.3-8)$$

Neglecting the term $dx\,dL$ and rearranging,

$$\frac{dL}{L} = \frac{dx}{y - x} \qquad (11.3-9)$$

Integrating,

$$\int_{L_2}^{L_1} \frac{dL}{L} = \ln\frac{L_1}{L_2} = \int_{x_2}^{x_1} \frac{dx}{y - x} \qquad (11.3-10)$$

where L_1 is the original moles charged, L_2 the moles left in the still, x_1 the original composition, and x_2 the final composition of liquid.

The integration of Eq. (11.3-10) can be done graphically by plotting $1/(y - x)$ versus x and getting the area under the curve between x_1 and x_2. The equilibrium curve gives the relationship between y and x. Equation (11.3-10) is known as the *Rayleigh equation*. The average composition of total material distilled, y_{av}, can be obtained by a material balance.

$$L_1 x_1 - L_2 x_2 + (L_1 - L_2)y_{av} \qquad (11.3-11)$$

EXAMPLE 11.3-2.   Simple Differential Distillation
A mixture of 100 mol containing 50 mol % *n*-pentane and 50 mol % *n*-heptane is distilled under differential conditions at 101.3 kPa until 40 mol is distilled. What is the average composition of the total vapor distilled and the composition of the liquid left? The equilibrium data are as follows, where x and y are mole fractions of *n*-pentane.

x	y	x	y	x	y
1.000	1.000	0.398	0.836	0.059	0.271
0.867	0.984	0.254	0.701	0	0
0.594	0.925	0.145	0.521		

Solution: The given values to be used in Eq. (11.3-10) are $L_1 = 100$ mol, $x_1 = 0.50$, $L_2 = 60$ mol, V (moles distilled) $= 40$ mol. Substituting into Eq. (11.3-10),

$$\ln \frac{100}{60} = 0.510 = \int_{x_2}^{x_1 = 0.5} \frac{dx}{y - x} \qquad (11.3\text{-}12)$$

The unknown is x_2, the composition of the liquid L_2 at the end of the differential distillation. To do the graphical integration a plot of $1/(y - x)$ versus x is made in Fig. 11.3-3 as follows. For $x = 0.594$, the equilibrium value of $y = 0.925$. Then $1/(y - x) = 1/(0.925 - 0.594) = 3.02$. The point $1/(y - x) = 3.02$ and $x = 0.594$ is plotted. In a similar manner, other points are plotted.

To determine the value of x_2, the area of Eq. (11.3-12) is obtained under the curve from $x_1 = 0.5$ to x_2 such that the area $= 0.510$. Hence, $x_2 = 0.277$. Substituting into Eq. (11.3-11) and solving for the average composition of the 40 mol distilled,

$$100(0.50) = 60(0.277) + 40(y_{av})$$

$$y_{av} = 0.835$$

11.3E Simple Steam Distillation

At atmospheric pressure high-boiling liquids cannot be purified by distillation since the components of the liquid may decompose at the high temperatures required. Often the high-boiling substances are essentially insoluble in water, so a separation at lower temperatures can be obtained by *simple steam distillation*. This method is often used to separate a high-boiling component from small amounts of nonvolatile impurities.

If a layer of liquid water (A) and an immiscible high-boiling component (B) such as a hydrocarbon are boiled at 101.3 kPa abs pressure, then, by the phase rule, Eq. (10.2-1), for three phases and two components,

$$F = 2 - 3 + 2 = 1 \text{ degree of freedom}$$

Hence, if the total pressure is fixed, the system is fixed. Since there are two liquid phases, each will exert its own vapor pressure at the prevailing temperature and cannot be influenced by the presence of the other. When the sum of the separate vapor pressures equals the total pressure, the mixture boils and

$$P_A + P_B = P \qquad (11.3\text{-}13)$$

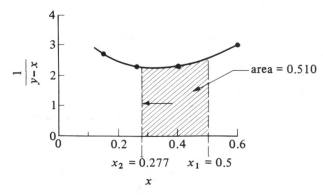

FIGURE 11.3-3. *Graphical integration for Example 11.3-2.*

where P_A is vapor pressure of pure water A and P_B is vapor pressure of pure B. Then the vapor composition is

$$y_A = \frac{P_A}{P} \qquad y_B = \frac{P_B}{P} \tag{11.3-14}$$

As long as the two liquid phases are present, the mixture will boil at the same temperature, giving a vapor of constant composition y_A. The temperature is found by using the vapor-pressure curves of pure A and pure B.

Note that by steam distillation, as long as liquid water is present, the high-boiling component B vaporizes at a temperature well below its normal boiling point without using a vacuum. The vapors of water (A) and high-boiling component (B) are usually condensed in a condenser and the resulting two immiscible liquid phases separated. This method has the disadvantage that large amounts of heat must be used to simultaneously evaporate the water with the high-boiling compound.

The ratio moles of B distilled to moles of A distilled is

$$\frac{n_B}{n_A} = \frac{P_B}{P_A} \tag{11.3-15}$$

Steam distillation is sometimes used in the food industry for the removal of volatile taints and flavors from edible fats and oils. In many cases vacuum distillation is used instead of steam distillation to purify high-boiling materials. The total pressure is quite low so that the vapor pressure of the system reaches the total pressure at relatively low temperatures.

Van Winkle (V1) derives equations for steam distillation where an appreciable amount of a nonvolatile component is present with the high-boiling component. This involves a three-component system. He also considers other cases for binary batch, continuous, and multicomponent batch steam distillation.

11.4 DISTILLATION WITH REFLUX AND McCABE–THIELE METHOD

11.4A Introduction to Distillation with Reflux

Rectification (fractionation) or stage distillation with reflux, from a simplified point of view, can be considered to be a process in which a series of flash-vaporization stages are arranged in a series in such a manner that the vapor and liquid products from each stage flow countercurrently to each other. The liquid in a stage is conducted or flows to the stage below and the vapor from a stage flows upward to the stage above. Hence, in each stage a vapor stream V and a liquid stream L enter, are mixed and equilibrated, and a vapor and a liquid stream leave in equilibrium. This process flow diagram was shown in Fig. 10.3-1 for a single stage and an example given in Example 11.2-1 for a benzene–toluene mixture.

For the countercurrent contact with multiple stages in Fig. 10.3-2, the material-balance or operating-line equation (10.3-13) was derived which relates the concentrations of the vapor and liquid streams passing each other in each stage. In a distillation column the stages (referred to as *sieve plates* or *trays*) in a distillation tower are arranged vertically, as shown schematically in Fig. 11.4-1.

The feed enters the column in Fig. 11.4-1 somewhere in the middle of the column. If the feed is liquid, it flows down to a sieve tray or stage. Vapor enters the tray and bubbles through the liquid on this tray as the entering liquid flows across. The vapor and liquid leaving the tray are essentially in equilibrium. The vapor continues up to the next tray or stage, where it is again contacted with a downflowing liquid. In this case the concentration of the more volatile component (the lower-boiling component A) is being increased in the vapor from each stage going upward and decreased in the liquid from each stage going downward. The final vapor product coming overhead is condensed in a condenser and a portion of the liquid product (distillate) is removed, which contains a high concentration of A. The remaining liquid from the condenser is returned (refluxed) as a liquid to the top tray.

The liquid leaving the bottom tray enters a reboiler, where it is partially vaporized, and the remaining liquid, which is lean in A or rich in B, is withdrawn as liquid product. The vapor from the reboiler is sent back to the bottom stage or tray. Only three trays are shown in the tower of Fig. 11.4-1. In most cases the number of trays is much greater. In the sieve tray the vapor enters through an opening and bubbles up through the liquid to give intimate contact of the liquid and vapor on the tray. In a theoretical tray the vapor and liquid leaving are in equilibrium. The reboiler can be considered as a theoretical stage or tray.

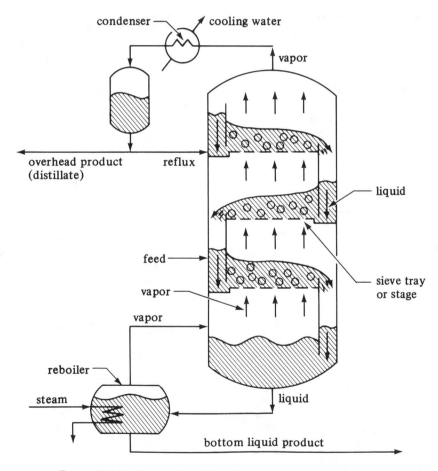

FIGURE 11.4-1.   *Process flow of a fractionating tower containing sieve trays.*

11.4B McCabe–Thiele Method of Calculation for Number of Theoretical Stages

1. Introduction and assumptions. A mathematical–graphical method for determining the number of theoretical trays or stages needed for a given separation of a binary mixture of A and B has been developed by McCabe and Thiele. The method uses material balances around certain parts of the tower, which give operating lines somewhat similar to Eq. (10.3-13), and the xy equilibrium curve for the system.

The main assumption made in the McCabe–Thiele method is that there must be equimolar overflow through the tower between the feed inlet and the top tray and the feed inlet and bottom tray. This can be shown in Fig. 11.4-2, where liquid and vapor streams enter a tray, are equilibrated, and leave. A total material balance gives

$$V_{n+1} + L_{n-1} = V_n + L_n \tag{11.4-1}$$

A component balance on A gives

$$V_{n+1}y_{n+1} + L_{n-1}x_{n-1} = V_n y_n + L_n x_n \tag{11.4-2}$$

where V_{n+1} is mol/h of vapor from tray $n + 1$, L_n is mol/h liquid from tray n, y_{n+1} is mole fraction of A in V_{n+1}, and so on. The compositions y_n and x_n are in equilibrium and the temperature of the tray n is T_n. If T_n is taken as a datum, it can be shown by a heat balance that the sensible heat differences in the four streams are quite small if heats of solution are negligible. Hence, only the latent heats in stream V_{n+1} and V_n are important. Since molar latent heats for chemically similar compounds are almost the same, $V_{n+1} = V_n$ and $L_n = L_{n-1}$. Therefore, we have constant molal overflow in the tower.

2. Equations for enriching section. In Fig. 11.4-3 a continuous distillation column is shown with feed being introduced to the column at an intermediate point and an overhead distillate product and a bottoms product being withdrawn. The upper part of the tower above the feed entrance is called the *enriching section*, since the entering feed of binary components A and B is enriched in this section, so that the distillate is richer in A than the feed. The tower is at steady state.

An overall material balance around the entire column in Fig. 11.4-3 states that the entering feed of F mol/h must equal the distillate D in mol/h plus the bottoms W in mol/h.

$$F = D + W \tag{11.4-3}$$

A total material balance on component A gives

$$Fx_F = Dx_D + Wx_W \tag{11.4-4}$$

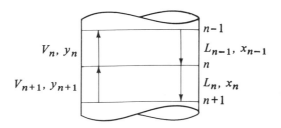

FIGURE 11.4-2. *Vapor and liquid flows entering and leaving a tray.*

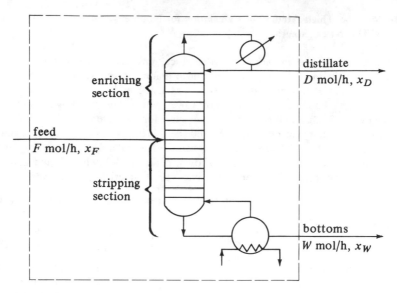

FIGURE 11.4-3. *Distillation column showing material-balance sections for McCabe–Thiele method.*

In Fig. 11.4-4a the distillation tower section above the feed, the enriching section, is shown schematically. The vapor from the top tray having a composition y_1 passes to the condenser, where it is condensed so that the resulting liquid is at the boiling point. The reflux stream L mol/h and distillate D mol/h have the same composition, so $y_1 = x_D$. Since equimolal overflow is assumed, $L_1 = L_2 = L_n$ and $V_1 = V_2 = V_n = V_{n+1}$.

Making a total material balance over the dashed-line section in Fig. 11.4-4a,

$$V_{n+1} = L_n + D \qquad (11.4\text{-}5)$$

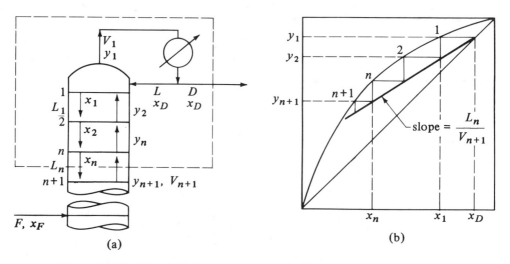

FIGURE 11.4-4. *Material balance and operating line for enriching section : (a) schematic of tower, (b) operating and equilibrium lines.*

Making a balance on component A,

$$V_{n+1} y_{n+1} = L_n x_n + D x_D \tag{11.4-6}$$

Solving for y_{n+1}, the enriching-section operating line is

$$y_{n+1} = \frac{L_n}{V_{n+1}} x_n + \frac{D x_D}{V_{n+1}} \tag{11.4-7}$$

Since $V_{n+1} = L_n + D$, $L_n/V_{n+1} = R/(R+1)$ and Eq. (11.4-7) becomes

$$y_{n+1} = \frac{R}{R+1} x_n + \frac{x_D}{R+1} \tag{11.4-8}$$

where $R = L_n/D$ = reflux ratio = constant. Equation (11.4-7) is a straight line on a plot of vapor composition versus liquid composition. It relates the compositions of two streams passing each other and is plotted in Fig. 11.4-4b. The slope is L_n/V_{n+1} or $R/(R+1)$, as given in Eq. (11.4-8). It intersects the $y = x$ line (45° diagonal line) at $x = x_D$. The intercept of the operating line at $x = 0$ is $y = x_D/(R+1)$.

The theoretical stages are determined by starting at x_D and stepping off the first plate to x_1. Then y_2 is the composition of the vapor passing the liquid x_1. In a similar manner, the other theoretical trays are stepped off down the tower in the enriching section to the feed tray.

3. *Equations for stripping section.*   Making a total material balance over the dashed-line section in Fig. 11.4-5a for the stripping section of the tower below the feed entrance,

$$V_{m+1} = L_m - W \tag{11.4-9}$$

Making a balance on component A,

$$V_{m+1} y_{m+1} = L_m x_m - W x_W \tag{11.4-10}$$

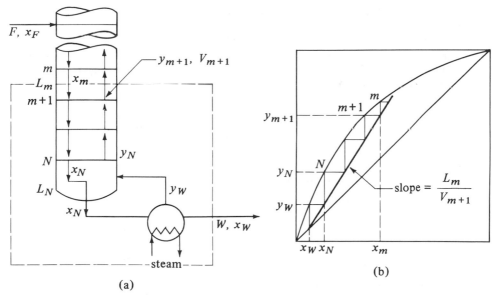

FIGURE 11.4-5.   *Material balance and operating line for stripping section: (a) schematic of tower, (b) operating and equilibrium lines.*

Solving for y_{m+1}, the stripping-section operating line is

$$y_{m+1} = \frac{L_m}{V_{m+1}} x_m - \frac{Wx_W}{V_{m+1}} \tag{11.4-11}$$

Again, since equimolal flow is assumed, $L_m = L_N = $ constant and $V_{m+1} = V_N = $ constant. Equation (11.4-11) is a straight line when plotted as y versus x in Fig. 11.4-5b, with a slope of L_m/V_{m+1}. It intersects the $y = x$ line at $x = x_W$. The intercept at $x = 0$ is $y = -Wx_W/V_{m+1}$.

Again the theoretical stages for the stripping section are determined by starting at x_W, going up to y_W, and then across to the operating line, etc.

4. Effect of feed conditions. The condition of the feed stream F entering the tower determines the relation between the vapor V_m in the stripping section and V_n in the enriching section and also between L_m and L_n. If the feed is part liquid and part vapor, the vapor will add to V_m to give V_n.

For convenience, we represent the condition of the feed by the quantity q, which is defined as

$$q = \frac{\text{heat needed to vaporize 1 mol of feed at entering conditions}}{\text{molar latent heat of vaporization of feed}} \tag{11.4-12}$$

If the feed enters at its boiling point, the numerator of Eq. (11.4-12), is the same as the denominator and $q = 1.0$. Equation (11.4-12) can also be written in terms of enthalpies.

$$q = \frac{H_V - H_F}{H_V - H_L} \tag{11.4-13}$$

where H_V is the enthalpy of the feed at the dew point, H_L the enthalpy of the feed at the boiling point (bubble point), and H_F the enthalpy of the feed at its entrance conditions. If the feed enters as vapor at the dew point, $q = 0$. For cold liquid feed $q > 1.0$, for superheated vapor $q < 0$, and for the feed being part liquid and part vapor, q is the fraction of feed that is liquid.

We can look at q also as the number of moles of saturated liquid produced on the feed plate by each mole of feed added to the tower. In Fig. 11.4-6 a diagram shows the relationship between flows above and below the feed extrance. From the definition of q, the following equations hold:

$$L_m = L_n + qF \tag{11.4-14}$$

$$V_n = V_m + (1 - q)F \tag{11.4-15}$$

FIGURE 11.4-6. *Relationship between flows above and below the feed entrance.*

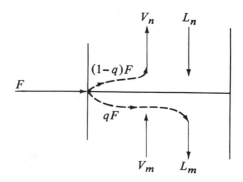

The point of intersection of the enriching and the stripping operating-line equations on an xy plot can be derived as follows. Rewriting Eqs. (11.4-6) and (11.4-10) as follows without the tray subscripts:

$$V_n y = L_n x + D x_D \tag{11.4-16}$$

$$V_m y = L_m x - W x_W \tag{11.4-17}$$

where the y and x values are the point of intersection of the two operating lines. Subtracting Eq. (11.4-16) from (11.4-17),

$$(V_m - V_n)y = (L_m - L_n)x - (D x_D + W x_W) \tag{11.4-18}$$

Substituting Eqs. (11.4-4), (11.4-14), and (11.4-15) into Eq. (11.4-18) and rearranging,

$$y = \frac{q}{q-1} x - \frac{x_F}{q-1} \tag{11.4-19}$$

This equation is the q-line equation and is the locus of the intersection of the two operating lines. Setting $y = x$ in Eq. (11.4-19), the intersection of the q-line equation with the 45° line is $y = x = x_F$, where x_F is the overall composition of the feed.

In Fig. 11.4-7 the q line is plotted for various feed conditions given below the figure. The slope of the q line is $q/(q-1)$. For example, for the liquid below the boiling point, $q > 1$, and the slope is > 1.0, as shown. The enriching and operating lines are plotted for the case of a feed of part liquid and part vapor and the two lines intersect on the q line. A convenient way to locate the stripping operating line is to first plot the enriching operating line and the q line. Then draw the stripping line between the intersection of the q line and enriching operating line and the point $y = x = x_W$.

5. *Location of the feed tray in a tower and number of trays.* To determine the number of theoretical trays needed in a tower, the stripping and operating lines are drawn to intersect on the q line as shown in Fig. 11.4-8. Starting at the top at x_D, the trays are stepped off. For trays 2 and 3, the steps can go to the enriching operating line, as shown

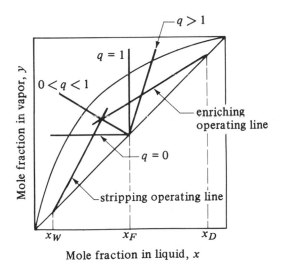

FIGURE 11.4-7. *Location of the q line for various feed conditions : liquid below boiling point (q > 1), liquid at boiling point (q = 1), liquid + vapor (0 < q < 1), saturated vapor (q = 0).*

in Fig. 11.4-8a. At step 4 the step goes to the stripping line. A total of about 4.6 theoretical steps are needed. The feed enters on tray 4.

For the correct method, the shift is made on step 2 to the stripping line, as shown in Fig. 11.4-8b. A total of only about 3.7 steps are needed with the feed on tray 2. To keep the number of trays to a minimum, the shift from the enriching to the stripping operating line should be made at the first opportunity after passing the operating-line intersection.

In Fig. 11.4-8b the feed is part liquid and part vapor since $0 < q < 1$. Hence, in adding the feed to tray 2, the vapor portion of the feed is separated and added beneath plate 2 and the liquid added to the liquid from above entering tray 2. If the feed is all liquid, it should be added to the liquid flowing to tray 2 from the tray above. If the feed is all vapor, it should be added below tray 2 and joins the vapor rising from the plate below.

Since a reboiler is considered a theoretical step when the vapor y_W is in equilibrium with x_W as in Fig. 11.4-5b, the number of theoretical trays in a tower is equal to the number of theoretical steps minus one.

EXAMPLE 11.4-1. Rectification of a Benzene–Toluene Mixture

A liquid mixture of benzene–toluene is to be distilled in a fractionating tower at 101.3 kPa pressure. The feed of 100 kg mol/h is liquid and it contains 45 mol % benzene and 55 mol % toluene and enters at 327.6 K (130°F). A distillate containing 95 mol % benzene and 5 mol % toluene and a bottoms containing 10 mol % benzene and 90 mol % toluene are to be obtained. The reflux ratio is 4:1. The average heat capacity of the feed is 159 kJ/kg mol·K (38 btu/lb mol·°F) and the average latent heat 32099 kJ/kg mol (13800 btu/lb mol). Equilibrium data for this system are given in Table 11.1-1 and in Fig. 11.1-1. Calculate the kg moles per hour distillate, kg moles per hour bottoms, and the number of theoretical trays needed.

Solution: The given data are $F = 100$ kg mol/h, $x_F = 0.45$, $x_D = 0.95$, $x_W = 0.10$, and $R = L_n/D = 4$. For the overall material balance substituting into Eq. (11.4-3),

$$F = D + W \qquad (11.4\text{-}3)$$

$$100 = D + W$$

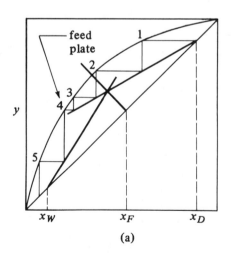

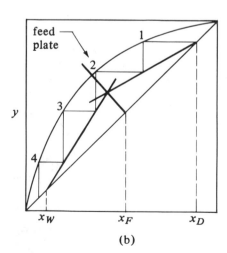

(a)

(b)

FIGURE 11.4-8. *Method of stepping off number of theoretical trays and location of feed plate : (a) improper location of feed on tray 4, (b) proper location of feed on tray 2 to give minimum number of steps.*

Substituting into Eq. (11.4-4) and solving for D and W,

$$Fx_F = Dx_D + Wx_W \qquad \text{(11.4-4)}$$

$$100(0.45) = D(0.95) + (100 - D)(0.10)$$

$$D = 41.2 \text{ kg mol/h} \qquad W = 58.8 \text{ kg mol/h}$$

For the enriching operating line, using Eq. (11.4-8),

$$y_{n+1} = \frac{R}{R+1} x_n + \frac{x_D}{R+1} = \frac{4}{4+1} x_n + \frac{0.95}{4+1} = 0.800 x_n + 0.190$$

The equilibrium data from Table 11.1-1 and the enriching operating line above are plotted in Fig. 11.4-9.

Next, the value of q is calculated. From the boiling-point diagram, Fig. 11.1-1, for $x_F = 0.45$, the boiling point of the feed is 93.5°C or 366.7 K (200.3°F). From Eq. (11.4-13),

$$q = \frac{H_V - H_F}{H_V - H_L} \qquad \text{(11.4-13)}$$

The value of $H_V - H_L$ = latent heat = 32 099 kJ/kg mol. The numerator of Eq. (11.4-13) is

$$H_V - H_F = (H_V - H_L) + (H_L - H_F) \qquad \text{(11.4-20)}$$

Also,

$$H_L - H_F = c_{pL}(T_B - T_F) \qquad \text{(11.4-21)}$$

where the heat capacity of the liquid feed c_{pL} = 159 kJ/kg mol · K, T_B = 366.7 K (boiling point of feed), and T_F = 327.6 K (inlet feed temperature). Substituting Eqs. (11.4-20) and (11.4-21) into (11.4-13),

$$q = \frac{(H_V - H_L) + c_{pL}(T_B - T_F)}{H_V - H_L} \qquad \text{(11.4-22)}$$

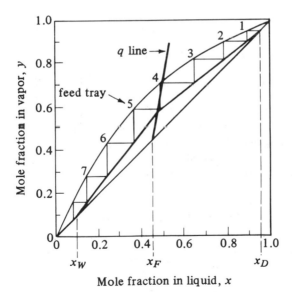

FIGURE 11.4-9. McCabe-Thiele diagram for distillation of benzene–toluene for Example 11.4-1.

Substituting the known values into Eq. (11.4-22),

$$q = \frac{32\,099 + 159(366.7 - 327.6)}{32\,099} = 1.195 \qquad \text{(SI)}$$

$$q = \frac{13\,800 + 38(200.3 - 130)}{13\,800} = 1.195 \qquad \text{(English)}$$

From Eq. (11.4-19), the slope of the q line is

$$\frac{q}{q - 1} = \frac{1.195}{1.195 - 1} = 6.12$$

The q line is plotted in Fig. 11.4-9 starting at the point $y = x_F = 0.45$ with a slope of 6.12.

The stripping operating line is drawn connecting the point $y = x = x_W = 0.10$ with the intersection of the q line and the enriching operating line. Starting at the point $y = x = x_D$, the theoretical steps are drawn in as shown in Fig. 11.4-9. The number of theoretical steps is 7.6 or 7.6 steps minus a reboiler, which gives 6.6 theoretical trays. The feed is introduced on tray 5 from the top.

11.4C Total and Minimum Reflux Ratio for McCabe–Thiele Method

1. *Total reflux.*   In distillation of a binary mixture A and B the feed conditions, distillate composition, and bottoms composition are usually specified and the number of theoretical trays are to be calculated. However, the number of theoretical trays needed depends upon the operating lines. To fix the operating lines, the reflux ratio $R = L_n/D$ at the top of the column must be set.

One of the limiting values of reflux ratio is that of total reflux, or $R = \infty$. Since $R = L_n/D$ and, by Eq. (11.4-5),

$$V_{n+1} = L_n + D \tag{11.4-5}$$

then L_n is very large, as is the vapor flow V_n. This means that the slope $R/(R + 1)$ of the enriching operating line becomes 1.0 and the operating lines of both sections of the column coincide with the 45° diagonal line, as shown in Fig. 11.4-10.

The number of theoretical trays required is obtained as before by stepping off the trays from the distillate to the bottoms. This gives the minimum number of trays that can possibly be used to obtain the given separation. In actual practice, this condition can be realized by returning all the overhead condensed vapor V_1 from the top of the tower back to the tower as reflux, i.e., total reflux. Also, all the liquid in the bottoms is reboiled. Hence, all the products distillate and bottoms are reduced to zero flow, as is the fresh feed to the tower.

This condition of total reflux can also be interpreted as requiring infinite sizes of condenser, reboiler, and tower diameter for a given feed rate.

If the relative volatility α of the binary mixture is approximately constant, the following analytical expression by Fenske can be used to calculate the minimum number of theoretical steps N_m when a total condenser is used.

$$N_m = \frac{\log\left(\dfrac{x_D}{1 - x_D} \dfrac{1 - x_W}{x_W}\right)}{\log \alpha_{\text{av}}} \tag{11.4-23}$$

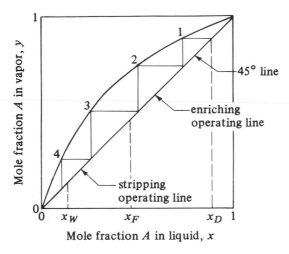

FIGURE 11.4-10. *Total reflux and minimum number of trays by McCabe–Thiele method.*

For small variations in α, $\alpha_{av} = (\alpha_1 \alpha_W)^{1/2}$, where α_1 is the relative volatility of the overhead vapor and α_W is the relative volatility of the bottoms liquid.

2. Minimum reflux ratio. The minimum reflux ratio can be defined as the reflux ratio R_m that will require an infinite number of trays for the given separation desired of x_D and x_W. This corresponds to the minimum vapor flow in the tower, and hence the minimum reboiler and condenser sizes. This case is shown in Fig. 11.4-11. If R is decreased, the slope of the enriching operating line $R/(R + 1)$ is decreased, and the intersection of this line and the stripping line with the q line moves farther from the 45° line and closer to the

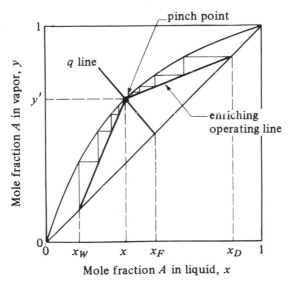

FIGURE 11.4-11. *Minimum reflux ratio and infinite number of trays by McCabe–Thiele method.*

equilibrium line. As a result, the number of steps required to give a fixed x_D and x_W increases. When the two operating lines touch the equilibrium line, a "pinch point" at y' and x' occurs where the number of steps required becomes infinite. The slope of the enriching operating line is as follows from Fig. 11.4-11, since the line passes through the points x', y', and x_D ($y = x_D$).

$$\frac{R_m}{R_m + 1} = \frac{x_D - y'}{x_D - x'} \tag{11.4-24}$$

In some cases, where the equilibrium line has an inflection in it as shown in Fig. 11.4-12, the operating line at minimum reflux will be tangent to the equilibrium line.

3. Operating and optimum reflux ratio. For the case of total reflux, the number of plates is a minimum, but the tower diameter is infinite. This corresponds to an infinite cost of tower and steam and cooling water. This is one limit in the tower operation. Also, for minimum reflux, the number of trays is infinite, which again gives an infinite cost. These are the two limits in operation of the tower.

The actual operating reflux ratio to use is in between these two limits. To select the proper value of R requires a complete economic balance on the fixed costs of the tower and operating costs. The optimum reflux ratio to use for lowest total cost per year is between the minimum R_m and total reflux. This has been shown for many cases to be at an operating reflux ratio between $1.2R_m$ to $1.5R_m$.

EXAMPLE 11.4-2. Minimum Reflux Ratio and Total Reflux in Rectification

For the rectification in Example 11.4-1, where a benzene–toluene feed is being distilled to give a distillate composition of $x_D = 0.95$ and a bottoms composition of $x_W = 0.10$, calculate the following.
  (a) Minimum reflux ratio R_m.
  (b) Minimum number of theoretical plates at total reflux.

Solution: For part (a) the equilibrium line is plotted in Fig. 11.4-13 and the q-line equation is also shown for $x_F = 0.45$. Using the same x_D and x_W as in Example 11.4-1, the enriching operating line for minimum reflux is plotted as a dashed line and intersects the equilibrium line at the same point at

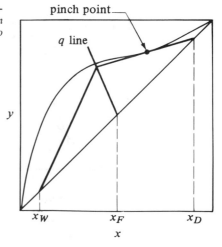

FIGURE 11.4-12. *Minimum reflux ratio and infinite number of trays when operating line is tangent to equilibrium line.*

which the q line intersects. Reading off the values of $x' = 0.49$ and $y' = 0.702$, substituting into Eq. (11.4-24), and solving for R_m,

$$\frac{R_m}{R_m + 1} = \frac{x_D - y'}{x_D - x'} = \frac{0.95 - 0.702}{0.95 - 0.49}$$

Hence, the minimum reflux ratio $R_m = 1.17$.

For the case of total reflux in part (b), the theoretical steps are drawn as shown in Fig. 11.4-13. The minimum number of theoretical steps is 5.8, which gives 4.8 theoretical trays plus a reboiler.

11.4D Special Cases for Rectification Using McCabe–Thiele Method

1. Stripping-column distillation.   In some cases the feed to be distilled is not supplied to an intermediate point in a column but is added to the top of the stripping column as shown in Fig. 11.4-14a. The feed is usually a saturated liquid at the boiling point and the overhead product V_D is the vapor rising from the top plate, which goes to a condenser with no reflux or liquid returned back to the tower.

The bottoms product W usually has a high concentration of the less volatile component B. Hence, the column operates as a stripping tower with the vapor removing the more volatile A from the liquid as it flows downward. Assuming constant molar flow rates, a material balance of the more volatile component A around the dashed line in Fig. 11.4-14a gives, on rearrangement,

$$y_{m+1} = \frac{L_m}{V_{m+1}} x_m - \frac{W x_W}{V_{m+1}} \tag{11.4-25}$$

This stripping-line equation is the same as the stripping-line equation for a complete tower given as Eq. (11.4-11). It intersects the $y = x$ line at $x = x_W$, and the slope is constant at L_m/V_{m+1}.

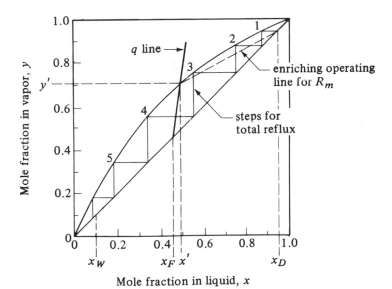

FIGURE 11.4-13.   *Graphical solution for minimum reflux ratio R_m and total reflux for Example 11.4-2.*

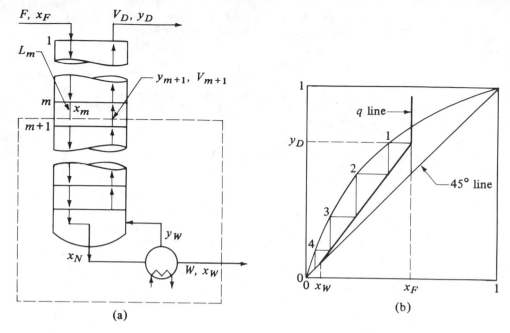

FIGURE 11.4-14.   *Material balance and operating line for stripping tower : (a) flows in tower, (b) operating and equilibrium line.*

If the feed is saturated liquid, then $L_m = F$. If the feed is cold liquid below the boiling point, the q line should be used and $q > 1$.

$$L_m = qF \qquad (11.4\text{-}26)$$

In Fig. 11.4-14 the stripping operating-line equation (11.4-25) is plotted and the q line, Eq. (11.4-19), is also shown for $q = 1.0$. Starting at x_F, the steps are drawn down the tower.

EXAMPLE 11.4-3.   *Number of Trays in Stripping Tower*
A liquid feed at the boiling point of 400 kg mol/h containing 70 mol % benzene (A) and 30 mol % toluene (B) is fed to a stripping tower at 101.3 kPa pressure. The bottoms product flow is to be 60 kg mol/h containing only 10 mol % A and the rest B. Calculate the kg mol/h overhead vapor, its composition, and the number of theoretical steps required.

Solution: Referring to Fig. 11.4-14a, the known values are $F = 400$ kg mol/h, $x_F = 0.70$, $W = 60$ kg mol/h, and $x_W = 0.10$. The equilibrium data from Table 11.1-1 are plotted in Fig. 11.4-15. Making an overall material balance,

$$F = W + V_D$$

$$400 = 60 + V_D$$

Solving, $V_D = 340$ kg mol/h. Making a component A balance and solving,

$$Fx_F = Wx_W + V_D y_D$$

$$400(0.70) = 60(0.10) + 340(y_D)$$

$$y_D = 0.806$$

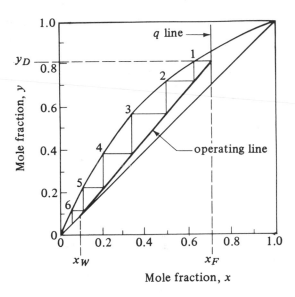

FIGURE 11.4-15. *Stripping tower for Example 11.4-3.*

For a saturated liquid, the q line is vertical and is plotted in Fig. 11.4-15. The operating line is plotted through the point $y = x_W = 0.10$ and the intersection of $y_D = 0.806$ with the q line. Alternatively, Eq. (11.4-25) can be used with a slope of $L_m/V_{m+1} = 400/340$. Stepping off the trays from the top, 5.3 theoretical steps or 4.3 theoretical trays plus a reboiler are needed.

2. Enriching-column distillation. Enriching towers are also sometimes used, where the feed enters the bottom of the tower as a vapor. The overhead distillate is produced in the same manner as in a complete fractionating tower and is usually quite rich in the more volatile component A. The liquid bottoms is usually comparable to the feed in composition, being slightly leaner in component A. If the feed is saturated vapor, the vapor in the tower $V_n = F$. Enriching-line equation (11.4-7) holds, as does the q-line equation (11.4-19).

3. Rectification with direct steam injection Generally, the heat to a distillation tower is applied to one side of a heat exchanger (reboiler) and the steam does not directly contact the boiling solution, as shown in Fig. 11.4-5. However, when an aqueous solution of more volatile A and water B is being distilled, the heat required may be provided by the use of open steam injected directly at the bottom of the tower. The reboiler exchanger is then not needed.

The steam is injected as small bubbles into the liquid in the tower bottom, as shown in Fig. 11.4-16a. The vapor leaving the liquid is then in equilibrium with the liquid if sufficient contact is obtained. Making an overall balance on the tower and a balance on A,

$$F + S = D + W \tag{11.4-27}$$

$$Fx_F + Sy_S = Dx_D + Wx_W \tag{11.4-28}$$

where S = mol/h of steam and $y_S = 0$ = mole fraction of A in steam. The enriching operating-line equation is the same as for indirect steam.

For the stripping-line equation, an overall balance and a balance on component A are as follows:

$$L_m + S = V_{m+1} + W \qquad (11.4\text{-}29)$$

$$L_m x_m + S(0) = V_{m+1} y_{m+1} + W x_W \qquad (11.4\text{-}30)$$

Solving for y_{m+1} in Eq. (11.4-30),

$$y_{m+1} = \frac{L_m}{V_{m+1}} x_m - \frac{W x_W}{V_{m+1}} \qquad (11.4\text{-}31)$$

For saturated steam entering, $S = V_{m+1}$ and hence, by Eq. (11.4-29), $L_m = W$. Substituting into Eq. (11.4-31), the stripping operating line is

$$y_{m+1} = \frac{W}{S} x_m - \frac{W}{S} x_W \qquad (11.4\text{-}32)$$

When $y = 0$, $x = x_W$. Hence, the stripping line passes through the point $y = 0, x = x_W$, as shown in Fig. 11.4-16b, and is continued to the x axis. Also, for the intersection of the stripping line with the 45° line, when $y = x$ in Eq. (11.4-32), $x = W x_W/(W - S)$.

For a given reflux ratio and overhead distillate composition, the use of open steam rather than closed requires an extra fraction of a stage, since the bottom step starts below the $y = x$ line (Fig. 11.4-16b). The advantage of open steam lies in simpler construction of the heater, which is a sparger.

4. *Rectification tower with side stream.* In certain situations, intermediate product or

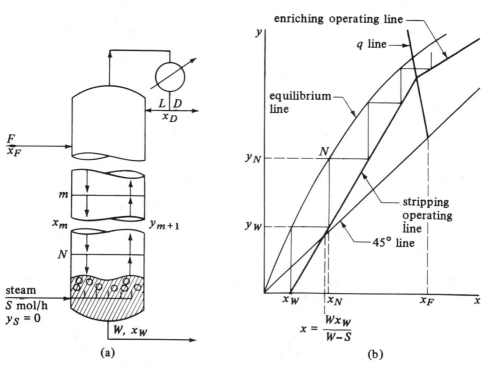

FIGURE 11.4-16. *Use of direct steam injection: (a) schematic of tower, (b) operating and equilibrium lines.*

side streams are removed from sections of the tower between the distillate and the bottoms. The side stream may be vapor or liquid and may be removed at a point above the feed entrance or below depending on the composition desired.

The flows for a column with a liquid side stream removed above the feed inlet are shown in Fig. 11.4-17. The top enriching operating line above the liquid side stream and the stripping operating line below the feed are found in the usual way. The equation of the q line is also unaffected by the side stream and is found as before. The liquid side stream alters the liquid rate below it, and hence the material balance or operating line in the middle portion between the feed and liquid side stream plates.

Making a total material balance on the top portion of the tower as shown in the dashed-line box in Fig. 11.4-17,

$$V_{S+1} = L_S + O + D \tag{11.4-33}$$

where O is mol/h saturated liquid removed as a side stream. Since the liquid side stream is saturated,

$$L_n = L_S + O \tag{11.4-34}$$

$$V_{S+1} = V_{n+1} \tag{11.4-35}$$

Making a balance on the most volatile component,

$$V_{S+1}y_{S+1} = L_S x_S + O x_O + D x_D \tag{11.4-36}$$

Solving for y_{S+1}, the operating line for the region between the side stream and the feed is

$$y_{S+1} = \frac{L_S}{V_{S+1}} x_S + \frac{O x_O + D x_D}{V_{S+1}} \tag{11.4-37}$$

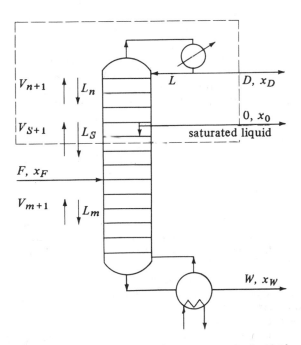

FIGURE 11.4-17. *Process flow for a rectification tower with a liquid side stream.*

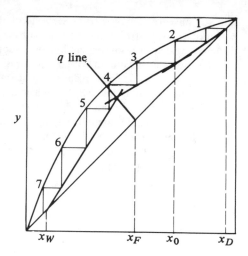

The slope of this line is L_S/V_{S+1}. The line can be located as shown in Fig. 11.4-18 by the q line, which determines the intersection of the stripping line and Eq. (11.4-37), or it may be fixed by the specification of x_O, the side-stream composition. The step on the McCabe–Thiele diagram must actually be at the intersection of the two operating lines at x_O in an actual tower. If this does not occur, the reflux ratio can be altered slightly to change the steps.

5. Partial condensers. In a few cases it may be desired to remove the overhead distillate product as a vapor instead of a liquid. This can also occur when the low boiling point of the distillate makes complete condensation difficult. The liquid condensate in a partial condenser is returned to the tower as reflux and the vapor removed as product as shown in Fig. 11.4-19.

If the time of contact between the vapor product and the liquid is sufficient, the partial condenser is a theoretical stage. Then the composition x_R of the liquid reflux is in equilibrium with the vapor composition y_D, where $y_D = x_D$. If the cooling in the condenser is rapid and the vapor and liquid do not reach equilibrium, only a partial stage separation is obtained.

11.5 DISTILLATION AND ABSORPTION TRAY EFFICIENCIES

11.5A Introduction

In all the previous discussions on theoretical trays or stages in distillation, we assumed that the vapor leaving a tray was in equilibrium with the liquid leaving. However, if the time of contact and the degree of mixing on the tray is insufficient, the streams will not be in equilibrium. As a result the efficiency of the stage or tray is not 100%. This means that we must use more actual trays for a given separation than the theoretical number of trays determined by calculation. The discussions in this section apply to both absorption and distillation tray towers.

Three types of tray or plate efficiency are used: *overall tray efficiency E_O*, *Murphree tray efficiency E_M*, and point or local tray efficiency E_{MP} (sometimes called *Murphree point efficiency*). These will be considered individually.

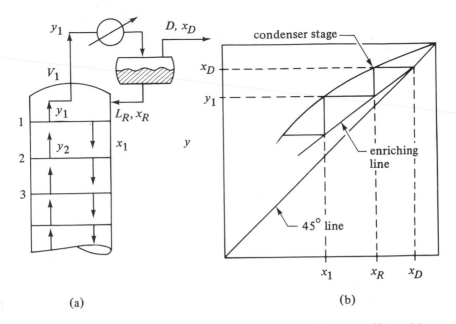

y_1

D, x_D

condenser stage

V_1

x_D

y_1

L_R, x_R

y_1

1

y_2

x_1

y

2

enriching line

3

$45°$ line

x_1 x_R x_D

(a)

(b)

FIGURE 11.4-19. *Partial condenser where the vapor and liquid leave in equilibrium : (a) process flow diagram, (b) McCabe–Thiele plot.*

11.5B Types of Tray Efficiencies

1. Overall tray efficiency. The *overall tray* or *plate efficiency* E_O concerns the entire tower and is simple to use but is the least fundamental. It is defined as the ratio of the number of theoretical or ideal trays needed in an entire tower to the number of actual trays used.

$$E_O = \frac{\text{number of ideal trays}}{\text{number of actual trays}} \qquad \text{(11.5-1)}$$

For example, if eight theoretical steps are needed and the overall efficiency is 60%, the number of theoretical trays is eight minus a reboiler, or seven trays. The actual number of trays is 7/0.60, or 11.7 trays.

Two empirical correlations for absorption and distillation overall tray efficiencies in commercial towers are available for standard tray designs (O1). For hydrocarbon distillation these values range from about 50 to 85% and for hydrocarbon absorption from about 10 to 50%. These correlations should only be used for approximate estimates.

2. Murphree tray efficiency. The *Murphree tray efficiency* E_M is defined as follows:

$$E_M = \frac{y_n - y_{n+1}}{y_n^* - y_{n+1}} \qquad \text{(11.5-2)}$$

where y_n is the average actual concentration of the mixed vapor leaving the tray n shown in Fig. 11.5-1, y_{n+1} the average actual concentration of the mixed vapor entering tray n, and y_n^* the concentration of the vapor that would be in equilibrium with the liquid of concentration x_n leaving the tray to the downcomer.

The liquid entering the tray has a concentration of x_{n-1}, and as it travels across the tray, its concentration drops to x_n at the outlet. Hence, there is a concentration gradient in the liquid as it flows across the tray. The vapor entering the tray contacts liquid of different concentrations, and the outlet vapor will not be uniform in concentration.

3. Point efficiency. The *point* or *local efficiency* E_{MP} on a tray is defined as

$$E_{MP} = \frac{y_n' - y_{n+1}'}{y_n^* - y_{n+1}'} \tag{11.5-3}$$

where y_n' is the concentration of the vapor at a specific point in plate n as shown in Fig. 11.5-1, y_{n+1}' the concentration of the vapor entering the plate n at the same point, and $y_n'^*$ the concentration of the vapor that would be in equilibrium with x_n' at the same point. Since y_n' cannot be greater than $y_n'^*$, the local efficiency cannot be greater than 1.00 or 100%.

In small-diameter towers the vapor flow sufficiently agitates the liquid so that it is uniform on the tray. Then the concentration of the liquid leaving is the same as that on the tray. Then $y_n' = y_n$, $y_{n+1}' = y_{n+1}$, and $y_n'^* = y_n^*$. The point efficiency then equals the Murphree tray efficiency or $E_M = E_{MP}$.

In large-diameter columns incomplete mixing of the liquid occurs on the trays. Some vapor will contact the entering liquid x_{n-1}, which is richer in component A than x_n. This will give a richer vapor at this point than at the exit point, where x_n leaves. Hence, the tray efficiency E_M will be greater than the point efficiency E_{MP}. The value of E_M can be related to E_{MP} by the integration of E_{MP} over the entire tray.

11.5C Relationship Between Efficiencies

The relationship between E_{MP} and E_M can be derived mathematically if the amount of liquid mixing is specified and the amount of vapor mixing is also set. Derivations for three different sets of assumptions are given by Robinson and Gilliland (R1). However, experimental data are usually needed to obtain amounts of mixing. Semitheoretical methods to predict E_{MP} and E_M are summarized in detail by Van Winkle (V1).

When the Murphree tray efficiency E_M is known or can be predicted, the overall tray

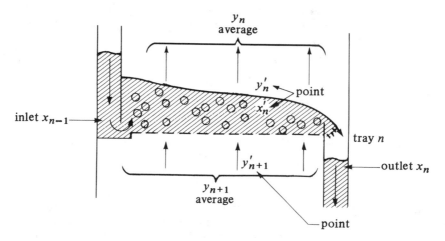

FIGURE 11.5-1. *Vapor and liquid compositions on a sieve tray and tray efficiency.*

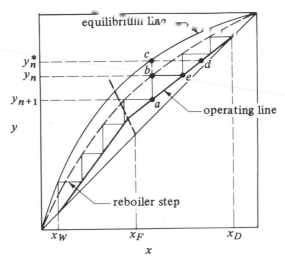

FIGURE 11.5-2.   *Use of Murphree plate efficiency to determine actual number of trays.*

efficiency E_O can be related to E_M by several methods. In the first method an analytical expression is as follows when the slope m of the equilibrium line is constant and also the slope L/V of the operating line:

$$E_O = \frac{\log\,[1 + E_M(mV/L - 1)]}{\log(mV/L)} \tag{11.5-4}$$

If the equilibrium and operating lines of the tower are not straight, a graphical method in the McCabe–Thiele diagram can be used to determine the actual number of trays when the Murphree tray efficiency is known. In Fig. 11.5-2 a diagram is given for an actual plate as compared with an ideal plate. The triangle acd represents an ideal plate and the smaller triangle abe the actual plate. For the case shown, the Murphree efficiency $E_M = 0.60 = ba/ca$. The dashed line going through point b is drawn so that ba/ca for each tray is 0.60. The trays are stepped off using this efficiency, and the total number of steps gives the actual number of trays needed. The reboiler is considered to be one theoretical tray, so the true equilibrium curve is used for this tray as shown. In Fig. 11.5-2, 6.0 actual trays plus a reboiler are obtained.

11.6   FRACTIONAL DISTILLATION USING ENTHALPY–CONCENTRATION METHOD

11.6A   Enthalpy–Concentration Data

1. Introduction.   In Section 11.4B the McCabe–Thiele method was used to calculate the number of theoretical steps or trays needed for a given separation of a binary mixture of A and B by rectification or fractional distillation. The main assumptions in the method are that the latent heats are equal, sensible heat differences are negligible, and constant molal overflow occurs in each section of the distillation tower. In this section we shall consider fractional distillation using enthalpy–concentration data where the molal overflow rates are not necessarily constant. The analysis will be made using enthalpy as well as material balances.

2. Enthalpy-concentration data.

An enthalpy–concentration diagram for a binary vapor–liquid mixture of A and B takes into account latent heats, heats of solution or mixing, and sensible heats of the components of the mixture. The following data are needed to construct such a diagram at a constant pressure: (1) heat capacity of the liquid as a function of temperature, composition, and pressure; (2) heat of solution as a function of temperature and composition; (3) latent heats of vaporization as a function of composition and pressure or temperature; and (4) boiling point as a function of pressure, composition, and temperature.

The diagram at a given constant pressure is based on arbitrary reference states of liquid and temperature, such a 273 K (32°F). The saturated liquid line in enthalpy h kJ/kg (btu/lb$_m$) or kJ/kg mol is calculated by

$$h = x_A c_{pA}(T - T_0) + (1 - x_A)c_{pB}(T - T_0) + \Delta H_{sol} \tag{11.6-1}$$

where x_A is wt or mole fraction A, T is boiling point of the mixture in K (°F) or °C, T_0 is reference temperature, K, c_{pA} is the liquid heat capacity of the component A in kJ/kg·K (btu/lb$_m$·°F) or kJ/kg mol·K, c_{pB} is heat capacity of B, and ΔH_{sol} is heat of solution at T_0 in kJ/kg (btu/lb$_m$) or kJ/kg mol. If heat is evolved on mixing, the ΔH_{sol} will be a negative value in Eq. (11.6-1). Often, the heats of solution are small, as in hydrocarbon mixtures, and are neglected.

The saturated vapor enthalpy line of H kJ/kg (btu/lb$_m$) or kJ/kg mol of a vapor composition y_A is calculated by

$$H = y_A[\lambda_A + c_{py_A}(T - T_0)] + (1 - y_A)[\lambda_B + c_{pyB}(T - T_0)] \tag{11.6-2}$$

where c_{py_A} is the vapor heat capacity of A and c_{py_B} for B. The latent heats λ_A and λ_B are the values at the reference temperature T_0. Generally, the latent heat is given as λ_{Ab} at the normal boiling point T_{bA} of the pure component A and λ_{Bb} for B. Then to correct this to the reference temperature T_0 to use in Eq. (11.6-2),

$$\lambda_A = c_{pA}(T_{bA} - T_0) + \lambda_{Ab} - c_{py_A}(T_{bA} - T_0) \tag{11.6-3}$$

$$\lambda_B = c_{pB}(T_{bB} - T_0) + \lambda_{Bb} - c_{py_B}(T_{bB} - T_0) \tag{11.6-4}$$

In Eq. (11.6-3) the pure liquid is heated from T_0 to T_{bA}, vaporized at T_{bA}, and then cooled as a vapor to T_0. Similarly, Eq. (11.6-4) also holds for λ_B. For convenience, the reference temperature T_0 is often taken as equal to the boiling point of the lower-boiling component A. This means $\lambda_A = \lambda_{Ab}$. Hence, only λ_{Bb} must be corrected to λ_B.

EXAMPLE 11.6-1. Enthalpy-Concentration Plot for Benzene-Toluene

Prepare an enthalpy-concentration plot for benzene-toluene at 1 atm pressure. Equilibrium data are given in Table 11.1-1 and Figs. 11.1-1 and 11.1-2. Physical property data are given in Table 11.6-1.

TABLE 11.6-1. *Physical Property Data for Benzene and Toluene*

Component	Boiling Point, (°C)	Heat Capacity, (kJ/kg mol · K) Liquid	Heat Capacity, (kJ/kg mol · K) Vapor	Latent Heat of Vaporization (kJ/kg mol)
Benzene (A)	80.1	138.2	96.3	30 820
Toluene (B)	110.6	167.5	138.2	33 330

Solution: A reference temperature of $T_0 = 80.1°C$ will be used for convenience so that the liquid enthalpy of pure benzene ($x_A = 1.0$) at the boiling point will be zero. For the first point we will select pure toluene ($x_A = 0$). For liquid toluene at the boiling point of 110.6°C using Eq. (11.6-1) with zero heat of solution and data from Table 11.6-1,

$$h = x_A c_{pA}(T - 80.1) + (1 - x_A)c_{pB}(T - 80.1) + 0 \qquad \textbf{(11.6-5)}$$

$$h = 0 + (1 - 0)(167.5)(110.6 - 80.1) = 5109 \text{ kJ/kg mol}$$

For the saturated vapor enthalpy line, Eq. (11.6-2) is used. However, we first must calculate λ_B at the reference temperature $T_0 = 80.1°C$ using Eq. (11.6-4).

$$\lambda_B = c_{pB}(T_{bB} - T_0) + \lambda_{Bb} - c_{pyB}(T_{bB} - T_0) \qquad \textbf{(11.6-4)}$$

$$= 167.5(110.6 - 80.1) + 33\,330 - 138.2(110.6 - 80.1)$$

$$= 34\,224 \text{ kJ/kg mol}$$

To calculate H, Eq. (11.6-2) is used and $y_A = 0$.

$$H = y_A[\lambda_A + c_{pyA}(T - T_0)] + (1 - y_A)[\lambda_B + c_{pyB}(T - T_0)] \qquad \textbf{(11.6-2)}$$

$$= 0 + (1.0 - 0)[34\,224 + 138.2(110.6 - 80.1)]$$

$$= 38\,439 \text{ kJ/kg mol}$$

For pure benzene, $x_A = 1.0$ and $y_A = 1.0$. Using Eq. (11.6-5), since $T = T_0 = 80.1$, $h = 0$. For the saturated vapor enthalpy, using Eq. (11.6-2) and $T = 80.1$,

$$H = 1.0[30\,820 + 96.3(80.1 - 80.1)] + 0 = 30\,820$$

Selecting $x_A = 0.50$, the boiling point $T_b = 92°C$ and the temperature of saturated vapor for $y_A = 0.50$ is 98.8°C from Fig. 11.1-1. Using Eq. (11.6-5) for the saturated liquid enthalpy at the boiling point,

$$h = 0.5(138.2)(92 - 80.1) + (1 - 0.5)(167.5)(92 - 80.1) = 1820$$

Using Eq. (11.6-2) for $y_A = 0.5$, the saturated vapor enthalpy at 98.8°C is

$$H = 0.5[30\,820 + 96.3(98.8 - 80.1)] + (1 - 0.5)[34\,224$$

$$+ 138.2(98.8 - 80.1)] = 34\,716$$

Selecting $x_A = 0.30$ and $y_A = 0.30$, $h = 2920$ and $H = 36\,268$. Also, for $x_A = 0.80$ and $y_A = 0.80$, $h = 562$ and $H = 32\,380$. These values are tabulated in Table 11.6-2 and plotted in Fig. 11.6-1.

Some properties of the enthalpy-concentration plot are as follows. The region in between the saturated vapor line and the saturated liquid line is the two-phase liquid-vapor region. From Table 11.1-1 for $x_A = 0.411$, the vapor in equilibrium is $y_A = 0.632$. These two points are plotted in Fig. 11.6-1 and this tie line represents the

Saturated Liquid		Saturated Vapor	
Mole fraction, x_A	*Enthalpy, h, (kJ/kg mol)*	*Mole fraction, y_A*	*enthalpy, H, (kJ/kg mol)*
0	5109	0	38 439
0.30	2920	0.30	36 268
0.50	1820	0.50	34 716
0.80	562	0.80	32 380
1.00	0	1.00	30 820

enthalpies and compositions of the liquid and vapor phases in equilibrium. Other tie lines can be drawn in a similar manner. The region below the h versus x_A line represents liquid below the boiling point.

11.6B   Distillation in Enriching Section of Tower

To analyze the enriching section of a fractionating tower using enthalpy-concentration data, we make an overall and a component balance in Fig. 11.6-2.

$$V_{n+1} = L_n + D \qquad \qquad \textbf{(11.6-6)}$$

$$V_{n+1} y_{n+1} = L_n x_n + D x_D \qquad \qquad \textbf{(11.6-7)}$$

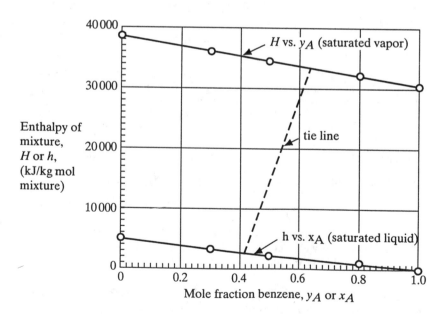

FIGURE 11.6-1. *Enthalpy-concentration plot for benzene-toluene mixture at 1.0 atm abs.*

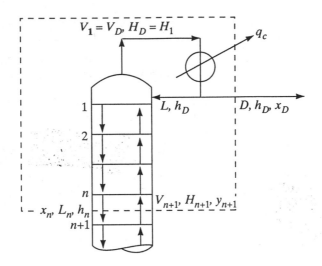

FIGURE 11.6-2. *Enriching section of distillation tower.*

Equation (11.6-7) can be rearranged to give the enriching-section operating line.

$$y_{n+1} = \frac{L_n}{V_{n+1}} x_n + \frac{Dx_D}{V_{n+1}} \qquad (11.6\text{-}8)$$

This is the same as Eq. (11.4-7) for the McCabe-Thiele method, but now the liquid and vapor flow rates V_{n+1} and L_n may vary throughout the tower and Eq. (11.6-8) will not be a straight line on an xy plot.

Making an enthalpy balance,

$$V_{n+1}H_{n+1} = L_n h_n + Dh_D + q_c \qquad (11.6\text{-}9)$$

where q_c is the condenser duty, kJ/h or kW (btu/h). An enthalpy balance can be made just around the condenser.

$$q_c = V_1 H_1 - Lh_D - Dh_D \qquad (11.6\text{-}10)$$

By combining Eqs. (11.6-9) and (11.6-10) to eliminate q_c, an alternative form is obtained.

$$V_{n+1}H_{n+1} = L_n h_n + V_1 H_1 - Lh_D \qquad (11.6\text{-}11)$$

Substituting the value of L_n from Eq. (11.6-6) into (11.6-11),

$$V_{n+1}H_{n+1} = (V_{n+1} - D) h_n + V_1 H_1 - Lh_D \qquad (11.6\text{-}12)$$

Equations (11.6-8) and (11.6-12) are the final working equations for the enriching section.

In order to plot the operating line Eq. (11.6-8), the terms V_{n+1} and L_n must be determined from Eq. (11.6-12). If the reflux ratio is set, V_1 and L are known. The values H_1 and h_D can be determined by Eqs. (11.6-1) and (11.6-2) or from an enthalpy-concentration plot. If a value of x_n is selected, it is a trial-and-error solution to obtain H_{n+1} since y_{n+1} is not known. The steps to follow are given below.

1. Select a value of x_n. Assume $V_{n+1} = V_1 = L + D$ and $L_n = L$. Then using these values in Eq. (11.6-8), calculate an approximate value of y_{n+1}. This assumes a straight operating line.
2. Using this y_{n+1}, obtain H_{n+1} and also obtain h_n using x_n. Substitute these values into Eq. (11.6-12) and solve for V_{n+1}. Obtain L_n from Eq. (11.6-6).
3. Substitute into Eq. (11.6-8) and solve for y_{n+1}.
4. If the calculated value of y_{n+1} does not equal the assumed value of y_{n+1}, repeat steps 2–3. Generally, a second trial is not needed. Assume another value of x_n and repeat steps 1–4.
5. Plot the curved operating line for the enriching section. Generally, only a few values of the flows L_n and V_{n+1} are needed to determine the operating line, which is slightly curved.

11.6C Distillation in Stripping Section of Tower

To analyze the stripping section of a distillation tower, an overall and a component material balance are made on Fig. 11.4-5a.

$$L_m = W + V_{m+1} \tag{11.6-13}$$

$$L_m x_m = W x_W + V_{m+1} y_{m+1} \tag{11.6-14}$$

$$y_{m+1} = \frac{L_m}{V_{m+1}} x_m - \frac{W x_W}{V_{m+1}} \tag{11.6-15}$$

Making an enthalpy balance with q_R kJ/h or kW(btu/h) entering the reboiler in Fig. 11.4-5a and substituting $(V_{m+1} + W)$ for L_m from Eq. (11.6-13),

$$V_{m+1} H_{m+1} = (V_{m+1} + W) h_m + q_R - W h_W \tag{11.6-16}$$

Making an overall enthalpy balance in Fig. 11.4-3,

$$q_R = D h_D + W h_W + q_c - F h_F \tag{11.6-17}$$

The final working equations to use are Eqs. (11.6-15)-(11.6-17).

Using a method similar to that for the enriching section to solve the equations, select a value of y_{m+1} and calculate an approximate value of x_m from Eq. (11.6-15) assuming constant molal overflow. Then calculate V_{m+1} and L_m from Eqs. (11.6-16) and (11.6-13). Then use Eq. (11.6-15) to determine x_m. Compare this calculated value of x_m with the assumed value.

EXAMPLE 11.6-2. Distillation Using Enthalpy-Concentration Method
A liquid mixture of benzene-toluene is being distilled using the same conditions as in Example 11.4-1 except that a reflux ratio of 1.5 times the minimum reflux ratio is to be used. The value of $R_m = 1.17$ from Example 11.4-2 will be used. Use enthalpy balances to calculate the flow rates of the liquid and vapor at various points in the tower and plot the curved operating lines. Determine the number of theoretical stages needed.

Solution: The given data are as follows: $F = 100$ kg mol/h, $x_F = 0.45$, $x_D = 0.95$, $x_W = 0.10$, $R = 1.5 R_m = 1.5(1.17) = 1.755$, $D = 41.2$ kg mol/h, $W = 58.8$ kg mol/h. The feed enters at 54.4°C and $q = 1.195$. The flows at the top of the tower are calculated as follows.

$$\frac{L}{D} = 1.755; \qquad L = 1.755(41.2) = 72.3; \qquad V_1 = L + D = 72.3$$

$$+ 41.2 = 113.5$$

The saturation temperature at the top of the tower for $y_1 = x_D = 0.95$ is 82.3°C from Fig. 11.1-1. Using Eq. (11.6-2),

$$H_1 = 0.95[30\ 820 + 96.3(82.3 - 80.1)] + (1 - 0.95)[34\ 224$$

$$+ 138.2(82.3 - 80.1)] = 31\ 206$$

This value of 31 206 could also have been obtained from the enthalpy-concentration plot, Fig. 11.6-1. The boiling point of the distillate D is obtained from Fig. 11.1-1 and is 81.1°C. Using Eq. (11.6-5),

$$h_D = 0.95(138.2)(81.1 - 80.1) + (1 - 0.95)(167.5)(81.1 - 80.1) = 139$$

Again this value could have been obtained from Fig. 11.6-1.

Following the procedure outlined for the enriching section for step 1, a value of $x_n = 0.55$ is selected. Assuming a straight operating line for Eq. (11.6-8), an approximate value of y_{n+1} is obtained.

$$y_{n+1} = \frac{72.3}{113.5} x_n + \frac{41.2}{113.5}(0.95) = 0.637(x_n) + 0.345$$

$$= 0.637(0.55) + 0.345 = 0.695$$

Starting with step 2 and using Fig. 11.6-1, for $x_n = 0.55$, $h_n = 1590$ and for $y_{n+1} = 0.695$, $H_{n+1} = 33\ 240$. Substituting into Eq. (11.6-12) and solving,

$$V_{n+1}(33\ 240) = (V_{n+1} - 41.2)1590 + 113.5(31\ 206) - 72.3(139)$$

$$V_{n+1} = 109.5$$

Using Eq. (11.6-6),

$$109.5 = L_n + 41.2 \quad \text{or} \quad L_n = 68.3$$

For step 3, substituting into Eq. (11.6-8),

$$y_{n+1} = \frac{68.3}{109.5}(0.55) + \frac{41.2}{109.5}(0.95) = 0.700$$

This calculated value of $y_{n+1} = 0.700$ is sufficiently close to the approximate value of 0.695 so that no further trials are needed.

Selecting another value for $x_n = 0.70$ and using Eq. (11.6-8), an approximate value of y_{n+1} is calculated.

$$y_{n+1} = \frac{72.3}{113.5}(0.70) + \frac{41.2}{113.5}(0.95) = 0.791$$

Using Fig. 11.6-1 for $x_n = 0.70$, $h_n = 1000$, and for $y_{n+1} = 0.791$, $H_{n+1} = 32\,500$. Substituting into Eq. (11.6-12) and solving,

$$V_{n+1}(32\,500) = (V_{n+1} - 41.2)1000 + 113.5(31\,206) - 72.3(139)$$

$$V_{n+1} = 110.8$$

Using Eq. (11.6-6),

$$L_n = 110.8 - 41.2 = 69.6$$

Substituting into Eq. 11.6-8),

$$y_{n+1} = \frac{69.6}{110.8}(0.70) + \frac{41.2}{110.8}(0.95) = 0.793$$

In Fig. 11.6-3, the points for the curved operating line in the enriching section are plotted. This line is approximately straight and is very slightly above that for constant molal overflow.

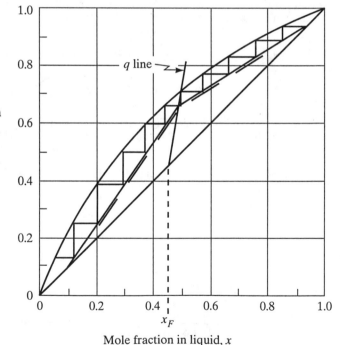

Mole fraction in vapor, y

Mole fraction in liquid, x

FIGURE 11.6-3. *Plot of curved operating lines using enthalpy-concentration method for Example 11.6-2. Solid lines are for enthalpy-concentration method and dashed lines for constant molal overflow.*

Using Eq. (11.6-10), the condenser duty is calculated.

$$q_c = 113.5(31\ 206) - 72.3(139) - 41.2(139)$$

$$= 3\ 526\ 100\ kJ/h$$

To obtain the reboiler duty q_R, values for h_W and h_F are needed. Using Fig. 11.6-1 for $x_W = 0.10$, $h_W = 4350$. The feed is at 54.5°C. Using Eq. (11.6-5),

$$h_F = 0.45(138.2)(54.5 - 80.1) + (1 - 0.45)(167.5)(54.5 - 80.1)$$

$$= -3929$$

Using Eq. (11.6-17),

$$q_R = 41.2(139) + 58.8(4350) + 3\ 526\ 100 - 100(-3929)$$

$$= 4\ 180\ 500 kJ/h$$

Using Fig. 11.4-5 and making a material balance below the bottom tray and around the reboiler,

$$L_N = W + V_W \tag{11.6-18}$$

Rewriting Eq. (11.6-16) for this bottom section,

$$V_W H_W = (V_W + W)h_N + q_R - W h_W \tag{11.6-19}$$

From the equilibrium diagram, Fig. 11.1-2, for $x_W = 0.10$, $y_W = 0.207$, which is the vapor composition leaving the reboiler.

For equimolal overflow in the stripping section using Eqs. (11.4-14) and (11.4-15),

$$L_m = L_n + qF = 72.3 + 1.195(100) = 191.8 \tag{11.4-14}$$

$$V_{m+1} = V_{n+1} - (1 - q)F$$

$$= 113.5 - (1 - 1.195)100 = 133.0 \tag{11.4-15}$$

Selecting $y_{m+1} = y_W = 0.207$, and using Eq. (11.6-15), an approximate value of $x_m = x_N$ is obtained.

$$y_{m+1} = \frac{L_m}{V_{m+1}} x_m - \frac{W x_W}{V_{m+1}} \tag{11.6-15}$$

$$0.207 = \frac{191.8}{133.0}(x_N) - \frac{58.8}{133.0}(0.10)$$

Solving, $x_N = 0.174$. From Fig. (11.6-1) for $x_N = 0.174$, $h_N = 3800$, and for $y_W = 0.207$, $H_W = 37\ 000$. Substituting into Eq. (11.6-19),

$$V_W(37\ 000) = (V_W + 58.8)(3800) + 4\ 180\ 500 - 58.8(4350)$$

Solving, $V_W = 125.0$. Using Eq. (11.6-18), $L_N = 183.8$. Substituting into Eq. (11.6-15) and solving for x_N,

$$0.207 = \frac{183.8}{125.0}(x_N) - \frac{58.8}{125.0}(0.10)$$

$$x_N = 0.173$$

This value of 0.173 is quite close to the approximate value of 0.174.

Assuming a value of $y_{m+1} = 0.55$ and using Eq. (11.6-15), an approximate value of x_m is obtained.

$$y_{m+1} = 0.55 = \frac{191.8}{133.0}(x_m) - \frac{58.8}{133.0}(0.10)$$

$$x_m = 0.412$$

From Fig. (11.6-1) for $x_m = 0.412$, $h_m = 2300$, and for $y_{m+1} = 0.55$, $H_m = 34\ 400$. Substituting into Eq. (11.6-16),

$$V_{m+1}(34\ 400) = (V_{m+1} + 58.8)(2300) + 4\ 180\ 500 - 58.8(4350)$$

Solving, $V_{m+1} = 126.5$. Using Eq. (11.6-13),

$$L_m = W + V_{m+1} = 58.8 + 126.5 = 185.3$$

Substituting into Eq. (11.6-15) and solving for x_m,

$$y_{m+1} = 0.55 = \frac{185.3}{126.5}x_m - \frac{58.8}{126.5}(0.1)$$

$$x_m = 0.407$$

This value of 0.407 is sufficiently close to the approximate value of 0.412 so that no further trials are needed. The two points calculated for the stripping section are plotted in Fig. 11.6-3. This stripping line is also approximately straight and is very slightly above the operating line for constant molal overflow.

Using the operating line for the enthalpy balance method, the number of theoretical steps is 10.4. For the equimolal method 9.9 steps are obtained. This difference would be larger if the reflux ratio of 1.5 times R_m were decreased to, say, 1.2 or 1.3. At larger reflux ratios, this difference in number of steps would be less.

Note that in Example 11.6-2 in the stripping section the vapor flow increases slightly from 125.0 to 126.5 in going from the reboiler to near the feed tray. These values are lower than the value of 133.0 obtained assuming equimolal overflow. Similar conclusions hold for the enriching section. The enthalpy-concentration method is useful in calculating the internal vapor and liquid flows at any point in the column. These data are then used in sizing the trays. Also, calculations of q_c and q_R are used in designing the condenser and reboiler. This method is very applicable for design using a computer solution for binary and multicomponent mixtures to make tray to tray mass and enthalpy balances for the whole tower. A more restrictive Ponchon-Savarit graphical method for only binary mixtures is available (K3, T2).

11.7 DISTILLATION OF MULTICOMPONENT MIXTURES

11.7A Introduction to Multicomponent Distillation

In industry many of the distillation processes involve the separation of more than two components. The general principles of design of multicomponent distillation towers are the same in many respects as those described for binary systems. There is one mass balance for each component in the multicomponent mixture. Enthalpy or heat balances are made which are similar to those for the binary case. Equilibrium data are used to calculate boiling points and dew points. The concepts of minimum reflux and total reflux as limiting cases are also used.

1. Number of distillation towers needed. In binary distillation one tower was used to separate the two components *A* and *B* into relatively pure components with *A* in the overhead and *B* in the bottoms. However, in a multicomponent mixture of *n* components, *n* − 1 fractionators will be required for separation. For example, for a three-component system of components *A*, *B*, and *C*, where *A* is the most volatile and *C* the least volatile, two columns will be needed, as shown in Fig. 11.7-1. The feed of *A*, *B*, and *C* is distilled in column 1 and *A* and *B* are removed in the overhead and *C* in the bottoms. Since the separation in this column is between *B* and *C*, the bottoms containing *C* will contain a small amount of *B* and often a negligible amount of *A* (often called trace component). The amount of the trace component *A* in the bottoms can often be neglected if the relative volatilities are reasonably large. In column 2 the feed of *A* and *B* is distilled with *A* in the distillate containing a small amount of component *B* and a much smaller amount of *C*. The bottoms containing *B* will also be contaminated with a small amount of *C* and *A*. Alternately, column 1 could be used to remove *A* overhead with *B* plus *C* being fed to column 2 for separation of *B* and *C*.

2. Design calculation methods. In multicomponent distillation, as in binary, ideal stages or trays are assumed in the stage-to-stage calculations. Using equilibrium data, equilibrium calculations are used to obtain the boiling point and equilibrium vapor composition from a given liquid composition or the dew point and liquid composition from a given vapor composition. Material balances and heat balances similar to those described in Section 11.6 are then used to calculate the flows to and from the adjacent stages. These stage-to-stage design calculations involve trial-and-error calculations, and high-speed digital computers are generally used to provide rigorous solutions.

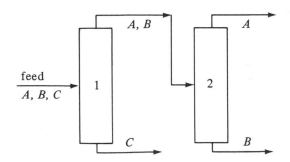

FIGURE 11.7-1. *Separation of a ternary system of A, B, and C.*

In a design the conditions of the feed are generally known or specified (temperature, pressure, composition, flow rate). Then, in most cases, the calculation procedure follows either of two general methods. In the first method, the desired separation or split between two of the components is specified and the number of theoretical trays are calculated for a selected reflux ratio. It is clear that with more than two components in the feed the complete compositions of the distillate and bottoms are not then known and trial-and-error procedures must be used. In the second method, the number of stages in the enriching section and in the stripping section and the reflux ratio are specified or assumed and the separation of the components is calculated using assumed liquid and vapor flows and temperatures for the first trial. This approach is often preferred for computer calculations (H2, P1). In the trial-and-error procedures, the design method of Thiele and Geddes (P1, S1, T1), which is a reliable procedure, is often used to calculate resulting distillate and bottoms compositions and tray temperatures and compositions. Various combinations and variations of the above rigorous calculation methods are available in the literature (H2, P1, S1) and are not considered further.

The variables in the design of a distillation column are all interrelated, and there are only a certain number of these which may be fixed in the design. For a more detailed discussion of the specification of these variables, see Kwauk (K2).

3. Shortcut calculation methods. In the remainder of this chapter, shortcut calculation methods for the approximate solution of multicomponent distillation are considered. These methods are quite useful to study a large number of cases rapidly to help orient the designer, to determine approximate optimum conditions, or to provide information for a cost estimate. Before discussing these methods, equilibrium relationships and calculation methods of bubble point, dew point, and flash vaporization for multicomponent systems are covered.

11.7B Equilibrium Data in Multicomponent Distillation

For multicomponent systems which can be considered ideal, Raoult's law can be used to determine the composition of the vapor in equilibrium with the liquid. For example, for a system composed of four components, A, B, C, and D,

$$p_A = P_A x_A, \qquad p_B = P_B x_B, \qquad p_C = P_C x_C, \qquad p_D = P_D x_D \qquad \text{(11.7-1)}$$

$$y_A = \frac{p_A}{P} = \frac{P_A}{P} x_A, \qquad y_B = \frac{P_B}{P} x_B, \qquad y_C = \frac{P_C}{P} x_C, \qquad y_D = \frac{P_D}{P} x_D \qquad \text{(11.7-2)}$$

In hydrocarbon systems, because of nonidealities, the equilibrium data are often represented by

$$y_A = K_A x_A, \qquad y_B = K_B x_B, \qquad y_C = K_C x_C, \qquad y_D = K_D x_D \qquad \text{(11.7-3)}$$

where K_A is the vapor–liquid equilibrium constant or distribution coefficient for component A. These K values for light hydrocarbon systems (methane to decane) have been determined semiempirically and each K is a function of temperature and pressure. Convenient K factor charts are available by Depriester (D1) and Hadden and Grayson (H1). For light hydrocarbon systems K is generally assumed not to be a function of composition, which is sufficiently accurate for most engineering calculations. Note that for an ideal system, $K_A = P_A/P$, and so on. As an example, data for the hydrocarbons *n*-butane, *n*-pentane, *n*-hexane, and *n*-heptane are plotted in Fig. 11.7-2 at 405.3 kPa (4.0 atm) absolute (D1, H1).

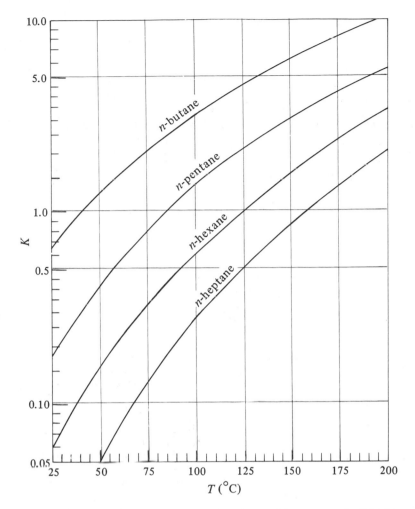

FIGURE 11.7-2. *Equilibrium K values for light hydrocarbon systems at 405.3 kPa (4.0 atm) absolute.*

The relative volatility α_i for each individual component in a multicomponent mixture can be defined in a manner similar to that for a binary mixture. If component C in a mixture of A, B, C, and D is selected as the base component,

$$\alpha_i = \frac{K_i}{K_C}, \qquad \alpha_A = \frac{K_A}{K_C}, \qquad \alpha_B = \frac{K_B}{K_C}, \qquad \alpha_C = \frac{K_C}{K_C} = 1.0, \qquad \alpha_D = \frac{K_D}{K_C} \quad (11.7\text{-}4)$$

The values of K_i will be a stronger function of temperature than the α_i values since the K_i lines in Fig. 11.7-2 all increase with temperature in a similar manner.

11.7C Boiling Point, Dew Point, and Flash Distillation

1. Boiling point. At a specified pressure, the boiling point or bubble point of a given multicomponent mixture must satisfy the relation $\sum y_i = 1.0$. For a mixture of A, B, C, and D with C as the base component,

$$\sum y_i = \sum K_i x_i = K_C \sum \alpha_i x_i = 1.0 \tag{11.7-5}$$

The calculation is a trial-and-error process, as follows. First a temperature is assumed and the values of α_i are calculated from the values of K_i at this temperature. Then the value of K_C is calculated from $K_C = 1.0/\sum \alpha_i x_i$. The temperature corresponding to the calculated value of $K_C = 1.0/\sum \alpha_i x_i$. The temperature corresponding to the calculated value of K_C is compared to the assumed temperature. If the values differ, the calculated temperature is used for the next iteration. After the final temperature is known, the vapor composition is calculated from

$$y_i = \frac{\alpha_i x_i}{\sum (\alpha_i x_i)} \tag{11.7-6}$$

2. *Dew point.* For the dew point calculation, which is also trial and error,

$$\sum x_i = \sum \left(\frac{y_i}{K_i}\right) = \left(\frac{1}{K_C}\right) \sum \left(\frac{y_i}{\alpha_i}\right) = 1.0 \tag{11.7-7}$$

The value of K_C is calculated from $K_C = \sum (y_i/\alpha_i)$. After the final temperature is known, the liquid composition is calculated from

$$x_i = \frac{y_i/\alpha_i}{\sum (y_i/\alpha_i)} \tag{11.7-8}$$

EXAMPLE 11.7-1. *Boiling Point of a Multicomponent Liquid*
A liquid feed to a distillation tower at 405.3 kPa abs is fed to a distillation tower. The composition in mol fractions is as follows: n-butane ($x_A = 0.40$), n-pentane ($x_B = 0.25$), n-hexane ($x_C = 0.20$), n-heptane ($x_D = 0.15$). Calculate the boiling point and the vapor in equilibrium with the liquid.

Solution: First a temperature of 65°C is assumed and the values of K obtained from Fig. 11.7-2. Using component C (n-hexane) as the base component, the following values are calculated using Eq. (11.7-5) for the first trial:

		Trial 1 (65°C)			Trial 3 (70°C) Final			
Comp.	x_i	K_i	$\dfrac{K_i}{K_C} = \alpha_i$	$\alpha_i x_i$	K_i	α_i	$\alpha_i x_i$	y_i
A	0.40	1.68	6.857	2.743	1.86	6.607	2.643	0.748
B	0.25	0.63	2.571	0.643	0.710	2.522	0.631	0.178
C	0.20	0.245	1.000	0.200	0.2815	1.000	0.200	0.057
D	0.15	0.093	0.380	0.057	0.110	0.391	0.059	0.017
	1.00			$\sum \alpha_i x_i = 3.643$			$\sum \alpha_i x_i = 3.533$	1.000

$$K_C = 1/\sum \alpha_i x_i = 1/3.643 = 0.2745 \ (69°\text{C}) \quad K_C = 1/3.533 = 0.2830 \ (70°\text{C})$$

The calculated value of K_C is 0.2745, which corresponds to 69°C, Fig. 11.7-2. Using 69°C for trial 2, a temperature of 70°C was obtained. Using 70°C for trial 3, the calculations shown in the table give a final calculated value of 70°C, which is the bubble point. Values of y_i are calculated from Eq. (11.7-6).

3. *Flash distillation of multicomponent mixture.* For flash distillation, the process flow diagram is shown in Fig. 11.3-1. Defining $f = V/F$ as the fraction of the feed vaporized

and $(1 - f) = L/F$ as the fraction of the feed remaining as liquid and making a component i balance as in Eq. (11.3-6), the following is obtained:

$$y_i = \frac{f-1}{f} x_i + \frac{x_{iF}}{f} \tag{11.7-9}$$

where y_i is the composition of component i in the vapor, which is in equilibrium with x_i in the liquid after vaporization. Also, for equilibrium, $y_i = K_i x_i = K_C \alpha_i x_i$, where $\alpha_i = K_i/K_C$. Then Eq. (11.7-9) becomes

$$y_i = K_C \alpha_i x_i = \frac{f-1}{f} x_i + \frac{x_{iF}}{f} \tag{11.7-10}$$

Solving for x_i and summing for all components,

$$\sum x_i = \sum \frac{x_{iF}}{f(K_C \alpha_i - 1) + 1} = 1.0 \tag{11.7-11}$$

This equation is solved by trial and error by first assuming a temperature if the fraction f vaporized has been set. When the $\sum x_i$ values add up to 1.0, the proper temperature has been chosen. The composition of the vapor y_i can be obtained from $y_i = K_C \alpha_i x_i$ or by a material balance.

11.7D Key Components in Multicomponent Distillation

Fractionation of a multicomponent mixture in a distillation tower will allow separation only between two components. For a mixture of A, B, C, D, and so on, a separation in one tower can only be made between A and B, or B and C, and so on. The components separated are called the *light key*, which is the more volatile (identified by the subscript L), and the *heavy key* (H). The components more volatile than the light key are called *light components* and will be present in the bottoms in small amounts. The components less volatile than the heavy key are called *heavy components* and are present in the distillate in small amounts. The two key components are present in significant amounts in both the distillate and bottoms.

11.7E Total Reflux for Multicomponent Distillation

1. Minimum stages for total reflux.   Just as in binary distillation, the minimum number of theoretical stages or steps, N_m, can be determined for multicomponent distillation for total reflux. The Fenske equation (11.4-23) also applies to any two components in a multicomponent system. When applied to the heavy key H and the light key L, it becomes

$$N_m = \frac{\log [(x_{LD} D/x_{HD} D)(x_{HW} W/x_{LW} W)]}{\log (\alpha_{L, av})} \tag{11.7-12}$$

where x_{LD} is mole fraction of light key in distillate, x_{LW} is mole fraction in bottoms, x_{HD} is mole fraction of heavy key in distillate, and x_{HW} is mole fraction in bottoms. The average value of α_L of the light key is calculated from the α_{LD} at the top temperature (dew point) of the tower and α_{LW} at the bottoms temperature.

$$\alpha_{L, av} = \sqrt{\alpha_{LD} \alpha_{LW}} \tag{11.7-13}$$

Note that the distillate dew-point and bottoms boiling-point estimation is partially trial and error, since the distribution of the other components in the distillate and bottoms is not known and can affect these values.

2. *Distribution of other components.* To determine the distribution or concentration of other components in the distillate and the bottoms at total reflux, Eq. (11.7-12) can be rearranged and written for any other component i as follows:

$$\frac{x_{iD} D}{x_{iW} W} = (\alpha_{i,\,av})^{N_m} \frac{x_{HD} D}{x_{HW} W} \qquad (11.7\text{-}14)$$

These concentrations of the other components determined at total reflux can be used as approximations with finite and minimum reflux ratios. More accurate methods for finite and minimum reflux are available elsewhere (H2, S1, V1).

EXAMPLE 11.7-2. Calculation of Top and Bottom Temperatures and Total Reflux

The liquid feed of 100 mol/h at the boiling point given in Example 11.7-1 is fed to a distillation tower at 405.3 kPa and is to be fractionated so that 90% of the *n*-pentane (*B*) is recovered in the distillate and 90% of the *n*-hexane (*C*) in the bottoms. Calculate the following.

    (a) Moles per hour and composition of distillate and bottoms.
    (b) Top temperature (dew point) of distillate and boiling point of bottoms.
    (c) Minimum stages for total reflux and distribution of other components in the distillate and bottoms.

Solution: For part (a) material balances are made for each component, with component *n*-pentane (*B*) being the light key (*L*) and *n*-hexane (*C*) the heavy key (*H*). For the overall balance,

$$F = D + W \qquad (11.7\text{-}15)$$

For component *B*, the light key,

$$x_{BF} F = 0.25(100) = 25.0 = y_{BD} D + x_{BW} W \qquad (11.7\text{-}16)$$

Since 90% of *B* is in the distillate, $y_{BD} D = (0.90)(25) = 22.5$. Hence, $x_{BW} W = 2.5$. For component *C*, the heavy key,

$$x_{CF} F = 0.20(100) = 20.0 = y_{CD} D + x_{CW} W \qquad (11.7\text{-}17)$$

Also, 90% of *C* is in the bottoms and $x_{CW} W = 0.90(20) = 18.0$. Then, $y_{CD} D = 2.0$. For the first trial, it is assumed that no component *D* (heavier than the heavy key *C*) is in the distillate and no light *A* in the bottoms. Hence, moles *A* in distillate $= y_{AD} D = 0.40(100) = 40.0$. Also, moles *D* in bottoms $= x_{DW} W = 0.15(100) = 15.0$. These values are tabulated below.

Comp.	Feed. F		Distillate, D		Bottoms, W	
	x_F	$x_F F$	$y_D = x_D$	$y_D D$	x_W	$x_W W$
A	0.40	40.0	0.620	40.0	0	0
B (lt key L)	0.25	25.0	0.349	22.5	0.070	2.5
C (hy key H)	0.20	20.0	0.031	2.0	0.507	18.0
D	0.15	15.0	0	0	0.423	15.0
	1.00	$F = 100.0$	1.000	$D = 64.5$	1.000	$W = 35.5$

For the dew point of the distillate (top temperature) in part (b), a value of 67°C will be estimated for the first trial. The K values are read from Fig. 11.7-2 and the α values calculated. Using Eqs. (11.7-7) and (11.7-8), the following values are calculated:

Comp.	y_{iD}	$K_i(67°C)$	α_i	y_i/α_i	x_i
A	0.620	1.75	6.73	0.0921	0.351
B (L)	0.349	0.65	2.50	0.1396	0.531
C (H)	0.031	0.26	1.00	0.0310	0.118
D	0	0.10	0.385	0	0
	1.000			$\sum y_i/\alpha_i = 0.2627$	1.000

$$K_C = \sum y_i/\alpha_i = 0.2627$$

The calculated value of K_C is 0.2627, which corresponds very closely to 67°C, which is the final temperature of the dew point.

For the bubble point of the bottoms, a temperature of 135°C is assumed for trial 1 and Eqs. (11.7-5) and (11.7-6) used for the calculations. A second trial using 132°C gives the final temperature as shown below.

Comp.	x_{iW}	K_i	α_i	$\alpha_i x_i$	y_i
A	0	5.00	4.348	0	0
B (L)	0.070	2.35	2.043	0.1430	0.164
C (H)	0.507	1.15	1.000	0.5070	0.580
D	0.423	0.61	0.530	0.2242	0.256
	1.000			$\sum \alpha_i x_i = 0.8742$	1.000

$$K_C = 1/0.8742 = 1.144$$

The calculated value of K_C is 1.144, which is close to the value at 132°C.

For part (c) the proper α values of the light key L (n-pentane) to use in Eq. (11.7-13) are as follows:

$$\alpha_{LD} = 2.50 \quad (t = 67°C \text{ at the column top})$$

$$\alpha_{LW} = 2.04 \quad (t = 132°C \text{ at the column bottom})$$

$$\alpha_{L,\text{av}} = \sqrt{\alpha_{LD}\alpha_{LW}} = \sqrt{2.50(2.04)} = 2.258 \quad \textbf{(11.7-13)}$$

Substituting into Eq. (11.7-12),

$$N_m = \frac{\log\,[(0.349 \times 64.5/0.031 \times 64.5)(0.507 \times 35.5/0.070 \times 35.5)]}{\log\,(2.258)}$$

$$= 5.404 \text{ theoretical stages (4.404 theoretical trays)}$$

The distribution or compositions of the other components can be calculated using Eq. (11.7-14). For component A, the average α value to use is

$$\alpha_{A,\text{av}} = \sqrt{\alpha_{AD}\alpha_{AW}} = \sqrt{6.73 \times 4.348} = 5.409$$

$$\frac{x_{AD} D}{x_{AW} W} = (\alpha_{A,\text{av}})^{N_m} \frac{x_{HD} D}{x_{HW} W} = (5.409)^{5.404} \frac{0.031(64.5)}{0.507(35.5)} = 1017 \quad \textbf{(11.7-14)}$$

Making an overall balance on A,

$$x_{AF} F = 40.0 = x_{AD} D + x_{AW} W \quad \textbf{(11.7-18)}$$

Substituting $x_{AD} D = 1017 x_{AW} W$ from Eq. (11.7-14) into (11.7-18) and solving,

$$x_{AW} W = 0.039, \qquad x_{AD} D = 39.961$$

For the distribution of component D, $\alpha_{D, \text{av}} = \sqrt{0.385 \times 0.530} = 0.452$.

$$\frac{x_{DD} D}{x_{DW} W} = (\alpha_{D, \text{av}})^{N_m} \frac{x_{HD} D}{x_{HW} W} = (0.452)^{5.404} \frac{0.031(645)}{0.507(355)} = 0.001521$$

$$x_{DF} F = 15.0 = x_{DD} D + x_{DW} W$$

Solving, $x_{DD} D = 0.023$, $x_{DW} W = 14.977$.
 The revised distillate and bottoms compositions are as follows.

	Distillate, D			Bottoms, W	
Comp.	$y_D = x_D$	$x_D D$	x_W		$x_W W$
A	0.6197	39.961	0.0011		0.039
B (L)	0.3489	22.500	0.0704		2.500
C (H)	0.0310	2.000	0.5068		18.000
D	0.0004	0.023	0.4217		14.977
	1.0000	$D = 64.484$	1.0000		$W = 35.516$

Hence, the moles of D in the distillate are quite small, as are the moles of A in the bottoms.
 Using the new distillate composition, a recalculation of the dew point assuming 67°C gives a calculated value of $K_C = 0.2637$. This is very close to that of 0.2627 obtained when the trace amount of D in the distillate was assumed as zero. Hence, the dew point is 67°C. Repeating the bubble-point calculation for the bottoms assuming 132°C, a calculated value of $K_C = 1.138$, which is close to the value at 132°C. Hence, the bubble point remains at 132°C. If either the bubble- or dew-point temperatures had changed, the new values would then be used in a recalculation of N_m.

11.7F Shortcut Method for Minimum Reflux Ratio for Multi-component Distillation

As in the case for binary distillation, the minimum reflux ratio R_m is that reflux ratio that will require an infinite number of trays for the given separation of the key components.
 For binary distillation only one "pinch point" occurs where the number of steps become infinite, and this is usually at the feed tray. For multicomponent distillation two pinch points or zones of constant composition occur: one in the section above the feed plate and another below the feed tray. The rigorous plate-by-plate stepwise procedure to calculate R_m is trial and error and can be extremely tedious for hand calculations.
 Underwood's shortcut method to calculate R_m (U1, U2) uses constant average α values and also assumes constant flows in both sections of the tower. This method provides a reasonably accurate value. The two equations to be solved to determine the minimum reflux ratio are

$$1 - q = \sum \frac{\alpha_i x_{iF}}{\alpha_i - \theta} \qquad (11.7\text{-}19)$$

$$R_m + 1 = \sum \frac{\alpha_i x_{iD}}{\alpha_i - \theta} \qquad (11.7\text{-}20)$$

The values of x_{iD} for each component in the distillate in Eq. (11.7-20) are supposed to be the values at the minimum reflex. However, as an approximation, the values obtained using the Fenski total reflux equation are used. Since each α_i may vary with temperature, the average value of α_i to use in the preceeding equations is approximated by using α_i at the average temperature of the top and bottom of the tower. Some (P1, S1) have used the average α which is used in the Fenske equation or the α at the entering feed temperature. To solve for R_m, the value of θ in Eq. (11.7-19) is first obtained by trial and error. This value of θ lies between the α value of the light key and α value of the heavy key, which is 1.0. Using this value of θ in Eq. (11.7-20), the value of R_m is obtained directly. When distributed components appear between the key components, modified methods described by others (S1, T2, V1) can be used.

11.7G   Shortcut Method for Number of Stages at Operating Reflux Ratio

1. Number of stages at operating reflux ratio. The determination of the minimum number of stages for total reflux in Section 11.7E and the minimum reflux ratio in Section 11.7F are useful for setting the allowable ranges of number of stages and flow conditions. These ranges are helpful for selecting the particular operating conditions for a design calculation. The relatively complex rigorous procedures for doing a stage-by-stage calculation at any operating reflux ratio have been discussed in Section 11.7A.

An important shortcut method to determine the theoretical number of stages required for an operating reflux ratio R is the empirical correlation of Erbar and Maddox (E1) given in Fig. 11.7-3. This correlation is somewhat similar to a correlation by Gilliland (G1) and should be considered as an approximate method. In Fig. 11.7-3 the operating reflux ratio R (for flow rates at the column top) is correlated with the minimum R_m obtained using the Underwood method, the minimum number of stages N_m obtained by the Fenske method, and the number of stages N at the operating R.

2. Estimate of feed plate location. Kirkbride (K1) has devised an approximate method to estimate the number of theoretical stages above and below the feed which can be used to estimate the feed stage location. This empirical relation is as follows:

$$\log \frac{N_e}{N_s} = 0.206 \log \left[\left(\frac{x_{HF}}{x_{LF}} \right) \frac{W}{D} \left(\frac{x_{LW}}{x_{HD}} \right)^2 \right] \qquad (11.7\text{-}21)$$

where N_e is the number of theoretical stages above the feed plate and N_s the number of theoretical stages below the feed plate.

EXAMPLE 11.7-3.   Minimum Reflux Ratio and Number of Stages at Operating Reflux Ratio
Using the conditions and results given in Example 11.7-2, calculate the following.
    (a) Minimum reflux ratio using the Underwood method.
    (b) Number of theoretical stages at an operating reflux ratio R of $1.5R_m$ using the Erbar–Maddox correlation.
    (c) Location of feed tray using the method of Kirkbride.

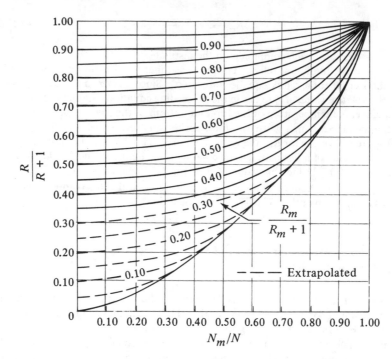

FIGURE 11.7-3.   *Erbar–Maddox correlation between reflux ratio and number of stages (R$_m$ based on Underwood method.) [From J. H. Erbar, R. N. Maddox, Petrol. Refiner.* **40** *(5), 183 (1961). With permission.]*

Solution: For part (a) the temperature to use for determining the values of α_i is the average between the top of 67°C and the bottom of 132°C (from Example 11.7-2) and is $(67 + 132)/2$ or 99.5°C. The K_i values obtained from Fig. 11.7-2 and the α_i values and distillate and feed compositions to use in Eqs. (11.7-19) and (11.7-20) are as follows:

Comp.	x_{iF}	x_{iD}	$K_i(99.5°C)$	$\alpha_i(99.5°C)$	x_{iW}
A	0.40	0.6197	3.12	5.20	0.0011
B (L)	0.25	0.3489	1.38	2.30	0.0704
C (H)	0.20	0.0310	0.60	1.00	0.5068
D	0.15	0.0004	0.28	0.467	0.4217
	$\overline{1.00}$	$\overline{1.0000}$			$\overline{1.0000}$

Substituting into Eq. (11.7-19) with $q = 1.0$ for feed at the boiling point,

$$1 - q = 1 - 1 = 0 = \frac{5.20(0.40)}{5.20 - \theta} + \frac{2.30(0.25)}{2.30 - \theta}$$

$$+ \frac{1.00(0.20)}{1.00 - \theta} + \frac{0.467(0.15)}{0.467 - \theta} \qquad \textbf{(11.7-22)}$$

This is trial and error, so a value of $\theta = 1.210$ will be used for the first trial (θ must be between 2.30 and 1.00). This and other trials are shown below.

θ (Assumed)	$\dfrac{2.08}{5.2 - \theta}$	$\dfrac{0.575}{2.3 - \theta}$	$\dfrac{0.200}{1.0 - \theta}$	$\dfrac{0.070}{0.467 - \theta}$	$\sum$ (Sum)
1.210	0.5213	0.5275	-0.9524	-0.0942	$+0.0022$
1.200	0.5200	0.5227	-1.0000	-0.0955	-0.0528
1.2096	0.5213	0.5273	-0.9542	-0.0943	$+0.0001$

The final value of $\theta = 1.2096$ is substituted into Eq. (11.7-20) to solve for R_m.

$$R_m + 1 = \frac{5.20(0.6197)}{5.20 - 1.2096} + \frac{2.30(0.3489)}{2.30 - 1.2096}$$

$$+ \frac{1.00(0.031)}{1.00 - 1.2096} + \frac{0.467(0.0004)}{0.467 - 1.2096}$$

Solving, $R_m = 0.395$.

For part (b) the following values are calculated. $R = 1.5 R_m = 1.5(0.395) = 0.593$, $R/(R + 1) = 0.593/(0.593 + 1.0) = 0.3723$, $R_m/(R_m + 1) = 0.395/(0.395 + 1.0) = 0.2832$. From Fig. 11.7-3, $N_m/N = 0.49$. Hence, $N_m/N = 0.49 = 5.40/N$. Solving, $N = 11.0$ theoretical stages in the tower. This gives $11.0 - 1.0$ (reboiler) or 10.0 theoretical trays.

For the location of the feed tray in part (c) using Eq. (11.7-21),

$$\log \frac{N_e}{N_s} = 0.206 \log \left[\left(\frac{0.20}{0.25} \right) \frac{35.516}{64.484} \left(\frac{0.0704}{0.0310} \right)^2 \right] = 0.07344$$

Hence, $N_e/N_s = 1.184$. Also, $N_e + N_s = 1.184 N_s + N_s = N = 11.0$ stages. Solving, $N_s = 5.0$ and $N_e = 6.0$. This means that the feed tray is 6.0 trays from the top.

PROBLEMS

11.1-1. Phase Rule for a Vapor System. For the system NH_3–water and only a vapor phase present, calculate the number of degrees of freedom. What variables can be fixed?

Ans. $F = 3$ degrees of freedom; variables T, P, y_A

11.1-2. Boiling Point and Raoult's Law. For the system benzene–toluene, do as follows using the data of Table 11.1-1.
(a) At 378.2 K, calculate y_A and x_A using Raoult's law.
(b) If a mixture has a composition of $x_A = 0.40$ and is at 358.2 K and 101.32 kPa pressure, will it boil? If not, at what temperature will it boil and what will be the composition of the vapor first coming off?

11.1-3. Boiling-Point-Diagram Calculation. The vapor-pressure data are given below for the system hexane–octane.

		Vapor Pressure			
		n-Hexane		n-Octane	
$T(°F)$	$T(°C)$	kPa	mm Hg	kPa	mm Hg
155.7	68.7	101.3	760	16.1	121
175	79.4	136.7	1025	23.1	173
200	93.3	197.3	1480	37.1	278
225	107.2	284.0	2130	57.9	434
258.2	125.7	456.0	3420	101.3	760

(a) Using Raoult's law, calculate and plot the xy data at a total pressure of 101.32 kPa.

(b) Plot the boiling-point diagram.

11.2-1 Single-Stage Contact of Vapor–Liquid System. A mixture of 100 mol containing 60 mol % n-pentane and 40 mol % n-heptane is vaporized at 101.32 kPa abs pressure until 40 mol of vapor and 60 mol of liquid in equilibrium with each other are produced. This occurs in a single-stage system and the vapor and liquid are kept in contact with each other until the vaporization is complete. The equilibrium data are given in Example 11.3-2. Calculate the composition of the vapor and the liquid.

11.3-1. Relative Volatility of a Binary System. Using the equilibrium data for the n-pentane–n-heptane system given in Example 11.3-2, calculate the relative volatility for each concentration and plot α versus the liquid composition x_A.

11.3-2. Comparison of Differential and Flash Distillation. A mixture of 100 kg mol which contains 60 mol % n-pentane (A) and 40 mol % n-heptane (B) is vaporized at 101.32 kPa pressure under differential conditions until 40 kg mol are distilled. Use equilibrium data from Example 11.3-2.

(a) What is the average composition of the total vapor distilled and the composition of the liquid left?

(b) If this same vaporization is done in an equilibrium or flash distillation and 40 kg mol are distilled, what is the composition of the vapor distilled and of the liquid left?

Ans. (a) $x_2 = 0.405$, $y_{av} = 0.892$; (b) $x_2 = 0.430$, $y_2 = 0.854$

11.3-3. Differential Distillation of Benzene–Toluene. A mixture containing 70 mol % benzene and 30 mol % toluene is distilled under differential conditions at 101.32 kPa (1 atm). A total of one third of the moles in the feed is vaporized. Calculate the average composition of the distillate and the composition of the remaining liquid. Use equilibrium data in Table 11.1-1.

11.3-4. Steam Distillation of Ethylaniline. A mixture contains 100 kg of H_2O and 100 kg of ethylaniline (mol wt = 121.1 kg/kg mol), which is immiscible with water. A very slight amount of nonvolatile impurity is dissolved in the organic. To purify the ethylaniline it is steam-distilled by bubbling saturated steam into the mixture at a total pressure of 101.32 kPa (1 atm). Determine the boiling point of the mixture and the composition of the vapor. The vapor pressure of each of the pure compounds is as follows (T1):

Temperature		P_A(water) (kPa)	P_B(ethylaniline) (kPa)
K	°C		
353.8	80.6	48.5	1.33
369.2	96.0	87.7	2.67
372.3	99.15	98.3	3.04
386.4	113.2	163.3	5.33

11.3-5. Steam Distillation of Benzene. A mixture of 50 g mol of liquid benzene and 50 g mol of water is boiling at 101.32 kPa pressure. Liquid benzene is immiscible in water. Determine the boiling point of the mixture and the composition of the vapor. Which component will first be removed completely from the still? Vapor-pressure data of the pure components are as follows:

Temperature		P_{water} (mm Hg)	$P_{benzene}$ (mm Hg)
K	°C		
308.5	35.3	43	150
325.9	52.7	106	300
345.8	72.6	261	600
353.3	80.1	356	760

11.4-1. Distillation Using McCabe–Thiele Method. A rectification column is fed 100 kg mol/h of a mixture of 50 mol % benzene and 50 mol % toluene at 101.32 kPa abs pressure. The feed is liquid at the boiling point. The distillate is to contain 90 mol % benzene and the bottoms 10 mol % benzene. The reflux ratio is 4.52 : 1. Calculate the kg mol/h distillate, kg mol/h bottoms, and the number of theoretical trays needed using the McCabe-Thiele method.

Ans. $D = 50$ kg mol/h, $W = 50$ kg mol/h, 4.9 theoretical trays plus reboiler

11.4-2. Rectification of a Heptane–Ethyl Benzene Mixture. A saturated liquid feed of 200 mol/h at the boiling point containing 42 mol % heptane and 58% ethyl benzene is to be fractionated at 101.32 kPa abs to give a distillate containing 97 mol % heptane and a bottoms containing 1.1 mol % heptane. The reflux ratio used is 2.5 : 1. Calculate the mol/h distillate, mol/h bottoms, theoretical number of trays, and the feed tray number. Equilibrium data are given below at 101.32 kPa abs pressure for the mole fraction n-heptane x_H and y_H.

Temperature		x_H	y_H	Temperature		x_H	y_H
K	°C			K	°C		
409.3	136.1	0	0	383.8	110.6	0.485	0.730
402.6	129.4	0.08	0.230	376.0	102.8	0.790	0.904
392.6	119.4	0.250	0.514	371.5	98.3	1.000	1.000

Ans. $D = 85.3$ mol/h, $W = 114.7$ mol/h, 9.5 trays + reboiler, feed on tray 6 from top

11.4-3. *Graphical Solution for Minimum Reflux Ratio and Total Reflux.* For the rectification given in Problem 11.4-1, where an equimolar liquid feed of benzene and toluene is being distilled to give a distillate of composition $x_D = 0.90$ and a bottoms of composition $x_w = 0.10$, calculate the following using graphical methods.

(a) Minimum reflux ratio R_m.

(b) Minimum number of theoretical plates at total reflux.

Ans. (a) $R_m = 0.91$; (b) 4.0 theoretical trays plus a reboiler

11.4-4. *Minimum Number of Theoretical Plates and Minimum Reflux Ratio.* Determine the minimum reflux ration R_m and the minimum number of theoretical plates at total reflux for the rectification of a mixture of heptane and ethyl benzene as given in Problem 11.4-2. Do this by the graphical methods of McCabe–Thiele.

11.4-5. *Rectification Using a Partially Vaporized Feed.* A total feed of 200 mol/h having an overall composition of 42 mol % heptane and 58 mol % ethyl benzene is to be fractionated at 101.3 kPa pressure to give a distillate containing 97 mol % heptane and a bottoms containing 1.1 mol % heptane. The feed enters the tower partially vaporized so that 40 mol % is liquid and 60 mol % vapor. Equilibrium data are given in Problem 11.4-2. Calculate the following.

(a) Moles per hour distillate and bottoms.

(b) Minimum reflux ratio R_m.

(c) Minimum steps and theoretical trays at total reflux.

(d) Theoretical number of trays required for an operating reflux ratio of 2.5 : 1. Compare with the results of Problem 11.4-2, which uses a saturated liquid feed.

11.4-6. *Distillation Using a Vapor Feed.* Repeat Problem 11.4-1 but use a feed that is saturated vapor at the dew point. Calculate the following.

(a) Minimum reflux ratio R_m.

(b) Minimum number of theoretical plates at total reflux.

(c) Theoretical number of trays at an operating reflux ratio of $1.5(R_m)$.

11.4-7 *Enriching Tower for Benzene–Toluene.* An enriching tower is fed 100 kg mol/h of a saturated vapor feed containing 40 mol % benzene (A) and 60 mol % toluene (B) at 101.32 kPa abs. The distillate is to contain 90 mol % benzene. The reflux ratio is set at 4.0 : 1. Calculate the kg mol/h distillate D and bottoms W and their compositions. Also, calculate the number of theoretical plates required.

Ans. $D = 20$ kg mol/h, $W = 80$ kg mol/h, $x_w = 0.275$

11.4-8. *Stripping Tower.* A liquid mixture containing 10 mol % *n*-heptane and 90 mol % *n*-octane is fed at its boiling point to the top of a stripping tower at 101.32 kPa abs. The bottoms are to contain 98 mol % *n*-octane. For every 3 mol of feed, 2 mol of vapor is withdrawn as product. Calculate the composition of the vapor and the number of theoretical plates required. The equilibrium data below are given as mole fraction *n*-heptane.

x	y	x	y
0.284	0.459	0.039	0.078
0.097	0.184	0.012	0.025
0.067	0.131		

11.4-9. *Stripping Tower and Direct Steam Injection.* A liquid feed at the boiling point contains 3.3 mol % ethanol and 96.7 mol % water and enters the top tray of a stripping tower. Saturated steam is injected directly into liquid in the bottom of the tower. The overhead vapor which is withdrawn contains 99% of the alcohol in the feed. Assume equimolar overflow for this problem. Equilibrium data for mole fraction of alcohol are as follows at 101.32 kPa abs pressure (1 atm abs).

x	y	x	y
0	0	0.0296	0.250
0.0080	0.0750	0.033	0.270
0.020	0.175		

(a) For an infinite number of theoretical steps, calculate the minimum moles of steam needed per mole of feed. (*Note :* Be sure and plot the q line.)

(b) Using twice the minimum moles of steam, calculate the number of theoretical steps needed, the composition of the overhead vapor, and the bottoms composition.

Ans. (a) 0.121 mol steam/mol feed;
(b) 5.0 theoretical steps, $x_D = 0.135$, $x_W = 0.00033$

11.5-1. *Murphree Efficiency and Actual Number of Trays.* For the distillation of heptane and ethyl benzene in Problem 11.4-2, the Murphree tray efficiency is estimated as 0.55. Determine the actual number of trays needed by stepping off the trays using the tray efficiency of 0.55. Also, calculate the overall tray efficiency E_O.

11.6-1. *Use of Enthalpy-Concentration Method to Distill an Ethanol-Water Solution.* A mixture of 50 wt % ethanol and 50 wt % water which is saturated liquid at the boiling point is to be distilled at 101.3 kPa pressure to give a distillate containing 85 wt % ethanol and a bottoms containing 3 wt % ethanol. The feed rate is 453.6 kg/h and a reflux ratio of 1.5 is to be used. Use equilibrium and enthalpy data from Appendix A.3. Note that the data are given in wt fraction and kJ/kg. Use these consistent units in plotting the enthalpy-concentration data and equilibrium data. Do as follows.

(a) Calculate the amount of distillate and bottoms.
(b) Calculate the number of theoretical trays needed.
(c) Calculate the condenser and reboiler loads.

Ans. (a) $D = 260.0$ kg/h, $W = 193.6$ kg/h
(b) 3.9 trays plus a reboiler
(c) $q_c = 698\ 750$ kJ/h, $q_R = 704\ 770$ kJ/h

11.6-2. *Distillation of Ethanol-Water Solution Using Enthalpy-Concentration Method.* Repeat Problem 11.6-1 but use a reflux ratio of 2.0 instead of 1.5.
Ans. 3.6 theoretical trays plus reboiler

11.6-3. *Minimum Reflux and Theoretical Number of Trays.* A feed of ethanol-water containing 60 wt % ethanol is to be distilled at 101.32 kPa pressure to give a distillate containing 85 wt % ethanol and a bottoms containing 2 wt % ethanol. The feed rate is 10 000 kg/h and its enthalpy is 116.3 kJ/kg (50 btu/lb$_m$). Use consistent units of kg/h, weight fraction, and kJ/kg.

(a) Calculate the amount of distillate and bottoms.
(b) Determine the minimum reflux ratio using enthalpy-concentration data from Appendix A.3.
(c) Using 2.0 times the minimum reflux ratio, determine the theoretical number of trays needed.
(d) Calculate the condenser and reboiler heat loads.
(e) Determine the minimum number of theoretical plates at total reflux.

Ans. (b) $R_m = 0.373$
(c) 4.4 theoretical trays plus reboiler
(d) $q_c = 3\ 634$ kW, $q_R = 4\ 096$ kW
(e) 2.8 theoretical trays plus reboiler

11.6-4. Distillation of Benzene-Toluene Feed Using Enthalpy-Concentration Method. A liquid feed of 100 kg mol/h of benzene-toluene at the boiling point contains 55 mol % benzene and 45 mol % toluene. It is being distilled at 101.32 kPa pressure to give a distillate with $x_D = 0.98$ and a bottoms of $x_W = 0.04$. Using a reflux ratio of 1.3 times the minimum and the enthalpy-concentration method do as follows.
(a) Determine the theoretical number of trays needed.
(b) Calculate the condenser and reboiler heat loads.
(c) Determine the minimum number of theoretical trays at total reflux.

11.6-5. Use of Enthalpy-Concentration Plot. For the system benzene-toluene do as follows.
(a) Plot the enthalpy-concentration data using values from Table 11.6-2. For a value of $x = 0.60 = y$, calculate the saturated liquid enthalpy h and the saturated vapor enthalpy H and plot these data on the graph.
(b) A mixture contains 60 mol of benzene and 40 mol of toluene. This mixture is heated so that 30 mol of vapor are produced. The mixture is in equilibrium. Determine the enthalpy of this overall mixture and plot this point on the enthalpy-concentration diagram.

11.7-1. Flash Vaporization of Multicomponent Feed. For the feed to the distillation tower of Example 11.1-1, calculate the following.
(a) Dew point of feed and composition of liquid in equilibrium. (*Note*: The boiling point of 70°C has already been calculated.)
(b) The temperature and composition of both phases when 40% of the feed is vaporized in a flash distillation.

  Ans. (a) 107°C, $x_A = 0.114$, $x_B = 0.158$, $x_C = 0.281$, $x_D = 0.447$;
  (b) 82°C, $x_A = 0.260$, $x_B = 0.254$, $x_C = 0.262$, $x_D = 0.224$;
  $y_A = 0.610$, $y_B = 0.244$, $y_C = 0.107$, $y_D = 0.039$

11.7-2. Boiling Point, Dew Point, and Flash Vaporization. Following is the composition of a liquid feed in mole fraction: n-butane ($x_A = 0.35$), n-pentane ($x_B = 0.20$), n-hexane, ($x_C = 0.25$), n-heptane ($x_D = 0.20$). At a pressure of 405.3 kPa calculate the following.
(a) Boiling point and composition of the vapor in equilibrium.
(b) Dew point and composition of the liquid in equilibrium.
(c) The temperature and composition of both phases when 60% of the feed is vaporized in a flash distillation.

11.7-3. Vaporization of Multicomponent Alcohol Mixture. The vapor-pressure data are given below for the following alcohols.

	Vapor Pressure (mm Hg)			
$T(°C)$	*Methanol*	*Ethanol*	*n-Propanol*	*n-Butanol*
50	415	220.0	88.9	33.7
60	629	351.5	148.9	59.2
65	767	438	190.1	77.7
70	929	542	240.6	99.6
75	1119	665	301.9	131.3
80	1339	812	376.0	165.0
85	1593	984	465	206.1
90	1884	1185	571	225.9
100	2598	1706	843	387.6

Following is the composition of a liquid alcohol mixture to be fed to a distillation tower at 101.32 kPa: methyl alcohol ($x_A = 0.30$), ethyl alcohol ($x_B = 0.20$), n-propyl alcohol ($x_C = 0.15$), and n-butyl alcohol ($x_D = 0.35$). Calculate the following assuming that the mixture follows Raoult's law.
(a) Boiling point and composition of vapor in equilibrium.
(b) Dew point and composition of liquid in equilibrium.
(c) The temperature and composition of both phases when 40% of the feed is vaporized in a flash distillation.

Ans. (a) 83°C, $y_A = 0.589$, $y_B = 0.241$, $y_C = 0.084$, $y_D = 0.086$;
(b) 100°C, $x_A = 0.088$, $x_B = 0.089$, $x_C = 0.136$, $x_D = 0.687$

11.7-4. Total Reflux, Minimum Reflux, Number of Stages. The following feed of 100 mol/h at the boiling point and 405.3 kPa pressure is fed to a fractionating tower: n-butane ($x_A = 0.40$), n-pentane ($x_B = 0.25$), n-hexane ($x_C = 0.20$), n-heptane ($x_D = 0.15$). This feed is distilled so that 95% of the n-pentane is recovered in the distillate and 95% of the n-hexane in the bottoms. Calculate the following.
(a) Moles per hour and composition of distillate and bottoms.
(b) Top and bottom temperature of tower.
(c) Minimum stages for total reflux and distribution of other components (trace components) in the distillate and bottoms, i.e., moles and mole fractions. [Also correct the compositions and moles in part (a) for the traces.]
(d) Minimum reflux ratio using the Underwood method.
(e) Number of theoretical stages at an operating reflux ratio of 1.3 times the minimum using the Erbar–Maddox correlation.
(f) Location of the feed tray using the Kirkbride method.

Ans. (a) $D = 64.75$ mol/h, $x_{AD} = 0.6178$, $x_{BD} = 0.3668$, $x_{CD} = 0.0154$, $x_{DD} = 0$;
$W = 35.25$ mol/h, $x_{AW} = 0$, $x_{BW} = 0.0355$, $x_{CW} = 0.5390$, $x_{DW} = 0.4255$;
(b) top, 66°C; bottom, 134°C; $N_m = 7.14$ stages; trace compositions,
$x_{AW} = 1.2 \times 10^{-4}$, $x_{DD} = 4.0 \times 10^{-5}$;
(d) $R_m = 0.504$;
(e) $N = 16.8$ stages;
(f) $N_e = 9.1$ stages, $N_s = 7.7$ stages, feed 9.1 stages from top

11.7-5. Shortcut Design of Multicomponent Distillation Tower. A feed of part liquid and part vapor ($q = 0.30$) at 405.4 kPa is fed at the rate of 1000 mol/h to a distillation tower. The overall composition of the feed is n-butane ($x_A = 0.35$), n-pentane ($x_B = 0.30$), n-hexane ($x_C = 0.20$) and n-heptane ($x_D = 0.15$). The feed is distilled so that 97% of the n-pentane is recovered in the distillate and 85% of the n-hexane in the bottoms. Calculate the following.
(a) Amount and composition of products and top and bottom tower temperatures.
(b) Number of stages at total reflux and distribution of other components in the products.
(c) Minimum reflux ratio, number of stages at $1.2R_m$, and feed tray location.

11.7-6. Distillation of Multicomponent Alcohol Mixture. A feed of 30 mol % methanol (A), 20% ethanol (B), 15% n-propanol (C), and 35% n-butanol (D) is distilled at 101.32 kPa absolute pressure to give a distillate composition containing 95.0 mole % methanol and a residue composition containing 5.0% methanol and the other components as calculated. The feed is below the boiling point, so that $q = 1.1$. The operating reflux ratio is 3.0. Assume that Raoult's law applies and use vapor-pressure data from Problem 11.7-3. Calculate the following.
(a) Composition and amounts of distillate and bottoms for a feed of 100 mol/h.
(b) Top and bottom temperatures and number of stages at total reflux. (Also, calculate the distribution of the other components.)

      (c) Minimum reflux ratio, number of stages at $R = 3.00$, and the feed tray location.

Ans. (a) $D = 27.778$ mol/h, $x_{AD} = 0.95$, $x_{BD} = 0.05$, $x_{CD} = 0$, $x_{DD} = 0$;
$W = 72.222$ mol/h, $x_{AW} = 0.0500$, $x_{BW} = 0.2577$, $x_{CW} = 0.2077$, $x_{DW} = 0.4846$;
(b) 65.5°C top temperature, 94.3°C bottom, $N_m = 9.21$ stages,
$x_{CD} = 3.04 \times 10^{-5}$, $x_{DD} = 8.79 \times 10^{-7}$ (trace compositions);
(c) $R_m = 2.20$, $N = 16.2$ stages, $N_s = 7.6$, $N_e = 8.6$, feed on stage 8.6 from top

11.7-7. Shortcut Design Method for Distillation of Ternary Mixture. A liquid feed at its bubble point is to be distilled in a tray tower to produce the distillate and bottoms as follows. Feed, $x_{AF} = 0.047$, $x_{BF} = 0.072$, $x_{CF} = 0.881$; distillate, $x_{AD} = 0.1260$, $x_{BD} = 0.1913$, $x_{CD} = 0.6827$; bottoms, $x_{AW} = 0$, $x_{BW} = 0.001$, $x_{CW} = 0.999$. Average α values to use are $\alpha_A = 4.19$, $\alpha_B = 1.58$, $\alpha_C = 1.00$.
    (a) For a feed rate of 100 mol/h, calculate D and W, number of stages at total reflux, and distribution (concentration) of A in the bottoms.
    (b) Calculate R_m and the number of stages at $1.25R_m$.

REFERENCES

(D1)   DEPRIESTER, C. L. *Chem. Eng. Progr. Symp. Ser.*, **49**(7), 1 (1953).

(E1)   ERBAR, J. H., and MADDOX, R. N. *Petrol. Refiner*, **40**(5), 183 (1961).

(G1)   GILLILAND, E. R. *Ind. Eng. Chem.*, **32,** 1220 (1940).

(H1)   HADDEN, S. T., and GRAYSON, H. G. *Petrol. Refiner*, **40**(9), 207 (1961).

(H2)   HOLLAND, C. D. *Multicomponent Distillation.* Englewood Cliffs, N. J.: Prentice-Hall, 1963.

(K1)   KIRKBRIDE, C. G., *Petrol. Refiner*, **23**, 32(1944).

(K2)   KWAUK, M. *A.I.Ch.E.J.*, **2**, 240 (1956).

(K3)   KING, C. J., *Separation Processes*, 2nd ed. New York: McGraw-Hill Book Company, 1980.

(O1)   O'CONNELL, H. E. *Trans. A.I.Ch.E.*, **42**, 741 (1946).

(P1)   PERRY, R. H., and GREEN, D. *Perry's Chemical Engineers' Handbook*, 6th ed. New York: McGraw-Hill Book Company, 1984.

(R1)   ROBINSON, C. S., and GILLILAND, E. R. *Elements of Fractional Distillation*, 4th ed. New York: McGraw-Hill Book Company, 1950.

(S1)   SMITH, B. D. *Design of Equilibrium Stage Processes.* New York: McGraw-Hill Book Company, 1963.

(T1)   THIELE, E. W., and GEDDES, R. L. *Ind. Eng. Chem.*, **25,** 289 (1933).

(T2)   TREYBAL, R. E. *Mass Transfer Operations*, 3rd ed. New York: McGraw-Hill Book Company, 1980.

(U1)   UNDERWOOD, A. J. V. *Chem. Eng. Progr.*, **44**, 603 (1948); **45**, 609 (1949).

(U2)   UNDERWOOD, A. J. V. *J. Inst. Petrol.*, **32**, 614 (1946).

(V1)   VAN WINKLE, M. *Distillation.* New York: McGraw-Hill Book Company, 1967.

CHAPTER 12

Liquid–Liquid and Fluid–Solid Separation Processes

12.1 INTRODUCTION TO ADSORPTION PROCESSES

12.1A Introduction

In adsorption processes one or more components of a gas or liquid stream are adsorbed on the surface of a solid adsorbent and a separation is accomplished. In commercial processes, the adsorbent is usually in the form of small particles in a fixed bed. The fluid is passed through the bed and the solid particles adsorb components from the fluid. When the bed is almost saturated, the flow in this bed is stopped and the bed is regenerated thermally or by other methods, so desorption occurs. The adsorbed material (adsorbate) is thus recovered and the solid adsorbent is ready for another cycle of adsorption.

Applications of liquid-phase adsorption include removal of organic compounds from water or organic solutions, colored impurities from organics, and various fermentation products from fermentor effluents. Separations include paraffins from aromatics and fructose from glucose using zeolites.

Applications of gas-phase adsorption include removal of water from hydrocarbon gases, sulfur compounds from natural gas, solvents from air and other gases, and odors from air.

12.1B Physical Properties of Adsorbents

Many adsorbents have been developed for a wide range of separations. Typically, the adsorbents are in the form of small pellets, beads, or granules ranging from about 0.1 mm to 12 mm in size with the larger particles being used in packed beds. A particle of adsorbent has a very porous structure with many fine pores and pore volumes up to 50% of total particle volume. The adsorption often occurs as a monolayer on the surface of the fine pores. However, several layers sometimes occur. Physical adsorption, or van der Waals adsorption, usually occurs between the adsorbed molecules and the solid internal pore surface and is readily reversible.

The overall adsorption process consists of a series of steps in series. When the fluid is flowing past the particle in a fixed bed, the solute first diffuses from the bulk fluid to the gross exterior surface of the particle. Then the solute diffuses inside the pore to the surface of the pore. Finally, the solute is adsorbed on the surface. Hence, the overall adsorption process is a series of steps.

There are a number of commercial adsorbents and some of the main ones are described below. All are characterized by very large pore surface areas of 100 to over 2000 m^2/g.

1. *Activated carbon*. This is a microcrystalline material made by thermal decomposition of wood, vegetable shells, coal, etc., and has surface areas of 300 to 1200 m^2/g with average pore diameters of 10 to 60 Å. Organics are generally adsorbed by activated carbon.

2. *Silica gel*. This adsorbent is made by acid treatment of sodium silicate solution and then dried. It has a surface area of 600 to 800 m^2/g and average pore diameters of 20 to 50 Å. It is primarily used to dehydrate gases and liquids and to fractionate hydrocarbons.

3. *Activated alumina*. To prepare this material, hydrated aluminum oxide is activated by heating to drive off the water. It is used mainly to dry gases and liquids. Surface areas range from 200 to 500 m^2/g, with average pore diameters of 20 to 140 Å.

4. *Molecular sieve zeolites*. These zeolites are porous crystalline aluminosilicates that form an open crystal lattice containing precisely uniform pores. Hence, the uniform pore size is different from other types of adsorbents which have a range of pore sizes. Different zeolites have pore sizes from about 3 to 10 Å. Zeolites are used for drying, separation of hydrocarbons, mixtures, and many other applications.

5. *Synthetic polymers or resins*. These are made by polymerizing two major types of monomers. Those made from aromatics such as styrene and divinylbenzene are used to adsorb nonpolar organics from aqueous solutions. Those made from acrylic esters are usable with more polar solutes in aqueous solutions.

12.1C Equilibrium Relations for Adsorbents

The equilibrium between the concentration of a solute in the fluid phase and its concentration on the solid resembles somewhat the equilibrium solubility of a gas in a liquid. Data are plotted as adsorption isotherms as shown in Fig. 12.1-1. The concentration in the solid phase is expressed as q, kg adsorbate(solute)/kg adsorbent (solid), and in the fluid phase (gas or liquid) as c, kg adsorbate/m^3 fluid.

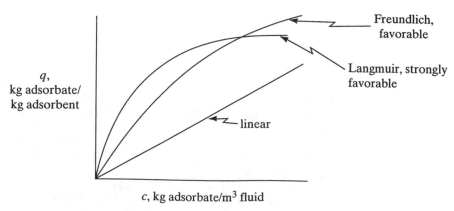

FIGURE 12.1-1.   *Some common types of adsorption isotherms.*

Data that follow a linear law can be expressed by an equation similar to Henry's law.

$$q = Kc \qquad (12.1\text{-}1)$$

where K is a constant determined experimentally, m^3/kg adsorbent. This linear isotherm is not common, but in the dilute region it can be used to approximate data of many systems.

The Freundlich isotherm equation, which is empirical, often approximates data for many physical adsorption systems and is particularly useful for liquids.

$$q = Kc^n \qquad (12.1\text{-}2)$$

where K and n are constants and must be determined experimentally. If a log-log plot of q versus c is made, the slope is the dimensionless exponent n. The dimensions of K depend on the value of n. This equation is sometimes used to correlate data for hydrocarbon gases on activated carbon.

The Langmuir isotherm has a theoretical basis and is given by the following, where q_o and K are empirical constants

$$q = \frac{q_o c}{K + c} \qquad (12.1\text{-}3)$$

where q_o is a constant, kg adsorbate/kg solid; and K is a constant, kg/m^3. The equation was derived assuming that there are only a fixed number of active sites available for adsorption, only a monolayer is formed, and the adsorption is reversible and reaches an equilibrium condition. By plotting $1/q$ versus $1/c$, the slope is K/q_o and the intercept is $1/q_o$.

Almost all adsorption systems show that as temperature is increased the amount adsorbed by the adsorbent decreases strongly. This is useful since adsorption is normally at room temperatures and desorption can be attained by raising the temperature.

EXAMPLE 12.1-1. Adsorption Isotherm for Phenol in Wastewater
Batch tests were performed in the laboratory using solutions of phenol in water and particles of granular activated carbon (R5). The equilibrium data at room temperature are shown in Table 12.1-1. Determine the isotherm that fits the data.

TABLE 12.1-1. Equilibrium Data for Example 12.1-1 (R5)

c, $\left(\dfrac{kg\ phenol}{m^3\ solution}\right)$	q, $\left(\dfrac{kg\ phenol}{kg\ carbon}\right)$
0.322	0.150
0.117	0.122
0.039	0.094
0.0061	0.059
0.0011	0.045

Solution: Plotting the data as $1/q$ versus $1/c$, the results are not a straight line and do not follow the Langmuir equation (12.1-3). A plot of log q versus log c in Fig. 12.1-2 gives a straight line and, hence, follow the Freundlich isotherm Eq. (12.1-2). The slope n is 0.229 and the constant K is 0.199, to give

$$q = 0.199c^{0.229}$$

12.2 BATCH ADSORPTION

Batch adsorption is often used to adsorb solutes from liquid solutions when the quantities treated are small in amount, such as in the pharmaceutical industry or other industries. As in many other processes, an equilibrium relation such as the Freundlich or Langmuir isotherms and a material balance are needed. The initial feed concentration is c_F and the final equilibrium concentration is c. Also, the initial concentration of the solute adsorbed on the solid is q_F and the final equilibrium value is q. The material balance on the adsorbate is

$$q_F M + c_F S = qM + cS \qquad (12.2-1)$$

where M is the amount of adsorbent, kg; and S is the volume of feed solution, m^3.

When the variable q in Eq. (12.2-1) is plotted versus c, the result is a straight line. If the equilibrium isotherm is also plotted on the same graph, the intersection of both lines gives the final equilibrium values of q and c.

EXAMPLE 12.2-1. Batch Adsorption on Activated Carbon
A wastewater solution having a volume of 1.0 m^3 contains 0.21 kg phenol/m^3 of solution (0.21 g/L). A total of 1.40 kg of fresh granular activated carbon is added to the solution, which is then mixed thoroughly to reach equilibrium. Using the isotherm from Example 12.1-1, what are the final equilibrium values, and what percent of phenol is extracted?

Solution: The given values are $c_F = 0.21$ kg phenol/m^3, $S = 1.0$ m^3, $M = 1.40$ kg carbon, and q_F is assumed as zero. Substituting into Eq. (12.2-1)

$$0(1.40) + 0.21(1.0) = q(1.40) + c(1.0)$$

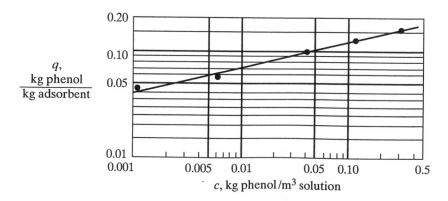

FIGURE 12.1-2. *Plot of data for Example 12.1-1.*

This straight-line equation is plotted in Fig. 12.2-1. The isotherm from Example 12.1-1 is also plotted in Fig. 12.2-1. At the intersection, $q = 0.106$ kg phenol/kg carbon and $c = 0.062$ kg phenol/m³. The percent of phenol extracted is

$$\% \text{ extracted} = \frac{c_F - c}{c_F}(100) = \frac{0.210 - 0.062}{0.210}(100) = 70.5$$

12.3 DESIGN OF FIXED-BED ADSORPTION COLUMNS

12.3A Introduction and Concentration Profiles

A widely used method for adsorption of solutes from liquid or gases employs a fixed bed of granular particles. The fluid to be treated is usually passed down through the packed bed at a constant flow rate. The situation is more complex than that for a simple stirred-tank batch process which reaches equilibrium. Mass-transfer resistances are important in the fixed-bed process and the process is unsteady state. The overall dynamics of the system determines the efficiency of the process rather than just the equilibrium considerations.

The concentrations of the solute in the fluid phase and of the solid adsorbent phase change with time and also with position in the fixed bed as adsorption proceeds. At the inlet to the bed the solid is assumed to contain no solute at the start of the process. As the fluid first contacts the inlet of the bed, most of the mass transfer and adsorption takes place here. As the fluid passes through the bed, the concentration in this fluid drops very rapidly with distance in the bed to zero way before the end of the bed is reached. The concentration profile at the start at time t_1 is shown in Fig. 12.3-1a, where the concentration ratio c/c_o is plotted versus bed length. The fluid concentration c_o is the feed concentration and c is the fluid concentration at a point in the bed.

After a short time, the solid near the entrance to the tower is almost saturated, and most of the mass transfer and adsorption now takes place at a point slightly farther from the inlet. At a later time t_2 the profile or mass-transfer zone where most of the

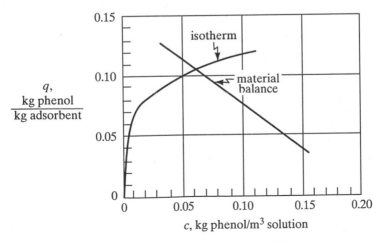

FIGURE 12.2-1. *Solution to Example 12.2-1.*

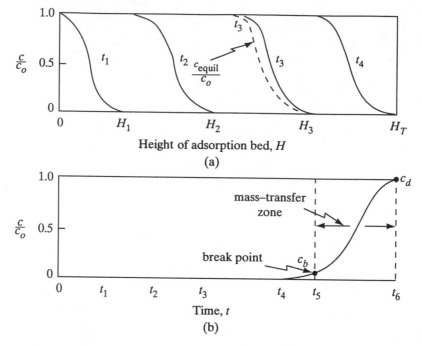

FIGURE 12.3-1. *Concentration profiles for adsorption in a fixed bed: (a) profiles at various positions and times in the bed, (b) breakthrough concentration profile in the fluid at outlet of bed.*

concentration change takes place has moved farther down the bed. The concentration profiles shown are for the fluid phase. Concentration profiles for the concentration of adsorbates on the solid would be similar. The solid at the entrance would be nearly saturated and this concentration would remain almost constant down to the mass-transfer zone, where it would drop off rapidly to almost zero. The dashed line for time t_3 shows the concentration in the fluid phase in equilibrium with the solid. The difference in concentrations is the driving force for mass transfer.

12.3B Breakthrough Concentration Curve

As seen in Fig. 12.3-1a, the major part of the adsorption at any time takes place in a relatively narrow adsorption or mass-transfer zone. As the solution continues to flow, this mass-transfer zone, which is S-shaped, moves down the column. At a given time t_3 in Fig. 12.3-1a when almost half of the bed is saturated with solute, the outlet concentration is still approximately zero, as shown in Fig. 12.3-1b. This outlet concentration remains near zero until the mass-transfer zone starts to reach the tower outlet at time t_4. Then the outlet concentration starts to rise and at t_5 the outlet concentration has risen to c_b, which is called the *break point*.

After the break-point time is reached, the concentration c rises very rapidly up to point c_d, which is the end of the breakthrough curve where the bed is judged ineffective. The break-point concentration represents the maximum that can be discarded and is often taken as 0.01 to 0.05 for c_b/c_o. The value c_d/c_o is taken as the point where c_d is approximately equal to c_o.

For a narrow mass-transfer zone, the breakthrough curve is very steep and most of the bed capacity is used at the break point. This makes efficient use of the adsorbent and lowers energy costs for regeneration.

12.3C Capacity of Column and Scale-Up Design Method

The mass-transfer zone width and shape depends on the adsorption isotherm, flow rate, mass-transfer rate to the particles, and diffusion in the pores. A number of theoretical methods have been published which predict the mass-transfer zone and concentration profiles in the bed. The predicted results may be inaccurate because of many uncertainties due to flow patterns and correlations to predict diffusion and mass transfer. Hence, experiments in laboratory scale are needed in order to scale up the results.

The total or stoichiometric capacity of the packed-bed tower, if the entire bed comes to equilibrium with the feed, can be shown to be proportional to the area between the curve and a line at $c/c_o = 1.0$ as shown in Fig. 12.3-2. The total shaded area represents the total or stoichiometric capacity of the bed as follows (R6):

$$t_t = \int_0^\infty \left(1 - \frac{c}{c_o}\right) dt \tag{12.3-1}$$

where t_t is the time equivalent to the total or stoichiometric capacity. The usable capacity of the bed up to the break-point time t_b is the crosshatched area.

$$t_u = \int_0^{t_b} \left(1 - \frac{c}{c_o}\right) dt \tag{12.3-2}$$

where t_u is the time equivalent to the usable capacity or the time at which the effluent concentration reaches its maximum permissible level. The value of t_u is usually very close to that of t_b.

The ratio t_u/t_t is the fraction of the total bed capacity or length utilized up to the break point (C3, L1, M1). Hence, for a total bed length of H_T m, H_B is the length of bed used up to the break point,

$$H_B = \frac{t_u}{t_t} H_T \tag{12.3-3}$$

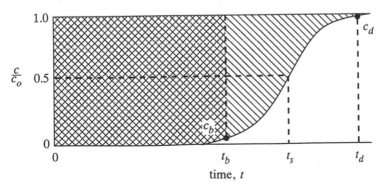

FIGURE 12.3-2. *Determination of capacity of column from breakthrough curve.*

The length of unused bed H_{UNB} in m is then the unused fraction times the total length.

$$H_{UNB} = \left(1 - \frac{t_u}{t_t}\right) H_T \tag{12.3-4}$$

The H_{UNB} represents the mass-transfer section or zone. It depends on the fluid velocity and is essentially independent of total length of the column. The value of H_{UNB} may, therefore, be measured at the design velocity in a small-diameter laboratory column packed with the desired adsorbent. Then the full-scale adsorber bed can be designed simply by first calculating the length of bed needed to achieve the required usable capacity, H_B, at the break point. The value of H_B is directly proportional to t_b. Then the length H_{UNB} of the mass-transfer section is simply added to the length H_B needed to obtain the total length, H_T.

$$H_T = H_{UNB} + H_B \tag{12.3-5}$$

This design procedure is widely used and its validity depends on the conditions in the laboratory column being similar to those for the full-scale unit. The small-diameter unit must be well insulated to be similar to the large-diameter tower, which operates adiabatically. The mass velocity in both units must be the same and the bed of sufficient length to contain a steady-state mass transfer zone (L1). Axial dispersion or axial mixing may not be exactly the same in both towers, but if caution is exercised, this method is a useful design method.

An approximate alternative procedure to use instead of integrating and obtaining areas is to assume that the breakthrough curve in Fig. 12.3-2 is symmetrical at $c/c_o = 0.5$ and t_s. Then the value of t_t in Eq. (12.3-1) is simply t_s. This assumes that the area below the curve between t_b and t_s is equal to the area above the curve between t_s and t_d.

EXAMPLE 12.3-1. Scale-Up of Laboratory Adsorption Column

A waste stream of alcohol vapor in air from a process was adsorbed by activated carbon particles in a packed bed having a diameter of 4 cm and length of 14 cm containing 79.2 g of carbon. The inlet gas stream having a concentration c_o of 600 ppm and a density of 0.00115 g/cm^3 entered the bed at a flow rate of 754 cm^3/s. Data in Table 12.3-1 give the concentrations of the breakthrough curve. The break-point concentration is set at $c/c_o = 0.01$. Do as follows.

(a) Determine the break-point time, the fraction of total capacity used up to the break point, and the length of the unused bed. Also determine the saturation loading capacity of the carbon.

TABLE 12.3-1. *Breakthrough Concentration for Example 12.3-1*

Time, h	c/c_o	Time, h	c/c_o
0	0	5.5	0.658
3	0	6.0	0.903
3.5	0.002	6.2	0.933
4	0.030	6.5	0.975
4.5	0.155	6.8	0.993
5	0.396		

(b) If the break-point time required for a new column is 6.0 h, what is the new total length of the column required?

Solution: The data from Table 12.3-1 are plotted in Fig. 12.3-3. For part (a), for $c/c_o = 0.01$, the break-point time is $t_b = 3.65$ h from the graph. The value of t_d is approximately 6.95 h. Graphically integrating, the areas are $A_1 = 3.65$ h and $A_2 = 1.51$ h. Then from Eq. (12.3-1), the time equivalent to the total or stoichiometric capacity of the bed is

$$t_t = \int_0^\infty \left(1 - \frac{c}{c_o}\right) dt = A_1 + A_2 = 3.65 + 1.51 = 5.16 \text{ h}$$

The time equivalent to the usable capacity of the bed up to the break-point time is, using Eq. (12.3-2),

$$t_u = \int_0^{t_b = 3.65} \left(1 - \frac{c}{c_o}\right) dt = A_1 = 3.65 \text{ h}$$

Hence, the fraction of total capacity used up to the break point is $t_u/t_t = 3.65/5.16 = 0.707$. From Eq. (12.3-3) the length of the used bed is $H_B = 0.707(14) = 9.9$ cm. To calculate the length of the unused bed from Eq. (12.3-4),

$$H_{UNB} = \left(1 - \frac{t_u}{t_t}\right) H_T = (1 - 0.707)14 = 4.1 \text{ cm}$$

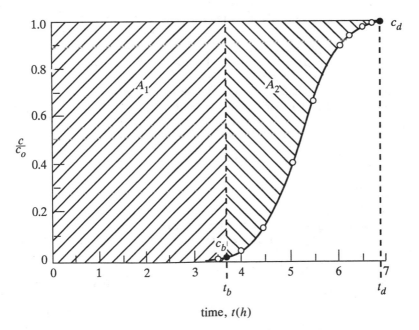

FIGURE 12.3-3. *Breakthrough curve for Example 12.3-1.*

For part (b) for a new t_b of 6.0 h, the new H_B is obtained simply from the ratio of the break-point times multiplied by the old H_B.

$$H_B = \frac{6.0}{3.65} (9.9) = 16.3 \text{ cm}$$

$$H_T = H_B + H_{UNB} = 16.3 + 4.1 = 20.4 \text{ cm}$$

We determine the saturation capacity of the carbon.

$$\text{Air flow rate} = (754 \text{ cm}^3/\text{s})(3600 \text{ s})(0.00115 \text{ g/cm}^3) = 3122 \text{ g air/h}$$

$$\text{Total alcohol adsorbed} = \left(\frac{600 \text{ g alcohol}}{10^6 \text{ g air}}\right)(3122 \text{ g air/h})(5.16 \text{ h})$$

$$= 9.67 \text{ g alcohol}$$

$$\text{Saturation capacity} = 9.67 \text{ g alcohol}/79.2 \text{ g carbon}$$

$$= 0.1220 \text{ g alcohol/g carbon}$$

The fraction of the new bed used up to the break point is now 16.3/20.4, or 0.799.

In the scale-up it may be necessary not only to change the column height, but also, the actual throughput of fluid might be different from that used in the experimental laboratory unit. Since the mass velocity in the bed must remain constant for scale-up, the diameter of the bed should be adjusted to keep this constant.

Typical gas adsorption systems use heights of fixed beds of about 0.3 m to 1.5 m with downflow of the gas. Low superficial gas velocities of 15 to 50 cm/s (0.5 to 1.7 ft/s) are used. Adsorbent particle sizes range from about 4 to 50 mesh (0.3 to 5 mm). Pressure drops are low and are only a few inches of water per foot of bed. The adsorption time is about 0.5 h up to 8 h. For liquids the superficial velocity of the liquid in the bed is about 0.3 to 0.7 cm/s (4 to 10 gpm/ft^2).

12.3D   Basic Models to Predict Adsorption

Adsorption in fixed beds is the most important method used for this process. A fixed or packed bed consists of a vertical cylindrical pipe filled or packed with the adsorbent particles. Adsorbers are mainly designed using laboratory data and the methods described in Section 12.3C. In this section the basic equations are described for isothermal adsorption so that the fundamentals involved in this process can be better understood.

An unsteady-state solute material balance in the fluid is as follows for a section dz length of bed.

$$\varepsilon \frac{\partial c}{\partial t} + (1 - \varepsilon)\rho_p \frac{\partial q}{\partial t} = -v \frac{\partial c}{\partial z} + E \frac{\partial^2 c}{\partial z^2} \qquad \text{(12.3-6)}$$

where ε is the external void fraction of the bed; v is superficial velocity in the empty bed, m/s; ρ_p is density of particle, kg/m^3; and E is an axial dispersion coefficient, m^2/s. The first term represents accumulation of solute in the liquid. The second term is accumulation of solute in the solid. The third term represents the amount of solute flowing in by convection to the section dz of the bed minus that flowing out. The last term represents axial dispersion of solute in the bed, which leads to mixing of the solute and solvent.

The second differential equation needed to describe this process relates the second term of Eq. (12.3-6) for accumulation of solute in the solid to the rate of external mass transfer of the solute from the bulk solution to the particle and the diffusion and adsorption on the internal surface area. The actual physical adsorption is very rapid. The third equation is the equilibrium isotherm.

There are many solutions to these three equations which are nonlinear and coupled. These solutions often do not fit experimental results very well and are not discussed here.

12.3E Processing Variables and Adsorption Cycles

Large-scale adsorption processes can be divided into two broad classes. The first and most important is the cyclic batch system, in which the adsorption fixed bed is alternately saturated and then regenerated in a cyclic manner. The second is a continuous flow system which involves a continuous flow of adsorbent countercurrent to a flow of feed.

There are four basic methods in common use for the cyclic batch adsorption system using fixed beds. These methods differ from each other mainly in the means used to regenerate the adsorbent after the adsorption cycle. In general, these four basic methods operate with two or sometimes three fixed beds in parallel, one in the adsorption cycle, and the other one or two in a desorbing cycle to provide continuity of flow. After a bed has completed the adsorption cycle, the flow is switched to the second newly regenerated bed for adsorption. The first bed is then regenerated by any of the following methods.

1. *Temperature-swing cycle*. This is also called the thermal-swing cycle. The spent adsorption bed is regenerated by heating it with embedded stream coils or with a hot purge gas stream to remove the adsorbate. Finally, the bed must be cooled so that it can be used for adsorption in the next cycle. The time for regeneration is generally a few hours or more.
2. *Pressure-swing cycle*. In this case the bed is desorbed by reducing the pressure at essentially constant temperature and then purging the bed at this low pressure with a small fraction of the product stream. This process for gases uses a very short cycle time for regeneration compared to that for the temperature-swing cycle.
3. *Inert-purge gas stripping cycle*. In this cycle the adsorbate is removed by passing a nonadsorbing or inert gas through the bed. This lowers the partial pressure or concentration around the particles and desorption occurs. Regeneration cycle times are usually only a few minutes.
4. *Displacement-purge cycle*. The pressure and temperature are kept essentially constant as in purge-gas stripping, but a gas or liquid is used that is adsorbed more

strongly than the adsorbate and displaces the adsorbate. Again, cycle times are usually only a few minutes.

Steam stripping is often used in regeneration of solvent recovery systems using activated carbon adsorbent. This can be considered as a combination of the temperature-swing cycle and the displacement-purge cycle.

12.4 ION-EXCHANGE PROCESSES

12.1A Introduction and Ion-Exchange Materials

Ion-exchange processes are basically chemical reactions between ions in solution and ions in an insoluble solid phase. The techniques used in ion exchange so closely resemble those used in adsorption that for the majority of engineering purposes ion exchange can be considered as a special case of adsorption.

In ion exchange certain ions are removed by the ion-exchange solid. Since electroneutrality must be maintained, the solid releases replacement ions to the solution. The first ion-exchange materials were natural-occurring porous sands called zeolites and are cation exchangers. Positively charged ions in solution such as Ca^{2+} diffuse into the pores of the solid and exchange with the Na^+ ions in the mineral.

$$Ca^{2+} \; + \; Na_2R \; \rightleftharpoons \; CaR \; + \; 2Na^+$$
$$\text{(solution)} \quad \text{(solid)} \quad \text{(solid)} \quad \text{(solution)}$$

$$(12.4\text{-}1)$$

where the R represents the solid. This is the basis for "softening" of water. To regenerate, a solution of NaCl is added which drives the reversible to reaction above the left. Almost all of these inorganic ion-exchange solids exchange only cations.

Most present-day ion-exchange solids are synthetic resins or polymers. Certain synthetic polymeric resins contain sulfonic, carboxylic, or phenolic groups. These anionic groups can exchange cations.

$$Na^+ \; + \; HR \; \rightleftharpoons \; NaR \; + \; H^+$$
$$\text{(solution)} \quad \text{(solid)} \quad \text{(solid)} \quad \text{(solution)}$$

$$(12.4\text{-}2)$$

Here the R represents the solid resin. The Na^+ in the solid resin can be exchanged with H^+ or other cations.

Similar synthetic resins containing amine groups can be used to exchange anions and OH^- in solution.

$$Cl^- \; + \; RNH_3OH \; \rightleftharpoons \; RNH_3Cl \; + \; OH^-$$
$$\text{(solution)} \quad \text{(solid)} . \quad \text{(solid)} \quad \text{(solution)}$$

$$(12.4\text{-}3)$$

12.4B Equilibrium Relations in Ion Exchange

The ion-exchange isotherms have been developed using the law of mass action. For example, for the case of a simple ion-exchange reaction such as Eq. (12.4-2), HR and

NaR represent the ion-exchange sites on the resin filled with a proton H^+ and a sodium ion Na^+. It is assumed that all of the fixed number of sites are filled with H^+ or Na^+. At equilibrium,

$$K = \frac{[NaR][H^+]}{[Na^+][HR]} \tag{12.4-4}$$

Since the total concentration of the ionic groups $[\overline{R}]$ on the resin is fixed (B7),

$$[\overline{R}] = \text{constant} = [NaR] + [HR] \tag{12.4-5}$$

Combining Eqs. (12.4-4) and (12.4-5),

$$[NaR] = \frac{K[\overline{R}][Na^+]}{[H^+] + K[Na^+]} \tag{12.4-6}$$

If the solution is buffered so $[H^+]$ is constant, the equation above for sodium exchange or adsorption is similar to the Langmuir isotherm.

12.4C Design of Fixed-Bed Ion-Exchange Columns

The rate of ion exchange depends on mass transfer of ions from the bulk solution to the particle surface, diffusion of the ions in the pores of the solid to the surface, exchange of the ions at the surface, and diffusion of the exchange ions back to the bulk solution. This is similar to adsorption. The differential equations derived are also very similar to those for adsorption. The design methods used for ion exchange and adsorption are similar and are described in Section 12.3 for adsorption processes.

12.5 SINGLE-STAGE LIQUID–LIQUID EXTRACTION PROCESSES

12.5A Introduction to Extraction Processes

In order to separate one or more of the components in a mixture, the mixture is contacted with another phase. The two-phase pair can be gas–liquid, which was discussed in Chapter 10; vapor–liquid, which was covered in Chapter 11; liquid–liquid; or fluid–solid. In this section *liquid–liquid extraction separation processes* are considered first. Alternative terms are *liquid extraction* or *solvent extraction*.

In distillation the liquid is partially vaporized to create another phase, which is a vapor. The separation of the components depends on the relative vapor pressures of the substances. The vapor and liquid phases are similar chemically. In liquid–liquid extraction the two phases are chemically quite different, which leads to a separation of the components according to physical and chemical properties.

Solvent extraction can sometimes be used as an alternative to separation by distillation or evaporation. For example, acetic acid can be removed from water by distillation or by solvent extraction using an organic solvent. The resulting organic solvent–acetic acid solution is then distilled. Choice of distillation or solvent extraction would depend on relative costs (C7). In another example high-molecular-weight fatty acids can be separated from vegetable oils by extraction with liquid propane or by high-vacuum distillation, which is more expensive.

In the pharmaceutical industry products such as penicillin occur in fermentation mixtures that are quite complex, and liquid extraction can be used to separate the penicillin. Many metal separations are being done commercially by extraction of aqueous solutions, such as copper–iron, uranium–vanadium, and tantalum–columbium.

12.5B   Equilibrium Relations in Extraction

1. *Phase rule.*   Generally in a liquid–liquid system we have three components, A, B, and C, and two phases in equilibrium. Substituting into the phase rule, Eq. (10.2-1), the number of degrees of freedom is 3. The variables are temperature, pressure, and four concentrations. Four concentrations occur because only two of the three mass fraction concentrations in a phase can be specified. The third must make the total mass fractions total to 1.0, $x_A + x_B + x_C = 1.0$. If pressure and temperature are set, which is the usual case, then, at equilibrium, setting one concentration in either phase fixes the system.

2. *Triangular coordinates and equilibrium data.*   Equilateral triangular coordinates are often used to represent the equilibrium data of a three-component system, since there are three axes. This is shown in Fig. 12.5-1. Each of the three corners represents a pure component, A, B, or C. The point M represents a mixture of A, B, and C. The perpendicular distance from the point M to the base AB represents the mass fraction x_C of C in the mixture at M, the distance to base CB the mass fraction x_A of A, and the distance to base AC the mass fraction x_B of B. Thus,

$$x_A + x_B + x_C = 0.40 + 0.20 + 0.40 = 1.0 \tag{12.5-1}$$

A common phase diagram where a pair of components A and B are partially miscible is shown in Fig. 12.5-2. Typical examples are methyl isobutyl ketone (A)–water

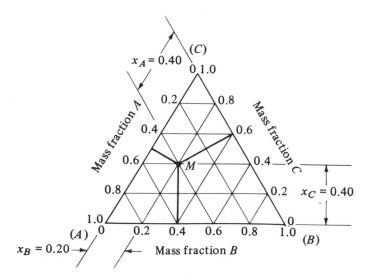

FIGURE 12.5-1.   *Coordinates for a triangular diagram.*

(B)–acetone (C), water (A)–chloroform (B)–acetone (C), and benzene (A)–water (B)–acetic acid (C). Referring to Fig. 12.5-2, liquid C dissolves completely in A or in B. Liquid A is only slightly soluble in B and B slightly soluble in A. The two-phase region is included inside below the curved envelope. An original mixture of composition M will separate into two phases a and b which are on the equilibrium tie line through point M. Other tie lines are also shown. The two phases are identical at point P, the *Plait point*.

3. Equilibrium data on rectangular coordinates. Since triangular diagrams have some disadvantages because of the special coordinates, a more useful method of plotting the three component data is to use rectangular coordinates. This is shown in Fig. 12.5-3 for the system acetic acid (A)–water (B)–isopropyl ether solvent (C). Data are from the Appendix A.3 for this system. The solvent pair B and C are partially miscible. The concentration of the component C is plotted on the vertical axis and that of A on the horizontal axis. The concentration of component B is obtained by difference from Eqs. (12.5-2) or (12.5-3).

$$x_B = 1.0 - x_A - x_C \qquad (12.5\text{-}2)$$

$$y_B = 1.0 - y_A - y_C \qquad (12.5\text{-}3)$$

The two-phase region in Fig. 12.5-3 is inside the envelope and the one-phase region outside. A tie line gi is shown connecting the water-rich layer i, called the *raffinate layer*, and the ether-rich solvent layer g, called the *extract layer*. The raffinate composition is designated by x and the extract by y. Hence, the mass fraction of C is designated as y_C in the extract layer and as x_C in the raffinate layer. To construct the tie line gi using the equilibrium $y_A - x_A$ plot below the phase diagram, vertical lines to g and i are drawn.

EXAMPLE 12.5-1. Material Balance for Equilibrium Layers
An original mixture weighing 100 kg and containing 30 kg of isopropyl ether (C), 10 kg of acetic acid (A), and 60 kg water (B) is equilibrated and the equilibrium phases separated. What are the compositions of the two equilibrium phases?

Solution: The composition of the original mixture is $x_C = 0.30$, $x_A = 0.10$, and $x_B = 0.60$. This composition of $x_C = 0.30$ and $x_A = 0.10$ is plotted as point h on Fig. 12.5-3. The tie line gi is drawn through point h by trial and error. The composition of the extract (ether) layer at g is $y_A = 0.04$, $y_C = 0.94$, and $y_B = 1.00 - 0.04 - 0.94 = 0.02$ mass fraction. The raffinate (water) layer composition at i is $x_A = 0.12$, $x_C = 0.02$, and $x_B = 1.00 - 0.12 - 0.02 = 0.86$.

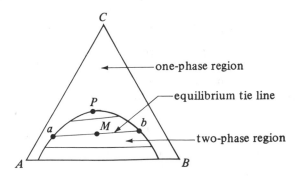

FIGURE 12.5-2. *Liquid–liquid phase diagram where components A and B are partially miscible.*

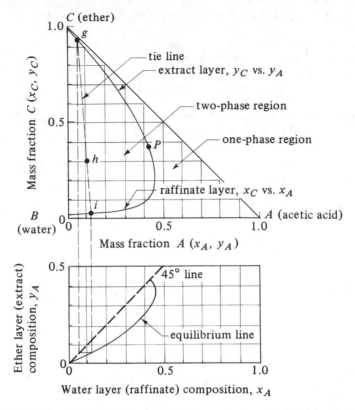

FIGURE 12.5-3. *Acetic acid (A)–water (B)–isopropyl ether (C) liquid–liquid phase diagram at 293 K (20°C).*

Another common type of phase diagram is shown in Fig. 12.5-4, where the solvent pairs B and C and also A and C are partially miscible. Examples are the system styrene (A)–ethylbenzene (B)–diethylene glycol (C) and the system chlorobenzene (A)–methyl ethyl ketone (B)–water (C).

12.5C   Single-Stage Equilibrium Extraction

1. Derivation of lever-arm rule for graphical addition.   This will be derived for use in the rectangular extraction-phase-diagram charts. In Fig. 12.5-5a two streams, L kg and V kg, containing components A, B, and C, are mixed (added) to give a resulting mixture stream M kg total mass. Writing an overall mass balance and a balance on A,

$$V + L = M \tag{12.5-4}$$

$$V y_A + L x_A = M x_{AM} \tag{12.5-5}$$

where x_{AM} is the mass fraction of A in the M stream. Writing a balance for component C,

$$V y_C + L x_C = M x_{CM} \tag{12.5-6}$$

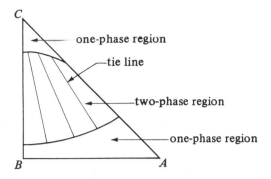

FIGURE 12.5-4. *Phase diagram where the solvent pairs B–C and A–C are partially miscible.*

Combining Eqs. (12.5-4) and (12.5-5),

$$\frac{L}{V} = \frac{y_A - x_{AM}}{x_{AM} - x_A} \tag{12.5-7}$$

Combining Eqs. (12.5-4) and (12.5-6),

$$\frac{L}{V} = \frac{y_C - x_{CM}}{x_{CM} - x_C} \tag{12.5-8}$$

Equating Eqs. (12.5-7) and (12.5-8) and rearranging,

$$\frac{x_C - x_{CM}}{x_A - x_{AM}} = \frac{x_{CM} - y_C}{x_{AM} - y_A} \tag{12.5-9}$$

This shows that points L, M, and V must lie on a straight line. By using the properties of similar right triangles,

$$\frac{L \ (\text{kg})}{V \ (\text{kg})} = \frac{\overline{VM}}{\overline{LM}} \tag{12.5-10}$$

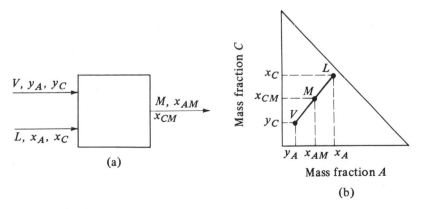

FIGURE 12.5-5. *Graphical addition and lever-arm rule : (a) process flow, (b) graphical addition.*

This is the lever-arm rule and states that kg L/kg V is equal to the length of line $\overline{VM}$/length of line $\overline{LM}$. Also,

$$\frac{L\ (\text{kg})}{M\ (\text{kg})} = \frac{\overline{VM}}{\overline{LV}} \tag{12.5-11}$$

These same equations also hold for kg mol and mol frac, lb_m, and so on.

EXAMPLE 12.5-2. *Amounts of Phases in Solvent Extraction*
The compositions of the two equilibrium layers in Example 12.5-1 are for the extract layer (V) $y_A = 0.04$, $y_B = 0.02$, and $y_C = 0.94$, and for the raffinate layer (L) $x_A = 0.12$, $x_B = 0.86$, and $x_C = 0.02$. The original mixture contained 100 kg and $x_{AM} = 0.10$. Determine the amounts of V and L.

Solution: Substituting into Eq. (12.5-4),

$$V + L = M = 100$$

Substituting into Eq. (12.5-5), where $M = 100$ kg and $x_{AM} = 0.10$,

$$V(0.04) + L(0.12) = 100(0.10)$$

Solving the two equations simultaneously, $L = 75.0$ and $V = 25.0$. Alternatively, using the lever-arm rule, the distance hg in Fig. 12.5-3 is measured as 4.2 units and gi as 5.8 units. Then by Eq. (12.5-11),

$$\frac{L}{M} = \frac{L}{100} = \frac{\overline{hg}}{\overline{gi}} = \frac{4.2}{5.8}$$

Solving, $L = 72.5$ kg and $V = 27.5$ kg, which is a reasonably close check on the material-balance method.

2. *Single-stage equilibrium extraction.* We now study the separation of A from a mixture of A and B by a solvent C in a single equilibrium stage. The process is shown in Fig. 12.5-6a, where the solvent, as stream V_2, enters and also the stream L_0. The streams are mixed and equilibrated and the exit streams L_1 and V_1 leave in equilibrium with each other.

The equations for this process are the same as those given in Section 10.3 for a single

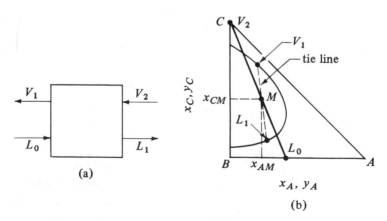

FIGURE 12.5-6. *Single-stage equilibrium liquid–liquid extraction: (a) process flow diagram, (b) plot on phase diagram.*

equilibrium stage where y represents the composition of the V streams and x the L streams.

$$L_0 + V_2 = L_1 + V_1 = M \tag{12.5-12}$$

$$L_0 x_{A0} + V_2 y_{A2} = L_1 x_{A1} + V_1 y_{A1} = M x_{AM} \tag{12.5-13}$$

$$L_0 x_{C0} + V_2 y_{C2} = L_1 x_{C1} + V_1 y_{C1} = M x_{CM} \tag{12.5-14}$$

Since $x_A + x_B + x_C = 1.0$, an equation for B is not needed. To solve the three equations, the equilibrium-phase diagram in Fig. 12.5-6b is used. Since the amounts and compositions of L_0 and V_2 are known, we can calculate values of M, x_{AM}, and x_{CM} from Eqs. (12.5-12)–(12.5-14). The points L_0, V_2, and M can be plotted as shown in Fig. 12.5-6b. Then using trial and error a tie line is drawn through the point M, which locates the compositions of L_1 and V_1. The amounts of L_1 and V_1 can be determined by substitution into Eqs. (12.5-12)–(12.5-14) or by using the lever-arm rule.

12.6 EQUIPMENT FOR LIQUID–LIQUID EXTRACTION

12.6A Introduction and Equipment Types

As in the separation processes of absorption and distillation, the two phases in liquid–liquid extraction must be brought into intimate contact with a high degree of turbulence in order to obtain high mass-transfer rates. After this contact of the two phases, they must be separated. In absorption and in distillation, this separation is rapid and easy because of the large difference in density between the gas or vapor phase and the liquid phase. In solvent extraction the density difference between the two phases is not large and separation is more difficult.

There are two main classes of solvent-extraction equipment, vessels in which mechanical agitation is provided for mixing, and vessels in which the mixing is done by the flow of the fluids themselves. The extraction equipment can be operated batchwise or operated continuously as in absorption and in distillation.

12.6B Mixer–Settlers for Extraction

To provide efficient mass transfer, a mechanical mixer is often used to provide intimate contact of the two liquid phases. One phase is usually dispersed into the other in the form of small droplets. Sufficient time of contact should be provided for the extraction to take place. Small droplets produce large interfacial areas and faster extraction. However, the droplets must not be so small that the subsequent settling time in the settler is too large.

The design and power requirements of baffled agitators or mixers have been discussed in detail in Section 3.4. In Fig. 12.6-1a a typical mixer–settler is shown, where the mixer or agitator is entirely separate from the settler. The feed of aqueous phase and organic phase are mixed in the mixer, and then the mixed phases are separated in the settler. In Fig. 12.6-1b a combined mixer–settler is shown, which is sometimes used in extraction of uranium salts or copper salts from aqueous solutions. Both types of mixer–settlers can be used in series for countercurrent or multiple-stage extraction.

12.6C Plate and Agitated Tower Contactors for Extraction

As discussed in Section 10.6 for plate absorption and distillation towers, somewhat similar types of devices are used for liquid–liquid contacting. In Fig. 12.6-2a a

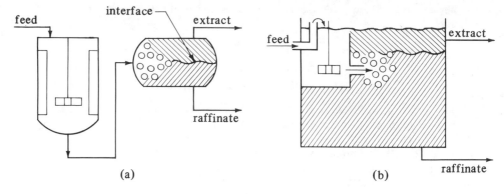

FIGURE 12.6-1. *Typical mixer–settlers for extraction: (a) separate mixer–settler, (b) combined mixer–settler.*

perforated-plate or sieve-tray extraction tower is shown wherein the rising droplets of the light solvent liquid are dispersed. The dispersed droplets coalesce below each tray and are then re-formed on each tray by passing through the perforations. The heavy aqueous liquid flows across each plate, where it is contacted by the rising droplets and then passes through the downcomer to the plate below.

In Fig. 12.6-2b an agitated extraction tower is shown. A series of paddle agitators mounted on a central rotating shaft provides the agitation for the two phases. Each agitator is separated from the next agitator by a calming section of wire mesh to encourage coalescence of the droplets and phase separation. This apparatus is essentially a series of mixer–settlers one above the other (C8, P1, T1). Another type is the Karr reciprocating-plate column, which contains a series of sieve trays with a large open area of 60% where the plates are moved up and down (C6, C8, L3). This is one of the few types of extraction towers that can be scaled up with reasonable accuracy (C6, K3).

12.6D Packed and Spray Extraction Towers

Packed and spray tower extractors give differential contacts, where mixing and settling proceed continuously and simultaneously (C8). In the plate-type towers or mixer–settler contactors, the extraction and settling proceeds in definite stages. In Fig. 12.6-3 the heavy liquid enters the top of the spray tower, fills the tower as the continuous phase, and flows out through the bottom. The light liquid enters through a nozzle distributor at the bottom which disperses or sprays the droplets upward. The light liquid coalesces at the top and flows out. In some cases the heavy liquid is sprayed downward into a rising light continuous phase.

A more effective type of tower is to pack the column with packing such as Raschig rings or Berl saddles, which cause the droplets to coalesce and redisperse at frequent intervals throughout the tower. Detailed discussions of flooding and construction of packed towers are given elsewhere (T1, P1).

12.7 CONTINUOUS MULTISTAGE COUNTERCURRENT EXTRACTION

12.7A Introduction

In Section 12.5 single-stage equilibrium contact was used to transfer the solute A from one liquid to the other liquid phase. To transfer more solute, the single-stage contact can

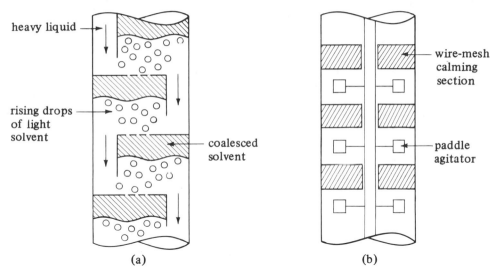

FIGURE 12.6-2. *Extraction towers: (a) perforated-plate or sieve-tray tower, (b) agitated extraction tower.*

be repeated by contacting the exit L_1 stream with fresh solvent V_2 in Fig. 12.5-6. In this way a greater percentage removal of the solute A is obtained. However, this is wasteful of the solvent stream and also gives a dilute product of A in the outlet solvent extract streams. In order to use less solvent and to obtain a more concentrated exit extract stream, countercurrent multistage contacting is often used.

Many of the fundamental equations of countercurrent gas absorption and of rectification are the same or similar to those used in countercurrent extraction. Because of the frequently high solubility of the two liquid phases in each other, the equilibrium relationships in extraction are more complicated than in absorption and distillation.

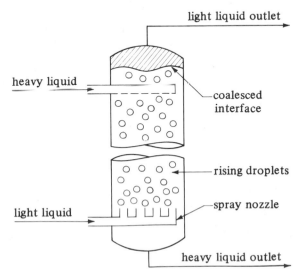

FIGURE 12.6-3. *Spray-type extraction tower.*

12.7B Continuous Multistage Countercurrent Extraction

1. Countercurrent process and overall balance. The process flow for this extraction process is the same as given previously in Fig. 10.3-2 and is shown in Fig. 12.7-1. The feed stream containing the solute A to be extracted enters at one end of the process and the solvent stream enters at the other end. The extract and raffinate streams flow countercurrently from stage to stage, and the final products are the extract stream V_1 leaving stage 1 and the raffinate stream L_N leaving stage N.

Making an overall balance on all the N stages,

$$L_0 + V_{N+1} = L_N + V_1 = M \tag{12.7-1}$$

where M represents total kg/h (lb$_m$/h) and is a constant, L_0 the inlet feed flow rate in kg/h, V_{N+1} the inlet solvent flow rate in kg/h, V_1 the exit extract stream, and L_N the exit raffinate stream. Making an overall component balance on component C,

$$L_0 x_{C0} + V_{N+1} y_{CN+1} = L_N x_{CN} + V_1 y_{C1} = M x_{CM} \tag{12.7-2}$$

Combining Eqs. (12.7-1) and (12.7-2) and rearranging,

$$x_{CM} = \frac{L_0 x_{C0} + V_{N+1} y_{CN+1}}{L_0 + V_{N+1}} = \frac{L_N x_{CN} + V_1 y_{C1}}{L_N + V_1} \tag{12.7-3}$$

A similar balance on component A gives

$$x_{AM} = \frac{L_0 x_{A0} + V_{N+1} y_{AN+1}}{L_0 + V_{N+1}} = \frac{L_N x_{AN} + V_1 y_{A1}}{L_N + V_1} \tag{12.7-4}$$

Equations (12.7-3) and (12.7-4) can be used to calculate the coordinates of point M on the phase diagram that ties together the two entering streams L_0 and V_{N+1} and the two exit streams V_1 and L_N. Usually, the flows and compositions of L_0 and V_{N+1} are known and the desired exit composition x_{AN} is set. If we plot points L_0, V_{N+1}, and M as in Fig. 12.7-2, a straight line must connect these three points. Then L_N, M, and V_1 must lie on one line. Also, L_N and V_1 must also lie on the phase envelope, as shown. These balances also hold for lb$_m$ and mass fraction, kg mol and mol fractions, and so on.

EXAMPLE 12.7-1. *Material Balance for Countercurrent Stage Process*
Pure solvent isopropyl ether at the rate of $V_{N+1} = 600$ kg/h is being used to extract an aqueous solution of $L_0 = 200$ kg/h containing 30 wt % acetic acid (A) by countercurrent multistage extraction. The desired exit acetic acid concentration in the aqueous phase is 4%. Calculate the compositions and amounts of the ether extract V_1 and the aqueous raffinate L_N. Use equilibrium data from Appendix A.3.

Solution: The given values are $V_{N+1} = 600$, $y_{AN+1} = 0$, $y_{CN+1} = 1.0$, $L_0 = 200$, $x_{A0} = 0.30$, $x_{B0} = 0.70$, $x_{C0} = 0$, and $x_{AN} = 0.04$. In Fig. 12.7-3, V_{N+1}

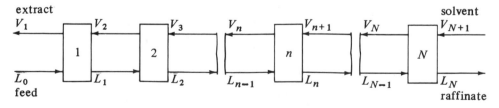

FIGURE 12.7-1. *Countercurrent-multistage-extraction-process flow diagram.*

FIGURE 12.7-2. *Use of the mixture point M for overall material balance in countercurrent solvent extraction.*

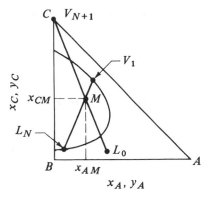

and L_0 are plotted. Also, since L_N is on the phase boundary, it can be plotted at $x_{AN} = 0.04$. For the mixture point M, substituting into Eqs. (12.7-3) and (12.7-4),

$$x_{CM} = \frac{L_0 x_{C0} + V_{N+1} y_{CN+1}}{L_0 + V_{N+1}} = \frac{200(0) + 600(1.0)}{200 + 600} = 0.75 \qquad \textbf{(12.7-3)}$$

$$x_{AM} = \frac{L_0 x_{A0} + V_{N+1} y_{AN+1}}{L_0 + V_{N+1}} = \frac{200(0.30) + 600(0)}{200 + 600} = 0.075 \qquad \textbf{(12.7-4)}$$

Using these coordinates, the point M is plotted in Fig. 12.7-3. We locate V_1 by drawing a line from L_N through M and extending it until it intersects the phase boundary. This gives $y_{A1} = 0.08$ and $y_{C1} = 0.90$. For L_N a value of $x_{CN} = 0.017$ is obtained. By substituting into Eqs. (12.7-1) and (12.7-2) and solving, $L_N = 136$ kg/h and $V_1 = 664$ kg/h.

2. Stage-to-stage calculations for countercurrent extraction. The next step after an overall balance has been made is to go stage by stage to determine the concentrations at each stage and the total number of stages N needed to reach L_N in Fig. 12.7-1.

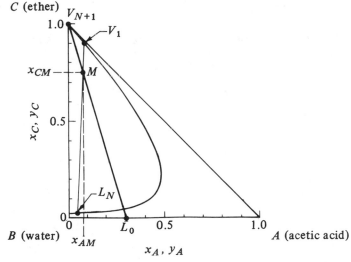

FIGURE 12.7-3. *Method to perform overall material balance for Example 12.7-1.*

Making a total balance on stage 1,

$$L_0 + V_2 = L_1 + V_1 \tag{12.7-5}$$

Making a similar balance on stage n,

$$L_{n-1} + V_{n+1} = L_n + V_n \tag{12.7-6}$$

Rearranging Eq. (12.7-5) to obtain the difference Δ in flows,

$$L_0 - V_1 = L_1 - V_2 = \Delta \tag{12.7-7}$$

This value of Δ in kg/h is constant and for all stages,

$$\Delta = L_0 - V_1 = L_n - V_{n+1} = L_N - V_{N+1} = \cdots \tag{12.7-8}$$

This also holds for a balance on component A, B, or C.

$$\Delta x_\Delta = L_0 x_0 - V_1 y_1 = L_n x_n - V_{n+1} y_{n+1} = L_N x_N - V_{N+1} y_{N+1} = \cdots \tag{12.7-9}$$

Combining Eqs. (12.7-8) and (12.7-9) and solving for x_Δ,

$$x_\Delta = \frac{L_0 x_0 - V_1 y_1}{L_0 - V_1} = \frac{L_n x_n - V_{n+1} y_{n+1}}{L_n - V_{n+1}} = \frac{L_N x_N - V_{N+1} y_{N+1}}{L_N - V_{N+1}} \tag{12.7-10}$$

where x_Δ is the x coordinate of point Δ.

Equations (12.7-7) and (12.7-8) can be written as

$$L_0 = \Delta + V_1 \qquad L_n = \Delta + V_{n+1} \qquad L_N = \Delta + V_{N+1} \tag{12.7-11}$$

From Eq. (12.7-11), we see that L_0 is on a line through Δ and V_1, L_n is on a line through Δ and V_{n+1}, and so on. This means Δ is a point common to all streams passing each other, such as L_0 and V_1, L_n and V_{n+1}, L_N and V_{N+1}, and so on. The coordinates to locate this Δ operating point are given for $x_{C\Delta}$ and $x_{A\Delta}$ in Eq. (12.7-10). Since the end points V_{N+1}, L_N or V_1, and L_0 are known, x_Δ can be calculated and point Δ located. Alternatively, the Δ point is located graphically in Fig. 12.7-4 as the intersection of lines $L_0 V_1$ and $L_N V_{N+1}$.

In order to step off the number of stages using Eq. (12.7-11) we start at L_0 and draw the line $L_0 \Delta$, which locates V_1 on the phase boundary. Next a tie line through V_1 locates

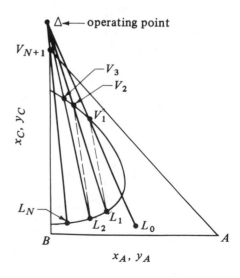

FIGURE 12.7-4. *Operating point Δ and number of theoretical stages needed for countercurrent extraction.*

Chap. 12   Liquid–Liquid and Fluid–Solid Separation Processes

L_1, which is in equilibrium with V_1. Then line $L_1 \Delta$ is drawn giving V_2. The tie line $V_2 L_2$ is drawn. This stepwise procedure is repeated until the desired raffinate composition L_N is reached. The number of stages N is obtained to perform the extraction.

EXAMPLE 12.7-2. Number of Stages in Countercurrent Extraction
Pure isopropyl ether of 450 kg/h is being used to extract an aqueous solution of 150 kg/h with 30 wt % acetic acid (A) by countercurrent multistage extraction. The exit acid concentration in the aqueous phase is 10 wt %. Calculate the number of stages required.

Solution: The known values are $V_{N+1} = 450$, $y_{AN+1} = 0$, $y_{CN+1} = 1.0$, $L_0 = 150$, $x_{A0} = 0.30$, $x_{B0} = 0.70$, $x_{C0} = 0$, and $x_{AN} = 0.10$. The points V_{N+1}, L_0, and L_N are plotted in Fig. 12.7-5. For the mixture point M, substituting into Eqs. (12.7-3) and (12.7-4), $x_{CM} = 0.75$ and $x_{AM} = 0.075$. The point M is plotted and V_1 is located at the intersection of line $L_N M$ with the phase boundary to give $y_{A1} = 0.072$ and $y_{C1} = 0.895$. This construction is not shown. (See Example 12.7-1 for construction of lines.)

The lines $L_0 V_1$ and $L_N V_{N+1}$ are drawn and the intersection is the operating point Δ as shown. Alternatively, the coordinates of Δ can be calculated from Eq. (12.7-10) to locate point Δ. Starting at L_0 we draw line $L_0 \Delta$, which locates V_1. Then a tie line through V_1 locates L_1 in equilibrium with V_1. (The tie-line data are obtained from an enlarged plot such as the bottom of Fig. 12.5-3.) Line $L_1 \Delta$ is next drawn locating V_2. A tie line through V_2 gives L_2. A line $L_2 \Delta$ gives V_3. A final tie line gives L_3, which has gone beyond the desired L_N. Hence, about 2.5 theoretical stages are needed.

3. *Minimum solvent rate.* If a solvent rate V_{N+1} is selected at too low a value a limiting

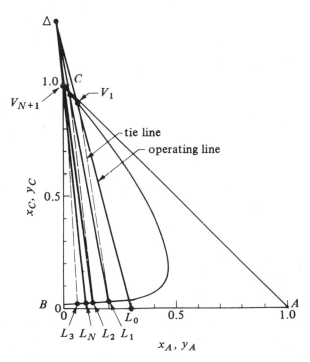

FIGURE 12.7-5. *Graphical solution for countercurrent extraction in Example 12.7-2.*

case will be reached with a line through Δ and a tie line being the same. Then an infinite number of stages will be needed to reach the desired separation. The minimum amount of solvent is reached. For actual operation a greater amount of solvent must be used.

The procedure to obtain this minimum is as follows. A tie line is drawn through point L_0 (Fig. 12.7-4) to intersect the extension of line $L_N V_{N+1}$. Other tie lines to the left of this tie line are drawn including one through L_N to intersect the line $L_N V_{N+1}$. The intersection of a tie line on line $L_N V_{N+1}$ which is nearest to V_{N+1} represents the Δ_{min} point for minimum solvent. The actual position of Δ used must be closer to V_{N+1} than Δ_{min} for a finite number of stages. This means that more solvent must be used. Usually, the tie line through L_0 represents the Δ_{min}.

12.7C Countercurrent-Stage Extraction with Immiscible Liquids

If the solvent stream V_{N+1} contains components A and C and the feed stream L_0 contains A and B and components B and C are relatively immiscible in each other, the stage calculations are made more easily. The solute A is relatively dilute and is being transferred from L_0 to V_{N+1}.

Referring to Fig. 12.7-1 and making an overall balance for A over the whole system and then over the first n stages,

$$L'\left(\frac{x_0}{1-x_0}\right) + V'\left(\frac{y_{N+1}}{1-y_{N+1}}\right) = L'\left(\frac{x_N}{1-x_N}\right) + V'\left(\frac{y_1}{1-y_1}\right) \qquad (12.7\text{-}12)$$

$$L'\left(\frac{x_0}{1-x_0}\right) + V'\left(\frac{y_{n+1}}{1-y_{n+1}}\right) = L'\left(\frac{x_n}{1-x_n}\right) + V'\left(\frac{y_1}{1-y_1}\right) \qquad (12.7\text{-}13)$$

where $L' = $ kg inert B/h, $V' = $ kg inert C/h, $y = $ mass fraction A in V stream, and $x = $ mass fraction A in L stream. This Eq. (12.7-13) is an operating-line equation whose slope $\cong L'/V'$. If y and x are quite dilute, the line will be straight when plotted on an xy diagram.

The number of stages are stepped off as shown previously in cases in distillation and absorption.

If the equilibrium line is relatively dilute, then since the operating line is essentially straight, the analytical Eqs. (10.3-21)–(10.3-26) given in Section 10.3D can be used to calculate the number of stages.

EXAMPLE 12.7-3. Extraction of Nicotine with Immiscible Liquids.
An inlet water solution of 100 kg/h containing 0.010 wt fraction nicotine (A) in water is stripped with a kerosene stream of 200 kg/h containing 0.0005 wt fraction nicotine in a countercurrent stage tower. The water and kerosene are essentially immiscible in each other. It is desired to reduce the concentration of the exit water to 0.0010 wt fraction nicotine. Determine the theoretical number of stages needed. The equilibrium data are as follows (C5), with x the weight fraction of nicotine in the water solution and y in the kerosene.

x	y	x	y
0.001010	0.000806	0.00746	0.00682
0.00246	0.001959	0.00988	0.00904
0.00500	0.00454	0.0202	0.0185

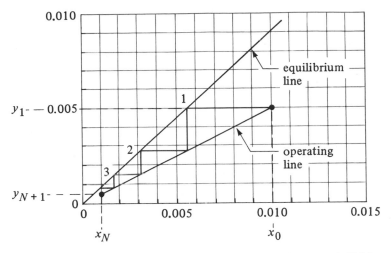

FIGURE 12.7-6. *Solution for extraction with immiscible liquids in Example 12.7-3.*

Solution: The given values are $L_0 = 100$ kg/h, $x_0 = 0.010$, $V_{N+1} = 200$ kg/h, $y_{N+1} = 0.0005$, $x_N = 0.0010$. The inert streams are

$$L' = L(1 - x) = L_0(1 - x_0) = 100(1 - 0.010) = 99.0 \text{ kg water/hr}$$

$$V' = V(1 - y) = V_{N+1}(1 - y_{N+1}) = 200(1 - 0.0005) = 199.9 \text{ kg kerosene/hr}$$

Making an overall balance on A using Eq. (12.7-12) and solving, $y_1 = 0.00497$. These end points on the operating line are plotted in Fig. 12.7-6. Since the solutions are quite dilute, the line is straight. The equilibrium line is also shown. The number of stages are stepped off, giving $N = 3.8$ theoretical stages.

12.8 INTRODUCTION AND EQUIPMENT FOR LIQUID–SOLID LEACHING

12.8A Leaching Processes

1. Introduction. Many biological and inorganic and organic substances occur in a mixture of different components in a solid. In order to separate the desired solute constituent or remove an undesirable solute component from the solid phase, the solid is contacted with a liquid phase. The two phases are in intimate contact and the solute or solutes can diffuse from the solid to the liquid phase, which causes a separation of the components originally in the solid. This process is called *liquid–solid leaching* or simply *leaching*. The term *extraction* is also used to describe this unit operation, although it also refers to liquid–liquid extraction. In leaching when an undesirable component is removed from a solid with water, the process is called *washing*.

2. Leaching processes for biological substances. In the biological and food processing industries, many products are separated from their original natural structure by liquid–solid leaching. An important process is the leaching of sugar from sugar beets with hot water. In the production of vegetable oils, organic solvents such as hexane, acetone, and ether are used to extract the oil from peanuts, soybeans, flax seeds, castor beans, sunflower seeds, cotton seeds, tung meal, and halibut livers. In the pharmaceutical industry, many different pharmaceutical products are obtained by leaching plant roots, leaves, and stems. For the production of soluble "instant" coffee, ground roasted coffee is leached with fresh water. Soluble tea is produced by water leaching of tea leaves. Tannin is removed from tree barks by leaching with water.

3. Leaching processes for inorganic and organic materials. Large uses of leaching processes occur in the metals processing industries. The useful metals usually occur in mixtures with very large amounts of undesirable constituents, and leaching is used to remove the metals as soluble salts. Copper salts are dissolved or leached from ground ores containing other minerals by sulfuric acid or ammoniacal solutions. Cobalt and nickel salts are leached from their ores by sulfuric acid–ammonia–oxygen mixtures. Gold is leached from its ore using an aqueous sodium cyanide solution. Sodium hydroxide is leached from a slurry of calcium carbonate and sodium hydroxide prepared by reacting Na_2CO_3 with $Ca(OH)_2$.

12.8B Preparation of Solids for Leaching

1. Inorganic and organic materials. The method of preparation of the solid depends to a large extent upon the proportion of the soluble constituent present, its distribution throughout the original solid, the nature of the solid—i.e., whether it is composed of plant cells or whether the soluble material is completely surrounded by a matrix of insoluble matter—and the original particle size.

   If the soluble material is surrounded by a matrix of insoluble matter, the solvent must diffuse inside to contact and dissolve the soluble material and then diffuse out. This occurs in many hydrometallurgical processes where metal salts are leached from mineral ores. In these cases crushing and grinding of the ores is used to increase the rate of leaching since the soluble portions are made more accessible to the solvent. If the soluble substance is in solid solution in the solid or is widely distributed throughout the whole solid, the solvent leaching action could form small channels. The passage of additional solvent is then made easier, and grinding to very small sizes may not be needed. Grinding of the particles is not necessary if the soluble material is dissolved in solution adhering to the solid. Then simple washing can be used as in washing of chemical precipitates.

2. Animal and vegetable materials. Biological materials are cellular in structure and the soluble constituents are generally found inside the cells. The rate of leaching may be comparatively slow because the cell walls provide another resistance to diffusion. However, to grind the biological material sufficiently small to expose the contents of individual cells is impractical. Sugar beets are cut into thin wedge-shaped slices for leaching so that the distance required for the water solvent to diffuse to reach individual cells is reduced. The cells of the sugar beet are kept essentially intact so that sugar will diffuse through the semipermeable cell walls, while the undesirable albuminous and colloidal components cannot pass through the walls.

   For the leaching of pharmaceutical products from leaves, stems, and roots, drying of the material before extraction helps rupture the cell walls. Thus, the solvent can directly dissolve the solute. The cell walls of soybeans and many vegetable seeds are largely ruptured when the original materials are reduced in size to about 0.1 mm to 0.5 mm by rolling or flaking. Cells are smaller in size, but the walls are ruptured and the vegetable oil is easily accessible to the solvent.

12.8C Rates of Leaching

1. Introduction and general steps. In the leaching of soluble materials from inside a particle by a solvent, the following general steps can occur in the overall process. The solvent must be transferred from the bulk solvent solution to the surface of the solid. Next, the solvent must penetrate or diffuse into the solid. The solute dissolves into the solvent. The solute then diffuses through the solid solvent mixture to the surface of the particle. Finally, the solute is transferred to the bulk solution. The many different phenomena encountered make it almost impracticable or impossible to apply any one theory to the leaching action.

In general, the rate of transfer of the solvent from the bulk solution to the solid surface is quite rapid, and the rate of transfer of the solvent into the solid can be somewhat rapid or slow. These are not, in many cases, the rate-limiting steps in the overall leaching process. This solvent transfer usually occurs initially when the particles are first contacted with the solvent. The dissolving of the solute into the solvent inside the solid may be simply a physical dissolution process or an actual chemical reaction that frees the solute for dissolution. Our knowledge of the dissolution process is limited and the mechanism may be different in each solid (K1).

The rate of diffusion of the solute through the solid and solvent to the surface of the solid is often the controlling resistance in the overall leaching process and can depend on a number of different factors. If the solid is made up of an inert porous solid structure with the solute and solvent in the pores in the solid, the diffusion through the porous solid can be described by an effective diffusivity. The void fraction and tortuosity are needed. This is described in Section 6.5C for diffusion in porous solids.

In biological or natural substances, additional complexity occurs because of the cells present. In the leaching of thin sugar beet slices, about one-fifth of the cells are ruptured in the slicing of the beets. The leaching of the sugar is then similar to a washing process (Y1). In the remaining cells, sugar must diffuse out through the cell walls. The net result of the two transfer processes does not follow the simple diffusion law with a constant effective diffusivity.

With soybeans, whole beans cannot be leached effectively. The rolling and flaking of the soybeans ruptures cell walls so that the solvent can more easily penetrate by capillary action. The rate of diffusion of the soybean oil solute from the soybean flakes does not permit simple interpretation. A method to design large-scale extractors is given by using small-scale laboratory experiments (O2) with flakes.

The resistance to mass transfer of the solute from the solid surface to the bulk solvent is in general quite small compared to the resistance to diffusion within the solid (O1) itself. This has been found for leaching soybeans where the degree of agitation of the external solvent has no appreciable effect on the extraction rate (O3, Y1).

2. Rate of leaching when dissolving a solid. When a material is being dissolved from the solid to the solvent solution, however, the rate of mass transfer from the solid surface to the liquid is the controlling factor. There is essentially no resistance in the solid phase if it is a pure material. The equation for this can be derived as follows for a batch system. The following can also be used for the case when diffusion in the solid is very rapid compared to the diffusion from the particle.

The rate of mass transfer of the solute A being dissolved to the solution of volume V m^3 is

$$\frac{\bar{N}_A}{A} = k_L(c_{AS} - c_A) \tag{12.8-1}$$

where $\bar{N}_A$ is kg mol of A dissolving to the solution/s, A is surface area of particles in m^2, k_L is a mass-transfer coefficient in m/s, c_{AS} is the saturation solubility of the solid solute A in the solution in kg mol/m^3, and c_A is the concentration of A in the solution at time t sec in kg mol/m^3. By a material balance, the rate of accumulation of A in the solution is equal to Eq. (12.8-1) times the area A.

$$\frac{V\ dc_A}{dt} = \bar{N}_A = Ak_L(c_{AS} - c_A) \tag{12.8-2}$$

Integrating from $t = 0$ and $c_A = c_{A0}$ to $t = t$ and $c_A = c_A$,

$$\int_{c_{A0}}^{c_A} \frac{dc_A}{c_{AS} - c_A} = \frac{Ak_L}{V} \int_{t=0}^{t} dt \tag{12.8-3}$$

$$\frac{c_{AS} - c_A}{c_{AS} - c_{A0}} = e^{-(k_L A/V)t} \tag{12.8-4}$$

The solution approaches a saturated condition exponentially. Often the interfacial area A will increase during the extraction if the external surface becomes very irregular. If the soluble material forms a high proportion of the total solid, disintegration of the particles may occur. If the solid is completely dissolving, the interfacial area changes markedly. Also, the mass-transfer coefficient may then change.

If the particles are very small, the mass-transfer coefficient to the particle in an agitated system can be predicted by using equations given in Section 7.4. For larger particles which are usually present in leaching, equations to predict the mass-transfer coefficient k_L in agitated mixing vessels are given in Section 7.4 and reference (B1).

3. Rate of leaching when diffusion in solid controls. In the case where unsteady-state diffusion in the solid is the controlling resistance in the leaching of the solute by an external solvent, the following approximations can be used. If the average diffusivity $D_{A\ eff}$ of the solute A is approximately constant, then for extraction in a batch process, unsteady-state mass-transfer equations can be used as discussed in Section 7.1. If the particle is approximately spherical, Fig. 5.3-13 can be used.

> **EXAMPLE 12.8-1.** *Prediction of Time for Batch Leaching*
> Particles having an average diameter of approximately 2.0 mm are leached in a batch-type apparatus with a large volume of solvent. The concentration of the solute A in the solvent is kept approximately constant. A time of 3.11 h is needed to leach 80% of the available solute from the solid. Assuming that diffusion in the solid is controlling and the effective diffusivity is constant, calculate the time of leaching if the particle size is reduced to 1.5 mm.
>
> **Solution:** For 80% extraction, the fraction unextracted E_s is 0.20. Using Fig. 5.3-13 for a sphere, for $E_s = 0.20$, a value of $D_{A\ eff}\ t/a^2 = 0.112$ is obtained, where $D_{A\ eff}$ is the effective diffusivity in mm^2/s, t is time in s, and a is radius in mm. For the same fraction E_s, the value of $D_{A\ eff}\ t/a^2$ is constant for a different size. Hence,
>
> $$t_2 = \frac{t_1 a_2^2}{a_1^2} \tag{12.8-5}$$
>
> where t_2 is time for leaching with a particle size a_2. Substituting into Eq. (12.8-5),
>
> $$t_2 = (3.11)\frac{(1.5/2)^2}{(2.0/2)^2} = 1.75 \text{ h}$$

4. Methods of operation in leaching.   There are a number of general methods of operation in the leaching of solids. The operations can be carried out in batch or unsteady-state conditions as well as in continuous or steady-state conditions. Both continuous and stagewise types of equipment are used in steady or unsteady-state operation.

In unsteady-state leaching a common method used in the mineral industries is *in-place leaching*, where the solvent is allowed to percolate through the actual ore body. In other cases the leach liquor is pumped over a pile of crushed ore and collected at the ground level as it drains from the heap. Copper is leached by sulfuric acid solutions from sulfide ores in this manner.

Crushed solids are often leached by percolation through stationary solid beds in a vessel with a perforated bottom to permit drainage of the solvent. The solids should not be too fine or a high resistance to flow is encountered. Sometimes a number of tanks are used in series, called an *extraction battery*, and fresh solvent is fed to the solid that is most nearly extracted. The tanks can be open tanks or closed tanks called *diffusers*. The solvent flows through the tanks in series, being withdrawn from the freshly charged tank. This simulates a continuous countercurrent stage operation. As a tank is completely leached, a fresh charge is added to the tank at the other end. Multiple piping is used so tanks do not have to be moved for countercurrent operation. This is often called the *Shanks system*. It is used widely in leaching sodium nitrate from ore, recovering tannins from barks and woods, in the mineral industries, in the sugar industry, and in other processes.

In some processes the crushed solid particles are moved continuously by bucket-type conveyors or a screw conveyor. The solvent flows countercurrently to the moving bed.

Finely ground solids may be leached in agitated vessels or in thickeners. The process can be unsteady-state batch or the vessels can be arranged in a series to obtain a countercurrent stage process.

12.8D   Types of Equipment for Leaching

1. Fixed-bed leaching.   *Fixed-bed leaching* is used in the beet sugar industry and is also used for the extraction of tanning extracts from tanbark, for the extraction of pharmaceuticals from barks and seeds, and in other processes. In Fig. 12.8-1 a typical sugar beet diffuser or extractor is shown. The cover is removable, so sugar beet slices called *cossettes* can be dumped into the bed. Heated water at 344 K (71°C) to 350 K (77°C) flows into the bed to leach out the sugar. The leached sugar solution flows out the

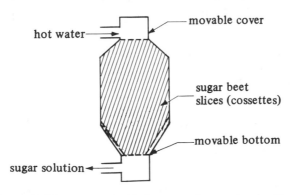

FIGURE 12.8-1.   *Typical fixed-bed apparatus for sugar beet leaching.*

bottom onto the next tank in series. Countercurrent operation is used in the Shanks system. The top and bottom covers are removable so that the leached beets can be removed and a fresh charge added. About 95% of the sugar in the beets is leached to yield an outlet solution from the system of about 12 wt %.

2. Moving-bed leaching. There are a number of devices for stagewise countercurrent leaching where the bed or stage moves instead of being stationary. These are used widely in extracting oil from vegetable seeds such as cottonseeds, peanuts, and soybeans. The seeds are usually dehulled first, sometimes precooked, often partially dried, and rolled or flaked. Sometimes preliminary removal of oil is accomplished by expression. The solvents are usually petroleum products, such as hexane. The final solvent–vegetable solution, called *miscella*, may contain some finely divided solids.

In Fig. 12.8-2a an enclosed moving-bed bucket elevator device is shown. This is called the *Bollman extractor*. Dry flakes or solids are added at the upper right side to a perforated basket or bucket. As the buckets on the right side descend, they are leached by a dilute solution of oil in solvent called *half miscella*. This liquid percolates downward through the moving buckets and is collected at the bottom as the strong solution or *full miscella*. The buckets moving upward on the left are leached countercurrently by fresh solvent sprayed on the top bucket. The wet flakes are dumped as shown and removed continuously.

The *Hildebrandt extractor* in Fig. 12.8-2b consists of three screw conveyors arranged in a U shape. The solids are charged at the top right, conveyed downward, across the bottom, and then up the other leg. The solvent flows countercurrently.

3. Agitated solid leaching. When the solid can be ground fine to about 200 mesh (0.074 mm), it can be kept in suspension by small amounts of agitation. Continuous countercurrent leaching can be accomplished by placing a number of agitators in series with settling tanks or thickeners between each agitator.

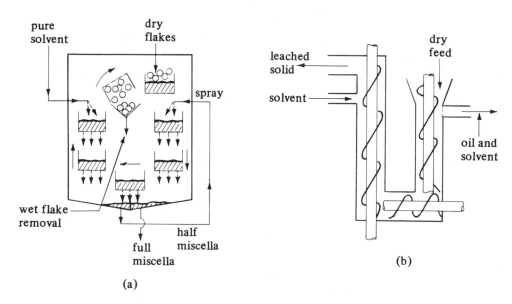

(a)

(b)

FIGURE 12.8-2. *Equipment for moving-bed leaching : (a) Bollman bucket-type extractor, (b) Hildebrandt screw-conveyor extractor.*

Sometimes the thickeners themselves are used as combination contactor-agitators and settlers as shown in Fig. 12.8-3. In this countercurrent stage system, fresh solvent enters as shown to the first stage thickener. The clear settled liquid leaves and flows from stage to stage. The feed solids enter the last stage, where they are contacted with solvent from the previous stage and then enter the settler. The slowly rotating rake moves the solids to the bottom discharge. The solids with some liquid are pumped as a slurry to the next tank. If the contact is insufficient, a mixer can be installed between each settler.

12.9 EQUILIBRIUM RELATIONS AND SINGLE-STAGE LEACHING

12.9A Equilibrium Relations in Leaching

1. Introduction. To analyze single-stage and countercurrent-stage leaching, an operating-line equation or material-balance relation and the equilibrium relations between the two streams are needed as in liquid–liquid extraction. It is assumed that the solute-free solid is insoluble in the solvent. In leaching, assuming there is sufficient solvent present so that all the solute in the entering solid can be dissolved into the liquid, equilibrium is reached when the solute is dissolved. Hence, all the solute is completely dissolved in the first stage. There usually is sufficient time for this to occur in the first stage.

It is also assumed that there is no adsorption of the solute by the solid in the leaching. This means that the solution in the liquid phase leaving a stage is the same as the solution that remains with the solid matrix in the settled slurry leaving the stage. In the settler in a stage it is not possible or feasible to separate all the liquid from the solid. Hence, the settled solid leaving a stage always contains some liquid in which dissolved solute is present. This solid–liquid stream is called the *underflow* or *slurry stream*. Consequently, the concentration of oil or solute in the liquid or *overflow stream* is equal to the concentration of solute in the liquid solution accompanying the slurry or underflow stream. Hence, on an xy plot the equilibrium line is on the 45° line.

The amount of solution retained with the solids in the settling portion of each stage may depend upon the viscosity and density of the liquid in which the solid is suspended.

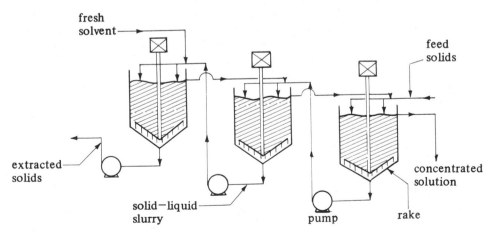

FIGURE 12.8-3. *Countercurrent leaching using thickeners.*

This, in turn, depends upon the concentration of the solute in the solution. Hence, experimental data showing the variation of the amount and composition of solution retained in the solids as a function of the solute composition are obtained. These data should be obtained under conditions of concentrations, time, and temperature similar to those in the process for which the stage calculations are to be made.

2. Equilibrium diagrams for leaching. The equilibrium data can be plotted on the rectangular diagram as wt fraction for the three components: solute (A), inert or leached solid (B), and solvent (C). The two phases are the overflow (liquid) phase and the underflow (slurry) phase. This method is discussed elsewhere (B2). Another convenient method of plotting the equilibrium data will be used, instead, which is similar to the method discussed in the enthalpy–concentration plots in Section 11.6.

The concentration of inert or insoluble solid B in the solution mixture or the slurry mixture can be expressed in kg (lb_m) units.

$$N = \frac{\text{kg } B}{\text{kg } A + \text{kg } C} = \frac{\text{kg solid}}{\text{kg solution}} = \frac{\text{lb solid}}{\text{lb solution}} \tag{12.9-1}$$

There will be a value of N for the overflow where $N = 0$ and for the underflow N will have different values, depending on the solute concentration in the liquid. The compositions of solute A in the liquid will be expressed as wt fractions.

$$x_A = \frac{\text{kg } A}{\text{kg } A + \text{kg } C} = \frac{\text{kg solute}}{\text{kg solution}} \quad \text{(overflow liquid)} \tag{12.9-2}$$

$$y_A = \frac{\text{kg } A}{\text{kg } A + \text{kg } C} = \frac{\text{kg solute}}{\text{kg solution}} \quad \text{(liquid in slurry)} \tag{12.9-3}$$

where x_A is the wt fraction of solute A in the overflow liquid and y_A is the wt fraction of A on a solid B free basis in the liquid associated with the slurry or underflow. For the entering solid feed to be leached, N is kg inert solid/kg solute A and $y_A = 1.0$. For pure entering solvent $N = 0$ and $x_A = 0$.

In Fig. 12.9-1a a typical equilibrium diagram is shown where solute A is infinitely soluble in solvent C, which would occur in the system of soybean oil (A)–soybean inert solid meal (B)–hexane solvent (C). The upper curve of N versus y_A for the slurry underflow represents the separated solid under experimental conditions similar to the actual stage process. The bottom line of N versus x_A, where $N = 0$ on the axis, represents the overflow liquid composition where all the solid has been removed. In some cases small amounts of solid may remain in the overflow. The tie lines are vertical, and on a yx diagram, the equilibrium line is $y_A = x_A$ on the 45° line. In Fig. 12.9-1b the tie lines are not vertical, which can result from insufficient contact time, so that all the solute is not dissolved; adsorption of solute A on the solid; or the solute being soluble in the solid B.

If the underflow line of N versus y is straight and horizontal, the amount of liquid associated with the solid in the slurry is constant for all concentrations. This would mean that the underflow liquid rate is constant throughout the various stages as well as the overflow stream. This is a special case which is sometimes approximated in practice.

12.9B   Single-Stage Leaching

In Fig. 12.9-2a a single-stage leaching process is shown where V is kg/h (lb_m/h) of overflow solution with composition x_A and L is kg/h of liquid in the slurry solution with composition y_A based on a given flow rate B kg/h of dry solute-free solid. The material-

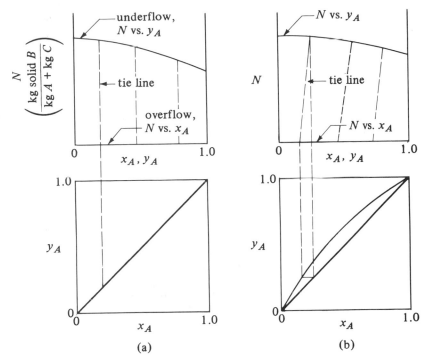

FIGURE 12.9-1. *Several typical equilibrium diagrams : (a) case for vertical tie lines and*
$y_A = x_A$, *(b) case where* $y_A \neq x_A$ *for tie lines.*

balance equations are almost identical to Eqs. (12.5-12)–(12.5-14) for single-stage liquid–liquid extraction and are as follows for a total solution balance (solute A + solvent C), a component balance on A, and a solids balance on B, respectively.

$$L_0 + V_2 = L_1 + V_1 = M \tag{12.9-4}$$

$$L_0 y_{A0} + V_2 x_{A2} = L_1 y_{A1} + V_1 x_{A1} = M x_{AM} \tag{12.9-5}$$

$$B = N_0 L_0 + 0 = N_1 L_1 + 0 = N_M M \tag{12.9-6}$$

where M is the total flow rate in kg A + C/h and x_{AM} and N_M are the coordinates of this point M. A balance on C is not needed, since $x_A + x_C = 1.0$ and $y_A + y_C = 1.0$. As shown before, $L_1 M V_1$ must lie on a straight line and $L_0 M V_2$ must also lie on a straight line. This is shown in Fig. 12.9-2b. Also, L_1 and V_1 must lie on the vertical tie line. The point M is the intersection of the two lines. If L_0 entering is the fresh solid feed to be leached with no solvent C present, it would be located above the N versus y line in Fig. 12.9-2b.

EXAMPLE 12.9-1. *Single-Stage Leaching of Flaked Soybeans*

In a single-stage leaching of soybean oil from flaked soybeans with hexane, 100 kg of soybeans containing 20 wt % oil is leached with 100 kg of fresh hexane solvent. The value of N for the slurry underflow is essentially constant at 1.5 kg insoluble solid/kg solution retained. Calculate the amounts and compositions of the overflow V_1 and the underflow slurry L_1 leaving the stage.

Solution: The process flow diagram is the same as given in Fig. 12.9-2a. The known process variables are as follows.

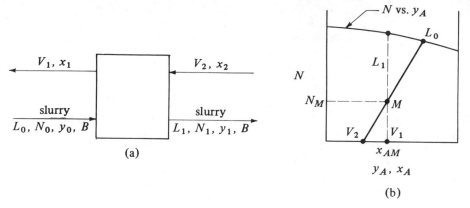

FIGURE 12.9-2. *Process flow and material balance for single-stage leaching: (a) process flow, (b) material balance.*

The entering solvent flow $V_2 = 100$ kg, $x_{A2} = 0$, $x_{C2} = 1.0$. For the entering slurry stream, $B = 100(1.0 - 0.2) = 80$ kg insoluble solid, $L_0 = 100(1.0 - 0.8) = 20$ kg A, $N_0 = 80/20 = 4.0$ kg solid/kg solution, $y_{A0} = 1.0$.

To calculate the location of M, substituting into Eqs. (12.9-4), (12.9-5), and (12.9-6) and solving,

$$L_0 + V_2 = 20 + 100 = 120 \text{ kg} = M$$

$$L_0 y_{A0} + V_2 x_{A2} = 20(1.0) + 100(0) = 120 x_{AM}$$

Hence, $x_{AM} = 0.167$.

$$B = N_0 L_0 = 4.0(20) = 80 = N_M(120)$$

$$N_M = 0.667$$

The point M is plotted in Fig. 12.9-3 along with V_2 and L_0. The vertical tie line is drawn locating L_1 and V_1 in equilibrium with each other. Then $N_1 = 1.5$, $y_{A1} = 0.167$, $x_{A1} = 0.167$. Substituting into Eqs. (12.9-4) and (12.9-6) and solving or using the lever-arm rule, $L_1 = 53.3$ kg and $V_1 = 66.7$ kg.

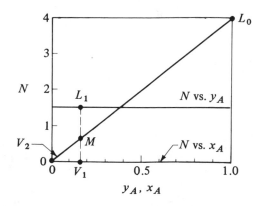

FIGURE 12.9-3. *Graphical solution of single-stage leaching for Example 12.9-1.*

Chap. 12   Liquid–Liquid and Fluid–Solid Separation Processes

12.10 COUNTERCURRENT MULTISTAGE LEACHING

12.10A Introduction and Operating Line for Countercurrent Leaching

A process flow sheet for countercurrent multistage leaching is shown in Fig. 12.10-1 and is similar to Fig. 12.7-1 for liquid–liquid extraction. The ideal stages are numbered in the direction of the solids or underflow stream. The solvent (C)–solute (A) phase or V phase is the liquid phase that overflows continuously from stage to stage countercurrently to the solid phase, and it dissolves solute as it moves along. The slurry phase L composed of inert solids (B) and a liquid phase of A and C is the continuous underflow from each stage. Note that the composition of the V phase is denoted by x and the composition of the L phase by y, which is the reverse of that for liquid–liquid extraction.

It is assumed that the solid B is insoluble and is not lost in the liquid V phase. The flow rate of the solids is constant throughout the cascade of stages. As in the single-stage leaching V is kg/h (lb$_m$/h) of overflow solution and L is kg/h of liquid solution in the slurry retained by the solid.

In order to derive the operating-line equation, an overall balance and a component balance on solute A is made over the first n stages.

$$V_{n+1} + L_0 = V_1 + L_n \tag{12.10-1}$$

$$V_{n+1}x_{n+1} + L_0 y_0 = V_1 x_1 + L_n y_n \tag{12.10-2}$$

Solving for x_{n+1} and eliminating V_{n+1},

$$x_{n+1} = \frac{1}{1 + (V_1 - L_0)/L_n} y_n + \frac{V_1 x_1 - L_0 x_0}{L_n + V_1 - L_0} \tag{12.10-3}$$

The operating-line equation (12.10-3) when plotted on an xy plot passes through the terminal points x_1, y_0 and x_{N+1}, y_N.

In the leaching process, if the viscosity and density of the solution changes appreciably with the solute (A) concentration, the solids from the lower-numbered stages where solute concentrations are high may retain more liquid solution than the solids from the higher-numbered stages, where the solute concentration is dilute. Then L_n, the liquid retained in the solids underflow, will vary and the slope of Eq. (12.10-3) will vary from stage to stage. This condition of variable underflow will be considered first. The overflow will also vary. If the amount of solution L_n retained by the solid is constant and independent of concentration, then constant underflow occurs. This simplifies somewhat the stage-to-stage calculations. This case will be considered later.

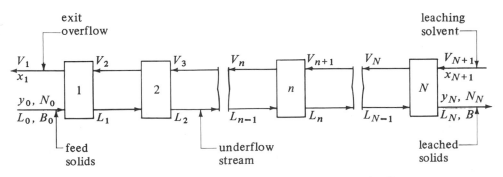

FIGURE 12.10-1. *Process flow for countercurrent multistage leaching.*

12.10B Variable Underflow in Countercurrent Multistage Leaching

The methods in this section are very similar to those used in Section 12.7B for countercurrent solvent extraction, where the L and V flow rates varied from stage to stage. Making an overall total solution (solute A + solvent C) balance on the process of Fig. 12.10-1,

$$L_0 + V_{N+1} = L_N + V_1 = M \qquad (12.10\text{-}4)$$

where M is the total mixture flow rate in kg A + C/h. Next making a component balance on A,

$$L_0 y_{A0} + V_{N+1} x_{AN+1} = L_N y_{AN} + V_1 x_{A1} = M x_{AM} \qquad (12.10\text{-}5)$$

Making a total solids balance on B,

$$B = N_0 L_0 = N_N L_N = N_M M \qquad (12.10\text{-}6)$$

where N_M and x_{AM} are the coordinates of point M shown in Fig. 12.10-2, which is the operating diagram for the process. As shown previously $L_0 M V_{N+1}$ must lie on a straight line and $V_1 M L_N$ must be on a straight line. Usually the flows and compositions of L_0 and V_{N+1} are known and the desired exit concentration y_{AN} is set. Then the coordinates N_M and x_{AM} can be calculated from Eqs. (12.10-4)–(12.10-6) and point M plotted. Then L_N, M, and V_1 must lie on one line as shown in Fig. 12.10-2.

In order to go stage by stage on Fig. 12.10-2, we must derive the operating-point

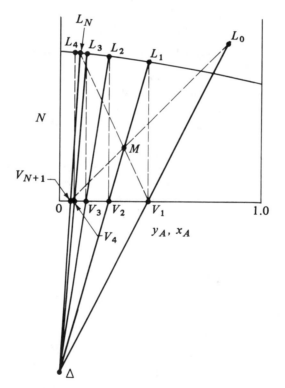

FIGURE 12.10-2.   *Number of stages for multistage countercurrent leaching.*

equation. Making a total balance on stage 1 and than on stage n,

$$L_0 + V_2 = L_1 + V_1 \qquad (12.10\text{-}7)$$

$$L_{n-1} + V_{n+1} = L_n + V_n \qquad (12.10\text{-}8)$$

Rearranging Eq. (12.10-7) for the difference flows Δ in kg/h,

$$L_0 - V_1 = L_1 - V_2 = \Delta \qquad (12.10\text{-}9)$$

This value Δ is constant and also holds for Eq. (12.10-8) rearranged and for all stages.

$$\Delta = L_0 - V_1 = L_n - V_{n+1} = L_N - V_{N+1} = \cdots \qquad (12.10\text{-}10)$$

This can also be written for a balance on solute A to give

$$x_{A\Delta} = \frac{L_0 y_{A0} - V_1 x_{A1}}{L_0 - V_1} = \frac{L_N y_{AN} - V_{N+1} x_{AN+1}}{L_N - V_{N+1}} \qquad (12.10\text{-}11)$$

where $x_{A\Delta}$ is the x coordinate of the operating point Δ. A balance given on solids gives

$$N_\Delta = \frac{B}{L_0 - V_1} = \frac{N_0 L_0}{L_0 - V_1} \qquad (12.10\text{-}12)$$

where N_Δ is the N coordinate of the operating point Δ.

As shown in Section 12.7B, Δ is the operating point. This point Δ is located graphically in Fig. 12.10-2 as the intersection of lines $L_0 V_1$ and $L_N V_{N+1}$. From Eq. (12.10-10) we see that V_1 is on a line between L_0 and Δ, V_2 is on a line between L_1 and Δ, V_{n+1} is on a line between L_n and Δ, and so on.

To graphically determine the number of stages, we start at L_0 and draw line $L_0 \Delta$ to locate V_1. A tie line through V_1 locates L_1. Line $L_1 \Delta$ is drawn given V_2. A tie line gives L_2. This is continued until the desired L_N is reached. In Fig. 12.10-2, about 3.5 stages are required.

EXAMPLE 12.10-1. Countercurrent Leaching of Oil from Meal
A continuous countercurrent multistage system is to be used to leach oil from meal by benzene solvent (B3). The process is to treat 2000 kg/h of inert solid meal (B) containing 800 kg oil (A) and also 50 kg benzene (C). The inlet flow per hour of fresh solvent mixture contains 1310 kg benzene and 20 kg oil. The leached solids are to contain 120 kg oil. Settling experiments similar to those in the actual extractor show that the solution retained depends upon the concentration of oil in the solution. The data (B3) are tabulated below as N kg inert solid B/kg solution and y_A kg oil A/kg solution.

N	y_A	N	y_A
2.00	0	1.82	0.4
1.98	0.1	1.75	0.5
1.94	0.2	1.68	0.6
1.89	0.3	1.61	0.7

Calculate the amounts and concentrations of the stream leaving the process and the number of stages required.

Solution: The underflow data from the table are plotted in Fig. 12.10-3 as N versus y_A. For the inlet solution with the untreated solid, $L_0 = 800 +$

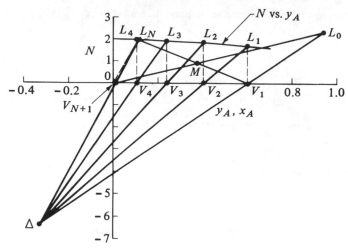

FIGURE 12.10-3. *Graphical construction for number of stages for Example 12.10-1.*

$50 = 850$ kg/h, $y_{A0} = 800/(800 + 50) = 0.941$, $B = 2000$ kg/h, $N_0 = 2000/(800 + 50) = 2.36$. For the inlet leaching solvent, $V_{N+1} = 1310 + 20 = 1330$ kg/h and $x_{AN+1} = 20/1330 = 0.015$. The points V_{N+1} and L_0 are plotted.

The point L_N lies on the N versus y_A line in Fig. 12.10-3. Also for this point, the ratio $N_N/y_{AN} =$ (kg solid/kg solution)/(kg oil/kg solution) = kg solid/kg oil $= 2000/120 = 16.67$. Hence, a dashed line through the origin at $y_A = 0$ and $N = 0$ is plotted with a slope of 16.67, which intersects the N versus y_A line at L_N. The coordinates of L_N at this intersection are $N_N = 1.95$ kg solid/kg solution and $y_{AN} = 0.118$ kg oil/kg solution.

Making an overall balance by substituting into Eq. (12.10-4) to determine point M,

$$L_0 + V_{N+1} = 850 + 1330 = 2180 \text{ kg/h} = M$$

Substituting into Eq. (12.10-5) and solving,

$$L_0 y_{A0} + V_{N+1}x_{AN+1} = 850(0.941) + 1330(0.015) = 2180x_{AM}$$

$$x_{AM} = 0.376$$

Substituting into Eq. (12.10-6) and solving,

$$B = 2000 = N_M M = N_M(2180) \qquad N_M = 0.918$$

The point M is plotted with the coordinates $x_{AM} = 0.376$ and $N_M = 0.918$ in Fig. 12.10-3. The line $V_{N+1}ML_0$ is drawn, as is line $L_N M$, which intersects the abscissa at point V_1 where $x_{A1} = 0.600$.

The amounts of streams V_1 and L_N are calculated by substituting into Eqs. (12.10-4) and (12.10-5), and solving simultaneously:

$$L_N + V_1 = M = 2180$$

$$L_N y_{AN} + V_1 x_{A1} = L_N(0.118) + V_1(0.600) = 2180(0.376)$$

Hence, $L_N = 1016$ kg solution/h in the outlet underflow stream and $V_1 = 1164$ kg solution/h in the exit overflow stream. Alternatively, the amounts could have been calculated using the lever-arm rule.

The operating point Δ is obtained as the intersection of lines $L_0 V_1$ and

$L_N V_{N+1}$ in Fig. 12.10-3. Its coordinates can also be calculated from Eqs. (12.10-11) and (12.10-12). The stages are stepped off as shown. The fourth stage for L_4 is slightly past the desired L_N. Hence, about 3.9 stages are required.

12.10C Constant Underflow in Countercurrent Multistage Leaching

In this case the liquid L_n retained in the underflow solids is constant from stage to stage. This means that a plot of N versus y_A is a horizontal straight line and N is constant. Then the operating-line equation (12.10-3) is a straight line when plotted as y_A versus x_A. The equilibrium line can also be plotted on the same diagram. In many cases the equilibrium line may also be straight with $y_A = x_A$. Special treatment must be given the first stage, however, since L_0 is generally not equal to L_n, since it contains little or no solvent. A separate material and equilibrium balance is made on stage 1 to obtain L_1 and V_2 (see Fig. 12.10-1). Then the straight operating line can be used and the McCabe–Thiele method used to step off the number of stages.

Since this procedure for constant underflow requires almost as many calculations as the general case for variable underflow, the general procedure can be used for constant underflow by simply using a horizontal line of N versus y_A in Fig. 12.10-2 and stepping off the stages with the Δ point.

12.11 INTRODUCTION AND EQUIPMENT FOR CRYSTALLIZATION

12.11A Crystallization and Types of Crystals

1. Introduction. Separation processes for gas–liquid and liquid–liquid systems have been treated in this and previous chapters. Also, the separation process of leaching was discussed for a solid–liquid system. Crystallization is also a solid–liquid separation process in which mass transfer occurs of a solute from the liquid solution to a pure solid crystalline phase. An important example is in the production of sucrose from sugar beet, where the sucrose is crystallized out from an aqueous solution.

Crystallization is a process where solid particles are formed from a homogeneous phase. This process can occur in the freezing of water to form ice, in the formation of snow particles from a vapor, in the formation of solid particles from a liquid melt, or in the formation of solid crystals from a liquid solution. The last process mentioned, crystallization from a solution, is the most important one commercially and will be treated in the present discussion. In crystallization the solution is concentrated and usually cooled until the solute concentration becomes greater than its solubility at that temperature. Then the solute comes out of the solution forming crystals of approximately pure solute.

In commercial crystallization the yield and purity of crystals are not only important but also the sizes and shapes of the crystals. It is often desirable that crystals be uniform in size. Size uniformity is desirable to minimize caking in the package, for ease of pouring, for ease in washing and filtering, and for uniform behavior when used. Sometimes large crystals are requested by the purchaser, even though smaller crystals are just as useful. Also, crystals of a certain shape are sometimes required, such as needles rather than cubes.

2. *Types of crystal geometry.* A crystal can be defined as a solid composed of atoms, ions, or molecules, which are arranged in an orderly and repetitive manner. It is a highly organized type of matter. The atoms, ions, or molecules are located in three-dimensional arrays or space lattices. The interatomic distances in a crystal between these imaginary planes or space lattices are measured by x-ray diffraction as are the angles between these planes. The pattern or arrangement of these space lattices is repeated in all directions.

Crystals appear as polyhedrons having flat faces and sharp corners. The relative sizes of the faces and edges of different crystals of the same material may differ greatly. However, the angles between the corresponding faces of all crystals of the same material are equal and are characteristic of that particular material. Crystals are thus classified on the basis of these interfacial angles.

There are seven classes of crystals, depending upon the arrangement of the axes to which the angles are referred:

1. Cubic system. Three equal axes at right angles to each other.
2. Tetragonal system. Three axes at right angles to each other, one axis longer than the other two.
3. Orthorhombic system. Three axes at right angles to each other, all of different lengths.
4. Hexagonal system. Three equal axes in one plane at 60° to each other, and a fourth axis at right angles to this plane and not necessarily the same length.
5. Monoclinic system. Three unequal axes, two at right angles in a plane and a third at some angle to this plane.
6. Triclinic system. Three unequal axes at unequal angles to each other and not 30, 60, or 90°.
7. Trigonal system. Three equal and equally inclined axes.

The relative development of different types of faces of a crystal may differ for a given solute crystallizing. Sodium chloride crystallizes from aqueous solutions with cubic faces only. In another case, if sodium chloride crystallizes from an aqueous solution with a given slight impurity present, the crystals will have octahedral faces. Both types of crystals are in the cubic system but differ in crystal habit. The crystallization in overall shapes of plates or needles has no relation to crystal habit or crystal system and usually depends upon the process conditions under which the crystals are grown.

12.11B Equilibrium Solubility in Crystallization

In crystallization equilibrium is attained when the solution or mother liquor is saturated. This is represented by a *solubility curve*. Solubility is dependent mainly upon temperature. Pressure has a negligible effect on solubility. Solubility data are given in the form of curves where solubilities in some convenient units are plotted versus temperature. Tables of solubilities are given in many chemical handbooks (P1). Solubility curves for some typical salts in water were given in Fig. 8.1-1. In general, the solubilities of most salts increase slightly or markedly with increasing temperature.

A very common type of curve is shown in Fig. 8.1-1 for KNO_3, where the solubility increases markedly with temperature and there are no hydrates. Over the whole range of temperatures, the solid phase is KNO_3. The solubility of NaCl is marked by its small change with temperature. In solubility plots the solubility data are ordinarily given as parts by weight of anhydrous material per 100 parts by weight of total solvent (i.e., water in many cases).

In Fig. 12.11-1 the solubility curve is shown for sodium thiosulfate, $Na_2S_2O_3$. The solubility increases rapidly with temperature, but there are definite breaks in the curve which indicate different hydrates. The stable phase up to 48.2°C is the pentahydrate $Na_2S_2O_3 \cdot 5H_2O$. This means that at concentrations above the solubility line (up to 48.2°C), the solid crystals formed are $Na_2S_2O_3 \cdot 5H_2O$. At concentrations below the solubility line, only a solution exists. From 48.2 to about 65°C, the stable phase is $Na_2S_2O_3 \cdot 2H_2O$. A half-hydrate is present between 65 to 70°C, and the anhydrous salt is the stable phase above 70°C.

12.11C Yields and Heat and Material Balances in Crystallization

1. Yields and material balances in crystallization. In most of the industrial crystallization processes, the solution (mother liquor) and the solid crystals are in contact for a long enough time to reach equilibrium. Hence, the mother liquor is saturated at the final temperature of the process, and the final concentration of the solute in the solution can be obtained from the solubility curve. The yield of crystals from a crystallization process can then be calculated knowing the initial concentration of solute, the final temperature, and the solubility at this temperature.

In some instances in commercial crystallization, the rate of crystal growth may be quite slow, due to a very viscous solution or a small surface of crystals exposed to the solution. Hence, some supersaturation may still exist, giving a lower yield of crystals than predicted.

In making the material balances, the calculations are straightforward when the solute crystals are anhydrous. Simple water and solute material balances are made. When the crystals are hydrated, some of the water in the solution is removed with the crystals as a hydrate.

> **EXAMPLE 12.11-1. *Yield of a Crystallization Process***
> A salt solution weighing 10 000 kg with 30 wt % Na_2CO_3 is cooled to 293 K (20°C). The salt crystallizes as the decahydrate. What will be the yield of $Na_2CO_3 \cdot 10H_2O$ crystals if the solubility is 21.5 kg anhydrous $Na_2CO_3/100$ kg of total water? Do this for the following cases.
> (a) Assume that no water is evaporated.
> (b) Assume that 3% of the total weight of the solution is lost by evaporation of water in cooling.

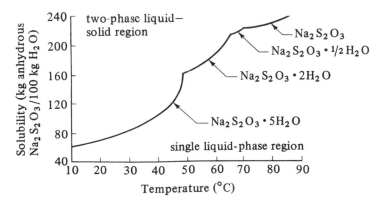

FIGURE 12.11-1. *Solubility of sodium thiosulfate, $Na_2S_2O_3$, in water.*

Solution: The molecular weights are 106.0 for Na_2CO_3, 180.2 for $10H_2O$, and 286.2 for $Na_2CO_3 \cdot 10H_2O$. The process flow diagram is shown in Fig. 12.11-2, with W being kg H_2O evaporated, S kg solution (mother liquor), and C kg crystals of $Na_2CO_3 \cdot 10H_2O$. Making a material balance around the dashed-line box for water for part (a), where $W = 0$,

$$0.70(10\,000) = \frac{100}{100 + 21.5}(S) + \frac{180.2}{286.2}(C) + 0 \qquad \textbf{(12.11-1)}$$

where $(180.2)/(286.2)$ is wt fraction of water in the crystals. Making a balance for Na_2CO_3,

$$0.30(10\,000) = \frac{21.5}{100 + 21.5}(S) + \frac{106.0}{286.2}(C) + 0 \qquad \textbf{(12.11-2)}$$

Solving the two equations simultaneously, $C = 6370$ kg of $Na_2CO_3 \cdot 10H_2O$ crystals and $S = 3630$ kg solution.

For part (b), $W = 0.03(10\,000) = 300$ kg H_2O. Equation (12.11-1) becomes

$$0.70(10\,000) = \frac{100}{100 + 21.5}(S) + \frac{180.2}{286.2}(C) + 300 \qquad \textbf{(12.11-3)}$$

Equation (12.11-2) does not change, since no salt is in the W stream. Solving Eqs. (12.11-2) and (12.11-3) simultaneously, $C = 6630$ kg of $Na_2CO_3 \cdot 10H_2O$ crystals and $S = 3070$ kg solution.

2. Heat effects and heat balances in crystallization. When a compound whose solubility increases as temperature increases dissolves, there is an absorption of heat, called the *heat of solution.* An evolution of heat occurs when a compound dissolves whose solubility decreases as temperature increases. For compounds dissolving whose solubility does not change with temperature, there is no heat evolution on dissolution. Most data on heats of solution are given as the change in enthalpy in kJ/kg mol (kcal/g mol) of solute occurring with the dissolution of 1 kg mol of the solid in a large amount of solvent at essentially infinite dilution.

In crystallization the opposite of dissolution occurs. At equilibrium the heat of crystallization is equal to the negative of the heat of solution at the same concentration in solution. If the heat of dilution from saturation in the solution to infinite dilution is small, this can be neglected, and the negative of the heat of solution at infinite dilution can be used for the heat of crystallization. With many materials this heat of dilution is small

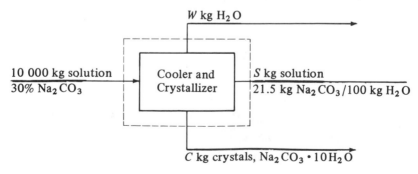

FIGURE 12.11-2. *Process flow for crystallization in Example 12.11-1.*

compared with the heat of solution, and this approximation is reasonably accurate. Heat of solution data are available in several references (P1, N1).

Probably the most satisfactory method of calculating heat effects during a crystallization process is to use the enthalpy–concentration chart for the solution and the various solid phases which are present for the system. However, only a few such charts are available, including the following systems: calcium chloride–water (H1), magnesium sulfate–water (P2), and ferrous sulfate–water (K2). When such a chart is available, the following procedure is used. The enthalpy H_1 of the entering solution at the initial temperature is read off the chart, where H_1 is kJ (btu) for the total feed. The enthalpy H_2 of the final mixture of crystals and mother liquor at the final temperature is also read off the chart. If some evaporation occurs, the enthalpy H_V of the water vapor is obtained from the steam tables. Then the total heat absorbed q in kJ is

$$q = (H_2 + H_V) - H_1 \qquad \text{(12.11-4)}$$

If q is positive, heat must be added to the system. If it is negative, heat is evolved or given off.

EXAMPLE 12.11-2. Heat Balance in Crystallization

A feed solution of 2268 kg at 327.6 K (54.4°C) containing 48.2 kg $MgSO_4$/100 kg total water is cooled to 293.2 K (20°C), where $MgSO_4 \cdot 7H_2O$ crystals are removed. The solubility of the salt is 35.5 kg $MgSO_4$/100 kg total water (P1). The average heat capacity of the feed solution can be assumed as 2.93 kJ/kg·K (H1). The heat of solution at 291.2 K (18°C) is -13.31×10^3 kJ/kg mol $MgSO_4 \cdot 7H_2O$ (P1). Calculate the yield of crystals and make a heat balance to determine the total heat absorbed, q, assuming that no water is vaporized.

Solution: Making a water balance and a balance for $MgSO_4$ using equations similar to (12.11-1) and (12.11-2) in Example 12.11-1, $C - 616.9$ kg $MgSO_4 \cdot 7H_2O$ crystals and $S = 1651.1$ kg solution.

To make a heat balance, a datum of 293.2 K (20°C) will be used. The molecular weight of $MgSO_4 \cdot 7H_2O$ is 246.49. The enthalpy of the feed is H_1.

$$H_1 = 2268(327.6 - 293.2)(2.93) = 228\,600 \text{ kJ}$$

The heat of solution is $-(13.31 \times 10^3)/246.49 = -54.0$ kJ/kg crystals. Then the heat of crystallization is $-(-54.0) = +54.0$ kJ/kg crystals, or $54.0(616.9) = 33\,312$ kJ. This assumes that the value at 291.2 K is the same as at 293.2 K. The total heat absorbed, q, is

$$q = -228\,600 - 33\,312 = -261\,912 \text{ kJ } (-248\,240 \text{ btu})$$

Since q is negative, heat is given off and must be removed.

12.11D Equipment for Crystallization

1. Introduction and classification of crystallizers. Crystallizers may be classified according to whether they are batch or continuous in operation. Batch operation is done for certain special applications. Continuous operation of crystallizers is generally preferred.

Crystallization cannot occur without *supersaturation*. A main function of any crystallizer is to cause a supersaturated solution to form. A classification of crystallizing equipment can be made based on the methods used to bring about supersaturation as follows: (1) supersaturation produced by cooling the solution with negligible evaporation—tank and batch-type crystallizers; (2) supersaturation produced by evaporation of the solvent with little or no cooling—evaporator–crystallizers and crystallizing evaporators; (3) supersaturation by combined cooling and evaporation in adiabatic evaporator—vacuum crystallizers.

In crystallizers producing supersaturation by cooling the substances must have a solubility curve that decreases markedly with temperature. This occurs for many substances, and this method is commonly used. When the solubility curve changes little with temperature, such as for common salt, evaporation of the solvent to produce supersaturation is often used. Sometimes evaporation with some cooling will also be used. In the method of cooling adiabatically in a vacuum, a hot solution is introduced into a vacuum, where the solvent flashes or evaporates and the solution is cooled adiabatically. This method to produce supersaturation is the most important one for large-scale production.

In another method of classification of crystallizers, the equipment is classified according to the method of suspending the growing product crystals. Examples are crystallizers where the suspension is agitated in a tank, is circulated by a heat exchanger, or is circulated in a scraped surface exchanger.

An important difference in many commercial crystallizers is the manner in which the supersaturated liquid contacts the growing crystals. In one method, called the *circulating magma method*, the entire magma of crystals and supersaturated liquid is circulated through both the supersaturation and crystallization steps without separating the solid from the liquid into two streams. Crystallization and supersaturation are occurring together in the presence of the crystals. In the second method, called the *circulating liquid method*, a separate stream of supersaturated liquid is passed through a fluidized bed of crystals, where the crystals grow and new ones form by nucleation. Then the saturated liquid is passed through an evaporating or cooling region to produce supersaturation again for recycling.

2. Tank crystallizers. In tank crystallization, which is an old method still used in some specialized cases, hot saturated solutions are allowed to cool in open tanks. After a period of time the mother liquor is drained and the crystals removed. Nucleation and the size of crystals are difficult to control. Crystals contain considerable amounts of occluded mother liquor. Labor costs are very high. In some cases the tank is cooled by coils or a jacket and an agitator used to improve the heat-transfer rate. However, crystals often build up on these surfaces. This type has limited application and sometimes is used to produce some fine chemicals and pharmaceutical products.

3. Scraped surface crystallizers. One type of scraped surface crystallizer is the Swenson–Walker crystallizer, which consists of an open trough 0.6 m wide with a semicircular bottom having a cooling jacket outside. A slow-speed spiral agitator rotates and suspends the growing crystals on turning. The blades pass close to the wall and break off any deposits of crystals on the cooled wall. The product generally has a somewhat wide crystal-size distribution.

In the double-pipe scraped surface crystallizer, cooling water passes in the annular space. An internal agitator is fitted with spring-loaded scrapers that wipe the wall and provide good heat-transfer coefficients. This type is called a *votator* and is used in crystallizing ice cream and plasticizing margarine. A sketch is shown in Fig. 4.13-2.

4. Circulating-liquid evaporator–crystallizer. In a combination evaporator–crystallizer shown in Fig. 12.11-3a, supersaturation is generated by evaporation. The circulating liquid is drawn by the screw pump down inside the tube side of the condensing steam heater. The heated liquid then flows into the vapor space, where flash evaporation occurs giving some supersaturation. The vapor leaving is condensed. The supersaturated liquid flows down the downflow tube and then up through the bed of fluidized and agitated crystals, which are growing in size. The leaving saturated liquid then goes back as a recycle stream to the heater, where it is joined by the entering feed. The larger crystals settle out and a slurry of crystals and mother liquor is withdrawn as product. This type is also called the *Oslo crystallizer*.

5. Circulating-magma vacuum crystallizer. In this circulating-magma vacuum-type crystallizer in Fig. 12.11-3b, the magma or suspension of crystals is circulated out of the main body through a circulating pipe by a screw pump. The magma flows through a heater, where its temperature is raised 2 to 6 K. The heated liquor then mixes with body slurry and boiling occurs at the liquid surface. This causes supersaturation in the swirling liquid near the surface, which causes deposits on the swirling suspended crystals until they leave again via the circulating pipe. The vapors leave through the top. A steam-jet ejector provides the vacuum.

12.12   CRYSTALLIZATION THEORY

12.12A   Introduction and Nucleation Theories

1. Introduction. When crystallization occurs in a homogeneous mixture, a new solid phase is created. An understanding of the mechanisms by which crystals form and then grow is helpful in designing and operating crystallizers. Much experimental and theoretical work has been done to help understand crystallization. However, the differences between predicted and actual performance in commercial crystallizers are still often quite large.

The overall process of crystallization from a supersaturated solution is considered to consist of the basic steps of nucleus formation or nucleation and of crystal growth. If the solution is free of all solid particles, foreign or of the crystallizing substance, then nucleus formation must first occur before crystal growth starts. New nuclei may continue to form

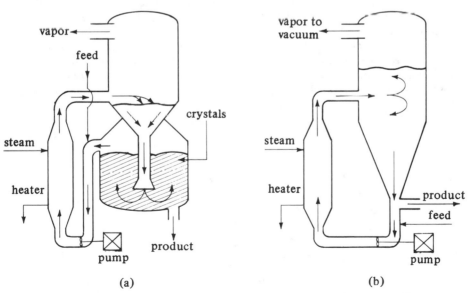

FIGURE 12.11-3.   *Types of crystallizers : (a) circulating-liquid evaporator–crystallizer,*
(b) circulating-magma vacuum crystallizer.

while the nuclei present are growing. The driving force for the nucleation step and the growth step is supersaturation. These two steps do not occur in a saturated or undersaturated solution.

2. Nucleation theories. *Primary nucleation* is a result of rapid local fluctuations on a molecular scale in a homogeneous phase. Particles or molecules of solute happen to come together and form clusters. More solute molecules may be added to one or more clusters so they grow, whereas others may break up and revert to individual molecules. The growing clusters become crystals and continue to absorb solute molecules from the solution.

This type of nucleation is called *homogeneous* or *primary nucleation*. The larger the crystal, the smaller its solubility. The solubility of small crystals in the micrometer size range is greater than that of a large crystal. The ordinary solubility data apply to large crystals. Hence, in a supersaturated solution a small crystal can be in equilibrium. If a large crystal is also present, the larger crystal will grow and the smaller one will dissolve. This effect of particle size is an important factor in nucleation. In magma crystallization, primary nucleation happens to a small degree.

An early qualitative explanation of crystallization by Miers attempts to explain formation of nuclei and crystals in an unseeded solution. This theory is shown in Fig. 12.12-1, where line *AB* is the normal solubility curve. If a sample of solution at point *a* is cooled, it first crosses the solubility curve. The sample will not crystallize until it has supercooled to some point *b* where crystallization begins, and the concentration drops to point *c* if no further cooling is done. The curve *CD*, called the *supersolubility curve*, represents the limit at which nucleus formation starts spontaneously and hence where crystallizaton can start. Any crystal in the metastable region will grow. The present tendency is to regard the supersolubility curve as a zone where the nucleation rate increases sharply. However, the value of Miers' explanation is that it points out that the greater the degree of supersaturation, the greater the chance of more nuclei forming.

Secondary or *contact nucleation*, which is the most effective method of nucleation, occurs when crystals collide with each other, with the impellors in mixing, or with the walls of the pipe or container. This type of nucleation is, of course, affected by the intensity of agitation. It occurs at low supersaturation, where the crystal growth rate is at the optimum for good crystals. This type of contact nucleation has been isolated and demonstrated experimentally. It is the most effective and common method of nucleation in magma crystallization. The precise mechanisms of contact nucleation are not known, nor is a complete theory available to predict these rates.

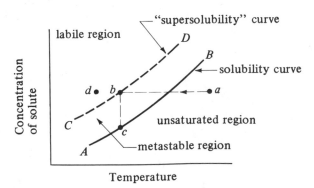

FIGURE 12.12-1. *Miers' qualitative explanation of crystallization: solubility curve (AB) and "supersolubility" curve (CD).*

12.12B   Rate of Crystal Growth and ΔL Law

1. Rate of crystal growth and growth coefficients.   The rate of growth of a crystal face is the distance moved per unit time in a direction that is perpendicular to the face. The crystal growth is a layer-by-layer process, and since the growth can occur only at the outer face of the crystal, the solute material must be transported to that face from the bulk of the solution. The solute molecules reach the face by diffusion through the liquid phase. The usual mass-transfer coefficient k_y applies in this case. At the surface the resistance to integration of the molecules into the space lattice at the face must be considered. This reaction at the surface occurs at a finite rate, and the overall process consists of two resistances in series. The solution must be supersaturated for the diffusion and interfacial steps to proceed.

The equation for mass transfer of solute A from the bulk solution of supersaturation concentration y_A, mole fraction of A, to the surface where the concentration is y'_A, is

$$\frac{\bar{N}_A}{A_i} = k_y(y_A - y'_A) \tag{12.12-1}$$

where k_y is the mass-transfer coefficient in kg mol/s $\cdot$ m$^2 \cdot$ mol frac, $\bar{N}_A$ is rate in kg mol A/s, and A_i is area in m^2 of surface i. Assuming that the rate of reaction at the crystal surface is also dependent on the concentration difference,

$$\frac{\bar{N}_A}{A_i} = k_s(y'_A - y_{Ae}) \tag{12.12-2}$$

where k_s is a surface reaction coefficient in kg mol/s $\cdot$ m$^2 \cdot$ mol frac and y_{Ae} is the saturation concentration. Combining Eqs. (12.12-1) and (12.12-2),

$$\frac{\bar{N}_A}{A_i} = \frac{y_A - y_{Ae}}{1/k_y + 1/k_s} = K(y_A - y_{Ae}) \tag{12.12-3}$$

where K is the overall transfer coefficient.

The mass-transfer coefficient k_y can be predicted by methods given in Section 7.4 for convective mass-transfer coefficients. The correlation for mass transfer through fixed and fluidized beds of solids can be used. The velocity obtained from the terminal settling velocity of particles can be used in the correlation or the velocity of the solution relative to the crystals in the suspension. The Schmidt number of the saturated solution is also needed for the prediction.

When the mass-transfer coefficient k_y is very large, the surface reaction is controlling and $1/k_y$ is negligible. Conversely, when the mass-transfer coefficient is very small, diffusional resistance is controlling. Surface reaction coefficients and overall transfer coefficients have been measured and reported on a number of systems (B4, H2, P3, V1). Much of the information in the literature is not directly applicable, because the conditions of measurement differ greatly from those in a commercial crystallizer. Also, the velocities and the level of supersaturation in a system are difficult to determine, and vary with position of the circulating magma in the crystallizer.

2. The ΔL law of crystal growth.   McCabe (M1) has shown that all crystals that are geometrically similar and of the same material in the same solution grow at the same rate. The growth is measured as the increase in length ΔL, in mm, in linear dimension of one crystal. This increase in length is for geometrically corresponding distances on all crystals. This increase is independent of the initial size of the initial crystals, provided that all the crystals are subject to the same environmental conditions. This law follows from Eq. (12.12-3), where the overall transfer coefficient is the same for each face of all crystals.

Mathematically, this can be written

$$\frac{\Delta L}{\Delta t} = G \tag{12.12-4}$$

where Δt is time in h and growth rate G is a constant in mm/h. Hence, if D_1 is the linear dimension of a given crystal at time t_1 and D_2 at time t_2,

$$\Delta L = D_2 - D_1 = G(t_2 - t_1) \tag{12.12-5}$$

The total growth $(D_2 - D_1)$ or ΔL is the same for all crystals.

The ΔL law fails in cases where the crystals are given any different treatment based on size. It has been found to hold for many materials, particularly when the crystals are under 50 mesh in size (0.3 mm). Even though this law is not applicable in all cases, it is reasonably accurate in many situations.

12.12C Particle-Size Distribution of Crystals

An important factor in the design of crystallization equipment is the expected particle-size distribution of the crystals obtained. Usually, the dried crystals are screened to determine the particle sizes. The percent retained on different-sized screens is recorded. The screens or sieves used are the Tyler standard screens, whose sieve or clear openings in mm are given in Appendix A.5.

The data are plotted as particle diameter (sieve opening in screen) in mm versus the cumulative percent retained at that size on arithmetic probability paper. Data for urea particles from a typical crystallizer (B5) are shown in Fig. 12.12-2. Many types of such data will show an approximate straight line for a large portion of the plot.

A common parameter used to characterize the size distribution is the coefficient of variation, CV, as a percent.

$$CV = 100 \frac{PD_{16\%} - PD_{84\%}}{2PD_{50\%}} \tag{12.12-6}$$

where $PD_{16\%}$ is the particle diameter at 16 percent retained. By giving the coefficient of variation and mean particle diameter, a description of the particle-size distribution is obtained if the line is approximately straight between 90 and 10%. For a product removed from a mixed-suspension crystallizer, the CV value is about 50% (R1). In a

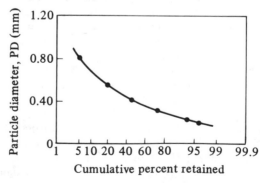

FIGURE 12.12-2. *Typical particle-size distribution from a crystallizer, [From R. C. Bennett and M. Van Buren, Chem. Eng. Progr. Symp., 65(95), 46 (1969).]*

Chap. 12   Liquid–Liquid and Fluid–Solid Separation Processes

mixed-suspension system, the crystallizer is at steady state and contains a well-mixed-suspension magma with no product classification and no solids entering with the feed.

12.12D Models for Crystallizers

In order to analyze data from a mixed-suspension crystallizer, an overall theory combining the effects of nucleation rate, growth rate, heat balance, and material balance is needed. Some progress has been made and an idealized model has been investigated by Randolph and Larson and their co-workers (R1, R2, R3, M2, S2, P1, P3, C2). Their equations are rather complicated but allow determination of some fundamental factors of growth rate and nucleation rate from experimental data.

A crystal product sample from the actual crystallizer is first obtained and a screen analysis run. The slurry density and retention time in the crystallizer are also needed. By converting the size analysis to a population density of crystals of various sizes and plotting the data, the nucleation rate and the growth rate in mm/h can be obtained for the actual conditions tested in the crystallizer. Then experiments can be conducted to determine the effects of operation effects on growth rate and nucleation rate. A sample calculation for this method is given elsewhere (P1). A contact nucleation model for the design of magma crystallizers has been developed (B6) based on single-particle-contact nucleation experiments (C4, M3). Larson (L2, P1) gives examples of design of crystallizing systems.

PROBLEMS

12.1-1. *Equilibrium Isotherm for Glucose Adsorption.* Equilibrium isotherm data for adsorption of glucose from an aqueous solution to activated alumina are as follows (H3):

c (g/cm^3)	0.0040	0.0087	0.019	0.027	0.094	0.195
q (g solute/g alumina)	0.026	0.053	0.075	0.082	0.123	0.129

Determine the isotherm that fits the data and give the constants of the equation using the given units.

Ans. Langmuir isotherm, $q = 0.145c/(0.0174 + c)$

12.2-1. *Batch Adsorption for Phenol Solution.* A wastewater solution having a volume of 2.5 m^3 contains 0.25 kg phenol/m^3 of solution. This solution is mixed thoroughly in a batch process with 3.0 kg of granular activated carbon until equilibrium is reached. Use the isotherm from Example 12.2-1 and calculate the final equilibrium values and the percent phenol extracted.

12.3-1. *Scale-Up of Laboratory Adsorption Column Data.* Using the break-point time and other results from Example 12.3-1, do as follows:
   (a) The break-point time for a new column is to be 8.5 h. Calculate the new total length of the column required, column diameter, and the fraction of total capacity used up to the break point. The flow rate is to remain constant at 754 cm^3/s.
   (b) Use the same conditions as part (a), but the flow rate is to be increased to 2000 cm^3/s.

Ans. (a) $H_T = 27.2$ cm, 0.849 fraction; (b) $D = 6.52$ cm

12.3-2. *Drying of Nitrogen and Scale-Up of Column.* Using molecular sieves, water vapor was removed from nitrogen gas in a packed bed (C3) at 28.3°C. The column height was 0.268 m, with the bulk density of the solid bed being equal to 712.8 kg/m^3. The initial water concentration in the solid was 0.01 kg water/kg solid and the mass velocity of the nitrogen gas was 4052 kg/m$^2 \cdot$ h.

The initial water concentration in the gas was $c_o = 926 \times 10^{-6}$ kg water/kg nitrogen. The breakthrough data are as follows.

t (h)	0	9	9.2	9.6	10	10.4
c (kg H_2O/kg $N_2 \times 10^6$)	<0.6	0.6	2.6	21	91	235
t (h)	10.8	11.25	11.5	12.0	12.5	12.8
c (kg H_2O/kg $N_2 \times 10^6$)	418	630	717	855	906	926

A value of $c/c_o = 0.02$ is desired at the break point. Do as follows.

(a) Determine the break-point time, the fraction of total capacity used up to the break point, the length of the unused bed, and the saturation loading capacity of the solid.

(b) For a proposed column length $H_T = 0.40$ m, calculate the break-point time and fraction of total capacity used.

Ans. (a) $t_b = 9.58$ h, fraction used = 0.878

12.4-1. *Scale-Up of Ion-Exchange Column.* An ion-exchange process using a resin to remove copper ions from aqueous solution is conducted in a 1.0-in.-diameter column 1.2 ft high. The flow rate is 1.5 gph and the break point occurred at 7.0 min. Integrating the breakthrough curve gives a ratio of usable capacity to total capacity of 0.60. Design a new tower that will be 3.0 ft high and operating at 4.5 gph. Calculate the new tower size and break-point time.

Ans. $t_b = 24.5$ min, $D = 1.732$ in.

12.4-2. *Height of Tower in Ion-Exchange.* In a given run using a flow rate of 0.2 m^3/h in an ion-exchange tower with a column height of 0.40 m, the break point occurred at 8.0 min. The ratio of usable capacity to total equilibrium capacity is 0.65. What is the height of a similar column operating for 13.0 min to the break point at the same flow rate?

12.4-3. *Ion Exchange of Copper in Column.* An ion-exchange column containing 99.3 g of amberlite ion-exchange resin was used to remove Cu^{2+} from a solution where $c_o = 0.18$ M $CuSO_4$. The tower height = 30.5 cm and the diameter = 2.59 cm. The flow rate was 1.37 cm^3 solution/s to the tower. The breakthrough data are shown below.

t (s)	420	480	510	540	600	660
c (g mol Cu/L)	0	0.0033	0.0075	0.0157	0.0527	0.1063
t (s)	720	780	810	870	900	
c (g mol Cu/L)	0.1433	0.1634	0.1722	0.1763	0.180	

The concentration desired at the break point is $c/c_o = 0.010$. Determine the break-point time, fraction of total capacity used up to the break point, length of unused bed, and the saturation loading capacity of the solid.

12.5-1. *Composition of Two Liquid Phases in Equilibrium.* An original mixture weighing 200 kg and containing 50 kg of isopropyl ether, 20 kg of acetic acid, and 130 kg of water is equilibrated in a mixer–settler and the phases separated. Determine the amounts and compositions of the raffinate and extract layers. Use equilibrium data from Appendix A.3.

12.5-2. *Single-Stage Extraction.* A single-stage extraction is performed in which 400 kg of a solution containing 35 wt % acetic acid in water is contacted with 400 kg of pure isopropyl ether. Calculate the amounts and compositions of the extract and raffinate layers. Solve for the amounts both algebraically and by the lever-arm rule. What percent of the acetic acid is removed? Use equilibrium data from Appendix A.3.

Ans. $L_1 = 358$ kg, $x_{B1} = 0.715$, $x_{C1} = 0.03$, $V_1 = 442$ kg, $y_{A1} = 0.11$, $y_{C1} = 0.86$, 34.7% removed.

12.5-3. Single-Stage Extraction with Unknown Composition. A feed mixture weighing 200 kg of unknown composition containing water, acetic acid, and isopropyl ether is contacted in a single stage with 280 kg of a mixture containing 40 wt % acetic acid, 10 wt % water, and 50 wt % isopropyl ether. The resulting raffinate layer weighs 320 kg and contains 29.5 wt % acetic acid, 66.5 wt % water, and 4.0 wt % isopropyl ether. Determine the original composition of the feed mixture and the composition of the resulting extract layer. Use equilibrium data from Appendix A.3.

Ans. $x_{A0} = 0.030$, $x_{B0} = 0.955$, $y_{A1} = 0.15$

12.5-4. Extraction of Acetone in a Single Stage. A mixture weighing 1000 kg contains 23.5 wt % acetone and 76.5 wt % water and is to be extracted by 500 kg methyl isobutyl ketone in a single-stage extraction. Determine the amounts and compositions of the extract and raffinate phases. Use equilibrium data from Appendix A.3.

12.7-1. Multiple-Stage Extraction with Fresh Solvent in Each Stage. Pure water is to be used to extract acetic acid from 400 kg of a feed solution containing 25 wt % acetic acid in isopropyl ether. Use equilibrium data from Appendix A.3.

(a) If 400 kg of water is used, calculate the percent recovery in the water solution in a one-stage process.

(b) If a multiple four-stage system is used and 100 kg fresh water is used in each stage, calculate the overall percent recovery of the acid in the total outlet water. (*Hint:* First, calculate the outlet extract and raffinate streams for the first stage using 400 kg of feed solution and 100 kg of water. For the second stage, 100 kg of water contacts the outlet organic phase from the first stage. For the third stage, 100 kg of water contacts the outlet organic phase from the second stage, and so on.)

12.7-2. Overall Balance in Countercurrent Stage Extraction. An aqueous feed of 200 kg/h containing 25 wt % acetic acid is being extracted by pure isopropyl ether at the rate of 600 kg/h in a countercurrent multistage system. The exit acid concentration in the aqueous phase is to contain 3.0 wt % acetic acid. Calculate the compositions and amounts of the exit extract and raffinate streams. Use equilibrium data from Appendix A.3.

12.7-3. Minimum Solvent and Countercurrent Extraction of Acetone. An aqueous feed solution of 1000 kg/h containing 23.5 wt % acetone and 76.5 wt % water is being extracted in a countercurrent multistage extraction system using pure methyl isobutyl ketone solvent at 298–299 K. The outlet water raffinate will contain 2.5 wt % acetone. Use equilibrium data from Appendix A.3.

(a) Calculate the minimum solvent that can be used. [*Hint:* In this case the tie line through the feed L_0 represents the condition for minimum solvent flow rate. This gives $V_{1\,min}$. Then draw lines $L_N V_{1\,min}$ and $L_0 V_{N+1}$ to give the mixture point M_{min} and the coordinate $x_{AM\,min}$. Using Eq. (12.7-4), solve for $V_{N+1\,min}$, the minimum value of the solvent flow rate V_{N+1}.]

(b) Using a solvent flow rate of 1.5 times the minimum, calculate the number of theoretical stages.

12.7-4. Countercurrent Extraction of Acetic Acid and Minimum Solvent. An aqueous feed solution of 1000 kg/h of acetic acid–water solution contains 30.0 wt % acetic acid and is to be extracted in a countercurrent multistage process with pure isopropyl ether to reduce the acid concentration to 2.0 wt % acid in the final raffinate. Use equilibrium data in Appendix A.3.

(a) Calculate the minimum solvent flow rate that can be used. (*Hint:* See Problem 12.7-3 for the method to use.)

(b) If 2500 kg/h of ether solvent is used, determine the number of theoretical stages required. (*Note*: It may be necessary to replot on an expanded scale the concentrations at the dilute end.)

Ans. (a) Minimum solvent flow rate $V_{N+1} = 1630$ kg/h; (b) 7.5 stages

12.7-5. Number of Stages in Countercurrent Extraction. Repeat Example 12.7-2 but use an exit acid concentration in the aqueous phase of 4.0 wt %.

12.7-6. Extraction with Immiscible Solvents. A water solution of 1000 kg/h containing 1.5 wt % nicotine in water is stripped with a kerosene stream of 2000 kg/h containing 0.05 wt % nicotine in a countercurrent stage tower. The exit water is to contain only 10% of the original nicotine, i.e., 90% is removed. Use equilibrium data from Example 12.7-3. Calculate the number of theoretical stages needed.

Ans. 3.7 stages

12.7-7. Analytical Equation for Number of Stages. Example 12.7-3 gives data for extraction of nicotine from water by kerosene where the two solvents are immiscible. A total of 3.8 theoretical stages were needed. Use the analytical equations (10.3-21)–(10.3-26) to calculate the number of theoretical stages and compare with the value obtained graphically.

12.7-8. Minimum Solvent Rate with Immiscible Solvents. Determine the minimum solvent kerosene rate to perform the desired extraction in Example 12.7-3. Using 1.25 times this minimum rate, determine the number of theoretical stages needed graphically and also by using Eqs. (10.3-21)–(10.3-26).

12.7-9. Stripping Nicotine from Kerosene. A kerosene flow of 100 kg/h contains 1.4 wt % nicotine and is to be stripped with pure water in a countercurrent multistage tower. It is desired to remove 90% of the nicotine. Using a water rate of 1.50 times the minimum, determine the number of theoretical stages required. (Use the equilibrium data from Example 12.7-3.)

12.8-1. Effective Diffusivity in Leaching Particles. In Example 12.8-1 a time of leaching of the solid particle of 3.11 h is needed to remove 80% of the solute. Do the following calculations.
(a) Using the experimental data, calculate the effective diffusivity, $D_{A\,eff}$.
(b) Predict the time to leach 90% of the solute from the 2.0 mm particle.

Ans. (a) $D_{A\,eff} = 1.0 \times 10^{-5}$ mm^2/s; (b) $t = 5.00$ h

12.9-1. Leaching of Oil from Soybeans in a Single Stage. Repeat Example 12.9-1 for single-stage leaching of oil from soybeans. The 100 kg of soybeans contains 22 wt % oil and the solvent feed is 80 kg of solvent containing 3 wt % soybean oil.

Ans. $L_1 = 52.0$ kg, $y_{A1} = 0.239$, $V_1 = 50.0$ kg, $x_{A1} = 0.239$, $N_1 = 1.5$

12.9-2. Leaching a Soybean Slurry in a Single Stage. A slurry of flaked soybeans weighing a total of 100 kg contains 75 kg of inert solids and 25 kg of solution with 10 wt % oil and 90 wt % solvent hexane. This slurry is contacted with 100 kg of pure hexane in a single stage so that the value of N for the outlet underflow is 1.5 kg insoluble solid/kg solution retained. Calculate the amounts and compositions of the overflow V_1 and the underflow L_1 leaving the stage.

12.10-1. Constant Underflow in Leaching Oil from Meal. Use the same conditions as given in Example 12.10-1, but assume constant underflow of $N = 1.85$ kg solid/kg solution. Calculate the exit flows and compositions and the number of stages required. Compare with Example 12.10-1.

Ans. $y_{AN} = 0.111$, $x_{A1} = 0.623$, 4.3 stages

12.10-2. Effect of Less Solvent Flow in Leaching Oil from Meal. Use the same conditions as given in Example 12.10-1, but the inlet fresh solvent mixture flow rate per hour is decreased by 10%, to 1179 kg of benzene and 18 kg of oil. Calculate the number of stages needed.

12.10-3. Countercurrent Multistage Washing of Ore. A treated ore containing inert solid gangue and copper sulfate is to be leached in a countercurrent multistage extractor using pure water to leach the $CuSO_4$. The solid charge rate per hour consists of 10 000 kg of inert gangue (B), 1200 kg of $CuSO_4$ (solute A), and 400 kg of water (C). The exit wash solution is to contain 92 wt % water and 8 wt % $CuSO_4$. A total of 95% of the $CuSO_4$ in the inlet ore is to be recovered. The underflow is constant at $N = 0.5$ kg inert gangue solid/kg aqueous solution. Calculate the number of stages required.

12.10-4. *Countercurrent Multistage Leaching of Halibut Livers.* Fresh halibut livers containing 25.7 wt % oil are to be extracted with pure ethyl ether to remove 95% of the oil in a countercurrent multistage leaching process. The feed rate is 1000 kg of fresh livers per hour. The final exit overflow solution is to contain 70 wt % oil. The retention of solution by the inert solids (oil-free liver) of the liver varies as follows (C1), where N is kg inert solid/kg solution retained and y_A is kg oil/kg solution.

N	y_A	N	y_A
4.88	0	1.67	0.6
3.50	0.2	1.39	0.81
2.47	0.4		

Calculate the amounts and compositions of the exit streams and the total number of theoretical stages.

12.10-5. *Countercurrent Leaching of Flaked Soybeans.* Soybean flakes containing 22 wt % oil are to be leached in a countercurrent multistage process to contain 0.8 kg oil/100 kg inert solid using fresh and pure hexane solvent. For every 1000 kg soybeans, 1000 kg hexane is used. Experiments (S1) give the following retention of solution with the solids in the underflow, where N is kg inert solid/kg solution retained and y_A is wt fraction of oil in solution.

N	y_A
1.73	0
1.52	0.20
1.43	0.30

Calculate the exit flows and compositions and the number of theoretical stages needed.

12.11-1. *Crystallization of $Ba(NO_3)_2$.* A hot solution of $Ba(NO_3)_2$ from an evaporator contains 30.6 kg $Ba(NO_3)_2$/100 kg H_2O and goes to a crystallizer where the solution is cooled and $Ba(NO_3)_2$ crystallizes. On cooling, 10% of the original water present evaporates. For a feed solution of 100 kg total, calculate the following.
(a) The yield of crystals if the solution is cooled to 290 K (17°C), where the solubility is 8.6 kg $Ba(NO_3)_2$/100 kg total water.
(b) The yield if cooled instead to 283 K, where the solubility is 7.0 kg $Ba(NO_3)_2$/100 kg total water.

Ans. (a) 17.47 kg $Ba(NO_3)_2$ crystals

12.11-2. *Dissolving and Subsequent Crystallization.* A batch of 1000 kg of KCl is dissolved in sufficient water to make a saturated solution at 363 K, where the solubility is 35 wt % KCl in water. The solution is cooled to 293 K, at which temperature its solubility is 25.4 wt %.

(a) What is the weight of water required for solution and the weight of crystals of KCl obtained?

(b) What is the weight of crystals obtained if 5% of the original water evaporates on cooling?

Ans. (a) 1857 kg water, 368 kg crystals; (b) 399 kg crystals

12.11-3. *Crystallization of $MgSO_4 \cdot 7H_2O$.* A hot solution containing 1000 kg of $MgSO_4$ and water having a concentration of 30 wt % $MgSO_4$ is cooled to 288.8 K, where crystals of $MgSO_4 \cdot 7H_2O$ are precipitated. The solubility at 288.8 K is 24.5 wt % anhydrous $MgSO_4$ in the solution. Calculate the yield of crystals obtained if 5% of the original water in the system evaporates on cooling.

12.11-4. *Heat Balance in Crystallization.* A feed solution of 10 000 lb$_m$ at 130°F containing 47.0 lb $FeSO_4$/100 lb total water is cooled to 80°F, where $FeSO_4 \cdot 7H_2O$ crystals are removed. The solubility of the salt is 30.5 lb $FeSO_4$/100 lb total water (P1). The average heat capacity of the feed solution is 0.70 btu/lb$_m$ · °F. The heat of solution at 18°C is -4.4 kcal/g mol (-18.4 kJ/g mol) $FeSO_4 \cdot 7H_2O$ (P1). Calculate the yield of crystals and make a heat balance. Assume that no water is vaporized.

Ans. 2750 lb$_m$ $FeSO_4 \cdot 7H_2O$ crystals, $q = -428\ 300$ btu ($-451\ 900$ kJ)

12.11-5. *Effect of Temperature on Yield and Heat Balance in Crystallization.* Use the conditions of Example 12.11-2, but the solution is cooled instead to 283.2 K, where the solubility is 30.9 kg $MgSO_4$/100 kg total water (P1). Calculate the effect on yield and the heat absorbed by using 283.2 K instead of 293.2 K for the crystallization.

REFERENCES

(B1) BLAKEBROUGH, N. *Biochemical and Biological Engineering Science*, Vol. 1. New York: Academic Press, Inc., 1968.

(B2) BADGER, W. L., and BANCHERO, J. T. *Introduction to Chemical Engineering.* New York: McGraw-Hill Book Company, 1955.

(B3) BADGER, W. L., and MCCABE, W. L. *Elements of Chemical Engineering.* New York: McGraw-Hill Book Company, 1936.

(B4) BUCKLEY, H. E. *Crystal Growth.* New York: John Wiley & Sons, Inc., 1951.

(B5) BENNETT, R. C., and VAN BUREN, M. *Chem. Eng. Progr. Symp.*, **65**(95), 46 (1969).

(B6) BENNETT, R. C., FIELDELMAN, H., and RANDOLPH, A. D. *Chem. Eng. Progr.*, **69**(7), 86 (1972).

(B7) BELTER, P. A., CUSSLER, E. L., and HU, W. S. *Bioseparations.* New York: John Wiley & Sons, Inc., 1988.

(C1) CHARM, S. E. *The Fundamentals of Food Engineering*, 2nd ed. Westport, Conn.: Avi Publishing Co., Inc., 1971.

(C2) *Crystallization from Solutions and Melts, Chem. Eng. Progr. Symp.*, **65**(95), (1969).

(C3) COLLINS, J. J. *Chem. Eng. Progr. Symp. Series*, **63**(74), 31 (1967).

(C4) CLONTZ, N. A., and MCCABE, W. L. *A.I.Ch.E. Symp. Ser. No. 110*, **67**, 6 (1971).

(C5) CLAFFEY, J. B., BADGETT, C. O., SKALAMERA, C. D., and PHILLIPS, G. W. *Ind. Eng. Chem.*, **42**, 166 (1950).

(C6) CUSACK, R. W., AND KARR, A. E. *Chem. Eng.*, **98**(4), 112 (1991).

(C7) CUSACK, R. W., FREMEAUX, P., AND GLATZ, D. *Chem. Eng.*, **98**(2), 66 (1991).

(C8) CUSACK, R. W., AND FREMEAUX, P. *Chem. Eng.*, **98**(3), 132 (1991).

(H1) HOUGEN, O. A., WATSON, K. M., and RAGATZ, R. A. *Chemical Process Principles*, Part I, 2nd ed. New York: John Wiley & Sons, Inc., 1954.

(H2) HIXSON, A. W., and KNOX, K. L. *Ind. Eng. Chem.*, **43**, 2144 (1951).

(H3) HU, M. C., HAERING, E. R., and GEANKOPLIS, C. J. *Chem. Eng. Sci.*, **40**, 2241 (1985).

(K1) KARNOFSKY, G. *J. Am. Oil Chemist's Soc.*, **26**, 564 (1949).

(K2) KOBE, K. A., and COUCH, E. J., JR. *Ind. Eng. Chem.*, **46**, 377 (1954).

(K3) KARR, A. E. AND LO, T. C., *Proc. Int. Solv. Ext. Conf. (ISEC)*, **1**, 299 (1971).

(L1) LUKCHIS, G. M. *Chem. Eng.*, **80**(June 11), 111 (1973).

(L2) LARSON, M. A. *Chem. Eng.*, **85**(Feb. 13), 90 (1978).

(L3) LO, T. C., AND KARR, A. E., *I.E.C. Proc. Des. Dev.*, **11**(4), 495 (1972).

(M1) MICHAELS, A. S. *Ind. Eng. Chem.*, **44**(8), 1922 (1952).

(M2) MURRAY, D. C., and LARSON, M. A. *A.I.Ch.E. J.*, **11**, 728 (1965).

(M3) McCABE, W. L., SMITH, J. C., and HARRIOTT, P. *Unit Operations of Chemical Engineering*, 4th ed. New York: McGraw-Hill Book Company, 1985.

(N1) NATIONAL RESEARCH COUNCIL, *International Critical Tables*, Vol. 5. New York: McGraw-Hill Book Company, 1929.

(O1) OSBURN, J. Q., and KATZ, D. L. *Trans. A.I.Ch.E.*, **40**, 511 (1944).

(O2) OTHMER, D. F., and JAATINEN, W. A. *Ind. Eng. Chem.*, **51**, 543 (1959).

(O3) OTHMER, D. F., and AGARWAL, J. C. *Ind. Eng. Chem.*, **51**, 372 (1955).

(P1) PERRY, R. H., and GREEN, D. *Perry's Chemical Engineers' Handbook*, 6th ed. New York: McGraw-Hill Book Company, 1984.

(P2) PERRY, J. H. *Chemical Engineers' Handbook*, 4th ed. New York: McGraw-Hill Book Company, 1963.

(P3) PERRY, R. H., AND CHILTON, C. H. *Chemical Engineers' Handbook*, 5th ed. New York: McGraw-Hill Book Company, 1973.

(R1) RANDOLPH, A. D., and LARSON, M. A. *Theory of Particulate Systems*, New York: Academic Press, Inc., 1971.

(R2) RANDOLPH, A. D., and LARSON, M. A. *A.I.Ch.E. J.*, **8**, 639 (1962).

(R3) RANDOLPH, A. D. *A.I.Ch.E. J.*, **11**, 424 (1965).

(R4) ROUSSEAU, R. W. (ed.). *Handbook of Separation Process Technology*, New York: John Wiley & Sons, Inc., 1987.

(R5) REYNOLDS, T. D. *Unit Operations and Processes in Environmental Engineering*, Boston: PWS Publishers, 1982.

(R6) RUTHVEN, D. M. *Principles of Adsorption and Adsorption Processes*, New York: John Wiley & Sons, Inc., 1984.

(S1) SMITH, C. T. *J. Am. Oil Chemists' Soc.*, **28**, 274 (1951).

(S2) SAEMAN, W. C. *A.I.Ch.E. J.*, **2**, 107 (1956).

(T1) TREYBAL, R. E. *Mass Transfer Operations*, 3rd ed. New York: McGraw-Hill Book Company, 1980.

(V1) VAN HOOK, A. *Crystallization, Theory and Practice*, New York: John Wiley & Sons, Inc., 1951.

(Y1) YANG, H. H., and BRIER, J. C. *A.I.Ch.E. J.*, **4**, 453 (1958).

CHAPTER 13
Membrane Separation Processes

13.1 INTRODUCTION AND TYPES OF MEMBRANE SEPARATION PROCESSES

13.1A Introduction

Separations by the use of membranes are becoming increasingly important in the process industries. In this relatively new unit operation, the membrane acts as a semipermeable barrier and separation occurs by the membrane controlling the rate of movement of various molecules between two liquid phases, two gas phases, or a liquid and gas phase. The two fluid phases are usually miscible and the membrane barrier prevents actual, ordinary hydrodynamic flow. A classification of the main types of membrane separation is as follows.

13.1B Classification of Membrane Processes

1. Gas diffusion in porous solid. In this type a gas phase is present on both sides of the membrane, which is a microporous solid. The rates of molecular diffusion of the various gas molecules depend on the pore sizes and the molecular weights. This type of diffusion in the molecular, transition, and Knudsen regions was discussed in detail in Section 7.6.

2. Gas permeation in a membrane. The membrane in this process is usually a polymer such as rubber, polyamide, and so on, and is not a porous solid. The solute gas first dissolves in the membrane and then diffuses in the solid to the other gas phase. This was discussed in detail in Section 6.5 for solutes following Fick's law and is considered again for the case where resistances are present in Section 13.3. Examples are hydrogen diffusing through rubber and helium being separated from natural gas by permeation through a fluorocarbon polymer. Separation of a gas mixture occurs since each type of molecule diffuses at a different rate through the membrane.

3. Liquid permeation or dialysis. In this case the small solutes in one liquid phase diffuse readily because of concentration differences through a porous membrane to the second liquid phase (or vapor phase). Passage of large molecules through the membrane is more difficult. This membrane process has been applied in chemical processing separations such as separation of H_2SO_4 from nickel and copper sulfates in aqueous solutions, food processing, and artificial kidneys and is covered in detail in Section 13.2. In electrodialysis, separation of ions occurs by imposing an emf difference across the membrane.

4. Reverse osmosis. A membrane, which impedes the passage of a low-molecular-weight solute, is placed between a solute–solvent solution and a pure solvent. The solvent diffuses into the solution by osmosis. In reverse osmosis a reverse pressure difference is imposed which causes the flow of solvent to reverse as in the desalination of seawater. This process also is used to separate other low-molecular-weight solutes, such as salts, sugars, and simple acids from a solvent (usually water). This process is covered in detail in Sections 13.9 and 13.10.

5. Ultrafiltration membrane process. In this process, pressure is used to obtain a separation of molecules by a semipermeable polymeric membrane (M2). The membrane discriminates on the basis of molecular size, shape, or chemical structure and separates relatively high molecular weight solutes such as proteins, polymers, colloidal materials such as minerals, and so on. The osmotic pressure is usually negligible because of the high molecular weights. This is covered in Section 13.11.

6. Gel permeation chromotography. The porous gel retards diffusion of the high-molecular-weight solutes. The driving force is concentration. This process is quite useful in analyzing complex chemical solutions and purification of very specialized and/or valuable components.

13.2 LIQUID PERMEATION MEMBRANE PROCESSES OR DIALYSIS

13.2A Series Resistances in Membrane Processes

In membrane processes with liquids, the solute molecules must first be transported or diffuse through the liquid film of the first liquid phase on one side of the solid membrane, through the membrane itself, and then through the film of the second liquid phase. This is shown in Fig. 13.2-1a, where c_1 is the bulk liquid-phase concentration of the diffusing

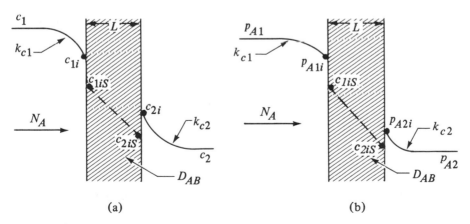

(a)   (b)

FIGURE 13.2-1. *Concentration profiles for membrane processes: (a) two liquid films and a solid, (b) two gas films and a solid.*

solute A in kg mol A/m^3, c_{1i} is the concentration of A in the fluid just adjacent to the solid, and c_{1iS} is the concentration of A in the solid at the surface and is in equilibrium with C_{1i}. The mass-transfer coefficients are k_{c1} and k_{c2} in m/s. The equilibrium distribution coefficient K' is defined as

$$K' = \frac{c_S}{c_L} = \frac{c_{1iS}}{c_{1i}} = \frac{c_{2iS}}{c_{2i}} \qquad (13.2\text{-}1)$$

Note that K' is the inverse of K defined in Eq. (7.1-16).

The flux equations through each phase are all equal to each other at steady state and are as follows:

$$N_A = k_{c1}(c_1 - c_{1i}) = \frac{D_{AB}}{L}(c_{1iS} - c_{2iS}) = k_{c2}(c_{2i} - c_2) \qquad (13.2\text{-}2)$$

Substituting $c_{1iS} = K'c_{1i}$ and $c_{2iS} = K'c_{2i}$ into Eq. (13.2-2),

$$N_A = k_{c1}(c_1 - c_{1i}) = \frac{D_{AB}K'}{L}(c_{1i} - c_{2i}) = p_M(c_{1i} - c_{2i}) = k_{c2}(c_{2i} - c_2) \quad (13.2\text{-}3)$$

$$p_M = \frac{D_{AB}K'}{L} \qquad (13.2\text{-}4)$$

where p_M is the permeability in the solid in m/s, L is the thickness in m, and D_{AB} is the diffusivity of A in the solid in m^2/s. Note that the permeability p_M in Eq. (13.2-4) is different from the permeability P_M defined in Eq. (6.5-9). Also, the value of p_M is inversely proportional to the thickness L. Instead of determining D_{AB} and K' in two separate experiments, it is more convenient to determine p_M in one separate diffusion experiment. Solving each of the parts in Eq. (13.2-3) for the concentration difference,

$$c_1 - c_{1i} = \frac{N_A}{k_{c1}} \qquad c_{1i} - c_{2i} = \frac{N_A}{p_M} \qquad c_{2i} - c_2 = \frac{N_A}{k_{c2}} \qquad (13.2\text{-}5)$$

Adding the equations, the internal concentrations c_{1i} and c_{2i} drop out, and the final equation is

$$N_A = \frac{c_1 - c_2}{1/k_{c1} + 1/p_M + 1/k_{c2}} \qquad (13.2\text{-}6)$$

In some cases, the resistances in the two liquid films are quite small compared to that of the membrane resistance, which controls the permeation rate.

EXAMPLE 13.2-1. Membrane Diffusion and Liquid Film Resistances

A liquid containing dilute solute A at a concentration $c_1 = 3 \times 10^{-2}$ kg mol/m^3 is flowing rapidly by a membrane of thickness $L = 3.0 \times 10^{-5}$ m. The distribution coefficient $K' = 1.5$ and $D_{AB} = 7.0 \times 10^{-11}$ m^2/s in the membrane. The solute diffuses through the membrane and its concentration on the other side is $c_2 = 0.50 \times 10^{-2}$ kg mol/m^3. The mass-transfer coefficient k_{c1} is large and can be considered as infinite and $k_{c2} = 2.02 \times 10^{-5}$ m/s.

(a) Derive the equation to calculate the steady-state flux N_A and make a sketch.

(b) Calculate the flux and the concentrations at the membrane interfaces.

Solution: For part (a) the sketch is shown in Fig. (13.2-2). Note that the concentration profile on the left side is flat ($k_{c1} = \infty$) and $c_1 = c_{1i}$. The

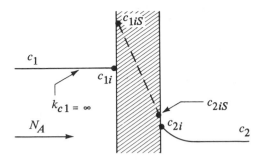

FIGURE 13.2-2. *Concentrations for Example 13.2-1.*

derivation is the same as for Eq. (13.2-6) but $1/k_{c1} = 0$ to give

$$N_A = \frac{c_1 - c_2}{1/p_M + 1/k_{c2}} \qquad (13.2\text{-}7)$$

For part (b) to calculate the flux using Eqs. (13.2-4) and (13.2-7),

$$p_M = \frac{D_{AB} K'}{L} = \frac{7.0 \times 10^{-11}(1.5)}{3.0 \times 10^{-5}} = 3.5 \times 10^{-6} \text{m/s}$$

$$N_A = \frac{c_1 - c_2}{1/p_M + 1/k_{c2}} = \frac{3.0 \times 10^{-2} - 0.5 \times 10^{-2}}{1/3.5 \times 10^{-6} + 1/2.02 \times 10^{-5}}$$

$$= 7.458 \times 10^{-8} \text{ kg mol/s} \cdot \text{m}^2$$

To calculate c_{2i},

$$N_A = 7.458 \times 10^{-8} = k_{c2}(C_{2i} - c_2) = 2.02 \times 10^{-5}(c_{2i} - 0.5 \times 10^{-2})$$

Solving, $c_{2i} = 0.869 \times 10^{-2}$ kg mol/m^3. Also, using Eq. (13.2-1),

$$K' = 1.5 = \frac{c_{2iS}}{c_{2i}} = \frac{c_{2iS}}{0.869 \times 10^{-2}}$$

Solving, $c_{2iS} = 1.304 \times 10^{-2}$ kg mol/m^3.

13.2B Dialysis Processes

Dialysis uses a semipermeable membrane to separate species by virtue of their different diffusion rates in the membrane. The feed solution or dialyzate, which contains the solutes to be separated, flows on one side of the membrane and the solvent or diffusate stream on the other side. Some solvent may also diffuse across the membrane in the opposite direction, which reduces the performance by diluting the dialyzate.

In practice it is used to separate species that differ appreciably in size, which have a reasonably large difference in diffusion rates. Solute fluxes depend on the concentration gradient in the membrane. Hence, dialysis is characterized by low flux rates, in comparison to other membrane processes, such as reverse osmosis and ultrafiltration, membrane processes that depend on applied pressure.

In general, dialysis is used with aqueous solutions on both sides of the membrane. The film resistances can be appreciable compared to the membrane resistance. Applications include recovery of sodium hydroxide in cellulose processing, recovery of acids from metallurgical liquors, removal of products from a culture solution in fermentation, desalting of cheese whey solids, and reduction of alcohol content of beer. Many small-scale applications are used in the pharmaceutical industry.

13.2C   Types of Equipment for Dialysis

Various types of geometrical configurations are used in liquid membrane processes. A common type is one similar to a filter press where the membrane is a flat plate. Vertical solid membranes are placed in between alternate liquor and solvent feed frames, with the liquor to be dialyzed being fed to the bottom and the solvent to the top of these frames. The dialyzate and the diffusate are removed through channels located at the top and bottom of the frames, respectively. The most important type consists of many small tubes or very fine hollow fibers arranged in a bundle like a heat exchanger. This type of unit has a very high ratio of membrane area to volume of the unit.

13.2D   Hemodialysis in Artificial Kidney

An important example of liquid permeation processes is dialysis with an artificial kidney in the biomedical field. In this application for purifying human blood, the principal solutes removed are the small solutes urea, uric acid, creatinine, phosphates, and excess amounts of chloride. A typical membrane used is cellophane about 0.025 mm thick, which allows small solutes to diffuse but retains the large proteins of the blood. During the hemodialysis, blood is passed on one side of the membrane while an aqueous dialyzing fluid flows on the other side. Solutes such as urea, uric acid, NaCl, and so on, which have elevated concentrations in the blood, diffuse across the membrane to the dialyzing aqueous solution, which contains certain concentrations of solutes such as potassium salts, and so on, to ensure that concentrations in the blood do not drop below certain levels. In one configuration the membranes are stacked in the form of a multilayered sandwich with blood flowing by one side of the membrane in a narrow channel and the dialyzing fluid by the other side in alternate channels.

> *EXAMPLE 13.2-2.   Dialysis to Remove Urea from Blood*
> Calculate the flux and the rate of removal of urea at steady state in g/h from blood in a cuprophane (cellophane) membrane dialyzer at 37°C. The membrane is 0.025 mm thick and has an area of 2.0 m². The mass-transfer coefficient on the blood side is estimated as $k_{c1} = 1.25 \times 10^{-5}$ m/s and that on the aqueous side is 3.33×10^{-5} m/s. The permeability of the membrane is 8.73×10^{-6} m/s (B2). The concentration of urea in the blood is 0.02 g urea/100 mL and that in the dialyzing fluid will be assumed as 0.
>
> *Solution:*   The concentration $c_1 = 0.02/100 = 2.0 \times 10^{-4}$ g/mL $= 200$ g/m³ and $c_2 = 0$. Substituting into Eq. (13.2-6),
>
> $$N_A = \frac{c_1 - c_2}{1/k_{c1} + 1/p_M + 1/k_{c2}}$$
>
> $$= \frac{200 - 0}{1/1.25 \times 10^{-5} + 1/8.73 \times 10^{-6} + 1/3.33 \times 10^{-5}}$$
>
> $$= 8.91 \times 10^{-4} \text{ g/s} \cdot \text{m}^2$$

For a time of 1 h and an area of 2.0 m²,

$$\text{rate of removal} = 8.91 \times 10^{-4}(3600)(2.0) = 6.42 \text{ g urea/h}$$

13.3 GAS PERMEATION MEMBRANE PROCESSES

13.3A Series Resistances in Membrane Processes

In membrane processes with two gas phases and a solid membrane, similar equations can be written for the case shown in Fig. 13.2-1b. The equilibrium relation between the solid and gas phases is given by

$$H = \frac{S}{22.414} = \frac{c_S}{p_A} = \frac{c_{1iS}}{p_{A1i}} = \frac{c_{2iS}}{p_{A2i}} \tag{13.3-1}$$

where S is the solubility of A in m^3 (STP)/atm · m^3 solid, as shown in Eq. (6.5-5), and H is the equilibrium relation in kg mol/m^3 · atm. This is similar to Henry's law. The flux equations in each phase are as follows:

$$N_A = \frac{k_{c1}}{RT}(p_{A1} - p_{A1i}) = \frac{D_{AB}}{L}(c_{1iS} - c_{2iS})$$

$$= \frac{D_{AB}H}{L}(p_{A1i} - p_{A2i})$$

$$= \frac{k_{c2}}{RT}(p_{A2i} - p_{A2}) \tag{13.3-2}$$

The permeability P_m in kg mol/s · m · atm is given by

$$P_m = D_{AB}H = \frac{D_{AB}S}{22.414} \tag{13.3-3}$$

Note that this P_m differs from the P_M defined in Eq. (6.5-9) as $D_{AB}S$. Eliminating the interfacial concentrations as before,

$$N_A = \frac{p_{A1} - p_{A2}}{1/(k_{c1}/RT) + 1/(P_m/L) + 1/(k_{c2}/RT)} \tag{13.3-4}$$

In the case where pure A (p_{A1}) is on the left side of the membrane, there is no diffusional resistance in the gas phase and k_{c1} can be considered to be infinite. Note that $k_{G1} = k_{c1}/RT$.

An example of gas permeation in a membrane is use of a polymeric membrane as an oxygenator for a heart–lung machine to oxygenate blood. In this biomedical application, pure O_2 gas is on one side of a thin membrane and blood is on the other side. Oxygen diffuses through the membrane into the blood and CO_2 diffuses in a reverse direction into the gas stream.

13.3B Types of Membranes and Permeabilities for Separation of Gases

1. Types of membranes. Early membranes were limited in their use because of low-selectivities in separating two gases and quite low permeation fluxes. This low-flux problem was due to the fact that the membranes had to be relatively thick (1 mil or 1/1000 of an inch or greater) in order to avoid tiny holes which reduced the separation by allowing viscous or Knudsen flow of the feed. Development of silicone polymers (1 mil thickness) increased the permeability by factors of 10 to 20 or so.

Some newer asymmetric membranes include a very thin but dense skin on one side of the membrane supported by a porous substructure (R1). The dense skin has a thickness of about 1000 Å and the porous support thickness is about 25–100 μm. The flux increase of these membranes is thousands of times higher than the 1-mil-thick original membranes. Some typical materials of present membranes are a composite of polysulfone coated with silicone rubber, cellulose acetate and modified cellulose acetates, aromatic polyamides or aromatic polyimides, and silicone-polycarbonate copolymer on a porous support.

2. Permeability of membranes. The accurate prediction of permeabilities of gases in membranes is generally not possible, and experimental values are needed. Experimental data for common gases in some typical membranes are given in Table 13.3-1. Note that there are wide differences among the permeabilities of various gases in a given membrane. Silicone rubber exhibits very high permeabilities for the gases in the table.

For the effect of temperature T in K, the ln P'_A is approximately a linear function of $1/T$ and increases with T. However, operation at high temperatures can often degrade the membranes. When a mixture of gases is present, reductions of permability of an individual component of up to 10% or so can often occur. In a few cases much larger reductions have been observed (R1). Hence, when using a mixture of gases, experimental data should be obtained to determine if there is any interaction between the gases. The presence of water vapor can also have similar effects on the permeabilities and can also possibly damage the membranes.

13.3C Types of Equipment for Gas Permeation Membrane Processes

1. Flat membranes. Flat membranes are mainly used for experimental use to characterize the permeability of the membrane. The modules are easy to fabricate and use and the areas of the membranes are well defined. In some cases modules are stacked together like a multilayer sandwich or plate-and-frame filter press. The major drawback of this type is the very small membrane area per unit separator volume.

TABLE 13.3-1. *Permeabilities of Various Gases in Membranes*

Material	Temp. (°C)	Permeability, P'_A, $\dfrac{cm^3(STP)\cdot cm}{s\cdot cm^2\cdot cm\,Hg}\times 10^{10}$						Ref.
		He	H_2	CH_4	CO_2	O_2	N_2	
Silicone rubber	25	300	550	800	2700	500	250	(S2)
Natural rubber	25	31	49	30	131	24	8.1	(S2)
Polycarbonate (Lexane)	25–30	15	12		5.6,10	1.4		(S2)
Nylon 66	25	1.0			0.17	0.034	0.008	(S2)
Polyester (Permasep)	—		1.65	0.035	0.31		0.031	(H1)
Silicone– polycarbonate copolymer (57% silicone)	25		210		970	160	70	(W2)
Teflon FEP	30	62		1.4			2.5	(S1)
Ethyl cellulose	30	35.7	49.2	7.47	47.5	11.2	3.29	(W3)
Polystyrene	30	40.8	56.0	2.72	23.3		7.47	2.55	(W3)

Small commercial flat membranes are used for producing oxygen-enriched air for individual medical applications.

2. Spiral-wound membranes. This configuration maintains the simplicity of fabricating flat membranes while increasing markedly the membrane area per unit separator volume up to 100 ft^2/ft^3 (328 m^2/m^3) while decreasing pressure drops (R1). The assembly consists of a sandwich of four sheets wrapped around a central core of a perforated collecting tube. The four sheets consist of a top sheet of an open separator grid for the feed channel, a membrane, a porous felt backing for the permeate channel, and another membrane as shown in Fig. 13.3-1. The spiral-wound element is 100 to 200 mm in diameter and is about 1 to 1.5 m long in the axial direction. The flat sheets before rolling are about 1 to 1.5 m by about 2 to 2.5 m. The space between the membranes (open grid for feed) is about 1 mm and the thickness of the porous backing (for permeate) is about 0.2 mm.

The whole spiral-wound element is located inside a metal shell. The feed gas enters at the left end of the shell, enters the feed channel, and flows through this channel in the axial direction of the spiral to the right end of the assembly (Fig. 13.3-1). Then the exit residue gas leaves the shell at this point. The feed stream, which is in the feed channel, permeates perpendicularly through the membrane. This permeate then flows through the permeate channel in a direction perpendicular to the feed stream toward the perforated collecting tube, where it leaves the apparatus at one end. This is illustrated in Fig. 13.3-2, where the local gas flow paths are shown for a small element of the assembly.

3. Hollow-fiber membranes. The membranes are in the shape of very small diameter hollow fibers. The inside diameter of the fibers is in the range of 100 to 500 μm and the outside 200 to 1000 μm with the length up to 3 to 5 m. The module resembles a shell-and-tube heat exchanger. Thousands of fine tubes are bound together at each end

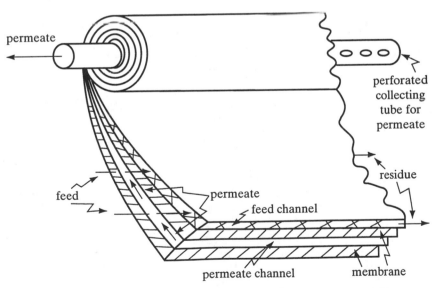

FIGURE 13.3-1. *Spiral-wound elements and assembly.* [From R. I. Berry, Chem. Eng., **88** (July 13), 63 (1981). With permission.]

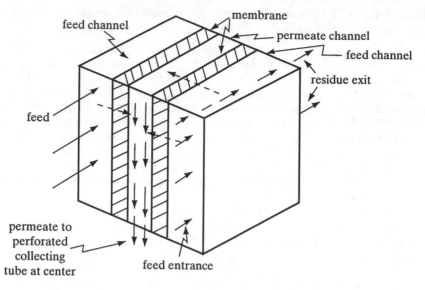

FIGURE 13.3-2.   *Local gas flow paths for spiral-wound separator.*

into a tube sheet that is surrounded by a metal shell having a diameter of 0.1 to 0.2 m, so that the membrane area per unit volume is up to 10 000 m^2/m^3 as in Fig. 13.3-3.

Typically, the high-pressure feed enters into the shell side at one end and leaves at the other end. The hollow fibers are closed at one end of the tube bundles. The permeate gas inside the fibers flows countercurrently to the shell-side flow and is collected in a chamber where the open ends of the fibers terminate. Then the permeate exits the device.

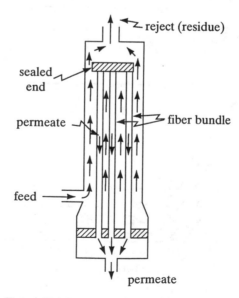

FIGURE 13.3-3.   *Hollow-fiber separator assembly.*

13.3D Introduction to Types of Flow in Gas Permeation

1. Types of flow and diffusion gradients. In a membrane process high-pressure feed gas is supplied to one side of the membrane and permeates normal to the membrane. The permeate leaves in a direction normal to the membrane, accumulating on the low-pressure side. Because of the very high diffusion coefficient in gases, concentration gradients in the gas phase in the direction normal to the surface of the membrane are quite small. Hence, gas film resistances compared to the membrane resistance can be neglected. This means that the concentration in the gas phase in a direction perpendicular to the membrane is essentially uniform whether or not the gas stream is flowing parallel to the surface or is not flowing.

If the gas stream is flowing parallel to the membrane in essentially plug flow, a concentration gradient occurs in this direction. Hence, several cases can occur in the operation of a membrane module. The permeate side of the membrane can be operated so that the phase is completely mixed (uniform concentration) or where the phase is in plug flow. The high-pressure feed side can also be completely mixed or in plug flow. Countercurrent or cocurrent flow can be used when both sides are in plug flow. Hence, separate theoretical models must be derived for these different types of operation as given in Sections 13.4 to 13.7.

2. Assumptions used and ideal flow patterns. In deriving theoretical models for gas separation by membranes, isothermal conditions and negligible pressure drop in the feed stream and permeate stream are generally assumed. It is also assumed that the effects of total pressure and/or composition of the gas are negligible and that the permeability of each component is constant (i.e., no interactions between different components).

Since there are a number of idealized flow patterns, the important types are summarized in Fig. 13.3-4. In Fig. 13.3-4a complete mixing is assumed for the feed chamber and the permeate chamber. Similar to a continuous-stirred tank, the reject or residue and the product or permeate compositions are equal to their respective uniform compositions in the chambers.

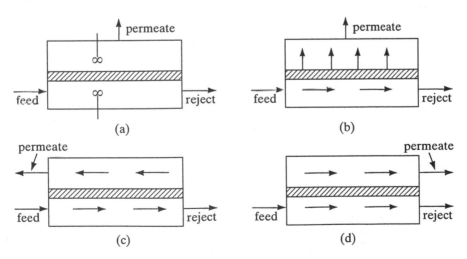

FIGURE 13.3-4. *Ideal flow patterns in a membrane separator for gases: (a) complete mixing, (b) cross-flow, (c) countercurrent flow, (d) cocurrent flow.*

An ideal cross-flow pattern is given in Fig. 13.3-4b, where the feed stream is in plug flow and the permeate flows in a normal direction away from the membrane without mixing. Since the feed composition varies along its flow path, the local permeate concentration also varies along the membrane path.

In Fig. 13.3-4c both the feed stream and permeate stream are in plug flow countercurrent to each other. The composition of each stream varies along its flow path. Cocurrent flow of the feed and permeate streams is shown in Fig. 13.3-4d.

13.4   COMPLETE-MIXING MODEL FOR GAS SEPARATION BY MEMBRANES

13.4A   Basic Equations Used

In Fig. 13.4-1 a detailed process flow diagram is shown for complete mixing. When a separator element is operated at a low recovery (i.e., where the permeate flow rate is a small fraction of the entering feed rate), there is a minimal change in composition. Then the results derived using the complete-mixing model provide reasonable estimates of permeate purity. This case was derived by Weller and Steiner (W4).

The overall material balance (Fig. 13.4-1) is as follows:

$$q_f = q_o + q_p \tag{13.4-1}$$

where q_f is total feed flow rate in cm^3(STP)/s; q_o is outlet reject flow rate, cm^3(STP)/s; and q_p is outlet permeate flow rate, cm^3(STP)/s. The cut or fraction of feed permeated, θ, is given as

$$\theta = \frac{q_p}{q_f} \tag{13.4-2}$$

The rate of diffusion or permeation of species A (in a binary of A and B) is given below by an equation similar to Eq. (6.5-8) but which uses cm^3(STP)/s as rate of permeation rather than flux in kg mol/s $\cdot$ cm^2.

$$\frac{q_A}{A_m} = \frac{q_p y_p}{A_m} = \left(\frac{P'_A}{t}\right)(p_h x_o - p_l y_p) \tag{13.4-3}$$

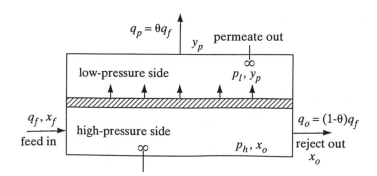

FIGURE 13.4-1.   *Process flow for complete mixing case.*

where P'_A is permeability of A in the membrane, $cm^3(STP) \cdot cm/(s \cdot cm^2 \cdot cm\ Hg)$; q_A is flow rate of A in permeate, $cm^3(STP)/s$; A_m is membrane area, cm^2; t is membrane thickness, cm; p_h is total pressure in the high-pressure (feed) side, $cm\ Hg$; p_l is total pressure in the low-pressure or permeate side, $cm\ Hg$; x_o is mole fraction of A in reject side; x_f is mole fraction of A in feed; and y_p is mole fraction of A in permeate. Note that $p_h x_o$ is the partial pressure of A in the reject gas phase.

A similar equation can be written for component B.

$$\frac{q_B}{A_m} = \frac{q_p(1 - y_p)}{A_m} = \left(\frac{P'_B}{t}\right)[p_h(1 - x_o) - p_l(1 - y_p)] \qquad (13.4\text{-}4)$$

where P'_B is permeability of B, $cm^3(STP) \cdot cm/(s \cdot cm^2 \cdot cm\ Hg)$. Dividing Eq. (13.4-3) by (13.4-4)

$$\frac{y_p}{1 - y_p} = \frac{\alpha^*[x_o - (p_l/p_h)y_p]}{(1 - x_o) - (p_l/p_h)(1 - y_p)} \qquad (13.4\text{-}5)$$

This equation relates y_p, the permeate composition, to x_o, the reject composition, and the ideal separation factor α^* is defined as

$$\alpha^* = \frac{P'_A}{P'_B} \qquad (13.4\text{-}6)$$

Making an overall material balance on component A

$$q_f x_f = q_o x_o + q_p y_p \qquad (13.4\text{-}7)$$

Dividing by q_f and solving for the outlet reject composition,

$$x_o = \frac{x_f - \theta y_p}{(1 - \theta)} \quad \text{or} \quad y_p = \frac{x_f - x_o(1 - \theta)}{\theta} \qquad (13.4\text{-}8)$$

Substituting $q_p = \theta q_f$ from Eq. (13.4-2) into Eq. (13.4-3) and solving for the membrane area, A_m,

$$A_m = \frac{\theta q_f y_p}{(P'_A/t)(p_h x_o - p_l y_p)} \qquad (13.4\text{-}9)$$

13.4B Solution of Equations for Design of Complete-Mixing Case

For design of a system there are seven variables in the complete-mixing model (H1), x_f, x_o, y_p, θ, α^*, p_l/p_h, and A_m, four of which are independent variables. Two commonly occurring cases are considered here.

Case 1. This is the simplest case where x_f, x_o, α^*, and p_l/p_h are given and y_p, θ, and A_m are to be determined by solution of the equations. By use of the quadratic equation formula, Eq. (13.4-5) is solved for the permeate composition y_p in terms of x_o.

$$y_p = \frac{-b + \sqrt{b^2 - 4ac}}{2a} \qquad (13.4\text{-}10)$$

where

$$a = 1 - \alpha^*$$

$$b = \frac{p_h}{p_l}(1 - x_o) - 1 + \alpha^* \frac{p_h}{p_l} x_o + \alpha^*$$

$$c = -\alpha^* \frac{p_h}{p_l} x_o$$

Hence, to solve this case, y_p is first calculated using Eq. (13.4-10). Then the fraction of feed permeated, θ, is calculated using Eq. (13.4-8) and the membrane area, A_m from Eq. (13.4-9).

EXAMPLE 13.4-1. *Design of a Membrane Unit for Complete Mixing*
A membrane is to be used to separate a gaseous mixture of A and B whose feed flow rate is $q_f = 1 \times 10^4$ cm^3(STP)/s and feed composition of A is $x_f = 0.50$ mole fraction. The desired composition of the reject is $x_o = 0.25$. The membrane thickness $t = 2.54 \times 10^{-3}$ cm, the pressure on the feed side is $p_h = 80$ cm Hg and on the permeate side is $p_l = 20$ cm Hg. The permeabilities are $P'_A = 50 \times 10^{-10}$ cm^3(STP)·cm/(s·cm^2· cm Hg) and $P'_B = 5 \times 10^{-10}$. Assuming the complete-mixing model, calculate the permeate composition, y_p, the fraction permeated, θ, and the membrane area, A_m.

Solution: Substituting into Eq. (13.4-6),

$$\alpha^* = \frac{P'_A}{P'_B} = \frac{50 \times 10^{-10}}{(5 \times 10^{-10})} = 10$$

Using Eq. (13.4-10),

$$a = 1 - \alpha^* = 1 - 10 = -9$$

$$b = \frac{p_h}{p_l}(1 - x_o) - 1 + \alpha^* \frac{p_h}{p_l} x_o + \alpha^*$$

$$= \frac{80}{20}(1 - 0.25) - 1 + 10 \left(\frac{80}{20}\right)(0.25) + 10 = 22.0$$

$$c = \alpha^* \frac{p_h}{p_l} x_o = -10 \left(\frac{80}{20}\right)(0.25) = -10$$

$$y_p = \frac{-b + \sqrt{b^2 - 4ac}}{2a}$$

$$= \frac{-22.0 + \sqrt{(22.0)^2 - 4(-9)(-10)}}{2(-9)} = 0.604$$

Using the material balance equation (13.4-8),

$$x_o = \frac{x_f - \theta y_p}{1 - \theta}; \qquad 0.25 = \frac{0.50 - \theta\,(0.604)}{1 - \theta}$$

Solving, $\theta = 0.706$. Also, using Eq. (13.4-9),

$$A_m = \frac{\theta q_f y_p}{(P'_A/t)(p_h x_o - p_l y_p)}$$

$$= \frac{0.706(1 \times 10^4)(0.604)}{[50 \times 10^{-10}/(2.54 \times 10^{-3})](80 \times 0.25 - 20 \times 0.604)}$$

$$= 2.735 \times 10^8 \text{ cm}^2 \ (2.735 \times 10^4 \text{ m}^2)$$

Case 2. In this case x_f, θ, α^*, and p_l/p_h are given and y_p, x_o, and A_m are to be determined. Equation (13.4-5) cannot be solved for y_p since x_o is unknown. Hence, x_o from Eq. (13.4-8) is substituted into Eq. (13.4-5) and the resulting equation solved for y_p using the quadratic equation to give

$$y_p = \frac{-b_1 + \sqrt{b_1^2 - 4a_1 c_1}}{2a_1} \tag{13.4-11}$$

where

$$a_1 = \theta + \frac{p_l}{p_h} - \frac{p_l}{p_h}\theta - \alpha^*\theta - \alpha^*\frac{p_l}{p_h} + \alpha^*\frac{p_l}{p_h}\theta$$

$$b_1 = 1 - \theta - x_f - \frac{p_l}{p_h} + \frac{p_l}{p_h}\theta + \alpha^*\theta + \alpha^*\frac{p_l}{p_h} - \alpha^*\frac{p_l}{p_h}\theta + \alpha^*x_f$$

$$c_1 = -\alpha^* x_f$$

After solving for y_p, the value of x_o is calculated from Eq. (13.4-8) and A_m from Eq. (13.4-9).

EXAMPLE 13.4-2. Membrane Design for Separation of Air

It is desired to determine the membrane area needed to separate an air stream using a membrane 1 mil thick with an oxygen permeability of $P'_A = 500 \times 10^{-10} \text{ cm}^3(\text{STP}) \cdot \text{cm}/(\text{s} \cdot \text{cm}^2 \cdot \text{cm Hg})$. An $\alpha^* = 10$ for oxygen permeability divided by nitrogen permeability (S6) will be used. The feed rate is $q_f = 1 \times 10^6 \text{ cm}^3(\text{STP})/\text{s}$ and the fraction cut $\theta = 0.20$. The pressure selected for use are $p_h = 190$ cm Hg and $p_l = 19$ cm Hg. Again, assuming the complete-mixing model, calculate the permeate composition, the reject composition, and the area.

Solution: Using Eq. (13.4-11) for a feed composition of $x_f = 0.209$,

$$a_1 = \theta + \frac{p_l}{p_h} - \frac{p_l}{p_h}\theta - \alpha^*\theta - \alpha^*\frac{p_l}{p_h} + \alpha^*\frac{p_l}{p_h}\theta$$

$$= 0.2 + \frac{19}{190} - \frac{19}{190}(0.2) - 10(0.2) - 10\frac{(19)}{190} + 10\left(\frac{19}{190}\right)(0.2) = -2.52$$

$$b_1 = 1 - \theta - x_f - \frac{p_l}{p_h} + \frac{p_l}{p_h}\theta + \alpha^*\theta + \alpha^*\frac{p_l}{p_h} - \alpha^*\frac{p_l}{p_h}\theta + \alpha^*x_f =$$

$$= 1 - 0.2 - 0.209 - \frac{19}{190} + \frac{19}{190}(0.2) + 10(0.2)$$

$$+ 10\left(\frac{19}{190}\right) - 10\left(\frac{19}{190}\right)(0.2)$$

$$+ 10(0.209) = 5.401$$

$$c_1 = -\alpha^*x_f = -10(0.209) = -2.09$$

$$y_p = \frac{-b_1 + \sqrt{b_1^2 - 4a_1c_1}}{2a_1}$$

$$= \frac{-5.401 + \sqrt{(5.401)^2 - 4(-2.52)(-2.09)}}{2(-2.52)}$$

$$= 0.5067$$

Substituting into Eq. (13.4-8),

$$x_o = \frac{x_f - \theta y_p}{(1 - \theta)} = \frac{0.209 - 0.2(0.5067)}{(1 - 0.2)} = 0.1346$$

Finally, using Eq. (13.4-9) to find the area

$$A_m = \frac{\theta q_f y_p}{(P_A'/t)(p_h x_o - p_l y_p)}$$

$$= \frac{0.2(1 \times 10^6)(0.5067)}{(500 \times 10^{-10}/2.54 \times 10^{-3})(190 \times 0.1346 - 19 \times 0.5067)}$$

$$= 3.228 \times 10^8 \text{cm}^2$$

13.4C Minimum Concentration of Reject Stream

If all of the feed is permeated, then $\theta = 1$ and the feed composition $x_f = y_p$. For all values of $\theta < 1$, the permeate composition $y_p > x_f$ (H1). Substituting the value $x_f = y_p$ into Eq. (13.4-5) and solving, the minimum reject composition x_{oM} for a given x_f value is obtained as

$$x_{oM} = \frac{x_f\left[1 + (\alpha^* - 1)\frac{p_l}{p_h}(1 - x_f)\right]}{\alpha^*(1 - x_f) + x_f} \tag{13.4-12}$$

Hence, a feed of x_f concentration cannot be stripped lower than a value of x_{oM} even with an infinitely large membrane area for a completely mixed system. To strip beyond this limiting value a cascade-type system could be used. However, a single unit could be used which is not completely mixed but is designed for plug flow.

EXAMPLE 13.4-3. *Effect of Feed Composition on Minimum Reject Concentration*

Calculate the minimum reject concentration for Example 13.4-1 where the feed concentration is $x_f = 0.50$. Also, what is the effect of raising the feed purity to $x_f = 0.65$?

Solution: Substituting $x_f = 0.50$ into Eq. (13.4-12),

$$x_{oM} = \frac{x_f \left[1 + (\alpha^* - 1) \frac{p_l}{p_h} (1 - x_f) \right]}{\alpha^*(1 - x_f) + x_f}$$

$$= \frac{0.50 \left[1 + (10 - 1) \left(\frac{20}{80} \right) (1 - 0.50) \right]}{10(1 - 0.50) + 0.50}$$

$$= 0.1932$$

For an $x_f = 0.65$,

$$x_{oM} = \frac{0.65 \left[1 + (10 - 1) \left(\frac{20}{80} \right) (1 - 0.65) \right]}{10(1 - 0.65) + 0.65} = 0.2780$$

13.5 COMPLETE-MIXING MODEL FOR MULTICOMPONENT MIXTURES

13.5A Derivation of Equations

When multicomponent mixtures are present, the iteration method by Stern et al. (S1) is quite useful. This method will be derived for a ternary mixture of components A, B, and C. The process flow diagram is the same as Fig. 13.4-1, where the feed composition x_f is x_{fA}, x_{fB}, and x_{fC}. The known values are

$$x_{fA}, \; x_{fB}, \; x_{fC}; \; q_f; \; \theta; \; p_h, \; p_l; \; P'_A, \; P'_R, \; P'_C; \; \text{and } t$$

The unknown values to be determined are

$$y_{pA}, \; y_{pB}, \; y_{pC}; \; x_{oA}, \; x_{oB}, \; x_{oC}; \; q_p \text{ or } q_o; \text{ and } A_m$$

These eight unknowns can be obtained by solving a set of eight simultaneous equations using an iteration method. Three rate of permeation equations similar to Eq. (13.4-3) are as follows for components A, B, and C.

$$q_p y_{pA} = \frac{P'_A}{t} A_m (p_h x_{oA} - p_l y_{pA}) \tag{13.5-1}$$

$$q_p y_{pB} = \frac{P'_B}{t} A_m (p_h x_{oB} - p_l y_{pB}) \tag{13.5-2}$$

$$q_p y_{pC} = \frac{P'_C}{t} A_m (p_h x_{oC} - p_l y_{pC}) \tag{13.5-3}$$

The three material balance equations similar to Eq. (13.4-8) are written for components A, B, and C.

$$x_{oA} = \frac{1}{1 - \theta} x_{fA} - \frac{\theta}{1 - \theta} y_{pA} \tag{13.5-4}$$

$$x_{oB} = \frac{1}{1 - \theta} x_{fB} - \frac{\theta}{1 - \theta} y_{pB} \tag{13.5-5}$$

$$x_{oC} = \frac{1}{1 - \theta} x_{fC} - \frac{\theta}{1 - \theta} y_{pC} \tag{13.5-6}$$

Also, two final equations can be written as

$$\sum_n y_{pn} = y_{pA} + y_{pB} + y_{pC} = 1.0 \tag{13.5-7}$$

$$\sum_n x_{on} = x_{oA} + x_{oB} + x_{oC} = 1.0 \tag{13.5-8}$$

Substituting x_{oA} from Eq. (13.5-4) into Eq. (13.5-1) and solving for A_m,

$$A_m = \frac{q_p y_{pA} t}{P'_A \left[\dfrac{p_h}{1 - \theta} (x_{fA} - \theta y_{pA}) - p_l y_{pA} \right]} \tag{13.5-9}$$

For component B, Eq. (13.5-5) is substituted into Eq. (13.5-2), giving

$$A_m = \frac{q_p y_{pB} t}{P'_B \left[\dfrac{p_h}{1 - \theta} (x_{fB} - \theta y_{pB}) - p_l y_{pB} \right]} \tag{13.5-10}$$

Rearranging Eq. (13.5-10) and solving for y_{pB},

$$y_{pB} = \frac{p_h x_{fB}/(1 - \theta)}{q_p t/(P'_B A_m) + \theta p_h/(1 - \theta) + p_l} \tag{13.5-11}$$

In a similar manner Eq. (13.5-12) is derived for y_{pC}.

$$y_{pC} = \frac{p_h x_{fC}/(1 - \theta)}{q_p t/(P'_C A_m) + \theta p_h/(1 - \theta) + p_l} \tag{13.5-12}$$

13.5B   Iteration Solution Procedure for Multicomponent Mixtures

The following iteration or trial-and-error procedure can be used to solve the equations above.

1. A value of y_{pA} is assumed where $y_{pA} > x_{fA}$.
2. Using Eq. (13.4-2) and the known value of θ, q_p is calculated.
3. The membrane area is calculated from Eq. (13.5-9).

4. Values of y_{pB} and y_{pC} are calculated from Eqs. (13.5-11) and (13.5-12).
5. The sum $\Sigma_n \, y_{pn}$ is calculated from Eq. (13.5-7). If this sum is not equal to 1.0, steps 1 through 5 are repeated until the sum is 1.0.
6. Finally, x_{oA}, x_{oB}, and x_{oC} are calculated from Eqs. (13.5-4), (13.5-5), and (13.5-6).

EXAMPLE 13.5-1. *Design of Membrane Unit for Multicomponent Mixture*
A multicomponent gaseous mixture having a composition of $x_{fA} = 0.25$, $x_{fB} = 0.55$, and $x_{fC} = 0.20$ is to be separated by a membrane with a thickness of 2.54×10^{-3} cm using the complete-mixing model. The feed flow rate is 1.0×10^4 cm^3 (STP)/s and the permeabilities are $P'_A = 200 \times 10^{-10}$ cm^3(STP)·cm/(s·cm^2·cm Hg), $P'_B = 50 \times 10^{-10}$, and $P'_C = 25 \times 10^{-10}$. The pressure on the feed side is 300 cm Hg and 30 cm Hg on the permeate side. The fraction permeated will be 0.25. Calculate the permeate composition, the reject composition, and the membrane area using the complete-mixing model.

Solution: Following the iteration procedure, a value of $y_{pA} = 0.50$ is assumed. Substituting into Eq. (13.4-2) for step 2,

$$q_p = \theta q_f = 0.25 \times 1.0 \times 10^4 = 0.25 \times 10^4 \text{ cm}^3\text{(STP)/s}$$

Using Eq. (13.5-9), the membrane area for step (3) is

$$A_m = \frac{q_p y_{pA} t}{P'_A \left[\dfrac{p_h}{1 - \theta} (x_{fA} - \theta y_{pA}) - p_l y_{pA} \right]}$$

$$= \frac{0.25 \times 10^4 (0.50)(2.54 \times 10^{-3})}{200 \times 10^{-10}} \Bigg/$$

$$\left[\frac{300}{1 - 0.25} (0.25 - 0.25 \times 0.50) - 30(0.50) \right]$$

$$= 4.536 \times 10^6 \text{ cm}^2$$

Following step 4, the values y_{pB} and y_{pC} are calculated using Eqs. (13.5-11) and (13.5-12).

$$y_{pB} = \frac{p_h \, x_{fB}/(1 - \theta)}{q_p t/(P'_B A_m) + \theta p_h/(1 - \theta) + p_l}$$

$$= \frac{300 \times 0.55/(1 - 0.25)}{0.25 \times 10^4 \times 2.54 \times 10^{-3}/(50 \times 10^{-10} \times 4.536 \times 10^6) + 0.25 \times 300/(1 - 0.25) + 30}$$

$$= 0.5366$$

$$y_{pC} = \frac{p_h \, x_{fC}/(1 - \theta)}{q_p t/(P'_C A_m) + \theta p_h/(1 - \theta) + p_l}$$

$$= \frac{300 \times 0.20/(1 - 0.25)}{0.25 \times 10^4 \times 2.54 \times 10^{-3}/(25 \times 10^{-10} \times 4.536 \times 10^6) + 0.25 \times 300/(1 - 0.25) + 30}$$

$$= 0.1159$$

Substituting into Eq. (13.5-7),

$$\sum_n y_{pn} = y_{pA} + y_{pB} + y_{pC} = 0.5000 + 0.5366 + 0.1159 = 1.1525$$

For the second iteration, assuming that $y_{pA} = 0.45$, the following values are calculated:

$$A_m = 3.546 \times 10^6 \text{ cm}^2 \qquad y_{pB} = 0.4410 \qquad y_{pC} = 0.0922$$

$$\sum_n y_{pn} = 0.9832$$

The final iteration values are $A_m = 3.536 \times 10^6 \text{ cm}^2$; $y_{pA} = 0.4555$, $y_{pB} = 0.4502$, and $y_{pC} = 0.0943$. Substituting into Eqs. (13.5-4), (13.5-5), and (13.5-6)

$$x_{oA} = \frac{1}{1-\theta} x_{fA} - \frac{\theta}{1-\theta} y_{pA} = \frac{1}{1-0.25}(0.25)$$

$$- \frac{0.25}{1-0.25}(0.4555) = 0.1815$$

$$x_{oB} = \frac{1}{1-\theta} x_{fB} - \frac{\theta}{1-\theta} y_{pB} = \frac{1}{1-0.25}(0.55)$$

$$- \frac{0.25}{1-0.25}(0.4502) = 0.5833$$

$$x_{oC} = \frac{1}{1-\theta} x_{fC} - \frac{\theta}{1-\theta} y_{pC} = \frac{1}{1-0.25}(0.20)$$

$$- \frac{0.25}{1-0.25}(0.0943) = 0.2352$$

13.6. CROSS-FLOW MODEL FOR GAS SEPARATION BY MEMBRANES

13.6A Derivation of Basic Equations

A detailed flow diagram for the cross-flow model derived by Weller and Steiner (W3, W4) is shown in Fig. 13.6-1. In this case the longitudinal velocity of the high-pressure or reject stream is large enough so that this gas stream is in plug flow and flows parallel to the membrane. On the low-pressure side the permeate stream is almost pulled into vacuum, so that the flow is essentially perpendicular to the membrane.

This model assumes no mixing in the permeate side and also no mixing on the high-pressure side. Hence, the permeate composition at any point along the membrane is determined by the relative rates of permeation of the feed components at that point. This cross-flow pattern approximates that in an actual spiral-wound membrane

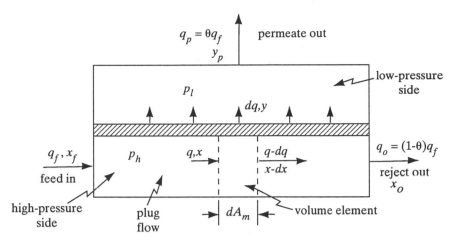

FIGURE 13.6-1. *Process flow diagram for cross-flow model.*

separator (Fig. 13.3-1) with a high-flux asymmetric membrane resting on a porous felt support (P2, R1).

Referring to Fig. 13.6-1, the local permeation rate over a differential membrane area dA_m at any point in the stage is

$$-y\,dq = \frac{P'_A}{t}[p_h x - p_l y]dA_m \tag{13.6-1}$$

$$-(1-y)\,dq = \frac{P'_B}{t}[p_h(1-x) - p_l(1-y)]dA_m \tag{13.6-2}$$

where dq is the total flow rate permeating through the area dA_m. Dividing Eq. (13.6-1) by (13.6-2) gives

$$\frac{y}{1-y} = \frac{\alpha^*[x - (p_l/p_h)y]}{(1-x) - (p_l/p_h)(1-y)} \tag{13.6-3}$$

This equation relates the permeate composition y to the reject composition x at a point along the path. It is similar to Eq. (13.4-5) for complete mixing. Hwang and Kammermeyer (H1) give a computer program for the solution of the above system of differential equations by numerical methods.

Weller and Steiner (W3, W4) used some ingenious transformations and were able to obtain an analytical solution to the three equations as follows:

$$\frac{(1 - \theta^*)(1-x)}{(1-x_f)} = \left(\frac{u_f - E/D}{u - E/D}\right)^R \left(\frac{u_f - \alpha^* + F}{u - \alpha^* + F}\right)^S \left(\frac{u_f - F}{u - F}\right)^T \tag{13.6-4}$$

where

$$\theta^* = 1 - \frac{q}{q_f}$$

$$i = \frac{x}{1-x}$$

$$u = -Di + (D^2i^2 + 2Ei + F^2)^{0.5}$$

$$D = 0.5 \left[\frac{(1 - \alpha^*)p_l}{p_h} + \alpha^* \right]$$

$$E = \frac{\alpha^*}{2} - DF$$

$$F = -0.5 \left[\frac{(1 - \alpha^*)p_l}{p_h} - 1 \right]$$

$$R = \frac{1}{2D - 1}$$

$$S = \frac{\alpha^*(D - 1) + F}{(2D - 1)(\alpha^*/2 - F)}$$

$$T = \frac{1}{1 - D - (E/F)}$$

The term u_f is the value of u at $i = i_f = x_f/(1 - x_f)$. The value of θ^* is the fraction permeated up to the value of x in Fig. 13.6-1. At the outlet where $x = x_o$, the value of θ^* is equal to θ, the total fraction permeated. The composition of the exit permeate stream is y_p and is calculated from the overall material balance, Eq. (13.4-8).

The total membrane area was obtained by Weller and Steiner (W3, W4) using some additional transformations of Eqs. (13.6-1) to (13.6-3) to give

$$A_m = \frac{tq_f}{p_h P'_B} \int_{i_0}^{i_f} \frac{(1 - \theta^*)(1 - x) \, di}{(f_i - i) \left[\frac{1}{1 + i} - \frac{p_l}{p_h} \left(\frac{1}{1 + f_i} \right) \right]} \tag{13.6-5}$$

where

$$f_i = (Di - F) + (D^2i^2 + 2Ei + F^2)^{0.5}$$

Values of θ^* in the integral can be obtained from Eq. (13.6-4). The integral can be calculated numerically. The term i_f is the value of i at the feed x_f and i_o is the value of i at the outlet x_o. A shortcut approximation of the area without using a numerical integration is available by Weller and Steiner (W3) which has a maximum error of about 20%.

13.6B Procedure for Design of Cross-Flow Case

In the design for the complete-mixing model there are seven variables and two of the most common cases were discussed in Section 13.4B. Similarly, for the cross-flow model these same common cases occur.

Case 1. The values of x_f, x_o, α^*, and p_l/p_h are given and y_p, θ, and A_m are to be determined. The value of θ^* or θ can be calculated directly from Eq. (13.6-4) since all other values in this equation are known. Then y_p is calculated from Eq. (13.4-8). To calculate the area A_m, a series of values of x less than the feed x_f and greater than the reject outlet x_o are substituted into Eq. (13.6-4) to give a series of θ^* values. These values are then used to numerically or graphically integrate Eq. (13.6-5) to obtain the area A_m.

Case 2. In this case the values of x_f, θ, α^*, and p_l/p_h are given and y_p, x_o, and A_m are to be determined. This is trial and error, where values of x_o are substituted into Eq. (13.6-4) to solve the equation. The membrane area is calculated as in Case 1.

EXAMPLE 13.6-1. Design of a Membrane Unit Using Cross-Flow

The same conditions for the separation of an air stream as given in Example 13.4-2 for complete mixing are to be used in this example. The process flow streams will be in cross-flow. The given values are $x_f = 0.209$, $\theta = 0.20$, $\alpha^* = 10$, $p_h = 190$ cm Hg, $p_l = 19$ cm Hg, $q_f = 1 \times 10^6$ cm^3(STP)/s, $P'_A = 500 \times 10^{-10}$ cm^3(STP)·cm/(s·cm^2·cm Hg), and $t = 2.54 \times 10^{-3}$ cm. Do as follows:
(a) Calculate y_p, x_o, and A_m.
(b) Compare the results with Example 13.4-2.

Solution: Since this is the same as Case 2, a value of $x_o = 0.1642$ will be used for the first trial for part (a). Substituting into Eq. (13.6-4)

$$i = i_f = \frac{x_f}{1 - x_f} = \frac{0.209}{1 - 0.209} = 0.2642$$

$$i = \frac{0.1642}{1 - 0.1642} = 0.1965$$

$$D = 0.5 \left[\frac{(1 - \alpha^*)p_l}{p_h} + \alpha^* \right]$$

$$= 0.5 \left[\frac{(1 - 10)19}{190} + 10 \right] = 4.550$$

$$F = -0.5 \left[\frac{(1 - \alpha^*)p_l}{p_h} - 1 \right]$$

$$= -0.5 \left[\frac{(1 - 10)19}{190} - 1 \right] = 0.950$$

$$E = \frac{\alpha^*}{2} - DF = \frac{10}{2} - 4.550(0.950) = 0.6775$$

$$R = \frac{1}{2D - 1} = \frac{1}{2(4.550) - 1} = 0.12346$$

$$S = \frac{\alpha^*(D - 1) + F}{(2D - 1)(\alpha^*/2 - F)}$$

$$= \frac{10(4.550 - 1) + 0.950}{(2 \times 4.550 - 1)(10/2 - 0.950)} = 1.1111$$

$$T = \frac{1}{1 - D - (E/F)}$$

$$= \frac{1}{1 - 4.550 - 0.6775/0.950} = -0.2346$$

$$u_f = -Di + (D^2 i^2 + 2Ei + F^2)^{0.5}$$

$$= -(4.550)(0.2642) + [(4.550)^2(0.2642)^2$$

$$+ 2(0.6775)(0.2642) + (0.950)^2]^{0.5}$$

$$= 0.4427$$

$$u = -(4.550)(0.1965) + [(4.550)^2(0.1965)^2$$

$$+ 2(0.6775)(0.1965) + (0.950)^2]^{0.5}$$

$$= 0.5089$$

$$\frac{(1 - \theta^*)(1 - x)}{(1 - x_f)} = \frac{(1 - \theta^*)(1 - 0.1642)}{(1 - 0.209)}$$

$$= \left(\frac{0.4427 - 0.6775/4.550}{0.5089 - 0.6775/4.550}\right)^{0.12346}$$

$$\left(\frac{0.4427 - 10 + 0.950}{0.5089 - 10 + 0.950}\right)^{1.1111}$$

$$\left(\frac{0.4427 - 0.950}{0.5089 - 0.950}\right)^{-0.2346}$$

Solving $\theta^* = 0.0992$. This value of 0.0992 does not check the given value of $\theta = 0.200$. However, these values can be used later to solve Eq. (13.6-5).

For the second iteration, a value of $x_o = 0.142$ is assumed and it is used again to solve for θ^* in Eq. (13.6-4), which results in $\theta^* = 0.1482$. For the final iteration, $x_o = 0.1190$ and $\theta^* = \theta = 0.2000$. Several more values are calculated for later use and are for $x_o = 0.187$, $\theta^* = 0.04876$, and for $x_o = 0.209$, $\theta^* = 0$. These values are tabulated in Table 13.6-1.

TABLE 13.6-1.   Calculated Values for
Example 13.6-1

θ^*	x	y_p	F_i
0	0.209	0.6550	0.6404
0.04876	0.1870	0.6383	0.7192
0.0992	0.1642	0.6158	0.8246
0.1482	0.1420	0.5940	0.9603
0.2000	0.1190	0.5690	1.1520

Using the material-balance equation (13.4-8) to calculate y_p,

$$y_p = \frac{x_f - x_o(1 - \theta)}{\theta} = \frac{0.209 - 0.1190(1 - 0.2000)}{0.2000} = 0.5690$$

To calculate y_p at $\theta^* = 0$, Eqs. (13.6-3) and (13.4-10) must be used and give $y_p = 0.6550$.

To solve for the area, Eq. (13.6-5) can be written as

$$A_m = \frac{tq_f}{P_h P_B'} \int_{i_o}^{i_f} \left[\frac{(1 - \theta^*)(1 - x)}{(f_i - i)\left[\dfrac{1}{1+i} - \dfrac{p_l}{p_h}\left(\dfrac{1}{1+f_i}\right)\right]} \right] di = \frac{tq_f}{P_h P_B'} \int_{i_o}^{i_f} F_i \, di$$

(13.6-6)

where the function F_i is defined as above. Values of F_i will be calculated for different values of i in order to integrate the equation. For $\theta^* = 0.200$, $x_o = 0.119$ and from Eq. (13.6-4),

$$i = i_o = \frac{x}{(1 - x)} = \frac{0.119}{(1 - 0.119)} = 0.1351$$

From Eq. (13.6-5),

$$f_i = (Di - F) + (D^2 i^2 + 2Ei + F^2)^{0.5}$$

$$= (4.55 \times 0.1351 - 0.950) + [(4.55)^2 (0.1351)^2$$

$$+ 2(0.6775)(0.1351) + (0.95)^2]^{0.5}$$

$$= 0.8744$$

Using the definition of F_i from Eq. (13.6-6),

$$F_i = \frac{(1 - \theta^*)(1 - x)}{(f_i - i)\left[\dfrac{1}{1+i} - \dfrac{p_l}{p_h}\left(\dfrac{1}{1+f_i}\right)\right]}$$

$$= \frac{(1 - 0.200)(1 - 0.119)}{(0.8744 - 0.1351)\left[\dfrac{1}{1 + 0.1351} - \dfrac{19}{190}\left(\dfrac{1}{1 + 0.8744}\right)\right]}$$

$$= 1.1520$$

Other values of F_i are calculated for the remaining values of θ^* and are tabulated in Table 13.6-1. The integral of Eq. (13.6-6) is obtained by using the values from Table 13.6-1 and plotting F_i versus i to give an area of 0.1082. Finally, substituting into Eq. (13.6-6)

$$A_m = \frac{tq_f}{P_h P_B'} \int_{i_o}^{i_f} F_i \, di = \frac{2.54 \times 10^{-3}(1 \times 10^6)}{190(50 \times 10^{-10})/10}(0.1082)$$

$$= 2.893 \times 10^8 \text{ cm}^2$$

For part (b), from Example 13.4-2, $y_p = 0.5067$ and $A_m = 3.228 \times 10^8$ cm². Hence, the cross-flow model yields a higher y_p of 0.5690 compared to 0.5067 for the complete-mixing model. Also, the area for the cross-flow model is 10% less than for the complete-mixing model.

13.7 COUNTERCURRENT-FLOW MODEL FOR GAS SEPARATION BY MEMBRANES

13.7A Derivation of Basic Equations

A flow diagram for the countercurrent-flow model is given in Fig. 13.7-1, where both streams are in plug flow. The derivation follows that given by Walawender and Stern (W5). Others (B1, P4) have also derived equations for this case.

Making a total and a component balance for A over the volume element and the reject,

$$q = q_o + q' \tag{13.7-1}$$

$$qx = q_o x_o + q'y \tag{13.7-2}$$

Differentiating Eq.(13.7-2)

$$d(qx) = 0 + d(q'y) \tag{13.7-3}$$

A balance for component A on the high- and low-pressure side of the volume element gives

$$qx = (q - dq)(x - dx) + y\, dq \tag{13.7-4}$$

Simplifying, we have

$$y\, dq = q\, dx + x\, dq = d(qx) \tag{13.7-5}$$

The local flux out of the element with area dA_m is

$$-y\, dq = \frac{P'_A}{t}(p_h x - p_l y)\, dA_m \tag{13.7-6}$$

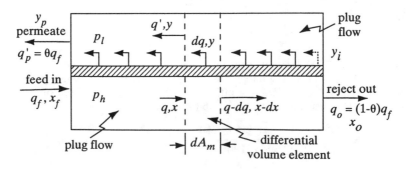

FIGURE 13.7-1. *Flow diagram for the countercurrent-flow model.*

Combining Eqs. (13.7-3), (13.7-5), and (13.7-6)

$$-d(q'y) = -d(qx) = \frac{P'_A}{t}(p_h x - p_l y)\, dA_m \qquad \text{(13.7-7)}$$

Similarly, for component B,

$$-d[q'(1-y)] = -d[q(1-x)] = \frac{P'_B}{t}[p_h(1-x) - p_l(1-y)]\, dA_m \qquad \text{(13.7-8)}$$

Combining Eq. (13.7-1) with (13.7-2) to eliminate q' and multiplying by dx,

$$q_o\, dx = \left(\frac{x-y}{y-x_o}\right)(-q\, dx) \qquad \text{(13.7-9)}$$

It can be shown that Eq. (13.7-10) is valid.

$$-q\, dx = -(1-x)\, d(qx) + x\, d[q(1-x)] \qquad \text{(13.7-10)}$$

Substituting $-q\, dx$ from Eq. (13.7-10), $-d(qx)$ from Eq. (13.7-7), and $d[q(1-x)]$ from Eq. (13.7-8) into Eq. (13.7-9) gives

$$\left(\frac{q_o t}{p_l P'_B}\right)\frac{dx}{dA_m} = \left(\frac{x-y}{y-x_o}\right)\{(1-x)\alpha^*(rx-y) - x[r(1-x) - (1-y)]\} \qquad \text{(13.7-11)}$$

where $r = p_h/p_l$ and $\alpha^* = P'_A/P'_B$.

Equation (13.7-12) can also be derived using the same methods as those used for Eq. (13.7-9).

$$q_o\, dy = \left(\frac{x-y}{x-x_o}\right)(-q'dy) \qquad \text{(13.7-12)}$$

It can also be shown that Eq. (13.7-13) is valid, which is similar to Eq. (13.7-10).

$$q'\, dy = (1-y)d(q'y) - y\, d[q'(1-y)] \qquad \text{(13.7-13)}$$

Substituting $q'\, dy$ from Eq. (13.7-13), $d(q'y)$ from Eq. (13.7-7), and $d[q'(1-y)]$ from Eq. (13.7-8) into Eq. (13.7-12),

$$\left(\frac{q_o t}{p_l P'_B}\right)\frac{dy}{dA_m} = \left(\frac{x-y}{x-x_o}\right)\{(1-y)\alpha^*(rx-y) - y[r(1-x) - (1-y)]\} \qquad \text{(13.7-14)}$$

At the outlet of the residue stream of composition x_o, the permeate $y = y_i$ and x_o are related by Eq. (13.4-5), which is given as Eq. (13.7-15).

$$\frac{y_i}{1-y_i} = \frac{\alpha^*[x_o - (p_l/p_h)y_i]}{(1-x_o) - (p_l/p_h)(1-y_i)} \qquad \text{(13.7-15)}$$

The solution of this quadratic equation is identical to Eq. (13.4-10).

13.7B Solution of Equations for Countercurrent Flow

Equations (13.7-11) and (13.7-14) are solved simultaneously by numerical methods starting at the high-pressure outlet stream of composition x_o. The area A_m can be arbitrarily set equal to zero at this outlet and a negative area will be obtained whose sign must be reversed. Equation (13.7-14) along with Eq. (13.7-15) is indeterminate at the high-pressure outlet. Using L'Hôpital's rule for $A_m \to 0$,

$$\left(\frac{dy}{dA_m}\right)_{A_m = 0}$$

$$= \frac{(x_o - y_i)r[\alpha^* - y_i(\alpha^* - 1)]}{q_ot/(p_lP'_B) - \{(x_o - y_i)[(\alpha^* - 1)(2y_i - rx_o - 1) - r]\}/(dx/dA_m)_{A_m = 0}} \qquad (13.7\text{-}16)$$

For Eq. (13.7-11),

$$\left(\frac{dx}{dA_m}\right)_{A_m = 0} = \frac{p_lP'_B}{q_ot}\left[\frac{\alpha^*(rx_o - y_i)(x_o - y_i)}{y_i}\right] \qquad (13.7\text{-}17)$$

It is more convenient to solve Eqs. (13.7-11) and (13.7-14) in terms of x as the independent variable. So dividing Eq. (13.7-14) by (13.7-11),

$$\frac{dy}{dx} = \frac{(y - x_o)}{(x - x_o)}\frac{\{(1 - y)\alpha^*(rx - y) - y[r(1 - x) - (1 - y)]\}}{\{(1 - x)\alpha^*(rx - y) - x[r(1 - x) - (1 - y)]\}} \qquad (13.7\text{-}18)$$

Inverting Eq. (13.7-11)

$$\frac{dA_m}{dx} = \frac{q_ot}{p_lP'_B}\frac{[(y - x_o)/(x - y)]}{\{(1 - x)\alpha^*(rx - y) - x[r(1 - x) - (1 - y)]\}} \qquad (13.7\text{-}19)$$

Since Eq. (13.7-18) is indeterminate at the high-pressure outlet, it can be evaluated using Eqs. (13.7-16) and (13.7-17) as follows

$$\left(\frac{dy}{dx}\right)_{A_m = 0} = \frac{(dy/dA_m)_{A_m = 0}}{(dx/dA_m)_{A_m = 0}} \qquad (13.7\text{-}20)$$

Assuming that the stage cut θ is specified, the procedure is trial and error. First a value of x_o at the high-pressure outlet is assumed, and by solving the two Eqs. (13.7-18) and (13.7-19) numerically, the value of the area A_m and the value of the permeate y_p at the outlet are obtained. Using this y_p and the material balance equation (13.4-8), x_o is calculated. If the assumed and the calculated values of x_o do not agree, another value is assumed and the procedure repeated. Further details and computer programs are given elsewhere (H1, R1, W5). For cocurrent flow, the equations are quite similar and are given by others (B1, H1, R1, W5).

13.8. EFFECTS OF PROCESSING VARIABLES ON GAS SEPARATION BY MEMBRANES

13.8A Effects of Pressure Ratio and Separation Factor on Recovery

Using the Weller–Steiner equation (13.4-5) for the compete-mixing model, the effects of pressure ratio, p_h/p_l, and separation factor, α^*, on permeate purity can be determined for a fixed feed composition. Figure 13.8-1 is a plot of this equation for a

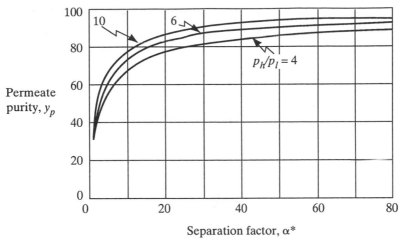

FIGURE 13.8-1. *Effects of separation factor and pressure ratio on permeate purity.* (Feed x_f = 0.30.) [From "Membranes Separate Gases Selectively," by D. J. Stookey, C. J. Patton, and G. L. Malcolm, *Chem. Eng. Progr.*, **82**(11), 36 (1986). Reproduced by permission of the American Institute of Chemical Engineers, 1986.]

feed concentration of 30% (S7). This equation can be expected to provide estimates of product purity and trends for conditions of low to modest recovery in all types of models, such as complete-mixing, cross-flow, and countercurrent.

Figure 13.8-1 shows that above an α^* of 20, the product purity is not greatly affected. Also, above a pressure ratio of about 6, this ratio has a diminishing effect on product purity. Some typical separation factors for commercial separators are given in Table 13.8-1.

If liquids are present in the gas separation process, a liquid film can increase the membrane resistance markedly. Liquids can also damage the membrane by chemical action or by swelling or softening. If water vapor is present in the gas streams, the dew point may be reached in the residue product and liquid condensed. Also, condensation of hydrocarbons must be avoided.

TABLE 13.8-1. *Typical Separation Factors for Some Industrial Membranes*

Gases Separated	Separation factor	Refs.
H_2O/CH_4	500	(M4)
He/CH_4	5–44	(M4), (S1), (W3)
H_2/CO	35–80	(K3), (M4)
H_2/N_2	3–200	(H1), (K3), (M4), (W2), (W3)
H_2/O_2	4–12	(M4), (S2), (W3)
H_2/CH_4	6–200	(H1), (K3), (M4), (W3)
O_2/N_2	2–12	(M4), (K3), (S2), (W2), (W3)
CO_2/CH_4	3–50	(H1), (M4), (S2), (W3)
CO_2/O_2	3–6	(M4), (S2), (W2), (W3)
CH_4/C_2H_6	2	(M4)

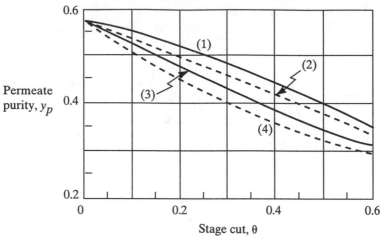

FIGURE 13.8-2. *Effect of stage cut and flow pattern on permeate purity. Operating conditions for air are as follows:* $x_f = 0.209$, $\alpha^* = 10$, $p_h/p_l = 380$ cm Hg/76 cm Hg $= 5$. $P'_A = 500 \times 10^{-10}$ $cm^3(STP) \cdot cm/s \cdot cm^2 \cdot cm$ Hg. *(1) countercurrent flow, (2) cross-flow, (3) cocurrent flow, (4) complete mixing (W5). [Reprinted from W. P. Walawender and S. A. Stern, Sep. Sci., 7, 553 (1972). By courtesy of Marcel Dekker, Inc.]*

13.8B Effects of Process Flow Patterns on Separation and Area

Detailed parametric studies have been done by various investigators (P4, P5, W5) for binary systems. They compared the four flow patterns of complete mixing, cross-flow, cocurrent, and countercurrent flow. In Fig. 13.8-2 (W5) the permeate concentration is shown plotted versus stage cut, θ, for a feed of air ($x_f = 0.209$ for oxygen) with $\alpha^* = 10$ and $p_h/p_l = 5$. It is shown that, as expected, the countercurrent flow pattern gives the best separation. The other patterns of cross-flow, cocurrent, and complete mixing give lower separations in descending order. Note that when the stage cut $\theta = 0$, all flow patterns are equivalent to the complete mixing model and give the same permeate composition. Also, at $\theta = 1.00$, all patterns again give the same value of $y_p = 0.209$, which is also the feed composition.

   The required membrane areas for the same process conditions and air feed versus stage cut were also determined (W5). The areas for all four types of flow patterns were shown to be within about 10% of each other. The countercurrent and cross-flow flow patterns give the lowest area required.

   In general, it has been concluded by many parametric studies that at the same operating conditions the countercurrent flow pattern yields the best separation and requires the lowest membrane area. The order of efficiency is as follows: countercurrent > cross-flow > cocurrent > complete mixing.

13.9 REVERSE-OSMOSIS MEMBRANE PROCESSES

13.9A Introduction

1. Introduction.   To be useful for separation of different species, a membrane must allow passage of certain molecules and exclude or greatly restrict passage of other molecules. In osmosis, a spontaneous transport of solvent occurs from a dilute solute or salt solution to a concentrated solute or salt solution across a semipermeable membrane which allows passage of the solvent but impedes passage of the salt solutes. In Fig. 13.9-1a the solvent water normally flows through the semipermeable membrane to the salt solution. The levels of both liquids are the same as shown. The solvent flow can be reduced by exerting a pressure on the salt-solution side and membrane, as shown in Fig. 13.9-1b, until at a certain pressure, called the osmotic pressure π of the salt solution, equilibrium is reached and the amount of the solvent passing in opposite directions is equal. The chemical potentials of the solvent on both sides of the membrane are equal. The property of the solution determines only the value of the osmotic pressure, not the membrane, provided that it is truly semipermeable. To reverse the flow of the water so that it flows from the salt solution to the fresh solvent as in Fig. 13.9-1c, the pressure is increased above the osmotic pressure on the solution side.

This phenomenon, called *reverse osmosis*, is used in a number of processes. An important commercial use is in the desalination of seawater or brackish water to produce fresh water. Unlike distillation and freezing processes used to remove solvents, reverse osmosis can operate at ambient temperature without phase change. This process is quite useful for processing of thermally and chemically unstable products. Applications include concentration of fruit juices and milk, recovery of protein and sugar from cheese whey, and concentration of enzymes.

2. Osmotic pressure of solutions. Experimental data show that the osmotic pressure π of a solution is proportional to the concentration of the solute and temperature T. Van't Hoff originally showed that the relationship is similar to that for pressure of an ideal gas. For example, for dilute water solutions,

$$\pi = \frac{n}{V_m} RT \qquad (13.9\text{-}1)$$

where n is the number of kg mol of solute, V_m the volume of pure solvent water in m^3 associated with n kg mol of solute, R the gas law constant 82.057×10^{-3} $m^3 \cdot atm/kg$ $mol \cdot K$, and T is temperature in K. If a solute exists as two or more ions in solution, n represents the total number of ions. For more concentrated solutions Eq. (13.9-1) is modified using the osmotic coefficient ϕ, which is the ratio of the actual osmotic pressure

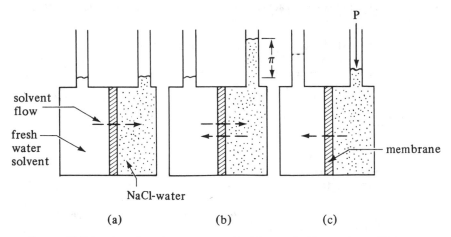

FIGURE 13.9-1. *Osmosis and reverse osmosis: (a) osmosis, (b) osmotic equilibrium, (c) reverse osmosis.*

π to the ideal π calculated from the equation. For very dilute solutions ϕ has a value of unity and usually decreases as concentration increases. In Table 13.9-1 some experimental values of π are given for NaCl solutions, sucrose solutions, and seawater solutions (S3, S5).

> **EXAMPLE 13.9-1.** *Calculation of Osmotic Pressure of Salt Solution*
> Calculate the osmotic pressure of a solution containing 0.10 g mol NaCl/1000 g H_2O at 25°C.
>
> *Solution:* From Table A.2-3, the density of water = 997.0 kg/m³. Then, $n = 2 \times 0.10 \times 10^{-3} = 2.00 \times 10^{-4}$ kg mol (NaCl gives two ions). Also, the volume of the pure solvent water $V_m = 1.00$ kg/(997.0 kg/m³). Substituting into Eq. (13.9-1),
>
> $$\pi = \frac{n}{V_m} RT = \frac{2.00 \times 10^{-4}(82.057 \times 10^{-3})(298.15)}{1.000/997.0} = 4.88 \text{ atm}$$
>
> This compares with the experimental value in Table 13.9-1 of 4.56 atm.

3. Types of membranes for reverse osmosis.   One of the more important membranes for reverse-osmosis desalination and many other reverse-osmosis processes is the cellulose acetate membrane. The asymmetric membrane is made as a composite film in which a thin dense layer about 0.1 to 10 μm thick of extremely fine pores supported upon a much thicker (50 to 125 μm) layer of microporous sponge with little resistance to permeation. The thin, dense layer has the ability to block the passage of quite small solute molecules. In desalination the membrane rejects the salt solute and allows the solvent water to pass through. Solutes which are most effectively excluded by the cellulose acetate membrane are the salts NaCl, NaBr, $CaCl_2$, and Na_2SO_4; sucrose; and tetralkyl ammonium salts. The main limitations of the cellulose acetate membrane are that it can only be used mainly in aqueous solutions and that it must be used below about 60°C.

Another important membrane useful for seawater, wastewater, nickel-plating rinse solutions, and other solutes is the synthetic aromatic polyamide membrane "Permasep," made in the form of very fine hollow fibers (L1, P3). This type of membrane used industrially withstands continued operation at pH values of 10 to 11 (S4). Many other anisotropic membranes have also been synthesized of synthetic polymers, some of which can be used in organic solvents, at higher temperatures, and at high or low pH (M2, R1).

TABLE 13.9-1.   *Osmotic Pressure of Various Aqueous Solutions at 25°C (P1, S3, S5)*

Sodium Chloride Solutions			Sea Salt Solutions		Sucrose Solutions	
g mol NaCl kg H_2O	Density (kg/m³)	Osmotic Pressure (atm)	Wt. % Salts	Osmotic Pressure (atm)	Solute Mol. Frac. $\times 10^3$	Osmotic Pressure (atm)
0	997.0	0	0	0	0	0
0.01	997.4	0.47	1.00	7.10	1.798	2.48
0.10	1001.1	4.56	3.45*	25.02	5.375	7.48
0.50	1017.2	22.55	7.50	58.43	10.69	15.31
1.00	1036.2	45.80	10.00	82.12	17.70	26.33
2.00	1072.3	96.2				

* Value for standard seawater.

13.9B Flux Equations for Reverse Osmosis

1. Basic models for membrane processes.   There are two basic types of mass-transport mechanisms which can take place in membranes. In the first basic type, using tight membranes, which are capable of retaining solutes of about 10 Å in size or less, diffusion-type transport mainly occurs. Both the solute and the solvent migrate by molecular or Fickian diffusion in the polymer, driven by concentration gradients set up in the membrane by the applied pressure difference. In the second basic type, using loose, microporous membranes which retain particles larger than 10 Å, a sieve-type mechanism occurs where the solvent moves through the micropores in essentially viscous flow and the solute molecules small enough to pass through the pores are carried by convection with the solvent. For details of the second type, see (M2, W1).

2. Diffusion-type model.   For diffusion-type membranes, the steady-state equations governing the transport of solvent and of solute are to a first approximation as follows (M2, M3). For the diffusion of the solvent through the membrane as shown in Fig. 13.9-2,

$$N_w = \frac{P_w}{L_m}(\Delta P - \Delta \pi) = A_w(\Delta P - \Delta \pi) \qquad \text{(13.9-2)}$$

$$P_w = \frac{D_w \bar{c}_w V_w}{RT} \qquad \text{(13.9-3)}$$

$$A_w = \frac{P_w}{L_m} \qquad \text{(13.9-4)}$$

where N_w is the solvent (water) flux in kg/s·m²; P_w the solvent membrane permeability, kg solvent/s·m·atm; L_m the membrane thickness, m; A_w the solvent permeability constant, kg solvent/s·m²·atm; $\Delta P = P_1 - P_2$ (hydrostatic pressure difference with P_1 pressure exerted on feed and P_2 on product solution), atm; $\Delta \pi = \pi_1 - \pi_2$ (osmotic pressure of feed solution − osmotic pressure of product solution), atm; D_w is the diffusivity of solvent in membrane, m²/s; $\bar{c}_w$ the mean concentration of solvent in membrane, kg solvent/m³; V_w the molar volume of solvent, m³/kg mol solvent; R the gas law constant, 82.057×10^{-3} m³·atm/kg mol·K; and T the temperature, K. Note that subscript 1 is the feed or upstream side of the membrane and 2 the product or downstream side of the membrane.

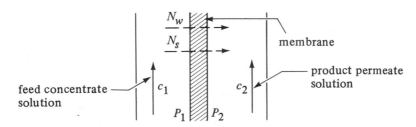

FIGURE 13.9-2. *Concentrations and fluxes in reverse-osmosis process.*

For the diffusion of the solute through the membrane, an approximation for the flux of the solute is (C1, M1)

$$N_s = \frac{D_s K_s}{L_m}(c_1 - c_2) = A_s(c_1 - c_2) \tag{13.9-5}$$

$$A_s = \frac{D_s K_s}{L_m} \tag{13.9-6}$$

where N_s is the solute (salt) flux in kg solute/s·m²; D_s the diffusivity of solute in membrane, m²/s; $K_s = c_m/c$ (distribution coefficient), concentration of solute in membrane/concentration of solute in solution; A_s is the solute permeability constant, m/s; c_1 the solute concentration in upstream or feed (concentrate) solution, kg solute/m³; and c_2 the solute concentration in downstream or product (permeate) solution, kg solute/m³. The distribution coefficient K_s is approximately constant over the membrane.

Making a material balance at steady state for the solute, the solute diffusing through the membrane must equal the amount of solute leaving in the downstream or product (permeate) solution.

$$N_s = \frac{N_w c_2}{c_{w2}} \tag{13.9-7}$$

where c_{w2} is the concentration of solvent in stream 2 (permeate), kg solvent/m³. If the stream 2 is dilute in solute, c_{w2} is approximately the density of the solvent. In reverse osmosis, the solute rejection R is defined as the ratio concentration difference across the membrane divided by the bulk concentration on the feed or concentrate side (fraction of solute remaining in the feed stream).

$$R = \frac{c_1 - c_2}{c_1} = 1 - \frac{c_2}{c_1} \tag{13.9-8}$$

This can be related to the flux equations as follows by first substituting Eqs. (13.9-2) and (13.9-5) into (13.9-7) to eliminate N_w and N_s in Eq. (13.9-7). Then solving for c_2/c_1 and substituting this result into Eq. (13.9-8),

$$R = \frac{B(\Delta P - \Delta \pi)}{1 + B(\Delta P - \Delta \pi)} \tag{13.9-9}$$

$$B = \frac{P_w}{D_s K_s c_{w2}} = \frac{A_w}{A_s c_{w2}} \tag{13.9-10}$$

where B is in atm^{-1}. Note that B is composed of the various physical properties P_w, D_s, and K_s of the membrane and must be determined experimentally for each membrane. Usually, the product $D_s K_s$ is determined, not the values of D_s and K_s separately. Also, many of the data reported in the literature give values of (P_w/L_m) or A_w in kg solvent/s·m²·atm and $(D_s K_s/L_m)$ or A_s in m/s and not separate values of L_m, P_w, and so on.

EXAMPLE 13.9-2. *Experimental Determination of Membrane Permeability*

Experiments at 25°C were performed to determine the permeabilities of a cellulose acetate membrane (A1, W1). The laboratory test section shown in Fig. 13.9-3 has membrane area $A = 2.00 \times 10^{-3}$ m². The inlet feed solution concentration of NaCl is $c_1 = 10.0$ kg NaCl/m³ solution (10.0g NaCl/L, $\rho_1 = 1004$ kg solution/m³). The water recovery is assumed low so that the

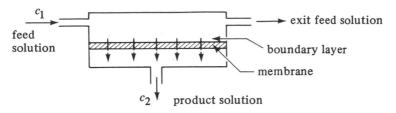

FIGURE 13.9-3. *Process flow diagram of experimental reverse-osmosis laboratory unit.*

concentration c_1 in the entering feed solution flowing past the membrane and the concentration of the exit feed solution are essentially equal. The product solution contains $c_2 = 0.39$ kg $NaCl/m^3$ solution ($\rho_2 = 997$ kg solution/m^3) and its measured flow rate is 1.92×10^{-8} m^3 solution/s. A pressure differential of 5514 kPa (54.42 atm) is used. Calculate the permeability constants of the membrane and the solute rejection R.

Solution: Since c_2 is very low (dilute solution), the value of c_{w2} can be assumed as the density of water (Table 13.9-1) or $c_{w2} = 997$ kg solvent/m^3. To convert the product flow rate to water flux, N_w, using an area of 2.00×10^{-3} m^2,

$$N_w = (1.92 \times 10^{-8} \text{ m}^3/\text{s})(997 \text{ kg solvent/m}^3)(1/2.00 \times 10^{-3} \text{ m}^2)$$

$$= 9.57 \times 10^{-3} \text{ kg solvent/s} \cdot \text{m}^2$$

Substituting into Eq. (13.9-7),

$$N_s = \frac{N_w c_2}{c_{w2}} = \frac{(9.57 \times 10^{-3})(0.39)}{997}$$

$$= 3.744 \times 10^{-6} \text{ kg solute NaCl/s} \cdot \text{m}^2$$

To determine the osmotic pressures from Table 13.9-1, the concentrations are converted as follows. For c_1, 10 kg NaCl is in 1004 kg solution/m^3 ($\rho_1 = 1004$). Then, $1004 - 10 = 994$ kg H_2O in 1 m^3 solution. Hence, in the feed solution where the molecular weight of NaCl = 58.45, $(10.00 \times 1000)/(994 \times 58.45) = 0.1721$g mol NaCl/kg H_2O. From Table 13.9-1, $\pi_1 = 7.80$ atm by linear interpolation. Substituting into Eq. (13.9-1), the predicted $\pi_1 = 8.39$ atm, which is higher than the experimental value. For the product solution, $997 - 0.39 = 996.6$ kg H_2O. Hence, $(0.39 \times 1000)/(996.6 \times 58.45) = 0.00670$g mol NaCl/kg H_2O. From Table 13.9-1, $\pi_2 = 0.32$ atm. Then, $\Delta\pi = \pi_1 - \pi_2 = 7.80 - 0.32 = 7.48$ atm and $\Delta P = 54.42$ atm.

Substituting into Eq. (13.9-2),

$$N_w = 9.57 \times 10^{-3} = \frac{P_w}{L_m}(\Delta P - \Delta\pi) = \frac{P_w}{L_m}(54.42 - 7.48)$$

Solving $(P_w/L_m) = A_w = 2.039 \times 10^{-4}$ kg solvent/s $\cdot$ $m^2 \cdot$ atm. Substituting into Eq. (13.9-5),

$$N_s = 3.744 \times 10^{-6} = \frac{D_s K_s}{L_m}(c_1 - c_2) = \frac{D_s K_s}{L_m}(10.00 - 0.39)$$

Solving, $(D_s K_s/L_m) = A_s = 3.896 \times 10^{-7}$ m/s.

To calculate the solute rejection R by substituting into Eq. (13.9-8),

$$R = \frac{c_1 - c_2}{c_1} = \frac{10.00 - 0.39}{10.00} = 0.961$$

Also substituting into Eq. (13.9-10) and then Eq. (13.9-9),

$$B = \frac{P_w/L_m}{(D_s K_s/L_m)c_{w2}} = \frac{2.039 \times 10^{-4}}{(3.896 \times 10^{-7})997} = 0.5249 \text{ atm}^{-1}$$

$$R = \frac{B(\Delta P - \Delta \pi)}{1 + B(\Delta P - \Delta \pi)} = \frac{0.5249(54.42 - 7.48)}{1 + 0.5249(54.42 - 7.48)} = 0.961$$

13.10 APPLICATIONS, EQUIPMENT, AND MODELS FOR REVERSE OSMOSIS

13.10A Effects of Operating Variables

In many commercial units operating pressures in reverse osmosis range from about 1035 up to 10 350 kPa (150 up to 1500 psi). Comparison of Eq. (13.9-2) for solvent flux and Eq. (13.9-5) for solute flux shows that the solvent flux N_w depends only on the net pressure difference, while the solute flux N_s depends only on the concentration difference. Hence, as the feed pressure is increased, solvent or water flow through the membrane increases and the solute flow remains approximately constant, giving lower solute concentration in the product solution.

At a constant applied pressure, increasing the feed solute concentration increases the product solute concentration. This is caused by the increase in the feed osmotic pressure, since as more solvent is extracted from the feed solution (as water recovery increases), the solute concentration becomes higher and the water flux decreases. Also, the amount of solute present in the product solution increases because of the higher feed concentration.

If a reverse-osmosis unit has a large membrane area (as in a commercial unit), and the path between the feed inlet and outlet is long, the outlet feed concentration can be considerably higher than the inlet feed c_1. Then the salt flux will be greater at the outlet feed compared to the inlet (K2). Many manufacturers use the feed solute or salt concentration average between inlet and outlet to calculate the solute or salt rejection R in Eq. (13.9-8).

EXAMPLE 13.10-1. Prediction of Performance in a Reverse-Osmosis Unit

A reverse-osmosis membrane to be used at 25°C for a NaCl feed solution containing 2.5 g NaCl/L (2.5 kg NaCl/m³, $\rho = 999$ kg/m³) has a water permeability constant $A_w = 4.81 \times 10^{-4}$ kg/s·m²·atm and a solute (NaCl) permeability constant $A_s = 4.42 \times 10^{-7}$ m/s (A1). Calculate the water flux and solute flux through the membrane using a $\Delta P = 27.20$ atm and the solute rejection R. Also calculate c_2 of the product solution.

Solution: In the feed solution, $c_1 = 2.5$ kg NaCl/m³ and $\rho_1 = 999$ kg solution/m³. Hence, for the feed, $999 - 2.5 = 996.5$ kg H_2O in 1.0 m³ solution; also for the feed, $(2.50 \times 1000)/(996.5 \times 58.45) = 0.04292$ g mol NaCl/kg H_2O. From Table 13.9-1, $\pi_1 = 1.97$ atm. Since the product solution c_2 is unknown, a value of $c_2 = 0.1$ kg NaCl/m³ will be assumed. Also, since this is quite dilute, $\rho_2 = 997$ kg solution/m³ and $C_w^2 = 997$ kg solvent/m³. Then for the product solution, $(0.10 \times 1000)/(996.9 \times 58.45) = 0.00172$ g mol NaCl/kg H_2O and $\pi_2 = 0.08$ atm. Also, $\Delta \pi = \pi_1 - \pi_2 = 1.97 - 0.08 = 1.89$ atm.

Substituting into Eq. (13.9-2),

$$N_w = A_w(\Delta P - \Delta \pi) = 4.81 \times 10^{-4}(27.20 - 1.89)$$
$$= 1.217 \times 10^{-2} \text{ kg H}_2\text{O/s} \cdot \text{m}^2$$

For calculation of R, substituting first into Eq. (13.9-10),

$$B = \frac{A_w}{A_s c_{w2}} = \frac{4.81 \times 10^{-4}}{4.42 \times 10^{-7} \times 997} = 1.092 \text{ atm}^{-1}$$

Next substituting into Eq. (13.9-9),

$$R = \frac{B(\Delta P - \Delta \pi)}{1 + B(\Delta P - \Delta \pi)} = \frac{1.092(27.20 - 1.89)}{1 + 1.092(27.20 - 1.89)} = 0.965$$

Using this value of R in Eq. (13.9-8),

$$R = 0.965 = \frac{c_1 - c_2}{c_1} = \frac{2.50 - c_2}{2.50}$$

Solving, $c_2 = 0.0875$ kg NaCl/m^3 for the product solution. This is close enough to the assumed value of $c_2 = 0.10$ so that π_2 will not change significantly on a second trial. Hence, the final value of c_2 is 0.0875 kg NaCl/m^3(0.0875 g NaCl/L).

Substituting into Eq. (13.9-5),

$$N_s = A_s(c_1 - c_2) = 4.42 \times 10^{-7}(2.50 - 0.0875)$$
$$= 1.066 \times 10^{-6} \text{ kg NaCl/s} \cdot \text{m}^2$$

13.10B Concentration Polarization in Reverse-Osmosis Diffusion Model

In desalination, localized concentrations of solute build up at the point where the solvent leaves the solution and enters the membrane. The solute accumulates in a relatively stable boundary layer (Fig. 13.9-3) next to the membrane. Concentration polarization, β, is defined as the ratio of the salt concentration at the membrane surface to the salt concentration in the bulk feed stream c_1. Concentration polarization causes the water flux to decrease since the osmotic pressure π_1 increases as the boundary layer concentration increases and the overall driving force $(\Delta P - \Delta \pi)$ decreases. Also, the solute flux increases since the solute concentration increases at the boundary. Hence, often the ΔP must be increased to compensate which gives higher power costs (K2).

The effect of the concentration polarization β can be included approximately by modifying the value of $\Delta \pi$ in Eqs. (13.9-2) and (13.9-9) as follows (P6):

$$\Delta \pi = \beta \pi_1 - \pi_2 \qquad (13.10\text{-}1)$$

It is assumed that the osmotic pressure π_1 is directly proportional to the concentration, which is approximately correct. Also, Eq. (13.9-5) can be modified as

$$N_s = A_s(\beta c_1 - c_2) \qquad (13.10\text{-}2)$$

The usual concentration polarization ratio (K3) is 1.2 to 2.0, i.e., the concentration in the boundary layer is 1.2 to 2.0 times c_1 in the bulk feed solution. This ratio is often difficult to predict. In desalination of seawater using values of about 1000 psia = ΔP, π_1 can be large. Increasing this π_1 by a factor of 1.2 to 2.0 can appreciably reduce the solvent flux. For brackish waters containing 2 to 10 g/L and using ΔP values of 17 to 55 atm abs, the value of π_1 is low and concentration polarization is not important.

The boundary layer can be reduced by increasing the turbulence using higher feed solution velocities. However, this extra flow results in a smaller ratio of product solution to feed. Also, screens can be put in the flow path to induce turbulence.

13.10C Permeability Constants of Reverse-Osmosis Membranes

Permeability constants for membranes must be determined experimentally for the particular type of membrane to be used. For cellulose acetate membranes, typical water permeability constants A_w range from about 1×10^{-4} to 5×10^{-4} kg solvents/s $\cdot m^2 \cdot atm$ (A1, M3, W1). Values for other types of membranes can differ widely. Generally, the water permeability constant for a particular membrane does not depend upon the solute present. For the solute permeability constants A_s of cellulose acetate membranes, some relative typical values are as follows, assuming a value of $A_s = 4 \times 10^{-7}$ m/s for NaCl: 1.6×10^{-7} m/s ($BaCl_2$), 2.2×10^{-7} ($MgCl_2$), 2.4×10^{-7} ($CaCl_2$), 4.0×10^{-7} (Na_2SO_4), 6.0×10^{-7} (KCl), 6.0×10^{-7} (NH_4Cl) (A1).

13.10D Types of Equipment for Reverse Osmosis

The equipment for reverse osmosis is quite similar to that for gas permeation membrane processes described in Section 13.3C. In the plate-and-frame type unit, thin plastic support plates with thin grooves are covered on both sides with membranes as in a filter press. Pressurized feed solution flows between the closely spaced membranes (L1). Solvent permeates through the membrane and flows in the grooves to an outlet. In the tubular-type unit, membranes in the form of tubes are inserted inside porous-tube casings, which serve as a pressure vessel. These tubes are then arranged in bundles like a heat exchanger.

In the spiral-wound type, a planar membrane is used and a flat, porous support material is sandwiched between the membranes. Then the membranes, support, and a mesh feed-side spacer are wrapped in a spiral around a tube. In the hollow-fiber type, fibers of 100 to 200 μm diameter with walls about 25 μm thick are arranged in a bundle similar to a heat exchanger (L1, R1).

13.10E Complete-Mixing Model for Reverse Osmosis

The process flow diagram for the complete-mixing model is shown in Fig. 13.10-1. The model is a simplified one for use with low concentrations of salt of about 1% or so such

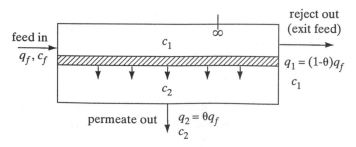

FIGURE 13.10-1. *Process flow for complete-mixing model for reverse osmosis.*

as occur in brackish waters. Also, a relatively low recovery of solvent occurs and the effects of concentration polarization are small. Since the concentration of the permeate is very low, the permeate side acts as though it were completely mixed.

For the overall material balance for dilute solutions,

$$q_f = q_1 + q_2 \tag{13.10-3}$$

where q_f is volumetric flow rate of feed, m^3/s; q_2 is flow rate of permeate, m^3/s; and q_1 is flow rate of residue or exit, m^3/s. Making a solute balance,

$$c_f q_f = c_1 q_1 + c_2 q_2 \tag{13.10-4}$$

Defining the cut or fraction of solvent recovered as $\theta = q_2/q_f$, Eq. (13.10-4) becomes

$$c_f = (1 - \theta)c_1 + \theta c_2 \tag{13.10-5}$$

The previous equations derived for the fluxes and rejection are useful in this case and are as follows:

$$N_w = A_2(\Delta P - \Delta \pi) \tag{13.9-2}$$

$$N_s = A_s(c_1 - c_2) \tag{13.9-5}$$

$$R = \frac{c_1 - c_2}{c_1} \tag{13.9-8}$$

$$R = \frac{B(\Delta P - \Delta \pi)}{1 + B(\Delta P - \Delta \pi)} \tag{13.9-9}$$

When the cut or fraction recovered, θ, is specified the solution is trial and error. Since the permeate and reject concentrations c_1 and c_2 are unknown, a value of c_2 is assumed. Then c_1 is calculated from Eq. (13.10-5). Next N_w is obtained from Eq. (13.9-2) and c_2 from Eqs. (13.9-8) and (13.9-9). If the calculated value of c_2 does not equal the assumed value, the procedure is repeated.

When concentration polarization effects are present an estimated value of β can be used to make an approximate correction for this effect. This is used in Eq. (13.10-1) to obtain a value of $\Delta \pi$ for use in Eqs. (13.9-2) and (13.9-9). Also, Eq. (13.10-2) will replace Eq. (13.9-5). A more detailed analysis of this complete mixing model is given by others (H1, K1) in which the mass transfer coefficient in the concentration polarization boundary layer is used.

The cross-flow model for reverse osmosis is similar to that for gas separation by membranes discussed in Section 13.6. Because of the small solute concentration, the permeate side acts as if completely mixed. Hence, even if the module is designed for countercurrent or cocurrent flow, the cross-flow model is valid. This is discussed in detail elsewhere (H1).

13.11 ULTRAFILTRATION MEMBRANE PROCESSES

13.11A Introduction

Ultrafiltration is a membrane process that is quite similar to reverse osmosis. It is a pressure-driven process where the solvent and, when present, small solute molecules pass through the membrane and are collected as a permeate. Larger solute molecules do not pass through the membrane and are recovered in a concentrated solution. The solutes or molecules to be separated generally have molecular weights greater than 500 and up to 1 000 000 or more, such as macromolecules of proteins, polymers, and starches and also colloidal dispersions of clays, latex particles, and microorganisms.

Unlike reverse osmosis, ultrafiltration membranes are too porous to be used for desalting. The rejection, often called retention, is also given by Eq. (13.9-8), which is defined for reverse osmosis. Ultrafiltration is also used to separate a mixture of different molecular weight proteins. The molecular weight cut-off of the membrane is defined as the molecular weight of globular proteins, which are 90% retained by the membrane.

Ultrafiltration is used in many different processes at the present time. Some of these are separation of oil–water emulsions, concentration of latex particles, processing of blood and plasma, fractionation or separation of proteins, recovery of whey proteins in cheese manufacturing, removal of bacteria and other particles to sterilize wine, and clarification of fruit juices.

Membranes for ultrafiltration are in general similar to those for reverse osmosis and are commonly asymmetric and more porous. The membrane consists of a very thin dense skin supported by a relatively porous layer for strength. Membranes are made from aromatic polyamides, cellulose acetate, cellulose nitrate, polycarbonate, polyimides, polysulfone, etc. (M2, P6, R1).

13.11B Types of Equipment for Ultrafiltration

The equipment for ultrafiltration is similar to that used for reverse osmosis and gas separation processes described in Sections 13.3C and 13.10D. The tubular type unit is less prone to foul and is more easily cleaned than any of the other three types. However, this type is relatively costly.

Flat sheet membranes in a plate-and-frame unit offer the greatest versatility but at the highest capital cost (P6). Membranes can easily be cleaned or replaced by disassembly of the unit. Spiral-wound modules provide relatively low costs per unit membrane area. These units are more prone to foul than tubular units but are more resistant to fouling than hollow-fiber units. Hollow-fiber modules are the least resistant to fouling when compared to the three other types. However, the hollow-fiber configuration has the highest ratio of membrane area per unit volume.

13.11C Flux Equations for Ultrafiltration

The flux equation for diffusion of solvent through the membrane is the same as Eq. (13.9-2) for reverse osmosis:

$$N_w = A_w(\Delta P - \Delta \pi) \tag{13.9-2}$$

In ultrafiltration the membrane does not allow the passage of the solute, which is generally a macromolecule. The concentration in moles/liter of the large solute molecules is usually small. Hence, the osmotic pressure is very low and is neglected. Then Eq. (13.9-2) becomes

$$N_w = A_w(\Delta P) \tag{13.11-1}$$

Ultrafiltration units operate at about 5 to 100 psi pressure drop compared to 400 to 2000 for reverse osmosis. For low-pressure drops of, say, 5 to 10 psi and dilute solutions of up to 1 wt % or so, Eq. (13.11-1) predicts the performance reasonably well for well-stirred systems.

Since the solute is rejected by the membrane, it accumulates and starts to build up at the surface of the membrane. As pressure drop is increased and/or concentration of the solute is increased, concentration polarization occurs, which is much more severe than in reverse osmosis. This is shown in Fig. 13.11-1a, where c_1 is the concentration of the solute in the bulk solution, kg solute/m^3, and c_s is the concentration of the solute at the surface of the membrane.

As the pressure drop increases, this increases the solvent flux N_w to and through the membrane. This gives a higher convective transport of the solute to the membrane, i.e., the solvent carries with it more solute. The concentration c_s increases and gives a larger back molecular diffusion of solute from the membrane to the bulk solution. At steady state the convective flux equals the diffusion flux,

$$\frac{N_w c}{\rho} = -D_{AB} \frac{dc}{dx} \qquad (13.11\text{-}2)$$

where $N_w c/\rho = $ [kg solvent/(s · m^2)](kg solute/m^3)/(kg solvent/m^3) = kg solute/s · m^2; D_{AB} is diffusivity of solute in solvent, m^2/s; and x is distance, m. Integrating this equation between the limits of $x = 0$ and $c = c_s$ and $x = \delta$ and $c = c_1$,

$$\frac{N_w}{\rho} = \left(\frac{D_{Ab}}{\delta}\right) \ln\left(\frac{c_s}{c_1}\right) = k_c \ln\left(\frac{c_s}{c_1}\right) \qquad (13.11\text{-}3)$$

where k_c is the mass-transfer coefficient, m/s. Further increases in pressure drop increase the value of c_s to a limiting concentration where the accumulated solute forms a semisolid gel where $c_s = c_g$, as shown in Fig. 13.11-1b.

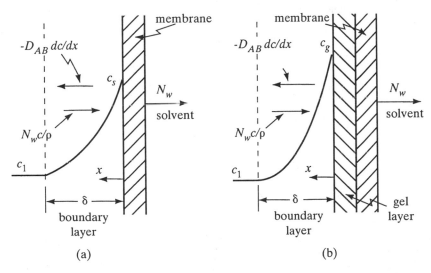

FIGURE 13.11-1. *Concentration polarization in ultrafiltration: (a) concentration profile before gel formation, (b) concentration profile with a gel layer formed at membrane surface.*

Still further increases in pressure drop do not change c_g and the membrane is said to be "gel polarized." Then Eq. (13.11-3) becomes (P1, P6, R1)

$$\frac{N_w}{\rho} = k_c \ln \left(\frac{c_g}{c_1}\right) \tag{13.11-4}$$

With increases in pressure drop, the gel layer increases in thickness and this causes the solvent flux to decrease because of the added gel layer resistance. Finally, the net flux of solute by convective transfer becomes equal to the back diffusion of solute into the bulk solution because of the polarized concentration gradient as given by Eq. (13.11-4).

The added gel layer resistance next to the membrane causes an increased resistance to solvent flux as given by

$$N_w = \frac{\Delta P}{1/A_w + R_g} \tag{13.11-5}$$

where $1/A_w$ is the membrane resistance and R_g is the variable gel layer resistance, $(s \cdot m^2 \cdot atm)/kg$ solvent. The solvent flux in this gel-polarized regime is independent of pressure difference and is determined by Eq. (13.11-4) for back diffusion. Experimental data confirm the use of Eq. (13.11-4) for a large number of macromolecular solutions, such as proteins, etc., and colloidal suspensions, such as latex particles, etc. (P1, P6).

13.11D Effects of Processing Variables in Ultrafiltration

A plot of typical experimental data of flux versus pressure difference is shown in Fig. 13.11-2 (H1, P6). At low pressure differences and/or low solute concentrations the data typically follow Eq. (13.11-1). For a given bulk concentration, c_1, the flux approaches a constant value at high pressure differences as shown in Eq. (13.11-4). Also, more dilute protein concentrations give higher flux rates as expected from Eq. (13.11-4). Most commercial applications are flux limited by concentration polarization and operate in the region where the flux is approximately independent of pressure difference (R1).

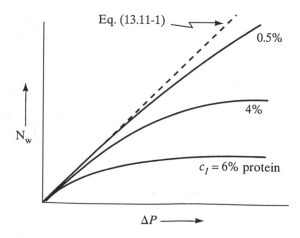

FIGURE 13.11-2. *Effect of pressure difference on solvent flux.*

Using experimental data, a plot of N_w/ρ versus $\ln c_1$ is a straight line with the negative slope of k_c, the mass-transfer coefficient, as shown by Eq. (13.11-4). These plots also give the value of c_g, the gel concentration. Data (P1) show that the gel concentration for many macromolecular solutions is about 25 wt %, with a range of 5 to 50%. For colloidal dispersions it is about 65 wt %, with a range of 50 to 75%.

The concentration polarization effects for hollow fibers is often quite small because of the low solvent flux. Hence, Eq. (13.11-1) describes the flux. In order to increase the ultrafiltration solvent flux, cross-flow of fluid past the membrane can be used to sweep away part of the polarized layer, thereby increasing k_c in Eq. (13.11-4). Higher velocities and other methods are used to increase turbulence, and hence, k_c. In most cases the solvent flux is too small to operate in a single-pass mode. It is necessary to recirculate the feed by the membrane with recirculation rates of 10/1 to 100/1 often used.

Methods to predict the mass-transfer coefficient k_c in Eq. (13.11-4) are given by others (P1, P6) for known geometries such as channels, etc. Predictions of flux in known geometries using these methods and experimental values of c_g in Eq. (13.11-4) in the gel polarization regime compare with experimental values for macromolecular solutions within about 25 to 30%. However, for colloidal dispersions the experimental flux is higher than the theoretical by factors of 20 to 30 for laminar flow and 8 to 10 for turbulent flow. Hence, Eq. (13.11-4) is not useful for predicting the solvent flux accurately. Generally, for design of commercial units it is necessary to obtain experimental data on single modules.

PROBLEMS

13.2-1. Diffusion Through Liquids and a Membrane. A membrane process is being designed to recover solute A from a dilute solution where $c_1 - 2.0 \times 10^{-2}$ kg mol A/m^3 by dialysis through a membrane to a solution where $c_2 = 0.3 \times 10^{-2}$. The membrane thickness is 1.59×10^{-5} m, the distribution coefficient $K' = 0.75$, $D_{AB} = 3.5 \times 10^{-11}$ m^2/s in the membrane, the mass-transfer coefficient in the dilute solution is $k_{c1} = 3.5 \times 10^{-5}$ m/s, and $k_{c2} - 2.1 \times 10^{-5}$.
  (a) Calculate the individual resistances, total resistance, and the total percent resistance of the two films.
  (b) Calculate the flux at steady state and the total area in m^2 for a transfer of 0.01 kg mol solute/h.
  (c) Increasing the velocity of both liquid phases flowing by the surface of the membrane will increase the mass-transfer coefficients, which are approximately proportional to $v^{0.6}$, where v is velocity. If the velocities are doubled, calculate the total percent resistance of the two films and the percent increase in flux.
  Ans. (a) Total resistance $= 6.823 \times 10^5$ s/m, 11.17% resistance, (b) $N_A = 2.492 \times 10^{-8}$ kg mol $A/s \cdot m^2$, area $= 111.5$ m^2

13.2-2. Suitability of a Membrane for Hemodialysis. Experiments are being conducted to determine the suitability of a cellophane membrane 0.029 mm thick for use in an artificial kidney device. In an experiment at 37°C using NaCl as the diffusing solute, the membrane separates two components containing stirred aqueous solutions of NaCl, where $c_1 = 1.0 \times 10^{-4}$ g mol/cm^3 (100 g mol/m^3) and $c_2 = 5.0 \times 10^{-7}$. The mass-transfer coefficients on either side of the membrane have been estimated as $k_{c1} = k_{c2} = 5.24 \times 10^{-5}$ m/s. Experimental data obtained gave a flux $N_A = 8.11 \times 10^{-4}$ g mol NaCl/s $\cdot$ m^2 at pseudo-steady-state conditions.
  (a) Calculate the permeability p_M in m/s and $D_{AB} K'$ in m^2/s.
  (b) Calculate the percent resistance to diffusion in the liquid films.

13.3-1. Gas-Permeation Membrane for Oxygenation. To determine the suitability of silicone rubber for its use as a membrane for a heart–lung machine to oxygenate blood, an experimental value of the permeability at 30°C of oxygen was obtained where $P''_M = 6.50 \times 10^{-7}$ cm^3 O$_2$ (STP)/(s·cm^2·cm Hg/mm).
(a) Predict the maximum flux of O$_2$ in kg mol/s·m^2 with an O$_2$ pressure of 700 mm Hg on one side of the membrane and an equivalent pressure in the blood film side of 50 mm. The membrane is 0.165 mm thick. Since the gas film is pure oxygen, the gas film resistance is zero. Neglect the blood film resistance in this case.
(b) Assuming a maximum requirement for an adult of 300 cm^3 O$_2$ (STP) per minute, calculate the membrane surface area required in m^2. (*Note:* The actual area needed should be considerably larger since the blood film resistance, which must be determined by experiment, can be appreciable.)

Ans.   (b) 1.953 m^2

13.4-1. Derivation of Equation for Permeate Concentration. Derive Eq. (13.4-11) for Case 2 for complete mixing. Note that x_o from Eq. (13.4-8) must first be substituted into Eq. (13.4-5) before multiplying out the equation and solving for y_p.

13.4-2. Use of Complete-Mixing Model for Membrane Design. A membrane having a thickness of 2×10^{-3} cm, a permeability $P'_A = 400 \times 10^{-10}$ cm^3(STP)·cm/ (s·cm^2·cm Hg), and an $\alpha^* = 10$ is to be used to separate a gas mixture of A and B. The feed flow rate is $q_f = 2 \times 10^3$ cm^3(STP)/s and its composition is $x_f = 0.413$. The feed-side pressure is 80 cm Hg and the permeate-side pressure is 20 cm Hg. The reject composition is to be $x_o = 0.30$. Using the complete-mixing model, calculate the permeate composition, the fraction of feed permeated, and the membrane area.

Ans. $y_p = 0.678$

13.4-3. Design Using Complete-Mixing Model. A gaseous feed stream having a composition $x_f = 0.50$ and a flow rate of 2×10^3 cm^3(STP)/s is to be separated in a membrane unit. The feed-side pressure is 40 cm Hg and the permeate is 10 cm Hg. The membrane has a thickness of 1.5×10^{-3} cm, a permeability $P'_A = 40 \times 10^{-10}$ cm^3(STP)·cm/(s·cm^2·cm Hg), and an $\alpha^* = 10$. The fraction of feed permeated is 0.529.
(a) Use the complete-mixing model to calculate the permeate composition, the reject composition, and the membrane area.
(b) Calculate the minimum reject concentration.
(c) If the feed composition is increased to $x_f = 0.60$, what is this new minimum reject concentration?

Ans.   (a) $A_m = 5.153 \times 10^7$ cm^2
(c) $x_{oM} = 0.2478$

13.4-4. Effect of Permeabilities on Minimum Reject Concentration. For the conditions of Problem 13.4-2, $x_f = 0.413$, $\alpha^* = 10$, $p_l = 20$ cm Hg, $p_h = 80$ cm Hg, and $x_o = 0.30$. Calculate the minimum reject concentration for the following cases.
(a) Calculate x_{oM} for the given conditions.
(b) Calculate the effect on x_{oM} if the permeability of B increases so that α^* decreases to 5.
(c) Calculate the limiting value of x_{oM} when α^* is lowered to its minimum value. Make a plot of x_{oM} versus α^* for these three cases.

13.4-5. Minimum Reject Concentration and Pressure Effect. For Example 13.4-2 for separation of air, do as follows.
(a) Calculate the minimum reject concentration.
(b) If the pressure on the feed side is reduced by one-half, calculate the effect on x_{oM}.

Ans.   (b) $x_{oM} = 0.0624$

13.5-1. *Separation of Multicomponent Gas Mixtures.* Using the same feed composition and flow rate, pressures, and membrane as in Example 13.5-1, do the following using the complete-mixing model.

(a) Calculate the permeate composition, the reject composition, and the membrane area for a fraction permeated of 0.50 instead of 0.25.

(b) Repeat part (a) but for $\theta = 0.90$.

(c) Make a plot of permeate composition y_{pA} versus θ and also of area A_m versus θ using the calculated values for $\theta = 0.25, 0.50,$ and 0.90.

13.5-2. *Separation of Helium from Natural Gas.* A typical composition of a natural gas (S1) is 0.5% He (A), 17.0% N_2 (B), 76.5% CH_4 (C), and 6.0% higher hydrocarbons (D). The membrane proposed to separate helium has a thickness of 2.54×10^{-3} cm and the permeabilities are $P'_A = 60 \times 10^{-10}$ cm^3(STP)·cm/ (s·cm^2·cm Hg), $P'_B = 3.0 \times 10^{-10}$, and $P'_C = 1.5 \times 10^{-10}$. It is assumed that the higher hydrocarbons are essentially non-permeable ($P'_D \cong 0$). The feed flow rate is 2.0×10^5 cm^3(STP)/s. The feed pressure $p_h = 500$ cm Hg and the permeate pressure $p_l = 20$ cm Hg.

(a) For a fraction permeated of 0.2, calculate the permeate composition, the reject composition, and the membrane area using the complete mixing model.

(b) Use the permeate from part (a) as feed to a completely mixed second stage. The pressure $p_h = 500$ cm Hg and $p_l = 20$ cm. For a fraction permeated of 0.20, calculate the permeate composition and the membrane area.

13.6-1. *Design Using Cross-Flow Model for Membrane.* Use the same conditions for the separation of an air stream as given in Example 13.6-1. These given values are $x_f = 0.209$, $\alpha^* = 10$, $p_h = 190$ cm Hg, $p_l = 19$ cm Hg, $q_f = 1 \times 10^6$ cm^3(STP)/s, $P'_A = 500 \times 10^{-10}$ cm^3(STP)·cm/(s·cm^2·cm Hg), and $t = 2.54 \times 10^{-3}$cm. Do as follows using the cross-flow model.

(a) Calculate y_p, x_o, and A_m for $\theta = 0.40$.

(b) Calculate y_p and x_o for $\theta = 0$.

Ans. (a) $y_p = 0.452$, $x_o = 0.0303$, $A_m = 6.94 \times 10^8$ cm^2 (S6)
(b) $y_p = 0.655$, $x_o = 0.209$

13.7-1. *Equations for Countercurrent-Flow Model.* For the derivation of the equations for countercurrent flow in a gas separation using a membrane do the following.

(a) Obtain Eq. (13.7-5) from Eq. (13.7-4).

(b) Show that Eq. (13.7-10) is valid.

(c) Obtain Eq. (13.7-12).

13.7-2. *Design Using Countercurrent-Flow Model for Membrane.* Use the same conditions as given in Example 13.6-1 for the separation of an air stream. The given values are $x_f = 0.209$, $\alpha^* = 10$, $p_h = 190$ cm Hg, $p_l = 19$ cm Hg, $q_f = 1 \times 10^6$ cm^3(STP)/s, $P'_A = 500 \times 10^{-10}$ cm^3(STP)·cm/(s·cm^2·cm Hg), and $t = 2.54 \times 10^{-3}$ cm. Using the countercurrent-flow model, calculate y_p, x_o, and A_m for $\theta = 0.40$. (Note that this problem involves a trial-and-error procedure along with the numerical solution of two differential equations.)

13.9-1. *Osmotic Pressure of Salt and Sugar Solutions.* Calculate the osmotic pressure of the following solutions at 25°C and compare with the experimental values.

(a) Solution of 0.50 g mol NaCl/kg H_2O. (See Table 13.9-1 for the experimental value.)

(b) Solution of 1.0 g sucrose/kg H_2O. (Experimental value = 0.0714 atm.)

(c) Solution of 1.0 g $MgCl_2$/kg H_2O. (Experimental value = 0.660 atm.)

Ans. (a) $\pi = 24.39$ atm, (b) $\pi = 0.0713$ atm,
(c) $\pi = 0.768$ atm

13.9-2. *Determination of Permeability Constants for Reverse Osmosis.* A cellulose-acetate membrane with an area of 4.0×10^{-3} m^2 is used at 25°C to determine the permeability constants for reverse osmosis of a feed salt solution containing 12.0 kg $NaCl/m^3$ ($\rho = 1005.5$ kg/m^3). The product solution has a concentration of 0.468 kg $NaCl/m^3$ ($\rho = 997.3$ kg/m^3). The measured product flow rate is 3.84×10^{-8} m^3/s and the pressure difference used is 56.0 atm. Calculate the permeability constants and the solute rejection R.

Ans. $A_w = 2.013 \times 10^{-4}$ kg solvent/s·m^2·atm, $R = 0.961$

13.9-3. *Performance of a Laboratory Reverse-Osmosis Unit.* A feed solution at 25°C contains 3500 mg NaCl/L ($\rho = 999.5$ kg/m^3). The permeability constant $A_w = 3.50 \times 10^{-4}$ kg solvent/s·m^2·atm and $A_s = 2.50 \times 10^{-7}$ m/s. Using a $\Delta P = 35.50$ atm, calculate the fluxes, solute rejection R, and the product solution concentration in mg NaCl/L. Repeat, but using a feed solution of 3500 mg BaCl$_2$/L. Use the same value of A_w but $A_s = 1.00 \times 10^{-7}$ m/s (A1).

13.10-1. *Effect of Pressure on Performance of Reverse-Osmosis Unit.* Using the same conditions and permeability constants as in Example 13.10-1, calculate the fluxes, solute rejection R, and the product concentration c_2 for ΔP pressures of 17.20, 27.20, and 37.20 atm. (*Note:* The values for 27.20 atm have already been calculated.) Plot the fluxes, R, and c_2 versus the pressure.

13.10-2. *Effect of Concentration Polarization on Reverse Osmosis.* Repeat Example 13.10-1 but use a concentration polarization of 1.5. (*Note:* The flux equations and the solute rejection R should be calculated using this new value of c_1.)

Ans. $N_w = 1.170 \times 10^{-2}$ kg solvent/s·m^2, $c_2 = 0.1361$ kg NaCl/m^3

13.10-3. *Performance of a Complete-Mixing Model for Reverse Osmosis.* Use the same feed conditions and pressures given in Example 13.10-1. Assume that the cut or fraction recovered of the solvent water will be 0.10 instead of the very low water recovery assumed in Example 13.10-1. Hence, the concentration of the entering feed solution and the exit feed will not be the same. The flow rate q_2 of the permeate water solution is 100 gal/h. Calculate c_1 and c_2 in kg NaCl/m^3 and the membrane area in m^2.

Ans. $c_1 = 2.767$ kg/m^3, $c_2 = 0.0973$ kg/m^3, area = 8.68 m^2

13.11-1. *Flux for Ultrafiltration.* A solution containing 0.9 wt % protein is to undergo ultrafiltration using a pressure difference of 5 psi. The membrane permeability is $A_w = 1.37 \times 10^{-2}$ kg/s·m^2·atm. Assuming no effects of polarization, predict the flux in kg/s·m^2 and in units of gal/ft^2·day which are often used in industry.

Ans. 9.88 gal/ft^2·day

13.11-2. *Time for Ultrafiltration Using Recirculation.* It is desired to use ultrafiltration for 800 kg of a solution containing 0.05 wt % of a protein to obtain a solution of 1.10 wt %. The feed is recirculated by the membrane with a surface area of 9.90 m^2. The permeability of the membrane is $A_w = 2.50 \times 10^{-2}$ kg/s·m^2·atm. Neglecting the effects of concentration polarization, if any, calculate the final amount of solution and the time to perform this using a pressure difference of 0.50 atm.

REFERENCES

(A1)   AGRAWAL, J. P., and SOURIRAJAN, S. *Ind. Eng. Chem.*, **69**(11), 62 (1969).

(B1)   BLAISDELL, C. T., and KAMMERMEYER, K. *Chem. Eng. Sci.*, **28**, 1249 (1973).

(B2)   BABB, A. L., MAURER, C. J., FRY, D. L., POPOVICH, R. P., and MCKEE, R. E. *Chem. Eng. Progr. Symp.*, **64**(84), 59 (1968).

(B3)   BERRY, R. I. *Chem. Eng.*, **88**(July 13), 63 (1981)

(C1)   CLARK, W. E. *Science,* **138**, 148 (1962).

(H1)   HWANG, S. T., and KAMMERMEYER, K. *Membranes in Separations.* New York: John Wiley & Sons, Inc., 1975.

(K1)   KIMURA, S., and SOURIRAJAN, S. *A.I.Ch.E. J.,* **13**, 497 (1967).

(K2)   KAUP, E. C. *Chem. Eng.,* **80**(Apr. 2), 46 (1973).

(K3)   KURZ, J. E., and NARAYAN, R. S., "New Developments and Applications in Membrane Technology."

(L1)   LACEY, R. E. *Chem. Eng.,* **79**(Sept. 4), 57 (1972).

(M1)   MCCABE, W. L. *Ind. Eng. Chem.,* **21**, 112 (1929).

(M2)   MICHAELS, A. S. *Chem. Eng. Progr.,* **64**(12), 31 (1968).

(M3)   MERTEN, U. (ED.). *Desalination by Reverse Osmosis.* Cambridge, Mass.: The MIT Press, 1966.

(M4)   MAZUR, W. H., AND MARTIN, C. C. *Chem. Eng. Progr.,* **78**(10), 38 (1982).

(P1)   PERRY, R. H., and GREEN, D. *Perry's Chemical Engineers' Handbook,* 6th ed. New York: McGraw-Hill Book Company, 1984.

(P2)   PAN, C. Y. *A.I.Ch.E. J.,* **29**, 545 (1983).

(P3)   PERMASEP PERMEATORS, *E. I. duPont Tech. Bull.,* 401, 403, 405 (1972).

(P4)   PAN, C. Y., and HABGOOD, H. W. *Ind. Eng. Chem. Fund.,* **13**, 323 (1974).

(P5)   PAN, C. Y., and HABGOOD, H. W. *Can. J. Chem. Eng.,* **56** 197, 210 (1978).

(P6)   PORTER, M. C. (ed.). *Handbook of Industrial Membrane Technology.* Park Ridge, N.J.: Noyes Publications, 1990.

(R1)   ROUSSEAU, R. W. (ed.). *Handbook of Separation Process Technology.* New York: John Wiley & Sons, Inc., 1987.

(S1)   STERN, S. A., SINCLAIR, T. F., GAREIS, P. J., VAHLDIECK, N. P., and MOHR, P. H. *Ind. Eng. Chem.,* **57**, 49 (1965).

(S2)   STANNETT, V. T., KOROS, W. J., PAUL, D. R., LONSDALE, H. K., and BAKER, R. W. *Adv. Polym. Sci.,* **32**, 69 (1979).

(S3)   STOUGHTEN, R. W., and LIETZKE, M. H. JR. *Chem. Eng. Data,* **10**, 254 (1965).

(S4)   SCHROEDER, E. D. *Water and Wastewater Treatment.* New York: McGraw-Hill Book Company, 1977.

(S5)   SOURIRAJAN, S. *Reverse Osmosis.* New York: Academic Press, Inc., 1970.

(S6)   STERN, S. A., and WALAWENDER, W. P. *Sep. Sci.,* **4**, 129 (1969).

(S7)   STOOKEY, D. J., PATTON, C. J., and MALCOLM, G. L. *Chem. Eng. Progr.,* **82**(11), 36 (1986).

(W1)   WEBER, W. J., JR. *Physicochemical Processes for Water Quality Control.* New York: Wiley-Interscience, 1972.

(W2)   WARD, W. J., BROWAL, W. R., and SALEMME, R. M. *J. Membr. Sci.,* **1**, 99 (1976).

(W3)   WELLER, S., and STEINER, W. A. *Chem. Eng. Progr.,* **46**, 585 (1950).

(W4)   WELLER, S., and STEINER, W. A. *J. Appl. Phys.,* **21**, 279 (1950).

(W5)   WALAWENDER, W. P., AND STERN, S. A. *Sep. Sci.,* **7**, 553 (1972).

CHAPTER 14

Mechanical–Physical
Separation Processes

14.1 INTRODUCTION AND CLASSIFICATION OF MECHANICAL–PHYSICAL SEPARATION PROCESSES

14.1A Introduction

In Chapters 10 and 11 gas–liquid and vapor–liquid separation processes were considered. The separation processes depended on molecules diffusing or vaporizing from one distinct phase to another phase. In Chapter 12 liquid–liquid separation processes were discussed. The two liquid phases are quite different chemically, which leads to a separation on a molecular scale according to physical–chemical properties. Also, in Chapter 12 we considered liquid–solid leaching and adsorption separation processes. Again differences in the physical–chemical properties of the molecules lead to separation on a molecular scale. In Chapter 13 we discussed membrane separation processes where the separation also depends on physical–chemical properties.

All the separation processes considered so far have been based upon physical–chemical differences in the molecules themselves and on mass transfer of the molecules. In this way individual molecules were separated into two phases because of these molecular differences. In the present chapter a group of separation processes will be considered where the separation is not accomplished on a molecular scale nor is it due to the differences among the various molecules. The separation will be accomplished using mechanical–physical forces and not molecular or chemical forces and diffusion. These mechanical–physical forces will be acting on particles, liquids, or mixtures of particles and liquids themselves and not necessarily on the individual molecules.

The mechanical–physical forces include gravitational and centrifugal, actual mechanical, and kinetic forces arising from flow. Particles and/or fluid streams are separated because of the different effects produced on them by these forces.

14.1B Classification of Mechanical–Physical Separation Processes

These mechanical–physical separation processes are considered under the following classifications.

1. Filtration. The general problem of the separation of solid particles from liquids can be solved by using a wide variety of methods, depending on the type of solids, the proportion of solid to liquid in the mixture, viscosity of the solution, and other factors. In filtration a pressure difference is set up and causes the fluid to flow through small holes of a screen or cloth which block the passage of the large solid particles, which, in turn, build up on the cloth as a porous cake.

2. Settling and sedimentation. In settling and sedimentation the particles are separated from the fluid by gravitational forces acting on the various size and density particles.

3. Centrifugal settling and sedimentation. In centrifugal separations the particles are separated from the fluid by centrifugal forces acting on the various size and density particles. Two general types of separation processes are used. In the first type of process, centrifugal settling or sedimentation occurs.

4. Centrifugal filtration. In this second type of centrifugal separation process, centrifugal filtration occurs which is similar to ordinary filtration where a bed or cake of solids builds up on a screen, but centrifugal force is used to cause the flow instead of a pressure difference.

5. Mechanical size reduction and separation. In mechanical size reduction the solid particles are broken mechanically into smaller particles and separated according to size.

14.2 FILTRATION IN SOLID–LIQUID SEPARATION

14.2A Introduction

In filtration, suspended solid particles in a fluid of liquid or gas are physically or mechanically removed by using a porous medium that retains the particles as a separate phase or cake and passes the clear filtrate. Commerical filtrations cover a very wide range of applications. The fluid can be a gas or a liquid. The suspended solid particles can be very fine (in the micrometer range) or much larger, very rigid or plastic particles, spherical or very irregular in shape, aggregates of particles or individual particles. The valuable product may be the clear filtrate from the filtration or the solid cake. In some cases complete removal of the solid particles is required and in other cases only partial removal.

The feed or slurry solution may carry a heavy load of solid particles or a very small amount. When the concentration is very low, the filters can operate for very long periods of time before the filter needs cleaning. Because of the wide diversity of filtration problems, a multitude of types of filters have been developed.

Industrial filtration equipment differs from laboratory filtration equipment only in the amount of material handled and in the necessity for low-cost operation. A typical laboratory filtration apparatus is shown in Fig. 14.2-1, which is similar to a Büchner funnel. The liquid is caused to flow through the filter cloth or paper by a vacuum on the exit end. The slurry consists of the liquid and the suspended particles. The passage of the particles is blocked by the small openings in the pores of the filter cloth. A support with relatively large holes is used to hold the filter cloth. The solid particles build up in the form of a porous filter cake as the filtration proceeds. This cake itself also acts as a filter for the suspended particles. As the cake builds up, resistance to flow also increases.

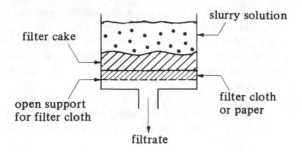

FIGURE 14.2-1. *Simple laboratory filtration apparatus.*

In the present section 14.2 the ordinary type of filtration will be considered where a pressure difference is used to force the liquid through the filter cloth and the filter cake that builds up.

In Section 14.4E centrifugal filtration will be discussed, where centrifugal force is used instead of a pressure difference. In many filtration applications, ordinary filters and centrifugal filters are often competitive and either type can be used.

14.2B Types of Filtration Equipment

1. Classification of filters. There are a number of ways to classify types of filtration equipment, and it is not possible to make a simple classification that includes all types of filters. In one classification filters are classified according to whether the filter cake is the desired product or whether the clarified filtrate or outlet liquid is desired. In either case the slurry can have a relatively large percentage of solids so that a cake is formed, or have just a trace of suspended particles.

Filters can be classified by operating cycle. Filters can be operated as batch, where the cake is removed after a run, or continuous, where the cake is continuously removed. In another classification, filters can be of the gravity type, where the liquid simply flows by a hydrostatic head, or pressure or vacuum can be used to increase the flow rates. An important method of classification depends upon the mechanical arrangement of the filter media. The filter cloth can be in a series arrangement as flat plates in an enclosure, as individual leaves dipped in the slurry, or on rotating-type rolls in the slurry. In the following sections only the most important types of filters will be described. For more details, see references (B1, P1).

2. Bed filters. The simplest type of filter is the bed filter shown schematically in Fig. 14.2-2. This type is useful mainly in cases where relatively small amounts of solids are to be removed from large amounts of water in clarifying the liquid. Often the bottom layer is composed of coarse pieces of gravel resting on a perforated or slotted plate. Above the gravel is fine sand, which acts as the actual filter medium. Water is introduced at the top onto a baffle which spreads the water out. The clarified liquid is drawn out at the bottom.

The filtration continues until the precipitate of filtered particles has clogged the sand so that the flow rate drops. Then the flow is stopped and water introduced in the reverse direction so that it flows upward, backwashing the bed and carrying the precipitated solid away. This apparatus can only be used on precipitates that do not adhere strongly to the sand and can be easily removed by backwashing. Open tank filters are used in filtering municipal water supplies.

3. Plate-and-frame filter presses.   One of the important types of filters is the plate-and-frame filter press, which is shown diagrammatically in Fig. 14.2-3a. These filters consist of plates and frames assembled alternatively with a filter cloth over each side of the plates. The plates have channels cut in them so that clear filtrate liquid can drain down along each plate. The feed slurry is pumped into the press and flows through the duct into each of the open frames so that slurry fills the frames. The filtrate flows through the filter cloth and the solids build up as a cake on the frame side of the cloth. The filtrate flows between the filter cloth and the face of the plate through the channels to the outlet.

The filtration proceeds until the frames are completely filled with solids. In Fig. 14.2-3a all the discharge outlets go to a common header. In many cases the filter press will have a separate discharge to the open for each frame. Then visual inspection can be made to see if the filtrate is running clear. If one is running cloudy because of a break in the filter cloth or other factors, it can be shut off separately. When the frames are completely full, the frames and plates are separated and the cake removed. Then the filter is reassembled and the cycle is repeated.

If the cake is to be washed, the cake is left in the plates and through washing is performed, as shown in Fig. 14.2-3b. In this press a separate channel is provided for the wash water inlet. The wash water enters the inlet, which has ports opening behind the filter cloths at every other plate of the filter press. The wash water then flows through the filter cloth, through the entire cake (not half the cake as in filtration), through the filter cloth at the other side of the frames, and out the discharge channel. It should be noted that there are two kinds of plates in Fig. 14.2-3b: those having ducts to admit wash water behind the filter cloth, alternating with those without such ducts.

The plate-and-frame presses suffer from the disadvantages common to batch processes. The cost of labor for removing the cakes and reassembling plus the cost of fixed charges for downtime can be an appreciable part of the total operating cost. Some newer types of plate-and-frame presses have duplicate sets of frames mounted on a rotating shaft. Half of the frames are in use while the others are being cleaned, saving downtime and labor costs. Other advances in automation have been applied to these types of filters.

Filter presses are used in batch processes but cannot be employed for high-throughput processes. They are simple to operate, very versatile and flexible in operation, and can be used at high pressures, when necessary, if viscous solutions are being used or the filter cake has a high resistance.

4. Leaf filters.   The filter press is useful for many purposes but is not economical for handling large quantities of sludge or for efficient washing with a small amount of wash

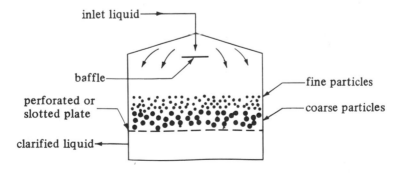

FIGURE 14.2-2.   *Bed filter of solid particles.*

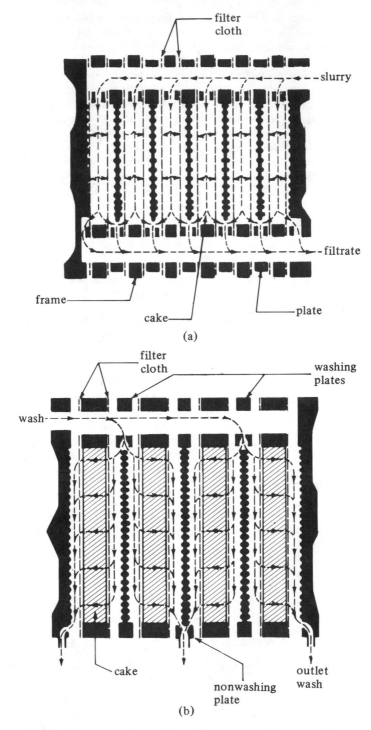

FIGURE 14.2-3. *Diagrams of plate-and-frame filter presses: (a) filtration of slurry with closed delivery, (b) through washing in a press with open delivery.*

water. The wash water often channels in the cake and large volumes of wash water may be needed. The leaf filter shown in Fig. 14.2-4 was developed for larger volumes of slurry and more efficient washing. Each leaf is a hollow wire framework covered by a sack of filter cloth.

A number of these leaves are hung in parallel in a closed tank. The slurry enters the tank and is forced under pressure through the filter cloth, where the cake deposits on the outside of the leaf. The filtrate flows inside the hollow framework and out a header. The wash liquid follows the same path as the slurry. Hence, the washing is more efficient than the through washing in plate-and-frame filter presses. To remove the cake, the shell is opened. Sometimes air is blown in the reverse direction into the leaves to help in dislodging the cake. If the solids are not wanted, water jets can be used to simply wash away the cakes without opening the filter.

Leaf filters also suffer from the disadvantage of batch operation. They can be automated for the filtering, washing, and cleaning cycle. However, they are still cyclical and are used for batch processes and relatively modest throughput processes.

5. *Continuous rotary filters.* The plate-and-frame filters suffer from the disadvantages common to all batch processes and cannot be used for large capacity processes. A number of continuous-type filters are available as discussed below.

(a). *Continuous rotary vacuum-drum filter.* This filter shown in Fig. 14.2-5 filters, washes, and discharges the cake in a continuous repeating sequence. The drum is covered with a suitable filtering medium. The drum rotates and an automatic valve in the center serves to activate the filtering, drying, washing, and cake discharge functions in the cycle. The filtrate leaves through the axle of the filter.

The automatic valve provides separate outlets for the filtrate and the wash liquid. Also, if needed, a connection for compressed air blowback just before discharge can be used to help in cake removal by the knife scraper. The maximum pressure differential for the vacuum filter is only 1 atm. Hence, this type is not suitable for viscous liquids or for

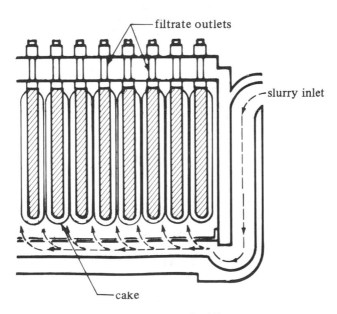

FIGURE 14.2-4. *Leaf filter.*

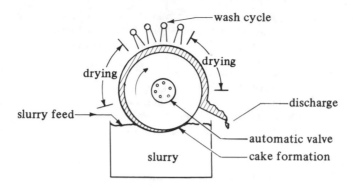

FIGURE 14.2-5. *Schematic of continuous rotary-drum filter.*

liquids that must be enclosed. If the drum is enclosed in a shell, pressures above atmospheric can be used. However, the cost of a pressure type is about two times that of a vacuum-type rotary drum filter (P2).

Modern, high-capacity processes use continuous filters. The important advantages are that the filters are continuous and automatic and labor costs are relatively low. However, the capital cost is relatively high.

(*b*). *Continuous rotary disk filter.* This filter consists of concentric vertical disks mounted on a horizontal rotating shaft. The filter operates on the same principle as the vacuum rotary drum filter. Each disk is hollow and covered with a filter cloth and is partly submerged in the slurry. The cake is washed, dried, and scraped off when the disk is in the upper half of its rotation. Washing is less efficient than with a rotating drum type.

(*c*). *Continuous rotary horizontal filter.* This type is a vacuum filter with the rotating annular filtering surface divided into sectors. As the horizontal filter rotates it successively receives slurry, is washed, dried, and the cake is scraped off. The washing efficiency is better than with the rotary disk filter. This filter is widely used in ore extraction processes, pulp washing, and other large-capacity processes.

14.2C Filter Media and Filter Aids

1. Filter media. The filter medium for industrial filtration must fulfill a number of requirements. First and foremost, it must remove the solids to be filtered from the slurry and give a clear filtrate. Also, the pores should not become plugged so that the rate of filtration becomes too slow. The filter medium must allow the filter cake to be removed easily and cleanly. Obviously, it must have sufficient strength to not tear and must be chemically resistant to the solutions used.

Some widely used filter media are twill or duckweave heavy cloth, other types of woven heavy cloth, woolen cloth, glass cloth, paper, felted pads of cellulose, metal cloth, nylon cloth, Dacron cloth, and other synthetic cloths. The ragged fibers of natural materials are more effective in removing fine particles than the smooth plastic or metal fibers. Sometimes the filtrate may come through somewhat cloudy at first before the first layers of particles, which help filter the subsequent slurry, are deposited. This filtrate can be recycled for refiltration.

2. Filter aids. Certain filter aids may be used to aid filtration. These are often incompressible diatomaceous earth or kieselguhr, which is composed primarily of silica. Also used are wood cellulose, asbestos, and other inert porous solids.

These filter aids can be used in a number of ways. They can be used as a precoat before the slurry is filtered. This will prevent gelatinous-type solids from plugging the filter medium and also give a clearer filtrate. They can also be added to the slurry before filtration. This increases the pororsity of the cake and reduces resistance of the cake during filtration. In a rotary filter the filter aid can be applied as a precoat, and subsequently thin slices of this layer are sliced off with the cake.

The use of filter aids is usually limited to cases where the cake is discarded or to cases where the precipitate can be separated chemically from the filter aid.

14.2D Basic Theory of Filtration

1. Pressure drop of fluid through filter cake. Figure 14.2-6 is a section through a filter cake and filter medium at a definite time t s from the start of the flow of filtrate. At this time the thickness of the cake is L m (ft). The filter cross-sectional area is A m^2 (ft^2), and the linear velocity of the filtrate in the L direction is v m/s (ft/s) based on the filter area of A m^2.

The flow of the filtrate through the packed bed of cake can be described by an equation similar to Poiseuille's law, assuming laminar flow occurs in the filter channels. Equation (2.10-2) gives Poiseuille's equation for laminar flow in a straight tube, which can be written

$$-\frac{\Delta p}{L} = \frac{32\mu v}{D^2} \qquad \text{(SI)}$$

$$-\frac{\Delta p}{L} = \frac{32\mu v}{g_c D^2} \qquad \text{(English)}$$

(14.2-1)

where Δp is pressure drop in N/m^2 (lb$_f$/ft^2), v is open-tube velocity in m/s (ft/s), D is diameter in m (ft), L is length in m (ft), μ is viscosity in Pa·s or kg/m·s (lb$_m$/ft·s), and g_c is 32.174 lb$_m$·ft/lb$_f$·s^2.

For laminar flow in a packed bed of particles, the *Carman–Kozeny relation* is similar to Eq. (14.2-1) and to the Blake–Kozeny equation (3.1-17) and has been shown to apply to filtration.

$$-\frac{\Delta p_c}{L} = \frac{k_1 \mu v (1-\varepsilon)^2 S_0^2}{\varepsilon^3}$$

(14.2-2)

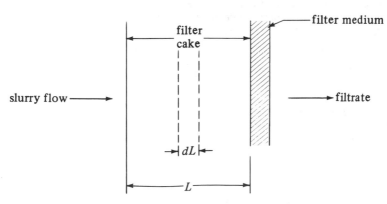

FIGURE 14.2-6. *Section through a filter cake.*

where k_1 is a constant and equals 4.17 for random particles of definite size and shape, μ is viscosity of filtrate in Pa $\cdot$ s (lb$_m$/ft $\cdot$ s), v is linear velocity based on filter area in m/s (ft/s), ε is void fraction or porosity of cake, L is thickness of cake in m (ft), S_0 is specific surface area of particle in m^2 (ft^2) of particle area per m^3 (ft^3) volume of solid particle, and Δp_c is pressure drop in the cake in N/m^2 (lb$_f$/ft^2). For English units, the right-hand side of Eq. (14.2-2) is divided by g_c. The linear velocity is based on the empty cross-sectional area and is

$$v = \frac{dV/dt}{A} \tag{14.2-3}$$

where A is filter area in m^2 (ft^2) and V is total m^3 (ft^3) of filtrate collected up to time t s. The cake thickness L may be related to the volume of filtrate V by a material balance. If c_S is kg solids/m^3 (lb$_m$/ft^3) of filtrate, a material balance gives

$$LA(1 - \varepsilon)\rho_p = c_S(V + \varepsilon LA) \tag{14.2-4}$$

where ρ_p is density of solid particles in the cake in kg/m^3 (lb$_m$/ft^3) solid. The final term of Eq. (14.2-4) is the volume of filtrate held in the cake. This is usually small and will be neglected.

Substituting Eq. (14.2-3) into (14.2-2) and using Eq. (14.2-4) to eliminate L, we obtain the final equation as

$$\frac{dV}{A\,dt} = \frac{-\Delta p_c}{\dfrac{k_1(1 - \varepsilon)S_0^2}{\rho_p \varepsilon^3}\dfrac{\mu c_S V}{A}} = \frac{-\Delta p_c}{\alpha\dfrac{\mu c_S V}{A}} \tag{14.2-5}$$

where α is the specific cake resistance in m/kg (ft/lb$_m$), defined as

$$\alpha = \frac{k_1(1 - \varepsilon)S_0^2}{\rho_p \varepsilon^3} \tag{14.2-6}$$

For the filter medium resistance, we can write, by analogy with Eq. (14.2-5),

$$\frac{dV}{A\,dt} = \frac{-\Delta p_f}{\mu R_m} \tag{14.2-7}$$

where R_m is the resistance of the filter medium to filtrate flow in m^{-1} (ft^{-1}) and Δp_f is the pressure drop. When R_m is treated as an empirical constant, it includes the resistance to flow of the piping leads to and from the filter and the filter medium resistance.

Since the resistances of the cake and the filter medium are in series, Eqs. (14.2-5) and (14.2-7) can be combined, and become

$$\frac{dV}{A\,dt} = \frac{-\Delta p}{\mu\left(\dfrac{\alpha c_s V}{A} + R_m\right)} \tag{14.2-8}$$

where $\Delta p = \Delta p_c + \Delta p_f$. Sometimes Eq. (14.2-8) is modified as follows:

$$\frac{dV}{A\,dt} = \frac{-\Delta p}{\dfrac{\mu \alpha c_S}{A}(V + V_e)} \tag{14.2-9}$$

where V_e is a volume of filtrate necessary to build up a fictitious filter cake whose resistance is equal to R_m.

The volume of filtrate V can also be related to W, the kg of accumulated dry cake solids, as follows:

$$W = c_S V = \frac{\rho c_x}{1 - mc_x} V \tag{14.2-10}$$

where c_x is mass fraction of solids in the slurry, m is mass ratio of wet cake to dry cake, and ρ is density of filtrate in kg/m^3 (lb_m/ft^3).

2. Specific cake resistance. From Eq. (14.2-6) we see that the specific cake resistance is a function of void fraction ε and S_0. It also is a function of pressure, since pressure can affect ε. By conducting constant-pressure experiments at various pressure drops, the variation of α with Δp can be found.

Alternatively, compression–permeability experiments can be performed. A filter cake at a low pressure drop and atm pressure is built up by gravity filtering in a cylinder with a porous bottom. A piston is loaded on top and the cake compressed to a given pressure. Then filtrate is fed to the cake and α is determined by a differential form of Eq. (14.2-9). This is then repeated for other compression pressures (G1).

If α is independent of $-\Delta p$, the sludge is incompressible. Usually, α increases with $-\Delta p$, since most cakes are somewhat compressible. An empirical equation often used is

$$\alpha = \alpha_0(-\Delta p)^s \tag{14.2-11}$$

where α_0 and s are empirical constants. The compressibility constant s is zero for incompressible sludges or cakes. The constant s usually falls between 0.1 to 0.8. Sometimes the following is used.

$$\alpha = \alpha_0'[1 + \beta(-\Delta p)^{s'}] \tag{14.2-12}$$

where α_0', β, and s' are empirical constants. Experimental data for various sludges are given by Grace (G1).

The data obtained from filtration experiments often do not have a high degree of reproducibility. The state of agglomeration of the particles in the slurry can vary and have an effect on the specific cake resistance.

14.2E Filtration Equations for Constant-Pressure Filtration

1. Basic equations for filtration rate in batch process. Often a filtration is done for conditions of constant pressure. Equation (14.2-8) can be inverted and rearranged to give

$$\frac{dt}{dV} = \frac{\mu \alpha c_S}{A^2(-\Delta p)} V + \frac{\mu}{A(-\Delta p)} R_m = K_p V + B \tag{14.2-13}$$

where K_p is in s/m^6 (s/ft^6) and B in s/m^3 (s/ft^3).

$$K_p = \frac{\mu \alpha c_S}{A^2(-\Delta p)} \quad \text{(SI)}$$

$$K_p = \frac{\mu \alpha c_S}{A^2(-\Delta p)g_c} \quad \text{(English)} \tag{14.2-14}$$

$$B = \frac{\mu R_m}{A(-\Delta p)} \quad \text{(SI)}$$

$$B = \frac{\mu R_m}{A(-\Delta p)g_c} \quad \text{(English)} \tag{14.2-15}$$

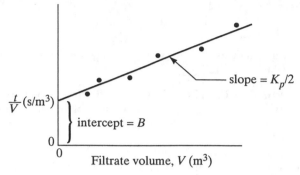

FIGURE 14.2-7.   *Determination of constants in a constant-pressure filtration run.*

For constant pressure, constant α, and incompressible cake, V and t are the only variables in Eq. (14.2-13). Integrating to obtain the time of filtration in t s,

$$\int_0^t dt = \int_0^V (K_p V + B)dV \tag{14.2-16}$$

$$t = \frac{K_p}{2} V^2 + BV \tag{14.2-17}$$

Dividing by V

$$\frac{t}{V} = \frac{K_p V}{2} + B \tag{14.2-18}$$

where V is total volume of filtrate in m^3 (ft^3) collected to t s.

 To evaluate Eq. (14.2-17) it is necessary to know α and R_m. This can be done by using Eq. (14.2-18). Data of V collected at different times t are obtained. Then the experimental data are plotted as t/V versus V as in Fig. 14.2-7. Often, the first point on the graph does not fall on the line and is omitted. The slope of the line is $K_p/2$ and the intercept B. Then, using Eqs. (14.2-14) and (14.2-15), values of α and R_m can be determined.

EXAMPLE 14.2-1.   *Evaluation of Filtration Constants for Constant-Pressure Filtration*

Data for the laboratory filtration of $CaCO_3$ slurry in water at 298.2 K (25°C) are reported as follows at a constant pressure $(-\Delta p)$ of 338 kN/m² (7060 lb$_f$/ft²) (R1, R2, M1). The filter area of the plate-and-frame press was $A = 0.0439$ m² (0.473 ft²) and the slurry concentration was $c_S = 23.47$ kg/m³ (1.465 lb$_m$/ft³). Calculate the constants α and R_m from the experimental data given, where t is time in s and V is filtrate volume collected in m³.

t	V	t	V	t	V
4.4	0.498×10^{-3}	34.7	2.498×10^{-3}	73.6	4.004×10^{-3}
9.5	1.000×10^{-3}	46.1	3.002×10^{-3}	89.4	4.502×10^{-3}
16.3	1.501×10^{-3}	59.0	3.506×10^{-3}	107.3	5.009×10^{-3}
24.6	2.000×10^{-3}				

Solution: First, the data are calculated as t/V and tabulated in Table 14.2-1. The data are plotted as t/V versus V in Fig. 14.2-8 and the intercept is determined as $B = 6400$ s/m^3 (181 s/ft^3) and the slope as $K_p/2 = 3.00 \times 10^6$ s/m^6. Hence, $K_p = 6.00 \times 10^6$ s/m^6 (4820 s/ft^6).

At 298.2 K the viscosity of water is 8.937×10^{-4} Pa·s $= 8.937 \times 10^{-4}$ kg/m·s. Substituting known values into Eq. (14.2-14) and solving

$$K_p = 6.00 \times 10^6 = \frac{\mu \alpha c_s}{A^2(-\Delta p)} = \frac{(8.937 \times 10^{-4})(\alpha)(23.47)}{(0.0439)^2(338 \times 10^3)}$$

$$\alpha = 1.863 \times 10^{11} \text{ m/kg } (2.77 \times 10^{11} \text{ ft/lb}_m)$$

Substituting into Eq. (14.2-15) and solving,

$$B = 6400 = \frac{\mu R_m}{A(-\Delta p)} = \frac{(8.937 \times 10^{-4})(R_m)}{0.0439(338 \times 10^3)}$$

$$R_m = 10.63 \times 10^{10} \text{ m}^{-1} \ (3.24 \times 10^{10} \text{ ft}^{-1})$$

EXAMPLE 14.2-2. Time Required to Perform a Filtration
The same slurry used in Example 14.2-1 is to be filtered in a plate-and-frame press having 20 frames and 0.873 m^2 (9.4 ft^2) area per frame. The same pressure will be used in constant-pressure filtration. Assuming the same filter cake properties and filter cloth, calculate the time to recover 3.37 m^3 (119 ft^3) filtrate.

Solution: In Example 14.2-1, the area $A = 0.0439$ m^2, $K_p = 6.00 \times 10^6$ s/m^6, and $B = 6400$ s/m^3. Since the α and R_m will be the same as before, K_p can be corrected. From Eq. (14.2-14), K_p is proportional to $1/A^2$. The

TABLE 14.2-1. *Values of t/V for Example 14.2-1 (t = s, V = m^3)*

t	$V \times 10^3$	$(t/V) \times 10^{-3}$
0	0	
4.4	0.498	8.84
9.5	1.000	9.50
16.3	1.501	10.86
24.6	2.000	12.30
34.7	2.498	13.89
46.1	3.002	15.36
59.0	3.506	16.83
73.6	4.004	18.38
89.4	4.502	19.86
107.3	5.009	21.42

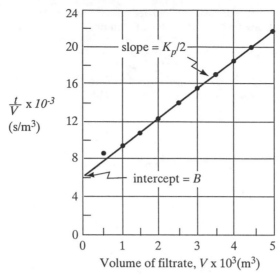

FIGURE 14.2-8. *Determination of constants for Example 14.2-1.*

new area is $A = 0.873(20) = 17.46 \text{ m}^2$ (188 ft^2). The new K_p is

$$K_p = 6.00 \times 10^6 (0.0439/17.46)^2 = 37.93 \text{ s/m}^6 \ (0.03042 \text{ s/ft}^6)$$

The new B is proportional to $1/A$ from Eq. (14.2-15).

$$B = (6400) \frac{0.0439}{17.46} = 16.10 \text{ s/m}^3 \ (0.456 \text{ s/ft}^3)$$

Substituting into Eq. (14.2-17),

$$t = \frac{K_p}{2} V^2 + BV = \frac{37.93}{2} (3.37)^2 + (16.10)(3.37) = 269.7 \text{ s}$$

Using English units,

$$t = \frac{K_p}{2} V^2 + BV = \frac{0.03042}{2} (119)^2 + (0.456)(119) = 269.7 \text{ s}$$

2. *Equations for washing of filter cakes and total cycle time.*   The washing of a cake after the filtration cycle takes place by displacement of the filtrate and by diffusion. The amount of wash liquid should be sufficient to give the desired washing effect. To calculate washing rates, it is assumed that the conditions during washing are the same as those that existed at the end of the filtration. It is assumed that the cake structure is not affected when wash liquid replaces the slurry liquid in the cake.

In filters where the wash liquid follows the flow path, similar to that during filtration as in leaf filters, the final filtering rate gives the predicted washing rate. For constant-pressure filtration using the same pressure in washing as in filtering, the final filtering rate is the reciprocal of Eq. (14.2-13).

$$\left(\frac{dV}{dt} \right)_f = \frac{1}{K_p V_f + B} \tag{14.2-19}$$

where $(dV/dt)_f$ = rate of washing in m^3/s (ft^3/s) and V_f is the total volume of filtrate for the entire period at the end of filtration in m^3 (ft^3).

For plate-and-frame filter presses, the wash liquid travels through a cake twice as thick and an area only half as large as in filtering, so the predicted washing rate is $\frac{1}{4}$ of the final filtration rate.

$$\left(\frac{dV}{dt}\right)_f = \frac{1}{4}\frac{1}{K_p V_f + B} \tag{14.2-20}$$

In actual experience the washing rate may be less than predicted because of cake consolidation, channeling, and formation of cracks. Washing rates in a small plate-and-frame filter were found to be from 70 to 92% of that predicted (M1).

After washing is completed, additional time is needed to remove the cake, clean the filter, and reassemble the filter. The total filter cycle time is the sum of the filtration time, plus the washing time, plus the cleaning time.

EXAMPLE 14.2-3. Rate of Washing and Total Filter Cycle Time
At the end of the filtration cycle in Example 14.2-2, a total filtrate volume of 3.37 m³ is collected in a total time of 269.7 s. The cake is to be washed by through washing in the plate-and-frame press using a volume of wash water equal to 10% of the filtrate volume. Calculate the time of washing and the total filter cycle time if cleaning the filter takes 20 min.

Solution: For this filter, Eq. (14.2-20) holds. Substituting $K_p = 37.93$ s/m⁶, $B = 16.10$ s/m³, and $V_f = 3.37$ m³, the washing rate is as follows:

$$\left(\frac{dV}{dt}\right)_f = \frac{1}{4}\frac{1}{(37.93)(3.37) + 16.10} = 1.737 \times 10^{-3} \text{ m}^3\text{/s } (0.0613 \text{ ft}^3\text{/s})$$

The time of washing is then as follows for 0.10 (3.37), or 0.337 m³ of wash water.

$$t = \frac{0.337 \text{ m}^3}{1.737 \times 10^{-3} \text{ m}^3\text{/s}} = 194.0$$

The total filtration cycle is

$$\frac{269.7}{60} + \frac{194.0}{60} + 20 = 27.73 \text{ min}$$

3. Equations for continuous filtration. In a filter that is continuous in operation, such as a rotary-drum vacuum type, the feed, filtrate, and the cake move at steady, continuous rates. In a rotary drum the pressure drop is held constant for the filtration. The cake formation involves a continual change in conditions. In continuous filtration the resistance of the filter medium is generally negligible compared with the cake resistance. So in Eq. (14.2-13), $B = 0$.

Integrating Eq. (14.2-13) with $B = 0$,

$$\int_0^t dt = K_p \int_0^V V \, dV \tag{14.2-21}$$

$$t = K_p \frac{V^2}{2} \tag{14.2-22}$$

where t is the time required for formation of the cake. In a rotary-drum filter, the filter time t is less than the total cycle time t_c by

$$t = f t_c \tag{14.2-23}$$

where f is the fraction of the cycle used for cake formation. In the rotary drum, f is the fraction submergence of the drum surface in the slurry.

Next, substituting Eq. (14.2-14) and Eq. (14.2-23) into (14.2-22) and rearranging,

$$\text{flow rate} = \frac{V}{At_c} = \left[\frac{2f(-\Delta p)}{t_c\, \mu\alpha c_S}\right]^{1/2} \tag{14.2-24}$$

If the specific cake resistance varies with pressure, the constants in Eq. (14.2-11) are needed to predict the value of α to be used in Eq. (14.2-24). Experimental verification of Eq. (14.2-24) shows that the flow rate varies inversely with the square root of the viscosity and the cycle time (N1).

When short cycle times are used in continuous filtration and/or the filter medium resistance is relatively large, the filter resistance term B must be included, and Eq. (14.2-13) becomes

$$t = ft_c = K_p \frac{V^2}{2} + BV \tag{14.2-25}$$

Then Eq. (14.2-25) becomes

$$\text{flow rate} = \frac{V}{At_c} = \frac{-R_m/t_c + [R_m^2/t_c^2 + 2c_S\alpha(-\Delta p)f/(\mu t_c)]^{1/2}}{\alpha c_S} \tag{14.2-26}$$

EXAMPLE 14.2-4. Filtration in a Continuous Rotary Drum Filter
A rotary vacuum drum filter having a 33% submergence of the drum in the slurry is to be used to filter a $CaCO_3$ slurry as given in Example 14.2-1 using a pressure drop of 67.0 kPa. The solids concentration in the slurry is $c_x = 0.191$ kg solid/kg slurry and the filter cake is such that the kg wet cake/kg dry cake $= m = 2.0$. The density and viscosity of the filtrate can be assumed as that of water at 298.2 K. Calculate the filter area needed to filter 0.778 kg slurry/s. The filter cycle time is 250 s. The specific cake resistance can be represented by $\alpha = (4.37 \times 10^9)(-\Delta p)^{0.3}$, where $-\Delta p$ is in Pa and α in m/kg.

Solution: From Appendix A.2 for water, $\rho = 996.9$ kg/m^3, $\mu = 0.8937 \times 10^{-3}$ Pa·s. From Eq. (14.2-10),

$$c_S = \frac{\rho c_x}{1 - mc_x} = \frac{996.9\,(0.191)}{1 - (2.0)(0.191)} = 308.1 \text{ kg solids/m}^3 \text{ filtrate}$$

Solving for α, $\alpha = (4.37 \times 10^9)(67.0 \times 10^3)^{0.3} = 1.225 \times 10^{11}$ m/kg. To calculate the flow rate of the filtrate,

$$\frac{V}{t_c} = 0.778\ (c_x)/(c_s)$$

$$= \left(0.778\ \frac{\text{kg slurry}}{\text{s}}\right)\left(0.191\ \frac{\text{kg solid}}{\text{kg slurry}}\right)\bigg/\left(\frac{1}{308.1 \text{ kg solid/m}^3 \text{ filtrate}}\right)$$

$$= 4.823 \times 10^{-4} \text{ m}^3 \text{ filtrate/s}$$

Substituting into Eq. (14.2-24), neglecting and setting $B = 0$, and solving,

$$\frac{V}{At_c} = \frac{4.823 \times 10^{-4}}{A} = \left[\frac{2(0.33)(67.0 \times 10^3)}{250\,(0.8937 \times 10^{-3})(1.225 \times 10^{11})(308.1)}\right]^{1/2}$$

Hence, $A = 6.60$ m^2.

14.2F Filtration Equations for Constant-Rate Filtration

In some cases filtration runs are made under conditions of constant rate rather than constant pressure. This occurs if the slurry is fed to the filter by a positive displacement pump. Equation (14.2-8) can be rearranged to give the following for a constant rate (dV/dt) m³/s.

$$-\Delta p = \left(\frac{\mu \alpha c_s}{A^2}\frac{dV}{dt}\right)V + \left(\frac{\mu R_m}{A}\frac{dV}{dt}\right) = K_V V + C \qquad (14.2\text{-}27)$$

where

$$K_V = \left(\frac{\mu \alpha c_s}{A^2}\frac{dV}{dt}\right) \qquad \text{(SI)}$$

$$K_V = \left(\frac{\mu \alpha c_s}{A^2 g_c}\frac{dV}{dt}\right) \qquad \text{(English)} \qquad (14.2\text{-}28)$$

$$C = \left(\frac{\mu R_m}{A}\frac{dV}{dt}\right) \qquad \text{(SI)}$$

$$C = \left(\frac{\mu R_m}{A g_c}\frac{dV}{dt}\right) \qquad \text{(English)} \qquad (14.2\text{-}29)$$

K_V is in N/m⁵ (lb$_f$/ft⁵) and C is in N/m² (lb$_f$/ft²).

Assuming that the cake is incompressible, K_V and C are constants characteristic of the slurry, cake, rate of filtrate flow, and so on. Hence, a plot of pressure, $-\Delta p$, versus the total volume of filtrate collected, V, gives a straight line for a constant rate dV/dt. The slope of the line is K_V and the intercept is C. The pressure increases as the cake thickness increases and the volume of filtrate collected increases.

The equations can also be rearranged in terms of $-\Delta p$ and time t as variables. At any moment during the filtration, the total volume V is related to the rate and total time t as follows:

$$V = t \frac{dV}{dt} \qquad (14.2\text{-}30)$$

Substituting Eq. (14.2-30) into Eq. (14.2-27),

$$-\Delta p = \left[\frac{\mu \alpha c_s}{A^2}\left(\frac{dV}{dt}\right)^2\right]t + \left(\frac{\mu R_m}{A}\frac{dV}{dt}\right) \qquad (14.2\text{-}31)$$

For the case where the specific cake resistance α is not constant but varies as in Eq. (14.2-11), this can be substituted for α in Eq. (14.2-27) to obtain a final equation.

14.3 SETTLING AND SEDIMENTATION IN PARTICLE–FLUID SEPARATION

14.3A Introduction

In filtration the solid particles are removed from the slurry by forcing the fluid through a filter medium, which blocks the passage of the solid particles and allows the filtrate to pass through. In settling and sedimentation the particles are separated from the fluid by gravitational forces acting on the particles.

Applications of settling and sedimentation include removal of solids from liquid sewage wastes, settling of crystals from the mother liquor, separation of liquid–liquid mixture from a solvent-extraction stage in a settler, settling of solid food particles from a liquid food, and settling of a slurry from a soybean leaching process. The particles can be solid particles or liquid drops. The fluid can be a liquid or gas and it may be at rest or in motion.

In some processes of settling and sedimentation the purpose is to remove the particles from the fluid stream so that the fluid is free of particle contaminants. In other processes the particles are recovered as the product, as in recovery of the dispersed phase in liquid–liquid extraction. In some cases the particles are suspended in fluids so that the particles can be separated into fractions differing in size or in density.

When a particle is at a sufficient distance from the walls of the container and from other particles so that its fall is not affected by them, the process is called *free settling*. Interference is less than 1% if the ratio of the particle diameter to the container diameter is less than 1:200 or if the particle concentration is less than 0.2 vol % in the solution. When the particles are crowded, they settle at a lower rate and the process is called *hindered settling*. The separation of a dilute slurry or suspension by gravity settling into a clear fluid and a slurry of higher solids content is called *sedimentation*.

14.3B Theory of Particle Movement Through a Fluid

1. Derivation of basic equations for rigid spheres. Whenever a particle is moving through a fluid, a number of forces will be acting on the particle. First, a density difference is needed between the particle and the fluid. An external force of gravity is needed to impart motion to the particle. If the densities of the fluid and particle are equal, the buoyant force on the particle will counterbalance the external force and the particle will not move relative to the fluid.

For a rigid particle moving in a fluid, there are three forces acting on the body: gravity acting downward, buoyant force acting upward, and resistance or drag force acting in opposite direction to the particle motion.

We will consider a particle of mass m kg falling at a velocity v m/s relative to the fluid. The density of the solid particle is ρ_p kg/m^3 solid and that of the liquid is ρ kg/m^3 liquid. The buoyant force F_b in N on the particle is

$$F_b = \frac{m\rho g}{\rho_p} = V_p \rho g \tag{14.3-1}$$

where m/ρ_p is the volume V_p in m^3 of the particle and g is the gravitational acceleration in m/s^2.

The gravitation or external force F_g in N on the particle is

$$F_g = mg \tag{14.3-2}$$

The drag force F_D on a body in N may be derived from the fact that, like in flow of fluids, the drag force or frictional resistance is proportional to the velocity head $v^2/2$ of the fluid displaced by the moving body. This must be multiplied by the density of the fluid and by a significant area A, such as the projected area of the particle. This was defined previously in Eq. (3.1-1).

$$F_D = C_D \frac{v^2}{2} \rho A \tag{14.3-3}$$

where the drag coefficient C_D is the proportionality constant and is dimensionless.

The resultant force on the body is then $F_g - F_b - F_D$. This resultant force must equal the force due to acceleration.

$$m \frac{dv}{dt} = F_g - F_b - F_D \tag{14.3-4}$$

Substituting Eqs. (14.3-1)–(14.3-3) into (14.3-4),

$$m \frac{dv}{dt} = mg - \frac{m\rho g}{\rho_p} - \frac{C_D v^2 \rho A}{2} \tag{14.3-5}$$

If we start from the moment the body is released from its position of rest, the falling of the body consists of two periods: the period of accelerated fall and the period of constant velocity fall. The initial acceleration period is usually very short, of the order of a tenth of a second or so. Hence, the period of constant velocity fall is the important one. The velocity is called the *free settling velocity* or *terminal velocity v_t*.

To solve for the terminal velocity in Eq. (14.3-5), $dv/dt = 0$ and the equation becomes

$$v_t = \sqrt{\frac{2g(\rho_p - \rho)m}{A\rho_p C_D \rho}} \tag{14.3-6}$$

For spherical particles $m = \pi D_p^3 \rho_p / 6$ and $A = \pi D_p^2 / 4$. Substituting these into Eq. (14.3-6), we obtain, for spherical particles

$$v_t = \sqrt{\frac{4(\rho_p - \rho)g D_p}{3 C_D \rho}} \tag{14.3-7}$$

where v_t is m/s (ft/s), ρ is kg/m³ (lb$_m$/ft³), g is 9.80665 m/s² (32.174 ft/s²), and D_p is m (ft).

2. *Drag coefficient for rigid spheres.* The drag coefficient for rigid spheres has been shown to be a function of the Reynolds number $D_p v\rho/\mu$ of the sphere and is shown in Fig. 14.3-1. In the laminar flow region, called the Stokes' law region for $N_{Re} < 1$, as discussed in Section 3.1B, the drag coefficient is

$$C_D = \frac{24}{D_p v\rho/\mu} = \frac{24}{N_{Re}} \tag{14.3-8}$$

where μ is the viscosity of the liquid in Pa·s or kg/m·s (lb$_m$/ft·s). Substituting this into Eq. (14.3-7) for laminar flow,

$$v_t = \frac{g D_p^2 (\rho_p - \rho)}{18\mu} \tag{14.3-9}$$

For other shapes of particles, drag coefficients will differ from those given in Fig. 14.3-1 and data are given in Fig. 3.1-2 and elsewhere (B2, L2, P1). In the turbulent Newton's law region above a Reynolds number of about 1000 to 2.0 × 10⁵, the drag coefficient is approximately constant at $C_D = 0.44$.

Solution of Eq. (14.3-7) is trial and error when the particle diameter is known and the terminal velocity is to be obtained. This occurs because C_D also depends upon the velocity v_t.

If the particles are quite small, Brownian motion is present. *Brownian movement* is the random motion imparted to the particle by collisions between the molecules of the fluid surrounding the particle and the particle. This movement of the particles in random directions tends to suppress the effect of gravity, so settling of the particles may occur

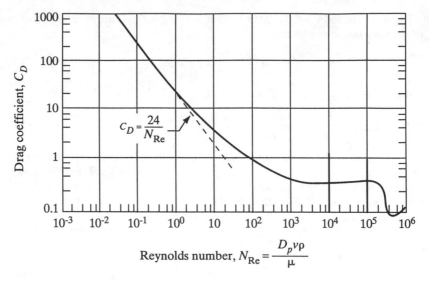

FIGURE 14.3-1. *Drag coefficient for a rigid sphere.*

more slowly or not at all. At particle sizes of a few micrometers, the Brownian effect becomes appreciable and at sizes of less than 0.1 μm, the effect predominates. In very small particles, application of centrifugal force helps reduce the effect of Brownian movement.

EXAMPLE 14.3-1.   Settling Velocity of Oil Droplets
Oil droplets having a diameter of 20 μm (0.020 mm) are to be settled from air at an air temperature of 37.8°C (311 K) at 101.3 kPa pressure. The density of the oil is 900 kg/m³. Calculate the terminal settling velocity of the drops.

Solution: The various knowns are $D_p = 2.0 \times 10^{-5}$ m, $\rho_p = 900$ kg/m³. From Appendix A.3 for air at 37.8°C, $\rho = 1.137$ kg/m³, $\mu = 1.90 \times 10^{-5}$ Pa · s. The drop will be assumed to be a rigid sphere.
    The solution is trial and error since the velocity is unknown. Hence, C_D cannot be directly evaluated. The Reynolds number is as follows:

$$N_{Re} = \frac{D_p v_t \rho}{\mu} = \frac{(2.0 \times 10^{-5})(v_t)(1.137)}{1.90 \times 10^{-5}} = 1.197 v_t \qquad \textbf{(14.3-10)}$$

For the first trial, assume that $v_t = 0.305$ m/s. Then $N_{Re} = 1.197(0.305) = 0.365$. Substituting into Eq. (14.3-7) and solving for C_D,

$$v_t = \sqrt{\frac{4(\rho_p - \rho)g D_p}{3 C_D \rho}} = \sqrt{\frac{4(900 - 1.137)(9.8066)(2.0 \times 10^{-5})}{(3) C_D (1.137)}}$$

$$C_D = \frac{0.2067}{v_t^2} \qquad \textbf{(14.3-11)}$$

Using $v_t = 0.305$ m/s, $C_D = 0.2067/(0.305)^2 = 2.22$.
    Assuming that $v_t = 0.0305$ m/s, $N_{Re} = 0.0365$ from Eq. (14.3-10) and $C_D = 222$ from Eq. (14.3-11). For the third trial, assuming that $v_t = 0.00305$ m/s, $N_{Re} = 0.00365$ and $C_D = 22\,200$. These three points of N_{Re} and

values of C_D calculated are plotted on a plot similar to Fig. 14.3-1 and shown in Fig. 14.3-2. It can be shown that the line through these points is a straight line. The intersection of this line and the drag-coefficient correlation line is the solution to the problem at $N_{Re} = 0.012$. The velocity can be calculated from the Reynolds number in Eq. (14.3-10).

$$N_{Re} = 0.012 = 1.197 v_t$$

$$v_t = 0.0100 \text{ m/s } (0.0328 \text{ ft/s})$$

The particle is in the Reynolds number range less than 1, which is the laminar Stokes' law region. Alternatively, the velocity can be calculated by substituting into Eq. (14.3-9).

$$v_t = \frac{9.8066(2.0 \times 10^{-5})^2(900 - 1.137)}{18(1.90 \times 10^{-5})} = 0.0103 \text{ m/s}$$

Note that Eq. (14.3-9) could not be used until it was determined that the particle fall was in the laminar region.

For particles that are rigid but nonspherical, the drag depends upon the shape of the particle and the orientation of the particle with respect to its motion. Correlations of drag coefficients for particles of different shapes are given in a number of references (B2, C1, P1).

3. Drag coefficients for nonrigid spheres. When particles are nonrigid, internal circulation inside the particle and particle deformation can occur. Both of these effects affect the drag coefficient and terminal velocity. Drag coefficients for air bubbles rising in water are given in Perry and Green (P1), and for a Reynolds number less than about 50, the curve is the same as for rigid spheres in water.

For liquid drops in gases, the same drag relationship as for solid spherical particles is obtained up to a Reynolds number of about 100 (H1). Large drops will deform with an increase in drag. Small liquid drops in immiscible liquids behave like rigid spheres and the drag coefficient curve follows that for rigid spheres up to a Reynolds number of about 10. Above this and up to a Reynolds number of 500, the terminal velocity is greater than that for solids because of internal circulation in the drop.

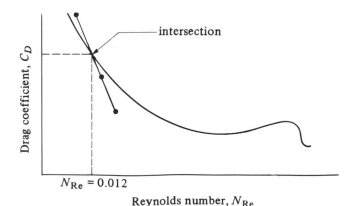

FIGURE 14.3-2. *Solution of Example 14.3-1 for settling velocity of a particle.*

14.3C Hindered Settling

For many cases in settling, a large number of particles are present and the surrounding particles interfere with the motion of individual particles. The velocity gradients surrounding each particle are affected by the close presence of other particles. The particles in settling in the liquid displace the liquid, and an appreciable upward velocity of the liquid is generated. Hence, the velocity of the liquid is appreciably greater with respect to the particle than with respect to the apparatus itself.

For this hindered flow the settling velocity is less than would be calculated from Eq. (14.3-9) for Stokes' law. The true drag force is greater in the suspension because of the interference of the other particles. This higher effective viscosity of the mixture μ_m is equal to the actual viscosity of the liquid itself, μ, divided by an empirical correction factor, ψ_p, which depends upon ε, the volume fraction of the slurry mixture occupied by the liquid (S1).

$$\mu_m = \frac{\mu}{\psi_p} \tag{14.3-12}$$

where ψ_p is dimensionless and is as follows (S1):

$$\psi_p = \frac{1}{10^{1.82(1-\varepsilon)}} \tag{14.3-13}$$

The density of the fluid phase becomes effectively the bulk density of the slurry ρ_m, which is as follows:

$$\rho_m = \varepsilon\rho + (1 - \varepsilon)\rho_p \tag{14.3-14}$$

where ρ_m is density of slurry in kg solid + liquid/m^3. The density difference is now

$$\rho_p - \rho_m = \rho_p - [\varepsilon\rho + (1 - \varepsilon)\rho_p] = \varepsilon(\rho_p - \rho) \tag{14.3-15}$$

The settling velocity v_t with respect to the apparatus is ε times the velocity calculated by Stokes' law.

Substituting mixture properties of μ_m from Eq. (14.3-12) for μ in Eq. (14.3-9), $(\rho_p - \rho_m)$ from Eq. (14.3-15) for $(\rho_p - \rho)$, and multiplying the result by ε for the relative velocity effect, Eq. (14.3-9) becomes, for laminar settling,

$$v_t = \frac{gD_p^2(\rho_p - \rho)}{18\mu}\,(\varepsilon^2\psi_p) \tag{14.3-16}$$

This is the velocity calculated from Eq. (14.3-9), multiplied by the correction factor $(\varepsilon^2\psi_p)$.

The Reynolds number is then based on the velocity relative to the fluid and is

$$N_{Re} = \frac{D_p v_t \rho_m}{\mu_m \varepsilon} = \frac{D_p^3 g(\rho_p - \rho)\rho_m \varepsilon\psi_p^2}{18\mu^2} \tag{14.3-17}$$

When the Reynolds number is less than 1, the settling is in the Stokes' law range. For Reynolds numbers above 1.0, see (P1). The effect of concentration is greater for non-spherical particles and angular particles (S1).

EXAMPLE 14.3-2. Hindered Settling of Glass Spheres
Calculate the settling velocity of glass spheres having a diameter of 1.554×10^{-4} m (5.10×10^{-4} ft) in water at 293.2 K (20°C). The slurry contains 60 wt% solids. The density of the glass spheres is $\rho_p = 2467$ kg/m^3 (154 lb$_m$/ft^3).

Solution: Density of water $\rho = 998$ kg/m³ (62.3 lb$_m$/ft³), and viscosity of water $\mu = 1.005 \times 10^{-3}$ Pa·s (6.72 × 10^{-4} lb$_m$/ft·s). To calculate the volume fraction ε of the liquid,

$$\varepsilon = \frac{40/998}{40/998 + 60/2467} = 0.622$$

The bulk density of the slurry ρ_m by Eq. (14.3-14) is

$$\rho_m = \varepsilon\rho + (1 - \varepsilon)\rho_p = 0.622(998) + (1 - 0.622)(2467)$$

$$= 1553 \text{ kg/m}^3 \text{ (96.9 lb}_m/\text{ft}^3)$$

Substituting into Eq. (14.3-13),

$$\psi_p = \frac{1}{10^{1.82(1-\varepsilon)}} = \frac{1}{10^{1.82(1-0.622)}} = 0.205$$

Substituting into Eq. (14.3-16), using SI and English units,

$$v_t = \frac{9.807(1.554 \times 10^{-4})^2(2467 - 998)(0.622^2 \times 0.205)}{18(1.005 \times 10^{-3})}$$

$$= 1.525 \times 10^{-3} \text{ m/s}$$

$$v_t = \frac{32.174(5.1 \times 10^{-4})^2(154 - 62.3)(0.622^2 \times 0.205)}{18(6.72 \times 10^{-4})}$$

$$= 5.03 \times 10^{-3} \text{ ft/s}$$

The Reynolds number is obtained by substituting into Eq. (14.3-17),

$$N_{Re} = \frac{D_p v_t \rho_m}{\mu_m \varepsilon} = \frac{D_p v_t \rho_m}{(\mu/\psi_p)\varepsilon} = \frac{(1.554 \times 10^{-4})(1.525 \times 10^{-3})1553}{(1.005 \times 10^{-3}/0.205)0.622}$$

$$= 0.121$$

Hence, the settling is in the laminar range.

14.3D Wall Effect on Free Settling

When the diameter of D_p of the particle becomes appreciable with respect to the diameter D_W of the container in which the settling is occurring, a retarding effect known as the *wall effect* is exerted on the particle. The terminal settling velocity is reduced. In the case of settling in the Stokes' law regime, the computed terminal velocity can be multiplied by the following to allow for the wall effect (Z1) for $D_p/D_W < 0.05$.

$$k_W = \frac{1}{1 + 2.1(D_p/D_W)} \tag{14.3-18}$$

For the completely turbulent regime, the correction factor is

$$k'_W = \frac{1 - (D_p/D_W)^2}{[1 + (D_p/D_W)^4]^{1/2}} \tag{14.3-19}$$

14.3E Differential Settling and Separation of Solids in Classification

1. Sink-and-float methods. Devices for the separation of solid particles into several fractions based upon their rates of flow or settling through fluids are known as *classifiers*.

There are several separation methods to accomplish this by sink-and-float and differential settling. In the *sink-and-float method*, a liquid is used whose density is intermediate between that of the heavy or high-density material and the light-density material. In this method the heavy particles will not float but settle from the medium, and the light particles float.

This method is independent of the sizes of the particles and depends only upon the relative densities of the two materials. This means liquids used must have densities greater than water, since most solids have high densities. Unfortunately, few such liquids exist that are cheap and noncorrosive. As a result, pseudoliquids are used, consisting of a suspension in water of very fine solid materials with high specific gravities such as galena (specific gravity = 7.5) and magnetite (specific gravity = 5.17).

Hindered settling is used and the bulk density of the medium can be varied widely by varying the amount of the fine solid materials in the medium. Common applications of this technique are concentrating ore materials and cleaning coal. The fine solid materials in the medium are so small in diameter that their settling velocity is negligible, giving a relatively stable suspension.

2. Differential settling methods.   The separation of solid particles into several size fractions based upon the settling velocities in a medium is called *differential settling* or *classification*. The density of the medium is less than that of either of the two substances to be separated.

In differential settling both light and heavy materials settle through the medium. A disadvantage of this method if the light and heavy materials both have a range of particle sizes is that the smaller, heavy particles settle at the same terminal velocity as the larger, light particles.

Suppose that we consider two different materials: heavy-density material A (such as galena, with a specific gravity $\rho_A = 7.5$) and light-density material B (such as quartz, with a specific gravity $\rho_B = 2.65$). The terminal settling velocity of components A and B from Eq. (14.3-7) can be written

$$v_{tA} = \left[\frac{4(\rho_{pA} - \rho)gD_{pA}}{3C_{DA}\rho}\right]^{1/2} \tag{14.3-20}$$

$$v_{tB} = \left[\frac{4(\rho_{pB} - \rho)gD_{pB}}{3C_{DB}\rho}\right]^{1/2} \tag{14.3-21}$$

For particles of equal settling velocities, $v_{tA} = v_{tB}$ and we obtain, by equating Eq. (14.3-20) to (14.3-21), canceling terms, and squaring both sides,

$$\frac{(\rho_{pA} - \rho)D_{pA}}{\rho C_{DA}} = \frac{(\rho_{pB} - \rho)D_{pB}}{\rho C_{DB}} \tag{14.3-22}$$

or

$$\frac{D_{pA}}{D_{pB}} = \frac{\rho_{pB} - \rho}{\rho_{pA} - \rho} \cdot \frac{C_{DA}}{C_{DB}} \tag{14.3-23}$$

For particles that are essentially spheres at very high Reynolds numbers in the turbulent Newton's law region, C_D is constant and $C_{DA} = C_{DB}$, giving

$$\frac{D_{pA}}{D_{pB}} = \left(\frac{\rho_{pB} - \rho}{\rho_{pA} - \rho}\right)^{1.0} \tag{14.3-24}$$

For laminar Stokes' law settling,

$$C_{DA} = \frac{24\mu}{D_{pA} v_{tA} \rho} \qquad C_{DB} = \frac{24\mu}{D_{pB} v_{tB} \rho} \qquad (14.3\text{-}25)$$

Substituting Eq. (14.3-25) into (14.3-23) and rearranging for Stokes' law settling, where $v_{tA} = v_{tB}$,

$$\frac{D_{pA}}{D_{pB}} = \left(\frac{\rho_{pB} - \rho}{\rho_{pA} - \rho}\right)^{0.5} \qquad (14.3\text{-}26)$$

For transition flow between laminar and turbulent flow,

$$\frac{D_{pA}}{D_{pB}} = \left(\frac{\rho_{pB} - \rho}{\rho_{pA} - \rho}\right)^{n} \qquad \text{where } \tfrac{1}{2} < n < 1 \qquad (14.3\text{-}27)$$

For particles settling in the turbulent range, Eq. (14.3-24) holds for equal settling velocities. For particles where $D_{pA} = D_{pB}$ and settling is in the turbulent Newton's law region, combining Eqs. (14.3-20) and (14.3-21),

$$\frac{v_{tA}}{v_{tB}} = \left(\frac{\rho_{pA} - \rho}{\rho_{pB} - \rho}\right)^{1/2} \qquad (14.3\text{-}28)$$

If both A and B particles are settling in the same medium, then Eqs. (14.3-24) and (14.3-28) can be used to make the plots given in Fig. 14.3-3 for the relation of velocity to diameter for A and B.

First, we consider a mixture of particles of materials A and B with a size range of D_{p1} to D_{pA} for both types of material. In the size range D_{p1} to D_{p2} in Fig. 14.3-3, a pure fraction of substance B can be obtained since no particles of A settle as slowly. In the size range D_{p3} to D_{p4}, a pure fraction of A can be obtained since no B particles settle as fast as the A particles in this size range. In the size range D_{p1} to D_{p3}, A particles settle as rapidly as B particles in the size range D_{p2} to D_{p4}, forming a mixed fraction of A and B.

Increasing the density ρ of the medium in Eq. (14.3-24), the numerator becomes smaller proportionately faster than the denominator, and the spread between D_{pA} and D_{pB} is increased. Somewhat similar curves are obtained in the Stokes' law region.

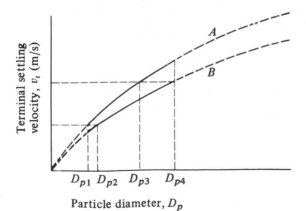

FIGURE 14.3-3. *Settling and separation of two materials A and B in Newton's law region.*

EXAMPLE 14.3-3. *Separation of a Mixture of Silica and Galena*
A mixture of silica (*B*) and galena (*A*) solid particles having a size range of 5.21×10^{-6} m to 2.50×10^{-5} m is to be separated by hydraulic classification using free settling conditions in water at 293.2 K (B1). The specific gravity of silica is 2.65 and that of galena is 7.5. Calculate the size range of the various fractions obtained in the settling. If the settling is in the laminar region, the drag coefficients will be reasonably close to that for spheres.

Solution: The particle-size range is $D_p = 5.21 \times 10^{-6}$ m to $D_p = 2.50 \times 10^{-5}$ m. Densities are $\rho_{pA} = 7.5(1000) = 7500$ kg/m^3, $\rho_{pB} = 2.65(1000) = 2650$ kg/m^3, $\rho = 998$ kg/m^3 for water at 293.2 K (20°C). The water viscosity $\mu = 1.005 \times 10^{-3}$ Pa·s $= 1.005 \times 10^{-3}$ kg/m·s.
Assuming Stokes' law settling, Eq. (14.3-9) becomes as follows:

$$v_{tA} = \frac{gD_{pA}^2(\rho_{pA} - \rho)}{18\mu} \tag{14.3-29}$$

The largest Reynolds number occurs for the largest particle and the biggest density, where $D_{pA} = 2.50 \times 10^{-5}$ m and $\rho_{pA} = 7500$. Substituting into Eq. (14.3-29),

$$v_{tA} = \frac{9.807(2.50 \times 10^{-5})^2(7500 - 998)}{18(1.005 \times 10^{-3})} = 2.203 \times 10^{-3} \text{ m/s}$$

Substituting into the Reynolds number equation,

$$N_{Re} = \frac{D_{pA}v_{tA}\rho}{\mu} \tag{14.3-30}$$

$$= \frac{(2.50 \times 10^{-5})(2.203 \times 10^{-3})998}{1.005 \times 10^{-3}} = 0.0547$$

Hence, the settling is in the Stokes' law region.
Referring to Fig. 14.3-3 and using the same nomenclature, the largest size is $D_{p4} = 2.50 \times 10^{-5}$ m. The smallest size is $D_{p1} = 5.21 \times 10^{-6}$ m. The pure fraction of *A* consists of $D_{pA4} = 2.50 \times 10^{-5}$ m to D_{pA3}. The particles, having diameters D_{pA3} and D_{pB4}, are related by having equal settling velocities in Eq. (14.3-26). Substituting $D_{pB4} = 2.50 \times 10^{-5}$ m into Eq. (14.3-26) and solving,

$$\frac{D_{pA3}}{2.50 \times 10^{-5}} = \left(\frac{2650 - 998}{7500 - 998}\right)^{1/2}$$

$$D_{pA3} = 1.260 \times 10^{-5} \text{ m}$$

The size range of pure *B* fraction is D_{pB2} to $D_{pB1} = 5.21 \times 10^{-6}$ m. The diameter D_{pB2} is related to $D_{pA1} = 5.21 \times 10^{-6}$ by Eq. (14.3-26) at equal settling velocities.

$$\frac{5.21 \times 10^{-6}}{D_{pB2}} = \left(\frac{2650 - 998}{7500 - 998}\right)^{1/2}$$

$$D_{pB2} = 1.033 \times 10^{-5} \text{ m}$$

The three fractions recovered are as follows.

1. The size range of the first fraction of pure *A* (galena) is as follows:

$$D_{pA3} = 1.260 \times 10^{-5} \text{ m} \quad \text{to} \quad D_{pA4} = 2.50 \times 10^{-5} \text{ m}$$

2. The mixed-fraction size range is as follows:

$$D_{pB2} = 1.033 \times 10^{-5} \text{ m} \quad \text{to} \quad D_{pB4} = 2.50 \times 10^{-5} \text{ m}$$

$$D_{pA1} = 5.21 \times 10^{-6} \text{ m} \quad \text{to} \quad D_{pA3} = 1.260 \times 10^{-5} \text{ m}$$

3. The size range of the third fraction of pure B (silica) is as follows:

$$D_{pB1} = 5.21 \times 10^{-6} \text{ m} \quad \text{to} \quad D_{pB2} = 1.033 \times 10^{-5} \text{ m}$$

14.3F Sedimentation and Thickening

1. Mechanisms of sedimentation.   When a dilute slurry is settled by gravity into a clear fluid and a slurry of higher solids concentration, the process is called *sedimentation* or sometimes *thickening*. To illustrate the method of determining settling velocities and the mechanisms of settling, a batch settling test is carried out by placing a uniform concentration of the slurry in a graduated cylinder. At the start, as shown in Fig. 14.3-4a, all the particles settle by free settling in suspension zone B. The particles in zone B settle at a uniform rate at the start and a clear liquid zone A appears in Fig. 14.3-4b. The height z drops at a constant rate. Also, zone D begins to appear, which contains the settled particles at the bottom. Zone C is a transition layer whose solids content varies from that in zone B to zone D. After further settling, zones B and C disappear, as shown in Fig. 14.3-4c. Then compression first appears, and this moment is called the *critical point*. During compression liquid is expelled upward from zone D and the thickness of zone D decreases.

2. Determination of settling velocity.   In Fig. 14.3-4d the height z of the clear liquid interface height is plotted versus time. As shown, the velocity of settling, which is the slope of the line, is constant at first. The critical point is shown at point C. Since sludges vary greatly in their rates, experimental settling rates of each sludge is necessary. Kynch (K1) and Talmage and Fitch (T1) describe a method to predict thickener sizes from the batch settling test.

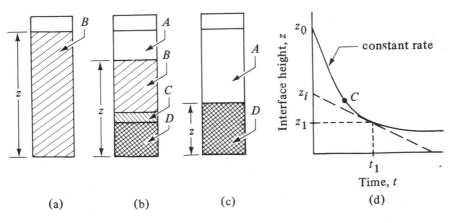

(a)   (b)   (c)   (d)

FIGURE 14.3-4.   *Batch sedimentation results : (a) original uniform suspension, (b) zones of settling after a given time, (c) compression of zone D after zones B and C disappear, (d) clear liquid interface height z versus time of settling.*

The settling velocity v is determined by drawing a tangent to the curve in Fig. 14.3-4d at a given time t_1 and the slope $-dz/dt = v_1$. At this point the height is z_1 and z_i is the intercept of the tangent to the curve. Then,

$$v_1 = \frac{z_i - z_1}{t_1 - 0} \tag{14.3-31}$$

The concentration c_1 is, therefore, the average concentration of the suspension if z_i is the height of this slurry. This is calculated by

$$c_1 z_i = c_0 z_0 \quad \text{or} \quad c_1 = \left(\frac{z_0}{z_i}\right) c_0 \tag{14.3-32}$$

where c_0 is the original slurry concentration in kg/m^3 at z_0 height and $t = 0$. This is repeated for other times and a plot of settling velocity versus concentration is made. Further details of this and other methods to design the thickener are given elsewhere (C1, F1, F2, T1, P1). These and other methods in the literature are highly empirical and care should be exercised in their use.

14.3G Equipment for Settling and Sedimentation

1. Simple gravity settling tank.   In Fig. 14.3-5a a simple gravity settler is shown for removing by settling a dispersed liquid phase from another phase. The velocity horizontally to the right must be slow enough to allow time for the smallest droplets to rise from the bottom to the interface or from the top down to the interface and coalesce.

In Fig. 14.3-5b a gravity settling chamber is shown schematically. Dust-laden air enters at one end of a large boxlike chamber. Particles settle toward the floor at their terminal settling velocities. The air must remain in the chamber a sufficient length of time (residence time) so that the particles reach the floor of the chamber. Knowing the throughput of the air stream through the chamber and the chamber size, the residence time of the air in the chamber can be calculated. The vertical height of the chamber must be small enough so that this height, divided by the settling velocity, gives a time less than the residence time of the air.

2. Equipment for classification.   The simplest type of classifier is one in which a large

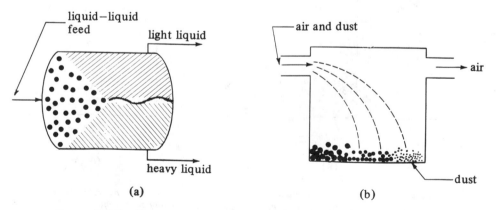

FIGURE 14.3-5.   *Gravity settling tanks : (a) settler for liquid–liquid dispersion, (b) dust-settling chamber.*

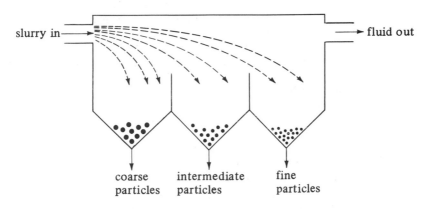

FIGURE 14.3-6. *Simple gravity settling classifier.*

tank is subdivided into several sections, as shown in Fig. 14.3-6. A liquid slurry feed enters the tank containing a size range of solid particles. The larger, faster-settling particles settle to the bottom close to the entrance and the slower-settling particles settle to the bottom close to the exit. The linear velocity of the entering feed decreases as a result of the enlargement of the cross-sectional area at the entrance. The vertical baffles in the tank allow for the collection of several fractions. The settling-velocity equations derived in this section hold.

3. Spitzkasten classifier. Another type of gravity settling chamber is the *Spitzkasten,* shown in Fig. 14.3-7, which consists of a series of conical vessels of increasing diameter in the direction of flow. The slurry enters the first vessel, where the largest and faster-settling particles are separated. The overflow goes to the next vessel, where another separation occurs. This continues in the succeeding vessel or vessels. In each vessel the velocity of upflowing inlet water is controlled to give the desired size range for each vessel.

4. Sedimentation thickener. The separation of a dilute slurry by gravity settling into a clear fluid and a slurry of higher solids concentration is called *sedimentation.* Industrially, sedimentation operations are often carried out continuously in equipment called *thick-*

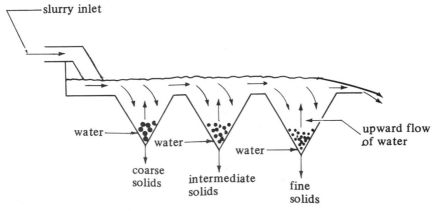

FIGURE 14.3-7. *Spitzkasten gravity settling chamber.*

Sec. 14.3   Settling and Sedimentation in Particle–Fluid Separation   **827**

eners. A continuous thickener with a slowly revolving rake for removing the sludge or thickened slurry is shown in Fig. 14.3-8.

The slurry in Fig. 14.3-8 is fed at the center of the tank several feet below the surface of the liquid. Around the top edge of the tank is a clear liquid overflow outlet. The rake serves to scrape the sludge toward the center of the bottom for removal. This gentle stirring aids in removing water from the sludge.

In the thickener the entering slurry spreads radially through the cross section of the thickener and the liquid flows upward and out the overflow. The solids settle in the upper zone by free settling. Below this dilute settling zone is the transition zone, in which the concentration of solids increases rapidly, and then the compression zone. A clear overflow can be obtained if the upward velocity of the fluid in the dilute zone is less than the minimal terminal settling velocity of the solids in this zone.

The settling rates are quite slow in the thickened zone, which consists of a compression of the solids with liquid being forced upward through the solids. This is an extreme case of hindered settling. Equation (14.3-16) may be used to estimate the settling velocities, but the results can be in considerable error because of agglomeration of particles. As a result, laboratory settling or sedimentation data must be used in the design of a thickener as discussed previously in Section 14.3F.

14.4 CENTRIFUGAL SEPARATION PROCESSES

14.4A Introduction

1. Centrifugal settling or sedimentation. In Section 14.3 were discussed the processing methods of settling and sedimentation where particles are separated from a fluid by gravitational forces acting on the particles. The particles were solid, gas, or liquid and the fluid was a liquid or a gas. In the present section we discuss settling or separation of particles from a fluid by centrifugal forces acting on the particles.

Use of centrifuges increases the forces on particles manyfold. Hence, particles that will not settle readily or at all in gravity settlers can often be separated from fluids by centrifugal force. The high settling force means that practical rates of settling can be obtained with much smaller particles than in gravity settlers. These high centrifugal

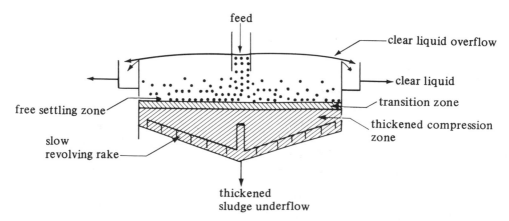

FIGURE 14.3-8. *Continuous thickener.*

forces do not change the relative settling velocities of small particles, but these forces do overcome the disturbing effects of Brownian motion and free convection currents.

Sometimes gravity separation may be too slow because of the closeness of the densities of the particle and the fluid, or because of association forces holding the components together, as in emulsions. An example in the dairy industry is the separation of cream from whole milk, giving skim milk. Gravity separation takes hours, while centrifugal separation is accomplished in minutes in a cream separator. Centrifugal settling or separation is employed in many food industries, such as breweries, vegetable oil processing, fish protein concentrate processing, fruit juice processing to remove cellular materials, and so on. Centrifugal separation is also used in drying crystals and for separating emulsions into their constituent liquids or solid–liquid. The principles of centrifugal sedimentation are discussed in Sections 14.4B and 14.4C.

2. Centrifugal filtration. Centrifuges are also used in centrifugal filtration where a centrifugal force is used instead of a pressure difference to cause the flow of slurry in a filter where a cake of solids builds up on a screen. The cake of granular solids from the slurry is deposited on a filter medium held in a rotating basket, washed, and then spun "dry." Centrifuges and ordinary filters are competitive in most solid–liquid separation problems. The principles of centrifugal filtration are discussed in Section 14.4E.

14.4B   Forces Developed in Centrifugal Separation

1. Introduction. Centrifugal separators make use of the common principle that an object whirled about an axis or center point at a constant radial distance from the point is acted on by a force. The object being whirled about an axis is constantly changing direction and is thus accelerating, even though the rotational speed is constant. This centripetal force acts in a direction toward the center of rotation.

If the object being rotated is a cylindrical container, the contents of fluid and solids exert an equal and opposite force, called *centrifugal force*, outward to the walls of the container. This is the force that causes settling or sedimentation of particles through a layer of liquid or filtration of a liquid through a bed of filter cake held inside a perforated rotating chamber.

In Fig. 14.4-1a a cylindrical bowl is shown rotating with a slurry feed of solid particles and liquid being admitted at the center. The feed enters and is immediately thrown outward to the walls of the container, as in Fig. 14.4-1b. The liquid and solids are now acted upon by the vertical gravitational force and the horizontal centrifugal force. The centrifugal force is usually so large that the force of gravity may be neglected. The liquid layer then assumes the equilibrium position with the surface almost vertical. The particles settle horizontally outward and press against the vertical bowl wall.

In Fig. 14.4-1c two liquids having different densities are being separated by the centrifuge. The more dense fluid will occupy the outer periphery, since the centrifugal force is greater on the more dense fluid.

2. Equations for centrifugal force. In circular motion the acceleration from the centrifugal force is

$$a_e = r\omega^2 \tag{14.4-1}$$

where a_e is the acceleration from a centrifugal force in m/s² (ft/s²), r is radial distance from center of rotation in m (ft), and ω is angular velocity in rad/s.

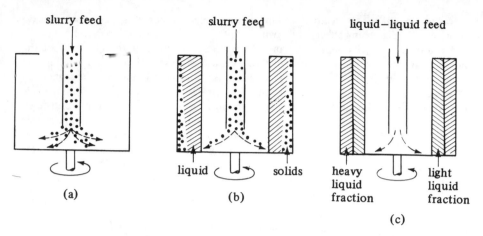

slurry feed   slurry feed   liquid–liquid feed

liquid   solids   heavy   light
liquid liquid
fraction fraction

(a)   (b)   (c)

FIGURE 14.4-1. *Sketch of centrifugal separation : (a) initial slurry feed entering, (b) settling of solids from a liquid, (c) separation of two liquid fractions.*

The centrifugal force F_c in N (lb$_f$) acting on the particle is given by

$$F_c = ma_e = mr\omega^2 \qquad \text{(SI)}$$

$$F_c = \frac{mr\omega^2}{g_c} \qquad \text{(English)}$$

(14.4-2)

where $g_c = 32.174\ \text{lb}_m \cdot \text{ft/lb}_f \cdot \text{s}^2$.

Since $\omega = v/r$, where v is the tangential velocity of the particle in m/s (ft/s),

$$F_c = mr\left(\frac{v}{r}\right)^2 = \frac{mv^2}{r}$$

(14.4-3)

Often rotational speeds are given as N rev/min and

$$\omega = \frac{2\pi N}{60}$$

(14.4-4)

$$N = \frac{60v}{2\pi r}$$

(14.4-5)

Substituting Eq. (14.4-4) into Eq. (14.4-2),

$$F_c \ \text{newton} = mr\left(\frac{2\pi N}{60}\right)^2 = 0.01097mrN^2 \qquad \text{(SI)}$$

$$F_c \ \text{lb}_f = \frac{mr}{g_c}\left(\frac{2\pi N}{60}\right)^2 = 0.000341mrN^2 \qquad \text{(English)}$$

(14.4-6)

By Eq. (14.3-2), the gravitational force on a particle is

$$F_g = mg$$

(14.3-2)

where g is the acceleration of gravity and is $9.80665\ \text{m/s}^2$. In terms of gravitational force, the centrifugal force is as follows by combining Eqs. (14.3-2), (14.4-2), and (14.4-3).

$$\frac{F_c}{F_g} = \frac{r\omega^2}{g} = \frac{v^2}{rg} = \frac{r}{g}\left(\frac{2\pi N}{60}\right)^2 = 0.001118rN^2 \quad \text{(SI)}$$

$$(14.4\text{-}7)$$

$$\frac{F_c}{F_g} = 0.000341rN^2 \qquad\qquad\qquad\qquad \text{(English)}$$

Hence, the force developed in a centrifuge is $r\omega^2/g$ or v^2/rg times as large as the gravity force. This is often expressed as equivalent to so many g forces.

EXAMPLE 14.4-1. Force in a Centrifuge

A centrifuge having a radius of the bowl of 0.1016 m (0.333 ft) is rotating at $N = 1000$ rev/min.

    (a) Calculate the centrifugal force developed in terms of gravity forces.
    (b) Compare this force to that for a bowl with a radius of 0.2032 m rotating at the same rev/min.

Solution: For part (a), $r = 0.1016$ m and $N = 1000$. Substituting into Eq. (14.4-7),

$$\frac{F_c}{F_g} = 0.001118rN^2 = 0.001118(0.1016)(1000)^2 \quad \text{(SI)}$$

$$= 113.6 \text{ gravities or } g\text{'s}$$

$$\frac{F_c}{F_g} = 0.000341(0.333)(1000)^2 = 113.6 \qquad\qquad \text{(English)}$$

For part (b), $r = 0.2032$ m. Substituting into Eq. (14.4-7),

$$\frac{F_c}{F_g} = 0.001118(0.2032)(1000)^2 = 227.2 \text{ gravities or } g\text{'s}$$

14.4C Equations for Rates of Settling in Centrifuges

1. General equation for settling. If a centrifuge is used for sedimentation (removal of particles by settling), a particle of a given size can be removed from the liquid in the bowl if sufficient residence time of the particle in the bowl is available for the particle to reach the wall. For a particle moving radially at its terminal settling velocity, the diameter of the smallest particle removed can be calculated.

In Fig. 14.4-2 a schematic of a tubular-bowl centrifuge is shown. The feed enters at the bottom and it is assumed all the liquid moves upward at a uniform velocity, carrying solid particles with it. The particle is assumed to be moving radially at its terminal settling velocity v_t. The trajectory or path of the particle is shown in Fig. 14.4-2. A particle of a given size is removed from the liquid if sufficient residence time is available for the particle to reach the wall of the bowl, where it is held. The length of the bowl is b m.

At the end of the residence time of the particle in the fluid, the particle is at a distance r_B m from the axis of rotation. If $r_B < r_2$, then the particle leaves the bowl with the fluid. If $r_B = r_2$, it is deposited on the wall of the bowl and effectively removed from the liquid.

For settling in the Stokes' law range, the terminal settling velocity at a radius r is obtained by substituting Eq. (14.4-1) for the acceleration g into Eq. (14.3-9).

$$v_t = \frac{\omega^2 r D_p^2(\rho_p - \rho)}{18\mu} \qquad\qquad (14.4\text{-}8)$$

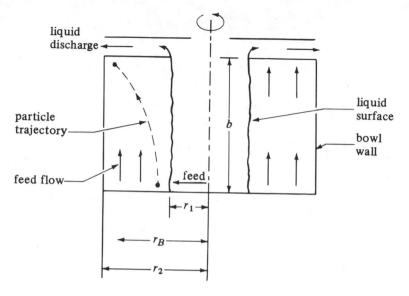

FIGURE 14.4-2. *Particle settling in sedimenting tubular-bowl centrifuge.*

where v_t is settling velocity in the radial direction in m/s, D_p is particle diameter in m, ρ_p is particle density in kg/m³, ρ is liquid density in kg/m³, and μ is liquid viscosity in Pa · s. If hindered settling occurs, the right-hand side of Eq. (14.4-8) is multiplied by the factor $(\varepsilon^2 \psi_p)$ given in Eq. (14.3-16).

Since $v_t = dr/dt$, then Eq. (14.4-8) becomes

$$dt = \frac{18\mu}{\omega^2(\rho_p - \rho)D_p^2} \frac{dr}{r} \tag{14.4-9}$$

Integrating between the limits $r = r_1$ at $t = 0$ and $r = r_2$ at $t = t_T$,

$$t_T = \frac{18\mu}{\omega^2(\rho_p - \rho)D_p^2} \ln \frac{r_2}{r_1} \tag{14.4-10}$$

The residence time t_T is equal to the volume of liquid V m³ in the bowl divided by the feed volumetric flow rate q in m³/s. The volume $V = \pi b(r_2^2 - r_1^2)$. Substituting into Eq. (14.4-10) and solving for q,

$$q = \frac{\omega^2(\rho_p - \rho)D_p^2}{18\mu \ln (r_2/r_1)} (V) = \frac{\omega^2(\rho_p - \rho)D_p^2}{18\mu \ln (r_2/r_1)} [\pi b(r_2^2 - r_1^2)] \tag{14.4-11}$$

Particles having diameters smaller than that calculated from Eq. (14.4-11) will not reach the wall of the bowl and will go out with the exit liquid. Larger particles will reach the wall and be removed from the liquid.

A cut point or critical diameter D_{pc} can be defined as the diameter of a particle that reaches $\frac{1}{2}$ the distance between r_1 and r_2. This particle moves a distance of half the liquid layer or $(r_2 - r_1)/2$ during the time this particle is in the centrifuge. The integration is then between $r = (r_1 + r_2)/2$ at $t = 0$ and $r = r_2$ at $t = t_T$. Then we obtain

$$q_c = \frac{\omega^2(\rho_p - \rho)D_{pc}^2}{18\mu \ln [2r_2/(r_1 + r_2)]} (V) = \frac{\omega^2(\rho_p - \rho)D_{pc}}{18\mu \ln [2r_2/(r_1 + r_2)]} [\pi b(r_2^2 - r_1^2)] \tag{14.4-12}$$

At this flow rate q_c, particles with a diameter greater than D_{pc} will predominantly settle to the wall and most smaller particles will remain in the liquid.

2. *Special case for settling.* For the special case where the thickness of the liquid layer is small compared to the radius, Eq. (14.4-8) can be written for a constant $r \cong r_2$ and $D_p = D_{pc}$ as follows:

$$v_t = \frac{\omega^2 r_2 D_{pc}^2(\rho_p - \rho)}{18\mu} \tag{14.4-13}$$

The time of settling t_T is then as follows for the critical D_{pc} case.

$$t_T = \frac{V}{q_c} = \frac{(r_2 - r_1)/2}{v_t} \tag{14.4-14}$$

Substituting Eq. (14.4-13) into (14.4-14) and rearranging,

$$q_c = \frac{\omega^2 r_2 D_{pc}^2(\rho_p - \rho)V}{18\mu[(r_2 - r_1)/2]} \tag{14.4-15}$$

The volume V can be expressed as

$$V \cong 2\pi r_2(r_2 - r_1)b \tag{14.4-16}$$

Combining Eqs. (14.4-15) and (14.4-16),

$$q_c = \frac{2\pi b r_2^2 \omega^2 D_{pc}^2(\rho_p - \rho)}{9\mu} \tag{14.4-17}$$

The analysis above is somewhat simplified. The pattern of flow of the fluid is actually more complicated. These equations can also be used for liquid–liquid systems where droplets of liquid migrate according to the equations and coalesce in the other liquid phase.

EXAMPLE 14.4-2. Settling in a Centrifuge
A viscous solution containing particles with a density $\rho_p = 1461$ kg/m^3 is to be clarified by centrifugation. The solution density $\rho = 801$ kg/m^3, and its viscosity is 100 cp. The centrifuge has a bowl with $r_2 = 0.02225$ m, $r_1 = 0.00716$ m, and height $b = 0.1970$ m. Calculate the critical particle diameter of the largest particles in the exit stream if $N = 23\ 000$ rev/min and the flow rate $q = 0.002832$ m^3/h.

Solution: Using Eq. (14.4-4),

$$\omega = \frac{2\pi N}{60} = \frac{2\pi(23\ 000)}{60} = 2410 \text{ rad/s}$$

The bowl volume V is

$$V = \pi b(r_2^2 - r_1^2)$$
$$= \pi(0.1970)[(0.02225)^2 - (0.00716)^2] = 2.747 \times 10^{-4} \text{ m}^3$$

Viscosity $\mu = 100 \times 10^{-3} = 0.100$ Pa·s $= 0.100$ kg/m·s. The flow rate q_c is

$$q_c = \frac{0.002832}{3600} = 7.87 \times 10^{-7} \text{ m}^3/\text{s}$$

Substituting into Eq. (14.4-12) and solving for D_{pc},

$$q_c = 7.87 \times 10^{-7}$$

$$= \frac{(2410)^2(1461 - 801)D_{pc}^2(2.747 \times 10^{-4})}{18(0.100) \ln [2 \times 0.02225/(0.00716 + 0.02225)]}$$

$$D_{pc} = 0.746 \times 10^{-6} \text{ m} \quad \text{or} \quad 0.746 \ \mu\text{m}$$

Substituting into Eq. (14.4-13) to obtain v_t and then calculating the Reynolds number, the settling is in the Stokes' law range.

3. Sigma values and scale-up of centrifuges. A useful physical characteristic of a tubular-bowl centrifuge can be derived by multiplying and dividing Eq. (14.4-12) by $2g$ and then substituting Eq. (14.3-9) written for D_{pc} into Eq. (14.4-12) to obtain

$$q_c = 2 \frac{(\rho_p - \rho)g D_{pc}^2}{18 \ \mu} \frac{\omega^2 V}{2g \ \ln [2r_2/(r_1 + r_2)]} = 2v_t \cdot \Sigma \qquad \textbf{(14.4-18)}$$

where v_t is the terminal settling velocity of the particle in a gravitational field and

$$\Sigma = \frac{\omega^2 V}{2g \ \ln [2r_2/(r_1 + r_2)]} = \frac{\omega^2 [\pi b(r_2^2 - r_1^2)]}{2g \ \ln [2r_2/(r_1 + r_2)]} \qquad \textbf{(14.4-19)}$$

where Σ is a physical characteristic of the centrifuge and not of the fluid–particle system being separated. Using Eq. (14.4-17) for the special case for settling for a thin layer,

$$\Sigma = \frac{\omega^2 \pi b 2 r_2^2}{g} \qquad \textbf{(14.4-20)}$$

The value of Σ is really the area in m^2 of a gravitational settler that will have the same sedimentation characteristics as the centrifuge for the same feed rate. To scale up from a laboratory test of q_1 and Σ_1 to q_2 (for $v_{t1} = v_{t2}$),

$$\frac{q_1}{\Sigma_1} = \frac{q_2}{\Sigma_2} \qquad \textbf{(14.4-21)}$$

This scale-up procedure is dependable for similar type and geometry centrifuges and if the centrifugal forces are within a factor of 2 from each other. If different configurations are used, efficiency factors E should be used where $q_1/\Sigma_1 E_1 = q_2/\Sigma_2 E_2$. These efficiencies are determined experimentally and values for different types of centrifuges are given elsewhere (F1, P1).

4. Separation of liquids in a centrifuge. Liquid–liquid separations in which the liquids are immiscible but finely dispersed as an emulsion are common operations in the food and other industries. An example is the dairy industry, in which the emulsion of milk is separated into skim milk and cream. In these liquid–liquid separations, the position of the outlet overflow weir in the centrifuge is quite important, not only in controlling the volumetric holdup V in the centrifuge but also in determining whether a separation is actually made.

In Fig. 14.4-3 a tubular-bowl centrifuge is shown in which the centrifuge is separating two liquid phases, one a heavy liquid with density ρ_H kg/m^3 and the second a light liquid with density ρ_L. The distances shown are as follows: r_1 is radius to surface of light liquid layer, r_2 is radius to liquid–liquid interface, and r_4 is radius to surface of heavy liquid downstream.

To locate the interface, a balance must be made of the pressures in the two layers.

The force on the fluid at distance r is, by Eq. (14.4-2),

$$F_c = mr\omega^2 \qquad (14.4-2)$$

The differential force across a thickness dr is

$$dF_c = dmr\omega^2 \qquad (14.4-22)$$

But,

$$dm = [(2\pi rb)\, dr]\rho \qquad (14.4-23)$$

where b is the height of the bowl in m and $(2\pi rb)\, dr$ is the volume of fluid. Substituting Eq. (14.4-23) in (14.4-22) and dividing both sides by the area $A = 2\pi rb$,

$$dP = \frac{dF_c}{A} = \omega^2 \rho r\, dr \qquad (14.4-24)$$

where P is pressure in N/m^2 (lb$_f$/ft^2).

Integrating Eq. (14.4-24) between r_1 and r_2,

$$P_2 - P_1 = \frac{\rho\omega^2}{2}(r_2^2 - r_1^2) \qquad (14.4-25)$$

Applying Eq. (14.4-25) to Fig. 14.4-3 and equating the pressure exerted by the light phase of thickness $r_2 - r_1$ to the pressure exerted by the heavy phase of thickness $r_2 - r_4$ at the liquid–liquid interface at r_2,

$$\frac{\rho_H \omega^2}{2}(r_2^2 - r_4^2) = \frac{\rho_L \omega^2}{2}(r_2^2 - r_1^2) \qquad (14.4-26)$$

Solving for r_2^2, the interface position,

$$r_2^2 = \frac{\rho_H r_4^2 - \rho_L r_1^2}{\rho_H - \rho_L} \qquad (14.4-27)$$

The interface at r_2 must be located at a radius smaller than r_3 in Fig. 14.4-3.

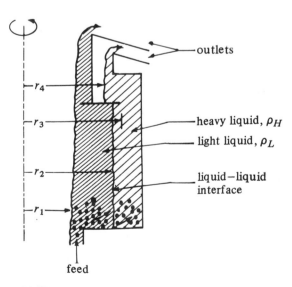

FIGURE 14.4-3. *Tubular bowl centrifuge for separating two liquid phases.*

EXAMPLE 14.4-3. Location of Interface in Centrifuge

In a vegetable-oil-refining process, an aqueous phase is being separated from the oil phase in a centrifuge. The density of the oil is 919.5 kg/m³ and that of the aqueous phase is 980.3 kg/m³. The radius r_1 for overflow of the light liquid has been set at 10.160 mm and the outlet for the heavy liquid at 10.414 mm. Calculate the location of the interface in the centrifuge.

Solution: The densities are $\rho_L = 919.5$ and $\rho_H = 980.3$ kg/m³. Substituting into Eq. (14.4-27) and solving for r_2,

$$r_2^2 = \frac{980.3(10.414)^2 - 919.5(10.160)^2}{980.3 - 919.5}$$

$$r_2 = 13.75 \text{ mm}$$

13.4D   Centrifuge Equipment for Sedimentation

1. Tubular centrifuge.   A schematic of a *tubular bowl centrifuge* is shown in Fig. 14.4-3. The bowl is tall and has a narrow diameter, 100 to 150 mm. Such centrifuges, known as *supercentrifuges*, develop a force about 13 000 times the force of gravity. Some narrow centrifuges having a diameter of 75 mm and very high speeds of 60 000 or so rev/min are known as *ultracentrifuges*. These supercentrifuges are often used to separate liquid–liquid emulsions.

2. Disk bowl centrifuge.   The *disk bowl centrifuge* shown in Fig. 14.4-4 is often used in liquid–liquid separations. The feed enters the actual compartment at the bottom and travels upward through vertically spaced feed holes, filling the spaces between the disks. The holes divide the vertical assembly into an inner section, where mostly light liquid is present, and an outer section, where mainly heavy liquid is present. This dividing line is similar to an interface in a tubular centrifuge.

The heavy liquid flows beneath the underside of a disk to the periphery of the bowl. The light liquid flows over the upper side of the disks and toward the inner outlet. Any

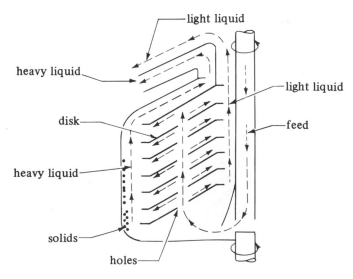

FIGURE 14.4-4.   *Schematic of disk bowl centrifuge.*

small amount of heavy solids is thrown to the outer wall. Periodic cleaning is required to remove solids deposited. Disk bowl centrifuges are used in starch–gluten separation, concentration of rubber latex, and cream separation. Details are given elsewhere (P1, L1).

14.4E Centrifugal Filtration

1. Theory for centrifugal filtration. Theoretical prediction of filtration rates in centrifugal filters have not been too successful. The filtration in centrifuges is more complicated than for ordinary filtration using pressure differences, since the area for flow and driving force increase with distance from the axis and the specific cake resistance may change markedly. Centrifuges for filtering are generally selected by scale-up from tests on a similar-type laboratory centrifuge using the slurry to be processed.

The theory of constant-pressure filtration discussed in Section 14.2E can be modified and used where the centrifugal force causes the flow instead of the impressed pressure difference. The equation will be derived for the case where a cake has already been deposited as shown in Fig. 14.4-5. The inside radius of the basket is r_2, r_i is the inner radius of the face of the cake, and r_1 is the inner radius of the liquid surface. We will assume that the cake is nearly incompressible so that an average value of α can be used for the cake. Also, the flow is laminar. If we assume a thin cake in a large-diameter centrifuge, then the area A for flow is approximately constant. The velocity of the liquid is

$$v = \frac{q}{A} = \frac{dV}{A\,dt} \tag{14.4-28}$$

where q is the filtrate flow rate in m³/s and v the velocity. Substituting Eq. (14.4-28) into (14.2-8),

$$-\Delta p = q\mu\left(\frac{m_c\,\alpha}{A^2} + \frac{R_m}{A}\right) \tag{14.4-29}$$

where $m_c = c_S V$, mass of cake in kg deposited on the filter.

For a hydraulic head of dz m, the pressure drop is

$$dp = \rho g\,dz \tag{14.4-30}$$

In a centrifugal field, g is replaced by $r\omega^2$ from Eq. (14.4-1) and dz by dr. Then,

$$dp = \rho r\omega^2\,dr \tag{14.4-31}$$

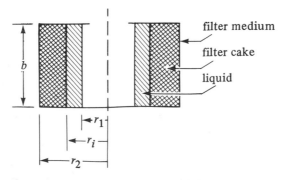

FIGURE 14.4-5. *Physical arrangement for centrifugal filtration.*

Integrating between r_1 and r_2,

$$-\Delta p = \frac{\rho \omega^2 (r_2^2 - r_1^2)}{2}$$

(14.4-32)

Combining Eqs. (14.4-29) and (14.4-32) and solving for q,

$$q = \frac{\rho \omega^2 (r_2^2 - r_1^2)}{2\mu \left(\dfrac{m_c \alpha}{A^2} + \dfrac{R_m}{A} \right)}$$

(14.4-33)

For the case where the flow area A varies considerably with the radius, the following has been derived (G1).

$$q = \frac{\rho \omega^2 (r_2^2 - r_1^2)}{2\mu \left(\dfrac{m_c \alpha}{\bar{A}_L \bar{A}_a} + \dfrac{R_m}{A_2} \right)}$$

(14.4-34)

where $A_2 = 2\pi r_2 b$ (area of filter medium), $\bar{A}_L = 2\pi b(r_2 - r_i)/\ln (r_2/r_i)$ (logarithmic cake area), and $\bar{A}_a = (r_i + r_2)\pi b$ (arithmetic mean cake area). This equation holds for a cake of a given mass at a given time. It is not an integrated equation covering the whole filtration cycle.

2. Equipment for centrifugal filtration. In a centrifugal filter, slurry is fed continuously to a rotating basket which has a perforated wall and is covered with a filter cloth. The cake builds up on the surface of the filter medium to the desired thickness. Then at the end of the filtration cycle, feed is stopped, and wash liquid is added or sprayed on to the cake. Then the wash liquid is stopped and the cake is spun as dry as possible. The motor is then shut off or slowed and the basked allowed to rotate while the solids are discharged by a scraper knife so the solids drop through an opening in the basket floor. Finally, the filtyer medium is rinsed clean to complete the cycle. Usually, the batch cycle is completely automated. Automatic batch centrifugals have basket sizes up to about 1.2 m in diameter and usually rotate below 4000 rpm.

Continuous centrifugal filters are available with capacities up to about 25 000 kg solids/h. Intermittently, the cake deposited on the filter medium is removed by being pushed toward the discharge end by a pusher, which then retreats again, allowing the cake to build up once more. As the cake is being pushed, it passes through a wash region. The filtrate and wash liquid are kept separate by partitions in the collector. Details of different types of centrifugal filters are available (P1).

14.4F Gas–Solid Cyclone Separators

1. Introduction and equipment. For separation of small solid particles or mist from gases, the most widely used type of equipment is the cyclone separator, shown in Fig. 14.4-6. The cyclone consists of a vertical cylinder with a conical bottom. The gas–solid particle mixture enters in a tangential inlet near the top. This gas–solid mixture enters in a rotating motion, and the vortex formed develops centrifugal force which throws the particles radially toward the wall.

On entering, the air in the cyclone flows downward in a spiral or vortex adjacent to the wall. When the air reaches near the bottom of the cone, it spirals upward in a smaller spiral in the center of the cone and cylinder. Hence, a double vortex is present. The downward and upward spirals are in the same direction.

The particles are thrown toward the wall and fall downward, leaving out the bottom of the cone. A cyclone is a settling device in which the outward force on the particles at high tangential velocities is many times the force of gravity. Hence, cyclones accomplish much more effective separation than gravity settling chambers.

The centrifugal force in a cyclone ranges from about 5 times gravity in large, low-velocity units to 2500 times gravity in small, high-resistance units. These devices are used often in many applications, such as in spray drying of foods, where the dried particles are removed by cyclones; in cleaning dust-laden air; and in removing mist droplets from gases. Cyclones offer one of the least expensive means of gas–particle separation. They are generally applicable in removing particles over 5 μm in diameter from gases. For particles over 200 μm in size, gravity settling chambers are often used. Wet scrubber cyclones are sometimes used where water is sprayed inside, helping to remove the solids.

2. Theory for cyclone separators. It is assumed that particles on entering a cyclone quickly reach their terminal settling velocities. Particle sizes are usually so small that Stokes' law is considered valid. For centrifugal motion, the terminal radial velocity v_{tR} is given by Eq. (14.4-8), with v_{tR} being used for v_t.

$$v_{tR} = \frac{\omega^2 r D_p^2 (\rho_p - \rho)}{18\mu} \tag{14.4-35}$$

Since $\omega = v_{\tan}/r$, where $v_{\tan}$ is tangential velocity of the particle at radius r, Eq. (14.4-35) becomes

$$v_{tR} = \frac{D_p^2 g(\rho_p - \rho)}{18\mu} \frac{v_{\tan}^2}{gr} = v_t \frac{v_{\tan}^2}{gr} \tag{14.4-36}$$

where v_t is the gravitational terminal settling velocity v_t in Eq. (14.3-9).

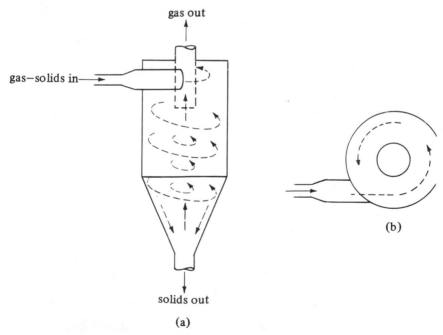

FIGURE 14.4-6. *Gas–solid cyclone separator : (a) side view, (b) top view.*

The higher the terminal velocity v_t, the greater the radial velocity v_{tR} and the easier it should be to "settle" the particle at the walls. However, the evaluation of the radial velocity is difficult, since it is a function of gravitational terminal velocity, tangential velocity, and position radially and axially in the cyclone. Hence, the following empirical equation is often used (S2).

$$v_{tR} = \frac{b_1 D_p^2 (\rho_p - \rho)}{18\mu r^n}$$

(14.4-37)

where b_1 and n are empirical constants.

3. *Efficiency of collection of cyclones.* Smaller particles have smaller settling velocities by Eq. (14.4-37) and do not have time to reach the wall to be collected. Hence, they leave with the exit air in a cyclone. Larger particles are more readily collected. The efficiency of separation for a given particle diameter is defined as the mass fraction of the size particles that are collected.

A typical collection efficiency plot for a cyclone shows that the efficiency rises rapidly with particle size. The cut diameter D_{pc} is the diameter for which one half of the mass of the entering particles is retained.

14.5   MECHANICAL SIZE REDUCTION

14.5A   Introduction

Many solid materials occur in sizes that are too large to be used and must be reduced. Often the solids are reduced in size so that the separation of various ingredients can be carried out. In general, the terms *crushing* and *grinding* are used to signify the subdividing of large solid particles to smaller particles.

In the food processing industry, a large number of food products are subjected to size reduction. Roller mills are used to grind wheat and rye to flour and corn. Soybeans are rolled, pressed, and ground to produce oil and flour. Hammer mills are often used to produce potato flour, tapioca, and other flours. Sugar is ground to a finer product.

Grinding operations are very extensive in the ore processing and cement industries. Examples are copper ores, nickel and cobalt ores, and iron ores being ground before chemical processing. Limestone, marble, gypsum, and dolomite are ground to use as fillers in paper, paint, and rubber. Raw materials for the cement industry, such as lime, alumina, and silica, are ground on a very large scale.

Solids may be reduced in size by a number of methods. *Compression* or *crushing* is generally used for reduction of hard solids to coarse sizes. *Impact* gives coarse, medium, or fine sizes. *Attrition* or *rubbing* yields fine products. *Cutting* is used to give definite sizes.

14.5B   Particle-Size Measurement

The feed-to-size reduction processes and the product are defined in terms of the particle-size distribution. One common way to plot particle sizes is to plot particle diameter (sieve opening in screen) in mm or μm versus the cumulative percent retained at that size. (Openings for various screen sizes are given in Appendix A.5.) Such a plot was given on arithmetic probability paper in Fig. 12.12-2.

Often the plot is made, instead, as the cumulative amount as percent smaller than the stated size versus particle size as shown in Fig. 14.5-1a. In Fig. 14.5-1b the same data are plotted as a particle distribution curve. The ordinate is obtained by taking the slopes of the 5-μm intervals of Fig. 14.5-1a and converting to percent by weight per μm. Complete particle-size analysis is necessary for most comparisons and calculations.

14.5C   Energy and Power Required in Size Reduction

1. Introduction.   In size reduction of solids, feed materials of solid are reduced to a smaller size by mechanical action. The materials are fractured. The particles of feed are first distorted and strained by the action of the size-reduction machine. This work to strain the particles is first stored temporarily in the solid as strain energy. As additional force is added to the stressed particles, the strain energy exceeds a certain level, and the material fractures into smaller pieces.

When the material fractures, new surface area is created. Each new unit area of surface requires a certain amount of energy. Some of the energy added is used to create the new surface, but a large portion of it appears as heat. The energy required for fracture is a complicated function of the type of material, size, hardness, and other factors.

The magnitude of the mechanical force applied; the duration; the type of force, such as compression, shear, and impact; and other factors affect the extent and efficiency of the size-reduction process. The important factors in the size-reduction process are the amount of energy or power used and the particle size and new surface formed.

2. Power required in size reduction.   The various theories or laws proposed for predicting power requirements for size reduction of solids do not apply well in practice. The most important ones will be discussed briefly. Part of the problem in the theories is that of estimating the theoretical amount of energy required to fracture and create new surface area. Approximate calculations give actual efficiencies of about 0.1 to 2%.

The theories derived depend upon the assumption that the energy E required to produce a change dX in a particle of size X is a power function of X.

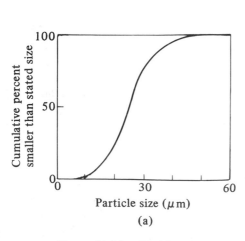

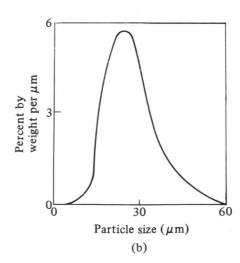

FIGURE 14.5-1.   *Particle-size-distribution curves : (a) cumulative percent versus particle size, (b) percent by weight versus particle size. (From R. H. Perry and C. H. Chilton, Chemical Engineers' Handbook, 5th ed. New York : McGraw-Hill Book Company, 1973. With permission.)*

Sec. 14.5   Mechanical Size Reduction   **841**

$$\frac{dE}{dX} = -\frac{C}{X^n} \tag{14.5-1}$$

where X is size or diameter of particle in mm, and n and C are constants depending upon type and size of material and type of machine.

Rittinger proposed a law which states that the work in crushing is proportional to the new surface created. This leads to a value of $n = 2$ in Eq. (14.5-1), since area is proportional to length squared. Integrating Eq. (14.5-1),

$$E = \frac{C}{n-1}\left(\frac{1}{X_2^{n-1}} - \frac{1}{X_1^{n-1}}\right) \tag{14.5-2}$$

where X_1 is mean diameter of feed and X_2 is mean diameter of product. Since $n = 2$ for Rittinger's equation, we obtain

$$E = K_R\left(\frac{1}{X_2} - \frac{1}{X_1}\right) \tag{14.5-3}$$

where E is work to reduce a unit mass of feed from X_1 to X_2 and K_R is a constant. The law implies that the same amount of energy is needed to reduce a material from 100 mm to 50 mm as is needed to reduce the same material from 50 mm to 33.3 mm. It has been found experimentally that this law has some validity in grinding fine powders.

Kick assumed that the energy required to reduce a material in size was directly proportional to the size-reduction ratio. This implies $n = 1$ in Eq. (14.5-1), giving

$$E = C \ln \frac{X_1}{X_2} = K_K \log \frac{X_1}{X_2} \tag{14.5-4}$$

where K_K is a constant. This law implies that the same amount of energy is required to reduce a material from 100 mm to 50 mm as is needed to reduce the same material from 50 mm to 25 mm.

Recent data by Bond (B3) on correlating extensive experimental data suggest that the work required using a large-size feed is proportional to the square root of the surface/volume ratio of the product. This corresponds to $n = 1.5$ in Eq. (14.5-1), giving

$$E = K_B \frac{1}{\sqrt{X_2}} \tag{14.5-5}$$

where K_B is a constant. To use Eq. (14.4-5), Bond proposed a work index E_i as the work in $kW \cdot h/ton$ required to reduce a unit weight from a very large size to 80% passing a 100-μm screen. Then the work E is the gross work required to reduce a unit weight of feed with 80% passing a diameter X_F μm to a product with 80% passing X_P μm.

Bond's final equation in terms of English units is

$$\frac{P}{T} = 1.46 E_i\left(\frac{1}{\sqrt{D_P}} - \frac{1}{\sqrt{D_F}}\right) \tag{14.5-6}$$

where P is hp, T is feed rate in tons/min, D_F is size of feed in ft, and D_P is product size in ft. Typical values of E_i for various types of materials are given in Perry and Green (P1) and by Bond (B3). Some typical values are bauxite ($E_i = 9.45$), coal (11.37), potash salt (8.23), shale (16.4), and granite (14.39). These values should be multiplied by 1.34 for dry grinding.

EXAMPLE 14.5-1. *Power to Crush Iron Ore by Bond's Theory*
It is desired to crush 10 ton/h of iron ore hematite. The size of the feed is

such that 80% passes a 3-in. (76.2-mm) screen and 80% of the product is to pass a $\frac{1}{8}$-in. (3.175-mm) screen. Calculate the gross power required. Use a work index E_i for iron ore hematite of 12.68 (P1).

Solution: The feed size is $D_F = \frac{3}{12} = 0.250$ ft (76.2 mm) and the product size is $D_P = \frac{1}{8}/12 = 0.0104$ ft (3.175 mm). The feed rate is $T = 10/60 = 0.167$ ton/min. Substituting into Eq. (14.5-6) and solving for P,

$$\frac{P}{0.167} = (1.46)(12.68)\left(\frac{1}{\sqrt{0.0104}} - \frac{1}{\sqrt{0.250}}\right)$$

$$P = 24.1 \text{ hp } (17.96 \text{ kW})$$

14.5D   Equipment for Size Reduction

1. Introduction and classification.   Size-reduction equipment may be classified according to the way the forces are applied as follows: between two surfaces, as in crushing and shearing; at one solid surface, as in impact; and by action of the surrounding medium, as in a colloid mill. A more practical classification is to divide the equipment into crushers, grinders, fine grinders, and cutters.

2. Jaw crushers.   Equipment for coarse reduction of large amounts of solids consists of slow-speed machines called *crushers*. Several types are in common use. In the first type, a jaw crusher, the material is fed between two heavy jaws or flat plates. As shown in the *Dodge crusher* in Fig. 14.5-2a, one jaw is fixed and the other reciprocating and movable on a pivot point at the bottom. The jaw swings back and forth, pivoting at the bottom of the V. The material is gradually worked down into a narrower space, being crushed as it moves.

The *Blake crusher* in Fig. 14.5-2b is more commonly used, and the pivot point is at the top of the movable jaw. The reduction ratios average about 8 : 1 in the Blake crusher. Jaw crushers are used mainly for primary crushing of hard materials and are usually followed by other types of crushers.

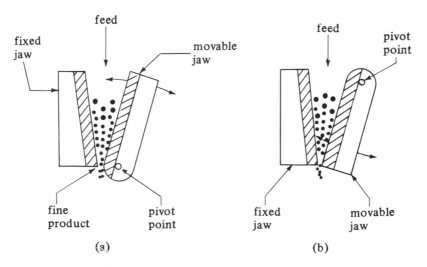

FIGURE 14.5-2.   *Types of jaw crushers : (a) Dodge type, (b) Blake type.*

3. *Gyratory crushers.* The *gyratory crusher* shown in Fig. 14.5-3a has taken over to a large extent in the field of large hard-ore and mineral crushing applications. Basically it is like a mortar-and-pestle crusher. The movable crushing head is shaped like an inverted truncated cone and is inside a truncated cone casing. The crushing head rotates eccentrically and the material being crushed is trapped between the outer fixed cone and the inner gyrating cone.

4. *Roll crushers.* In Fig. 14.5-3b a typical smooth *roll crusher* is shown. The rolls are rotated toward each other at the same or different speeds. Wear of the rolls is a serious problem. The reduction ratio varies from about 4 : 1 to 2.5 : 1. Single rolls are often used, rotating against a fixed surface, and corrugated and toothed rolls are also used. Many food products that are not hard materials, such as flour, soybeans, and starch, are ground on rolls.

5. *Hammer mill grinders.* *Hammer mill devices* are used to reduce intermediate-sized material to small sizes or powder. Often the product from jaw and gyratory crushers is the feed to the hammer mill. In the hammer mill a high-speed rotor turns inside a cylindrical casing. Sets of hammers are attched to pivot points at the outside of the rotor. The feed enters the top of the casing and the particles are broken as they fall through the cylinder. The material is broken by the impact of the hammers and pulverized into powder between the hammers and casing. The powder then passes through a grate or screen at the discharge end.

6. *Revolving grinding mills.* For intermediate and fine reduction of materials *revolving grinding mills* are often used. In such mills a cylindrical or conical shell rotating on a horizontal axis is charged with a grinding medium such as steel, flint, or porcelain balls, or with steel rods. The size reduction is effected by the tumbling of the balls or rods on the material between them. In the revolving mill the grinding elements are carried up the side of the shell and fall on the particles underneath. These mills may operate wet or dry.

   Equipment for very fine grinding is very specialized. In some cases two flat disks are used where one or both disks rotate and grind the material caught between the disks (P1).

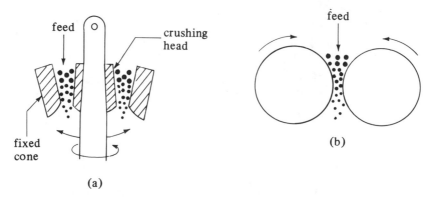

FIGURE 14.5-3.   *Types of size-reduction equipment : (a) gyratory crusher, (b) roll crusher.*

PROBLEMS

14.2-1 *Constant-Pressure Filtration and Filtration Constants.* Data for the filtration of $CaCO_3$ slurry in water at 298.2 K (25°C) are reported as follows (R1, R2, M1) at a constant pressure $(-\Delta p)$ of 46.2 kN/m^2 (6.70 psia). The area of the plate-and-frame press was 0.0439 m^2 (0.473 ft^2) and the slurry concentration was 23.47 kg solid/m^3 filtrate. Calculate the constants α and R_m. Data are given as t = time in s and V = volume of filtrate collected in m^3.

$V \times 10^3$	t	$V \times 10^3$	t	$V \times 10^3$	t
0.5	17.3	1.5	72.0	2.5	152.0
1.0	41.3	2.0	108.3	3.0	201.7

Ans. $\alpha = 1.106 \times 10^{11}$ m/kg (1.65×10^{11} ft/lb$_m$), $R_m = 6.40 \times 10^{10}$ m^{-1} (1.95×10^{10} ft^{-1})

14.2-2. *Filtration Constants for Constant-Pressure Filtration.* Data for constant-pressure filtration at 194.4 kN/m^2 are reported for the same slurry and press as in Problem 14.2-1 as follows, where t is in s and V in m^3:

$V \times 10^3$	t	$V \times 10^3$	t	$V \times 10^3$	t
0.5	6.3	2.5	51.7	4.5	134.0
1.0	14.0	3.0	69.0	5.0	160.0
1.5	24.2	3.5	88.8		
2.0	37.0	4.0	110.0		

Calculate the constants α and R_m.

Ans. $\alpha = 1.61 \times 10^{11}$ m/kg

14.2-3. *Compressibility of Filter Cake.* Use the data for specific cake resistance α from Example 14.2-1 and Problems 14.2-1 and 14.2-2 and determine the compressibility constant s in Eq. (14.2-11). Plot the ln of α versus the ln of $-\Delta p$ and determine the slope s.

14.2-4. *Prediction of Filtration Time and Washing Time.* The slurry used in Problem 14.2-1 is to be filtered in a plate-and-frame press having 30 frames and 0.873 m^2 area per frame. The same pressure, 46.2 kN/m^2, will be used in constant-pressure filtration. Assume the same filter cake properties and filter cloth, and calculate the time to recover 2.26 m^3 of filtrate. At the end, using through washing and 0.283 m^3 of wash water, calculate the time of washing and the total filter cycle time if cleaning the press takes 30 min.

14.2-5. *Constants in Constant-Pressure Filtration.* McMillen and Webber (M2), using a filter press with an area of 0.0929 m^2, performed constant-pressure filtration at 34.5 kPa of a 13.9 wt % $CaCO_3$ solids in water slurry at 300 K. The mass ratio of wet cake to dry cake was 1.59. The dry cake density was 1017 kg/m^3. The data obtained are as follows, where W = kg filtrate and t = time in s:

W	t	W	t	W	t
0.91	24	3.63	244	6.35	690
1.81	71	4.54	372	7.26	888
2.72	146	5.44	524	8.16	1188

Calculate the values of α and R_m.

14.2-6. Constant-Pressure Filtration and Washing in a Leaf Filter. An experimental filter press having an area of 0.0414 m^2 (R1) is used to filter an aqueous $BaCO_3$ slurry at a constant pressure of 267 kPa. The filtration equation obtained was

$$\frac{t}{V} = 10.25 \times 10^6 V + 3.4 \times 10^3$$

where t is in s and V in m^3.
(a) If the same slurry and conditions are used in a leaf press having an area of 6.97 m^2, how long will it take to obtain 1.00 m^3 of filtrate?
(b) After the filtration the cake is to be washed with 0.100 m^3 of water. Calculate the time of washing.

Ans. (a) $t = 381.8$ s

14.2-7. Constant-Rate Filtration of Incompressible Cake. The filtration equation for filtration at a constant pressure of 38.7 psia (266.8 kPa) is

$$\frac{t}{V} = 6.10 \times 10^{-5} V + 0.01$$

where t is in s, $-\Delta p$ in psia, and V in liters. The specific resistance of the cake is independent of pressure. If the filtration is run at a constant rate of 10 liters/s, how long will it take to reach 50 psia?

14.2-8. Effect of Filter Medium Resistance on Continuous Rotary-Drum Filter. Repeat Example 14.2-4 for the continuous rotary-drum vacuum filter but do not neglect the constant R_m, which is the filter medium resistance to flow. Compare with results of Example 14.2-4.

Ans. $A = 7.78 \text{ m}^2$

14.2-9. Throughput in Continuous Rotary Drum Filter. A rotary drum filter having an area of 2.20 m^2 is to be used to filter the $CaCO_3$ slurry given in Example 14.2-4. The drum has a 28% submergence and the filter cycle time is 300 s. A pressure drop of 62.0 kN/m^2 is to be used. Calculate the slurry feed rate in kg slurry/s for the following cases.
(a) Neglect the filter medium resistance.
(b) Do not neglect the value of B.

14.3-1. Settling Velocity of a Coffee Extract Particle. Solid spherical particles of coffee extract (F1) from a dryer having a diameter of 400 μm are falling through air at a temperature of 422 K. The density of the particles is 1030 kg/m^3. Calculate the terminal settling velocity and the distance of fall in 5 s. The pressure is 101.32 kPa.

Ans. $v_t = 1.49$ m/s, 7.45 m fall

14.3-2. Terminal Settling Velocity of Dust Particles. Calculate the terminal settling velocity of dust particles having a diameter of 60 μm in air at 294.3 K and 101.32 kPa. The dust particles can be considered spherical with a density of 1280 kg/m^3.

Ans. $v_t = 0.1372$ m/s

14.3-3. Settling Velocity of Liquid Particles. Oil droplets having a diameter of 200 μm are settling from still air at 294.3 K and 101.32 kPa. The density of the oil is 900 kg/m^3. A settling chamber is 0.457 m high. Calculate the terminal settling velocity. How long will it take the particles to settle? (*Note :* If the Reynolds number is above about 100, the equations and form drag correlation for rigid spheres cannot be used.)

14.3-4. Settling Velocity of Quartz Particles in Water. Solid quartz particles having a diameter of 1000 μm are settling from water at 294.3 K. The density of the spherical particles is 2650 kg/m^3. Calculate the terminal settling velocity of these particles.

14.3-5. *Hindered Settling of Solid Particles.* Solid spherical particles having a diameter of 0.090 mm and a solid density of 2002 kg/m^3 are settling in a solution of water at 26.7°C. The volume fraction of the solids in the water is 0.45. Calculate the settling velocity and the Reynolds number.

14.3-6. *Settling of Quartz Particles in Hindered Settling.* Particles of quartz having a diameter of 0.127 mm and a specific gravity of 2.65 are settling in water at 293.2 K. The volume fraction of the particles in the slurry mixture of quartz and water is 0.25. Calculate the hindered settling velocity and the Reynolds number.

14.3-7. *Density Effect on Settling Velocity and Diameter.* Calculate the terminal settling velocity of a glass sphere 0.080 mm in diameter having a density of 2469 kg/m^3 in air at 300 K and 101.32 kPa. Also calculate the diameter of a sphalerite sphere having a specific gravity of 4.00 with the same terminal settling velocity.

14.3-8. *Differential Settling of Particles.* Repeat Example 14.3-3 for particles having a size range of 1.27×10^{-2} mm to 5.08×10^{-2} mm. Calculate the size range of the various fractions obtained using free settling conditions. Also calculate the value of the largest Reynolds number occurring.

14.3-9. *Separation by Settling.* A mixture of galena and silica particles has a size range of 0.075–0.65 mm and is to be separated by a rising stream of water at 293.2 K. Use specific gravities from Example 14.3-3.
   (a) To obtain an uncontaminated product of galena, what velocity of water flow is needed and what is the size range of the pure product?
   (b) If another liquid, such as benzene, having a specific gravity of 0.85 and a viscosity of 6.50×10^{-4} Pa·s is used, what velocity is needed and what is the size range of the pure product?

14.3-10. *Separation by Sink-and-Float Method.* Quartz having a specific gravity of 2.65 and hematite having a specific gravity of 5.1 are present in a mixture of particles. It is desired to separate them by a sink-and-float method using a suspension of fine particles of ferrosilicon having a specific gravity of 6.7 in water. At what consistency in vol % ferrosilicon solids in water should the medium be maintained for the separation?

14.3-11. *Batch Settling and Sedimentation Velocities.* A batch settling test on a slurry gave the following results, where the height z in meters between the clear liquid and the suspended solids is given at time t hours.

t (h)	z (m)	t (h)	z (m)	t (h)	z (m)
0	0.360	1.75	0.150	12.0	0.102
0.50	0.285	3.00	0.125	20.0	0.090
1.00	0.211	5.00	0.113		

The original slurry concentration is 250 kg/m^3 of slurry. Determine the velocities of settling and concentrations and make a plot of velocity versus concentration.

14.4-1. *Comparison of Forces in Centrifuges.* Two centrifuges rotate at the same peripheral velocity of 53.34 m/s. The first bowl has a radius of $r_1 = 76.2$ mm and the second $r_2 = 305$ mm. Calculate the rev/min and the centrifugal forces developed in each bowl.
   Ans. $N_1 = 6684$ rev/min,
   $N_2 = 1670$ rev/min, 3806 g's in bowl 1, 951 g's in bowl 2

14.4-2. *Forces in a Centrifuge.* A centrifuge bowl is spinning at a constant 2000 rev/min. What radius bowl is needed for the following?

(a) A force of 455 g's.

(b) A force four times that in part (a).

Ans. (a) $r = 0.1017$ m

14.4-3. *Effect of Varying Centrifuge Dimensions and Speed.* Repeat Example 14.4-2 but with the following changes.

(a) Reduce the rev/min to 10 000 and double the outer-bowl radius r_2 to 0.0445 m, keeping $r_1 = 0.00716$ m.

(b) Keep all variables as in Example 14.4-2 but double the throughput.

Ans. (b) $D_p = 1.747 \times 10^{-6}$ m

14.4-4. *Centrifuging to Remove Food Particles.* A dilute slurry contains small solid food particles having a diameter of 5×10^{-2} mm which are to be removed by centrifuging. The particle density is 1050 kg/m^3 and the solution density is 1000 kg/m^3. The viscosity of the liquid is 1.2×10^{-3} Pa·s. A centrifuge at 3000 rev/min is to be used. The bowl dimensions are $b = 100.1$ mm, $r_1 = 5.00$ mm, and $r_2 = 30.0$ mm. Calculate the expected flow rate in m^3/s just to remove these particles.

14.4-5. *Effect of Oil Density on Interface Location.* Repeat Example 14.4-3, but for the case where the vegetable oil density has been decreased to 914.7 kg/m^3.

14.4-6. *Interface in Cream Separator.* A cream separator centrifuge has an outlet discharge radius $r_1 = 50.8$ mm and outlet radius $r_4 = 76.2$ mm. The density of the skim milk is 1032 kg/m^3 and that of the cream is 865 kg/m^3 (E1). Calculate the radius of the interface neutral zone.

Ans. $r_2 = 150$ mm

14.4-7. *Scale-Up and Σ Values of Centrifuges.* For the conditions given in Example 14.4-2, do as follows.

(a) Calculate the Σ value.

(b) A new centrifuge having the following dimensions is to be used. $r_2 = 0.0445$ m, $r_1 = 0.01432$ m, $b = 0.394$ m, and $N = 26\,000$ rev/min. Calculate the new Σ value and scale-up the flow rate using the same solution.

Ans. (a) $\Sigma = 196.3$ m^2

14.4-8. *Centrifugal Filtration Process.* A batch centrifugal filter similar to Fig. 14.4-5 has a bowl height $b = 0.457$ m and $r_2 = 0.381$ m and operates at 33.33 rev/s at 25.0°C. The filtrate is essentially water. At a given time in the cycle the slurry and cake formed have the following properties. $c_S = 60.0$ kg solids/m^3 filtrate, $\varepsilon = 0.82$, $\rho_p = 2002$ kg solids/m^3, cake thickness $= 0.152$ m, $\alpha = 6.38 \times 10^{10}$ m/kg, $R_m = 8.53 \times 10^{10}$ m^{-1}, $r_1 = 0.2032$ m. Calculate the rate of filtrate flow.

Ans. $q = 6.11 \times 10^{-4}$ m^3/s

14.5-1. *Change in Power Requirements in Crushing.* In crushing a certain ore, the feed is such that 80% is less than 50.8 mm in size and the product size is such that 80% is less than 6.35 mm. The power required is 89.5 kW. What will be the power required using the same feed so that 80% is less than 3.18 mm? Use the Bond equation. (*Hint :* The work index E_i is unknown, but it can be determined using the original experimental data in terms of T. In the equation for the new size, the same unknowns appear. Dividing one equation by the other will eliminate these unknowns.)

Ans. 146.7 kW

14.5-2. *Crushing of Phosphate Rock.* It is desired to crush 100 ton/h of phosphate rock from a feed size where 80% is less than 4 in. to a product where 80% is less than $\frac{1}{8}$ in. The work index is 10.13 (P1).

(a) Calculate the power required.

(b) Calculate the power required to crush the product further where 80% is less than 1000 μm.

REFERENCES

(B1)   BADGER, W. L., and BANCHERO, J. T. *Introduction to Chemical Engineering.* New York: McGraw-Hill Book Company, 1955.

(B2)   BECKER, H. A. *Can. J. Chem. Eng.*, **37**, 85 (1951).

(B3)   BOND, F. C. *Trans. A.I.M.E.*, **193**, 484 (1952).

(C1)   COULSON, J. M., and RICHARDSON, J. F. *Chemical Engineering*, Vol. 2, 3rd ed. New York: Pergamon Press, Inc., 1978.

(E1)   EARLE, R. L. *Unit Operations in Food Processing.* Oxford: Pergamon Press, Inc., 1966.

(F1)   FOUST, A. S., et al. *Principles of Unit Operations*, 2nd ed. New York: John Wiley & Sons, Inc., 1980.

(F2)   FITCH, B. *Ind. Eng. Chem.*, **58** (10), 18 (1966).

(G1)   GRACE, H. P. *Chem. Eng. Progr.*, **46**, 467 (1950); **49**, 303, 367, 427 (1953); *A.I.Ch.E. J.*, **2**, 307 (1956).

(H1)   HUGHES, R. R., and GILLILAND, E. R. *Chem. Eng. Progr.*, **48**, 497 (1952).

(K1)   KYNCH, G. J. *Trans. Faraday Soc.*, **48**, 166 (1952).

(L1)   LARIAN, M. G. *Fundamentals of Chemical Engineering Operations.* Englewood Cliffs, N.J.: Prentice-Hall, Inc., 1958.

(L2)   LAPPLE, C. E., and SHEPHERD, C. B. *Ind. Eng. Chem.*, **32**, 605 (1940).

(M1)   MCCABE, W. L., and SMITH, J. C. *Unit Operations of Chemical Engineering*, 3rd ed. New York: McGraw-Hill Book Company, 1976.

(M2)   MCMILLEN, E. L., and WEBBER, H. A. *Trans. A.I.Ch.E.*, **34**, 213 (1938).

(N1)   NICKOLAUS, N., and DAHLSTROM, D. A. *Chem. Eng. Progr.*, **52**(3), 87M (1956).

(P1)   PERRY, R. H., and GREEN, D. *Perry's Chemical Engineers' Handbook*, 6th ed. New York: McGraw-Hill Book Company, 1984.

(P2)   PERRY, R. H., and CHILTON, C. H. *Chemical Engineers' Handbook*, 5th ed. New York: McGraw-Hill Book Company, 1973.

(R1)   RUTH, B. F., and KEMPE, L. L. *Trans. A.I.Ch.E.*, **34**, 213 (1938).

(R2)   RUTH, B. F. *Ind. Eng. Chem.*, **25**, 157 (1933).

(S1)   STEINOUR, H. H. *Ind. Eng. Chem.*, **36**, 618, 840 (1944).

(S2)   SHEPHERD, C. B., and LAPPLE, C. E. *Ind. Eng. Chem.*, **31**, 972 (1939); **32**, 1246 (1940).

(T1)   TALMAGE, W. P., and FITCH, E. B. *Ind. Eng. Chem.*, **47**(1), 38 (1955).

(Z1)   ZENZ, F. A., and OTHMER, D. F. *Fluidization and Fluid-Particle Systems.* New York: Reinhold Publishing Co., Inc., 1960.

APPENDIX A.1

Fundamental Constants
and Conversion Factors

A.1-1   Gas Law Constant R

Numerical Value	Units
1.9872	g cal/g mol $\cdot$ K
1.9872	btu/lb mol $\cdot$ °R
82.057	cm$^3 \cdot$ atm/g mol $\cdot$ K
8314.34	J/kg mol $\cdot$ K
82.057 × 10^{-3}	m$^3 \cdot$ atm/kg mol $\cdot$ K
8314.34	kg $\cdot$ m^2/s$^2 \cdot$ kg mol $\cdot$ K
10.731	ft$^3 \cdot$ lb$_f$/in.$^2 \cdot$ lb mol $\cdot$ °R
0.7302	ft$^3 \cdot$ atm/lb mol $\cdot$ °R
1545.3	ft $\cdot$ lb$_f$/lb mol $\cdot$ °R
8314.34	m$^3 \cdot$ Pa/kg mol $\cdot$ K

A.1-2   Volume and Density

1 g mol ideal gas at 0°C, 760 mm Hg = 22.4140 liters = 22 414 cm^3
1 lb mol ideal gas at 0°C, 760 mm Hg = 359.05 ft^3
1 kg mol ideal gas at 0°C, 760 mm Hg = 22.414 m^3
Density of dry air at 0°C, 760 mm Hg = 1.2929 g/liter
$\qquad\qquad\qquad\qquad\qquad\qquad\quad$ = 0.080711 lb$_m$/ft^3
Molecular weight of air = 28.97 lb$_m$/lb mol = 28.97 g/g mol
1 g/cm^3 = 62.43 lb$_m$/ft^3 = 1000 kg/m^3
1 g/cm^3 = 8.345 lb$_m$/U.S. gal
1 lb$_m$/ft^3 = 16.0185 kg/m^3

A.1-3   Length

1 in. = 2.540 cm
100 cm = 1 m (meter)

1 micron $= 10^{-6}$ m $= 10^{-4}$ cm $= 10^{-3}$ mm $= 1\ \mu$m (micrometer)
1 Å (angstrom) $= 10^{-10}$ m $= 10^{-4}\ \mu$m
1 mile $= 5280$ ft
1 m $= 3.2808$ ft $= 39.37$ in.

A.1-4 Mass

$1\ \text{lb}_m = 453.59$ g $= 0.45359$ kg
$1\ \text{lb}_m = 16$ oz $= 7000$ grains
1 kg $= 1000$ g $= 2.2046\ \text{lb}_m$
1 ton (short) $= 2000\ \text{lb}_m$
1 ton (long) $= 2240\ \text{lb}_m$
1 ton (metric) $= 1000$ kg

A.1-5 Standard Acceleration of Gravity

$g = 9.80665$ m/s^2
$g = 980.665$ cm/s^2
$g = 32.174$ ft/s^2
g_c (gravitational conversion factor) $= 32.1740\ \text{lb}_m \cdot \text{ft/lb}_f \cdot \text{s}^2$
$$= 980.665\ \text{g}_m \cdot \text{cm/g}_f \cdot \text{s}^2$$

A.1-6 Volume

1 L (liter) $= 1000$ cm^3	1 m^3 $= 1000$ L (liter)
1 in.^{3} $= 16.387$ cm^3	1 U.S. gal $= 4$ qt
1 ft^3 $= 28.317$ L (liter)	1 U.S. gal $= 3.7854$ L (liter)
1 ft^3 $= 0.028317$ m^3	1 U.S. gal $= 3785.4$ cm^3
1 ft^3 $= 7.481$ U.S. gal	1 British gal $= 1.20094$ U.S. gal
1 m^3 $= 264.17$ U.S. gal	1 m^3 $= 35.313$ ft^3

A.1-7 Force

$1\ \text{g} \cdot \text{cm/s}^2$ (dyn) $= 10^{-5}$ kg $\cdot$ m/s^2 $= 10^{-5}$ N (newton)
$1\ \text{g} \cdot \text{cm/s}^2 = 7.2330 \times 10^{-5}\ \text{lb}_m \cdot \text{ft/s}^2$ (poundal)
1 kg $\cdot$ m/s^2 $= 1$ N (newton)
$1\ \text{lb}_f = 4.4482$ N
$1\text{g} \cdot \text{cm/s}^2 = 2.2481 \times 10^{-6}\ \text{lb}_f$

A.1-8 Pressure

1 bar $= 1 \times 10^5$ Pa (pascal) $= 1 \times 10^5$ N/m^2
1 psia $= 1\ \text{lb}_f$/in.^{2}
1 psia $= 2.0360$ in. Hg at 0°C
1 psia $= 2.311$ ft H$_2$O at 70°F
1 psia $= 51.715$ mm Hg at 0°C $(\rho_{Hg} = 13.5955$ g/cm$^3)$
1 atm $= 14.696$ psia $= 1.01325 \times 10^5$ N/m^2 $= 1.01325$ bar
1 atm $= 760$ mm Hg at 0°C $= 1.01325 \times 10^5$ Pa
1 atm $= 29.921$ in. Hg at 0°C
1 atm $= 33.90$ ft H$_2$O at 4°C

$1 \text{ psia} = 6.89476 \times 10^4 \text{ g/cm} \cdot \text{s}^2$

$1 \text{ psia} = 6.89476 \times 10^4 \text{ dyn/cm}^2$

$1 \text{ dyn/cm}^2 = 2.0886 \times 10^{-3} \text{ lb}_f/\text{ft}^2$

$1 \text{ psia} = 6.89476 \times 10^3 \text{ N/m}^2 = 6.89476 \times 10^3 \text{ Pa}$

$1 \text{ lb}_f/\text{ft}^2 = 4.7880 \times 10^2 \text{ dyn/cm}^2 = 47.880 \text{ N/m}^2$

$1 \text{ mm Hg} (0°\text{C}) = 1.333224 \times 10^2 \text{ N/m}^2 = 0.1333224 \text{ kPa}$

A.1-9 Power

$1 \text{ hp} = 0.74570 \text{ kW}$ $1 \text{ watt (W)} = 14.340 \text{ cal/min}$

$1 \text{ hp} = 550 \text{ ft} \cdot \text{lb}_f/\text{s}$ $1 \text{ btu/h} = 0.29307 \text{ W (watt)}$

$1 \text{ hp} = 0.7068 \text{ btu/s}$ $1 \text{ J/s (joule/s)} = 1 \text{ W}$

A.1-10 Heat, Energy, Work

$1 \text{ J} = 1 \text{ N} \cdot \text{m} = 1 \text{ kg} \cdot \text{m}^2/\text{s}^2$

$1 \text{ kg} \cdot \text{m}^2/\text{s}^2 = 1 \text{ J (joule)} = 10^7 \text{ g} \cdot \text{cm}^2/\text{s}^2 \text{ (erg)}$

$1 \text{ btu} = 1055.06 \text{ J} = 1.05506 \text{ kJ}$

$1 \text{ btu} = 252.16 \text{ cal (thermochemical)}$

$1 \text{ kcal (thermochemical)} = 1000 \text{ cal} = 4.1840 \text{ kJ}$

$1 \text{ cal (thermochemical)} = 4.1840 \text{ J}$

$1 \text{ cal (IT)} = 4.1868 \text{ J}$

$1 \text{ btu} = 251.996 \text{ cal (IT)}$

$1 \text{ btu} = 778.17 \text{ ft} \cdot \text{lb}_f$

$1 \text{ hp} \cdot \text{h} = 0.7457 \text{ kW} \cdot \text{h}$

$1 \text{ hp} \cdot \text{h} = 2544.5 \text{ btu}$

$1 \text{ ft} \cdot \text{lb}_f = 1.35582 \text{ J}$

$1 \text{ ft} \cdot \text{lb}_f/\text{lb}_m = 2.9890 \text{ J/kg}$

A.1-11 Thermal Conductivity

$1 \text{ btu/h} \cdot \text{ft} \cdot °\text{F} = 4.1365 \times 10^{-3} \text{ cal/s} \cdot \text{cm} \cdot °\text{C}$

$1 \text{ btu/h} \cdot \text{ft} \cdot °\text{F} = 1.73073 \text{ W/m} \cdot \text{K}$

A.1-12 Heat-Transfer Coefficient

$1 \text{ btu/h} \cdot \text{ft}^2 \cdot °\text{F} = 1.3571 \times 10^{-4} \text{ cal/s} \cdot \text{cm}^2 \cdot °\text{C}$

$1 \text{ btu/h} \cdot \text{ft}^2 \cdot °\text{F} = 5.6783 \times 10^{-4} \text{ W/cm}^2 \cdot °\text{C}$

$1 \text{ btu/h} \cdot \text{ft}^2 \cdot °\text{F} = 5.6783 \text{ W/m}^2 \cdot \text{K}$

$1 \text{ kcal/h} \cdot \text{m}^2 \cdot °\text{F} = 0.2048 \text{ btu/h} \cdot \text{ft}^2 \cdot °\text{F}$

A.1-13 Viscosity

$1 \text{ cp} = 10^{-2} \text{ g/cm} \cdot \text{s (poise)}$

$1 \text{ cp} = 2.4191 \text{ lb}_m/\text{ft} \cdot \text{h}$

$1 \text{ cp} = 6.7197 \times 10^{-4} \text{ lb}_m/\text{ft} \cdot \text{s}$

$1 \text{ cp} = 10^{-3} \text{ Pa} \cdot \text{s} = 10^{-3} \text{ kg/m} \cdot \text{s} = 10^{-3} \text{ N} \cdot \text{s/m}^2$

$1 \text{ cp} = 2.0886 \times 10^{-5} \text{ lb}_f \cdot \text{s/ft}^2$

$1 \text{ Pa} \cdot \text{s} = 1 \text{ N} \cdot \text{s/m}^2 = 1 \text{ kg/m} \cdot \text{s} = 1000 \text{ cp} = 0.67197 \text{ lb}_m/\text{ft} \cdot \text{s}$

A.1-14 Diffusivity

$1 \text{ cm}^2/\text{s} = 3.875 \text{ ft}^2/\text{h}$ $1 \text{ m}^2/\text{s} = 3.875 \times 10^4 \text{ ft}^2/\text{h}$

$1 \text{ cm}^2/\text{s} = 10^{-4} \text{ m}^2/\text{s}$ $1 \text{ centistoke} = 10^{-2} \text{ cm}^2/\text{s}$

$1 \text{ m}^2/\text{h} = 10.764 \text{ ft}^2/\text{h}$

A.1-15 Mass Flux and Molar Flux

$1 \text{ g/s} \cdot \text{cm}^2 = 7.3734 \times 10^3 \text{ lb}_m/\text{h} \cdot \text{ft}^2$

$1 \text{ g mol/s} \cdot \text{cm}^2 = 7.3734 \times 10^3 \text{ lb mol/h} \cdot \text{ft}^2$

$1 \text{ g mol/s} \cdot \text{cm}^2 = 10 \text{ kg mol/s} \cdot \text{m}^2 = 1 \times 10^4 \text{ g mol/s} \cdot \text{m}^2$

$1 \text{ lb mol/h} \cdot \text{ft}^2 = 1.3562 \times 10^{-3} \text{ kg mol/s} \cdot \text{m}^2$

A.1-16 Heat Flux and Heat Flow

$1 \text{ btu/h} \cdot \text{ft}^2 = 3.1546 \text{ W/m}^2$

$1 \text{ btu/h} = 0.29307 \text{ W}$

$1 \text{ cal/h} - 1.1622 \times 10^{-3} \text{ W}$

A.1-17 Heat Capacity and Enthalpy

$1 \text{ btu/lb}_m \cdot {}^\circ\text{F} = 4.1868 \text{ kJ/kg} \cdot \text{K}$

$1 \text{ btu/lb}_m \cdot {}^\circ\text{F} = 1.000 \text{ cal/g} \cdot {}^\circ\text{C}$

$1 \text{ btu/lb}_m = 2326.0 \text{ J/kg}$

$1 \text{ ft} \cdot \text{lb}_f/\text{lb}_m = 2.9890 \text{ J/kg}$

$1 \text{ cal (IT)/g} \cdot {}^\circ\text{C} = 4.1868 \text{ kJ/kg} \cdot \text{K}$

$1 \text{ kcal/g mol} = 4.1840 \times 10^3 \text{ kJ/kg mol}$

A.1-18 Mass-Transfer Coefficient

$1 \, k_c \text{ cm/s} = 10^{-2} \text{ m/s}$

$1 \, k_c \text{ ft/h} = 8.4668 \times 10^{-5} \text{ m/s}$

$1 \, k_x \text{ g mol/s} \cdot \text{cm}^2 \cdot \text{mol frac} = 10 \text{ kg mol/s} \cdot \text{m}^2 \cdot \text{mol frac}$

$1 \, k_x \text{ g mol/s} \cdot \text{cm}^2 \cdot \text{mol frac} = 1 \times 10^4 \text{ g mol/s} \cdot \text{m}^2 \cdot \text{mol frac}$

$1 \, k_x \text{ lb mol/h} \cdot \text{ft}^2 \cdot \text{mol frac} = 1.3562 \times 10^{-3} \text{ kg mol/s} \cdot \text{m}^2 \cdot \text{mol frac}$

$1 \, k_x a \text{ lb mol/h} \cdot \text{ft}^3 \cdot \text{mol frac} = 4.449 \times 10^{-3} \text{ kg mol/s} \cdot \text{m}^3 \cdot \text{mol frac}$

$1 \, k_G \text{ kg mol/s} \cdot \text{m}^2 \cdot \text{atm} = 0.98692 \times 10^{-5} \text{ kg mol/s} \cdot \text{m}^2 \cdot \text{Pa}$

$1 \, k_G a \text{ kg mol/s} \cdot \text{m}^3 \cdot \text{atm} = 0.98692 \times 10^{-5} \text{ kg mol/s} \cdot \text{m}^3 \cdot \text{Pa}$

A.2-1 Latent Heat of Water at 273.15 K (0°C)

$$\text{Latent heat of fusion} = 1436.3 \text{ cal/g mol}$$

$$= 79.724 \text{ cal/g}$$

$$= 2585.3 \text{ btu/lb mol}$$

$$= 6013.4 \text{ kJ/kg mol}$$

Source: O. A. Hougen, K. M. Watson, and R. A. Ragatz, *Chemical Process Principles*, Part I, 2nd ed. New York: John Wiley & Sons, Inc., 1954.

Latent heat of vaporization at 298.15 K (25°C)

Pressure (mm Hg)	*Latent Heat*
23.75	44 020 kJ/kg mol, 10.514 kcal/g mol, 18 925 btu/lb mol
760	44 045 kJ/kg mol, 10.520 kcal/g mol, 18 936 btu/lb mol

Source: National Bureau of Standards, *Circular 500*.

A.2-2 Vapor Pressure of Water

Temperature		*Vapor Pressure*		*Temperature*		*Vapor Pressure*	
K	*°C*	*kPa*	*mm Hg*	*K*	*°C*	*kPa*	*mm Hg*
273.15	0	0.611	4.58	323.15	50	12.333	92.51
283.15	10	1.228	9.21	333.15	60	19.92	149.4
293.15	20	2.338	17.54	343.15	70	31.16	233.7
298.15	25	3.168	23.76	353.15	80	47.34	355.1
303.15	30	4.242	31.82	363.15	90	70.10	525.8
313.15	40	7.375	55.32	373.15	100	101.325	760.0

Source: Physikalish-technishe, Reichsansalt, Holborn, Scheel, and Henning, *Warmetabellen*. Brunswick, Germany: Friedrich Viewig and Son, 1909.

A.2-3 Density of Liquid Water

Temperature		Density		Temperature		Density	
K	°C	g/cm^3	kg/m^3	K	°C	g/cm^3	kg/m^3
273.15	0	0.99987	999.87	323.15	50	0.98807	988.07
277.15	4	1.00000	1000.00	333.15	60	0.98324	983.24
283.15	10	0.99973	999.73	343.15	70	0.97781	977.81
293.15	20	0.99823	998.23	353.15	80	0.97183	971.83
298.15	25	0.99708	997.08	363.15	90	0.96534	965.34
303.15	30	0.99568	995.68	373.15	100	0.95838	958.38
313.15	40	0.99225	992.25				

Source: R. H. Perry and C. H. Chilton, *Chemical Engineers' Handbook*, 5th ed. New York: McGraw-Hill Book Company, 1973. With permission.

A.2-4 Viscosity of Liquid Water

Temperature		Viscosity $[(Pa \cdot s) 10^3,$ $(kg/m \cdot s) 10^3,$ or cp]	Temperature		Viscosity $[(Pa \cdot s) 10^3,$ $(kg/m \cdot s) 10^3,$ or cp]
K	°C		K	°C	
273.15	0	1.7921	323.15	50	0.5494
275.15	2	1.6728	325.15	52	0.5315
277.15	4	1.5674	327.15	54	0.5146
279.15	6	1.4728	329.15	56	0.4985
281.15	8	1.3860	331.15	58	0.4832
283.15	10	1.3077	333.15	60	0.4688
285.15	12	1.2363	335.15	62	0.4550
287.15	14	1.1709	337.15	64	0.4418
289.15	16	1.1111	339.15	66	0.4293
291.15	18	1.0559	341.15	68	0.4174
293.15	20	1.0050	343.15	70	0.4061
293.35	20.2	1.0000	345.15	72	0.3952
295.15	22	0.9579	347.15	74	0.3849
297.15	24	0.9142	349.15	76	0.3750
298.15	25	0.8937	351.15	78	0.3655
299.15	26	0.8737	353.15	80	0.3565
301.15	28	0.8360	355.15	82	0.3478
303.15	30	0.8007	357.15	84	0.3395
305.15	32	0.7679	359.15	86	0.3315
307.15	34	0.7371	361.15	88	0.3239
309.15	36	0.7085	363.15	90	0.3165
311.15	38	0.6814	365.15	92	0.3095
313.15	40	0.6560	367.15	94	0.3027
315.15	42	0.6321	369.15	96	0.2962
317.15	44	0.6097	371.15	98	0.2899
319.15	46	0.5883	373.15	100	0.2838
321.15	48	0.5683			

Source: Bingham, *Fluidity and Plasticity*. New York: McGraw-Hill Book Company, 1922. With permission.

A.2-5 Heat Capacity of Liquid Water at 101.325 kPa (1 Atm)

Temperature		Heat Capacity, c_p		Temperature		Heat Capacity, c_p	
°C	K	cal/g · °C	kJ/kg · K	°C	K	cal/g · °C	kJ/kg · K
0	273.15	1.0080	4.220	50	323.15	0.9992	4.183
10	283.15	1.0019	4.195	60	333.15	1.0001	4.187
20	293.15	0.9995	4.185	70	343.15	1.0013	4.192
25	298.15	0.9989	4.182	80	353.15	1.0029	4.199
30	303.15	0.9987	4.181	90	363.15	1.0050	4.208
40	313.15	0.9987	4.181	100	373.15	1.0076	4.219

Source : N. S. Osborne, H. F. Stimson, and D. C. Ginnings, *Bur. Standards J. Res.,* **23**, 197 (1939).

A.2-6 Thermal Conductivity of Liquid Water

Temperature			Thermal Conductivity	
°C	°F	K	btu/h · ft · °F	W/m · K
0	32	273.15	0.329	0.569
37.8	100	311.0	0.363	0.628
93.3	200	366.5	0.393	0.680
148.9	300	422.1	0.395	0.684
215.6	420	588.8	0.376	0.651
326.7	620	599.9	0.275	0.476

Source : D. L. Timrot and N. B. Vargaftik, *J. Tech. Phys. (U.S.S.R.),* **10**, 1063 (1940); 6th International Conference on the Properties of Steam, Paris, 1964.

A.2-7 Vapor Pressure of Saturated Ice–Water Vapor and Heat of Sublimation

Temperature			Vapor Pressure			Heat of Sublimation	
K	°F	°C	kPa	psia	mm Hg	btu/lb$_m$	kJ/kg
273.2	32	0	6.107×10^{-1}	8.858×10^{-2}	4.581	1218.6	2834.5
266.5	20	−6.7	3.478×10^{-1}	5.045×10^{-2}	2.609	1219.3	2836.1
261.0	10	−12.2	2.128×10^{-1}	3.087×10^{-2}	1.596	1219.7	2837.0
255.4	0	−17.8	1.275×10^{-1}	1.849×10^{-2}	0.9562	1220.1	2838.0
249.9	−10	−23.3	7.411×10^{-2}	1.082×10^{-2}	0.5596	1220.3	2838.4
244.3	−20	−28.9	3.820×10^{-2}	6.181×10^{-3}	0.3197	1220.5	2838.9
238.8	−30	−34.4	2.372×10^{-2}	3.440×10^{-3}	0.1779	1220.5	2838.9
233.2	−40	−40.0	1.283×10^{-2}	1.861×10^{-3}	0.09624	1220.5	2838.9

Source : ASHRAE, *Handbook of Fundamentals.* New York: ASHRAE, 1972.

A.2-8 Heat Capacity of Ice

Temperature		c_p		Temperature		c_p	
°F	K	$btu/lb_m \cdot °F$	$kJ/kg \cdot K$	°F	K	$btu/lb_m \cdot °F$	$kJ/kg \cdot K$
32	273.15	0.500	2.093	−10	249.85	0.461	1.930
20	266.45	0.490	2.052	−20	244.25	0.452	1.892
10	260.95	0.481	2.014	−30	238.75	0.442	1.850
0	255.35	0.472	1.976	−40	233.15	0.433	1.813

Source. Adapted from ASHRAE, *Handbook of Fundamentals.* New York: ASHRAE, 1972.

A.2-9 Properties of Saturated Steam and Water (Steam Table), SI Units

Temper-ature (°C)	Vapor Pressure (kPa)	Specific Volume (m^3/kg)		Enthalpy (kJ/kg)		Entropy ($kJ/kg \cdot K$)	
		Liquid	Sat'd Vapor	Liquid	Sat'd Vapor	Liquid	Sat'd Vapor
0.01	0.6113	0.0010002	206.136	0.00	2501.4	0.0000	9.1562
3	0.7577	0.0010001	168.132	12.57	2506.9	0.0457	9.0773
6	0.9349	0.0010001	137.734	25.20	2512.4	0.0912	9.0003
9	1.1477	0.0010003	113.386	37.80	2517.9	0.1362	8.9253
12	1.4022	0.0010005	93.784	50.41	2523.4	0.1806	8.8524
15	1.7051	0.0010009	77.926	62.99	2528.9	0.2245	8.7814
18	2.0640	0.0010014	65.038	75.58	2534.4	0.2679	8.7123
21	2.487	0.0010020	54.514	88.14	2539.9	0.3109	8.6450
24	2.985	0.0010027	45.883	100.70	2545.4	0.3534	8.5794
25	3.169	0.0010029	43.360	104.89	2547.2	0.3674	8.5580
27	3.567	0.0010035	38.774	113.25	2550.8	0.3954	8.5156
30	4.246	0.0010043	32.894	125.79	2556.3	0.4369	8.4533
33	5.034	0.0010053	28.011	138.33	2561.7	0.4781	8.3927
36	5.947	0.0010063	23.940	150.86	2567.1	0.5188	8.3336
40	7.384	0.0010078	19.523	167.57	2574.3	0.5725	8.2570
45	9.593	0.0010099	15.258	188.45	2583.2	0.6387	8.1648
50	12.349	0.0010121	12.032	209.33	2592.1	0.7038	8.0763
55	15.758	0.0010146	9.568	230.23	2600.9	0.7679	7.9913
60	19.940	0.0010172	7.671	251.13	2609.6	0.8312	7.9096
65	25.03	0.0010199	6.197	272.06	2618.3	0.8935	7.8310
70	31.19	0.0010228	5.042	292.98	2626.8	0.9549	7.7553
75	38.58	0.0010259	4.131	313.93	2635.3	1.0155	7.6824
80	47.39	0.0010291	3.407	334.91	2643.7	1.0753	7.6122
85	57.83	0.0010325	2.828	355.90	2651.9	1.1343	7.5445
90	70.14	0.0010360	2.361	376.92	2660.1	1.1925	7.4791
95	84.55	0.0010397	1.9819	397.96	2668.1	1.2500	7.4159
100	101.35	0.0010435	1.6729	419.04	2676.1	1.3069	7.3549

Temper- ature (°C)	Vapor Pressure (kPa)	Specific Volume (m³/kg)		Enthalpy (kJ/kg)		Entropy (kJ/kg · K)	
		Liquid	Sat'd Vapor	Liquid	Sat'd Vapor	Liquid	Sat'd Vapor
105	120.82	0.0010475	1.4194	440.15	2683.8	1.3630	7.2958
110	143.27	0.0010516	1.2102	461.30	2691.5	1.4185	7.2387
115	169.06	0.0010559	1.0366	482.48	2699.0	1.4734	7.1833
120	198.53	0.0010603	0.8919	503.71	2706.3	1.5276	7.1296
125	232.1	0.0010649	0.7706	524.99	2713.5	1.5813	7.0775
130	270.1	0.0010697	0.6685	546.31	2720.5	1.6344	7.0269
135	313.0	0.0010746	0.5822	567.69	2727.3	1.6870	6.9777
140	316.3	0.0010797	0.5089	589.13	2733.9	1.7391	6.9299
145	415.4	0.0010850	0.4463	610.63	2740.3	1.7907	6.8833
150	475.8	0.0010905	0.3928	632.20	2746.5	1.8418	6.8379
155	543.1	0.0010961	0.3468	653.84	2752.4	1.8925	6.7935
160	617.8	0.0011020	0.3071	675.55	2758.1	1.9427	6.7502
165	700.5	0.0011080	0.2727	697.34	2763.5	1.9925	6.7078
170	791.7	0.0011143	0.2428	719.21	2768.7	2.0419	6.6663
175	892.0	0.0011207	0.2168	741.17	2773.6	2.0909	6.6256
180	1002.1	0.0011274	0.19405	763.22	2778.2	2.1396	6.5857
190	1254.4	0.0011414	0.15654	807.62	2786.4	2.2359	6.5079
200	1553.8	0.0011565	0.12736	852.45	2793.2	2.3309	6.4323
225	2548	0.0011992	0.07849	966.78	2803.3	2.5639	6.2503
250	3973	0.0012512	0.05013	1085.36	2801.5	2.7927	6.0730
275	5942	0.0013168	0.03279	1210.07	2785.0	3.0208	5.8938
300	8581	0.0010436	0.02167	1344.0	2749.0	3.2534	5.7045

Source: Abridged from J. H. Keenan, F. G. Keyes, P. G. Hill, and J. G. Moore, *Steam Tables—Metric Units.* New New York: John Wiley & Sons, Inc., 1969. Reprinted by permission of John Wiley & Sons, Inc.

A.2-9   Properties of Saturated Steam and Water (Steam Table), English Units

Temper- ature (°F)	Vapor Pressure (psia)	Specific Volume (ft³/lb$_m$)		Enthalpy (btu/lb$_m$)		Entropy (btu/lb$_m$ · °F)	
		Liquid	Sat'd Vapor	Liquid	Sat'd Vapor	Liquid	Sat'd Vapor
32.02	0.08866	0.016022	3302	0.00	1075.4	0.000	2.1869
35	0.09992	0.016021	2948	3.00	1076.7	0.00607	2.1764
40	0.12166	0.016020	2445	8.02	1078.9	0.01617	2.1592
45	0.14748	0.016021	2037	13.04	1081.1	0.02618	2.1423
50	0.17803	0.016024	1704.2	18.06	1083.3	0.03607	2.1259
55	0.2140	0.016029	1431.4	23.07	1085.5	0.04586	2.1099

Temper-ature (°F)	Vapor Pressure (psia)	Specific Volume (ft³/lbₘ)		Enthalpy (btu/lbₘ)		Entropy (btu/lbₘ·°F)	
		Liquid	Sat'd Vapor	Liquid	Sat'd Vapor	Liquid	Sat'd Vapor
60	0.2563	0.016035	1206.9	28.08	1087.7	0.05555	2.0943
65	0.3057	0.016042	1021.5	33.09	1089.9	0.06514	2.0791
70	0.3622	0.016051	867.7	38.09	1092.0	0.07463	2.0642
75	0.4300	0.016061	739.7	43.09	1094.2	0.08402	2.0497
80	0.5073	0.016073	632.8	48.09	1096.4	0.09332	2.0356
85	0.5964	0.016085	543.1	53.08	1098.6	0.10252	2.0218
90	0.6988	0.016099	467.7	58.07	1100.7	0.11165	2.0083
95	0.8162	0.016114	404.0	63.06	1102.9	0.12068	1.9951
100	0.9503	0.016130	350.0	68.05	1105.0	0.12963	1.9822
110	1.2763	0.016166	265.1	78.02	1109.3	0.14730	1.9574
120	1.6945	0.016205	203.0	88.00	1113.5	0.16465	1.9336
130	2.225	0.016247	157.17	97.98	1117.8	0.18172	1.9109
140	2.892	0.016293	122.88	107.96	1121.9	0.19851	1.8892
150	3.722	0.016343	96.99	117.96	1126.1	0.21503	1.8684
160	4.745	0.016395	77.23	127.96	1130.1	0.23130	1.8484
170	5.996	0.016450	62.02	137.97	1134.2	0.24732	1.8293
180	7.515	0.016509	50.20	147.99	1138.2	0.26311	1.8109
190	9.343	0.016570	40.95	158.03	1142.1	0.27866	1.7932
200	11.529	0.016634	33.63	168.07	1145.9	0.29400	1.7762
210	14.125	0.016702	27.82	178.14	1149.7	0.30913	1.7599
212	14.698	0.016716	26.80	180.16	1150.5	0.31213	1.7567
220	17.188	0.016772	23.15	188.22	1153.5	0.32406	1.7441
230	20.78	0.016845	19.386	198.32	1157.1	0.33880	1.7289
240	24.97	0.016922	16.327	208.44	1160.7	0.35335	1.7143
250	29.82	0.017001	13.826	218.59	1164.2	0.36772	1.7001
260	35.42	0.017084	11.768	228.76	1167.6	0.38193	1.6864
270	41.85	0.017170	10.066	238.95	1170.9	0.39597	1.6731
280	49.18	0.017259	8.650	249.18	1174.1	0.40986	1.6602
290	57.33	0.017352	7.467	259.44	1177.2	0.42360	1.6477
300	66.98	0.017448	6.472	269.73	1180.2	0.43720	1.6356
310	77.64	0.017548	5.632	280.06	1183.0	0.45067	1.6238
320	89.60	0.017652	4.919	290.43	1185.8	0.46400	1.6123
330	103.00	0.017760	4.312	300.84	1188.4	0.47722	1.6010
340	117.93	0.017872	3.792	311.30	1190.8	0.49031	1.5901
350	134.53	0.017988	3.346	321.80	1193.1	0.50329	1.5793
360	152.92	0.018108	2.961	332.35	1195.2	0.51617	1.5688
370	173.23	0.018233	2.628	342.96	1197.2	0.52894	1.5585
380	195.60	0.018363	2.339	353.62	1199.0	0.54163	1.5483
390	220.2	0.018498	2.087	364.34	1200.6	0.55422	1.5383
400	247.1	0.018638	1.8661	375.12	1202.0	0.56672	1.5284
410	276.5	0.018784	1.6726	385.97	1203.1	0.57916	1.5187
450	422.1	0.019433	1.1011	430.2	1205.6	0.6282	1.4806

Source : Abridged from J. H. Keenan, F. G. Keyes, P. G. Hill, and J. G. Moore, *Steam Tables—English Units.* New York: John Wiley & Sons, Inc., 1969. Reprinted by permission of John Wiley & Sons, Inc.

A.2-10 Properties of Superheated Steam (Steam Table), SI Units (v, specific volume, m^3/kg; H, enthalpy, kJ/kg; s, entropy, kJ/kg · K)

Absolute Pressure, kPa (Sat. Temp., °C)		Temperature (°C)							
		100	150	200	250	300	360	420	500
10 (45.81)	v	17.196	19.512	21.825	24.136	26.445	29.216	31.986	35.679
	H	2687.5	2783.0	2879.5	2977.3	3076.5	3197.6	3320.9	3489.1
	s	8.4479	8.6882	8.9038	9.1002	9.2813	9.4821	9.6682	9.8978
50 (81.33)	v	3.418	3.889	4.356	4.820	5.284	5.839	6.394	7.134
	H	2682.5	2780.1	2877.7	2976.0	3075.5	3196.8	3320.4	3488.7
	s	7.6947	7.9401	8.1580	8.3556	8.5373	8.7385	8.9249	9.1546
75 (91.78)	v	2.270	2.587	2.900	3.211	3.520	3.891	4.262	4.755
	H	2679.4	2778.2	2876.5	2975.2	3074.9	3196.4	3320.0	3488.4
	s	7.5009	7.7496	7.9690	8.1673	8.3493	8.5508	8.7374	8.9672
100 (99.63)	v	1.6958	1.9364	2.172	2.406	2.639	2.917	3.195	3.565
	H	2672.2	2776.4	2875.3	2974.3	3074.3	3195.9	3319.6	3488.1
	s	7.3614	7.6134	7.8343	8.0333	8.2158	8.4175	8.6042	8.8342
150 (111.37)	v		1.2853	1.4443	1.6012	1.7570	1.9432	2.129	2.376
	H		2772.6	2872.9	2972.7	3073.1	3195.0	3318.9	3487.6
	s		7.4193	7.6433	7.8438	8.0720	8.2293	8.4163	8.6466
400 (143.63)	v		0.4708	0.5342	0.5951	0.6548	0.7257	0.7960	0.8893
	H		2752.8	2860.5	2964.2	3066.8	3190.3	3315.3	3484.9
	s		6.9299	7.1706	7.3789	7.5662	7.7712	7.9598	8.1913
700 (164.97)	v			0.2999	0.3363	0.3714	0.4126	0.4533	0.5070
	H			2844.8	2953.6	3059.1	3184.7	3310.9	3481.7
	s			6.8865	7.1053	7.2979	7.5063	7.6968	7.9299
1000 (179.91)	v			0.2060	0.2327	0.2579	0.2873	0.3162	0.3541
	H			2827.9	2942.6	3051.2	3178.9	3306.5	3478.5
	s			6.6940	6.9247	7.1229	7.3349	7.5275	7.7622
1500 (198.32)	v			0.13248	0.15195	0.16966	0.18988	0.2095	0.2352
	H			2796.8	2923.3	3037.6	3.1692	3299.1	3473.1
	s			6.4546	6.7090	6.9179	7.1363	7.3323	7.5698
2000 (212.42)	v				0.11144	0.12547	0.14113	0.15616	0.17568
	H				2902.5	3023.5	3159.3	3291.6	3467.6
	s				6.5453	6.7664	6.9917	7.1915	7.4317
2500 (223.99)	v				0.08700	0.09890	0.11186	0.12414	0.13998
	H				2880.1	3008.8	3149.1	3284.0	3462.1
	s				6.4085	6.6438	6.8767	7.0803	7.3234
3000 (233.90)	v				0.07058	0.08114	0.09233	0.10279	0.11619
	H				2855.8	2993.5	3138.7	3276.3	3456.5
	s				6.2872	6.5390	6.7801	6.9878	7.2338

Source: Abridged from J. H. Keenan, F. G. Keyes, P. G. Hill, and J. G. Moore, *Steam Tables—Metric Units.* New York: John Wiley & Sons, Inc., 1969. Reprinted by permission of John Wiley & Sons, Inc.

A.2-10 Properties of Superheated Steam (Steam Table), English Units (v, specific volume, ft^3/lb$_m$; H, enthalpy, btu/lb$_m$; s, entropy, btu/lb$_m \cdot$ °F)

Absolute Pressure, psia (Sat. Temp., °F)		Temperature (°F)								
		200	300	400	500	600	700	800	900	1000
1.0 (101.70)	v	392.5	452.3	511.9	571.5	631.1	690.7	750.3	809.9	869.5
	H	1150.1	1195.7	1241.8	1288.5	1336.1	1384.5	1433.7	1483.8	1534.8
	s	2.0508	2.1150	2.1720	2.2235	2.2706	2.3142	2.3550	2.3932	2.4294
5.0 (162.21)	v	78.15	90.24	102.24	114.20	126.15	138.08	150.01	161.94	173.86
	H	1148.6	1194.8	1241.2	1288.2	1335.8	1384.3	1433.5	1483.7	1534.7
	s	1.8715	1.9367	1.9941	2.0458	2.0930	2.1367	2.1775	2.2158	2.2520
10.0 (193.19)	v	38.85	44.99	51.03	57.04	63.03	69.01	74.98	80.95	86.91
	H	1146.6	1193.7	1240.5	1287.7	1335.5	1384.0	1433.3	1483.5	1534.6
	s	1.7927	1.8592	1.9171	1.9690	2.0164	2.0601	2.1009	2.1393	2.1755
14.696 (211.99)	v		30.52	34.67	38.77	42.86	46.93	51.00	55.07	59.13
	H		1192.6	1239.9	1287.3	1335.2	1383.8	1433.1	1483.4	1534.5
	s		1.8157	1.8741	1.9263	1.9737	2.0175	2.0584	2.0967	2.1330
20.0 (227.96)	v		22.36	25.43	28.46	31.47	34.77	37.46	40.45	43.44
	H		1191.5	1239.2	1286.8	1334.8	1383.5	1432.9	1483.2	1534.3
	s		1.7805	1.8395	1.8919	1.9395	1.9834	2.0243	2.0627	2.0989
60.0 (292.73)	v		7.260	8.353	9.399	10.425	11.440	12.448	13.452	14.454
	H		1181.9	1233.5	1283.0	1332.1	1381.4	1431.2	1481.8	1533.2
	s		1.6496	1.7134	1.7678	1.8165	1.8609	1.9022	1.9408	1.9773
100.0 (327.86)	v			4.934	5.587	6.216	6.834	7.445	8.053	8.657
	H			1227.5	1279.1	1329.3	1379.2	1429.6	1480.5	1532.1
	s			1.6517	1.7085	1.7582	1.8033	1.8449	1.8838	1.9204
150.0 (358.48)	v			3.221	3.679	4.111	4.531	4.944	5.353	5.759
	H			1219.5	1274.1	1325.7	1376.6	1427.5	1478.8	1530.7
	s			1.5997	1.6598	1.7110	1.7568	1.7989	1.8381	1.8750
200.0 (381.86)	v			2.361	2.724	3.058	3.379	3.693	4.003	4.310
	H			1210.8	1268.8	1322.1	1373.8	1425.3	1477.1	1529.3
	s			1.5600	1.6239	1.6767	1.7234	1.7660	1.8055	1.8425
250.0 (401.04)	v				2.150	2.426	2.688	2.943	3.193	3.440
	H				1263.3	1318.3	1371.1	1423.2	1475.3	1527.9
	s				1.5948	1.6494	1.6970	1.7401	1.7799	1.8172
300.0 (417.43)	v				1.766	2.004	2.227	2.442	2.653	2.860
	H				1257.5	1314.5	1368.3	1421.0	1473.6	1526.5
	s				1.5701	1.6266	1.6751	1.7187	1.7589	1.7964
400 (444.70)	v				1.2843	1.4760	1.6503	1.8163	1.9776	2.136
	H				1245.2	1306.6	1362.5	1416.6	1470.1	1523.6
	s				1.5282	1.5892	1.6397	1.6884	1.7252	1.7632

Source: Abridged from J. H. Keenan, F. G. Keyes, P. G. Hill, and J. G. Moore, *Steam Tables—Metric Units.* New York: John Wiley & Sons, Inc., 1969. Reprinted by permission of John Wiley & Sons, Inc.

A.2-11 Heat-Transfer Properties of Liquid Water, SI Units

T (°C)	T (K)	ρ (kg/m³)	c_p (kJ/kg·K)	$\mu \times 10^3$ (Pa·s, or kg/m·s)	k (W/m·K)	N_{Pr}	$\beta \times 10^4$ (1/K)	$(g\beta\rho^2/\mu^2)$ $\times 10^{-8}$ (1/K·m³)
0	273.2	999.6	4.229	1.786	0.5694	13.3	−0.630	
15.6	288.8	998.0	4.187	1.131	0.5884	8.07	1.44	10.93
26.7	299.9	996.4	4.183	0.860	0.6109	5.89	2.34	30.70
37.8	311.0	994.7	4.183	0.682	0.6283	4.51	3.24	68.0
65.6	338.8	981.9	4.187	0.432	0.6629	2.72	5.04	256.2
93.3	366.5	962.7	4.229	0.3066	0.6802	1.91	6.66	642
121.1	394.3	943.5	4.271	0.2381	0.6836	1.49	8.46	1300
148.9	422.1	917.9	4.312	0.1935	0.6836	1.22	10.08	2231
204.4	477.6	858.6	4.522	0.1384	0.6611	0.950	14.04	5308
260.0	533.2	784.9	4.982	0.1042	0.6040	0.859	19.8	11 030
315.6	588.8	679.2	6.322	0.0862	0.5071	1.07	31.5	19 260

A.2-11 Heat-Transfer Properties of Liquid Water, English Units

T (°F)	ρ $\left(\dfrac{lb_m}{ft^3}\right)$	c_p $\left(\dfrac{btu}{lb_m \cdot °F}\right)$	$\mu \times 10^3$ $\left(\dfrac{lb_m}{ft \cdot s}\right)$	k $\left(\dfrac{btu}{h \cdot ft \cdot °F}\right)$	N_{Pr}	$\beta \times 10^4$ (1/°R)	$(g\beta\rho^2/\mu^2)$ $\times 10^{-6}$ (1/°R·ft³)
32	62.4	1.01	1.20	0.329	13.3	−0.350	
60	62.3	1.00	0.760	0.340	8.07	0.800	17.2
80	62.2	0.999	0.578	0.353	5.89	1.30	48.3
100	62.1	0.999	0.458	0.363	4.51	1.80	107
150	61.3	1.00	0.290	0.383	2.72	2.80	403
200	60.1	1.01	0.206	0.393	1.91	3.70	1010
250	58.9	1.02	0.160	0.395	1.49	4.70	2045
300	57.3	1.03	0.130	0.395	1.22	5.60	3510
400	53.6	1.08	0.0930	0.382	0.950	7.80	8350
500	49.0	1.19	0.0700	0.349	0.859	11.0	17 350
600	42.4	1.51	0.0579	0.293	1.07	17.5	30 300

A.2-12 Heat-Transfer Properties of Water Vapor (Steam) at 101.32 kPa (1 Atm Abs), SI Units

T ($°C$)	T (K)	ρ (kg/m^3)	c_p ($kJ/kg \cdot K$)	$\mu \times 10^5$ ($Pa \cdot s$, or $kg/m \cdot s$)	k ($W/m \cdot K$)	N_{Pr}	$\beta \times 10^3$ ($1/K$)	$g\beta\rho^2/\mu^2$ ($1/K \cdot m^3$)
100.0	373.2	0.596	1.888	1.295	0.02510	0.96	2.68	0.557×10^8
148.9	422.1	0.525	1.909	1.488	0.02960	0.95	2.38	0.292×10^8
204.4	477.6	0.461	1.934	1.682	0.03462	0.94	2.09	0.154×10^8
260.0	533.2	0.413	1.968	1.883	0.03946	0.94	1.87	0.0883×10^8
315.6	588.8	0.373	1.997	2.113	0.04448	0.94	1.70	52.1×10^5
371.1	644.3	0.341	2.030	2.314	0.04985	0.93	1.55	33.1×10^5
426.7	699.9	0.314	2.068	2.529	0.05556	0.92	1.43	21.6×10^5

A.2-12 Heat-Transfer Properties of Water Vapor (Steam) at 101.32 kPa (1 Atm Abs), English Units

T ($°F$)	ρ $\left(\dfrac{lb_m}{ft^3}\right)$	c_p $\left(\dfrac{btu}{lb_m \cdot °F}\right)$	$\mu \times 10^5$ $\left(\dfrac{lb_m}{ft \cdot s}\right)$	k $\left(\dfrac{btu}{h \cdot ft \cdot °F}\right)$	N_{Pr}	$\beta \times 10^3$ ($1/°R$)	$g\beta\rho^2/\mu^2$ ($1/°R \cdot ft^3$)
212	0.0372	0.451	0.870	0.0145	0.96	1.49	0.877×10^6
300	0.0328	0.456	1.000	0.0171	0.95	1.32	0.459×10^6
400	0.0288	0.462	1.130	0.0200	0.94	1.16	0.243×10^6
500	0.0258	0.470	1.265	0.0228	0.94	1.04	0.139×10^6
600	0.0233	0.477	1.420	0.0257	0.94	0.943	82×10^3
700	0.0213	0.485	1.555	0.0288	0.93	0.862	52.1×10^3
800	0.0196	0.494	1.700	0.0321	0.92	0.794	34.0×10^3

Source: D. L. Timrot and N. B. Vargaftik, *J, Tech. Phys (U.S.S.R.),* **10,** 1063 (1940), R. H. Perry and C. H. Chilton, *Chemical Engineers' Handbook,* 5th ed. New York: McGraw-Hill Book Company, 1973; J. H. Keenan, F. G. Keyes, P. G. Hill, and J. G. Moore, *Steam Tables.* New York: John Wiley & Sons, Inc., 1969; National Research Council, *International Critical Tables.* New York: McGraw-Hill Book Company, 1929; L. S. Marks, *Mechanical Engineers' Handbook,* 5th ed. New York: McGraw-Hill Book Company, 1951.

APPENDIX A.3

Physical Properties
of Inorganic
and Organic Compounds

A.3-1   Standard Heats of Formation at 298.15 K (25°C) and 101.325 kPa (1 Atm Abs), (c) = crystalline, (g) = gas, (l) = liquid

Compound	ΔH_f°		Compound	ΔH_f°	
	$(kJ/kg\,mol)10^{-3}$	$kcal/g\,mol$		$(kJ/kg\,mol)10^{-3}$	$kcal/g\,mol$
$NH_3(g)$	-46.19	-11.04	$CaCO_3(c)$	-1206.87	-288.45
$NO(g)$	$+90.374$	$+21.600$	$CaO(c)$	-635.5	-151.9
$H_2O(l)$	-285.840	-68.3174	$CO(g)$	-110.523	-26.4157
$H_2O(g)$	-241.826	-57.7979	$CO_2(g)$	-393.513	-94.0518
$HCN(g)$	$+130.1$	$+31.1$	$CH_4(g)$	-74.848	-17.889
$HCl(g)$	-92.312	-22.063	$C_2H_6(g)$	-84.667	-20.236
$H_2SO_4(l)$	-811.32	-193.91	$C_3H_8(g)$	-103.847	-24.820
$H_3PO_4(c)$	-1281.1	-306.2	$CH_3OH(l)$	-238.66	-57.04
$NaCl(c)$	-411.003	-98.232	$CH_3CH_2OH(l)$	-277.61	-66.35
$NH_4Cl(c)$	-315.39	-75.38			

Source: J. H. Perry and C. H. Chilton, *Chemical Engineers' Handbook*, 5th ed. New York: McGraw-Hill Book Company, 1973; and O. A. Hougen, K. M. Watson, and R. A. Ragatz, *Chemical Process Principles*, Part I, 2nd ed. New York: John Wiley & Sons, Inc., 1954.

A.3-2 Standard Heats of Combustion at 298.15 K (25°C) and 101.325 kPa (1 Atm Abs)
(g) = gas, (l) = liquid, (s) = solid

Compound	Combustion Reaction	ΔH_c° kcal/g mol	ΔH_c° (kJ/kg mol)10^{-3}
C(s)	$C(s) + \frac{1}{2}O_2(g) \rightarrow CO(g)$	−26.4157	−110.523
CO(g)	$CO(g) + \frac{1}{2}O_2(g) \rightarrow CO_2(g)$	−67.6361	−282.989
C(s)	$C(s) + O_2(g) \rightarrow CO_2(g)$	−94.0518	−393.513
$H_2(g)$	$H_2(g) + \frac{1}{2}O_2(g) \rightarrow H_2O(l)$	−68.3174	−285.840
$H_2(g)$	$H_2(g) + \frac{1}{2}O_2(g) \rightarrow H_2O(g)$	−57.7979	−241.826
$CH_4(g)$	$CH_4(g) + 2O_2(g) \rightarrow CO_2(g) + 2H_2O(l)$	−212.798	−890.346
$C_2H_6(g)$	$C_2H_6(g) + \frac{7}{2}O_2(g) \rightarrow 2CO_2(g) + 3H_2O(l)$	−372.820	−1559.879
$C_3H_8(g)$	$C_3H_8(g) + 5O_2(g) \rightarrow 3CO_2(g) + 4H_2O(l)$	−530.605	−2220.051
d-Glucose (dextrose) $C_6H_{12}O_6(s)$	$C_6H_{12}O_6(s) + 6O_2(g) \rightarrow 6CO_2(g) + 6H_2O(l)$	−673	−2816
Lactose (anhydrous) $C_{12}H_{22}O_{11}(s)$	$C_{12}H_{22}O_{11}(s) + 12O_2(g) \rightarrow 12CO_2(g) + 11H_2O(l)$	−1350.1	−5648.8
Sucrose $C_{12}H_{22}O_{11}(s)$	$C_{12}H_{22}O_{11}(s) + 12O_2(g) \rightarrow 12CO_2(g) + 11H_2O(l)$	−1348.9	−5643.8

Source : R. H. Perry and C. H. Chilton, *Chemical Engineers' Handbook*, 5th ed. New York: McGraw-Hill Book Company, 1973; and O. A. Hougen, K. M. Watson, and R. A. Ragatz, *Chemical Process Principles*, Part I, 2nd ed. New York: John Wiley & Sons, Inc., 1954.

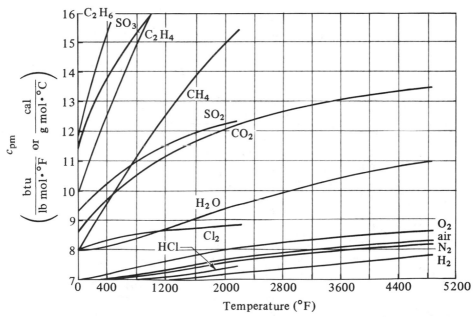

FIGURE A.3-1. *Mean molar heat capacities from 77°F (25°C) to t°F at constant pressure of 101.325 kPa (1 atm abs). (From O. A. Hougen, K. M. Watson, and R. A. Ragatz, Chemical Process Principles, Part I, 2nd ed. New York: John Wiley & Sons, Inc., 1954. With permission.)*

A.3-3 Physical Properties of Air at 101.325 kPa (1 Atm Abs), SI Units

T (°C)	T (K)	ρ (kg/m³)	c_p (kJ/kg·K)	$\mu \times 10^5$ (Pa·s, or kg/m·s)	k (W/m·K)	N_{Pr}	$\beta \times 10^3$ (1/K)	$g\beta\rho^2/\mu^2$ (1/K·m³)
−17.8	255.4	1.379	1.0048	1.62	0.02250	0.720	3.92	2.79×10^8
0	273.2	1.293	1.0048	1.72	0.02423	0.715	3.65	2.04×10^8
10.0	283.2	1.246	1.0048	1.78	0.02492	0.713	3.53	1.72×10^8
37.8	311.0	1.137	1.0048	1.90	0.02700	0.705	3.22	1.12×10^8
65.6	338.8	1.043	1.0090	2.03	0.02925	0.702	2.95	0.775×10^8
93.3	366.5	0.964	1.0090	2.15	0.03115	0.694	2.74	0.534×10^8
121.1	394.3	0.895	1.0132	2.27	0.03323	0.692	2.54	0.386×10^8
148.9	422.1	0.838	1.0174	2.37	0.03531	0.689	2.38	0.289×10^8
176.7	449.9	0.785	1.0216	2.50	0.03721	0.687	2.21	0.214×10^8
204.4	477.6	0.740	1.0258	2.60	0.03894	0.686	2.09	0.168×10^8
232.2	505.4	0.700	1.0300	2.71	0.04084	0.684	1.98	0.130×10^8
260.0	533.2	0.662	1.0341	2.80	0.04258	0.680	1.87	0.104×10^8

A.3-3 Physical Properties of Air at 101.325 kPa (1 Atm Abs), English Units

T (°F)	ρ $\left(\frac{lb_m}{ft^3}\right)$	c_p $\left(\frac{btu}{lb_m \cdot °F}\right)$	μ (centipoise)	k $\left(\frac{btu}{h \cdot ft \cdot °F}\right)$	N_{pr}	$\beta \times 10^3$ (1/°R)	$g\beta\rho^2/\mu^2$ (1/°R·ft³)
0	0.0861	0.240	0.0162	0.0130	0.720	2.18	4.39×10^6
32	0.0807	0.240	0.0172	0.0140	0.715	2.03	3.21×10^6
50	0.0778	0.240	0.0178	0.0144	0.713	1.96	2.70×10^6
100	0.0710	0.240	0.0190	0.0156	0.705	1.79	1.76×10^6
150	0.0651	0.241	0.0203	0.0169	0.702	1.64	1.22×10^6
200	0.0602	0.241	0.0215	0.0180	0.694	1.52	0.840×10^6
250	0.0559	0.242	0.0227	0.0192	0.692	1.41	0.607×10^6
300	0.0523	0.243	0.0237	0.0204	0.689	1.32	0.454×10^6
350	0.0490	0.244	0.0250	0.0215	0.687	1.23	0.336×10^6
400	0.0462	0.245	0.0260	0.0225	0.686	1.16	0.264×10^6
450	0.0437	0.246	0.0271	0.0236	0.674	1.10	0.204×10^6
500	0.0413	0.247	0.0280	0.0246	0.680	1.04	0.163×10^6

Source: National Bureau of Standards, *Circular* 461C, 1947; **564**, 1955; NBS–NACA, *Tables of Thermal Properties of Gases*, 1949; F. G. Keyes, *Trans.* A.S.M.E., **73**, 590, 597 (1951); **74**, 1303 (1952); D. D. Wagman, *Selected Values of Chemical Thermodynamic Properties.* Washington, D.C.: National Bureau of Standards, 1953.

App. A.3   *Physical Properties of Inorganic and Organic Compounds*

A.3-4 Viscosity of Gases at 101.325 kPa (1 Atm Abs) [Viscosity in (Pa · s) 10^3, (kg/m · s) 10^3, or cp]

Temperature			H_2	O_2	N_2	CO	CO_2
K	°F	°C					
255.4	0	−17.8	0.00800	0.0181	0.0158	0.0156	0.0128
273.2	32	0	0.00840	0.0192	0.0166	0.0165	0.0137
283.2	50	10.0	0.00862	0.0197	0.0171	0.0169	0.0141
311.0	100	37.8	0.00915	0.0213	0.0183	0.0183	0.0154
338.8	150	65.6	0.00960	0.0228	0.0196	0.0195	0.0167
366.5	200	93.3	0.0101	0.0241	0.0208	0.0208	0.0179
394.3	250	121.1	0.0106	0.0256	0.0220	0.0220	0.0191
422.1	300	148.9	0.0111	0.0267	0.0230	0.0231	0.0203
449.9	350	176.7	0.0115	0.0282	0.0240	0.0242	0.0215
477.6	400	204.4	0.0119	0.0293	0.0250	0.0251	0.0225
505.4	450	232.2	0.0124	0.0307	0.0260	0.0264	0.0236
533.2	500	260.0	0.0128	0.0315	0.0273	0.0276	0.0247

Source: National Bureau of Standards, *Circular* **461C**, 1947; **564**, 1955; NBS–NACA, *Tables of Thermal Properties of Gases*, 1949; F. G. Keyes, *Trans. A.S.M.E.*, **73**, 590, 597 (1951); **74**, 1303 (1952); D. D. Wagman, *Selected Values of Chemical Thermodynamic Properties*. Washington, D.C.: National Bureau of Standards, 1953.

A.3-5 Thermal Conductivity of Gases at 101.325 kPa (1 Atm Abs)

Temperature			H_2		O_2		N_2		CO		CO_2	
K	°C	°F	$W/m \cdot K$	$\frac{btu}{h \cdot ft \cdot °F}$	$W/m \cdot K$	$\frac{btu}{h \cdot ft \cdot °F}$	$W/m \cdot K$	$\frac{btu}{h \cdot ft \cdot °F}$	$W/m \cdot K$	$\frac{btu}{h \cdot ft \cdot °F}$	$W/m \cdot K$	$\frac{btu}{h \cdot ft \cdot °F}$
255.4	−17.8	0	0.1592	0.0920	0.0228	0.0132	0.0228	0.0132	0.0222	0.0128	0.0132	0.0076
273.2	0	32	0.1667	0.0963	0.0246	0.0142	0.0239	0.0138	0.0233	0.0135	0.0145	0.0084
283.2	10.0	50	0.1720	0.0994	0.0253	0.0146	0.0248	0.0143	0.0239	0.0138	0.0152	0.0088
311.0	37.8	100	0.1852	0.107	0.0277	0.0160	0.0267	0.0154	0.0260	0.0150	0.0173	0.0100
338.8	65.6	150	0.1990	0.115	0.0299	0.0173	0.0287	0.0166	0.0279	0.0161	0.0190	0.0110
366.5	93.3	200	0.2111	0.122	0.0320	0.0185	0.0303	0.0175	0.0296	0.0171	0.0216	0.0125
394.3	121.1	250	0.2233	0.129	0.0343	0.0198	0.0329	0.0190	0.0318	0.0184	0.0239	0.0138
422.1	148.9	300	0.2353	0.136	0.0363	0.0210	0.0348	0.0201	0.0338	0.0195	0.0260	0.0150
449.9	176.7	350	0.2458	0.142	0.0382	0.0221	0.0365	0.0211	0.0355	0.0205	0.0286	0.0165
477.6	204.4	400	0.2579	0.149	0.0398	0.0230	0.0382	0.0221	0.0369	0.0213	0.0308	0.0178
505.4	232.2	450	0.2683	0.155	0.0422	0.0244	0.0400	0.0231	0.0384	0.0222	0.0334	0.0193
533.2	260.0	500	0.2786	0.161	0.0438	0.0253	0.0419	0.0242	0.0407	0.0235	0.0355	0.0205

Source: National Bureau of Standards, Circular 461C, 1947; 564, 1955; NBS–NACA, Table of Thermal Properties of Gases, 1949; F. G. Keyes, Trans. A.S.M.E., 73, 590, 597 (1951); 74, 1303 (1952); D. D. Wagman, Selected Values of Chemical Thermodynamic Properties. Washington, D.C.: National Bureau of Standards, 1953.

Heat Capacity of Gases at Constant Pressure at 101.325 kPa (1 Atm Abs)

Temperature			H_2		O_2		N_2		CO		CO_2	
K	°C	°F	kJ/ kg·K	btu/ lb_m·°F	kJ/ kg·K	btu/ lb_m·°F	kJ/ kg·K	btu/ lb_m·°F	kJ/ kg·K	btu/ lb_m·°F	kJ/ kg·K	btu/ lb_m·°F
255.4	−17.8	0	14.07	3.36	0.909	0.217	1.034	0.247	1.034	0.247	0.800	0.191
273.2	0	32	14.19	3.39	0.913	0.218	1.038	0.248	1.038	0.248	0.816	0.195
283.2	10.0	50	14.19	3.39	0.917	0.219	1.038	0.248	1.038	0.248	0.825	0.197
311.0	37.8	100	14.32	3.42	0.921	0.220	1.038	0.248	1.043	0.249	0.854	0.204
338.8	65.6	150	14.36	3.43	0.925	0.221	1.038	0.248	1.043	0.249	0.883	0.211
366.5	93.3	200	14.40	3.44	0.929	0.222	1.043	0.249	1.047	0.250	0.904	0.216
394.3	121.1	250	14.44	3.45	0.938	0.224	1.043	0.249	1.047	0.250	0.929	0.222
422.1	148.9	300	14.49	3.46	0.946	0.226	1.047	0.250	1.051	0.251	0.950	0.227
449.9	176.7	350	14.49	3.46	0.955	0.228	1.047	0.250	1.055	0.252	0.976	0.233
477.6	204.4	400	14.49	3.46	0.963	0.230	1.051	0.251	1.059	0.253	0.996	0.238
505.4	232.2	450	14.52	3.47	0.971	0.232	1.055	0.252	1.063	0.254	1.017	0.243
533.2	260.0	500	14.52	3.47	0.976	0.233	1.059	0.253	1.068	0.255	1.030	0.246

Source: National Bureau of Standards, *Circular* **461**C, 1947; **564**, 1955; NBS–NACA, *Tables of Thermal Properties of Gases*, 1949; F. G. Keyes, *Trans. A.S.M.E.*, **73**, 590, 597 (1951); **74**, 1303 (1952); D. D. Wagman, *Selected Values of Chemical Thermodynamic Properties*. Washington, D.C.: National Bureau of Standards, 1953.

A.3-7 Prandtl Number of Gases at 101.325 kPa (1 Atm Abs)

Temperature							
°C	°F	K	H_2	O_2	N_2	CO	CO_2
−17.8	0	255.4	0.720	0.720	0.720	0.740	0.775
0	32	273.2	0.715	0.711	0.720	0.738	0.770
10.0	50	283.2	0.710	0.710	0.717	0.735	0.769
37.8	100	311.0	0.700	0.707	0.710	0.731	0.764
65.6	150	338.8	0.700	0.706	0.700	0.727	0.755
93.3	200	366.5	0.694	0.703	0.700	0.724	0.752
121.1	250	394.3	0.688	0.703	0.696	0.720	0.746
148.9	300	422.1	0.683	0.703	0.690	0.720	0.738
176.6	350	449.9	0.677	0.704	0.689	0.720	0.734
204.4	400	477.6	0.670	0.706	0.688	0.720	0.725
232.2	450	505.4	0.668	0.702	0.688	0.720	0.716
260.0	500	533.2	0.666	0.700	0.688	0.720	0.702

Source: National Bureau of Standards, *Circular* **461C**, 1947; **564**, 1955; NBS–NACA, *Tables of Thermal Properties of Gases*, 1949; F. G. Keyes, *Trans. A.S.M.E.*, **73**, 590, 597 (1951); **74**, 1303 (1952); D. D. Wagman, *Selected Values of Chemical Thermodynamic Properties.* Washington, D.C.: National Bureau of Standards, 1953.

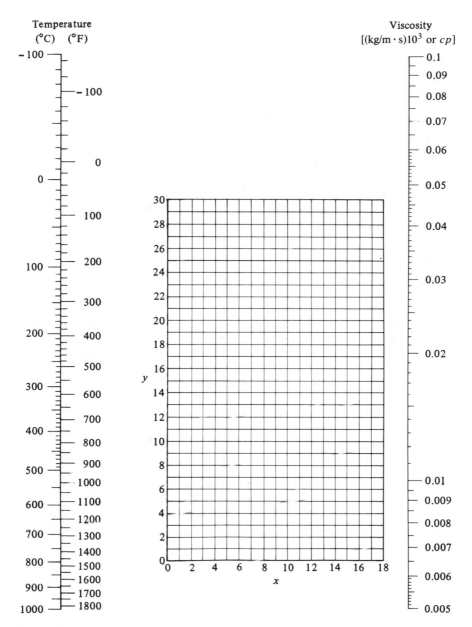

Temperature
(°C) (°F)

Viscosity
[(kg/m · s)10³ or cp]

FIGURE A.3-2.   *Viscosities of gases at 101.325 kPa (1 atm abs). (From R. H. Perry and C. H. Chilton, Chemical Engineers' Handbook, 5th ed. New York: McGraw-Hill Book Company, 1973. With permission.) See Table A.3-8 for coordinates for use with Fig. A.3-2.*

No.	Gas	X	Y	No.	Gas	X	Y
1	Acetic acid	7.7	14.3	29	Freon-113	11.3	14.0
2	Acetone	8.9	13.0	30	Helium	10.9	20.5
3	Acetylene	9.8	14.9	31	Hexane	8.6	11.8
4	Air	11.0	20.0	32	Hydrogen	11.2	12.4
5	Ammonia	8.4	16.0	33	$3H_2 + 1N_2$	11.2	17.2
6	Argon	10.5	22.4	34	Hydrogen bromide	8.8	20.9
7	Benzene	8.5	13.2	35	Hydrogen chloride	8.8	18.7
8	Bromine	8.9	19.2	36	Hydrogen cyanide	9.8	14.9
9	Butene	9.2	13.7	37	Hydrogen iodide	9.0	21.3
10	Butylene	8.9	13.0	38	Hydrogen sulfide	8.6	18.0
11	Carbon dioxide	9.5	18.7	39	Iodine	9.0	18.4
12	Carbon disulfide	8.0	16.0	40	Mercury	5.3	22.9
13	Carbon monoxide	11.0	20.0	41	Methane	9.9	15.5
14	Chlorine	9.0	18.4	42	Methyl alcohol	8.5	15.6
15	Chloroform	8.9	15.7	43	Nitric oxide	10.9	20.5
16	Cyanogen	9.2	15.2	44	Nitrogen	10.6	20.0
17	Cyclohexane	9.2	12.0	45	Nitrosyl chloride	8.0	17.6
18	Ethane	9.1	14.5	46	Nitrous oxide	8.8	19.0
19	Ethyl acetate	8.5	13.2	47	Oxygen	11.0	21.3
20	Ethyl alcohol	9.2	14.2	48	Pentane	7.0	12.8
21	Ethyl chloride	8.5	15.6	49	Propane	9.7	12.9
22	Ethyl ether	8.9	13.0	50	Propyl alcohol	8.4	13.4
23	Ethylene	9.5	15.1	51	Propylene	9.0	13.8
24	Fluorine	7.3	23.8	52	Sulfur dioxide	9.6	17.0
25	Freon-11	10.6	15.1	53	Toluene	8.6	12.4
26	Freon-12	11.1	16.0	54	2,3,3-Trimethylbutane	9.5	10.5
27	Freon-21	10.8	15.3	55	Water	8.0	16.0
28	Freon-22	10.1	17.0	56	Xenon	9.3	23.0

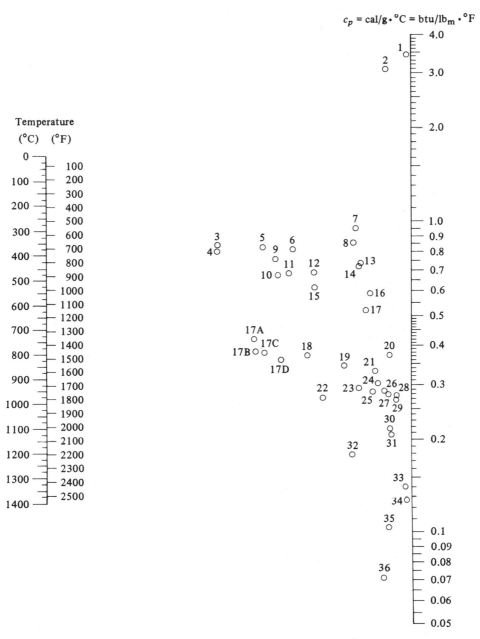

$c_p = \text{cal/g} \cdot {}^\circ\text{C} = \text{btu/lb}_m \cdot {}^\circ\text{F}$

FIGURE A.3-3. *Heat capacity of gases at constant pressure at 101.325 kPa (1 atm abs). (From R. H. Perry and C. H. Chilton, Chemical Engineers' Handbook, 5th ed. New York: McGraw-Hill Book Company, 1973. With permission.) See Table A.3-9 for use with Fig. A.3-3.*

A.3-9 Heat Capacity of Gases at Constant Pressure (for Use with Fig. A.3-3)

No.	Gas	Range (°C)
10	Acetylene	0–200
15	Acetylene	200–400
16	Acetylene	400–1400
27	Air	0–1400
12	Ammonia	0–600
14	Ammonia	600–1400
18	Carbon dioxide	0–400
24	Carbon dioxide	400–1400
26	Carbon monoxide	0–1400
32	Chlorine	0–200
34	Chlorine	200–1400
3	Ethane	0–200
9	Ethane	200–600
8	Ethane	600–1400
4	Ethylene	0–200
11	Ethylene	200–600
13	Ethylene	600–1400
17B	Freon-11 (CCl_3F)	0–150
17C	Freon-21 ($CHCl_2F$)	0–150
17A	Freon-22 ($CHClF_2$)	0–150
17D	Freon-113 ($CCl_2F–CClF_2$)	0–150
1	Hydrogen	0–600
2	Hydrogen	600–1400
35	Hydrogen bromide	0–1400
30	Hydrogen chloride	0–1400
20	Hydrogen fluoride	0–1400
36	Hydrogen iodide	0–1400
19	Hydrogen sulfide	0–700
21	Hydrogen sulfide	700–1400
5	Methane	0–300
6	Methane	300–700
7	Methane	700–1400
25	Nitric oxide	0–700
28	Nitric oxide	700–1400
26	Nitrogen	0–1400
23	Oxygen	0–500
29	Oxygen	500–1400
33	Sulfur	300–1400
22	Sulfur dioxide	0–400
31	Sulfur dioxide	400–1400
17	Water	0–1400

A.3-10 Thermal Conductivities of Gases and Vapors at 101.325 kPa (1 Atm Abs); $k = W/m \cdot K$

Gas or Vapor	K	k	Gas or Vapor	K	k
Acetone[1]	273	0.0099	Ethane[5, 6]	239	0.0149
	319	0.0130		273	0.0183
	373	0.0171		373	0.0303
	457	0.0254	Ethyl alcohol[1]	293	0.0154
Ammonia[2]	273	0.0218		373	0.0215
	373	0.0332	Ethyl ether[1]	273	0.0133
	473	0.0484		319	0.0171
Butane[3]	273	0.0135		373	0.0227
	373	0.0234	Ethylene[6]	273	0.0175
Carbon monoxide[2]	173	0.0152		323	0.0227
	273	0.0232		373	0.0279
	373	0.0305	n-Hexane[3]	273	0.0125
Chlorine[4]	273	0.00744		293	0.0138
			Sulfur dioxide[7]	273	0.0087
				373	0.0119

Source : (1) Moser, dissertation, Berlin, 1913; (2) F. G. Keyes, *Tech. Rept. 37*, Project Squid, Apr. 1, 1952; (3) W. B. Mann and B. G. Dickens, *Proc. Roy. Soc. (London)*, **A134**, 77 (1931); (4) *International Critical Tables*. New York: McGraw-Hill Book Company, 1929; (5) T. H. Chilton and R. P. Genereaux, personal communication, 1946; (6) A. Eucken, *Physik, Z.*, **12**, 1101 (1911); **14**, 324 (1913); (7) B. G. Dickens, *Proc. Roy. Soc. (London)*, **A143**, 517 (1934).

A.3-11 Heat Capacities of Liquids ($c_p = kJ/kg \cdot K$)

Liquid	K	c_p	Liquid	K	c_p
Acetic acid	273	1.959	Hydrochloric acid (20 mol %)	273	2.43
	311	2.240		293	2.474
Acetone	273	2.119	Mercury	293	0.01390
	293	2.210	Methyl alcohol	293	2.512
Aniline	273	2.001		313	2.583
	323	2.181	Nitrobenzene	283	1.499
Benzene	293	1.700		303	1.419
	333	1.859		363	1.436
Butane	273	2.300	Sodium chloride (9.1 mol %)	293	3.39
i-Butyl alcohol	303	2.525		330	3.43
Ethyl alcohol	273	2.240	Sulfuric acid (100%)	293	1.403
	298	2.433	Toluene	273	1.616
Formic acid	273	1.825		323	1.763
	289	2.131	o-Xylene	303	1.721
Glycerol	288	2.324			
	305	2.412			

Source : N. A. Lange, *Handbook of Chemistry*, 10th ed. New York: McGraw-Hill Book Company, 1967; National Research Council, *International Critical Tables*, Vol. V. New York: McGraw-Hill Book Company, 1929; R. H. Perry and C. H. Chilton, *Chemical Engineers' Handbook*, 5th ed. New York: McGraw-Hill Book Company, 1973.

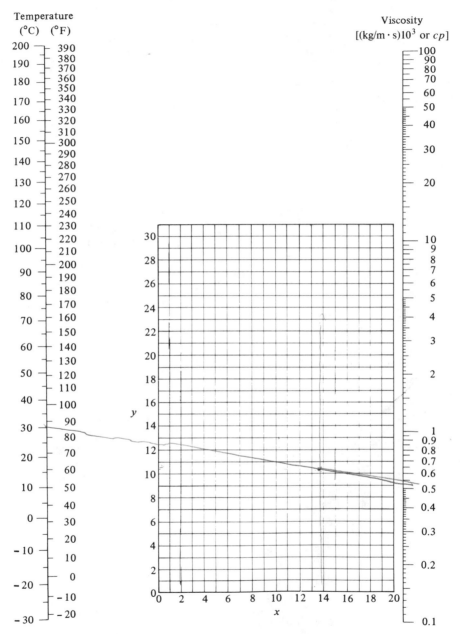

FIGURE A.3-4. *Viscosities of liquids. (From R. H. Perry and C. H. Chilton, Chemical Engineers' Handbook, 5th ed. New York : McGraw-Hill Book Company, 1973. With permission.) See Table A.3-12 for use with Fig. A.3-4.*

A.3-12 Viscosities of Liquids (Coordinates for Use with Fig. A.3-4)

Liquid	X	Y	Liquid	X	Y
Acetaldehyde	15.2	4.8	Cyclohexanol	2.9	24.3
Acetic acid, 100%	12.1	14.2	Cyclohexane	9.8	12.9
Acetic acid, 70%	9.5	17.0	Dibromomethane	12.7	15.8
Acetic anhydride	12.7	12.8	Dichloroethane	13.2	12.2
Acetone, 100%	14.5	7.2	Dichloromethane	14.6	8.9
Acetone, 35%	7.9	15.0	Diethyl ketone	13.5	9.2
Acetonitrile	14.4	7.4	Diethyl oxalate	11.0	16.4
Acrylic acid	12.3	13.9	Diethylene glycol	5.0	24.7
Allyl alcohol	10.2	14.3	Diphenyl	12.0	18.3
Allyl bromide	14.4	9.6	Dipropyl ether	13.2	8.6
Allyl iodide	14.0	11.7	Dipropyl oxalate	10.3	17.7
Ammonia, 100%	12.6	2.0	Ethyl acetate	13.7	9.1
Ammonia, 26%	10.1	13.9	Ethyl acrylate	12.7	10.4
Amyl acetate	11.8	12.5	Ethyl alcohol, 100%	10.5	13.8
Amyl alcohol	7.5	18.4	Ethyl alcohol, 95%	9.8	14.3
Aniline	8.1	18.7	Ethyl alcohol, 40%	6.5	16.6
Anisole	12.3	13.5	Ethyl benzene	13.2	11.5
Arsenic trichloride	13.9	14.5	Ethyl bromide	14.5	8.1
Benzene	12.5	10.9	2-Ethyl butyl acrylate	11.2	14.0
Brine, $CaCL_2$, 25%	6.6	15.9	Ethyl chloride	14.8	6.0
Brine, NaCl, 25%	10.2	16.6	Ethyl ether	14.5	5.3
Bromine	14.2	13.2	Ethyl formate	14.2	8.4
Bromotoluene	20.0	15.9	2-Ethyl hexyl acrylate	9.0	15.0
Butyl acetate	12.3	11.0	Ethyl iodide	14.7	10.3
Butyl acrylate	11.5	12.6	Ethyl propionate	13.2	9.9
Butyl alcohol	8.6	17.2	Ethyl propyl ether	14.0	7.0
Butyric acid	12.1	15.3	Ethyl sulfide	13.8	8.9
Carbon dioxide	11.6	0.3	Ethylene bromide	11.9	15.7
Carbon disulfide	16.1	7.5	Ethylene chloride	12.7	12.2
Carbon tetrachloride	12.7	13.1	Ethylene glycol	6.0	23.6
Chlorobenzene	12.3	12.4	Ethylidene chloride	14.1	8.7
Chloroform	14.4	10.2	Fluorobenzene	13.7	10.4
Chlorosulfonic acid	11.2	18.1	Formic acid	10.7	15.8
Chlorotoluene, ortho	13.0	13.3	Freon-11	14.4	9.0
Chlorotoluene, meta	13.3	12.5	Freon-12	16.8	15.6
Chlorotoluene, para	13.3	12.5	Freon-21	15.7	7.5
Cresol, meta	2.5	20.8	Freon-22	17.2	4.7

Liquid	X	Y	Liquid	X	Y
Freon-113	12.5	11.4	Octyl alcohol	6.6	21.1
Glycerol, 100%	2.0	30.0	Pentachloroethane	10.9	17.3
Glycerol, 50%	6.9	19.6	Pentane	14.9	5.2
Heptane	14.1	8.4	Phenol	6.9	20.8
Hexane	14.7	7.0	Phosphorus tribromide	13.8	16.7
Hydrochloric acid, 31.5%	13.0	16.6	Phosphorus trichloride	16.2	10.9
Iodobenzene	12.8	15.9	Propionic acid	12.8	13.8
Isobutyl alcohol	7.1	18.0	Propyl acetate	13.1	10.3
Isobutyric acid	12.2	14.4	Propyl alcohol	9.1	16.5
Isopropyl alcohol	8.2	16.0	Propyl bromide	14.5	9.6
Isopropyl bromide	14.1	9.2	Propyl chloride	14.4	7.5
Isopropyl chloride	13.9	7.1	Propyl formate	13.1	9.7
Isopropyl iodide	13.7	11.2	Propyl iodide	14.1	11.6
Kerosene	10.2	16.9	Sodium	16.4	13.9
Linseed oil, raw	7.5	27.2	Sodium hydroxide, 50%	3.2	25.8
Mercury	18.4	16.4	Stannic chloride	13.5	12.8
Methanol, 100%	12.4	10.5	Succinonitrile	10.1	20.8
Methanol, 90%	12.3	11.8	Sulfur dioxide	15.2	7.1
Methanol, 40%	7.8	15.5	Sulfuric acid, 110%	7.2	27.4
Methyl acetate	14.2	8.2	Sulfuric acid, 100%	8.0	25.1
Methyl acrylate	13.0	9.5	Sulfuric acid, 98%	7.0	24.8
Methyl *i*-butyrate	12.3	9.7	Sulfuric acid, 60%	10.2	21.3
Methyl *n*-butyrate	13.2	10.3	Sulfuryl chloride	15.2	12.4
Methyl chloride	15.0	3.8	Tetrachloroethane	11.9	15.7
Methyl ethyl ketone	13.9	8.6	Thiophene	13.2	11.0
Methyl formate	14.2	7.5	Titanium tetrachloride	14.4	12.3
Methyl iodide	14.3	9.3	Toluene	13.7	10.4
Methyl propionate	13.5	9.0	Trichloroethylene	14.8	10.5
Methyl propyl ketone	14.3	9.5	Triethylene glycol	4.7	24.8
Methyl sulfide	15.3	6.4	Turpentine	11.5	14.9
Naphthalene	7.9	18.1	Vinyl acetate	14.0	8.8
Nitric acid, 95%	12.8	13.8	Vinyl toluene	13.4	12.0
Nitric acid, 60%	10.8	17.0	Water	10.2	13.0
Nitrobenzene	10.6	16.2	Xylene, ortho	13.5	12.1
Nitrogen dioxide	12.9	8.6	Xylene, meta	13.9	10.6
Nitrotoluene	11.0	17.0	Xylene, para	13.9	10.9
Octane	13.7	10.0			

$$c_p = \text{cal/g} \cdot {}^\circ\text{C} = \text{btu/lb}_m \cdot {}^\circ\text{F}$$

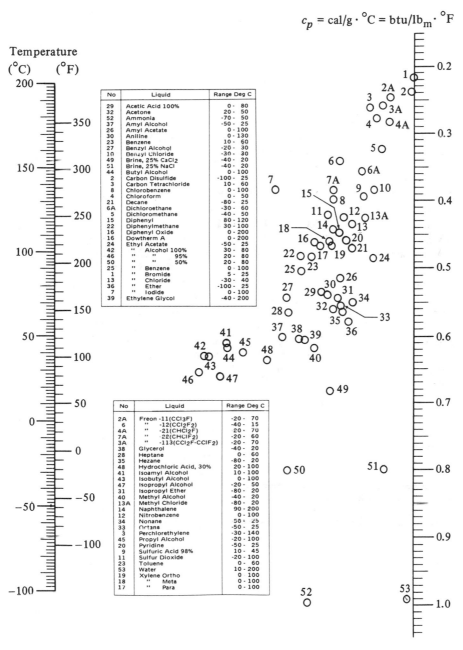

Temperature
(°C)   (°F)

No	Liquid	Range Deg C
29	Acetic Acid 100%	0 - 80
32	Acetone	20 - 50
52	Ammonia	-70 - 50
37	Amyl Alcohol	-50 - 25
26	Amyl Acetate	0 - 100
30	Aniline	0 - 130
23	Benzene	10 - 60
27	Benzyl Alcohol	-20 - 30
10	Benzyl Chloride	-30 - 30
49	Brine, 25% CaCl2	-40 - 20
51	Brine, 25% NaCl	-40 - 20
44	Butyl Alcohol	0 - 100
2	Carbon Disulfide	-100 - 25
3	Carbon Tetrachloride	10 - 60
8	Chlorobenzene	0 - 100
4	Chloroform	0 - 50
21	Decane	-80 - 25
6A	Dichloroethane	-30 - 60
5	Dichloromethane	-40 - 50
15	Diphenyl	80 - 120
22	Diphenylmethane	30 - 100
16	Diphenyl Oxide	0 - 200
16	Dowtherm A	0 - 200
24	Ethyl Acetate	-50 - 25
42	" Alcohol 100%	30 - 80
46	" " 95%	20 - 80
50	" " 50%	20 - 80
25	" Benzene	0 - 100
1	" Bromide	5 - 25
13	" Chloride	-30 - 40
36	" Ether	-100 - 25
7	" Iodide	0 - 100
39	Ethylene Glycol	-40 - 200

No	Liquid	Range Deg C
2A	Freon -11(CCl3F)	-20 - 70
6	" -12(CCl2F2)	-40 - 15
4A	" -21(CHCl2F)	20 - 70
7A	" -22(CHClF2)	-20 - 60
3A	" -113(CCl2F-CClF2)	-20 - 70
38	Glycerol	-40 - 20
28	Heptane	0 - 60
35	Hexane	-80 - 20
48	Hydrochloric Acid, 30%	20 - 100
41	Isoamyl Alcohol	10 - 100
43	Isobutyl Alcohol	0 - 100
47	Isopropyl Alcohol	-20 - 50
31	Isopropyl Ether	-80 - 20
40	Methyl Alcohol	-40 - 20
13A	Methyl Chloride	-80 - 20
14	Naphthalene	90 - 200
12	Nitrobenzene	0 - 100
34	Nonane	50 - 25
33	Octane	-50 - 25
3	Perchlorethylene	-30 - 140
45	Propyl Alcohol	-20 - 100
20	Pyridine	-50 - 25
9	Sulfuric Acid 98%	10 - 45
11	Sulfur Dioxide	-20 - 100
23	Toluene	0 - 60
53	Water	10 - 200
19	Xylene Ortho	0 - 100
18	" Meta	0 - 100
17	" Para	0 - 100

FIGURE A.3-5.   Heat capacity of liquids. (From R. H. Perry and C. H. Chilton, Chemical Engineers' Handbook, 5th ed. New York : McGraw-Hill Book Company, 1973. With permission.)

A.3-13   Thermal Conductivities of Liquids (k = W/m · K)*

Liquid	K	k	Liquid	K	k
Acetic acid			Ethylene glycol	273	0.265
100%	293	0.171	Glycerol, 100%	293	0.284
50%	293	0.346	n-Hexane	303	0.138
Ammonia	243–258	0.502		333	0.135
n-Amyl alcohol	303	0.163	Kerosene	293	0.149
	373	0.154		348	0.140
Benzene	303	0.159	Methyl alcohol		
	333	0.151	100%	293	0.215
Carbon tetrachloride	273	0.185	60%	293	0.329
	341	0.163	20%	293	0.492
n-Decane	303	0.147	100%	323	0.197
	333	0.144	n-Octane	303	0.144
Ethyl acetate	293	0.175		333	0.140
Ethyl alcohol			NaCl brine		
100%	293	0.182	25%	303	0.571
60%	293	0.305	12.5%	303	0.589
20%	293	0.486	Sulfuric acid		
100%	323	0.151	90%	303	0.364
			60%	303	0.433
			Vaseline	332	0.183

* A linear variation with temperature may be assumed between the temperature limits given.
Source: R. H. Perry and C. H. Chilton, *Chemical Engineers' Handbook*, 5th ed. New York: McGraw-Hill Book Company, 1973. With permission.

A.3-14 Heat Capacities of Solids ($c_p = kJ/kg \cdot K$)

Solid	K	c_p	Solid	K	c_p
Alumina	373	0.84	Benzene	273	1.570
	1773	1.147	Benzoic acid	293	1.243
Asbestos		1.05	Camphene	308	1.591
Asphalt		0.92	Caprylic acid	271	2.629
Brick, fireclay	373	0.829	Dextrin	273	1.218
	1773	1.248	Formic acid	273	1.800
Cement, portland		0.779	Glycerol	273	1.382
Clay		0.938	Lactose	293	1.202
Concrete		0.63	Oxalic acid	323	1.612
Corkboard	303	0.167	Tartaric acid	309	1.202
Glass		0.84	Urea	293	1.340
Magnesia	373	0.980			
	1773	0.787			
Oak		2.39			
Pine, yellow	298	2.81			
Porcelain	293–373	0.775			
Rubber, vulcanized		2.01			
Steel		0.50			
Wool		1.361			

Source: R. H. Perry and C. H. Chilton, *Chemical Engineers' Handbook*, 5th ed. New York: McGraw-Hill Book Company, 1973; National Research Council, *International Critical Tables*, Vol. V. New York: McGraw-Hill Book Company, 1929; L. S. Marks, *Mechanical Engineers' Handbook*, 5th ed. New York: McGraw-Hill Book Company, 1951; F. Kreith, *Principles of Heat Transfer*, 2nd ed. Scranton, Pa.: International Textbook Co, 1965.

A.3-15 Thermal Conductivities of Building and Insulating Materials

Material	$\rho \left(\dfrac{kg}{m^3}\right)$	t^* (°C)	$k(W/m \cdot K)$		
Asbestos	577		0.151 (0°C)	0.168 (37.8°C)	0.190 (93.3°C)
Asbestos sheets	889	51	0.166		
Brick, building		20	0.69		
Brick, fireclay			1.00 (200°C)	1.47 (600°C)	1.64 (1000°C)
Clay soil, 4% H_2O	1666	4.5	0.57		
Concrete, 1:4 dry			0.762		
Corkboard	160.2	30	0.0433		
Cotton	80.1		0.055 (0°C)	0.061 (37.8°C)	0.068 (93.3°C)
Felt, wool	330	30	0.052		
Fiber insulation board	237	21	0.048		
Glass, window			0.52–1.06		
Glass wool	64.1	30	0.0310 (-6.7°C)	0.0414 (37.8°C)	0.0549 (93.3°C)
Ice	921	0	2.25		
Magnesia, 85%	271		0.068 (37.8°C)	0.071 (93.3°C)	0.080 (204.4°C)
	208		0.059 (37.8°C)	0.062 (93.3°C)	0.066 (148.9°C)
Oak, across grain	825	15	0.208		
Pine, across grain	545	15	0.151		
Paper			0.130		
Rock wool	192		0.0317 (-6.7°C)	0.0391 (37.8°C)	0.0486 (93.3°C)
	128		0.0296 (-6.7°C)	0.0395 (37.8°C)	0.0518 (93.3°C)
Rubber, hard	1198	0	0.151		
Sand soil					
4% H_2O	1826	4.5	1.51		
10% H_2O	1922	4.5	2.16		
Sandstone	2243	40	1.83		
Snow	559	0	0.47		
Wool	110.5	30	0.036		

* Room temperature when none is noted.

Source: L. S. Marks, *Mechanical Engineers' Handbook*, 5th ed. New York: McGraw-Hill Book Company, 1951; W. H. McAdams, *Heat Transmission*, 3rd ed. New York: McGraw-Hill Book Company, 1954; F. H. Norton, *Refractories*. New York: McGraw-Hill Book Company, 1949; National Research Council, *International Critical Tables*. New York: McGraw-Hill Book Company, 1929; M. S. Kersten, *Univ. Minn. Eng. Ex. Sta., Bull. 28,* June 1949; R. H. Heilman, *Ind. Eng. Chem.,* 28, 782 (1936).

A.3-16 Thermal Conductivities, Densities, and Heat Capacities of Metals

Material	t ($°C$)	ρ $\left(\dfrac{kg}{m^3}\right)$	c_p $\left(\dfrac{kJ}{kg \cdot K}\right)$		$k(W/m \cdot K)$	
Aluminum	20	2707	0.896	202 (0°C) 230 (300°C)	206 (100°C)	215 (200°C)
Brass (70–30)	20	8522	0.385	97 (0°C)	104 (100°C)	109 (200°C)
Cast iron	20	7593	0.465	55 (0°C)	52 (100°C)	48 (200°C)
Copper	20	8954	0.383	388 (0°C)	377 (100°C)	372 (200°C)
Lead	20	11 370	0.130	35 (0°C)	33 (100°C)	31 (200°C)
Steel 1%C	20	7801	0.473	45.3 (18°C) 43 (300°C)	45 (100°C)	45 (200°C)
308 stainless	20	7849	0.461	15.2 (100°C)	21.6 (500°C)	
304 stainless	0	7817	0.461	13.8 (0°C)	16.3 (100°C)	18.9 (300°C)
Tin	20	7304	0.227	62 (0°C)	59 (100°C)	57 (200°C)

Source: L. S. Marks, *Mechanical Engineers' Handbook*, 5th ed. New York: McGraw-Hill Book Company, 1951; E. R. G. Eckert and R. M. Drake, *Heat and Mass Transfer*, 2nd ed. New York: McGraw-Hill Book Company, 1959; R. H. Perry and C. H. Chilton, *Chemical Engineers' Handbook*, 5th ed. New York: McGraw-Hill Book Company, 1973; National Research Council, *International Critical Tables*. New York: McGraw-Hill Book Company, 1929.

A.3-17 Normal Total Emmissivities of Surfaces

Surface	K	ε	Surface	K	ε
Aluminum			Lead, unoxidized	400	0.057
highly oxidized	366	0.20	Nickel, polished	373	0.072
highly polished	500	0.039	Nickel oxide	922	0.59
	850	0.057	Oak, planed	294	0.90
Aluminum oxide	550	0.63	Paint		
Asbestos board	296	0.96	aluminum	373	0.52
Brass, highly	520	0.028	oil (16 different,		
			all colors)	373	0.92–0.96
polished	630	0.031	Paper	292	0.924
Chromium,					
polished	373	0.075	Roofing paper	294	0.91
Copper			Rubber (hard, glossy)	296	0.94
oxidized	298	0.78	Steel		
polished	390	0.023	oxidized at 867 K	472	0.79
Glass, smooth	295	0.94	polished stainless	373	0.074
Iron			304 stainless	489	0.44
oxidized	373	0.74	Water	273	0.95
tin-plated	373	0.07		373	0.963
Iron oxide	772	0.85			

Source: R. H. Perry and C. H. Chilton, *Chemical Engineers' Handbook*, 5th ed. New York: McGraw-Hill Book Company, 1973; W. H. McAdams, *Heat Transmission*, 3rd ed. New York: McGraw-Hill Book Company, 1954; E. Schmidt, *Gesundh.-Ing. Beiheft*, **20**, Reihe 1, 1 (1927).

A.3-18 Henry's Law Constants for Gases in Water ($H \times 10^{-4}$)*

T											
K	°C	CO_2	CO	C_2H_6	C_2H_4	He	H_2	H_2S	CH_4	N_2	O_2
273.2	0	0.0728	3.52	1.26	0.552	12.9	5.79	0.0268	2.24	5.29	2.55
283.2	10	0.104	4.42	1.89	0.768	12.6	6.36	0.0367	2.97	6.68	3.27
293.2	20	0.142	5.36	2.63	1.02	12.5	6.83	0.0483	3.76	8.04	4.01
303.2	30	0.186	6.20	3.42	1.27	12.4	7.29	0.0609	4.49	9.24	4.75
313.2	40	0.233	6.96	4.23		12.1	7.51	0.0745	5.20	10.4	5.35

* $p_A = Hx_A$, p_A = partial pressure of A in gas in atm, x_A = mole fraction of A in liquid, H = Henry's law constant in atm/mole frac.
Source: National Research Council, *International Critical Tables*, Vol. III. New York: McGraw-Hill Book Company, 1929.

A.3-19 Equilibrium Data for SO_2–Water System

Mole Fraction SO_2 in Liquid, x_A	Partial Pressure of SO_2 in Vapor, p_A (mm Hg)		Mole Fraction SO_2 in Vapor, y_A; $P = 1$ Atm	
	20°C (293 K)	30°C (303 K)	20°C	30°C
0	0	0	0	0
0.0000562	0.5	0.6	0.000658	0.000790
0.0001403	1.2	1.7	0.00158	0.00223
0.000280	3.2	4.7	0.00421	0.00619
0.000422	5.8	8.1	0.00763	0.01065
0.000564	8.5	11.8	0.01120	0.0155
0.000842	14.1	19.7	0.01855	0.0259
0.001403	26.0	36	0.0342	0.0473
0.001965	39.0	52	0.0513	0.0685
0.00279	59	79	0.0775	0.1040
0.00420	92	125	0.121	0.1645
0.00698	161	216	0.212	0.284
0.01385	336	452	0.443	0.594
0.0206	517	688	0.682	0.905
0.0273	698		0.917	

Source : T. K. Sherwood, *Ind. Eng. Chem.*, **17**, 745 (1925).

A.3-20 Equilibrium Data for Methanol–Water System

Mole Fraction Methanol in Liquid, x_A	Partial Pressure of Methanol in Vapor, p_A (mm Hg)	
	39.9°C (313.1 K)	59.4°C (332.6K)
0	0	0
0.05	25.0	50
0.10	46.0	102
0.15	66.5	151

Source : National Research Council, *International Critical Tables*, Vol. III. New York: McGraw-Hill Book Company, 1929.

A.3-21 Equilibrium Data for Acetone–Water System at 20°C (293 K)

Mole Fraction Acetone in Liquid, x_A	Partial Pressure of Acetone in Vapor, p_A (mm Hg)
0	0
0.0333	30.0
0.0720	62.8
0.117	85.4
0.171	103

Source: T. K. Sherwood, *Absorption and Extraction*. New York: McGraw-Hill Book Company, 1937. With permission.

A.3-22 Equilibrium Data for Ammonia–Water System

Mole Fraction NH_3 in Liquid, x_A	Partial Pressure of NH_3 in Vapor, p_A (mm Hg)		Mole Fraction NH_3 in Vapor, y_A; $P = 1$ Atm	
	20°C (293 K)	30°C (303 K)	20°C	30°C
0	0	0	0	0
0.0126		11.5		0.0151
0.0167		15.3		0.0201
0.0208	12	19.3	0.0158	0.0254
0.0258	15	24.4	0.0197	0.0321
0.0309	18.2	29.6	0.0239	0.0390
0.0405	24.9	40.1	0.0328	0.0527
0.0503	31.7	51.0	0.0416	0.0671
0.0737	50.0	79.7	0.0657	0.105
0.0960	69.6	110	0.0915	0.145
0.137	114	179	0.150	0.235
0.175	166	260	0.218	0.342
0.210	227	352	0.298	0.463
0.241	298	454	0.392	0.597
0.297	470	719	0.618	0.945

Source: J. H. Perry, *Chemical Engineers' Handbook*, 4th ed. New York: McGraw-Hill Book Company, 1963. With permission.

A.3-23 Equilibrium Data for Ethanol–Water System at 101.325 kPa (1 Atm)*

Temperature °C	°F	x_A	y_A
100.0	212	0	0
98.1	208.5	0.020	0.192
95.2	203.4	0.050	0.377
91.8	197.2	0.100	0.527
87.3	189.2	0.200	0.656
84.7	184.5	0.300	0.713
83.2	181.7	0.400	0.746
82.0	179.6	0.500	0.771

Vapor–Liquid Equilibria, Mass Fraction Ethanol

Temperature °C	°F	x_A	y_A
81.0	177.8	0.600	0.794
80.1	176.2	0.700	0.822
79.1	174.3	0.800	0.858
78.3	173.0	0.900	0.912
78.2	172.8	0.940	0.942
78.1	172.7	0.960	0.959
78.2	172.8	0.980	0.978
78.3	173.0	1.00	1.00

Temperature °C	°F	Mass Fraction	Enthalpy (btu/lb$_m$ of mixture) Liquid	Vapor	Enthalpy (kJ/kg of mixture) Liquid	Vapor
100.0	212	0	180.1	1150	418.9	2675
91.8	197.2	0.1	159.8	1082	371.7	2517
84.7	184.5	0.3	135.0	943	314.0	2193
82.0	179.6	0.5	122.9	804	285.9	1870
80.1	176.2	0.7	111.1	664	258.4	1544
78.3	173.0	0.9	96.6	526	224.7	1223
78.3	173.0	1.0	89.0	457.5	207.0	1064

* Reference state for enthalpy is pure liquid at 273 K or 0°C.

Source: Data from L. W. Cornell and R. E. Montonna, Ind. Eng. Chem., **25**, 1331 (1933); and W. A. Noyes and R. R. Warfel, J. Am. Chem. Soc., **23**, 463 (1901), as given by G. G. Brown, Unit Operations. New York: John Wiley & Sons, Inc., 1950. With permission.

A.3-24 Acetic Acid–Water–Isopropyl Ether System, Liquid–Liquid Equilibria at 293 K or 20°C

Water Layer (wt %)			Isopropyl Ether Layer (wt %)		
Acetic Acid	Water	Isopropyl Ether	Acetic Acid	Water	Isopropyl Ether
0	98.8	1.2	0	0.6	99.4
0.69	98.1	1.2	0.18	0.5	99.3
1.41	97.1	1.5	0.37	0.7	98.9
2.89	95.5	1.6	0.79	0.8	98.4
6.42	91.7	1.9	1.93	1.0	97.1
13.30	84.4	2.3	4.82	1.9	93.3
25.50	71.1	3.4	11.40	3.9	84.7
36.70	58.9	4.4	21.60	6.9	71.5
44.30	45.1	10.6	31.10	10.8	58.1
46.40	37.1	16.5	36.20	15.1	48.7

Source: Trans. A.I.Ch.E., 36, 601, 628 (1940). With permission.

A.3-25 Liquid–Liquid Equilibrium Data for Acetone–Water–Methyl Isobutyl Ketone (MIK) System at 298–299 K or 25–26°C

Composition Data (wt %)			Acetone Distribution Data (wt %)	
MIK	Acetone	Water	Water Phase	MIK Phase
98.0	0	2.00	2.5	4.5
93.2	4.6	2.33	5.5	10.0
77.3	18.95	3.86	7.5	13.5
71.0	24.4	4.66	10.0	17.5
65.5	28.9	5.53	12.5	21.3
54.7	37.6	7.82	15.5	25.5
46.2	43.2	10.7	17.5	28.2
12.4	42.7	45.0	20.0	31.2
5.01	30.9	64.2	22.5	34.0
3.23	20.9	75.8	25.0	36.5
2.12	3.73	94.2	26.0	37.5
2.20	0	97.8		

Source: Reprinted with permission from D. F. Othmer, R. E. White, and E. Trueger, Ind. Eng. Chem., 33, 1240 (1941). Copyright by the American Chemical Society.

Physical Properties
of Foods and
Biological Materials

A.4-1 Heat Capacities of Foods (Average c_p 273–373 K or 0–100°C)

Material	H_2O (wt %)	c_p $(kJ/kg \cdot K)$
Apples	75–85	3.73–4.02
Apple sauce		4.02*
Asparagus		
Fresh	93	3.94†
Frozen	93	2.01‡
Bacon, lean	51	3.43
Banana purée		3.66§
Beef, lean	72	3.43
Bread, white	44–45	2.72–2.85
Butter	15	2.30¶
Cantaloupe	92.7	3.94†
Cheese, Swiss	55	2.68†
Corn, sweet		
Fresh		3.32†
Frozen		1.77‡
Cream, 45–60% fat	57–73	3.06–3.27
Cucumber	97	4.10
Eggs		
Fresh		3.18†
Frozen		1.68‡
Fish, cod		
Fresh	70	3.18
Frozen	70	1.72‡
Flour	12–13.5	1.80–1.88
Ice	100	1.958‖

Material	H_2O (wt %)	c_p (kJ/kg·K)
Ice cream		
Fresh	58–66	3.27†
Frozen	58–66	1.88‡
Lamb	70	3.18*
Macaroni	12.5–13.5	1.84–1.88
Milk, cows'		
Whole	87.5	3.85
Skim	91	3.98–4.02
Olive oil		2.01**
Oranges		
Fresh	87.2	3.77†
Frozen	87.2	1.93‡
Peas, air-dried	14	1.84
Peas, green		
Fresh	74.3	3.31†
Frozen	74.3	1.76‡
Pea soup		4.10
Plums	75–78	3.52
Pork		
Fresh	60	2.85†
Frozen	60	1.34‡
Potatoes	75	3.52
Poultry		
Fresh	74	3.31†
Frozen	74	1.55‡
Sausage, franks		
Fresh	60	3.60†
Frozen	60	2.35‡
String beans		
Fresh	88.9	3.81†
Frozen	88.9	1.97‡
Tomatoes	95	3.98†
Veal	63	3.22
Water	100	4.185**

* 32.8°C.
† Above freezing.
‡ Below freezing.
§ 24.4°C.
¶ 4.4°C.
‖ −20°C.
** 20°C.

Source : W. O. Ordinanz, *Food Ind.*, **18**, 101 (1946); G. A. Reidy, Department of Food Science, Michigan State University, 1968; S. E. Charm, *The Fundamentals of Food Engineering*, 2nd ed. Westport, Conn.: Avi Publishing Co., Inc., 1971; R. L. Earle, *Unit Operations in Food Processing.* Oxford: Pergamon Press, Inc., 1966; ASHRAE, *Handbook of Fundamentals.* New York: ASHRAE, 1972, 1967; H. C. Mannheim, M. P. Steinberg, and A. I. Nelson, *Food Technol.*, **9**, 556 (1955).

A.4-2 Thermal Conductivities, Densities, and Viscosities of Foods

Material	H_2O (wt %)	Temperature (K)	k (W/m·K)	ρ (kg/m³)	μ [(Pa·s)10³, or cp]
Apple sauce		295.7	0.692		
Butter	15	277.6	0.197	998	
Cantaloupe			0.571		
Fish					
Fresh		273.2	0.431		
Frozen		263.2	1.22		
Flour, wheat	8.8		0.450		
Honey	12.6	275.4	0.50		
Ice	100	273.2	2.25		
	100	253.2	2.42		
Lamb	71	278.8	0.415		
Milk					
Whole		293.2		1030	2.12
Skim		274.7	0.538		
		298.2		1041	1.4
Oil					
Cod liver		298.2		924	
Corn		288.2		921	
Olive		293.2	0.168	919	84
Peanut		277.1	0.168		
Soybean		303.2		919	40
Oranges	61.2	303.5	0.431		
Pears		281.9	0.595		
Pork, lean					
Fresh	74	275.4	0.460		
Frozen		258.2	1.109		
Potatoes					
Raw			0.554		
Frozen		260.4	1.09	977	
Salmon					
Fresh	67	277.1	0.50		
Frozen	67	248.2	1.30		
Sucrose solution	80	294.3		1073	1.92
Turkey					
Fresh	74	276.0	0.502		
Frozen		248.2	1.675		
Veal					
Fresh	75	335.4	0.485		
Frozen	75	263.6	1.30		
Water	100	293.2	0.602		
	100	273.2	0.569		

Source: R. C. Weast, Handbook of Chemistry and Physics, 48th ed. Cleveland: Chemical Rubber Co., Inc., 1967; C. P. Lentz, Food Technol., 15, 243 (1961); G. A. Reidy, Department of Food Science, Michigan State University, 1968; S. E. Charm, The Fundamentals of Food Engineering, 2nd ed. Westport, Conn.: Avi Publishing Co., Inc., 1971; R. Earle, Unit Operations in Food Processing, Oxford: Pergamon Press, 1966; R. H. Perry and C. H. Chilton, Chemical Engineers' Handbook, 5th ed. New York: McGraw-Hill Book Company, 1973; V. E. Sweat, J. Food Sci., 39, 1080 (1974).

Properties of Pipes, Tubes, and Screens

A.5-1 Dimensions of Standard Steel Pipe

Nominal Pipe Size (in.)	Outside Diameter		Sched-ule Number	Wall Thickness		Inside Diameter		Inside Cross-Sectional Area	
	in.	mm		in.	mm	in.	mm	ft^2	$m^2 \times 10^4$
$\frac{1}{8}$	0.405	10.29	40	0.068	1.73	0.269	6.83	0.00040	0.3664
			80	0.095	2.41	0.215	5.46	0.00025	0.2341
$\frac{1}{4}$	0.540	13.72	40	0.088	2.24	0.364	9.25	0.00072	0.6720
			80	0.119	3.02	0.302	7.67	0.00050	0.4620
$\frac{3}{8}$	0.675	17.15	40	0.091	2.31	0.493	12.52	0.00133	1.231
			80	0.126	3.20	0.423	10.74	0.00098	0.9059
$\frac{1}{2}$	0.840	21.34	40	0.109	2.77	0.622	15.80	0.00211	1.961
			80	0.147	3.73	0.546	13.87	0.00163	1.511
$\frac{3}{4}$	1.050	26.67	40	0.113	2.87	0.824	20.93	0.00371	3.441
			80	0.154	3.91	0.742	18.85	0.00300	2.791
1	1.315	33.40	40	0.133	3.38	1.049	26.64	0.00600	5.574
			80	0.179	4.45	0.957	24.31	0.00499	4.641
$1\frac{1}{4}$	1.660	42.16	40	0.140	3.56	1.380	35.05	0.01040	9.648
			80	0.191	4.85	1.278	32.46	0.00891	8.275
$1\frac{1}{2}$	1.900	48.26	40	0.145	3.68	1.610	40.89	0.01414	13.13
			80	0.200	5.08	1.500	38.10	0.01225	11.40
2	2.375	60.33	40	0.154	3.91	2.067	52.50	0.02330	21.65
			80	0.218	5.54	1.939	49.25	0.02050	19.05
$2\frac{1}{2}$	2.875	73.03	40	0.203	5.16	2.469	62.71	0.03322	30.89
			80	0.276	7.01	2.323	59.00	0.02942	27.30
3	3.500	88.90	40	0.216	5.49	3.068	77.92	0.05130	47.69
				0.300	7.62	2.900	73.66	0.04587	42.61
$3\frac{1}{2}$	4.000	101.6	40	0.226	5.74	3.548	90.12	0.06870	63.79
			80	0.318	8.08	3.364	85.45	0.06170	57.35
4	4.500	114.3	40	0.237	6.02	4.026	102.3	0.08840	82.19
			80	0.337	8.56	3.826	97.18	0.07986	74.17
5	5.563	141.3	40	0.258	6.55	5.047	128.2	0.1390	129.1
			80	0.375	9.53	4.813	122.3	0.1263	117.5
6	6.625	168.3	40	0.280	7.11	6.065	154.1	0.2006	186.5
			80	0.432	10.97	5.761	146.3	0.1810	168.1
8	8.625	219.1	40	0.322	8.18	7.981	202.7	0.3474	322.7
			80	0.500	12.70	7.625	193.7	0.3171	294.7

A.5-2 Dimensions of Heat-Exchanger Tubes

Outside Diameter		BWG Number	Wall Thickness		Inside Diameter		Inside Cross-Sectional Area	
in.	mm		in.	mm	in.	mm	ft^2	$m^2 \times 10^4$
$\frac{5}{8}$	15.88	12	0.109	2.77	0.407	10.33	0.000903	0.8381
		14	0.083	2.11	0.459	11.66	0.00115	1.068
		16	0.065	1.65	0.495	12.57	0.00134	1.241
		18	0.049	1.25	0.527	13.39	0.00151	1.408
$\frac{3}{4}$	19.05	12	0.109	2.77	0.532	13.51	0.00154	1.434
		14	0.083	2.11	0.584	14.83	0.00186	1.727
		16	0.065	1.65	0.620	15.75	0.00210	1.948
		18	0.049	1.25	0.652	16.56	0.00232	2.154
$\frac{7}{8}$	22.23	12	0.109	2.77	0.657	16.69	0.00235	2.188
		14	0.083	2.11	0.709	18.01	0.00274	2.548
		16	0.065	1.65	0.745	18.92	0.00303	2.811
		18	0.049	1.25	0.777	19.74	0.00329	3.060
1	25.40	10	0.134	3.40	0.732	18.59	0.00292	2.714
		12	0.109	2.77	0.782	19.86	0.00334	3.098
		14	0.083	2.11	0.834	21.18	0.00379	3.523
		16	0.065	1.65	0.870	22.10	0.00413	3.836
$1\frac{1}{4}$	31.75	10	0.134	3.40	0.982	24.94	0.00526	4.885
		12	0.109	2.77	1.032	26.21	0.00581	5.395
		14	0.083	2.11	1.084	27.53	0.00641	5.953
		16	0.065	1.65	1.120	28.45	0.00684	6.357
$1\frac{1}{2}$	38.10	10	0.134	3.40	1.232	31.29	0.00828	7.690
		12	0.109	2.77	1.282	32.56	0.00896	8.326
		14	0.083	2.11	1.334	33.88	0.00971	9.015
2	50.80	10	0.134	3.40	1.732	43.99	0.0164	15.20
		12	0.109	2.77	1.782	45.26	0.0173	16.09

A.5-3 Tyler Standard Screen Scale

	Sieve Opening		Nominal Wire Diameter		
mm	in. (approx. equivalents)	mm	in. (approx. equivalents)	Tyler Equivalent Designation
26.9	1.06	3.90	0.1535	1.050 in.
25.4	1.00	3.80	0.1496	
22.6	0.875	3.50	0.1378	0.883 in.
19.0	0.750	3.30	0.1299	0.742 in.
16.0	0.625	3.00	0.1181	0.624 in.
13.5	0.530	2.75	0.1083	0.525 in.
12.7	0.500	2.67	0.1051	
11.2	0.438	2.45	0.0965	0.441 in.
9.51	0.375	2.27	0.0894	0.371 in.
8.00	0.312	2.07	0.0815	$2\frac{1}{2}$ mesh
6.73	0.265	1.87	0.0736	3 mesh
6.35	0.250	1.82	0.0717	
5.66	0.223	1.68	0.0661	$3\frac{1}{2}$ mesh
4.76	0.187	1.54	0.0606	4 mesh
4.00	0.157	1.37	0.0539	5 mesh
3.36	0.132	1.23	0.0484	6 mesh
2.83	0.111	1.10	0.0430	7 mesh
2.38	0.0937	1.00	0.0394	8 mesh
2.00	0.0787	0.900	0.0354	9 mesh
1.68	0.0661	0.810	0.0319	10 mesh
1.41	0.0555	0.725	0.0285	12 mesh
1.19	0.0469	0.650	0.0256	14 mesh
1.00	0.0394	0.580	0.0228	16 mesh
0.841	0.0331	0.510	0.0201	20 mesh
0.707	0.0278	0.450	0.0177	24 mesh
0.595	0.0234	0.390	0.0154	28 mesh
0.500	0.0197	0.340	0.0134	32 mesh
0.420	0.0165	0.290	0.0114	35 mesh
0.354	0.0139	0.247	0.0097	42 mesh
0.297	0.0117	0.215	0.0085	48 mesh
0.250	0.0098	0.180	0.0071	60 mesh
0.210	0.0083	0.152	0.0060	65 mesh
0.177	0.0070	0.131	0.0052	80 mesh
0.149	0.0059	0.110	0.0043	100 mesh
0.125	0.0049	0.091	0.0036	115 mesh
0.105	0.0041	0.076	0.0030	150 mesh
0.088	0.0035	0.064	0.0025	170 mesh
0.074	0.0029	0.053	0.0021	200 mesh
0.063	0.0025	0.044	0.0017	250 mesh
0.053	0.0021	0.037	0.0015	270 mesh
0.044	0.0017	0.030	0.0012	325 mesh
0.037	0.0015	0.025	0.0010	400 mesh

Notation

SI units are given first (followed by English and/or cgs units).

a	particle radius, m (ft)
a	area, m^2 (ft^2); also m^2 area/m^3 volume bed or packing (ft^2/ft^3)
a_e	acceleration from centrifugal force, m/s^2 (ft/s^2)
a_v	specific surface area of particle, m^{-1} (ft^{-1})
A	cross-sectional area, m^2 (ft^2, cm^2); also area, m^2 (ft^2)
A	absorption factor $= L/mV$, dimensionless; also filter area, m^2 (ft^2)
A_m	membrane area, cm^2, m^2 (ft^2)
A_w	solvent permeability constant, kg/s $\cdot$ m^2 $\cdot$ atm
A_s	solute permeability constant, m/s (ft/h)
b	length, m (ft, cm)
B	flow rate of dry solid, kg/h (lb$_m$/h); also filtration constant, s/m^3 (s/ft^3)
B	physical property of membrane, atm^{-1}
c	concentration, kg/m^3, kg mol/m^3 (lb$_m$/ft^3, g mol/cm^3)
c	concentration of absorbate in fluid, kg/m^3 (lb$_m$/ft^3)
c_A	concentration of A, kg mol/m^3 (lb mol/ft^3, g mol/cm^3)
c_b	break-point concentration, kg/m^3 (lb$_m$/ft^3)
c_g	concentration of solute in gel, kg solute/m^3 (lb$_m$/ft^3, g/cm^3)
c_s	concentration of solute at surface of membrane, kg solute/m^3 (lb$_m$/ft^3, g/cm^3)
c_S	concentration of A in solid, kg mol A/m^3 (lb$_m$/ft^3, g mol/cm^3)
c_S	concentration of solids in slurry, kg/m^3 (lb$_m$/ft^3)
c_S	humid heat, kJ/kg dry air $\cdot$ K (btu/lb$_m$ dry air $\cdot$ °F)
c_x	concentration of solids in slurry, mass frac
c_p	heat capacity at constant pressure, J/kg $\cdot$ K, kJ/kg $\cdot$ K, kJ/kg mol $\cdot$ K (btu/lb$_m$ $\cdot$ °F, cal/g $\cdot$ °C)
$\bar{c}_w$	mean solvent concentration in membrane, kg/m^3 (lb$_m$/ft^3)
c'_A	deviation of concentration from mean concentration $\bar{c}_A$, kg mol/m^3
c_P	concentration of P, kg P/m^3
c_v	heat capacity at constant volume, J/kg $\cdot$ K
C	filtration constant, N/m^2 (lb$_f$ ft^2); also, number of components
C	fluid heat capacity, W/K (btu/h $\cdot$ °F)
C	height of bottom of agitator above tank bottom, m (ft)
C_p	pitot tube coefficient, dimensionless

C_D	drag coefficient, dimensionless
C_v, C_0	Venturi coefficient, orifice coefficient, dimensionless
D	molecular diffusivity, m^2/s (ft^2/h, cm^2/s); also diameter, m (ft)
D	decimal reduction time, min; also distillate flow rate, kg/h, kg mol/h (lb_m/h)
D_{AB}	molecular diffusivity, m^2/s (ft^2/h, cm^2/s)
D_p	particle diameter, m (ft)
D_{KA}	Knudsen diffusivity m^2/s (ft^2/h, cm^2/s)
D_{NA}	transition-region diffusivity, m^2/s (ft^2/h, cm^2/s)
$D_{A\,eff}$	effective diffusivity, m^2/s (ft^2/h, cm^2/s)
$D_{p,\,m}$	effective mean diameter for mixture, m (ft)
D_a	diameter of agitator, m (ft)
D_t	diameter of tank, m (ft)
D_{pc}	critical diameter, m (ft)
D_{AP}	diffusivity of A in protein solution, m^2/s
E	activation energy, J/kg mol (cal/g mol)
E	energy for size reduction, $kW \cdot h/ton$; also tray efficiency, dimensionless
E	total energy, J/kg ($ft \cdot lb_f/lb_m$)
E	fraction unaccomplished change, dimensionless
E	emitted radiation energy, W/m^2 ($btu/h \cdot ft^2$)
E	axial dispersion coefficient, m^2/s (ft^2/h)
$E_{B\lambda}$	monochromatic emissive power, W/m^3 ($btu/h \cdot ft^3$)
f	fraction of feed vaporized; also cycle fraction, dimensionless
f	Fanning friction factor, dimensionless
f_t	mixing factor, dimensionless
F	number of degrees of freedom
F	frictional loss, J/kg ($ft \cdot lb_f/lb_m$)
F	flow rate, kg/h, kg mol/h (lb_m/h)
F	force, N (lb_f, dyn)
F_T	correction factor for temperature difference, dimensionless
F_0	process time at $121.1°C$ ($250°F$), min
$\mathscr{F}_{12}$	geometric view factor for gray surfaces, dimensionless
F_{12}	geometric view factor, dimensionless
g	standard acceleration of gravity (see Appendix A.1)
g_c	gravitional conversion factor (see Appendix A.1)
G	mass velocity $= v\rho$, $kg/s \cdot m^2$, $kg/h \cdot m^2$ ($lb_m/h \cdot ft^2$)
G	growth constant, mm/h
G'	mass velocity $= v'\rho$, $kg/s \cdot m^2$, $kg/h \cdot m^2$ ($lb_m/h \cdot ft^2$)
G	irradiation on a body, W/m^2 ($btu/h \cdot ft^2$)
h	constant spacing in x for Simpson's rule
h	head, J/kg ($ft \cdot lb_f/lb_m$); also height of fluid, m (ft)
h	heat-transfer coefficient, $W/m^2 \cdot K$ ($btu/h \cdot ft^2 \cdot °F$)
h	enthalpy of liquid, J/kg, kJ/kg, kJ/kg mol (btu/lb_m)
h_{fg}	latent heat of vaporization, J/kg, kJ/kg (btu/lb_m)
h_c	contact resistance coefficient, $W/m^2 \cdot K$ ($btu/h \cdot ft^2 \cdot °F$)
H	distance, m (ft)
H	Henry's law constant, atm/mol frac
H	head, J/kg ($ft \cdot lb_f/lb_m$); also height of fluid, m (ft)
H	enthalpy, J/kg, kJ/kg, kJ/kg mol (btu/lb_m); also enthalpy kJ/kg dry air (btu/lb_m dry air)

H	humidity, kg water vapor/kg dry air (lb water vapor/lb dry air)
H	enthalpy of vapor, J/kg, kJ/kg, kJ/kg mol (btu/lb$_m$)
H	equilibrium relation, kg mol/m^3 · atm
H	bed length, m (ft); also effective length of spindle, m (ft)
H'	enthalpy, kJ/kg dry solid (btu/lb$_m$); also enthalpy, kJ/kg dry air (btu/lb$_m$ dry air)
H_B	length of bed used up to break point, m (ft)
$H_G, H_L,$ H_{OG}, H_{OL}	height of transfer unit, m (ft)
H_P, H_R	percentage humidity, percentage relative humidity, respectively
H_T	total bed length, m (ft)
H_{UNB}	length of unused bed, m (ft)
$\mathbf{i}$	unit vector along x axis
I	intensity of turbulence, dimensionless; also current, amp
I_B	radiation intensity of black body, W/m^2 · sr (btu/h · ft^2 · sr)
I_λ	intensity of radiation, W/m^2 (btu/h · ft^2)
$\mathbf{j}$	unit vector along y axis
J	width of baffle, m (ft)
j_A	mass flux of A relative to mass average velocity, kg/s · m^2
j_A^*	mass flux of A relative to molar average velocity, kg/s · m^2
J_A	molar flux of A relative to mass average velocity, kg mol/s · m^2
$\mathbf{J}_A^*$	molar flux vector of A relative to molar average velocity, kg mol/s · m^2 (lb mol/h · ft^2, g mol/s · cm^2)
J_D, J_H	mass-transfer and heat-transfer factors, dimensionless
$\mathbf{k}$	unit vector along z axis
k, k'	reaction velocity constant, h^{-1}, min^{-1}, or s^{-1}
k	thermal conductivity, W/m · K (btu/h · ft · °F)
k	reaction velocity constant, h^{-1}, min^{-1}, or s^{-1}
k_1'	first-order heterogeneous reaction velocity constant, m/s (cm/s)
$k_c', k_c, k_G, k_x, \ldots$	mass-transfer coefficient, kg mol/s · m^2 · conc diff (lb mol/h · ft^2 · conc diff, g mol/s · cm^2 · conc diff); kg mol/s · m^2 · Pa, kg mol/s · m^2 · atm (lb mol/h · ft^2 · atm). (See Table 7.2-1.)
k_c	mass-transfer coefficient, m/s (ft/h, cm/s)
$k_G a, k_x a, k_y a, \ldots$	volumetric mass-transfer coefficient, kg mol/s · m^3 · conc diff (lb mol/h · ft^3 · conc diff, g mol/s · cm^3 · conc diff); kg mol/s · m^3 · Pa
k_s	surface reaction coefficient, kg mol/s · m^2 · mol frac
k_w, k_w'	wall constants, dimensionless
K_y', K_x'	overall mass-transfer coefficient, kg mol/s · m^2 · mol frac (lb mol/h · ft^2 · mol frac, g mol/s · cm^2 · mol frac)
K	consistency index, N · s^n/m^2 (lb$_f$ · s^n/ft^2)
K_1	constant in Eq. (3.1-39), m/s (ft/s)
K', K_s	equilibrium distribution coefficient, dimensionless
K'	consistency index, N · s$^{n'}$/m^2 (lb$_f$ · s$^{n'}$/ft^2)
K	equilibrium distribution coefficient, dimensionless
K_p	filtration constant, s/m^6 (s/ft^6)
K_V	filtration constant, N/m^5 (lb$_f$/ft^5)
K_c, K_{ex}, K_f	contraction, expansion, fitting loss coefficient, dimensionless
L	length, m (ft); also amount, kg, kg mol (lb$_m$); also kg/h · m^2
L	liquid flow rate, kg/h, kg mol/h (lb$_m$/h)

L	Prandtl mixing length, m (ft)
L	mean beam length, m (ft)
L_S	dry solid weight, kg dry solid (lb_m dry solid)
L'	flow rate, kg inert/h, kg mol inert/h (lb_m inert/h)
m, m', m''	slope of equilibrium line, dimensionless
m	flow rate, kg/s, kg/h (lb_m/s)
m	dimensionless ratio $= k/hx_1$; also, position m
m	parameter in Table 7.1-1, dimensionless; also position parameter
m	ratio, kg wet cake/kg dry cake (lb wet cake/lb dry cake)
M	molecular weight, kg/kg mol (lb_m/lb mol)
M	total mass, kg (lb_m); also parameter $= (\Delta x)^2/\alpha \Delta t$, dimensionless
M	modulus $= (\Delta x)^2/D_{AB} \Delta t$, dimensionless
M	flow rate, kg/h, kg mol/h (lb_m/h)
M	amount of adsorbent, kg(lb_m)
n	exponent, dimensionless; also flow behavior index, dimensionless
n'	slope of line for power-law fluid, dimensionless
n	position parameter; also dimensionless ratio $= x/x_1$
n	total amount, kg mol (lb mol, g mol)
n_A	flux of A relative to stationary coordinates, kg/s $\cdot$ m^2
N	rpm or rps; also number of radiation shields
N	parameter $= h \Delta x/k$, dimensionless; also number of viable organisms, dimensionless
N	total flux relative to stationary coordinates, kg mol/s $\cdot$ m^2 (lb mol/h $\cdot$ ft^2, g mol/s $\cdot$ cm^2)
N	modulus $= k_c \Delta x/D_{AB}$, dimensionless; also number of stages
N	concentration of solid B, kg solid B/kg solution (lb solid B/lb solution)
N	number of equal temperature subdivisions, dimensionless
N_s	solute flux, kg/s $\cdot$ m^2 (lb_m/h $\cdot$ ft^2)
N_w	solvent flux, kg solvent/s $\cdot$ m^2 (lb_m/h $\cdot$ ft^2)
$\bar{N}_{n, m}$	residual defined by Eq. (6.6-3), kg mol/m^3 (lb mol/ft^3)
N_A	molar flux vector of A relative to stationary coordinates, kg mol/s $\cdot$ m^2 (lb mol/h $\cdot$ ft^2, g mol/s $\cdot$ cm^2)
$\bar{N}_A$	mass transfer of A relative to stationary coordinates, kg mol/s (lb mol/h, g mol/s)
N_G, N_L, N_{OG}, N_{OL}	number of transfer units, dimensionless
N_{Gr}	Grashof number defined in Eq. (4.7-4), dimensionless
N_{Bi}	Biot number $= hx_1/k$, dimensionless
N_{Gz}	Graetz number defined in Eq. (4.12-3), dimensionless
$N_{Gr, \delta}$	Grashof number defined in Eq. (4.7-5), dimensionless
N_{Ma}	Mach number defined by Eq. (2.11-15), dimensionless
N_{Pe}	Peclet number $= N_{Re} N_{Pr}$, dimensionless
$N_{Nu, \delta}$	Nusselt number $= h\delta/k$, dimensionless
N_{Fr}	Froude number $= v^2/gL$, dimensionless
N_{Kn}	Knudsen number $= \lambda/2\bar{r}$, dimensionless
N_{Nu}	Nusselt number $= hL/k$, dimensionless
$N_{Nu, x}$	Nusselt number $= h_x x/k$, dimensionless
N_{Pr}	Prandtl number $= c_p \mu/k$, dimensionless
N_{Re}	Reynolds number $= Dv\rho/\mu$, dimensionless
N'_{Re}	Reynolds number $= D_a^2 N\rho/\mu$, dimensionless
$N_{Re, gen}$	Reynolds number defined by Eq. (3.5-11), dimensionless

$N_{\text{Re}, L}$	Reynolds number $= Lv_\infty \rho/\mu$, dimensionless
$N_{\text{Re}, x}$	Reynolds number $= xv_\infty \rho/\mu$, dimensionless
$N'_{\text{Re}, n}$	Reynolds number defined by Eq. (3.5-20), dimensionless
$N_{\text{Re}, p}$	Reynolds number defined by Eq. (3.1-15), dimensionless
$N_{\text{Re}, mf}$	Reynolds number at minimum fluidization defined by Eq. (3.1-35), dimensionless
N_{Eu}	Euler number $= p/\rho v^2$, dimensionless
N_{Sc}	Schmidt number $= \mu/\rho D_{\text{AB}}$, dimensionless
N_{Sh}	Sherwood number $= k'_c D/D_{\text{AB}}$, dimensionless
N_P	power number defined by Eq. (3.4-2), dimensionless
N_Q	flow number, dimensionless
N_{St}	Stanton number $= k'_c/v$, dimensionless
NTU	number of transfer units, dimensionless
O	flow rate, kg/h, kg mol/h (lb$_\text{m}$/h)
p	pressure, N/m^2, Pa (lb$_\text{f}$/ft^2, atm, psia, mm Hg)
p_A	partial pressure of A, N/m^2, Pa (lb$_\text{f}$/ft^2, atm, psia, mm Hg)
p_{BM}	log mean inert partial pressure of B in Eq. (6.2-21), N/m^2, Pa (lb$_\text{f}$/ft^2, atm, psia, mm Hg)
p_h	total pressure in high-pressure side (feed), cm Hg, Pa (atm)
p_l	total pressure in low-pressure side (permeate) cm Hg, Pa (atm)
p_M	permeability in solid, m/s (ft/h)
P	parameter in Eq. (5.5-12), dimensionless
P	total pressure, N/m^2, Pa (lb$_\text{f}$/ft^2, atm, psia, mm H)
P	power, W (ft $\cdot$ lb$_\text{f}$/s, hp)
P	flow rate, kg/h, kg/min; also number of phases at equilibrium
$\mathbf{P}$	momentum vector, kg $\cdot$ m/s (lb$_\text{m}$ $\cdot$ ft/s)
P_A	vapor pressure of pure A, N/m^2, Pa (lb$_\text{f}$/ft^2, atm, psia, mm Hg)
P'_A	permeability of A, cm^3 (STP) $\cdot$ cm/s $\cdot$ cm^2 $\cdot$ cm Hg
P_w	solvent membrane permeability, kg solvent/s $\cdot$ m $\cdot$ atm (lb$_\text{m}$/h $\cdot$ ft $\cdot$ atm)
P_m	permeability, kg mol/s $\cdot$ m $\cdot$ atm (lb mol/h $\cdot$ ft $\cdot$ atm)
P_M	permeability, m^3(STP)/(s $\cdot$ m^2 C.S. $\cdot$ atm/m)
P'_M	permeability, cm^3(STP)/(s $\cdot$ cm^2 C.S. $\cdot$ atm/cm)
P''_M	permeability, cm^3(STP)/(s $\cdot$ cm^2 C.S. $\cdot$ cm Hg/mm)
q	heat-transfer rate, W (btu/h); also net energy added to system, W (btu/h); also J (btu)
q	flow rate, m^3/s (ft^3/s)
$\mathbf{q}$	heat flux vector, W/m^2 (btu/h $\cdot$ ft^2)
$\dot{q}$	rate of heat generation, W/m^3 (btu/h $\cdot$ ft^3)
q	feed condition defined by Eq. (11.4-12)
q'	flow rate in Darcy's law, cm^3/s
q	kg adsorbate/kg adsorbent (lb$_\text{m}$/lb$_\text{m}$)
q_1	flow rate of residue, m^3/s (ft^3/h, cm^3/s)
q_2	flow rate of permeate, m^3/s (ft^3/h, cm^3/s)
q_A	flow rate of A in permeate, cm^3 (STP)/s, m^3/s (ft^3/h)
q_c	condenser duty, kJ/h, kW (btu/h)
q_f	feed flow rate, cm^3 (STP)/s, m^3/s (ft^3/h)
$\bar{q}_{n,m}$	residual defined by Eq. (4.5-11), K (°F)
q_o	reject flow rate, cm^3 (STP)/s, m^3/s (ft^3/h)
q_p	permeate flow rate, cm^3 (STP)/s, m^3/s (ft^3/h)
q_R	reboiler duty, kJ/h, kW (btu/h)

Q	circulation rate, m^3/s (ft^3/h)
Q	amount absorbed, kg mol/m^2; also heat loss, W (btu/h); also heat absorbed, J/kg (btu/lb_m, ft · lb_f/lb_m)
r	radius, m (ft)
r_A	rate of generation, kg $A/s \cdot m^3$ (lb_m $A/h \cdot ft^3$)
r_H	hydraulic radius, m (ft)
$(r_2)_{cr}$	critical value of radius, m (ft)
R	rate of drying, kg/h · m^2 (lb_m/h · ft^2)
R	solute rejection, dimensionless
R	scaleup ratio, dimensionless
R	radius, m (ft); also resistance, K/W (h · °F/btu)
R	parameter in Eq. (5.5-12), dimensionless; also resistance, ohms
R	gas constant (see Appendix A.1); also reflux ratio = L_n/D, dimensionless
R_A	rate of generation, kg mol $A/s \cdot m^3$ (lb mol $A/h \cdot ft^3$)
R_b	radius of spindle, m (ft)
R_c	radius of outer cylinder, m (ft)
R_c	contact resistance, K/W (h · °F/btu)
R_g	gel layer resistance, s · m^2 · atm/kg
R_m	resistance of filter medium, m^{-1} (ft^{-1})
R_i	rate of generation of i, kg/s (lb_m/h)
R_x	x component of force, N (lb_f, dyn)
s	compressibility constant, dimensionless
s	mean surface renewal factor, s^{-1}
S	conduction shape factor, m (ft)
S	solubility of a gas, m^3 solute(STP)/m^3 solid · atm [cc solute(STP)/cc solid · atm]
S	distance between centers, m (ft); also steam flow rate, kg/h, kg mol/h (lb_m/h)
S	volume of feed solution, m^3 (ft^3)
S	cross-sectional area of tower, m^2 (ft^2); also stripping factor, $1/A$
S_0	specific surface area, m^2/m^3 volume (ft^2/ft^3 volume)
S_p	surface area of particle, m^2 (ft^2)
t	fin thickness, m (ft)
t	time, s, min, or h
t	temperature, K, °C (°F)
t	membrane thickness, cm, m (ft)
t_b	break-point time, h
t_t	time equivalent to total capacity, h
t_T	mixing time, s
t_u	time equivalent to usable capacity up to break-point time, h
T	torque, kg · m^2/s^2
T	temperature, K, °C (°F, °R); also feed rate, ton/min
T'	deviation of temperature from mean temperature $\bar{T}$, K (°F)
$\bar{u}$	average velocity, m/s (ft/s)
U	overall heat-transfer coefficient, W/m^2 · K (btu/h · ft^2 · °F)
U	internal energy, J/kg (btu/lb_m)
v	velocity, m/s (ft/s)
$\mathbf{v}$	velocity vector, m/s (ft/s)

v_A	velocity of A relative to stationary coordinates, m/s (ft/s, cm/s)
v_{Ad}	diffusion velocity of A relative to molar average velocity, m/s (ft/s, cm/s)
v_M	molar average velocity of stream relative to stationary coordinates, m/s (ft/s, cm/s)
v_H	humid volume, m^3/kg dry air (ft^3/lb_m dry air)
v_t	terminal settling velocity, m/s (ft/s)
v^+	dimensionless velocity defined by Eq. (3.10-34)
v^*	velocity defined by Eq. (3.10-42), dimensionless
v'_x	deviation of velocity in x direction from mean velocity $\bar{v}_x$, m/s (ft/s)
v'	superficial velocity based on cross section of empty tube, m/s (ft/s, cm/s)
v'_{mf}	velocity at minimum fluidization, m/s (ft/s)
v'_t	terminal settling velocity, m/s (ft/s)
V_A	solute molar volume, m^3/kg mol
V	flow rate, kg/h, kg mol/h, m^3/s (lb_m/h, ft^3/s)
V	volume, m^3 (ft^3, cm^3); also specific volume, m^3/kg (ft^3/lb_m)
V	velocity, m/s (ft/s); also total amount, kg, kg mol (lb_m)
V'	inert flow rate, kg/h, kg/s, kg mol/h (lb_m/h)
w_A	mass fraction of A
$\dot{W}$	work done on surroundings, W (ft · lb_f/s)
W	free water, kg (lb_m)
$\dot{W}_s$	mechanical shaft work done on surroundings, W (ft · lb_f /s)
W	flow rate, kg/h, kg mol/h (lb_m/h)
W	work done on surroundings, J/kg (ft · lb_f /lb_m)
W	weight of wet solid, kg (lb_m)
W	height or width, m (ft); also power, W (hp); also mass dry cake, kg (lb_m)
W	width of paddle, m (ft)
W_S	mechanical shaft work done on surroundings, J/kg (ft · lb_f/lb_m)
W_p	shaft work delivered to pump, J/kg (ft · lb_f/lb_m)
x	distance in x direction, m (ft)
x	mass fraction or mole fraction; also fraction remaining of original free moisture
x_A	mole fraction of A, dimensionless
x'_B	inert mole ratio, mol B/mol inert
x_f	mole fraction of A in feed, dimensionless
x_o	mole fraction of A in reject, dimensionless
x_{oM}	mole fraction of minimum reject concentration, dimensionless
x_{BM}	log mean mole fraction of inert or stagnant B given in Eq. (6.3-4)
$(1 - x_A)_{*M}$	log mean inert mole fraction defined by Eq. (10.4-27)
$(1 - x_A)_{iM}$	log mean inert mole fraction defined by Eq. (10.4-7)
X	particle size, m (ft); also parameter $= \alpha t/x_1^2$, dimensionless
X	parameter $= Dt/x^2$, dimensionless; also free moisture, kg water/kg dry solid (lb water/lb dry solid)
y	distance in y direction, m (ft); also mole fraction
y_A	mass fraction of A or mole fraction of A; also kg A/kg solution (lb A/lb solution)
y_{BM}	log mean mole fraction of inert or stagnant B given in Eq. (7.2-11)
y^+	dimensionless number defined by Eq. (3.10-35)

y_i	mole fraction of A in permeate at outlet of residue stream, dimensionless
y_p	mole fraction of A in permeate, dimensionless
$(y - y_i)_M$	log mean driving force defined by Eq. (10.6-24)
$(1 - y_A)_{iM}$	log mean inert mole fraction defined by Eq. (10.4-6)
Y	temperature ratio defined by Eq. (4.10-2), dimensionless
Y	fraction of unaccomplished change in Table 5.3-1 or Eq. (7.1-12)
Y	expansion correction factor defined in Eqs. (3.2-9) and (3.2-11)
z	distance in z direction, m (ft); also tower height, m (ft)
z	temperature range for 10:1 change in D_T, °C (°F)
Z	height, m (ft); also temperature ratio in Eq. (4.10-2)

Greek letters

α	correction factor = 1.0, turbulent flow and $\frac{1}{2}$, laminar flow
α	absorptivity, dimensionless; also flux ratio = $1 + N_B/N_A$
α	thermal diffusivity = $k/\rho c_p$, m²/s (ft²/h, cm²/s)
α, α_{AB}	relative volatility of A with respect to B, dimensionless
α	specific cake resistance, m/kg (ft/lb$_m$)
α	angle, rad
α	kinetic-energy velocity correction factor, dimensionless
α^*	ideal separation factor = P'_A/P'_B, dimensionless
α_G	gas absorptivity, dimensionless
α_T	eddy thermal diffusivity, m²/s (ft²/h)
β	concentration polarization, ratio of salt concentration at membrane surface to the salt concentration in bulk feed stream, dimensionless
β	momentum velocity correction factor, dimensionless
β	volumetric coefficient of expansion, 1/K (1/°R)
γ	viscosity coefficient, N · s^n/m² (lb$_m$/ft · s^{2-n})
γ	ratio of heat capacities = c_p/c_v, dimensionless
Γ	flow rate, kg/s · m (lb$_m$/h · ft)
Γ	concentration of property, amount of property/m³
δ	molecular diffusivity, m²/s
δ	boundary-layer thickness, m (ft); also distance, m (ft)
δ	constant in Eq. (4.12-2), dimensionless
Δ	difference; also difference operating-point flow rate, kg/h(lb$_m$/h)
ΔT_a	arithmetic temperature drop, K, °C (°F)
ΔT_{1m}	log mean temperature driving force, K, °C(°F)
ΔH	enthalpy change, J/kg, kJ/kg, kJ/kg mol (btu/lb$_m$, btu/lb mol)
∇	design criterion for sterilization, dimensionless
Δp	pressure drop, N/m², Pa (lb$_f$/ft²)
ε	roughness parameter, m (ft); or void fraction, dimensionless
ε	emissivity, dimensionless; also volume fraction, dimensionless
ε_M	mass eddy diffusivity m²/s (ft²/h, cm²/s)
ε	heat-exchanger effectiveness, dimensionless
ε_G	gas emissivity, dimensionless
ε_t	momentum eddy diffusivity, m²/s (ft²/s)
ε_{mf}	void fraction at minimum fluidization, dimensionless
η_f	fin efficiency, dimensionless
η_t	turbulent eddy viscosity, Pa · s, kg/m · s (lb$_m$/ft · s)
η	efficiency, dimensionless
θ	angle, rad

θ	parameter in Eq. (11.7-19)
θ	cut or fraction of feed permeated, dimensionless
θ^*	fraction permeated up to a value of $x = 1 - q/q_f$, dimensionless
λ	wavelength, m (ft)
λ	latent heat, J/kg, kJ/kg (btu/lb$_m$); also mean free path, m (ft)
λ_{Ab}	latent heat of A at normal boiling point, J/kg, kJ/kg mol (btu/lb$_m$)
μ	viscosity, Pa $\cdot$ s, kg/m $\cdot$ s, N $\cdot$ s/m^2 (lb$_m$/ft $\cdot$ s, lb$_m$/ft $\cdot$ h, cp)
μ_a	apparent viscosity, Pa $\cdot$ s, kg/m $\cdot$ s (lb$_m$/ft $\cdot$ s)
ν	momentum diffusivity μ/ρ, m^2/s (ft^2/s, cm^2/s)
π_1	dimensionless group, dimensionless
π	osmotic pressure, Pa, N/m^2 (lb$_f$/ft^2, atm)
ρ	density, kg/m^3 (lb$_m$/ft^3); also reflectivity, dimensionless
σ	constant, 5.676×10^{-8} W/m$^2 \cdot$ K^4 (0.1714×10^{-8} btu/h $\cdot$ ft$^2 \cdot$ $^\circ$R^4); also collision diameter, Å
Σ	sigma value for centrifuge, m^2 (ft^2)
τ_{zx}	flux of x-directed momentum in z direction, (kg $\cdot$ m/s)/s $\cdot$ m^2 or N/m^2 (lb$_f$/ft^2, dyn/cm^2)
τ	shear stress, N/m^2 (lb$_f$/ft^2, dyn/cm^2)
τ	tortuosity, dimensionless
ϕ	velocity potential, m^2/s (ft^2/h)
ϕ	angle, rad; also association parameter, dimensionless
ϕ	osmotic coefficient, dimensionless
ϕ_S	shape factor of particle, dimensionless
Ψ_z	flux of property, amount of property/s $\cdot$ m^2
ψ	correction factor, dimensionless
ψ	stream function, m^2/s (ft^2/h)
ψ_p	parameter defined by Eq. (14.3-13), dimensionless
ω	angular velocity, rad/s
ω	solid angle, sr
$\Omega_{D, AB}$	collision integral, dimensionless

Index

A

Absorption (*see also* Humidification
    processes; Stage processes)
  absorption factors, 593
  design methods for packed towers
    for concentrated gas mixtures,
      627–628
    for dilute gas mixtures, 619–621
    general method, 615–619
    transfer unit method, 624–625
  equipment for, 610–613
  gas-liquid equilibrium, 586–587
  interface compositions, 595–597
  interphase mass transfer, 594–602
  introduction to, 584–585
  Kremser analytical equations, 592–
    593
  log mean driving force, 620
  minimum liquid to gas ratio, 616–
    617
  number of theoretical trays, 614
  operating lines, 613–616
  optimum liquid to gas ratio, 616–617
  packing mass transfer coefficients
    film, 595–597, 617–618, 632–633
    overall, 599–601, 618
  stage calculations, 587–592
  stripping, 616–617
Absorptivity, 276–277, 283–284
Adiabatic compressible flow, 103–104

Adiabatic saturation temperature, 530–
    531
Adsorbents (*see* Adsorption)
Adsorption
  adsorbents
    applications of, 697
    physical properties of, 697–698
    types of, 698
  by batch process, 700
  equilibrium in
    Freundlich isotherm, 698–699
    Langmuir isotherm, 698–699
    linear isotherm, 698–699
  by fixed-bed process
    adsorption cycles, 707–708
    basic model, 706–707
    breakthrough curves, 702–703
    concentration profiles, 701–702
    design method, 703–706
  ion exchange (*see* Ion exchange
    processes)
Agitation (*see also* Mixing)
  baffles for, 143–144
  circulation rate in, 151
  equipment for, 141–145
  flow number in, 151
  flow patterns in, 143–144
  heat transfer in, 300–302
  mixing time in, 149–151
  motionless mixers, 152
  of non-Newtonian fluids, 163–164
  power consumption, 144–147

purposes of, 140–141
scale-up in, 147–149
special agitation systems, 151–152
standard agitation design, 144
types of agitators, 141–144
Air, physical properties of, 866
Analogies
in boundary layer, 370–375
between momentum, heat, and mass
transfer, 42–43, 375, 381–382,
430, 438–440, 477–478
Arithmetic mean temperature drop,
238
Azeotrope, 642–643

B

Batch distillation, 646–647
Bernoulli equation, 67–68
Bingham-plastic fluids, 154
Biological diffusion
in gels, 406–407
in solutions, 403–406
Biological materials
chilling of, 360–361
diffusion of, 403–407
equilibrium moisture content, 533–
535
evaporation of
fruit juices, 513
paper pulp liquors, 514
sugar solutions, 513–514
freeze drying of, 566–569
freezing of, 362–365
grain dryer, 524
heat of respiration, 229
leaching of, 723–729
physical properties of, 889–891
sterilization of, 569–577
Biot number, 332–333
Blake crusher, 843
Blasius theory, 193, 201
Blowers, 138–139
Boiling
film, 261
natural convection, 259–260
nucleate, 260–261
physical mechanisms of, 259–261
temperature of, 9, 681–683
Boiling point diagrams, 640–642

Boiling point rise, 499–500
Bollman extractor, 728
Bond's crushing law, 842
Boundary layers (*see also* Turbulent
flow)
continuity equation for, 192
energy equation for, 370–373
laminar flow in, 192–193
mass transfer equation for, 475–477
momentum equation for, 199–201
separation of, 115, 191–192
theory for mass transfer in, 478–479
wakes in, 115, 191–192
Bound moisture, 534–535
Bourdon gage, 38
Brownian motion, 817–818
Bubble trays
efficiencies, 666–669
types of, 611
Buckingham pi theorem, 203–204,
308–310, 474–475

C

Capillaries (*see* Porous solids)
Capillary flow, in drying, 540, 553–555
Centrifugal filtration
equipment, 838
theory for, 837–838
Centrifugal pumps, 134–136
Centrifugal settling and sedimentation
(*see also* Centrifugal filtration;
Centrifuges; Cyclones)
critical diameter, 832–833
equations for centrifugal force, 829–
831
purpose of, 829
separation of liquids, 834–835
settling of particles, 831–834
Centrifuges (*see also* Centrifugal
settling and sedimentation)
disk bowl, 836–837
scale-up of, 834
sigma value of, 834
tubular, 836
Cgs system of units, 4
Chapman-Enskog theory, 394–395
Chemical reaction and diffusion, 456–
460

Chilling of food and biological
   materials, 360–361
Chilton-Colburn analogies, 440
Circular pipes and tubes
   compressible flow in, 101–104
   dimensions of, 892–893
   friction factors in, 86–92
   heat transfer coefficients in, 238–243
   laminar flow in, 78–80, 83–87
   mass transfer in, 440–443
   turbulent flow in, 83–84, 87–91
   universal velocity distribution in,
     197–199
Classifiers (*see also* Settling and
    sedimentation)
   centrifugal, 831–833
   differential settling, 821–823, 826–
    827
Comminution (*see* Mechanical size
    reduction)
Compressible flow of gases
   adiabatic conditions, 103–104
   basic differential equation, 101
   isothermal conditions, 101–103
   Mach number for, 104
   maximum flow conditions, 102–104
Compressors
   equations for, 138–139
   equipment, 138
Concentration units, 7
Condensation
   derivation of equation for, 263–267
   mechanisms of, 263
   outside horizontal cylinders, 266–
    267
   outside vertical surfaces, 263–265
Condensers for evaporation
   calculation methods, 512
   direct-contact type, 511–512
   surface type, 511
Conduction heat transfer (*see also*
    Unsteady-state heat transfer)
   combined conduction and
    convection, 227–228
   contact resistance at interface, 233
   critical thickness of insulation, 231–
    232
   in cylinders, 221, 224–226
   effect of variable thermal
    conductivity, 220
   equations for

   in cylindrical coordinates, 368
   in rectangular coordinates, 368
   in spherical coordinates, 368
   Fourier's law, 43, 214, 216–217, 368
   graphical curvilinear square method,
    233–235
   with heat generation, 229–231
   in hollow sphere, 222–223
   through materials in parallel, 226–
    227
   mechanism of, 215
   shape factors in, 235–236
   steady-state in two dimensions
    Laplace equation, 311
    numerical method, 312–317
    with other boundary conditions,
     316–317
   through a wall, 220
   through walls in series, 223–225
Continuity equation
   for binary mixture, 454–456
   for boundary layer, 192–193
   for pure fluid, 50–54, 164–165, 167–
    169, 176
Control volume, 52–54
Convection heat transfer (*see also*
    Heat transfer coefficients;
    Natural convection)
   general discussion of, 215–216, 219,
    227–228
   physical mechanism of, 236–238
Convective heat transfer coefficients
    (*see* Heat transfer coefficients)
Convective mass transfer coefficients
    (*see* Mass transfer coefficients)
Conversion factors, 850–853
Cooling towers (*see* Humidification
    processes)
Coriolis force, 175
Countercurrent processes (*see also*
    Absorption; Distillation;
    Leaching; Liquid-liquid
    extraction; Stage processes)
   analytical equations, 592–593
   in heat transfer, 244–245
   multiple stages, 589–591
Creeping flow, 189–190
Critical diameter, 832–833
Critical moisture content, 538–539
Critical thickness of insulation, 231–
   232

Crushing and grinding (*see* Mechanical size reduction)
Crystallization (*see also* Crystallizers)
  crystal geometry, 737–738
  crystal particle-size distribution, 746–747
  crystal size effect on solubility, 744
  heat balances, 740–741
  heat of solution, 740–741
  mass transfer in, 745
  material balances, 739
  McCabe ΔL law, 745–746
  Miers' qualitative theory, 744
  models for, 747
  nucleation theories, 744, 747
  purpose of, 737
  rate of crystal growth, 743–746
  solubility (phase equilibria), 490, 738–739
  supersaturation, 741–744
Crystallizers (*see also* Crystallization)
  circulating-liquid evaporator-crystallizer, 742–743
  circulating-magma vacuum crystallizer, 743
  classification of, 741–743
  scraped-surface crystallizer, 742
  tank crystallizer, 742
Crystals (*see also* Crystallization)
  particle-size distribution, 746
  rate of growth, 745–746
  types of, 738
Cubes, mass transfer to, 448
Curvilinear squares method, 233–235
Cyclones
  equipment, 838–839
  theory, 839–840
Cylinders
  drag coefficient for, 116–117
  flow across banks of, 250–251
  heat transfer coefficients for, 249–251, 559
  mass transfer coefficients for, 450
Cylindrical coordinates, 169, 174, 369

D

Dalton's law, 8
Darcy's law, 123
Dehumidification, 525, 602–603

Density
  of foods, 891
  of metals, 883
  of solids, 882
  of water, 855
Dew point temperature, 527, 682
Dialysis
  equipment for, 758
  hemodialysis, 758
  theory for, 755–756
  use of, 754–755, 757–758
Differential operations
  time derivatives, 165
  with scalars, 166–167
  with vectors, 166–167
Differential settling, 821–823, 826–827
Diffusion (*see also* Steady-state diffusion; Unsteady-state diffusion)
  of A through stagnant B, 388–390, 456
  in biological gels, 406–407
  of biological solutes, 403–406
  and blockage by proteins, 405–406
  in capillaries, 462–468
  with chemical reaction, 456–460
  and convection, 387–389
  in drying, 539–540, 552–553
  and equation of continuity, 454–456
  equimolar counterdiffusion, 385–386, 456
  Fick's law, 43, 383–384, 453–454, 464
  in gases, 385–397
  general case for A and B, 387–388
  introduction to, 42–43, 381–383
  Knudsen, 463–464
  in leaching, 725–726
  in liquids, 397–399
  molecular, 464
  multicomponent
    gases, 461–462
    liquids, 402
  in porous solids, 412–413, 462, 468
  similarity of mass, heat, and momentum transfer, 39–43, 381–382
  in solids
    classification of, 408
    Fick's law, 408–409
    permeability equations, 410–411

Diffusion (*cont.*)
  to a sphere, 391–392
  steady-state, in two dimensions, 413–416
  transition, 464–466
  with variable cross-sectional area, 390–393, 408–409
  velocities in, 387
Diffusivity (mass)
  in biological gels, 406–407
  of biological solutes, 404–406
  effective
    in capillaries, 464–466
    in porous solids, 412, 468
  estimation of
    for biological solutes, 405–406
    in gases, 394–396
    in liquids, 400–402
  experimental determination
    in biological gels, 406
    in biological solutions, 404
    in gases, 390, 393–394
    in liquids, 399–400
  experimental values
    for biological gels, 406–407
    for biological solutes, 404–405
    for gases, 394–395
    for liquids, 400–401
    for solids, 410–411
  Knudsen, 463–464
  molecular and eddy, 374–375
  transition, 464–466
Diffusivity (momentum), 374–375, 381–382
Diffusivity (thermal), 331, 364–365, 374–375, 381–382
Dilitant fluids, 155
Dimensional analysis
  Buckingham pi theorem, 203–204, 308–310, 474
  in differential equations, 202–203
  in heat transfer, 308–310
  in mass transfer, 474–475
  in momentum transfer, 202–204
Dimensional homogeneity, 5
Disk, drag coefficient for, 116–117
Distillation (*see also* Multicomponent distillation; Vapor-liquid equilibrium)
  constant molal overflow, 651–652
  direct steam injection, 663–664

enriching operating line, 651–653
enriching tower, 663
enthalpy-concentration method
  benzene-toluene data, 672
  construction of plot for, 669–672
  enriching equations, 672–674
  equilibrium data, 672, 887
  example of, 674–678
  number of stages, 676, 678
  stripping equations, 674
equilibrium or flash, 682–683
feed condition and location, 654–656, 687–688
Fenske equation, 658–659, 683
introduction and process flow, 644, 649–650
McCabe-Thiele method, 651–666
packed towers, 612–613
partial condensers, 666
Ponchon-Savarit method, 678
reflux ratio
  minimum, 659–660, 686–687
  operating, 660
  total, 658–659, 683
relative volatility, 644–645, 681
sidestream, 664–665
simple batch or differential, 646–647
simple stream, 648–649
single stage contact, 642
stripping operating line, 653–654
stripping tower, 661–662
tray efficiencies
  introduction, 666
  Murphree, 667–669
  overall, 667–669
  point, 668
  types of trays, 611–612
Distribution coefficient, in mass transfer, 429–430, 470–471
Dodge crusher, 843
Drag coefficient
  definition of, 115–116, 201
  for disk, 116–118
  for flat plate, 115–116, 192–193
  form drag, 115–117, 190
  for long cylinder, 116–117
  skin drag, 114–116
  for sphere, 115–118, 190, 816–819, 822–823
Dryers (*see also* Drying)
  continuous tunnel, 522

drum, 523
fixed bed, 556–559
grain, 524
rotary, 523
spray, 523–524
tray, 521, 561
vacuum shelf, 521–522
Drying (*see also* Dryers)
  air recirculation in, 563–564
  of biological materials, 533–535, 540
  bound moisture, 534–535
  capillary movement theory, 540,
    553–555
  through circulation, 556–559
  constant-drying conditions
    constant-rate period, 538, 540–545
    effect of air velocity, 544
    effect of humidity, 544–545
    effect of solid thickness, 545
    effect of temperature, 545
    falling-rate period, 538–540, 545–
      547
    prediction for constant-rate
      period, 541–545
    rate-of-drying curve, 536–538
    time for, 540–543
    use of drying curve, 540–541
  continuous countercurrent, 564–566
  critical moisture, 538
  effect of shrinkage, 540
  equilibrium moisture, 533–535
  experimental methods, 536
  free moisture, 535
  freeze drying, 566–569
  heat balance in continuous dryers,
    561–562
  introduction to, 520–521
  liquid diffusion theory, 539–540,
    551–555
  in packed beds, 556–559
  in trays with varying air conditions,
    561
  unbound moisture, 534–535
Dühring lines, 499–500

E

Emissivity
  definition of, 277–278, 283–284
  values of, 884

Energy balances
  for boundary layer, 370–373
  in compressible flow, 101–104
  differential, 365–368
  mechanical, 63–67
  overall, 56–61
English system of units, 4
Enthalpy, 15, 57
Enthalpy of air-water vapor mixtures,
  528, 606–607
Enthalpy-concentration diagram
  in distillation, 669–672
  in evaporation, 500–501
Enthalpy-concentration method (*see*
  Distillation)
Equilibrium or flash distillation, 645–
  646
Equilibrium moisture, 533–535
Equivalent diameter
  in fluid flow, 98–99
  in heat transfer, 241
Erbar-Maddox correlation, 687–688
Euler equations, 185–186
Euler number, 202
Evaporation (*see also* Evaporators)
  of biological materials, 513–514
  boiling point rise in, 499–500
  capacity in multiple-effect, 504
  Dühring lines, 499–500
  effects of processing variables on,
    489–490, 498–499
  enthalpy-concentration charts for,
    500–501
  heat transfer coefficients for, 495–
    496
  methods of operation
    backward-feed multiple-effect,
      494–495
    forward-feed multiple-effect, 494
    parallel-feed multiple-effect, 495
    single-effect, 493–494
  multiple-effect calculations, 502–505
  single-effect calculations, 496–497
  steam economy, 494
  temperature drops in, 503–504
  vapor recompression
    mechanical recompression, 514–
      515
    thermal recompression, 515
Evaporators (*see also* Evaporation)
  agitated film, 493

Evaporators (*cont.*)
  condensers for, 511–512
  falling-film, 492
  forced-circulation, 492–493
  horizontal-tube natural circulation, 491
  long-tube vertical, 491
  open kettle, 491
  short-tube, 491
  vertical-type natural circulation, 491
Extraction (*see* Leaching; Liquid-liquid extraction)

F

Falling film
  diffusion in, 441–442
  shell momentum balance, 80–82
  velocity profile, 82
Fans, 137
Fenske equation, 658–659, 683
Fermentation, mass transfer in, 450–453
Fick's law (*see also* Diffusion)
  forms of, 453–454
  for steady-state, 383–384
  for unsteady-state, 426–427
Film temperature, 248
Film theory, 478
Filter aids, 806–807
Filter media, 806
Filters (*see also* Filtration)
  bed, 802
  classification of, 802
  continuous rotary, 805–806
  leaf, 803–805
  plate-and-frame, 803
Filtration (*see also* Centrifugal filtration; Filters)
  basic theory, 809–810
  compressible cake, 809
  constant pressure, 809–810
  constant rate, 815
  continuous, 813–814
  filter aids, 806–807
  filter cycle time, 812–813
  filter media, 806
  pressure drop, 807–808
  purpose of, 801
  specific cake resistance, 808–809

washing, 803–805, 812–813
Finned-surface exchangers
  efficiency of, 304–306
  overall coefficients for, 307–308
  types of, 303–304
First law of thermodynamics, 56–57
Fixed-bed extractor, 727–728
Flat plate
  boundary layer equations for
    heat transfer, 370–373
    mass transfer, 475–477
    total drag, 190–193, 199–201
  drag coefficient for, 114–117, 193
  heat transfer to, 248, 543
  mass transfer to, 444
Flooding velocities, 613
Flow meters (*see* Orifice meter; Pitot tube meter; Venturi meter; Weirs)
Flow separation, 191–192
Fluid friction
  chart for Newtonian fluids, 88
  chart for non-Newtonian fluids, 159–160
  effect of heat transfer on, 92
  in entrance section of pipe, 99–100
  in fittings and valves, 92–94
  for flow of gases, 91
  friction factor in pipes, 86–92
  for noncircular channels, 98–99
  for non-Newtonian fluids, 153–161
  roughness effect on, 89
  from sudden contraction, 93
  from sudden expansion, 75–76, 92–93
Fluidized beds
  expansion of, 126
  heat transfer in, 253
  mass transfer in, 448
  minimum fluidization velocity in, 123–125
  minimum porosity in, 123–124
Fluid statics, 32–39
Flux (mass)
  conversion factors for, 853
  ratios, 464, 466
  types of, 453–454
Foods, physical properties of, 889–891
Form drag, 114–117, 190
Fouling factors, 275–276

Fourier's law, 43, 214, 216–217, 368, 382 (*see also* Conduction heat transfer)

Fractionation (*see* Distillation)

Free moisture, 535

Free settling (*see* Settling and sedimentation)

Freeze drying, 566–569

Freezing of food and biological materials, 362–365

Freundlich isotherm (*see* Adsorption)

Friction factor (*see* Fluid friction)

Froude number, 202

Fuller et al. equation, 396

Fundamental constants, 850–853

G

Gases
  equations for, 7–9
  physical properties of, 864–875, 884–886

Gas law constant, R, 7, 850

Gas law, ideal, 7–9

Gas-liquid equilibrium
  acetone-water, 886
  ammonia-water, 886
  enthalpy-temperature, 606–607
  Henry's law, 586–587, 884
  methanol-water, 885
  phase rule, 586
  sulfur dioxide-water, 586–587, 885

Gas-liquid separation processes (*see* Absorption; Humidification processes; Stage processes)

Gas permeation membrane processes
  equipment for, 760–762
  minimum reject concentration, 768
  permeability in, 759–760
  processing variables for, 780–782
  separation factor in, 765
  series resistances in, 759
  theory of
    cocurrent model, 780
    completely-mixed model, 764–768
    countercurrent model, 778–780
    cross-flow model, 772–775
    introduction to, 763–764
    multicomponent mixture, 769–771
  types of membranes, 759–760

Gas radiation, 293–296

Gas-solid equilibrium, 409–411

General molecular transport equation
  for heat transfer, 214–216
  similarity between momentum, heat, and mass, 39–43, 381–382
  for steady state, 39–40
  for unsteady state, 41–42

General property balance, 39–42

Graphical methods
  integration by, 23–24
  two-dimensional conduction, 233–235

Grashof number, 254

Gravitational constant, 851

Gravity separator, 38–39

Grinding (*see* Mechanical size reduction)

Gurney-Lurie charts, 340, 343, 345

Gyratory crusher, 844

H

Hagen-Poiseuille equation, 80, 87, 180

Heat balances
  in crystallization, 740–741
  in drying, 561–562
  in evaporation, 496–497, 505
  principles of, 19–22

Heat capacity
  data for foods, 889–890
  data for gases, 16, 866, 869, 873–874
  data for liquids, 875, 879
  data for solids, 881, 883
  data for water, 856 857
  discussion of, 14–15

Heat exchangers (*see also* Heat transfer coefficients)
  cross-flow, 268–269, 271
  double-pipe, 267
  effectiveness of, 272–274
  extended-surface, 303–308
  fouling factors for, 275–276
  log mean temperature difference, 244–245, 269–271
  scraped-surface, 302–303
  shell and tube, 267–268
  temperature correction factors, 269–271

Heat of reaction, 17–18

Heat of solution, 740–741
Heat transfer coefficients
  for agitated vessels, 300–302
  approximate values of, 219
  average coefficient, 238
  to banks of tubes, 250–251
  for condensation, 263–267
  definition of, 219, 237
  entrance region effect on, 242–243
  in evaporators, 495–496
  for film boiling, 261
  for finned-surface exchangers, 303–
    308
  for flow parallel to flat plate, 248
  for flow past a cylinder, 249
  for fluidized bed, 253
  fouling factors, 275–276
  for laminar flow in pipes, 238
  for liquid metals, 243
  for natural convection, 253–259
  in noncircular conducts, 241
  for non-Newtonian fluids, 297–299
  for nucleate boiling, 260–261
  for other geometries, 253
  overall, 227–228, 275–276
  for packed beds, 252–253, 447–448
  for radiation and convection, 279–
    281
  for scraped-surface exchangers, 302–
    303
  for spheres, 249
  for transition flow in pipes, 240–241
  for turbulent flow in pipes, 239–240
Heat transfer mechanisms, 215–216
Height of a transfer unit, 609–610,
    624–625, 632–633
Heisler charts, 341, 344, 346
Hemodialysis, 758
Henry's law
  data for gases, 884
  equation, 586–587
Hildebrandt extractor, 728
Hindered settling, 820–822
Humid heat, 527
Humidification processes (see also
    Humidity)
  adiabatic saturation temperature,
    530–531
  definition of, 525
  dehumidification, 525, 590–591, 602–
    603

equilibrium relations, 606–607
equipment for, 602–603
water-cooling
  calculation methods for, 603–610
  minimum air flow, 609
  operating lines, 604–605
  packed tower height, 607, 609–
    610
Humidity (see also Humidification
    processes)
  adiabatic saturation temperature,
    530–531
  chart for, 528–529
  definition of, 526
  dew point temperature, 527
  equations for, 526
  humid heat, 527
  humid volume, 527
  percentage, 526
  relative, 526
  saturated, 526
  total enthalpy, 528
  wet bulb temperature, 531–532
Humid volume, 527–528

I

Ice, properties of, 856–857, 889
Ideal fluids, 185–189
Ideal gas volume, 850
Ideal tray (see Bubble trays)
Integral analysis of boundary layers
  for energy balance, 373
  for momentum balance, 199–201
Intensity of turbulence, 194–195
Interface contact resistance, 233
Internal energy, 16, 57
Interphase mass transfer
  interface compositions, 595–597
  introduction to, 594–595
  use of film coefficients, 595–597,
    605–606, 617–621
  use of overall coefficients, 599–601,
    607, 610
Inverse lever-arm rule, 712–713
Inviscid flow (see Ideal fluids)
Ion exchange processes (see also
    Adsorption)
  design of, 709
  equilibrium relations in, 708–709

introduction to, 708
types of, 708
Isothermal compressible flow, 101–103

J

J-factor, 429, 438, 440

K

K factors (distillation), 680–681
Kick's crushing law, 842
Kinetic energy
  velocity correction factor for, 59–60, 159
  definition of, 57
Kirchhoff's law, 277–278, 283
Kirkbride method, 687
Knudsen diffusion, 463–464
Knudsen number, 464
Kremser equations for stage
  processes, 592–593

L

Laminar flow
  boundary layer for, 192–193, 199–201
  definition of, 47–49
  on flat plate, 190–193
  Hagen-Poiseuille equation for, 80, 87, 180
  kinetic energy correction factor for, 58–60, 159
  mass transfer in, 440–443
  momentum correction factor for, 72–73
  non-Newtonian fluids in, 155–159
  pressure drop in, 84–87
  Reynolds number for, 49, 86
  velocity profile in tubes for, 80, 83–84
Langmuir isotherm (*see* Adsorption)
Laplace's equation
  in heat transfer, 310–311
  for potential flow, 187
  for stream function, 187
Latent heat

discussion of, 16–17
  of ice, 854, 856
  of water, 854, 857–859
Leaching
  countercurrent multistage
    constant underflow, 737
    number of stages, 734–735
    operating line, 733
    variable underflow, 734–735
  equilibrium relations, 729–730
  equipment for
    agitated tanks, 728–729
    Bollman extractor, 728
    fixed-bed (Shanks), 727
    Hildebrandt extractor, 728
    thickeners, 728–729
  preparation of solids, 724–725
  processing methods, 727
  purpose of, 723
  rates of
    when diffusion in solid controls, 726
    when dissolving a solid, 725–726
    introduction to, 725
  single-stage contact, 730–731
  washing, 723
Le Bas molar volumes, 401–402
Lennard-Jones function, 394–395
Lever-arm rule, 712–713
Liquid-liquid equilibrium
  acetic acid-water-isopropyl ether, 711–712, 888
  acetone-water-methyl isobutyl ketone, 888
Liquid-liquid extraction
  countercurrent multistage
    immiscible liquids, 722
    minimum solvent rate, 721–722
    number of stages, 719–722
    overall material balance, 718
  equilibrium relations
    acetic acid-water-isopropyl ether, 831, 711–712, 888
    acetone-water-methyl isobutyl ketone, 888
    phase rule, 710
    rectangular coordinates, 711
    triangular coordinates, 710–711
    types of phase diagrams, 710–712
  equipment for
    agitated tower, 715–716

Liquid-liquid extraction (*cont.*)
    Karr column, 716
    mixer-settler, 715
    packed tower, 716
    plate tower, 715–716
    spray tower, 716
    type and classes of operation,
      715–716
  lever-arm rule, 712–713
  purpose of, 709–710
  single-stage equilibrium contact,
    714–715
Liquid-metals heat transfer
  coefficients, 243
Liquid-metals mass transfer, 450
Liquid-solid equilibrium, 729–730
Liquid-solid leaching (*see* Leaching)
Liquid-solid separation (*see*
    Crystallization; Settling and
    sedimentation)
Log-mean value
  area, 225
  concentration difference, 448–449,
    620
  corrections for heat exchanger, 269–
    271
  temperature difference, 244–245,
    269–271
Lost work, 63
Lumped capacity analysis, 332–334

M

Mach number, 104
Manometers, 36–38
Mass balance overall, 50–56
Mass transfer, boundary conditions in,
    456
Mass transfer coefficients
  analogies for, 438–440
  for A through stagnant B, 435–436
  for crystallization, 745
  for cubes in packed bed, 448
  for cylinder, 450
  for cylinders in packed bed, 448
  definition of, 429, 433–434
  dimensionless numbers for, 437–438
  for equimolar counterdiffusion, 434–
    435

  experimental determination of, 437,
    632
  for falling film, 441–442
  film and overall, 595–597, 599–601,
    606–610, 632–633
  for flat plate, 444
  for fluidized bed, 448
  inside pipes, 440–443
  introduction to, 385, 432–433
  for liquid metals, 450
  models for, 478–479
  for packed bed, 447–449, 556–557
  for packed tower, 632–633
  to small particle suspensions, 450–
    453
  for sphere, 445–446
  types of, 433–436
  for wetted-wall tower, 441–443
Mass transfer fluxes, 453–454
Mass transfer between phases
  concentration profiles, 594–595
  film coefficients, 595–597, 606–607
  overall coefficients, 599–601, 607,
    610, 617
Mass transfer models, 478–479
Mass units, 6
Mass velocity, 51
Material balances
  chemical reaction and, 12–13
  for crystallization, 739
  for evaporation, 496–497, 504–505
  methods of calculation, 10–11
  overall, 50–56
  recycle and, 11
McCabe ΔL Law of crystal growth,
    745–746
McCabe-Thiele method, 651–666
Mean free path, 462
Mechanical energy balance, 63–67
Mechanical-physical separation
    processes (*see also* Centrifugal
    filtration; Centrifugal settling
    and sedimentation; Cyclones;
    Filtration; Mechanical size
    reduction; Settling and
    sedimentation)
  classification of, 800–801
  methods of separation, 800–801
Mechanical size reduction
  equipment for
    Blake crusher, 843

Dodge crusher, 843
  gyratory crusher, 844
  jaw crushers, 843
  revolving grinding mill, 844
  roll crusher, 844
particle size measurement, 840–841
power required for
  Bond's law, 842
  general theory, 841–842
  Kick's law, 842
  Rittinger's law, 842
purpose of, 840
Membrane processes (*see also*
    Dialysis; Gas permeation
    membrane processes; Reverse
    osmosis; Ultrafiltration)
  equipment for, 758, 760–762, 790,
    792
  permeability in, 410–411, 756, 785–
    786, 790
  series resistances in, 755–756, 759
  types of, 585, 754–755
Metals, properties of, 883–884
Miers' theory, 744
Mixer-settlers, 715, 728–729
Mixing (*see also* Agitation)
  discussion of, 140–141, 152–153
  equipment for, 152–153
  with pastes, 152–153
  with powders, 152
Mixing time (*see* Agitation)
Moisture (*see also* Drying; Humidity)
  bound and unbound, 534–535
  capillary flow of, 540, 553–555
  diffusion of, 539–540, 551–555
  equilibrium, for air-solids, 533–535
  free moisture, 535
Molar volumes, 400–402
Molecular diffusion (*see* Diffusion)
Molecular transport (*see* Diffusion)
Mole units, 6
Momentum
  definition of, 46, 69–70
  introduction to, 31
  velocity correction factor for, 72–73
Momentum balance
  applications of, 175–184
  for boundary layer, 192–193, 199–
    201
  in circular tube, 179–181
  Coriolis force, 175

differential equations for, 164–165,
    170–175
  Euler equations, 185–186
  ideal fluids, 185–186
  for jet striking a vane, 76–78
  Laplace's equation in, 187
  for Newtonian fluids, 172–175
  overall, 69–78
  between parallel plates, 175–178
  potential flow, 186–189
  in rotating cylinder, 181–184
  shell, 78–82
  stream function, 185
  transfer of momentum, 46, 79
Motionless mixers (*see* Agitation)
Multicomponent diffusion, 402, 461–
    462
Multicomponent distillation
  boiling point, 681–682
  dew point, 682
  distribution of other components,
    684
  equilibrium data for, 680–681
  flash distillation, 682–683
  introduction to, 679–680
  key components, 683
  minimum reflux (Underwood
    equation), 686–687
  number of stages
    feed tray location, 687
    by short-cut method, 687–688
  number of towers in, 679
  total reflux (Fenske equation), 683

N

Natural convection heat transfer
  derivation of equation (vertical
    plate), 253–254
  equations for various geometries,
    254–259
  introduction to, 253
Navier-Stokes equations, 173–175
Newtonian fluids, 46–47, 153–154
Newton's law
  of momentum transfer, 42–45
  second law (momentum balance),
    69–70
  of settling, 817
  of viscosity, 42–45

Non-Newtonian fluids
  agitation of, 163–164
  flow-property constants, 156–157
  friction loss in fittings, 159
  heat transfer for, 297–299
  laminar flow for, 155–159
  rotational viscometer, 161–163
  turbulent flow for, 159–160
  types of, 153–155
  velocity profiles for, 161
Notation, 895–903
Nucleation (*see also* Crystallization)
  primary, 744
  secondary, 744
Number of transfer units
  for heat exchangers, 274
  for humidification, 609–610
  for mass transfer, 624–625
Numerical methods
  for integration, by Simpson's
    method, 24
  for steady-state conduction
    with other boundary conditions,
      316–317
    in two dimensions, 310–317
  for steady-state diffusion
    with other boundary conditions,
      414–416
    in two dimensions, 413–414
  for unsteady-state heat transfer
    boundary conditions, 351–353
    in a cylinder, 358–359
    implicit method, 359–360
    Schmidt method, 351
    in a slab, 350–353
  for unsteady-state mass transfer
    boundary conditions, 470–471
    Schmidt method, 469–470
    in a slab, 468–471
Nusselt equation, 263–265
Nusselt number, 238

O

Operating lines (*see* Absorption;
    Distillation; Humidification
    processes; Leaching)
Orifice meter, 131–132
Osmotic pressure, 783–784

Overall energy balance (*see* Energy
    balances)

P

Packed beds
  Darcy's law for flow in, 123
  drying in, 556–559
  heat transfer coefficients for, 252–
    253, 447–448, 558–559
  mass transfer coefficients for, 447–
    448
  mass transfer in, 448–449
  tortuosity in, 412, 468
  pressure drop in
    laminar flow, 118–120, 123
    turbulent flow, 120–121
  shape factors for particles, 121–122
  surface area in, 118, 448, 558–559
Packed towers, 602–603, 612–613, 716
Packing, 612–613
Particle-size measurement
  in crystallization, 746–747
  in mechanical size reduction, 840–
    841
  standard screens for, 894
Pasteurization, 576
Penetration theory, 442, 478–479
Permeability of solids, 410–411, 756,
    760, 764–765, 785–786, 790
Phase rule, 586, 640, 648, 710
Physical properties of compounds,
    854–891
Pipes
  dimensions of, 892–893
  schedule number of, 83, 892
  size selection of, 100
Pitot tube meter, 127–128
Planck's law, 281–282
Plate tower, 611, 715–716
Pohlhausen boundary layer relation,
    372
Poiseuille equation (*see* Hagen-
    Poiseuille equation)
Porous solids
  effective diffusivity of, 412, 468
  introduction to, 408, 412–413, 462
  Knudsen diffusion in, 463–464
  molecular diffusion in, 412–413, 464
  transition diffusion in, 464–466

Potential energy, 57
Potential flow, 186–189
Prandtl analogy, 439
Prandtl mixing length (*see* Turbulent flow)
Prandtl number
  definition of, 237–238
  of gases, 866, 870
Pressure
  conversion factors for, 851–852
  devices to measure, 36–39
  head and, 35
  units of, 7, 32–34
Pressure drop (*see also* Fluid friction)
  in compressible flow, 91, 101–104
  in laminar flow, 84–87
  in packed beds, 118–121
  in pipes, 84–94
  in turbulent flow, 87–90
Pseudoplastic fluids, 154
Psychometric ratio, 532
Pumps
  centrifugal, 134–135
  developed head of, 134–136
  efficiency of, 133–136
  positive displacement, 136–137
  power requirements for, 133–136
  suction lift of, 134

R

Radiation heat transfer
  in absorbing gases, 293–296
  absorptivity, 277
  black body, 277–278
  combined radiation and convection, 279–281
  emissivity, 277–278, 283, 884
  gray body, 278, 283–284
  heat transfer coefficient, 279–280
  introduction to, 216, 276–278, 281
  Kirchhoff's law, 277, 283
  Planck's law for emissive power, 281–282
  shields, 286
  Stefan-Boltzmann law, 278, 283
  to small object, 278–279
  view factors
    between black bodies, 284–291
    general equation for, 286–288

    between gray bodies, 292–293
Raoult's law, 640–641, 680
Reflux ratio, 658–660, 686–687
Relative volatility, 644–645, 681
Reverse osmosis
  complete-mixing model, 790–791
  concentration polarization in, 789–790
  introduction to, 782–783
  membranes in, 784
  operating variables in, 788
  osmotic pressure in, 783–784
  solute rejection, 786
  theory of, 785–786, 789–791
Reynolds analogy, 438–439
Reynolds number
  in agitation, 144–145
  for condensation, 265
  definition of, 49, 202, 437
  for flat plate, 191, 193
  for flow in tube, 49, 238, 437
  for non-Newtonian fluids, 157
Reynolds stresses, 195–196
Rheopectic fluids, 155
Rittinger's crushing law, 842
Rotational viscometer, 161–163
Roughness in pipes, 87–89

S

Schmidt method
  for heat transfer, 351
  for mass transfer, 469–470
Schmidt number, 396–397, 437–438
Scraped-surface heat exchangers, 302–303
Screen analyses
  in crystallization, 746–747
  particle size measurement, 746
  Tyler screen table, 894
Sedimentation (*see* Settling and sedimentation)
Separation (*see also* Settling and sedimentation)
  by differential settling, 821–823, 826–827
  of particles from gases, 838–840
  of particles from liquids, 815–823, 831–834
  of two liquids, 834–835

Separation processes, types of, 584–585

Settlers (*see also* Settling and sedimentation)
gravity chamber, 826
gravity classifier, 826–827
gravity tank, 826
Spitzkasten classifier, 827
thickener, 718, 728–729, 825–829

Settling and sedimentation (*see also* Centrifugal settling and sedimentation; Settlers; Thickeners)
Brownian movement, 817–818
classification, 821–823, 826–827
differential settling, 821–827
drag coefficient for sphere, 816–819, 822–823
hindered settling, 820, 822
Newton's law, 817
purpose of, 815–816
sedimentation and thickening, 825–828
Stoke's law, 116, 189–190, 817
terminal settling velocity for sphere, 817
theory for rigid sphere, 816–819
wall effect, 821

Shape factors, in conduction, 235–236

Shape factors, for particles, 121–122, 124

Shear stress (*see also* Momentum balance)
components of, 172–173
definition of, 44–45
Euler equations, 185–186
ideal fluids, 185–186
in laminar flow on flat plate, 193
in laminar flow in tube, 79
normal, 170
potential flow, 186–189
stream function, 185

Sherwood number, 438

Sieve (perforated) plate tower, 611, 667–668, 716–717

Sigma value (centrifuge), 834

Simple batch or differential distillation, 646–647

Simpson's numerical integration method, 24

SI system of units, 3–4

Size reduction (*see* Mechanical size reduction)

Skin friction, 114–115, 440

Solid-liquid equilibrium, 729–731

Solids, properties of, 881–884

Solubility of gases
in liquids, 586–587, 884–886
in solids, 409–411

Solubility of salts
sodium thiosulfate, 738–739
typical solubility curves, 490

Solvent extraction (*see* Liquid-liquid extraction)

Spheres
diffusion to, 391–392
drag coefficient for, 114–117, 190
heat transfer to, 249
mass transfer to, 445–446, 450–453
Newton's law for, 817
settling velocity of, 817
Stoke's law for, 116, 190, 817

Spherical coordinates, 169, 368

Spitzkasten classifier, 827

Spray tower, 716

Stage processes (*see also* Absorption; Distillation; Liquid-liquid extraction; Leaching)
absorption, 613–614
analytical equations for, 592–593
countercurrent multistage, 589–591
distillation, 649–688
leaching, 730–735, 737
liquid-liquid extraction, 715–722
single stage, 587–588, 642, 712–715, 730–731

Standard heat of combustion, 18, 865

Standard heat of formation, 18, 864

Standard screen sizes, 894

Stanton number, 438

Steady-state diffusion (*see also* Diffusion; Diffusivity; Numerical methods)
in biological solutions and gels, 403–407
in gases, 385–393
in liquids, 397–398
in two dimensions, 413–416
in solids, 408–413

Steady-state heat transfer (*see* Conduction heat transfer;

Convective heat transfer;
Numerical methods)
Steam distillation, 648–649
Steam table, 16–17, 857–861
Stefan-Boltzmann radiation law, 278, 283
Sterilization of biological materials
effects on food, 577
introduction to, 569–570
pasteurization, 576
thermal death rate kinetics, 570–571, 575–576
thermal process time, 571–574
Stoke's law, 116, 190, 817
Stream function, 185
Streamlined body, 115
Stripping (see Absorption)
Supersaturation, 741–746 (see also Crystallization)
Suspensions, mass transfer to, 450–453
Système International (SI)
system of basic units, 3–4
table of values and conversion factors, 850–853

T

Temperature scales, 5
Thermal conductivity
definition of, 217–219
of foods, 891
of gases, 217–218, 866, 868, 875
of liquids, 218, 880
of solids, 218–219, 882–883
of water, 856, 862–863
Thickeners, 728–729, 825–828 (see also Settling and sedimentation)
Thixotropic fluids, 155
Tortuosity factor, 412–413, 468
Total energy, 56–57
Transfer unit method, 609–610, 624–625
Transition region diffusion, 464–466
Transport processes
in boundary layers, 370–373
classification of, 2
similarity of, 39–43, 381–382, 426, 430, 438–440
Tray efficiency

introduction to, 666
Murphree, 667–669
overall, 667–669
point, 668
Tray towers, 611, 613–614, 715–716
Triangular coordinates, 710–711
Tubes, sizes of, 893 (see also Circular pipes and tubes)
Turboblowers, 138
Turbulence, 194–195
Turbulent flow (see also Boundary layers)
boundary layer separation in, 191–192
boundary layer theory for
heat transfer, 373
mass transfer, 475–477
momentum transfer, 192–193, 199–201
deviating velocities in, 194–195
discussion of, 48–49, 194–195
on flat plate, 190–191, 199–201
maximum velocity in, 83–84
Prandtl mixing length
in heat transfer, 374–375
in mass transfer, 477–478
in momentum transfer, 196–197, 374–375
Reynolds number for, 49
in tubes, 49, 83–84, 87–89
turbulent shear in, 195–196
turbulent diffusion in, 382
velocity profile in, 83–84, 197–199
Two-dimensional conduction, 233–236
Tyler standard screens, 894

U

Ultrafiltration
comparison to reverse osmosis, 791–792
concentration polarization in, 789–791
equipment for, 792
introduction to, 791–792
membranes for, 792
processing variables in, 794–795
theory of, 792–794
Unbound moisture, 534–535

Underwood, minimum reflux ratio, 686–687
Unit operations, classification of, 1–2
Units
cgs system, 4
and dimensions, 5
English system, 4
SI system, 3–4
table of, 850–853
Universal velocity distribution, 197–199
Unsteady-state diffusion (see also Numerical methods)
analytical equations for flat plate, 428–429
basic equation for, 426–427
boundary conditions for, 428–430
charts for various geometries, 431
and chemical reaction, 460
in leaching, 724–726
mass and heat transfer parameters for, 430
in three directions, 432
Unsteady-state heat transfer (conduction) (see also Numerical methods)
analytical equations for, 334–336
average temperature for, 348–349
in biological materials, 360–365
in cylinder, 342–344
derivation of equation for, 214–215, 330–332
to flat plate, 334–336, 338–341
heat and mass transfer parameters for, 339, 430
lumped capacity method for, 332–334
method for negligible internal resistance, 332–334
in semiinfinite solid, 336–337
in sphere, 343, 345–346
in three directions, 345, 347–349

V

Valve trays, 611
Vapor-liquid equilibrium
azeotropic mixtures, 642
benzene-toluene, 641–642
boiling point diagrams, 640–642
calculations using Raoult's law, 640–641
enthalpy-concentration, 669–672
ethanol-water, 887
n-heptane-ethylbenzene, 691
n-heptane-n-octane, 692
n-hexane-n-octane, 690
multicomponent, 680–682
n-pentane-n-heptane, 647
phase rule, 640, 648
water-benzene, 691
water-ethylaniline, 691
xy plots, 641–643
Vapor-liquid separation processes (see Distillation)
Vapor pressure
data for organic compounds, 642, 690–692
data for water, 525–526, 854, 856–859
discussion of, 9
Vapor recompression (see Evaporation)
Vectors, 165–167
Velocity
average, 55, 80, 82
interstitial, 437
maximum, 83–84
profile in laminar flow, 80, 82–84
profile in turbulent flow, 83–84
relation between velocities, 83–84
representative values in pipes, 100
superficial, 119, 437
types of, in mass transfer, 387, 753–754
universal, in pipes, 197–199
Venturi meter, 129–131
View factors in radiation, 284–293
Viscoelastic fluids, 155
Viscosity
discussion of, 43–47
of foods, 891
of gases, 866–867, 871–872
of liquids, 876–878
Newton's law of, 43–45, 381–382
of water, 855, 863
Void fraction (packed and fluidized beds), 118–119, 123–124, 447
Von Kármán analogy, 439–440

W

Washing (*see* Filtration; Leaching)
Water
  physical states of, 525
  properties of, 854–863
Water cooling (*see* Humidification
    processes)

Weirs, 132–133
Wet-bulb temperature
  relation to adiabatic saturation
    temperature, 532
  theory of, 531–532
Wilke-Change correlation, 401–402
Work, 57–58, 63
Work index for crushing, 842